KU-256-265

CHAMBERS

SHORTER SIX-FIGURE

MATHEMATICAL TABLES

CHAMBERS SHORTER SIX-FIGURE

MATHEMATICAL TABLES

BY

L. J. COMRIE, M.A., PH.D., F.R.S.

W. & R. CHAMBERS LTD.

11 THISTLE STREET, EDINBURGH 2 : 6 DEAN STREET, LONDON, W.1

1972

SBN 550 77701 6

Printed in Great Britain
by T. and A. Constable Ltd., Hopetoun Street,
Printers to the University of Edinburgh

PREFACE

In 1944, the centenary year of the first publication of *Chambers's Mathematical Tables*,* the publishers asked me to report on that work with a view to bringing it thoroughly abreast of modern requirements. The two volumes whose contents are described on pages 378-379 were compiled and published on the basis of that report. The general principles underlying this undertaking are detailed in the Preface to each volume; those that are relevant to this volume are given below.

Chambers's Shorter Six-Figure Mathematical Tables meet the demand for a single volume of about 400 pages, rather than two volumes of 1200 pages. Many pages have been taken from one or other of the parent volumes; others have been reset with a wider interval of argument. Naturally, many tables of Volumes I and II have had to be omitted—among them the trigonometrical functions in degrees and decimals, both natural and logarithmic.

The following points are mostly taken from the Preface to the more extended volumes.

(1) Six-figure Values instead of Seven-figure. It is well known that Briggs, in his first tables of logarithms, gave fourteen decimals. Vlacq, shortly afterwards, was content with ten. John Newton, in the same century, dropped to eight. Then in the eighteenth century various authors, conspicuous among whom was Hutton, saw that not more than seven were needed for most practical problems. Hence Bell, under the spell of Hutton, gave seven in the first *Chambers's Tables*. About that time 6-figure and even 5-figure tables appeared; their position was firmly and permanently established, a few years after Bell's volume, by the 6-figure tables of Bremiker.

The tendency towards fewer figures for popular tables continued, until to-day schools are equipped with 4-figure tables, which are ten times as accurate as the common 10-inch slide rule with which the great majority of engineering calculations are done. There is no doubt that 4-figure tables are now used more than all others put together. When four figures fail, usually five or six suffice; rarely does the need arise for 7-figure tables, which are 1000 times as accurate as 4-figure tables. It is true that in the past a great deal of the most laborious computation arose in astronomy, which shares with surveying and geodesy the need for high accuracy; but to-day these subjects must be classed as specialist, and are well provided for.

I contend that good 4-figure tables and good 6-figure tables will cover 95 per cent of computational requirements. Where five figures suffice, one figure can be dropped from the 6-figure tables, with the advantages that the smaller interval of argument gives more tabular values and that the effect of second differences is usually negligible. Another consideration is that 6-figure tables can be made linear (or almost so) in a reasonable compass. The number of decimals required for different degrees of accuracy in angles is discussed on page ix.†

(2) Contents. Since nautical, astronomical, surveying, and other specialised requirements are now amply met by a multiplicity of tables, many of them excellent, it seemed to me advisable to concentrate on a collection of general mathematical functions.

Neither Bell nor Pryde is to be blamed for not including exponential and hyperbolic functions. Not to do so to-day would be unpardonable. They represent the frontier of elementary functions, beyond which lies the vast range of so-called higher mathematical functions.

Inverse circular and hyperbolic functions, although obtainable by inverse interpolation of the direct tables, are now regarded so much as functions in themselves that they have established their right to an independent existence. For this reason a short table has been included here, mostly at ten times the interval of the more extended tables in Volume II. The frequency with which the

* Chambers's *Mathematical Tables, consisting of logarithmic and other tables required in the various branches of practical mathematics* (pp. xxxi+316, 8vo; 1844) were compiled by Andrew Bell, at one time mathematics master at Dollar Academy, as an accompaniment to his *Treatise of Practical Mathematics* (1842) in "Chambers's Educational Course". Later the original book was enlarged by the addition of nautical and other tables to form a "necessary accompaniment" to James Pryde's *Treatise on Navigation* (1867) in the same series. In 1878, when a new and reset edition was issued, James Pryde's name appeared on the title-page. James Pryde, one of the original Fellows of the Educational Institute of Scotland, was lecturer in mathematics at the Edinburgh School of Arts, the forerunner of the Heriot-Watt College.

† In accordance with my views, I have also produced for Messrs. Chambers two 4-figure tables, one of which, with 64 pages, is a miniature version of these tables, while the other, with 32 pages, contains only the essentials needed for schools where the teaching of mathematics is not pursued beyond matriculation standard.

tabular values themselves can be used as they stand without interpolation would in itself justify the table maker in not insisting that the user should get every inverse function by inverse interpolation.

Once logarithmic computation is replaced by mechanical computation, tables of roots and powers become necessary. It is true that good tables of these functions exist, but our tables, being a comprehensive collection, would have suffered if roots and powers had not been included. Similarly, reciprocals are often used by calculating machine computers to turn division into multiplication.

Great care has been expended on the provision of means for accurate interpolation. In many places this has taken the form of expanding parts of the tables (at the expense of more pages) to a finer interval of argument.

(3) TYPOGRAPHY. I hope it will be evident without any words of mine that great attention has been paid to typography. It was Hutton who introduced the so-called " modern face " or equal height figures (1 2 3 4 5 6 7 8 9 0) into mathematical tables. But De Morgan clearly demonstrated the superiority of " old style " figures with heads and tails (1 2 3 4 5 6 7 8 9 0). In this he has been followed by the best editors (especially Peters) in Germany and (latterly) in this country. It is only in the U.S.A. that the modern face figure persists to any extent. A very convincing testimony to the superior legibility of old style figures came from the compositors and proof readers of these tables, who emphatically agreed that the figures here used were less fatiguing and so less liable to misreading than modern face figures.

Legibility depends not only on the size and shape of the figures used but also on the proper relation of black and white. Figures have to be read one by one, not in groups like words. To help the eye to separate them, we have introduced between the figures and between lines extra spaces of a few thousandths of an inch.

The sparing use of rules will also be noticed. As usually overdone, they give a heavy appearance to the page. White spaces can be just as effective, and are more restful to the eye. The partial suppression of leading figures that change very slowly also enables relief from black to be given. At the same time every precaution has been taken not to mislead the user by overdoing this suppression; a further safeguard is the space separating third and fourth decimals.

(4) ACCURACY. Proofs have been read against a great variety of published tables where appropriate ones exist. It is confidently believed that the cases where the error exceeds $\pm 0{\cdot}51$ units of the last decimal could be counted on the fingers of one hand; those that are known to exist form an uncomfortable trap for any would-be plagiarist. Our aim has been the same as that of Bell a hundred years ago, from whom we quote: " As accuracy is, next to the principles of construction, the most important element in tables intended for practical application, the greatest care has been bestowed in collating the whole . . . with others of established character."

ACKNOWLEDGMENTS

I cannot close these remarks without mentioning some of those specially deserving of praise and thanks. The tasks of copy making and of many proof readings, which fell to my staff, were no light ones, but they were faithfully, cheerfully and diligently accomplished.

The printers, Messrs. T. and A. Constable Ltd., deserve every possible praise. It would be unfair not to mention by name Mr. Peter Wilson, in charge of the composition, although he will wish to share this tribute with his fellow compositors and readers. No author or publisher could have wished for more helpful co-operation in meeting exacting typographical demands or for a more intelligent grasp of the whole problem of table presentation.

Sincere thanks are also due to the Monotype Corporation Ltd. for their provision, under conditions of great difficulty, of a large number of special figures, all to extremely fine specifications.

Finally, it is a great pleasure to express my admiration for Mr. T. C. Collocott, of Messrs. W. and R. Chambers Ltd., who shepherded the tables through the press on behalf of his firm. Throughout he has been an ideal team-mate. After all, these tables, which may fairly be described as one of the greatest of all British table-making enterprises, are the work of three teams—my staff, the publishers and the printers.

L. J. COMRIE

CONTENTS

CONTENTS

EXPLANATION

Introduction

No systematic attempt will be made here to explain points that should be known by every table user; it suffices to cover points that may be unfamiliar, or which are peculiar to these tables. Those that are general will be dealt with first, followed by a detailed description of each table.

How Accurate are Six-Figure Tables?

The tendency of inexperienced computers is to use too many decimals. As a guide to the number of decimals *normally* required when working to any specified units, and conversely the units one *normally* works to with any specified number of decimals, we may consult the table below. It will be found that this table is equivalent to saying that the number of decimals in the function and the number of significant figures required to express the angle (assumed for the purpose of this statement to be over 10°) are the same. Thus 12°·3456 or 12° 34′·56 or 12° 34′ 56″ or 0^r·123 456 are all expressed with six figures, and need 6-figure trigonometrical functions.

Working unit	Value of unit in radians	Decimals required
0°·01	0·0002	4
1′	0·0003	4
0^r·0001	0·0001	4
0°·001	0·00002	5
0′·1	0·00003	5
10″	0·00005	5
0^r·00001	0·00001	5
0°·0001	0·000 002	6
0′·01	0·000 003	6
1″	0·000 005	6
0^r·000 001	0·000 001	6

This can be continued by adding another decimal each time the working unit is divided by ten. Thus an 8-figure table permits working to six decimals of a degree, four decimals of a minute, two decimals of a second, or eight decimals of a radian.

It should be emphasised, however, that this table is not for extreme cases. With small angles, fewer decimals would be needed in the logarithms of sines, tangents and cotangents, or in natural cotangents and cosecants; there are corresponding anomalies for large angles. In logarithmic working the safest function for determining an angle that may lie anywhere in the quadrant is the log tan (or log cot), whose weakest point is at 45°. A small angle can be determined $1/M$ times (or slightly more than twice) as accurately from its natural cosine as from its logarithmic cosine. As the natural sine is at its best for small angles, the user of natural tables (with a calculating machine) is at an advantage over the logarithmic user as far as both ease and accuracy are concerned.

The general principles to bear in mind are:

(1) With rounded-off values of a function, which are subject to a possible error of half a unit in the last decimal retained, there will be uncertainty in the last decimal of an interpolate if the difference between consecutive tabular values of the function, in units of its last decimal, greatly exceeds the difference between consecutive values of the argument, expressed in units of the last working (not tabular) decimal. Thus an interpolated function of an angle rounded off to 0′·01 could not be relied on to several units in the last decimal if obtained from a table at interval 1′ at a point where the tabular differences are of the order 300-1000, or more. The fault is not with the tables; the angle is insufficiently defined to take full advantage of the tables.

(2) In inverse use, i.e. when finding an angle from a function, the interval between consecutive arguments cannot be divided *accurately* into more parts than there are units of the last decimal in the difference between consecutive tabular values of the function. Thus a difference of 12 would entitle the user to one decimal of the interval, but not to two. Similarly, a difference of 444 would be safe for two decimals of the interval but might yield a result 0·002 or 0·003 in error if three decimals were written. Provided the user is fully conscious of the limitations of his accuracy, there is no reason why he should not, in a case like this, write three decimals in order to reduce cumulative rounding-off errors when this angle is used in subsequent calculations. But it serves no useful purpose to write values that may be 10 units or more in error.

Leading Figures

Where successive tabular values are in horizontal lines, a change in the leading figures is indicated by printing the entire group of following figures in a contrasting type—a bold equal-height figure—that cannot fail to be noticed when using such an entry. The leading figures to be annexed are then to be found at the beginning of the next lower line.

Where successive tabular values are in columns, leading figures are to be found always at the beginning of each block of five lines (or on every tenth line of some of the trigonometrical tables). If a change occurs, the last line with the old leading figures and the first line with the new are given in full. Thus, *whichever* way the user looks for the missing figures (i.e. up or down), he will find them. In the example below, the lines for $x = 1{\cdot}385$ and $1{\cdot}390$, being at beginnings of blocks of five lines, are given in full. The line for $x = 1{\cdot}386$, being the last in which the integral part is 3, is given in full. The line for $x = 1{\cdot}387$, being the first in which the integral part is 4, is also given in full.

x	e^x
1·385	3·994 826
·386	3·998 823
·387	4·002 824
·388	·006 828
·389	·010 837
1·390	4·014 850

Critical Tables

A function is normally tabulated at equal intervals of the independent variable or argument. If, however, the function is changing slowly in its last figure, it is possible to write down *all* the values that the function may assume, and then to show the limiting values of the independent variable that lead to each function value. Suppose we wanted sines to one decimal from angles in the first quadrant known to the nearest degree. The necessary critical table is shown alongside.

x	$\sin x$
°	
0	
	0·0
2	
	0·1
8	
	0·2
14	
	0·3
20	
	0·4
26	
	0·5
33	
	0·6
40	
	0·7
48	
	0·8
58	
	0·9
71	
	1·0
90	

In this form the argument (here x) is on the left and the respondent (here $\sin x$) on the right as usual, but on lines half way between those of the argument. Thus any angle *between* 20° and 26° has a sine of 0·4. By a convention that is now well established, the respondent to adopt when the argument corresponds to a tabular (or critical) value is the value half a line higher on the page. Thus 20° has a sine of 0·3 and 26° of 0·4. This is often denoted by the legend: *In critical cases ascend.*

The method of preparing such a table is to compute the values of x corresponding to half-way values of $\sin x$, as shown in the table alongside. The argument in critical tables is never rounded *off*. It is rounded *down* if it is increasing numerically, and *up* if it is decreasing numerically, regardless of whether the respondent is increasing or decreasing. Thus the table shows 2° rather than 3° for its first significant entry, and use of the table gives 0·1 for sin 3° to one decimal, which is correct. Hence the meaning of the working table is that for angles from 3° to 8° inclusive $\sin x$ ranges from a little over 0·05 to a little under 0·15, and so correctly rounds off to 0·1.

$\sin x$	x
	°
0·05	2·9
0·15	8·6
0·25	14·5
0·35	20·5
0·45	26·7
0·55	33·4
0·65	40·5
0·75	48·6
0·85	58·2
0·95	71·8

The advantages of critical tables are:

(1) They are used without interpolation.

(2) They are correct to within approximately half a unit of the last decimal, whereas the error in an ordinary table after interpolation may be as much as a unit in this decimal.

Their use is limited by the fact that *every* possible answer must be printed; nevertheless, when they are applicable, the fact that they eliminate interpolation represents a considerable advantage. They have been used throughout this volume wherever appropriate.

Critical tables are not intended to be used inversely, although the knowledge of their method of construction informs us that the value of the argument corresponding to a given respondent must lie approximately half way between the two tabular arguments above and below, with a preference in doubtful cases for *descending* in the column of arguments. Thus the angle whose sine is 0·6 is approximately $\frac{1}{2}(33° + 40°)$ or $36\frac{1}{2}°$, for which we take 37°. But this method of inverse use of critical tables breaks down when the argument ceases to be linear. In fact, inverse use of critical tables should be avoided unless the computer is really expert.

Ranges and Intervals of Tables

There is an established abbreviation for indicating the range and intervals of a table, which we follow. Thus 0°(1″)1°(10″)10°(1′)45° means that a table begins at 0°, proceeds at interval 1″ to 1°, then changes the interval to 10″, and at 10° changes again to interval 1′, which continues to 45°, where the table ends.

Arrangement of Trigonometrical Tables

All tables based on the 90° quadrant are arranged in the usual semi-quadrantal manner, which has the great advantage of presenting all the functions of any one angle on the same line. The method of dealing with angles not in the first quadrant is described on page 1.

Left	Right
0	60
1	59
2	58
3	57
4	56
5	55
6	54
7	53
8	52
9	51

These trigonometrical tables do not follow the usual arrangement of blocks of five lines, since that would introduce a lack of symmetry in the position of the terminal decimals of the argument. Thus in left-hand columns the middle of blocks would end in 2 or 7, while in right-hand columns they would end in 3 or 8. To avoid this we have used the Bremiker arrangement, in which multiples of 10 stand on lines by themselves and the intervening nine lines are divided into three groups of three each, as shown alongside. Thus in both columns of arguments the middle of the group always ends in 2, 5 or 8. This advantage will be appreciated by the constant user.

Left	Right
0	60
1	59
2	58
3	57
4	56
5	55
6	54
7	53
8	52
9	51
10	50

Differences

To facilitate interpolation, interlinear differences (in units of the last decimal) are given where necessary. No signs are given. The printing of these differences in italics is a warning that second differences are not negligible. In general, italic is used when the second difference exceeds 4 or 5, but there are a few exceptions, especially where the error caused by neglecting second differences is not likely to be of any real consequence.

Use of Proportional Parts

The fact that differences have been provided has already been mentioned. The user who does not have a calculating machine (for the use of which see Volume II of *Chambers's Six-Figure Mathematical Tables*) will require other means of doing the multiplications needed for interpolation. Most of these are within the capacity of a slide rule. Another convenient method is the use of a multiplication table. That of Crelle gives all 3 × 3 products; that of Zimmermann gives 3 × 2 products and that of Peters gives 4 × 2 products. The full titles of these tables are:

Crelle, A. L. *Calculating Tables*. Berlin: de Gruyter.

Peters, J. *New Calculating Tables . . .* (*Neue Rechentafeln . . .*). Berlin and Leipzig: de Gruyter.

Zimmermann, H. *Rechentafeln . . .* Berlin: Ernst.

Tables of so-called proportional parts are really 2 × 1 or 3 × 1 multiplication tables. They are less bulky than the tables just mentioned, but on the other hand need an entry for each digit of the multiplier. Special tables of proportional parts will be described in conjunction with the tables with which they are given. For general use, Table XII (pages 358-377), giving multiples of all numbers less than 1000 by ·1, ·2, ·3 . . . ·9, has been provided.

Example. Multiply 456 × 0·789.

From the table for 456 on page 367

$$456 \times 0{\cdot}7 = 319{\cdot}2$$
$$456 \times 0{\cdot}08 = 36{\cdot}5$$
$$456 \times 0{\cdot}009 = 4{\cdot}1$$
$$456 \times 0{\cdot}789 = 359{\cdot}8$$

360 to the nearest integer.

The responsibility of determining whether the proportional part is to be added or subtracted lies with the user. He should at least take the precaution, after interpolating, of verifying that his interpolate lies *between* the two function values between which he has interpolated. Exceptions to this rule can only occur when the function passes through a maximum or a minimum between the two tabular values that are being used.

Interpolation by Machine

For the owner of a calculating machine interpolation has no terrors. The machine technique that the writer considers *safest* will now be described. A knowledge of the use of the machine is assumed, but the nomenclature to be used for the various parts must first be stated.

A calculating machine contains setting levers or a keyboard for setting numbers to be added or subtracted, or multiplicands or divisors. This will be abbreviated S.L. (setting levers) in deference to the Brunsviga, the machine that is best known among British scientists. Most machines have a setting register showing the amount set on the levers or keyboard.

A multiplier register (M.R.) records the number of turns or revolutions that the machine has made, thus showing multipliers and quotients. A product register (P.R.) shows the results of additions and subtractions, and products. It is mounted in a carriage, which enables it to be displaced relatively to the S.L. Turns of the handle or motor are recorded in the units, tens, etc., place of the M.R. according to the position of the P.R.

The terms "upper dials" and "lower dials", denoting M.R. and P.R. (or vice versa), which are to be found in calculating machine trade literature, are deprecated, as they are positional only (and thus peculiar to any one machine) rather than functional.

If we have two consecutive tabular values f_0 and f_1, and their difference $\Delta(=f_1-f_0)$, the problem of interpolating linearly to a fraction n of the interval between them may be expressed as

$$\begin{matrix} f_0 & \\ & \Delta \\ f_1 & \end{matrix}$$

$$f_n = f_0 + n\Delta$$

This may be mechanised thus: Set f_0 on the S.L., and multiply by 1 followed by as many ciphers as there are decimals in n. Clear S.L. and M.R. Set Δ and turn the handle once (without moving the carriage), forwards if f is increasing numerically and vice versa. It must be provided that the M.R. shows 1 (followed by ciphers) at this stage; this provision will depend on the particular machine in use. See that the P.R. now shows f_1. This check guards against quite a number of sins that might have been committed up to this stage, e.g. a misprint in f_0 or f_1, wrong setting of f_0, use of an incorrect Δ, wrong setting of Δ, wrong direction of turning, or displacement of the carriage between the two turnings.

The machine now says that

$$f_1 = f_0 + 1 \times \Delta$$

but our question is: what is $f_0 + n \times \Delta$? So, without any further clearing or setting, the handle is turned until the M.R. shows not 1 but n. The product register will then show f_n. Provided the user verifies now that the M.R. actually does show n, he may have every confidence in his interpolate.

Example. Find y from the following entries in a table, when $x = 1{\cdot}666\ 873$.

Here $n = 0{\cdot}873$

x	y	Δ
1·666	0·123 456	
		789
1·667	0·124 245	

Set 123 456, multiply by 1·000, and clear S.L. and M.R. Set 789, turn once, and verify that the P.R. shows 124 245. To change the M.R. from 1·000 to 0·873 we can use the familiar process of "short cutting", i.e. turn back one turn, then three turns, then forward three turns, thus showing ·873 in the M.R. (machines without tens transmission in the M.R. are not recommended) and the interpolate 0·124 145 in the P.R.

The procedure for inverse interpolation is precisely the same up to the point where the P.R. shows f_1 and the M.R. shows 1. Then instead of turning until the M.R. shows the known value of n, we turn, watching the P.R., until it shows the known value of f_n. The desired value of n is then read from the M.R. The reader may study this from the same values of x, y and Δ, finding n for $f_n = 0{\cdot}124\ 145$. When he has finished this problem, the machine will be in exactly the same condition as when he finished the first, i.e. showing n in the M.R. and f_n in the P.R. This is precisely equivalent to the more familiar process of

$$n = \frac{f_n - f_0}{\Delta}$$

since that equation immediately transforms to our working equation

$$f_n = f_0 + n\Delta$$

Characteristics of Logarithms

When logarithms are taken to base 10, it is well known that it is advantageous to keep the mantissa positive, even when the number is less than unity, and its logarithm negative. There are then two ways of expressing the characteristic. The first is by writing a bar over it to denote that it is negative; e.g. $\bar{1}{\cdot}234\ 567$ denotes $-1 + 0{\cdot}234\ 567$ or $-0{\cdot}765\ 433$. The second method is to increase the characteristic (usually by 10) so that it is positive, but is subject to the subtraction of the amount that has been added. When this amount is conventionally 10, it is not usually written, but merely understood by the computer. Thus the logarithm just given is $9{\cdot}234\ 567\ -10$, normally written 9·234 567.

British teaching practice seems to favour the negative characteristic. Indeed, the present writer was so taught, but when he began to use logarithms professionally as a computer, he soon realised the superiority, for practical use, of positive characteristics, and since then has used them exclusively. Astronomers and surveyors have always used positive characteristics, which are to be found in all the best 6-, 7-, 8- and 10-figure tables. For that reason, and also because of a firm belief that they are better for serious use, they are adopted in the trigonometrical tables in this volume.

No confusion need arise with positive characteristics, but the user should beware of overlooking a factor of 10^{10} or 10^{-10}. In doubtful cases, the amount that has been added to the characteristic should be written after the logarithm. For instance, physical constants are often written as a number between 1 and 10 (whose logarithm has the characteristic 0) and a power of 10. The characteristic of the logarithm, whether positive or negative, is then the power of 10 required. Thus Planck's constant is $6{\cdot}62 \times 10^{-27}$ and its logarithm may be written $0{\cdot}821\ -27$. The logarithm of its square root is $\frac{1}{2}(1{\cdot}821 - 28)$ or $0{\cdot}910\ -14$.

S and *T* Functions

The difficulty of interpolating the log sin, log tan or log cot of a small angle (or their co-functions of a large angle) is well known. There are three ways in which relief may be obtained:

(1) By using fewer decimals, since such angles have one or more 0's immediately after the decimal, and there is no need for more decimals in the logarithms than there are *significant* figures in the function.

(2) By decreasing the interval of tabulation, in order to render interpolation more manageable.

(3) By the use of the auxiliary functions S and T.

All three of these devices are used in different parts of this volume—in some cases two (or even all three) together. The third steps in when the first two ultimately break down, but is often preferred to interpolation in direct tables.

The experienced table maker knows that when a function tends to become infinite, its direct tabulation becomes burdensome, but there is always some means of overcoming the difficulty by a suitable auxiliary function suggested by the known form of the function. Writing (for x in radians)

$$\sin x = x\left(1 - \frac{x^2}{3!} + \frac{x^4}{5!} - \cdots\right)$$

and taking logarithms of both sides, we find

$$\log \sin x = \log x + \log\left(1 - \tfrac{1}{6}x^2 + \cdots\right)$$

We see at once that, for small x, the trouble arises from the term $\log x$, since the logarithm of the other term on the right-hand side can never differ much from zero, and must vary slowly. Hence, since we already have provision for finding the logarithm of a number, however small, all that is necessary is to tabulate the second term, which is called S, so that the working equation becomes

$$\log \sin x = \log x + S$$

If x is not measured in radians, but in some other (smaller) units (such as seconds of arc), $\log x$ will be too great, and it becomes necessary to subtract the logarithm of the number of times the unit is contained in a radian. This correcting term, which is constant for any specified unit, may obviously be combined with S, and separate tables of S prepared for each of the different units in which x may be measured. Thus, using x'' (for instance) to denote the number of seconds of arc in the angle x, we write

$$\log \sin x = \log x'' + S$$

where it is now tacitly understood that S has been adjusted for a unit of a second of arc in x.

The T function is defined similarly from the equation

$$\log \tan x = \log x + \log (1 + \tfrac{1}{3}x^2 + \cdots)$$

It is seen that, for small x (with which alone we are concerned), T changes twice as rapidly as S, and in the opposite direction. Also $\log \cos x = S - T$, whatever the unit in which the angles are measured.

To find a small angle from its log sin, we may ascertain S, and then

$$\log x = \log \sin x - S$$

x will be in the same units as those in the table from which S is taken.

The τ and σ Functions

Note. Much of the phraseology of this paragraph is similar to that of the preceding paragraph on the S and T functions, but as certain readers may read one and not the other, it has been deemed advisable to make each complete in itself.

The difficulty of interpolating cotangents and cosecants of small angles, or their cofunctions of large angles, is well known. There are three ways in which relief may be obtained:

(1) By using fewer decimals, since the differences would otherwise be so large that, unless the angles are very much more precisely defined than is necessary for other angles, meaningless figures would be obtained.

(2) By decreasing the interval of tabulation, in order to render interpolation more manageable.

(3) By the use of the auxiliary functions τ and σ.

All three of these devices are used in this volume—often together. The third steps in when the first two ultimately break down, but is often preferred to interpolation in direct tables.

The experienced table maker knows that when a function tends to become infinite its direct tabulation becomes burdensome, but there is always some means of overcoming the difficulty by a suitable auxiliary function suggested by the known form of the function. Writing (for x in radians)

$$\cot x = \frac{1}{x}\left(1 - \frac{1}{3}x^2 - \frac{1}{45}x^4 - \cdots\right)$$

we see that, for small x, the trouble arises from the factor $1/x$, which tends to become infinite as x approaches zero, whereas the other factor can never differ much from unity, and must vary slowly. Hence, since we have in our calculating machine provision for finding the reciprocal of a number, however small, all that is necessary is to tabulate the second factor, which is called τ, so that the working equation becomes

$$\cot x = \frac{\tau}{x}$$

If x is not measured in radians, but in some other (smaller) units (such as seconds of arc), x will be too great, so it becomes necessary to multiply τ by the number of times the unit is contained in a radian. Hence separate tables of τ have been prepared for each of the different units in which x may be measured. Thus, using x'' (for instance) to denote the number of seconds of arc in an angle x, we write

$$\cot x = \frac{\tau}{x''}$$

where it is now tacitly assumed that τ has been adjusted for a unit of a second of arc in x.

The σ function is defined similarly. From the equation

$$\operatorname{cosec} x = \frac{1}{x}\left(1 + \frac{1}{6}x^2 + \frac{7}{360}x^4 + \cdots\right) = \frac{\sigma}{x}$$

it is seen that, for small x (with which alone we are concerned), τ changes twice as rapidly as σ, and in the opposite direction. Also $\cos x = \tau/\sigma$, whatever the unit in which the angles are measured.

To find a small angle from its cotangent, an excellent method is to form its reciprocal on the calculating machine and use a table of tangents inversely. Alternatively ascertain τ, and then

$$x = \frac{\tau}{\cot x}$$

x will be in the same units as those in the table from which τ is taken.

The functions τh and σh are used similarly to find hyperbolic cotangents and cosecants.

This volume gives these auxiliary functions for various units in the form of critical tables. The tables for angles in seconds of arc are on page 121 (with illustrations), for radians on page 171, and those for hyperbolic functions on page 205. In each case the left-hand argument is x, for finding τ or σ when the cotangent or cosecant of a known angle is required, and the right-hand argument is the corresponding function, to be used for finding τ or σ when seeking an angle from a function. Further illustrations are given later when describing each table.

TABLE I—LOGARITHMS OF NUMBERS

This table gives the mantissæ (without any decimal points) of logarithms of numbers from 1000 to 10,000. The characteristic (always positive for numbers greater than 1 and positive or negative—according to the preferred way of working—for numbers less than 1) must be prefixed according to rules that the user of a 6-figure table may be expected to have learned at an earlier stage of his computing career.

The leading figure of each number is given at the beginning of each block of five lines and at changes. When the leading figure increases in the middle of a line the number is printed in full a sufficient number of times to avoid any uncertainty on the part of the reader, who has simply to glance *either* up or down.

As the differences, which vary from 434 down to 43, are much more numerous in the early part of the table, it has been necessary to devote considerable portions of the early pages to proportional parts. Those on each page are in ascending order, and are so arranged that the proportional parts for any logarithm will always be found on the same opening as the logarithm itself. The only proportional parts omitted are those for 200, 300 and 400, for which the user may substitute 0's for the decimals in the proportional parts for 201, 301 and 401.

The column of small-type figures at the end of each line, marked Δ, gives the difference between column 9 of that line and column 0 of the next line. Thus any difference may be inferred by subtracting the final digits only of the consecutive logarithms, and glancing at this column.

Example. Required log 4567·89.

log 4567 (page 15)	3·659 631
P.P. for ·8 with $\Delta = 95$	76
P.P. for ·09 with $\Delta = 95$	9
Sum = log 4567·89	3·659 716

The user of a slide rule will see that the evaluation of the proportional part is simply the calculation $0{\cdot}89 \times 95$.

Example. Required log 23·4567.

log 23·45 (page 8)	1·370 143
P.P. for ·6 with $\Delta = 185$	111
P.P. for ·07 with $\Delta = 185$	13
Sum = log 23·4567	1·370 267

Example. Find the number whose logarithm is 0·123 456.

log 1·328 (page 4)	0·123 198
Δ at this point	327
Excess (i.e. 0·123 456 − 0·123 198)	258

The problem is to find two further decimals to be added to 1·328 in order to account for the excess 258. These are obviously found by the evaluation of $\frac{258}{327}$ to two decimals. Using the P.P. table for $\Delta = 327$ (page 5) it is seen that 228·9 is 0·7 of 327, leaving a residual excess of 29·1. 29·1 is 0·09 (nearly) of 327. Hence 258 is 0·79 of 327.

Therefore the number whose logarithm is 0·123 456 is 1·328 79.

Example. Find the number whose logarithm is 3·456 789.

log 2862 (page 10)	3·456 670
Δ at this point	151
Excess	119

Working as in the preceding example, the fifth figure for excess 106 is found to be 7, and the sixth figure for excess 13 is 9, so the required number is 2862·79.

The four fundamental rules for the practical application of logarithms, although doubtless familiar, may be summarised

Multiplication.

The logarithm of a product is the sum of the logarithms of its factors, i.e.

$$\log NM = \log N + \log M$$

Division.

The logarithm of a quotient is found as the logarithm of the numerator *minus* the logarithm of the denominator, i.e.

$$\log \frac{N}{M} = \log N - \log M$$

Involution.

The logarithm of a power is obtained by multiplying the logarithm of the exponent by the power to which it is raised, i.e.

$$\log N^M = M \log N$$

Evolution.

The logarithm of a root is obtained by dividing the logarithm of the radicand by the root to which it is taken, i.e.

$$\log N^{\frac{1}{M}} \text{ or } \sqrt[M]{N} = \frac{\log N}{M}$$

We append an example of each process, with parallel columns showing positive and negative characteristics. See note on positive characteristics on page xii.

Multiplication.

Find 123·456 × 3·141 59 × 0·000 137 246.

	With positive characteristics	With negative characteristics
log 123·456 (page 2)	2·091 512	2·091 512
log 3·141 59 (page 11)	0·497 149	0·497 149
log 0·000 137 246 (p. 4)	6·137 500	$\bar{4}$·137 500
	8·726 161	$\bar{2}$·726 161

From page 17, the required number is 0·053 2306.

Division.

Find 0·246 890 ÷ 0·434 294.

log 0·246 890 (page 8)	9·392 503	$\bar{1}$·392 503
log 0·434 294 (page 15)	9·637 784	$\bar{1}$·637 784
	9·754 719	$\bar{1}$·754 719

From page 18, the required number is 0·568 486.

Powers of Numbers or Involution.

Find $0{\cdot}765\ 432^{11}$.	With positive characteristics	With negative characteristics
log 0·765 432 (page 22)	9·883 906	$\bar{1}$·883 906
log $0{\cdot}765\ 432^{11}$ = 11 × 9·883 906	8·722 97	$\bar{2}$·722 97

From page 17, $0{\cdot}765\ 432^{11} = 0{\cdot}052\ 841$.

Caution. A six-figure logarithm, when multiplied by a factor of the order 10, loses a decimal; if the factor is of the order 100, two decimals are lost, and so on. It so happens, in this example, that the sixth decimal is regained by an answer to five significant figures, because the first is a cipher.

Roots of Numbers or Evolution.

Find $0{\cdot}000\ 246\ 789^{\frac{1}{7}}$ or $\sqrt[7]{0{\cdot}000\ 246\ 789}$.

log 0·000 246 789 (page 8)

$$6{\cdot}392\ 326 - 10 \text{ or } \bar{4}{\cdot}392\ 326$$

Using the logarithm with a positive characteristic, a sufficient multiple of 10 is added, numerically, to both components to render the negative component a multiple of 7. With a negative characteristic, the negative characteristic is made a multiple of 7 and the change is balanced by adding an equal positive characteristic to the mantissa. This gives

$$66{\cdot}392\ 326 - 70 \text{ or } \bar{7} + 3{\cdot}392\ 326.$$

Then, dividing by 7, log $0{\cdot}000\ 246\ 789^{\frac{1}{7}}$ is found to be 9·484 618 − 10 or $\bar{1}$·484 618.

From page 11, $0{\cdot}000\ 246\ 789^{\frac{1}{7}} = 0{\cdot}305\ 223$.

TABLE II—LOGARITHMS OF TRIGONOMETRICAL FUNCTIONS

(Angles in Degrees, Minutes and Seconds)

Table IIA (pages 28-31) gives, to five decimals, the log sin and log tan of angles up to 1° 20′ at interval 10″. Only the last three decimals of log tan are given, since the characteristic and the first two decimals can be annexed from the log sin. In the two cases (page 31) where these three decimals are printed in bold type, the second decimal must be increased by a unit; thus

$$\log \tan 1°\ 05'\ 30'' = 8{\cdot}28002$$
$$\log \tan 1°\ 13'\ 30'' = 8{\cdot}33008$$

The differences of log tan are the same as those of log sin or differ from them by a unit. Hence to find a log tan difference it is only necessary to subtract the end figures and glance at the printed difference for the log sin. Thus the difference between log tan 21′ 50″ and log tan 22′ 00″ (page 29) ends in 0 and so must be 330, since the difference for the corresponding values of log sin is 331.

In this range five decimals suffice even if dealing with angles to 0″·1. For the range up to 10′, for which no differences are given, or to obtain six decimals in any part of the range, the *S* and *T* functions given on the right of these pages may be used. The theory of these functions has been given on page xii; it remains to illustrate them, first recapitulating the formulæ:

$$\log \sin x = \log x'' + S \qquad \log x'' = \log \sin x - S$$
$$\log \tan x = \log x'' + T \qquad \log x'' = \log \tan x - T$$

Example. Find log sin 12′ 34″·56.

Reducing the angle to seconds with the aid of the table on the right of page 28, $x = 754''{\cdot}56$. Hence

log x'' (page 22)	2·877 694
S (page 29)	4·685 574
Sum = log sin x	7·563 268

Direct interpolation from page 28 yields 7·56326, the discrepancy of 1 unit in the fifth decimal being the maximum that can normally arise; it is due to the combined effect of the neglected sixth decimal, the neglected second difference and the inevitable rounding-off uncertainty. It is of no consequence if the working unit is not finer than 0″·1.

Now suppose that we had been given log sin x = 7·563 268 and it is desired to find x. *S* must now be obtained from the same table, but with argument log sin x. Hence

log sin x (given)	7·563 268
S (page 29)	4·685 574
Difference = log x''	2·877 694
x	754″·56
	= 12′ 34″·56

Example. Find log cot 7′ 24″·441.

log 444·441 (page 15)	2·647 814
T (page 30)	4·685 576
log tan 7′ 24″·441	7·333 390
log cot 7′ 24″·441	2·666 610

It is, of course, not difficult to write the log cot without writing the log tan, by adding the first two lines and building up to zero at the same time.

Example. Given log tan x = 2·666 610 to find x.

log tan (90° − x)	7·333 390
T (page 30)	4·685 576
log (90° − x)	2·647 814
90° − x (page 15)	444″·441
	= 7′ 24″·441
x	89° 52′ 35″·559

Example. Find log cos 89° 51′ 23″·4. This is the same as log sin (90° − 89° 51′ 23″·4) = log sin 8′ 36″·6 = log sin 516″·6.

log 516″·6 (page 17)	2·7132
S (page 29)	4·6856
Sum = log cos 89° 51′ 23″·4	7·3988

We have here assumed that the given angle is subject to the usual uncertainty of half a unit in the last decimal. In that case we are not entitled to more decimals in the logarithm than there are significant figures in $90° - x$.

Example. Find x when log cot x = 8·123 456.

log tan $(90° - x)$	8·123 456
T (page 30)	4·685 600
log $(90° - x)$	3·437 856
$90° - x$ (page 10)	2740″·66
	= 45′ 40″·66
x	89° 14′ 19″·34

We have here taken six significant figures in $90° - x$, to correspond with the six decimals in its logarithm.

Table IIB (pages 32-33) comprises two pages for dealing with the missing logarithms of cosines of angles up to 1° 20′, or of sines of angles greater than 88° 40′. Page 33 gives these values in a critical table. For determining an angle in this range from a given function value, the direct table on page 32 shows, for each such function value, the best value of the angle to adopt. It may perhaps prevent the unwary from writing angles to 1″ when they are not known to within a score of seconds.

Table IIC (pages 34-120) gives the logarithms of the four principal functions for arguments 1° 20′(10″)10°(1′)45°, the arrangement being semi-quadrantal. As the differences of log cot are precisely the same as those of log tan, they are not printed. In the early pages (up to $2\frac{1}{2}$°) it has not been possible to give all proportional parts; the omissions have been carefully selected so that, if not more than two decimals of a second are being used, it suffices to take the nearest proportional part given.

Example. Find log cot 1° 23′ 45″·67.

The difference is here 865, for which we use 866.

log cot 1° 23′ 40″ (page 34)		1·613 636
P.Ps. for	5″	−4330
	0″·6	− 520
	0″·07	− 61
log cot 1° 23′ 45″·67		1·613 145

When the proportional parts for interval 1′ (i.e. for angles greater than 10°) are given in the middle of an opening, they serve the entire opening. They are given for 6″(1″)10″(10″)50″ only, since those for smaller increments of angle can be obtained by dropping decimals from those printed.

Example. Find log tan 78° 23′ 34″·56.

The difference is 641, whose P.Ps. are not given, but we may adopt 640 or 642 (or a mean between them for increments of 10″ and over).

log tan 78° 23′ (page 87)		0·687 032
P.Ps. from page 86	30″	3205
	4″	427
	0″·5	53
	0″·06	6
log tan 78° 23′ 34″·56		0·687 401

Care is necessary to avoid confusion about the number of figures to be retained in the P.Ps., especially for increments less than 1″. If, however, no decimals of a second appear, the danger of error is much less. For most purposes the above example could be abbreviated thus:

log tan 78° 23′		0·687 032
P.Ps. for	30″	320
	4″	43
	0″·56	6
log tan 78° 23′ 34″·56		0·687 401

Example. Find the angle whose log sin is 9·345 678.

The tabular difference is here 555 (page 88).

Given log sin	9·345 678
log sin 12° 48′	9·345 469
Remainder	209
P.P. for 20″	185
Remainder	24
P.P. for 2″	18
Remainder	6
P.P. for 0″·6	6

Hence the angle is 12° 48′ 22″·6.

Example. Find the angle whose log cos is 9·967 891.

log cos 21° 45′	9·967 927
Given log cos	9·967 891
Remainder	36
P.P. for 40″	34
Remainder	2
P.P. for 2″	2

Hence the angle is 21° 45′ 42″. The tabular difference (51) is here too small to enable even one decimal of a second to be found.

TABLE III—NATURAL TRIGONOMETRICAL FUNCTIONS

(Degrees, Minutes and Seconds)

The derivation of the τ and σ functions in Table IIIA on page 121 has already been explained on page xiii.

Table IIIB (pages 122-166) gives the six principal functions throughout the quadrant at interval 1′, in the usual semi-quadrantal manner.

The user of a calculating machine may modify the procedure previously described (page xi) for decimal interpolation. The M.R. is first pointed off with the number of decimals to be used in the seconds. With this decimal pointing f_0 is multiplied by 100, and instead of setting Δ, set $\Delta/6$ to *one* decimal, with the last digit in the same position as the last digit of f_0. Multiplication by ± 60 produces f_1 in the P.R. The M.R. is then changed to n to give f_n.

Example. Find cot 30° 12′ 34″·5.

On page 152 we find

30° 12′	1·718 172	1149
30° 13′	1·717 023	

Set 1·718 172 and multiply by 100·0. Clear M.R. and S.L. Set 1915 and multiply by −60 to produce 1·717 023. By changing the M.R. to 34·5 we find the desired cot as 1·717 511

TABLE IV—CIRCULAR FUNCTIONS

(*Argument in Radians*)

Table IVA (page 167) gives sines and cosines for $0^r(0^r{\cdot}1)10^r{\cdot}4$ and $10^r(1^r)54^r$. The footnotes on these pages should be studied. The method described avoids interpolation entirely (up to 55^r) if there are only three decimals in x. In any event it provides a completely independent means of checking, if the elimination of accidental errors is vital.

Table IVB (pages 168-169) gives, with a full description of their use, the means of reducing an angle in radians to the first quadrant, so that functions may be found from the main table (Table IVE, on pages 172-203).

Table IVC (page 170) lists various formulæ connected with the circular functions. Similar formulæ for the hyperbolic functions are on pages 204 and 380.

Table IVD (page 171) gives critical tables of the auxiliary functions τ and σ (see page xiii) up to $0^r{\cdot}01$ as well as tabular values for $0^r(0^r{\cdot}001)0^r{\cdot}1$.

Example. Find cot $0^r{\cdot}006\ 54321$.

$$\cot 0^r{\cdot}006\ 54321 = \frac{0{\cdot}999\ 986}{0{\cdot}006\ 54321} = 152{\cdot}828$$

Example. Find x to five significant figures when $\cot x = 123{\cdot}456$.

$$x = \frac{0{\cdot}999\ 978}{123{\cdot}456} = 0^r{\cdot}008\ 0999$$

If six decimals only had been required, it would have sufficed to take the reciprocal of 123·456 or 0·008 100.

Table IVE (pages 172-203) gives the six principal functions throughout the first quadrant and a little beyond (to $1^r{\cdot}6$) at interval $0^r{\cdot}001$. As there is no complementary argument (since $\frac{1}{2}\pi$ is irrational) the opportunity has been taken of putting the most used functions—sine and cosine—nearest to the argument, and the least used functions on the extreme right. The user of this table may normally be expected to be working to six decimals of a radian. Beyond the range of Table IVD, six significant figures have been kept in the cotangents and cosecants of small angles and the tangents and secants of large angles, but at the expense of 4-figure first differences and the necessity for second-difference correction if all the decimals given are required.

TABLE V—EXPONENTIAL AND HYPERBOLIC FUNCTIONS

Table VA (page 204) lists various formulæ connected with the exponential and hyperbolic functions. Further formulæ, and similar formulæ for circular functions, are on pages 170 and 380.

Table VB (page 205) gives critical tables of the auxiliary functions τh and σh (see page xiii) up to 0·01, as well as tabular values for 0(0·001)0·1. Its use is similar to that of the table of τ and σ on page 171, which is described on this page.

Table VC (pages 206-265) gives the ascending and descending exponential and the four principal hyperbolic functions at interval 0·001 up to $x = 3$. It became necessary to drop a decimal from e^x at $x = 2$, but the curtailing of $\sinh x$ and $\cosh x$ was postponed till $x = 2{\cdot}5$. We have not followed the practice of some tables in maintaining a constant number of significant figures in e^{-x}; more decimals, if required, can be found by determining it as $1/e^x$.

Table VD (pages 266-269) continues these six functions at interval 0·01 from $x = 3$ to $x = 5$. Differences are no longer given for the ascending exponential, or for $\sinh x$ and $\cosh x$ (now approaching $\frac{1}{2}e^x$), since they are far from linear at this interval.

Table VE (pages 270-271) continues e^{-x}, $\tanh x$ and $\coth x$ at interval 0·01 to $x = 6$. At that point $\tanh x$ and $\coth x$ are each within 12 units in the sixth decimal of unity, so are completed on page 271 by critical tables. On the same page e^{-x} is continued for $x = 6(0{\cdot}01)7{\cdot}5(0{\cdot}1)12{\cdot}5$, and then its remaining four units in the sixth decimal are extinguished by a short critical table.

Table VF (pages 272-275) resumes the tabulation of e^x, and continues it for $x = 5(0{\cdot}01)10(0{\cdot}1)50$, always with six or seven significant figures. Sinh x and cosh x are not continued beyond $x = 5$. Since

$$\sinh x = \tfrac{1}{2}e^x - \tfrac{1}{2}e^{-x} \text{ and } \cosh x = \tfrac{1}{2}e^x + \tfrac{1}{2}e^{-x}$$

we have ample access to these functions. Beyond $x = 6$, where e^{-x} is only 0·001 and is decreasing, and where three (or fewer) decimals suffice in $\sinh x$ and $\cosh x$, we may use

$$\sinh x = \cosh x = \tfrac{1}{2}e^x$$

It will be noticed that no differences are given for e^x after $x = 3$, but there are several ways in which values for non-tabular arguments may be obtained. Of these the best is probably that in which x is split into two parts a (a tabular argument) and b, so that

$$e^x = e^{a+b} = e^a \times e^b$$

Example. Required e^x for $x = 8{\cdot}765\ 432$.

Let $a = 8{\cdot}76$ and $b = 0{\cdot}005\ 432$. Then $e^{8{\cdot}76}$ is found from page 273 to be 6374·11 and by interpolation from page 206 e^b is 1·005 447. The product of these is 6408·83.

Between $x = 3$ and $x = 10$, where b need not

exceed 0·01, we may avoid the looking up of e^b by using

$$e^b = 1 + b + \tfrac{1}{2}b^2$$

The term $\frac{1}{2}b^2$ (here 15 in the sixth decimal) is formed by squaring b on the machine and halving mentally, since under the conditions specified b^2 cannot exceed 100 units of the sixth decimal. Then set e^a, multiply by $1 + b$, clear the M.R. (only) and multiply by $\frac{1}{2}b^2$. The quantity e^b is never found as such, but as its components have been used as multipliers, the P.R. must show e^x.

It is not essential that a should be large and b small. What is essential is that their sum should be x. It may perhaps be considered that, with this easy means of getting e^x at our disposal, we have tabulated this function extravagantly. But there are four reasons for giving as many values as we have done. The first (and strongest) is that tabular values are often wanted for calculations in which interpolation can be avoided. The second is that the first four or five figures of an exponential table at interval 0·01 may be interpolated linearly. The third is that, at this interval, the looking up and interpolation of e^b can be avoided, as has been shown. Fourthly, the inverse use of this table, when the number of decimals is reduced to the point where the table is linear (i.e. five or six significant figures) is the least laborious way of getting natural logarithms to five decimals of numbers between 10 and 20,000. See the examples in the description of Table VI.

A second method of getting e^x is from

$$\log e^x = Mx = 0{\cdot}434\ 294\ 482x$$

Applying this to the same example as before, $\log e^x = 3{\cdot}806\ 779$, from which $e^x = 6408{\cdot}83$ (page 20).

Thirdly, since $\ln e^x = x$, we may use the table of natural logarithms (Table VI) inversely. Again using the same example,

x (given)	8·765 432
$\ln 10^3$ (Table VI)	6·907 755
Difference = natural log	1·857 677
Corresponding number (page 286)	6·408 83
e^x	6408·83

Obviously to get x from a given value of e^x we use the natural logarithm table directly, with argument e^x, since $\ln e^x = x$.

TABLE VI—NATURAL LOGARITHMS

$$\log = \log_{10} = \text{common logarithm}$$

$$\ln = \log_e = \text{natural logarithm}$$

Since the base e (=2·718 281 828 459 ...) of natural logarithms is irrational, the great advantage of common logarithms, namely that the mantissa is independent of the decimal point, is lost. But as they occur so frequently in mathematical analysis as the result of integrations, and in other ways, provision for obtaining them is essential. A common logarithm may always be converted to a natural logarithm by dividing by the modulus M, or better still by multiplying by its reciprocal, namely 2·302 585 092 994 . . . Conversely a natural logarithm may be converted to a common logarithm by multiplying by the modulus, i.e. by 0·434 294 481 903 ...

Table VI (pages 276-293) enables the natural logarithm of a number to be found directly. The range of direct tabulation is $x = 1(0{\cdot}001)10$, in which ln is positive and varies from 0 to 2·302 ... Any number may be expressed as a number in this range, multiplied by a power of 10. To obtain its natural logarithm, we first express it in this form, then ascertain from Table VI the natural logarithm of the first factor, and add the natural logarithm (which may be negative) of the second factor. The natural logarithms of positive powers of 10 up to the twelfth are given at the foot of each opening.

The last column Δ gives the difference between the last entry on its line and the first entry on the next line. The remaining differences on any line will usually lie between the Δ at the end of their line and that on the line above; hence a subtraction of the end figures of two consecutive logarithms, together with a glance at the Δ's concerned, suffices to determine the appropriate Δ in its entirety.

Example. Find ln 222·792. This must be treated (mentally) as $2{\cdot}22792 \times 10^2$. Then

ln 2·22792 (page 278)	0·801 068 ($\Delta = 449$)
$\ln 10^2$ (page 278)	4·605 170
Sum = ln 222·792	5·406 238

A natural logarithm may, of course, be found by using the exponential table inversely, since each of these functions is the inverse of the other. This is indeed an easy way if not pushed beyond the stage where linear inverse interpolation can be used.

Example. Find ln 456·78. From page 272, dropping a decimal in e^x,

x	e^x	
6·12	454·87	457
6·13	459·44	

from which ln 456·78 = 6·12418.

Since $\ln 10^{-x} = -\ln 10^x$, the natural logarithms of negative powers of 10 must be subtracted. The natural logarithm of a number less than 1 is negative, and so when written will be preceded by a minus sign. There is no point in making the mantissa positive, as there is with common logarithms.

Example. Find ln 0·000 123 456. The number is equivalent to $1{\cdot}23456 \times 10^{-4}$. From page 276,

ln 1·23456	0·210 715 ($\Delta = 810$)
$\ln 10^4$	9·210 340
ln 0·000 123 456	− 8·999 625

For numbers less than 1, but with no cipher after the decimal, the user may prefer the inverse use of our table of e^{-x}, prefixing a minus sign to the value of x thus obtained.

Example. Find ln 0·123 456. From page 247, the value of x when $e^{-x} = 0·123\ 456$ is 2·09187. Hence ln 0·123 456 = −2·09187. It will be noticed that at this point only five decimals are obtainable. The diminishing differences in e^{-x} limit the usefulness of this method if six decimals are required. But it is valuable if it gives sufficient decimals in exploratory calculations. Thus we see at sight that ln 0·000 123 is −9·00 to two decimals.

TABLE VII—INVERSE CIRCULAR AND HYPERBOLIC FUNCTIONS

Any table of integrals will show how frequently these functions arise in integration. It is not so very long since they had to be found by the inverse use of trigonometrical tables, followed by conversion of the resulting degrees, minutes and seconds to a common unit, and then by division by the number of those units in a radian. They could, of course, be found by inverse use of the direct tables of the circular and hyperbolic functions here provided. But their direct tabulation is easier for the user, and is particularly advantageous when the tabular values suffice, as may easily happen when preparing figures for a graph, or when compiling a table whose argument is a circular or a hyperbolic function. We have followed the British practice of writing $\sin^{-1} x$, etc., rather than that in vogue in some places of writing arc sin x, etc.

$$\sin^{-1} x + \cos^{-1} x = \tan^{-1} x + \cot^{-1} x = \sec^{-1} x + \text{cosec}^{-1} x = \tfrac{1}{2}\pi$$

Since the above holds, it was not strictly necessary to tabulate more than three inverse circular functions. But here again the user's convenience dictates that we should do so.

In Volume II of *Chambers's Six-Figure Mathematical Tables* we have given extended and interpolable tables, with differences, but the present much shorter table suffices to stimulate interest in these functions, for many of the purposes mentioned above and for rough checks. For non-tabular values, inverse interpolation of our direct tables should be used; for this reason, as well as because of the non-linearity of this illustrative table, differences are not given.

Pages 294-295 give the inverses of those functions that can have values less than 1. The functions $\sin^{-1} x$, $\cos^{-1} x$, $\tanh^{-1} x$ and $\text{sech}^{-1} x$ change very rapidly as x approaches 1. For this reason the first three are given on page 296 at interval 0·001 when x is greater than 0·9, and the last when x is greater than 0·95.

Pages 297-305 give the inverses of those functions that can have values greater than 1. Those that move rapidly when x is just greater than 1 are $\sec^{-1} x$, $\text{cosec}^{-1} x$, $\cosh^{-1} x$ and $\coth^{-1} x$. Page 297 gives these functions at interval 0·001 up to $x = 1·01$, except that $\coth^{-1} x$ ceases at 1·05.

TABLE VIII—POWERS, ROOTS, FACTORS AND RECIPROCALS

Table VIII (pages 306-345) gives the square, cube, fourth power, square root, reciprocal of square root, cube root, reciprocal, factorial, log factorial, and factors of each number up to 1000. For numbers up to 100, fifth powers and fourth and fifth roots also are given, as well as the reciprocals of squares or the squares of reciprocals. The functions of each number spread across a full opening.

The figures in each positive power are printed in groups, each of which contains as many figures as the exponent of the power; thus the cubes are printed in triads. The reason for this is that each movement of the decimal point by one position in x causes a movement of a positions in x^a. Thus $12·3^3$ is 1 860·867 and $1·23^4$ is 2·2888 6641. The value of this grouping will be even more apparent when finding roots by the inverse use of the table.

It will be noticed that no differences are given in this table, which is really a table for the integers concerned, or those integers multiplied by powers of 10. The two square root columns provide both $\sqrt{x}$ and $\sqrt{10x}$, and the three cube root columns give $\sqrt[3]{x}$, $\sqrt[3]{10x}$, and $\sqrt[3]{100x}$. Thus they enable the roots of any *sequence* of three figures to be found, regardless of the position of the decimal point. For instance, the square root of 1·23 (page 310) is 1·109 0537 and that of 12·3 is 3·507 1356. The cube root of 0·123 (page 311) is 0·497 3190; that of 0·0123 is 0·230 83502; and that of 0·00123 is 0·107 14413. The determination of roots for non-tabular values of x is considered later in this section.

Reciprocals of square roots (which are the same as square roots of reciprocals) will often turn a division into a multiplication. In statistical work they are frequently required for integers less than 1000, for which we provide completely. As with square roots, the position of the decimal point may be varied. Thus, from page 310, $\dfrac{1}{\sqrt{1·23}}$ is 0·901 670 and $\dfrac{1}{\sqrt{12·3}}$ is 0·285 133.

The chief application of reciprocals is in machine calculation, for converting division to multiplication, since $\dfrac{a}{b} = a \times \dfrac{1}{b}$. This is particularly advantageous when b is constant for several values of a, since then its reciprocal is set and multiplied by the various values of a in succession.

No.	%
47	22·2
58	27·4
2	0·9
29	13·7
76	35·8
212	100·0

Example. What percentage (to 0·1) of their total is each of the numbers in the left-hand column alongside? The sum of the numbers is 212, whose reciprocal (page 315) has the significant figures 471 698, which are set on the machine. It is recommended that this be multiplied firstly by 212 itself, thus checking the setting and fixing the position of the decimal point. Multiplication by the various numbers gives the percentages shown. In this case the total happens to be precisely 100·0, but accumulation of rounding-off errors could produce a small discrepancy either way; unless this exceeds $\sqrt{n}$ units of the last decimal, where n is the number of items, no error need be suspected.

A large part of the table of factorials, namely that for x greater than 200, is here published for the first time. 123!, for instance, is to be read (page 311) as $1{\cdot}214\,630 \times 10^{205}$.

This table also gives all the prime factors of all numbers up to 1000. Prime numbers are shown in the same bold type as the argument, and are thus a help in locating the factors of adjacent numbers.

Roots

If we know an approximation A to the pth root of a number N and find a quotient Q from

$$Q = \frac{N}{A^{p-1}}$$

we shall have another approximation to the desired root, from which a still better approximation B may be obtained. If $A = N^{1/p} + \epsilon$, where ϵ is small, we find, by expanding $NA^{-(p-1)}$ to terms of the second order in ϵ,

$$Q = N^{1/p} - (p-1)\epsilon + \frac{p(p-1)}{2A}\epsilon^2$$

From this we easily deduce

$$B = \frac{(p-1)A + Q}{p} = N^{1/p} + \frac{p-1}{2A}\epsilon^2$$

in which the term in ϵ has vanished.

This suggests a general method for finding pth roots. First find an approximation A, raise it to the power $p-1$, divide it into N, and take the weighted mean of $p-1$ times the approximation and the quotient. An equivalent process is

$$B = A + \frac{Q - A}{p}$$

Roughly speaking, this process yields twice as many correct figures as there are correct in the original approximation A, but from the presence of A in the denominator of the error term we see that it yields better results the larger the first figure of the approximation.

If we have a table showing, in successive columns, x, x^2, x^3, x^4, x^5, etc.

and find the first approximation A to a pth root from the column x after entering the column x^p with the number N, the desired divisor A^{p-1} is taken, without interpolation, from the same line x and the preceding column x^{p-1}. This generalised process will be used below when dealing with various roots.

Square Roots

How the approximation is found for a square root is immaterial. The number is divided by the approximation, and the mean of divisor and quotient taken. The algebra above shows that our process always tends to give an improved root that is too large; hence in getting a $2n$-figure root from an n-figure approximation, if the quotient ends in an odd figure, it is decreased by a unit before dividing by 2. If getting fewer than $2n$ figures, the quotient should be taken to the nearest *even* figure.

The reader will be familiar with the rules for pointing off the digits of the radicand in pairs from the decimal point in whichever direction the first significant figure occurs. This is still advisable here (although generally it need be done only mentally) for two reasons:

(1) To determine the first figure of the root and so avoid getting $\sqrt{10}$ or $\sqrt{0{\cdot}1}$ times the root.

(2) To determine the position of the decimal point in the root.

Example. Find the square root of 12·3456. By inspection, the root begins with 3·. From the column of 300's on page 320,

x	x^2
351	12 32 01
352	12 39 04

The approximation yielded by inverse use of the tables is therefore 3·51. Dividing,

$$\frac{12{\cdot}3456}{3{\cdot}51} = 3{\cdot}51726$$

We now have confirmation that 3·51 is a good approximation, and to get the desired root we have only to divide the new figures by 2 (making due allowance if the last figure of the divisor is replaced by a greater or a smaller figure in the quotient) to get the root 3·51363.

Now suppose we had started with an approximation 3·52—say from a slide rule.

$$\frac{12{\cdot}3456}{3{\cdot}52} = 3{\cdot}50727$$

from which, dropping the terminal 7 to 6, the root is once more 3·51363. Naturally, the better the original approximation, the more reliable is the answer, especially as its error is proportional to the square of the error of the approximation.

Still another method of doing square roots with the aid of a table of squares is to interpolate inversely. The difference between the squares of two consecutive integers is the sum of those integers. The second difference is always 2, which may be neglected, but the method will not yield (in general) more new figures in the root than there are figures in the first difference and may yield one fewer if the first figure of the first difference is small. Reverting to our first example, for 12·3456, we have $\Delta = 351 + 352 = 703$. The usual process of inverse interpolation yields 3·51363, as before.

A method often taught by vendors of calculating machines, which involves subtraction of various consecutive odd numbers, is not recommended. On the other hand the user of the first method here described must exercise caution when taking the mean of the divisor and quotient, in order to avoid errors of 5, 50, 500 or 5000 units in his answer.

Cube Roots

First triad of N	Root begins with
000	
	1
007	
	2
026	
	3
063	
	4
124	
	5
215	
	6
342	
	7
511	
	8
728	
	9
999	

The principles here are similar to those for square roots. The division of the number into triads from the decimal point in the direction of the first significant figure of the radicand provides the first figure of the cube root (with the aid of the critical table alongside, or the first ten lines of x^3 on page 306), as well as the position of the decimal point.

Example. Find the cube root of 123·456. Since this begins with 4·, we turn to the 400's on page 324 to find $A = 4{\cdot}98$ and $A^2 = 24{\cdot}8004$. A check that the user has selected the right portion of the table is that the x^3 selected begins with the first significant triad of the radicand and that this triad is isolated by a space from subsequent triads. Now

$$\frac{123{\cdot}456}{24{\cdot}8004} = 4{\cdot}97798$$

In forming (mentally) the mean of twice 4·98 and this quotient, the computer first writes 4·97 with a 2 to carry, and then continues dividing, obtaining 4·97933.

There is no satisfactory machine method of calculating cube roots that is independent of the help of tables of some kind.

Fourth Roots

A fourth root is of course the square root of a square root. The first significant tetrad of the radicand must be found to fix the first digit of the root (from the first ten lines of x^4 on page 306) and the position of the decimal point.

The number N is divided by A^3 to get Q, and then the improved root B is given by

$$B = \tfrac{1}{4}(3A+Q) = 0{\cdot}75A + 0{\cdot}25Q$$

If the first tetrad does not exceed 0016 (i.e. if the first figure of the root is 1) the number may be multiplied by $625 = 5^4$. Taking A and Q from $625N$

$$B = 0{\cdot}15A + 0{\cdot}05Q$$

Example. Find the fourth root of 12·34567.

12·34567 × 625 = 7716·04375, whose root begins with 9.

Turning to the table of 900's, we find, from page 342, $A = 9{\cdot}37$, and $A^3 = 822{\cdot}656\ 953$. Hence $Q = 9{\cdot}37942$ and $B = 1{\cdot}87447$.

Similarly, if the first figure of the root is 2, 3 or 4, A and Q may be found from $16N$, and then

$$B = 0{\cdot}375A + 0{\cdot}125Q$$

If the first figure of the root is 5 or more, no factor is needed to get a reliable sixth significant figure in the root. If only five figures are required, there is no need for any factor, whatever the number.

Higher Roots

Higher roots are perhaps best done by logarithms, although some are possible by combinations of the methods already illustrated; thus a sixth root is the cube root of the square root, or the square root of the cube root.

The British Association *Power Tables* (Volume IX, 1940) gives powers up to the 12th of integers up to 1100, and so may be used in the way described for getting fifth to twelfth roots to six significant figures.

Fractional Powers

The best way of obtaining fractional powers, e.g. $x = a^b$, where b is not a simple integer or its reciprocal, is the combined use of logarithms and a machine, together with the equation

$$\log x = b \log a$$

There is no satisfactory way of evaluating say $0{\cdot}1234^{0{\cdot}5678}$ by machine alone.

The table below shows the limits of x (always positive) for all ranges of a (always positive) and b.

b	a less than 1	a greater than 1
$-\infty$	∞	0
-1	$1/a$	$1/a$
0	1	1
$+1$	a	a
$+\infty$	0	∞

For example, if a is less than 1, and b is negative but less than 1, x lies between 1 and $1/a$. Similarly, if a is greater than 1, and b is positive and greater than 1, x is greater than a. This table may assist in avoiding or detecting confusion with negative logarithms, especially when working wholly by logarithms and the equation

$$\log \log x = \log \log a + \log b$$

TABLE IX—PRIME NUMBERS

Table IX (pages 346-347) gives the 1540 prime numbers (including 1) up to 12919. There are 55 in each column, so the user who wishes to count prime numbers in certain ranges may head the columns 0, 55, 110 ... 825 (page 347) ... 1485.

The factors of a composite number between 1000 and 12919 can be found, in the absence of any better method, by trying as divisors prime numbers less than the square root of the number until a factor is found.

TABLES XA AND XC
INTERPOLATION COEFFICIENTS

Linear interpolation with a calculating machine, both direct and inverse, has already been described on page xi. These tables make provision for non-linear interpolation, with special emphasis on cases where the effect of fourth differences is either negligible or small.

The notation used is:

Tabular values of function	Differences				
	First	Second	Third	Fourth	Fifth
f_{-2}					
	$\Delta'_{-1\frac{1}{2}}$				
f_{-1}		Δ''_{-1}			
	$\Delta'_{-\frac{1}{2}}$		$\Delta'''_{-\frac{1}{2}}$		
f_0		Δ''_0		Δ^{iv}_0	
	$\Delta'_{\frac{1}{2}}$		$\Delta'''_{\frac{1}{2}}$		$\Delta^{v}_{\frac{1}{2}}$
f_1		Δ''_1		Δ^{iv}_1	
	$\Delta'_{1\frac{1}{2}}$		$\Delta'''_{1\frac{1}{2}}$		
f_2		Δ''_2			
	$\Delta'_{2\frac{1}{2}}$				
f_3					

A difference in general is denoted by Δ. The roman index is the order of the difference, and the suffix is its level; thus all differences on the same level have the same suffix. With the modern preference for central differences, this old-established and common-sense notation appears to be superior to those in which advancing differences are denoted by Δ, backward differences by ∇, central differences by δ, and the order of the difference by an arabic figure that is easily mistaken for a power. It has also the advantage of an easy notation for a mean difference; thus

$$\Delta'_1 = \tfrac{1}{2}(\Delta'_{\frac{1}{2}} + \Delta'_{1\frac{1}{2}}) \text{ or } \Delta''_{\frac{1}{2}} = \tfrac{1}{2}(\Delta''_0 + \Delta''_1)$$

Differences are always to be taken in the sense defined by

$$\Delta^p_q = \Delta^{p-1}_{q+\frac{1}{2}} - \Delta^{p-1}_{q-\frac{1}{2}}$$

The two function values between which we are interpolating are always given the floating suffixes 0 and 1. The fraction of the interval between them at which the desired interpolate lies is denoted by n, and the interpolate itself by f_n.

In this volume only the formula of Bessel—the simplest for ordinary use—is described. Other formulæ are covered in Volume II of *Chambers's Six-Figure Mathematical Tables*.

The Bessel formula is

$$\begin{aligned} f_n &= f_0 + n\Delta'_{\frac{1}{2}} + B''(\Delta''_0 + \Delta''_1) + B'''\Delta'''_{\frac{1}{2}} \\ &\quad + B^{iv}(\Delta^{iv}_0 + \Delta^{iv}_1) \\ &= f_0 + n\Delta'_{\frac{1}{2}} + B''(M''_0 + M''_1) + B'''\Delta'''_{\frac{1}{2}} \end{aligned}$$

The coefficients required are symbolised by B (for Bessel) with roman indices for the order and suffixes for the level, as for differences; where there is no ambiguity about the level the suffix is sometimes dropped. They are defined thus:

$$B'' = \frac{n(n-1)}{2 \times 2!}$$

$$B''' = \frac{n(n-1)(n-\frac{1}{2})}{3!}$$

$$B^{iv} = \frac{(n+1)n(n-1)(n-2)}{2 \times 4!}$$

$$B^{v} = \frac{(n+1)n(n-1)(n-2)(n-\frac{1}{2})}{5!}$$

It will be noted that the even-order coefficients contain an extra factor $\frac{1}{2}$ in their definition, in order to avoid taking the mean of two adjacent even differences.

In rigorous interpolation, all differences whose effect may amount to half a unit of the last decimal must be taken into account. As a practical working guide, the limits below which the effect of a difference may be considered negligible are shown alongside.

Difference		Limit
Second	Δ''	4
Third	Δ'''	60
Fourth	Δ^{iv}	20
Fifth	Δ^{v}	500
Sixth	Δ^{vi}	100

A further saving in labour and printing is effected by the device known as the throw-back, which has been named and popularised by the present author. The terms

$$B''(\Delta''_0 + \Delta''_1) + B^{iv}(\Delta^{iv}_0 + \Delta^{iv}_1)$$

in Bessel's formula may be rewritten

$$\begin{aligned} &B''\{\Delta''_0 + \Delta''_1 - C(\Delta^{iv}_0 + \Delta^{iv}_1)\} \\ &\quad + (B^{iv} + CB'')(\Delta^{iv}_0 + \Delta^{iv}_1) \end{aligned}$$

The coefficient $B^{iv} + CB''$, otherwise written as T^{iv}, varies with the adopted value of C. Choosing the value that equalises the maximum positive and negative values of T^{iv}, we find $C = 0{\cdot}184$, for which T^{iv} lies between $\pm 0{\cdot}00023$. Hence, if the fourth differences do not exceed 1000, the effect of neglecting entirely the term in T^{iv} will not exceed half a unit of the last decimal. Thus with this limitation we may write

$$M'' = \Delta'' - 0{\cdot}184\,\Delta^{iv}$$

and in Bessel's formula replace Δ'' by M'', neglecting Δ^{iv} entirely. If the table maker tabulates M'' instead of Δ'', he confers a great boon on the user and saves valuable printing space.

Chambers's Six-Figure Mathematical Tables, Volume II, gives, in a critical table, multiples of 0·184. For fifth and sixth differences less than 10,000 we may use

$$M''' = \Delta''' - 0{\cdot}108\,\Delta^{v}$$

$$M^{iv} = \Delta^{iv} - 0{\cdot}207\,\Delta^{vi}$$

Table XA (page 348) gives, in critical tables, B'' to two decimals and B'', B''' and B^{iv} to three decimals, which suffice for differences less than 1000. For more decimals (if necessary) the user is referred to Volume II of *Chambers's Six-Figure Mathematical Tables*.

Table Xc (pages 350-351) contains a series of critical tables giving (without signs) the second difference correction $B''(\Delta''_0 + \Delta''_1)$ for values of $\Delta''_0 + \Delta''_1$ not exceeding 200, i.e. for values of Δ'' with not more than two figures. There is a table for every third value of $\Delta''_0 + \Delta''_1$; it suffices always to use the nearest table.

The number of decimals in an interpolation coefficient should not be less than the number of significant figures in the difference by which it is multiplied.

Direct Interpolation

The following examples have been constructed from values in this volume at wide intervals, in order that the results may be compared with values interpolated from linear tables.

Example A. Find sin 12° 25′ 55″·6. Here n is $\frac{1555 \cdot 6}{3600} = 0 \cdot 4321$, so we must interpolate for 12°·4321.

x	$f = \sin x$	Δ'	Δ''	Δ'''
°				
11	0·190 809			
		+17 103		
12	0·207 912		−64	
		+17 039		−4
13	0·224 951		−68	
		+16 971		
14	0·241 922			

Here Δ''' is negligible, and two decimals in B'' will suffice. From page 348, $B'' = -0 \cdot 06$, and since $\Delta''_0 + \Delta''_1 = -132$,

f_0	0·207 912
0·4321 × 17039	7 363
−0·06 × −132	+8
$f_n = \sin 12° \cdot 4321$	0·215 283

This agrees with direct interpolation from page 88 for 12° 25′·93. The second difference correction could have been obtained from the table for 132 on page 350. The calculating machine user would, of course, write nothing but the interpolate.

Example B. Find e^x for $x = 2 \cdot 78901$.

x	e^x	Δ'	Δ''	Δ'''	Δ^{iv}
		+	+	+	+
2·5	12·1825				
		12812			
2·6	13·4637		1348		
		14160		141	
2·7	14·8797		1489		16
		15649		157	
2·8	16·4446		1646		16
		17295		173	
2·9	18·1741		1819		
		19114			
3·0	20·0855				

Here Δ^{iv} is negligible and four decimals are required in B'', which is computed from its formula as −0·0245. B''' is taken from page 348.

f_0	14·8797
·8901 × 15649	1·39292
−·0245 × 3135	−768
−·006 × 157	− 9
Sum $= f_n = e^x$	16·2648

Interpolation from page 261 yields 16·2649.

Example C. Find sin 1ʳ·234 567 from the following values at interval 0ʳ·1, taken from page 167.

x	$\sin x$	Δ'	Δ''	Δ'''	Δ^{iv}	Δ^{v}
r		+	−	−	+	
1·0	0·841 471					
		49 736				
1·1	0·891 207		8904			
		40 832		409		
1·2	0·932 039		9313		95	
		31 519		314		−1
1·3	0·963 558		9627		94	
		21 892		220		
1·4	0·985 450		9847			
		12 045				
1·5	0·997 495					

As $\Delta''_0 + \Delta''_1$ has five significant figures we must here (in the absence of more extended tables) calculate B'' on the machine as −0·05655. B''' and B^{iv} are taken from page 348.

f_0	0·932 039
·34567 × +31519	+10 8952
−·05655 × −18940	+ 1 0711
+·006 × − 314	− 19
+·010 × + 189	+ 19
sin 1ʳ·234 567	0·944 005

Interpolation from page 196 also yields 0·944 005.

Using the throw-back,

$$M''_0 + M''_1 = -18940 - 35 = -18975$$

Hence

f_0	0·932 039
$n\Delta'$	+10 8952
−·05655 × −18975	+ 1 0730
+·006 × − 314	− 19
sin 1ʳ·234 567	0·944 005

Inverse Interpolation

To find the value of n corresponding to a given f_n when second or higher order differences have to be taken into account involves the solution of an equation of high degree. In practice, however, the small coefficients of the high-order differences permit convergent iterative processes to be used. That now presented was first published by the present writer in 1936, and has become standard practice in many computing circles. As here described it is limited to simple cases that can be dealt with by the tables in this volume.

Bessel's formula may be rewritten

$$f_n - B''(\Delta''_0 + \Delta''_1) = f_0 + n\Delta'_{\frac{1}{2}}$$

If we omit the terms in Δ'', we may determine n

by linear inverse interpolation, as described on page xii. Strictly speaking, we cannot introduce the omitted terms till n is known. But as B'' is not sensitive to n, we may first obtain an approximate n. We then introduce $-B''(\Delta''_0+\Delta''_1)$ with B'' based on the approximate n. This will lead to a revised n and hence a revised B''. This iterative process is continued until no further change occurs in B''.

As an example, we reverse the first of those previously given, remembering that we cannot divide the tabular interval into more parts than the number forming Δ' (in units of the last decimal) without introducing uncertainty in the last decimal of n.

Example A′. Find $\sin^{-1} 0{\cdot}215\ 283$ to four decimals of a degree. Here we must solve

$$215\ 283 + 132B'' = 207\ 912 + 17\ 039n$$

Using the calculating machine in the manner already described, we find $n = 0{\cdot}43$ to two decimals. From page 350, $132B'' = -8$, so that the left-hand side becomes 215 275 and n becomes 0·4321, which has the same value of $132B''$, so the process is completed, and the desired angle is 12°·4321.

We now examine the criteria for the rejection of differences. First write

$$n = \frac{f_n - f_0}{\Delta'_{\frac{1}{2}}} - \frac{B''(\Delta''_0+\Delta''_1)}{\Delta'_{\frac{1}{2}}} - \frac{B'''\Delta'''_{\frac{1}{2}}}{\Delta'_{\frac{1}{2}}} - \frac{B^{iv}(\Delta^{iv}_0+\Delta^{iv}_1)}{\Delta'_{\frac{1}{2}}}$$

Then, if we require the effect of a neglected difference to be less than $\frac{1}{2}$ unit of the Dth decimal of n, we see that

$$\text{Mean } \Delta'' \text{ must not exceed } \quad 4 \times 10^{-D}\Delta'_{\frac{1}{2}}$$
$$\Delta''' \text{ must not exceed } \quad 60 \times 10^{-D}\Delta'_{\frac{1}{2}}$$

If the differences are too large to prevent the processes here described for direct and inverse interpolation being applied, reference must be made to Volume II of *Chambers's Six-Figure Mathematical Tables*.

TABLE XB
NUMERICAL DIFFERENTIATION

Table XB (page 349) gives the coefficients used for determining numerically the derivatives of a tabulated function.

Example. From the data of *Example C* on the previous page, find the first and second derivative of $f = \sin x$ at $x = 1^r{\cdot}2$.

$$0{\cdot}1f' = \tfrac{1}{2}10^{-6}\{40832+31519+\tfrac{1}{6}(409+314)\}$$
$$f' = 0{\cdot}36236$$

This is the value of $\cos 1^r{\cdot}2$ correct to five decimals. As there are only five significant figures in Δ', we cannot (with the data of the example) get more in f'.

$$0{\cdot}01f'' = 10^{-6}(-9313-\tfrac{1}{12}95)$$
$$f'' = -0{\cdot}9321$$

The correct value is $-\sin 1^r{\cdot}2$ or $-0{\cdot}9320$. Using the formulæ for derivatives from five tabular points yields precisely the same results.

TABLE XD—NUMERICAL INTEGRATION OR QUADRATURE

Table XD (pages 352 - 353) contains the formulæ that constitute the necessary working tools for determining definite integrals from given numerical values of an integrand. The text with which the formulæ are interwoven should make them intelligible to those who do not wish to be overburdened with theory, because their need for quadrature is occasional or incidental only.

To illustrate the various formulæ, the following integral has been evaluated:

$$I = \int_0^{1{\cdot}2} \frac{dx}{1+x^2}$$

using the values of the integrand given in the table at the top of the opposite page.

The true value of this integral, here denoted by T, is $\tan^{-1} 1{\cdot}2 = 0{\cdot}876\ 058$. The values yielded by the various formulæ are:

Formula	w	n	I	$T-I$
Trapezoidal	1·2	1	0·845 902	30156
	0·6	2	0·864 127	11931
	0·4	3	0·870 697	5361
	0·3	4	0·873 039	3019
	0·2	6	0·874 715	1343
	0·1	12	0·875 722	336
Simpson	0·6	2	0·870 202	5856
	0·3	4	0·876 009	49
	0·2	6	0·876 054	4
	0·1	12	0·876 058	0
Three-eighths	0·4	3	0·873 797	2261
	0·2	6	0·876 038	20
	0·1	12	0·876 058	0
4-strip	0·3	4	0·876 396	−338
	0·1	12	0·876 058	0
6-strip	0·2	6	0·876 050	8
	0·1	12	0·876 058	0
Weddle	0·2	6	0·876 067	−9
	0·1	12	0·876 058	0

The weakness of the trapezoidal rule is abundantly evident, as is also the small reward (with this formula) for increasing the number of ordinates. The simple Weddle integral at interval 0·1 is the same as that of the 6-strip formula; that at interval 0·2 (where Δ^{vi} is +12103) requires a correction of

$$-\frac{0{\cdot}2}{140} \times 12103 = -17$$

units of the last decimal. Actually this interval is too wide for accurate integration without help from ordinates outside the range of integration.

x	$\frac{1}{1+x^2}$	Δ'	Δ''	Δ'''	Δ^{iv}	Δ^{v}	Δ^{vi}	Δ^{vii}	Δ^{viii}
	+								
−0·3	0·917 431								
		+44 107							
−0·2	0·961 538		−15 546						
		+28 561		−3114					
−0·1	0·990 099		−18 660		+1972				
		+ 9 901		−1142		+312			
0·0	1·000 000		−19 802		+2284		−624		
		− 9 901		+1142		−312		+141	
0·1	0·990 099		−18 660		+1972		−483		+217
		−28 561		+3114		−795		+358	
0·2	0·961 538		−15 546		+1177		−125		− 48
		−44 107		+4291		−920		+310	
0·3	0·917 431		−11 255		+ 257		+185		−162
		−55 362		+4548		−735		+148	
0·4	0·862 069		− 6 707		− 478		+333		−174
		−62 069		+4070		−402		− 26	
0·5	0·800 000		− 2 637		− 880		+307		− 79
		−64 706		+3190		− 95		−105	
0·6	0·735 294		+ 553		− 975		+202		− 14
		−64 153		+2215		+107		−119	
0·7	0·671 141		+ 2 768		− 868		+ 83		+ 46
		−61 385		+1347		+190		− 73	
0·8	0·609 756		+ 4 115		− 678		+ 10		+ 33
		−57 270		+ 669		+200		− 40	
0·9	0·552 486		+ 4 784		− 478		− 30		+ 31
		−52 486		+ 191		+170		− 9	
1·0	0·500 000		+ 4 975		− 308		− 39		+ 4
		−47 511		− 117		+131		− 5	
1·1	0·452 489		+ 4 858		− 177		− 44		
		−42 653		− 294		+ 87			
1·2	0·409 836		+ 4 564		− 90				
		−38 089		− 384					
1·3	0·371 747		+ 4 180						
		−33 909							
1·4	0·337 838								
...									
1·6	0·280 899								
1·8	0·235 849								

If D is the difference neglected:

Formula	w	D	I	$T-I$
Gregory	0·2	Δ'	0·874 715	1343
		Δ''	0·875 577	481
		Δ'''	0·875 922	136
		Δ^{iv}	0·876 080	−22
		Δ^{v}	0·876 139	−81
		Δ^{vi}	0·876 105	−47
		Δ^{vii}	0·876 050	8
	0·1	Δ'	0·875 722	336
		Δ''	0·875 995	63
		Δ'''	0·876 053	5
		Δ^{iv}	0·876 061	−3
		Δ^{v}	0·876 059	−1
		Δ^{vi}	0·876 058	0
Gauss	0·2	Δ'	0·874 715	1343
		Δ'''	0·876 066	−8
		Δ^{v}	0·876 059	−1
		Δ^{vii}	0·876 058	0
	0·1	Δ'	0·875 722	336
		Δ'''	0·876 058	0

The improvement brought about by having ordinates beyond the range of integration is amply demonstrated; whereas the Gregory formula is hunting for the true value of the integral, the Gauss formula converges rapidly to it. Comparison of the results obtained by taking $w = 0·2$ and 0·1 illustrates the advantage of a smaller interval, especially when no extra ordinates are known.

TABLE XI—CONVERSION TABLES

The table on page 354 enables degrees and any of the usual divisions thereof to be converted to radians. As a safeguard only, seven decimals of a radian are given, so that, after several items have been added, the rounded-off sixth decimal may still be reliable. The angles at the foot of the first column have been so chosen that they do not contribute to the rounding-off error, as their radian equivalents are correct to less than a unit of the tenth decimal.

The radian equivalent of the first pair of decimals of an angle expressed in degrees and decimals is found by rounding off the last two decimals of the equivalent for that number of degrees, and moving the remaining decimals two places to the right. Thus $0^\circ{\cdot}12 = 0^{r}{\cdot}002\ 0944$. Similarly for the second pair we drop the last four decimals, i.e. round off at the break in the printed equivalent.

The equivalents of minutes and seconds have been extended to 100 in order to make provision for decimals of these units.

Example. Convert $12345^\circ{\cdot}67891$ to radians.

First note that $12345 = 10655 + 1690$.

°	r
10655	185·964 8318
1600	27·925 2680
90	1·570 7963
0·67	11 6937
0·00 89	1553
0·00 00 1	2
12345·67891	215·472 745

Example. Convert $12^\circ\ 34'\ 56''{\cdot}78$ to radians.

12°	$0^{r}{\cdot}209\ 4395$
34′	9 8902
56″	2715
0″·78	38
12° 34′ 56″·78	$0^{r}{\cdot}219\ 605$

The table on page 355 converts radians to degrees, minutes and seconds.

Example. Convert $12^{r}{\cdot}345\ 678$ to degrees, minutes, seconds and decimals.

r	° ′ ″
12	687 32 57·7
0·34	19 28 50·0
0·00 56	19 15·1
0·00 00 78	16·1
12·345 678	707 21 18·9

It must be remembered that a result produced by adding four, five or six lines, each of which is correct to half a unit in the last decimal, may be as much as two units in error in that decimal.

In astronomical work, where hour angles are measured in time, conversions from time to arc and (less frequently) from arc to time arise. Page 356 provides for the former conversion, which is illustrated.

The other conversion frequently arising in astronomical determinations of latitude, longitude or azimuth is from mean to sidereal time; the converse conversion is seldom required in modern practice. The so-called acceleration of sidereal time on mean time is given and illustrated.

The tables on page 357 for converting from arc to time give three means of converting the seconds of arc:

(1) Conversion of exact seconds of arc to two decimals of a second of time.

(2) Exact equivalents of multiples of 3″, followed by a critical table for converting arcs up to 3″ to time with two decimals of a second of arc.

(3) A critical table for converting angles to $0''{\cdot}1$ to time to $0^{s}{\cdot}1$.

The user will be able to select from these alternatives according to his requirements.

Example. Convert $76^\circ\ 54'\ 32''{\cdot}1$ to time to $0^{s}{\cdot}1$.

	h	m	s
76°	5	04	
54′		3	36
32″·1			2·1
76° 54′ 32″·1	5	07	38·1

TABLE XII—PROPORTIONAL PARTS

These 20 pages (358-377) constitute a multiplication table up to 999×9 that can be used for multiplication or division by 2- or 3-figure numbers.

FORMULÆ

On page 380 are given lists of formulæ for circular and hyperbolic functions, with comparable formulæ or groups of formulæ placed near one another for easy comparison. These are followed (pages 381-384) by formulæ connected with derivatives and integrals and by a number of useful expansions in series. Other formulæ connected with circular and hyperbolic functions are given on pages 170 and 204.

DIFFERENCES IN ITALIC

The printing of differences in italic is a warning that the effect of second differences is not negligible. See also page xi.

TRIGONOMETRICAL FUNCTIONS IN THE DIFFERENT QUADRANTS

The functions of first quadrant angles here tabulated enable angles in any quadrant to be obtained. To obtain a function in any other quadrant, first subtract the multiple of 90° that

Quadrant	I	II	III	IV
Range	0°–90°	90°–180°	180°–270°	270°–360°
sin	+sin	+cos	−sin	−cos
cos	+cos	−sin	−cos	+sin
tan	+tan	−cot	+tan	−cot
cot	+cot	−tan	+cot	−tan
sec	+sec	−cosec	−sec	+cosec
cosec	+cosec	+sec	−cosec	−sec

marks the beginning of that quadrant. The table above then shows, for each of the six principal functions, the function of the remaining first quadrant angle that must be looked up, and the sign that must be prefixed. Thus sin 100° = +cos 10°; sin 200° = −sin 20°; sin 300° = −cos 30°. Angles should never be subtracted from a higher multiple of 90°; this is unnecessary and dangerous.

When finding an angle in a quadrant other than the first from a given function, we consult the trigonometrical tables for the function named in the table above (which may be the given function, or its co-function) to find an angle less than 90°, which is then added to the multiple of 90° that marks the beginning of the desired quadrant. Thus $\tan^{-1} - 0{\cdot}123\,456$, if in the fourth quadrant, is $270° + \cot^{-1} 0{\cdot}123\,456$. As before, methods that involve subtraction of an acute angle from a higher multiple of 90° should be avoided.

The user who wishes to be independent of this table will have no difficulty in remembering that when 0° or 180° (i.e. an even multiple of 90°) is to be subtracted or added, there is no change of function; when 90° or 270° (i.e. an odd multiple of 90°) is to be subtracted or added, the function is replaced by its co-function. A function changes sign when it passes through zero or infinity, but not at a maximum or minimum. A diagram that may help for giving signs quickly is shown alongside. The other three functions are reciprocals of those shown; every function has the same sign as its reciprocal.

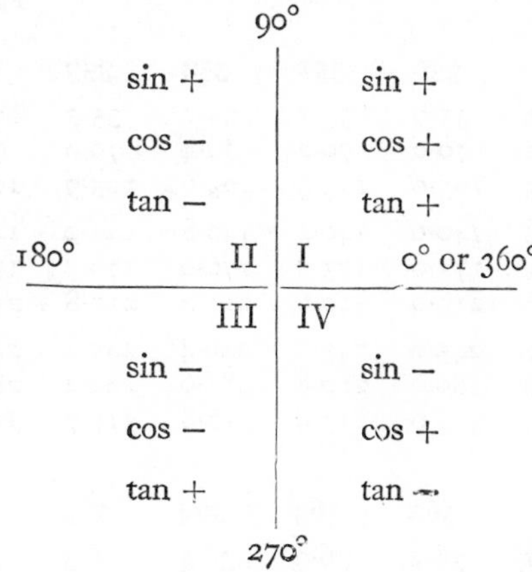

TABLE I—LOGARITHMS OF NUMBERS

	0	1	2	3	4	5	6	7	8	9	Δ
100	000 000	000 434	000 868	001 301	001 734	002 166	002 598	003 029	003 461	003 891	430
101	04 321	04 751	05 181	05 609	06 038	06 466	06 894	07 321	07 748	08 174	426
102	08 600	09 026	09 451	09 876	10 300	10 724	11 147	11 570	11 993	12 415	422
103	12 837	13 259	13 680	14 100	14 521	14 940	15 360	15 779	16 197	16 616	417
104	17 033	17 451	17 868	18 284	18 700	19 116	19 532	19 947	20 361	20 775	414
105	021 189	021 603	022 016	022 428	022 841	023 252	023 664	024 075	024 486	024 896	410
106	25 306	25 715	26 125	26 533	26 942	27 350	27 757	28 164	28 571	28 978	406
107	29 384	29 789	30 195	30 600	31 004	31 408	31 812	32 216	32 619	33 021	403
108	33 424	33 826	34 227	34 628	35 029	35 430	35 830	36 230	36 629	37 028	398
109	37 426	37 825	38 223	38 620	39 017	39 414	39 811	40 207	40 602	40 998	395
110	041 393	041 787	042 182	042 576	042 969	043 362	043 755	044 148	044 540	044 932	391
111	45 323	45 714	46 105	46 495	46 885	47 275	47 664	48 053	48 442	48 830	388
112	49 218	49 606	49 993	50 380	50 766	51 153	51 538	51 924	52 309	52 694	384
113	53 078	53 463	53 846	54 230	54 613	54 996	55 378	55 760	56 142	56 524	381
114	56 905	57 286	57 666	58 046	58 426	58 805	59 185	59 563	59 942	60 320	378
115	060 698	061 075	061 452	061 829	062 206	062 582	062 958	063 333	063 709	064 083	375
116	64 458	64 832	65 206	65 580	65 953	66 326	66 699	67 071	67 443	67 815	371
117	68 186	68 557	68 928	69 298	69 668	70 038	70 407	70 776	71 145	71 514	368
118	71 882	72 250	72 617	72 985	73 352	73 718	74 085	74 451	74 816	75 182	365
119	75 547	75 912	76 276	76 640	77 004	77 368	77 731	78 094	78 457	78 819	362
120	079 181	079 543	079 904	080 266	080 626	080 987	081 347	081 707	082 067	082 426	359
121	82 785	83 144	83 503	83 861	84 219	84 576	84 934	85 291	85 647	86 004	356
122	86 360	86 716	87 071	87 426	87 781	88 136	88 490	88 845	89 198	89 552	353
123	89 905	90 258	90 611	90 963	91 315	91 667	92 018	92 370	92 721	93 071	351
124	93 422	93 772	94 122								
	0	**1**	**2**	**3**	**4**	**5**	**6**	**7**	**8**	**9**	

PROPORTIONAL PARTS

	350	**351**	**352**	**353**	**354**	**355**	**356**	**357**	**358**	**359**	**360**	**361**
1	35·0	35·1	35·2	35·3	35·4	35·5	35·6	35·7	35·8	35·9	36·0	36·1
2	70·0	70·2	70·4	70·6	70·8	71·0	71·2	71·4	71·6	71·8	72·0	72·2
3	105·0	105·3	105·6	105·9	106·2	106·5	106·8	107·1	107·4	107·7	108·0	108·3
4	140·0	140·4	140·8	141·2	141·6	142·0	142·4	142·8	143·2	143·6	144·0	144·4
5	175·0	175·5	176·0	176·5	177·0	177·5	178·0	178·5	179·0	179·5	180·0	180·5
6	210·0	210·6	211·2	211·8	212·4	213·0	213·6	214·2	214·8	215·4	216·0	216·6
7	245·0	245·7	246·4	247·1	247·8	248·5	249·2	249·9	250·6	251·3	252·0	252·7
8	280·0	280·8	281·6	282·4	283·2	284·0	284·8	285·6	286·4	287·2	288·0	288·8
9	315·0	315·9	316·8	317·7	318·6	319·5	320·4	321·3	322·2	323·1	324·0	324·9

	362	**363**	**364**	**365**	**366**	**367**	**368**	**369**	**370**	**371**	**372**	**373**
1	36·2	36·3	36·4	36·5	36·6	36·7	36·8	36·9	37·0	37·1	37·2	37·3
2	72·4	72·6	72·8	73·0	73·2	73·4	73·6	73·8	74·0	74·2	74·4	74·6
3	108·6	108·9	109·2	109·5	109·8	110·1	110·4	110·7	111·0	111·3	111·6	111·9
4	144·8	145·2	145·6	146·0	146·4	146·8	147·2	147·6	148·0	148·4	148·8	149·2
5	181·0	181·5	182·0	182·5	183·0	183·5	184·0	184·5	185·0	185·5	186·0	186·5
6	217·2	217·8	218·4	219·0	219·6	220·2	220·8	221·4	222·0	222·6	223·2	223·8
7	253·4	254·1	254·8	255·5	256·2	256·9	257·6	258·3	259·0	259·7	260·4	261·1
8	289·6	290·4	291·2	292·0	292·8	293·6	294·4	295·2	296·0	296·8	297·6	298·4
9	325·8	326·7	327·6	328·5	329·4	330·3	331·2	332·1	333·0	333·9	334·8	335·7

PROPORTIONAL PARTS, *continued*

	374	375	376	377	378	379	380	381	382	383	384	385
1	37·4	37·5	37·6	37·7	37·8	37·9	38·0	38·1	38·2	38·3	38·4	38·5
2	74·8	75·0	75·2	75·4	75·6	75·8	76·0	76·2	76·4	76·6	76·8	77·0
3	112·2	112·5	112·8	113·1	113·4	113·7	114·0	114·3	114·6	114·9	115·2	115·5
4	149·6	150·0	150·4	150·8	151·2	151·6	152·0	152·4	152·8	153·2	153·6	154·0
5	187·0	187·5	188·0	188·5	189·0	189·5	190·0	190·5	191·0	191·5	192·0	192·5
6	224·4	225·0	225·6	226·2	226·8	227·4	228·0	228·6	229·2	229·8	230·4	231·0
7	261·8	262·5	263·2	263·9	264·6	265·3	266·0	266·7	267·4	268·1	268·8	269·5
8	299·2	300·0	300·8	301·6	302·4	303·2	304·0	304·8	305·6	306·4	307·2	308·0
9	336·6	337·5	338·4	339·3	340·2	341·1	342·0	342·9	343·8	344·7	345·6	346·5

	386	387	388	389	390	391	392	393	394	395	396	397
1	38·6	38·7	38·8	38·9	39·0	39·1	39·2	39·3	39·4	39·5	39·6	39·7
2	77·2	77·4	77·6	77·8	78·0	78·2	78·4	78·6	78·8	79·0	79·2	79·4
3	115·8	116·1	116·4	116·7	117·0	117·3	117·6	117·9	118·2	118·5	118·8	119·1
4	154·4	154·8	155·2	155·6	156·0	156·4	156·8	157·2	157·6	158·0	158·4	158·8
5	193·0	193·5	194·0	194·5	195·0	195·5	196·0	196·5	197·0	197·5	198·0	198·5
6	231·6	232·2	232·8	233·4	234·0	234·6	235·2	235·8	236·4	237·0	237·6	238·2
7	270·2	270·9	271·6	272·3	273·0	273·7	274·4	275·1	275·8	276·5	277·2	277·9
8	308·8	309·6	310·4	311·2	312·0	312·8	313·6	314·4	315·2	316·0	316·8	317·6
9	347·4	348·3	349·2	350·1	351·0	351·9	352·8	353·7	354·6	355·5	356·4	357·3

	398	399	401	402	403	404	405	406	407	408	409	410
1	39·8	39·9	40·1	40·2	40·3	40·4	40·5	40·6	40·7	40·8	40·9	41·0
2	79·6	79·8	80·2	80·4	80·6	80·8	81·0	81·2	81·4	81·6	81·8	82·0
3	119·4	119·7	120·3	120·6	120·9	121·2	121·5	121·8	122·1	122·4	122·7	123·0
4	159·2	159·6	160·4	160·8	161·2	161·6	162·0	162·4	162·8	163·2	163·6	164·0
5	199·0	199·5	200·5	201·0	201·5	202·0	202·5	203·0	203·5	204·0	204·5	205·0
6	238·8	239·4	240·6	241·2	241·8	242·4	243·0	243·6	244·2	244·8	245·4	246·0
7	278·6	279·3	280·7	281·4	282·1	282·8	283·5	284·2	284·9	285·6	286·3	287·0
8	318·4	319·2	320·8	321·6	322·4	323·2	324·0	324·8	325·6	326·4	327·2	328·0
9	358·2	359·1	360·9	361·8	362·7	363·6	364·5	365·4	366·3	367·2	368·1	369·0

	411	412	413	414	415	416	417	418	419	420	421	422
1	41·1	41·2	41·3	41·4	41·5	41·6	41·7	41·8	41·9	42·0	42·1	42·2
2	82·2	82·4	82·6	82·8	83·0	83·2	83·4	83·6	83·8	84·0	84·2	84·4
3	123·3	123·6	123·9	124·2	124·5	124·8	125·1	125·4	125·7	126·0	126·3	126·6
4	164·4	164·8	165·2	165·6	166·0	166·4	166·8	167·2	167·6	168·0	168·4	168·8
5	205·5	206·0	206·5	207·0	207·5	208·0	208·5	209·0	209·5	210·0	210·5	211·0
6	246·6	247·2	247·8	248·4	249·0	249·6	250·2	250·8	251·4	252·0	252·6	253·2
7	287·7	288·4	289·1	289·8	290·5	291·2	291·9	292·6	293·3	294·0	294·7	295·4
8	328·8	329·6	330·4	331·2	332·0	332·8	333·6	334·4	335·2	336·0	336·8	337·6
9	369·9	370·8	371·7	372·6	373·5	374·4	375·3	376·2	377·1	378·0	378·9	379·8

	423	424	425	426	427	428	429	430	431	432	433	434
1	42·3	42·4	42·5	42·6	42·7	42·8	42·9	43·0	43·1	43·2	43·3	43·4
2	84·6	84·8	85·0	85·2	85·4	85·6	85·8	86·0	86·2	86·4	86·6	86·8
3	126·9	127·2	127·5	127·8	128·1	128·4	128·7	129·0	129·3	129·6	129·9	130·2
4	169·2	169·6	170·0	170·4	170·8	171·2	171·6	172·0	172·4	172·8	173·2	173·6
5	211·5	212·0	212·5	213·0	213·5	214·0	214·5	215·0	215·5	216·0	216·5	217·0
6	253·8	254·4	255·0	255·6	256·2	256·8	257·4	258·0	258·6	259·2	259·8	260·4
7	296·1	296·8	297·5	298·2	298·9	299·6	300·3	301·0	301·7	302·4	303·1	303·8
8	338·4	339·2	340·0	340·8	341·6	342·4	343·2	344·0	344·8	345·6	346·4	347·2
9	380·7	381·6	382·5	383·4	384·3	385·2	386·1	387·0	387·9	388·8	389·7	390·6

TABLE I—LOGARITHMS OF NUMBERS

	0	1	2	3	4	5	6	7	8	9	Δ
124			094 122	094 471	094 820	095 169	095 518	095 866	096 215	096 562	348
125	096 910	097 257	097 604	097 951	098 298	098 644	098 990	099 335	099 681	100 026	345
126	100 371	100 715	101 059	101 403	101 747	102 091	102 434	102 777	103 119	03 462	342
127	03 804	04 146	04 487	04 828	05 169	05 510	05 851	06 191	06 531	06 871	339
128	07 210	07 549	07 888	08 227	08 565	08 903	09 241	09 579	09 916	10 253	337
129	10 590	10 926	11 263	11 599	11 934	12 270	12 605	12 940	13 275	13 609	334
130	113 943	114 277	114 611	114 944	115 278	115 611	115 943	116 276	116 608	116 940	331
131	17 271	17 603	17 934	18 265	18 595	18 926	19 256	19 586	19 915	20 245	329
132	20 574	20 903	21 231	21 560	21 888	22 216	22 544	22 871	23 198	23 525	327
133	23 852	24 178	24 504	24 830	25 156	25 481	25 806	26 131	26 456	26 781	324
134	27 105	27 429	27 753	28 076	28 399	28 722	29 045	29 368	29 690	30 012	322
135	130 334	130 655	130 977	131 298	131 619	131 939	132 260	132 580	132 900	133 219	320
136	33 539	33 858	34 177	34 496	34 814	35 133	35 451	35 769	36 086	36 403	318
137	36 721	37 037	37 354	37 671	37 987	38 303	38 618	38 934	39 249	39 564	315
138	39 879	40 194	40 508	40 822	41 136	41 450	41 763	42 076	42 389	42 702	313
139	43 015	43 327	43 639	43 951	44 263	44 574	44 885	45 196	45 507	45 818	310
140	146 128	146 438	146 748	147 058	147 367	147 676	147 985	148 294	148 603	148 911	308
141	49 219	49 527	49 835	50 142	50 449	50 756	51 063	51 370	51 676	51 982	306
142	52 288	52 594	52 900	53 205	53 510	53 815	54 120	54 424	54 728	55 032	304
143	55 336	55 640	55 943	56 246	56 549	56 852	57 154	57 457	57 759	58 061	301
144	58 362	58 664	58 965	59 266	59 567	59 868	60 168	60 469	60 769	61 068	300
145	161 368	161 667	161 967	162 266	162 564	162 863	163 161	163 460	163 758	164 055	298
146	64 353	64 650	64 947	65 244	65 541	65 838	66 134	66 430	66 726	67 022	295
147	67 317	67 613	67 908	68 203	68 497	68 792	69 086	69 380	69 674	69 968	294
148	70 262	70 555	70 848	71 141	71 434	71 726	72 019	72 311	72 603	72 895	291
149	73 186	73 478	73 769	74 060	74 351	74 641	74 932	75 222	75 512	75 802	289
150	176 091	176 381	176 670	176 959	177 248	177 536	177 825	178 113	178 401	178 689	288
151	78 977	79 264	79 552	79 839	80 126	80 413	80 699	80 986	81 272	81 558	286
152	81 844	82 129	82 415	82 700	82 985	83 270	83 555	83 839	84 123	84 407	284
153	84 691	84 975	85 259	85 542	85 825	86 108	86 391	86 674	86 956	87 239	282
154	87 521	87 803	88 084	88 366	88 647	88 928	89 209	89 490	89 771	90 051	281
155	190 332	190 612	190 892	191 171	191 451	191 730	192 010	192 289	192 567	192 846	279
156	93 125	93 403	93 681	93 959	94 237	94 514	94 792	95 069	95 346	95 623	277
157	95 900	96 176	96 453								
	0	1	2	3	4	5	6	7	8	9	

PROPORTIONAL PARTS

	277	278	279	280	281	282	283	284	285	286	287	288
1	27·7	27·8	27·9	28·0	28·1	28·2	28·3	28·4	28·5	28·6	28·7	28·8
2	55·4	55·6	55·8	56·0	56·2	56·4	56·6	56·8	57·0	57·2	57·4	57·6
3	83·1	83·4	83·7	84·0	84·3	84·6	84·9	85·2	85·5	85·8	86·1	86·4
4	110·8	111·2	111·6	112·0	112·4	112·8	113·2	113·6	114·0	114·4	114·8	115·2
5	138·5	139·0	139·5	140·0	140·5	141·0	141·5	142·0	142·5	143·0	143·5	144·0
6	166·2	166·8	167·4	168·0	168·6	169·2	169·8	170·4	171·0	171·6	172·2	172·8
7	193·9	194·6	195·3	196·0	196·7	197·4	198·1	198·8	199·5	200·2	200·9	201·6
8	221·6	222·4	223·2	224·0	224·8	225·6	226·4	227·2	228·0	228·8	229·6	230·4
9	249·3	250·2	251·1	252·0	252·9	253·8	254·7	255·6	256·5	257·4	258·3	259·2

PROPORTIONAL PARTS, *continued*

	289	290	291	292	293	294	295	296	297	298	299	301
1	28·9	29·0	29·1	29·2	29·3	29·4	29·5	29·6	29·7	29·8	29·9	30·1
2	57·8	58·0	58·2	58·4	58·6	58·8	59·0	59·2	59·4	59·6	59·8	60·2
3	86·7	87·0	87·3	87·6	87·9	88·2	88·5	88·8	89·1	89·4	89·7	90·3
4	115·6	116·0	116·4	116·8	117·2	117·6	118·0	118·4	118·8	119·2	119·6	120·4
5	144·5	145·0	145·5	146·0	146·5	147·0	147·5	148·0	148·5	149·0	149·5	150·5
6	173·4	174·0	174·6	175·2	175·8	176·4	177·0	177·6	178·2	178·8	179·4	180·6
7	202·3	203·0	203·7	204·4	205·1	205·8	206·5	207·2	207·9	208·6	209·3	210·7
8	231·2	232·0	232·8	233·6	234·4	235·2	236·0	236·8	237·6	238·4	239·2	240·8
9	260·1	261·0	261·9	262·8	263·7	264·6	265·5	266·4	267·3	268·2	269·1	270·9

	302	303	304	305	306	307	308	309	310	311	312	313
1	30·2	30·3	30·4	30·5	30·6	30·7	30·8	30·9	31·0	31·1	31·2	31·3
2	60·4	60·6	60·8	61·0	61·2	61·4	61·6	61·8	62·0	62·2	62·4	62·6
3	90·6	90·9	91·2	91·5	91·8	92·1	92·4	92·7	93·0	93·3	93·6	93·9
4	120·8	121·2	121·6	122·0	122·4	122·8	123·2	123·6	124·0	124·4	124·8	125·2
5	151·0	151·5	152·0	152·5	153·0	153·5	154·0	154·5	155·0	155·5	156·0	156·5
6	181·2	181·8	182·4	183·0	183·6	184·2	184·8	185·4	186·0	186·6	187·2	187·8
7	211·4	212·1	212·8	213·5	214·2	214·9	215·6	216·3	217·0	217·7	218·4	219·1
8	241·6	242·4	243·2	244·0	244·8	245·6	246·4	247·2	248·0	248·8	249·6	250·4
9	271·8	272·7	273·6	274·5	275·4	276·3	277·2	278·1	279·0	279·9	280·8	281·7

	314	315	316	317	318	319	320	321	322	323	324	325
1	31·4	31·5	31·6	31·7	31·8	31·9	32·0	32·1	32·2	32·3	32·4	32·5
2	62·8	63·0	63·2	63·4	63·6	63·8	64·0	64·2	64·4	64·6	64·8	65·0
3	94·2	94·5	94·8	95·1	95·4	95·7	96·0	96·3	96·6	96·9	97·2	97·5
4	125·6	126·0	126·4	126·8	127·2	127·6	128·0	128·4	128·8	129·2	129·6	130·0
5	157·0	157·5	158·0	158·5	159·0	159·5	160·0	160·5	161·0	161·5	162·0	162·5
6	188·4	189·0	189·6	190·2	190·8	191·4	192·0	192·6	193·2	193·8	194·4	195·0
7	219·8	220·5	221·2	221·9	222·6	223·3	224·0	224·7	225·4	226·1	226·8	227·5
8	251·2	252·0	252·8	253·6	254·4	255·2	256·0	256·8	257·6	258·4	259·2	260·0
9	282·6	283·5	284·4	285·3	286·2	287·1	288·0	288·9	289·8	290·7	291·6	292·5

	326	327	328	329	330	331	332	333	334	335	336	337
1	32·6	32·7	32·8	32·9	33·0	33·1	33·2	33·3	33·4	33·5	33·6	33·7
2	65·2	65·4	65·6	65·8	66·0	66·2	66·4	66·6	66·8	67·0	67·2	67·4
3	97·8	98·1	98·4	98·7	99·0	99·3	99·6	99·9	100·2	100·5	100·8	101·1
4	130·4	130·8	131·2	131·6	132·0	132·4	132·8	133·2	133·6	134·0	134·4	134·8
5	163·0	163·5	164·0	164·5	165·0	165·5	166·0	166·5	167·0	167·5	168·0	168·5
6	195·6	196·2	196·8	197·4	198·0	198·6	199·2	199·8	200·4	201·0	201·6	202·2
7	228·2	228·9	229·6	230·3	231·0	231·7	232·4	233·1	233·8	234·5	235·2	235·9
8	260·8	261·6	262·4	263·2	264·0	264·8	265·6	266·4	267·2	268·0	268·8	269·6
9	293·4	294·3	295·2	296·1	297·0	297·9	298·8	299·7	300·6	301·5	302·4	303·3

	338	339	340	341	342	343	344	345	346	347	348	349
1	33·8	33·9	34·0	34·1	34·2	34·3	34·4	34·5	34·6	34·7	34·8	34·9
2	67·6	67·8	68·0	68·2	68·4	68·6	68·8	69·0	69·2	69·4	69·6	69·8
3	101·4	101·7	102·0	102·3	102·6	102·9	103·2	103·5	103·8	104·1	104·4	104·7
4	135·2	135·6	136·0	136·4	136·8	137·2	137·6	138·0	138·4	138·8	139·2	139·6
5	169·0	169·5	170·0	170·5	171·0	171·5	172·0	172·5	173·0	173·5	174·0	174·5
6	202·8	203·4	204·0	204·6	205·2	205·8	206·4	207·0	207·6	208·2	208·8	209·4
7	236·6	237·3	238·0	238·7	239·4	240·1	240·8	241·5	242·2	242·9	243·6	244·3
8	270·4	271·2	272·0	272·8	273·6	274·4	275·2	276·0	276·8	277·6	278·4	279·2
9	304·2	305·1	306·0	306·9	307·8	308·7	309·6	310·5	311·4	312·3	313·2	314·1

TABLE I—LOGARITHMS OF NUMBERS

	0	1	2	3	4	5	6	7	8	9	Δ
157	195 900	196 176	196 453	196 729	197 005	197 281	197 556	197 832	198 107	198 382	275
158	198 657	198 932	199 206	199 481	199 755	200 029	200 303	200 577	200 850	201 124	273
159	201 397	201 670	201 943	202 216	202 488	02 761	03 033	03 305	03 577	03 848	272
160	204 120	204 391	204 663	204 934	205 204	205 475	205 746	206 016	206 286	206 556	270
161	06 826	07 096	07 365	07 634	07 904	08 173	08 441	08 710	08 979	09 247	268
162	09 515	09 783	10 051	10 319	10 586	10 853	11 121	11 388	11 654	11 921	267
163	12 188	12 454	12 720	12 986	13 252	13 518	13 783	14 049	14 314	14 579	265
164	14 844	15 109	15 373	15 638	15 902	16 166	16 430	16 694	16 957	17 221	263
165	217 484	217 747	218 010	218 273	218 536	218 798	219 060	219 323	219 585	219 846	262
166	20 108	20 370	20 631	20 892	21 153	21 414	21 675	21 936	22 196	22 456	260
167	22 716	22 976	23 236	23 496	23 755	24 015	24 274	24 533	24 792	25 051	258
168	25 309	25 568	25 826	26 084	26 342	26 600	26 858	27 115	27 372	27 630	257
169	27 887	28 144	28 400	28 657	28 913	29 170	29 426	29 682	29 938	30 193	256
170	230 449	230 704	230 960	231 215	231 470	231 724	231 979	232 234	232 488	232 742	254
171	32 996	33 250	33 504	33 757	34 011	34 264	34 517	34 770	35 023	35 276	252
172	35 528	35 781	36 033	36 285	36 537	36 789	37 041	37 292	37 544	37 795	251
173	38 046	38 297	38 548	38 799	39 049	39 299	39 550	39 800	40 050	40 300	249
174	40 549	40 799	41 048	41 297	41 546	41 795	42 044	42 293	42 541	42 790	248
175	243 038	243 286	243 534	243 782	244 030	244 277	244 525	244 772	245 019	245 266	247
176	45 513	45 759	46 006	46 252	46 499	46 745	46 991	47 237	47 482	47 728	245
177	47 973	48 219	48 464	48 709	48 954	49 198	49 443	49 687	49 932	50 176	244
178	50 420	50 664	50 908	51 151	51 395	51 638	51 881	52 125	52 368	52 610	243
179	52 853	53 096	53 338	53 580	53 822	54 064	54 306	54 548	54 790	55 031	242
180	255 273	255 514	255 755	255 996	256 237	256 477	256 718	256 958	257 198	257 439	240
181	57 679	57 918	58 158	58 398	58 637	58 877	59 116	59 355	59 594	59 833	238
182	60 071	60 310	60 548	60 787	61 025	61 263	61 501	61 739	61 976	62 214	237
183	62 451	62 688	62 925	63 162	63 399	63 636	63 873	64 109	64 346	64 582	236
184	64 818	65 054	65 290	65 525	65 761	65 996	66 232	66 467	66 702	66 937	235
185	267 172	267 406	267 641	267 875	268 110	268 344	268 578	268 812	269 046	269 279	234
186	69 513	69 746	69 980	70 213	70 446	70 679	70 912	71 144	71 377	71 609	233
187	71 842	72 074	72 306	72 538	72 770	73 001	73 233	73 464	73 696	73 927	231
188	74 158	74 389	74 620	74 850	75 081	75 311	75 542	75 772	76 002	76 232	230
189	76 462	76 692	76 921	77 151	77 380	77 609	77 838	78 067	78 296	78 525	229
190	278 754	278 982	279 211	279 439	279 667	279 895	280 123	280 351	280 578	280 806	227
191	81 033	81 261	81 488	81 715	81 942	82 169	82 396	82 622	82 849	83 075	226
192	83 301	83 527	83 753	83 979	84 205	84 431	84 656	84 882	85 107	85 332	225
193	85 557	85 782	86 007	86 232	86 456	86 681	86 905	87 130	87 354	87 578	224
194	87 802	88 026	88 249	88 473	88 696	88 920	89 143	89 366	89 589	89 812	223
195	290 035	290 257	290 480	290 702	290 925	291 147	291 369	291 591	291 813	292 034	222
196	92 256	92 478	92 699	92 920	93 141	93 363	93 584	93 804	94 025	94 246	220
197	94 466	94 687	94 907	95 127	95 347	95 567	95 787	96 007	96 226	96 446	219
198	96 665	96 884	97 104	97 323	97 542	97 761	297 979	298 198	298 416	298 635	218
199	98 853	99 071	99 289	99 507	99.725	299 943	300 161	300 378	300 595	300 813	217
	0	1	2	3	4	5	6	7	8	9	

Proportional Parts will be found on the facing page.

TABLE I—LOGARITHMS OF NUMBERS

PROPORTIONAL PARTS

	217	218	219	220	221	222	223	224	225	226	227	228
1	21·7	21·8	21·9	22·0	22·1	22·2	22·3	22·4	22·5	22·6	22·7	22·8
2	43·4	43·6	43·8	44·0	44·2	44·4	44·6	44·8	45·0	45·2	45·4	45·6
3	65·1	65·4	65·7	66·0	66·3	66·6	66·9	67·2	67·5	67·8	68·1	68·4
4	86·8	87·2	87·6	88·0	88·4	88·8	89·2	89·6	90·0	90·4	90·8	91·2
5	108·5	109·0	109·5	110·0	110·5	111·0	111·5	112·0	112·5	113·0	113·5	114·0
6	130·2	130·8	131·4	132·0	132·6	133·2	133·8	134·4	135·0	135·6	136·2	136·8
7	151·9	152·6	153·3	154·0	154·7	155·4	156·1	156·8	157·5	158·2	158·9	159·6
8	173·6	174·4	175·2	176·0	176·8	177·6	178·4	179·2	180·0	180·8	181·6	182·4
9	195·3	196·2	197·1	198·0	198·9	199·8	200·7	201·6	202·5	203·4	204·3	205·2

	229	230	231	232	233	234	235	236	237	238	239	240
1	22·9	23·0	23·1	23·2	23·3	23·4	23·5	23·6	23·7	23·8	23·9	24·0
2	45·8	46·0	46·2	46·4	46·6	46·8	47·0	47·2	47·4	47·6	47·8	48·0
3	68·7	69·0	69·3	69·6	69·9	70·2	70·5	70·8	71·1	71·4	71·7	72·0
4	91·6	92·0	92·4	92·8	93·2	93·6	94·0	94·4	94·8	95·2	95·6	96·0
5	114·5	115·0	115·5	116·0	116·5	117·0	117·5	118·0	118·5	119·0	119·5	120·0
6	137·4	138·0	138·6	139·2	139·8	140·4	141·0	141·6	142·2	142·8	143·4	144·0
7	160·3	161·0	161·7	162·4	163·1	163·8	164·5	165·2	165·9	166·6	167·3	168·0
8	183·2	184·0	184·8	185·6	186·4	187·2	188·0	188·8	189·6	190·4	191·2	192·0
9	206·1	207·0	207·9	208·8	209·7	210·6	211·5	212·4	213·3	214·2	215·1	216·0

	241	242	243	244	245	246	247	248	249	250	251	252
1	24·1	24·2	24·3	24·4	24·5	24·6	24·7	24·8	24·9	25·0	25·1	25·2
2	48·2	48·4	48·6	48·8	49·0	49·2	49·4	49·6	49·8	50·0	50·2	50·4
3	72·3	72·6	72·9	73·2	73·5	73·8	74·1	74·4	74·7	75·0	75·3	75·6
4	96·4	96·8	97·2	97·6	98·0	98·4	98·8	99·2	99·6	100·0	100·4	100·8
5	120·5	121·0	121·5	122·0	122·5	123·0	123·5	124·0	124·5	125·0	125·5	126·0
6	144·6	145·2	145·8	146·4	147·0	147·6	148·2	148·8	149·4	150·0	150·6	151·2
7	168·7	169·4	170·1	170·8	171·5	172·2	172·9	173·6	174·3	175·0	175·7	176·4
8	192·8	193·6	194·4	195·2	196·0	196·8	197·6	198·4	199·2	200·0	200·8	201·6
9	216·9	217·8	218·7	219·6	220·5	221·4	222·3	223·2	224·1	225·0	225·9	226·8

	253	254	255	256	257	258	259	260	261	262	263	264
1	25·3	25·4	25·5	25·6	25·7	25·8	25·9	26·0	26·1	26·2	26·3	26·4
2	50·6	50·8	51·0	51·2	51·4	51·6	51·8	52·0	52·2	52·4	52·6	52·8
3	75·9	76·2	76·5	76·8	77·1	77·4	77·7	78·0	78·3	78·6	78·9	79·2
4	101·2	101·6	102·0	102·4	102·8	103·2	103·6	104·0	104·4	104·8	105·2	105·6
5	126·5	127·0	127·5	128·0	128·5	129·0	129·5	130·0	130·5	131·0	131·5	132·0
6	151·8	152·4	153·0	153·6	154·2	154·8	155·4	156·0	156·6	157·2	157·8	158·4
7	177·1	177·8	178·5	179·2	179·9	180·6	181·3	182·0	182·7	183·4	184·1	184·8
8	202·4	203·2	204·0	204·8	205·6	206·4	207·2	208·0	208·8	209·6	210·4	211·2
9	227·7	228·6	229·5	230·4	231·3	232·2	233·1	234·0	234·9	235·8	236·7	237·6

	265	266	267	268	269	270	271	272	273	274	275	276
1	26·5	26·6	26·7	26·8	26·9	27·0	27·1	27·2	27·3	27·4	27·5	27·6
2	53·0	53·2	53·4	53·6	53·8	54·0	54·2	54·4	54·6	54·8	55·0	55·2
3	79·5	79·8	80·1	80·4	80·7	81·0	81·3	81·6	81·9	82·2	82·5	82·8
4	106·0	106·4	106·8	107·2	107·6	108·0	108·4	108·8	109·2	109·6	110·0	110·4
5	132·5	133·0	133·5	134·0	134·5	135·0	135·5	136·0	136·5	137·0	137·5	138·0
6	159·0	159·6	160·2	160·8	161·4	162·0	162·6	163·2	163·8	164·4	165·0	165·6
7	185·5	186·2	186·9	187·6	188·3	189·0	189·7	190·4	191·1	191·8	192·5	193·2
8	212·0	212·8	213·6	214·4	215·2	216·0	216·8	217·6	218·4	219·2	220·0	220·8
9	238·5	239·4	240·3	241·2	242·1	243·0	243·9	244·8	245·7	246·6	247·5	248·4

TABLE I—LOGARITHMS OF NUMBERS

	0	1	2	3	4	5	6	7	8	9	Δ
200	301 030	301 247	301 464	301 681	301 898	302 114	302 331	302 547	302 764	302 980	216
201	03 196	03 412	03 628	03 844	04 059	04 275	04 491	04 706	04 921	05 136	215
202	05 351	05 566	05 781	05 996	06 211	06 425	06 639	06 854	07 068	07 282	214
203	07 496	07 710	07 924	08 137	08 351	08 564	08 778	08 991	09 204	09 417	213
204	09 630	09 843	10 056	10 268	10 481	10 693	10 906	11 118	11 330	11 542	212
205	311 754	311 966	312 177	312 389	312 600	312 812	313 023	313 234	313 445	313 656	211
206	13 867	14 078	14 289	14 499	14 710	14 920	15 130	15 340	15 551	15 760	210
207	15 970	16 180	16 390	16 599	16 809	17 018	17 227	17 436	17 646	17 854	209
208	18 063	18 272	18 481	18 689	18 898	19 106	19 314	19 522	19 730	19 938	208
209	20 146	20 354	20 562	20 769	20 977	21 184	21 391	21 598	21 805	22 012	207
210	322 219	322 426	322 633	322 839	323 046	323 252	323 458	323 665	323 871	324 077	205
211	24 282	24 488	24 694	24 899	25 105	25 310	25 516	25 721	25 926	26 131	205
212	26 336	26 541	26 745	26 950	27 155	27 359	27 563	27 767	27 972	28 176	204
213	28 380	28 583	28 787	28 991	29 194	29 398	29 601	29 805	30 008	30 211	203
214	30 414	30 617	30 819	31 022	31 225	31 427	31 630	31 832	32 034	32 236	202
215	332 438	332 640	332 842	333 044	333 246	333 447	333 649	333 850	334 051	334 253	201
216	34 454	34 655	34 856	35 057	35 257	35 458	35 658	35 859	36 059	36 260	200
217	36 460	36 660	36 860	37 060	37 260	37 459	37 659	37 858	38 058	38 257	199
218	38 456	38 656	38 855	39 054	39 253	39 451	39 650	39 849	40 047	40 246	198
219	40 444	40 642	40 841	41 039	41 237	41 435	41 632	41 830	42 028	42 225	198
220	342 423	342 620	342 817	343 014	343 212	343 409	343 606	343 802	343 999	344 196	196
221	44 392	44 589	44 785	44 981	45 178	45 374	45 570	45 766	45 962	46 157	196
222	46 353	46 549	46 744	46 939	47 135	47 330	47 525	47 720	47 915	48 110	195
223	48 305	48 500	48 694	48 889	49 083	49 278	49 472	49 666	49 860	50 054	194
224	50 248	50 442	50 636	50 829	51 023	51 216	51 410	51 603	51 796	51 989	194
225	352 183	352 375	352 568	352 761	352 954	353 147	353 339	353 532	353 724	353 916	192
226	54 108	54 301	54 493	54 685	54 876	55 068	55 260	55 452	55 643	55 834	192
227	56 026	56 217	56 408	56 599	56 790	56 981	57 172	57 363	57 554	57 744	191
228	57 935	58 125	58 316	58 506	58 696	58 886	59 076	59 266	59 456	59 646	189
229	59 835	60 025	60 215	60 404	60 593	60 783	60 972	61 161	61 350	61 539	189
230	361 728	361 917	362 105	362 294	362 482	362 671	362 859	363 048	363 236	363 424	188
231	63 612	63 800	63 988	64 176	64 363	64 551	64 739	64 926	65 113	65 301	187
232	65 488	65 675	65 862	66 049	66 236	66 423	66 610	66 796	66 983	67 169	187
233	67 356	67 542	67 729	67 915	68 101	68 287	68 473	68 659	68 845	69 030	186
234	69 216	69 401	69 587	69 772	69 958	70 143	70 328	70 513	70 698	70 883	185
235	371 068	371 253	371 437	371 622	371 806	371 991	372 175	372 360	372 544	372 728	184
236	72 912	73 096	73 280	73 464	73 647	73 831	74 015	74 198	74 382	74 565	183
237	74 748	74 932	75 115	75 298	75 481	75 664	75 846	76 029	76 212	76 394	183
238	76 577	76 759	76 942	77 124	77 306	77 488	77 670	77 852	78 034	78 216	182
239	78 398	78 580	78 761	78 943	79 124	79 306	79 487	79 668	79 849	80 030	181
240	380 211	380 392	380 573	380 754	380 934	381 115	381 296	381 476	381 656	381 837	180
241	82 017	82 197	82 377	82 557	82 737	82 917	83 097	83 277	83 456	83 636	179
242	83 815	83 995	84 174	84 353	84 533	84 712	84 891	85 070	85 249	85 428	178
243	85 606	85 785	85 964	86 142	86 321	86 499	86 677	86 856	87 034	87 212	178
244	87 390	87 568	87 746	87 923	88 101	88 279	88 456	88 634	88 811	88 989	177
245	389 166	389 343	389 520	389 698	389 875	390 051	390 228	390 405	390 582	390 759	176
246	90 935	91 112	91 288	91 464	91 641	91 817	91 993	92 169	92 345	92 521	176
247	92 697	92 873	93 048	93 224	93 400	93 575	93 751	93 926	94 101	94 277	175
248	94 452	94 627	94 802	94 977	95 152	95 326	95 501	95 676	95 850	96 025	174
249	96 199	96 374	96 548	96 722	96 896	97 071	97 245	97 419	97 592	97 766	174
	0	**1**	**2**	**3**	**4**	**5**	**6**	**7**	**8**	**9**	

TABLE I—LOGARITHMS OF NUMBERS

	0	1	2	3	4	5	6	7	8	9	Δ
250	397 940	398 114	398 287	398 461	398 634	398 808	398 981	399 154	399 328	399 501	173
251	399 674	399 847	400 020	400 192	400 365	400 538	400 711	400 883	401 056	401 228	173
252	401 401	401 573	01 745	01 917	02 089	02 261	02 433	02 605	02 777	02 949	172
253	03 121	03 292	03 464	03 635	03 807	03 978	04 149	04 320	04 492	04 663	171
254	04 834	05 005	05 176	05 346	05 517	05 688	05 858	06 029	06 199	06 370	170
255	406 540	406 710	406 881	407 051	407 221	407 391	407 561	407 731	407 901	408 070	170
256	08 240	08 410	08 579	08 749	08 918	09 087	09 257	09 426	09 595	09 764	169

PROPORTIONAL PARTS

	169	**170**	**171**	**172**	**173**	**174**	**175**	**176**	**177**	**178**	**179**	**180**
1	16·9	17·0	17·1	17·2	17·3	17·4	17·5	17·6	17·7	17·8	17·9	18·0
2	33·8	34·0	34·2	34·4	34·6	34·8	35·0	35·2	35·4	35·6	35·8	36·0
3	50·7	51·0	51·3	51·6	51·9	52·2	52·5	52·8	53·1	53·4	53·7	54·0
4	67·6	68·0	68·4	68·8	69·2	69·6	70·0	70·4	70·8	71·2	71·6	72·0
5	84·5	85·0	85·5	86·0	86·5	87·0	87·5	88·0	88·5	89·0	89·5	90·0
6	101·4	102·0	102·6	103·2	103·8	104·4	105·0	105·6	106·2	106·8	107·4	108·0
7	118·3	119·0	119·7	120·4	121·1	121·8	122·5	123·2	123·9	124·6	125·3	126·0
8	135·2	136·0	136·8	137·6	138·4	139·2	140·0	140·8	141·6	142·4	143·2	144·0
9	152·1	153·0	153·9	154·8	155·7	156·6	157·5	158·4	159·3	160·2	161·1	162·0

	181	**182**	**183**	**184**	**185**	**186**	**187**	**188**	**189**	**190**	**191**	**192**
1	18·1	18·2	18·3	18·4	18·5	18·6	18·7	18·8	18·9	19·0	19·1	19·2
2	36·2	36·4	36·6	36·8	37·0	37·2	37·4	37·6	37·8	38·0	38·2	38·4
3	54·3	54·6	54·9	55·2	55·5	55·8	56·1	56·4	56·7	57·0	57·3	57·6
4	72·4	72·8	73·2	73·6	74·0	74·4	74·8	75·2	75·6	76·0	76·4	76·8
5	90·5	91·0	91·5	92·0	92·5	93·0	93·5	94·0	94·5	95·0	95·5	96·0
6	108·6	109·2	109·8	110·4	111·0	111·6	112·2	112·8	113·4	114·0	114·6	115·2
7	126·7	127·4	128·1	128·8	129·5	130·2	130·9	131·6	132·3	133·0	133·7	134·4
8	144·8	145·6	146·4	147·2	148·0	148·8	149·6	150·4	151·2	152·0	152·8	153·6
9	162·9	163·8	164·7	165·6	166·5	167·4	168·3	169·2	170·1	171·0	171·9	172·8

	193	**194**	**195**	**196**	**197**	**198**	**199**	**201**	**202**	**203**	**204**	**205**
1	19·3	19·4	19·5	19·6	19·7	19·8	19·9	20·1	20·2	20·3	20·4	20·5
2	38·6	38·8	39·0	39·2	39·4	39·6	39·8	40·2	40·4	40·6	40·8	41·0
3	57·9	58·2	58·5	58·8	59·1	59·4	59·7	60·3	60·6	60·9	61·2	61·5
4	77·2	77·6	78·0	78·4	78·8	79·2	79·6	80·4	80·8	81·2	81·6	82·0
5	96·5	97·0	97·5	98·0	98·5	99·0	99·5	100·5	101·0	101·5	102·0	102·5
6	115·8	116·4	117·0	117·6	118·2	118·8	119·4	120·6	121·2	121·8	122·4	123·0
7	135·1	135·8	136·5	137·2	137·9	138·6	139·3	140·7	141·4	142·1	142·8	143·5
8	154·4	155·2	156·0	156·8	157·6	158·4	159·2	160·8	161·6	162·4	163·2	164·0
9	173·7	174·6	175·5	176·4	177·3	178·2	179·1	180·9	181·8	182·7	183·6	184·5

	206	**207**	**208**	**209**	**210**	**211**	**212**	**213**	**214**	**215**	**216**	**217**
1	20·6	20·7	20·8	20·9	21·0	21·1	21·2	21·3	21·4	21·5	21·6	21·7
2	41·2	41·4	41·6	41·8	42·0	42·2	42·4	42·6	42·8	43·0	43·2	43·4
3	61·8	62·1	62·4	62·7	63·0	63·3	63·6	63·9	64·2	64·5	64·8	65·1
4	82·4	82·8	83·2	83·6	84·0	84·4	84·8	85·2	85·6	86·0	86·4	86·8
5	103·0	103·5	104·0	104·5	105·0	105·5	106·0	106·5	107·0	107·5	108·0	108·5
6	123·6	124·2	124·8	125·4	126·0	126·6	127·2	127·8	128·4	129·0	129·6	130·2
7	144·2	144·9	145·6	146·3	147·0	147·7	148·4	149·1	149·8	150·5	151·2	151·9
8	164·8	165·6	166·4	167·2	168·0	168·8	169·6	170·4	171·2	172·0	172·8	173·6
9	185·4	186·3	187·2	188·1	189·0	189·9	190·8	191·7	192·6	193·5	194·4	195·3

TABLE I—LOGARITHMS OF NUMBERS

	0	1	2	3	4	5	6	7	8	9	Δ
257	409 933	410 102	410 271	410 440	410 609	410 777	410 946	411 114	411 283	411 451	169
258	11 620	11 788	11 956	12 124	12 293	12 461	12 629	12 796	12 964	13 132	168
259	13 300	13 467	13 635	13 803	13 970	14 137	14 305	14 472	14 639	14 806	167
260	414 973	415 140	415 307	415 474	415 641	415 808	415 974	416 141	416 308	416 474	167
261	16 641	16 807	16 973	17 139	17 306	17 472	17 638	17 804	17 970	18 135	166
262	18 301	18 467	18 633	18 798	18 964	19 129	19 295	19 460	19 625	19 791	165
263	19 956	20 121	20 286	20 451	20 616	20 781	20 945	21 110	21 275	21 439	165
264	21 604	21 768	21 933	22 097	22 261	22 426	22 590	22 754	22 918	23 082	164
265	423 246	423 410	423 574	423 737	423 901	424 065	424 228	424 392	424 555	424 718	164
266	24 882	25 045	25 208	25 371	25 534	25 697	25 860	26 023	26 186	26 349	162
267	26 511	26 674	26 836	26 999	27 161	27 324	27 486	27 648	27 811	27 973	162
268	28 135	28 297	28 459	28 621	28 783	28 944	29 106	29 268	29 429	29 591	161
269	29 752	29 914	30 075	30 236	30 398	30 559	30 720	30 881	31 042	31 203	161
270	431 364	431 525	431 685	431 846	432 007	432 167	432 328	432 488	432 649	432 809	160
271	32 969	33 130	33 290	33 450	33 610	33 770	33 930	34 090	34 249	34 409	160
272	34 569	34 729	34 888	35 048	35 207	35 367	35 526	35 685	35 844	36 004	159
273	36 163	36 322	36 481	36 640	36 799	36 957	37 116	37 275	37 433	37 592	159
274	37 751	37 909	38 067	38 226	38 384	38 542	38 701	38 859	39 017	39 175	158
275	439 333	439 491	439 648	439 806	439 964	440 122	440 279	440 437	440 594	440 752	157
276	40 909	41 066	41 224	41 381	41 538	41 695	41 852	42 009	42 166	42 323	157
277	42 480	42 637	42 793	42 950	43 106	43 263	43 419	43 576	43 732	43 889	156
278	44 045	44 201	44 357	44 513	44 669	44 825	44 981	45 137	45 293	45 449	155
279	45 604	45 760	45 915	46 071	46 226	46 382	46 537	46 692	46 848	47 003	155
280	447 158	447 313	447 468	447 623	447 778	447 933	448 088	448 242	448 397	448 552	154
281	48 706	48 861	49 015	49 170	49 324	49 478	49 633	49 787	49 941	50 095	154
282	50 249	50 403	50 557	50 711	50 865	51 018	51 172	51 326	51 479	51 633	153
283	51 786	51 940	52 093	52 247	52 400	52 553	52 706	52 859	53 012	53 165	153
284	53 318	53 471	53 624	53 777	53 930	54 082	54 235	54 387	54 540	54 692	153
285	454 845	454 997	455 150	455 302	455 454	455 606	455 758	455 910	456 062	456 214	152
286	56 366	56 518	56 670	56 821	56 973	57 125	57 276	57 428	57 579	57 731	151
287	57 882	58 033	58 184	58 336	58 487	58 638	58 789	58 940	59 091	59 242	150
288	59 392	59 543	59 694	59 845	59 995	60 146	60 296	60 447	60 597	60 748	150
289	60 898	61 048	61 198	61 348	61 499	61 649	61 799	61 948	62 098	62 248	150
290	462 398	462 548	462 697	462 847	462 997	463 146	463 296	463 445	463 594	463 744	149
291	63 893	64 042	64 191	64 340	64 490	64 639	64 788	64 936	65 085	65 234	149
292	65 383	65 532	65 680	65 829	65 977	66 126	66 274	66 423	66 571	66 719	149
293	66 868	67 016	67 164	67 312	67 460	67 608	67 756	67 904	68 052	68 200	147
294	68 347	68 495	68 643	68 790	68 938	69 085	69 233	69 380	69 527	69 675	147
295	469 822	469 969	470 116	470 263	470 410	470 557	470 704	470 851	470 998	471 145	147
296	71 292	71 438	71 585	71 732	71 878	72 025	72 171	72 318	72 464	72 610	146
297	72 756	72 903	73 049	73 195	73 341	73 487	73 633	73 779	73 925	74 071	145
298	74 216	74 362	74 508	74 653	74 799	74 944	75 090	75 235	75 381	75 526	145
299	75 671	75 816	75 962	76 107	76 252	76 397	76 542	76 687	76 832	76 976	145
300	477 121	477 266	477 411	477 555	477 700	477 844	477 989	478 133	478 278	478 422	144
301	78 566	78 711	78 855	78 999	79 143	79 287	79 431	79 575	79 719	79 863	144
302	80 007	80 151	80 294	80 438	80 582	80 725	80 869	81 012	81 156	81 299	144
303	81 443	81 586	81 729	81 872	82 016	82 159	82 302	82 445	82 588	82 731	143
304	82 874	83 016	83 159	83 302	83 445	83 587	83 730	83 872	84 015	84 157	143
	0	1	2	3	4	5	6	7	8	9	

Proportional Parts will be found on the facing page.

TABLE I—LOGARITHMS OF NUMBERS

	0	1	2	3	4	5	6	7	8	9	Δ
305	484 300	484 442	484 585	484 727	484 869	485 011	485 153	485 295	485 437	485 579	142
306	85 721	85 863	86 005	86 147	86 289	86 430	86 572	86 714	86 855	86 997	141
307	87 138	87 280	87 421	87 563	87 704	87 845	87 986	88 127	88 269	88 410	141
308	88 551	88 692	88 833	88 974	89 114	89 255	89 396	89 537	89 677	89 818	140
309	89 958	90 099	90 239	90 380	90 520	90 661	90 801	90 941	91 081	91 222	140
310	491 362	491 502	491 642	491 782	491 922	492 062	492 201	492 341	492 481	492 621	139
311	92 760	92 900	93 040	93 179	93 319	93 458	93 597	93 737	93 876	94 015	140
312	94 155	94 294	94 433	94 572	94 711	94 850	94 989	95 128	95 267	95 406	138
313	95 544	95 683	95 822	95 960	96 099	96 238	96 376	96 515	96 653	96 791	139
314	96 930	97 068	97 206	97 344	97 483	97 621	97 759	97 897	98 035	98 173	138
315	498 311	498 448	498 586	498 724	498 862	498 999	499 137	499 275	499 412	499 550	137
316	499 687	499 824	499 962	500 099	500 236	500 374	500 511	500 648	500 785	500 922	137
317	501 059	501 196	501 333	01 470	01 607	01 744	01 880	02 017	02 154	02 291	136
318	02 427	02 564	02 700	02 837	02 973	03 109	03 246	03 382	03 518	03 655	136
319	03 791	03 927	04 063	04 199	04 335	04 471	04 607	04 743	04 878	05 014	136
320	505 150	505 286	505 421	505 557	505 693	505 828	505 964	506 099	506 234	506 370	135
321	06 505	06 640	06 776	06 911	07 046	07 181	07 316	07 451	07 586	07 721	135
322	07 856	07 991	08 126	08 260	08 395	08 530	08 664	08 799	08 934	09 068	135
323	09 203	09 337	09 471	09 606	09 740	09 874	10 009	10 143	10 277	10 411	134
	0	1	2	3	4	5	6	7	8	9	

PROPORTIONAL PARTS

	134	135	136	137	138	139	140	141	142	143	144	145
1	13·4	13·5	13·6	13·7	13·8	13·9	14·0	14·1	14·2	14·3	14·4	14·5
2	26·8	27·0	27·2	27·4	27·6	27·8	28·0	28·2	28·4	28·6	28·8	29·0
3	40·2	40·5	40·8	41·1	41·4	41·7	42·0	42·3	42·6	42·9	43·2	43·5
4	53·6	54·0	54·4	54·8	55·2	55·6	56·0	56·4	56·8	57·2	57·6	58·0
5	67·0	67·5	68·0	68·5	69·0	69·5	70·0	70·5	71·0	71·5	72·0	72·5
6	80·4	81·0	81·6	82·2	82·8	83·4	84·0	84·6	85·2	85·8	86·4	87·0
7	93·8	94·5	95·2	95·9	96·6	97·3	98·0	98·7	99·4	100·1	100·8	101·5
8	107·2	108·0	108·8	109·6	110·4	111·2	112·0	112·8	113·6	114·4	115·2	116·0
9	120·6	121·5	122·4	123·3	124·2	125·1	126·0	126·9	127·8	128·7	129·6	130·5

	146	147	148	149	150	151	152	153	154	155	156	157
1	14·6	14·7	14·8	14·9	15·0	15·1	15·2	15·3	15·4	15·5	15·6	15·7
2	29·2	29·4	29·6	29·8	30·0	30·2	30·4	30·6	30·8	31·0	31·2	31·4
3	43·8	44·1	44·4	44·7	45·0	45·3	45·6	45·9	46·2	46·5	46·8	47·1
4	58·4	58·8	59·2	59·6	60·0	60·4	60·8	61·2	61·6	62·0	62·4	62·8
5	73·0	73·5	74·0	74·5	75·0	75·5	76·0	76·5	77·0	77·5	78·0	78·5
6	87·6	88·2	88·8	89·4	90·0	90·6	91·2	91·8	92·4	93·0	93·6	94·2
7	102·2	102·9	103·6	104·3	105·0	105·7	106·4	107·1	107·8	108·5	109·2	109·9
8	116·8	117·6	118·4	119·2	120·0	120·8	121·6	122·4	123·2	124·0	124·8	125·6
9	131·4	132·3	133·2	134·1	135·0	135·9	136·8	137·7	138·6	139·5	140·4	141·3

	158	159	160	161	162	163	164	165	166	167	168	169
1	15·8	15·9	16·0	16·1	16·2	16·3	16·4	16·5	16·6	16·7	16·8	16·9
2	31·6	31·8	32·0	32·2	32·4	32·6	32·8	33·0	33·2	33·4	33·6	33·8
3	47·4	47·7	48·0	48·3	48·6	48·9	49·2	49·5	49·8	50·1	50·4	50·7
4	63·2	63·6	64·0	64·4	64·8	65·2	65·6	66·0	66·4	66·8	67·2	67·6
5	79·0	79·5	80·0	80·5	81·0	81·5	82·0	82·5	83·0	83·5	84·0	84·5
6	94·8	95·4	96·0	96·6	97·2	97·8	98·4	99·0	99·6	100·2	100·8	101·4
7	110·6	111·3	112·0	112·7	113·4	114·1	114·8	115·5	116·2	116·9	117·6	118·3
8	126·4	127·2	128·0	128·8	129·6	130·4	131·2	132·0	132·8	133·6	134·4	135·2
9	142·2	143·1	144·0	144·9	145·8	146·7	147·6	148·5	149·4	150·3	151·2	152·1

TABLE I—LOGARITHMS OF NUMBERS

	0	1	2	3	4	5	6	7	8	9	Δ
324	510 545	510 679	510 813	510 947	511 081	511 215	511 349	511 482	511 616	511 750	133
325	511 883	512 017	512 151	512 284	512 418	512 551	512 684	512 818	512 951	513 084	134
326	13 218	13 351	13 484	13 617	13 750	13 883	14 016	14 149	14 282	14 415	133
327	14 548	14 681	14 813	14 946	15 079	15 211	15 344	15 476	15 609	15 741	133
328	15 874	16 006	16 139	16 271	16 403	16 535	16 668	16 800	16 932	17 064	132
329	17 196	17 328	17 460	17 592	17 724	17 855	17 987	18 119	18 251	18 382	132
330	518 514	518 646	518 777	518 909	519 040	519 171	519 303	519 434	519 566	519 697	131
331	19 828	19 959	20 090	20 221	20 353	20 484	20 615	20 745	20 876	21 007	131
332	21 138	21 269	21 400	21 530	21 661	21 792	21 922	22 053	22 183	22 314	130
333	22 444	22 575	22 705	22 835	22 966	23 096	23 226	23 356	23 486	23 616	130
334	23 746	23 876	24 006	24 136	24 266	24 396	24 526	24 656	24 785	24 915	130
335	525 045	525 174	525 304	525 434	525 563	525 693	525 822	525 951	526 081	526 210	129
336	26 339	26 469	26 598	26 727	26 856	26 985	27 114	27 243	27 372	27 501	129
337	27 630	27 759	27 888	28 016	28 145	28 274	28 402	28 531	28 660	28 788	129
338	28 917	29 045	29 174	29 302	29 430	29 559	29 687	29 815	29 943	30 072	128
339	30 200	30 328	30 456	30 584	30 712	30 840	30 968	31 096	31 223	31 351	128
340	531 479	531 607	531 734	531 862	531 990	532 117	532 245	532 372	532 500	532 627	127
341	32 754	32 882	33 009	33 136	33 264	33 391	33 518	33 645	33 772	33 899	127
342	34 026	34 153	34 280	34 407	34 534	34 661	34 787	34 914	35 041	35 167	127
343	35 294	35 421	35 547	35 674	35 800	35 927	36 053	36 180	36 306	36 432	126
344	36 558	36 685	36 811	36 937	37 063	37 189	37 315	37 441	37 567	37 693	126
345	537 819	537 945	538 071	538 197	538 322	538 448	538 574	538 699	538 825	538 951	125
346	39 076	39 202	39 327	39 452	39 578	39 703	39 829	39 954	40 079	40 204	125
347	40 329	40 455	40 580	40 705	40 830	40 955	41 080	41 205	41 330	41 454	125
348	41 579	41 704	41 829	41 953	42 078	42 203	42 327	42 452	42 576	42 701	124
349	42 825	42 950	43 074	43 199	43 323	43 447	43 571	43 696	43 820	43 944	124
350	544 068	544 192	544 316	544 440	544 564	544 688	544 812	544 936	545 060	545 183	124
351	45 307	45 431	45 555	45 678	45 802	45 925	46 049	46 172	46 296	46 419	124
352	46 543	46 666	46 789	46 913	47 036	47 159	47 282	47 405	47 529	47 652	123
353	47 775	47 898	48 021	48 144	48 267	48 389	48 512	48 635	48 758	48 881	122
354	49 003	49 126	49 249	49 371	49 494	49 616	49 739	49 861	49 984	50 106	122
355	550 228	550 351	550 473	550 595	550 717	550 840	550 962	551 084	551 206	551 328	122
356	51 450	51 572	51 694	51 816	51 938	52 060	52 181	52 303	52 425	52 547	121
357	52 668	52 790	52 911	53 033	53 155	53 276	53 398	53 519	53 640	53 762	121
358	53 883	54 004	54 126	54 247	54 368	54 489	54 610	54 731	54 852	54 973	121
359	55 094	55 215	55 336	55 457	55 578	55 699	55 820	55 940	56 061	56 182	121
360	556 303	556 423	556 544	556 664	556 785	556 905	557 026	557 146	557 267	557 387	120
361	57 507	57 627	57 748	57 868	57 988	58 108	58 228	58 349	58 469	58 589	120
362	58 709	58 829	58 948	59 068	59 188	59 308	59 428	59 548	59 667	59 787	120
363	59 907	60 026	60 146	60 265	60 385	60 504	60 624	60 743	60 863	60 982	119
364	61 101	61 221	61 340	61 459	61 578	61 698	61 817	61 936	62 055	62 174	119
365	562 293	562 412	562 531	562 650	562 769	562 887	563 006	563 125	563 244	563 362	119
366	63 481	63 600	63 718	63 837	63 955	64 074	64 192	64 311	64 429	64 548	118
367	64 666	64 784	64 903	65 021	65 139	65 257	65 376	65 494	65 612	65 730	118
368	65 848	65 966	66 084	66 202	66 320	66 437	66 555	66 673	66 791	66 909	117
369	67 026	67 144	67 262	67 379	67 497	67 614	67 732	67 849	67 967	68 084	118
	0	**1**	**2**	**3**	**4**	**5**	**6**	**7**	**8**	**9**	

Proportional Parts will be found on the facing page.

TABLE I—LOGARITHMS OF NUMBERS

	0	1	2	3	4	5	6	7	8	9	Δ
370	568 202	568 319	568 436	568 554	568 671	568 788	568 905	569 023	569 140	569 257	117
371	69 374	69 491	69 608	69 725	69 842	69 959	70 076	70 193	70 309	70 426	117
372	70 543	70 660	70 776	70 893	71 010	71 126	71 243	71 359	71 476	71 592	117
373	71 709	71 825	71 942	72 058	72 174	72 291	72 407	72 523	72 639	72 755	117
374	72 872	72 988	73 104	73 220	73 336	73 452	73 568	73 684	73 800	73 915	116
375	574 031	574 147	574 263	574 379	574 494	574 610	574 726	574 841	574 957	575 072	116
376	75 188	75 303	75 419	75 534	75 650	75 765	75 880	75 996	76 111	76 226	115
377	76 341	76 457	76 572	76 687	76 802	76 917	77 032	77 147	77 262	77 377	115
378	77 492	77 607	77 722	77 836	77 951	78 066	78 181	78 295	78 410	78 525	114
379	78 639	78 754	78 868	78 983	79 097	79 212	79 326	79 441	79 555	79 669	115
380	579 784	579 898	580 012	580 126	580 241	580 355	580 469	580 583	580 697	580 811	114
381	80 925	81 039	81 153	81 267	81 381	81 495	81 608	81 722	81 836	81 950	113
382	82 063	82 177	82 291	82 404	82 518	82 631	82 745	82 858	82 972	83 085	114
383	83 199	83 312	83 426	83 539	83 652	83 765	83 879	83 992	84 105	84 218	113
384	84 331	84 444	84 557	84 670	84 783	84 896	85 009	85 122	85 235	85 348	113
385	585 461	585 574	585 686	585 799	585 912	586 024	586 137	586 250	586 362	586 475	112
386	86 587	86 700	86 812	86 925	87 037	87 149	87 262	87 374	87 486	87 599	112
387	87 711	87 823	87 935	88 047	88 160	88 272	88 384	88 496	88 608	88 720	112
388	88 832	88 944	89 056	89 167	89 279	89 391	89 503	89 615	89 726	89 838	112
389	89 950	90 061	90 173	90 284	90 396	90 507	90 619	90 730	90 842	90 953	112
390	591 065	591 176	591 287	591 399	591 510	591 621	591 732	591 843	591 955	592 066	111
	0	**1**	**2**	**3**	**4**	**5**	**6**	**7**	**8**	**9**	

PROPORTIONAL PARTS

	111	112	113	114	115	116	117	118	119	120	121	122
1	11·1	11·2	11·3	11·4	11·5	11·6	11·7	11·8	11·9	12·0	12·1	12·2
2	22·2	22·4	22·6	22·8	23·0	23·2	23·4	23·6	23·8	24·0	24·2	24·4
3	33·3	33·6	33·9	34·2	34·5	34·8	35·1	35·4	35·7	36·0	36·3	36·6
4	44·4	44·8	45·2	45·6	46·0	46·4	46·8	47·2	47·6	48·0	48·4	48·8
5	55·5	56·0	56·5	57·0	57·5	58·0	58·5	59·0	59·5	60·0	60·5	61·0
6	66·6	67·2	67·8	68·4	69·0	69·6	70·2	70·8	71·4	72·0	72·6	73·2
7	77·7	78·4	79·1	79·8	80·5	81·2	81·9	82·6	83·3	84·0	84·7	85·4
8	88·8	89·6	90·4	91·2	92·0	92·8	93·6	94·4	95·2	96·0	96·8	97·6
9	99·9	100·8	101·7	102·6	103·5	104·4	105·3	106·2	107·1	108·0	108·9	109·8

	123	124	125	126	127	128	129	130	131	132	133	134
1	12·3	12·4	12·5	12·6	12·7	12·8	12·9	13·0	13·1	13·2	13·3	13·4
2	24·6	24·8	25·0	25·2	25·4	25·6	25·8	26·0	26·2	26·4	26·6	26·8
3	36·9	37·2	37·5	37·8	38·1	38·4	38·7	39·0	39·3	39·6	39·9	40·2
4	49·2	49·6	50·0	50·4	50·8	51·2	51·6	52·0	52·4	52·8	53·2	53·6
5	61·5	62·0	62·5	63·0	63·5	64·0	64·5	65·0	65·5	66·0	66·5	67·0
6	73·8	74·4	75·0	75·6	76·2	76·8	77·4	78·0	78·6	79·2	79·8	80·4
7	86·1	86·8	87·5	88·2	88·9	89·6	90·3	91·0	91·7	92·4	93·1	93·8
8	98·4	99·2	100·0	100·8	101·6	102·4	103·2	104·0	104·8	105·6	106·4	107·2
9	110·7	111·6	112·5	113·4	114·3	115·2	116·1	117·0	117·9	118·8	119·7	120·6

TABLE I—LOGARITHMS OF NUMBERS

	0	1	2	3	4	5	6	7	8	9	Δ
390	591 065	591 176	591 287	591 399	591 510	591 621	591 732	591 843	591 955	592 066	111
391	92 177	92 288	92 399	92 510	92 621	92 732	92 843	92 954	93 064	93 175	111
392	93 286	93 397	93 508	93 618	93 729	93 840	93 950	94 061	94 171	94 282	111
393	94 393	94 503	94 614	94 724	94 834	94 945	95 055	95 165	95 276	95 386	110
394	95 496	95 606	95 717	95 827	95 937	96 047	96 157	96 267	96 377	96 487	110
395	596 597	596 707	596 817	596 927	597 037	597 146	597 256	597 366	597 476	597 586	109
396	97 695	97 805	97 914	98 024	98 134	98 243	98 353	98 462	98 572	98 681	110
397	98 791	98 900	599 009	599 119	599 228	599 337	599 446	599 556	599 665	599 774	109
398	599 883	599 992	600 101	600 210	600 319	600 428	600 537	600 646	600 755	600 864	109
399	600 973	601 082	01 191	01 299	01 408	01 517	01 625	01 734	01 843	01 951	109
400	602 060	602 169	602 277	602 386	602 494	602 603	602 711	602 819	602 928	603 036	108
401	03 144	03 253	03 361	03 469	03 577	03 686	03 794	03 902	04 010	04 118	108
402	04 226	04 334	04 442	04 550	04 658	04 766	04 874	04 982	05 089	05 197	108
403	05 305	05 413	05 521	05 628	05 736	05 844	05 951	06 059	06 166	06 274	107
404	06 381	06 489	06 596	06 704	06 811	06 919	07 026	07 133	07 241	07 348	107
405	607 455	607 562	607 669	607 777	607 884	607 991	608 098	608 205	608 312	608 419	107
406	08 526	08 633	08 740	08 847	08 954	09 061	09 167	09 274	09 381	09 488	106
407	09 594	09 701	09 808	09 914	10 021	10 128	10 234	10 341	10 447	10 554	106
408	10 660	10 767	10 873	10 979	11 086	11 192	11 298	11 405	11 511	11 617	106
409	11 723	11 829	11 936	12 042	12 148	12 254	12 360	12 466	12 572	12 678	106
410	612 784	612 890	612 996	613 102	613 207	613 313	613 419	613 525	613 630	613 736	106
411	13 842	13 947	14 053	14 159	14 264	14 370	14 475	14 581	14 686	14 792	105
412	14 897	15 003	15 108	15 213	15 319	15 424	15 529	15 634	15 740	15 845	105
413	15 950	16 055	16 160	16 265	16 370	16 476	16 581	16 686	16 790	16 895	105
414	17 000	17 105	17 210	17 315	17 420	17 525	17 629	17 734	17 839	17 943	105
415	618 048	618 153	618 257	618 362	618 466	618 571	618 676	618 780	618 884	618 989	104
416	19 093	19 198	19 302	19 406	19 511	19 615	19 719	19 824	19 928	20 032	104
417	20 136	20 240	20 344	20 448	20 552	20 656	20 760	20 864	20 968	21 072	104
418	21 176	21 280	21 384	21 488	21 592	21 695	21 799	21 903	22 007	22 110	104
419	22 214	22 318	22 421	22 525	22 628	22 732	22 835	22 939	23 042	23 146	103
420	623 249	623 353	623 456	623 559	623 663	623 766	623 869	623 973	624 076	624 179	103
421	24 282	24 385	24 488	24 591	24 695	24 798	24 901	25 004	25 107	25 210	102
422	25 312	25 415	25 518	25 621	25 724	25 827	25 929	26 032	26 135	26 238	102
423	26 340	26 443	26 546	26 648	26 751	26 853	26 956	27 058	27 161	27 263	103
424	27 366	27 468	27 571	27 673	27 775	27 878	27 980	28 082	28 185	28 287	102
425	628 389	628 491	628 593	628 695	628 797	628 900	629 002	629 104	629 206	629 308	102
426	29 410	29 512	29 613	29 715	29 817	29 919	30 021	30 123	30 224	30 326	102
427	30 428	30 530	30 631	30 733	30 835	30 936	31 038	31 139	31 241	31 342	102
428	31 444	31 545	31 647	31 748	31 849	31 951	32 052	32 153	32 255	32 356	101
429	32 457	32 559	32 660	32 761	32 862	32 963	33 064	33 165	33 266	33 367	101
	0	1	2	3	4	5	6	7	8	9	

	101	**102**	**103**	**104**	**105**	**106**	**107**	**108**	**109**	**110**	**111**	**112**
1	10·1	10·2	10·3	10·4	10·5	10·6	10·7	10·8	10·9	11·0	11·1	11·2
2	20·2	20·4	20·6	20·8	21·0	21·2	21·4	21·6	21·8	22·0	22·2	22·4
3	30·3	30·6	30·9	31·2	31·5	31·8	32·1	32·4	32·7	33·0	33·3	33·6
4	40·4	40·8	41·2	41·6	42·0	42·4	42·8	43·2	43·6	44·0	44·4	44·8
5	50·5	51·0	51·5	52·0	52·5	53·0	53·5	54·0	54·5	55·0	55·5	56·0
6	60·6	61·2	61·8	62·4	63·0	63·6	64·2	64·8	65·4	66·0	66·6	67·2
7	70·7	71·4	72·1	72·8	73·5	74·2	74·9	75·6	76·3	77·0	77·7	78·4
8	80·8	81·6	82·4	83·2	84·0	84·8	85·6	86·4	87·2	88·0	88·8	89·6
9	90·9	91·8	92·7	93·6	94·5	95·4	96·3	97·2	98·1	99·0	99·9	100·8

TABLE I—LOGARITHMS OF NUMBERS

	0	1	2	3	4	5	6	7	8	9	Δ
430	633 468	633 569	633 670	633 771	633 872	633 973	634 074	634 175	634 276	634 376	101
431	34 477	34 578	34 679	34 779	34 880	34 981	35 081	35 182	35 283	35 383	101
432	35 484	35 584	35 685	35 785	35 886	35 986	36 087	36 187	36 287	36 388	100
433	36 488	36 588	36 688	36 789	36 889	36 989	37 089	37 189	37 290	37 390	100
434	37 490	37 590	37 690	37 790	37 890	37 990	38 090	38 190	38 290	38 389	100
435	638 489	638 589	638 689	638 789	638 888	638 988	639 088	639 188	639 287	639 387	99
436	39 486	39 586	39 686	39 785	39 885	39 984	40 084	40 183	40 283	40 382	99
437	40 481	40 581	40 680	40 779	40 879	40 978	41 077	41 177	41 276	41 375	99
438	41 474	41 573	41 672	41 771	41 871	41 970	42 069	42 168	42 267	42 366	99
439	42 465	42 563	42 662	42 761	42 860	42 959	43 058	43 156	43 255	43 354	99
440	643 453	643 551	643 650	643 749	643 847	643 946	644 044	644 143	644 242	644 340	99
441	44 439	44 537	44 636	44 734	44 832	44 931	45 029	45 127	45 226	45 324	98
442	45 422	45 521	45 619	45 717	45 815	45 913	46 011	46 110	46 208	46 306	98
443	46 404	46 502	46 600	46 698	46 796	46 894	46 992	47 089	47 187	47 285	98
444	47 383	47 481	47 579	47 676	47 774	47 872	47 969	48 067	48 165	48 262	98
445	648 360	648 458	648 555	648 653	648 750	648 848	648 945	649 043	649 140	649 237	98
446	49 335	49 432	49 530	49 627	49 724	49 821	49 919	50 016	50 113	50 210	98
447	50 308	50 405	50 502	50 599	50 696	50 793	50 890	50 987	51 084	51 181	97
448	51 278	51 375	51 472	51 569	51 666	51 762	51 859	51 956	52 053	52 150	96
449	52 246	52 343	52 440	52 536	52 633	52 730	52 826	52 923	53 019	53 116	97
450	653 213	653 309	653 405	653 502	653 598	653 695	653 791	653 888	653 984	654 080	97
451	54 177	54 273	54 369	54 465	54 562	54 658	54 754	54 850	54 946	55 042	96
452	55 138	55 235	55 331	55 427	55 523	55 619	55 715	55 810	55 906	56 002	96
453	56 098	56 194	56 290	56 386	56 482	56 577	56 673	56 769	56 864	56 960	96
454	57 056	57 152	57 247	57 343	57 438	57 534	57 629	57 725	57 820	57 916	95
455	658 011	658 107	658 202	658 298	658 393	658 488	658 584	658 679	658 774	658 870	95
456	58 965	59 060	59 155	59 250	59 346	59 441	59 536	59 631	59 726	59 821	95
457	59 916	60 011	60 106	60 201	60 296	60 391	60 486	60 581	60 676	60 771	94
458	60 865	60 960	61 055	61 150	61 245	61 339	61 434	61 529	61 623	61 718	95
459	61 813	61 907	62 002	62 096	62 191	62 286	62 380	62 475	62 569	62 663	95
460	662 758	662 852	662 947	663 041	663 135	663 230	663 324	663 418	663 512	663 607	94
461	63 701	63 795	63 889	63 983	64 078	64 172	64 266	64 360	64 454	64 548	94
462	64 642	64 736	64 830	64 924	65 018	65 112	65 206	65 299	65 393	65 487	94
463	65 581	65 675	65 769	65 862	65 956	66 050	66 143	66 237	66 331	66 424	94
464	66 518	66 612	66 705	66 799	66 892	66 986	67 079	67 173	67 266	67 360	93
465	667 453	667 546	667 640	667 733	667 826	667 920	668 013	668 106	668 199	668 293	93
466	68 386	68 479	68 572	68 665	68 759	68 852	68 945	69 038	69 131	69 224	93
467	69 317	69 410	69 503	69 596	69 689	69 782	69 875	69 967	70 060	70 153	93
468	70 246	70 339	70 431	70 524	70 617	70 710	70 802	70 895	70 988	71 080	93
469	71 173	71 265	71 358	71 451	71 543	71 636	71 728	71 821	71 913	72 005	93
	0	**1**	**2**	**3**	**4**	**5**	**6**	**7**	**8**	**9**	

	92	93	94	95	96	97	98	99	100	101
1	9·2	9·3	9·4	9·5	9·6	9·7	9·8	9·9	10·0	10·1
2	18·4	18·6	18·8	19·0	19·2	19·4	19·6	19·8	20·0	20·2
3	27·6	27·9	28·2	28·5	28·8	29·1	29·4	29·7	30·0	30·3
4	36·8	37·2	37·6	38·0	38·4	38·8	39·2	39·6	40·0	40·4
5	46·0	46·5	47·0	47·5	48·0	48·5	49·0	49·5	50·0	50·5
6	55·2	55·8	56·4	57·0	57·6	58·2	58·8	59·4	60·0	60·6
7	64·4	65·1	65·8	66·5	67·2	67·9	68·6	69·3	70·0	70·7
8	73·6	74·4	75·2	76·0	76·8	77·6	78·4	79·2	80·0	80·8
9	82·8	83·7	84·6	85·5	86·4	87·3	88·2	89·1	90·0	90·9

TABLE I—LOGARITHMS OF NUMBERS

	0	1	2	3	4	5	6	7	8	9	Δ
470	672 098	672 190	672 283	672 375	672 467	672 560	672 652	672 744	672 836	672 929	92
471	73 021	73 113	73 205	73 297	73 390	73 482	73 574	73 666	73 758	73 850	92
472	73 942	74 034	74 126	74 218	74 310	74 402	74 494	74 586	74 677	74 769	92
473	74 861	74 953	75 045	75 137	75 228	75 320	75 412	75 503	75 595	75 687	91
474	75 778	75 870	75 962	76 053	76 145	76 236	76 328	76 419	76 511	76 602	92
475	676 694	676 785	676 876	676 968	677 059	677 151	677 242	677 333	677 424	677 516	91
476	77 607	77 698	77 789	77 881	77 972	78 063	78 154	78 245	78 336	78 427	91
477	78 518	78 609	78 700	78 791	78 882	78 973	79 064	79 155	79 246	79 337	91
478	79 428	79 519	79 610	79 700	79 791	79 882	79 973	80 063	80 154	80 245	91
479	80 336	80 426	80 517	80 607	80 698	80 789	80 879	80 970	81 060	81 151	90
480	681 241	681 332	681 422	681 513	681 603	681 693	681 784	681 874	681 964	682 055	90
481	82 145	82 235	82 326	82 416	82 506	82 596	82 686	82 777	82 867	82 957	90
482	83 047	83 137	83 227	83 317	83 407	83 497	83 587	83 677	83 767	83 857	90
483	83 947	84 037	84 127	84 217	84 307	84 396	84 486	84 576	84 666	84 756	89
484	84 845	84 935	85 025	85 114	85 204	85 294	85 383	85 473	85 563	85 652	90
485	685 742	685 831	685 921	686 010	686 100	686 189	686 279	686 368	686 458	686 547	89
486	86 636	86 726	86 815	86 904	86 994	87 083	87 172	87 261	87 351	87 440	89
487	87 529	87 618	87 707	87 796	87 886	87 975	88 064	88 153	88 242	88 331	89
488	88 420	88 509	88 598	88 687	88 776	88 865	88 953	89 042	89 131	89 220	89
489	89 309	89 398	89 486	89 575	89 664	89 753	89 841	89 930	90 019	90 107	89
490	690 196	690 285	690 373	690 462	690 550	690 639	690 728	690 816	690 905	690 993	88
491	91 081	91 170	91 258	91 347	91 435	91 524	91 612	91 700	91 789	91 877	88
492	91 965	92 053	92 142	92 230	92 318	92 406	92 494	92 583	92 671	92 759	88
493	92 847	92 935	93 023	93 111	93 199	93 287	93 375	93 463	93 551	93 639	88
494	93 727	93 815	93 903	93 991	94 078	94 166	94 254	94 342	94 430	94 517	88
495	694 605	694 693	694 781	694 868	694 956	695 044	695 131	695 219	695 307	695 394	88
496	95 482	95 569	95 657	95 744	95 832	95 919	96 007	96 094	96 182	96 269	87
497	96 356	96 444	96 531	96 618	96 706	96 793	96 880	96 968	97 055	97 142	87
498	97 229	97 317	97 404	97 491	97 578	97 665	97 752	97 839	97 926	98 014	87
499	98 101	98 188	98 275	98 362	98 449	98 535	98 622	98 709	98 796	98 883	87
500	698 970	699 057	699 144	699 231	699 317	699 404	699 491	699 578	699 664	699 751	87
501	699 838	699 924	700 011	700 098	700 184	700 271	700 358	700 444	700 531	700 617	87
502	700 704	700 790	00 877	00 963	01 050	01 136	01 222	01 309	01 395	01 482	86
503	01 568	01 654	01 741	01 827	01 913	01 999	02 086	02 172	02 258	02 344	87
504	02 431	02 517	02 603	02 689	02 775	02 861	02 947	03 033	03 119	03 205	86
505	703 291	703 377	703 463	703 549	703 635	703 721	703 807	703 893	703 979	704 065	86
506	04 151	04 236	04 322	04 408	04 494	04 579	04 665	04 751	04 837	04 922	86
507	05 008	05 094	05 179	05 265	05 350	05 436	05 522	05 607	05 693	05 778	86
508	05 864	05 949	06 035	06 120	06 206	06 291	06 376	06 462	06 547	06 632	86
509	06 718	06 803	06 888	06 974	07 059	07 144	07 229	07 315	07 400	07 485	85
	0	1	2	3	4	5	6	7	8	9	

	85	**86**	**87**	**88**	**89**	**90**	**91**	**92**	**93**
1	8·5	8·6	8·7	8·8	8·9	9·0	9·1	9·2	9·3
2	17·0	17·2	17·4	17·6	17·8	18·0	18·2	18·4	18·6
3	25·5	25·8	26·1	26·4	26·7	27·0	27·3	27·6	27·9
4	34·0	34·4	34·8	35·2	35·6	36·0	36·4	36·8	37·2
5	42·5	43·0	43·5	44·0	44·5	45·0	45·5	46·0	46·5
6	51·0	51·6	52·2	52·8	53·4	54·0	54·6	55·2	55·8
7	59·5	60·2	60·9	61·6	62·3	63·0	63·7	64·4	65·1
8	68·0	68·8	69·6	70·4	71·2	72·0	72·8	73·6	74·4
9	76·5	77·4	78·3	79·2	80·1	81·0	81·9	82·8	83·7

TABLE I—LOGARITHMS OF NUMBERS

	0	1	2	3	4	5	6	7	8	9	Δ
510	707 570	707 655	707 740	707 826	707 911	707 996	708 081	708 166	708 251	708 336	85
511	08 421	08 506	08 591	08 676	08 761	08 846	08 931	09 015	09 100	09 185	85
512	09 270	09 355	09 440	09 524	09 609	09 694	09 779	09 863	09 948	10 033	84
513	10 117	10 202	10 287	10 371	10 456	10 540	10 625	10 710	10 794	10 879	84
514	10 963	11 048	11 132	11 217	11 301	11 385	11 470	11 554	11 639	11 723	84
515	711 807	711 892	711 976	712 060	712 144	712 229	712 313	712 397	712 481	712 566	84
516	12 650	12 734	12 818	12 902	12 986	13 070	13 154	13 238	13 323	13 407	84
517	13 491	13 575	13 659	13 742	13 826	13 910	13 994	14 078	14 162	14 246	84
518	14 330	14 414	14 497	14 581	14 665	14 749	14 833	14 916	15 000	15 084	83
519	15 167	15 251	15 335	15 418	15 502	15 586	15 669	15 753	15 836	15 920	83
520	716 003	716 087	716 170	716 254	716 337	716 421	716 504	716 588	716 671	716 754	84
521	16 838	16 921	17 004	17 088	17 171	17 254	17 338	17 421	17 504	17 587	84
522	17 671	17 754	17 837	17 920	18 003	18 086	18 169	18 253	18 336	18 419	83
523	18 502	18 585	18 668	18 751	18 834	18 917	19 000	19 083	19 165	19 248	83
524	19 331	19 414	19 497	19 580	19 663	19 745	19 828	19 911	19 994	20 077	82
525	720 159	720 242	720 325	720 407	720 490	720 573	720 655	720 738	720 821	720 903	83
526	20 986	21 068	21 151	21 233	21 316	21 398	21 481	21 563	21 646	21 728	83
527	21 811	21 893	21 975	22 058	22 140	22 222	22 305	22 387	22 469	22 552	82
528	22 634	22 716	22 798	22 881	22 963	23 045	23 127	23 209	23 291	23 374	82
529	23 456	23 538	23 620	23 702	23 784	23 866	23 948	24 030	24 112	24 194	82
530	724 276	724 358	724 440	724 522	724 604	724 685	724 767	724 849	724 931	725 013	82
531	25 095	25 176	25 258	25 340	25 422	25 503	25 585	25 667	25 748	25 830	82
532	25 912	25 993	26 075	26 156	26 238	26 320	26 401	26 483	26 564	26 646	81
533	26 727	26 809	26 890	26 972	27 053	27 134	27 216	27 297	27 379	27 460	81
534	27 541	27 623	27 704	27 785	27 866	27 948	28 029	28 110	28 191	28 273	81
535	728 354	728 435	728 516	728 597	728 678	728 759	728 841	728 922	729 003	729 084	81
536	29 165	29 246	29 327	29 408	29 489	29 570	29 651	29 732	29 813	29 893	81
537	29 974	30 055	30 136	30 217	30 298	30 378	30 459	30 540	30 621	30 702	80
538	30 782	30 863	30 944	31 024	31 105	31 186	31 266	31 347	31 428	31 508	81
539	31 589	31 669	31 750	31 830	31 911	31 991	32 072	32 152	32 233	32 313	81
540	732 394	732 474	732 555	732 635	732 715	732 796	732 876	732 956	733 037	733 117	80
541	33 197	33 278	33 358	33 438	33 518	33 598	33 679	33 759	33 839	33 919	80
542	33 999	34 079	34 160	34 240	34 320	34 400	34 480	34 560	34 640	34 720	80
543	34 800	34 880	34 960	35 040	35 120	35 200	35 279	35 359	35 439	35 519	80
544	35 599	35 679	35 759	35 838	35 918	35 998	36 078	36 157	36 237	36 317	80
545	736 397	736 476	736 556	736 635	736 715	736 795	736 874	736 954	737 034	737 113	80
546	37 193	37 272	37 352	37 431	37 511	37 590	37 670	37 749	37 829	37 908	79
547	37 987	38 067	38 146	38 225	38 305	38 384	38 463	38 543	38 622	38 701	80
548	38 781	38 860	38 939	39 018	39 097	39 177	39 256	39 335	39 414	39 493	79
549	39 572	39 651	39 731	39 810	39 889	39 968	40 047	40 126	40 205	40 284	79
	0	**1**	**2**	**3**	**4**	**5**	**6**	**7**	**8**	**9**	

	79	80	81	82	83	84	85	86
1	7·9	8·0	8·1	8·2	8·3	8·4	8·5	8·6
2	15·8	16·0	16·2	16·4	16·6	16·8	17·0	17·2
3	23·7	24·0	24·3	24·6	24·9	25·2	25·5	25·8
4	31·6	32·0	32·4	32·8	33·2	33·6	34·0	34·4
5	39·5	40·0	40·5	41·0	41·5	42·0	42·5	43·0
6	47·4	48·0	48·6	49·2	49·8	50·4	51·0	51·6
7	55·3	56·0	56·7	57·4	58·1	58·8	59·5	60·2
8	63·2	64·0	64·8	65·6	66·4	67·2	68·0	68·8
9	71·1	72·0	72·9	73·8	74·7	75·6	76·5	77·4

TABLE I—LOGARITHMS OF NUMBERS

	0	1	2	3	4	5	6	7	8	9	Δ
550	740 363	740 442	740 521	740 600	740 678	740 757	740 836	740 915	740 994	741 073	79
551	41 152	41 230	41 309	41 388	41 467	41 546	41 624	41 703	41 782	41 860	79
552	41 939	42 018	42 096	42 175	42 254	42 332	42 411	42 489	42 568	42 647	78
553	42 725	42 804	42 882	2 961	43 039	43 118	43 196	43 275	43 353	43 431	79
554	43 510	43 588	43 667	43 745	43 823	43 902	43 980	44 058	44 136	44 215	78
555	744 293	744 371	744 449	744 528	744 606	744 684	744 762	744 840	744 919	744 997	78
556	45 075	45 153	45 231	45 309	45 387	45 465	45 543	45 621	45 699	45 777	78
557	45 855	45 933	46 011	46 089	46 167	46 245	46 323	46 401	46 479	46 556	78
558	46 634	46 712	46 790	46 868	46 945	47 023	47 101	47 179	47 256	47 334	78
559	47 412	47 489	47 567	47 645	47 722	47 800	47 878	47 955	48 033	48 110	78
560	748 188	748 266	748 343	748 421	748 498	748 576	748 653	748 731	748 808	748 885	78
561	48 963	49 040	49 118	49 195	49 272	49 350	49 427	49 504	49 582	49 659	77
562	49 736	49 814	49 891	49 968	50 045	50 123	50 200	50 277	50 354	50 431	77
563	50 508	50 586	50 663	50 740	50 817	50 894	50 971	51 048	51 125	51 202	77
564	51 279	51 356	51 433	51 510	51 587	51 664	51 741	51 818	51 895	51 972	76
565	752 048	752 125	752 202	752 279	752 356	752 433	752 509	752 586	752 663	752 740	76
566	52 816	52 893	52 970	53 047	53 123	53 200	53 277	53 353	53 430	53 506	77
567	53 583	53 660	53 736	53 813	53 889	53 966	54 042	54 119	54 195	54 272	76
568	54 348	54 425	54 501	54 578	54 654	54 730	54 807	54 883	54 960	55 036	76
569	55 112	55 189	55 265	55 341	55 417	55 494	55 570	55 646	55 722	55 799	76
570	755 875	755 951	756 027	756 103	756 180	756 256	756 332	756 408	756 484	756 560	76
571	56 636	56 712	56 788	56 864	56 940	57 016	57 092	57 168	57 244	57 320	76
572	57 396	57 472	57 548	57 624	57 700	57 775	57 851	57 927	58 003	58 079	76
573	58 155	58 230	58 306	58 382	58 458	58 533	58 609	58 685	58 761	58 836	76
574	58 912	58 988	59 063	59 139	59 214	59 290	59 366	59 441	59 517	59 592	76
575	759 668	759 743	759 819	759 894	759 970	760 045	760 121	760 196	760 272	760 347	75
576	60 422	60 498	60 573	60 649	60 724	60 799	60 875	60 950	61 025	61 101	75
577	61 176	61 251	61 326	61 402	61 477	61 552	61 627	61 702	61 778	61 853	75
578	61 928	62 003	62 078	62 153	62 228	62 303	62 378	62 453	62 529	62 604	75
579	62 679	62 754	62 829	62 904	62 978	63 053	63 128	63 203	63 278	63 353	75
580	763 428	763 503	763 578	763 653	763 727	763 802	763 877	763 952	764 027	764 101	75
581	64 176	64 251	64 326	64 400	64 475	64 550	64 624	64 699	64 774	64 848	75
582	64 923	64 998	65 072	65 147	65 221	65 296	65 370	65 445	65 520	65 594	75
583	65 669	65 743	65 818	65 892	65 966	66 041	66 115	66 190	66 264	66 338	75
584	66 413	66 487	66 562	66 636	66 710	66 785	66 859	66 933	67 007	67 082	74
585	767 156	767 230	767 304	767 379	767 453	767 527	767 601	767 675	767 749	767 823	75
586	67 898	67 972	68 046	68 120	68 194	68 268	68 342	68 416	68 490	68 564	74
587	68 638	68 712	68 786	68 860	68 934	69 008	69 082	69 156	69 230	69 303	74
588	69 377	69 451	69 525	69 599	69 673	69 746	69 820	69 894	69 968	70 042	73
589	70 115	70 189	70 263	70 336	70 410	70 484	70 557	70 631	70 705	70 778	74
590	770 852	770 926	770 999	771 073	771 146	771 220	771 293	771 367	771 440	771 514	73
591	71 587	71 661	71 734	71 808	71 881	71 955	72 028	72 102	72 175	72 248	74
592	72 322	72 395	72 468	72 542	72 615	72 688	72 762	72 835	72 908	72 981	74
593	73 055	73 128	73 201	73 274	73 348	73 421	73 494	73 567	73 640	73 713	73
594	73 786	73 860	73 933	74 006	74 079	74 152	74 225	74 298	74 371	74 444	73
595	774 517	774 590	774 663	774 736	774 809	774 882	774 955	775 028	775 100	775 173	73
596	75 246	75 319	75 392	75 465	75 538	75 610	75 683	75 756	75 829	75 902	72
597	75 974	76 047	76 120	76 193	76 265	76 338	76 411	76 483	76 556	76 629	72
598	76 701	76 774	76 846	76 919	76 992	77 064	77 137	77 209	77 282	77 354	73
599	77 427	77 499	77 572	77 644	77 717	77 789	77 862	77 934	78 006	78 079	72
	0	1	2	3	4	5	6	7	8	9	

Proportional Parts will be found on the facing page.

TABLE I—LOGARITHMS OF NUMBERS

	0	**1**	**2**	**3**	**4**	**5**	**6**	**7**	**8**	**9**	*Δ*
600	778 151	778 224	778 296	778 368	778 441	778 513	778 585	778 658	778 730	778 802	72
601	78 874	78 947	79 019	79 091	79 163	79 236	79 308	79 380	79 452	79 524	72
602	79 596	79 669	79 741	79 813	79 885	79 957	80 029	80 101	80 173	80 245	72
603	80 317	80 389	80 461	80 533	80 605	80 677	80 749	80 821	80 893	80 965	72
604	81 037	81 109	81 181	81 253	81 324	81 396	81 468	81 540	81 612	81 684	71
605	781 755	781 827	781 899	781 971	782 042	782 114	782 186	782 258	782 329	782 401	72
606	82 473	82 544	82 616	82 688	82 759	82 831	82 902	82 974	83 046	83 117	72
607	83 189	83 260	83 332	83 403	83 475	83 546	83 618	83 689	83 761	83 832	72
608	83 904	83 975	84 046	84 118	84 189	84 261	84 332	84 403	84 475	84 546	71
609	84 617	84 689	84 760	84 831	84 902	84 974	85 045	85 116	85 187	85 259	71
610	785 330	785 401	785 472	785 543	785 615	785 686	785 757	785 828	785 899	785 970	71
611	86 041	86 112	86 183	86 254	86 325	86 396	86 467	86 538	86 609	86 680	71
612	86 751	86 822	86 893	86 964	87 035	87 106	87 177	87 248	87 319	87 390	70
613	87 460	87 531	87 602	87 673	87 744	87 815	87 885	87 956	88 027	88 098	70
614	88 168	88 239	88 310	88 381	88 451	88 522	88 593	88 663	88 734	88 804	71
615	788 875	788 946	789 016	789 087	789 157	789 228	789 299	789 369	789 440	789 510	71
616	89 581	89 651	89 722	89 792	89 863	89 933	90 004	90 074	90 144	90 215	70
617	90 285	90 356	90 426	90 496	90 567	90 637	90 707	90 778	90 848	90 918	70
618	90 988	91 059	91 129	91 199	91 269	91 340	91 410	91 480	91 550	91 620	71
619	91 691	91 761	91 831	91 901	91 971	92 041	92 111	92 181	92 252	92 322	70
620	792 392	792 462	792 532	792 602	792 672	792 742	792 812	792 882	792 952	793 022	70
621	93 092	93 162	93 231	93 301	93 371	93 441	93 511	93 581	93 651	93 721	69
622	93 790	93 860	93 930	94 000	94 070	94 139	94 209	94 279	94 349	94 418	70
623	94 488	94 558	94 627	94 697	94 767	94 836	94 906	94 976	95 045	95 115	70
624	95 185	95 254	95 324	95 393	95 463	95 532	95 602	95 672	95 741	95 811	69
625	795 880	795 949	796 019	796 088	796 158	796 227	796 297	796 366	796 436	796 505	69
626	96 574	96 644	96 713	96 782	96 852	96 921	96 990	97 060	97 129	97 198	70
627	97 268	97 337	97 406	97 475	97 545	97 614	97 683	97 752	97 821	97 890	70
628	97 960	98 029	98 098	98 167	98 236	98 305	98 374	98 443	98 513	98 582	69
629	98 651	98 720	98 789	98 858	98 927	98 996	99 065	99 134	99 203	99 272	69
630	799 341	799 409	799 478	799 547	799 616	799 685	799 754	799 823	799 892	799 961	68
631	800 029	800 098	800 167	800 236	800 305	800 373	800 442	800 511	800 580	800 648	69
632	00 717	00 786	00 854	00 923	00 992	01 061	01 129	01 198	01 266	01 335	69
633	01 404	01 472	01 541	01 609	01 678	01 747	01 815	01 884	01 952	02 021	68
634	02 089	02 158	02 226	02 295	02 363	02 432	02 500	02 568	02 637	02 705	69
635	802 774	802 842	802 910	802 979	803 047	803 116	803 184	803 252	803 321	803 389	68
636	03 457	03 525	03 594	03 662	03 730	03 798	03 867	03 935	04 003	04 071	68
637	04 139	04 208	04 276	04 344	04 412	04 480	04 548	04 616	04 685	04 753	68
638	04 821	04 889	04 957	05 025	05 093	05 161	05 229	05 297	05 365	05 433	68
639	05 501	05 569	05 637	05 705	05 773	05 841	05 908	05 976	06 044	06 112	68
	0	**1**	**2**	**3**	**4**	**5**	**6**	**7**	**8**	**9**	

	67	**68**	**69**	**70**	**71**	**72**	**73**	**74**	**75**	**76**	**77**	**78**	**79**
1	6·7	6·8	6·9	7·0	7·1	7·2	7·3	7·4	7·5	7·6	7·7	7·8	7·9
2	13·4	13·6	13·8	14·0	14·2	14·4	14·6	14·8	15·0	15·2	15·4	15·6	15·8
3	20·1	20·4	20·7	21·0	21·3	21·6	21·9	22·2	22·5	22·8	23·1	23·4	23·7
4	26·8	27·2	27·6	28·0	28·4	28·8	29·2	29·6	30·0	30·4	30·8	31·2	31·6
5	33·5	34·0	34·5	35·0	35·5	36·0	36·5	37·0	37·5	38·0	38·5	39·0	39·5
6	40·2	40·8	41·4	42·0	42·6	43·2	43·8	44·4	45·0	45·6	46·2	46·8	47·4
7	46·9	47·6	48·3	49·0	49·7	50·4	51·1	51·8	52·5	53·2	53·9	54·6	55·3
8	53·6	54·4	55·2	56·0	56·8	57·6	58·4	59·2	60·0	60·8	61·6	62·4	63·2
9	60·3	61·2	62·1	63·0	63·9	64·8	65·7	66·6	67·5	68·4	69·3	70·2	71·1

TABLE I—LOGARITHMS OF NUMBERS

	0	1	2	3	4	5	6	7	8	9	Δ
640	806 180	806 248	806 316	806 384	806 451	806 519	806 587	806 655	806 723	806 790	68
641	06 858	06 926	06 994	07 061	07 129	07 197	07 264	07 332	07 400	07 467	68
642	07 535	07 603	07 670	07 738	07 806	07 873	07 941	08 008	08 076	08 143	68
643	08 211	08 279	08 346	08 414	08 481	08 549	08 616	08 684	08 751	08 818	68
644	08 886	08 953	09 021	09 088	09 156	09 223	09 290	09 358	09 425	09 492	68
645	809 560	809 627	809 694	809 762	809 829	809 896	809 964	810 031	810 098	810 165	68
646	10 233	10 300	10 367	10 434	10 501	10 569	10 636	10 703	10 770	10 837	67
647	10 904	10 971	11 039	11 106	11 173	11 240	11 307	11 374	11 441	11 508	67
648	11 575	11 642	11 709	11 776	11 843	11 910	11 977	12 044	12 111	12 178	67
649	12 245	12 312	12 379	12 445	12 512	12 579	12 646	12 713	12 780	12 847	66
650	812 913	812 980	813 047	813 114	813 181	813 247	813 314	813 381	813 448	813 514	67
651	13 581	13 648	13 714	13 781	13 848	13 914	13 981	14 048	14 114	14 181	67
652	14 248	14 314	14 381	14 447	14 514	14 581	14 647	14 714	14 780	14 847	66
653	14 913	14 980	15 046	15 113	15 179	15 246	15 312	15 378	15 445	15 511	67
654	15 578	15 644	15 711	15 777	15 843	15 910	15 976	16 042	16 109	16 175	66
655	816 241	816 308	816 374	816 440	816 506	816 573	816 639	816 705	816 771	816 838	66
656	16 904	16 970	17 036	17 102	17 169	17 235	17 301	17 367	17 433	17 499	66
657	17 565	17 631	17 698	17 764	17 830	17 896	17 962	18 028	18 094	18 160	66
658	18 226	18 292	18 358	18 424	18 490	18 556	18 622	18 688	18 754	18 820	65
659	18 885	18 951	19 017	19 083	19 149	19 215	19 281	19 346	19 412	19 478	66
660	819 544	819 610	819 676	819 741	819 807	819 873	819 939	820 004	820 070	820 136	65
661	20 201	20 267	20 333	20 399	20 464	20 530	20 595	20 661	20 727	20 792	66
662	20 858	20 924	20 989	21 055	21 120	21 186	21 251	21 317	21 382	21 448	66
663	21 514	21 579	21 645	21 710	21 775	21 841	21 906	21 972	22 037	22 103	65
664	22 168	22 233	22 299	22 364	22 430	22 495	22 560	22 626	22 691	22 756	66
665	822 822	822 887	822 952	823 018	823 083	823 148	823 213	823 279	823 344	823 409	65
666	23 474	23 539	23 605	23 670	23 735	23 800	23 865	23 930	23 996	24 061	65
667	24 126	24 191	24 256	24 321	24 386	24 451	24 516	24 581	24 646	24 711	65
668	24 776	24 841	24 906	24 971	25 036	25 101	25 166	25 231	25 296	25 361	65
669	25 426	25 491	25 556	25 621	25 686	25 751	25 815	25 880	25 945	26 010	65
670	826 075	826 140	826 204	826 269	826 334	826 399	826 464	826 528	826 593	826 658	65
671	26 723	26 787	26 852	26 917	26 981	27 046	27 111	27 175	27 240	27 305	64
672	27 369	27 434	27 499	27 563	27 628	27 692	27 757	27 821	27 886	27 951	64
673	28 015	28 080	28 144	28 209	28 273	28 338	28 402	28 467	28 531	28 595	65
674	28 660	28 724	28 789	28 853	28 918	28 982	29 046	29 111	29 175	29 239	65
675	829 304	829 368	829 432	829 497	829 561	829 625	829 690	829 754	829 818	829 882	65
676	29 947	30 011	30 075	30 139	30 204	30 268	30 332	30 396	30 460	30 525	64
677	30 589	30 653	30 717	30 781	30 845	30 909	30 973	31 037	31 102	31 166	64
678	31 230	31 294	31 358	31 422	31 486	31 550	31 614	31 678	31 742	31 806	64
679	31 870	31 934	31 998	32 062	32 126	32 189	32 253	32 317	32 381	32 445	64
680	832 509	832 573	832 637	832 700	832 764	832 828	832 892	832 956	833 020	833 083	64
681	33 147	33 211	33 275	33 338	33 402	33 466	33 530	33 593	33 657	33 721	63
682	33 784	33 848	33 912	33 975	34 039	34 103	34 166	34 230	34 294	34 357	64
683	34 421	34 484	34 548	34 611	34 675	34 739	34 802	34 866	34 929	34 993	63
684	35 056	35 120	35 183	35 247	35 310	35 373	35 437	35 500	35 564	35 627	64
685	835 691	835 754	835 817	835 881	835 944	836 007	836 071	836 134	836 197	836 261	63
686	36 324	36 387	36 451	36 514	36 577	36 641	36 704	36 767	36 830	36 894	63
687	36 957	37 020	37 083	37 146	37 210	37 273	37 336	37 399	37 462	37 525	63
688	37 588	37 652	37 715	37 778	37 841	37 904	37 967	38 030	38 093	38 156	63
689	38 219	38 282	38 345	38 408	38 471	38 534	38 597	38 660	38 723	38 786	63
	0	1	2	3	4	5	6	7	8	9	

Proportional Parts will be found on the facing page.

TABLE I—LOGARITHMS OF NUMBERS

	0	1	2	3	4	5	6	7	8	9	Δ
690	838 849	838 912	838 975	839 038	839 101	839 164	839 227	839 289	839 352	839 415	63
691	39 478	39 541	39 604	39 667	39 729	39 792	39 855	39 918	39 981	40 043	63
692	40 106	40 169	40 232	40 294	40 357	40 420	40 482	40 545	40 608	40 671	62
693	40 733	40 796	40 859	40 921	40 984	41 046	41 109	41 172	41 234	41 297	62
694	41 359	41 422	41 485	41 547	41 610	41 672	41 735	41 797	41 860	41 922	63
695	841 985	842 047	842 110	842 172	842 235	842 297	842 360	842 422	842 484	842 547	62
696	42 609	42 672	42 734	42 796	42 859	42 921	42 983	43 046	43 108	43 170	63
697	43 233	43 295	43 357	43 420	43 482	43 544	43 606	43 669	43 731	43 793	62
698	43 855	43 918	43 980	44 042	44 104	44 166	44 229	44 291	44 353	44 415	62
699	44 477	44 539	44 601	44 664	44 726	44 788	44 850	44 912	44 974	45 036	62
700	845 098	845 160	845 222	845 284	845 346	845 408	845 470	845 532	845 594	845 656	62
701	45 718	45 780	45 842	45 904	45 966	46 028	46 090	46 151	46 213	46 275	62
702	46 337	46 399	46 461	46 523	46 585	46 646	46 708	46 770	46 832	46 894	61
703	46 955	47 017	47 079	47 141	47 202	47 264	47 326	47 388	47 449	47 511	62
704	47 573	47 634	47 696	47 758	47 819	47 881	47 943	48 004	48 066	48 128	61
705	848 189	848 251	848 312	848 374	848 435	848 497	848 559	848 620	848 682	848 743	62
706	48 805	48 866	48 928	48 989	49 051	49 112	49 174	49 235	49 297	49 358	61
707	49 419	49 481	49 542	49 604	49 665	49 726	49 788	49 849	49 911	49 972	61
708	50 033	50 095	50 156	50 217	50 279	50 340	50 401	50 462	50 524	50 585	61
709	50 646	50 707	50 769	50 830	50 891	50 952	51 014	51 075	51 136	51 197	61
710	851 258	851 320	851 381	851 442	851 503	851 564	851 625	851 686	851 747	851 809	61
711	51 870	51 931	51 992	52 053	52 114	52 175	52 236	52 297	52 358	52 419	61
712	52 480	52 541	52 602	52 663	52 724	52 785	52 846	52 907	52 968	53 029	61
713	53 090	53 150	53 211	53 272	53 333	53 394	53 455	53 516	53 577	53 637	61
714	53 698	53 759	53 820	53 881	53 941	54 002	54 063	54 124	54 185	54 245	61
715	854 306	854 367	854 428	854 488	854 549	854 610	854 670	854 731	854 792	854 852	61
716	54 913	54 974	55 034	55 095	55 156	55 216	55 277	55 337	55 398	55 459	60
717	55 519	55 580	55 640	55 701	55 761	55 822	55 882	55 943	56 003	56 064	60
718	56 124	56 185	56 245	56 306	56 366	56 427	56 487	56 548	56 608	56 668	61
719	56 729	56 789	56 850	56 910	56 970	57 031	57 091	57 152	57 212	57 272	60
720	857 332	857 393	857 453	857 513	857 574	857 634	857 694	857 755	857 815	857 875	60
721	57 935	57 995	58 056	58 116	58 176	58 236	58 297	58 357	58 417	58 477	60
722	58 537	58 597	58 657	58 718	58 778	58 838	58 898	58 958	59 018	59 078	60
723	59 138	59 198	59 258	59 318	59 379	59 439	59 499	59 559	59 619	59 679	60
724	59 739	59 799	59 859	59 918	59 978	60 038	60 098	60 158	60 218	60 278	60
725	860 338	860 398	860 458	860 518	860 578	860 637	860 697	860 757	860 817	860 877	60
726	60 937	60 996	61 056	61 116	61 176	61 236	61 295	61 355	61 415	61 475	59
727	61 534	61 594	61 654	61 714	61 773	61 833	61 893	61 952	62 012	62 072	59
728	62 131	62 191	62 251	62 310	62 370	62 430	62 489	62 549	62 608	62 668	60
729	62 728	62 787	62 847	62 906	62 966	63 025	63 085	63 144	63 204	63 263	60
	0	1	2	3	4	5	6	7	8	9	

	59	60	61	62	63	64	65	66	67	68
1	5·9	6·0	6·1	6·2	6·3	6·4	6·5	6·6	6·7	6·8
2	11·8	12·0	12·2	12·4	12·6	12·8	13·0	13·2	13·4	13·6
3	17·7	18·0	18·3	18·6	18·9	19·2	19·5	19·8	20·1	20·4
4	23·6	24·0	24·4	24·8	25·2	25·6	26·0	26·4	26·8	27·2
5	29·5	30·0	30·5	31·0	31·5	32·0	32·5	33·0	33·5	34·0
6	35·4	36·0	36·6	37·2	37·8	38·4	39·0	39·6	40·2	40·8
7	41·3	42·0	42·7	43·4	44·1	44·8	45·5	46·2	46·9	47·6
8	47·2	48·0	48·8	49·6	50·4	51·2	52·0	52·8	53·6	54·4
9	53·1	54·0	54·9	55·8	56·7	57·6	58·5	59·4	60·3	61·2

TABLE I—LOGARITHMS OF NUMBERS

	0	1	2	3	4	5	6	7	8	9	Δ
730	863 323	863 382	863 442	863 501	863 561	863 620	863 680	863 739	863 799	863 858	59
731	63 917	63 977	64 036	64 096	64 155	64 214	64 274	64 333	64 392	64 452	59
732	64 511	64 570	64 630	64 689	64 748	64 808	64 867	64 926	64 985	65 045	59
733	65 104	65 163	65 222	65 282	65 341	65 400	65 459	65 519	65 578	65 637	59
734	65 696	65 755	65 814	65 874	65 933	65 992	66 051	66 110	66 169	66 228	59
735	866 287	866 346	866 405	866 465	866 524	866 583	866 642	866 701	866 760	866 819	59
736	66 878	66 937	66 996	67 055	67 114	67 173	67 232	67 291	67 350	67 409	58
737	67 467	67 526	67 585	67 644	67 703	67 762	67 821	67 880	67 939	67 998	58
738	68 056	68 115	68 174	68 233	68 292	68 350	68 409	68 468	68 527	68 586	58
739	68 644	68 703	68 762	68 821	68 879	68 938	68 997	69 056	69 114	69 173	59
740	869 232	869 290	869 349	869 408	869 466	869 525	869 584	869 642	869 701	869 760	58
741	69 818	69 877	69 935	69 994	70 053	70 111	70 170	70 228	70 287	70 345	59
742	70 404	70 462	70 521	70 579	70 638	70 696	70 755	70 813	70 872	70 930	59
743	70 989	71 047	71 106	71 164	71 223	71 281	71 339	71 398	71 456	71 515	58
744	71 573	71 631	71 690	71 748	71 806	71 865	71 923	71 981	72 040	72 098	58
745	872 156	872 215	872 273	872 331	872 389	872 448	872 506	872 564	872 622	872 681	58
746	72 739	72 797	72 855	72 913	72 972	73 030	73 088	73 146	73 204	73 262	59
747	73 321	73 379	73 437	73 495	73 553	73 611	73 669	73 727	73 785	73 844	58
748	73 902	73 960	74 018	74 076	74 134	74 192	74 250	74 308	74 366	74 424	58
749	74 482	74 540	74 598	74 656	74 714	74 772	74 830	74 888	74 945	75 003	58
750	875 061	875 119	875 177	875 235	875 293	875 351	875 409	875 466	875 524	875 582	58
751	75 640	75 698	75 756	75 813	75 871	75 929	75 987	76 045	76 102	76 160	58
752	76 218	76 276	76 333	76 391	76 449	76 507	76 564	76 622	76 680	76 737	58
753	76 795	76 853	76 910	76 968	77 026	77 083	77 141	77 199	77 256	77 314	57
754	77 371	77 429	77 487	77 544	77 602	77 659	77 717	77 774	77 832	77 889	58
755	877 947	878 004	878 062	878 119	878 177	878 234	878 292	878 349	878 407	878 464	58
756	78 522	78 579	78 637	78 694	78 752	78 809	78 866	78 924	78 981	79 039	57
757	79 096	79 153	79 211	79 268	79 325	79 383	79 440	79 497	79 555	79 612	57
758	79 669	79 726	79 784	79 841	79 898	79 956	80 013	80 070	80 127	80 185	57
759	80 242	80 299	80 356	80 413	80 471	80 528	80 585	80 642	80 699	80 756	58
760	880 814	880 871	880 928	880 985	881 042	881 099	881 156	881 213	881 271	881 328	57
761	81 385	81 442	81 499	81 556	81 613	81 670	81 727	81 784	81 841	81 898	57
762	81 955	82 012	82 069	82 126	82 183	82 240	82 297	82 354	82 411	82 468	57
763	82 525	82 581	82 638	82 695	82 752	82 809	82 866	82 923	82 980	83 037	56
764	83 093	83 150	83 207	83 264	83 321	83 377	83 434	83 491	83 548	83 605	56
765	883 661	883 718	883 775	883 832	883 888	883 945	884 002	884 059	884 115	884 172	57
766	84 229	84 285	84 342	84 399	84 455	84 512	84 569	84 625	84 682	84 739	56
767	84 795	84 852	84 909	84 965	85 022	85 078	85 135	85 192	85 248	85 305	56
768	85 361	85 418	85 474	85 531	85 587	85 644	85 700	85 757	85 813	85 870	56
769	85 926	85 983	86 039	86 096	86 152	86 209	86 265	86 321	86 378	86 434	57
770	886 491	886 547	886 604	886 660	886 716	886 773	886 829	886 885	886 942	886 998	56
771	87 054	87 111	87 167	87 223	87 280	87 336	87 392	87 449	87 505	87 561	56
772	87 617	87 674	87 730	87 786	87 842	87 898	87 955	88 011	88 067	88 123	56
773	88 179	88 236	88 292	88 348	88 404	88 460	88 516	88 573	88 629	88 685	56
774	88 741	88 797	88 853	88 909	88 965	89 021	89 077	89 134	89 190	89 246	56
775	889 302	889 358	889 414	889 470	889 526	889 582	889 638	889 694	889 750	889 806	56
776	89 862	89 918	89 974	90 030	90 086	90 141	90 197	90 253	90 309	90 365	56
777	90 421	90 477	90 533	90 589	90 645	90 700	90 756	90 812	90 868	90 924	56
778	90 980	91 035	91 091	91 147	91 203	91 259	91 314	91 370	91 426	91 482	55
779	91 537	91 593	91 649	91 705	91 760	91 816	91 872	91 928	91 983	92 039	56
	0	**1**	**2**	**3**	**4**	**5**	**6**	**7**	**8**	**9**	

Proportional Parts will be found on the facing page.

TABLE I—LOGARITHMS OF NUMBERS

	0	1	2	3	4	5	6	7	8	9	Δ
780	892 095	892 150	892 206	892 262	892 317	892 373	892 429	892 484	892 540	892 595	56
781	92 651	92 707	92 762	92 818	92 873	92 929	92 985	93 040	93 096	93 151	56
782	93 207	93 262	93 318	93 373	93 429	93 484	93 540	93 595	93 651	93 706	56
783	93 762	93 817	93 873	93 928	93 984	94 039	94 094	94 150	94 205	94 261	55
784	94 316	94 371	94 427	94 482	94 538	94 593	94 648	94 704	94 759	94 814	56
785	894 870	894 925	894 980	895 036	895 091	895 146	895 201	895 257	895 312	895 367	56
786	95 423	95 478	95 533	95 588	95 644	95 699	95 754	95 809	95 864	95 920	55
787	95 975	96 030	96 085	96 140	96 195	96 251	96 306	96 361	96 416	96 471	55
788	96 526	96 581	96 636	96 692	96 747	96 802	96 857	96 912	96 967	97 022	55
789	97 077	97 132	97 187	97 242	97 297	97 352	97 407	97 462	97 517	97 572	55
790	897 627	897 682	897 737	897 792	897 847	897 902	897 957	898 012	898 067	898 122	54
791	98 176	98 231	98 286	98 341	98 396	98 451	98 506	98 561	98 615	98 670	55
792	98 725	98 780	98 835	98 890	98 944	98 999	99 054	99 109	99 164	99 218	55
793	99 273	99 328	99 383	99 437	899 492	899 547	899 602	899 656	899 711	899 766	55
794	899 821	899 875	899 930	899 985	900 039	900 094	900 149	900 203	900 258	900 312	55
795	900 367	900 422	900 476	900 531	900 586	900 640	900 695	900 749	900 804	900 859	54
796	00 913	00 968	01 022	01 077	01 131	01 186	01 240	01 295	01 349	01 404	54
797	01 458	01 513	01 567	01 622	01 676	01 731	01 785	01 840	01 894	01 948	55
798	02 003	02 057	02 112	02 166	02 221	02 275	02 329	02 384	02 438	02 492	55
799	02 547	02 601	02 655	02 710	02 764	02 818	02 873	02 927	02 981	03 036	54
800	903 090	903 144	903 199	903 253	903 307	903 361	903 416	903 470	903 524	903 578	55
801	03 633	03 687	03 741	03 795	03 849	03 904	03 958	04 012	04 066	04 120	54
802	04 174	04 229	04 283	04 337	04 391	04 445	04 499	04 553	04 607	04 661	55
803	04 716	04 770	04 824	04 878	04 932	04 986	05 040	05 094	05 148	05 202	54
804	05 256	05 310	05 364	05 418	05 472	05 526	05 580	05 634	05 688	05 742	54
805	905 796	905 850	905 904	905 958	906 012	906 066	906 119	906 173	906 227	906 281	54
806	06 335	06 389	06 443	06 497	06 551	06 604	06 658	06 712	06 766	06 820	54
807	06 874	06 927	06 981	07 035	07 089	07 143	07 196	07 250	07 304	07 358	53
808	07 411	07 465	07 519	07 573	07 626	07 680	07 734	07 787	07 841	07 895	54
809	07 949	08 002	08 056	08 110	08 163	08 217	08 270	08 324	08 378	08 431	54
810	908 485	908 539	908 592	908 646	908 699	908 753	908 807	908 860	908 914	908 967	54
811	09 021	09 074	09 128	09 181	09 235	09 289	09 342	09 396	09 449	09 503	53
812	09 556	09 610	09 663	09 716	09 770	09 823	09 877	09 930	09 984	10 037	54
813	10 091	10 144	10 197	10 251	10 304	10 358	10 411	10 464	10 518	10 571	53
814	10 624	10 678	10 731	10 784	10 838	10 891	10 944	10 998	11 051	11 104	54
815	911 158	911 211	911 264	911 317	911 371	911 424	911 477	911 530	911 584	911 637	53
816	11 690	11 743	11 797	11 850	11 903	11 956	12 009	12 063	12 116	12 169	53
817	12 222	12 275	12 328	12 381	12 435	12 488	12 541	12 594	12 647	12 700	53
818	12 753	12 806	12 859	12 913	12 966	13 019	13 072	13 125	13 178	13 231	53
819	13 284	13 337	13 390	13 443	13 496	13 549	13 602	13 655	13 708	13 761	53
	0	**1**	**2**	**3**	**4**	**5**	**6**	**7**	**8**	**9**	

	53	**54**	**55**	**56**	**57**	**58**	**59**	**60**
1	5·3	5·4	5·5	5·6	5·7	5·8	5·9	6·0
2	10·6	10·8	11·0	11·2	11·4	11·6	11·8	12·0
3	15·9	16·2	16·5	16·8	17·1	17·4	17·7	18·0
4	21·2	21·6	22·0	22·4	22·8	23·2	23·6	24·0
5	26·5	27·0	27·5	28·0	28·5	29·0	29·5	30·0
6	31·8	32·4	33·0	33·6	34·2	34·8	35·4	36·0
7	37·1	37·8	38·5	39·2	39·9	40·6	41·3	42·0
8	42·4	43·2	44·0	44·8	45·6	46·4	47·2	48·0
9	47·7	48·6	49·5	50·4	51·3	52·2	53·1	54·0

TABLE I—LOGARITHMS OF NUMBERS

	0	1	2	3	4	5	6	7	8	9	Δ
820	913 814	913 867	913 920	913 973	914 026	914 079	914 132	914 184	914 237	914 290	53
821	14 343	14 396	14 449	14 502	14 555	14 608	14 660	14 713	14 766	14 819	53
822	14 872	14 925	14 977	15 030	15 083	15 136	15 189	15 241	15 294	15 347	53
823	15 400	15 453	15 505	15 558	15 611	15 664	15 716	15 769	15 822	15 875	52
824	15 927	15 980	16 033	16 085	16 138	16 191	16 243	16 296	16 349	16 401	53
825	916 454	916 507	916 559	916 612	916 664	916 717	916 770	916 822	916 875	916 927	53
826	16 980	17 033	17 085	17 138	17 190	17 243	17 295	17 348	17 400	17 453	53
827	17 506	17 558	17 611	17 663	17 716	17 768	17 820	17 873	17 925	17 978	52
828	18 030	18 083	18 135	18 188	18 240	18 293	18 345	18 397	18 450	18 502	53
829	18 555	18 607	18 659	18 712	18 764	18 816	18 869	18 921	18 973	19 026	52
830	919 078	919 130	919 183	919 235	919 287	919 340	919 392	919 444	919 496	919 549	52
831	19 601	19 653	19 706	19 758	19 810	19 862	19 914	19 967	20 019	20 071	52
832	20 123	20 176	20 228	20 280	20 332	20 384	20 436	20 489	20 541	20 593	52
833	20 645	20 697	20 749	20 801	20 853	20 906	20 958	21 010	21 062	21 114	52
834	21 166	21 218	21 270	21 322	21 374	21 426	21 478	21 530	21 582	21 634	52
835	921 686	921 738	921 790	921 842	921 894	921 946	921 998	922 050	922 102	922 154	52
836	22 206	22 258	22 310	22 362	22 414	22 466	22 518	22 570	22 622	22 674	51
837	22 725	22 777	22 829	22 881	22 933	22 985	23 037	23 089	23 140	23 192	52
838	23 244	23 296	23 348	23 399	23 451	23 503	23 555	23 607	23 658	23 710	52
839	23 762	23 814	23 865	23 917	23 969	24 021	24 072	24 124	24 176	24 228	51
840	924 279	924 331	924 383	924 434	924 486	924 538	924 589	924 641	924 693	924 744	52
841	24 796	24 848	24 899	24 951	25 003	25 054	25 106	25 157	25 209	25 261	51
842	25 312	25 364	25 415	25 467	25 518	25 570	25 621	25 673	25 725	25 776	52
843	25 828	25 879	25 931	25 982	26 034	26 085	26 137	26 188	26 240	26 291	51
844	26 342	26 394	26 445	26 497	26 548	26 600	26 651	26 702	26 754	26 805	52
845	926 857	926 908	926 959	927 011	927 062	927 114	927 165	927 216	927 268	927 319	51
846	27 370	27 422	27 473	27 524	27 576	27 627	27 678	27 730	27 781	27 832	51
847	27 883	27 935	27 986	28 037	28 088	28 140	28 191	28 242	28 293	28 345	51
848	28 396	28 447	28 498	28 549	28 601	28 652	28 703	28 754	28 805	28 857	51
849	28 908	28 959	29 010	29 061	29 112	29 163	29 215	29 266	29 317	29 368	51
850	929 419	929 470	929 521	929 572	929 623	929 674	929 725	929 776	929 827	929 879	51
851	29 930	29 981	30 032	30 083	30 134	30 185	30 236	30 287	30 338	30 389	51
852	30 440	30 491	30 542	30 592	30 643	30 694	30 745	30 796	30 847	30 898	51
853	30 949	31 000	31 051	31 102	31 153	31 204	31 254	31 305	31 356	31 407	51
854	31 458	31 509	31 560	31 610	31 661	31 712	31 763	31 814	31 865	31 915	51
855	931 966	932 017	932 068	932 118	932 169	932 220	932 271	932 322	932 372	932 423	51
856	32 474	32 524	32 575	32 626	32 677	32 727	32 778	32 829	32 879	32 930	51
857	32 981	33 031	33 082	33 133	33 183	33 234	33 285	33 335	33 386	33 437	50
858	33 487	33 538	33 589	33 639	33 690	33 740	33 791	33 841	33 892	33 943	50
859	33 993	34 044	34 094	34 145	34 195	34 246	34 296	34 347	34 397	34 448	50
860	934 498	934 549	934 599	934 650	934 700	934 751	934 801	934 852	934 902	934 953	50
861	35 003	35 054	35 104	35 154	35 205	35 255	35 306	35 356	35 406	35 457	50
862	35 507	35 558	35 608	35 658	35 709	35 759	35 809	35 860	35 910	35 960	51
863	36 011	36 061	36 111	36 162	36 212	36 262	36 313	36 363	36 413	36 463	51
864	36 514	36 564	36 614	36 665	36 715	36 765	36 815	36 865	36 916	36 966	50
865	937 016	937 066	937 117	937 167	937 217	937 267	937 317	937 367	937 418	937 468	50
866	37 518	37 568	37 618	37 668	37 718	37 769	37 819	37 869	37 919	37 969	50
867	38 019	38 069	38 119	38 169	38 219	38 269	38 320	38 370	38 420	38 470	50
868	38 520	38 570	38 620	38 670	38 720	38 770	38 820	38 870	38 920	38 970	50
869	39 020	39 070	39 120	39 170	39 220	39 270	39 320	39 369	39 419	39 469	50
	0	1	2	3	4	5	6	7	8	9	

Proportional Parts will be found on the facing page.

TABLE I—LOGARITHMS OF NUMBERS

	0	1	2	3	4	5	6	7	8	9	Δ
870	939 519	939 569	939 619	939 669	939 719	939 769	939 819	939 869	939 918	939 968	50
871	40 018	40 068	40 118	40 168	40 218	40 267	40 317	40 367	40 417	40 467	49
872	40 516	40 566	40 616	40 666	40 716	40 765	40 815	40 865	40 915	40 964	50
873	41 014	41 064	41 114	41 163	41 213	41 263	41 313	41 362	41 412	41 462	49
874	41 511	41 561	41 611	41 660	41 710	41 760	41 809	41 859	41 909	41 958	50
875	942 008	942 058	942 107	942 157	942 207	942 256	942 306	942 355	942 405	942 455	49
876	42 504	42 554	42 603	42 653	42 702	42 752	42 801	42 851	42 901	42 950	50
877	43 000	43 049	43 099	43 148	43 198	43 247	43 297	43 346	43 396	43 445	50
878	43 495	43 544	43 593	43 643	43 692	43 742	43 791	43 841	43 890	43 939	50
879	43 989	44 038	44 088	44 137	44 186	44 236	44 285	44 335	44 384	44 433	50
880	944 483	944 532	944 581	944 631	944 680	944 729	944 779	944 828	944 877	944 927	49
881	44 976	45 025	45 074	45 124	45 173	45 222	45 272	45 321	45 370	45 419	50
882	45 469	45 518	45 567	45 616	45 665	45 715	45 764	45 813	45 862	45 912	49
883	45 961	46 010	46 059	46 108	46 157	46 207	46 256	46 305	46 354	46 403	49
884	46 452	46 501	46 551	46 600	46 649	46 698	46 747	46 796	46 845	46 894	49
885	946 943	946 992	947 041	947 090	947 140	947 189	947 238	947 287	947 336	947 385	49
886	47 434	47 483	47 532	47 581	47 630	47 679	47 728	47 777	47 826	47 875	49
887	47 924	47 973	48 022	48 070	48 119	48 168	48 217	48 266	48 315	48 364	49
888	48 413	48 462	48 511	48 560	48 609	48 657	48 706	48 755	48 804	48 853	49
889	48 902	48 951	48 999	49 048	49 097	49 146	49 195	49 244	49 292	49 341	49
890	949 390	949 439	949 488	949 536	949 585	949 634	949 683	949 731	949 780	949 829	49
891	49 878	49 926	49 975	50 024	50 073	50 121	50 170	50 219	50 267	50 316	49
892	50 365	50 414	50 462	50 511	50 560	50 608	50 657	50 706	50 754	50 803	48
893	50 851	50 900	50 949	50 997	51 046	51 095	51 143	51 192	51 240	51 289	49
894	51 338	51 386	51 435	51 483	51 532	51 580	51 629	51 677	51 726	51 775	48
895	951 823	951 872	951 920	951 969	952 017	952 066	952 114	952 163	952 211	952 260	48
896	52 308	52 356	52 405	52 453	52 502	52 550	52 599	52 647	52 696	52 744	48
897	52 792	52 841	52 889	52 938	52 986	53 034	53 083	53 131	53 180	53 228	48
898	53 276	53 325	53 373	53 421	53 470	53 518	53 566	53 615	53 663	53 711	49
899	53 760	53 808	53 856	53 905	53 953	54 001	54 049	54 098	54 146	54 194	49
900	954 243	954 291	954 339	954 387	954 435	954 484	954 532	954 580	954 628	954 677	48
901	54 725	54 773	54 821	54 869	54 918	54 966	55 014	55 062	55 110	55 158	49
902	55 207	55 255	55 303	55 351	55 399	55 447	55 495	55 543	55 592	55 640	48
903	55 688	55 736	55 784	55 832	55 880	55 928	55 976	56 024	56 072	56 120	48
904	56 168	56 216	56 265	56 313	56 361	56 409	56 457	56 505	56 553	56 601	48
905	956 649	956 697	956 745	956 793	956 840	956 888	956 936	956 984	957 032	957 080	48
906	57 128	57 176	57 224	57 272	57 320	57 368	57 416	57 464	57 512	57 559	48
907	57 607	57 655	57 703	57 751	57 799	57 847	57 894	57 942	57 990	58 038	48
908	58 086	58 134	58 181	58 229	58 277	58 325	58 373	58 421	58 468	58 516	48
909	58 564	58 612	58 659	58 707	58 755	58 803	58 850	58 898	58 946	58 994	47
	0	1	2	3	4	5	6	7	8	9	

	47	48	49	50	51	52	53
1	4·7	4·8	4·9	5·0	5·1	5·2	5·3
2	9·4	9·6	9·8	10·0	10·2	10·4	10·6
3	14·1	14·4	14·7	15·0	15·3	15·6	15·9
4	18·8	19·2	19·6	20·0	20·4	20·8	21·2
5	23·5	24·0	24·5	25·0	25·5	26·0	26·5
6	28·2	28·8	29·4	30·0	30·6	31·2	31·8
7	32·9	33·6	34·3	35·0	35·7	36·4	37·1
8	37·6	38·4	39·2	40·0	40·8	41·6	42·4
9	42·3	43·2	44·1	45·0	45·9	46·8	47·7

TABLE I—LOGARITHMS OF NUMBERS

	0	1	2	3	4	5	6	7	8	9	Δ
910	959 041	959 089	959 137	959 185	959 232	959 280	959 328	959 375	959 423	959 471	47
911	59 518	59 566	59 614	59 661	59 709	59 757	59 804	59 852	59 900	59 947	48
912	59 995	60 042	60 090	60 138	60 185	60 233	60 280	60 328	60 376	60 423	48
913	60 471	60 518	60 566	60 613	60 661	60 709	60 756	60 804	60 851	60 899	47
914	60 946	60 994	61 041	61 089	61 136	61 184	61 231	61 279	61 326	61 374	47
915	961 421	961 469	961 516	961 563	961 611	961 658	961 706	961 753	961 801	961 848	47
916	61 895	61 943	61 990	62 038	62 085	62 132	62 180	62 227	62 275	62 322	47
917	62 369	62 417	62 464	62 511	62 559	62 606	62 653	62 701	62 748	62 795	48
918	62 843	62 890	62 937	62 985	63 032	63 079	63 126	63 174	63 221	63 268	48
919	63 316	63 363	63 410	63 457	63 504	63 552	63 599	63 646	63 693	63 741	47
920	963 788	963 835	963 882	963 929	963 977	964 024	964 071	964 118	964 165	964 212	48
921	64 260	64 307	64 354	64 401	64 448	64 495	64 542	64 590	64 637	64 684	47
922	64 731	64 778	64 825	64 872	64 919	64 966	65 013	65 061	65 108	65 155	47
923	65 202	65 249	65 296	65 343	65 390	65 437	65 484	65 531	65 578	65 625	47
924	65 672	65 719	65 766	65 813	65 860	65 907	65 954	66 001	66 048	66 095	47
925	966 142	966 189	966 236	966 283	966 329	966 376	966 423	966 470	966 517	966 564	47
926	66 611	66 658	66 705	66 752	66 799	66 845	66 892	66 939	66 986	67 033	47
927	67 080	67 127	67 173	67 220	67 267	67 314	67 361	67 408	67 454	67 501	47
928	67 548	67 595	67 642	67 688	67 735	67 782	67 829	67 875	67 922	67 969	47
929	68 016	68 062	68 109	68 156	68 203	68 249	68 296	68 343	68 390	68 436	47
930	968 483	968 530	968 576	968 623	968 670	968 716	968 763	968 810	968 856	968 903	47
931	68 950	68 996	69 043	69 090	69 136	69 183	69 229	69 276	69 323	69 369	47
932	69 416	69 463	69 509	69 556	69 602	69 649	69 695	69 742	69 789	69 835	47
933	69 882	69 928	69 975	70 021	70 068	70 114	70 161	70 207	70 254	70 300	47
934	70 347	70 393	70 440	70 486	70 533	70 579	70 626	70 672	70 719	70 765	47
935	970 812	970 858	970 904	970 951	970 997	971 044	971 090	971 137	971 183	971 229	47
936	71 276	71 322	71 369	71 415	71 461	71 508	71 554	71 601	71 647	71 693	47
937	71 740	71 786	71 832	71 879	71 925	71 971	72 018	72 064	72 110	72 157	46
938	72 203	72 249	72 295	72 342	72 388	72 434	72 481	72 527	72 573	72 619	47
939	72 666	72 712	72 758	72 804	72 851	72 897	72 943	72 989	73 035	73 082	46
940	973 128	973 174	973 220	973 266	973 313	973 359	973 405	973 451	973 497	973 543	47
941	73 590	73 636	73 682	73 728	73 774	73 820	73 866	73 913	73 959	74 005	46
942	74 051	74 097	74 143	74 189	74 235	74 281	74 327	74 374	74 420	74 466	46
943	74 512	74 558	74 604	74 650	74 696	74 742	74 788	74 834	74 880	74 926	46
944	74 972	75 018	75 064	75 110	75 156	75 202	75 248	75 294	75 340	75 386	46
945	975 432	975 478	975 524	975 570	975 616	975 662	975 707	975 753	975 799	975 845	46
946	75 891	75 937	75 983	76 029	76 075	76 121	76 167	76 212	76 258	76 304	46
947	76 350	76 396	76 442	76 488	76 533	76 579	76 625	76 671	76 717	76 763	45
948	76 808	76 854	76 900	76 946	76 992	77 037	77 083	77 129	77 175	77 220	46
949	77 266	77 312	77 358	77 403	77 449	77 495	77 541	77 586	77 632	77 678	46
950	977 724	977 769	977 815	977 861	977 906	977 952	977 998	978 043	978 089	978 135	46
951	78 181	78 226	78 272	78 317	78 363	78 409	78 454	78 500	78 546	78 591	46
952	78 637	78 683	78 728	78 774	78 819	78 865	78 911	78 956	79 002	79 047	46
953	79 093	79 138	79 184	79 230	79 275	79 321	79 366	79 412	79 457	79 503	45
954	79 548	79 594	79 639	79 685	79 730	79 776	79 821	79 867	79 912	79 958	45
955	980 003	980 049	980 094	980 140	980 185	980 231	980 276	980 322	980 367	980 412	46
956	80 458	80 503	80 549	80 594	80 640	80 685	80 730	80 776	80 821	80 867	45
957	80 912	80 957	81 003	81 048	81 093	81 139	81 184	81 229	81 275	81 320	46
958	81 366	81 411	81 456	81 501	81 547	81 592	81 637	81 683	81 728	81 773	46
959	81 819	81 864	81 909	81 954	82 000	82 045	82 090	82 135	82 181	82 226	45
	0	1	2	3	4	5	6	7	8	9	

Proportional Parts will be found on the facing page.

TABLE I—LOGARITHMS OF NUMBERS

	0	1	2	3	4	5	6	7	8	9	Δ
960	982 271	982 316	982 362	982 407	982 452	982 497	982 543	982 588	982 633	982 678	45
961	82 723	82 769	82 814	82 859	82 904	82 949	82 994	83 040	83 085	83 130	45
962	83 175	83 220	83 265	83 310	83 356	83 401	83 446	83 491	83 536	83 581	45
963	83 626	83 671	83 716	83 762	83 807	83 852	83 897	83 942	83 987	84 032	45
964	84 077	84 122	84 167	84 212	84 257	84 302	84 347	84 392	84 437	84 482	45
965	984 527	984 572	984 617	984 662	984 707	984 752	984 797	984 842	984 887	984 932	45
966	84 977	85 022	85 067	85 112	85 157	85 202	85 247	85 292	85 337	85 382	44
967	85 426	85 471	85 516	85 561	85 606	85 651	85 696	85 741	85 786	85 830	45
968	85 875	85 920	85 965	86 010	86 055	86 100	86 144	86 189	86 234	86 279	45
969	86 324	86 369	86 413	86 458	86 503	86 548	86 593	86 637	86 682	86 727	45
970	986 772	986 817	986 861	986 906	986 951	986 996	987 040	987 085	987 130	987 175	44
971	87 219	87 264	87 309	87 353	87 398	87 443	87 488	87 532	87 577	87 622	44
972	87 666	87 711	87 756	87 800	87 845	87 890	87 934	87 979	88 024	88 068	45
973	88 113	88 157	88 202	88 247	88 291	88 336	88 381	88 425	88 470	88 514	45
974	88 559	88 604	88 648	88 693	88 737	88 782	88 826	88 871	88 916	88 960	45
975	989 005	989 049	989 094	989 138	989 183	989 227	989 272	989 316	989 361	989 405	45
976	89 450	89 494	89 539	89 583	89 628	89 672	89 717	89 761	89 806	89 850	45
977	89 895	89 939	89 983	90 028	90 072	90 117	90 161	90 206	90 250	90 294	45
978	90 339	90 383	90 428	90 472	90 516	90 561	90 605	90 650	90 694	90 738	45
979	90 783	90 827	90 871	90 916	90 960	91 004	91 049	91 093	91 137	91 182	44
980	991 226	991 270	991 315	991 359	991 403	991 448	991 492	991 536	991 580	991 625	44
981	91 669	91 713	91 758	91 802	91 846	91 890	91 935	91 979	92 023	92 067	44
982	92 111	92 156	92 200	92 244	92 288	92 333	92 377	92 421	92 465	92 509	45
983	92 554	92 598	92 642	92 686	92 730	92 774	92 819	92 863	92 907	92 951	44
984	92 995	93 039	93 083	93 127	93 172	93 216	93 260	93 304	93 348	93 392	44
985	993 436	993 480	993 524	993 568	993 613	993 657	993 701	993 745	993 789	993 833	44
986	93 877	93 921	93 965	94 009	94 053	94 097	94 141	94 185	94 229	94 273	44
987	94 317	94 361	94 405	94 449	94 493	94 537	94 581	94 625	94 669	94 713	44
988	94 757	94 801	94 845	94 889	94 933	94 977	95 021	95 065	95 108	95 152	44
989	95 196	95 240	95 284	95 328	95 372	95 416	95 460	95 504	95 547	95 591	44
990	995 635	995 679	995 723	995 767	995 811	995 854	995 898	995 942	995 986	996 030	44
991	96 074	96 117	96 161	96 205	96 249	96 293	96 337	96 380	96 424	96 468	44
992	96 512	96 555	96 599	96 643	96 687	96 731	96 774	96 818	96 862	96 906	43
993	96 949	96 993	97 037	97 080	97 124	97 168	97 212	97 255	97 299	97 343	43
994	97 386	97 430	97 474	97 517	97 561	97 605	97 648	97 692	97 736	97 779	44
995	997 823	997 867	997 910	997 954	997 998	998 041	998 085	998 129	998 172	998 216	43
996	98 259	98 303	98 347	98 390	98 434	98 477	98 521	98 564	98 608	98 652	43
997	98 695	98 739	98 782	98 826	98 869	98 913	98 956	99 000	99 043	99 087	44
998	99 131	99 174	99 218	99 261	99 305	99 348	99 392	99 435	99 479	99 522	43
999	99 565	99 609	99 652	99 696	99 739	99 783	99 826	99 870	99 913	99 957	43
	0	1	2	3	4	5	6	7	8	9	

	43	44	45	46	47	48
1	4·3	4·4	4·5	4·6	4·7	4·8
2	8·6	8·8	9·0	9·2	9·4	9·6
3	12·9	13·2	13·5	13·8	14·1	14·4
4	17·2	17·6	18·0	18·4	18·8	19·2
5	21·5	22·0	22·5	23·0	23·5	24·0
6	25·8	26·4	27·0	27·6	28·2	28·8
7	30·1	30·8	31·5	32·2	32·9	33·6
8	34·4	35·2	36·0	36·8	37·6	38·4
9	38·7	39·6	40·5	41·4	42·3	43·2

TABLE IIA—LOGARITHMS OF TRIGONOMETRICAL FUNCTIONS

0° ′ ″	sin	tan	′ ″
00 00	—	—	60 00
10	5·68557	557	59 50
20	5·98660	660	40
30	6·16270	270	30
40	·28763	763	20
50	·38454	454	10
01 00	6·46373	373	59 00
10	·53067	067	58 50
20	·58866	866	40
30	·63982	982	30
40	·68557	557	20
50	·72697	697	10
02 00	6·76476	476	58 00
10	·79952	952	57 50
20	·83170	170	40
30	·86167	167	30
40	·88969	969	20
50	·91602	602	10
03 00	6·94085	085	57 00
10	·96433	433	56 50
20	6·98660	660	40
30	7·00779	779	30
40	·02800	800	20
50	·04730	730	10
04 00	7·06579	579	56 00
10	·08351	352	55 50
20	·10055	055	40
30	·11694	694	30
40	·13273	273	20
50	·14797	797	10
05 00	7·16270	270	55 00
10	·17694	694	54 50
20	·19072	073	40
30	·20409	409	30
40	·21705	705	20
50	·22964	964	10
06 00	7·24188	188	54 00
10	·25378	378	53 50
20	·26536	536	40
30	·27664	664	30
40	·28763	764	20
50	·29836	836	10
07 00	7·30882	882	53 00
10	·31904	904	52 50
20	·32903	903	40
30	·33879	879	30
40	·34833	833	20
50	·35767	767	10
08 00	7·36682	682	52 00
10	·37577	577	51 50
20	·38454	455	40
30	·39314	315	30
40	·40158	158	20
50	·40985	985	10
09 00	7·41797	797	51 00
10	·42594	594	50 50
20	·43376	376	40
30	·44145	145	30
40	·44900	900	20
50	·45643	643	10
10 00	7·46373	373	50 00
	cos	cot	89°

0° ′ ″	sin	diff.	tan	′ ″
10 00	7·46373	717	373	50 00
10	·47090	707	091	49 50
20	·47797	694	797	40
30	·48491	684	492	30
40	·49175	674	176	20
50	·49849	663	849	10
11 00	7·50512	653	512	49 00
10	·51165	643	165	48 50
20	·51808	634	809	40
30	·52442	625	443	30
40	·53067	616	067	20
50	·53683	608	683	10
12 00	7·54291	599	291	48 00
10	·54890	591	890	47 50
20	·55481	583	481	40
30	·56064	575	064	30
40	·56639	567	639	20
50	·57206	561	207	10
13 00	7·57767	553	767	47 00
10	·58320	546	320	46 50
20	·58866	540	867	40
30	·59406	533	406	30
40	·59939	526	939	20
50	·60465	520	466	10
14 00	7·60985	514	986	46 00
10	·61499	508	500	45 50
20	·62007	502	008	40
30	·62509	497	510	30
40	·63006	490	006	20
50	·63496	486	497	10
15 00	7·63982	479	982	45 00
10	·64461	475	462	44 50
20	·64936	470	937	40
30	·65406	464	406	30
40	·65870	460	871	20
50	·66330	454	330	10
16 00	7·66784	451	785	44 00
10	·67235	445	235	43 50
20	·67680	441	680	40
30	·68121	436	121	30
40	·68557	432	558	20
50	·68989	428	990	10
17 00	7·69417	424	418	43 00
10	·69841	420	842	42 50
20	·70261	415	261	40
30	·70676	412	677	30
40	·71088	408	088	20
50	·71496	404	496	10
18 00	7·71900	400	900	42 00
10	·72300	397	301	41 50
20	·72697	393	697	40
30	·73090	389	090	30
40	·73479	386	480	20
50	·73865	383	866	10
19 00	7·74248	379	248	41 00
10	·74627	376	628	40 50
20	·75003	373	004	40
30	·75376	369	377	30
40	·75745	367	746	20
50	·76112	363	113	10
20 00	7·76475		476	40 00
	cos		cot	89°

CONVERSION TO SECONDS

′	0° ″	1° ″	2° ″
0	0	3600	7200
1	60	3660	7260
2	120	3720	7320
3	180	3780	7380
4	240	3840	7440
5	300	3900	7500
6	360	3960	7560
7	420	4020	7620
8	480	4080	7680
9	540	4140	7740
10	600	4200	7800
11	660	4260	7860
12	720	4320	7920
13	780	4380	7980
14	840	4440	8040
15	900	4500	8100
16	960	4560	8160
17	1020	4620	8220
18	1080	4680	8280
19	1140	4740	8340
20	1200	4800	8400
21	1260	4860	8460
22	1320	4920	8520
23	1380	4980	8580
24	1440	5040	8640
25	1500	5100	8700
26	1560	5160	8760
27	1620	5220	8820
28	1680	5280	8880
29	1740	5340	8940
30	1800	5400	9000
31	1860	5460	9060
32	1920	5520	9120
33	1980	5580	9180
34	2040	5640	9240
35	2100	5700	9300
36	2160	5760	9360
37	2220	5820	9420
38	2280	5880	9480
39	2340	5940	9540
40	2400	6000	9600
41	2460	6060	9660
42	2520	6120	9720
43	2580	6180	9780
44	2640	6240	9840
45	2700	6300	9900
46	2760	6360	9960
47	2820	6420	10020
48	2880	6480	10080
49	2940	6540	10140
50	3000	6600	10200
51	3060	6660	10260
52	3120	6720	10320
53	3180	6780	10380
54	3240	6840	10440
55	3300	6900	10500
56	3360	6960	10560
57	3420	7020	10620
58	3480	7080	10680
59	3540	7140	10740
60	3600	7200	10800

0° ′ ″	sin		tan	′ ″
20 00	7·76475	361	476	40 00
10	·76836	357	837	39 50
20	·77193	355	194	40
30	·77548	351	549	30
40	·77899	349	900	20
50	·78248	346	249	10
21 00	7·78594	344	595	39 00
10	·78938	340	938	38 50
20	·79278	338	279	40
30	·79616	336	617	30
40	·79952	332	952	20
50	·80284	331	285	10
22 00	7·80615	327	615	38 00
10	·80942	326	943	37 50
20	·81268	323	269	40
30	·81591	320	591	30
40	·81911	318	912	20
50	·82229	316	230	10
23 00	7·82545	314	546	37 00
10	·82859	311	860	36 50
20	·83170	309	171	40
30	·83479	307	480	30
40	·83786	305	787	20
50	·84091	302	092	10
24 00	7·84393	301	394	36 00
10	·84694	298	695	35 50
20	·84992	297	993	40
30	·85289	294	290	30
40	·85583	293	584	20
50	·85876	290	877	10
25 00	7·86166	289	167	35 00
10	·86455	286	456	34 50
20	·86741	285	743	40
30	·87026	283	027	30
40	·87309	281	310	20
50	·87590	280	591	10
26 00	7·87870	277	871	34 00
10	·88147	276	148	33 50
20	·88423	274	424	40
30	·88697	272	698	30
40	·88969	271	970	20
50	·89240	269	241	10
27 00	7·89509	267	510	33 00
10	·89776	265	777	32 50
20	·90041	264	043	40
30	·90305	263	307	30
40	·90568	261	569	20
50	·90829	259	830	10
28 00	7·91088	258	089	32 00
10	·91346	256	347	31 50
20	·91602	255	603	40
30	·91857	253	858	30
40	·92110	252	111	20
50	·92362	250	363	10
29 00	7·92612	249	613	31 00
10	·92861	247	862	30 50
20	·93108	246	110	40
30	·93354	245	356	30
40	·93599	243	601	20
50	·93842	242	844	10
30 00	7·94084		086	30 00
	cos		cot	89°

0° ′ ″	sin		tan	′ ″
30 00	7·94084	241	086	30 00
10	·94325	239	326	29 50
20	·94564	238	566	40
30	·94802	237	804	30
40	·95039	235	040	20
50	·95274	234	276	10
31 00	7·95508	233	510	29 00
10	·95741	232	743	28 50
20	·95973	230	974	40
30	·96203	229	205	30
40	·96432	228	434	20
50	·96660	227	662	10
32 00	7·96887	226	889	28 00
10	·97113	224	114	27 50
20	·97337	223	339	40
30	·97560	222	562	30
40	·97782	221	784	20
50	·98003	220	005	10
33 00	7·98223	219	225	27 00
10	·98442	218	444	26 50
20	·98660	216	662	40
30	·98876	216	878	30
40	·99092	214	094	20
50	·99306	214	308	10
34 00	7·99520	212	522	26 00
10	·99732	211	734	25 50
20	7·99943	211	946	40
30	8·00154	209	156	30
40	·00363	208	365	20
50	·00571	208	574	10
35 00	8·00779	206	781	25 00
10	·00985	205	987	24 50
20	·01190	205	193	40
30	·01395	203	397	30
40	·01598	203	600	20
50	·01801	201	803	10
36 00	8·02002	201	004	24 00
10	·02203	199	205	23 50
20	·02402	199	405	40
30	·02601	198	604	30
40	·02799	197	801	20
50	·02996	196	998	10
37 00	8·03192	195	194	23 00
10	·03387	194	390	22 50
20	·03581	194	584	40
30	·03775	192	777	30
40	·03967	192	970	20
50	·04159	191	162	10
38 00	8·04350	190	353	22 00
10	·04540	189	543	21 50
20	·04729	189	732	40
30	·04918	187	921	30
40	·05105	187	108	20
50	·05292	186	295	10
39 00	8·05478	185	481	21 00
10	·05663	185	666	20 50
20	·05848	183	851	40
30	·06031	183	034	30
40	·06214	182	217	20
50	·06396	182	399	10
40 00	8·06578		581	20 00
	cos		cot	89°

x ″	S	sin x
0		—
	4·685 575	
464		7·352 411
	4·685 574	
896		7·638 038
	4·685 573	
1179		7·757 263
	4·685 572	
1406		7·833 789
	4·685 571	
1602		7·890 261
	4·685 570	
1776		7·935 036
	4·685 569	
1934		7·972 137
	4·685 568	
2080		8·003 815
	4·685 567	
2217		8·031 454
	4·685 566	
2346		8·055 969
	4·685 565	
2468		8·077 994
	4·685 564	
2584		8·097 990
	4·685 563	
2696		8·116 298
	4·685 562	
2802		8·133 182
	4·685 561	
2905		8·148 848
	4·685 560	
3005		8·163 458
	4·685 559	
3101		8·177 147
	4·685 558	
3194		8·190 024
	4·685 557	
3285		8·202 180
	4·685 556	
3373		8·213 691
	4·685 555	
3459		8·224 622
	4·685 554	
3543		8·235 029
	4·685 553	
3625		8·244 961
	4·685 552	
3706		8·254 457
	4·685 551	
3784		8·263 556
	4·685 550	
3861		8·272 288
	4·685 549	
3936		8·280 683
	4·685 548	
4010		8·288 765
	4·685 547	
4083		8·296 557
	4·685 546	
4154		8·304 079
	4·685 545	
4224		8·311 349
	4·685 544	
4293		8·318 384
	4·685 543	
4361		8·325 197
	4·685 542	
4428		8·331 804
	4·685 541	
4494		8·338 215
	4·685 540	
4559		8·344 442
	4·685 539	
4623		8·350 496
	4·685 538	
4686		8·356 385
	4·685 537	
4748		8·362 119
	4·685 536	
4810		8·367 705
	4·685 535	
4870		8·373 151
	4·685 534	
4930		8·378 464
	4·685 533	
4990		8·383 649
	4·685 532	
5048		8·388 714
	4·685 531	
5106		8·393 663
	4·685 530	
5163		8·398 502
	4·685 529	
5220		8·403 236
	4·685 528	
5276		8·407 868
	4·685 527	
5331		8·412 404
	4·685 526	
5386		8·416 846
	4·685 525	
5440		8·421 200

$\log \sin x = \log x'' + S$

$\log x'' = \log \sin x - S$

0° ′ ″	sin		tan	′ ″
40 00	8·06578	180	581	20 00
10	·06758	180	761	19 50
20	·06938	179	941	40
30	·07117	178	120	30
40	·07295	178	298	20
50	·07473	177	476	10
41 00	8·07650	176	653	19 00
10	·07826	176	829	18 50
20	·08002	174	005	40
30	·08176	174	180	30
40	·08350	174	354	20
50	·08524	172	527	10
42 00	8·08696	172	700	18 00
10	·08868	172	872	17 50
20	·09040	170	043	40
30	·09210	170	214	30
40	·09380	170	384	20
50	·09550	168	553	10
43 00	8·09718	168	722	17 00
10	·09886	168	890	16 50
20	·10054	166	057	40
30	·10220	166	224	30
40	·10386	166	390	20
50	·10552	165	555	10
44 00	8·10717	164	720	16 00
10	·10881	163	884	15 50
20	·11044	163	048	40
30	·11207	163	211	30
40	·11370	161	373	20
50	·11531	162	535	10
45 00	8·11693	160	696	15 00
10	·11853	160	857	14 50
20	·12013	159	017	40
30	·12172	159	176	30
40	·12331	158	335	20
50	·12489	158	493	10
46 00	8·12647	157	651	14 00
10	·12804	157	808	13 50
20	·12961	156	965	40
30	·13117	155	121	30
40	·13272	155	276	20
50	·13427	154	431	10
47 00	8·13581	154	585	13 00
10	·13735	153	739	12 50
20	·13888	153	892	40
30	·14041	152	045	30
40	·14193	151	197	20
50	·14344	151	348	10
48 00	8·14495	151	500	12 00
10	·14646	150	650	11 50
20	·14796	149	800	40
30	·14945	149	950	30
40	·15094	149	099	20
50	·15243	148	247	10
49 00	8·15391	147	395	11 00
10	·15538	147	543	10 50
20	·15685	147	690	40
30	·15832	146	836	30
40	·15978	145	982	20
50	·16123	145	128	10
50 00	8·16268		273	10 00
	cos		cot	89°

0° ′ ″	sin		tan	′ ″
50 00	8·16268	145	273	10 00
10	·16413	144	417	09 50
20	·16557	143	561	40
30	·16700	143	705	30
40	·16843	143	848	20
50	·16986	142	991	10
51 00	8·17128	142	133	09 00
10	·17270	141	275	08 50
20	·17411	141	416	40
30	·17552	140	557	30
40	·17692	140	697	20
50	·17832	139	837	10
52 00	8·17971	139	976	08 00
10	·18110	139	115	07 50
20	·18249	138	254	40
30	·18387	137	392	30
40	·18524	138	530	20
50	·18662	136	667	10
53 00	8·18798	137	804	07 00
10	·18935	136	940	06 50
20	·19071	135	076	40
30	·19206	135	211	30
40	·19341	135	347	20
50	·19476	134	481	10
54 00	8·19610	134	616	06 00
10	·19744	133	749	05 50
20	·19877	133	883	40
30	·20010	133	016	30
40	·20143	132	149	20
50	·20275	132	281	10
55 00	8·20407	131	413	05 00
10	·20538	131	544	04 50
20	·20669	131	675	40
30	·20800	130	806	30
40	·20930	130	936	20
50	·21060	129	066	10
56 00	8·21189	130	195	04 00
10	·21319	128	324	03 50
20	·21447	129	453	40
30	·21576	127	581	30
40	·21703	128	709	20
50	·21831	127	837	10
57 00	8·21958	127	964	03 00
10	·22085	126	091	02 50
20	·22211	126	217	40
30	·22337	126	343	30
40	·22463	125	469	20
50	·22588	125	595	10
58 00	8·22713	125	720	02 00
10	·22838	124	844	01 50
20	·22962	124	968	40
30	·23086	124	092	30
40	·23210	123	216	20
50	·23333	123	339	10
59 00	8·23456	122	462	01 00
10	·23578	122	585	00 50
20	·23700	122	707	40
30	·23822	122	829	30
40	·23944	121	950	20
50	·24065	121	071	10
60 00	8·24186		192	00 00
	cos		cot	89°

x ″	T	tan x
0		—
	4·685 575	
431		7·320 431
	4·685 576	
692		7·526 186
	4·685 577	
879		7·629 910
	4·685 578	
1033		7·699 814
	4·685 579	
1166		7·752 610
	4·685 580	
1286		7·795 048
	4·685 581	
1396		7·830 533
	4·685 582	
1497		7·861 026
	4·685 583	
1592		7·887 759
	4·685 584	
1682		7·911 559
	4·685 585	
1767		7·933 006
	4·685 586	
1849		7·952 525
	4·685 587	
1926		7·970 432
	4·685 588	
2001		7·986 975
	4·685 589	
2073		8·002 347
	4·685 590	
2143		8·016 702
	4·685 591	
2210		8·030 166
	4·685 592	
2276		8·042 845
	4·685 593	
2340		8·054 824
	4·685 594	
2402		8·066 176
	4·685 595	
2462		8·076 964
	4·685 596	
2521		8·087 242
	4·685 597	
2579		8·097 055
	4·685 598	
2635		8·106 444
	4·685 599	
2690		8·115 444
	4·685 600	
2744		8·124 086
	4·685 601	
2797		8·132 397
	4·685 602	
2849		8·140 401
	4·685 603	
2900		8·148 121
	4·685 604	
2951		8·155 576
	4·685 605	
3000		8·162 784
	4·685 606	
3048		8·169 760
	4·685 607	
3096		8·176 519
	4·685 608	
3143		8·183 074
	4·685 609	
3190		8·189 437
	4·685 610	
3235		8·195 618
	4·685 611	
3281		8·201 629
	4·685 612	
3325		8·207 478
	4·685 613	
3369		8·213 173
	4·685 614	
3412		8·218 723
	4·685 615	
3455		8·224 135
	4·685 616	
3497		8·229 415
	4·685 617	
3539		8·234 570
	4·685 618	
3580		8·239 605
	4·685 619	
3621		8·244 526
	4·685 620	
3661		8·249 338
	4·685 621	
3701		8·254 046
	4·685 622	
3741		8·258 654
	4·685 623	
3780		8·263 166
	4·685 624	
3819		8·267 586

$\log \tan x = \log x'' + T$

$\log x'' = \log \tan x - T$

1° ′ ″	sin	d	tan	′ ″
00 00	8·24186	120	192	60 00
10	·24306	120	313	59 50
20	·24426	120	433	40
30	·24546	119	553	30
40	·24665	120	672	20
50	·24785	118	791	10
01 00	8·24903	119	910	59 00
10	·25022	118	029	58 50
20	·25140	118	147	40
30	·25258	117	265	30
40	·25375	118	382	20
50	·25493	116	500	10
02 00	8·25609	117	616	58 00
10	·25726	116	733	57 50
20	·25842	116	849	40
30	·25958	116	965	30
40	·26074	115	081	20
50	·26189	115	196	10
03 00	8·26304	115	312	57 00
10	·26419	114	426	56 50
20	·26533	115	541	40
30	·26648	113	655	30
40	·26761	114	769	20
50	·26875	113	882	10
04 00	8·26988	113	996	56 00
10	·27101	113	109	55 50
20	·27214	112	221	40
30	·27326	112	334	30
40	·27438	112	446	20
50	·27550	111	558	10
05 00	8·27661	112	669	55 00
10	·27773	110	780	54 50
20	·27883	111	891	40
30	·27994	110	002	30
40	·28104	111	112	20
50	·28215	109	223	10
06 00	8·28324	110	332	54 00
10	·28434	109	442	53 50
20	·28543	109	551	40
30	·28652	109	660	30
40	·28761	108	769	20
50	·28869	108	877	10
07 00	8·28977	108	986	53 00
10	·29085	108	094	52 50
20	·29193	107	201	40
30	·29300	107	309	30
40	·29407	107	416	20
50	·29514	107	523	10
08 00	8·29621	106	629	52 00
10	·29727	106	736	51 50
20	·29833	106	842	40
30	·29939	105	947	30
40	·30044	106	053	20
50	·30150	105	158	10
09 00	8·30255	104	263	51 00
10	·30359	105	368	50 50
20	·30464	104	473	40
30	·30568	104	577	30
40	·30672	104	681	20
50	·30776	103	785	10
10 00	8·30879		888	50 00
	cos		cot	88°

1° ′ ″	sin	d	tan	′ ″
10 00	8·30879	104	888	50 00
10	·30983	103	992	49 50
20	·31086	102	095	40
30	·31188	103	198	30
40	·31291	102	300	20
50	·31393	102	403	10
11 00	8·31495	102	505	49 00
10	·31597	102	606	48 50
20	·31699	101	708	40
30	·31800	101	809	30
40	·31901	101	911	20
50	·32002	101	012	10
12 00	8·32103	100	112	48 00
10	·32203	100	213	47 50
20	·32303	100	313	40
30	·32403	100	413	30
40	·32503	99	513	20
50	·32602	100	612	10
13 00	8·32702	99	711	47 00
10	·32801	98	810	46 50
20	·32899	99	909	40
30	·32998	98	008	30
40	·33096	99	106	20
50	·33195	97	205	10
14 00	8·33292	98	302	46 00
10	·33390	98	400	45 50
20	·33488	97	498	40
30	·33585	97	595	30
40	·33682	97	692	20
50	·33779	96	789	10
15 00	8·33875	97	886	45 00
10	·33972	96	982	44 50
20	·34068	96	078	40
30	·34164	96	174	30
40	·34260	95	270	20
50	·34355	95	366	10
16 00	8·34450	96	461	44 00
10	·34546	94	556	43 50
20	·34640	95	651	40
30	·34735	95	746	30
40	·34830	94	840	20
50	·34924	94	935	10
17 00	8·35018	94	029	43 00
10	·35112	94	123	42 50
20	·35206	93	217	40
30	·35299	93	310	30
40	·35392	93	403	20
50	·35485	93	497	10
18 00	8·35578	93	590	42 00
10	·35671	93	682	41 50
20	·35764	92	775	40
30	·35856	92	867	30
40	·35948	92	959	20
50	·36040	91	051	10
19 00	8·36131	92	143	41 00
10	·36223	91	235	40 50
20	·36314	91	326	40
30	·36405	91	417	30
40	·36496	91	508	20
50	·36587	91	599	10
20 00	8·36678		689	40 00
	cos		cot	88°

x ″	T	$\tan x$
3819	4·685 625	8·267 586
3857	4·685 626	8·271 918
3895	4·685 627	8·276 166
3932	4·685 628	8·280 332
3970	4·685 629	8·284 420
4006	4·685 630	8·288 432
4043	4·685 631	8·292 371
4079	4·685 632	8·296 240
4115	4·685 633	8·300 042
4150	4·685 634	8·303 778
4186	4·685 635	8·307 451
4221	4·685 636	8·311 063
4255	4·685 637	8·314 615
4290	4·685 638	8·318 111
4324	4·685 639	8·321 551
4358	4·685 640	8·324 938
4391	4·685 641	8·328 272
4425	4·685 642	8·331 557
4458	4·685 643	8·334 792
4490	4·685 644	8·337 980
4523	4·685 645	8·341 121
4555	4·685 646	8·344 218
4588	4·685 647	8·347 272
4619	4·685 648	8·350 283
4651	4·685 649	8·353 253
4683	4·685 650	8·356 182
4714	4·685 651	8·359 073
4745	4·685 652	8·361 926
4776	4·685 653	8·364 742
4806	4·685 654	8·367 522
4837	4·685 655	8·370 267
4867	4·685 656	8·372 977
4897	4·685 657	8·375 654
4927	4·685 658	8·378 299
4957	4·685 659	8·380 912
4986	4·685 660	8·383 493
5016	4·685 661	8·386 045
5045	4·685 662	8·388 566
5074	4·685 663	8·391 059
5103	4·685 664	8·393 524
5132	4·685 665	8·395 961
5160	4·685 666	8·398 370
5189	4·685 667	8·400 754
5217	4·685 668	8·403 111
5245	4·685 669	8·405 444
5273	4·685 670	8·407 751
5301	4·685 671	8·410 034
5328	4·685 672	8·412 294
5356	4·685 673	8·414 530
5383	4·685 674	8·416 744
5410		8·418 935

$\log \tan x = \log x'' + T$

$\log x'' = \log \tan x - T$

TABLE IIB—LOGARITHMS OF TRIGONOMETRICAL FUNCTIONS

(Degrees, Minutes and Seconds)

Angles 0° to 1° 20′ and 88° 40′ to 90°

For angles less than 1° 20′, log sines (to 5 decimals) and log tangents (last 3 decimals) at interval 10″ are given on the four pages (28-31) immediately preceding. These same values represent log cosines and log cotangents of angles greater than 88° 40′. It will be noted that the differences of log tan are the same as those of log sin, or differ by a unit. Bold type in the log tan column indicates an increase in the second decimal; thus log tan 1° 13′ 30″ = 8·33008. In this range 5-decimal values suffice for working to 0″·1, but 6-decimal values, if required, may be obtained from the critical tables of S and T given on the right of these pages. The logarithm of a cotangent is the cologarithm of a tangent, and vice versa; i.e. log cot = 10 − log tan or log tan = 10 − log cot.

Log cosines of small angles or log sines of large angles may be obtained from the critical table on the facing page. The table below gives the best value of such an angle whose function is given; in the first two triple columns angles are given to the nearest 10″, in the next to ½′, and in the last to 1′; greater accuracy is not warranted.

If S and T are known for a small angle x, then log cos $x = S - T$.

	Angle from			Angle from			Angle from			Angle from	
log	cos	sin	log	cos	sin	log	cos	sin	log	cos	sin
9·999	° ′ ″	° ′ ″	9·999	° ′ ″	° ′ ″	9·999	′	° ′	9·999	′	° ′
870	1 24 10	88 35 50	**905**	1 11 50	88 48 10	**935**	59½	89 00½	**965**	44	89 16
871	1 23 50	88 36 10	**906**	1 11 30	88 48 30	**936**	59	89 01	**966**	43	89 17
872	1 23 30	88 36 30	**907**	1 11 10	88 48 50	**937**	58½	89 01½	**967**	42	89 18
873	1 23 10	88 36 50	**908**	1 10 50	88 49 10	**938**	58	89 02	**968**	42	89 18
874	1 22 50	88 37 10	**909**	1 10 20	88 49 40	**939**	57½	89 02½	**969**	41	89 19
875	1 22 30	88 37 30	**910**	1 10 00	88 50 00	**940**	57	89 03	**970**	40	89 20
876	1 22 10	88 37 50	**911**	1 09 40	88 50 20	**941**	56½	89 03½	**971**	40	89 20
877	1 21 50	88 38 10	**912**	1 09 10	88 50 50	**942**	56	89 04	**972**	39	89 21
878	1 21 30	88 38 30	**913**	1 08 50	88 51 10	**943**	55½	89 04½	**973**	38	89 22
879	1 21 10	88 38 50	**914**	1 08 20	88 51 40	**944**	55	89 05	**974**	38	89 22
880	1 20 50	88 39 10	**915**	1 08 00	88 52 00	**945**	54½	89 05½	**975**	37	89 23
881	1 20 30	88 39 30	**916**	1 07 40	88 52 20	**946**	54	89 06	**976**	36	89 24
882	1 20 10	88 39 50	**917**	1 07 10	88 52 50	**947**	53½	89 06½	**977**	35	89 25
883	1 19 50	88 40 10	**918**	1 06 50	88 53 10	**948**	53	89 07	**978**	35	89 25
884	1 19 30	88 40 30	**919**	1 06 20	88 53 40	**949**	52½	89 07½	**979**	34	89 26
885	1 19 10	88 40 50	**920**	1 06 00	88 54 00	**950**	52	89 08	**980**	33	89 27
886	1 18 50	88 41 10	**921**	1 05 30	88 54 30	**951**	51½	89 08½	**981**	32	89 28
887	1 18 30	88 41 30	**922**	1 05 10	88 54 50	**952**	51	89 09	**982**	31	89 29
888	1 18 00	88 42 00	**923**	1 04 40	88 55 20	**953**	50½	89 09½	**983**	30	89 30
889	1 17 40	88 42 20	**924**	1 04 20	88 55 40	**954**	50	89 10	**984**	30	89 30
890	1 17 20	88 42 40	**925**	1 03 50	88 56 10	**955**	49½	89 10½	**985**	29	89 31
891	1 17 00	88 43 00	**926**	1 03 30	88 56 30	**956**	49	89 11	**986**	28	89 32
892	1 16 40	88 43 20	**927**	1 03 00	88 57 00	**957**	48½	89 11½	**987**	27	89 33
893	1 16 20	88 43 40	**928**	1 02 40	88 57 20	**958**	48	89 12	**988**	26	89 34
894	1 16 00	88 44 00	**929**	1 02 10	88 57 50	**959**	47	89 13	**989**	24	89 36
895	1 15 40	88 44 20	**930**	1 01 40	88 58 20	**960**	46½	89 13½	**990**	23	89 37
896	1 15 10	88 44 50	**931**	1 01 20	88 58 40	**961**	46	89 14	**991**	22	89 38
897	1 14 50	88 45 10	**932**	1 00 50	88 59 10	**962**	45½	89 14½	**992**	21	89 39
898	1 14 30	88 45 30	**933**	1 00 20	88 59 40	**963**	45	89 15	**993**	20	89 40
899	1 14 10	88 45 50	**934**	1 00 00	89 00 00	**964**	44½	89 15½	**994**	18	89 42
900	1 13 50	88 46 10	**935**	59 30	89 00 30	**965**	43½	89 16½	**995**	16	89 44
901	1 13 20	88 46 40	**936**	59 00	89 01 00	**966**	43	89 17	**996**	15	89 45
902	1 13 00	88 47 00	**937**	58 30	89 01 30	**967**	42½	89 17½	**997**	13	89 47
903	1 12 40	88 47 20	**938**	58 10	89 01 50	**968**	41½	89 18½	**998**	10	89 50
904	1 12 20	88 47 40	**939**	57 40	89 02 20	**969**	41	89 19	**999**	7	89 53
	To nearest 10″			To nearest 10″			To nearest ½′			To nearest 1′	

TABLE IIb—LOGARITHMS OF TRIGONOMETRICAL FUNCTIONS

(Degrees, Minutes and Seconds)

This critical table gives log cosines of small angles and log sines of large angles. It must not be used inversely; if the log sine or log cosine is greater than 9·999 870, the corresponding angle can be found from the table on the previous page.

For angles less than 5′ 13″, log cos = 0·000 000.

For angles greater than 89° 54′ 47″, log sin = 0·000 000.

Angle	cos	Angle	cos	Angle	cos	Angle	sin	Angle	sin	Angle	sin
′ ″	9·999	° ′ ″	9·999	° ′ ″	9·999	° ′ ″	9·999	° ′ ″	9·999	° ′ ″	9·999
5 12		46 56		1 06 11		88 39 01		88 53 48		89 13 03	
	999		959		919		880		920		960
9 02		47 31		1 06 35		88 39 21		88 54 13		89 13 38	
	998		958		918		881		921		961
11 39		48 05		1 07 00		88 39 41		88 54 38		89 14 13	
	997		957		917		882		922		962
13 48		48 39		1 07 24		88 40 02		88 55 03		89 14 49	
	996		956		916		883		923		963
15 38		49 12		1 07 48		88 40 22		88 55 28		89 15 25	
	995		955		915		884		924		964
17 18		49 45		1 08 12		88 40 43		88 55 54		89 16 02	
	994		954		914		885		925		965
18 48		50 18		1 08 36		88 41 03		88 56 19		89 16 40	
	993		953		913		886		926		966
20 12		50 50		1 09 00		88 41 24		88 56 45		89 17 18	
	992		952		912		887		927		967
21 30		51 22		1 09 23		88 41 45		88 57 11		89 17 56	
	991		951		911		888		928		968
22 44		51 54		1 09 47		88 42 06		88 57 37		89 18 35	
	990		950		910		889		929		969
23 54		52 25		1 10 10		88 42 27		88 58 03		89 19 15	
	989		949		909		890		930		970
25 01		52 56		1 10 33		88 42 48		88 58 29		89 19 55	
	988		948		908		891		931		971
26 04		53 27		1 10 56		88 43 09		88 58 56		89 20 36	
	987		947		907		892		932		972
27 06		53 57		1 11 19		88 43 30		88 59 23		89 21 18	
	986		946		906		893		933		973
28 05		54 27		1 11 42		88 43 52		88 59 50		89 22 01	
	985		945		905		894		934		974
29 02		54 57		1 12 05		88 44 13		89 00 17		89 22 44	
	984		944		904		895		935		975
29 57		55 27		1 12 28		88 44 35		89 00 45		89 23 29	
	983		943		903		896		936		976
30 51		55 56		1 12 50		88 44 57		89 01 12		89 24 14	
	982		942		902		897		937		977
31 43		56 25		1 13 12		88 45 18		89 01 40		89 25 00	
	981		941		901		898		938		978
32 34		56 54		1 13 35		88 45 40		89 02 08		89 25 47	
	980		940		900		899		939		979
33 24		57 22		1 13 57		88 46 02		89 02 37		89 26 35	
	979		939		899		900		940		980
34 12		57 51		1 14 19		88 46 24		89 03 05		89 27 25	
	978		938		898		901		941		981
34 59		58 19		1 14 41		88 46 47		89 03 34		89 28 16	
	977		937		897		902		942		982
35 45		58 47		1 15 02		88 47 09		89 04 03		89 29 08	
	976		936		896		903		943		983
36 30		59 14		1 15 24		88 47 31		89 04 32		89 30 02	
	975		935		895		904		944		984
37 15		59 42		1 15 46		88 47 54		89 05 02		89 30 57	
	974		934		894		905		945		985
37 58		1 00 09		1 16 07		88 48 17		89 05 32		89 31 54	
	973		933		893		906		946		986
38 41		1 00 36		1 16 29		88 48 40		89 06 02		89 32 53	
	972		932		892		907		947		987
39 23		1 01 03		1 16 50		88 49 03		89 06 32		89 33 55	
	971		931		891		908		948		988
40 04		1 01 30		1 17 11		88 49 26		89 07 03		89 34 58	
	970		930		890		909		949		989
40 44		1 01 56		1 17 32		88 49 49		89 07 34		89 36 05	
	969		929		889		910		950		990
41 24		1 02 22		1 17 53		88 50 12		89 08 05		89 37 15	
	968		928		888		911		951		991
42 03		1 02 48		1 18 14		88 50 36		89 08 37		89 38 29	
	967		927		887		912		952		992
42 41		1 03 14		1 18 35		88 50 59		89 09 09		89 39 47	
	966		926		886		913		953		993
43 19		1 03 40		1 18 56		88 51 23		89 09 41		89 41 11	
	965		925		885		914		954		994
43 57		1 04 05		1 19 16		88 51 47		89 10 14		89 42 41	
	964		924		884		915		955		995
44 34		1 04 31		1 19 37		88 52 11		89 10 47		89 44 21	
	963		923		883		916		956		996
45 10		1 04 56		1 19 57		88 52 35		89 11 20		89 46 11	
	962		922		882		917		957		997
45 46		1 05 21		1 20 18		88 52 59		89 11 54		89 48 20	
	961		921		881		918		958		998
46 21		1 05 46		1 20 38		88 53 24		89 12 28		89 50 57	
	960		920		880		919		959		999
46 56		1 06 11		1 20 58		88 53 48		89 13 03		89 54 47	

TABLE IIc—LOGARITHMS OF TRIGONOMETRICAL FUNCTIONS

(Degrees, Minutes and Seconds)

1°

′	″	sin	d	tan	d	cot	cos	′	″
20	00	8·366 777	904	8·366 895	904	1·633 105	9·999 882	**40**	00
	10	·367 681	901	·367 799	902	·632 201	·999 882	**39**	50
	20	·368 582	900	·368 701	900	·631 299	·999 881		40
	30	·369 482	898	·369 601	899	·630 399	·999 881		30
	40	·370 380	897	·370 500	897	·629 500	·999 880		20
	50	·371 277	894	·371 397	895	·628 603	·999 880		10
21	00	8·372 171	892	8·372 292	893	1·627 708	9·999 879	**39**	00
	10	·373 063	891	·373 185	891	·626 815	·999 879	**38**	50
	20	·373 954	889	·374 076	889	·625 924	·999 878		40
	30	·374 843	887	·374 965	888	·625 035	·999 878		30
	40	·375 730	885	·375 853	885	·624 147	·999 877		20
	50	·376 615	884	·376 738	884	·623 262	·999 877		10
22	00	8·377 499	881	8·377 622	882	1·622 378	9·999 876	**38**	00
	10	·378 380	880	·378 504	881	·621 496	·999 876	**37**	50
	20	·379 260	878	·379 385	878	·620 615	·999 875		40
	30	·380 138	877	·380 263	877	·619 737	·999 875		30
	40	·381 015	874	·381 140	875	·618 860	·999 874		20
	50	·381 889	873	·382 015	874	·617 985	·999 874		10
23	00	8·382 762	871	8·382 889	871	1·617 111	9·999 873	**37**	00
	10	·383 633	869	·383 760	870	·616 240	·999 873	**36**	50
	20	·384 502	868	·384 630	868	·615 370	·999 872		40
	30	·385 370	866	·385 498	866	·614 502	·999 872		30
	40	·386 236	864	·386 364	865	·613 636	·999 871		20
	50	·387 100	862	·387 229	863	·612 771	·999 871		10
24	00	8·387 962	861	8·388 092	861	1·611 908	9·999 870	**36**	00
	10	·388 823	859	·388 953	860	·611 047	·999 870	**35**	50
	20	·389 682	857	·389 813	857	·610 187	·999 869		40
	30	·390 539	856	·390 670	856	·609 330	·999 869		30
	40	·391 395	854	·391 526	855	·608 474	·999 868		20
	50	·392 249	852	·392 381	853	·607 619	·999 868		10
25	00	8·393 101	850	8·393 234	851	1·606 766	9·999 867	**35**	00
	10	·393 951	849	·394 085	849	·605 915	·999 867	**34**	50
	20	·394 800	847	·394 934	848	·605 066	·999 866		40
	30	·395 647	846	·395 782	846	·604 218	·999 866		30
	40	·396 493	844	·396 628	844	·603 372	·999 865		20
	50	·397 337	842	·397 472	843	·602 528	·999 865		10
26	00	8·398 179	841	8·398 315	841	1·601 685	9·999 864	**34**	00
	10	·399 020	839	·399 156	840	·600 844	·999 864	**33**	50
	20	·399 859	837	·399 996	838	·600 004	·999 863		40
	30	·400 696	836	·400 834	836	·599 166	·999 863		30
	40	·401 532	834	·401 670	835	·598 330	·999 862		20
	50	·402 366	833	·402 505	833	·597 495	·999 861		10
27	00	8·403 199	831	8·403 338	832	1·596 662	9·999 861	**33**	00
	10	·404 030	829	·404 170	830	·595 830	·999 860	**32**	50
	20	·404 859	828	·405 000	828	·595 000	·999 860		40
	30	·405 687	827	·405 828	827	·594 172	·999 859		30
	40	·406 514	824	·406 655	825	·593 345	·999 859		20
	50	·407 338	823	·407 480	824	·592 520	·999 858		10
28	00	8·408 161	822	8·408 304	822	1·591 696	9·999 858	**32**	00
	10	·408 983	820	·409 126	820	·590 874	·999 857	**31**	50
	20	·409 803	818	·409 946	819	·590 054	·999 857		40
	30	·410 621	817	·410 765	818	·589 235	·999 856		30
	40	·411 438	816	·411 583	816	·588 417	·999 856		20
	50	·412 254	814	·412 399	814	·587 601	·999 855		10
29	00	8·413 068	812	8·413 213	813	1·586 787	9·999 854	**31**	00
	10	·413 880	811	·414 026	811	·585 974	·999 854	**30**	50
	20	·414 691	809	·414 837	810	·585 163	·999 853		40
	30	·415 500	808	·415 647	809	·584 353	·999 853		30
	40	·416 308	806	·416 456	807	·583 544	·999 852		20
	50	·417 114	805	·417 263	805	·582 737	·999 852		10
30	00	8·417 919		8·418 068		1·581 932	9·999 851	**30**	00
		cos		cot		tan	sin	**88°**	

PROPORTIONAL PARTS

″	805	808	811	814	818
1	80·5	80·8	81·1	81·4	81·8
2	161·0	161·6	162·2	162·8	163·6
3	241·5	242·4	243·3	244·2	245·4
4	322·0	323·2	324·4	325·6	327·2
5	402·5	404·0	405·5	407·0	409·0
6	483·0	484·8	486·6	488·4	490·8
7	563·5	565·6	567·7	569·8	572·6
8	644·0	646·4	648·8	651·2	654·4
9	724·5	727·2	729·9	732·6	736·2

″	824	828	831	834	837
1	82·4	82·8	83·1	83·4	83·7
2	164·8	165·6	166·2	166·8	167·4
3	247·2	248·4	249·3	250·2	251·1
4	329·6	331·2	332·4	333·6	334·8
5	412·0	414·0	415·5	417·0	418·5
6	494·4	496·8	498·6	500·4	502·2
7	576·8	579·6	581·7	583·8	585·9
8	659·2	662·4	664·8	667·2	669·6
9	741·6	745·2	747·9	750·6	753·3

″	840	843	847	850	853
1	84·0	84·3	84·7	85·0	85·3
2	168·0	168·6	169·4	170·0	170·6
3	252·0	252·9	254·1	255·0	255·9
4	336·0	337·2	338·8	340·0	341·2
5	420·0	421·5	423·5	425·0	426·5
6	504·0	505·8	508·2	510·0	511·8
7	588·0	590·1	592·9	595·0	597·1
8	672·0	674·4	677·6	680·0	682·4
9	756·0	758·7	762·3	765·0	767·7

″	856	860	863	866	868
1	85·6	86·0	86·3	86·6	86·8
2	171·2	172·0	172·6	173·2	173·6
3	256·8	258·0	258·9	259·8	260·4
4	342·4	344·0	345·2	346·4	347·2
5	428·0	430·0	431·5	433·0	434·0
6	513·6	516·0	517·8	519·6	520·8
7	599·2	602·0	604·1	606·2	607·6
8	684·8	688·0	690·4	692·8	694·4
9	770·4	774·0	776·7	779·4	781·2

″	871	874	878	881	884
1	87·1	87·4	87·8	88·1	88·4
2	174·2	174·8	175·6	176·2	176·8
3	261·3	262·2	263·4	264·3	265·2
4	348·4	349·6	351·2	352·4	353·6
5	435·5	437·0	439·0	440·5	442·0
6	522·6	524·4	526·8	528·6	530·4
7	609·7	611·8	614·6	616·7	618·8
8	696·8	699·2	702·4	704·8	707·2
9	783·9	786·6	790·2	792·9	795·6

″	888	893	898	901	904
1	88·8	89·3	89·8	90·1	90·4
2	177·6	178·6	179·6	180·2	180·8
3	266·4	267·9	269·4	270·3	271·2
4	355·2	357·2	359·2	360·4	361·6
5	444·0	446·5	449·0	450·5	452·0
6	532·8	535·8	538·8	540·6	542·4
7	621·6	625·1	628·6	630·7	632·8
8	710·4	714·4	718·4	720·8	723·2
9	799·2	803·7	808·2	810·9	813·6

TABLE IIc—LOGARITHMS OF TRIGONOMETRICAL FUNCTIONS

(Degrees, Minutes and Seconds)

1°

′ ″	sin	diff.	tan	diff.	cot	cos	′ ″
30 00	8·417 919	803	8·418 068	804	1·581 932	9·999 851	30 00
10	·418 722	802	·418 872	802	·581 128	·999 851	29 50
20	·419 524	801	·419 674	801	·580 326	·999 850	40
30	·420 325	798	·420 475	799	·579 525	·999 849	30
40	·421 123	798	·421 274	798	·578 726	·999 849	20
50	·421 921	796	·422 072	797	·577 928	·999 848	10
31 00	8·422 717	794	8·422 869	795	1·577 131	9·999 848	29 00
10	·423 511	793	·423 664	794	·576 336	·999 847	28 50
20	·424 304	792	·424 458	792	·575 542	·999 847	40
30	·425 096	790	·425 250	791	·574 750	·999 846	30
40	·425 886	789	·426 041	789	·573 959	·999 846	20
50	·426 675	787	·426 830	788	·573 170	·999 845	10
32 00	8·427 462	786	8·427 618	786	1·572 382	9·999 844	28 00
10	·428 248	784	·428 404	785	·571 596	·999 844	27 50
20	·429 032	783	·429 189	784	·570 811	·999 843	40
30	·429 815	782	·429 973	782	·570 027	·999 843	30
40	·430 597	780	·430 755	781	·569 245	·999 842	20
50	·431 377	779	·431 536	779	·568 464	·999 842	10
33 00	8·432 156	778	8·432 315	778	1·567 685	9·999 841	27 00
10	·432 934	776	·433 093	777	·566 907	·999 840	26 50
20	·433 710	774	·433 870	775	·566 130	·999 840	40
30	·434 484	773	·434 645	774	·565 355	·999 839	30
40	·435 257	772	·435 419	772	·564 581	·999 839	20
50	·436 029	771	·436 191	771	·563 809	·999 838	10
34 00	8·436 800	769	8·436 962	770	1·563 038	9·999 838	26 00
10	·437 569	768	·437 732	768	·562 268	·999 837	25 50
20	·438 337	766	·438 500	767	·561 500	·999 836	40
30	·439 103	765	·439 267	766	·560 733	·999 836	30
40	·439 868	764	·440 033	764	·559 967	·999 835	20
50	·440 632	762	·440 797	763	·559 203	·999 835	10
35 00	8·441 394	762	8·441 560	762	1·558 440	9·999 834	25 00
10	·442 156	759	·442 322	760	·557 678	·999 834	24 50
20	·442 915	759	·443 082	759	·556 918	·999 833	40
30	·443 674	757	·443 841	758	·556 159	·999 832	30
40	·444 431	755	·444 599	756	·555 401	·999 832	20
50	·445 186	755	·445 355	755	·554 645	·999 831	10
36 00	8·445 941	753	8·446 110	754	1·553 890	9·999 831	24 00
10	·446 694	752	·446 864	752	·553 136	·999 830	23 50
20	·447 446	750	·447 616	752	·552 384	·999 829	40
30	·448 196	750	·448 368	749	·551 632	·999 829	30
40	·448 946	748	·449 117	749	·550 883	·999 828	20
50	·449 694	746	·449 866	747	·550 134	·999 828	10
37 00	8·450 440	746	8·450 613	746	1·549 387	9·999 827	23 00
10	·451 186	744	·451 359	745	·548 641	·999 827	22 50
20	·451 930	743	·452 104	743	·547 896	·999 826	40
30	·452 673	741	·452 847	742	·547 153	·999 825	30
40	·453 414	740	·453 589	741	·546 411	·999 825	20
50	·454 154	739	·454 330	740	·545 670	·999 824	10
38 00	8·454 893	738	8·455 070	738	1·544 930	9·999 824	22 00
10	·455 631	737	·455 808	737	·544 192	·999 823	21 50
20	·456 368	735	·456 545	736	·543 455	·999 822	40
30	·457 103	734	·457 281	735	·542 719	·999 822	30
40	·457 837	733	·458 016	733	·541 984	·999 821	20
50	·458 570	731	·458 749	732	·541 251	·999 820	10
39 00	8·459 301	731	8·459 481	731	1·540 519	9·999 820	21 00
10	·460 032	729	·460 212	730	·539 788	·999 819	20 50
20	·460 761	728	·460 942	728	·539 058	·999 819	40
30	·461 489	726	·461 670	728	·538 330	·999 818	30
40	·462 215	726	·462 398	726	·537 602	·999 817	20
50	·462 941	724	·463 124	725	·536 876	·999 817	10
40 00	8·463 665		8·463 849		1·536 151	9·999 816	20 00
	cos		cot		tan	sin	**88°**

PROPORTIONAL PARTS

″	724	726	728	731	734
1	72·4	72·6	72·8	73·1	73·4
2	144·8	145·2	145·6	146·2	146·8
3	217·2	217·8	218·4	219·3	220·2
4	289·6	290·4	291·2	292·4	293·6
5	362·0	363·0	364·0	365·5	367·0
6	434·4	435·6	436·8	438·6	440·4
7	506·8	508·2	509·6	511·7	513·8
8	579·2	580·8	582·4	584·8	587·2
9	651·6	653·4	655·2	657·9	660·6

″	737	740	743	746	749
1	73·7	74·0	74·3	74·6	74·9
2	147·4	148·0	148·6	149·2	149·8
3	221·1	222·0	222·9	223·8	224·7
4	294·8	296·0	297·2	298·4	299·6
5	368·5	370·0	371·5	373·0	374·5
6	442·2	444·0	445·8	447·6	449·4
7	515·9	518·0	520·1	522·2	524·3
8	589·6	592·0	594·4	596·8	599·2
9	663·3	666·0	668·7	671·4	674·1

″	752	755	757	759	762
1	75·2	75·5	75·7	75·9	76·2
2	150·4	151·0	151·4	151·8	152·4
3	225·6	226·5	227·1	227·7	228·6
4	300·8	302·0	302·8	303·6	304·8
5	376·0	377·5	378·5	379·5	381·0
6	451·2	453·0	454·2	455·4	457·2
7	526·4	528·5	529·9	531·3	533·4
8	601·6	604·0	605·6	607·2	609·6
9	676·8	679·5	681·3	683·1	685·8

″	765	768	771	774	777
1	76·5	76·8	77·1	77·4	77·7
2	153·0	153·6	154·2	154·8	155·4
3	229·5	230·4	231·3	232·2	233·1
4	306·0	307·2	308·4	309·6	310·8
5	382·5	384·0	385·5	387·0	388·5
6	459·0	460·8	462·6	464·4	466·2
7	535·5	537·6	539·7	541·8	543·9
8	612·0	614·4	616·8	619·2	621·6
9	688·5	691·2	693·9	696·6	699·3

″	780	782	784	786	789
1	78·0	78·2	78·4	78·6	78·9
2	156·0	156·4	156·8	157·2	157·8
3	234·0	234·6	235·2	235·8	236·7
4	312·0	312·8	313·6	314·4	315·6
5	390·0	391·0	392·0	393·0	394·5
6	468·0	469·2	470·4	471·6	473·4
7	546·0	547·4	548·8	550·2	552·3
8	624·0	625·6	627·2	628·8	631·2
9	702·0	703·8	705·6	707·4	710·1

″	792	795	798	802	804
1	79·2	79·5	79·8	80·2	80·4
2	158·4	159·0	159·6	160·4	160·8
3	237·6	238·5	239·4	240·6	241·2
4	316·8	318·0	319·2	320·8	321·6
5	396·0	397·5	399·0	401·0	402·0
6	475·2	477·0	478·8	481·2	482·4
7	554·4	556·5	558·6	561·4	562·8
8	633·6	636·0	638·4	641·6	643·2
9	712·8	715·5	718·2	721·8	723·6

TABLE IIc—LOGARITHMS OF TRIGONOMETRICAL FUNCTIONS

(Degrees, Minutes and Seconds)

1°

′ ″	sin		tan		cot	cos	′ ″
40 00	8·463 665	723	8·463 849	723	1·536 151	9·999 816	20 00
10	·464 388	722	·464 572	723	·535 428	·999 816	19 50
20	·465 110	720	·465 295	721	·534 705	·999 815	40
30	·465 830	720	·466 016	720	·533 984	·999 814	30
40	·466 550	718	·466 736	719	·533 264	·999 814	20
50	·467 268	717	·467 455	717	·532 545	·999 813	10
41 00	8·467 985	716	8·468 172	717	1·531 828	9·999 813	19 00
10	·468 701	715	·468 889	715	·531 111	·999 812	18 50
20	·469 416	713	·469 604	714	·530 396	·999 811	40
30	·470 129	712	·470 318	713	·529 682	·999 811	30
40	·470 841	712	·471 031	712	·528 969	·999 810	20
50	·471 553	710	·471 743	711	·528 257	·999 809	10
42 00	8·472 263	708	8·472 454	709	1·527 546	9·999 809	18 00
10	·472 971	708	·473 163	709	·526 837	·999 808	17 50
20	·473 679	707	·473 872	707	·526 128	·999 808	40
30	·474 386	705	·474 579	706	·525 421	·999 807	30
40	·475 091	704	·475 285	705	·524 715	·999 806	20
50	·475 795	703	·475 990	703	·524 010	·999 806	10
43 00	8·476 498	702	8·476 693	703	1·523 307	9·999 805	17 00
10	·477 200	701	·477 396	701	·522 604	·999 804	16 50
20	·477 901	700	·478 097	701	·521 903	·999 804	40
30	·478 601	698	·478 798	699	·521 202	·999 803	30
40	·479 299	698	·479 497	698	·520 503	·999 803	20
50	·479 997	696	·480 195	697	·519 805	·999 802	10
44 00	8·480 693	695	8·480 892	696	1·519 108	9·999 801	16 00
10	·481 388	695	·481 588	695	·518 412	·999 801	15 50
20	·482 083	693	·482 283	693	·517 717	·999 800	40
30	·482 776	691	·482 976	693	·517 024	·999 799	30
40	·483 467	691	·483 669	691	·516 331	·999 799	20
50	·484 158	690	·484 360	690	·515 640	·999 798	10
45 00	8·484 848	688	8·485 050	690	1·514 950	9·999 797	15 00
10	·485 536	688	·485 740	688	·514 260	·999 797	14 50
20	·486 224	686	·486 428	687	·513 572	·999 796	40
30	·486 910	686	·487 115	686	·512 885	·999 795	30
40	·487 596	684	·487 801	685	·512 199	·999 795	20
50	·488 280	683	·488 486	684	·511 514	·999 794	10
46 00	8·488 963	682	8·489 170	682	1·510 830	9·999 794	14 00
10	·489 645	681	·489 852	682	·510 148	·999 793	13 50
20	·490 326	680	·490 534	681	·509 466	·999 792	40
30	·491 006	679	·491 215	679	·508 785	·999 792	30
40	·491 685	678	·491 894	679	·508 106	·999 791	20
50	·492 363	677	·492 573	677	·507 427	·999 790	10
47 00	8·493 040	675	8·493 250	677	1·506 750	9·999 790	13 00
10	·493 715	675	·493 927	675	·506 073	·999 789	12 50
20	·494 390	674	·494 602	674	·505 398	·999 788	40
30	·495 064	672	·495 276	673	·504 724	·999 788	30
40	·495 736	672	·495 949	673	·504 051	·999 787	20
50	·496 408	670	·496 622	671	·503 378	·999 786	10
48 00	8·497 078	670	8·497 293	670	1·502 707	9·999 786	12 00
10	·497 748	668	·497 963	669	·502 037	·999 785	11 50
20	·498 416	668	·498 632	668	·501 368	·999 784	40
30	·499 084	666	·499 300	667	·500 700	·999 784	30
40	·499 750	666	·499 967	666	·500 033	·999 783	20
50	·500 416	664	·500 633	665	·499 367	·999 782	10
49 00	8·501 080	663	8·501 298	664	1·498 702	9·999 782	11 00
10	·501 743	662	·501 962	663	·498 038	·999 781	10 50
20	·502 405	662	·502 625	662	·497 375	·999 780	40
30	·503 067	660	·503 287	661	·496 713	·999 780	30
40	·503 727	659	·503 948	660	·496 052	·999 779	20
50	·504 386	659	·504 608	659	·495 392	·999 778	10
50 00	8·505 045		8·505 267		1·494 733	9·999 778	10 00
	cos		cot		tan	sin	88°

PROPORTIONAL PARTS

″	659	660	662	664	666
1	65·9	66·0	66·2	66·4	66·6
2	131·8	132·0	132·4	132·8	133·2
3	197·7	198·0	198·6	199·2	199·8
4	263·6	264·0	264·8	265·6	266·4
5	329·5	330·0	331·0	332·0	333·0
6	395·4	396·0	397·2	398·4	399·6
7	461·3	462·0	463·4	464·8	466·2
8	527·2	528·0	529·6	531·2	532·8
9	593·1	594·0	595·8	597·6	599·4

″	668	670	672	674	677
1	66·8	67·0	67·2	67·4	67·7
2	133·6	134·0	134·4	134·8	135·4
3	200·4	201·0	201·6	202·2	203·1
4	267·2	268·0	268·8	269·6	270·8
5	334·0	335·0	336·0	337·0	338·5
6	400·8	402·0	403·2	404·4	406·2
7	467·6	469·0	470·4	471·8	473·9
8	534·4	536·0	537·6	539·2	541·6
9	601·2	603·0	604·8	606·6	609·3

″	679	682	684	686	688
1	67·9	68·2	68·4	68·6	68·8
2	135·8	136·4	136·8	137·2	137·6
3	203·7	204·6	205·2	205·8	206·4
4	271·6	272·8	273·6	274·4	275·2
5	339·5	341·0	342·0	343·0	344·0
6	407·4	409·2	410·4	411·6	412·8
7	475·3	477·4	478·8	480·2	481·6
8	543·2	545·6	547·2	548·8	550·4
9	611·1	613·8	615·6	617·4	619·2

″	690	691	693	695	698
1	69·0	69·1	69·3	69·5	69·8
2	138·0	138·2	138·6	139·0	139·6
3	207·0	207·3	207·9	208·5	209·4
4	276·0	276·4	277·2	278·0	279·2
5	345·0	345·5	346·5	347·5	349·0
6	414·0	414·6	415·8	417·0	418·8
7	483·0	483·7	485·1	486·5	488·6
8	552·0	552·8	554·4	556·0	558·4
9	621·0	621·9	623·7	625·5	628·2

″	701	703	705	707	709
1	70·1	70·3	70·5	70·7	70·9
2	140·2	140·6	141·0	141·4	141·8
3	210·3	210·9	211·5	212·1	212·7
4	280·4	281·2	282·0	282·8	283·6
5	350·5	351·5	352·5	353·5	354·5
6	420·6	421·8	423·0	424·2	425·4
7	490·7	492·1	493·5	494·9	496·3
8	560·8	562·4	564·0	565·6	567·2
9	630·9	632·7	634·5	636·3	638·1

″	712	715	717	720	723
1	71·2	71·5	71·7	72·0	72·3
2	142·4	143·0	143·4	144·0	144·6
3	213·6	214·5	215·1	216·0	216·9
4	284·8	286·0	286·8	288·0	289·2
5	356·0	357·5	358·5	360·0	361·5
6	427·2	429·0	430·2	432·0	433·8
7	498·4	500·5	501·9	504·0	506·1
8	569·6	572·0	573·6	576·0	578·4
9	640·8	643·5	645·3	648·0	650·7

TABLE IIc—LOGARITHMS OF TRIGONOMETRICAL FUNCTIONS

(Degrees, Minutes and Seconds)

1° ′	″	sin		tan		cot	cos	′	″
50	00	8·505 045	657	8·505 267	658	1·494 733	9·999 778	10	00
	10	·505 702	656	·505 925	657	·494 075	·999 777	09	50
	20	·506 358	656	·506 582	656	·493 418	·999 776		40
	30	·507 014	654	·507 238	655	·492 762	·999 776		30
	40	·507 668	653	·507 893	654	·492 107	·999 775		20
	50	·508 321	653	·508 547	653	·491 453	·999 774		10
51	00	8·508 974	651	8·509 200	652	1·490 800	9·999 774	09	00
	10	·509 625	650	·509 852	651	·490 148	·999 773	08	50
	20	·510 275	650	·510 503	650	·489 497	·999 772		40
	30	·510 925	648	·511 153	649	·488 847	·999 772		30
	40	·511 573	648	·511 802	649	·488 198	·999 771		20
	50	·512 221	646	·512 451	647	·487 549	·999 770		10
52	00	8·512 867	646	8·513 098	646	1·486 902	9·999 769	08	00
	10	·513 513	644	·513 744	645	·486 256	·999 769	07	50
	20	·514 157	644	·514 389	645	·485 611	·999 768		40
	30	·514 801	643	·515 034	643	·484 966	·999 767		30
	40	·515 444	642	·515 677	643	·484 323	·999 767		20
	50	·516 086	640	·516 320	641	·483 680	·999 766		10
53	00	8·516 726	640	8·516 961	641	1·483 039	9·999 765	07	00
	10	·517 366	639	·517 602	639	·482 398	·999 765	06	50
	20	·518 005	638	·518 241	639	·481 759	·999 764		40
	30	·518 643	637	·518 880	638	·481 120	·999 763		30
	40	·519 280	636	·519 518	636	·480 482	·999 763		20
	50	·519 916	635	·520 154	636	·479 846	·999 762		10
54	00	8·520 551	635	8·520 790	635	1·479 210	9·999 761	06	00
	10	·521 186	633	·521 425	634	·478 575	·999 760	05	50
	20	·521 819	632	·522 059	633	·477 941	·999 760		40
	30	·522 451	632	·522 692	632	·477 308	·999 759		30
	40	·523 083	630	·523 324	632	·476 676	·999 758		20
	50	·523 713	630	·523 956	630	·476 044	·999 758		10
55	00	8·524 343	629	8·524 586	629	1·475 414	9·999 757	05	00
	10	·524 972	627	·525 215	629	·474 785	·999 756	04	50
	20	·525 599	627	·525 844	628	·474 156	·999 756		40
	30	·526 226	626	·526 472	626	·473 528	·999 755		30
	40	·526 852	625	·527 098	626	·472 902	·999 754		20
	50	·527 477	625	·527 724	625	·472 276	·999 753		10
56	00	8·528 102	623	8·528 349	624	1·471 651	9·999 753	04	00
	10	·528 725	622	·528 973	623	·471 027	·999 752	03	50
	20	·529 347	622	·529 596	622	·470 404	·999 751		40
	30	·529 969	621	·530 218	622	·469 782	·999 751		30
	40	·530 590	619	·530 840	620	·469 160	·999 750		20
	50	·531 209	619	·531 460	620	·468 540	·999 749		10
57	00	8·531 828	618	8·532 080	618	1·467 920	9·999 748	03	00
	10	·532 446	617	·532 698	618	·467 302	·999 748	02	50
	20	·533 063	616	·533 316	617	·466 684	·999 747		40
	30	·533 679	616	·533 933	616	·466 067	·999 746		30
	40	·534 295	614	·534 549	615	·465 451	·999 746		20
	50	·534 909	614	·535 164	615	·464 836	·999 745		10
58	00	8·535 523	613	8·535 779	613	1·464 221	9·999 744	02	00
	10	·536 136	611	·536 392	613	·463 608	·999 743	01	50
	20	·536 747	611	·537 005	612	·462 995	·999 743		40
	30	·537 358	611	·537 617	610	·462 383	·999 742		30
	40	·537 969	609	·538 227	610	·461 773	·999 741		20
	50	·538 578	608	·538 837	610	·461 163	·999 740		10
59	00	8·539 186	608	8·539 447	608	1·460 553	9·999 740	01	00
	10	·539 794	607	·540 055	607	·459 945	·999 739	00	50
	20	·540 401	606	·540 662	607	·459 338	·999 738		40
	30	·541 007	605	·541 269	606	·458 731	·999 738		30
	40	·541 612	604	·541 875	605	·458 125	·999 737		20
	50	·542 216	603	·542 480	604	·457 520	·999 736		10
60	00	8·542 819		8·543 084		1·456 916	9·999 735	00	00
		cos		cot		tan	sin	88°	

PROPORTIONAL PARTS

″	603	605	607	608	610
1	60·3	60·5	60·7	60·8	61·0
2	120·6	121·0	121·4	121·6	122·0
3	180·9	181·5	182·1	182·4	183·0
4	241·2	242·0	242·8	243·2	244·0
5	301·5	302·5	303·5	304·0	305·0
6	361·8	363·0	364·2	364·8	366·0
7	422·1	423·5	424·9	425·6	427·0
8	482·4	484·0	485·6	486·4	488·0
9	542·7	544·5	546·3	547·2	549·0

″	611	613	614	616	618
1	61·1	61·3	61·4	61·6	61·8
2	122·2	122·6	122·8	123·2	123·6
3	183·3	183·9	184·2	184·8	185·4
4	244·4	245·2	245·6	246·4	247·2
5	305·5	306·5	307·0	308·0	309·0
6	366·6	367·8	368·4	369·6	370·8
7	427·7	429·1	429·8	431·2	432·6
8	488·8	490·4	491·2	492·8	494·4
9	549·9	551·7	552·6	554·4	556·2

″	620	622	625	626	629
1	62·0	62·2	62·5	62·6	62·9
2	124·0	124·4	125·0	125·2	125·8
3	186·0	186·6	187·5	187·8	188·7
4	248·0	248·8	250·0	250·4	251·6
5	310·0	311·0	312·5	313·0	314·5
6	372·0	373·2	375·0	375·6	377·4
7	434·0	435·4	437·5	438·2	440·3
8	496·0	497·6	500·0	500·8	503·2
9	558·0	559·8	562·5	563·4	566·1

″	630	632	633	635	636
1	63·0	63·2	63·3	63·5	63·6
2	126·0	126·4	126·6	127·0	127·2
3	189·0	189·6	189·9	190·5	190·8
4	252·0	252·8	253·2	254·0	254·4
5	315·0	316·0	316·5	317·5	318·0
6	378·0	379·2	379·8	381·0	381·6
7	441·0	442·4	443·1	444·5	445·2
8	504·0	505·6	506·4	508·0	508·8
9	567·0	568·8	569·7	571·5	572·4

″	639	641	643	644	646
1	63·9	64·1	64·3	64·4	64·6
2	127·8	128·2	128·6	128·8	129·2
3	191·7	192·3	192·9	193·2	193·8
4	255·6	256·4	257·2	257·6	258·4
5	319·5	320·5	321·5	322·0	323·0
6	383·4	384·6	385·8	386·4	387·6
7	447·3	448·7	450·1	450·8	452·2
8	511·2	512·8	514·4	515·2	516·8
9	575·1	576·9	578·7	579·6	581·4

″	648	650	653	656	658
1	64·8	65·0	65·3	65·6	65·8
2	129·6	130·0	130·6	131·2	131·6
3	194·4	195·0	195·9	196·8	197·4
4	259·2	260·0	261·2	262·4	263·2
5	324·0	325·0	326·5	328·0	329·0
6	388·8	390·0	391·8	393·6	394·8
7	453·6	455·0	457·1	459·2	460·6
8	518·4	520·0	522·4	524·8	526·4
9	583·2	585·0	587·7	590·4	592·2

2°

′ ″	sin	d	tan	d	cot	cos	′ ″
00 00	8·542 819	603	8·543 084	603	1·456 916	9·999 735	60 00
10	·543 422	601	·543 687	602	·456 313	·999 735	59 50
20	·544 023	601	·544 289	602	·455 711	·999 734	40
30	·544 624	600	·544 891	601	·455 109	·999 733	30
40	·545 224	599	·545 492	600	·454 508	·999 732	20
50	·545 823	599	·546 092	599	·453 908	·999 732	10
01 00	8·546 422	597	8·546 691	598	1·453 309	9·999 731	59 00
10	·547 019	597	·547 289	598	·452 711	·999 730	58 50
20	·547 616	596	·547 887	596	·452 113	·999 729	40
30	·548 212	595	·548 483	596	·451 517	·999 729	30
40	·548 807	594	·549 079	595	·450 921	·999 728	20
50	·549 401	594	·549 674	594	·450 326	·999 727	10
02 00	8·549 995	592	8·550 268	594	1·449 732	9·999 726	58 00
10	·550 587	592	·550 862	592	·449 138	·999 726	57 50
20	·551 179	591	·551 454	592	·448 546	·999 725	40
30	·551 770	591	·552 046	591	·447 954	·999 724	30
40	·552 361	589	·552 637	590	·447 363	·999 723	20
50	·552 950	589	·553 227	590	·446 773	·999 723	10
03 00	8·553 539	587	8·553 817	588	1·446 183	9·999 722	57 00
10	·554 126	587	·554 405	588	·445 595	·999 721	56 50
20	·554 713	587	·554 993	587	·445 007	·999 720	40
30	·555 300	585	·555 580	586	·444 420	·999 720	30
40	·555 885	585	·556 166	586	·443 834	·999 719	20
50	·556 470	584	·556 752	584	·443 248	·999 718	10
04 00	8·557 054	583	8·557 336	584	1·442 664	9·999 717	56 00
10	·557 637	582	·557 920	583	·442 080	·999 717	55 50
20	·558 219	582	·558 503	582	·441 497	·999 716	40
30	·558 801	580	·559 085	582	·440 915	·999 715	30
40	·559 381	580	·559 667	581	·440 333	·999 714	20
50	·559 961	579	·560 248	580	·439 752	·999 714	10
05 00	8·560 540	579	8·560 828	579	1·439 172	9·999 713	55 00
10	·561 119	577	·561 407	578	·438 593	·999 712	54 50
20	·561 696	577	·561 985	578	·438 015	·999 711	40
30	·562 273	576	·562 563	577	·437 437	·999 711	30
40	·562 849	576	·563 140	576	·436 860	·999 710	20
50	·563 425	574	·563 716	575	·436 284	·999 709	10
06 00	8·563 999	574	8·564 291	575	1·435 709	9·999 708	54 00
10	·564 573	573	·564 866	574	·435 134	·999 707	53 50
20	·565 146	573	·565 440	573	·434 560	·999 707	40
30	·565 719	571	·566 013	572	·433 987	·999 706	30
40	·566 290	571	·566 585	572	·433 415	·999 705	20
50	·566 861	570	·567 157	570	·432 843	·999 704	10
07 00	8·567 431	569	8·567 727	571	1·432 273	9·999 704	53 00
10	·568 000	569	·568 298	569	·431 702	·999 703	52 50
20	·568 569	568	·568 867	568	·431 133	·999 702	40
30	·569 137	567	·569 435	568	·430 565	·999 701	30
40	·569 704	566	·570 003	567	·429 997	·999 700	20
50	·570 270	566	·570 570	567	·429 430	·999 700	10
08 00	8·570 836	565	8·571 137	565	1·428 863	9·999 699	52 00
10	·571 401	564	·571 702	565	·428 298	·999 698	51 50
20	·571 965	563	·572 267	565	·427 733	·999 697	40
30	·572 528	563	·572 832	563	·427 168	·999 697	30
40	·573 091	562	·573 395	563	·426 605	·999 696	20
50	·573 653	561	·573 958	562	·426 042	·999 695	10
09 00	8·574 214	560	8·574 520	561	1·425 480	9·999 694	51 00
10	·574 774	560	·575 081	561	·424 919	·999 693	50 50
20	·575 334	559	·575 642	559	·424 358	·999 693	40
30	·575 893	558	·576 201	559	·423 799	·999 692	30
40	·576 451	558	·576 760	559	·423 240	·999 691	20
50	·577 009	557	·577 319	558	·422 681	·999 690	10
10 00	8·577 566		8·577 877		1·422 123	9·999 689	50 00
	cos		cot		tan	sin	87°

PROPORTIONAL PARTS

″	557	558	559	561	563
1	55·7	55·8	55·9	56·1	56·3
2	111·4	111·6	111·8	112·2	112·6
3	167·1	167·4	167·7	168·3	168·9
4	222·8	223·2	223·6	224·4	225·2
5	278·5	279·0	279·5	280·5	281·5
6	334·2	334·8	335·4	336·6	337·8
7	389·9	390·6	391·3	392·7	394·1
8	445·6	446·4	447·2	448·8	450·4
9	501·3	502·2	503·1	504·9	506·7

″	565	567	568	569	571
1	56·5	56·7	56·8	56·9	57·1
2	113·0	113·4	113·6	113·8	114·2
3	169·5	170·1	170·4	170·7	171·3
4	226·0	226·8	227·2	227·6	228·4
5	282·5	283·5	284·0	284·5	285·5
6	339·0	340·2	340·8	341·4	342·6
7	395·5	396·9	397·6	398·3	399·7
8	452·0	453·6	454·4	455·2	456·8
9	508·5	510·3	511·2	512·1	513·9

″	573	574	576	577	579
1	57·3	57·4	57·6	57·7	57·9
2	114·6	114·8	115·2	115·4	115·8
3	171·9	172·2	172·8	173·1	173·7
4	229·2	229·6	230·4	230·8	231·6
5	286·5	287·0	288·0	288·5	289·5
6	343·8	344·4	345·6	346·2	347·4
7	401·1	401·8	403·2	403·9	405·3
8	458·4	459·2	460·8	461·6	463·2
9	515·7	516·6	518·4	519·3	521·1

″	580	582	584	585	587
1	58·0	58·2	58·4	58·5	58·7
2	116·0	116·4	116·8	117·0	117·4
3	174·0	174·6	175·2	175·5	176·1
4	232·0	232·8	233·6	234·0	234·8
5	290·0	291·0	292·0	292·5	293·5
6	348·0	349·2	350·4	351·0	352·2
7	406·0	407·4	408·8	409·5	410·9
8	464·0	465·6	467·2	468·0	469·6
9	522·0	523·8	525·6	526·5	528·3

″	588	589	591	592	594
1	58·8	58·9	59·1	59·2	59·4
2	117·6	117·8	118·2	118·4	118·8
3	176·4	176·7	177·3	177·6	178·2
4	235·2	235·6	236·4	236·8	237·6
5	294·0	294·5	295·5	296·0	297·0
6	352·8	353·4	354·6	355·2	356·4
7	411·6	412·3	413·7	414·4	415·8
8	470·4	471·2	472·8	473·6	475·2
9	529·2	530·1	531·9	532·8	534·6

″	596	597	599	601	603
1	59·6	59·7	59·9	60·1	60·3
2	119·2	119·4	119·8	120·2	120·6
3	178·8	179·1	179·7	180·3	180·9
4	238·4	238·8	239·6	240·4	241·2
5	298·0	298·5	299·5	300·5	301·5
6	357·6	358·2	359·4	360·6	361·8
7	417·2	417·9	419·3	420·7	422·1
8	476·8	477·6	479·2	480·8	482·4
9	536·4	537·3	539·1	540·9	542·7

TABLE IIc—LOGARITHMS OF TRIGONOMETRICAL FUNCTIONS

(Degrees, Minutes and Seconds)

2°

′	″	sin	diff.	tan	diff.	cot	cos	′	″
10	00	8·577 566	556	8·577 877	557	1·422 123	9·999 689	50	00
	10	·578 122	556	·578 434	556	·421 566	·999 689	49	50
	20	·578 678	554	·578 990	555	·421 010	·999 688		40
	30	·579 232	554	·579 545	555	·420 455	·999 687		30
	40	·579 786	554	·580 100	554	·419 900	·999 686		20
	50	·580 340	552	·580 654	554	·419 346	·999 685		10
11	00	8·580 892	552	8·581 208	552	1·418 792	9·999 685	49	00
	10	·581 444	551	·581 760	552	·418 240	·999 684	48	50
	20	·581 995	551	·582 312	552	·417 688	·999 683		40
	30	·582 546	550	·582 864	550	·417 136	·999 682		30
	40	·583 096	549	·583 414	550	·416 586	·999 681		20
	50	·583 645	548	·583 964	550	·416 036	·999 681		10
12	00	8·584 193	548	8·584 514	548	1·415 486	9·999 680	48	00
	10	·584 741	547	·585 062	548	·414 938	·999 679	47	50
	20	·585 288	546	·585 610	547	·414 390	·999 678		40
	30	·585 834	546	·586 157	547	·413 843	·999 677		30
	40	·586 380	545	·586 704	545	·413 296	·999 677		20
	50	·586 925	544	·587 249	546	·412 751	·999 676		10
13	00	8·587 469	544	8·587 795	544	1·412 205	9·999 675	47	00
	10	·588 013	543	·588 339	544	·411 661	·999 674	46	50
	20	·588 556	542	·588 883	543	·411 117	·999 673		40
	30	·589 098	542	·589 426	542	·410 574	·999 672		30
	40	·589 640	541	·589 968	542	·410 032	·999 672		20
	50	·590 181	540	·590 510	541	·409 490	·999 671		10
14	00	8·590 721	539	8·591 051	540	1·408 949	9·999 670	46	00
	10	·591 260	539	·591 591	540	·408 409	·999 669	45	50
	20	·591 799	539	·592 131	539	·407 869	·999 668		40
	30	·592 338	537	·592 670	538	·407 330	·999 668		30
	40	·592 875	537	·593 208	538	·406 792	·999 667		20
	50	·593 412	536	·593 746	537	·406 254	·999 666		10
15	00	8·593 948	536	8·594 283	537	1·405 717	9·999 665	45	00
	10	·594 484	535	·594 820	535	·405 180	·999 664	44	50
	20	·595 019	534	·595 355	535	·404 645	·999 663		40
	30	·595 553	534	·595 890	535	·404 110	·999 663		30
	40	·596 087	532	·596 425	534	·403 575	·999 662		20
	50	·596 619	533	·596 959	533	·403 041	·999 661		10
16	00	8·597 152	531	8·597 492	532	1·402 508	9·999 660	44	00
	10	·597 683	531	·598 024	532	·401 976	·999 659	43	50
	20	·598 214	531	·598 556	531	·401 444	·999 658		40
	30	·598 745	529	·599 087	531	·400 913	·999 658		30
	40	·599 274	529	·599 618	529	·400 382	·999 657		20
	50	·599 803	529	·600 147	530	·399 853	·999 656		10
17	00	8·600 332	527	8·600 677	528	1·399 323	9·999 655	43	00
	10	·600 859	528	·601 205	528	·398 795	·999 654	42	50
	20	·601 387	526	·601 733	527	·398 267	·999 653		40
	30	·601 913	526	·602 260	527	·397 740	·999 653		30
	40	·602 439	525	·602 787	526	·397 213	·999 652		20
	50	·602 964	525	·603 313	526	·396 687	·999 651		10
18	00	8·603 489	523	8·603 839	524	1·396 161	9·999 650	42	00
	10	·604 012	524	·604 363	524	·395 637	·999 649	41	50
	20	·604 536	522	·604 887	524	·395 113	·999 648		40
	30	·605 058	522	·605 411	523	·394 589	·999 647		30
	40	·605 580	522	·605 934	522	·394 066	·999 647		20
	50	·606 102	521	·606 456	522	·393 544	·999 646		10
19	00	8·606 623	520	8·606 978	521	1·393 022	9·999 645	41	00
	10	·607 143	519	·607 499	520	·392 501	·999 644	40	50
	20	·607 662	519	·608 019	520	·391 981	·999 643		40
	30	·608 181	518	·608 539	519	·391 461	·999 642		30
	40	·608 699	518	·609 058	518	·390 942	·999 641		20
	50	·609 217	517	·609 576	518	·390 424	·999 641		10
20	00	8·609 734		8·610 094		1·389 906	9·999 640	40	00
		cos		cot		tan	sin		

87°

PROPORTIONAL PARTS

″	517	518	519	520	522
1	51·7	51·8	51·9	52·0	52·2
2	103·4	103·6	103·8	104·0	104·4
3	155·1	155·4	155·7	156·0	156·6
4	206·8	207·2	207·6	208·0	208·8
5	258·5	259·0	259·5	260·0	261·0
6	310·2	310·8	311·4	312·0	313·2
7	361·9	362·6	363·3	364·0	365·4
8	413·6	414·4	415·2	416·0	417·6
9	465·3	466·2	467·1	468·0	469·8

″	523	524	526	527	528
1	52·3	52·4	52·6	52·7	52·8
2	104·6	104·8	105·2	105·4	105·6
3	156·9	157·2	157·8	158·1	158·4
4	209·2	209·6	210·4	210·8	211·2
5	261·5	262·0	263·0	263·5	264·0
6	313·8	314·4	315·6	316·2	316·8
7	366·1	366·8	368·2	368·9	369·6
8	418·4	419·2	420·8	421·6	422·4
9	470·7	471·6	473·4	474·3	475·2

″	529	531	532	534	535
1	52·9	53·1	53·2	53·4	53·5
2	105·8	106·2	106·4	106·8	107·0
3	158·7	159·3	159·6	160·2	160·5
4	211·6	212·4	212·8	213·6	214·0
5	264·5	265·5	266·0	267·0	267·5
6	317·4	318·6	319·2	320·4	321·0
7	370·3	371·7	372·4	373·8	374·5
8	423·2	424·8	425·6	427·2	428·0
9	476·1	477·9	478·8	480·6	481·5

″	537	538	539	540	542
1	53·7	53·8	53·9	54·0	54·2
2	107·4	107·6	107·8	108·0	108·4
3	161·1	161·4	161·7	162·0	162·6
4	214·8	215·2	215·6	216·0	216·8
5	268·5	269·0	269·5	270·0	271·0
6	322·2	322·8	323·4	324·0	325·2
7	375·9	376·6	377·3	378·0	379·4
8	429·6	430·4	431·2	432·0	433·6
9	483·3	484·2	485·1	486·0	487·8

″	543	544	546	547	548
1	54·3	54·4	54·6	54·7	54·8
2	108·6	108·8	109·2	109·4	109·6
3	162·9	163·2	163·8	164·1	164·4
4	217·2	217·6	218·4	218·8	219·2
5	271·5	272·0	273·0	273·5	274·0
6	325·8	326·4	327·6	328·2	328·8
7	380·1	380·8	382·2	382·9	383·6
8	434·4	435·2	436·8	437·6	438·4
9	488·7	489·6	491·4	492·3	493·2

″	550	552	554	556	557
1	55·0	55·2	55·4	55·6	55·7
2	110·0	110·4	110·8	111·2	111·4
3	165·0	165·6	166·2	166·8	167·1
4	220·0	220·8	221·6	222·4	222·8
5	275·0	276·0	277·0	278·0	278·5
6	330·0	331·2	332·4	333·6	334·2
7	385·0	386·4	387·8	389·2	389·9
8	440·0	441·6	443·2	444·8	445·6
9	495·0	496·8	498·6	500·4	501·3

6–III–3

TABLE IIc—LOGARITHMS OF TRIGONOMETRICAL FUNCTIONS

(Degrees, Minutes and Seconds)

2°

′	″	sin		tan		cot	cos	′	″
20	**00**	8·609 734	517	8·610 094	518	1·389 906	9·999 640	**40**	**00**
	10	·610 251	515	·610 612	516	·389 388	·999 639	**39**	50
	20	·610 766	516	·611 128	516	·388 872	·999 638		40
	30	·611 282	514	·611 644	516	·388 356	·999 637		30
	40	·611 796	514	·612 160	515	·387 840	·999 636		20
	50	·612 310	513	·612 675	514	·387 325	·999 635		10
21	**00**	8·612 823	513	8·613 189	513	1·386 811	9·999 635	**39**	**00**
	10	·613 336	512	·613 702	513	·386 298	·999 634	**38**	50
	20	·613 848	512	·614 215	513	·385 785	·999 633		40
	30	·614 360	511	·614 728	512	·385 272	·999 632		30
	40	·614 871	510	·615 240	511	·384 760	·999 631		20
	50	·615 381	510	·615 751	511	·384 249	·999 630		10
22	**00**	8·615 891	509	8·616 262	510	1·383 738	9·999 629	**38**	**00**
	10	·616 400	509	·616 772	509	·383 228	·999 629	**37**	50
	20	·616 909	508	·617 281	509	·382 719	·999 628		40
	30	·617 417	507	·617 790	508	·382 210	·999 627		30
	40	·617 924	507	·618 298	508	·381 702	·999 626		20
	50	·618 431	506	·618 806	507	·381 194	·999 625		10
23	**00**	8·618 937	505	8·619 313	506	1·380 687	9·999 624	**37**	**00**
	10	·619 442	505	·619 819	506	·380 181	·999 623	**36**	50
	20	·619 947	505	·620 325	505	·379 675	·999 622		40
	30	·620 452	504	·620 830	505	·379 170	·999 622		30
	40	·620 956	503	·621 335	504	·378 665	·999 621		20
	50	·621 459	503	·621 839	504	·378 161	·999 620		10
24	**00**	8·621 962	502	8·622 343	503	1·377 657	9·999 619	**36**	**00**
	10	·622 464	501	·622 846	502	·377 154	·999 618	**35**	50
	20	·622 965	501	·623 348	502	·376 652	·999 617		40
	30	·623 466	500	·623 850	501	·376 150	·999 616		30
	40	·623 966	500	·624 351	501	·375 649	·999 615		20
	50	·624 466	499	·624 852	500	·375 148	·999 614		10
25	**00**	8·624 965	499	8·625 352	499	1·374 648	9·999 614	**35**	**00**
	10	·625 464	498	·625 851	499	·374 149	·999 613	**34**	50
	20	·625 962	497	·626 350	499	·373 650	·999 612		40
	30	·626 459	497	·626 849	497	·373 151	·999 611		30
	40	·626 956	497	·627 346	498	·372 654	·999 610		20
	50	·627 453	495	·627 844	496	·372 156	·999 609		10
26	**00**	8·627 948	496	8·628 340	496	1·371 660	9·999 608	**34**	**00**
	10	·628 444	494	·628 836	496	·371 164	·999 607	**33**	50
	20	·628 938	494	·629 332	495	·370 668	·999 606		40
	30	·629 432	494	·629 827	494	·370 173	·999 606		30
	40	·629 926	493	·630 321	494	·369 679	·999 605		20
	50	·630 419	492	·630 815	493	·369 185	·999 604		10
27	**00**	8·630 911	492	8·631 308	493	1·368 692	9·999 603	**33**	**00**
	10	·631 403	491	·631 801	492	·368 199	·999 602	**32**	50
	20	·631 894	491	·632 293	492	·367 707	·999 601		40
	30	·632 385	490	·632 785	491	·367 215	·999 600		30
	40	·632 875	490	·633 276	490	·366 724	·999 599		20
	50	·633 365	489	·633 766	490	·366 234	·999 598		10
28	**00**	8·633 854	488	8·634 256	490	1·365 744	9·999 597	**32**	**00**
	10	·634 342	488	·634 746	489	·365 254	·999 597	**31**	50
	20	·634 830	487	·635 235	488	·364 765	·999 596		40
	30	·635 317	487	·635 723	488	·364 277	·999 595		30
	40	·635 804	487	·636 211	487	·363 789	·999 594		20
	50	·636 291	485	·636 698	486	·363 302	·999 593		10
29	**00**	8·636 776	486	8·637 184	487	1·362 816	9·999 592	**31**	**00**
	10	·637 262	484	·637 671	485	·362 329	·999 591	**30**	50
	20	·637 746	484	·638 156	485	·361 844	·999 590		40
	30	·638 230	484	·638 641	485	·361 359	·999 589		30
	40	·638 714	483	·639 126	484	·360 874	·999 588		20
	50	·639 197	483	·639 610	483	·360 390	·999 587		10
30	**00**	8·639 680		8·640 093		1·359 907	9·999 586	**30**	**00**
		cos		cot		tan	sin	**87°**	

PROPORTIONAL PARTS

″	483	484	485	487	488
1	48·3	48·4	48·5	48·7	48·8
2	96·6	96·8	97·0	97·4	97·6
3	144·9	145·2	145·5	146·1	146·4
4	193·2	193·6	194·0	194·8	195·2
5	241·5	242·0	242·5	243·5	244·0
6	289·8	290·4	291·0	292·2	292·8
7	338·1	338·8	339·5	340·9	341·6
8	386·4	387·2	388·0	389·6	390·4
9	434·7	435·6	436·5	438·3	439·2

″	490	491	492	493	494
1	49·0	49·1	49·2	49·3	49·4
2	98·0	98·2	98·4	98·6	98·8
3	147·0	147·3	147·6	147·9	148·2
4	196·0	196·4	196·8	197·2	197·6
5	245·0	245·5	246·0	246·5	247·0
6	294·0	294·6	295·2	295·8	296·4
7	343·0	343·7	344·4	345·1	345·8
8	392·0	392·8	393·6	394·4	395·2
9	441·0	441·9	442·8	443·7	444·6

″	496	497	499	500	501
1	49·6	49·7	49·9	50·0	50·1
2	99·2	99·4	99·8	100·0	100·2
3	148·8	149·1	149·7	150·0	150·3
4	198·4	198·8	199·6	200·0	200·4
5	248·0	248·5	249·5	250·0	250·5
6	297·6	298·2	299·4	300·0	300·6
7	347·2	347·9	349·3	350·0	350·7
8	396·8	397·6	399·2	400·0	400·8
9	446·4	447·3	449·1	450·0	450·9

″	502	503	504	505	506
1	50·2	50·3	50·4	50·5	50·6
2	100·4	100·6	100·8	101·0	101·2
3	150·6	150·9	151·2	151·5	151·8
4	200·8	201·2	201·6	202·0	202·4
5	251·0	251·5	252·0	252·5	253·0
6	301·2	301·8	302·4	303·0	303·6
7	351·4	352·1	352·8	353·5	354·2
8	401·6	402·4	403·2	404·0	404·8
9	451·8	452·7	453·6	454·5	455·4

″	507	508	509	510	511
1	50·7	50·8	50·9	51·0	51·1
2	101·4	101·6	101·8	102·0	102·2
3	152·1	152·4	152·7	153·0	153·3
4	202·8	203·2	203·6	204·0	204·4
5	253·5	254·0	254·5	255·0	255·5
6	304·2	304·8	305·4	306·0	306·6
7	354·9	355·6	356·3	357·0	357·7
8	405·6	406·4	407·2	408·0	408·8
9	456·3	457·2	458·1	459·0	459·9

″	512	513	514	516	518
1	51·2	51·3	51·4	51·6	51·8
2	102·4	102·6	102·8	103·2	103·6
3	153·6	153·9	154·2	154·8	155·4
4	204·8	205·2	205·6	206·4	207·2
5	256·0	256·5	257·0	258·0	259·0
6	307·2	307·8	308·4	309·6	310·8
7	358·4	359·1	359·8	361·2	362·6
8	409·6	410·4	411·2	412·8	414·4
9	460·8	461·7	462·6	464·4	466·2

TABLE IIc—LOGARITHMS OF TRIGONOMETRICAL FUNCTIONS

(Degrees, Minutes and Seconds)

2°

′ ″	sin	d	tan	d	cot	cos	′ ″
30 00	8·639 680	482	8·640 093	483	1·359 907	9·999 586	30 00
10	·640 162	481	·640 576	482	·359 424	·999 586	29 50
20	·640 643	481	·641 058	482	·358 942	·999 585	40
30	·641 124	480	·641 540	481	·358 460	·999 584	30
40	·641 604	480	·642 021	481	·357 979	·999 583	20
50	·642 084	479	·642 502	480	·357 498	·999 582	10
31 00	8·642 563	479	8·642 982	480	1·357 018	9·999 581	29 00
10	·643 042	478	·643 462	479	·356 538	·999 580	28 50
20	·643 520	478	·643 941	479	·356 059	·999 579	40
30	·643 998	477	·644 420	478	·355 580	·999 578	30
40	·644 475	477	·644 898	478	·355 102	·999 577	20
50	·644 952	476	·645 376	477	·354 624	·999 576	10
32 00	8·645 428	476	8·645 853	476	1·354 147	9·999 575	28 00
10	·645 904	475	·646 329	476	·353 671	·999 574	27 50
20	·646 379	475	·646 805	476	·353 195	·999 573	40
30	·646 854	474	·647 281	475	·352 719	·999 573	30
40	·647 328	473	·647 756	474	·352 244	·999 572	20
50	·647 801	473	·648 230	474	·351 770	·999 571	10
33 00	8·648 274	473	8·648 704	474	1·351 296	9·999 570	27 00
10	·648 747	472	·649 178	473	·350 822	·999 569	26 50
20	·649 219	471	·649 651	472	·350 349	·999 568	40
30	·649 690	471	·650 123	472	·349 877	·999 567	30
40	·650 161	471	·650 595	472	·349 405	·999 566	20
50	·650 632	470	·651 067	470	·348 933	·999 565	10
34 00	8·651 102	469	8·651 537	471	1·348 463	9·999 564	26 00
10	·651 571	469	·652 008	470	·347 992	·999 563	25 50
20	·652 040	468	·652 478	469	·347 522	·999 562	40
30	·652 508	468	·652 947	469	·347 053	·999 561	30
40	·652 976	468	·653 416	468	·346 584	·999 560	20
50	·653 444	467	·653 884	468	·346 116	·999 559	10
35 00	8·653 911	466	8·654 352	468	1·345 648	9·999 558	25 00
10	·654 377	466	·654 820	466	·345 180	·999 557	24 50
20	·654 843	465	·655 286	467	·344 714	·999 557	40
30	·655 308	465	·655 753	466	·344 247	·999 556	30
40	·655 773	465	·656 219	465	·343 781	·999 555	20
50	·656 238	464	·656 684	465	·343 316	·999 554	10
36 00	8·656 702	463	8·657 149	464	1·342 851	9·999 553	24 00
10	·657 165	463	·657 613	464	·342 387	·999 552	23 50
20	·657 628	462	·658 077	464	·341 923	·999 551	40
30	·658 090	462	·658 541	463	·341 459	·999 550	30
40	·658 552	462	·659 004	462	·340 996	·999 549	20
50	·659 014	461	·659 466	462	·340 534	·999 548	10
37 00	8·659 475	460	8·659 928	461	1·340 072	9·999 547	23 00
10	·659 935	460	·660 389	461	·339 611	·999 546	22 50
20	·660 395	460	·660 850	461	·339 150	·999 545	40
30	·660 855	459	·661 311	460	·338 689	·999 544	30
40	·661 314	458	·661 771	459	·338 229	·999 543	20
50	·661 772	458	·662 230	459	·337 770	·999 542	10
38 00	8·662 230	458	8·662 689	459	1·337 311	9·999 541	22 00
10	·662 688	457	·663 148	458	·336 852	·999 540	21 50
20	·663 145	457	·663 606	457	·336 394	·999 539	40
30	·663 602	456	·664 063	457	·335 937	·999 538	30
40	·664 058	455	·664 520	457	·335 480	·999 537	20
50	·664 513	455	·664 977	456	·335 023	·999 536	10
39 00	8·664 968	455	8·665 433	456	1·334 567	9·999 535	21 00
10	·665 423	454	·665 889	455	·334 111	·999 534	20 50
20	·665 877	454	·666 344	455	·333 656	·999 533	40
30	·666 331	453	·666 799	454	·333 201	·999 532	30
40	·666 784	453	·667 253	454	·332 747	·999 531	20
50	·667 237	452	·667 707	453	·332 293	·999 530	10
40 00	8·667 689		8·668 160		1·331 840	9·999 529	20 00
	cos		cot		tan	sin	**87°**

PROPORTIONAL PARTS

″	453	454	455	456	457
1	45·3	45·4	45·5	45·6	45·7
2	90·6	90·8	91·0	91·2	91·4
3	135·9	136·2	136·5	136·8	137·1
4	181·2	181·6	182·0	182·4	182·8
5	226·5	227·0	227·5	228·0	228·5
6	271·8	272·4	273·0	273·6	274·2
7	317·1	317·8	318·5	319·2	319·9
8	362·4	363·2	364·0	364·8	365·6
9	407·7	408·6	409·5	410·4	411·3

″	458	459	460	461	462
1	45·8	45·9	46·0	46·1	46·2
2	91·6	91·8	92·0	92·2	92·4
3	137·4	137·7	138·0	138·3	138·6
4	183·2	183·6	184·0	184·4	184·8
5	229·0	229·5	230·0	230·5	231·0
6	274·8	275·4	276·0	276·6	277·2
7	320·6	321·3	322·0	322·7	323·4
8	366·4	367·2	368·0	368·8	369·6
9	412·2	413·1	414·0	414·9	415·8

″	463	464	465	466	467
1	46·3	46·4	46·5	46·6	46·7
2	92·6	92·8	93·0	93·2	93·4
3	138·9	139·2	139·5	139·8	140·1
4	185·2	185·6	186·0	186·4	186·8
5	231·5	232·0	232·5	233·0	233·5
6	277·8	278·4	279·0	279·6	280·2
7	324·1	324·8	325·5	326·2	326·9
8	370·4	371·2	372·0	372·8	373·6
9	416·7	417·6	418·5	419·4	420·3

″	468	469	470	471	472
1	46·8	46·9	47·0	47·1	47·2
2	93·6	93·8	94·0	94·2	94·4
3	140·4	140·7	141·0	141·3	141·6
4	187·2	187·6	188·0	188·4	188·8
5	234·0	234·5	235·0	235·5	236·0
6	280·8	281·4	282·0	282·6	283·2
7	327·6	328·3	329·0	329·7	330·4
8	374·4	375·2	376·0	376·8	377·6
9	421·2	422·1	423·0	423·9	424·8

″	473	474	475	476	477
1	47·3	47·4	47·5	47·6	47·7
2	94·6	94·8	95·0	95·2	95·4
3	141·9	142·2	142·5	142·8	143·1
4	189·2	189·6	190·0	190·4	190·8
5	236·5	237·0	237·5	238·0	238·5
6	283·8	284·4	285·0	285·6	286·2
7	331·1	331·8	332·5	333·2	333·9
8	378·4	379·2	380·0	380·8	381·6
9	425·7	426·6	427·5	428·4	429·3

″	478	479	480	481	482
1	47·8	47·9	48·0	48·1	48·2
2	95·6	95·8	96·0	96·2	96·4
3	143·4	143·7	144·0	144·3	144·6
4	191·2	191·6	192·0	192·4	192·8
5	239·0	239·5	240·0	240·5	241·0
6	286·8	287·4	288·0	288·6	289·2
7	334·6	335·3	336·0	336·7	337·4
8	382·4	383·2	384·0	384·8	385·6
9	430·2	431·1	432·0	432·9	433·8

TABLE IIc—LOGARITHMS OF TRIGONOMETRICAL FUNCTIONS

(Degrees, Minutes and Seconds)

2°

′	″	sin		tan		cot	cos	′	″
40	00	8·667 689	452	8·668 160	453	1·331 840	9·999 529	**20**	00
	10	·668 141	451	·668 613	452	·331 387	·999 528	**19**	50
	20	·668 592	451	·669 065	452	·330 935	·999 527		40
	30	·669 043	451	·669 517	451	·330 483	·999 527		30
	40	·669 494	450	·669 968	451	·330 032	·999 526		20
	50	·669 944	449	·670 419	451	·329 581	·999 525		10
41	00	8·670 393	449	8·670 870	450	1·329 130	9·999 524	**19**	00
	10	·670 842	449	·671 320	449	·328 680	·999 523	**18**	50
	20	·671 291	448	·671 769	449	·328 231	·999 522		40
	30	·671 739	448	·672 218	449	·327 782	·999 521		30
	40	·672 187	447	·672 667	448	·327 333	·999 520		20
	50	·672 634	446	·673 115	448	·326 885	·999 519		10
42	00	8·673 080	447	8·673 563	447	1·326 437	9·999 518	**18**	00
	10	·673 527	445	·674 010	447	·325 990	·999 517	**17**	50
	20	·673 972	446	·674 457	446	·325 543	·999 516		40
	30	·674 418	445	·674 903	446	·325 097	·999 515		30
	40	·674 863	444	·675 349	445	·324 651	·999 514		20
	50	·675 307	444	·675 794	445	·324 206	·999 513		10
43	00	8·675 751	443	8·676 239	445	1·323 761	9·999 512	**17**	00
	10	·676 194	444	·676 684	444	·323 316	·999 511	**16**	50
	20	·676 638	442	·677 128	444	·322 872	·999 510		40
	30	·677 080	442	·677 572	443	·322 428	·999 509		30
	40	·677 522	442	·678 015	442	·321 985	·999 508		20
	50	·677 964	441	·678 457	443	·321 543	·999 507		10
44	00	8·678 405	441	8·678 900	441	1·321 100	9·999 506	**16**	00
	10	·678 846	440	·679 341	442	·320 659	·999 505	**15**	50
	20	·679 286	440	·679 783	441	·320 217	·999 504		40
	30	·679 726	440	·680 224	440	·319 776	·999 503		30
	40	·680 166	439	·680 664	440	·319 336	·999 502		20
	50	·680 605	438	·681 104	440	·318 896	·999 501		10
45	00	8·681 043	438	8·681 544	439	1·318 456	9·999 500	**15**	00
	10	·681 481	438	·681 983	439	·318 017	·999 499	**14**	50
	20	·681 919	437	·682 422	438	·317 578	·999 498		40
	30	·682 356	437	·682 860	438	·317 140	·999 497		30
	40	·682 793	437	·683 298	437	·316 702	·999 496		20
	50	·683 230	435	·683 735	437	·316 265	·999 495		10
46	00	8·683 665	436	8·684 172	436	1·315 828	9·999 493	**14**	00
	10	·684 101	435	·684 608	436	·315 392	·999 492	**13**	50
	20	·684 536	435	·685 044	436	·314 956	·999 491		40
	30	·684 971	434	·685 480	435	·314 520	·999 490		30
	40	·685 405	433	·685 915	435	·314 085	·999 489		20
	50	·685 838	434	·686 350	434	·313 650	·999 488		10
47	00	8·686 272	433	8·686 784	434	1·313 216	9·999 487	**13**	00
	10	·686 705	432	·687 218	434	·312 782	·999 486	**12**	50
	20	·687 137	432	·687 652	433	·312 348	·999 485		40
	30	·687 569	432	·688 085	432	·311 915	·999 484		30
	40	·688 001	431	·688 517	433	·311 483	·999 483		20
	50	·688 432	431	·688 950	431	·311 050	·999 482		10
48	00	8·688 863	430	8·689 381	432	1·310 619	9·999 481	**12**	00
	10	·689 293	430	·689 813	431	·310 187	·999 480	**11**	50
	20	·689 723	429	·690 244	430	·309 756	·999 479		40
	30	·690 152	429	·690 674	430	·309 326	·999 478		30
	40	·690 581	429	·691 104	430	·308 896	·999 477		20
	50	·691 010	428	·691 534	429	·308 466	·999 476		10
49	00	8·691 438	428	8·691 963	429	1·308 037	9·999 475	**11**	00
	10	·691 866	427	·692 392	428	·307 608	·999 474	**10**	50
	20	·692 293	427	·692 820	428	·307 180	·999 473		40
	30	·692 720	426	·693 248	427	·306 752	·999 472		30
	40	·693 146	426	·693 675	428	·306 325	·999 471		20
	50	·693 572	426	·694 103	426	·305 897	·999 470		10
50	00	8·693 998		8·694 529		1·305 471	9·999 469	**10**	00
		cos		cot		tan	sin	**87°**	

PROPORTIONAL PARTS

″	**426**	**427**	**428**	**429**	**430**
1	42·6	42·7	42·8	42·9	43·0
2	85·2	85·4	85·6	85·8	86·0
3	127·8	128·1	128·4	128·7	129·0
4	170·4	170·8	171·2	171·6	172·0
5	213·0	213·5	214·0	214·5	215·0
6	255·6	256·2	256·8	257·4	258·0
7	298·2	298·9	299·6	300·3	301·0
8	340·8	341·6	342·4	343·2	344·0
9	383·4	384·3	385·2	386·1	387·0

″	**431**	**432**	**433**	**434**	**435**
1	43·1	43·2	43·3	43·4	43·5
2	86·2	86·4	86·6	86·8	87·0
3	129·3	129·6	129·9	130·2	130·5
4	172·4	172·8	173·2	173·6	174·0
5	215·5	216·0	216·5	217·0	217·5
6	258·6	259·2	259·8	260·4	261·0
7	301·7	302·4	303·1	303·8	304·5
8	344·8	345·6	346·4	347·2	348·0
9	387·9	388·8	389·7	390·6	391·5

″	**436**	**437**	**438**	**439**	**440**
1	43·6	43·7	43·8	43·9	44·0
2	87·2	87·4	87·6	87·8	88·0
3	130·8	131·1	131·4	131·7	132·0
4	174·4	174·8	175·2	175·6	176·0
5	218·0	218·5	219·0	219·5	220·0
6	261·6	262·2	262·8	263·4	264·0
7	305·2	305·9	306·6	307·3	308·0
8	348·8	349·6	350·4	351·2	352·0
9	392·4	393·3	394·2	395·1	396·0

″	**441**	**442**	**443**	**444**	**445**
1	44·1	44·2	44·3	44·4	44·5
2	88·2	88·4	88·6	88·8	89·0
3	132·3	132·6	132·9	133·2	133·5
4	176·4	176·8	177·2	177·6	178·0
5	220·5	221·0	221·5	222·0	222·5
6	264·6	265·2	265·8	266·4	267·0
7	308·7	309·4	310·1	310·8	311·5
8	352·8	353·6	354·4	355·2	356·0
9	396·9	397·8	398·7	399·6	400·5

″	**446**	**447**	**448**	**449**	**450**
1	44·6	44·7	44·8	44·9	45·0
2	89·2	89·4	89·6	89·8	90·0
3	133·8	134·1	134·4	134·7	135·0
4	178·4	178·8	179·2	179·6	180·0
5	223·0	223·5	224·0	224·5	225 0
6	267·6	268·2	268·8	269·4	270·0
7	312·2	312·9	313·6	314·3	315·0
8	356·8	357·6	358·4	359·2	360·0
9	401·4	402·3	403·2	404·1	405·0

″	**451**	**452**	**453**
1	45·1	45·2	45·3
2	90·2	90·4	90·6
3	135·3	135·6	135·9
4	180·4	180·8	181·2
5	225·5	226·0	226·5
6	270·6	271·2	271·8
7	315·7	316·4	317·1
8	360·8	361·6	362·4
9	405·9	406·8	407·7

TABLE IIc—LOGARITHMS OF TRIGONOMETRICAL FUNCTIONS

(Degrees, Minutes and Seconds)

2° ′	″	sin		tan		cot	cos	′	″
50	00	8·693 998	425	8·694 529	427	1·305 471	9·999 469	10	00
	10	·694 423	425	·694 956	425	·305 044	·999 468	09	50
	20	·694 848	424	·695 381	426	·304 619	·999 467		40
	30	·695 272	424	·695 807	425	·304 193	·999 466		30
	40	·695 696	424	·696 232	424	·303 768	·999 465		20
	50	·696 120	423	·696 656	425	·303 344	·999 464		10
51	00	8·696 543	423	8·697 081	423	1·302 919	9·999 463	09	00
	10	·696 966	422	·697 504	424	·302 496	·999 461	08	50
	20	·697 388	422	·697 928	423	·302 072	·999 460		40
	30	·697 810	422	·698 351	422	·301 649	·999 459		30
	40	·698 232	421	·698 773	422	·301 227	·999 458		20
	50	·698 653	420	·699 195	422	·300 805	·999 457		10
52	00	8·699 073	421	8·699 617	421	1·300 383	9·999 456	08	00
	10	·699 494	419	·700 038	421	·299 962	·999 455	07	50
	20	·699 913	420	·700 459	421	·299 541	·999 454		40
	30	·700 333	419	·700 880	420	·299 120	·999 453		30
	40	·700 752	419	·701 300	420	·298 700	·999 452		20
	50	·701 171	418	·701 720	419	·298 280	·999 451		10
53	00	8·701 589	418	8·702 139	419	1·297 861	9·999 450	07	00
	10	·702 007	417	·702 558	418	·297 442	·999 449	06	50
	20	·702 424	417	·702 976	419	·297 024	·999 448		40
	30	·702 841	417	·703 395	417	·296 605	·999 447		30
	40	·703 258	416	·703 812	418	·296 188	·999 446		20
	50	·703 674	416	·704 230	416	·295 770	·999 445		10
54	00	8·704 090	415	8·704 646	417	1·295 354	9·999 443	06	00
	10	·704 505	415	·705 063	416	·294 937	·999 442	05	50
	20	·704 920	415	·705 479	416	·294 521	·999 441		40
	30	·705 335	414	·705 895	415	·294 105	·999 440		30
	40	·705 749	414	·706 310	415	·293 690	·999 439		20
	50	·706 163	414	·706 725	415	·293 275	·999 438		10
55	00	8·706 577	413	8·707 140	414	1·292 860	9·999 437	05	00
	10	·706 990	412	·707 554	413	·292 446	·999 436	04	50
	20	·707 402	413	·707 967	414	·292 033	·999 435		40
	30	·707 815	411	·708 381	413	·291 619	·999 434		30
	40	·708 226	412	·708 794	412	·291 206	·999 433		20
	50	·708 638	411	·709 206	412	·290 794	·999 432		10
56	00	8·709 049	411	8·709 618	412	1·290 382	9·999 431	04	00
	10	·709 460	410	·710 030	412	·289 970	·999 430	03	50
	20	·709 870	410	·710 442	411	·289 558	·999 428		40
	30	·710 280	410	·710 853	410	·289 147	·999 427		30
	40	·710 690	409	·711 263	411	·288 737	·999 426		20
	50	·711 099	408	·711 674	409	·288 326	·999 425		10
57	00	8·711 507	409	8·712 083	410	1·287 917	9·999 424	03	00
	10	·711 916	408	·712 493	409	·287 507	·999 423	02	50
	20	·712 324	407	·712 902	409	·287 098	·999 422		40
	30	·712 731	408	·713 311	408	·286 689	·999 421		30
	40	·713 139	407	·713 719	408	·286 281	·999 420		20
	50	·713 546	406	·714 127	407	·285 873	·999 419		10
58	00	8·713 952	406	8·714 534	408	1·285 466	9·999 418	02	00
	10	·714 358	406	·714 942	406	·285 058	·999 416	01	50
	20	·714 764	405	·715 348	407	·284 652	·999 415		40
	30	·715 169	405	·715 755	406	·284 245	·999 414		30
	40	·715 574	405	·716 161	406	·283 839	·999 413		20
	50	·715 979	404	·716 567	405	·283 433	·999 412		10
59	00	8·716 383	404	8·716 972	405	1·283 028	9·999 411	01	00
	10	·716 787	403	·717 377	404	·282 623	·999 410	00	50
	20	·717 190	403	·717 781	405	·282 219	·999 409		40
	30	·717 593	403	·718 186	403	·281 814	·999 408		30
	40	·717 996	402	·718 589	404	·281 411	·999 407		20
	50	·718 398	402	·718 993	403	·281 007	·999 406		10
60	00	8·718 800		8·719 396		1·280 604	9·999 404	00	00
		cos		cot		tan	sin	87°	

PROPORTIONAL PARTS

″	402	403	404	405	406
1	40·2	40·3	40·4	40·5	40·6
2	80·4	80·6	80·8	81·0	81·2
3	120·6	120·9	121·2	121·5	121·8
4	160·8	161·2	161·6	162·0	162·4
5	201·0	201·5	202·0	202·5	203·0
6	241·2	241·8	242·4	243·0	243·6
7	281·4	282·1	282·8	283·5	284·2
8	321·6	322·4	323·2	324·0	324·8
9	361·8	362·7	363·6	364·5	365·4

″	407	408	409	410	411
1	40·7	40·8	40·9	41·0	41·1
2	81·4	81·6	81·8	82·0	82·2
3	122·1	122·4	122·7	123·0	123·3
4	162·8	163·2	163·6	164·0	164·4
5	203·5	204·0	204·5	205·0	205·5
6	244·2	244·8	245·4	246·0	246·6
7	284·9	285·6	286·3	287·0	287·7
8	325·6	326·4	327·2	328·0	328·8
9	366·3	367·2	368·1	369·0	369·9

″	412	413	414	415	416
1	41·2	41·3	41·4	41·5	41·6
2	82·4	82·6	82·8	83·0	83·2
3	123·6	123·9	124·2	124·5	124·8
4	164·8	165·2	165·6	166·0	166·4
5	206·0	206·5	207·0	207·5	208·0
6	247·2	247·8	248·4	249·0	249·6
7	288·4	289·1	289·8	290·5	291·2
8	329·6	330·4	331·2	332·0	332·8
9	370·8	371·7	372·6	373·5	374·4

″	417	418	419	420	421
1	41·7	41·8	41·9	42·0	42·1
2	83·4	83·6	83·8	84·0	84·2
3	125·1	125·4	125·7	126·0	126·3
4	166·8	167·2	167·6	168·0	168·4
5	208·5	209·0	209·5	210·0	210·5
6	250·2	250·8	251·4	252·0	252·6
7	291·9	292·6	293·3	294·0	294·7
8	333·6	334·4	335·2	336·0	336·8
9	375·3	376·2	377·1	378·0	378·9

″	422	423	424	425	426
1	42·2	42·3	42·4	42·5	42·6
2	84·4	84·6	84·8	85·0	85·2
3	126·6	126·9	127·2	127·5	127·8
4	168·8	169·2	169·6	170·0	170·4
5	211·0	211·5	212·0	212·5	213·0
6	253·2	253·8	254·4	255·0	255·6
7	295·4	296·1	296·8	297·5	298·2
8	337·6	338·4	339·2	340·0	340·8
9	379·8	380·7	381·6	382·5	383·4

TABLE IIc—LOGARITHMS OF TRIGONOMETRICAL FUNCTIONS

(Degrees, Minutes and Seconds)

3°

′ ″	sin		tan		cot	cos	′ ″
00 00	8·718 800	402	8·719 396	402	1·280 604	9·999 404	60 00
10	·719 202	401	·719 798	403	·280 202	·999 403	59 50
20	·719 603	401	·720 201	402	·279 799	·999 402	40
30	·720 004	400	·720 603	401	·279 397	·999 401	30
40	·720 404	400	·721 004	401	·278 996	·999 400	20
50	·720 804	400	·721 405	401	·278 595	·999 399	10
01 00	8·721 204	399	8·721 806	401	1·278 194	9·999 398	59 00
10	·721 603	399	·722 207	400	·277 793	·999 397	58 50
20	·722 002	399	·722 607	400	·277 393	·999 396	40
30	·722 401	398	·723 007	399	·276 993	·999 394	30
40	·722 799	398	·723 406	399	·276 594	·999 393	20
50	·723 197	398	·723 805	399	·276 195	·999 392	10
02 00	8·723 595	397	8·724 204	398	1·275 796	9·999 391	58 00
10	·723 992	397	·724 602	398	·275 398	·999 390	57 50
20	·724 389	396	·725 000	397	·275 000	·999 389	40
30	·724 785	396	·725 397	397	·274 603	·999 388	30
40	·725 181	396	·725 794	397	·274 206	·999 387	20
50	·725 577	395	·726 191	397	·273 809	·999 385	10
03 00	8·725 972	395	8·726 588	396	1·273 412	9·999 384	57 00
10	·726 367	395	·726 984	396	·273 016	·999 383	56 50
20	·726 762	394	·727 380	395	·272 620	·999 382	40
30	·727 156	394	·727 775	395	·272 225	·999 381	30
40	·727 550	393	·728 170	395	·271 830	·999 380	20
50	·727 943	394	·728 565	394	·271 435	·999 379	10
04 00	8·728 337	392	8·728 959	394	1·271 041	9·999 378	56 00
10	·728 729	393	·729 353	393	·270 647	·999 377	55 50
20	·729 122	392	·729 746	394	·270 254	·999 375	40
30	·729 514	392	·730 140	393	·269 860	·999 374	30
40	·729 906	391	·730 533	392	·269 467	·999 373	20
50	·730 297	391	·730 925	392	·269 075	·999 372	10
05 00	8·730 688	391	8·731 317	392	1·268 683	9·999 371	55 00
10	·731 079	390	·731 709	392	·268 291	·999 370	54 50
20	·731 469	390	·732 101	391	·267 899	·999 369	40
30	·731 859	390	·732 492	391	·267 508	·999 367	30
40	·732 249	389	·732 883	390	·267 117	·999 366	20
50	·732 638	389	·733 273	390	·266 727	·999 365	10
06 00	8·733 027	389	8·733 663	390	1·266 337	9·999 364	54 00
10	·733 416	388	·734 053	389	·265 947	·999 363	53 50
20	·733 804	388	·734 442	389	·265 558	·999 362	40
30	·734 192	387	·734 831	389	·265 169	·999 361	30
40	·734 579	388	·735 220	388	·264 780	·999 359	20
50	·734 967	387	·735 608	388	·264 392	·999 358	10
07 00	8·735 354	386	8·735 996	388	1·264 004	9·999 357	53 00
10	·735 740	386	·736 384	387	·263 616	·999 356	52 50
20	·736 126	386	·736 771	387	·263 229	·999 355	40
30	·736 512	386	·737 158	387	·262 842	·999 354	30
40	·736 898	385	·737 545	386	·262 455	·999 353	20
50	·737 283	384	·737 931	386	·262 069	·999 351	10
08 00	8·737 667	385	8·738 317	386	1·261 683	9·999 350	52 00
10	·738 052	384	·738 703	385	·261 297	·999 349	51 50
20	·738 436	384	·739 088	385	·260 912	·999 348	40
30	·738 820	383	·739 473	385	·260 527	·999 347	30
40	·739 203	383	·739 858	384	·260 142	·999 346	20
50	·739 586	383	·740 242	384	·259 758	·999 344	10
09 00	8·739 969	383	8·740 626	383	1·259 374	9·999 343	51 00
10	·740 352	382	·741 009	384	·258 991	·999 342	50 50
20	·740 734	381	·741 393	383	·258 607	·999 341	40
30	·741 115	382	·741 776	382	·258 224	·999 340	30
40	·741 497	381	·742 158	382	·257 842	·999 339	20
50	·741 878	381	·742 540	382	·257 460	·999 338	10
10 00	8·742 259		8·742 922		1·257 078	9·999 336	50 00
	cos		cot		tan	sin	86°

PROPORTIONAL PARTS

″	381	382	383	384
1	38·1	38·2	38·3	38·4
2	76·2	76·4	76·6	76·8
3	114·3	114·6	114·9	115·2
4	152·4	152·8	153·2	153·6
5	190·5	191·0	191·5	192·0
6	228·6	229·2	229·8	230·4
7	266·7	267·4	268·1	268·8
8	304·8	305·6	306·4	307·2
9	342·9	343·8	344·7	345·6

″	385	386	387	388
1	38·5	38·6	38·7	38·8
2	77·0	77·2	77·4	77·6
3	115·5	115·8	116·1	116·4
4	154·0	154·4	154·8	155·2
5	192·5	193·0	193·5	194·0
6	231·0	231·6	232·2	232·8
7	269·5	270·2	270·9	271·6
8	308·0	308·8	309·6	310·4
9	346·5	347·4	348·3	349·2

″	389	390	391	392
1	38·9	39·0	39·1	39·2
2	77·8	78·0	78·2	78·4
3	116·7	117·0	117·3	117·6
4	155·6	156·0	156·4	156·8
5	194·5	195·0	195·5	196·0
6	233·4	234·0	234·6	235·2
7	272·3	273·0	273·7	274·4
8	311·2	312·0	312·8	313·6
9	350·1	351·0	351·9	352·8

″	393	394	395	396
1	39·3	39·4	39·5	39·6
2	78·6	78·8	79·0	79·2
3	117·9	118·2	118·5	118·8
4	157·2	157·6	158·0	158·4
5	196·5	197·0	197·5	198·0
6	235·8	236·4	237·0	237·6
7	275·1	275·8	276·5	277·2
8	314·4	315·2	316·0	316·8
9	353·7	354·6	355·5	356·4

″	397	398	399	400
1	39·7	39·8	39·9	40·0
2	79·4	79·6	79·8	80·0
3	119·1	119·4	119·7	120·0
4	158·8	159·2	159·6	160·0
5	198·5	199·0	199·5	200·0
6	238·2	238·8	239·4	240·0
7	277·9	278·6	279·3	280·0
8	317·6	318·4	319·2	320·0
9	357·3	358·2	359·1	360·0

″	401	402	403
1	40·1	40·2	40·3
2	80·2	80·4	80·6
3	120·3	120·6	120·9
4	160·4	160·8	161·2
5	200·5	201·0	201·5
6	240·6	241·2	241·8
7	280·7	281·4	282·1
8	320·8	321·6	322·4
9	360·9	361·8	362·7

TABLE IIc—LOGARITHMS OF TRIGONOMETRICAL FUNCTIONS

(Degrees, Minutes and Seconds)

3°

′ ″	sin	d	tan	d	cot	cos	′ ″
10 00	8·742 259	380	8·742 922	382	1·257 078	9·999 336	**50 00**
10	·742 639	380	·743 304	381	·256 696	·999 335	**49** 50
20	·743 019	380	·743 685	381	·256 315	·999 334	40
30	·743 399	379	·744 066	381	·255 934	·999 333	30
40	·743 778	379	·744 447	380	·255 553	·999 332	20
50	·744 157	379	·744 827	380	·255 173	·999 331	10
11 00	8·744 536	378	8·745 207	379	1·254 793	9·999 329	**49 00**
10	·744 914	379	·745 586	380	·254 414	·999 328	**48** 50
20	·745 293	377	·745 966	378	·254 034	·999 327	40
30	·745 670	378	·746 344	379	·253 656	·999 326	30
40	·746 048	377	·746 723	378	·253 277	·999 325	20
50	·746 425	377	·747 101	378	·252 899	·999 323	10
12 00	8·746 802	376	8·747 479	378	1·252 521	9·999 322	**48 00**
10	·747 178	376	·747 857	377	·252 143	·999 321	**47** 50
20	·747 554	376	·748 234	377	·251 766	·999 320	40
30	·747 930	375	·748 611	377	·251 389	·999 319	30
40	·748 305	375	·748 988	376	·251 012	·999 318	20
50	·748 680	375	·749 364	376	·250 636	·999 316	10
13 00	8·749 055	375	8·749 740	376	1·250 260	9·999 315	**47 00**
10	·749 430	374	·750 116	375	·249 884	·999 314	**46** 50
20	·749 804	374	·750 491	375	·249 509	·999 313	40
30	·750 178	373	·750 866	375	·249 134	·999 312	30
40	·750 551	373	·751 241	374	·248 759	·999 310	20
50	·750 924	373	·751 615	374	·248 385	·999 309	10
14 00	8·751 297	373	8·751 989	374	1·248 011	9·999 308	**46 00**
10	·751 670	372	·752 363	373	·247 637	·999 307	**45** 50
20	·752 042	372	·752 736	373	·247 264	·999 306	40
30	·752 414	372	·753 109	373	·246 891	·999 305	30
40	·752 786	371	·753 482	373	·246 518	·999 303	20
50	·753 157	371	·753 855	372	·246 145	·999 302	10
15 00	8·753 528	370	8·754 227	372	1·245 773	9·999 301	**45 00**
10	·753 898	371	·754 599	371	·245 401	·999 300	**44** 50
20	·754 269	370	·754 970	371	·245 030	·999 299	40
30	·754 639	369	·755 341	371	·244 659	·999 297	30
40	·755 008	370	·755 712	371	·244 288	·999 296	20
50	·755 378	369	·756 083	370	·243 917	·999 295	10
16 00	8·755 747	369	8·756 453	370	1·243 547	9·999 294	**44 00**
10	·756 116	368	·756 823	370	·243 177	·999 293	**43** 50
20	·756 484	368	·757 193	369	·242 807	·999 291	40
30	·756 852	368	·757 562	369	·242 438	·999 290	30
40	·757 220	367	·757 931	369	·242 069	·999 289	20
50	·757 587	368	·758 300	368	·241 700	·999 288	10
17 00	8·757 955	366	8·758 668	368	1·241 332	9·999 287	**43 00**
10	·758 321	367	·759 036	368	·240 964	·999 285	**42** 50
20	·758 688	366	·759 404	367	·240 596	·999 284	40
30	·759 054	366	·759 771	368	·240 229	·999 283	30
40	·759 420	366	·760 139	366	·239 861	·999 282	20
50	·759 786	365	·760 505	367	·239 495	·999 280	10
18 00	8·760 151	365	8·760 872	366	1·239 128	9·999 279	**42 00**
10	·760 516	365	·761 238	366	·238 762	·999 278	**41** 50
20	·760 881	364	·761 604	366	·238 396	·999 277	40
30	·761 245	364	·761 970	365	·238 030	·999 276	30
40	·761 609	364	·762 335	365	·237 665	·999 274	20
50	·761 973	364	·762 700	365	·237 300	·999 273	10
19 00	8·762 337	363	8·763 065	364	1·236 935	9·999 272	**41 00**
10	·762 700	363	·763 429	364	·236 571	·999 271	**40** 50
20	·763 063	362	·763 793	364	·236 207	·999 270	40
30	·763 425	362	·764 157	363	·235 843	·999 268	30
40	·763 787	362	·764 520	364	·235 480	·999 267	20
50	·764 149	362	·764 884	362	·235 116	·999 266	10
20 00	8·764 511		8·765 246		1·234 754	9·999 265	**40 00**
	cos		cot		tan	sin	**86°**

PROPORTIONAL PARTS

″	362	363	364	365
1	36·2	36·3	36·4	36·5
2	72·4	72·6	72·8	73·0
3	108·6	108·9	109·2	109·5
4	144·8	145·2	145·6	146·0
5	181·0	181·5	182·0	182·5
6	217·2	217·8	218·4	219·0
7	253·4	254·1	254·8	255·5
8	289·6	290·4	291·2	292·0
9	325·8	326·7	327·6	328·5
″	**366**	**367**	**368**	**369**
1	36·6	36·7	36·8	36·9
2	73·2	73·4	73·6	73·8
3	109·8	110·1	110·4	110·7
4	146·4	146·8	147·2	147·6
5	183·0	183·5	184·0	184·5
6	219·6	220·2	220·8	221·4
7	256·2	256·9	257·6	258·3
8	292·8	293·6	294·4	295·2
9	329·4	330·3	331·2	332·1
″	**370**	**371**	**372**	**373**
1	37·0	37·1	37·2	37·3
2	74·0	74·2	74·4	74·6
3	111·0	111·3	111·6	111·9
4	148·0	148·4	148·8	149·2
5	185·0	185·5	186·0	186·5
6	222·0	222·6	223·2	223·8
7	259·0	259·7	260·4	261·1
8	296·0	296·8	297·6	298·4
9	333·0	333·9	334·8	335·7
″	**374**	**375**	**376**	**377**
1	37·4	37·5	37·6	37·7
2	74·8	75·0	75·2	75·4
3	112·2	112·5	112·8	113·1
4	149·6	150·0	150·4	150·8
5	187·0	187·5	188·0	188·5
6	224·4	225·0	225·6	226·2
7	261·8	262·5	263·2	263·9
8	299·2	300·0	300·8	301·6
9	336·6	337·5	338·4	339·3
″	**378**	**379**	**380**	**381**
1	37·8	37·9	38·0	38·1
2	75·6	75·8	76·0	76·2
3	113·4	113·7	114·0	114·3
4	151·2	151·6	152·0	152·4
5	189·0	189·5	190·0	190·5
6	226·8	227·4	228·0	228·6
7	264·6	265·3	266·0	266·7
8	302·4	303·2	304·0	304·8
9	340·2	341·1	342·0	342·9

(Degrees, Minutes and Seconds)

3°

′	″	sin		tan		cot	cos	′	″
20	00	8·764 511	361	8·765 246	363	1·234 754	9·999 265	40	00
	10	·764 872	362	·765 609	362	·234 391	·999 263	39	50
	20	·765 234	360	·765 971	362	·234 029	·999 262		40
	30	·765 594	361	·766 333	362	·233 667	·999 261		30
	40	·765 955	360	·766 695	361	·233 305	·999 260		20
	50	·766 315	360	·767 056	361	·232 944	·999 258		10
21	00	8·766 675	359	8·767 417	361	1·232 583	9·999 257	39	00
	10	·767 034	360	·767 778	361	·232 222	·999 256	38	50
	20	·767 394	358	·768 139	360	·231 861	·999 255		40
	30	·767 752	359	·768 499	360	·231 501	·999 254		30
	40	·768 111	358	·768 859	359	·231 141	·999 252		20
	50	·768 469	359	·769 218	360	·230 782	·999 251		10
22	00	8·768 828	357	8·769 578	359	1·230 422	9·999 250	38	00
	10	·769 185	358	·769 937	358	·230 063	·999 249	37	50
	20	·769 543	357	·770 295	359	·229 705	·999 247		40
	30	·769 900	357	·770 654	358	·229 346	·999 246		30
	40	·770 257	356	·771 012	358	·228 988	·999 245		20
	50	·770 613	357	·771 370	357	·228 630	·999 244		10
23	00	8·770 970	356	8·771 727	358	1·228 273	9·999 242	37	00
	10	·771 326	355	·772 085	357	·227 915	·999 241	36	50
	20	·771 681	356	·772 442	356	·227 558	·999 240		40
	30	·772 037	355	·772 798	357	·227 202	·999 239		30
	40	·772 392	355	·773 155	356	·226 845	·999 237		20
	50	·772 747	354	·773 511	355	·226 489	·999 236		10
24	00	8·773 101	355	8·773 866	356	1·226 134	9·999 235	36	00
	10	·773 456	354	·774 222	355	·225 778	·999 234	35	50
	20	·773 810	353	·774 577	355	·225 423	·999 232		40
	30	·774 163	354	·774 932	355	·225 068	·999 231		30
	40	·774 517	353	·775 287	354	·224 713	·999 230		20
	50	·774 870	353	·775 641	354	·224 359	·999 229		10
25	00	8·775 223	352	8·775 995	354	1·224 005	9·999 227	35	00
	10	·775 575	352	·776 349	353	·223 651	·999 226	34	50
	20	·775 927	352	·776 702	354	·223 298	·999 225		40
	30	·776 279	352	·777 056	353	·222 944	·999 224		30
	40	·776 631	351	·777 409	352	·222 591	·999 222		20
	50	·776 982	351	·777 761	353	·222 239	·999 221		10
26	00	8·777 333	351	8·778 114	352	1·221 886	9·999 220	34	00
	10	·777 684	351	·778 466	351	·221 534	·999 219	33	50
	20	·778 035	350	·778 817	352	·221 183	·999 217		40
	30	·778 385	350	·779 169	351	·220 831	·999 216		30
	40	·778 735	350	·779 520	351	·220 480	·999 215		20
	50	·779 085	349	·779 871	351	·220 129	·999 213		10
27	00	8·779 434	349	8·780 222	350	1·219 778	9·999 212	33	00
	10	·779 783	349	·780 572	350	·219 428	·999 211	32	50
	20	·780 132	348	·780 922	350	·219 078	·999 210		40
	30	·780 480	349	·781 272	350	·218 728	·999 208		30
	40	·780 829	348	·781 622	349	·218 378	·999 207		20
	50	·781 177	347	·781 971	349	·218 029	·999 206		10
28	00	8·781 524	348	8·782 320	349	1·217 680	9·999 205	32	00
	10	·781 872	347	·782 669	348	·217 331	·999 203	31	50
	20	·782 219	347	·783 017	348	·216 983	·999 202		40
	30	·782 566	346	·783 365	348	·216 635	·999 201		30
	40	·782 912	347	·783 713	348	·216 287	·999 199		20
	50	·783 259	346	·784 061	347	·215 939	·999 198		10
29	00	8·783 605	346	8·784 408	347	1·215 592	9·999 197	31	00
	10	·783 951	345	·784 755	347	·215 245	·999 196	30	50
	20	·784 296	345	·785 102	346	·214 898	·999 194		40
	30	·784 641	345	·785 448	346	·214 552	·999 193		30
	40	·784 986	345	·785 794	346	·214 206	·999 192		20
	50	·785 331	344	·786 140	346	·213 860	·999 190		10
30	00	8·785 675		8·786 486		1·213 514	9·999 189	30	00
		cos		cot		tan	sin	86°	

PROPORTIONAL PARTS

″	344	345	346	347
1	34·4	34·5	34·6	34·7
2	68·8	69·0	69·2	69·4
3	103·2	103·5	103·8	104·1
4	137·6	138·0	138·4	138·8
5	172·0	172·5	173·0	173·5
6	206·4	207·0	207·6	208·2
7	240·8	241·5	242·2	242·9
8	275·2	276·0	276·8	277·6
9	309·6	310·5	311·4	312·3

″	348	349	350	351
1	34·8	34·9	35·0	35·1
2	69·6	69·8	70·0	70·2
3	104·4	104·7	105·0	105·3
4	139·2	139·6	140·0	140·4
5	174·0	174·5	175·0	175·5
6	208·8	209·4	210·0	210·6
7	243·6	244·3	245·0	245·7
8	278·4	279·2	280·0	280·8
9	313·2	314·1	315·0	315·9

″	352	353	354	355
1	35·2	35·3	35·4	35·5
2	70·4	70·6	70·8	71·0
3	105·6	105·9	106·2	106·5
4	140·8	141·2	141·6	142·0
5	176·0	176·5	177·0	177·5
6	211·2	211·8	212·4	213·0
7	246·4	247·1	247·8	248·5
8	281·6	282·4	283·2	284·0
9	316·8	317·7	318·6	319·5

″	356	357	358	359
1	35·6	35·7	35·8	35·9
2	71·2	71·4	71·6	71·8
3	106·8	107·1	107·4	107·7
4	142·4	142·8	143·2	143·6
5	178·0	178·5	179·0	179·5
6	213·6	214·2	214·8	215·4
7	249·2	249·9	250·6	251·3
8	284·8	285·6	286·4	287·2
9	320·4	321·3	322·2	323·1

″	360	361	362	363
1	36·0	36·1	36·2	36·3
2	72·0	72·2	72·4	72·6
3	108·0	108·3	108·6	108·9
4	144·0	144·4	144·8	145·2
5	180·0	180·5	181·0	181·5
6	216·0	216·6	217·2	217·8
7	252·0	252·7	253·4	254·1
8	288·0	288·8	289·6	290·4
9	324·0	324·9	325·8	326·7

(Degrees, Minutes and Seconds)

3°

′	″	sin		tan		cot	cos	′	″
30	00	8·785 675	344	8·786 486	345	1·213 514	9·999 189	30	00
	10	·786 019	344	·786 831	346	·213 169	·999 188	29	50
	20	·786 363	344	·787 177	344	·212 823	·999 187		40
	30	·786 707	343	·787 521	345	·212 479	·999 185		30
	40	·787 050	343	·787 866	344	·212 134	·999 184		20
	50	·787 393	343	·788 210	344	·211 790	·999 183		10
31	00	8·787 736	342	8·788 554	344	1·211 446	9·999 181	29	00
	10	·788 078	343	·788 898	344	·211 102	·999 180	28	50
	20	·788 421	341	·789 242	343	·210 758	·999 179		40
	30	·788 762	342	·789 585	343	·210 415	·999 178		30
	40	·789 104	342	·789 928	343	·210 072	·999 176		20
	50	·789 446	341	·790 271	342	·209 729	·999 175		10
32	00	8·789 787	341	8·790 613	342	1·209 387	9·999 174	28	00
	10	·790 128	340	·790 955	342	·209 045	·999 172	27	50
	20	·790 468	340	·791 297	342	·208 703	·999 171		40
	30	·790 808	341	·791 639	341	·208 361	·999 170		30
	40	·791 149	339	·791 980	341	·208 020	·999 168		20
	50	·791 488	340	·792 321	341	·207 679	·999 167		10
33	00	8·791 828	339	8·792 662	341	1·207 338	9·999 166	27	00
	10	·792 167	339	·793 003	340	·206 997	·999 165	26	50
	20	·792 506	339	·793 343	340	·206 657	·999 163		40
	30	·792 845	338	·793 683	340	·206 317	·999 162		30
	40	·793 183	338	·794 023	339	·205 977	·999 161		20
	50	·793 521	338	·794 362	339	·205 638	·999 159		10
34	00	8·793 859	338	8·794 701	339	1·205 299	9·999 158	26	00
	10	·794 197	337	·795 040	339	·204 960	·999 157	25	50
	20	·794 534	338	·795 379	339	·204 621	·999 155		40
	30	·794 872	336	·795 718	338	·204 282	·999 154		30
	40	·795 208	337	·796 056	338	·203 944	·999 153		20
	50	·795 545	336	·796 394	337	·203 606	·999 151		10
35	00	8·795 881	337	8·796 731	338	1·203 269	9·999 150	25	00
	10	·796 218	335	·797 069	337	·202 931	·999 149	24	50
	20	·796 553	336	·797 406	337	·202 594	·999 147		40
	30	·796 889	335	·797 743	336	·202 257	·999 146		30
	40	·797 224	335	·798 079	337	·201 921	·999 145		20
	50	·797 559	335	·798 416	336	·201 584	·999 143		10
36	00	8·797 894	335	8·798 752	336	1·201 248	9·999 142	24	00
	10	·798 229	334	·799 088	335	·200 912	·999 141	23	50
	20	·798 563	334	·799 423	336	·200 577	·999 140		40
	30	·798 897	334	·799 759	335	·200 241	·999 138		30
	40	·799 231	333	·800 094	335	·199 906	·999 137		20
	50	·799 564	333	·800 429	334	·199 571	·999 136		10
37	00	8·799 897	333	8·800 763	335	1·199 237	9·999 134	23	00
	10	·800 230	333	·801 098	334	·198 902	·999 133	22	50
	20	·800 563	333	·801 432	333	·198 568	·999 132		40
	30	·800 896	332	·801 765	334	·198 235	·999 130		30
	40	·801 228	332	·802 099	333	·197 901	·999 129		20
	50	·801 560	332	·802 432	333	·197 568	·999 128		10
38	00	8·801 892	331	8·802 765	333	1·197 235	9·999 126	22	00
	10	·802 223	331	·803 098	333	·196 902	·999 125	21	50
	20	·802 554	331	·803 431	332	·196 569	·999 124		40
	30	·802 885	331	·803 763	332	·196 237	·999 122		30
	40	·803 216	330	·804 095	332	·195 905	·999 121		20
	50	·803 546	330	·804 427	331	·195 573	·999 120		10
39	00	8·803 876	330	8·804 758	332	1·195 242	9·999 118	21	00
	10	·804 206	330	·805 090	331	·194 910	·999 117	20	50
	20	·804 536	330	·805 421	330	·194 579	·999 115		40
	30	·804 866	329	·805 751	331	·194 249	·999 114		30
	40	·805 195	329	·806 082	330	·193 918	·999 113		20
	50	·805 524	328	·806 412	330	·193 588	·999 111		10
40	00	8·805 852		8·806 742		1·193 258	9·999 110	20	00
		cos		cot		tan	sin		

86°

PROPORTIONAL PARTS

″	328	329	330	331
1	32·8	32·9	33·0	33·1
2	65·6	65·8	66·0	66·2
3	98·4	98·7	99·0	99·3
4	131·2	131·6	132·0	132·4
5	164·0	164·5	165·0	165·5
6	196·8	197·4	198·0	198·6
7	229·6	230·3	231·0	231·7
8	262·4	263·2	264·0	264·8
9	295·2	296·1	297·0	297·9

″	332	333	334	335
1	33·2	33·3	33·4	33·5
2	66·4	66·6	66·8	67·0
3	99·6	99·9	100·2	100·5
4	132·8	133·2	133·6	134·0
5	166·0	166·5	167·0	167·5
6	199·2	199·8	200·4	201·0
7	232·4	233·1	233·8	234·5
8	265·6	266·4	267·2	268·0
9	298·8	299·7	300·6	301·5

″	336	337	338	339
1	33·6	33·7	33·8	33·9
2	67·2	67·4	67·6	67·8
3	100·8	101·1	101·4	101·7
4	134·4	134·8	135·2	135·6
5	168·0	168·5	169·0	169·5
6	201·6	202·2	202·8	203·4
7	235·2	235·9	236·6	237·3
8	268·8	269·6	270·4	271·2
9	302·4	303·3	304·2	305·1

″	340	341	342	343
1	34·0	34·1	34·2	34·3
2	68·0	68·2	68·4	68·6
3	102·0	102·3	102·6	102·9
4	136·0	136·4	136·8	137·2
5	170·0	170·5	171·0	171·5
6	204·0	204·6	205·2	205·8
7	238·0	238·7	239·4	240·1
8	272·0	272·8	273·6	274·4
9	306·0	306·9	307·8	308·7

″	344	345	346
1	34·4	34·5	34·6
2	68·8	69·0	69·2
3	103·2	103·5	103·8
4	137·6	138·0	138·4
5	172·0	172·5	173·0
6	206·4	207·0	207·6
7	240·8	241·5	242·2
8	275·2	276·0	276·8
9	309·6	310·5	311·4

6—III—3*

TABLE IIc—LOGARITHMS OF TRIGONOMETRICAL FUNCTIONS

(Degrees, Minutes and Seconds)

3°

′ ″	sin		tan		cot	cos	′ ″
40 00	8·805 852	329	8·806 742	330	1·193 258	9·999 110	20 00
10	·806 181	328	·807 072	330	·192 928	·999 109	19 50
20	·806 509	328	·807 402	329	·192 598	·999 107	40
30	·806 837	328	·807 731	329	·192 269	·999 106	30
40	·807 165	327	·808 060	329	·191 940	·999 105	20
50	·807 492	327	·808 389	328	·191 611	·999 103	10
41 00	8·807 819	327	8·808 717	329	1·191 283	9·999 102	19 00
10	·808 146	327	·809 046	328	·190 954	·999 101	18 50
20	·808 473	326	·809 374	327	·190 626	·999 099	40
30	·808 799	327	·809 701	328	·190 299	·999 098	30
40	·809 126	325	·810 029	327	·189 971	·999 097	20
50	·809 451	326	·810 356	327	·189 644	·999 095	10
42 00	8·809 777	326	8·810 683	327	1·189 317	9·999 094	18 00
10	·810 103	325	·811 010	327	·188 990	·999 092	17 50
20	·810 428	325	·811 337	326	·188 663	·999 091	40
30	·810 753	325	·811 663	326	·188 337	·999 090	30
40	·811 078	324	·811 989	326	·188 011	·999 088	20
50	·811 402	324	·812 315	326	·187 685	·999 087	10
43 00	8·811 726	324	8·812 641	325	1·187 359	9·999 086	17 00
10	·812 050	324	·812 966	325	·187 034	·999 084	16 50
20	·812 374	324	·813 291	325	·186 709	·999 083	40
30	·812 698	323	·813 616	325	·186 384	·999 082	30
40	·813 021	323	·813 941	324	·186 059	·999 080	20
50	·813 344	323	·814 265	324	·185 735	·999 079	10
44 00	8·813 667	322	8·814 589	324	1·185 411	9·999 077	16 00
10	·813 989	323	·814 913	324	·185 087	·999 076	15 50
20	·814 312	322	·815 237	323	·184 763	·999 075	40
30	·814 634	322	·815 560	324	·184 440	·999 073	30
40	·814 956	321	·815 884	323	·184 116	·999 072	20
50	·815 277	322	·816 207	322	·183 793	·999 071	10
45 00	8·815 599	321	8·816 529	323	1·183 471	9·999 069	15 00
10	·815 920	321	·816 852	322	·183 148	·999 068	14 50
20	·816 241	320	·817 174	322	·182 826	·999 066	40
30	·816 561	321	·817 496	322	·182 504	·999 065	30
40	·816 882	320	·817 818	322	·182 182	·999 064	20
50	·817 202	320	·818 140	321	·181 860	·999 062	10
46 00	8·817 522	319	8·818 461	321	1·181 539	9·999 061	14 00
10	·817 841	320	·818 782	321	·181 218	·999 059	13 50
20	·818 161	319	·819 103	320	·180 897	·999 058	40
30	·818 480	319	·819 423	321	·180 577	·999 057	30
40	·818 799	319	·819 744	320	·180 256	·999 055	20
50	·819 118	318	·820 064	320	·179 936	·999 054	10
47 00	8·819 436	319	8·820 384	319	1·179 616	9·999 053	13 00
10	·819 755	318	·820 703	320	·179 297	·999 051	12 50
20	·820 073	317	·821 023	319	·178 977	·999 050	40
30	·820 390	318	·821 342	319	·178 658	·999 048	30
40	·820 708	317	·821 661	319	·178 339	·999 047	20
50	·821 025	318	·821 980	318	·178 020	·999 046	10
48 00	8·821 343	316	8·822 298	319	1·177 702	9·999 044	12 00
10	·821 659	317	·822 617	318	·177 383	·999 043	11 50
20	·821 976	316	·822 935	318	·177 065	·999 041	40
30	·822 292	317	·823 253	317	·176 747	·999 040	30
40	·822 609	316	·823 570	318	·176 430	·999 039	20
50	·822 925	315	·823 888	317	·176 112	·999 037	10
49 00	8·823 240	316	8·824 205	317	1·175 795	9·999 036	11 00
10	·823 556	315	·824 522	316	·175 478	·999 034	10 50
20	·823 871	315	·824 838	317	·175 162	·999 033	40
30	·824 186	315	·825 155	316	·174 845	·999 032	30
40	·824 501	315	·825 471	316	·174 529	·999 030	20
50	·824 816	314	·825 787	316	·174 213	·999 029	10
50 00	8·825 130		8·826 103		1·173 897	9·999 027	10 00
	cos		cot		tan	sin	86°

PROPORTIONAL PARTS

″	314	315	316	317	318	319	320	321	322	323	324	325	326	327	328	329	330
1	31·4	31·5	31·6	31·7	31·8	31·9	32·0	32·1	32·2	32·3	32·4	32·5	32·6	32·7	32·8	32·9	33·0
2	62·8	63·0	63·2	63·4	63·6	63·8	64·0	64·2	64·4	64·6	64·8	65·0	65·2	65·4	65·6	65·8	66·0
3	94·2	94·5	94·8	95·1	95·4	95·7	96·0	96·3	96·6	96·9	97·2	97·5	97·8	98·1	98·4	98·7	99·0
4	125·6	126·0	126·4	126·8	127·2	127·6	128·0	128·4	128·8	129·2	129·6	130·0	130·4	130·8	131·2	131·6	132·0
5	157·0	157·5	158·0	158·5	159·0	159·5	160·0	160·5	161·0	161·5	162·0	162·5	163·0	163·5	164·0	164·5	165·0
6	188·4	189·0	189·6	190·2	190·8	191·4	192·0	192·6	193·2	193·8	194·4	195·0	195·6	196·2	196·8	197·4	198·0
7	219·8	220·5	221·2	221·9	222·6	223·3	224·0	224·7	225·4	226·1	226·8	227·5	228·2	228·9	229·6	230·3	231·0
8	251·2	252·0	252·8	253·6	254·4	255·2	256·0	256·8	257·6	258·4	259·2	260·0	260·8	261·6	262·4	263·2	264·0
9	282·6	283·5	284·4	285·3	286·2	287·1	288·0	288·9	289·8	290·7	291·6	292·5	293·4	294·3	295·2	296·1	297·0

(Degrees, Minutes and Seconds)

3°

′	″	sin	diff.	tan	diff.	cot	cos	′	″
50	00	8·825 130	314	8·826 103	315	1·173 897	9·999 027	10	00
	10	·825 444	314	·826 418	315	·173 582	·999 026	09	50
	20	·825 758	314	·826 733	316	·173 267	·999 024		40
	30	·826 072	313	·827 049	314	·172 951	·999 023		30
	40	·826 385	313	·827 363	315	·172 637	·999 022		20
	50	·826 698	313	·827 678	314	·172 322	·999 020		10
51	00	8·827 011	313	8·827 992	315	1·172 008	9·999 019	09	00
	10	·827 324	313	·828 307	314	·171 693	·999 017	08	50
	20	·827 637	312	·828 621	313	·171 379	·999 016		40
	30	·827 949	312	·828 934	314	·171 066	·999 015		30
	40	·828 261	312	·829 248	313	·170 752	·999 013		20
	50	·828 573	311	·829 561	313	·170 439	·999 012		10
52	00	8·828 884	312	8·829 874	313	1·170 126	9·999 010	08	00
	10	·829 196	311	·830 187	313	·169 813	·999 009	07	50
	20	·829 507	311	·830 500	312	·169 500	·999 007		40
	30	·829 818	311	·830 812	312	·169 188	·999 006		30
	40	·830 129	310	·831 124	312	·168 876	·999 005		20
	50	·830 439	310	·831 436	312	·168 564	·999 003		10
53	00	8·830 749	311	8·831 748	311	1·168 252	9·999 002	07	00
	10	·831 060	309	·832 059	312	·167 941	·999 000	06	50
	20	·831 369	310	·832 371	311	·167 629	·998 999		40
	30	·831 679	309	·832 682	310	·167 318	·998 997		30
	40	·831 988	310	·832 992	311	·167 008	·998 996		20
	50	·832 298	309	·833 303	310	·166 697	·998 995		10
54	00	8·832 607	308	8·833 613	311	1·166 387	9·998 993	06	00
	10	·832 915	309	·833 924	310	·166 076	·998 992	05	50
	20	·833 224	308	·834 234	309	·165 766	·998 990		40
	30	·833 532	308	·834 543	310	·165 457	·998 989		30
	40	·833 840	308	·834 853	309	·165 147	·998 987		20
	50	·834 148	308	·835 162	309	·164 838	·998 986		10
55	00	8·834 456	307	8·835 471	309	1·164 529	9·998 984	05	00
	10	·834 763	307	·835 780	309	·164 220	·998 983	04	50
	20	·835 070	307	·836 089	308	·163 911	·998 982		40
	30	·835 377	307	·836 397	308	·163 603	·998 980		30
	40	·835 684	307	·836 705	308	·163 295	·998 979		20
	50	·835 991	306	·837 013	308	·162 987	·998 977		10
56	00	8·836 297	306	8·837 321	308	1·162 679	9·998 976	04	00
	10	·836 603	306	·837 629	307	·162 371	·998 974	03	50
	20	·836 909	306	·837 936	307	·162 064	·998 973		40
	30	·837 215	305	·838 243	307	·161 757	·998 971		30
	40	·837 520	305	·838 550	307	·161 450	·998 970		20
	50	·837 825	305	·838 857	306	·161 143	·998 969		10
57	00	8·838 130	305	8·839 163	307	1·160 837	9·998 967	03	00
	10	·838 435	305	·839 470	306	·160 530	·998 966	02	50
	20	·838 740	304	·839 776	305	·160 224	·998 964		40
	30	·839 044	304	·840 081	306	·159 919	·998 963		30
	40	·839 348	304	·840 387	305	·159 613	·998 961		20
	50	·839 652	304	·840 692	306	·159 308	·998 960		10
58	00	8·839 956	304	8·840 998	305	1·159 002	9·998 958	02	00
	10	·840 260	303	·841 303	304	·158 697	·998 957	01	50
	20	·840 563	303	·841 607	305	·158 393	·998 955		40
	30	·840 866	303	·841 912	304	·158 088	·998 954		30
	40	·841 169	303	·842 216	305	·157 784	·998 953		20
	50	·841 472	302	·842 521	304	·157 479	·998 951		10
59	00	8·841 774	302	8·842 825	303	1·157 175	9·998 950	01	00
	10	·842 076	302	·843 128	304	·156 872	·998 948	00	50
	20	·842 378	302	·843 432	303	·156 568	·998 947		40
	30	·842 680	302	·843 735	303	·156 265	·998 945		30
	40	·842 982	301	·844 038	303	·155 962	·998 944		20
	50	·843 283	302	·844 341	303	·155 659	·998 942		10
60	00	8·843 585		8·844 644		1·155 356	9·998 941	00	00
		cos		cot		tan	sin		

86°

PROPORTIONAL PARTS

″	301	302	303
1	30·1	30·2	30·3
2	60·2	60·4	60·6
3	90·3	90·6	90·9
4	120·4	120·8	121·2
5	150·5	151·0	151·5
6	180·6	181·2	181·8
7	210·7	211·4	212·1
8	240·8	241·6	242·4
9	270·9	271·8	272·7

″	304	305	306
1	30·4	30·5	30·6
2	60·8	61·0	61·2
3	91·2	91·5	91·8
4	121·6	122·0	122·4
5	152·0	152·5	153·0
6	182·4	183·0	183·6
7	212·8	213·5	214·2
8	243·2	244·0	244·8
9	273·6	274·5	275·4

″	307	308	309
1	30·7	30·8	30·9
2	61·4	61·6	61·8
3	92·1	92·4	92·7
4	122·8	123·2	123·6
5	153·5	154·0	154·5
6	184·2	184·8	185·4
7	214·9	215·6	216·3
8	245·6	246·4	247·2
9	276·3	277·2	278·1

″	310	311	312
1	31·0	31·1	31·2
2	62·0	62·2	62·4
3	93·0	93·3	93·6
4	124·0	124·4	124·8
5	155·0	155·5	156·0
6	186·0	186·6	187·2
7	217·0	217·7	218·4
8	248·0	248·8	249·6
9	279·0	279·9	280·8

″	313	314	315
1	31·3	31·4	31·5
2	62·6	62·8	63·0
3	93·9	94·2	94·5
4	125·2	125·6	126·0
5	156·5	157·0	157·5
6	187·8	188·4	189·0
7	219·1	219·8	220·5
8	250·4	251·2	252·0
9	281·7	282·6	283·5

TABLE IIc—LOGARITHMS OF TRIGONOMETRICAL FUNCTIONS

(Degrees, Minutes and Seconds)

4°

′	″	sin		tan		cot	cos	′	″
00	00	8·843 585	301	8·844 644	302	1·155 356	9·998 941	60	00
	10	·843 886	300	·844 946	302	·155 054	·998 939	59	50
	20	·844 186	301	·845 248	303	·154 752	·998 938		40
	30	·844 487	300	·845 551	301	·154 449	·998 936		30
	40	·844 787	300	·845 852	302	·154 148	·998 935		20
	50	·845 087	300	·846 154	301	·153 846	·998 933		10
01	00	8·845 387	300	8·846 455	302	1·153 545	9·998 932	59	00
	10	·845 687	300	·846 757	301	·153 243	·998 930	58	50
	20	·845 987	299	·847 058	300	·152 942	·998 929		40
	30	·846 286	299	·847 358	301	·152 642	·998 927		30
	40	·846 585	299	·847 659	300	·152 341	·998 926		20
	50	·846 884	299	·847 959	301	·152 041	·998 925		10
02	00	8·847 183	298	8·848 260	300	1·151 740	9·998 923	58	00
	10	·847 481	299	·848 560	299	·151 440	·998 922	57	50
	20	·847 780	298	·848 859	300	·151 141	·998 920		40
	30	·848 078	298	·849 159	299	·150 841	·998 919		30
	40	·848 376	297	·849 458	300	·150 542	·998 917		20
	50	·848 673	298	·849 758	299	·150 242	·998 916		10
03	00	8·848 971	297	8·850 057	298	1·149 943	9·998 914	57	00
	10	·849 268	297	·850 355	299	·149 645	·998 913	56	50
	20	·849 565	297	·850 654	298	·149 346	·998 911		40
	30	·849 862	297	·850 952	298	·149 048	·998 910		30
	40	·850 159	296	·851 250	298	·148 750	·998 908		20
	50	·850 455	296	·851 548	298	·148 452	·998 907		10
04	00	8·850 751	296	8·851 846	298	1·148 154	9·998 905	56	00
	10	·851 047	296	·852 144	297	·147 856	·998 904	55	50
	20	·851 343	296	·852 441	297	·147 559	·998 902		40
	30	·851 639	295	·852 738	297	·147 262	·998 901		30
	40	·851 934	295	·853 035	297	·146 965	·998 899		20
	50	·852 229	296	·853 332	296	·146 668	·998 898		10
05	00	8·852 525	294	8·853 628	297	1·146 372	9·998 896	55	00
	10	·852 819	295	·853 925	296	·146 075	·998 895	54	50
	20	·853 114	294	·854 221	296	·145 779	·998 893		40
	30	·853 408	295	·854 517	296	·145 483	·998 892		30
	40	·853 703	294	·854 813	295	·145 187	·998 890		20
	50	·853 997	294	·855 108	295	·144 892	·998 889		10
06	00	8·854 291	293	8·855 403	296	1·144 597	9·998 887	54	00
	10	·854 584	294	·855 699	294	·144 301	·998 886	53	50
	20	·854 878	293	·855 993	295	·144 007	·998 884		40
	30	·855 171	293	·856 288	295	·143 712	·998 883		30
	40	·855 464	293	·856 583	294	·143 417	·998 881		20
	50	·855 757	292	·856 877	294	·143 123	·998 880		10
07	00	8·856 049	293	8·857 171	294	1·142 829	9·998 878	53	00
	10	·856 342	292	·857 465	294	·142 535	·998 877	52	50
	20	·856 634	292	·857 759	294	·142 241	·998 875		40
	30	·856 926	292	·858 053	293	·141 947	·998 873		30
	40	·857 218	292	·858 346	293	·141 654	·998 872		20
	50	·857 510	291	·858 639	293	·141 361	·998 870		10
08	00	8·857 801	291	8·858 932	293	1·141 068	9·998 869	52	00
	10	·858 092	291	·859 225	292	·140 775	·998 867	51	50
	20	·858 383	291	·859 517	293	·140 483	·998 866		40
	30	·858 674	291	·859 810	292	·140 190	·998 864		30
	40	·858 965	290	·860 102	292	·139 898	·998 863		20
	50	·859 255	291	·860 394	292	·139 606	·998 861		10
09	00	8·859 546	290	8·860 686	291	1·139 314	9·998 860	51	00
	10	·859 836	290	·860 977	292	·139 023	·998 858	50	50
	20	·860 126	289	·861 269	291	·138 731	·998 857		40
	30	·860 415	290	·861 560	291	·138 440	·998 855		30
	40	·860 705	289	·861 851	291	·138 149	·998 854		20
	50	·860 994	289	·862 142	291	·137 858	·998 852		10
10	00	8·861 283		8·862 433		1·137 567	9·998 851	50	00
		cos		cot		tan	sin	85°	

PROPORTIONAL PARTS

″	289	290	291
1	28·9	29·0	29·1
2	57·8	58·0	58·2
3	86·7	87·0	87·3
4	115·6	116·0	116·4
5	144·5	145·0	145·5
6	173·4	174·0	174·6
7	202·3	203·0	203·7
8	231·2	232·0	232·8
9	260·1	261·0	261·9

″	292	293	294
1	29·2	29·3	29·4
2	58·4	58·6	58·8
3	87·6	87·9	88·2
4	116·8	117·2	117·6
5	146·0	146·5	147·0
6	175·2	175·8	176·4
7	204·4	205·1	205·8
8	233·6	234·4	235·2
9	262·8	263·7	264·6

″	295	296	297
1	29·5	29·6	29·7
2	59·0	59·2	59·4
3	88·5	88·8	89·1
4	118·0	118·4	118·8
5	147·5	148·0	148·5
6	177·0	177·6	178·2
7	206·5	207·2	207·9
8	236·0	236·8	237·6
9	265·5	266·4	267·3

″	298	299	300
1	29·8	29·9	30·0
2	59·6	59·8	60·0
3	89·4	89·7	90·0
4	119·2	119·6	120·0
5	149·0	149·5	150·0
6	178·8	179·4	180·0
7	208·6	209·3	210·0
8	238·4	239·2	240·0
9	268·2	269·1	270·0

″	301	302	303
1	30·1	30·2	30·3
2	60·2	60·4	60·6
3	90·3	90·6	90·9
4	120·4	120·8	121·2
5	150·5	151·0	151·5
6	180·6	181·2	181·8
7	210·7	211·4	212·1
8	240·8	241·6	242·4
9	270·9	271·8	272·7

TABLE IIc—LOGARITHMS OF TRIGONOMETRICAL FUNCTIONS

(Degrees, Minutes and Seconds)

4° ′	″	sin	d	tan	d	cot	cos	′	″
10	00	8·861 283	289	8·862 433	290	1·137 567	9·998 851	50	00
	10	·861 572	289	·862 723	290	·137 277	·998 849	49	50
	20	·861 861	288	·863 013	290	·136 987	·998 848		40
	30	·862 149	289	·863 303	290	·136 697	·998 846		30
	40	·862 438	288	·863 593	290	·136 407	·998 844		20
	50	·862 726	288	·863 883	290	·136 117	·998 843		10
11	00	8·863 014	288	8·864 173	289	1·135 827	9·998 841	49	00
	10	·863 302	287	·864 462	289	·135 538	·998 840	48	50
	20	·863 589	288	·864 751	289	·135 249	·998 838		40
	30	·863 877	287	·865 040	289	·134 960	·998 837		30
	40	·864 164	287	·865 329	288	·134 671	·998 835		20
	50	·864 451	287	·865 617	289	·134 383	·998 834		10
12	00	8·864 738	286	8·865 906	288	1·134 094	9·998 832	48	00
	10	·865 024	287	·866 194	288	·133 806	·998 831	47	50
	20	·865 311	286	·866 482	287	·133 518	·998 829		40
	30	·865 597	286	·866 769	288	·133 231	·998 827		30
	40	·865 883	286	·867 057	287	·132 943	·998 826		20
	50	·866 169	286	·867 344	288	·132 656	·998 824		10
13	00	8·866 455	285	8·867 632	287	1·132 368	9·998 823	47	00
	10	·866 740	285	·867 919	287	·132 081	·998 821	46	50
	20	·867 025	285	·868 206	286	·131 794	·998 820		40
	30	·867 310	285	·868 492	287	·131 508	·998 818		30
	40	·867 595	285	·868 779	286	·131 221	·998 817		20
	50	·867 880	285	·869 065	286	·130 935	·998 815		10
14	00	8·868 165	284	8·869 351	286	1·130 649	9·998 813	46	00
	10	·868 449	284	·869 637	286	·130 363	·998 812	45	50
	20	·868 733	284	·869 923	285	·130 077	·998 810		40
	30	·869 017	284	·870 208	286	·129 792	·998 809		30
	40	·869 301	284	·870 494	285	·129 506	·998 807		20
	50	·869 585	283	·870 779	285	·129 221	·998 806		10
15	00	8·869 868	283	8·871 064	285	1·128 936	9·998 804	45	00
	10	·870 151	283	·871 349	284	·128 651	·998 803	44	50
	20	·870 434	283	·871 633	285	·128 367	·998 801		40
	30	·870 717	283	·871 918	284	·128 082	·998 799		30
	40	·871 000	282	·872 202	284	·127 798	·998 798		20
	50	·871 282	283	·872 486	284	·127 514	·998 796		10
16	00	8·871 565	282	8·872 770	284	1·127 230	9·998 795	44	00
	10	·871 847	282	·873 054	283	·126 946	·998 793	43	50
	20	·872 129	281	·873 337	283	·126 663	·998 792		40
	30	·872 410	282	·873 620	284	·126 380	·998 790		30
	40	·872 692	281	·873 904	283	·126 096	·998 788		20
	50	·872 973	282	·874 187	282	·125 813	·998 787		10
17	00	8·873 255	281	8·874 469	283	1·125 531	9·998 785	43	00
	10	·873 536	281	·874 752	282	·125 248	·998 784	42	50
	20	·873 817	280	·875 034	283	·124 966	·998 782		40
	30	·874 097	281	·875 317	282	·124 683	·998 781		30
	40	·874 378	280	·875 599	282	·124 401	·998 779		20
	50	·874 658	280	·875 881	281	·124 119	·998 777		10
18	00	8·874 938	280	8·876 162	282	1·123 838	9·998 776	42	00
	10	·875 218	280	·876 444	281	·123 556	·998 774	41	50
	20	·875 498	279	·876 725	281	·123 275	·998 773		40
	30	·875 777	280	·877 006	281	·122 994	·998 771		30
	40	·876 057	279	·877 287	281	·122 713	·998 769		20
	50	·876 336	279	·877 568	281	·122 432	·998 768		10
19	00	8·876 615	279	8·877 849	280	1·122 151	9·998 766	41	00
	10	·876 894	278	·878 129	280	·121 871	·998 765	40	50
	20	·877 172	279	·878 409	280	·121 591	·998 763		40
	30	·877 451	278	·878 689	280	·121 311	·998 762		30
	40	·877 729	278	·878 969	280	·121 031	·998 760		20
	50	·878 007	278	·879 249	280	·120 751	·998 758		10
20	00	8·878 285		8·879 529		1·120 471	9·998 757	40	00
		cos		cot		tan	sin	85°	

PROPORTIONAL PARTS

″	278	279	280
1	27·8	27·9	28·0
2	55·6	55·8	56·0
3	83·4	83·7	84·0
4	111·2	111·6	112·0
5	139·0	139·5	140·0
6	166·8	167·4	168·0
7	194·6	195·3	196·0
8	222·4	223·2	224·0
9	250·2	251·1	252·0

″	281	282	283
1	28·1	28·2	28·3
2	56·2	56·4	56·6
3	84·3	84·6	84·9
4	112·4	112·8	113·2
5	140·5	141·0	141·5
6	168·6	169·2	169·8
7	196·7	197·4	198·1
8	224·8	225·6	226·4
9	252·9	253·8	254·7

″	284	285	286
1	28·4	28·5	28·6
2	56·8	57·0	57·2
3	85·2	85·5	85·8
4	113·6	114·0	114·4
5	142·0	142·5	143·0
6	170·4	171·0	171·6
7	198·8	199·5	200·2
8	227·2	228·0	228·8
9	255·6	256·5	257·4

″	287	288	289
1	28·7	28·8	28·9
2	57·4	57·6	57·8
3	86·1	86·4	86·7
4	114·8	115·2	115·6
5	143·5	144·0	144·5
6	172·2	172·8	173·4
7	200·9	201·6	202·3
8	229·6	230·4	231·2
9	258·3	259·2	260·1

″	290
1	29·0
2	58·0
3	87·0
4	116·0
5	145·0
6	174·0
7	203·0
8	232·0
9	261·0

(Degrees, Minutes and Seconds)

4°

′ ″	sin	d	tan	d	cot	cos	′ ″
20 00	8·878 285	278	8·879 529	279	1·120 471	9·998 757	40 00
10	·878 563	278	·879 808	279	·120 192	·998 755	39 50
20	·878 841	277	·880 087	279	·119 913	·998 754	40
30	·879 118	277	·880 366	279	·119 634	·998 752	30
40	·879 395	277	·880 645	279	·119 355	·998 750	20
50	·879 672	277	·880 924	278	·119 076	·998 749	10
21 00	8·879 949	277	8·881 202	278	1·118 798	9·998 747	39 00
10	·880 226	277	·881 480	279	·118 520	·998 746	38 50
20	·880 503	276	·881 759	278	·118 241	·998 744	40
30	·880 779	276	·882 037	277	·117 963	·998 742	30
40	·881 055	276	·882 314	278	·117 686	·998 741	20
50	·881 331	276	·882 592	277	·117 408	·998 739	10
22 00	8·881 607	276	8·882 869	278	1·117 131	9·998 738	38 00
10	·881 883	275	·883 147	277	·116 853	·998 736	37 50
20	·882 158	275	·883 424	277	·116 576	·998 734	40
30	·882 433	275	·883 701	276	·116 299	·998 733	30
40	·882 708	275	·883 977	277	·116 023	·998 731	20
50	·882 983	275	·884 254	276	·115 746	·998 729	10
23 00	8·883 258	275	8·884 530	277	1·115 470	9·998 728	37 00
10	·883 533	274	·884 807	276	·115 193	·998 726	36 50
20	·883 807	274	·885 083	275	·114 917	·998 725	40
30	·884 081	274	·885 358	276	·114 642	·998 723	30
40	·884 355	274	·885 634	276	·114 366	·998 721	20
50	·884 629	274	·885 910	275	·114 090	·998 720	10
24 00	8·884 903	274	8·886 185	275	1·113 815	9·998 718	36 00
10	·885 177	273	·886 460	275	·113 540	·998 717	35 50
20	·885 450	273	·886 735	275	·113 265	·998 715	40
30	·885 723	273	·887 010	275	·112 990	·998 713	30
40	·885 996	273	·887 285	274	·112 715	·998 712	20
50	·886 269	273	·887 559	274	·112 441	·998 710	10
25 00	8·886 542	272	8·887 833	275	1·112 167	9·998 708	35 00
10	·886 814	273	·888 108	274	·111 892	·998 707	34 50
20	·887 087	272	·888 382	273	·111 618	·998 705	40
30	·887 359	272	·888 655	274	·111 345	·998 704	30
40	·887 631	272	·888 929	273	·111 071	·998 702	20
50	·887 903	271	·889 202	274	·110 798	·998 700	10
26 00	8·888 174	272	8·889 476	273	1·110 524	9·998 699	34 00
10	·888 446	271	·889 749	273	·110 251	·998 697	33 50
20	·888 717	271	·890 022	273	·109 978	·998 695	40
30	·888 988	271	·890 295	272	·109 705	·998 694	30
40	·889 259	271	·890 567	273	·109 433	·998 692	20
50	·889 530	271	·890 840	272	·109 160	·998 690	10
27 00	8·889 801	270	8·891 112	272	1·108 888	9·998 689	33 00
10	·890 071	270	·891 384	272	·108 616	·998 687	32 50
20	·890 341	271	·891 656	272	·108 344	·998 686	40
30	·890 612	270	·891 928	271	·108 072	·998 684	30
40	·890 882	269	·892 199	272	·107 801	·998 682	20
50	·891 151	270	·892 471	271	·107 529	·998 681	10
28 00	8·891 421	269	8·892 742	271	1·107 258	9·998 679	32 00
10	·891 690	270	·893 013	271	·106 987	·998 677	31 50
20	·891 960	269	·893 284	271	·106 716	·998 676	40
30	·892 229	269	·893 555	270	·106 445	·998 674	30
40	·892 498	269	·893 825	271	·106 175	·998 672	20
50	·892 767	268	·894 096	270	·105 904	·998 671	10
29 00	8·893 035	269	8·894 366	270	1·105 634	9·998 669	31 00
10	·893 304	268	·894 636	270	·105 364	·998 667	30 50
20	·893 572	268	·894 906	270	·105 094	·998 666	40
30	·893 840	268	·895 176	269	·104 824	·998 664	30
40	·894 108	268	·895 445	270	·104 555	·998 662	20
50	·894 376	267	·895 715	269	·104 285	·998 661	10
30 00	8·894 643		8·895 984		1·104 016	9·998 659	30 00
	cos		cot		tan	sin	85°

PROPORTIONAL PARTS

″	268	269	270	271	272	273	274	275	276	277	278	279
1	26·8	26·9	27·0	27·1	27·2	27·3	27·4	27·5	27·6	27·7	27·8	27·9
2	53·6	53·8	54·0	54·2	54·4	54·6	54·8	55·0	55·2	55·4	55·6	55·8
3	80·4	80·7	81·0	81·3	81·6	81·9	82·2	82·5	82·8	83·1	83·4	83·7
4	107·2	107·6	108·0	108·4	108·8	109·2	109·6	110·0	110·4	110·8	111·2	111·6
5	134·0	134·5	135·0	135·5	136·0	136·5	137·0	137·5	138·0	138·5	139·0	139·5
6	160·8	161·4	162·0	162·6	163·2	163·8	164·4	165·0	165·6	166·2	166·8	167·4
7	187·6	188·3	189·0	189·7	190·4	191·1	191·8	192·5	193·2	193·9	194·6	195·3
8	214·4	215·2	216·0	216·8	217·6	218·4	219·2	220·0	220·8	221·6	222·4	223·2
9	241·2	242·1	243·0	243·9	244·8	245·7	246·6	247·5	248·4	249·3	250·2	251·1

(Degrees, Minutes and Seconds)

4°

′ ″	sin		tan		cot	cos	′ ″
30 00	8·894 643	268	8·895 984	269	1·104 016	9·998 659	30 00
10	·894 911	267	·896 253	269	·103 747	·998 657	29 50
20	·895 178	267	·896 522	269	·103 478	·998 656	40
30	·895 445	267	·896 791	269	·103 209	·998 654	30
40	·895 712	267	·897 060	268	·102 940	·998 653	20
50	·895 979	267	·897 328	268	·102 672	·998 651	10
31 00	8·896 246	266	8·897 596	268	1·102 404	9·998 649	29 00
10	·896 512	266	·897 864	268	·102 136	·998 648	28 50
20	·896 778	266	·898 132	268	·101 868	·998 646	40
30	·897 044	266	·898 400	268	·101 600	·998 644	30
40	·897 310	266	·898 668	267	·101 332	·998 643	20
50	·897 576	266	·898 935	268	·101 065	·998 641	10
32 00	8·897 842	265	8·899 203	267	1·100 797	9·998 639	28 00
10	·898 107	266	·899 470	267	·100 530	·998 638	27 50
20	·898 373	265	·899 737	267	·100 263	·998 636	40
30	·898 638	265	·900 004	266	·099 996	·998 634	30
40	·898 903	265	·900 270	267	·099 730	·998 633	20
50	·899 168	264	·900 537	266	·099 463	·998 631	10
33 00	8·899 432	265	8·900 803	266	1·099 197	9·998 629	27 00
10	·899 697	264	·901 069	266	·098 931	·998 627	26 50
20	·899 961	264	·901 335	266	·098 665	·998 626	40
30	·900 225	264	·901 601	266	·098 399	·998 624	30
40	·900 489	264	·901 867	265	·098 133	·998 622	20
50	·900 753	264	·902 132	266	·097 868	·998 621	10
34 00	8·901 017	263	8·902 398	265	1·097 602	9·998 619	26 00
10	·901 280	264	·902 663	265	·097 337	·998 617	25 50
20	·901 544	263	·902 928	265	·097 072	·998 616	40
30	·901 807	263	·903 193	265	·096 807	·998 614	30
40	·902 070	263	·903 458	264	·096 542	·998 612	20
50	·902 333	263	·903 722	265	·096 278	·998 611	10
35 00	8·902 596	262	8·903 987	264	1·096 013	9·998 609	25 00
10	·902 858	263	·904 251	264	·095 749	·998 607	24 50
20	·903 121	262	·904 515	264	·095 485	·998 606	40
30	·903 383	262	·904 779	264	·095 221	·998 604	30
40	·903 645	262	·905 043	263	·094 957	·998 602	20
50	·903 907	262	·905 306	264	·094 694	·998 601	10
36 00	8·904 169	261	8·905 570	263	1·094 430	9·998 599	24 00
10	·904 430	262	·905 833	263	·094 167	·998 597	23 50
20	·904 692	261	·906 096	263	·093 904	·998 595	40
30	·904 953	261	·906 359	263	·093 641	·998 594	30
40	·905 214	261	·906 622	263	·093 378	·998 592	20
50	·905 475	261	·906 885	262	·093 115	·998 590	10
37 00	8·905 736	261	8·907 147	263	1·092 853	9·998 589	23 00
10	·905 997	260	·907 410	262	·092 590	·998 587	22 50
20	·906 257	260	·907 672	262	·092 328	·998 585	40
30	·906 517	261	·907 934	262	·092 066	·998 584	30
40	·906 778	260	·908 196	261	·091 804	·998 582	20
50	·907 038	259	·908 457	262	·091 543	·998 580	10
38 00	8·907 297	260	8·908 719	261	1·091 281	9·998 578	22 00
10	·907 557	260	·908 980	262	·091 020	·998 577	21 50
20	·907 817	259	·909 242	261	·090 758	·998 575	40
30	·908 076	259	·909 503	261	·090 497	·998 573	30
40	·908 335	260	·909 764	261	·090 236	·998 572	20
50	·908 595	258	·910 025	260	·089 975	·998 570	10
39 00	8·908 853	259	8·910 285	261	1·089 715	9·998 568	21 00
10	·909 112	259	·910 546	260	·089 454	·998 566	20 50
20	·909 371	258	·910 806	260	·089 194	·998 565	40
30	·909 629	259	·911 066	260	·088 934	·998 563	30
40	·909 888	258	·911 326	260	·088 674	·998 561	20
50	·910 146	258	·911 586	260	·088 414	·998 560	10
40 00	8·910 404		8·911 846		1·088 154	9·998 558	20 00
	cos		cot		tan	sin	85°

PROPORTIONAL PARTS

″	258	″	259
1	25·8	1	25·9
2	51·6	2	51·8
3	77·4	3	77·7
4	103·2	4	103·6
5	129·0	5	129·5
6	154·8	6	155·4
7	180·6	7	181·3
8	206·4	8	207·2
9	232·2	9	233·1

″	260	″	261
1	26·0	1	26·1
2	52·0	2	52·2
3	78·0	3	78·3
4	104·0	4	104·4
5	130·0	5	130·5
6	156·0	6	156·6
7	182·0	7	182·7
8	208·0	8	208·8
9	234·0	9	234·9

″	262	″	263
1	26·2	1	26·3
2	52·4	2	52·6
3	78·6	3	78·9
4	104·8	4	105·2
5	131·0	5	131·5
6	157·2	6	157·8
7	183·4	7	184·1
8	209·6	8	210·4
9	235·8	9	236·7

″	264	″	265
1	26·4	1	26·5
2	52·8	2	53·0
3	79·2	3	79·5
4	105·6	4	106·0
5	132·0	5	132·5
6	158·4	6	159·0
7	184·8	7	185·5
8	211·2	8	212·0
9	237·6	9	238·5

″	266	″	267
1	26·6	1	26·7
2	53·2	2	53·4
3	79·8	3	80·1
4	106·4	4	106·8
5	133·0	5	133·5
6	159·6	6	160·2
7	186·2	7	186·9
8	212·8	8	213·6
9	239·4	9	240·3

″	268	″	269
1	26·8	1	26·9
2	53·6	2	53·8
3	80·4	3	80·7
4	107·2	4	107·6
5	134·0	5	134·5
6	160·8	6	161·4
7	187·6	7	188·3
8	214·4	8	215·2
9	241·2	9	242·1

TABLE IIc—LOGARITHMS OF TRIGONOMETRICAL FUNCTIONS

(Degrees, Minutes and Seconds)

4°

′ ″	sin		tan		cot	cos	′ ″
40 00	8·910 404	258	8·911 846	260	1·088 154	9·998 558	20 00
10	·910 662	257	·912 106	259	·087 894	·998 556	19 50
20	·910 919	258	·912 365	259	·087 635	·998 554	40
30	·911 177	257	·912 624	259	·087 376	·998 553	30
40	·911 434	258	·912 883	259	·087 117	·998 551	20
50	·911 692	257	·913 142	259	·086 858	·998 549	10
41 00	8·911 949	257	8·913 401	259	1·086 599	9·998 548	19 00
10	·912 206	256	·913 660	258	·086 340	·998 546	18 50
20	·912 462	257	·913 918	259	·086 082	·998 544	40
30	·912 719	257	·914 177	258	·085 823	·998 542	30
40	·912 976	256	·914 435	258	·085 565	·998 541	20
50	·913 232	256	·914 693	258	·085 307	·998 539	10
42 00	8·913 488	256	8·914 951	258	1·085 049	9·998 537	18 00
10	·913 744	256	·915 209	257	·084 791	·998 535	17 50
20	·914 000	256	·915 466	258	·084 534	·998 534	40
30	·914 256	255	·915 724	257	·084 276	·998 532	30
40	·914 511	256	·915 981	257	·084 019	·998 530	20
50	·914 767	255	·916 238	257	·083 762	·998 529	10
43 00	8·915 022	255	8·916 495	257	1·083 505	9·998 527	17 00
10	·915 277	255	·916 752	257	·083 248	·998 525	16 50
20	·915 532	255	·917 009	256	·082 991	·998 523	40
30	·915 787	254	·917 265	257	·082 735	·998 522	30
40	·916 041	255	·917 522	256	·082 478	·998 520	20
50	·916 296	254	·917 778	256	·082 222	·998 518	10
44 00	8·916 550	255	8·918 034	256	1·081 966	9·998 516	16 00
10	·916 805	254	·918 290	256	·081 710	·998 515	15 50
20	·917 059	254	·918 546	255	·081 454	·998 513	40
30	·917 313	253	·918 801	256	·081 199	·998 511	30
40	·917 566	254	·919 057	255	·080 943	·998 509	20
50	·917 820	253	·919 312	256	·080 688	·998 508	10
45 00	8·918 073	254	8·919 568	255	1·080 432	9·998 506	15 00
10	·918 327	253	·919 823	255	·080 177	·998 504	14 50
20	·918 580	253	·920 078	254	·079 922	·998 502	40
30	·918 833	253	·920 332	255	·079 668	·998 501	30
40	·919 086	252	·920 587	254	·079 413	·998 499	20
50	·919 338	253	·920 841	255	·079 159	·998 497	10
46 00	8·919 591	252	8·921 096	254	1·078 904	9·998 495	14 00
10	·919 843	253	·921 350	254	·078 650	·998 494	13 50
20	·920 096	252	·921 604	254	·078 396	·998 492	40
30	·920 348	252	·921 858	254	·078 142	·998 490	30
40	·920 600	252	·922 112	253	·077 888	·998 488	20
50	·920 852	251	·922 365	254	·077 635	·998 487	10
47 00	8·921 103	252	8·922 619	253	1·077 381	9·998 485	13 00
10	·921 355	251	·922 872	253	·077 128	·998 483	12 50
20	·921 606	252	·923 125	253	·076 875	·998 481	40
30	·921 858	251	·923 378	253	·076 622	·998 479	30
40	·922 109	251	·923 631	253	·076 369	·998 478	20
50	·922 360	250	·923 884	252	·076 116	·998 476	10
48 00	8·922 610	251	8·924 136	253	1·075 864	9·998 474	12 00
10	·922 861	251	·924 389	252	·075 611	·998 472	11 50
20	·923 112	250	·924 641	252	·075 359	·998 471	40
30	·923 362	250	·924 893	252	·075 107	·998 469	30
40	·923 612	250	·925 145	252	·074 855	·998 467	20
50	·923 862	250	·925 397	252	·074 603	·998 465	10
49 00	8·924 112	250	8·925 649	251	1·074 351	9·998 464	11 00
10	·924 362	250	·925 900	252	·074 100	·998 462	10 50
20	·924 612	249	·926 152	251	·073 848	·998 460	40
30	·924 861	250	·926 403	251	·073 597	·998 458	30
40	·925 111	249	·926 654	251	·073 346	·998 456	20
50	·925 360	249	·926 905	251	·073 095	·998 455	10
50 00	8·925 609		8·927 156		1·072 844	9·998 453	10 00
	cos		cot		tan	sin	85°

PROPORTIONAL PARTS

″	249	250	251	252	253	254	255	256	257	258	259	260
1	24·9	25·0	25·1	25·2	25·3	25·4	25·5	25·6	25·7	25·8	25·9	26·0
2	49·8	50·0	50·2	50·4	50·6	50·8	51·0	51·2	51·4	51·6	51·8	52·0
3	74·7	75·0	75·3	75·6	75·9	76·2	76·5	76·8	77·1	77·4	77·7	78·0
4	99·6	100·0	100·4	100·8	101·2	101·6	102·0	102·4	102·8	103·2	103·6	104·0
5	124·5	125·0	125·5	126·0	126·5	127·0	127·5	128·0	128·5	129·0	129·5	130·0
6	149·4	150·0	150·6	151·2	151·8	152·4	153·0	153·6	154·2	154·8	155·4	156·0
7	174·3	175·0	175·7	176·4	177·1	177·8	178·5	179·2	179·9	180·6	181·3	182·0
8	199·2	200·0	200·8	201·6	202·4	203·2	204·0	204·8	205·6	206·4	207·2	208·0
9	224·1	225·0	225·9	226·8	227·7	228·6	229·5	230·4	231·3	232·2	233·1	234·0

4°

′ ″	sin		tan		cot	cos	′ ″
50 00	8·925 609	249	8·927 156	251	1·072 844	9·998 453	10 00
10	·925 858	249	·927 407	250	·072 593	·998 451	09 50
20	·926 107	248	·927 657	251	·072 343	·998 449	40
30	·926 355	249	·927 908	250	·072 092	·998 448	30
40	·926 604	248	·928 158	250	·071 842	·998 446	20
50	·926 852	248	·928 408	250	·071 592	·998 444	10
51 00	8·927 100	248	8·928 658	250	1·071 342	9·998 442	09 00
10	·927 348	248	·928 908	250	·071 092	·998 440	08 50
20	·927 596	248	·929 158	249	·070 842	·998 439	40
30	·927 844	248	·929 407	250	·070 593	·998 437	30
40	·928 092	247	·929 657	249	·070 343	·998 435	20
50	·928 339	248	·929 906	249	·070 094	·998 433	10
52 00	8·928 587	247	8·930 155	249	1·069 845	9·998 431	08 00
10	·928 834	247	·930 404	249	·069 596	·998 430	07 50
20	·929 081	247	·930 653	249	·069 347	·998 428	40
30	·929 328	247	·930 902	248	·069 098	·998 426	30
40	·929 575	246	·931 150	249	·068 850	·998 424	20
50	·929 821	247	·931 399	248	·068 601	·998 422	10
53 00	8·930 068	246	8·931 647	248	1·068 353	9·998 421	07 00
10	·930 314	246	·931 895	248	·068 105	·998 419	06 50
20	·930 560	246	·932 143	248	·067 857	·998 417	40
30	·930 806	246	·932 391	248	·067 609	·998 415	30
40	·931 052	246	·932 639	248	·067 361	·998 413	20
50	·931 298	246	·932 887	247	·067 113	·998 412	10
54 00	8·931 544	245	8·933 134	247	1·066 866	9·998 410	06 00
10	·931 789	246	·933 381	248	·066 619	·998 408	05 50
20	·932 035	245	·933 629	247	·066 371	·998 406	40
30	·932 280	245	·933 876	247	·066 124	·998 404	30
40	·932 525	245	·934 123	246	·065 877	·998 403	20
50	·932 770	245	·934 369	247	·065 631	·998 401	10
55 00	8·933 015	245	8·934 616	246	1·065 384	9·998 399	05 00
10	·933 260	244	·934 862	247	·065 138	·998 397	04 50
20	·933 504	245	·935 109	246	·064 891	·998 395	40
30	·933 749	244	·935 355	246	·064 645	·998 394	30
40	·933 993	244	·935 601	246	·064 399	·998 392	20
50	·934 237	244	·935 847	246	·064 153	·998 390	10
56 00	8·934 481	244	8·936 093	246	1·063 907	9·998 388	04 00
10	·934 725	244	·936 339	245	·063 661	·998 386	03 50
20	·934 969	243	·936 584	246	·063 416	·998 385	40
30	·935 212	244	·936 830	245	·063 170	·998 383	30
40	·935 456	243	·937 075	245	·062 925	·998 381	20
50	·935 699	243	·937 320	245	·062 680	·998 379	10
57 00	8·935 942	243	8·937 565	245	1·062 435	9·998 377	03 00
10	·936 185	243	·937 810	245	·062 190	·998 375	02 50
20	·936 428	243	·938 055	244	·061 945	·998 374	40
30	·936 671	243	·938 299	245	·061 701	·998 372	30
40	·936 914	242	·938 544	244	·061 456	·998 370	20
50	·937 156	242	·938 788	244	·061 212	·998 368	10
58 00	8·937 398	243	8·939 032	244	1·060 968	9·998 366	02 00
10	·937 641	242	·939 276	244	·060 724	·998 364	01 50
20	·937 883	242	·939 520	244	·060 480	·998 363	40
30	·938 125	241	·939 764	243	·060 236	·998 361	30
40	·938 366	242	·940 007	244	·059 993	·998 359	20
50	·938 608	242	·940 251	243	·059 749	·998 357	10
59 00	8·938 850	241	8·940 494	244	1·059 506	9·998 355	01 00
10	·939 091	241	·940 738	243	·059 262	·998 353	00 50
20	·939 332	241	·940 981	243	·059 019	·998 352	40
30	·939 573	241	·941 224	243	·058 776	·998 350	30
40	·939 814	241	·941 467	242	·058 533	·998 348	20
50	·940 055	241	·941 709	243	·058 291	·998 346	10
60 00	8·940 296		8·941 952		1·058 048	9·998 344	00 00
	cos		cot		tan	sin	85°

PROPORTIONAL PARTS

″	241	242	243	244	245	246	247	248	249	250	251
1	24·1	24·2	24·3	24·4	24·5	24·6	24·7	24·8	24·9	25·0	25·1
2	48·2	48·4	48·6	48·8	49·0	49·2	49·4	49·6	49·8	50·0	50·2
3	72·3	72·6	72·9	73·2	73·5	73·8	74·1	74·4	74·7	75·0	75·3
4	96·4	96·8	97·2	97·6	98·0	98·4	98·8	99·2	99·6	100·0	100·4
5	120·5	121·0	121·5	122·0	122·5	123·0	123·5	124·0	124·5	125·0	125·5
6	144·6	145·2	145·8	146·4	147·0	147·6	148·2	148·8	149·4	150·0	150·6
7	168·7	169·4	170·1	170·8	171·5	172·2	172·9	173·6	174·3	175·0	175·7
8	192·8	193·6	194·4	195·2	196·0	196·8	197·6	198·4	199·2	200·0	200·8
9	216·9	217·8	218·7	219·6	220·5	221·4	222·3	223·2	224·1	225·0	225·9

TABLE IIc—LOGARITHMS OF TRIGONOMETRICAL FUNCTIONS

(Degrees, Minutes and Seconds)

5°

′ ″	sin		tan		cot	cos	′ ″
00 00	8·940 296	241	8·941 952	242	1·058 048	9·998 344	60 00
10	·940 537	240	·942 194	243	·057 806	·998 342	59 50
20	·940 777	240	·942 437	242	·057 563	·998 341	40
30	·941 017	241	·942 679	242	·057 321	·998 339	30
40	·941 258	240	·942 921	242	·057 079	·998 337	20
50	·941 498	240	·943 163	241	·056 837	·998 335	10
01 00	8·941 738	239	8·943 404	242	1·056 596	9·998 333	59 00
10	·941 977	240	·943 646	242	·056 354	·998 331	58 50
20	·942 217	240	·943 888	241	·056 112	·998 329	40
30	·942 457	239	·944 129	241	·055 871	·998 328	30
40	·942 696	239	·944 370	241	·055 630	·998 326	20
50	·942 935	239	·944 611	241	·055 389	·998 324	10
02 00	8·943 174	239	8·944 852	241	1·055 148	9·998 322	58 00
10	·943 413	239	·945 093	241	·054 907	·998 320	57 50
20	·943 652	239	·945 334	240	·054 666	·998 318	40
30	·943 891	238	·945 574	241	·054 426	·998 316	30
40	·944 129	239	·945 815	240	·054 185	·998 315	20
50	·944 368	238	·946 055	240	·053 945	·998 313	10
03 00	8·944 606	238	8·946 295	240	1·053 705	9·998 311	57 00
10	·944 844	239	·946 535	240	·053 465	·998 309	56 50
20	·945 083	238	·946 775	240	·053 225	·998 307	40
30	·945 321	237	·947 015	240	·052 985	·998 305	30
40	·945 558	238	·947 255	239	·052 745	·998 303	20
50	·945 796	238	·947 494	240	·052 506	·998 302	10
04 00	8·946 034	237	8·947 734	239	1·052 266	9·998 300	56 00
10	·946 271	237	·947 973	239	·052 027	·998 298	55 50
20	·946 508	237	·948 212	239	·051 788	·998 296	40
30	·946 745	237	·948 451	239	·051 549	·998 294	30
40	·946 982	237	·948 690	239	·051 310	·998 292	20
50	·947 219	237	·948 929	239	·051 071	·998 290	10
05 00	8·947 456	237	8·949 168	238	1·050 832	9·998 289	55 00
10	·947 693	236	·949 406	238	·050 594	·998 287	54 50
20	·947 929	237	·949 644	239	·050 356	·998 285	40
30	·948 166	236	·949 883	238	·050 117	·998 283	30
40	·948 402	236	·950 121	238	·049 879	·998 281	20
50	·948 638	236	·950 359	238	·049 641	·998 279	10
06 00	8·948 874	236	8·950 597	237	1·049 403	9·998 277	54 00
10	·949 110	235	·950 834	238	·049 166	·998 275	53 50
20	·949 345	236	·951 072	237	·048 928	·998 273	40
30	·949 581	236	·951 309	238	·048 691	·998 272	30
40	·949 817	235	·951 547	237	·048 453	·998 270	20
50	·950 052	235	·951 784	237	·048 216	·998 268	10
07 00	8·950 287	235	8·952 021	237	1·047 979	9·998 266	53 00
10	·950 522	235	·952 258	237	·047 742	·998 264	52 50
20	·950 757	235	·952 495	237	·047 505	·998 262	40
30	·950 992	235	·952 732	236	·047 268	·998 260	30
40	·951 227	234	·952 968	237	·047 032	·998 258	20
50	·951 461	235	·953 205	236	·046 795	·998 257	10
08 00	8·951 696	234	8·953 441	236	1·046 559	9·998 255	52 00
10	·951 930	234	·953 677	236	·046 323	·998 253	51 50
20	·952 164	234	·953 913	236	·046 087	·998 251	40
30	·952 398	234	·954 149	236	·045 851	·998 249	30
40	·952 632	234	·954 385	236	·045 615	·998 247	20
50	·952 866	234	·954 621	235	·045 379	·998 245	10
09 00	8·953 100	233	8·954 856	236	1·045 144	9·998 243	51 00
10	·953 333	234	·955 092	235	·044 908	·998 241	50 50
20	·953 567	233	·955 327	235	·044 673	·998 239	40
30	·953 800	233	·955 562	235	·044 438	·998 238	30
40	·954 033	233	·955 797	235	·044 203	·998 236	20
50	·954 266	233	·956 032	235	·043 968	·998 234	10
10 00	8·954 499		8·956 267		1·043 733	9·998 232	50 00
	cos		cot		tan	sin	84°

PROPORTIONAL PARTS

″	233	″	234
1	23·3	1	23·4
2	46·6	2	46·8
3	69·9	3	70·2
4	93·2	4	93·6
5	116·5	5	117·0
6	139·8	6	140·4
7	163·1	7	163·8
8	186·4	8	187·2
9	209·7	9	210·6

″	235	″	236
1	23·5	1	23·6
2	47·0	2	47·2
3	70·5	3	70·8
4	94·0	4	94·4
5	117·5	5	118·0
6	141·0	6	141·6
7	164·5	7	165·2
8	188·0	8	188·8
9	211·5	9	212·4

″	237	″	238
1	23·7	1	23·8
2	47·4	2	47·6
3	71·1	3	71·4
4	94·8	4	95·2
5	118·5	5	119·0
6	142·2	6	142·8
7	165·9	7	166·6
8	189·6	8	190·4
9	213·3	9	214·2

″	239	″	240
1	23·9	1	24·0
2	47·8	2	48·0
3	71·7	3	72·0
4	95·6	4	96·0
5	119·5	5	120·0
6	143·4	6	144·0
7	167·3	7	168·0
8	191·2	8	192·0
9	215·1	9	216·0

″	241	″	242
1	24·1	1	24·2
2	48·2	2	48·4
3	72·3	3	72·6
4	96·4	4	96·8
5	120·5	5	121·0
6	144·6	6	145·2
7	168·7	7	169·4
8	192·8	8	193·6
9	216·9	9	217·8

″	243
1	24·3
2	48·6
3	72·9
4	97·2
5	121·5
6	145·8
7	170·1
8	194·4
9	218·7

TABLE IIc—LOGARITHMS OF TRIGONOMETRICAL FUNCTIONS

(Degrees, Minutes and Seconds)

5°

′ ″	sin		tan		cot	cos	′ ″
10 00	8·954 499	233	8·956 267	235	1·043 733	9·998 232	50 00
10	·954 732	233	·956 502	234	·043 498	·998 230	49 50
20	·954 965	232	·956 736	235	·043 264	·998 228	40
30	·955 197	232	·956 971	234	·043 029	·998 226	30
40	·955 429	233	·957 205	234	·042 795	·998 224	20
50	·955 662	232	·957 439	235	·042 561	·998 222	10
11 00	8·955 894	232	8·957 674	234	1·042 326	9·998 220	49 00
10	·956 126	232	·957 908	233	·042 092	·998 218	48 50
20	·956 358	232	·958 141	234	·041 859	·998 217	40
30	·956 590	231	·958 375	234	·041 625	·998 215	30
40	·956 821	232	·958 609	233	·041 391	·998 213	20
50	·957 053	231	·958 842	233	·041 158	·998 211	10
12 00	8·957 284	232	8·959 075	234	1·040 925	9·998 209	48 00
10	·957 516	231	·959 309	233	·040 691	·998 207	47 50
20	·957 747	231	·959 542	233	·040 458	·998 205	40
30	·957 978	231	·959 775	233	·040 225	·998 203	30
40	·958 209	231	·960 008	232	·039 992	·998 201	20
50	·958 440	230	·960 240	233	·039 760	·998 199	10
13 00	8·958 670	231	8·960 473	232	1·039 527	9·998 197	47 00
10	·958 901	230	·960 705	233	·039 295	·998 195	46 50
20	·959 131	231	·960 938	232	·039 062	·998 194	40
30	·959 362	230	·961 170	232	·038 830	·998 192	30
40	·959 592	230	·961 402	232	·038 598	·998 190	20
50	·959 822	230	·961 634	232	·038 366	·998 188	10
14 00	8·960 052	230	8·961 866	232	1·038 134	9·998 186	46 00
10	·960 282	229	·962 098	231	·037 902	·998 184	45 50
20	·960 511	230	·962 329	232	·037 671	·998 182	40
30	·960 741	229	·962 561	231	·037 439	·998 180	30
40	·960 970	230	·962 792	231	·037 208	·998 178	20
50	·961 200	229	·963 023	232	·036 977	·998 176	10
15 00	8·961 429	229	8·963 255	231	1·036 745	9·998 174	45 00
10	·961 658	229	·963 486	230	·036 514	·998 172	44 50
20	·961 887	229	·963 716	231	·036 284	·998 170	40
30	·962 116	228	·963 947	231	·036 053	·998 168	30
40	·962 344	229	·964 178	230	·035 822	·998 167	20
50	·962 573	228	·964 408	231	·035 592	·998 165	10
16 00	8·962 801	229	8·964 639	230	1·035 361	9·998 163	44 00
10	·963 030	228	·964 869	230	·035 131	·998 161	43 50
20	·963 258	228	·965 099	230	·034 901	·998 159	40
30	·963 486	228	·965 329	230	·034 671	·998 157	30
40	·963 714	228	·965 559	230	·034 441	·998 155	20
50	·963 942	228	·965 789	230	·034 211	·998 153	10
17 00	8·964 170	227	8·966 019	229	1·033 981	9·998 151	43 00
10	·964 397	228	·966 248	230	·033 752	·998 149	42 50
20	·964 625	227	·966 478	229	·033 522	·998 147	40
30	·964 852	228	·966 707	229	·033 293	·998 145	30
40	·965 080	227	·966 936	229	·033 064	·998 143	20
50	·965 307	227	·967 165	229	·032 835	·998 141	10
18 00	8·965 534	227	8·967 394	229	1·032 606	9·998 139	42 00
10	·965 761	226	·967 623	229	·032 377	·998 137	41 50
20	·965 987	227	·967 852	229	·032 148	·998 135	40
30	·966 214	227	·968 081	228	·031 919	·998 133	30
40	·966 441	226	·968 309	229	·031 691	·998 131	20
50	·966 667	226	·968 538	228	·031 462	·998 130	10
19 00	8·966 893	227	8·968 766	228	1·031 234	9·998 128	41 00
10	·967 120	226	·968 994	228	·031 006	·998 126	40 50
20	·967 346	226	·969 222	228	·030 778	·998 124	40
30	·967 572	225	·969 450	228	·030 550	·998 122	30
40	·967 797	226	·969 678	227	·030 322	·998 120	20
50	·968 023	226	·969 905	228	·030 095	·998 118	10
20 00	8·968 249		8·970 133		1·029 867	9·998 116	40 00
	cos		cot		tan	sin	84°

PROPORTIONAL PARTS

″	225	226	227	228	229	230	231	232	233	234	235
1	22·5	22·6	22·7	22·8	22·9	23·0	23·1	23·2	23·3	23·4	23·5
2	45·0	45·2	45·4	45·6	45·8	46·0	46·2	46·4	46·6	46·8	47·0
3	67·5	67·8	68·1	68·4	68·7	69·0	69·3	69·6	69·9	70·2	70·5
4	90·0	90·4	90·8	91·2	91·6	92·0	92·4	92·8	93·2	93·6	94·0
5	112·5	113·0	113·5	114·0	114·5	115·0	115·5	116·0	116·5	117·0	117·5
6	135·0	135·6	136·2	136·8	137·4	138·0	138·6	139·2	139·8	140·4	141·0
7	157·5	158·2	158·9	159·6	160·3	161·0	161·7	162·4	163·1	163·8	164·5
8	180·0	180·8	181·6	182·4	183·2	184·0	184·8	185·6	186·4	187·2	188·0
9	202·5	203·4	204·3	205·2	206·1	207·0	207·9	208·8	209·7	210·6	211·5

TABLE IIc—LOGARITHMS OF TRIGONOMETRICAL FUNCTIONS

(Degrees, Minutes and Seconds)

5°

′ ″	sin		tan		cot	cos	′ ″
20 00	8·968 249	225	8·970 133	227	1·029 867	9·998 116	**40 00**
10	·968 474	226	·970 360	228	·029 640	·998 114	**39 50**
20	·968 700	225	·970 588	227	·029 412	·998 112	**40**
30	·968 925	225	·970 815	227	·029 185	·998 110	**30**
40	·969 150	225	·971 042	227	·028 958	·998 108	**20**
50	·969 375	225	·971 269	227	·028 731	·998 106	**10**
21 00	8·969 600	225	8·971 496	227	1·028 504	9·998 104	**39 00**
10	·969 825	224	·971 723	226	·028 277	·998 102	**38 50**
20	·970 049	225	·971 949	227	·028 051	·998 100	**40**
30	·970 274	224	·972 176	226	·027 824	·998 098	**30**
40	·970 498	225	·972 402	226	·027 598	·998 096	**20**
50	·970 723	224	·972 628	227	·027 372	·998 094	**10**
22 00	8·970 947	224	8·972 855	226	1·027 145	9·998 092	**38 00**
10	·971 171	224	·973 081	226	·026 919	·998 090	**37 50**
20	·971 395	224	·973 307	225	·026 693	·998 088	**40**
30	·971 619	223	·973 532	226	·026 468	·998 086	**30**
40	·971 842	224	·973 758	226	·026 242	·998 084	**20**
50	·972 066	223	·973 984	225	·026 016	·998 082	**10**
23 00	8·972 289	224	8·974 209	226	1·025 791	9·998 080	**37 00**
10	·972 513	223	·974 435	225	·025 565	·998 078	**36 50**
20	·972 736	223	·974 660	225	·025 340	·998 076	**40**
30	·972 959	223	·974 885	225	·025 115	·998 074	**30**
40	·973 182	223	·975 110	225	·024 890	·998 072	**20**
50	·973 405	223	·975 335	225	·024 665	·998 070	**10**
24 00	8·973 628	223	8·975 560	224	1·024 440	9·998 068	**36 00**
10	·973 851	222	·975 784	225	·024 216	·998 066	**35 50**
20	·974 073	223	·976 009	224	·023 991	·998 064	**40**
30	·974 296	222	·976 233	225	·023 767	·998 062	**30**
40	·974 518	222	·976 458	224	·023 542	·998 060	**20**
50	·974 740	222	·976 682	224	·023 318	·998 058	**10**
25 00	8·974 962	222	8·976 906	224	1·023 094	9·998 056	**35 00**
10	·975 184	222	·977 130	224	·022 870	·998 054	**34 50**
20	·975 406	222	·977 354	224	·022 646	·998 052	**40**
30	·975 628	222	·977 578	223	·022 422	·998 050	**30**
40	·975 850	221	·977 801	224	·022 199	·998 048	**20**
50	·976 071	222	·978 025	223	·021 975	·998 046	**10**
26 00	8·976 293	221	8·978 248	224	1·021 752	9·998 044	**34 00**
10	·976 514	221	·978 472	223	·021 528	·998 042	**33 50**
20	·976 735	221	·978 695	223	·021 305	·998 040	**40**
30	·976 956	221	·978 918	223	·021 082	·998 038	**30**
40	·977 177	221	·979 141	223	·020 859	·998 036	**20**
50	·977 398	221	·979 364	222	·020 636	·998 034	**10**
27 00	8·977 619	220	8·979 586	223	1·020 414	9·998 032	**33 00**
10	·977 839	221	·979 809	223	·020 191	·998 030	**32 50**
20	·978 060	220	·980 032	222	·019 968	·998 028	**40**
30	·978 280	221	·980 254	222	·019 746	·998 026	**30**
40	·978 501	220	·980 476	223	·019 524	·998 024	**20**
50	·978 721	220	·980 699	222	·019 301	·998 022	**10**
28 00	8·978 941	220	8·980 921	222	1·019 079	9·998 020	**32 00**
10	·979 161	220	·981 143	221	·018 857	·998 018	**31 50**
20	·979 381	219	·981 364	222	·018 636	·998 016	**40**
30	·979 600	220	·981 586	222	·018 414	·998 014	**30**
40	·979 820	219	·981 808	221	·018 192	·998 012	**20**
50	·980 039	220	·982 029	222	·017 971	·998 010	**10**
29 00	8·980 259	219	8·982 251	221	1·017 749	9·998 008	**31 00**
10	·980 478	219	·982 472	221	·017 528	·998 006	**30 50**
20	·980 697	219	·982 693	221	·017 307	·998 004	**40**
30	·980 916	219	·982 914	221	·017 086	·998 002	**30**
40	·981 135	219	·983 135	221	·016 865	·998 000	**20**
50	·981 354	219	·983 356	221	·016 644	·997 998	**10**
30 00	8·981 573		8·983 577		1·016 423	9·997 996	**30 00**
	cos		cot		tan	sin	**84°**

PROPORTIONAL PARTS

″	**219**	″	**220**
1	21·9	**1**	22·0
2	43·8	**2**	44·0
3	65·7	**3**	66·0
4	87·6	**4**	88·0
5	109·5	**5**	110·0
6	131·4	**6**	132·0
7	153·3	**7**	154·0
8	175·2	**8**	176·0
9	197·1	**9**	198·0

″	**221**	″	**222**
1	22·1	**1**	22·2
2	44·2	**2**	44·4
3	66·3	**3**	66·6
4	88·4	**4**	88·8
5	110·5	**5**	111·0
6	132·6	**6**	133·2
7	154·7	**7**	155·4
8	176·8	**8**	177·6
9	198·9	**9**	199·8

″	**223**	″	**224**
1	22·3	**1**	22·4
2	44·6	**2**	44·8
3	66·9	**3**	67·2
4	89·2	**4**	89·6
5	111·5	**5**	112·0
6	133·8	**6**	134·4
7	156·1	**7**	156·8
8	178·4	**8**	179·2
9	200·7	**9**	201·6

″	**225**	″	**226**
1	22·5	**1**	22·6
2	45·0	**2**	45·2
3	67·5	**3**	67·8
4	90·0	**4**	90·4
5	112·5	**5**	113·0
6	135·0	**6**	135·6
7	157·5	**7**	158·2
8	180·0	**8**	180·8
9	202·5	**9**	203·4

″	**227**	″	**228**
1	22·7	**1**	22·8
2	45·4	**2**	45·6
3	68·1	**3**	68·4
4	90·8	**4**	91·2
5	113·5	**5**	114·0
6	136·2	**6**	136·8
7	158·9	**7**	159·6
8	181·6	**8**	182·4
9	204·3	**9**	205·2

TABLE IIc—LOGARITHMS OF TRIGONOMETRICAL FUNCTIONS

(Degrees, Minutes and Seconds)

5°

′ ″	sin	d	tan	d	cot	cos	′ ″
30 00	8·981 573	218	8·983 577	221	1·016 423	9·997 996	**30 00**
10	·981 791	219	·983 798	220	·016 202	·997 994	**29 50**
20	·982 010	218	·984 018	220	·015 982	·997 992	40
30	·982 228	219	·984 238	221	·015 762	·997 990	30
40	·982 447	218	·984 459	220	·015 541	·997 988	20
50	·982 665	218	·984 679	220	·015 321	·997 986	10
31 00	8·982 883	218	8·984 899	220	1·015 101	9·997 984	**29 00**
10	·983 101	218	·985 119	220	·014 881	·997 982	**28 50**
20	·983 319	217	·985 339	220	·014 661	·997 980	40
30	·983 536	218	·985 559	219	·014 441	·997 978	30
40	·983 754	218	·985 778	220	·014 222	·997 976	20
50	·983 972	217	·985 998	219	·014 002	·997 974	10
32 00	8·984 189	217	8·986 217	220	1·013 783	9·997 972	**28 00**
10	·984 406	217	·986 437	219	·013 563	·997 970	**27 50**
20	·984 623	217	·986 656	219	·013 344	·997 967	40
30	·984 840	217	·986 875	219	·013 125	·997 965	30
40	·985 057	217	·987 094	219	·012 906	·997 963	20
50	·985 274	217	·987 313	219	·012 687	·997 961	10
33 00	8·985 491	217	8·987 532	218	1·012 468	9·997 959	**27 00**
10	·985 708	216	·987 750	219	·012 250	·997 957	**26 50**
20	·985 924	217	·987 969	218	·012 031	·997 955	40
30	·986 141	216	·988 187	219	·011 813	·997 953	30
40	·986 357	216	·988 406	218	·011 594	·997 951	20
50	·986 573	216	·988 624	218	·011 376	·997 949	10
34 00	8·986 789	216	8·988 842	218	1·011 158	9·997 947	**26 00**
10	·987 005	216	·989 060	218	·010 940	·997 945	**25 50**
20	·987 221	216	·989 278	218	·010 722	·997 943	40
30	·987 437	215	·989 496	218	·010 504	·997 941	30
40	·987 652	216	·989 714	217	·010 286	·997 939	20
50	·987 868	215	·989 931	218	·010 069	·997 937	10
35 00	8·988 083	216	8·990 149	217	1·009 851	9·997 935	**25 00**
10	·988 299	215	·990 366	217	·009 634	·997 933	**24 50**
20	·988 514	215	·990 583	218	·009 417	·997 931	40
30	·988 729	215	·990 801	217	·009 199	·997 929	30
40	·988 944	215	·991 018	217	·008 982	·997 926	20
50	·989 159	215	·991 235	216	·008 765	·997 924	10
36 00	8·989 374	214	8·991 451	217	1·008 549	9·997 922	**24 00**
10	·989 588	215	·991 668	217	·008 332	·997 920	**23 50**
20	·989 803	214	·991 885	216	·008 115	·997 918	40
30	·990 017	215	·992 101	217	·007 899	·997 916	30
40	·990 232	214	·992 318	216	·007 682	·997 914	20
50	·990 446	214	·992 534	216	·007 466	·997 912	10
37 00	8·990 660	214	8·992 750	216	1·007 250	9·997 910	**23 00**
10	·990 874	214	·992 966	216	·007 034	·997 908	**22 50**
20	·991 088	214	·993 182	216	·006 818	·997 906	40
30	·991 302	214	·993 398	216	·006 602	·997 904	30
40	·991 516	213	·993 614	216	·006 386	·997 902	20
50	·991 729	214	·993 830	215	·006 170	·997 900	10
38 00	8·991 943	213	8·994 045	216	1·005 955	9·997 897	**22 00**
10	·992 156	214	·994 261	215	·005 739	·997 895	**21 50**
20	·992 370	213	·994 476	216	·005 524	·997 893	40
30	·992 583	213	·994 692	215	·005 308	·997 891	30
40	·992 796	213	·994 907	215	·005 093	·997 889	20
50	·993 009	213	·995 122	215	·004 878	·997 887	10
39 00	8·993 222	213	8·995 337	215	1·004 663	9·997 885	**21 00**
10	·993 435	212	·995 552	214	·004 448	·997 883	**20 50**
20	·993 647	213	·995 766	215	·004 234	·997 881	40
30	·993 860	212	·995 981	215	·004 019	·997 879	30
40	·994 072	213	·996 196	214	·003 804	·997 877	20
50	·994 285	212	·996 410	214	·003 590	·997 875	10
40 00	8·994 497		8·996 624		1·003 376	9·997 872	**20 00**
	cos		cot		tan	sin	**84°**

PROPORTIONAL PARTS

″	212	213	214	215	216	217	218	219	220	221
1	21·2	21·3	21·4	21·5	21·6	21·7	21·8	21·9	22·0	22·1
2	42·4	42·6	42·8	43·0	43·2	43·4	43·6	43·8	44·0	44·2
3	63·6	63·9	64·2	64·5	64·8	65·1	65·4	65·7	66·0	66·3
4	84·8	85·2	85·6	86·0	86·4	86·8	87·2	87·6	88·0	88·4
5	106·0	106·5	107·0	107·5	108·0	108·5	109·0	109·5	110·0	110·5
6	127·2	127·8	128·4	129·0	129·6	130·2	130·8	131·4	132·0	132·6
7	148·4	149·1	149·8	150·5	151·2	151·9	152·6	153·3	154·0	154·7
8	169·6	170·4	171·2	172·0	172·8	173·6	174·4	175·2	176·0	176·8
9	190·8	191·7	192·6	193·5	194·4	195·3	196·2	197·1	198·0	198·9

(Degrees, Minutes and Seconds)

5°

′ ″	sin		tan		cot	cos	′ ″
40 00	8·994 497	212	8·996 624	215	1·003 376	9·997 872	20 00
10	·994 709	212	·996 839	214	·003 161	·997 870	19 50
20	·994 921	212	·997 053	214	·002 947	·997 868	40
30	·995 133	212	·997 267	214	·002 733	·997 866	30
40	·995 345	211	·997 481	213	·002 519	·997 864	20
50	·995 556	212	·997 694	214	·002 306	·997 862	10
41 00	8·995 768	212	8·997 908	214	1·002 092	9·997 860	19 00
10	·995 980	211	·998 122	213	·001 878	·997 858	18 50
20	·996 191	211	·998 335	214	·001 665	·997 856	40
30	·996 402	212	·998 549	213	·001 451	·997 854	30
40	·996 614	211	·998 762	213	·001 238	·997 852	20
50	·996 825	211	·998 975	213	·001 025	·997 849	10
42 00	8·997 036	211	8·999 188	213	1·000 812	9·997 847	18 00
10	·997 247	210	·999 401	213	·000 599	·997 845	17 50
20	·997 457	211	·999 614	213	·000 386	·997 843	40
30	·997 668	211	8·999 827	213	1·000 173	·997 841	30
40	·997 879	210	9·000 040	212	0·999 960	·997 839	20
50	·998 089	210	·000 252	213	·999 748	·997 837	10
43 00	8·998 299	211	9·000 465	212	0·999 535	9·997 835	17 00
10	·998 510	210	·000 677	212	·999 323	·997 833	16 50
20	·998 720	210	·000 889	213	·999 111	·997 830	40
30	·998 930	210	·001 102	212	·998 898	·997 828	30
40	·999 140	210	·001 314	212	·998 686	·997 826	20
50	·999 350	210	·001 526	212	·998 474	·997 824	10
44 00	8·999 560	209	9·001 738	211	0·998 262	9·997 822	16 00
10	·999 769	210	·001 949	212	·998 051	·997 820	15 50
20	8·999 979	209	·002 161	212	·997 839	·997 818	40
30	9·000 188	210	·002 373	211	·997 627	·997 816	30
40	·000 398	209	·002 584	211	·997 416	·997 814	20
50	·000 607	209	·002 795	212	·997 205	·997 811	10
45 00	9·000 816	209	9·003 007	211	0·996 993	9·997 809	15 00
10	·001 025	209	·003 218	211	·996 782	·997 807	14 50
20	·001 234	209	·003 429	211	·996 571	·997 805	40
30	·001 443	209	·003 640	211	·996 360	·997 803	30
40	·001 652	208	·003 851	210	·996 149	·997 801	20
50	·001 860	209	·004 061	211	·995 939	·997 799	10
46 00	9·002 069	208	9·004 272	211	0·995 728	9·997 797	14 00
10	·002 277	209	·004 483	210	·995 517	·997 794	13 50
20	·002 486	208	·004 693	211	·995 307	·997 792	40
30	·002 694	208	·004 904	210	·995 096	·997 790	30
40	·002 902	208	·005 114	210	·994 886	·997 788	20
50	·003 110	208	·005 324	210	·994 676	·997 786	10
47 00	9·003 318	208	9·005 534	210	0·994 466	9·997 784	13 00
10	·003 526	207	·005 744	210	·994 256	·997 782	12 50
20	·003 733	208	·005 954	210	·994 046	·997 780	40
30	·003 941	208	·006 164	209	·993 836	·997 777	30
40	·004 149	207	·006 373	210	·993 627	·997 775	20
50	·004 356	207	·006 583	209	·993 417	·997 773	10
48 00	9·004 563	208	9·006 792	210	0·993 208	9·997 771	12 00
10	·004 771	207	·007 002	209	·992 998	·997 769	11 50
20	·004 978	207	·007 211	209	·992 789	·997 767	40
30	·005 185	207	·007 420	209	·992 580	·997 765	30
40	·005 392	207	·007 629	209	·992 371	·997 762	20
50	·005 599	206	·007 838	209	·992 162	·997 760	10
49 00	9·005 805	207	9·008 047	209	0·991 953	9·997 758	11 00
10	·006 012	206	·008 256	209	·991 744	·997 756	10 50
20	·006 218	207	·008 465	208	·991 535	·997 754	40
30	·006 425	206	·008 673	209	·991 327	·997 752	30
40	·006 631	206	·008 882	208	·991 118	·997 750	20
50	·006 837	207	·009 090	208	·990 910	·997 747	10
50 00	9·007 044		9·009 298		0·990 702	9·997 745	10 00
	cos		cot		tan	sin	84°

PROPORTIONAL PARTS

″	206	″	207
1	20·6	1	20·7
2	41·2	2	41·4
3	61·8	3	62·1
4	82·4	4	82·8
5	103·0	5	103·5
6	123·6	6	124·2
7	144·2	7	144·9
8	164·8	8	165·6
9	185·4	9	186·3

″	208	″	209
1	20·8	1	20·9
2	41·6	2	41·8
3	62·4	3	62·7
4	83·2	4	83·6
5	104·0	5	104·5
6	124·8	6	125·4
7	145·6	7	146·3
8	166·4	8	167·2
9	187·2	9	188·1

″	210	″	211
1	21·0	1	21·1
2	42·0	2	42·2
3	63·0	3	63·3
4	84·0	4	84·4
5	105·0	5	105·5
6	126·0	6	126·6
7	147·0	7	147·7
8	168·0	8	168·8
9	189·0	9	189·9

″	212	″	213
1	21·2	1	21·3
2	42·4	2	42·6
3	63·6	3	63·9
4	84·8	4	85·2
5	106·0	5	106·5
6	127·2	6	127·8
7	148·4	7	149·1
8	169·6	8	170·4
9	190·8	9	191·7

″	214	″	215
1	21·4	1	21·5
2	42·8	2	43·0
3	64·2	3	64·5
4	85·6	4	86·0
5	107·0	5	107·5
6	128·4	6	129·0
7	149·8	7	150·5
8	171·2	8	172·0
9	192·6	9	193·5

TABLE IIc—LOGARITHMS OF TRIGONOMETRICAL FUNCTIONS

(Degrees, Minutes and Seconds)

5°

′ ″	sin		tan		cot	cos	′ ″
50 00	9·007 044	206	9·009 298	209	0·990 702	9·997 745	**10 00**
10	·007 250	206	·009 507	208	·990 493	·997 743	**09 50**
20	·007 456	205	·009 715	208	·990 285	·997 741	**40**
30	·007 661	206	·009 923	208	·990 077	·997 739	**30**
40	·007 867	206	·010 131	207	·989 869	·997 737	**20**
50	·008 073	205	·010 338	208	·989 662	·997 735	**10**
51 00	9·008 278	206	9·010 546	208	0·989 454	9·997 732	**09 00**
10	·008 484	205	·010 754	207	·989 246	·997 730	**08 50**
20	·008 689	205	·010 961	208	·989 039	·997 728	**40**
30	·008 894	206	·011 169	207	·988 831	·997 726	**30**
40	·009 100	205	·011 376	207	·988 624	·997 724	**20**
50	·009 305	205	·011 583	207	·988 417	·997 722	**10**
52 00	9·009 510	205	9·011 790	207	0·988 210	9·997 719	**08 00**
10	·009 715	204	·011 997	207	·988 003	·997 717	**07 50**
20	·009 919	205	·012 204	207	·987 796	·997 715	**40**
30	·010 124	205	·012 411	207	·987 589	·997 713	**30**
40	·010 329	204	·012 618	206	·987 382	·997 711	**20**
50	·010 533	204	·012 824	207	·987 176	·997 709	**10**
53 00	9·010 737	205	9·013 031	206	0·986 969	9·997 706	**07 00**
10	·010 942	204	·013 237	207	·986 763	·997 704	**06 50**
20	·011 146	204	·013 444	206	·986 556	·997 702	**40**
30	·011 350	204	·013 650	206	·986 350	·997 700	**30**
40	·011 554	204	·013 856	206	·986 144	·997 698	**20**
50	·011 758	204	·014 062	206	·985 938	·997 696	**10**
54 00	9·011 962	203	9·014 268	206	0·985 732	9·997 693	**06 00**
10	·012 165	204	·014 474	206	·985 526	·997 691	**05 50**
20	·012 369	203	·014 680	206	·985 320	·997 689	**40**
30	·012 572	204	·014 886	205	·985 114	·997 687	**30**
40	·012 776	203	·015 091	206	·984 909	·997 685	**20**
50	·012 979	203	·015 297	205	·984 703	·997 682	**10**
55 00	9·013 182	203	9·015 502	205	0·984 498	9·997 680	**05 00**
10	·013 385	203	·015 707	206	·984 293	·997 678	**04 50**
20	·013 588	203	·015 913	205	·984 087	·997 676	**40**
30	·013 791	203	·016 118	205	·983 882	·997 674	**30**
40	·013 994	203	·016 323	205	·983 677	·997 672	**20**
50	·014 197	203	·016 528	204	·983 472	·997 669	**10**
56 00	9·014 400	202	9·016 732	205	0·983 268	9·997 667	**04 00**
10	·014 602	203	·016 937	205	·983 063	·997 665	**03 50**
20	·014 805	202	·017 142	204	·982 858	·997 663	**40**
30	·015 007	202	·017 346	205	·982 654	·997 661	**30**
40	·015 209	202	·017 551	204	·982 449	·997 658	**20**
50	·015 411	202	·017 755	204	·982 245	·997 656	**10**
57 00	9·015 613	202	9·017 959	205	0·982 041	9·997 654	**03 00**
10	·015 815	202	·018 164	204	·981 836	·997 652	**02 50**
20	·016 017	202	·018 368	204	·981 632	·997 650	**40**
30	·016 219	202	·018 572	204	·981 428	·997 647	**30**
40	·016 421	201	·018 776	203	·981 224	·997 645	**20**
50	·016 622	202	·018 979	204	·981 021	·997 643	**10**
58 00	9·016 824	201	9·019 183	204	0·980 817	9·997 641	**02 00**
10	·017 025	202	·019 387	203	·980 613	·997 639	**01 50**
20	·017 227	201	·019 590	204	·980 410	·997 636	**40**
30	·017 428	201	·019 794	203	·980 206	·997 634	**30**
40	·017 629	201	·019 997	203	·980 003	·997 632	**20**
50	·017 830	201	·020 200	203	·979 800	·997 630	**10**
59 00	9·018 031	201	9·020 403	203	0·979 597	9·997 628	**01 00**
10	·018 232	201	·020 606	203	·979 394	·997 625	**00 50**
20	·018 433	200	·020 809	203	·979 191	·997 623	**40**
30	·018 633	201	·021 012	203	·978 988	·997 621	**30**
40	·018 834	200	·021 215	203	·978 785	·997 619	**20**
50	·019 034	201	·021 418	202	·978 582	·997 617	**10**
60 00	9·019 235		9·021 620		0·978 380	9·997 614	**00 00**
	cos		cot		tan	sin	**84°**

PROPORTIONAL PARTS

″	200	″	201
1	20·0	1	20·1
2	40·0	2	40·2
3	60·0	3	60·3
4	80·0	4	80·4
5	100·0	5	100·5
6	120·0	6	120·6
7	140·0	7	140·7
8	160·0	8	160·8
9	180·0	9	180·9

″	202	″	203
1	20·2	1	20·3
2	40·4	2	40·6
3	60·6	3	60·9
4	80·8	4	81·2
5	101·0	5	101·5
6	121·2	6	121·8
7	141·4	7	142·1
8	161·6	8	162·4
9	181·8	9	182·7

″	204	″	205
1	20·4	1	20·5
2	40·8	2	41·0
3	61·2	3	61·5
4	81·6	4	82·0
5	102·0	5	102·5
6	122·4	6	123·0
7	142·8	7	143·5
8	163·2	8	164·0
9	183·6	9	184·5

″	206	″	207
1	20·6	1	20·7
2	41·2	2	41·4
3	61·8	3	62·1
4	82·4	4	82·8
5	103·0	5	103·5
6	123·6	6	124·2
7	144·2	7	144·9
8	164·8	8	165·6
9	185·4	9	186·3

″	208	″	209
1	20·8	1	20·9
2	41·6	2	41·8
3	62·4	3	62·7
4	83·2	4	83·6
5	104·0	5	104·5
6	124·8	6	125·4
7	145·6	7	146·3
8	166·4	8	167·2
9	187·2	9	188·1

6°

′ ″	sin	diff.	tan	diff.	cot	cos	′ ″
00 00	9·019 235	200	9·021 620	203	0·978 380	9·997 614	60 00
10	·019 435	200	·021 823	202	·978 177	·997 612	59 50
20	·019 635	200	·022 025	202	·977 975	·997 610	40
30	·019 835	200	·022 227	203	·977 773	·997 608	30
40	·020 035	200	·022 430	202	·977 570	·997 605	20
50	·020 235	200	·022 632	202	·977 368	·997 603	10
01 00	9·020 435	200	9·022 834	202	0·977 166	9·997 601	59 00
10	·020 635	199	·023 036	202	·976 964	·997 599	58 50
20	·020 834	200	·023 238	201	·976 762	·997 597	40
30	·021 034	199	·023 439	202	·976 561	·997 594	30
40	·021 233	200	·023 641	202	·976 359	·997 592	20
50	·021 433	199	·023 843	201	·976 157	·997 590	10
02 00	9·021 632	199	9·024 044	201	0·975 956	9·997 588	58 00
10	·021 831	199	·024 245	202	·975 755	·997 585	57 50
20	·022 030	199	·024 447	201	·975 553	·997 583	40
30	·022 229	199	·024 648	201	·975 352	·997 581	30
40	·022 428	199	·024 849	201	·975 151	·997 579	20
50	·022 627	198	·025 050	201	·974 950	·997 577	10
03 00	9·022 825	199	9·025 251	201	0·974 749	9·997 574	57 00
10	·023 024	199	·025 452	201	·974 548	·997 572	56 50
20	·023 223	198	·025 653	200	·974 347	·997 570	40
30	·023 421	198	·025 853	201	·974 147	·997 568	30
40	·023 619	199	·026 054	200	·973 946	·997 565	20
50	·023 818	198	·026 254	201	·973 746	·997 563	10
04 00	9·024 016	198	9·026 455	200	0·973 545	9·997 561	56 00
10	·024 214	198	·026 655	200	·973 345	·997 559	55 50
20	·024 412	198	·026 855	200	·973 145	·997 556	40
30	·024 610	197	·027 055	200	·972 945	·997 554	30
40	·024 807	198	·027 255	200	·972 745	·997 552	20
50	·025 005	198	·027 455	200	·972 545	·997 550	10
05 00	9·025 203	197	9·027 655	200	0·972 345	9·997 547	55 00
10	·025 400	198	·027 855	200	·972 145	·997 545	54 50
20	·025 598	197	·028 055	199	·971 945	·997 543	40
30	·025 795	197	·028 254	200	·971 746	·997 541	30
40	·025 992	197	·028 454	199	·971 546	·997 539	20
50	·026 189	197	·028 653	199	·971 347	·997 536	10
06 00	9·026 386	197	9·028 852	200	0·971 148	9·997 534	54 00
10	·026 583	197	·029 052	199	·970 948	·997 532	53 50
20	·026 780	197	·029 251	199	·970 749	·997 530	40
30	·026 977	197	·029 450	199	·970 550	·997 527	30
40	·027 174	196	·029 649	199	·970 351	·997 525	20
50	·027 370	197	·029 848	198	·970 152	·997 523	10
07 00	9·027 567	196	9·030 046	199	0·969 954	9·997 520	53 00
10	·027 763	197	·030 245	199	·969 755	·997 518	52 50
20	·027 960	196	·030 444	198	·969 556	·997 516	40
30	·028 156	196	·030 642	199	·969 358	·997 514	30
40	·028 352	196	·030 841	198	·969 159	·997 511	20
50	·028 548	196	·031 039	198	·968 961	·997 509	10
08 00	9·028 744	196	9·031 237	198	0·968 763	9·997 507	52 00
10	·028 940	196	·031 435	198	·968 565	·997 505	51 50
20	·029 136	196	·031 633	198	·968 367	·997 502	40
30	·029 332	195	·031 831	198	·968 169	·997 500	30
40	·029 527	196	·032 029	198	·967 971	·997 498	20
50	·029 723	195	·032 227	198	·967 773	·997 496	10
09 00	9·029 918	196	9·032 425	198	0·967 575	9·997 493	51 00
10	·030 114	195	·032 623	197	·967 377	·997 491	50 50
20	·030 309	195	·032 820	197	·967 180	·997 489	40
30	·030 504	195	·033 017	198	·966 983	·997 487	30
40	·030 699	195	·033 215	197	·966 785	·997 484	20
50	·030 894	195	·033 412	197	·966 588	·997 482	10
10 00	9·031 089		9·033 609		0·966 391	9·997 480	50 00
	cos		cot		tan	sin	83°

PROPORTIONAL PARTS

″	195	″	196
1	19·5	1	19·6
2	39·0	2	39·2
3	58·5	3	58·8
4	78·0	4	78·4
5	97·5	5	98·0
6	117·0	6	117·6
7	136·5	7	137·2
8	156·0	8	156·8
9	175·5	9	176·4

″	197	″	198
1	19·7	1	19·8
2	39·4	2	39·6
3	59·1	3	59·4
4	78·8	4	79·2
5	98·5	5	99·0
6	118·2	6	118·8
7	137·9	7	138·6
8	157·6	8	158·4
9	177·3	9	178·2

″	199	″	200
1	19·9	1	20·0
2	39·8	2	40·0
3	59·7	3	60·0
4	79·6	4	80·0
5	99·5	5	100·0
6	119·4	6	120·0
7	139·3	7	140·0
8	159·2	8	160·0
9	179·1	9	180·0

″	201	″	202
1	20·1	1	20·2
2	40·2	2	40·4
3	60·3	3	60·6
4	80·4	4	80·8
5	100·5	5	101·0
6	120·6	6	121·2
7	140·7	7	141·4
8	160·8	8	161·6
9	180·9	9	181·8

″	203
1	20·3
2	40·6
3	60·9
4	81·2
5	101·5
6	121·8
7	142·1
8	162·4
9	182·7

TABLE IIc—LOGARITHMS OF TRIGONOMETRICAL FUNCTIONS

(Degrees, Minutes and Seconds)

6°

′ ″	sin	d	tan	d	cot	cos	′ ″
10 00	9·031 089	195	9·033 609	197	0·966 391	9·997 480	50 00
10	·031 284	195	·033 806	197	·966 194	·997 477	49 50
20	·031 479	194	·034 003	197	·965 997	·997 475	40
30	·031 673	195	·034 200	197	·965 800	·997 473	30
40	·031 868	194	·034 397	197	·965 603	·997 471	20
50	·032 062	195	·034 594	197	·965 406	·997 468	10
11 00	9·032 257	194	9·034 791	196	0·965 209	9·997 466	49 00
10	·032 451	194	·034 987	197	·965 013	·997 464	48 50
20	·032 645	194	·035 184	196	·964 816	·997 461	40
30	·032 839	194	·035 380	196	·964 620	·997 459	30
40	·033 033	194	·035 576	197	·964 424	·997 457	20
50	·033 227	194	·035 773	196	·964 227	·997 455	10
12 00	9·033 421	194	9·035 969	196	0·964 031	9·997 452	48 00
10	·033 615	194	·036 165	196	·963 835	·997 450	47 50
20	·033 809	193	·036 361	196	·963 639	·997 448	40
30	·034 002	194	·036 557	196	·963 443	·997 445	30
40	·034 196	193	·036 753	195	·963 247	·997 443	20
50	·034 389	193	·036 948	196	·963 052	·997 441	10
13 00	9·034 582	194	9·037 144	195	0·962 856	9·997 439	47 00
10	·034 776	193	·037 339	196	·962 661	·997 436	46 50
20	·034 969	193	·037 535	195	·962 465	·997 434	40
30	·035 162	193	·037 730	196	·962 270	·997 432	30
40	·035 355	193	·037 926	195	·962 074	·997 429	20
50	·035 548	193	·038 121	195	·961 879	·997 427	10
14 00	9·035 741	192	9·038 316	195	0·961 684	9·997 425	46 00
10	·035 933	193	·038 511	195	·961 489	·997 423	45 50
20	·036 126	193	·038 706	195	·961 294	·997 420	40
30	·036 319	192	·038 901	194	·961 099	·997 418	30
40	·036 511	192	·039 095	195	·960 905	·997 416	20
50	·036 703	193	·039 290	195	·960 710	·997 413	10
15 00	9·036 896	192	9·039 485	194	0·960 515	9·997 411	45 00
10	·037 088	192	·039 679	195	·960 321	·997 409	44 50
20	·037 280	192	·039 874	194	·960 126	·997 406	40
30	·037 472	192	·040 068	194	·959 932	·997 404	30
40	·037 664	192	·040 262	194	·959 738	·997 402	20
50	·037 856	192	·040 456	195	·959 544	·997 399	10
16 00	9·038 048	191	9·040 651	194	0·959 349	9·997 397	44 00
10	·038 239	192	·040 845	194	·959 155	·997 395	43 50
20	·038 431	192	·041 039	193	·958 961	·997 393	40
30	·038 623	191	·041 232	194	·958 768	·997 390	30
40	·038 814	191	·041 426	194	·958 574	·997 388	20
50	·039 005	192	·041 620	193	·958 380	·997 386	10
17 00	9·039 197	191	9·041 813	194	0·958 187	9·997 383	43 00
10	·039 388	191	·042 007	193	·957 993	·997 381	42 50
20	·039 579	191	·042 200	194	·957 800	·997 379	40
30	·039 770	191	·042 394	193	·957 606	·997 376	30
40	·039 961	191	·042 587	193	·957 413	·997 374	20
50	·040 152	190	·042 780	193	·957 220	·997 372	10
18 00	9·040 342	191	9·042 973	193	0·957 027	9·997 369	42 00
10	·040 533	191	·043 166	193	·956 834	·997 367	41 50
20	·040 724	190	·043 359	193	·956 641	·997 365	40
30	·040 914	191	·043 552	193	·956 448	·997 362	30
40	·041 105	190	·043 745	192	·956 255	·997 360	20
50	·041 295	190	·043 937	193	·956 063	·997 358	10
19 00	9·041 485	190	9·044 130	192	0·955 870	9·997 355	41 00
10	·041 675	190	·044 322	193	·955 678	·997 353	40 50
20	·041 865	190	·044 515	192	·955 485	·997 351	40
30	·042 055	190	·044 707	192	·955 293	·997 348	30
40	·042 245	190	·044 899	193	·955 101	·997 346	20
50	·042 435	190	·045 092	192	·954 908	·997 344	10
20 00	9·042 625		9·045 284		0·954 716	9·997 341	40 00
	cos		cot		tan	sin	83°

PROPORTIONAL PARTS

″	190	″	191
1	19·0	1	19·1
2	38·0	2	38·2
3	57·0	3	57·3
4	76·0	4	76·4
5	95·0	5	95·5
6	114·0	6	114·6
7	133·0	7	133·7
8	152·0	8	152·8
9	171·0	9	171·9

″	192	″	193
1	19·2	1	19·3
2	38·4	2	38·6
3	57·6	3	57·9
4	76·8	4	77·2
5	96·0	5	96·5
6	115·2	6	115·8
7	134·4	7	135·1
8	153·6	8	154·4
9	172·8	9	173·7

″	194	″	195
1	19·4	1	19·5
2	38·8	2	39·0
3	58·2	3	58·5
4	77·6	4	78·0
5	97·0	5	97·5
6	116·4	6	117·0
7	135·8	7	136·5
8	155·2	8	156·0
9	174·6	9	175·5

″	196	″	197
1	19·6	1	19·7
2	39·2	2	39·4
3	58·8	3	59·1
4	78·4	4	78·8
5	98·0	5	98·5
6	117·6	6	118·2
7	137·2	7	137·9
8	156·8	8	157·6
9	176·4	9	177·3

TABLE IIc—LOGARITHMS OF TRIGONOMETRICAL FUNCTIONS
(Degrees, Minutes and Seconds)

6°

′ ″	sin		tan		cot	cos	′ ″
20 00	9·042 625	190	9·045 284	192	0·954 716	9·997 341	40 00
10	·042 815	189	·045 476	192	·954 524	·997 339	39 50
20	·043 004	190	·045 668	191	·954 332	·997 337	40
30	·043 194	189	·045 859	192	·954 141	·997 334	30
40	·043 383	189	·046 051	192	·953 949	·997 332	20
50	·043 572	190	·046 243	191	·953 757	·997 330	10
21 00	9·043 762	189	9·046 434	192	0·953 566	9·997 327	39 00
10	·043 951	189	·046 626	191	·953 374	·997 325	38 50
20	·044 140	189	·046 817	192	·953 183	·997 323	40
30	·044 329	189	·047 009	191	·952 991	·997 320	30
40	·044 518	189	·047 200	191	·952 800	·997 318	20
50	·044 707	188	·047 391	191	·952 609	·997 316	10
22 00	9·044 895	189	9·047 582	191	0·952 418	9·997 313	38 00
10	·045 084	189	·047 773	191	·952 227	·997 311	37 50
20	·045 273	188	·047 964	191	·952 036	·997 309	40
30	·045 461	189	·048 155	191	·951 845	·997 306	30
40	·045 650	188	·048 346	190	·951 654	·997 304	20
50	·045 838	188	·048 536	191	·951 464	·997 301	10
23 00	9·046 026	188	9·048 727	190	0·951 273	9·997 299	37 00
10	·046 214	188	·048 917	191	·951 083	·997 297	36 50
20	·046 402	188	·049 108	190	·950 892	·997 294	40
30	·046 590	188	·049 298	191	·950 702	·997 292	30
40	·046 778	188	·049 489	190	·950 511	·997 290	20
50	·046 966	188	·049 679	190	·950 321	·997 287	10
24 00	9·047 154	188	9·049 869	190	0·950 131	9·997 285	36 00
10	·047 342	187	·050 059	190	·949 941	·997 283	35 50
20	·047 529	188	·050 249	190	·949 751	·997 280	40
30	·047 717	187	·050 439	190	·949 561	·997 278	30
40	·047 904	187	·050 629	189	·949 371	·997 276	20
50	·048 091	188	·050 818	190	·949 182	·997 273	10
25 00	9·048 279	187	9·051 008	189	0·948 992	9·997 271	35 00
10	·048 466	187	·051 197	190	·948 803	·997 268	34 50
20	·048 653	187	·051 387	189	·948 613	·997 266	40
30	·048 840	187	·051 576	190	·948 424	·997 264	30
40	·049 027	187	·051 766	189	·948 234	·997 261	20
50	·049 214	186	·051 955	189	·948 045	·997 259	10
26 00	9·049 400	187	9·052 144	189	0·947 856	9·997 257	34 00
10	·049 587	187	·052 333	189	·947 667	·997 254	33 50
20	·049 774	186	·052 522	189	·947 478	·997 252	40
30	·049 960	187	·052 711	189	·947 289	·997 249	30
40	·050 147	186	·052 900	188	·947 100	·997 247	20
50	·050 333	186	·053 088	189	·946 912	·997 245	10
27 00	9·050 519	187	9·053 277	189	0·946 723	9·997 242	33 00
10	·050 706	186	·053 466	188	·946 534	·997 240	32 50
20	·050 892	186	·053 654	189	·946 346	·997 238	40
30	·051 078	186	·053 843	188	·946 157	·997 235	30
40	·051 264	186	·054 031	188	·945 969	·997 233	20
50	·051 450	185	·054 219	188	·945 781	·997 230	10
28 00	9·051 635	186	9·054 407	189	0·945 593	9·997 228	32 00
10	·051 821	186	·054 596	188	·945 404	·997 226	31 50
20	·052 007	185	·054 784	188	·945 216	·997 223	40
30	·052 192	186	·054 972	187	·945 028	·997 221	30
40	·052 378	185	·055 159	188	·944 841	·997 218	20
50	·052 563	186	·055 347	188	·944 653	·997 216	10
29 00	9·052 749	185	9·055 535	188	0·944 465	9·997 214	31 00
10	·052 934	185	·055 723	187	·944 277	·997 211	30 50
20	·053 119	185	·055 910	188	·944 090	·997 209	40
30	·053 304	185	·056 098	187	·943 902	·997 206	30
40	·053 489	185	·056 285	187	·943 715	·997 204	20
50	·053 674	185	·056 472	187	·943 528	·997 202	10
30 00	9·053 859		9·056 659		0·943 341	9·997 199	30 00
	cos		cot		tan	sin	83°

PROPORTIONAL PARTS

″	185	″	186
1	18·5	1	18·6
2	37·0	2	37·2
3	55·5	3	55·8
4	74·0	4	74·4
5	92·5	5	93·0
6	111·0	6	111·6
7	129·5	7	130·2
8	148·0	8	148·8
9	166·5	9	167·4

″	187	″	188
1	18·7	1	18·8
2	37·4	2	37·6
3	56·1	3	56·4
4	74·8	4	75·2
5	93·5	5	94·0
6	112·2	6	112·8
7	130·9	7	131·6
8	149·6	8	150·4
9	168·3	9	169·2

″	189	″	190
1	18·9	1	19·0
2	37·8	2	38·0
3	56·7	3	57·0
4	75·6	4	76·0
5	94·5	5	95·0
6	113·4	6	114·0
7	132·3	7	133·0
8	151·2	8	152·0
9	170·1	9	171·0

″	191	″	192
1	19·1	1	19·2
2	38·2	2	38·4
3	57·3	3	57·6
4	76·4	4	76·8
5	95·5	5	96·0
6	114·6	6	115·2
7	133·7	7	134·4
8	152·8	8	153·6
9	171·9	9	172·8

TABLE IIc—LOGARITHMS OF TRIGONOMETRICAL FUNCTIONS

(Degrees, Minutes and Seconds)

6°

′ ″	sin	d	tan	d	cot	cos	′ ″
30 00	9·053 859	185	9·056 659	188	0·943 341	9·997 199	30 00
10	·054 044	184	·056 847	187	·943 153	·997 197	29 50
20	·054 228	185	·057 034	187	·942 966	·997 194	40
30	·054 413	184	·057 221	187	·942 779	·997 192	30
40	·054 597	185	·057 408	186	·942 592	·997 190	20
50	·054 782	184	·057 594	187	·942 406	·997 187	10
31 00	9·054 966	184	9·057 781	187	0·942 219	9·997 185	29 00
10	·055 150	185	·057 968	187	·942 032	·997 182	28 50
20	·055 335	184	·058 155	186	·941 845	·997 180	40
30	·055 519	184	·058 341	187	·941 659	·997 178	30
40	·055 703	184	·058 528	186	·941 472	·997 175	20
50	·055 887	184	·058 714	186	·941 286	·997 173	10
32 00	9·056 071	183	9·058 900	186	0·941 100	9·997 170	28 00
10	·056 254	184	·059 086	187	·940 914	·997 168	27 50
20	·056 438	184	·059 273	186	·940 727	·997 166	40
30	·056 622	183	·059 459	186	·940 541	·997 163	30
40	·056 805	184	·059 645	186	·940 355	·997 161	20
50	·056 989	183	·059 831	185	·940 169	·997 158	10
33 00	9·057 172	184	9·060 016	186	0·939 984	9·997 156	27 00
10	·057 356	183	·060 202	186	·939 798	·997 154	26 50
20	·057 539	183	·060 388	185	·939 612	·997 151	40
30	·057 722	183	·060 573	186	·939 427	·997 149	30
40	·057 905	183	·060 759	185	·939 241	·997 146	20
50	·058 088	183	·060 944	186	·939 056	·997 144	10
34 00	9·058 271	183	9·061 130	185	0·938 870	9·997 141	26 00
10	·058 454	183	·061 315	185	·938 685	·997 139	25 50
20	·058 637	183	·061 500	185	·938 500	·997 137	40
30	·058 820	182	·061 685	185	·938 315	·997 134	30
40	·059 002	183	·061 870	185	·938 130	·997 132	20
50	·059 185	182	·062 055	185	·937 945	·997 129	10
35 00	9·059 367	183	9·062 240	185	0·937 760	9·997 127	25 00
10	·059 550	182	·062 425	185	·937 575	·997 124	24 50
20	·059 732	182	·062 610	185	·937 390	·997 122	40
30	·059 914	182	·062 795	184	·937 205	·997 120	30
40	·060 096	182	·062 979	185	·937 021	·997 117	20
50	·060 278	182	·063 164	184	·936 836	·997 115	10
36 00	9·060 460	182	9·063 348	185	0·936 652	9·997 112	24 00
10	·060 642	182	·063 533	184	·936 467	·997 110	23 50
20	·060 824	182	·063 717	184	·936 283	·997 107	40
30	·061 006	182	·063 901	184	·936 099	·997 105	30
40	·061 188	181	·064 085	184	·935 915	·997 102	20
50	·061 369	182	·064 269	184	·935 731	·997 100	10
37 00	9·061 551	181	9·064 453	184	0·935 547	9·997 098	23 00
10	·061 732	182	·064 637	184	·935 363	·997 095	22 50
20	·061 914	181	·064 821	184	·935 179	·997 093	40
30	·062 095	181	·065 005	183	·934 995	·997 090	30
40	·062 276	181	·065 188	184	·934 812	·997 088	20
50	·062 457	182	·065 372	184	·934 628	·997 085	10
38 00	9·062 639	181	9·065 556	183	0·934 444	9·997 083	22 00
10	·062 820	181	·065 739	183	·934 261	·997 080	21 50
20	·063 001	180	·065 922	184	·934 078	·997 078	40
30	·063 181	181	·066 106	183	·933 894	·997 076	30
40	·063 362	181	·066 289	183	·933 711	·997 073	20
50	·063 543	181	·066 472	183	·933 528	·997 071	10
39 00	9·063 724	180	9·066 655	183	0·933 345	9·997 068	21 00
10	·063 904	181	·066 838	183	·933 162	·997 066	20 50
20	·064 085	180	·067 021	183	·932 979	·997 063	40
30	·064 265	180	·067 204	183	·932 796	·997 061	30
40	·064 445	181	·067 387	183	·932 613	·997 058	20
50	·064 626	180	·067 570	182	·932 430	·997 056	10
40 00	9·064 806		9·067 752		0·932 248	9·997 053	20 00
	cos		cot		tan	sin	83°

PROPORTIONAL PARTS

″	180	″	181
1	18·0	1	18·1
2	36·0	2	36·2
3	54·0	3	54·3
4	72·0	4	72·4
5	90·0	5	90·5
6	108·0	6	108·6
7	126·0	7	126·7
8	144·0	8	144·8
9	162·0	9	162·9

″	182	″	183
1	18·2	1	18·3
2	36·4	2	36·6
3	54·6	3	54·9
4	72·8	4	73·2
5	91·0	5	91·5
6	109·2	6	109·8
7	127·4	7	128·1
8	145·6	8	146·4
9	163·8	9	164·7

″	184	″	185
1	18·4	1	18·5
2	36·8	2	37·0
3	55·2	3	55·5
4	73·6	4	74·0
5	92·0	5	92·5
6	110·4	6	111·0
7	128·8	7	129·5
8	147·2	8	148·0
9	165·6	9	166·5

″	186	″	187
1	18·6	1	18·7
2	37·2	2	37·4
3	55·8	3	56·1
4	74·4	4	74·8
5	93·0	5	93·5
6	111·6	6	112·2
7	130·2	7	130·9
8	148·8	8	149·6
9	167·4	9	168·3

″	188
1	18·8
2	37·6
3	56·4
4	75·2
5	94·0
6	112·8
7	131·6
8	150·4
9	169·2

6°

′ ″	sin		tan		cot	cos	′ ″
40 00	9·064 806	180	9·067 752	183	0·932 248	9·997 053	20 00
10	·064 986	180	·067 935	182	·932 065	·997 051	19 50
20	·065 166	180	·068 117	183	·931 883	·997 049	40
30	·065 346	180	·068 300	182	·931 700	·997 046	30
40	·065 526	179	·068 482	182	·931 518	·997 044	20
50	·065 705	180	·068 664	182	·931 336	·997 041	10
41 00	9·065 885	180	9·068 846	183	0·931 154	9·997 039	19 00
10	·066 065	179	·069 029	182	·930 971	·997 036	18 50
20	·066 244	180	·069 211	182	·930 789	·997 034	40
30	·066 424	179	·069 393	182	·930 607	·997 031	30
40	·066 603	180	·069 575	181	·930 425	·997 029	20
50	·066 783	179	·069 756	182	·930 244	·997 026	10
42 00	9·066 962	179	9·069 938	182	0·930 062	9·997 024	18 00
10	·067 141	179	·070 120	181	·929 880	·997 021	17 50
20	·067 320	179	·070 301	182	·929 699	·997 019	40
30	·067 499	179	·070 483	181	·929 517	·997 016	30
40	·067 678	179	·070 664	182	·929 336	·997 014	20
50	·067 857	179	·070 846	181	·929 154	·997 011	10
43 00	9·068 036	179	9·071 027	181	0·928 973	9·997 009	17 00
10	·068 215	178	·071 208	181	·928 792	·997 007	16 50
20	·068 393	179	·071 389	181	·928 611	·997 004	40
30	·068 572	179	·071 570	181	·928 430	·997 002	30
40	·068 751	178	·071 751	181	·928 249	·996 999	20
50	·068 929	178	·071 932	181	·928 068	·996 997	10
44 00	9·069 107	179	9·072 113	181	0·927 887	9·996 994	16 00
10	·069 286	178	·072 294	181	·927 706	·996 992	15 50
20	·069 464	178	·072 475	180	·927 525	·996 989	40
30	·069 642	178	·072 655	181	·927 345	·996 987	30
40	·069 820	178	·072 836	180	·927 164	·996 984	20
50	·069 998	178	·073 016	181	·926 984	·996 982	10
45 00	9·070 176	178	9·073 197	180	0·926 803	9·996 979	15 00
10	·070 354	178	·073 377	181	·926 623	·996 977	14 50
20	·070 532	177	·073 558	180	·926 442	·996 974	40
30	·070 709	178	·073 738	180	·926 262	·996 972	30
40	·070 887	178	·073 918	180	·926 082	·996 969	20
50	·071 065	177	·074 098	180	·925 902	·996 967	10
46 00	9·071 242	178	9·074 278	180	0·925 722	9·996 964	14 00
10	·071 420	177	·074 458	180	·925 542	·996 962	13 50
20	·071 597	177	·074 638	179	·925 362	·996 959	40
30	·071 774	177	·074 817	180	·925 183	·996 957	30
40	·071 951	177	·074 997	180	·925 003	·996 954	20
50	·072 128	178	·075 177	179	·924 823	·996 952	10
47 00	9·072 306	176	9·075 356	180	0·924 644	9·996 949	13 00
10	·072 482	177	·075 536	179	·924 464	·996 947	12 50
20	·072 659	177	·075 715	180	·924 285	·996 944	40
30	·072 836	177	·075 895	179	·924 105	·996 942	30
40	·073 013	177	·076 074	179	·923 926	·996 939	20
50	·073 190	176	·076 253	179	·923 747	·996 937	10
48 00	9·073 366	177	9·076 432	179	0·923 568	9·996 934	12 00
10	·073 543	176	·076 611	179	·923 389	·996 932	11 50
20	·073 719	177	·076 790	179	·923 210	·996 929	40
30	·073 896	176	·076 969	179	·923 031	·996 927	30
40	·074 072	176	·077 148	179	·922 852	·996 924	20
50	·074 248	176	·077 327	178	·922 673	·996 922	10
49 00	9·074 424	176	9·077 505	179	0·922 495	9·996 919	11 00
10	·074 600	177	·077 684	178	·922 316	·996 917	10 50
20	·074 777	175	·077 862	179	·922 138	·996 914	40
30	·074 952	176	·078 041	178	·921 959	·996 912	30
40	·075 128	176	·078 219	179	·921 781	·996 909	20
50	·075 304	176	·078 398	178	·921 602	·996 906	10
50 00	9·075 480		9·078 576		0·921 424	9·996 904	10 00
	cos		cot		tan	sin	83°

PROPORTIONAL PARTS

″	175	″	176
1	17·5	1	17·6
2	35·0	2	35·2
3	52·5	3	52·8
4	70·0	4	70·4
5	87·5	5	88·0
6	105·0	6	105·6
7	122·5	7	123·2
8	140·0	8	140·8
9	157·5	9	158·4

″	177	″	178
1	17·7	1	17·8
2	35·4	2	35·6
3	53·1	3	53·4
4	70·8	4	71·2
5	88·5	5	89·0
6	106·2	6	106·8
7	123·9	7	124·6
8	141·6	8	142·4
9	159·3	9	160·2

″	179	″	180
1	17·9	1	18·0
2	35·8	2	36·0
3	53·7	3	54·0
4	71·6	4	72·0
5	89·5	5	90·0
6	107·4	6	108·0
7	125·3	7	126·0
8	143·2	8	144·0
9	161·1	9	162·0

″	181	″	182
1	18·1	1	18·2
2	36·2	2	36·4
3	54·3	3	54·6
4	72·4	4	72·8
5	90·5	5	91·0
6	108·6	6	109·2
7	126·7	7	127·4
8	144·8	8	145·6
9	162·9	9	163·8

″	183
1	18·3
2	36·6
3	54·9
4	73·2
5	91·5
6	109·8
7	128·1
8	146·4
9	164·7

(Degrees, Minutes and Seconds)

6° ′ ″	sin		tan		cot	cos	′ ″
50 00	9·075 480	176	9·078 576	178	0·921 424	9·996 904	10 00
10	·075 656	175	·078 754	178	·921 246	·996 901	09 50
20	·075 831	176	·078 932	178	·921 068	·996 899	40
30	·076 007	175	·079 110	178	·920 890	·996 896	30
40	·076 182	176	·079 288	178	·920 712	·996 894	20
50	·076 358	175	·079 466	178	·920 534	·996 891	10
51 00	9·076 533	175	9·079 644	178	0·920 356	9·996 889	09 00
10	·076 708	175	·079 822	178	·920 178	·996 886	08 50
20	·076 883	175	·080 000	177	·920 000	·996 884	40
30	·077 058	175	·080 177	178	·919 823	·996 881	30
40	·077 233	175	·080 355	177	·919 645	·996 879	20
50	·077 408	175	·080 532	178	·919 468	·996 876	10
52 00	9·077 583	175	9·080 710	177	0·919 290	9·996 874	08 00
10	·077 758	175	·080 887	177	·919 113	·996 871	07 50
20	·077 933	174	·081 064	177	·918 936	·996 869	40
30	·078 107	175	·081 241	178	·918 759	·996 866	30
40	·078 282	175	·081 419	177	·918 581	·996 863	20
50	·078 457	174	·081 596	177	·918 404	·996 861	10
53 00	9·078 631	174	9·081 773	177	0·918 227	9·996 858	07 00
10	·078 805	175	·081 950	176	·918 050	·996 856	06 50
20	·078 980	174	·082 126	177	·917 874	·996 853	40
30	·079 154	174	·082 303	177	·917 697	·996 851	30
40	·079 328	174	·082 480	177	·917 520	·996 848	20
50	·079 502	174	·082 657	176	·917 343	·996 846	10
54 00	9·079 676	174	9·082 833	177	0·917 167	9·996 843	06 00
10	·079 850	174	·083 010	176	·916 990	·996 841	05 50
20	·080 024	174	·083 186	176	·916 814	·996 838	40
30	·080 198	174	·083 362	177	·916 638	·996 835	30
40	·080 372	173	·083 539	176	·916 461	·996 833	20
50	·080 545	174	·083 715	176	·916 285	·996 830	10
55 00	9·080 719	173	9·083 891	176	0·916 109	9·996 828	05 00
10	·080 892	174	·084 067	176	·915 933	·996 825	04 50
20	·081 066	173	·084 243	176	·915 757	·996 823	40
30	·081 239	174	·084 419	176	·915 581	·996 820	30
40	·081 413	173	·084 595	176	·915 405	·996 818	20
50	·081 586	173	·084 771	176	·915 229	·996 815	10
56 00	9·081 759	173	9·084 947	175	0·915 053	9·996 812	04 00
10	·081 932	173	·085 122	176	·914 878	·996 810	03 50
20	·082 105	173	·085 298	175	·914 702	·996 807	40
30	·082 278	173	·085 473	176	·914 527	·996 805	30
40	·082 451	173	·085 649	175	·914 351	·996 802	20
50	·082 624	173	·085 824	176	·914 176	·996 800	10
57 00	9·082 797	172	9·086 000	175	0·914 000	9·996 797	03 00
10	·082 969	173	·086 175	175	·913 825	·996 795	02 50
20	·083 142	172	·086 350	175	·913 650	·996 792	40
30	·083 314	173	·086 525	175	·913 475	·996 789	30
40	·083 487	172	·086 700	175	·913 300	·996 787	20
50	·083 659	173	·086 875	175	·913 125	·996 784	10
58 00	9·083 832	172	9·087 050	175	0·912 950	9·996 782	02 00
10	·084 004	172	·087 225	175	·912 775	·996 779	01 50
20	·084 176	172	·087 400	174	·912 600	·996 777	40
30	·084 348	172	·087 574	175	·912 426	·996 774	30
40	·084 520	172	·087 749	175	·912 251	·996 771	20
50	·084 692	172	·087 924	174	·912 076	·996 769	10
59 00	9·084 864	172	9·088 098	175	0·911 902	9·996 766	01 00
10	·085 036	172	·088 273	174	·911 727	·996 764	00 50
20	·085 208	172	·088 447	174	·911 553	·996 761	40
30	·085 380	171	·088 621	174	·911 379	·996 758	30
40	·085 551	172	·088 795	175	·911 205	·996 756	20
50	·085 723	171	·088 970	174	·911 030	·996 753	10
60 00	9·085 894		9·089 144		0·910 856	9·996 751	00 00
	cos		cot		tan	sin	83°

PROPORTIONAL PARTS

″	171	″	172
1	17·1	1	17·2
2	34·2	2	34·4
3	51·3	3	51·6
4	68·4	4	68·8
5	85·5	5	86·0
6	102·6	6	103·2
7	119·7	7	120·4
8	136·8	8	137·6
9	153·9	9	154·8

″	173	″	174
1	17·3	1	17·4
2	34·6	2	34·8
3	51·9	3	52·2
4	69·2	4	69·6
5	86·5	5	87·0
6	103·8	6	104·4
7	121·1	7	121·8
8	138·4	8	139·2
9	155·7	9	156·6

″	175	″	176
1	17·5	1	17·6
2	35·0	2	35·2
3	52·5	3	52·8
4	70·0	4	70·4
5	87·5	5	88·0
6	105·0	6	105·6
7	122·5	7	123·2
8	140·0	8	140·8
9	157·5	9	158·4

″	177	″	178
1	17·7	1	17·8
2	35·4	2	35·6
3	53·1	3	53·4
4	70·8	4	71·2
5	88·5	5	89·0
6	106·2	6	106·8
7	123·9	7	124·6
8	141·6	8	142·4
9	159·3	9	160·2

(Degrees, Minutes and Seconds)

7°

′ ″	sin	d	tan	d	cot	cos	′ ″
00 00	9·085 894	172	9·089 144	174	0·910 856	9·996 751	60 00
10	·086 066	171	·089 318	174	·910 682	·996 748	59 50
20	·086 237	172	·089 492	174	·910 508	·996 746	40
30	·086 409	171	·089 666	173	·910 334	·996 743	30
40	·086 580	171	·089 839	174	·910 161	·996 740	20
50	·086 751	171	·090 013	174	·909 987	·996 738	10
01 00	9·086 922	171	9·090 187	174	0·909 813	9·996 735	59 00
10	·087 093	171	·090 361	173	·909 639	·996 733	58 50
20	·087 264	171	·090 534	174	·909 466	·996 730	40
30	·087 435	171	·090 708	173	·909 292	·996 727	30
40	·087 606	171	·090 881	173	·909 119	·996 725	20
50	·087 777	170	·091 054	174	·908 946	·996 722	10
02 00	9·087 947	171	9·091 228	173	0·908 772	9·996 720	58 00
10	·088 118	170	·091 401	173	·908 599	·996 717	57 50
20	·088 288	171	·091 574	173	·908 426	·996 714	40
30	·088 459	170	·091 747	173	·908 253	·996 712	30
40	·088 629	171	·091 920	173	·908 080	·996 709	20
50	·088 800	170	·092 093	173	·907 907	·996 707	10
03 00	9·088 970	170	9·092 266	173	0·907 734	9·996 704	57 00
10	·089 140	170	·092 439	173	·907 561	·996 701	56 50
20	·089 310	170	·092 612	172	·907 388	·996 699	40
30	·089 480	171	·092 784	173	·907 216	·996 696	30
40	·089 651	169	·092 957	172	·907 043	·996 694	20
50	·089 820	170	·093 129	173	·906 871	·996 691	10
04 00	9·089 990	170	9·093 302	172	0·906 698	9·996 688	56 00
10	·090 160	170	·093 474	173	·906 526	·996 686	55 50
20	·090 330	170	·093 647	172	·906 353	·996 683	40
30	·090 500	169	·093 819	172	·906 181	·996 681	30
40	·090 669	170	·093 991	172	·906 009	·996 678	20
50	·090 839	169	·094 163	173	·905 837	·996 675	10
05 00	9·091 008	170	9·094 336	172	0·905 664	9·996 673	55 00
10	·091 178	169	·094 508	172	·905 492	·996 670	54 50
20	·091 347	169	·094 680	171	·905 320	·996 667	40
30	·091 516	169	·094 851	172	·905 149	·996 665	30
40	·091 685	170	·095 023	172	·904 977	·996 662	20
50	·091 855	169	·095 195	172	·904 805	·996 660	10
06 00	9·092 024	169	9·095 367	171	0·904 633	9·996 657	54 00
10	·092 193	169	·095 538	172	·904 462	·996 654	53 50
20	·092 362	168	·095 710	171	·904 290	·996 652	40
30	·092 530	169	·095 881	172	·904 119	·996 649	30
40	·092 699	169	·096 053	171	·903 947	·996 646	20
50	·092 868	169	·096 224	171	·903 776	·996 644	10
07 00	9·093 037	168	9·096 395	172	0·903 605	9·996 641	53 00
10	·093 205	169	·096 567	171	·903 433	·996 639	52 50
20	·093 374	168	·096 738	171	·903 262	·996 636	40
30	·093 542	169	·096 909	171	·903 091	·996 633	30
40	·093 711	168	·097 080	171	·902 920	·996 631	20
50	·093 879	168	·097 251	171	·902 749	·996 628	10
08 00	9·094 047	169	9·097 422	171	0·902 578	9·996 625	52 00
10	·094 216	168	·097 593	171	·902 407	·996 623	51 50
20	·094 384	168	·097 764	170	·902 236	·996 620	40
30	·094 552	168	·097 934	171	·902 066	·996 618	30
40	·094 720	168	·098 105	171	·901 895	·996 615	20
50	·094 888	168	·098 276	170	·901 724	·996 612	10
09 00	9·095 056	167	9·098 446	170	0·901 554	9·996 610	51 00
10	·095 223	168	·098 616	171	·901 384	·996 607	50 50
20	·095 391	168	·098 787	170	·901 213	·996 604	40
30	·095 559	167	·098 957	170	·901 043	·996 602	30
40	·095 726	168	·099 127	171	·900 873	·996 599	20
50	·095 894	168	·099 298	170	·900 702	·996 596	10
10 00	9·096 062		9·099 468		0·900 532	9·996 594	50 00
	cos		cot		tan	sin	82°

PROPORTIONAL PARTS

″	167	168	169	170	171	172	173	174
1	16·7	16·8	16·9	17·0	17·1	17·2	17·3	17·4
2	33·4	33·6	33·8	34·0	34·2	34·4	34·6	34·8
3	50·1	50·4	50·7	51·0	51·3	51·6	51·9	52·2
4	66·8	67·2	67·6	68·0	68·4	68·8	69·2	69·6
5	83·5	84·0	84·5	85·0	85·5	86·0	86·5	87·0
6	100·2	100·8	101·4	102·0	102·6	103·2	103·8	104·4
7	116·9	117·6	118·3	119·0	119·7	120·4	121·1	121·8
8	133·6	134·4	135·2	136·0	136·8	137·6	138·4	139·2
9	150·3	151·2	152·1	153·0	153·9	154·8	155·7	156·6

TABLE IIc—LOGARITHMS OF TRIGONOMETRICAL FUNCTIONS

(Degrees, Minutes and Seconds)

7°

′ ″	sin		tan		cot	cos	′ ″
10 00	9·096 062	167	9·099 468	170	0·900 532	9·996 594	**50 00**
10	·096 229	167	·099 638	170	·900 362	·996 591	**49** 50
20	·096 396	168	·099 808	170	·900 192	·996 588	40
30	·096 564	167	·099 978	170	·900 022	·996 586	30
40	·096 731	167	·100 148	169	·899 852	·996 583	20
50	·096 898	167	·100 317	170	·899 683	·996 580	10
11 00	9·097 065	167	9·100 487	170	0·899 513	9·996 578	**49 00**
10	·097 232	167	·100 657	170	·899 343	·996 575	**48** 50
20	·097 399	167	·100 827	169	·899 173	·996 573	40
30	·097 566	167	·100 996	170	·899 004	·996 570	30
40	·097 733	167	·101 166	169	·898 834	·996 567	20
50	·097 900	166	·101 335	169	·898 665	·996 565	10
12 00	9·098 066	167	9·101 504	170	0·898 496	9·996 562	**48 00**
10	·098 233	166	·101 674	169	·898 326	·996 559	**47** 50
20	·098 399	167	·101 843	169	·898 157	·996 557	40
30	·098 566	166	·102 012	169	·897 988	·996 554	30
40	·098 732	167	·102 181	169	·897 819	·996 551	20
50	·098 899	166	·102 350	169	·897 650	·996 549	10
13 00	9·099 065	166	9·102 519	169	0·897 481	9·996 546	**47 00**
10	·099 231	167	·102 688	169	·897 312	·996 543	**46** 50
20	·099 398	166	·102 857	169	·897 143	·996 541	40
30	·099 564	166	·103 026	168	·896 974	·996 538	30
40	·099 730	166	·103 194	169	·896 806	·996 535	20
50	·099 896	166	·103 363	169	·896 637	·996 533	10
14 00	9·100 062	165	9·103 532	168	0·896 468	9·996 530	**46 00**
10	·100 227	166	·103 700	169	·896 300	·996 527	**45** 50
20	·100 393	166	·103 869	168	·896 131	·996 525	40
30	·100 559	166	·104 037	168	·895 963	·996 522	30
40	·100 725	165	·104 205	169	·895 795	·996 519	20
50	·100 890	166	·104 374	168	·895 626	·996 517	10
15 00	9·101 056	165	9·104 542	168	0·895 458	9·996 514	**45 00**
10	·101 221	166	·104 710	168	·895 290	·996 511	**44** 50
20	·101 387	165	·104 878	168	·895 122	·996 508	40
30	·101 552	165	·105 046	168	·894 954	·996 506	30
40	·101 717	166	·105 214	168	·894 786	·996 503	20
50	·101 883	165	·105 382	168	·894 618	·996 500	10
16 00	9·102 048	165	9·105 550	168	0·894 450	9·996 498	**44 00**
10	·102 213	165	·105 718	167	·894 282	·996 495	**43** 50
20	·102 378	165	·105 885	168	·894 115	·996 492	40
30	·102 543	165	·106 053	168	·893 947	·996 490	30
40	·102 708	165	·106 221	167	·893 779	·996 487	20
50	·102 873	164	·106 388	168	·893 612	·996 484	10
17 00	9·103 037	165	9·106 556	167	0·893 444	9·996 482	**43 00**
10	·103 202	165	·106 723	167	·893 277	·996 479	**42** 50
20	·103 367	164	·106 890	168	·893 110	·996 476	40
30	·103 531	165	·107 058	167	·892 942	·996 474	30
40	·103 696	164	·107 225	167	·892 775	·996 471	20
50	·103 860	165	·107 392	167	·892 608	·996 468	10
18 00	9·104 025	164	9·107 559	167	0·892 441	9·996 465	**42 00**
10	·104 189	164	·107 726	167	·892 274	·996 463	**41** 50
20	·104 353	164	·107 893	167	·892 107	·996 460	40
30	·104 517	165	·108 060	167	·891 940	·996 457	30
40	·104 682	164	·108 227	167	·891 773	·996 455	20
50	·104 846	164	·108 394	166	·891 606	·996 452	10
19 00	9·105 010	164	9·108 560	167	0·891 440	9·996 449	**41 00**
10	·105 174	163	·108 727	167	·891 273	·996 447	**40** 50
20	·105 337	164	·108 894	166	·891 106	·996 444	40
30	·105 501	164	·109 060	167	·890 940	·996 441	30
40	·105 665	164	·109 227	166	·890 773	·996 438	20
50	·105 829	163	·109 393	166	·890 607	·996 436	10
20 00	9·105 992		9·109 559		0·890 441	9·996 433	**40 00**
	cos		cot		tan	sin	**82°**

PROPORTIONAL PARTS

″	163	″	164
1	16·3	1	16·4
2	32·6	2	32·8
3	48·9	3	49·2
4	65·2	4	65·6
5	81·5	5	82·0
6	97·8	6	98·4
7	114·1	7	114·8
8	130·4	8	131·2
9	146·7	9	147·6

″	165	″	166
1	16·5	1	16·6
2	33·0	2	33·2
3	49·5	3	49·8
4	66·0	4	66·4
5	82·5	5	83·0
6	99·0	6	99·6
7	115·5	7	116·2
8	132·0	8	132·8
9	148·5	9	149·4

″	167	″	168
1	16·7	1	16·8
2	33·4	2	33·6
3	50·1	3	50·4
4	66·8	4	67·2
5	83·5	5	84·0
6	100·2	6	100·8
7	116·9	7	117·6
8	133·6	8	134·4
9	150·3	9	151·2

″	169	″	170
1	16·9	1	17·0
2	33·8	2	34·0
3	50·7	3	51·0
4	67·6	4	68·0
5	84·5	5	85·0
6	101·4	6	102·0
7	118·3	7	119·0
8	135·2	8	136·0
9	152·1	9	153·0

TABLE IIc—LOGARITHMS OF TRIGONOMETRICAL FUNCTIONS

(Degrees, Minutes and Seconds)

7°

′ ″	sin	d	tan	d	cot	cos	′ ″
20 00	9·105 992	164	9·109 559	167	0·890 441	9·996 433	**40 00**
10	·106 156	163	·109 726	166	·890 274	·996 430	**39 50**
20	·106 319	164	·109 892	166	·890 108	·996 428	**40**
30	·106 483	163	·110 058	166	·889 942	·996 425	**30**
40	·106 646	164	·110 224	166	·889 776	·996 422	**20**
50	·106 810	163	·110 390	166	·889 610	·996 419	**10**
21 00	9·106 973	163	9·110 556	166	0·889 444	9·996 417	**39 00**
10	·107 136	163	·110 722	166	·889 278	·996 414	**38 50**
20	·107 299	163	·110 888	166	·889 112	·996 411	**40**
30	·107 462	163	·111 054	165	·888 946	·996 409	**30**
40	·107 625	163	·111 219	166	·888 781	·996 406	**20**
50	·107 788	163	·111 385	166	·888 615	·996 403	**10**
22 00	9·107 951	163	9·111 551	165	0·888 449	9·996 400	**38 00**
10	·108 114	163	·111 716	166	·888 284	·996 398	**37 50**
20	·108 277	162	·111 882	165	·888 118	·996 395	**40**
30	·108 439	163	·112 047	166	·887 953	·996 392	**30**
40	·108 602	163	·112 213	165	·887 787	·996 390	**20**
50	·108 765	162	·112 378	165	·887 622	·996 387	**10**
23 00	9·108 927	163	9·112 543	165	0·887 457	9·996 384	**37 00**
10	·109 090	162	·112 708	165	·887 292	·996 381	**36 50**
20	·109 252	162	·112 873	166	·887 127	·996 379	**40**
30	·109 414	163	·113 039	165	·886 961	·996 376	**30**
40	·109 577	162	·113 204	164	·886 796	·996 373	**20**
50	·109 739	162	·113 368	165	·886 632	·996 370	**10**
24 00	9·109 901	162	9·113 533	165	0·886 467	9·996 368	**36 00**
10	·110 063	162	·113 698	165	·886 302	·996 365	**35 50**
20	·110 225	162	·113 863	165	·886 137	·996 362	**40**
30	·110 387	162	·114 028	164	·885 972	·996 359	**30**
40	·110 549	162	·114 192	165	·885 808	·996 357	**20**
50	·110 711	162	·114 357	164	·885 643	·996 354	**10**
25 00	9·110 873	161	9·114 521	165	0·885 479	9·996 351	**35 00**
10	·111 034	162	·114 686	164	·885 314	·996 349	**34 50**
20	·111 196	162	·114 850	165	·885 150	·996 346	**40**
30	·111 358	161	·115 015	164	·884 985	·996 343	**30**
40	·111 519	162	·115 179	164	·884 821	·996 340	**20**
50	·111 681	161	·115 343	164	·884 657	·996 338	**10**
26 00	9·111 842	161	9·115 507	164	0·884 493	9·996 335	**34 00**
10	·112 003	162	·115 671	164	·884 329	·996 332	**33 50**
20	·112 165	161	·115 835	164	·884 165	·996 329	**40**
30	·112 326	161	·115 999	164	·884 001	·996 327	**30**
40	·112 487	161	·116 163	164	·883 837	·996 324	**20**
50	·112 648	161	·116 327	164	·883 673	·996 321	**10**
27 00	9·112 809	161	9·116 491	164	0·883 509	9·996 318	**33 00**
10	·112 970	161	·116 655	163	·883 345	·996 316	**32 50**
20	·113 131	161	·116 818	164	·883 182	·996 313	**40**
30	·113 292	161	·116 982	163	·883 018	·996 310	**30**
40	·113 453	160	·117 145	164	·882 855	·996 307	**20**
50	·113 613	161	·117 309	163	·882 691	·996 305	**10**
28 00	9·113 774	161	9·117 472	164	0·882 528	9·996 302	**32 00**
10	·113 935	160	·117 636	163	·882 364	·996 299	**31 50**
20	·114 095	161	·117 799	163	·882 201	·996 296	**40**
30	·114 256	160	·117 962	164	·882 038	·996 293	**30**
40	·114 416	161	·118 126	163	·881 874	·996 291	**20**
50	·114 577	160	·118 289	163	·881 711	·996 288	**10**
29 00	9·114 737	160	9·118 452	163	0·881 548	9·996 285	**31 00**
10	·114 897	160	·118 615	163	·881 385	·996 282	**30 50**
20	·115 057	161	·118 778	163	·881 222	·996 280	**40**
30	·115 218	160	·118 941	163	·881 059	·996 277	**30**
40	·115 378	160	·119 104	162	·880 896	·996 274	**20**
50	·115 538	160	·119 266	163	·880 734	·996 271	**10**
30 00	9·115 698		9·119 429		0·880 571	9·996 269	**30 00**
	cos		cot		tan	sin	**82°**

PROPORTIONAL PARTS

″	**160**	″	**161**
1	16·0	**1**	16·1
2	32·0	**2**	32·2
3	48·0	**3**	48·3
4	64·0	**4**	64·4
5	80·0	**5**	80·5
6	96·0	**6**	96·6
7	112·0	**7**	112·7
8	128·0	**8**	128·8
9	144·0	**9**	144·9

″	**162**	″	**163**
1	16·2	**1**	16·3
2	32·4	**2**	32·6
3	48·6	**3**	48·9
4	64·8	**4**	65·2
5	81·0	**5**	81·5
6	97·2	**6**	97·8
7	113·4	**7**	114·1
8	129·6	**8**	130·4
9	145·8	**9**	146·7

″	**164**	″	**165**
1	16·4	**1**	16·5
2	32·8	**2**	33·0
3	49·2	**3**	49·5
4	65·6	**4**	66·0
5	82·0	**5**	82·5
6	98·4	**6**	99·0
7	114·8	**7**	115·5
8	131·2	**8**	132·0
9	147·6	**9**	148·5

″	**166**	″	**167**
1	16·6	**1**	16·7
2	33·2	**2**	33·4
3	49·8	**3**	50·1
4	66·4	**4**	66·8
5	83·0	**5**	83·5
6	99·6	**6**	100·2
7	116·2	**7**	116·9
8	132·8	**8**	133·6
9	149·4	**9**	150·3

7°

′ ″	sin	d	tan	d	cot	cos	′ ″
30 00	9·115 698	160	9·119 429	163	0·880 571	9·996 269	30 00
10	·115 858	159	·119 592	162	·880 408	·996 266	29 50
20	·116 017	160	·119 754	163	·880 246	·996 263	40
30	·116 177	160	·119 917	162	·880 083	·996 260	30
40	·116 337	160	·120 079	163	·879 921	·996 257	20
50	·116 497	159	·120 242	162	·879 758	·996 255	10
31 00	9·116 656	160	9·120 404	163	0·879 596	9·996 252	29 00
10	·116 816	159	·120 567	162	·879 433	·996 249	28 50
20	·116 975	160	·120 729	162	·879 271	·996 246	40
30	·117 135	159	·120 891	162	·879 109	·996 244	30
40	·117 294	159	·121 053	162	·878 947	·996 241	20
50	·117 453	160	·121 215	162	·878 785	·996 238	10
32 00	9·117 613	159	9·121 377	162	0·878 623	9·996 235	28 00
10	·117 772	159	·121 539	162	·878 461	·996 232	27 50
20	·117 931	159	·121 701	162	·878 299	·996 230	40
30	·118 090	159	·121 863	162	·878 137	·996 227	30
40	·118 249	159	·122 025	162	·877 975	·996 224	20
50	·118 408	159	·122 187	161	·877 813	·996 221	10
33 00	9·118 567	159	9·122 348	162	0·877 652	9·996 219	27 00
10	·118 726	158	·122 510	161	·877 490	·996 216	26 50
20	·118 884	159	·122 671	162	·877 329	·996 213	40
30	·119 043	159	·122 833	161	·877 167	·996 210	30
40	·119 202	158	·122 994	162	·877 006	·996 207	20
50	·119 360	159	·123 156	161	·876 844	·996 205	10
34 00	9·119 519	158	9·123 317	161	0·876 683	9·996 202	26 00
10	·119 677	159	·123 478	162	·876 522	·996 199	25 50
20	·119 836	158	·123 640	161	·876 360	·996 196	40
30	·119 994	158	·123 801	161	·876 199	·996 193	30
40	·120 152	159	·123 962	161	·876 038	·996 191	20
50	·120 311	158	·124 123	161	·875 877	·996 188	10
35 00	9·120 469	158	9·124 284	161	0·875 716	9·996 185	25 00
10	·120 627	158	·124 445	161	·875 555	·996 182	24 50
20	·120 785	158	·124 606	160	·875 394	·996 179	40
30	·120 943	158	·124 766	161	·875 234	·996 177	30
40	·121 101	158	·124 927	161	·875 073	·996 174	20
50	·121 259	158	·125 088	161	·874 912	·996 171	10
36 00	9·121 417	157	9·125 249	160	0·874 751	9·996 168	24 00
10	·121 574	158	·125 409	161	·874 591	·996 165	23 50
20	·121 732	158	·125 570	160	·874 430	·996 162	40
30	·121 890	157	·125 730	161	·874 270	·996 160	30
40	·122 047	158	·125 891	160	·874 109	·996 157	20
50	·122 205	157	·126 051	160	·873 949	·996 154	10
37 00	9·122 362	158	9·126 211	160	0·873 789	9·996 151	23 00
10	·122 520	157	·126 371	161	·873 629	·996 148	22 50
20	·122 677	158	·126 532	160	·873 468	·996 146	40
30	·122 835	157	·126 692	160	·873 308	·996 143	30
40	·122 992	157	·126 852	160	·873 148	·996 140	20
50	·123 149	157	·127 012	160	·872 988	·996 137	10
38 00	9·123 306	157	9·127 172	160	0·872 828	9·996 134	22 00
10	·123 463	157	·127 332	160	·872 668	·996 131	21 50
20	·123 620	157	·127 492	159	·872 508	·996 129	40
30	·123 777	157	·127 651	160	·872 349	·996 126	30
40	·123 934	157	·127 811	160	·872 189	·996 123	20
50	·124 091	157	·127 971	159	·872 029	·996 120	10
39 00	9·124 248	156	9·128 130	160	0·871 870	9·996 117	21 00
10	·124 404	157	·128 290	159	·871 710	·996 115	20 50
20	·124 561	157	·128 449	160	·871 551	·996 112	40
30	·124 718	156	·128 609	159	·871 391	·996 109	30
40	·124 874	157	·128 768	160	·871 232	·996 106	20
50	·125 031	156	·128 928	159	·871 072	·996 103	10
40 00	9·125 187		9·129 087		0·870 913	9·996 100	20 00
	cos		cot		tan	sin	82°

PROPORTIONAL PARTS

″	156	″	157
1	15·6	1	15·7
2	31·2	2	31·4
3	46·8	3	47·1
4	62·4	4	62·8
5	78·0	5	78·5
6	93·6	6	94·2
7	109·2	7	109·9
8	124·8	8	125·6
9	140·4	9	141·3

″	158	″	159
1	15·8	1	15·9
2	31·6	2	31·8
3	47·4	3	47·7
4	63·2	4	63·6
5	79·0	5	79·5
6	94·8	6	95·4
7	110·6	7	111·3
8	126·4	8	127·2
9	142·2	9	143·1

″	160	″	161
1	16·0	1	16·1
2	32·0	2	32·2
3	48·0	3	48·3
4	64·0	4	64·4
5	80·0	5	80·5
6	96·0	6	96·6
7	112·0	7	112·7
8	128·0	8	128·8
9	144·0	9	144·9

″	162	″	163
1	16·2	1	16·3
2	32·4	2	32·6
3	48·6	3	48·9
4	64·8	4	65·2
5	81·0	5	81·5
6	97·2	6	97·8
7	113·4	7	114·1
8	129·6	8	130·4
9	145·8	9	146·7

TABLE IIc—LOGARITHMS OF TRIGONOMETRICAL FUNCTIONS
(Degrees, Minutes and Seconds)

7°

′ ″	sin		tan		cot	cos	′ ″
40 00	9·125 187	157	9·129 087	159	0·870 913	9·996 100	20 00
10	·125 344	156	·129 246	159	·870 754	·996 098	19 50
20	·125 500	156	·129 405	159	·870 595	·996 095	40
30	·125 656	156	·129 564	159	·870 436	·996 092	30
40	·125 812	157	·129 723	159	·870 277	·996 089	20
50	·125 969	156	·129 882	159	·870 118	·996 086	10
41 00	9·126 125	156	9·130 041	159	0·869 959	9·996 083	19 00
10	·126 281	156	·130 200	159	·869 800	·996 081	18 50
20	·126 437	156	·130 359	159	·869 641	·996 078	40
30	·126 593	155	·130 518	158	·869 482	·996 075	30
40	·126 748	156	·130 676	159	·869 324	·996 072	20
50	·126 904	156	·130 835	159	·869 165	·996 069	10
42 00	9·127 060	156	9·130 994	158	0·869 006	9·996 066	18 00
10	·127 216	155	·131 152	159	·868 848	·996 063	17 50
20	·127 371	156	·131 311	158	·868 689	·996 061	40
30	·127 527	155	·131 469	159	·868 531	·996 058	30
40	·127 682	156	·131 628	158	·868 372	·996 055	20
50	·127 838	155	·131 786	158	·868 214	·996 052	10
43 00	9·127 993	156	9·131 944	158	0·868 056	9·996 049	17 00
10	·128 149	155	·132 102	159	·867 898	·996 046	16 50
20	·128 304	155	·132 261	158	·867 739	·996 043	40
30	·128 459	155	·132 419	158	·867 581	·996 041	30
40	·128 614	156	·132 577	158	·867 423	·996 038	20
50	·128 770	155	·132 735	158	·867 265	·996 035	10
44 00	9·128 925	155	9·132 893	157	0·867 107	9·996 032	16 00
10	·129 080	155	·133 050	158	·866 950	·996 029	15 50
20	·129 235	155	·133 208	158	·866 792	·996 026	40
30	·129 390	154	·133 366	158	·866 634	·996 023	30
40	·129 544	155	·133 524	157	·866 476	·996 021	20
50	·129 699	155	·133 681	158	·866 319	·996 018	10
45 00	9·129 854	155	9·133 839	158	0·866 161	9·996 015	15 00
10	·130 009	154	·133 997	157	·866 003	·996 012	14 50
20	·130 163	155	·134 154	158	·865 846	·996 009	40
30	·130 318	154	·134 312	157	·865 688	·996 006	30
40	·130 472	155	·134 469	157	·865 531	·996 003	20
50	·130 627	154	·134 626	158	·865 374	·996 001	10
46 00	9·130 781	155	9·134 784	157	0·865 216	9·995 998	14 00
10	·130 936	154	·134 941	157	·865 059	·995 995	13 50
20	·131 090	154	·135 098	157	·864 902	·995 992	40
30	·131 244	154	·135 255	157	·864 745	·995 989	30
40	·131 398	154	·135 412	157	·864 588	·995 986	20
50	·131 552	154	·135 569	157	·864 431	·995 983	10
47 00	9·131 706	154	9·135 726	157	0·864 274	9·995 980	13 00
10	·131 860	154	·135 883	157	·864 117	·995 978	12 50
20	·132 014	154	·136 040	157	·863 960	·995 975	40
30	·132 168	154	·136 197	156	·863 803	·995 972	30
40	·132 322	154	·136 353	157	·863 647	·995 969	20
50	·132 476	154	·136 510	157	·863 490	·995 966	10
48 00	9·132 630	153	9·136 667	156	0·863 333	9·995 963	12 00
10	·132 783	154	·136 823	157	·863 177	·995 960	11 50
20	·132 937	154	·136 980	156	·863 020	·995 957	40
30	·133 091	153	·137 136	156	·862 864	·995 954	30
40	·133 244	154	·137 292	157	·862 708	·995 952	20
50	·133 398	153	·137 449	156	·862 551	·995 949	10
49 00	9·133 551	153	9·137 605	156	0·862 395	9·995 946	11 00
10	·133 704	154	·137 761	157	·862 239	·995 943	10 50
20	·133 858	153	·137 918	156	·862 082	·995 940	40
30	·134 011	153	·138 074	156	·861 926	·995 937	30
40	·134 164	153	·138 230	156	·861 770	·995 934	20
50	·134 317	153	·138 386	156	·861 614	·995 931	10
50 00	9·134 470		9·138 542		0·861 458	9·995 928	10 00
	cos		cot		tan	sin	82°

PROPORTIONAL PARTS

″	153	″	154
1	15·3	1	15·4
2	30·6	2	30·8
3	45·9	3	46·2
4	61·2	4	61·6
5	76·5	5	77·0
6	91·8	6	92·4
7	107·1	7	107·8
8	122·4	8	123·2
9	137·7	9	138·6

″	155	″	156
1	15·5	1	15·6
2	31·0	2	31·2
3	46·5	3	46·8
4	62·0	4	62·4
5	77·5	5	78·0
6	93·0	6	93·6
7	108·5	7	109·2
8	124·0	8	124·8
9	139·5	9	140·4

″	157	″	158
1	15·7	1	15·8
2	31·4	2	31·6
3	47·1	3	47·4
4	62·8	4	63·2
5	78·5	5	79·0
6	94·2	6	94·8
7	109·9	7	110·6
8	125·6	8	126·4
9	141·3	9	142·2

″	159
1	15·9
2	31·8
3	47·7
4	63·6
5	79·5
6	95·4
7	111·3
8	127·2
9	143·1

7°

′ ″	sin		tan		cot	cos	′ ″
50 00	9·134 470	153	9·138 542	156	0·861 458	9·995 928	10 00
10	·134 623	153	·138 698	156	·861 302	·995 926	09 50
20	·134 776	153	·138 854	155	·861 146	·995 923	40
30	·134 929	153	·139 009	156	·860 991	·995 920	30
40	·135 082	153	·139 165	156	·860 835	·995 917	20
50	·135 235	152	·139 321	155	·860 679	·995 914	10
51 00	9·135 387	153	9·139 476	156	0·860 524	9·995 911	09 00
10	·135 540	153	·139 632	156	·860 368	·995 908	08 50
20	·135 693	152	·139 788	155	·860 212	·995 905	40
30	·135 845	153	·139 943	155	·860 057	·995 902	30
40	·135 998	152	·140 098	156	·859 902	·995 899	20
50	·136 150	153	·140 254	155	·859 746	·995 897	10
52 00	9·136 303	152	9·140 409	155	0·859 591	9·995 894	08 00
10	·136 455	152	·140 564	156	·859 436	·995 891	07 50
20	·136 607	153	·140 720	155	·859 280	·995 888	40
30	·136 760	152	·140 875	155	·859 125	·995 885	30
40	·136 912	152	·141 030	155	·858 970	·995 882	20
50	·137 064	152	·141 185	155	·858 815	·995 879	10
53 00	9·137 216	152	9·141 340	155	0·858 660	9·995 876	07 00
10	·137 368	152	·141 495	155	·858 505	·995 873	06 50
20	·137 520	152	·141 650	155	·858 350	·995 870	40
30	·137 672	152	·141 805	154	·858 195	·995 867	30
40	·137 824	152	·141 959	155	·858 041	·995 864	20
50	·137 976	152	·142 114	155	·857 886	·995 862	10
54 00	9·138 128	151	9·142 269	155	0·857 731	9·995 859	06 00
10	·138 279	152	·142 424	154	·857 576	·995 856	05 50
20	·138 431	151	·142 578	155	·857 422	·995 853	40
30	·138 582	152	·142 733	154	·857 267	·995 850	30
40	·138 734	152	·142 887	155	·857 113	·995 847	20
50	·138 886	151	·143 042	154	·856 958	·995 844	10
55 00	9·139 037	151	9·143 196	154	0·856 804	9·995 841	05 00
10	·139 188	152	·143 350	154	·856 650	·995 838	04 50
20	·139 340	151	·143 504	155	·856 496	·995 835	40
30	·139 491	151	·143 659	154	·856 341	·995 832	30
40	·139 642	151	·143 813	154	·856 187	·995 829	20
50	·139 793	151	·143 967	154	·856 033	·995 826	10
56 00	9·139 944	152	9·144 121	154	0·855 879	9·995 823	04 00
10	·140 096	151	·144 275	154	·855 725	·995 821	03 50
20	·140 247	151	·144 429	154	·855 571	·995 818	40
30	·140 398	150	·144 583	154	·855 417	·995 815	30
40	·140 548	151	·144 737	153	·855 263	·995 812	20
50	·140 699	151	·144 890	154	·855 110	·995 809	10
57 00	9·140 850	151	9·145 044	154	0·854 956	9·995 806	03 00
10	·141 001	150	·145 198	153	·854 802	·995 803	02 50
20	·141 151	151	·145 351	154	·854 649	·995 800	40
30	·141 302	151	·145 505	154	·854 495	·995 797	30
40	·141 453	150	·145 659	153	·854 341	·995 794	20
50	·141 603	151	·145 812	154	·854 188	·995 791	10
58 00	9·141 754	150	9·145 966	153	0·854 034	9·995 788	02 00
10	·141 904	151	·146 119	153	·853 881	·995 785	01 50
20	·142 055	150	·146 272	153	·853 728	·995 782	40
30	·142 205	150	·146 425	154	·853 575	·995 779	30
40	·142 355	150	·146 579	153	·853 421	·995 776	20
50	·142 505	150	·146 732	153	·853 268	·995 773	10
59 00	9·142 655	151	9·146 885	153	0·853 115	9·995 771	01 00
10	·142 806	150	·147 038	153	·852 962	·995 768	00 50
20	·142 956	150	·147 191	153	·852 809	·995 765	40
30	·143 106	150	·147 344	153	·852 656	·995 762	30
40	·143 256	149	·147 497	153	·852 503	·995 759	20
50	·143 405	150	·147 650	153	·852 350	·995 756	10
60 00	9·143 555		9·147 803		0·852 197	9·995 753	00 00
	cos		cot		tan	sin	82°

PROPORTIONAL PARTS

″	149	″	150
1	14·9	1	15·0
2	29·8	2	30·0
3	44·7	3	45·0
4	59·6	4	60·0
5	74·5	5	75·0
6	89·4	6	90·0
7	104·3	7	105·0
8	119·2	8	120·0
9	134·1	9	135·0

″	151	″	152
1	15·1	1	15·2
2	30·2	2	30·4
3	45·3	3	45·6
4	60·4	4	60·8
5	75·5	5	76·0
6	90·6	6	91·2
7	105·7	7	106·4
8	120·8	8	121·6
9	135·9	9	136·8

″	153	″	154
1	15·3	1	15·4
2	30·6	2	30·8
3	45·9	3	46·2
4	61·2	4	61·6
5	76·5	5	77·0
6	91·8	6	92·4
7	107·1	7	107·8
8	122·4	8	123·2
9	137·7	9	138·6

″	155	″	156
1	15·5	1	15·6
2	31·0	2	31·2
3	46·5	3	46·8
4	62·0	4	62·4
5	77·5	5	78·0
6	93·0	6	93·6
7	108·5	7	109·2
8	124·0	8	124·8
9	139·5	9	140·4

(Degrees, Minutes and Seconds)

8°

′ ″	sin		tan		cot	cos	′ ″
00 00	9·143 555	150	9·147 803	152	0·852 197	9·995 753	60 00
10	·143 705	150	·147 955	153	·852 045	·995 750	59 50
20	·143 855	150	·148 108	153	·851 892	·995 747	40
30	·144 005	149	·148 261	152	·851 739	·995 744	30
40	·144 154	150	·148 413	153	·851 587	·995 741	20
50	·144 304	149	·148 566	152	·851 434	·995 738	10
01 00	9·144 453	150	9·148 718	153	0·851 282	9·995 735	59 00
10	·144 603	149	·148 871	152	·851 129	·995 732	58 50
20	·144 752	150	·149 023	152	·850 977	·995 729	40
30	·144 902	149	·149 175	153	·850 825	·995 726	30
40	·145 051	149	·149 328	152	·850 672	·995 723	20
50	·145 200	149	·149 480	152	·850 520	·995 720	10
02 00	9·145 349	149	9·149 632	152	0·850 368	9·995 717	58 00
10	·145 498	150	·149 784	152	·850 216	·995 714	57 50
20	·145 648	149	·149 936	152	·850 064	·995 711	40
30	·145 797	149	·150 088	152	·849 912	·995 708	30
40	·145 946	149	·150 240	152	·849 760	·995 705	20
50	·146 095	148	·150 392	152	·849 608	·995 702	10
03 00	9·146 243	149	9·150 544	152	0·849 456	9·995 699	57 00
10	·146 392	149	·150 696	152	·849 304	·995 696	56 50
20	·146 541	149	·150 848	151	·849 152	·995 693	40
30	·146 690	149	·150 999	152	·849 001	·995 690	30
40	·146 839	148	·151 151	152	·848 849	·995 687	20
50	·146 987	149	·151 303	151	·848 697	·995 684	10
04 00	9·147 136	148	9·151 454	152	0·848 546	9·995 681	56 00
10	·147 284	149	·151 606	151	·848 394	·995 678	55 50
20	·147 433	148	·151 757	152	·848 243	·995 675	40
30	·147 581	149	·151 909	151	·848 091	·995 672	30
40	·147 730	148	·152 060	151	·847 940	·995 670	20
50	·147 878	148	·152 211	152	·847 789	·995 667	10
05 00	9·148 026	148	9·152 363	151	0·847 637	9·995 664	55 00
10	·148 174	149	·152 514	151	·847 486	·995 661	54 50
20	·148 323	148	·152 665	151	·847 335	·995 658	40
30	·148 471	148	·152 816	151	·847 184	·995 655	30
40	·148 619	148	·152 967	151	·847 033	·995 652	20
50	·148 767	148	·153 118	151	·846 882	·995 649	10
06 00	9·148 915	148	9·153 269	151	0·846 731	9·995 646	54 00
10	·149 063	148	·153 420	151	·846 580	·995 643	53 50
20	·149 211	147	·153 571	151	·846 429	·995 640	40
30	·149 358	148	·153 722	151	·846 278	·995 637	30
40	·149 506	148	·153 873	150	·846 127	·995 634	20
50	·149 654	148	·154 023	151	·845 977	·995 631	10
07 00	9·149 802	147	9·154 174	151	0·845 826	9·995 628	53 00
10	·149 949	148	·154 325	150	·845 675	·995 625	52 50
20	·150 097	147	·154 475	151	·845 525	·995 622	40
30	·150 244	148	·154 626	150	·845 374	·995 619	30
40	·150 392	147	·154 776	150	·845 224	·995 616	20
50	·150 539	147	·154 926	151	·845 074	·995 613	10
08 00	9·150 686	148	9·155 077	150	0·844 923	9·995 610	52 00
10	·150 834	147	·155 227	150	·844 773	·995 607	51 50
20	·150 981	147	·155 377	151	·844 623	·995 604	40
30	·151 128	147	·155 528	150	·844 472	·995 601	30
40	·151 275	147	·155 678	150	·844 322	·995 597	20
50	·151 422	147	·155 828	150	·844 172	·995 594	10
09 00	9·151 569	147	9·155 978	150	0·844 022	9·995 591	51 00
10	·151 716	147	·156 128	150	·843 872	·995 588	50 50
20	·151 863	147	·156 278	150	·843 722	·995 585	40
30	·152 010	147	·156 428	150	·843 572	·995 582	30
40	·152 157	147	·156 578	150	·843 422	·995 579	20
50	·152 304	147	·156 728	149	·843 272	·995 576	10
10 00	9·152 451		9·156 877		0·843 123	9·995 573	50 00
	cos		cot		tan	sin	81°

PROPORTIONAL PARTS

″	147	″	148
1	14·7	1	14·8
2	29·4	2	29·6
3	44·1	3	44·4
4	58·8	4	59·2
5	73·5	5	74·0
6	88·2	6	88·8
7	102·9	7	103·6
8	117·6	8	118·4
9	132·3	9	133·2

″	149	″	150
1	14·9	1	15·0
2	29·8	2	30·0
3	44·7	3	45·0
4	59·6	4	60·0
5	74·5	5	75·0
6	89·4	6	90·0
7	104·3	7	105·0
8	119·2	8	120·0
9	134·1	9	135·0

″	151	″	152
1	15·1	1	15·2
2	30·2	2	30·4
3	45·3	3	45·6
4	60·4	4	60·8
5	75·5	5	76·0
6	90·6	6	91·2
7	105·7	7	106·4
8	120·8	8	121·6
9	135·9	9	136·8

″	153
1	15·3
2	30·6
3	45·9
4	61·2
5	76·5
6	91·8
7	107·1
8	122·4
9	137·7

8°

′ ″	sin		tan		cot	cos	′ ″
10 00	9·152 451	146	9·156 877	150	0·843 123	9·995 573	50 00
10	·152 597	147	·157 027	150	·842 973	·995 570	49 50
20	·152 744	147	·157 177	149	·842 823	·995 567	40
30	·152 891	146	·157 326	150	·842 674	·995 564	30
40	·153 037	147	·157 476	149	·842 524	·995 561	20
50	·153 184	146	·157 625	150	·842 375	·995 558	10
11 00	9·153 330	146	9·157 775	149	0·842 225	9·995 555	49 00
10	·153 476	147	·157 924	150	·842 076	·995 552	48 50
20	·153 623	146	·158 074	149	·841 926	·995 549	40
30	·153 769	146	·158 223	149	·841 777	·995 546	30
40	·153 915	146	·158 372	149	·841 628	·995 543	20
50	·154 061	147	·158 521	150	·841 479	·995 540	10
12 00	9·154 208	146	9·158 671	149	0·841 329	9·995 537	48 00
10	·154 354	146	·158 820	149	·841 180	·995 534	47 50
20	·154 500	146	·158 969	149	·841 031	·995 531	40
30	·154 646	146	·159 118	149	·840 882	·995 528	30
40	·154 792	146	·159 267	149	·840 733	·995 525	20
50	·154 938	145	·159 416	149	·840 584	·995 522	10
13 00	9·155 083	146	9·159 565	148	0·840 435	9·995 519	47 00
10	·155 229	146	·159 713	149	·840 287	·995 516	46 50
20	·155 375	146	·159 862	149	·840 138	·995 513	40
30	·155 521	145	·160 011	149	·839 989	·995 510	30
40	·155 666	146	·160 160	148	·839 840	·995 507	20
50	·155 812	145	·160 308	149	·839 692	·995 504	10
14 00	9·155 957	146	9·160 457	148	0·839 543	9·995 501	46 00
10	·156 103	145	·160 605	149	·839 395	·995 497	45 50
20	·156 248	146	·160 754	148	·839 246	·995 494	40
30	·156 394	145	·160 902	149	·839 098	·995 491	30
40	·156 539	145	·161 051	148	·838 949	·995 488	20
50	·156 684	146	·161 199	148	·838 801	·995 485	10
15 00	9·156 830	145	9·161 347	149	0·838 653	9·995 482	45 00
10	·156 975	145	·161 496	148	·838 504	·995 479	44 50
20	·157 120	145	·161 644	148	·838 356	·995 476	40
30	·157 265	145	·161 792	148	·838 208	·995 473	30
40	·157 410	145	·161 940	148	·838 060	·995 470	20
50	·157 555	145	·162 088	148	·837 912	·995 467	10
16 00	9·157 700	145	9·162 236	148	0·837 764	9·995 464	44 00
10	·157 845	145	·162 384	148	·837 616	·995 461	43 50
20	·157 990	145	·162 532	148	·837 468	·995 458	40
30	·158 135	144	·162 680	148	·837 320	·995 455	30
40	·158 279	145	·162 828	147	·837 172	·995 452	20
50	·158 424	145	·162 975	148	·837 025	·995 449	10
17 00	9·158 569	144	9·163 123	148	0·836 877	9·995 446	43 00
10	·158 713	145	·163 271	147	·836 729	·995 442	42 50
20	·158 858	144	·163 418	148	·836 582	·995 439	40
30	·159 002	145	·163 566	147	·836 434	·995 436	30
40	·159 147	144	·163 713	148	·836 287	·995 433	20
50	·159 291	144	·163 861	147	·836 139	·995 430	10
18 00	9·159 435	145	9·164 008	148	0·835 992	9·995 427	42 00
10	·159 580	144	·164 156	147	·835 844	·995 424	41 50
20	·159 724	144	·164 303	147	·835 697	·995 421	40
30	·159 868	144	·164 450	148	·835 550	·995 418	30
40	·160 012	144	·164 598	147	·835 402	·995 415	20
50	·160 156	145	·164 745	147	·835 255	·995 412	10
19 00	9·160 301	144	9·164 892	147	0·835 108	9·995 409	41 00
10	·160 445	144	·165 039	147	·834 961	·995 406	40 50
20	·160 589	143	·165 186	147	·834 814	·995 403	40
30	·160 732	144	·165 333	147	·834 667	·995 399	30
40	·160 876	144	·165 480	147	·834 520	·995 396	20
50	·161 020	144	·165 627	147	·834 373	·995 393	10
20 00	9·161 164		9·165 774		0·834 226	9·995 390	40 00
	cos		cot		tan	sin	81°

PROPORTIONAL PARTS

″	143	″	144
1	14·3	1	14·4
2	28·6	2	28·8
3	42·9	3	43·2
4	57·2	4	57·6
5	71·5	5	72·0
6	85·8	6	86·4
7	100·1	7	100·8
8	114·4	8	115·2
9	128·7	9	129·6

″	145	″	146
1	14·5	1	14·6
2	29·0	2	29·2
3	43·5	3	43·8
4	58·0	4	58·4
5	72·5	5	73·0
6	87·0	6	87·6
7	101·5	7	102·2
8	116·0	8	116·8
9	130·5	9	131·4

″	147	″	148
1	14·7	1	14·8
2	29·4	2	29·6
3	44·1	3	44·4
4	58·8	4	59·2
5	73·5	5	74·0
6	88·2	6	88·8
7	102·9	7	103·6
8	117·6	8	118·4
9	132·3	9	133·2

″	149	″	150
1	14·9	1	15·0
2	29·8	2	30·0
3	44·7	3	45·0
4	59·6	4	60·0
5	74·5	5	75·0
6	89·4	6	90·0
7	104·3	7	105·0
8	119·2	8	120·0
9	134·1	9	135·0

TABLE IIc—LOGARITHMS OF TRIGONOMETRICAL FUNCTIONS

(Degrees, Minutes and Seconds)

8°

′	″	sin		tan		cot	cos	′	″
20	00	9·161 164	144	9·165 774	146	0·834 226	9·995 390	40	00
	10	·161 308	143	·165 920	147	·834 080	·995 387	39	50
	20	·161 451	144	·166 067	147	·833 933	·995 384		40
	30	·161 595	143	·166 214	147	·833 786	·995 381		30
	40	·161 738	144	·166 361	146	·833 639	·995 378		20
	50	·161 882	143	·166 507	147	·833 493	·995 375		10
21	00	9·162 025	144	9·166 654	146	0·833 346	9·995 372	39	00
	10	·162 169	143	·166 800	147	·833 200	·995 369	38	50
	20	·162 312	144	·166 947	146	·833 053	·995 365		40
	30	·162 456	143	·167 093	147	·832 907	·995 362		30
	40	·162 599	143	·167 240	146	·832 760	·995 359		20
	50	·162 742	143	·167 386	146	·832 614	·995 356		10
22	00	9·162 885	143	9·167 532	146	0·832 468	9·995 353	38	00
	10	·163 028	144	·167 678	147	·832 322	·995 350	37	50
	20	·163 172	143	·167 825	146	·832 175	·995 347		40
	30	·163 315	143	·167 971	146	·832 029	·995 344		30
	40	·163 458	142	·168 117	146	·831 883	·995 341		20
	50	·163 600	143	·168 263	146	·831 737	·995 338		10
23	00	9·163 743	143	9·168 409	146	0·831 591	9·995 334	37	00
	10	·163 886	143	·168 555	146	·831 445	·995 331	36	50
	20	·164 029	143	·168 701	146	·831 299	·995 328		40
	30	·164 172	142	·168 847	145	·831 153	·995 325		30
	40	·164 314	143	·168 992	146	·831 008	·995 322		20
	50	·164 457	143	·169 138	146	·830 862	·995 319		10
24	00	9·164 600	142	9·169 284	146	0·830 716	9·995 316	36	00
	10	·164 742	143	·169 430	145	·830 570	·995 313	35	50
	20	·164 885	142	·169 575	146	·830 425	·995 310		40
	30	·165 027	143	·169 721	145	·830 279	·995 307		30
	40	·165 170	142	·169 866	146	·830 134	·995 303		20
	50	·165 312	142	·170 012	145	·829 988	·995 300		10
25	00	9·165 454	143	9·170 157	146	0·829 843	9·995 297	35	00
	10	·165 597	142	·170 303	145	·829 697	·995 294	34	50
	20	·165 739	142	·170 448	145	·829 552	·995 291		40
	30	·165 881	142	·170 593	146	·829 407	·995 288		30
	40	·166 023	142	·170 739	145	·829 261	·995 285		20
	50	·166 165	142	·170 884	145	·829 116	·995 282		10
26	00	9·166 307	142	9·171 029	145	0·828 971	9·995 278	34	00
	10	·166 449	142	·171 174	145	·828 826	·995 275	33	50
	20	·166 591	142	·171 319	145	·828 681	·995 272		40
	30	·166 733	142	·171 464	145	·828 536	·995 269		30
	40	·166 875	142	·171 609	145	·828 391	·995 266		20
	50	·167 017	142	·171 754	145	·828 246	·995 263		10
27	00	9·167 159	141	9·171 899	145	0·828 101	9·995 260	33	00
	10	·167 300	142	·172 044	144	·827 956	·995 257	32	50
	20	·167 442	142	·172 188	145	·827 812	·995 253		40
	30	·167 584	141	·172 333	145	·827 667	·995 250		30
	40	·167 725	142	·172 478	145	·827 522	·995 247		20
	50	·167 867	141	·172 623	144	·827 377	·995 244		10
28	00	9·168 008	142	9·172 767	145	0·827 233	9·995 241	32	00
	10	·168 150	141	·172 912	144	·827 088	·995 238	31	50
	20	·168 291	141	·173 056	145	·826 944	·995 235		40
	30	·168 432	142	·173 201	144	·826 799	·995 232		30
	40	·168 574	141	·173 345	144	·826 655	·995 228		20
	50	·168 715	141	·173 489	145	·826 511	·995 225		10
29	00	9·168 856	141	9·173 634	144	0·826 366	9·995 222	31	00
	10	·168 997	141	·173 778	144	·826 222	·995 219	30	50
	20	·169 138	141	·173 922	145	·826 078	·995 216		40
	30	·169 279	141	·174 067	144	·825 933	·995 213		30
	40	·169 420	141	·174 211	144	·825 789	·995 210		20
	50	·169 561	141	·174 355	144	·825 645	·995 206		10
30	00	9·169 702		9·174 499		0·825 501	9·995 203	30	00
		cos		cot		tan	sin	81°	

PROPORTIONAL PARTS

″	141	″	142
1	14·1	1	14·2
2	28·2	2	28·4
3	42·3	3	42·6
4	56·4	4	56·8
5	70·5	5	71·0
6	84·6	6	85·2
7	98·7	7	99·4
8	112·8	8	113·6
9	126·9	9	127·8

″	143	″	144
1	14·3	1	14·4
2	28·6	2	28·8
3	42·9	3	43·2
4	57·2	4	57·6
5	71·5	5	72·0
6	85·8	6	86·4
7	100·1	7	100·8
8	114·4	8	115·2
9	128·7	9	129·6

″	145	″	146
1	14·5	1	14·6
2	29·0	2	29·2
3	43·5	3	43·8
4	58·0	4	58·4
5	72·5	5	73·0
6	87·0	6	87·6
7	101·5	7	102·2
8	116·0	8	116·8
9	130·5	9	131·4

″	147
1	14·7
2	29·4
3	44·1
4	58·8
5	73·5
6	88·2
7	102·9
8	117·6
9	132·3

TABLE IIc—LOGARITHMS OF TRIGONOMETRICAL FUNCTIONS

(Degrees, Minutes and Seconds)

8°

′ ″	sin	d	tan	d	cot	cos	′ ″
30 00	9·169 702	141	9·174 499	144	0·825 501	9·995 203	30 00
10	·169 843	141	·174 643	144	·825 357	·995 200	29 50
20	·169 984	141	·174 787	144	·825 213	·995 197	40
30	·170 125	140	·174 931	144	·825 069	·995 194	30
40	·170 265	141	·175 075	143	·824 925	·995 191	20
50	·170 406	141	·175 218	144	·824 782	·995 188	10
31 00	9·170 547	140	9·175 362	144	0·824 638	9·995 184	29 00
10	·170 687	141	·175 506	144	·824 494	·995 181	28 50
20	·170 828	140	·175 650	143	·824 350	·995 178	40
30	·170 968	141	·175 793	144	·824 207	·995 175	30
40	·171 109	140	·175 937	143	·824 063	·995 172	20
50	·171 249	140	·176 080	144	·823 920	·995 169	10
32 00	9·171 389	141	9·176 224	143	0·823 776	9·995 165	28 00
10	·171 530	140	·176 367	144	·823 633	·995 162	27 50
20	·171 670	140	·176 511	143	·823 489	·995 159	40
30	·171 810	140	·176 654	143	·823 346	·995 156	30
40	·171 950	140	·176 797	144	·823 203	·995 153	20
50	·172 090	140	·176 941	143	·823 059	·995 150	10
33 00	9·172 230	140	9·177 084	143	0·822 916	9·995 146	27 00
10	·172 370	140	·177 227	143	·822 773	·995 143	26 50
20	·172 510	140	·177 370	143	·822 630	·995 140	40
30	·172 650	140	·177 513	143	·822 487	·995 137	30
40	·172 790	140	·177 656	144	·822 344	·995 134	20
50	·172 930	140	·177 800	142	·822 200	·995 131	10
34 00	9·173 070	140	9·177 942	143	0·822 058	9·995 127	26 00
10	·173 210	139	·178 085	143	·821 915	·995 124	25 50
20	·173 349	140	·178 228	143	·821 772	·995 121	40
30	·173 489	140	·178 371	143	·821 629	·995 118	30
40	·173 629	139	·178 514	143	·821 486	·995 115	20
50	·173 768	140	·178 657	142	·821 343	·995 112	10
35 00	9·173 908	139	9·178 799	143	0·821 201	9·995 108	25 00
10	·174 047	140	·178 942	143	·821 058	·995 105	24 50
20	·174 187	139	·179 085	142	·820 915	·995 102	40
30	·174 326	139	·179 227	143	·820 773	·995 099	30
40	·174 465	140	·179 370	142	·820 630	·995 096	20
50	·174 605	139	·179 512	143	·820 488	·995 092	10
36 00	9·174 744	139	9·179 655	142	0·820 345	9·995 089	24 00
10	·174 883	139	·179 797	142	·820 203	·995 086	23 50
20	·175 022	139	·179 939	143	·820 061	·995 083	40
30	·175 161	139	·180 082	142	·819 918	·995 080	30
40	·175 300	139	·180 224	142	·819 776	·995 077	20
50	·175 439	139	·180 366	142	·819 634	·995 073	10
37 00	9·175 578	139	9·180 508	142	0·819 492	9·995 070	23 00
10	·175 717	139	·180 650	142	·819 350	·995 067	22 50
20	·175 856	139	·180 792	142	·819 208	·995 064	40
30	·175 995	139	·180 934	142	·819 066	·995 061	30
40	·176 134	139	·181 076	142	·818 924	·995 057	20
50	·176 273	138	·181 218	142	·818 782	·995 054	10
38 00	9·176 411	139	9·181 360	142	0·818 640	9·995 051	22 00
10	·176 550	138	·181 502	142	·818 498	·995 048	21 50
20	·176 688	139	·181 644	142	·818 356	·995 045	40
30	·176 827	139	·181 786	141	·818 214	·995 041	30
40	·176 966	138	·181 927	142	·818 073	·995 038	20
50	·177 104	138	·182 069	142	·817 931	·995 035	10
39 00	9·177 242	139	9·182 211	141	0·817 789	9·995 032	21 00
10	·177 381	138	·182 352	142	·817 648	·995 029	20 50
20	·177 519	138	·182 494	141	·817 506	·995 025	40
30	·177 657	139	·182 635	142	·817 365	·995 022	30
40	·177 796	138	·182 777	141	·817 223	·995 019	20
50	·177 934	138	·182 918	141	·817 082	·995 016	10
40 00	9·178 072		9·183 059		0·816 941	9·995 013	20 00
	cos		cot		tan	sin	81°

PROPORTIONAL PARTS

″	138	″	139
1	13·8	1	13·9
2	27·6	2	27·8
3	41·4	3	41·7
4	55·2	4	55·6
5	69·0	5	69·5
6	82·8	6	83·4
7	96·6	7	97·3
8	110·4	8	111·2
9	124·2	9	125·1

″	140	″	141
1	14·0	1	14·1
2	28·0	2	28·2
3	42·0	3	42·3
4	56·0	4	56·4
5	70·0	5	70·5
6	84·0	6	84·6
7	98·0	7	98·7
8	112·0	8	112·8
9	126·0	9	126·9

″	142	″	143
1	14·2	1	14·3
2	28·4	2	28·6
3	42·6	3	42·9
4	56·8	4	57·2
5	71·0	5	71·5
6	85·2	6	85·8
7	99·4	7	100·1
8	113·6	8	114·4
9	127·8	9	128·7

″	144
1	14·4
2	28·8
3	43·2
4	57·6
5	72·0
6	86·4
7	100·8
8	115·2
9	129·6

8°

′ ″	sin		tan		cot	cos	′ ″
40 00	9·178 072	138	9·183 059	142	0·816 941	9·995 013	20 00
10	·178 210	138	·183 201	141	·816 799	·995 009	19 50
20	·178 348	138	·183 342	141	·816 658	·995 006	40
30	·178 486	138	·183 483	142	·816 517	·995 003	30
40	·178 624	138	·183 625	141	·816 375	·995 000	20
50	·178 762	138	·183 766	141	·816 234	·994 997	10
41 00	9·178 900	138	9·183 907	141	0·816 093	9·994 993	19 00
10	·179 038	138	·184 048	141	·815 952	·994 990	18 50
20	·179 176	137	·184 189	141	·815 811	·994 987	40
30	·179 313	138	·184 330	141	·815 670	·994 984	30
40	·179 451	138	·184 471	141	·815 529	·994 980	20
50	·179 589	137	·184 612	140	·815 388	·994 977	10
42 00	9·179 726	138	9·184 752	141	0·815 248	9·994 974	18 00
10	·179 864	138	·184 893	141	·815 107	·994 971	17 50
20	·180 002	137	·185 034	141	·814 966	·994 968	40
30	·180 139	137	·185 175	140	·814 825	·994 964	30
40	·180 276	138	·185 315	141	·814 685	·994 961	20
50	·180 414	137	·185 456	141	·814 544	·994 958	10
43 00	9·180 551	138	9·185 597	140	0·814 403	9·994 955	17 00
10	·180 689	137	·185 737	141	·814 263	·994 951	16 50
20	·180 826	137	·185 878	140	·814 122	·994 948	40
30	·180 963	137	·186 018	140	·813 982	·994 945	30
40	·181 100	137	·186 158	141	·813 842	·994 942	20
50	·181 237	137	·186 299	140	·813 701	·994 938	10
44 00	9·181 374	137	9·186 439	140	0·813 561	9·994 935	16 00
10	·181 511	137	·186 579	141	·813 421	·994 932	15 50
20	·181 648	137	·186 720	140	·813 280	·994 929	40
30	·181 785	137	·186 860	140	·813 140	·994 926	30
40	·181 922	137	·187 000	140	·813 000	·994 922	20
50	·182 059	137	·187 140	140	·812 860	·994 919	10
45 00	9·182 196	137	9·187 280	140	0·812 720	9·994 916	15 00
10	·182 333	136	·187 420	140	·812 580	·994 913	14 50
20	·182 469	137	·187 560	140	·812 440	·994 909	40
30	·182 606	137	·187 700	140	·812 300	·994 906	30
40	·182 743	136	·187 840	140	·812 160	·994 903	20
50	·182 879	137	·187 980	140	·812 020	·994 900	10
46 00	9·183 016	136	9·188 120	139	0·811 880	9·994 896	14 00
10	·183 152	137	·188 259	140	·811 741	·994 893	13 50
20	·183 289	136	·188 399	140	·811 601	·994 890	40
30	·183 425	137	·188 539	139	·811 461	·994 887	30
40	·183 562	136	·188 678	140	·811 322	·994 883	20
50	·183 698	136	·188 818	140	·811 182	·994 880	10
47 00	9·183 834	137	9·188 958	139	0·811 042	9·994 877	13 00
10	·183 971	136	·189 097	139	·810 903	·994 874	12 50
20	·184 107	136	·189 236	140	·810 764	·994 870	40
30	·184 243	136	·189 376	139	·810 624	·994 867	30
40	·184 379	136	·189 515	140	·810 485	·994 864	20
50	·184 515	136	·189 655	139	·810 345	·994 861	10
48 00	9·184 651	136	9·189 794	139	0·810 206	9·994 857	12 00
10	·184 787	136	·189 933	139	·810 067	·994 854	11 50
20	·184 923	136	·190 072	140	·809 928	·994 851	40
30	·185 059	136	·190 212	139	·809 788	·994 848	30
40	·185 195	136	·190 351	139	·809 649	·994 844	20
50	·185 331	135	·190 490	139	·809 510	·994 841	10
49 00	9·185 466	136	9·190 629	139	0·809 371	9·994 838	11 00
10	·185 602	136	·190 768	139	·809 232	·994 834	10 50
20	·185 738	136	·190 907	139	·809 093	·994 831	40
30	·185 874	135	·191 046	138	·808 954	·994 828	30
40	·186 009	136	·191 184	139	·808 816	·994 825	20
50	·186 145	135	·191 323	139	·808 677	·994 821	10
50 00	9·186 280		9·191 462		0·808 538	9·994 818	10 00
	cos		cot		tan	sin	81°

PROPORTIONAL PARTS

″	135	″	136
1	13·5	1	13·6
2	27·0	2	27·2
3	40·5	3	40·8
4	54·0	4	54·4
5	67·5	5	68·0
6	81·0	6	81·6
7	94·5	7	95·2
8	108·0	8	108·8
9	121·5	9	122·4

″	137	″	138
1	13·7	1	13·8
2	27·4	2	27·6
3	41·1	3	41·4
4	54·8	4	55·2
5	68·5	5	69·0
6	82·2	6	82·8
7	95·9	7	96·6
8	109·6	8	110·4
9	123·3	9	124·2

″	139	″	140
1	13·9	1	14·0
2	27·8	2	28·0
3	41·7	3	42·0
4	55·6	4	56·0
5	69·5	5	70·0
6	83·4	6	84·0
7	97·3	7	98·0
8	111·2	8	112·0
9	125·1	9	126·0

″	141	″	142
1	14·1	1	14·2
2	28·2	2	28·4
3	42·3	3	42·6
4	56·4	4	56·8
5	70·5	5	71·0
6	84·6	6	85·2
7	98·7	7	99·4
8	112·8	8	113·6
9	126·9	9	127·8

(Degrees, Minutes and Seconds)

8°

′ ″	sin		tan		cot	cos	′ ″
50 00	9·186 280	136	9·191 462	139	0·808 538	9·994 818	**10 00**
10	·186 416	135	·191 601	138	·808 399	·994 815	**09 50**
20	·186 551	135	·191 739	139	·808 261	·994 812	40
30	·186 686	136	·191 878	139	·808 122	·994 808	30
40	·186 822	135	·192 017	138	·807 983	·994 805	20
50	·186 957	135	·192 155	139	·807 845	·994 802	10
51 00	9·187 092	136	9·192 294	138	0·807 706	9·994 798	**09 00**
10	·187 228	135	·192 432	139	·807 568	·994 795	**08 50**
20	·187 363	135	·192 571	138	·807 429	·994 792	40
30	·187 498	135	·192 709	139	·807 291	·994 789	30
40	·187 633	135	·192 848	138	·807 152	·994 785	20
50	·187 768	135	·192 986	138	·807 014	·994 782	10
52 00	9·187 903	135	9·193 124	138	0·806 876	9·994 779	**08 00**
10	·188 038	135	·193 262	139	·806 738	·994 776	**07 50**
20	·188 173	135	·193 401	138	·806 599	·994 772	40
30	·188 308	134	·193 539	138	·806 461	·994 769	30
40	·188 442	135	·193 677	138	·806 323	·994 766	20
50	·188 577	135	·193 815	138	·806 185	·994 762	10
53 00	9·188 712	135	9·193 953	138	0·806 047	9·994 759	**07 00**
10	·188 847	134	·194 091	138	·805 909	·994 756	**06 50**
20	·188 981	135	·194 229	138	·805 771	·994 752	40
30	·189 116	134	·194 367	138	·805 633	·994 749	30
40	·189 250	135	·194 505	137	·805 495	·994 746	20
50	·189 385	134	·194 642	138	·805 358	·994 743	10
54 00	9·189 519	135	9·194 780	138	0·805 220	9·994 739	**06 00**
10	·189 654	134	·194 918	138	·805 082	·994 736	**05 50**
20	·189 788	135	·195 056	137	·804 944	·994 733	40
30	·189 923	134	·195 193	138	·804 807	·994 729	30
40	·190 057	134	·195 331	137	·804 669	·994 726	20
50	·190 191	134	·195 468	138	·804 532	·994 723	10
55 00	9·190 325	135	9·195 606	137	0·804 394	9·994 720	**05 00**
10	·190 460	134	·195 743	138	·804 257	·994 716	**04 50**
20	·190 594	134	·195 881	137	·804 119	·994 713	40
30	·190 728	134	·196 018	138	·803 982	·994 710	30
40	·190 862	134	·196 156	137	·803 844	·994 706	20
50	·190 996	134	·196 293	137	·803 707	·994 703	10
56 00	9·191 130	134	9·196 430	137	0·803 570	9·994 700	**04 00**
10	·191 264	134	·196 567	138	·803 433	·994 696	**03 50**
20	·191 398	134	·196 705	137	·803 295	·994 693	40
30	·191 532	133	·196 842	137	·803 158	·994 690	30
40	·191 665	134	·196 979	137	·803 021	·994 686	20
50	·191 799	134	·197 116	137	·802 884	·994 683	10
57 00	9·191 933	133	9·197 253	137	0·802 747	9·994 680	**03 00**
10	·192 066	134	·197 390	137	·802 610	·994 676	**02 50**
20	·192 200	134	·197 527	137	·802 473	·994 673	40
30	·192 334	133	·197 664	137	·802 336	·994 670	30
40	·192 467	134	·197 801	137	·802 199	·994 667	20
50	·192 601	133	·197 938	136	·802 062	·994 663	10
58 00	9·192 734	134	9·198 074	137	0·801 926	9·994 660	**02 00**
10	·192 868	133	·198 211	137	·801 789	·994 657	**01 50**
20	·193 001	133	·198 348	136	·801 652	·994 653	40
30	·193 134	134	·198 484	137	·801 516	·994 650	30
40	·193 268	133	·198 621	137	·801 379	·994 647	20
50	·193 401	133	·198 758	136	·801 242	·994 643	10
59 00	9·193 534	133	9·198 894	137	0·801 106	9·994 640	**01 00**
10	·193 667	133	·199 031	136	·800 969	·994 637	**00 50**
20	·193 800	133	·199 167	137	·800 833	·994 633	40
30	·193 933	133	·199 304	136	·800 696	·994 630	30
40	·194 066	133	·199 440	136	·800 560	·994 627	20
50	·194 199	133	·199 576	137	·800 424	·994 623	10
60 00	9·194 332		9·199 713		0·800 287	9·994 620	**00 00**
	cos		cot		tan	sin	**81°**

PROPORTIONAL PARTS

″	133	″	134
1	13·3	1	13·4
2	26·6	2	26·8
3	39·9	3	40·2
4	53·2	4	53·6
5	66·5	5	67·0
6	79·8	6	80·4
7	93·1	7	93·8
8	106·4	8	107·2
9	119·7	9	120·6

″	135	″	136
1	13·5	1	13·6
2	27·0	2	27·2
3	40·5	3	40·8
4	54·0	4	54·4
5	67·5	5	68·0
6	81·0	6	81·6
7	94·5	7	95·2
8	108·0	8	108·8
9	121·5	9	122·4

″	137	″	138
1	13·7	1	13·8
2	27·4	2	27·6
3	41·1	3	41·4
4	54·8	4	55·2
5	68·5	5	69·0
6	82·2	6	82·8
7	95·9	7	96·6
8	109·6	8	110·4
9	123·3	9	124·2

″	139
1	13·9
2	27·8
3	41·7
4	55·6
5	69·5
6	83·4
7	97·3
8	111·2
9	125·1

TABLE IIc—LOGARITHMS OF TRIGONOMETRICAL FUNCTIONS

(Degrees, Minutes and Seconds)

9°

′ ″	sin		tan		cot	cos	′ ″
00 00	9·194 332	133	9·199 713	136	0·800 287	9·994 620	60 00
10	·194 465	133	·199 849	136	·800 151	·994 617	59 50
20	·194 598	133	·199 985	136	·800 015	·994 613	40
30	·194 731	133	·200 121	136	·799 879	·994 610	30
40	·194 864	133	·200 257	136	·799 743	·994 607	20
50	·194 997	132	·200 393	136	·799 607	·994 603	10
01 00	9·195 129	133	9·200 529	136	0·799 471	9·994 600	59 00
10	·195 262	133	·200 665	136	·799 335	·994 597	58 50
20	·195 395	132	·200 801	136	·799 199	·994 593	40
30	·195 527	133	·200 937	136	·799 063	·994 590	30
40	·195 660	132	·201 073	136	·798 927	·994 587	20
50	·195 792	133	·201 209	136	·798 791	·994 583	10
02 00	9·195 925	132	9·201 345	136	0·798 655	9·994 580	58 00
10	·196 057	132	·201 481	135	·798 519	·994 576	57 50
20	·196 189	133	·201 616	136	·798 384	·994 573	40
30	·196 322	132	·201 752	136	·798 248	·994 570	30
40	·196 454	132	·201 888	135	·798 112	·994 566	20
50	·196 586	133	·202 023	136	·797 977	·994 563	10
03 00	9·196 719	132	9·202 159	135	0·797 841	9·994 560	57 00
10	·196 851	132	·202 294	136	·797 706	·994 556	56 50
20	·196 983	132	·202 430	135	·797 570	·994 553	40
30	·197 115	132	·202 565	136	·797 435	·994 550	30
40	·197 247	132	·202 701	135	·797 299	·994 546	20
50	·197 379	132	·202 836	135	·797 164	·994 543	10
04 00	9·197 511	132	9·202 971	136	0·797 029	9·994 540	56 00
10	·197 643	132	·203 107	135	·796 893	·994 536	55 50
20	·197 775	132	·203 242	135	·796 758	·994 533	40
30	·197 907	131	·203 377	135	·796 623	·994 530	30
40	·198 038	132	·203 512	135	·796 488	·994 526	20
50	·198 170	132	·203 647	135	·796 353	·994 523	10
05 00	9·198 302	132	9·203 782	136	0·796 218	9·994 519	55 00
10	·198 434	131	·203 918	135	·796 082	·994 516	54 50
20	·198 565	132	·204 053	135	·795 947	·994 513	40
30	·198 697	131	·204 188	134	·795 812	·994 509	30
40	·198 828	132	·204 322	135	·795 678	·994 506	20
50	·198 960	131	·204 457	135	·795 543	·994 503	10
06 00	9·199 091	132	9·204 592	135	0·795 408	9·994 499	54 00
10	·199 223	131	·204 727	135	·795 273	·994 496	53 50
20	·199 354	132	·204 862	134	·795 138	·994 492	40
30	·199 486	131	·204 996	135	·795 004	·994 489	30
40	·199 617	131	·205 131	135	·794 869	·994 486	20
50	·199 748	131	·205 266	134	·794 734	·994 482	10
07 00	9·199 879	132	9·205 400	135	0·794 600	9·994 479	53 00
10	·200 011	131	·205 535	134	·794 465	·994 476	52 50
20	·200 142	131	·205 669	135	·794 331	·994 472	40
30	·200 273	131	·205 804	134	·794 196	·994 469	30
40	·200 404	131	·205 938	135	·794 062	·994 465	20
50	·200 535	131	·206 073	134	·793 927	·994 462	10
08 00	9·200 666	131	9·206 207	135	0·793 793	9·994 459	52 00
10	·200 797	131	·206 342	134	·793 658	·994 455	51 50
20	·200 928	131	·206 476	134	·793 524	·994 452	40
30	·201 059	130	·206 610	134	·793 390	·994 448	30
40	·201 189	131	·206 744	134	·793 256	·994 445	20
50	·201 320	131	·206 878	135	·793 122	·994 442	10
09 00	9·201 451	131	9·207 013	134	0·792 987	9·994 438	51 00
10	·201 582	130	·207 147	134	·792 853	·994 435	50 50
20	·201 712	131	·207 281	134	·792 719	·994 432	40
30	·201 843	130	·207 415	134	·792 585	·994 428	30
40	·201 973	131	·207 549	134	·792 451	·994 425	20
50	·202 104	130	·207 683	134	·792 317	·994 421	10
10 00	9·202 234		9·207 817		0·792 183	9·994 418	50 00
	cos		cot		tan	sin	80°

PROPORTIONAL PARTS

″	130	″	131
1	13·0	1	13·1
2	26·0	2	26·2
3	39·0	3	39·3
4	52·0	4	52·4
5	65·0	5	65·5
6	78·0	6	78·6
7	91·0	7	91·7
8	104·0	8	104·8
9	117·0	9	117·9

″	132	″	133
1	13·2	1	13·3
2	26·4	2	26·6
3	39·6	3	39·9
4	52·8	4	53·2
5	66·0	5	66·5
6	79·2	6	79·8
7	92·4	7	93·1
8	105·6	8	106·4
9	118·8	9	119·7

″	134	″	135
1	13·4	1	13·5
2	26·8	2	27·0
3	40·2	3	40·5
4	53·6	4	54·0
5	67·0	5	67·5
6	80·4	6	81·0
7	93·8	7	94·5
8	107·2	8	108·0
9	120·6	9	121·5

″	136
1	13·6
2	27·2
3	40·8
4	54·4
5	68·0
6	81·6
7	95·2
8	108·8
9	122·4

(Degrees, Minutes and Seconds)

9°

′	″	sin		tan		cot	cos	′	″
10	00	9·202 234	131	9·207 817	133	0·792 183	9·994 418	50	00
	10	·202 365	130	·207 950	134	·792 050	·994 415	49	50
	20	·202 495	131	·208 084	134	·791 916	·994 411		40
	30	·202 626	130	·208 218	134	·791 782	·994 408		30
	40	·202 756	130	·208 352	133	·791 648	·994 404		20
	50	·202 886	131	·208 485	134	·791 515	·994 401		10
11	00	9·203 017	130	9·208 619	134	0·791 381	9·994 398	49	00
	10	·203 147	130	·208 753	133	·791 247	·994 394	48	50
	20	·203 277	130	·208 886	134	·791 114	·994 391		40
	30	·203 407	130	·209 020	133	·790 980	·994 387		30
	40	·203 537	130	·209 153	134	·790 847	·994 384		20
	50	·203 667	130	·209 287	133	·790 713	·994 381		10
12	00	9·203 797	130	9·209 420	134	0·790 580	9·994 377	48	00
	10	·203 927	130	·209 554	133	·790 446	·994 374	47	50
	20	·204 057	130	·209 687	133	·790 313	·994 370		40
	30	·204 187	130	·209 820	134	·790 180	·994 367		30
	40	·204 317	130	·209 954	133	·790 046	·994 363		20
	50	·204 447	130	·210 087	133	·789 913	·994 360		10
13	00	9·204 577	129	9·210 220	133	0·789 780	9·994 357	47	00
	10	·204 706	130	·210 353	133	·789 647	·994 353	46	50
	20	·204 836	130	·210 486	133	·789 514	·994 350		40
	30	·204 966	129	·210 619	133	·789 381	·994 346		30
	40	·205 095	130	·210 752	133	·789 248	·994 343		20
	50	·205 225	129	·210 885	133	·789 115	·994 340		10
14	00	9·205 354	130	9·211 018	133	0·788 982	9·994 336	46	00
	10	·205 484	129	·211 151	133	·788 849	·994 333	45	50
	20	·205 613	130	·211 284	133	·788 716	·994 329		40
	30	·205 743	129	·211 417	133	·788 583	·994 326		30
	40	·205 872	130	·211 550	133	·788 450	·994 322		20
	50	·206 002	129	·211 683	132	·788 317	·994 319		10
15	00	9·206 131	129	9·211 815	133	0·788 185	9·994 316	45	00
	10	·206 260	129	·211 948	133	·788 052	·994 312	44	50
	20	·206 389	130	·212 081	132	·787 919	·994 309		40
	30	·206 519	129	·212 213	133	·787 787	·994 305		30
	40	·206 648	129	·212 346	132	·787 654	·994 302		20
	50	·206 777	129	·212 478	133	·787 522	·994 298		10
16	00	9·206 906	129	9·212 611	132	0·787 389	9·994 295	44	00
	10	·207 035	129	·212 743	133	·7 7 257	·994 292	43	50
	20	·207 164	129	·212 876	132	·787 124	·994 288		40
	30	·207 293	129	·213 008	133	·786 992	·994 285		30
	40	·207 422	129	·213 141	132	·786 859	·994 281		20
	50	·207 551	128	·213 273	132	·786 727	·994 278		10
17	00	9·207 679	129	9·213 405	132	0·786 595	9·994 274	43	00
	10	·207 808	129	·213 537	133	·786 463	·994 271	42	50
	20	·207 937	129	·213 670	132	·786 330	·994 267		40
	30	·208 066	128	·213 802	132	·786 198	·994 264		30
	40	·208 194	129	·213 934	132	·786 066	·994 261		20
	50	·208 323	129	·214 066	132	·785 934	·994 257		10
18	00	9·208 452	128	9·214 198	132	0·785 802	9·994 254	42	00
	10	·208 580	129	·214 330	132	·785 670	·994 250	41	50
	20	·208 709	128	·214 462	132	·785 538	·994 247		40
	30	·208 837	129	·214 594	132	·785 406	·994 243		30
	40	·208 966	128	·214 726	132	·785 274	·994 240		20
	50	·209 094	128	·214 858	131	·785 142	·994 236		10
19	00	9·209 222	129	9·214 989	132	0·785 011	9·994 233	41	00
	10	·209 351	128	·215 121	132	·784 879	·994 230	40	50
	20	·209 479	128	·215 253	132	·784 747	·994 226		40
	30	·209 607	128	·215 385	131	·784 615	·994 223		30
	40	·209 735	129	·215 516	132	·784 484	·994 219		20
	50	·209 864	128	·215 648	132	·784 352	·994 216		10
20	00	9·209 992		9·215 780		0·784 220	9·994 212	40	00
		cos		cot		tan	sin	80°	

PROPORTIONAL PARTS

″	128	″	129
1	12·8	1	12·9
2	25·6	2	25·8
3	38·4	3	38·7
4	51·2	4	51·6
5	64·0	5	64·5
6	76·8	6	77·4
7	89·6	7	90·3
8	102·4	8	103·2
9	115·2	9	116·1

″	130	″	131
1	13·0	1	13·1
2	26·0	2	26·2
3	39·0	3	39·3
4	52·0	4	52·4
5	65·0	5	65·5
6	78·0	6	78·6
7	91·0	7	91·7
8	104·0	8	104·8
9	117·0	9	117·9

″	132	″	133
1	13·2	1	13·3
2	26·4	2	26·6
3	39·6	3	39·9
4	52·8	4	53·2
5	66·0	5	66·5
6	79·2	6	79·8
7	92·4	7	93·1
8	105·6	8	106·4
9	118·8	9	119·7

″	134
1	13·4
2	26·8
3	40·2
4	53·6
5	67·0
6	80·4
7	93·8
8	107·2
9	120·6

(Degrees, Minutes and Seconds)

9°

′ ″	sin		tan		cot	cos	′ ″
20 00	9·209 992	128	9·215 780	131	0·784 220	9·994 212	**40 00**
10	·210 120	128	·215 911	132	·784 089	·994 209	**39 50**
20	·210 248	128	·216 043	131	·783 957	·994 205	**40**
30	·210 376	128	·216 174	131	·783 826	·994 202	**30**
40	·210 504	128	·216 305	132	·783 695	·994 198	**20**
50	·210 632	128	·216 437	131	·783 563	·994 195	**10**
21 00	9·210 760	128	9·216 568	132	0·783 432	9·994 191	**39 00**
10	·210 888	127	·216 700	131	·783 300	·994 188	**38 50**
20	·211 015	128	·216 831	131	·783 169	·994 184	**40**
30	·211 143	128	·216 962	131	·783 038	·994 181	**30**
40	·211 271	128	·217 093	132	·782 907	·994 178	**20**
50	·211 399	127	·217 225	131	·782 775	·994 174	**10**
22 00	9·211 526	128	9·217 356	131	0·782 644	9·994 171	**38 00**
10	·211 654	127	·217 487	131	·782 513	·994 167	**37 50**
20	·211 781	128	·217 618	131	·782 382	·994 164	**40**
30	·211 909	128	·217 749	131	·782 251	·994 160	**30**
40	·212 037	127	·217 880	131	·782 120	·994 157	**20**
50	·212 164	127	·218 011	131	·781 989	·994 153	**10**
23 00	9·212 291	128	9·218 142	131	0·781 858	9·994 150	**37 00**
10	·212 419	127	·218 273	130	·781 727	·994 146	**36 50**
20	·212 546	128	·218 403	131	·781 597	·994 143	**40**
30	·212 674	127	·218 534	131	·781 466	·994 139	**30**
40	·212 801	127	·218 665	131	·781 335	·994 136	**20**
50	·212 928	127	·218 796	130	·781 204	·994 132	**10**
24 00	9·213 055	127	9·218 926	131	0·781 074	9·994 129	**36 00**
10	·213 182	128	·219 057	131	·780 943	·994 125	**35 50**
20	·213 310	127	·219 188	130	·780 812	·994 122	**40**
30	·213 437	127	·219 318	131	·780 682	·994 118	**30**
40	·213 564	127	·219 449	130	·780 551	·994 115	**20**
50	·213 691	127	·219 579	131	·780 421	·994 111	**10**
25 00	9·213 818	127	9·219 710	130	0·780 290	9·994 108	**35 00**
10	·213 945	126	·219 840	131	·780 160	·994 104	**34 50**
20	·214 071	127	·219 971	130	·780 029	·994 101	**40**
30	·214 198	127	·220 101	130	·779 899	·994 097	**30**
40	·214 325	127	·220 231	130	·779 769	·994 094	**20**
50	·214 452	127	·220 361	131	·779 639	·994 090	**10**
26 00	9·214 579	126	9·220 492	130	0·779 508	9·994 087	**34 00**
10	·214 705	127	·220 622	130	·779 378	·994 083	**33 50**
20	·214 832	127	·220 752	130	·779 248	·994 080	**40**
30	·214 959	126	·220 882	130	·779 118	·994 076	**30**
40	·215 085	127	·221 012	130	·778 988	·994 073	**20**
50	·215 212	126	·221 142	130	·778 858	·994 069	**10**
27 00	9·215 338	127	9·221 272	130	0·778 728	9·994 066	**33 00**
10	·215 465	126	·221 402	130	·778 598	·994 062	**32 50**
20	·215 591	127	·221 532	130	·778 468	·994 059	**40**
30	·215 718	126	·221 662	130	·778 338	·994 055	**30**
40	·215 844	126	·221 792	130	·778 208	·994 052	**20**
50	·215 970	127	·221 922	130	·778 078	·994 048	**10**
28 00	9·216 097	126	9·222 052	130	0·777 948	9·994 045	**32 00**
10	·216 223	126	·222 182	129	·777 818	·994 041	**31 50**
20	·216 349	126	·222 311	130	·777 689	·994 038	**40**
30	·216 475	126	·222 441	130	·777 559	·994 034	**30**
40	·216 601	127	·222 571	129	·777 429	·994 031	**20**
50	·216 728	126	·222 700	130	·777 300	·994 027	**10**
29 00	9·216 854	126	9·222 830	129	0·777 170	9·994 024	**31 00**
10	·216 980	126	·222 959	130	·777 041	·994 020	**30 50**
20	·217 106	126	·223 089	129	·776 911	·994 017	**40**
30	·217 232	126	·223 218	130	·776 782	·994 013	**30**
40	·217 358	125	·223 348	129	·776 652	·994 010	**20**
50	·217 483	126	·223 477	130	·776 523	·994 006	**10**
30 00	9·217 609		9·223 607		0·776 393	9·994 003	**30 00**
	cos		cot		tan	sin	**80°**

PROPORTIONAL PARTS

″	125	″	126
1	12·5	1	12·6
2	25·0	2	25·2
3	37·5	3	37·8
4	50·0	4	50·4
5	62·5	5	63·0
6	75·0	6	75·6
7	87·5	7	88·2
8	100·0	8	100·8
9	112·5	9	113·4

″	127	″	128
1	12·7	1	12·8
2	25·4	2	25·6
3	38·1	3	38·4
4	50·8	4	51·2
5	63·5	5	64·0
6	76·2	6	76·8
7	88·9	7	89·6
8	101·6	8	102·4
9	114·3	9	115·2

″	129	″	130
1	12·9	1	13·0
2	25·8	2	26·0
3	38·7	3	39·0
4	51·6	4	52·0
5	64·5	5	65·0
6	77·4	6	78·0
7	90·3	7	91·0
8	103·2	8	104·0
9	116·1	9	117·0

″	131	″	132
1	13·1	1	13·2
2	26·2	2	26·4
3	39·3	3	39·6
4	52·4	4	52·8
5	65·5	5	66·0
6	78·6	6	79·2
7	91·7	7	92·4
8	104·8	8	105·6
9	117·9	9	118·8

TABLE IIc—LOGARITHMS OF TRIGONOMETRICAL FUNCTIONS

(Degrees, Minutes and Seconds)

9°

′ ″	sin	diff.	tan	diff.	cot	cos	′ ″
30 00	9·217 609	126	9·223 607	129	0·776 393	9·994 003	30 00
10	·217 735	126	·223 736	129	·776 264	·993 999	29 50
20	·217 861	126	·223 865	129	·776 135	·993 996	40
30	·217 987	125	·223 994	130	·776 006	·993 992	30
40	·218 112	126	·224 124	129	·775 876	·993 989	20
50	·218 238	125	·224 253	129	·775 747	·993 985	10
31 00	9·218 363	126	9·224 382	129	0·775 618	9·993 982	29 00
10	·218 489	126	·224 511	129	·775 489	·993 978	28 50
20	·218 615	125	·224 640	129	·775 360	·993 974	40
30	·218 740	126	·224 769	129	·775 231	·993 971	30
40	·218 866	125	·224 898	129	·775 102	·993 967	20
50	·218 991	125	·225 027	129	·774 973	·993 964	10
32 00	9·219 116	126	9·225 156	129	0·774 844	9·993 960	28 00
10	·219 242	125	·225 285	129	·774 715	·993 957	27 50
20	·219 367	125	·225 414	129	·774 586	·993 953	40
30	·219 492	126	·225 543	128	·774 457	·993 950	30
40	·219 618	125	·225 671	129	·774 329	·993 946	20
50	·219 743	125	·225 800	129	·774 200	·993 943	10
33 00	9·219 868	125	9·225 929	129	0·774 071	9·993 939	27 00
10	·219 993	125	·226 058	128	·773 942	·993 936	26 50
20	·220 118	125	·226 186	129	·773 814	·993 932	40
30	·220 243	125	·226 315	128	·773 685	·993 928	30
40	·220 368	125	·226 443	129	·773 557	·993 925	20
50	·220 493	125	·226 572	128	·773 428	·993 921	10
34 00	9·220 618	125	9·226 700	129	0·773 300	9·993 918	26 00
10	·220 743	125	·226 829	128	·773 171	·993 914	25 50
20	·220 868	125	·226 957	129	·773 043	·993 911	40
30	·220 993	125	·227 086	128	·772 914	·993 907	30
40	·221 118	124	·227 214	128	·772 786	·993 904	20
50	·221 242	125	·227 342	129	·772 658	·993 900	10
35 00	9·221 367	125	9·227 471	128	0·772 529	9·993 897	25 00
10	·221 492	124	·227 599	128	·772 401	·993 893	24 50
20	·221 616	125	·227 727	128	·772 273	·993 889	40
30	·221 741	125	·227 855	128	·772 145	·993 886	30
40	·221 866	124	·227 983	128	·772 017	·993 882	20
50	·221 990	125	·228 111	128	·771 889	·993 879	10
36 00	9·222 115	124	9·228 239	129	0·771 761	9·993 875	24 00
10	·222 239	125	·228 368	128	·771 632	·993 872	23 50
20	·222 364	124	·228 496	127	·771 504	·993 868	40
30	·222 488	124	·228 623	128	·771 377	·993 864	30
40	·222 612	125	·228 751	128	·771 249	·993 861	20
50	·222 737	124	·228 879	128	·771 121	·993 857	10
37 00	9·222 861	124	9·229 007	128	0·770 993	9·993 854	23 00
10	·222 985	124	·229 135	128	·770 865	·993 850	22 50
20	·223 109	125	·229 263	127	·770 737	·993 847	40
30	·223 234	124	·229 390	128	·770 610	·993 843	30
40	·223 358	124	·229 518	128	·770 482	·993 840	20
50	·223 482	124	·229 646	127	·770 354	·993 836	10
38 00	9·223 606	124	9·229 773	128	0·770 227	9·993 832	22 00
10	·223 730	124	·229 901	128	·770 099	·993 829	21 50
20	·223 854	124	·230 029	127	·769 971	·993 825	40
30	·223 978	124	·230 156	128	·769 844	·993 822	30
40	·224 102	124	·230 284	127	·769 716	·993 818	20
50	·224 226	123	·230 411	128	·769 589	·993 814	10
39 00	9·224 349	124	9·230 539	127	0·769 461	9·993 811	21 00
10	·224 473	124	·230 666	127	·769 334	·993 807	20 50
20	·224 597	124	·230 793	128	·769 207	·993 804	40
30	·224 721	124	·230 921	127	·769 079	·993 800	30
40	·224 845	123	·231 048	127	·768 952	·993 797	20
50	·224 968	124	·231 175	127	·768 825	·993 793	10
40 00	9·225 092		9·231 302		0·768 698	9·993 789	20 00
	cos		cot		tan	sin	80°

PROPORTIONAL PARTS

″	123	″	124
1	12·3	1	12·4
2	24·6	2	24·8
3	36·9	3	37·2
4	49·2	4	49·6
5	61·5	5	62·0
6	73·8	6	74·4
7	86·1	7	86·8
8	98·4	8	99·2
9	110·7	9	111·6

″	125	″	126
1	12·5	1	12·6
2	25·0	2	25·2
3	37·5	3	37·8
4	50·0	4	50·4
5	62·5	5	63·0
6	75·0	6	75·6
7	87·5	7	88·2
8	100·0	8	100·8
9	112·5	9	113·4

″	127	″	128
1	12·7	1	12·8
2	25·4	2	25·6
3	38·1	3	38·4
4	50·8	4	51·2
5	63·5	5	64·0
6	76·2	6	76·8
7	88·9	7	89·6
8	101·6	8	102·4
9	114·3	9	115·2

″	129	″	130
1	12·9	1	13·0
2	25·8	2	26·0
3	38·7	3	39·0
4	51·6	4	52·0
5	64·5	5	65·0
6	77·4	6	78·0
7	90·3	7	91·0
8	103·2	8	104·0
9	116·1	9	117·0

9°

′ ″	sin		tan		cot	cos	′ ″
40 00	9·225 092	123	9·231 302	128	0·768 698	9·993 789	20 00
10	·225 215	124	·231 430	127	·768 570	·993 786	19 50
20	·225 339	123	·231 557	127	·768 443	·993 782	40
30	·225 462	124	·231 684	127	·768 316	·993 779	30
40	·225 586	123	·231 811	127	·768 189	·993 775	20
50	·225 709	124	·231 938	127	·768 062	·993 771	10
41 00	9·225 833	123	9·232 065	127	0·767 935	9·993 768	19 00
10	·225 956	124	·232 192	127	·767 808	·993 764	18 50
20	·226 080	123	·232 319	127	·767 681	·993 761	40
30	·226 203	123	·232 446	127	·767 554	·993 757	30
40	·226 326	123	·232 573	126	·767 427	·993 753	20
50	·226 449	124	·232 699	127	·767 301	·993 750	10
42 00	9·226 573	123	9·232 826	127	0·767 174	9·993 746	18 00
10	·226 696	123	·232 953	127	·767 047	·993 743	17 50
20	·226 819	123	·233 080	126	·766 920	·993 739	40
30	·226 942	123	·233 206	127	·766 794	·993 735	30
40	·227 065	123	·233 333	127	·766 667	·993 732	20
50	·227 188	123	·233 460	126	·766 540	·993 728	10
43 00	9·227 311	123	9·233 586	127	0·766 414	9·993 725	17 00
10	·227 434	123	·233 713	126	·766 287	·993 721	16 50
20	·227 557	123	·233 839	127	·766 161	·993 717	40
30	·227 680	123	·233 966	126	·766 034	·993 714	30
40	·227 803	122	·234 092	127	·765 908	·993 710	20
50	·227 925	123	·234 219	126	·765 781	·993 707	10
44 00	9·228 048	123	9·234 345	126	0·765 655	9·993 703	16 00
10	·228 171	123	·234 471	127	·765 529	·993 699	15 50
20	·228 294	122	·234 598	126	·765 402	·993 696	40
30	·228 416	123	·234 724	126	·765 276	·993 692	30
40	·228 539	122	·234 850	126	·765 150	·993 689	20
50	·228 661	123	·234 976	127	·765 024	·993 685	10
45 00	9·228 784	122	9·235 103	126	0·764 897	9·993 681	15 00
10	·228 906	123	·235 229	126	·764 771	·993 678	14 50
20	·229 029	122	·235 355	126	·764 645	·993 674	40
30	·229 151	123	·235 481	126	·764 519	·993 670	30
40	·229 274	122	·235 607	126	·764 393	·993 667	20
50	·229 396	122	·235 733	126	·764 267	·993 663	10
46 00	9·229 518	123	9·235 859	126	0·764 141	9·993 660	14 00
10	·229 641	122	·235 985	126	·764 015	·993 656	13 50
20	·229 763	122	·236 111	126	·763 889	·993 652	40
30	·229 885	122	·236 237	125	·763 763	·993 649	30
40	·230 007	123	·236 362	126	·763 638	·993 645	20
50	·230 130	122	·236 488	126	·763 512	·993 641	10
47 00	9·230 252	122	9·236 614	126	0·763 386	9·993 638	13 00
10	·230 374	122	·236 740	125	·763 260	·993 634	12 50
20	·230 496	122	·236 865	126	·763 135	·993 631	40
30	·230 618	122	·236 991	126	·763 009	·993 627	30
40	·230 740	122	·237 117	125	·762 883	·993 623	20
50	·230 862	122	·237 242	126	·762 758	·993 620	10
48 00	9·230 984	122	9·237 368	125	0·762 632	9·993 616	12 00
10	·231 106	122	·237 493	126	·762 507	·993 612	11 50
20	·231 228	121	·237 619	125	·762 381	·993 609	40
30	·231 349	122	·237 744	126	·762 256	·993 605	30
40	·231 471	122	·237 870	125	·762 130	·993 601	20
50	·231 593	122	·237 995	125	·762 005	·993 598	10
49 00	9·231 715	121	9·238 120	126	0·761 880	9·993 594	11 00
10	·231 836	122	·238 246	125	·761 754	·993 591	10 50
20	·231 958	121	·238 371	125	·761 629	·993 587	40
30	·232 079	122	·238 496	125	·761 504	·993 583	30
40	·232 201	122	·238 621	126	·761 379	·993 580	20
50	·232 323	121	·238 747	125	·761 253	·993 576	10
50 00	9·232 444		9·238 872		0·761 128	9·993 572	10 00
	cos		cot		tan	sin	80°

PROPORTIONAL PARTS

″	121	″	122
1	12·1	1	12·2
2	24·2	2	24·4
3	36·3	3	36·6
4	48·4	4	48·8
5	60·5	5	61·0
6	72·6	6	73·2
7	84·7	7	85·4
8	96·8	8	97·6
9	108·9	9	109·8

″	123	″	124
1	12·3	1	12·4
2	24·6	2	24·8
3	36·9	3	37·2
4	49·2	4	49·6
5	61·5	5	62·0
6	73·8	6	74·4
7	86·1	7	86·8
8	98·4	8	99·2
9	110·7	9	111·6

″	125	″	126
1	12·5	1	12·6
2	25·0	2	25·2
3	37·5	3	37·8
4	50·0	4	50·4
5	62·5	5	63·0
6	75·0	6	75·6
7	87·5	7	88·2
8	100·0	8	100·8
9	112·5	9	113·4

″	127	″	128
1	12·7	1	12·8
2	25·4	2	25·6
3	38·1	3	38·4
4	50·8	4	51·2
5	63·5	5	64·0
6	76·2	6	76·8
7	88·9	7	89·6
8	101·6	8	102·4
9	114·3	9	115·2

(Degrees, Minutes and Seconds)

9°

′	″	sin		tan		cot	cos	′	″
50	00	9·232 444	121	9·238 872	125	0·761 128	9·993 572	10	00
	10	·232 565	122	·238 997	125	·761 003	·993 569	09	50
	20	·232 687	121	·239 122	125	·760 878	·993 565		40
	30	·232 808	122	·239 247	125	·760 753	·993 561		30
	40	·232 930	121	·239 372	125	·760 628	·993 558		20
	50	·233 051	121	·239 497	125	·760 503	·993 554		10
51	00	9·233 172	121	9·239 622	125	0·760 378	9·993 550	09	00
	10	·233 293	122	·239 747	125	·760 253	·993 547	08	50
	20	·233 415	121	·239 872	124	·760 128	·993 543		40
	30	·233 536	121	·239 996	125	·760 004	·993 539		30
	40	·233 657	121	·240 121	125	·759 879	·993 536		20
	50	·233 778	121	·240 246	125	·759 754	·993 532		10
52	00	9·233 899	121	9·240 371	124	0·759 629	9·993 528	08	00
	10	·234 020	121	·240 495	125	·759 505	·993 525	07	50
	20	·234 141	121	·240 620	125	·759 380	·993 521		40
	30	·234 262	121	·240 745	124	·759 255	·993 517		30
	40	·234 383	121	·240 869	125	·759 131	·993 514		20
	50	·234 504	121	·240 994	124	·759 006	·993 510		10
53	00	9·234 625	121	9·241 118	125	0·758 882	9·993 506	07	00
	10	·234 746	121	·241 243	124	·758 757	·993 503	06	50
	20	·234 867	120	·241 367	125	·758 633	·993 499		40
	30	·234 987	121	·241 492	124	·758 508	·993 495		30
	40	·235 108	121	·241 616	125	·758 384	·993 492		20
	50	·235 229	120	·241 741	124	·758 259	·993 488		10
54	00	9·235 349	121	9·241 865	124	0·758 135	9·993 484	06	00
	10	·235 470	121	·241 989	125	·758 011	·993 481	05	50
	20	·235 591	120	·242 114	124	·757 886	·993 477		40
	30	·235 711	121	·242 238	124	·757 762	·993 473		30
	40	·235 832	120	·242 362	124	·757 638	·993 470		20
	50	·235 952	121	·242 486	124	·757 514	·993 466		10
55	00	9·236 073	120	9·242 610	124	0·757 390	9·993 462	05	00
	10	·236 193	120	·242 734	124	·757 266	·993 459	04	50
	20	·236 313	121	·242 858	124	·757 142	·993 455		40
	30	·236 434	120	·242 982	124	·757 018	·993 451		30
	40	·236 554	120	·243 106	124	·756 894	·993 448		20
	50	·236 674	121	·243 230	124	·756 770	·993 444		10
56	00	9·236 795	120	9·243 354	124	0·756 646	9·993 440	04	00
	10	·236 915	120	·243 478	124	·756 522	·993 437	03	50
	20	·237 035	120	·243 602	124	·756 398	·993 433		40
	30	·237 155	120	·243 726	124	·756 274	·993 429		30
	40	·237 275	120	·243 850	124	·756 150	·993 426		20
	50	·237 395	120	·243 974	123	·756 026	·993 422		10
57	00	9·237 515	120	9·244 097	124	0·755 903	9·993 418	03	00
	10	·237 635	120	·244 221	124	·755 779	·993 414	02	50
	20	·237 755	120	·244 345	123	·755 655	·993 411		40
	30	·237 875	120	·244 468	124	·755 532	·993 407		30
	40	·237 995	120	·244 592	123	·755 408	·993 403		20
	50	·238 115	120	·244 715	124	·755 285	·993 400		10
58	00	9·238 235	120	9·244 839	123	0·755 161	9·993 396	02	00
	10	·238 355	119	·244 962	124	·755 038	·993 392	01	50
	20	·238 474	120	·245 086	123	·754 914	·993 389		40
	30	·238 594	120	·245 209	124	·754 791	·993 385		30
	40	·238 714	120	·245 333	123	·754 667	·993 381		20
	50	·238 834	119	·245 456	123	·754 544	·993 377		10
59	00	9·238 953	120	9·245 579	124	0·754 421	9·993 374	01	00
	10	·239 073	119	·245 703	123	·754 297	·993 370	00	50
	20	·239 192	120	·245 826	123	·754 174	·993 366		40
	30	·239 312	119	·245 949	123	·754 051	·993 363		30
	40	·239 431	120	·246 072	124	·753 928	·993 359		20
	50	·239 551	119	·246 196	123	·753 804	·993 355		10
60	00	9·239 670		9·246 319		0·753 681	9·993 351	00	00
		cos		cot		tan	sin	80°	

PROPORTIONAL PARTS

″	119	″	120
1	11·9	1	12·0
2	23·8	2	24·0
3	35·7	3	36·0
4	47·6	4	48·0
5	59·5	5	60·0
6	71·4	6	72·0
7	83·3	7	84·0
8	95·2	8	96·0
9	107·1	9	108·0

″	121	″	122
1	12·1	1	12·2
2	24·2	2	24·4
3	36·3	3	36·6
4	48·4	4	48·8
5	60·5	5	61·0
6	72·6	6	73·2
7	84·7	7	85·4
8	96·8	8	97·6
9	108·9	9	109·8

″	123	″	124
1	12·3	1	12·4
2	24·6	2	24·8
3	36·9	3	37·2
4	49·2	4	49·6
5	61·5	5	62·0
6	73·8	6	74·4
7	86·1	7	86·8
8	98·4	8	99·2
9	110·7	9	111·6

″	125
1	12·5
2	25·0
3	37·5
4	50·0
5	62·5
6	75·0
7	87·5
8	100·0
9	112·5

10°

′	sin	d	tan	d	cot	cos	d	′
00	9·239 670	716	9·246 319	738	0·753 681	9·993 351	22	60
01	·240 386	715	·247 057	737	·752 943	·993 329	22	59
02	·241 101	713	·247 794	736	·752 206	·993 307	23	58
03	·241 814	712	·248 530	734	·751 470	·993 284	22	57
04	·242 526	711	·249 264	734	·750 736	·993 262	22	56
05	·243 237	710	·249 998	732	·750 002	·993 240	23	55
06	·243 947	709	·250 730	731	·749 270	·993 217	22	54
07	·244 656	707	·251 461	730	·748 539	·993 195	23	53
08	·245 363	706	·252 191	729	·747 809	·993 172	23	52
09	·246 069	706	·252 920	728	·747 080	·993 149	22	51
10	9·246 775	703	9·253 648	726	0·746 352	9·993 127	23	50
11	·247 478	703	·254 374	726	·745 626	·993 104	23	49
12	·248 181	702	·255 100	724	·744 900	·993 081	22	48
13	·248 883	700	·255 824	723	·744 176	·993 059	23	47
14	·249 583	699	·256 547	722	·743 453	·993 036	23	46
15	·250 282	698	·257 269	721	·742 731	·993 013	23	45
16	·250 980	697	·257 990	720	·742 010	·992 990	23	44
17	·251 677	696	·258 710	719	·741 290	·992 967	23	43
18	·252 373	694	·259 429	717	·740 571	·992 944	23	42
19	·253 067	694	·260 146	717	·739 854	·992 921	23	41
20	9·253 761	692	9·260 863	715	0·739 137	9·992 898	23	40
21	·254 453	691	·261 578	714	·738 422	·992 875	23	39
22	·255 144	690	·262 292	713	·737 708	·992 852	23	38
23	·255 834	689	·263 005	712	·736 995	·992 829	23	37
24	·256 523	688	·263 717	711	·736 283	·992 806	23	36
25	·257 211	687	·264 428	710	·735 572	·992 783	24	35
26	·257 898	685	·265 138	709	·734 862	·992 759	23	34
27	·258 583	685	·265 847	708	·734 153	·992 736	23	33
28	·259 268	683	·266 555	706	·733 445	·992 713	23	32
29	·259 951	682	·267 261	706	·732 739	·992 690	24	31
30	9·260 633	681	9·267 967	704	0·732 033	9·992 666	23	30
31	·261 314	680	·268 671	704	·731 329	·992 643	24	29
32	·261 994	679	·269 375	702	·730 625	·992 619	23	28
33	·262 673	678	·270 077	702	·729 923	·992 596	24	27
34	·263 351	676	·270 779	700	·729 221	·992 572	23	26
35	·264 027	676	·271 479	699	·728 521	·992 549	24	25
36	·264 703	674	·272 178	698	·727 822	·992 525	24	24
37	·265 377	674	·272 876	697	·727 124	·992 501	23	23
38	·266 051	672	·273 573	696	·726 427	·992 478	24	22
39	·266 723	672	·274 269	695	·725 731	·992 454	24	21
40	9·267 395	670	9·274 964	694	0·725 036	9·992 430	24	20
41	·268 065	669	·275 658	693	·724 342	·992 406	24	19
42	·268 734	668	·276 351	692	·723 649	·992 382	23	18
43	·269 402	667	·277 043	691	·722 957	·992 359	24	17
44	·270 069	666	·277 734	690	·722 266	·992 335	24	16
45	·270 735	665	·278 424	689	·721 576	·992 311	24	15
46	·271 400	664	·279 113	688	·720 887	·992 287	24	14
47	·272 064	662	·279 801	687	·720 199	·992 263	24	13
48	·272 726	662	·280 488	686	·719 512	·992 239	25	12
49	·273 388	661	·281 174	684	·718 826	·992 214	24	11
50	9·274 049	659	9·281 858	684	0·718 142	9·992 190	24	10
51	·274 708	659	·282 542	683	·717 458	·992 166	24	09
52	·275 367	658	·283 225	682	·716 775	·992 142	24	08
53	·276 025	656	·283 907	681	·716 093	·992 118	25	07
54	·276 681	656	·284 588	680	·715 412	·992 093	24	06
55	·277 337	654	·285 268	679	·714 732	·992 069	25	05
56	·277 991	654	·285 947	677	·714 053	·992 044	24	04
57	·278 645	652	·286 624	677	·713 376	·992 020	24	03
58	·279 297	651	·287 301	676	·712 699	·991 996	25	02
59	·279 948	651	·287 977	675	·712 023	·991 971	24	01
60	9·280 599		9·288 652		0·711 348	9·991 947		00
	cos		cot		tan	sin		79°

PROPORTIONAL PARTS FOR SECONDS

″	22	23	24	25	26	27
6	2·2	2·3	2·4	2·5	2·6	2·7
7	2·6	2·7	2·8	2·9	3·0	3·2
8	2·9	3·1	3·2	3·3	3·5	3·6
9	3·3	3·4	3·6	3·8	3·9	4·0
10	3·7	3·8	4·0	4·2	4·3	4·5
20	7·3	7·7	8·0	8·3	8·7	9·0
30	11·0	11·5	12·0	12·5	13·0	13·5
40	14·7	15·3	16·0	16·7	17·3	18·0
50	18·3	19·2	20·0	20·8	21·7	22·5

″	595	598	601	605	608
6	59·5	59·8	60·1	60·5	60·8
7	69·4	69·8	70·1	70·6	70·9
8	79·3	79·7	80·1	80·7	81·1
9	89·2	89·7	90·2	90·8	91·2
10	99·2	99·7	100·2	100·8	101·3
20	198·3	199·3	200·3	201·7	202·7
30	297·5	299·0	300·5	302·5	304·0
40	396·7	398·7	400·7	403·3	405·3
50	495·8	498·3	500·8	504·2	506·7

″	611	615	619	622	624
6	61·1	61·5	61·9	62·2	62·4
7	71·3	71·8	72·2	72·6	72·8
8	81·5	82·0	82·5	82·9	83·2
9	91·6	92·2	92·8	93·3	93·6
10	101·8	102·5	103·2	103·7	104·0
20	203·7	205·0	206·3	207·3	208·0
30	305·5	307·5	309·5	311·0	312·0
40	407·3	410·0	412·7	414·7	416·0
50	509·2	512·5	515·8	518·3	520·0

″	627	629	631	634	636
6	62·7	62·9	63·1	63·4	63·6
7	73·2	73·4	73·6	74·0	74·2
8	83·6	83·9	84·1	84·5	84·8
9	94·0	94·4	94·6	95·1	95·4
10	104·5	104·8	105·2	105·7	106·0
20	209·0	209·7	210·3	211·3	212·0
30	313·5	314·5	315·5	317·0	318·0
40	418·0	419·3	420·7	422·7	424·0
50	522·5	524·2	525·8	528·3	530·0

″	638	640	642	644	646
6	63·8	64·0	64·2	64·4	64·6
7	74·4	74·7	74·9	75·1	75·4
8	85·1	85·3	85·6	85·9	86·1
9	95·7	96·0	96·3	96·6	96·9
10	106·3	106·7	107·0	107·3	107·7
20	212·7	213·3	214·0	214·7	215·3
30	319·0	320·0	321·0	322·0	323·0
40	425·3	426·7	428·0	429·3	430·7
50	531·7	533·3	535·0	536·7	538·3

″	649	651	654	656	658
6	64·9	65·1	65·4	65·6	65·8
7	75·7	76·0	76·3	76·5	76·8
8	86·5	86·8	87·2	87·5	87·7
9	97·4	97·6	98·1	98·4	98·7
10	108·2	108·5	109·0	109·3	109·7
20	216·3	217·0	218·0	218·7	219·3
30	324·5	325·5	327·0	328·0	329·0
40	432·7	434·0	436·0	437·3	438·7
50	540·8	542·5	545·0	546·7	548·3

PROPORTIONAL PARTS FOR SECONDS

″	659	662	664	667	669
6	65·9	66·2	66·4	66·7	66·9
7	76·9	77·2	77·5	77·8	78·0
8	87·9	88·3	88·5	88·9	89·2
9	98·8	99·3	99·6	100·0	100·4
10	109·8	110·3	110·7	111·2	111·5
20	219·7	220·7	221·3	222·3	223·0
30	329·5	331·0	332·0	333·5	334·5
40	439·3	441·3	442·7	444·7	446·0
50	549·2	551·7	553·3	555·8	557·5

″	672	674	676	679	682
6	67·2	67·4	67·6	67·9	68·2
7	78·4	78·6	78·9	79·2	79·6
8	89·6	89·9	90·1	90·5	90·9
9	100·8	101·1	101·4	101·8	102·3
10	112·0	112·3	112·7	113·2	113·7
20	224·0	224·7	225·3	226·3	227·3
30	336·0	337·0	338·0	339·5	341·0
40	448·0	449·3	450·7	452·7	454·7
50	560·0	561·7	563·3	565·8	568·3

″	685	688	691	694	697
6	68·5	68·8	69·1	69·4	69·7
7	79·9	80·3	80·6	81·0	81·3
8	91·3	91·7	92·1	92·5	92·9
9	102·8	103·2	103·6	104·1	104·6
10	114·2	114·7	115·2	115·7	116·2
20	228·3	229·3	230·3	231·3	232·3
30	342·5	344·0	345·5	347·0	348·5
40	456·7	458·7	460·7	462·7	464·7
50	570·8	573·3	575·8	578·3	580·8

″	699	703	706	709	712
6	69·9	70·3	70·6	70·9	71·2
7	81·6	82·0	82·4	82·7	83·1
8	93·2	93·7	94·1	94·5	94·9
9	104·8	105·4	105·9	106·4	106·8
10	116·5	117·2	117·7	118·2	118·7
20	233·0	234·3	235·3	236·3	237·3
30	349·5	351·5	353·0	354·5	356·0
40	466·0	468·7	470·7	472·7	474·7
50	582·5	585·8	588·3	590·8	593·3

″	715	717	720	723	726
6	71·5	71·7	72·0	72·3	72·6
7	83·4	83·6	84·0	84·4	84·7
8	95·3	95·6	96·0	96·4	96·8
9	107·2	107·6	108·0	108·4	108·9
10	119·2	119·5	120·0	120·5	121·0
20	238·3	239·0	240·0	241·0	242·0
30	357·5	358·5	360·0	361·5	363·0
40	476·7	478·0	480·0	482·0	484·0
50	595·8	597·5	600·0	602·5	605·0

″	729	731	734	736	738
6	72·9	73·1	73·4	73·6	73·8
7	85·0	85·3	85·6	85·9	86·1
8	97·2	97·5	97·9	98·1	98·4
9	109·4	109·6	110·1	110·4	110·7
10	121·5	121·8	122·3	122·7	123·0
20	243·0	243·7	244·7	245·3	246·0
30	364·5	365·5	367·0	368·0	369·0
40	486·0	487·3	489·3	490·7	492·0
50	607·5	609·2	611·7	613·3	615·0

11°

′	sin		tan		cot	cos		″
00	9·280 599	649	9·288 652	674	0·711 348	9·991 947	25	60
01	·281 248	649	·289 326	673	·710 674	·991 922	25	59
02	·281 897	647	·289 999	672	·710 001	·991 897	24	58
03	·282 544	646	·290 671	671	·709 329	·991 873	25	57
04	·283 190	646	·291 342	671	·708 658	·991 848	25	56
05	·283 836	644	·292 013	669	·707 987	·991 823	24	55
06	·284 480	644	·292 682	668	·707 318	·991 799	25	54
07	·285 124	642	·293 350	667	·706 650	·991 774	25	53
08	·285 766	642	·294 017	667	·705 983	·991 749	25	52
09	·286 408	640	·294 684	665	·705 316	·991 724	25	51
10	9·287 048	640	9·295 349	664	0·704 651	9·991 699	25	50
11	·287 688	638	·296 013	664	·703 987	·991 674	25	49
12	·288 326	638	·296 677	662	·703 323	·991 649	25	48
13	·288 964	636	·297 339	662	·702 661	·991 624	25	47
14	·289 600	636	·298 001	661	·701 999	·991 599	25	46
15	·290 236	634	·298 662	660	·701 338	·991 574	25	45
16	·290 870	634	·299 322	658	·700 678	·991 549	25	44
17	·291 504	633	·299 980	658	·700 020	·991 524	26	43
18	·292 137	631	·300 638	657	·699 362	·991 498	25	42
19	·292 768	631	·301 295	656	·698 705	·991 473	25	41
20	9·293 399	630	9·301 951	656	0·698 049	9·991 448	26	40
21	·294 029	629	·302 607	654	·697 393	·991 422	25	39
22	·294 658	628	·303 261	653	·696 739	·991 397	25	38
23	·295 286	627	·303 914	653	·696 086	·991 372	26	37
24	·295 913	626	·304 567	651	·695 433	·991 346	25	36
25	·296 539	625	·305 218	651	·694 782	·991 321	26	35
26	·297 164	624	·305 869	650	·694 131	·991 295	25	34
27	·297 788	624	·306 519	649	·693 481	·991 270	26	33
28	·298 412	622	·307 168	648	·692 832	·991 244	26	32
29	·299 034	621	·307 816	647	·692 184	·991 218	25	31
30	9·299 655	621	9·308 463	646	0·691 537	9·991 193	26	30
31	·300 276	619	·309 109	645	·690 891	·991 167	26	29
32	·300 895	619	·309 754	645	·690 246	·991 141	26	28
33	·301 514	618	·310 399	643	·689 601	·991 115	25	27
34	·302 132	616	·311 042	643	·688 958	·991 090	26	26
35	·302 748	616	·311 685	642	·688 315	·991 064	26	25
36	·303 364	615	·312 327	641	·687 673	·991 038	26	24
37	·303 979	614	·312 968	640	·687 032	·991 012	26	23
38	·304 593	614	·313 608	639	·686 392	·990 986	26	22
39	·305 207	612	·314 247	638	·685 753	·990 960	26	21
40	9·305 819	611	9·314 885	638	0·685 115	9·990 934	26	20
41	·306 430	611	·315 523	636	·684 477	·990 908	26	19
42	·307 041	609	·316 159	636	·683 841	·990 882	27	18
43	·307 650	609	·316 795	635	·683 205	·990 855	26	17
44	·308 259	608	·317 430	634	·682 570	·990 829	26	16
45	·308 867	607	·318 064	633	·681 936	·990 803	26	15
46	·309 474	606	·318 697	633	·681 303	·990 777	27	14
47	·310 080	605	·319 330	631	·680 670	·990 750	26	13
48	·310 685	604	·319 961	631	·680 039	·990 724	27	12
49	·311 289	604	·320 592	630	·679 408	·990 697	26	11
50	9·311 893	602	9·321 222	629	0·678 778	9·990 671	26	10
51	·312 495	602	·321 851	628	·678 149	·990 645	27	09
52	·313 097	601	·322 479	627	·677 521	·990 618	27	08
53	·313 698	599	·323 106	627	·676 894	·990 591	26	07
54	·314 297	600	·323 733	625	·676 267	·990 565	27	06
55	·314 897	598	·324 358	625	·675 642	·990 538	27	05
56	·315 495	597	·324 983	624	·675 017	·990 511	26	04
57	·316 092	597	·325 607	624	·674 393	·990 485	27	03
58	·316 689	595	·326 231	622	·673 769	·990 458	27	02
59	·317 284	595	·326 853	622	·673 147	·990 431	27	01
60	9·317 879		9·327 475		0·672 525	9·990 404		00
	cos		cot		tan	sin		78°

12°

′	sin	d	tan	d	cot	cos	d	′
00	9·317 879	594	9·327 475	620	0·672 525	9·990 404	26	60
01	·318 473	593	·328 095	620	·671 905	·990 378	27	59
02	·319 066	592	·328 715	619	·671 285	·990 351	27	58
03	·319 658	591	·329 334	619	·670 666	·990 324	27	57
04	·320 249	591	·329 953	617	·670 047	·990 297	27	56
05	·320 840	590	·330 570	617	·669 430	·990 270	27	55
06	·321 430	589	·331 187	616	·668 813	·990 243	28	54
07	·322 019	588	·331 803	615	·668 197	·990 215	27	53
08	·322 607	587	·332 418	615	·667 582	·990 188	27	52
09	·323 194	586	·333 033	613	·666 967	·990 161	27	51
10	9·323 780	586	9·333 646	613	0·666 354	9·990 134	27	50
11	·324 366	584	·334 259	612	·665 741	·990 107	28	49
12	·324 950	584	·334 871	611	·665 129	·990 079	27	48
13	·325 534	583	·335 482	611	·664 518	·990 052	27	47
14	·326 117	583	·336 093	609	·663 907	·990 025	28	46
15	·326 700	581	·336 702	609	·663 298	·989 997	27	45
16	·327 281	581	·337 311	608	·662 689	·989 970	28	44
17	·327 862	580	·337 919	608	·662 081	·989 942	27	43
18	·328 442	579	·338 527	606	·661 473	·989 915	28	42
19	·329 021	578	·339 133	606	·660 867	·989 887	27	41
20	9·329 599	577	9·339 739	605	0·660 261	9·989 860	28	40
21	·330 176	577	·340 344	604	·659 656	·989 832	28	39
22	·330 753	576	·340 948	604	·659 052	·989 804	27	38
23	·331 329	574	·341 552	603	·658 448	·989 777	28	37
24	·331 903	575	·342 155	602	·657 845	·989 749	28	36
25	·332 478	573	·342 757	601	·657 243	·989 721	28	35
26	·333 051	573	·343 358	600	·656 642	·989 693	28	34
27	·333 624	571	·343 958	600	·656 042	·989 665	28	33
28	·334 195	572	·344 558	599	·655 442	·989 637	27	32
29	·334 767	570	·345 157	598	·654 843	·989 610	28	31
30	9·335 337	569	9·345 755	598	0·654 245	9·989 582	29	30
31	·335 906	569	·346 353	596	·653 647	·989 553	28	29
32	·336 475	568	·346 949	596	·653 051	·989 525	28	28
33	·337 043	567	·347 545	596	·652 455	·989 497	28	27
34	·337 610	566	·348 141	594	·651 859	·989 469	28	26
35	·338 176	566	·348 735	594	·651 265	·989 441	28	25
36	·338 742	565	·349 329	593	·650 671	·989 413	28	24
37	·339 307	564	·349 922	592	·650 078	·989 385	29	23
38	·339 871	563	·350 514	592	·649 486	·989 356	28	22
39	340 434	562	·351 106	591	·648 894	·989 328	28	21
40	9·340 996	562	9·351 697	590	0·648 303	9·989 300	29	20
41	·341 558	561	·352 287	589	·647 713	·989 271	28	19
42	·342 119	560	·352 876	589	·647 124	·989 243	29	18
43	·342 679	560	·353 465	588	·646 535	·989 214	28	17
44	·343 239	558	·354 053	587	·645 947	·989 186	29	16
45	·343 797	558	·354 640	587	·645 360	·989 157	29	15
46	·344 355	557	·355 227	586	·644 773	·989 128	28	14
47	·344 912	557	·355 813	585	·644 187	·989 100	29	13
48	·345 469	555	·356 398	584	·643 602	·989 071	29	12
49	·346 024	555	·356 982	584	·643 018	·989 042	28	11
50	9·346 579	555	9·357 566	583	0·642 434	9·989 014	29	10
51	·347 134	553	·358 149	582	·641 851	·988 985	29	09
52	·347 687	553	·358 731	582	·641 269	·988 956	29	08
53	·348 240	552	·359 313	580	·640 687	·988 927	29	07
54	·348 792	551	·359 893	581	·640 107	·988 898	29	06
55	·349 343	550	·360 474	579	·639 526	·988 869	29	05
56	·349 893	550	·361 053	579	·638 947	·988 840	29	04
57	·350 443	549	·361 632	578	·638 368	·988 811	29	03
58	·350 992	548	·362 210	577	·637 790	·988 782	29	02
59	·351 540	548	·362 787	577	·637 213	·988 753	29	01
60	9·352 088		9·363 364		0·636 636	9·988 724		00
	cos		cot		tan	sin		77°

PROPORTIONAL PARTS FOR SECONDS

″	27	28	29	30	31	32
6	2·7	2·8	2·9	3·0	3·1	3·2
7	3·2	3·3	3·4	3·5	3·6	3·7
8	3·6	3·7	3·9	4·0	4·1	4·3
9	4·0	4·2	4·4	4·5	4·6	4·8
10	4·5	4·7	4·8	5·0	5·2	5·3
20	9·0	9·3	9·7	10·0	10·3	10·7
30	13·5	14·0	14·5	15·0	15·5	16·0
40	18·0	18·7	19·3	20·0	20·7	21·3
50	22·5	23·3	24·2	25·0	25·8	26·7

″	507	509	512	514	516
6	50·7	50·9	51·2	51·4	51·6
7	59·2	59·4	59·7	60·0	60·2
8	67·6	67·9	68·3	68·5	68·8
9	76·0	76·4	76·8	77·1	77·4
10	84·5	84·8	85·3	85·7	86·0
20	169·0	169·7	170·7	171·3	172·0
30	253·5	254·5	256·0	257·0	258·0
40	338·0	339·3	341·3	342·7	344·0
50	422·5	424·2	426·7	428·3	430·0

″	519	521	524	526	529
6	51·9	52·1	52·4	52·6	52·9
7	60·6	60·8	61·1	61·4	61·7
8	69·2	69·5	69·9	70·1	70·5
9	77·8	78·2	78·6	78·9	79·4
10	86·5	86·8	87·3	87·7	88·2
20	173·0	173·7	174·7	175·3	176·3
30	259·5	260·5	262·0	263·0	264·5
40	346·0	347·3	349·3	350·7	352·7
50	432·5	434·2	436·7	438·3	440·8

″	531	533	535	537	540
6	53·1	53·3	53·5	53·7	54·0
7	62·0	62·2	62·4	62·6	63·0
8	70·8	71·1	71·3	71·6	72·0
9	79·6	80·0	80·2	80·6	81·0
10	88·5	88·8	89·2	89·5	90·0
20	177·0	177·7	178·3	179·0	180·0
30	265·5	266·5	267·5	268·5	270·0
40	354·0	355·3	356·7	358·0	360·0
50	442·5	444·2	445·8	447·5	450·0

″	542	543	545	546	547
6	54·2	54·3	54·5	54·6	54·7
7	63·2	63·4	63·6	63·7	63·8
8	72·3	72·4	72·7	72·8	72·9
9	81·3	81·4	81·8	81·9	82·0
10	90·3	90·5	90·8	91·0	91·2
20	180·7	181·0	181·7	182·0	182·3
30	271·0	271·5	272·5	273·0	273·5
40	361·3	362·0	363·3	364·0	364·7
50	451·7	452·5	454·2	455·0	455·8

″	549	551	552	553	555
6	54·9	55·1	55·2	55·3	55·5
7	64·0	64·3	64·4	64·5	64·8
8	73·2	73·5	73·6	73·7	74·0
9	82·4	82·6	82·8	83·0	83·2
10	91·5	91·8	92·0	92·2	92·5
20	183·0	183·7	184·0	184·3	185·0
30	274·5	275·5	276·0	276·5	277·5
40	366·0	367·3	368·0	368·7	370·0
50	457·5	459·2	460·0	460·8	462·5

PROPORTIONAL PARTS FOR SECONDS

″	557	558	560	562	563
6	55·7	55·8	56·0	56·2	56·3
7	65·0	65·1	65·3	65·6	65·7
8	74·3	74·4	74·7	74·9	75·1
9	83·6	83·7	84·0	84·3	84·4
10	92·8	93·0	93·3	93·7	93·8
20	185·7	186·0	186·7	187·3	187·7
30	278·5	279·0	280·0	281·0	281·5
40	371·3	372·0	373·3	374·7	375·3
50	464·2	465·0	466·7	468·3	469·2

″	565	566	569	571	573
6	56·5	56·6	56·9	57·1	57·3
7	65·9	66·0	66·4	66·6	66·8
8	75·3	75·5	75·9	76·1	76·4
9	84·8	84·9	85·4	85·6	86·0
10	94·2	94·3	94·8	95·2	95·5
20	188·3	188·7	189·7	190·3	191·0
30	282·5	283·0	284·5	285·5	286·5
40	376·7	377·3	379·3	380·7	382·0
50	470·8	471·7	474·2	475·8	477·5

″	575	577	579	581	583
6	57·5	57·7	57·9	58·1	58·3
7	67·1	67·3	67·6	67·8	68·0
8	76·7	76·9	77·2	77·5	77·7
9	86·2	86·6	86·8	87·2	87·4
10	95·8	96·2	96·5	96·8	97·2
20	191·7	192·3	193·0	193·7	194·3
30	287·5	288·5	289·5	290·5	291·5
40	383·3	384·7	386·0	387·3	388·7
50	479·2	480·8	482·5	484·2	485·8

″	584	586	587	589	591
6	58·4	58·6	58·7	58·9	59·1
7	68·1	68·4	68·5	68·7	69·0
8	77·9	78·1	78·3	78·5	78·8
9	87·6	87·9	88·0	88·4	88·6
10	97·3	97·7	97·8	98·2	98·5
20	194·7	195·3	195·7	196·3	197·0
30	292·0	293·0	293·5	294·5	295·5
40	389·3	390·7	391·3	392·7	394·0
50	486·7	488·3	489·2	490·8	492·5

″	592	594	596	599	602
6	59·2	59·4	59·6	59·9	60·2
7	69·1	69·3	69·5	69·9	70·2
8	78·9	79·2	79·5	79·9	80·3
9	88·8	89·1	89·4	89·8	90·3
10	98·7	99·0	99·3	99·8	100·3
20	197·3	198·0	198·7	199·7	200·7
30	296·0	297·0	298·0	299·5	301·0
40	394·7	396·0	397·3	399·3	401·3
50	493·3	495·0	496·7	499·2	501·7

″	605	608	612	616	620
6	60·5	60·8	61·2	61·6	62·0
7	70·6	70·9	71·4	71·9	72·3
8	80·7	81·1	81·6	82·1	82·7
9	90·8	91·2	91·8	92·4	93·0
10	100·8	101·3	102·0	102·7	103·3
20	201·7	202·7	204·0	205·3	206·7
30	302·5	304·0	306·0	308·0	310·0
40	403·3	405·3	408·0	410·7	413·3
50	504·2	506·7	510·0	513·3	516·7

13°

′	sin	d	tan	d	cot	cos	d	′
00	9·352 088	547	9·363 364	576	0·636 636	9·988 724	29	60
01	·352 635	546	·363 940	575	·636 060	·988 695	29	59
02	·353 181	545	·364 515	575	·635 485	·988 666	30	58
03	·353 726	545	·365 090	574	·634 910	·988 636	29	57
04	·354 271	544	·365 664	573	·634 336	·988 607	29	56
05	·354 815	543	·366 237	573	·633 763	·988 578	30	55
06	·355 358	543	·366 810	572	·633 190	·988 548	29	54
07	·355 901	542	·367 382	571	·632 618	·988 519	30	53
08	·356 443	541	·367 953	571	·632 047	·988 489	29	52
09	·356 984	540	·368 524	570	·631 476	·988 460	30	51
10	9·357 524	540	9·369 094	569	0·630 906	9·988 430	29	50
11	·358 064	539	·369 663	569	·630 337	·988 401	30	49
12	·358 603	538	·370 232	567	·629 768	·988 371	29	48
13	·359 141	537	·370 799	568	·629 201	·988 342	30	47
14	·359 678	537	·371 367	566	·628 633	·988 312	30	46
15	·360 215	537	·371 933	566	·628 067	·988 282	30	45
16	·360 752	535	·372 499	565	·627 501	·988 252	29	44
17	·361 287	535	·373 064	565	·626 936	·988 223	30	43
18	·361 822	534	·373 629	564	·626 371	·988 193	30	42
19	·362 356	533	·374 193	563	·625 807	·988 163	30	41
20	9·362 889	533	9·374 756	563	0·625 244	9·988 133	30	40
21	·363 422	532	·375 319	562	·624 681	·988 103	30	39
22	·363 954	531	·375 881	561	·624 119	·988 073	30	38
23	·364 485	531	·376 442	561	·623 558	·988 043	30	37
24	·365 016	530	·377 003	560	·622 997	·988 013	30	36
25	·365 546	529	·377 563	559	·622 437	·987 983	30	35
26	·366 075	529	·378 122	559	·621 878	·987 953	31	34
27	·366 604	527	·378 681	558	·621 319	·987 922	30	33
28	·367 131	528	·379 239	558	·620 761	·987 892	30	32
29	·367 659	526	·379 797	557	·620 203	·987 862	30	31
30	9·368 185	526	9·380 354	556	0·619 646	9·987 832	31	30
31	·368 711	525	·380 910	556	·619 090	·987 801	30	29
32	·369 236	525	·381 466	554	·618 534	·987 771	31	28
33	·369 761	524	·382 020	555	·617 980	·987 740	30	27
34	·370 285	523	·382 575	554	·617 425	·987 710	31	26
35	·370 808	522	·383 129	553	·616 871	·987 679	30	25
36	·371 330	522	·383 682	552	·616 318	·987 649	31	24
37	·371 852	521	·384 234	552	·615 766	·987 618	30	23
38	·372 373	521	·384 786	551	·615 214	·987 588	31	22
39	·372 894	520	·385 337	551	·614 663	·987 557	31	21
40	9·373 414	519	9·385 888	550	0·614 112	9·987 526	30	20
41	·373 933	519	·386 438	549	·613 562	·987 496	31	19
42	·374 452	518	·386 987	549	·613 013	·987 465	31	18
43	·374 970	517	·387 536	548	·612 464	·987 434	31	17
44	·375 487	516	·388 084	547	·611 916	·987 403	31	16
45	·376 003	516	·388 631	547	·611 369	·987 372	31	15
46	·376 519	516	·389 178	546	·610 822	·987 341	31	14
47	·377 035	514	·389 724	546	·610 276	·987 310	31	13
48	·377 549	514	·390 270	545	·609 730	·987 279	31	12
49	·378 063	514	·390 815	545	·609 185	·987 248	31	11
50	9·378 577	512	9·391 360	543	0·608 640	9·987 217	31	10
51	·379 089	512	·391 903	544	·608 097	·987 186	31	09
52	·379 601	512	·392 447	542	·607 553	·987 155	31	08
53	·380 113	511	·392 989	542	·607 011	·987 124	32	07
54	·380 624	510	·393 531	542	·606 469	·987 092	31	06
55	·381 134	509	·394 073	541	·605 927	·987 061	31	05
56	·381 643	509	·394 614	540	·605 386	·987 030	32	04
57	·382 152	509	·395 154	540	·604 846	·986 998	31	03
58	·382 661	507	·395 694	539	·604 306	·986 967	31	02
59	·383 168	507	·396 233	538	·603 767	·986 936	32	01
60	9·383 675		9·396 771		0·603 229	9·986 904		00
	cos		cot		tan	sin		**76°**

14°

′	sin		tan		cot	cos		′
00	9·383 675	507	9·396 771	538	0·603 229	9·986 904	31	60
01	·384 182	505	·397 309	537	·602 691	·986 873	32	59
02	·384 687	505	·397 846	537	·602 154	·986 841	32	58
03	·385 192	505	·398 383	536	·601 617	·986 809	31	57
04	·385 697	504	·398 919	536	·601 081	·986 778	32	56
05	·386 201	503	·399 455	535	·600 545	·986 746	32	55
06	·386 704	503	·399 990	534	·600 010	·986 714	31	54
07	·387 207	502	·400 524	534	·599 476	·986 683	32	53
08	·387 709	501	·401 058	533	·598 942	·986 651	32	52
09	·388 210	501	·401 591	533	·598 409	·986 619	32	51
10	9·388 711	500	9·402 124	532	0·597 876	9·986 587	32	50
11	·389 211	500	·402 656	531	·597 344	·986 555	32	49
12	·389 711	499	·403 187	531	·596 813	·986 523	32	48
13	·390 210	498	·403 718	531	·596 282	·986 491	32	47
14	·390 708	498	·404 249	529	·595 751	·986 459	32	46
15	·391 206	497	·404 778	530	·595 222	·986 427	32	45
16	·391 703	496	·405 308	528	·594 692	·986 395	32	44
17	·392 199	496	·405 836	528	·594 164	·986 363	32	43
18	·392 695	496	·406 364	528	·593 636	·986 331	32	42
19	·393 191	494	·406 892	527	·593 108	·986 299	33	41
20	9·393 685	494	9·407 419	526	0·592 581	9·986 266	32	40
21	·394 179	494	·407 945	526	·592 055	·986 234	32	39
22	·394 673	493	·408 471	525	·591 529	·986 202	33	38
23	·395 166	492	·408 996	525	·591 004	·986 169	32	37
24	·395 658	492	·409 521	524	·590 479	·986 137	33	36
25	·396 150	491	·410 045	524	·589 955	·986 104	32	35
26	·396 641	491	·410 569	523	·589 431	·986 072	33	34
27	·397 132	489	·411 092	523	·588 908	·986 039	32	33
28	·397 621	490	·411 615	522	·588 385	·986 007	33	32
29	·398 111	489	·412 137	521	·587 863	·985 974	32	31
30	9·398 600	488	9·412 658	521	0·587 342	9·985 942	33	30
31	·399 088	487	·413 179	520	·586 821	·985 909	33	29
32	·399 575	487	·413 699	520	·586 301	·985 876	33	28
33	·400 062	487	·414 219	519	·585 781	·985 843	32	27
34	·400 549	486	·414 738	519	·585 262	·985 811	33	26
35	·401 035	485	·415 257	518	·584 743	·985 778	33	25
36	·401 520	485	·415 775	518	·584 225	·985 745	33	24
37	·402 005	484	·416 293	517	·583 707	·985 712	33	23
38	·402 489	483	·416 810	516	·583 190	·985 679	33	22
39	·402 972	483	·417 326	516	·582 674	·985 646	33	21
40	9·403 455	483	9·417 842	516	0·582 158	9·985 613	33	20
41	·403 938	482	·418 358	515	·581 642	·985 580	33	19
42	·404 420	481	·418 873	514	·581 127	·985 547	33	18
43	·404 901	481	·419 387	514	·580 613	·985 514	34	17
44	·405 382	480	·419 901	514	·580 099	·985 480	33	16
45	·405 862	479	·420 415	512	·579 585	·985 447	33	15
46	·406 341	479	·420 927	513	·579 073	·985 414	33	14
47	·406 820	479	·421 440	512	·578 560	·985 381	34	13
48	·407 299	478	·421 952	511	·578 048	·985 347	33	12
49	·407 777	477	·422 463	511	·577 537	·985 314	34	11
50	9·408 254	477	9·422 974	510	0·577 026	9·985 280	33	10
51	·408 731	476	·423 484	509	·576 516	·985 247	34	09
52	·409 207	475	·423 993	510	·576 007	·985 213	33	08
53	·409 682	475	·424 503	508	·575 497	·985 180	34	07
54	·410 157	475	·425 011	508	·574 989	·985 146	33	06
55	·410 632	474	·425 519	508	·574 481	·985 113	34	05
56	·411 106	473	·426 027	507	·573 973	·985 079	34	04
57	·411 579	473	·426 534	507	·573 466	·985 045	34	03
58	·412 052	472	·427 041	506	·572 959	·985 011	33	02
59	·412 524	472	·427 547	505	·572 453	·984 978	34	01
60	9·412 996		9·428 052		0·571 948	9·984 944		00
	cos		cot		tan	sin		75°

PROPORTIONAL PARTS FOR SECONDS

″	31	32	33	34	35	36
6	3·1	3·2	3·3	3·4	3·5	3·6
7	3·6	3·7	3·8	4·0	4·1	4·2
8	4·1	4·3	4·4	4·5	4·7	4·8
9	4·6	4·8	5·0	5·1	5·2	5·4
10	5·2	5·3	5·5	5·7	5·8	6·0
20	10·3	10·7	11·0	11·3	11·7	12·0
30	15·5	16·0	16·5	17·0	17·5	18·0
40	20·7	21·3	22·0	22·7	23·3	24·0
50	25·8	26·7	27·5	28·3	29·2	30·0

″	441	443	445	447	449
6	44·1	44·3	44·5	44·7	44·9
7	51·4	51·7	51·9	52·2	52·4
8	58·8	59·1	59·3	59·6	59·9
9	66·2	66·4	66·8	67·0	67·4
10	73·5	73·8	74·2	74·5	74·8
20	147·0	147·7	148·3	149·0	149·7
30	220·5	221·5	222·5	223·5	224·5
40	294·0	295·3	296·7	298·0	299·3
50	367·5	369·2	370·8	372·5	374·2

″	451	453	455	457	459
6	45·1	45·3	45·5	45·7	45·9
7	52·6	52·8	53·1	53·3	53·6
8	60·1	60·4	60·7	60·9	61·2
9	67·6	68·0	68·2	68·6	68·8
10	75·2	75·5	75·8	76·2	76·5
20	150·3	151·0	151·7	152·3	153·0
30	225·5	226·5	227·5	228·5	229·5
40	300·7	302·0	303·3	304·7	306·0
50	375·8	377·5	379·2	380·8	382·5

″	462	463	466	468	471
6	46·2	46·3	46·6	46·8	47·1
7	53·9	54·0	54·4	54·6	55·0
8	61·6	61·7	62·1	62·4	62·8
9	69·3	69·4	69·9	70·2	70·6
10	77·0	77·2	77·7	78·0	78·5
20	154·0	154·3	155·3	156·0	157·0
30	231·0	231·5	233·0	234·0	235·5
40	308·0	308·7	310·7	312·0	314·0
50	385·0	385·8	388·3	390·0	392·5

″	473	475	477	478	479
6	47·3	47·5	47·7	47·8	47·9
7	55·2	55·4	55·6	55·8	55·9
8	63·1	63·3	63·6	63·7	63·9
9	71·0	71·2	71·6	71·7	71·8
10	78·8	79·2	79·5	79·7	79·8
20	157·7	158·3	159·0	159·3	159·7
30	236·5	237·5	238·5	239·0	239·5
40	315·3	316·7	318·0	318·7	319·3
50	394·2	395·8	397·5	398·3	399·2

″	481	483	485	486	487
6	48·1	48·3	48·5	48·6	48·7
7	56·1	56·4	56·6	56·7	56·8
8	64·1	64·4	64·7	64·8	64·9
9	72·2	72·4	72·8	72·9	73·0
10	80·2	80·5	80·8	81·0	81·2
20	160·3	161·0	161·7	162·0	162·3
30	240·5	241·5	242·5	243·0	243·5
40	320·7	322·0	323·3	324·0	324·7
50	400·8	402·5	404·2	405·0	405·8

PROPORTIONAL PARTS FOR SECONDS

″	488	489	491	492	493
6	48·8	48·9	49·1	49·2	49·3
7	56·9	57·0	57·3	57·4	57·5
8	65·1	65·2	65·5	65·6	65·7
9	73·2	73·4	73·6	73·8	74·0
10	81·3	81·5	81·8	82·0	82·2
20	162·7	163·0	163·7	164·0	164·3
30	244·0	244·5	245·5	246·0	246·5
40	325·3	326·0	327·3	328·0	328·7
50	406·7	407·5	409·2	410·0	410·8

″	494	496	497	498	499
6	49·4	49·6	49·7	49·8	49·9
7	57·6	57·9	58·0	58·1	58·2
8	65·9	66·1	66·3	66·4	66·5
9	74·1	74·4	74·6	74·7	74·8
10	82·3	82·7	82·8	83·0	83·2
20	164·7	165·3	165·7	166·0	166·3
30	247·0	248·0	248·5	249·0	249·5
40	329·3	330·7	331·3	332·0	332·7
50	411·7	413·3	414·2	415·0	415·8

″	500	502	503	504	505
6	50·0	50·2	50·3	50·4	50·5
7	58·3	58·6	58·7	58·8	58·9
8	66·7	66·9	67·1	67·2	67·3
9	75·0	75·3	75·4	75·6	75·8
10	83·3	83·7	83·8	84·0	84·2
20	166·7	167·3	167·7	168·0	168·3
30	250·0	251·0	251·5	252·0	252·5
40	333·3	334·7	335·3	336·0	336·7
50	416·7	418·3	419·2	420·0	420·8

″	507	508	510	512	514
6	50·7	50·8	51·0	51·2	51·4
7	59·2	59·3	59·5	59·7	60·0
8	67·6	67·7	68·0	68·3	68·5
9	76·0	76·2	76·5	76·8	77·1
10	84·5	84·7	85·0	85·3	85·7
20	169·0	169·3	170·0	170·7	171·3
30	253·5	254·0	255·0	256·0	257·0
40	338·0	338·7	340·0	341·3	342·7
50	422·5	423·3	425·0	426·7	428·3

″	516	518	520	523	525
6	51·6	51·8	52·0	52·3	52·5
7	60·2	60·4	60·7	61·0	61·2
8	68·8	69·1	69·3	69·7	70·0
9	77·4	77·7	78·0	78·4	78·8
10	86·0	86·3	86·7	87·2	87·5
20	172·0	172·7	173·3	174·3	175·0
30	258·0	259·0	260·0	261·5	262·5
40	344·0	345·3	346·7	348·7	350·0
50	430·0	431·7	433·3	435·8	437·5

″	528	531	533	536	538
6	52·8	53·1	53·3	53·6	53·8
7	61·6	62·0	62·2	62·5	62·8
8	70·4	70·8	71·1	71·5	71·7
9	79·2	79·6	80·0	80·4	80·7
10	88·0	88·5	88·8	89·3	89·7
20	176·0	177·0	177·7	178·7	179·3
30	264·0	265·5	266·5	268·0	269·0
40	352·0	354·0	355·3	357·3	358·7
50	440·0	442·5	444·2	446·7	448·3

15°

′	sin		tan		cot	cos		′
00	9·412 996	471	9·428 052	506	0·571 948	9·984 944	34	60
01	·413 467	471	·428 558	504	·571 442	·984 910	34	59
02	·413 938	470	·429 062	504	·570 938	·984 876	34	58
03	·414 408	470	·429 566	504	·570 434	·984 842	34	57
04	·414 878	469	·430 070	503	·569 930	·984 808	34	56
05	·415 347	468	·430 573	502	·569 427	·984 774	34	55
06	·415 815	468	·431 075	502	·568 925	·984 740	34	54
07	·416 283	468	·431 577	502	·568 423	·984 706	34	53
08	·416 751	466	·432 079	501	·567 921	·984 672	34	52
09	·417 217	467	·432 580	500	·567 420	·984 638	35	51
10	9·417 684	466	9·433 080	500	0·566 920	9·984 603	34	50
11	·418 150	465	·433 580	500	·566 420	·984 569	34	49
12	·418 615	464	·434 080	499	·565 920	·984 535	35	48
13	·419 079	465	·434 579	499	·565 421	·984 500	34	47
14	·419 544	463	·435 078	498	·564 922	·984 466	34	46
15	·420 007	463	·435 576	497	·564 424	·984 432	35	45
16	·420 470	463	·436 073	497	·563 927	·984 397	34	44
17	·420 933	462	·436 570	497	·563 430	·984 363	35	43
18	·421 395	462	·437 067	496	·562 933	·984 328	34	42
19	·421 857	461	·437 563	496	·562 437	·984 294	35	41
20	9·422 318	460	9·438 059	495	0·561 941	9·984 259	35	40
21	·422 778	460	·438 554	494	·561 446	·984 224	34	39
22	·423 238	459	·439 048	495	·560 952	·984 190	35	38
23	·423 697	459	·439 543	493	·560 457	·984 155	35	37
24	·424 156	459	·440 036	493	·559 964	·984 120	35	36
25	·424 615	458	·440 529	493	·559 471	·984 085	35	35
26	·425 073	457	·441 022	492	·558 978	·984 050	35	34
27	·425 530	457	·441 514	492	·558 486	·984 015	34	33
28	·425 987	456	·442 006	491	·557 994	·983 981	35	32
29	·426 443	456	·442 497	491	·557 503	·983 946	35	31
30	9·426 899	455	9·442 988	491	0·557 012	9·983 911	36	30
31	·427 354	455	·443 479	489	·556 521	·983 875	35	29
32	·427 809	454	·443 968	490	·556 032	·983 840	35	28
33	·428 263	454	·444 458	489	·555 542	·983 805	35	27
34	·428 717	453	·444 947	488	·555 053	·983 770	35	26
35	·429 170	453	·445 435	488	·554 565	·983 735	35	25
36	·429 623	452	·445 923	488	·554 077	·983 700	36	24
37	·430 075	452	·446 411	487	·553 589	·983 664	35	23
38	·430 527	451	·446 898	486	·553 102	·983 629	35	22
39	·430 978	451	·447 384	486	·552 616	·983 594	36	21
40	9·431 429	450	9·447 870	486	0·552 130	9·983 558	35	20
41	·431 879	450	·448 356	485	·551 644	·983 523	36	19
42	·432 329	449	·448 841	485	·551 159	·983 487	35	18
43	·432 778	448	·449 326	484	·550 674	·983 452	36	17
44	·433 226	449	·449 810	484	·550 190	·983 416	35	16
45	·433 675	447	·450 294	483	·549 706	·983 381	36	15
46	·434 122	447	·450 777	483	·549 223	·983 345	36	14
47	·434 569	447	·451 260	483	·548 740	·983 309	36	13
48	·435 016	446	·451 743	482	·548 257	·983 273	35	12
49	·435 462	446	·452 225	481	·547 775	·983 238	36	11
50	9·435 908	445	9·452 706	481	0·547 294	9·983 202	36	10
51	·436 353	445	·453 187	481	·546 813	·983 166	36	09
52	·436 798	444	·453 668	480	·546 332	·983 130	36	08
53	·437 242	444	·454 148	480	·545 852	·983 094	36	07
54	·437 686	443	·454 628	479	·545 372	·983 058	36	06
55	·438 129	443	·455 107	479	·544 893	·983 022	36	05
56	·438 572	442	·455 586	478	·544 414	·982 986	36	04
57	·439 014	442	·456 064	478	·543 936	·982 950	36	03
58	·439 456	441	·456 542	477	·543 458	·982 914	36	02
59	·439 897	441	·457 019	477	·542 981	·982 878	36	01
60	9·440 338		9·457 496		0·542 504	9·982 842		00
	cos		cot		tan	sin		74°

16°

′	sin		tan		cot	cos		′
00	9·440 338	440	9·457 496	477	0·542 504	9·982 842	37	60
01	·440 778	440	·457 973	476	·542 027	·982 805	36	59
02	·441 218	440	·458 449	476	·541 551	·982 769	36	58
03	·441 658	438	·458 925	475	·541 075	·982 733	37	57
04	·442 096	439	·459 400	475	·540 600	·982 696	36	56
05	·442 535	438	·459 875	474	·540 125	·982 660	36	55
06	·442 973	437	·460 349	474	·539 651	·982 624	37	54
07	·443 410	437	·460 823	474	·539 177	·982 587	36	53
08	·443 847	437	·461 297	473	·538 703	·982 551	37	52
09	·444 284	436	·461 770	472	·538 230	·982 514	37	51
10	9·444 720	435	9·462 242	473	0·537 758	9·982 477	36	50
11	·445 155	435	·462 715	471	·537 285	·982 441	37	49
12	·445 590	435	·463 186	472	·536 814	·982 404	37	48
13	·446 025	434	·463 658	470	·536 342	·982 367	36	47
14	·446 459	434	·464 128	471	·535 872	·982 331	37	46
15	·446 893	433	·464 599	470	·535 401	·982 294	37	45
16	·447 326	433	·465 069	470	·534 931	·982 257	37	44
17	·447 759	432	·465 539	469	·534 461	·982 220	37	43
18	·448 191	432	·466 008	469	·533 992	·982 183	37	42
19	·448 623	431	·466 477	468	·533 523	·982 146	37	41
20	9·449 054	431	9·466 945	468	0·533 055	9·982 109	37	40
21	·449 485	430	·467 413	467	·532 587	·982 072	37	39
22	·449 915	430	·467 880	467	·532 120	·982 035	37	38
23	·450 345	430	·468 347	467	·531 653	·981 998	37	37
24	·450 775	429	·468 814	466	·531 186	·981 961	37	36
25	·451 204	428	·469 280	466	·530 720	·981 924	38	35
26	·451 632	428	·469 746	465	·530 254	·981 886	37	34
27	·452 060	428	·470 211	465	·529 789	·981 849	37	33
28	·452 488	427	·470 676	465	·529 324	·981 812	38	32
29	·452 915	427	·471 141	464	·528 859	·981 774	37	31
30	9·453 342	426	9·471 605	464	0·528 395	9·981 737	37	30
31	·453 768	426	·472 069	463	·527 931	·981 700	38	29
32	·454 194	425	·472 532	463	·527 468	·981 662	37	28
33	·454 619	425	·472 995	462	·527 005	·981 625	38	27
34	·455 044	425	·473 457	462	·526 543	·981 587	38	26
35	·455 469	424	·473 919	462	·526 081	·981 549	37	25
36	·455 893	423	·474 381	461	·525 619	·981 512	38	24
37	·456 316	423	·474 842	461	·525 158	·981 474	38	23
38	·456 739	423	·475 303	460	·524 697	·981 436	37	22
39	·457 162	422	·475 763	460	·524 237	·981 399	38	21
40	9·457 584	422	9·476 223	460	0·523 777	9·981 361	38	20
41	·458 006	421	·476 683	459	·523 317	·981 323	38	19
42	·458 427	421	·477 142	459	·522 858	·981 285	38	18
43	·458 848	420	·477 601	458	·522 399	·981 247	38	17
44	·459 268	420	·478 059	458	·521 941	·981 209	38	16
45	·459 688	420	·478 517	458	·521 483	·981 171	38	15
46	·460 108	419	·478 975	457	·521 025	·981 133	38	14
47	·460 527	419	·479 432	457	·520 568	·981 095	38	13
48	·460 946	418	·479 889	456	·520 111	·981 057	38	12
49	·461 364	418	·480 345	456	·519 655	·981 019	38	11
50	9·461 782	417	9·480 801	456	0·519 199	9·980 981	39	10
51	·462 199	417	·481 257	455	·518 743	·980 942	38	09
52	·462 616	416	·481 712	455	·518 288	·980 904	38	08
53	·463 032	416	·482 167	454	·517 833	·980 866	39	07
54	·463 448	416	·482 621	454	·517 379	·980 827	38	06
55	·463 864	415	·483 075	454	·516 925	·980 789	39	05
56	·464 279	415	·483 529	453	·516 471	·980 750	38	04
57	·464 694	414	·483 982	453	·516 018	·980 712	39	03
58	·465 108	414	·484 435	452	·515 565	·980 673	38	02
59	·465 522	413	·484 887	452	·515 113	·980 635	39	01
60	9·465 935		9·485 339		0·514 661	9·980 596		00
	cos		cot		tan	sin		73°

PROPORTIONAL PARTS FOR SECONDS

″	36	37	38	39	40	41
6	3·6	3·7	3·8	3·9	4·0	4·1
7	4·2	4·3	4·4	4·6	4·7	4·8
8	4·8	4·9	5·1	5·2	5·3	5·5
9	5·4	5·6	5·7	5·8	6·0	6·2
10	6·0	6·2	6·3	6·5	6·7	6·8
20	12·0	12·3	12·7	13·0	13·3	13·7
30	18·0	18·5	19·0	19·5	20·0	20·5
40	24·0	24·7	25·3	26·0	26·7	27·3
50	30·0	30·8	31·7	32·5	33·3	34·2

″	389	390	393	394	397
6	38·9	39·0	39·3	39·4	39·7
7	45·4	45·5	45·8	46·0	46·3
8	51·9	52·0	52·4	52·5	52·9
9	58·4	58·5	59·0	59·1	59·6
10	64·8	65·0	65·5	65·7	66·2
20	129·7	130·0	131·0	131·3	132·3
30	194·5	195·0	196·5	197·0	198·5
40	259·3	260·0	262·0	262·7	264·7
50	324·2	325·0	327·5	328·3	330·8

″	399	400	403	404	406
6	39·9	40·0	40·3	40·4	40·6
7	46·6	46·7	47·0	47·1	47·4
8	53·2	53·3	53·7	53·9	54·1
9	59·8	60·0	60·4	60·6	60·9
10	66·5	66·7	67·2	67·3	67·7
20	133·0	133·3	134·3	134·7	135·3
30	199·5	200·0	201·5	202·0	203·0
40	266·0	266·7	268·7	269·3	270·7
50	332·5	333·3	335·8	336·7	338·3

″	408	410	413	415	416
6	40·8	41·0	41·3	41·5	41·6
7	47·6	47·8	48·2	48·4	48·5
8	54·4	54·7	55·1	55·3	55·5
9	61·2	61·5	62·0	62·2	62·4
10	68·0	68·3	68·8	69·2	69·3
20	136·0	136·7	137·7	138·3	138·7
30	204·0	205·0	206·5	207·5	208·0
40	272·0	273·3	275·3	276·7	277·3
50	340·0	341·7	344·2	345·8	346·7

″	418	420	423	425	427
6	41·8	42·0	42·3	42·5	42·7
7	48·8	49·0	49·4	49·6	49·8
8	55·7	56·0	56·4	56·7	56·9
9	62·7	63·0	63·4	63·8	64·0
10	69·7	70·0	70·5	70·8	71·2
20	139·3	140·0	141·0	141·7	142·3
30	209·0	210·0	211·5	212·5	213·5
40	278·7	280·0	282·0	283·3	284·7
50	348·3	350·0	352·5	354·2	355·8

″	428	430	431	432	433
6	42·8	43·0	43·1	43·2	43·3
7	49·9	50·2	50·3	50·4	50·5
8	57·1	57·3	57·5	57·6	57·7
9	64·2	64·5	64·6	64·8	65·0
10	71·3	71·7	71·8	72·0	72·2
20	142·7	143·3	143·7	144·0	144·3
30	214·0	215·0	215·5	216·0	216·5
40	285·3	286·7	287·3	288·0	288·7
50	356·7	358·3	359·2	360·0	360·8

PROPORTIONAL PARTS FOR SECONDS

″	434	435	436	437	438
6	43·4	43·5	43·6	43·7	43·8
7	50·6	50·8	50·9	51·0	51·1
8	57·9	58·0	58·1	58·3	58·4
9	65·1	65·2	65·4	65·6	65·7
10	72·3	72·5	72·7	72·8	73·0
20	144·7	145·0	145·3	145·7	146·0
30	217·0	217·5	218·0	218·5	219·0
40	289·3	290·0	290·7	291·3	292·0
50	361·7	362·5	363·3	364·2	365·0

″	439	440	441	442	443
6	43·9	44·0	44·1	44·2	44·3
7	51·2	51·3	51·4	51·6	51·7
8	58·5	58·7	58·8	58·9	59·1
9	65·8	66·0	66·2	66·3	66·4
10	73·2	73·3	73·5	73·7	73·8
20	146·3	146·7	147·0	147·3	147·7
30	219·5	220·0	220·5	221·0	221·5
40	292·7	293·3	294·0	294·7	295·3
50	365·8	366·7	367·5	368·3	369·2

″	444	446	447	449	450
6	44·4	44·6	44·7	44·9	45·0
7	51·8	52·0	52·2	52·4	52·5
8	59·2	59·5	59·6	59·9	60·0
9	66·6	66·9	67·0	67·4	67·5
10	74·0	74·3	74·5	74·8	75·0
20	148·0	148·7	149·0	149·7	150·0
30	222·0	223·0	223·5	224·5	225·0
40	296·0	297·3	298·0	299·3	300·0
50	370·0	371·7	372·5	374·2	375·0

″	452	454	456	458	460
6	45·2	45·4	45·6	45·8	46·0
7	52·7	53·0	53·2	53·4	53·7
8	60·3	60·5	60·8	61·1	61·3
9	67·8	68·1	68·4	68·7	69·0
10	75·3	75·7	76·0	76·3	76·7
20	150·7	151·3	152·0	152·7	153·3
30	226·0	227·0	228·0	229·0	230·0
40	301·3	302·7	304·0	305·3	306·7
50	376·7	378·3	380·0	381·7	383·3

″	462	464	465	467	468
6	46·2	46·4	46·5	46·7	46·8
7	53·9	54·1	54·2	54·5	54·6
8	61·6	61·9	62·0	62·3	62·4
9	69·3	69·6	69·8	70·0	70·2
10	77·0	77·3	77·5	77·8	78·0
20	154·0	154·7	155·0	155·7	156·0
30	231·0	232·0	232·5	233·5	234·0
40	308·0	309·3	310·0	311·3	312·0
50	385·0	386·7	387·5	389·2	390·0

″	470	472	474	476	477
6	47·0	47·2	47·4	47·6	47·7
7	54·8	55·1	55·3	55·5	55·6
8	62·7	62·9	63·2	63·5	63·6
9	70·5	70·8	71·1	71·4	71·6
10	78·3	78·7	79·0	79·3	79·5
20	156·7	157·3	158·0	158·7	159·0
30	235·0	236·0	237·0	238·0	238·5
40	313·3	314·7	316·0	317·3	318·0
50	391·7	393·3	395·0	396·7	397·5

17°

′	sin		tan		cot	cos		′
00	9·465 935	413	9·485 339	452	0·514 661	9·980 596	38	60
01	·466 348	413	·485 791	451	·514 209	·980 558	39	59
02	·466 761	412	·486 242	451	·513 758	·980 519	39	58
03	·467 173	412	·486 693	450	·513 307	·980 480	38	57
04	·467 585	411	·487 143	450	·512 857	·980 442	39	56
05	·467 996	411	·487 593	450	·512 407	·980 403	39	55
06	·468 407	410	·488 043	449	·511 957	·980 364	39	54
07	·468 817	410	·488 492	449	·511 508	·980 325	39	53
08	·469 227	410	·488 941	449	·511 059	·980 286	39	52
09	·469 637	409	·489 390	448	·510 610	·980 247	39	51
10	9·470 046	409	9·489 838	448	0·510 162	9·980 208	39	50
11	·470 455	408	·490 286	447	·509 714	·980 169	39	49
12	·470 863	408	·490 733	447	·509 267	·980 130	39	48
13	·471 271	408	·491 180	447	·508 820	·980 091	39	47
14	·471 679	407	·491 627	446	·508 373	·980 052	40	46
15	·472 086	406	·492 073	446	·507 927	·980 012	39	45
16	·472 492	406	·492 519	446	·507 481	·979 973	39	44
17	·472 898	406	·492 965	445	·507 035	·979 934	39	43
18	·473 304	406	·493 410	444	·506 590	·979 895	40	42
19	·473 710	405	·493 854	445	·506 146	·979 855	39	41
20	9·474 115	404	9·494 299	444	0·505 701	9·979 816	40	40
21	·474 519	404	·494 743	443	·505 257	·979 776	39	39
22	·474 923	404	·495 186	444	·504 814	·979 737	40	38
23	·475 327	403	·495 630	443	·504 370	·979 697	39	37
24	·475 730	403	·496 073	442	·503 927	·979 658	40	36
25	·476 133	403	·496 515	442	·503 485	·979 618	39	35
26	·476 536	402	·496 957	442	·503 043	·979 579	40	34
27	·476 938	402	·497 399	442	·502 601	·979 539	40	33
28	·477 340	401	·497 841	441	·502 159	·979 499	40	32
29	·477 741	401	·498 282	440	·501 718	·979 459	39	31
30	9·478 142	400	9·498 722	441	0·501 278	9·979 420	40	30
31	·478 542	400	·499 163	440	·500 837	·979 380	40	29
32	·478 942	400	·499 603	439	·500 397	·979 340	40	28
33	·479 342	399	·500 042	439	·499 958	·979 300	40	27
34	·479 741	399	·500 481	439	·499 519	·979 260	40	26
35	·480 140	399	·500 920	439	·499 080	·979 220	40	25
36	·480 539	398	·501 359	438	·498 641	·979 180	40	24
37	·480 937	397	·501 797	438	·498 203	·979 140	40	23
38	·481 334	397	·502 235	437	·497 765	·979 100	41	22
39	·481 731	397	·502 672	437	·497 328	·979 059	40	21
40	9·482 128	397	9·503 109	437	0·496 891	9·979 019	40	20
41	·482 525	396	·503 546	436	·496 454	·978 979	40	19
42	·482 921	395	·503 982	436	·496 018	·978 939	41	18
43	·483 316	396	·504 418	436	·495 582	·978 898	40	17
44	·483 712	395	·504 854	435	·495 146	·978 858	41	16
45	·484 107	394	·505 289	435	·494 711	·978 817	40	15
46	·484 501	394	·505 724	435	·494 276	·978 777	40	14
47	·484 895	394	·506 159	434	·493 841	·978 737	41	13
48	·485 289	393	·506 593	434	·493 407	·978 696	41	12
49	·485 682	393	·507 027	433	·492 973	·978 655	40	11
50	9·486 075	392	9·507 460	433	0·492 540	9·978 615	41	10
51	·486 467	393	·507 893	433	·492 107	·978 574	41	09
52	·486 860	391	·508 326	433	·491 674	·978 533	40	08
53	·487 251	392	·508 759	432	·491 241	·978 493	41	07
54	·487 643	391	·509 191	431	·490 809	·978 452	41	06
55	·488 034	390	·509 622	432	·490 378	·978 411	41	05
56	·488 424	390	·510 054	431	·489 946	·978 370	41	04
57	·488 814	390	·510 485	431	·489 515	·978 329	41	03
58	·489 204	389	·510 916	430	·489 084	·978 288	41	02
59	·489 593	389	·511 346	430	·488 654	·978 247	41	01
60	9·489 982		9·511 776		0·488 224	9·978 206		00
	cos		cot		tan	sin		72°

18°

′	sin	d	tan	d	cot	cos	d	′
00	9·489 982	389	9·511 776	430	0·488 224	9·978 206	41	60
01	·490 371	388	·512 206	429	·487 794	·978 165	41	59
02	·490 759	388	·512 635	429	·487 365	·978 124	41	58
03	·491 147	388	·513 064	429	·486 936	·978 083	41	57
04	·491 535	387	·513 493	428	·486 507	·978 042	41	56
05	·491 922	386	·513 921	428	·486 079	·978 001	42	55
06	·492 308	387	·514 349	428	·485 651	·977 959	41	54
07	·492 695	386	·514 777	427	·485 223	·977 918	41	53
08	·493 081	385	·515 204	427	·484 796	·977 877	42	52
09	·493 466	385	·515 631	426	·484 369	·977 835	41	51
10	9·493 851	385	9·516 057	427	0·483 943	9·977 794	42	50
11	·494 236	385	·516 484	426	·483 516	·977 752	41	49
12	·494 621	384	·516 910	425	·483 090	·977 711	42	48
13	·495 005	383	·517 335	426	·482 665	·977 669	41	47
14	·495 388	384	·517 761	425	·482 239	·977 628	42	46
15	·495 772	382	·518 186	424	·481 814	·977 586	42	45
16	·496 154	383	·518 610	424	·481 390	·977 544	41	44
17	·496 537	382	·519 034	424	·480 966	·977 503	42	43
18	·496 919	382	·519 458	424	·480 542	·977 461	42	42
19	·497 301	381	·519 882	423	·480 118	·977 419	42	41
20	9·497 682	382	9·520 305	423	0·479 695	9·977 377	42	40
21	·498 064	380	·520 728	423	·479 272	·977 335	42	39
22	·498 444	381	·521 151	422	·478 849	·977 293	42	38
23	·498 825	379	·521 573	422	·478 427	·977 251	42	37
24	·499 204	380	·521 995	422	·478 005	·977 209	42	36
25	·499 584	379	·522 417	421	·477 583	·977 167	42	35
26	·499 963	379	·522 838	421	·477 162	·977 125	42	34
27	·500 342	379	·523 259	421	·476 741	·977 083	42	33
28	·500 721	378	·523 680	420	·476 320	·977 041	42	32
29	·501 099	377	·524 100	420	·475 900	·976 999	42	31
30	9·501 476	378	9·524 520	420	0·475 480	9·976 957	43	30
31	·501 854	377	·524 940	419	·475 060	·976 914	42	29
32	·502 231	376	·525 359	419	·474 641	·976 872	42	28
33	·502 607	377	·525 778	419	·474 222	·976 830	43	27
34	·502 984	376	·526 197	418	·473 803	·976 787	42	26
35	·503 360	375	·526 615	418	·473 385	·976 745	43	25
36	·503 735	375	·527 033	418	·472 967	·976 702	42	24
37	·504 110	375	·527 451	417	·472 549	·976 660	43	23
38	·504 485	375	·527 868	417	·472 132	·976 617	43	22
39	·504 860	374	·528 285	417	·471 715	·976 574	42	21
40	9·505 234	374	9·528 702	417	0·471 298	9·976 532	43	20
41	·505 608	373	·529 119	416	·470 881	·976 489	43	19
42	·505 981	373	·529 535	416	·470 465	·976 446	42	18
43	·506 354	373	·529 951	415	·470 049	·976 404	43	17
44	·506 727	372	·530 366	415	·469 634	·976 361	43	16
45	·507 099	372	·530 781	415	·469 219	·976 318	43	15
46	·507 471	372	·531 196	415	·468 804	·976 275	43	14
47	·507 843	371	·531 611	414	·468 389	·976 232	43	13
48	·508 214	371	·532 025	414	·467 975	·976 189	43	12
49	·508 585	371	·532 439	414	·467 561	·976 146	43	11
50	9·508 956	370	9·532 853	413	0·467 147	9·976 103	43	10
51	·509 326	370	·533 266	413	·466 734	·976 060	43	09
52	·509 696	369	·533 679	413	·466 321	·976 017	43	08
53	·510 065	369	·534 092	412	·465 908	·975 974	44	07
54	·510 434	369	·534 504	412	·465 496	·975 930	43	06
55	·510 803	369	·534 916	412	·465 084	·975 887	43	05
56	·511 172	368	·535 328	411	·464 672	·975 844	44	04
57	·511 540	367	·535 739	411	·464 261	·975 800	43	03
58	·511 907	368	·536 150	411	·463 850	·975 757	43	02
59	·512 275	367	·536 561	411	·463 439	·975 714	44	01
60	9·512 642		9·536 972		0·463 028	9·975 670		00
	cos		cot		tan	sin		**71°**

PROPORTIONAL PARTS FOR SECONDS

″	41	42	43	44	45	46
6	4·1	4·2	4·3	4·4	4·5	4·6
7	4·8	4·9	5·0	5·1	5·2	5·4
8	5·5	5·6	5·7	5·9	6·0	6·1
9	6·2	6·3	6·4	6·6	6·8	6·9
10	6·8	7·0	7·2	7·3	7·5	7·7
20	13·7	14·0	14·3	14·7	15·0	15·3
30	20·5	21·0	21·5	22·0	22·5	23·0
40	27·3	28·0	28·7	29·3	30·0	30·7
50	34·2	35·0	35·8	36·7	37·5	38·3

″	347	349	351	352	353
6	34·7	34·9	35·1	35·2	35·3
7	40·5	40·7	41·0	41·1	41·2
8	46·3	46·5	46·8	46·9	47·1
9	52·0	52·4	52·6	52·8	53·0
10	57·8	58·2	58·5	58·7	58·8
20	115·7	116·3	117·0	117·3	117·7
30	173·5	174·5	175·5	176·0	176·5
40	231·3	232·7	234·0	234·7	235·3
50	289·2	290·8	292·5	293·3	294·2

″	354	355	356	357	359
6	35·4	35·5	35·6	35·7	35·9
7	41·3	41·4	41·5	41·6	41·9
8	47·2	47·3	47·5	47·6	47·9
9	53·1	53·2	53·4	53·6	53·8
10	59·0	59·2	59·3	59·5	59·8
20	118·0	118·3	118·7	119·0	119·7
30	177·0	177·5	178·0	178·5	179·5
40	236·0	236·7	237·3	238·0	239·3
50	295·0	295·8	296·7	297·5	299·2

″	361	363	364	365	366
6	36·1	36·3	36·4	36·5	36·6
7	42·1	42·4	42·5	42·6	42·7
8	48·1	48·4	48·5	48·7	48·8
9	54·2	54·4	54·6	54·8	54·9
10	60·2	60·5	60·7	60·8	61·0
20	120·3	121·0	121·3	121·7	122·0
30	180·5	181·5	182·0	182·5	183·0
40	240·7	242·0	242·7	243·3	244·0
50	300·8	302·5	303·3	304·2	305·0

″	367	369	371	372	373
6	36·7	36·9	37·1	37·2	37·3
7	42·8	43·0	43·3	43·4	43·5
8	48·9	49·2	49·5	49·6	49·7
9	55·0	55·4	55·6	55·8	56·0
10	61·2	61·5	61·8	62·0	62·2
20	122·3	123·0	123·7	124·0	124·3
30	183·5	184·5	185·5	186·0	186·5
40	244·7	246·0	247·3	248·0	248·7
50	305·8	307·5	309·2	310·0	310·8

″	375	377	379	382	385
6	37·5	37·7	37·9	38·2	38·5
7	43·8	44·0	44·2	44·6	44·9
8	50·0	50·3	50·5	50·9	51·3
9	56·2	56·6	56·8	57·3	57·8
10	62·5	62·8	63·2	63·7	64·2
20	125·0	125·7	126·3	127·3	128·3
30	187·5	188·5	189·5	191·0	192·5
40	250·0	251·3	252·7	254·7	256·7
50	312·5	314·2	315·8	318·3	320·8

PROPORTIONAL PARTS FOR SECONDS

″	388	389	393	394	396
6	38·8	38·9	39·3	39·4	39·6
7	45·3	45·4	45·8	46·0	46·2
8	51·7	51·9	52·4	52·5	52·8
9	58·2	58·4	59·0	59·1	59·4
10	64·7	64·8	65·5	65·7	66·0
20	129·3	129·7	131·0	131·3	132·0
30	194·0	194·5	196·5	197·0	198·0
40	258·7	259·3	262·0	262·7	264·0
50	323·3	324·2	327·5	328·3	330·0

″	397	398	399	401	402
6	39·7	39·8	39·9	40·1	40·2
7	46·3	46·4	46·6	46·8	46·9
8	52·9	53·1	53·2	53·5	53·6
9	59·6	59·7	59·8	60·2	60·3
10	66·2	66·3	66·5	66·8	67·0
20	132·3	132·7	133·0	133·7	134·0
30	198·5	199·0	199·5	200·5	201·0
40	264·7	265·3	266·0	267·3	268·0
50	330·8	331·7	332·5	334·2	335·0

″	403	404	405	406	407
6	40·3	40·4	40·5	40·6	40·7
7	47·0	47·1	47·2	47·4	47·5
8	53·7	53·9	54·0	54·1	54·3
9	60·4	60·6	60·8	60·9	61·0
10	67·2	67·3	67·5	67·7	67·8
20	134·3	134·7	135·0	135·3	135·7
30	201·5	202·0	202·5	203·0	203·5
40	268·7	269·3	270·0	270·7	271·3
50	335·8	336·7	337·5	338·3	339·2

″	408	411	412	413	414
6	40·8	41·1	41·2	41·3	41·4
7	47·6	48·0	48·1	48·2	48·3
8	54·4	54·8	54·9	55·1	55·2
9	61·2	61·6	61·8	62·0	62·1
10	68·0	68·5	68·7	68·8	69·0
20	136·0	137·0	137·3	137·7	138·0
30	204·0	205·5	206·0	206·5	207·0
40	272·0	274·0	274·7	275·3	276·0
50	340·0	342·5	343·3	344·2	345·0

″	415	417	419	421	422
6	41·5	41·7	41·9	42·1	42·2
7	48·4	48·6	48·9	49·1	49·2
8	55·3	55·6	55·9	56·1	56·3
9	62·2	62·6	62·8	63·2	63·3
10	69·2	69·5	69·8	70·2	70·3
20	138·3	139·0	139·7	140·3	140·7
30	207·5	208·5	209·5	210·5	211·0
40	276·7	278·0	279·3	280·7	281·3
50	345·8	347·5	349·2	350·8	351·7

″	423	424	426	428	430
6	42·3	42·4	42·6	42·8	43·0
7	49·4	49·5	49·7	49·9	50·2
8	56·4	56·5	56·8	57·1	57·3
9	63·4	63·6	63·9	64·2	64·5
10	70·5	70·7	71·0	71·3	71·7
20	141·0	141·3	142·0	142·7	143·3
30	211·5	212·0	213·0	214·0	215·0
40	282·0	282·7	284·0	285·3	286·7
50	352·5	353·3	355·0	356·7	358·3

19° ′	sin		tan		cot	cos		′
00	9·512 642	367	9·536 972	410	0·463 028	9·975 670	43	60
01	·513 009	366	·537 382	410	·462 618	·975 627	44	59
02	·513 375	366	·537 792	410	·462 208	·975 583	44	58
03	·513 741	366	·538 202	409	·461 798	·975 539	43	57
04	·514 107	365	·538 611	409	·461 389	·975 496	44	56
05	·514 472	365	·539 020	409	·460 980	·975 452	44	55
06	·514 837	365	·539 429	408	·460 571	·975 408	43	54
07	·515 202	364	·539 837	408	·460 163	·975 365	44	53
08	·515 566	364	·540 245	408	·459 755	·975 321	44	52
09	·515 930	364	·540 653	408	·459 347	·975 277	44	51
10	9·516 294	363	9·541 061	407	0·458 939	9·975 233	44	50
11	·516 657	363	·541 468	407	·458 532	·975 189	44	49
12	·517 020	362	·541 875	406	·458 125	·975 145	44	48
13	·517 382	363	·542 281	407	·457 719	·975 101	44	47
14	·517 745	362	·542 688	406	·457 312	·975 057	44	46
15	·518 107	361	·543 094	405	·456 906	·975 013	44	45
16	·518 468	361	·543 499	406	·456 501	·974 969	44	44
17	·518 829	361	·543 905	405	·456 095	·974 925	45	43
18	·519 190	361	·544 310	405	·455 690	·974 880	44	42
19	·519 551	360	·544 715	404	·455 285	·974 836	44	41
20	9·519 911	360	9·545 119	405	0·454 881	9·974 792	44	40
21	·520 271	360	·545 524	404	·454 476	·974 748	45	39
22	·520 631	359	·545 928	403	·454 072	·974 703	44	38
23	·520 990	359	·546 331	404	·453 669	·974 659	45	37
24	·521 349	358	·546 735	403	·453 265	·974 614	44	36
25	·521 707	359	·547 138	402	·452 862	·974 570	45	35
26	·522 066	358	·547 540	403	·452 460	·974 525	44	34
27	·522 424	357	·547 943	402	·452 057	·974 481	45	33
28	·522 781	357	·548 345	402	·451 655	·974 436	45	32
29	·523 138	357	·548 747	402	·451 253	·974 391	44	31
30	9·523 495	357	9·549 149	401	0·450 851	9·974 347	45	30
31	·523 852	356	·549 550	401	·450 450	·974 302	45	29
32	·524 208	356	·549 951	401	·450 049	·974 257	45	28
33	·524 564	356	·550 352	400	·449 648	·974 212	45	27
34	·524 920	355	·550 752	401	·449 248	·974 167	45	26
35	·525 275	355	·551 153	399	·448 847	·974 122	45	25
36	·525 630	354	·551 552	400	·448 448	·974 077	45	24
37	·525 984	355	·551 952	399	·448 048	·974 032	45	23
38	·526 339	354	·552 351	399	·447 649	·973 987	45	22
39	·526 693	353	·552 750	399	·447 250	·973 942	45	21
40	9·527 046	354	9·553 149	399	0·446 851	9·973 897	45	20
41	·527 400	353	·553 548	398	·446 452	·973 852	45	19
42	·527 753	352	·553 946	398	·446 054	·973 807	46	18
43	·528 105	353	·554 344	397	·445 656	·973 761	45	17
44	·528 458	352	·554 741	398	·445 259	·973 716	45	16
45	·528 810	351	·555 139	397	·444 861	·973 671	46	15
46	·529 161	352	·555 536	397	·444 464	·973 625	45	14
47	·529 513	351	·555 933	396	·444 067	·973 580	45	13
48	·529 864	351	·556 329	396	·443 671	·973 535	46	12
49	·530 215	350	·556 725	396	·443 275	·973 489	45	11
50	9·530 565	350	9·557 121	396	0·442 879	9·973 444	46	10
51	·530 915	350	·557 517	396	·442 483	·973 398	46	09
52	·531 265	349	·557 913	395	·442 087	·973 352	45	08
53	·531 614	349	·558 308	395	·441 692	·973 307	46	07
54	·531 963	349	·558 703	394	·441 297	·973 261	46	06
55	·532 312	349	·559 097	394	·440 903	·973 215	46	05
56	·532 661	348	·559 491	394	·440 509	·973 169	45	04
57	·533 009	348	·559 885	394	·440 115	·973 124	46	03
58	·533 357	347	·560 279	394	·439 721	·973 078	46	02
59	·533 704	348	·560 673	393	·439 327	·973 032	46	01
60	9·534 052		9·561 066		0·438 934	9·972 986		00
	cos		cot		tan	sin		70°

20°

′	sin	d	tan	d	cot	cos	d	′
00	9·534 052	347	9·561 066	393	0·438 934	9·972 986	46	60
01	·534 399	346	·561 459	392	·438 541	·972 940	46	59
02	·534 745	347	·561 851	393	·438 149	·972 894	46	58
03	·535 092	346	·562 244	392	·437 756	·972 848	46	57
04	·535 438	345	·562 636	392	·437 364	·972 802	47	56
05	·535 783	346	·563 028	391	·436 972	·972 755	46	55
06	·536 129	345	·563 419	392	·436 581	·972 709	46	54
07	·536 474	344	·563 811	391	·436 189	·972 663	46	53
08	·536 818	345	·564 202	391	·435 798	·972 617	47	52
09	·537 163	344	·564 593	390	·435 407	·972 570	46	51
10	9·537 507	344	9·564 983	390	0·435 017	9·972 524	46	50
11	·537 851	343	·565 373	390	·434 627	·972 478	47	49
12	·538 194	344	·565 763	390	·434 237	·972 431	46	48
13	·538 538	342	·566 153	389	·433 847	·972 385	47	47
14	·538 880	343	·566 542	390	·433 458	·972 338	47	46
15	·539 223	342	·566 932	388	·433 068	·972 291	46	45
16	·539 565	342	·567 320	389	·432 680	·972 245	47	44
17	·539 907	342	·567 709	389	·432 291	·972 198	47	43
18	·540 249	341	·568 098	388	·431 902	·972 151	46	42
19	·540 590	341	·568 486	387	·431 514	·972 105	47	41
20	9·540 931	341	9·568 873	388	0·431 127	9·972 058	47	40
21	·541 272	341	·569 261	387	·430 739	·972 011	47	39
22	·541 613	340	·569 648	387	·430 352	·971 964	47	38
23	·541 953	340	·570 035	387	·429 965	·971 917	47	37
24	·542 293	339	·570 422	387	·429 578	·971 870	47	36
25	·542 632	339	·570 809	386	·429 191	·971 823	47	35
26	·542 971	339	·571 195	386	·428 805	·971 776	47	34
27	·543 310	339	·571 581	386	·428 419	·971 729	47	33
28	·543 649	338	·571 967	385	·428 033	·971 682	47	32
29	·543 987	338	·572 352	386	·427 648	·971 635	47	31
30	9·544 325	338	9·572 738	385	0·427 262	9·971 588	48	30
31	·544 663	337	·573 123	384	·426 877	·971 540	47	29
32	·545 000	338	·573 507	385	·426 493	·971 493	47	28
33	·545 338	336	·573 892	384	·426 108	·971 446	48	27
34	·545 674	337	·574 276	384	·425 724	·971 398	47	26
35	·546 011	336	·574 660	384	·425 340	·971 351	48	25
36	·546 347	336	·575 044	383	·424 956	·971 303	47	24
37	·546 683	336	·575 427	383	·424 573	·971 256	48	23
38	·547 019	335	·575 810	383	·424 190	·971 208	47	22
39	·547 354	335	·576 193	383	·423 807	·971 161	48	21
40	9·547 689	335	9·576 576	383	0·423 424	9·971 113	47	20
41	·548 024	335	·576 959	382	·423 041	·971 066	48	19
42	·548 359	334	·577 341	382	·422 659	·971 018	48	18
43	·548 693	334	·577 723	381	·422 277	·970 970	48	17
44	·549 027	333	·578 104	382	·421 896	·970 922	48	16
45	·549 360	333	·578 486	381	·421 514	·970 874	47	15
46	·549 693	333	·578 867	381	·421 133	·970 827	48	14
47	·550 026	333	·579 248	381	·420 752	·970 779	48	13
48	·550 359	333	·579 629	380	·420 371	·970 731	48	12
49	·550 692	332	·580 009	380	·419 991	·970 683	48	11
50	9·551 024	332	9·580 389	380	0·419 611	9·970 635	49	10
51	·551 356	331	·580 769	380	·419 231	·970 586	48	09
52	·551 687	331	·581 149	379	·418 851	·970 538	48	08
53	·552 018	331	·581 528	379	·418 472	·970 490	48	07
54	·552 349	331	·581 907	379	·418 093	·970 442	48	06
55	·552 680	330	·582 286	379	·417 714	·970 394	49	05
56	·553 010	331	·582 665	379	·417 335	·970 345	48	04
57	·553 341	329	·583 044	378	·416 956	·970 297	48	03
58	·553 670	330	·583 422	378	·416 578	·970 249	49	02
59	·554 000	329	·583 800	377	·416 200	·970 200	48	01
60	9·554 329		9·584 177		0·415 823	9·970 152		00
	cos		cot		tan	sin		′

69°

PROPORTIONAL PARTS FOR SECONDS

″	46	47	48	49	50	51
6	4·6	4·7	4·8	4·9	5·0	5·1
7	5·4	5·5	5·6	5·7	5·8	6·0
8	6·1	6·3	6·4	6·5	6·7	6·8
9	6·9	7·0	7·2	7·4	7·5	7·6
10	7·7	7·8	8·0	8·2	8·3	8·5
20	15·3	15·7	16·0	16·3	16·7	17·0
30	23·0	23·5	24·0	24·5	25·0	25·5
40	30·7	31·3	32·0	32·7	33·3	34·0
50	38·3	39·2	40·0	40·8	41·7	42·5

″	312	314	315	316	317
6	31·2	31·4	31·5	31·6	31·7
7	36·4	36·6	36·8	36·9	37·0
8	41·6	41·9	42·0	42·1	42·3
9	46·8	47·1	47·2	47·4	47·6
10	52·0	52·3	52·5	52·7	52·8
20	104·0	104·7	105·0	105·3	105·7
30	156·0	157·0	157·5	158·0	158·5
40	208·0	209·3	210·0	210·7	211·3
50	260·0	261·7	262·5	263·3	264·2

″	318	320	322	323	324
6	31·8	32·0	32·2	32·3	32·4
7	37·1	37·3	37·6	37·7	37·8
8	42·4	42·7	42·9	43·1	43·2
9	47·7	48·0	48·3	48·4	48·6
10	53·0	53·3	53·7	53·8	54·0
20	106·0	106·7	107·3	107·7	108·0
30	159·0	160·0	161·0	161·5	162·0
40	212·0	213·3	214·7	215·3	216·0
50	265·0	266·7	268·3	269·2	270·0

″	325	326	327	328	329
6	32·5	32·6	32·7	32·8	32·9
7	37·9	38·0	38·2	38·3	38·4
8	43·3	43·5	43·6	43·7	43·9
9	48·8	48·9	49·0	49·2	49·4
10	54·2	54·3	54·5	54·7	54·8
20	108·3	108·7	109·0	109·3	109·7
30	162·5	163·0	163·5	164·0	164·5
40	216·7	217·3	218·0	218·7	219·3
50	270·8	271·7	272·5	273·3	274·2

″	331	333	335	336	338
6	33·1	33·3	33·5	33·6	33·8
7	38·6	38·8	39·1	39·2	39·4
8	44·1	44·4	44·7	44·8	45·1
9	49·6	50·0	50·2	50·4	50·7
10	55·2	55·5	55·8	56·0	56·3
20	110·3	111·0	111·7	112·0	112·7
30	165·5	166·5	167·5	168·0	169·0
40	220·7	222·0	223·3	224·0	225·3
50	275·8	277·5	279·2	280·0	281·7

″	339	341	342	344	346
6	33·9	34·1	34·2	34·4	34·6
7	39·6	39·8	39·9	40·1	40·4
8	45·2	45·5	45·6	45·9	46·1
9	50·8	51·2	51·3	51·6	51·9
10	56·5	56·8	57·0	57·3	57·7
20	113·0	113·7	114·0	114·7	115·3
30	169·5	170·5	171·0	172·0	173·0
40	226·0	227·3	228·0	229·3	230·7
50	282·5	284·2	285·0	286·7	288·3

PROPORTIONAL PARTS FOR SECONDS

″	347	364	365	366	367
6	34·7	36·4	36·5	36·6	36·7
7	40·5	42·5	42·6	42·7	42·8
8	46·3	48·5	48·7	48·8	48·9
9	52·0	54·6	54·8	54·9	55·0
10	57·8	60·7	60·8	61·0	61·2
20	115·7	121·3	121·7	122·0	122·3
30	173·5	182·0	182·5	183·0	183·5
40	231·3	242·7	243·3	244·0	244·7
50	289·2	303·3	304·2	305·0	305·8

″	368	369	370	371	372
6	36·8	36·9	37·0	37·1	37·2
7	42·9	43·0	43·2	43·3	43·4
8	49·1	49·2	49·3	49·5	49·6
9	55·2	55·4	55·5	55·6	55·8
10	61·3	61·5	61·7	61·8	62·0
20	122·7	123·0	123·3	123·7	124·0
30	184·0	184·5	185·0	185·5	186·0
40	245·3	246·0	246·7	247·3	248·0
50	306·7	307·5	308·3	309·2	310·0

″	373	374	375	376	377
6	37·3	37·4	37·5	37·6	37·7
7	43·5	43·6	43·8	43·9	44·0
8	49·7	49·9	50·0	50·1	50·3
9	56·0	56·1	56·2	56·4	56·6
10	62·2	62·3	62·5	62·7	62·8
20	124·3	124·7	125·0	125·3	125·7
30	186·5	187·0	187·5	188·0	188·5
40	248·7	249·3	250·0	250·7	251·3
50	310·8	311·7	312·5	313·3	314·2

″	378	379	380	381	382
6	37·8	37·9	38·0	38·1	38·2
7	44·1	44·2	44·3	44·4	44·6
8	50·4	50·5	50·7	50·8	50·9
9	56·7	56·8	57·0	57·2	57·3
10	63·0	63·2	63·3	63·5	63·7
20	126·0	126·3	126·7	127·0	127·3
30	189·0	189·5	190·0	190·5	191·0
40	252·0	252·7	253·3	254·0	254·7
50	315·0	315·8	316·7	317·5	318·3

″	383	384	385	386	387
6	38·3	38·4	38·5	38·6	38·7
7	44·7	44·8	44·9	45·0	45·2
8	51·1	51·2	51·3	51·5	51·6
9	57·4	57·6	57·8	57·9	58·0
10	63·8	64·0	64·2	64·3	64·5
20	127·7	128·0	128·3	128·7	129·0
30	191·5	192·0	192·5	193·0	193·5
40	255·3	256·0	256·7	257·3	258·0
50	319·2	320·0	320·8	321·7	322·5

″	388	389	390	392	393
6	38·8	38·9	39·0	39·2	39·3
7	45·3	45·4	45·5	45·7	45·8
8	51·7	51·9	52·0	52·3	52·4
9	58·2	58·4	58·5	58·8	59·0
10	64·7	64·8	65·0	65·3	65·5
20	129·3	129·7	130·0	130·7	131·0
30	194·0	194·5	195·0	196·0	196·5
40	258·7	259·3	260·0	261·3	262·0
50	323·3	324·2	325·0	326·7	327·5

21°

′	sin	d.	tan	d.	cot	cos	d.	′
00	9·554 329	329	9·584 177	378	0·415 823	9·970 152	49	60
01	·554 658	329	·584 555	377	·415 445	·970 103	48	59
02	·554 987	328	·584 932	377	·415 068	·970 055	49	58
03	·555 315	328	·585 309	377	·414 691	·970 006	49	57
04	·555 643	328	·585 686	376	·414 314	·969 957	48	56
05	·555 971	328	·586 062	377	·413 938	·969 909	49	55
06	·556 299	327	·586 439	376	·413 561	·969 860	49	54
07	·556 626	327	·586 815	375	·413 185	·969 811	49	53
08	·556 953	327	·587 190	376	·412 810	·969 762	48	52
09	·557 280	326	·587 566	375	·412 434	·969 714	49	51
10	9·557 606	326	9·587 941	375	0·412 059	9·969 665	49	50
11	·557 932	326	·588 316	375	·411 684	·969 616	49	49
12	·558 258	325	·588 691	375	·411 309	·969 567	49	48
13	·558 583	326	·589 066	374	·410 934	·969 518	49	47
14	·558 909	325	·589 440	374	·410 560	·969 469	49	46
15	·559 234	324	·589 814	374	·410 186	·969 420	50	45
16	·559 558	325	·590 188	374	·409 812	·969 370	49	44
17	·559 883	324	·590 562	373	·409 438	·969 321	49	43
18	·560 207	324	·590 935	373	·409 065	·969 272	49	42
19	·560 531	324	·591 308	373	·408 692	·969 223	50	41
20	9·560 855	323	9·591 681	373	0·408 319	9·969 173	49	40
21	·561 178	323	·592 054	372	·407 946	·969 124	49	39
22	·561 501	323	·592 426	373	·407 574	·969 075	50	38
23	·561 824	322	·592 799	372	·407 201	·969 025	49	37
24	·562 146	322	·593 171	371	·406 829	·968 976	50	36
25	·562 468	322	·593 542	372	·406 458	·968 926	49	35
26	·562 790	322	·593 914	371	·406 086	·968 877	50	34
27	·563 112	321	·594 285	371	·405 715	·968 827	50	33
28	·563 433	322	·594 656	371	·405 344	·968 777	49	32
29	·563 755	320	·595 027	371	·404 973	·968 728	50	31
30	9·564 075	321	9·595 398	370	0·404 602	9·968 678	50	30
31	·564 396	320	·595 768	370	·404 232	·968 628	50	29
32	·564 716	320	·596 138	370	·403 862	·968 578	50	28
33	·565 036	320	·596 508	370	·403 492	·968 528	49	27
34	·565 356	320	·596 878	369	·403 122	·968 479	50	26
35	·565 676	319	·597 247	369	·402 753	·968 429	50	25
36	·565 995	319	·597 616	369	·402 384	·968 379	50	24
37	·566 314	318	·597 985	369	·402 015	·968 329	51	23
38	·566 632	319	·598 354	368	·401 646	·968 278	50	22
39	·566 951	318	·598 722	369	·401 278	·968 228	50	21
40	9·567 269	318	9·599 091	368	0·400 909	9·968 178	50	20
41	·567 587	317	·599 459	368	·400 541	·968 128	50	19
42	·567 904	318	·599 827	367	·400 173	·968 078	51	18
43	·568 222	317	·600 194	368	·399 806	·968 027	50	17
44	·568 539	317	·600 562	367	·399 438	·967 977	50	16
45	·568 856	316	·600 929	367	·399 071	·967 927	51	15
46	·569 172	316	·601 296	367	·398 704	·967 876	50	14
47	·569 488	316	·601 663	366	·398 337	·967 826	51	13
48	·569 804	316	·602 029	366	·397 971	·967 775	50	12
49	·570 120	315	·602 395	366	·397 605	·967 725	51	11
50	9·570 435	316	9·602 761	366	0·397 239	9·967 674	50	10
51	·570 751	315	·603 127	366	·396 873	·967 624	51	09
52	·571 066	314	·603 493	365	·396 507	·967 573	51	08
53	·571 380	315	·603 858	365	·396 142	·967 522	51	07
54	·571 695	314	·604 223	365	·395 777	·967 471	50	06
55	·572 009	314	·604 588	365	·395 412	·967 421	51	05
56	·572 323	313	·604 953	364	·395 047	·967 370	51	04
57	·572 636	314	·605 317	365	·394 683	·967 319	51	03
58	·572 950	313	·605 682	364	·394 318	·967 268	51	02
59	·573 263	312	·606 046	364	·393 954	·967 217	51	01
60	9·573 575		9·606 410		0·393 590	9·967 166		00
	cos		cot		tan	sin		68°

22°

′	sin	d	tan	d	cot	cos	d	′
00	9·573 575	313	9·606 410	363	0·393 590	9·967 166	51	60
01	·573 888	312	·606 773	364	·393 227	·967 115	51	59
02	·574 200	312	·607 137	363	·392 863	·967 064	51	58
03	·574 512	312	·607 500	363	·392 500	·967 013	52	57
04	·574 824	312	·607 863	362	·392 137	·966 961	51	56
05	·575 136	311	·608 225	363	·391 775	·966 910	51	55
06	·575 447	311	·608 588	362	·391 412	·966 859	51	54
07	·575 758	311	·608 950	362	·391 050	·966 808	52	53
08	·576 069	310	·609 312	362	·390 688	·966 756	51	52
09	·576 379	310	·609 674	362	·390 326	·966 705	52	51
10	9·576 689	310	9·610 036	361	0·389 964	9·966 653	51	50
11	·576 999	310	·610 397	362	·389 603	·966 602	52	49
12	·577 309	309	·610 759	361	·389 241	·966 550	51	48
13	·577 618	309	·611 120	360	·388 880	·966 499	52	47
14	·577 927	309	·611 480	361	·388 520	·966 447	52	46
15	·578 236	309	·611 841	360	·388 159	·966 395	51	45
16	·578 545	308	·612 201	360	·387 799	·966 344	52	44
17	·578 853	309	·612 561	360	·387 439	·966 292	52	43
18	·579 162	308	·612 921	360	·387 079	·966 240	52	42
19	·579 470	307	·613 281	360	·386 719	·966 188	52	41
20	9·579 777	308	9·613 641	359	0·386 359	9·966 136	51	40
21	·580 085	307	·614 000	359	·386 000	·966 085	52	39
22	·580 392	307	·614 359	359	·385 641	·966 033	52	38
23	·580 699	306	·614 718	359	·385 282	·965 981	52	37
24	·581 005	307	·615 077	358	·384 923	·965 929	53	36
25	·581 312	306	·615 435	358	·384 565	·965 876	52	35
26	·581 618	306	·615 793	358	·384 207	·965 824	52	34
27	·581 924	305	·616 151	358	·383 849	·965 772	52	33
28	·582 229	306	·616 509	358	·383 491	·965 720	52	32
29	·582 535	305	·616 867	357	·383 133	·965 668	53	31
30	9·582 840	305	9·617 224	358	0·382 776	9·965 615	52	30
31	·583 145	304	·617 582	357	·382 418	·965 563	52	29
32	·583 449	305	·617 939	356	·382 061	·965 511	53	28
33	·583 754	304	·618 295	357	·381 705	·965 458	52	27
34	·584 058	303	·618 652	356	·381 348	·965 406	53	26
35	·584 361	304	·619 008	356	·380 992	·965 353	52	25
36	·584 665	303	·619 364	356	·380 636	·965 301	53	24
37	·584 968	304	·619 720	356	·380 280	·965 248	53	23
38	·585 272	302	·620 076	356	·379 924	·965 195	52	22
39	·585 574	303	·620 432	355	·379 568	·965 143	53	21
40	9·585 877	302	9·620 787	355	0·379 213	9·965 090	53	20
41	·586 179	303	·621 142	355	·378 858	·965 037	53	19
42	·586 482	301	·621 497	355	·378 503	·964 984	53	18
43	·586 783	302	·621 852	355	·378 148	·964 931	52	17
44	·587 085	301	·622 207	354	·377 793	·964 879	53	16
45	·587 386	302	·622 561	354	·377 439	·964 826	53	15
46	·587 688	301	·622 915	354	·377 085	·964 773	53	14
47	·587 989	300	·623 269	354	·376 731	·964 720	54	13
48	·588 289	301	·623 623	353	·376 377	·964 666	53	12
49	·588 590	300	·623 976	354	·376 024	·964 613	53	11
50	9·588 890	300	9·624 330	353	0·375 670	9·964 560	53	10
51	·589 190	299	·624 683	353	·375 317	·964 507	53	09
52	·589 489	300	·625 036	352	·374 964	·964 454	54	08
53	·589 789	299	·625 388	353	·374 612	·964 400	53	07
54	·590 088	299	·625 741	352	·374 259	·964 347	53	06
55	·590 387	299	·626 093	352	·373 907	·964 294	54	05
56	·590 686	298	·626 445	352	·373 555	·964 240	53	04
57	·590 984	298	·626 797	352	·373 203	·964 187	54	03
58	·591 282	298	·627 149	352	·372 851	·964 133	53	02
59	·591 580	298	·627 501	351	·372 499	·964 080	54	01
60	9·591 878		9·627 852		0·372 148	9·964 026		00
	cos		cot		tan	sin		67°

PROPORTIONAL PARTS FOR SECONDS

″	51	52	53	54	55	56
6	5·1	5·2	5·3	5·4	5·5	5·6
7	6·0	6·1	6·2	6·3	6·4	6·5
8	6·8	6·9	7·1	7·2	7·3	7·5
9	7·6	7·8	8·0	8·1	8·2	8·4
10	8·5	8·7	8·8	9·0	9·2	9·3
20	17·0	17·3	17·7	18·0	18·3	18·7
30	25·5	26·0	26·5	27·0	27·5	28·0
40	34·0	34·7	35·3	36·0	36·7	37·3
50	42·5	43·3	44·2	45·0	45·8	46·7

″	284	285	286	287	288
6	28·4	28·5	28·6	28·7	28·8
7	33·1	33·2	33·4	33·5	33·6
8	37·9	38·0	38·1	38·3	38·4
9	42·6	42·8	42·9	43·0	43·2
10	47·3	47·5	47·7	47·8	48·0
20	94·7	95·0	95·3	95·7	96·0
30	142·0	142·5	143·0	143·5	144·0
40	189·3	190·0	190·7	191·3	192·0
50	236·7	237·5	238·3	239·2	240·0

″	289	290	291	292	293
6	28·9	29·0	29·1	29·2	29·3
7	33·7	33·8	34·0	34·1	34·2
8	38·5	38·7	38·8	38·9	39·1
9	43·4	43·5	43·6	43·8	44·0
10	48·2	48·3	48·5	48·7	48·8
20	96·3	96·7	97·0	97·3	97·7
30	144·5	145·0	145·5	146·0	146·5
40	192·7	193·3	194·0	194·7	195·3
50	240·8	241·7	242·5	243·3	244·2

″	294	295	296	297	298
6	29·4	29·5	29·6	29·7	29·8
7	34·3	34·4	34·5	34·6	34·8
8	39·2	39·3	39·5	39·6	39·7
9	44·1	44·2	44·4	44·6	44·7
10	49·0	49·2	49·3	49·5	49·7
20	98·0	98·3	98·7	99·0	99·3
30	147·0	147·5	148·0	148·5	149·0
40	196·0	196·7	197·3	198·0	198·7
50	245·0	245·8	246·7	247·5	248·3

″	299	300	301	302	303
6	29·9	30·0	30·1	30·2	30·3
7	34·9	35·0	35·1	35·2	35·4
8	39·9	40·0	40·1	40·3	40·4
9	44·8	45·0	45·2	45·3	45·4
10	49·8	50·0	50·2	50·3	50·5
20	99·7	100·0	100·3	100·7	101·0
30	149·5	150·0	150·5	151·0	151·5
40	199·3	200·0	200·7	201·3	202·0
50	249·2	250·0	250·8	251·7	252·5

″	304	305	306	307	308
6	30·4	30·5	30·6	30·7	30·8
7	35·5	35·6	35·7	35·8	35·9
8	40·5	40·7	40·8	40·9	41·1
9	45·6	45·8	45·9	46·0	46·2
10	50·7	50·8	51·0	51·2	51·3
20	101·3	101·7	102·0	102·3	102·7
30	152·0	152·5	153·0	153·5	154·0
40	202·7	203·3	204·0	204·7	205·3
50	253·3	254·2	255·0	255·8	256·7

PROPORTIONAL PARTS FOR SECONDS

″	309	310	311	312	313
6	30·9	31·0	31·1	31·2	31·3
7	36·0	36·2	36·3	36·4	36·5
8	41·2	41·3	41·5	41·6	41·7
9	46·4	46·5	46·6	46·8	47·0
10	51·5	51·7	51·8	52·0	52·2
20	103·0	103·3	103·7	104·0	104·3
30	154·5	155·0	155·5	156·0	156·5
40	206·0	206·7	207·3	208·0	208·7
50	257·5	258·3	259·2	260·0	260·8

″	340	341	342	343	344
6	34·0	34·1	34·2	34·3	34·4
7	39·7	39·8	39·9	40·0	40·1
8	45·3	45·5	45·6	45·7	45·9
9	51·0	51·2	51·3	51·4	51·6
10	56·7	56·8	57·0	57·2	57·3
20	113·3	113·7	114·0	114·3	114·7
30	170·0	170·5	171·0	171·5	172·0
40	226·7	227·3	228·0	228·7	229·3
50	283·3	284·2	285·0	285·8	286·7

″	345	346	347	348	349
6	34·5	34·6	34·7	34·8	34·9
7	40·2	40·4	40·5	40·6	40·7
8	46·0	46·1	46·3	46·4	46·5
9	51·8	51·9	52·0	52·2	52·4
10	57·5	57·7	57·8	58·0	58·2
20	115·0	115·3	115·7	116·0	116·3
30	172·5	173·0	173·5	174·0	174·5
40	230·0	230·7	231·3	232·0	232·7
50	287·5	288·3	289·2	290·0	290·8

″	350	351	352	353	354
6	35·0	35·1	35·2	35·3	35·4
7	40·8	41·0	41·1	41·2	41·3
8	46·7	46·8	46·9	47·1	47·2
9	52·5	52·6	52·8	53·0	53·1
10	58·3	58·5	58·7	58·8	59·0
20	116·7	117·0	117·3	117·7	118·0
30	175·0	175·5	176·0	176·5	177·0
40	233·3	234·0	234·7	235·3	236·0
50	291·7	292·5	293·3	294·2	295·0

″	355	356	357	358	359
6	35·5	35·6	35·7	35·8	35·9
7	41·4	41·5	41·6	41·8	41·9
8	47·3	47·5	47·6	47·7	47·9
9	53·2	53·4	53·6	53·7	53·8
10	59·2	59·3	59·5	59·7	59·8
20	118·3	118·7	119·0	119·3	119·7
30	177·5	178·0	178·5	179·0	179·5
40	236·7	237·3	238·0	238·7	239·3
50	295·8	296·7	297·5	298·3	299·2

″	360	361	362	363	364
6	36·0	36·1	36·2	36·3	36·4
7	42·0	42·1	42·2	42·4	42·5
8	48·0	48·1	48·3	48·4	48·5
9	54·0	54·2	54·3	54·4	54·6
10	60·0	60·2	60·3	60·5	60·7
20	120·0	120·3	120·7	121·0	121·3
30	180·0	180·5	181·0	181·5	182·0
40	240·0	240·7	241·3	242·0	242·7
50	300·0	300·8	301·7	302·5	303·3

23°

′	sin	d	tan	d	cot	cos	d	′
00	9·591 878	298	9·627 852	351	0·372 148	9·964 026	54	60
01	·592 176	297	628 203	351	·371 797	·963 972	53	59
02	·592 473	297	·628 554	351	·371 446	·963 919	54	58
03	·592 770	297	·628 905	350	·371 095	·963 865	54	57
04	·593 067	296	·629 255	351	·370 745	·963 811	54	56
05	·593 363	296	·629 606	350	·370 394	·963 757	53	55
06	·593 659	296	·629 956	350	·370 044	·963 704	54	54
07	·593 955	296	·630 306	350	·369 694	·963 650	54	53
08	·594 251	296	·630 656	349	·369 344	·963 596	54	52
09	·594 547	295	·631 005	350	·368 995	·963 542	54	51
10	9·594 842	295	9·631 355	349	0·368 645	9·963 488	54	50
11	·595 137	295	·631 704	349	·368 296	·963 434	55	49
12	·595 432	295	·632 053	349	·367 947	·963 379	54	48
13	·595 727	294	·632 402	348	·367 598	·963 325	54	47
14	·596 021	294	·632 750	349	·367 250	·963 271	54	46
15	·596 315	294	·633 099	348	·366 901	·963 217	54	45
16	·596 609	294	·633 447	348	·366 553	·963 163	55	44
17	·596 903	293	·633 795	348	·366 205	·963 108	54	43
18	·597 196	294	·634 143	347	·365 857	·963 054	55	42
19	·597 490	293	·634 490	348	·365 510	·962 999	54	41
20	9·597 783	292	9·634 838	347	0·365 162	9·962 945	55	40
21	·598 075	293	·635 185	347	·364 815	·962 890	54	39
22	·598 368	292	·635 532	347	·364 468	·962 836	55	38
23	·598 660	292	·635 879	347	·364 121	·962 781	54	37
24	·598 952	292	·636 226	346	·363 774	·962 727	55	36
25	·599 244	292	·636 572	347	·363 428	·962 672	55	35
26	·599 536	291	·636 919	346	·363 081	·962 617	55	34
27	·599 827	291	·637 265	346	·362 735	·962 562	54	33
28	·600 118	291	·637 611	345	·362 389	·962 508	55	32
29	·600 409	291	·637 956	346	·362 044	·962 453	55	31
30	9·600 700	290	9·638 302	345	0·361 698	9·962 398	55	30
31	·600 990	290	·638 647	345	·361 353	·962 343	55	29
32	·601 280	290	·638 992	345	·361 008	·962 288	55	28
33	·601 570	290	·639 337	345	·360 663	·962 233	55	27
34	·601 860	290	·639 682	345	·360 318	·962 178	55	26
35	·602 150	289	·640 027	344	·359 973	·962 123	56	25
36	·602 439	289	·640 371	345	·359 629	·962 067	55	24
37	·602 728	289	·640 716	344	·359 284	·962 012	55	23
38	·603 017	288	·641 060	344	·358 940	·961 957	55	22
39	·603 305	289	·641 404	343	·358 596	·961 902	56	21
40	9·603 594	288	9·641 747	344	0·358 253	9·961 846	55	20
41	·603 882	288	·642 091	343	·357 909	·961 791	56	19
42	·604 170	287	·642 434	343	·357 566	·961 735	55	18
43	·604 457	288	·642 777	343	·357 223	·961 680	56	17
44	·604 745	287	·643 120	343	·356 880	·961 624	55	16
45	·605 032	287	·643 463	343	·356 537	·961 569	56	15
46	·605 319	287	·643 806	342	·356 194	·961 513	55	14
47	·605 606	286	·644 148	342	·355 852	·961 458	56	13
48	·605 892	287	·644 490	342	·355 510	·961 402	56	12
49	·606 179	286	·644 832	342	·355 168	·961 346	56	11
50	9·606 465	286	9·645 174	342	0·354 826	9·961 290	55	10
51	·606 751	285	·645 516	341	·354 484	·961 235	56	09
52	·607 036	286	·645 857	342	·354 143	·961 179	56	08
53	·607 322	285	·646 199	341	·353 801	·961 123	56	07
54	·607 607	285	·646 540	341	·353 460	·961 067	56	06
55	·607 892	285	·646 881	341	·353 119	·961 011	56	05
56	·608 177	284	·647 222	340	·352 778	·960 955	56	04
57	·608 461	284	·647 562	341	·352 438	·960 899	56	03
58	·608 745	284	·647 903	340	·352 097	·960 843	57	02
59	·609 029	284	·648 243	340	·351 757	·960 786	56	01
60	9·609 313		9·648 583		0·351 417	9·960 730		00
	cos		cot		tan	sin		66°

24°

′	sin	d	tan	d	cot	cos	d	′
00	9·609 313	284	9·648 583	340	0·351 417	9·960 730	56	60
01	·609 597	283	·648 923	340	·351 077	·960 674	56	59
02	·609 880	284	·649 263	339	·350 737	·960 618	57	58
03	·610 164	283	·649 602	340	·350 398	·960 561	56	57
04	·610 447	282	·649 942	339	·350 058	·960 505	57	56
05	·610 729	283	·650 281	339	·349 719	·960 448	56	55
06	·611 012	282	·650 620	339	·349 380	·960 392	57	54
07	·611 294	282	·650 959	338	·349 041	·960 335	56	53
08	·611 576	282	·651 297	339	·348 703	·960 279	57	52
09	·611 858	282	·651 636	338	·348 364	·960 222	57	51
10	9·612 140	281	9·651 974	338	0·348 026	9·960 165	56	50
11	·612 421	281	·652 312	338	·347 688	·960 109	57	49
12	·612 702	281	·652 650	338	·347 350	·960 052	57	48
13	·612 983	281	·652 988	338	·347 012	·959 995	57	47
14	·613 264	281	·653 326	337	·346 674	·959 938	56	46
15	·613 545	280	·653 663	337	·346 337	·959 882	57	45
16	·613 825	280	·654 000	337	·346 000	·959 825	57	44
17	·614 105	280	·654 337	337	·345 663	·959 768	57	43
18	·614 385	280	·654 674	337	·345 326	·959 711	57	42
19	·614 665	279	·655 011	337	·344 989	·959 654	58	41
20	9·614 944	279	9·655 348	336	0·344 652	9·959 596	57	40
21	·615 223	279	·655 684	336	·344 316	·959 539	57	39
22	·615 502	279	·656 020	336	·343 980	·959 482	57	38
23	·615 781	279	·656 356	336	·343 644	·959 425	57	37
24	·616 060	278	·656 692	336	·343 308	·959 368	58	36
25	·616 338	278	·657 028	336	·342 972	·959 310	57	35
26	·616 616	278	·657 364	335	·342 636	·959 253	58	34
27	·616 894	278	·657 699	335	·342 301	·959 195	57	33
28	·617 172	278	·658 034	335	·341 966	·959 138	58	32
29	·617 450	277	·658 369	335	·341 631	·959 080	57	31
30	9·617 727	277	9·658 704	335	0·341 296	9·959 023	58	30
31	·618 004	277	·659 039	334	·340 961	·958 965	57	29
32	·618 281	277	·659 373	335	·340 627	·958 908	58	28
33	·618 558	276	·659 708	334	·340 292	·958 850	58	27
34	·618 834	276	·660 042	334	·339 958	·958 792	58	26
35	·619 110	276	·660 376	334	·339 624	·958 734	57	25
36	·619 386	276	·660 710	333	·339 290	·958 677	58	24
37	·619 662	276	·661 043	334	·338 957	·958 619	58	23
38	·619 938	275	·661 377	333	·338 623	·958 561	58	22
39	·620 213	275	·661 710	333	·338 290	·958 503	58	21
40	9·620 488	275	9·662 043	333	0·337 957	9·958 445	58	20
41	·620 763	275	·662 376	333	·337 624	·958 387	58	19
42	·621 038	275	·662 709	333	·337 291	·958 329	58	18
43	·621 313	274	·663 042	333	·336 958	·958 271	58	17
44	·621 587	274	·663 375	332	·336 625	·958 213	59	16
45	·621 861	274	·663 707	332	·336 293	·958 154	58	15
46	·622 135	274	·664 039	332	·335 961	·958 096	58	14
47	·622 409	273	·664 371	332	·335 629	·958 038	59	13
48	·622 682	274	·664 703	332	·335 297	·957 979	58	12
49	·622 956	273	·665 035	331	·334 965	·957 921	58	11
50	9·623 229	273	9·665 366	332	0·334 634	9·957 863	59	10
51	·623 502	272	·665 698	331	·334 302	·957 804	58	09
52	·623 774	273	·666 029	331	·333 971	·957 746	59	08
53	·624 047	272	·666 360	331	·333 640	·957 687	59	07
54	·624 319	272	·666 691	330	·333 309	·957 628	58	06
55	·624 591	272	·667 021	331	·332 979	·957 570	59	05
56	·624 863	272	·667 352	330	·332 648	·957 511	59	04
57	·625 135	271	·667 682	331	·332 318	·957 452	59	03
58	·625 406	271	·668 013	330	·331 987	·957 393	58	02
59	·625 677	271	·668 343	330	·331 657	·957 335	59	01
60	9·625 948		9·668 673		0·331 327	9·957 276		00
	cos		cot		tan	sin		′

65°

PROPORTIONAL PARTS FOR SECONDS

″	56	57	58	59
6	5·6	5·7	5·8	5·9
7	6·5	6·6	6·8	6·9
8	7·5	7·6	7·7	7·9
9	8·4	8·6	8·7	8·8
10	9·3	9·5	9·7	9·8
20	18·7	19·0	19·3	19·7
30	28·0	28·5	29·0	29·5
40	37·3	38·0	38·7	39·3
50	46·7	47·5	48·3	49·2

″	271	272	273	274	275
6	27·1	27·2	27·3	27·4	27·5
7	31·6	31·7	31·8	32·0	32·1
8	36·1	36·3	36·4	36·5	36·7
9	40·6	40·8	41·0	41·1	41·2
10	45·2	45·3	45·5	45·7	45·8
20	90·3	90·7	91·0	91·3	91·7
30	135·5	136·0	136·5	137·0	137·5
40	180·7	181·3	182·0	182·7	183·3
50	225·8	226·7	227·5	228·3	229·2

″	276	277	278	279	280
6	27·6	27·7	27·8	27·9	28·0
7	32·2	32·3	32·4	32·6	32·7
8	36·8	36·9	37·1	37·2	37·3
9	41·4	41·6	41·7	41·8	42·0
10	46·0	46·2	46·3	46·5	46·7
20	92·0	92·3	92·7	93·0	93·3
30	138·0	138·5	139·0	139·5	140·0
40	184·0	184·7	185·3	186·0	186·7
50	230·0	230·8	231·7	232·5	233·3

″	281	282	283	284	330
6	28·1	28·2	28·3	28·4	33·0
7	32·8	32·9	33·0	33·1	38·5
8	37·5	37·6	37·7	37·9	44·0
9	42·2	42·3	42·4	42·6	49·5
10	46·8	47·0	47·2	47·3	55·0
20	93·7	94·0	94·3	94·7	110·0
30	140·5	141·0	141·5	142·0	165·0
40	187·3	188·0	188·7	189·3	220·0
50	234·2	235·0	235·8	236·7	275·0

″	331	332	333	334	335
6	33·1	33·2	33·3	33·4	33·5
7	38·6	38·7	38·8	39·0	39·1
8	44·1	44·3	44·4	44·5	44·7
9	49·6	49·8	50·0	50·1	50·2
10	55·2	55·3	55·5	55·7	55·8
20	110·3	110·7	111·0	111·3	111·7
30	165·5	166·0	166·5	167·0	167·5
40	220·7	221·3	222·0	222·7	223·3
50	275·8	276·7	277·5	278·3	279·2

″	336	337	338	339	340
6	33·6	33·7	33·8	33·9	34·0
7	39·2	39·3	39·4	39·6	39·7
8	44·8	44·9	45·1	45·2	45·3
9	50·4	50·6	50·7	50·8	51·0
10	56·0	56·2	56·3	56·5	56·7
20	112·0	112·3	112·7	113·0	113·3
30	168·0	168·5	169·0	169·5	170·0
40	224·0	224·7	225·3	226·0	226·7
50	280·0	280·8	281·7	282·5	283·3

25° ′	sin		tan		cot	cos		′
00	9·625 948	271	9·668 673	329	0·331 327	9·957 276	59	60
01	·626 219	271	·669 002	330	·330 998	·957 217	59	59
02	·626 490	270	·669 332	329	·330 668	·957 158	59	58
03	·626 760	270	·669 661	330	·330 339	·957 099	59	57
04	·627 030	270	·669 991	329	·330 009	·957 040	59	56
05	·627 300	270	·670 320	329	·329 680	·956 981	60	55
06	·627 570	270	·670 649	328	·329 351	·956 921	59	54
07	·627 840	269	·670 977	329	·329 023	·956 862	59	53
08	·628 109	269	·671 306	329	·328 694	·956 803	59	52
09	·628 378	269	·671 635	328	·328 365	·956 744	60	51
10	9·628 647	269	9·671 963	328	0·328 037	9·956 684	59	50
11	·628 916	269	·672 291	328	·327 709	·956 625	59	49
12	·629 185	268	·672 619	328	·327 381	·956 566	60	48
13	·629 453	268	·672 947	327	·327 053	·956 506	59	47
14	·629 721	268	·673 274	328	·326 726	·956 447	60	46
15	·629 989	268	·673 602	327	·326 398	·956 387	60	45
16	·630 257	267	·673 929	328	·326 071	·956 327	59	44
17	·630 524	268	·674 257	327	·325 743	·956 268	60	43
18	·630 792	267	·674 584	327	·325 416	·956 208	60	42
19	·631 059	267	·674 911	326	·325 089	·956 148	59	41
20	9·631 326	267	9·675 237	327	0·324 763	9·956 089	60	40
21	·631 593	266	·675 564	326	·324 436	·956 029	60	39
22	·631 859	266	·675 890	327	·324 110	·955 969	60	38
23	·632 125	267	·676 217	326	·323 783	·955 909	60	37
24	·632 392	266	·676 543	326	·323 457	·955 849	60	36
25	·632 658	265	·676 869	325	·323 131	·955 789	60	35
26	·632 923	266	·677 194	326	·322 806	·955 729	60	34
27	·633 189	265	·677 520	326	·322 480	·955 669	60	33
28	·633 454	265	·677 846	325	·322 154	·955 609	61	32
29	·633 719	265	·678 171	325	·321 829	·955 548	60	31
30	9·633 984	265	9·678 496	325	0·321 504	9·955 488	60	30
31	·634 249	265	·678 821	325	·321 179	·955 428	60	29
32	·634 514	264	·679 146	325	·320 854	·955 368	61	28
33	·634 778	264	·679 471	324	·320 529	·955 307	60	27
34	·635 042	264	·679 795	325	·320 205	·955 247	61	26
35	·635 306	264	·680 120	324	·319 880	·955 186	60	25
36	·635 570	264	·680 444	324	·319 556	·955 126	61	24
37	·635 834	263	·680 768	324	·319 232	·955 065	60	23
38	·636 097	263	·681 092	324	·318 908	·955 005	61	22
39	·636 360	263	·681 416	324	·318 584	·954 944	61	21
40	9·636 623	263	9·681 740	323	0·318 260	9·954 883	60	20
41	·636 886	262	·682 063	324	·317 937	·954 823	61	19
42	·637 148	263	·682 387	323	·317 613	·954 762	61	18
43	·637 411	262	·682 710	323	·317 290	·954 701	61	17
44	·637 673	262	·683 033	323	·316 967	·954 640	61	16
45	·637 935	262	·683 356	323	·316 644	·954 579	61	15
46	·638 197	261	·683 679	322	·316 321	·954 518	61	14
47	·638 458	262	·684 001	323	·315 999	·954 457	61	13
48	·638 720	261	·684 324	322	·315 676	·954 396	61	12
49	·638 981	261	·684 646	322	·315 354	·954 335	61	11
50	9·639 242	261	9·684 968	322	0·315 032	9·954 274	61	10
51	·639 503	261	·685 290	322	·314 710	·954 213	61	09
52	·639 764	260	·685 612	322	·314 388	·954 152	62	08
53	·640 024	260	·685 934	321	·314 066	·954 090	61	07
54	·640 284	260	·686 255	322	·313 745	·954 029	61	06
55	·640 544	260	·686 577	321	·313 423	·953 968	62	05
56	·640 804	260	·686 898	321	·313 102	·953 906	61	04
57	·641 064	260	·687 219	321	·312 781	·953 845	62	03
58	·641 324	259	·687 540	321	·312 460	·953 783	61	02
59	·641 583	259	·687 861	321	·312 139	·953 722	62	01
60	9·641 842		9·688 182		0·311 818	9·953 660		00
	cos		cot		tan	sin		64°

PROPORTIONAL PARTS FOR SECONDS

″	59	60	61	62
6	5·9	6·0	6·1	6·2
7	6·9	7·0	7·1	7·2
8	7·9	8·0	8·1	8·3
9	8·8	9·0	9·2	9·3
10	9·8	10·0	10·2	10·3
20	19·7	20·0	20·3	20·7
30	29·5	30·0	30·5	31·0
40	39·3	40·0	40·7	41·3
50	49·2	50·0	50·8	51·7

″	259	260	261	262	263
6	25·9	26·0	26·1	26·2	26·3
7	30·2	30·3	30·4	30·6	30·7
8	34·5	34·7	34·8	34·9	35·1
9	38·8	39·0	39·2	39·3	39·4
10	43·2	43·3	43·5	43·7	43·8
20	86·3	86·7	87·0	87·3	87·7
30	129·5	130·0	130·5	131·0	131·5
40	172·7	173·3	174·0	174·7	175·3
50	215·8	216·7	217·5	218·3	219·2

″	264	265	266	267	268
6	26·4	26·5	26·6	26·7	26·8
7	30·8	30·9	31·0	31·2	31·3
8	35·2	35·3	35·5	35·6	35·7
9	39·6	39·8	39·9	40·0	40·2
10	44·0	44·2	44·3	44·5	44·7
20	88·0	88·3	88·7	89·0	89·3
30	132·0	132·5	133·0	133·5	134·0
40	176·0	176·7	177·3	178·0	178·7
50	220·0	220·8	221·7	222·5	223·3

″	269	270	271
6	26·9	27·0	27·1
7	31·4	31·5	31·6
8	35·9	36·0	36·1
9	40·4	40·5	40·6
10	44·8	45·0	45·2
20	89·7	90·0	90·3
30	134·5	135·0	135·5
40	179·3	180·0	180·7
50	224·2	225·0	225·8

″	321	322	323	324	325
6	32·1	32·2	32·3	32·4	32·5
7	37·4	37·6	37·7	37·8	37·9
8	42·8	42·9	43·1	43·2	43·3
9	48·2	48·3	48·4	48·6	48·8
10	53·5	53·7	53·8	54·0	54·2
20	107·0	107·3	107·7	108·0	108·3
30	160·5	161·0	161·5	162·0	162·5
40	214·0	214·7	215·3	216·0	216·7
50	267·5	268·3	269·2	270·0	270·8

″	326	327	328	329	330
6	32·6	32·7	32·8	32·9	33·0
7	38·0	38·2	38·3	38·4	38·5
8	43·5	43·6	43·7	43·9	44·0
9	48·9	49·0	49·2	49·4	49·5
10	54·3	54·5	54·7	54·8	55·0
20	108·7	109·0	109·3	109·7	110·0
30	163·0	163·5	164·0	164·5	165·0
40	217·3	218·0	218·7	219·3	220·0
50	271·7	272·5	273·3	274·2	275·0

26°

′	sin	d	tan	d	cot	cos	d	′
00	9·641 842	259	9·688 182	320	0·311 818	9·953 660	61	60
01	·642 101	259	·688 502	321	·311 498	·953 599	62	59
02	·642 360	258	·688 823	320	·311 177	·953 537	62	58
03	·642 618	259	·689 143	320	·310 857	·953 475	62	57
04	·642 877	258	·689 463	320	·310 537	·953 413	61	56
05	·643 135	258	·689 783	320	·310 217	·953 352	62	55
06	·643 393	257	·690 103	320	·309 897	·953 290	62	54
07	·643 650	258	·690 423	319	·309 577	·953 228	62	53
08	·643 908	257	·690 742	320	·309 258	·953 166	62	52
09	·644 165	258	·691 062	319	·308 938	·953 104	62	51
10	9·644 423	257	9·691 381	319	0·308 619	9·953 042	62	50
11	·644 680	256	·691 700	319	·308 300	·952 980	62	49
12	·644 936	257	·692 019	319	·307 981	·952 918	63	48
13	·645 193	257	·692 338	318	·307 662	·952 855	62	47
14	·645 450	256	·692 656	319	·307 344	·952 793	62	46
15	·645 706	256	·692 975	318	·307 025	·952 731	62	45
16	·645 962	256	·693 293	319	·306 707	·952 669	63	44
17	·646 218	256	·693 612	318	·306 388	·952 606	62	43
18	·646 474	255	·693 930	318	·306 070	·952 544	63	42
19	·646 729	255	·694 248	318	·305 752	·952 481	62	41
20	9·646 984	256	9·694 566	317	0·305 434	9·952 419	63	40
21	·647 240	254	·694 883	318	·305 117	·952 356	62	39
22	·647 494	255	·695 201	317	·304 799	·952 294	63	38
23	·647 749	255	·695 518	318	·304 482	·952 231	63	37
24	·648 004	254	·695 836	317	·304 164	·952 168	62	36
25	·648 258	254	·696 153	317	·303 847	·952 106	63	35
26	·648 512	254	·696 470	317	·303 530	·952 043	63	34
27	·648 766	254	·696 787	316	·303 213	·951 980	63	33
28	·649 020	254	·697 103	317	·302 897	·951 917	63	32
29	·649 274	253	·697 420	316	·302 580	·951 854	63	31
30	9·649 527	254	9·697 736	317	0·302 264	9·951 791	63	30
31	·649 781	253	·698 053	316	·301 947	·951 728	63	29
32	·650 034	253	·698 369	316	·301 631	·951 665	63	28
33	·650 287	252	·698 685	316	·301 315	·951 602	63	27
34	·650 539	253	·699 001	315	·300 999	·951 539	63	26
35	·650 792	252	·699 316	316	·300 684	·951 476	64	25
36	·651 044	253	·699 632	315	·300 368	·951 412	63	24
37	·651 297	252	·699 947	316	·300 053	·951 349	63	23
38	·651 549	251	·700 263	315	·299 737	·951 286	64	22
39	·651 800	252	·700 578	315	·299 422	·951 222	63	21
40	9·652 052	252	9·700 893	315	0·299 107	9·951 159	63	20
41	·652 304	251	·701 208	315	·298 792	·951 096	64	19
42	·652 555	251	·701 523	314	·298 477	·951 032	64	18
43	·652 806	251	·701 837	315	·298 163	·950 968	63	17
44	·653 057	251	·702 152	314	·297 848	·950 905	64	16
45	·653 308	250	·702 466	315	·297 534	·950 841	63	15
46	·653 558	250	·702 781	314	·297 219	·950 778	64	14
47	·653 808	251	·703 095	314	·296 905	·950 714	64	13
48	·654 059	250	·703 409	313	·296 591	·950 650	64	12
49	·654 309	249	·703 722	314	·296 278	·950 586	64	11
50	9·654 558	250	9·704 036	314	0·295 964	9·950 522	64	10
51	·654 808	250	·704 350	313	·295 650	·950 458	64	09
52	·655 058	249	·704 663	313	·295 337	·950 394	64	08
53	·655 307	249	·704 976	314	·295 024	·950 330	64	07
54	·655 556	249	·705 290	313	·294 710	·950 266	64	06
55	·655 805	249	·705 603	313	·294 397	·950 202	64	05
56	·656 054	248	·705 916	312	·294 084	·950 138	64	04
57	·656 302	249	·706 228	313	·293 772	·950 074	64	03
58	·656 551	248	·706 541	313	·293 459	·950 010	65	02
59	·656 799	248	·706 854	312	·293 146	·949 945	64	01
60	9·657 047		9·707 166		0·292 834	9·949 881		00
	cos		cot		tan	sin		63°

PROPORTIONAL PARTS FOR SECONDS

″	61	62	63	64	65
6	6·1	6·2	6·3	6·4	6·5
7	7·1	7·2	7·4	7·5	7·6
8	8·1	8·3	8·4	8·5	8·7
9	9·2	9·3	9·4	9·6	9·8
10	10·2	10·3	10·5	10·7	10·8
20	20·3	20·7	21·0	21·3	21·7
30	30·5	31·0	31·5	32·0	32·5
40	40·7	41·3	42·0	42·7	43·3
50	50·8	51·7	52·5	53·3	54·2

″	248	249	250	251	252
6	24·8	24·9	25·0	25·1	25·2
7	28·9	29·0	29·2	29·3	29·4
8	33·1	33·2	33·3	33·5	33·6
9	37·2	37·4	37·5	37·6	37·8
10	41·3	41·5	41·7	41·8	42·0
20	82·7	83·0	83·3	83·7	84·0
30	124·0	124·5	125·0	125·5	126·0
40	165·3	166·0	166·7	167·3	168·0
50	206·7	207·5	208·3	209·2	210·0

″	253	254	255	256	257
6	25·3	25·4	25·5	25·6	25·7
7	29·5	29·6	29·8	29·9	30·0
8	33·7	33·9	34·0	34·1	34·3
9	38·0	38·1	38·2	38·4	38·6
10	42·2	42·3	42·5	42·7	42·8
20	84·3	84·7	85·0	85·3	85·7
30	126·5	127·0	127·5	128·0	128·5
40	168·7	169·3	170·0	170·7	171·3
50	210·8	211·7	212·5	213·3	214·2

″	258	259
6	25·8	25·9
7	30·1	30·2
8	34·4	34·5
9	38·7	38·8
10	43·0	43·2
20	86·0	86·3
30	129·0	129·5
40	172·0	172·7
50	215·0	215·8

″	312	313	314	315	316
6	31·2	31·3	31·4	31·5	31·6
7	36·4	36·5	36·6	36·8	36·9
8	41·6	41·7	41·9	42·0	42·1
9	46·8	47·0	47·1	47·2	47·4
10	52·0	52·2	52·3	52·5	52·7
20	104·0	104·3	104·7	105·0	105·3
30	156·0	156·5	157·0	157·5	158·0
40	208·0	208·7	209·3	210·0	210·7
50	260·0	260·8	261·7	262·5	263·3

″	317	318	319	320
6	31·7	31·8	31·9	32·0
7	37·0	37·1	37·2	37·3
8	42·3	42·4	42·5	42·7
9	47·6	47·7	47·8	48·0
10	52·8	53·0	53·2	53·3
20	105·7	106·0	106·3	106·7
30	158·5	159·0	159·5	160·0
40	211·3	212·0	212·7	213·3
50	264·2	265·0	265·8	266·7

27°

′	sin	d.	tan	c.d.	cot	cos	d.	′
00	9·657 047	248	9·707 166	312	0·292 834	9·949 881	65	60
01	·657 295	247	·707 478	312	·292 522	·949 816	64	59
02	·657 542	248	·707 790	312	·292 210	·949 752	64	58
03	·657 790	247	·708 102	312	·291 898	·949 688	65	57
04	·658 037	247	·708 414	312	·291 586	·949 623	65	56
05	·658 284	247	·708 726	311	·291 274	·949 558	64	55
06	·658 531	247	·709 037	312	·290 963	·949 494	65	54
07	·658 778	247	·709 349	311	·290 651	·949 429	65	53
08	·659 025	246	·709 660	311	·290 340	·949 364	64	52
09	·659 271	246	·709 971	311	·290 029	·949 300	65	51
10	9·659 517	246	9·710 282	311	0·289 718	9·949 235	65	50
11	·659 763	246	·710 593	311	·289 407	·949 170	65	49
12	·660 009	246	·710 904	311	·289 096	·949 105	65	48
13	·660 255	246	·711 215	310	·288 785	·949 040	65	47
14	·660 501	245	·711 525	311	·288 475	·948 975	65	46
15	·660 746	245	·711 836	310	·288 164	·948 910	65	45
16	·660 991	245	·712 146	310	·287 854	·948 845	65	44
17	·661 236	245	·712 456	310	·287 544	·948 780	65	43
18	·661 481	245	·712 766	310	·287 234	·948 715	65	42
19	·661 726	244	·713 076	310	·286 924	·948 650	66	41
20	9·661 970	244	9·713 386	310	0·286 614	9·948 584	65	40
21	·662 214	245	·713 696	309	·286 304	·948 519	65	39
22	·662 459	244	·714 005	309	·285 995	·948 454	66	38
23	·662 703	243	·714 314	310	·285 686	·948 388	65	37
24	·662 946	244	·714 624	309	·285 376	·948 323	66	36
25	·663 190	243	·714 933	309	·285 067	·948 257	65	35
26	·663 433	244	·715 242	309	·284 758	·948 192	66	34
27	·663 677	243	·715 551	309	·284 449	·948 126	66	33
28	·663 920	243	·715 860	308	·284 140	·948 060	65	32
29	·664 163	243	·716 168	309	·283 832	·947 995	66	31
30	9·664 406	242	9·716 477	308	0·283 523	9·947 929	66	30
31	·664 648	243	·716 785	308	·283 215	·947 863	66	29
32	·664 891	242	·717 093	308	·282 907	·947 797	66	28
33	·665 133	242	·717 401	308	·282 599	·947 .31	66	27
34	·665 375	242	·717 709	308	·282 291	·947 665	65	26
35	·665 617	242	·718 017	308	·281 983	·947 600	67	25
36	·665 859	241	·718 325	308	·281 675	·947 533	66	24
37	·666 100	242	·718 633	307	·281 367	·947 467	66	23
38	·666 342	241	·718 940	308	·281 060	·947 401	66	22
39	·666 583	241	·719 248	307	·280 752	·947 335	66	21
40	9·666 824	241	9·719 555	307	0·280 445	9·947 269	66	20
41	·667 065	240	·719 862	307	·280 138	·947 203	67	19
42	·667 305	241	·720 169	307	·279 831	·947 136	66	18
43	·667 546	240	·720 476	307	·279 524	·947 070	66	17
44	·667 786	241	·720 783	306	·279 217	·947 004	67	16
45	·668 027	240	·721 089	307	·278 911	·946 937	66	15
46	·668 267	239	·721 396	306	·278 604	·946 871	67	14
47	·668 506	240	·721 702	307	·278 298	·946 804	66	13
48	·668 746	240	·722 009	306	·277 991	·946 738	67	12
49	·668 986	239	·722 315	306	·277 685	·946 671	67	11
50	9·669 225	239	9·722 621	306	0·277 379	9·946 604	66	10
51	·669 464	239	·722 927	305	·277 073	·946 538	67	09
52	·669 703	239	·723 232	306	·276 768	·946 471	67	08
53	·669 942	239	·723 538	306	·276 462	·946 404	67	07
54	·670 181	238	·723 844	305	·276 156	·946 337	67	06
55	·670 419	239	·724 149	305	·275 851	·946 270	67	05
56	·670 658	238	·724 454	306	·275 546	·946 203	67	04
57	·670 896	238	·724 760	305	·275 240	·946 136	67	03
58	·671 134	238	·725 065	305	·274 935	·946 069	67	02
59	·671 372	237	·725 370	304	·274 630	·946 002	67	01
60	9·671 609		9·725 674		0·274 326	9·945 935		00
	cos		cot		tan	sin		**62°**

PROPORTIONAL PARTS FOR SECONDS

″	64	65	66	67
6	6·4	6·5	6·6	6·7
7	7·5	7·6	7·7	7·8
8	8·5	8·7	8·8	8·9
9	9·6	9·8	9·9	10·0
10	10·7	10·8	11·0	11·2
20	21·3	21·7	22·0	22·3
30	32·0	32·5	33·0	33·5
40	42·7	43·3	44·0	44·7
50	53·3	54·2	55·0	55·8

″	237	238	239	240	241
6	23·7	23·8	23·9	24·0	24·1
7	27·6	27·8	27·9	28·0	28·1
8	31·6	31·7	31·9	32·0	32·1
9	35·6	35·7	35·8	36·0	36·2
10	39·5	39·7	39·8	40·0	40·2
20	79·0	79·3	79·7	80·0	80·3
30	118·5	119·0	119·5	120·0	120·5
40	158·0	158·7	159·3	160·0	160·7
50	197·5	198·3	199·2	200·0	200·8

″	242	243	244	245	246
6	24·2	24·3	24·4	24·5	24·6
7	28·2	28·4	28·5	28·6	28·7
8	32·3	32·4	32·5	32·7	32·8
9	36·3	36·4	36·6	36·8	36·9
10	40·3	40·5	40·7	40·8	41·0
20	80·7	81·0	81·3	81·7	82·0
30	121·0	121·5	122·0	122·5	123·0
40	161·3	162·0	162·7	163·3	164·0
50	201·7	202·5	203·3	204·2	205·0

″	247	248
6	24·7	24·8
7	28·8	28·9
8	32·9	33·1
9	37·0	37·2
10	41·2	41·3
20	82·3	82·7
30	123·5	124·0
40	164·7	165·3
50	205·8	206·7

″	304	305	306	307	308
6	30·4	30·5	30·6	30·7	30·8
7	35·5	35·6	35·7	35·8	35·9
8	40·5	40·7	40·8	40·9	41·1
9	45·6	45·8	45·9	46·0	46·2
10	50·7	50·8	51·0	51·2	51·3
20	101·3	101·7	102·0	102·3	102·7
30	152·0	152·5	153·0	153·5	154·0
40	202·7	203·3	204·0	204·7	205·3
50	253·3	254·2	255·0	255·8	256·7

″	309	310	311	312
6	30·9	31·0	31·1	31·2
7	36·0	36·2	36·3	36·4
8	41·2	41·3	41·5	41·6
9	46·4	46·5	46·6	46·8
10	51·5	51·7	51·8	52·0
20	103·0	103·3	103·7	104·0
30	154·5	155·0	155·5	156·0
40	206·0	206·7	207·3	208·0
50	257·5	258·3	259·2	260·0

28°

′	sin	d	tan	d	cot	cos	d	′
00	9·671 609	238	9·725 674	305	0·274 326	9·945 935	67	60
01	·671 847	237	·725 979	305	·274 021	·945 868	68	59
02	·672 084	237	·726 284	304	·273 716	·945 800	67	58
03	·672 321	237	·726 588	304	·273 412	·945 733	67	57
04	·672 558	237	·726 892	305	·273 108	·945 666	68	56
05	·672 795	237	·727 197	304	·272 803	·945 598	67	55
06	·673 032	236	·727 501	304	·272 499	·945 531	67	54
07	·673 268	237	·727 805	304	·272 195	·945 464	68	53
08	·673 505	236	·728 109	303	·271 891	·945 396	68	52
09	·673 741	236	·728 412	304	·271 588	·945 328	67	51
10	9·673 977	236	9·728 716	304	0·271 284	9·945 261	68	50
11	·674 213	235	·729 020	303	·270 980	·945 193	68	49
12	·674 448	236	·729 323	303	·270 677	·945 125	67	48
13	·674 684	235	·729 626	303	·270 374	·945 058	68	47
14	·674 919	236	·729 929	304	·270 071	·944 990	68	46
15	·675 155	235	·730 233	302	·269 767	·944 922	68	45
16	·675 390	234	·730 535	303	·269 465	·944 854	68	44
17	·675 624	235	·730 838	303	·269 162	·944 786	68	43
18	·675 859	235	·731 141	303	·268 859	·944 718	68	42
19	·676 094	234	·731 444	302	·268 556	·944 650	68	41
20	9·676 328	234	9·731 746	302	0·268 254	9·944 582	68	40
21	·676 562	234	·732 048	303	·267 952	·944 514	68	39
22	·676 796	234	·732 351	302	·267 649	·944 446	69	38
23	·677 030	234	·732 653	302	·267 347	·944 377	68	37
24	·677 264	234	·732 955	302	·267 045	·944 309	68	36
25	·677 498	233	·733 257	301	·266 743	·944 241	69	35
26	·677 731	233	·733 558	302	·266 442	·944 172	68	34
27	·677 964	233	·733 860	302	·266 140	·944 104	68	33
28	·678 197	233	·734 162	301	·265 838	·944 036	69	32
29	·678 430	233	·734 463	301	·265 537	·943 967	68	31
30	9·678 663	232	9·734 764	302	0·265 236	9·943 899	69	30
31	·678 895	233	·735 066	301	·264 934	·943 830	69	29
32	·679 128	232	·735 367	301	·264 633	·943 761	68	28
33	·679 360	232	·735 668	301	·264 332	·943 693	69	27
34	·679 592	232	·735 969	300	·264 031	·943 624	69	26
35	·679 824	232	·736 269	301	·263 731	·943 555	69	25
36	·680 056	232	·736 570	300	·263 430	·943 486	69	24
37	·680 288	231	·736 870	301	·263 130	·943 417	69	23
38	·680 519	231	·737 171	300	·262 829	·943 348	69	22
39	·680 750	232	·737 471	300	·262 529	·943 279	69	21
40	9·680 982	231	9·737 771	300	0·262 229	9·943 210	69	20
41	·681 213	230	·738 071	300	·261 929	·943 141	69	19
42	·681 443	231	·738 371	300	·261 629	·943 072	69	18
43	·681 674	231	·738 671	300	·261 329	·943 003	69	17
44	·681 905	230	·738 971	300	·261 029	·942 934	70	16
45	·682 135	230	·739 271	299	·260 729	·942 864	69	15
46	·682 365	230	·739 570	300	·260 430	·942 795	69	14
47	·682 595	230	·739 870	299	·260 130	·942 726	70	13
48	·682 825	230	·740 169	299	·259 831	·942 656	69	12
49	·683 055	229	·740 468	299	·259 532	·942 587	70	11
50	9·683 284	230	9·740 767	299	0·259 233	9·942 517	69	10
51	·683 514	229	·741 066	299	·258 934	·942 448	70	09
52	·683 743	229	·741 365	299	·258 635	·942 378	70	08
53	·683 972	229	·741 664	298	·258 336	·942 308	69	07
54	·684 201	229	·741 962	299	·258 038	·942 239	70	06
55	·684 430	228	·742 261	298	·257 739	·942 169	70	05
56	·684 658	229	·742 559	299	·257 441	·942 099	70	04
57	·684 887	228	·742 858	298	·257 142	·942 029	70	03
58	·685 115	228	·743 156	298	·256 844	·941 959	70	02
59	·685 343	228	·743 454	298	·256 546	·941 889	70	01
60	9·685 571		9·743 752		0·256 248	9·941 819		00
	cos		cot		tan	sin		61°

P.Ps. FOR SECONDS

″	67	68	69	70
6	6·7	6·8	6·9	7·0
7	7·8	7·9	8·0	8·2
8	8·9	9·1	9·2	9·3
9	10·0	10·2	10·4	10·5
10	11·2	11·3	11·5	11·7
20	22·3	22·7	23·0	23·3
30	33·5	34·0	34·5	35·0
40	44·7	45·3	46·0	46·7
50	55·8	56·7	57·5	58·3

″	228	229	230	231
6	22·8	22·9	23·0	23·1
7	26·6	26·7	26·8	27·0
8	30·4	30·5	30·7	30·8
9	34·2	34·4	34·5	34·6
10	38·0	38·2	38·3	38·5
20	76·0	76·3	76·7	77·0
30	114·0	114·5	115·0	115·5
40	152·0	152·7	153·3	154·0
50	190·0	190·8	191·7	192·5

″	232	233	234	235
6	23·2	23·3	23·4	23·5
7	27·1	27·2	27·3	27·4
8	30·9	31·1	31·2	31·3
9	34·8	35·0	35·1	35·2
10	38·7	38·8	39·0	39·2
20	77·3	77·7	78·0	78·3
30	116·0	116·5	117·0	117·5
40	154·7	155·3	156·0	156·7
50	193·3	194·2	195·0	195·8

″	236	237	238
6	23·6	23·7	23·8
7	27·5	27·6	27·8
8	31·5	31·6	31·7
9	35·4	35·6	35·7
10	39·3	39·5	39·7
20	78·7	79·0	79·3
30	118·0	118·5	119·0
40	157·3	158·0	158·7
50	196·7	197·5	198·3

″	298	299	300	301
6	29·8	29·9	30·0	30·1
7	34·8	34·9	35·0	35·1
8	39·7	39·9	40·0	40·1
9	44·7	44·8	45·0	45·2
10	49·7	49·8	50·0	50·2
20	99·3	99·7	100·0	100·3
30	149·0	149·5	150·0	150·5
40	198·7	199·3	200·0	200·7
50	248·3	249·2	250·0	250·8

″	302	303	304	305
6	30·2	30·3	30·4	30·5
7	35·2	35·4	35·5	35·6
8	40·3	40·4	40·5	40·7
9	45·3	45·4	45·6	45·8
10	50·3	50·5	50·7	50·8
20	100·7	101·0	101·3	101·7
30	151·0	151·5	152·0	152·5
40	201·3	202·0	202·7	203·3
50	251·7	252·5	253·3	254·2

29° ′	sin		tan		cot	cos		′
00	9·685 571	228	9·743 752	298	0·256 248	9·941 819	70	60
01	·685 799	228	·744 050	298	·255 950	·941 749	70	59
02	·686 027	227	·744 348	297	·255 652	·941 679	70	58
03	·686 254	228	·744 645	298	·255 355	·941 609	70	57
04	·686 482	227	·744 943	297	·255 057	·941 539	70	56
05	·686 709	227	·745 240	298	·254 760	·941 469	71	55
06	·686 936	227	·745 538	297	·254 462	·941 398	70	54
07	·687 163	226	·745 835	297	·254 165	·941 328	70	53
08	·687 389	227	·746 132	297	·253 868	·941 258	71	52
09	·687 616	227	·746 429	297	·253 571	·941 187	70	51
10	9·687 843	226	9·746 726	297	0·253 274	9·941 117	71	50
11	·688 069	226	·747 023	296	·252 977	·941 046	71	49
12	·688 295	226	·747 319	297	·252 681	·940 975	70	48
13	·688 521	226	·747 616	297	·252 384	·940 905	71	47
14	·688 747	225	·747 913	296	·252 087	·940 834	71	46
15	·688 972	226	·748 209	296	·251 791	·940 763	70	45
16	·689 198	225	·748 505	296	·251 495	·940 693	71	44
17	·689 423	225	·748 801	296	·251 199	·940 622	71	43
18	·689 648	225	·749 097	296	·250 903	·940 551	71	42
19	·689 873	225	·749 393	296	·250 607	·940 480	71	41
20	9·690 098	225	9·749 689	296	0·250 311	9·940 409	71	40
21	·690 323	225	·749 985	296	·250 015	·940 338	71	39
22	·690 548	224	·750 281	295	·249 719	·940 267	71	38
23	·690 772	224	·750 576	296	·249 424	·940 196	71	37
24	·690 996	224	·750 872	295	·249 128	·940 125	71	36
25	·691 220	224	·751 167	295	·248 833	·940 054	72	35
26	·691 444	224	·751 462	295	·248 538	·939 982	71	34
27	·691 668	224	·751 757	295	·248 243	·939 911	71	33
28	·691 892	223	·752 052	295	·247 948	·939 840	72	32
29	·692 115	224	·752 347	295	·247 653	·939 768	71	31
30	9·692 339	223	9·752 642	295	0·247 358	9·939 697	72	30
31	·692 562	223	·752 937	294	·247 063	·939 625	71	29
32	·692 785	223	·753 231	295	·246 769	·939 554	72	28
33	·693 008	223	·753 526	294	·246 474	·939 482	72	27
34	·693 231	222	·753 820	295	·246 180	·939 410	71	26
35	·693 453	223	·754 115	294	·245 885	·939 339	72	25
36	·693 676	222	·754 409	294	·245 591	·939 267	72	24
37	·693 898	222	·754 703	294	·245 297	·939 195	72	23
38	·694 120	222	·754 997	294	·245 003	·939 123	71	22
39	·694 342	222	·755 291	294	·244 709	·939 052	72	21
40	9·694 564	222	9·755 585	293	0·244 415	9·938 980	72	20
41	·694 786	221	·755 878	294	·244 122	·938 908	72	19
42	·695 007	222	·756 172	293	·243 828	·938 836	73	18
43	·695 229	221	·756 465	294	·243 535	·938 763	72	17
44	·695 450	221	·756 759	293	·243 241	·938 691	72	16
45	·695 671	221	·757 052	293	·242 948	·938 619	72	15
46	·695 892	221	·757 345	293	·242 655	·938 547	72	14
47	·696 113	221	·757 638	293	·242 362	·938 475	73	13
48	·696 334	220	·757 931	293	·242 069	·938 402	72	12
49	·696 554	221	·758 224	293	·241 776	·938 330	72	11
50	9·696 775	220	9·758 517	293	0·241 483	9·938 258	73	10
51	·696 995	220	·758 810	292	·241 190	·938 185	72	09
52	·697 215	220	·759 102	293	·240 898	·938 113	73	08
53	·697 435	219	·759 395	292	·240 605	·938 040	73	07
54	·697 654	220	·759 687	292	·240 313	·937 967	72	06
55	·697 874	220	·759 979	293	·240 021	·937 895	73	05
56	·698 094	219	·760 272	292	·239 728	·937 822	73	04
57	·698 313	219	·760 564	292	·239 436	·937 749	73	03
58	·698 532	219	·760 856	292	·239 144	·937 676	72	02
59	·698 751	219	·761 148	291	·238 852	·937 604	73	01
60	9·698 970		9·761 439		0·238 561	9·937 531		00
	cos		cot		tan	sin		60°

P.Ps. FOR SECONDS

″	70	71	72	73
6	7·0	7·1	7·2	7·3
7	8·2	8·3	8·4	8·5
8	9·3	9·5	9·6	9·7
9	10·5	10·6	10·8	11·0
10	11·7	11·8	12·0	12·2
20	23·3	23·7	24·0	24·3
30	35·0	35·5	36·0	36·5
40	46·7	47·3	48·0	48·7
50	58·3	59·2	60·0	60·8

″	219	220	221	222
6	21·9	22·0	22·1	22·2
7	25·6	25·7	25·8	25·9
8	29·2	29·3	29·5	29·6
9	32·8	33·0	33·2	33·3
10	36·5	36·7	36·8	37·0
20	73·0	73·3	73·7	74·0
30	109·5	110·0	110·5	111·0
40	146·0	146·7	147·3	148·0
50	182·5	183·3	184·2	185·0

″	223	224	225	226
6	22·3	22·4	22·5	22·6
7	26·0	26·1	26·2	26·4
8	29·7	29·9	30·0	30·1
9	33·4	33·6	33·8	33·9
10	37·2	37·3	37·5	37·7
20	74·3	74·7	75·0	75·3
30	111·5	112·0	112·5	113·0
40	148·7	149·3	150·0	150·7
50	185·8	186·7	187·5	188·3

″	227	228
6	22·7	22·8
7	26·5	26·6
8	30·3	30·4
9	34·0	34·2
10	37·8	38·0
20	75·7	76·0
30	113·5	114·0
40	151·3	152·0
50	189·2	190·0

″	291	292	293	294
6	29·1	29·2	29·3	29·4
7	34·0	34·1	34·2	34·3
8	38·8	38·9	39·1	39·2
9	43·6	43·8	44·0	44·1
10	48·5	48·7	48·8	49·0
20	97·0	97·3	97·7	98·0
30	145·5	146·0	146·5	147·0
40	194·0	194·7	195·3	196·0
50	242·5	243·3	244·2	245·0

″	295	296	297	298
6	29·5	29·6	29·7	29·8
7	34·4	34·5	34·6	34·8
8	39·3	39·5	39·6	39·7
9	44·2	44·4	44·6	44·7
10	49·2	49·3	49·5	49·7
20	98·3	98·7	99·0	99·3
30	147·5	148·0	148·5	149·0
40	196·7	197·3	198·0	198·7
50	245·8	246·7	247·5	248·3

30°

′	sin	d.	tan	d.c.	cot	cos	d.	′
00	9·698 970	219	9·761 439	292	0·238 561	9·937 531	73	60
01	·699 189	218	·761 731	292	·238 269	·937 458	73	59
02	·699 407	219	·762 023	291	·237 977	·937 385	73	58
03	·699 626	218	·762 314	292	·237 686	·937 312	74	57
04	·699 844	218	·762 606	291	·237 394	·937 238	73	56
05	·700 062	218	·762 897	291	·237 103	·937 165	73	55
06	·700 280	218	·763 188	291	·236 812	·937 092	73	54
07	·700 498	218	·763 479	291	·236 521	·937 019	73	53
08	·700 716	217	·763 770	291	·236 230	·936 946	74	52
09	·700 933	218	·764 061	291	·235 939	·936 872	73	51
10	9·701 151	217	9·764 352	291	0·235 648	9·936 799	74	50
11	·701 368	217	·764 643	290	·235 357	·936 725	73	49
12	·701 585	217	·764 933	291	·235 067	·936 652	74	48
13	·701 802	217	·765 224	290	·234 776	·936 578	73	47
14	·702 019	217	·765 514	291	·234 486	·936 505	74	46
15	·702 236	216	·765 805	290	·234 195	·936 431	74	45
16	·702 452	217	·766 095	290	·233 905	·936 357	73	44
17	·702 669	216	·766 385	290	·233 615	·936 284	74	43
18	·702 885	216	·766 675	290	·233 325	·936 210	74	42
19	·703 101	216	·766 965	290	·233 035	·936 136	74	41
20	9·703 317	216	9·767 255	290	0·232 745	9·936 062	74	40
21	·703 533	216	·767 545	289	·232 455	·935 988	74	39
22	·703 749	215	·767 834	290	·232 166	·935 914	74	38
23	·703 964	215	·768 124	290	·231 876	·935 840	74	37
24	·704 179	216	·768 414	289	·231 586	·935 766	74	36
25	·704 395	215	·768 703	289	·231 297	·935 692	74	35
26	·704 610	215	·768 992	289	·231 008	·935 618	75	34
27	·704 825	215	·769 281	290	·230 719	·935 543	74	33
28	·705 040	214	·769 571	289	·230 429	·935 469	74	32
29	·705 254	215	·769 860	288	·230 140	·935 395	75	31
30	9·705 469	214	9·770 148	289	0·229 852	9·935 320	74	30
31	·705 683	215	·770 437	289	·229 563	·935 246	75	29
32	·705 898	214	·770 726	289	·229 274	·935 171	74	28
33	·706 112	214	·771 015	288	·228 985	·935 097	75	27
34	·706 326	213	·771 303	289	·228 697	·935 022	74	26
35	·706 539	214	·771 592	288	·228 408	·934 948	75	25
36	·706 753	214	·771 880	288	·228 120	·934 873	75	24
37	·706 967	213	·772 168	289	·227 832	·934 798	75	23
38	·707 180	213	·772 457	288	·227 543	·934 723	74	22
39	·707 393	213	·772 745	288	·227 255	·934 649	75	21
40	9·707 606	213	9·773 033	288	0·226 967	9·934 574	75	20
41	·707 819	213	·773 321	287	·226 679	·934 499	75	19
42	·708 032	213	·773 608	288	·226 392	·934 424	75	18
43	·708 245	213	·773 896	288	·226 104	·934 349	75	17
44	·708 458	212	·774 184	287	·225 816	·934 274	75	16
45	·708 670	212	·774 471	288	·225 529	·934 199	76	15
46	·708 882	212	·774 759	287	·225 241	·934 123	75	14
47	·709 094	212	·775 046	287	·224 954	·934 048	75	13
48	·709 306	212	·775 333	288	·224 667	·933 973	75	12
49	·709 518	212	·775 621	287	·224 379	·933 898	76	11
50	9·709 730	211	9·775 908	287	0·224 092	9·933 822	75	10
51	·709 941	212	·776 195	287	·223 805	·933 747	76	09
52	·710 153	211	·776 482	286	·223 518	·933 671	75	08
53	·710 364	211	·776 768	287	·223 232	·933 596	76	07
54	·710 575	211	·777 055	287	·222 945	·933 520	75	06
55	·710 786	211	·777 342	286	·222 658	·933 445	76	05
56	·710 997	211	·777 628	287	·222 372	·933 369	76	04
57	·711 208	211	·777 915	286	·222 085	·933 293	76	03
58	·711 419	210	·778 201	287	·221 799	·933 217	76	02
59	·711 629	210	·778 488	286	·221 512	·933 141	75	01
60	9·711 839		9·778 774		0·221 226	9·933 066		00
	cos		cot		tan	sin		**59°**

P.Ps. FOR SECONDS

″	73	74	75	76
6	7·3	7·4	7·5	7·6
7	8·5	8·6	8·8	8·9
8	9·7	9·9	10·0	10·1
9	11·0	11·1	11·2	11·4
10	12·2	12·3	12·5	12·7
20	24·3	24·7	25·0	25·3
30	36·5	37·0	37·5	38·0
40	48·7	49·3	50·0	50·7
50	60·8	61·7	62·5	63·3

″	210	211	212	213
6	21·0	21·1	21·2	21·3
7	24·5	24·6	24·7	24·8
8	28·0	28·1	28·3	28·4
9	31·5	31·6	31·8	32·0
10	35·0	35·2	35·3	35·5
20	70·0	70·3	70·7	71·0
30	105·0	105·5	106·0	106·5
40	140·0	140·7	141·3	142·0
50	175·0	175·8	176·7	177·5

″	214	215	216	217
6	21·4	21·5	21·6	21·7
7	25·0	25·1	25·2	25·3
8	28·5	28·7	28·8	28·9
9	32·1	32·2	32·4	32·6
10	35·7	35·8	36·0	36·2
20	71·3	71·7	72·0	72·3
30	107·0	107·5	108·0	108·5
40	142·7	143·3	144·0	144·7
50	178·3	179·2	180·0	180·8

″	218	219
6	21·8	21·9
7	25·4	25·6
8	29·1	29·2
9	32·7	32·8
10	36·3	36·5
20	72·7	73·0
30	109·0	109·5
40	145·3	146·0
50	181·7	182·5

″	286	287	288	289
6	28·6	28·7	28·8	28·9
7	33·4	33·5	33·6	33·7
8	38·1	38·3	38·4	38·5
9	42·9	43·0	43·2	43·4
10	47·7	47·8	48·0	48·2
20	95·3	95·7	96·0	96·3
30	143·0	143·5	144·0	144·5
40	190·7	191·3	192·0	192·7
50	238·3	239·2	240·0	240·8

″	290	291	292
6	29·0	29·1	29·2
7	33·8	34·0	34·1
8	38·7	38·8	38·9
9	43·5	43·6	43·8
10	48·3	48·5	48·7
20	96·7	97·0	97·3
30	145·0	145·5	146·0
40	193·3	194·0	194·7
50	241·7	242·5	243·3

31°

′	sin	d	tan	d	cot	cos	d	′
00	9·711 839	211	9·778 774	286	0·221 226	9·933 066	76	60
01	·712 050	210	·779 060	286	·220 940	·932 990	76	59
02	·712 260	209	·779 346	286	·220 654	·932 914	76	58
03	·712 469	210	·779 632	286	·220 368	·932 838	76	57
04	·712 679	210	·779 918	285	·220 082	·932 762	77	56
05	·712 889	209	·780 203	286	·219 797	·932 685	76	55
06	·713 098	210	·780 489	286	·219 511	·932 609	76	54
07	·713 308	209	·780 775	285	·219 225	·932 533	76	53
08	·713 517	209	·781 060	286	·218 940	·932 457	77	52
09	·713 726	209	·781 346	285	·218 654	·932 380	76	51
10	9·713 935	209	9·781 631	285	0·218 369	9·932 304	76	50
11	·714 144	208	·781 916	285	·218 084	·932 228	77	49
12	·714 352	209	·782 201	285	·217 799	·932 151	76	48
13	·714 561	208	·782 486	285	·217 514	·932 075	77	47
14	·714 769	209	·782 771	285	·217 229	·931 998	77	46
15	·714 978	208	·783 056	285	·216 944	·931 921	76	45
16	·715 186	208	·783 341	285	·216 659	·931 845	77	44
17	·715 394	208	·783 626	284	·216 374	·931 768	77	43
18	·715 602	207	·783 910	285	·216 090	·931 691	77	42
19	·715 809	208	·784 195	284	·215 805	·931 614	77	41
20	9·716 017	207	9·784 479	285	0·215 521	9·931 537	77	40
21	·716 224	208	·784 764	284	·215 236	·931 460	77	39
22	·716 432	207	·785 048	284	·214 952	·931 383	77	38
23	·716 639	207	·785 332	284	·214 668	·931 306	77	37
24	·716 846	207	·785 616	284	·214 384	·931 229	77	36
25	·717 053	206	·785 900	284	·214 100	·931 152	77	35
26	·717 259	207	·786 184	284	·213 816	·931 075	77	34
27	·717 466	207	·786 468	284	·213 532	·930 998	77	33
28	·717 673	206	·786 752	284	·213 248	·930 921	78	32
29	·717 879	206	·787 036	283	·212 964	·930 843	77	31
30	9·718 085	206	9·787 319	284	0·212 681	9·930 766	78	30
31	·718 291	206	·787 603	283	·212 397	·930 688	77	29
32	·718 497	206	·787 886	284	·212 114	·930 611	78	28
33	·718 703	206	·788 170	283	·211 830	·930 533	77	27
34	·718 909	205	·788 453	283	·211 547	·930 456	78	26
35	·719 114	206	·788 736	283	·211 264	·930 378	78	25
36	·719 320	205	·789 019	283	·210 981	·930 300	77	24
37	·719 525	205	·789 302	283	·210 698	·930 223	78	23
38	·719 730	205	·789 585	283	·210 415	·930 145	78	22
39	·719 935	205	·789 868	283	·210 132	·930 067	78	21
40	9·720 140	205	9·790 151	283	0·209 849	9·929 989	78	20
41	·720 345	204	·790 434	282	·209 566	·929 911	78	19
42	·720 549	205	·790 716	283	·209 284	·929 833	78	18
43	·720 754	204	·790 999	282	·209 001	·929 755	78	17
44	·720 958	204	·791 281	282	·208 719	·929 677	78	16
45	·721 162	204	·791 563	283	·208 437	·929 599	78	15
46	·721 366	204	·791 846	282	·208 154	·929 521	79	14
47	·721 570	204	·792 128	282	·207 872	·929 442	78	13
48	·721 774	204	·792 410	282	·207 590	·929 364	78	12
49	·721 978	203	·792 692	282	·207 308	·929 286	79	11
50	9·722 181	204	9·792 974	282	0·207 026	9·929 207	78	10
51	·722 385	203	·793 256	282	·206 744	·929 129	79	09
52	·722 588	203	·793 538	281	·206 462	·929 050	78	08
53	·722 791	203	·793 819	282	·206 181	·928 972	79	07
54	·722 994	203	·794 101	282	·205 899	·928 893	78	06
55	·723 197	203	·794 383	281	·205 617	·928 815	79	05
56	·723 400	203	·794 664	282	·205 336	·928 736	79	04
57	·723 603	202	·794 946	281	·205 054	·928 657	79	03
58	·723 805	202	·795 227	281	·204 773	·928 578	79	02
59	·724 007	203	·795 508	281	·204 492	·928 499	79	01
60	9·724 210		9·795 789		0·204 211	9·928 420		00
	cos		cot		tan	sin		**58°**

P.Ps. FOR SECONDS

″	76	77	78	79
6	7·6	7·7	7·8	7·9
7	8·9	9·0	9·1	9·2
8	10·1	10·3	10·4	10·5
9	11·4	11·6	11·7	11·8
10	12·7	12·8	13·0	13·2
20	25·3	25·7	26·0	26·3
30	38·0	38·5	39·0	39·5
40	50·7	51·3	52·0	52·7
50	63·3	64·2	65·0	65·8

″	202	203	204	205
6	20·2	20·3	20·4	20·5
7	23·6	23·7	23·8	23·9
8	26·9	27·1	27·2	27·3
9	30·3	30·4	30·6	30·8
10	33·7	33·8	34·0	34·2
20	67·3	67·7	68·0	68·3
30	101·0	101·5	102·0	102·5
40	134·7	135·3	136·0	136·7
50	168·3	169·2	170·0	170·8

″	206	207	208	209
6	20·6	20·7	20·8	20·9
7	24·0	24·2	24·3	24·4
8	27·5	27·6	27·7	27·9
9	30·9	31·0	31·2	31·4
10	34·3	34·5	34·7	34·8
20	68·7	69·0	69·3	69·7
30	103·0	103·5	104·0	104·5
40	137·3	138·0	138·7	139·3
50	171·7	172·5	173·3	174·2

″	210	211
6	21·0	21·1
7	24·5	24·6
8	28·0	28·1
9	31·5	31·6
10	35·0	35·2
20	70·0	70·3
30	105·0	105·5
40	140·0	140·7
50	175·0	175·8

″	281	282	283	284
6	28·1	28·2	28·3	28·4
7	32·8	32·9	33·0	33·1
8	37·5	37·6	37·7	37·9
9	42·2	42·3	42·4	42·6
10	46·8	47·0	47·2	47·3
20	93·7	94·0	94·3	94·7
30	140·5	141·0	141·5	142·0
40	187·3	188·0	188·7	189·3
50	234·2	235·0	235·8	236·7

″	285	286
6	28·5	28·6
7	33·2	33·4
8	38·0	38·1
9	42·8	42·9
10	47·5	47·7
20	95·0	95·3
30	142·5	143·0
40	190·0	190·7
50	237·5	238·3

32° ′	sin		tan		cot	cos		
00	9·724 210	202	9·795 789	281	0·204 211	9·928 420	78	60′
01	·724 412	202	·796 070	281	·203 930	·928 342	79	59
02	·724 614	202	·796 351	281	·203 649	·928 263	80	58
03	·724 816	201	·796 632	281	·203 368	·928 183	79	57
04	·725 017	202	·796 913	281	·203 087	·928 104	79	56
05	·725 219	201	·797 194	280	·202 806	·928 025	79	55
06	·725 420	202	·797 474	281	·202 526	·927 946	79	54
07	·725 622	201	·797 755	281	·202 245	·927 867	80	53
08	·725 823	201	·798 036	280	·201 964	·927 787	79	52
09	·726 024	201	·798 316	280	·201 684	·927 708	79	51
10	9·726 225	201	9·798 596	281	0·201 404	9·927 629	80	50
11	·726 426	200	·798 877	280	·201 123	·927 549	79	49
12	·726 626	201	·799 157	280	·200 843	·927 470	80	48
13	·726 827	200	·799 437	280	·200 563	·927 390	80	47
14	·727 027	201	·799 717	280	·200 283	·927 310	79	46
15	·727 228	200	·799 997	280	·200 003	·927 231	80	45
16	·727 428	200	·800 277	280	·199 723	·927 151	80	44
17	·727 628	200	·800 557	279	·199 443	·927 071	80	43
18	·727 828	199	·800 836	280	·199 164	·926 991	80	42
19	·728 027	200	·801 116	280	·198 884	·926 911	80	41
20	9·728 227	200	9·801 396	279	0·198 604	9·926 831	80	40
21	·728 427	199	·801 675	280	·198 325	·926 751	80	39
22	·728 626	199	·801 955	279	·198 045	·926 671	80	38
23	·728 825	199	·802 234	279	·197 766	·926 591	80	37
24	·729 024	199	·802 513	279	·197 487	·926 511	80	36
25	·729 223	199	·802 792	280	·197 208	·926 431	80	35
26	·729 422	199	·803 072	279	·196 928	·926 351	81	34
27	·729 621	199	·803 351	279	·196 649	·926 270	80	33
28	·729 820	198	·803 630	279	·196 370	·926 190	80	32
29	·730 018	199	·803 909	278	·196 091	·926 110	81	31
30	9·730 217	198	9·804 187	279	0·195 813	9·926 029	80	30
31	·730 415	198	·804 466	279	·195 534	·925 949	81	29
32	·730 613	198	·804 745	278	·195 255	·925 868	80	28
33	·730 811	198	·805 023	279	·194 977	·925 788	81	27
34	·731 009	197	·805 302	278	·194 698	·925 707	81	26
35	·731 206	198	·805 580	279	·194 420	·925 626	81	25
36	·731 404	198	·805 859	278	·194 141	·925 545	80	24
37	·731 602	197	·806 137	278	·193 863	·925 465	81	23
38	·731 799	197	·806 415	278	·193 585	·925 384	81	22
39	·731 996	197	·806 693	278	·193 307	·925 303	81	21
40	9·732 193	197	9·806 971	278	0·193 029	9·925 222	81	20
41	·732 390	197	·807 249	278	·192 751	·925 141	81	19
42	·732 587	197	·807 527	278	·192 473	925 060	81	18
43	·732 784	196	·807 805	278	·192 195	·924 979	82	17
44	·732 980	197	·808 083	278	·191 917	·924 897	81	16
45	·733 177	196	·808 361	277	·191 639	·924 816	81	15
46	·733 373	196	·808 638	278	·191 362	·924 735	81	14
47	·733 569	196	·808 916	277	·191 084	·924 654	82	13
48	·733 765	196	·809 193	278	·190 807	·924 572	81	12
49	·733 961	196	·809 471	277	·190 529	·924 491	82	11
50	9·734 157	196	9·809 748	277	0·190 252	9·924 409	81	10
51	·734 353	196	·810 025	277	·189 975	·924 328	82	09
52	·734 549	195	·810 302	278	·189 698	·924 246	82	08
53	·734 744	195	·810 580	277	·189 420	·924 164	81	07
54	·734 939	196	·810 857	277	·189 143	·924 083	82	06
55	·735 135	195	·811 134	276	·188 866	·924 001	82	05
56	·735 330	195	·811 410	277	·188 590	·923 919	82	04
57	·735 525	194	·811 687	277	·188 313	·923 837	82	03
58	·735 719	195	·811 964	277	·188 036	·923 755	82	02
59	·735 914	195	·812 241	276	·187 759	·923 673	82	01
60	9·736 109		9·812 517		0·187 483	9·923 591		00
	cos		cot		tan	sin		57°

P.Ps. FOR SECONDS

″	78	79	80	81
6	7·8	7·9	8·0	8·1
7	9·1	9·2	9·3	9·4
8	10·4	10·5	10·7	10·8
9	11·7	11·8	12·0	12·2
10	13·0	13·2	13·3	13·5
20	26·0	26·3	26·7	27·0
30	39·0	39·5	40·0	40·5
40	52·0	52·7	53·3	54·0
50	65·0	65·8	66·7	67·5

″	82
6	8·2
7	9·6
8	10·9
9	12·3
10	13·7
20	27·3
30	41·0
40	54·7
50	68·3

″	194	195	196	197
6	19·4	19·5	19·6	19·7
7	22·6	22·8	22·9	23·0
8	25·9	26·0	26·1	26·3
9	29·1	29·2	29·4	29·6
10	32·3	32·5	32·7	32·8
20	64·7	65·0	65·3	65·7
30	97·0	97·5	98·0	98·5
40	129·3	130·0	130·7	131·3
50	161·7	162·5	163·3	164·2

″	198	199	200	201
6	19·8	19·9	20·0	20·1
7	23·1	23·2	23·3	23·4
8	26·4	26·5	26·7	26·8
9	29·7	29·8	30·0	30·2
10	33·0	33·2	33·3	33·5
20	66·0	66·3	66·7	67·0
30	99·0	99·5	100·0	100·5
40	132·0	132·7	133·3	134·0
50	165·0	165·8	166·7	167·5

″	202		276	277
6	20·2		27·6	27·7
7	23·6		32·2	32·3
8	26·9		36·8	36·9
9	30·3		41·4	41·6
10	33·7		46·0	46·2
20	67·3		92·0	92·3
30	101·0		138·0	138·5
40	134·7		184·0	184·7
50	168·3		230·0	230·8

″	278	279	280	281
6	27·8	27·9	28·0	28·1
7	32·4	32·6	32·7	32·8
8	37·1	37·2	37·3	37·5
9	41·7	41·8	42·0	42·2
10	46·3	46·5	46·7	46·8
20	92·7	93·0	93·3	93·7
30	139·0	139·5	140·0	140·5
40	185·3	186·0	186·7	187·3
50	231·7	232·5	233·3	234·2

33° ′	sin		tan		cot	cos		′
00	9·736 109	194	9·812 517	277	0·187 483	9·923 591	82	60
01	·736 303	195	·812 794	276	·187 206	·923 509	82	59
02	·736 498	194	·813 070	277	·186 930	·923 427	82	58
03	·736 692	194	·813 347	276	·186 653	·923 345	82	57
04	·736 886	194	·813 623	276	·186 377	·923 263	82	56
05	·737 080	194	·813 899	277	·186 101	·923 181	83	55
06	·737 274	193	·814 176	276	·185 824	·923 098	82	54
07	·737 467	194	·814 452	276	·185 548	·923 016	83	53
08	·737 661	194	·814 728	276	·185 272	·922 933	82	52
09	·737 855	193	·815 004	276	·184 996	·922 851	83	51
10	9·738 048	193	9·815 280	275	0·184 720	9·922 768	82	50
11	·738 241	193	·815 555	276	·184 445	·922 686	83	49
12	·738 434	193	·815 831	276	·184 169	·922 603	83	48
13	·738 627	193	·816 107	275	·183 893	·922 520	82	47
14	·738 820	193	·816 382	276	·183 618	·922 438	83	46
15	·739 013	193	·816 658	275	·183 342	·922 355	83	45
16	·739 206	192	·816 933	276	·183 067	·922 272	83	44
17	·739 398	192	·817 209	275	·182 791	·922 189	83	43
18	·739 590	193	·817 484	275	·182 516	·922 106	83	42
19	·739 783	192	·817 759	276	·182 241	·922 023	83	41
20	9·739 975	192	9·818 035	275	0·181 965	9·921 940	83	40
21	·740 167	192	·818 310	275	·181 690	·921 857	83	39
22	·740 359	191	·818 585	275	·181 415	·921 774	83	38
23	·740 550	192	·818 860	275	·181 140	·921 691	84	37
24	·740 742	192	·819 135	275	·180 865	·921 607	83	36
25	·740 934	191	·819 410	274	·180 590	·921 524	83	35
26	·741 125	191	·819 684	275	·180 316	·921 441	84	34
27	·741 316	192	·819 959	275	·180 041	·921 357	83	33
28	·741 508	191	·820 234	274	·179 766	·921 274	84	32
29	·741 699	190	·820 508	275	·179 492	·921 190	83	31
30	9·741 889	191	9·820 783	274	0·179 217	9·921 107	84	30
31	·742 080	191	·821 057	275	·178 943	·921 023	84	29
32	·742 271	191	·821 332	274	·178 668	·920 939	83	28
33	·742 462	190	·821 606	274	·178 394	·920 856	84	27
34	·742 652	190	·821 880	274	·178 120	·920 772	84	26
35	·742 842	191	·822 154	275	·177 846	·920 688	84	25
36	·743 033	190	·822 429	274	·177 571	·920 604	84	24
37	·743 223	190	·822 703	274	·177 297	·920 520	84	23
38	·743 413	189	·822 977	274	·177 023	·920 436	84	22
39	·743 602	190	·823 251	273	·176 749	·920 352	84	21
40	9·743 792	190	9·823 524	274	0·176 476	9·920 268	84	20
41	·743 982	189	·823 798	274	·176 202	·920 184	85	19
42	·744 171	190	·824 072	273	·175 928	·920 099	84	18
43	·744 361	189	·824 345	274	·175 655	·920 015	84	17
44	·744 550	189	·824 619	274	·175 381	·919 931	85	16
45	·744 739	189	·824 893	273	·175 107	·919 846	84	15
46	·744 928	189	·825 166	273	·174 834	·919 762	85	14
47	·745 117	189	·825 439	274	·174 561	·919 677	84	13
48	·745 306	188	·825 713	273	·174 287	·919 593	85	12
49	·745 494	189	·825 986	273	·174 014	·919 508	84	11
50	9·745 683	188	9·826 259	273	0·173 741	9·919 424	85	10
51	·745 871	189	·826 532	273	·173 468	·919 339	85	09
52	·746 060	188	·826 805	273	·173 195	·919 254	85	08
53	·746 248	188	·827 078	273	·172 922	·919 169	84	07
54	·746 436	188	·827 351	273	·172 649	·919 085	85	06
55	·746 624	188	·827 624	273	·172 376	·919 000	85	05
56	·746 812	187	·827 897	273	·172 103	·918 915	85	04
57	·746 999	188	·828 170	272	·171 830	·918 830	85	03
58	·747 187	187	·828 442	273	·171 558	·918 745	86	02
59	·747 374	188	·828 715	272	·171 285	·918 659	85	01
60	9·747 562		9·828 987		0·171 013	9·918 574		00
	cos		cot		tan	sin		56°

P.Ps. FOR SECONDS

″	82	83	84	85
6	8·2	8·3	8·4	8·5
7	9·6	9·7	9·8	9·9
8	10·9	11·1	11·2	11·3
9	12·3	12·4	12·6	12·8
10	13·7	13·8	14·0	14·2
20	27·3	27·7	28·0	28·3
30	41·0	41·5	42·0	42·5
40	54·7	55·3	56·0	56·7
50	68·3	69·2	70·0	70·8

″	86
6	8·6
7	10·0
8	11·5
9	12·9
10	14·3
20	28·7
30	43·0
40	57·3
50	71·7

″	187	188	189	190
6	18·7	18·8	18·9	19·0
7	21·8	21·9	22·0	22·2
8	24·9	25·1	25·2	25·3
9	28·0	28·2	28·4	28·5
10	31·2	31·3	31·5	31·7
20	62·3	62·7	63·0	63·3
30	93·5	94·0	94·5	95·0
40	124·7	125·3	126·0	126·7
50	155·8	156·7	157·5	158·3

″	191	192	193	194
6	19·1	19·2	19·3	19·4
7	22·3	22·4	22·5	22·6
8	25·5	25·6	25·7	25·9
9	28·6	28·8	29·0	29·1
10	31·8	32·0	32·2	32·3
20	63·7	64·0	64·3	64·7
30	95·5	96·0	96·5	97·0
40	127·3	128·0	128·7	129·3
50	159·2	160·0	160·8	161·7

″	195		272	273
6	19·5		27·2	27·3
7	22·8		31·7	31·8
8	26·0		36·3	36·4
9	29·2		40·8	41·0
10	32·5		45·3	45·5
20	65·0		90·7	91·0
30	97·5		136·0	136·5
40	130·0		181·3	182·0
50	162·5		226·7	227·5

″	274	275	276	277
6	27·4	27·5	27·6	27·7
7	32·0	32·1	32·2	32·3
8	36·5	36·7	36·8	36·9
9	41·1	41·2	41·4	41·6
10	45·7	45·8	46·0	46·2
20	91·3	91·7	92·0	92·3
30	137·0	137·5	138·0	138·5
40	182·7	183·3	184·0	184·7
50	228·3	229·2	230·0	230·8

34° ′	sin		tan		cot	cos		′
00	9·747 562	187	9·828 987	273	0·171 013	9·918 574	85	60
01	·747 749	187	·829 260	272	·170 740	·918 489	85	59
02	·747 936	187	·829 532	273	·170 468	·918 404	86	58
03	·748 123	187	·829 805	272	·170 195	·918 318	85	57
04	·748 310	187	·830 077	272	·169 923	·918 233	86	56
05	·748 497	186	·830 349	272	·169 651	·918 147	85	55
06	·748 683	187	·830 621	272	·169 379	·918 062	86	54
07	·748 870	186	·830 893	272	·169 107	·917 976	85	53
08	·749 056	187	·831 165	272	·168 835	·917 891	86	52
09	·749 243	186	·831 437	272	·168 563	·917 805	86	51
10	9·749 429	186	9·831 709	272	0·168 291	9·917 719	85	50
11	·749 615	186	·831 981	272	·168 019	·917 634	86	49
12	·749 801	186	·832 253	272	·167 747	·917 548	86	48
13	·749 987	185	·832 525	271	·167 475	·917 462	86	47
14	·750 172	186	·832 796	272	·167 204	·917 376	86	46
15	·750 358	185	·833 068	271	·166 932	·917 290	86	45
16	·750 543	186	·833 339	272	·166 661	·917 204	86	44
17	·750 729	185	·833 611	271	·166 389	·917 118	86	43
18	·750 914	185	·833 882	272	·166 118	·917 032	86	42
19	·751 099	185	·834 154	271	·165 846	·916 946	87	41
20	9·751 284	185	9·834 425	271	0·165 575	9·916 859	86	40
21	·751 469	185	·834 696	271	·165 304	·916 773	86	39
22	·751 654	185	·834 967	271	·165 033	·916 687	87	38
23	·751 839	184	·835 238	271	·164 762	·916 600	86	37
24	·752 023	185	·835 509	271	·164 491	·916 514	87	36
25	·752 208	184	·835 780	271	·164 220	·916 427	86	35
26	·752 392	184	·836 051	271	·163 949	·916 341	87	34
27	·752 576	184	·836 322	271	·163 678	·916 254	87	33
28	·752 760	184	·836 593	271	·163 407	·916 167	86	32
29	·752 944	184	·836 864	270	·163 136	·916 081	87	31
30	9·753 128	184	9·837 134	271	0·162 866	9·915 994	87	30
31	·753 312	183	·837 405	270	·162 595	·915 907	87	29
32	·753 495	184	·837 675	271	·162 325	·915 820	87	28
33	·753 679	183	·837 946	270	·162 054	·915 733	87	27
34	·753 862	184	·838 216	271	·161 784	·915 646	87	26
35	·754 046	183	·838 487	270	·161 513	·915 559	87	25
36	·754 229	183	·838 757	270	·161 243	·915 472	87	24
37	·754 412	183	·839 027	270	·160 973	·915 385	88	23
38	·754 595	183	·839 297	271	·160 703	·915 297	87	22
39	·754 778	182	·839 568	270	·160 432	·915 210	87	21
40	9·754 960	183	9·839 838	270	0·160 162	9·915 123	88	20
41	·755 143	183	·840 108	270	·159 892	·915 035	87	19
42	·755 326	182	·840 378	270	·159 622	·914 948	88	18
43	·755 508	182	·840 648	269	·159 352	·914 860	87	17
44	·755 690	182	·840 917	270	·159 083	·914 773	88	16
45	·755 872	182	·841 187	270	·158 813	·914 685	87	15
46	·756 054	182	·841 457	270	·158 543	·914 598	88	14
47	·756 236	182	·841 727	269	·158 273	·914 510	88	13
48	·756 418	182	·841 996	270	·158 004	·914 422	88	12
49	·756 600	182	·842 266	269	·157 734	·914 334	88	11
50	9·756 782	181	9·842 535	270	0·157 465	9·914 246	88	10
51	·756 963	181	·842 805	269	·157 195	·914 158	88	09
52	·757 144	182	·843 074	269	·156 926	·914 070	88	08
53	·757 326	181	·843 343	269	·156 657	·913 982	88	07
54	·757 507	181	·843 612	270	·156 388	·913 894	88	06
55	·757 688	181	·843 882	269	·156 118	·913 806	88	05
56	·757 869	181	·844 151	269	·155 849	·913 718	88	04
57	·758 050	180	·844 420	269	·155 580	·913 630	89	03
58	·758 230	181	·844 689	269	·155 311	·913 541	88	02
59	·758 411	180	·844 958	269	·155 042	·913 453	88	01
60	9·758 591		9·845 227		0·154 773	9·913 365		00
	cos		cot		tan	sin		55°

P.Ps. FOR SECONDS

″	85	86	87
6	8·5	8·6	8·7
7	9·9	10·0	10·2
8	11·3	11·5	11·6
9	12·8	12·9	13·0
10	14·2	14·3	14·5
20	28·3	28·7	29·0
30	42·5	43·0	43·5
40	56·7	57·3	58·0
50	70·8	71·7	72·5
″	88	89	180
6	8·8	8·9	18·0
7	10·3	10·4	21·0
8	11·7	11·9	24·0
9	13·2	13·4	27·0
10	14·7	14·8	30·0
20	29·3	29·7	60·0
30	44·0	44·5	90·0
40	58·7	59·3	120·0
50	73·3	74·2	150·0
″	181	182	183
6	18·1	18·2	18·3
7	21·1	21·2	21·4
8	24·1	24·3	24·4
9	27·2	27·3	27·4
10	30·2	30·3	30·5
20	60·3	60·7	61·0
30	90·5	91·0	91·5
40	120·7	121·3	122·0
50	150·8	151·7	152·5
″	184	185	186
6	18·4	18·5	18·6
7	21·5	21·6	21·7
8	24·5	24·7	24·8
9	27·6	27·8	27·9
10	30·7	30·8	31·0
20	61·3	61·7	62·0
30	92·0	92·5	93·0
40	122·7	123·3	124·0
50	153·3	154·2	155·0
″	187	269	270
6	18·7	26·9	27·0
7	21·8	31·4	31·5
8	24·9	35·9	36·0
9	28·0	40·4	40·5
10	31·2	44·8	45·0
20	62·3	89·7	90·0
30	93·5	134·5	135·0
40	124·7	179·3	180·0
50	155·8	224·2	225·0
″	271	272	273
6	27·1	27·2	27·3
7	31·6	31·7	31·8
8	36·1	36·3	36·4
9	40·6	40·8	41·0
10	45·2	45·3	45·5
20	90·3	90·7	91·0
30	135·5	136·0	136·5
40	180·7	181·3	182·0
50	225·8	226·7	227·5

35° ′	sin		tan		cot	cos		′
00	9·758 591	181	9·845 227	269	0·154 773	9·913 365	89	60
01	·758 772	180	·845 496	268	·154 504	·913 276	89	59
02	·758 952	180	·845 764	269	·154 236	·913 187	88	58
03	·759 132	180	·846 033	269	·153 967	·913 099	89	57
04	·759 312	180	·846 302	268	·153 698	·913 010	88	56
05	·759 492	180	·846 570	269	·153 430	·912 922	89	55
06	·759 672	180	·846 839	269	·153 161	·912 833	89	54
07	·759 852	179	·847 108	268	·152 892	·912 744	89	53
08	·760 031	180	·847 376	268	·152 624	·912 655	89	52
09	·760 211	179	·847 644	269	·152 356	·912 566	89	51
10	9·760 390	179	9·847 913	268	0·152 087	9·912 477	89	50
11	·760 569	179	·848 181	268	·151 819	·912 388	89	49
12	·760 748	179	·848 449	268	·151 551	·912 299	89	48
13	·760 927	179	·848 717	269	·151 283	·912 210	89	47
14	·761 106	179	·848 986	268	·151 014	·912 121	90	46
15	·761 285	179	·849 254	268	·150 746	·912 031	89	45
16	·761 464	178	·849 522	268	·150 478	·911 942	89	44
17	·761 642	179	·849 790	267	·150 210	·911 853	90	43
18	·761 821	178	·850 057	268	·149 943	·911 763	89	42
19	·761 999	178	·850 325	268	·149 675	·911 674	90	41
20	9·762 177	179	9·850 593	268	0·149 407	9·911 584	89	40
21	·762 356	178	·850 861	268	·149 139	·911 495	90	39
22	·762 534	178	·851 129	267	·148 871	·911 405	90	38
23	·762 712	177	·851 396	268	·148 604	·911 315	89	37
24	·762 889	178	·851 664	267	·148 336	·911 226	90	36
25	·763 067	178	·851 931	268	·148 069	·911 136	90	35
26	·763 245	177	·852 199	267	·147 801	·911 046	90	34
27	·763 422	178	·852 466	267	·147 534	·910 956	90	33
28	·763 600	177	·852 733	268	·147 267	·910 866	90	32
29	·763 777	177	·853 001	267	·146 999	·910 776	90	31
30	9·763 954	177	9·853 268	267	0·146 732	9·910 686	90	30
31	·764 131	177	·853 535	267	·146 465	·910 596	90	29
32	·764 308	177	·853 802	267	·146 198	·910 506	91	28
33	·764 485	177	·854 069	267	·145 931	·910 415	90	27
34	·764 662	176	·854 336	267	·145 664	·910 325	90	26
35	·764 838	177	·854 603	267	·145 397	·910 235	91	25
36	·765 015	176	·854 870	267	·145 130	·910 144	90	24
37	·765 191	176	·855 137	267	·144 863	·910 054	91	23
38	·765 367	177	·855 404	267	·144 596	·909 963	90	22
39	·765 544	176	·855 671	267	·144 329	·909 873	91	21
40	9·765 720	176	9·855 938	266	0·144 062	9·909 782	91	20
41	·765 896	176	·856 204	267	·143 796	·909 691	90	19
42	·766 072	175	·856 471	266	·143 529	·909 601	91	18
43	·766 247	176	·856 737	267	·143 263	·909 510	91	17
44	·766 423	175	·857 004	266	·142 996	·909 419	91	16
45	·766 598	176	·857 270	267	·142 730	·909 328	91	15
46	·766 774	175	·857 537	266	·142 463	·909 237	91	14
47	·766 949	175	·857 803	266	·142 197	·909 146	91	13
48	·767 124	176	·858 069	267	·141 931	·909 055	91	12
49	·767 300	175	·858 336	266	·141 664	·908 964	91	11
50	9·767 475	174	9·858 602	266	0·141 398	9·908 873	92	10
51	·767 649	175	·858 868	266	·141 132	·908 781	91	09
52	·767 824	175	·859 134	266	·140 866	·908 690	91	08
53	·767 999	174	·859 400	266	·140 600	·908 599	92	07
54	·768 173	175	·859 666	266	·140 334	·908 507	91	06
55	·768 348	174	·859 932	266	·140 068	·908 416	92	05
56	·768 522	175	·860 198	266	·139 802	·908 324	91	04
57	·768 697	174	·860 464	266	·139 536	·908 233	92	03
58	·768 871	174	·860 730	265	·139 270	·908 141	92	02
59	·769 045	174	·860 995	266	·139 005	·908 049	91	01
60	9·769 219		9·861 261		0·138 739	9·907 958		00
	cos		cot		tan	sin		54°

P.Ps. FOR SECONDS

″	89	90	91
6	8·9	9·0	9·1
7	10·4	10·5	10·6
8	11·9	12·0	12·1
9	13·4	13·5	13·6
10	14·8	15·0	15·2
20	29·7	30·0	30·3
30	44·5	45·0	45·5
40	59·3	60·0	60·7
50	74·2	75·0	75·8

″	92	174
6	9·2	17·4
7	10·7	20·3
8	12·3	23·2
9	13·8	26·1
10	15·3	29·0
20	30·7	58·0
30	46·0	87·0
40	61·3	116·0
50	76·7	145·0

″	175	176	177
6	17·5	17·6	17·7
7	20·4	20·5	20·6
8	23·3	23·5	23·6
9	26·2	26·4	26·6
10	29·2	29·3	29·5
20	58·3	58·7	59·0
30	87·5	88·0	88·5
40	116·7	117·3	118·0
50	145·8	146·7	147·5

″	178	179	180
6	17·8	17·9	18·0
7	20·8	20·9	21·0
8	23·7	23·9	24·0
9	26·7	26·8	27·0
10	29·7	29·8	30·0
20	59·3	59·7	60·0
30	89·0	89·5	90·0
40	118·7	119·3	120·0
50	148·3	149·2	150·0

″	181	265	266
6	18·1	26·5	26·6
7	21·1	30·9	31·0
8	24·1	35·3	35·5
9	27·2	39·8	39·9
10	30·2	44·2	44·3
20	60·3	88·3	88·7
30	90·5	132·5	133·0
40	120·7	176·7	177·3
50	150·8	220·8	221·7

″	267	268	269
6	26·7	26·8	26·9
7	31·2	31·3	31·4
8	35·6	35·7	35·9
9	40·0	40·2	40·4
10	44·5	44·7	44·8
20	89·0	89·3	89·7
30	133·5	134·0	134·5
40	178·0	178·7	179·3
50	222·5	223·3	224·2

36° ′	sin	d	tan	d	cot	cos	d	′
00	9·769 219	174	9·861 261	266	0·138 739	9·907 958	92	60
01	·769 393	173	·861 527	265	·138 473	·907 866	92	59
02	·769 566	174	·861 792	266	·138 208	·907 774	92	58
03	·769 740	173	·862 058	265	·137 942	·907 682	92	57
04	·769 913	174	·862 323	266	·137 677	·907 590	92	56
05	·770 087	173	·862 589	265	·137 411	·907 498	92	55
06	·770 260	173	·862 854	265	·137 146	·907 406	92	54
07	·770 433	173	·863 119	266	·136 881	·907 314	92	53
08	·770 606	173	·863 385	265	·136 615	·907 222	93	52
09	·770 779	173	·863 650	265	·136 350	·907 129	92	51
10	9·770 952	173	9·863 915	265	0·136 085	9·907 037	92	50
11	·771 125	173	·864 180	265	·135 820	·906 945	93	49
12	·771 298	172	·864 445	265	·135 555	·906 852	92	48
13	·771 470	173	·864 710	265	·135 290	·906 760	93	47
14	·771 643	172	·864 975	265	·135 025	·906 667	92	46
15	·771 815	172	·865 240	265	·134 760	·906 575	93	45
16	·771 987	172	·865 505	265	·134 495	·906 482	93	44
17	·772 159	172	·865 770	265	·134 230	·906 389	93	43
18	·772 331	172	·866 035	265	·133 965	·906 296	92	42
19	·772 503	172	·866 300	264	·133 700	·906 204	93	41
20	9·772 675	172	9·866 564	265	0·133 436	9·906 111	93	40
21	·772 847	171	·866 829	265	·133 171	·906 018	93	39
22	·773 018	172	·867 094	264	·132 906	·905 925	93	38
23	·773 190	171	·867 358	265	·132 642	·905 832	93	37
24	·773 361	172	·867 623	264	·132 377	·905 739	94	36
25	·773 533	171	·867 887	265	·132 113	·905 645	93	35
26	·773 704	171	·868 152	264	·131 848	·905 552	93	34
27	·773 875	171	·868 416	264	·131 584	·905 459	93	33
28	·774 046	171	·868 680	265	·131 320	·905 366	94	32
29	·774 217	171	·868 945	264	·131 055	·905 272	93	31
30	9·774 388	170	9·869 209	264	0·130 791	9·905 179	94	30
31	·774 558	171	·869 473	264	·130 527	·905 085	93	29
32	·774 729	170	·869 737	264	·130 263	·904 992	94	28
33	·774 899	171	·870 001	264	·129 999	·904 898	94	27
34	·775 070	170	·870 265	264	·129 735	·904 804	93	26
35	·775 240	170	·870 529	264	·129 471	·904 711	94	25
36	·775 410	170	·870 793	264	·129 207	·904 617	94	24
37	·775 580	170	·871 057	264	·128 943	·904 523	94	23
38	·775 750	170	·871 321	264	·128 679	·904 429	94	22
39	·775 920	170	·871 585	264	·128 415	·904 335	94	21
40	9·776 090	169	9·871 849	263	0·128 151	9·904 241	94	20
41	·776 259	170	·872 112	264	·127 888	·904 147	94	19
42	·776 429	169	·872 376	264	·127 624	·904 053	94	18
43	·776 598	170	·872 640	263	·127 360	·903 959	95	17
44	·776 768	169	·872 903	264	·127 097	·903 864	94	16
45	·776 937	169	·873 167	263	·126 833	·903 770	94	15
46	·777 106	169	·873 430	264	·126 570	·903 676	95	14
47	·777 275	169	·873 694	263	·126 306	·903 581	94	13
48	·777 444	169	·873 957	263	·126 043	·903 487	95	12
49	·777 613	168	·874 220	264	·125 780	·903 392	94	11
50	9·777 781	169	9·874 484	263	0·125 516	9·903 298	95	10
51	·777 950	169	·874 747	263	·125 253	·903 203	95	09
52	·778 119	168	·875 010	263	·124 990	·903 108	94	08
53	·778 287	168	·875 273	264	·124 727	·903 014	95	07
54	·778 455	169	·875 537	263	·124 463	·902 919	95	06
55	·778 624	168	·875 800	263	·124 200	·902 824	95	05
56	·778 792	168	·876 063	263	·123 937	·902 729	95	04
57	·778 960	168	·876 326	263	·123 674	·902 634	95	03
58	·779 128	167	·876 589	263	·123 411	·902 539	95	02
59	·779 295	168	·876 852	262	·123 148	·902 444	95	01
60	9·779 463		9·877 114		0·122 886	9·902 349		00
	cos		cot		tan	sin		53°

P.Ps. FOR SECONDS

″	92	93	94
6	9·2	9·3	9·4
7	10·7	10·8	11·0
8	12·3	12·4	12·5
9	13·8	14·0	14·1
10	15·3	15·5	15·7
20	30·7	31·0	31·3
30	46·0	46·5	47·0
40	61·3	62·0	62·7
50	76·7	77·5	78·3

″	95	167
6	9·5	16·7
7	11·1	19·5
8	12·7	22·3
9	14·2	25·0
10	15·8	27·8
20	31·7	55·7
30	47·5	83·5
40	63·3	111·3
50	79·2	139·2

″	168	169	170
6	16·8	16·9	17·0
7	19·6	19·7	19·8
8	22·4	22·5	22·7
9	25·2	25·4	25·5
10	28·0	28·2	28·3
20	56·0	56·3	56·7
30	84·0	84·5	85·0
40	112·0	112·7	113·3
50	140·0	140·8	141·7

″	171	172	173
6	17·1	17·2	17·3
7	20·0	20·1	20·2
8	22·8	22·9	23·1
9	25·6	25·8	26·0
10	28·5	28·7	28·8
20	57·0	57·3	57·7
30	85·5	86·0	86·5
40	114·0	114·7	115·3
50	142·5	143·3	144·2

″	174	262	263
6	17·4	26·2	26·3
7	20·3	30·6	30·7
8	23·2	34·9	35·1
9	26·1	39·3	39·4
10	29·0	43·7	43·8
20	58·0	87·3	87·7
30	87·0	131·0	131·5
40	116·0	174·7	175·3
50	145·0	218·3	219·2

″	264	265	266
6	26·4	26·5	26·6
7	30·8	30·9	31·0
8	35·2	35·3	35·5
9	39·6	39·8	39·9
10	44·0	44·2	44·3
20	88·0	88·3	88·7
30	132·0	132·5	133·0
40	176·0	176·7	177·3
50	220·0	220·8	221·7

37°

′	sin	d.	tan	d.	cot	cos	d.	′
00	9·779 463	168	9·877 114	263	0·122 886	9·902 349	96	60
01	·779 631	167	·877 377	263	·122 623	·902 253	95	59
02	·779 798	168	·877 640	263	·122 360	·902 158	95	58
03	·779 966	167	·877 903	262	·122 097	·902 063	96	57
04	·780 133	167	·878 165	263	·121 835	·901 967	95	56
05	·780 300	167	·878 428	263	·121 572	·901 872	96	55
06	·780 467	167	·878 691	262	·121 309	·901 776	95	54
07	·780 634	167	·878 953	263	·121 047	·901 681	96	53
08	·780 801	167	·879 216	262	·120 784	·901 585	95	52
09	·780 968	166	·879 478	263	·120 522	·901 490	96	51
10	9·781 134	167	9·879 741	262	0·120 259	9·901 394	96	50
11	·781 301	167	·880 003	262	·119 997	·901 298	96	49
12	·781 468	166	·880 265	263	·119 735	·901 202	96	48
13	·781 634	166	·880 528	262	·119 472	·901 106	96	47
14	·781 800	166	·880 790	262	·119 210	·901 010	96	46
15	·781 966	166	·881 052	262	·118 948	·900 914	96	45
16	·782 132	166	·881 314	263	·118 686	·900 818	96	44
17	·782 298	166	·881 577	262	·118 423	·900 722	96	43
18	·782 464	166	·881 839	262	·118 161	·900 626	97	42
19	·782 630	166	·882 101	262	·117 899	·900 529	96	41
20	9·782 796	165	9·882 363	262	0·117 637	9·900 433	96	40
21	·782 961	166	·882 625	262	·117 375	·900 337	97	39
22	·783 127	165	·882 887	261	·117 113	·900 240	96	38
23	·783 292	166	·883 148	262	·116 852	·900 144	97	37
24	·783 458	165	·883 410	262	·116 590	·900 047	96	36
25	·783 623	165	·883 672	262	·116 328	·899 951	97	35
26	·783 788	165	·883 934	262	·116 066	·899 854	97	34
27	·783 953	165	·884 196	261	·115 804	·899 757	97	33
28	·784 118	164	·884 457	262	·115 543	·899 660	96	32
29	·784 282	165	·884 719	261	·115 281	·899 564	97	31
30	9·784 447	165	9·884 980	262	0·115 020	9·899 467	97	30
31	·784 612	164	·885 242	262	·114 758	·899 370	97	29
32	·784 776	165	·885 504	261	·114 496	·899 273	97	28
33	·784 941	164	·885 765	261	·114 235	·899 176	98	27
34	·785 105	164	·886 026	262	·113 974	·899 078	97	26
35	·785 269	164	·886 288	261	·113 712	·898 981	97	25
36	·785 433	164	·886 549	262	·113 451	·898 884	97	24
37	·785 597	164	·886 811	261	·113 189	·898 787	98	23
38	·785 761	164	·887 072	261	·112 928	·898 689	97	22
39	·785 925	164	·887 333	261	·112 667	·898 592	98	21
40	9·786 089	163	9·887 594	261	0·112 406	9·898 494	97	20
41	·786 252	164	·887 855	261	·112 145	·898 397	98	19
42	·786 416	163	·888 116	262	·111 884	·898 299	97	18
43	·786 579	163	·888 378	261	·111 622	·898 202	98	17
44	·786 742	164	·888 639	261	·111 361	·898 104	98	16
45	·786 906	163	·888 900	261	·111 100	·898 006	98	15
46	·787 069	163	·889 161	260	·110 839	·897 908	98	14
47	·787 232	163	·889 421	261	·110 579	·897 810	98	13
48	·787 395	162	·889 682	261	·110 318	·897 712	98	12
49	·787 557	163	·889 943	261	·110 057	·897 614	98	11
50	9·787 720	163	9·890 204	261	0·109 796	9·897 516	98	10
51	·787 883	162	·890 465	260	·109 535	·897 418	98	09
52	·788 045	163	·890 725	261	·109 275	·897 320	98	08
53	·788 208	162	·890 986	261	·109 014	·897 222	99	07
54	·788 370	162	·891 247	260	·108 753	·897 123	98	06
55	·788 532	162	·891 507	261	·108 493	·897 025	99	05
56	·788 694	162	·891 768	260	·108 232	·896 926	98	04
57	·788 856	162	·892 028	261	·107 972	·896 828	99	03
58	·789 018	162	·892 289	260	·107 711	·896 729	98	02
59	·789 180	162	·892 549	261	·107 451	·896 631	99	01
60	9·789 342		9·892 810		0·107 190	9·896 532		00
	cos		cot		tan	sin		52°

P.Ps. for seconds

″	95	96	97
6	9·5	9·6	9·7
7	11·1	11·2	11·3
8	12·7	12·8	12·9
9	14·2	14·4	14·6
10	15·8	16·0	16·2
20	31·7	32·0	32·3
30	47·5	48·0	48·5
40	63·3	64·0	64·7
50	79·2	80·0	80·8

″	98	99
6	9·8	9·9
7	11·4	11·6
8	13·1	13·2
9	14·7	14·8
10	16·3	16·5
20	32·7	33·0
30	49·0	49·5
40	65·3	66·0
50	81·7	82·5

″	162	163	164
6	16·2	16·3	16·4
7	18·9	19·0	19·1
8	21·6	21·7	21·9
9	24·3	24·4	24·6
10	27·0	27·2	27·3
20	54·0	54·3	54·7
30	81·0	81·5	82·0
40	108·0	108·7	109·3
50	135·0	135·8	136·7

″	165	166	167
6	16·5	16·6	16·7
7	19·2	19·4	19·5
8	22·0	22·1	22·3
9	24·8	24·9	25·0
10	27·5	27·7	27·8
20	55·0	55·3	55·7
30	82·5	83·0	83·5
40	110·0	110·7	111·3
50	137·5	138·3	139·2

″	168	260
6	16·8	26·0
7	19·6	30·3
8	22·4	34·7
9	25·2	39·0
10	28·0	43·3
20	56·0	86·7
30	84·0	130·0
40	112·0	173·3
50	140·0	216·7

″	261	262	263
6	26·1	26·2	26·3
7	30·4	30·6	30·7
8	34·8	34·9	35·1
9	39·2	39·3	39·4
10	43·5	43·7	43·8
20	87·0	87·3	87·7
30	130·5	131·0	131·5
40	174·0	174·7	175·3
50	217·5	218·3	219·2

38°

′	sin		tan		cot	cos		′
00	9·789 342	162	9·892 810	260	0·107 190	9·896 532	99	60
01	·789 504	161	·893 070	261	·106 930	·896 433	98	59
02	·789 665	162	·893 331	260	·106 669	·896 335	99	58
03	·789 827	161	·893 591	260	·106 409	·896 236	99	57
04	·789 988	161	·893 851	260	·106 149	·896 137	99	56
05	·790 149	161	·894 111	261	·105 889	·896 038	99	55
06	·790 310	161	·894 372	260	·105 628	·895 939	99	54
07	·790 471	161	·894 632	260	·105 368	·895 840	99	53
08	·790 632	161	·894 892	260	·105 108	·895 741	100	52
09	·790 793	161	·895 152	260	·104 848	·895 641	99	51
10	9·790 954	161	9·895 412	260	0·104 588	9·895 542	99	50
11	·791 115	160	·895 672	260	·104 328	·895 443	100	49
12	·791 275	161	·895 932	260	·104 068	·895 343	99	48
13	·791 436	160	·896 192	260	·103 808	·895 244	99	47
14	·791 596	161	·896 452	260	·103 548	·895 145	100	46
15	·791 757	160	·896 712	259	·103 288	·895 045	100	45
16	·791 917	160	·896 971	260	·103 029	·894 945	99	44
17	·792 077	160	·897 231	260	·102 769	·894 846	100	43
18	·792 237	160	·897 491	260	·102 509	·894 746	100	42
19	·792 397	160	·897 751	259	·102 249	·894 646	100	41
20	9·792 557	159	9·898 010	260	0·101 990	9·894 546	100	40
21	·792 716	160	·898 270	260	·101 730	·894 446	100	39
22	·792 876	159	·898 530	259	·101 470	·894 346	100	38
23	·793 035	160	·898 789	260	·101 211	·894 246	100	37
24	·793 195	159	·899 049	259	·100 951	·894 146	100	36
25	·793 354	160	·899 308	260	·100 692	·894 046	100	35
26	·793 514	159	·899 568	259	·100 432	·893 946	100	34
27	·793 673	159	·899 827	260	·100 173	·893 846	101	33
28	·793 832	159	·900 087	259	·099 913	·893 745	100	32
29	·793 991	159	·900 346	259	·099 654	·893 645	101	31
30	9·794 150	158	9·900 605	259	0·099 395	9·893 544	100	30
31	·794 308	159	·900 864	260	·099 136	·893 444	101	29
32	·794 467	159	·901 124	259	·098 876	·893 343	100	28
33	·794 626	158	·901 383	259	·098 617	·893 243	101	27
34	·794 784	158	·901 642	259	·098 358	·893 142	101	26
35	·794 942	159	·901 901	259	·098 099	·893 041	101	25
36	·795 101	158	·902 160	260	·097 840	·892 940	101	24
37	·795 259	158	·902 420	259	·097 580	·892 839	100	23
38	·795 417	158	·902 679	259	·097 321	·892 739	101	22
39	·795 575	158	·902 938	259	·097 062	·892 638	102	21
40	9·795 733	158	9·903 197	259	0·096 803	9·892 536	101	20
41	·795 891	158	·903 456	258	·096 544	·892 435	101	19
42	·796 049	157	·903 714	259	·096 286	·892 334	101	18
43	·796 206	158	·903 973	259	·096 027	·892 233	101	17
44	·796 364	157	·904 232	259	·095 768	·892 132	102	16
45	·796 521	158	·904 491	259	·095 509	·892 030	101	15
46	·796 679	157	·904 750	258	·095 250	·891 929	102	14
47	·796 836	157	·905 008	259	·094 992	·891 827	101	13
48	·796 993	157	·905 267	259	·094 733	·891 726	102	12
49	·797 150	157	·905 526	259	·094 474	·891 624	101	11
50	9·797 307	157	9·905 785	258	0·094 215	9·891 523	102	10
51	·797 464	157	·906 043	259	·093 957	·891 421	102	09
52	·797 621	156	·906 302	258	·093 698	·891 319	102	08
53	·797 777	157	·906 560	259	·093 440	·891 217	102	07
54	·797 934	157	·906 819	258	·093 181	·891 115	102	06
55	·798 091	156	·907 077	259	·092 923	·891 013	102	05
56	·798 247	156	·907 336	258	·092 664	·890 911	102	04
57	·798 403	157	·907 594	259	·092 406	·890 809	102	03
58	·798 560	156	·907 853	258	·092 147	·890 707	102	02
59	·798 716	156	·908 111	258	·091 889	·890 605	102	01
60	9·798 872		9·908 369		0·091 631	9·890 503		00
	cos		cot		tan	sin		51°

P.Ps. FOR SECONDS

″	98	99	100
6	9·8	9·9	10·0
7	11·4	11·6	11·7
8	13·1	13·2	13·3
9	14·7	14·8	15·0
10	16·3	16·5	16·7
20	32·7	33·0	33·3
30	49·0	49·5	50·0
40	65·3	66·0	66·7
50	81·7	82·5	83·3

″	101	102
6	10·1	10·2
7	11·8	11·9
8	13·5	13·6
9	15·2	15·3
10	16·8	17·0
20	33·7	34·0
30	50·5	51·0
40	67·3	68·0
50	84·2	85·0

″	156	157	158
6	15·6	15·7	15·8
7	18·2	18·3	18·4
8	20·8	20·9	21·1
9	23·4	23·6	23·7
10	26·0	26·2	26·3
20	52·0	52·3	52·7
30	78·0	78·5	79·0
40	104·0	104·7	105·3
50	130·0	130·8	131·7

″	159	160	161
6	15·9	16·0	16·1
7	18·6	18·7	18·8
8	21·2	21·3	21·5
9	23·8	24·0	24·2
10	26·5	26·7	26·8
20	53·0	53·3	53·7
30	79·5	80·0	80·5
40	106·0	106·7	107·3
50	132·5	133·3	134·2

″	162	258
6	16·2	25·8
7	18·9	30·1
8	21·6	34·4
9	24·3	38·7
10	27·0	43·0
20	54·0	86·0
30	81·0	129·0
40	108·0	172·0
50	135·0	215·0

″	259	260	261
6	25·9	26·0	26·1
7	30·2	30·3	30·4
8	34·5	34·7	34·8
9	38·8	39·0	39·2
10	43·2	43·3	43·5
20	86·3	86·7	87·0
30	129·5	130·0	130·5
40	172·7	173·3	174·0
50	215·8	216·7	217·5

39°

′	sin	d	tan	d	cot	cos	d	′
00	9·798 872	156	9·908 369	259	0·091 631	9·890 503	103	60
01	·799 028	156	·908 628	258	·091 372	·890 400	102	59
02	·799 184	155	·908 886	258	·091 114	·890 298	103	58
03	·799 339	156	·909 144	258	·090 856	·890 195	102	57
04	·799 495	156	·909 402	258	·090 598	·890 093	103	56
05	·799 651	155	·909 660	258	·090 340	·889 990	102	55
06	·799 806	156	·909 918	259	·090 082	·889 888	103	54
07	·799 962	155	·910 177	258	·089 823	·889 785	103	53
08	·800 117	155	·910 435	258	·089 565	·889 682	103	52
09	·800 272	155	·910 693	258	·089 307	·889 579	102	51
10	9·800 427	155	9·910 951	258	0·089 049	9·889 477	103	50
11	·800 582	155	·911 209	258	·088 791	·889 374	103	49
12	·800 737	155	·911 467	258	·088 533	·889 271	103	48
13	·800 892	155	·911 725	257	·088 275	·889 168	104	47
14	·801 047	154	·911 982	258	·088 018	·889 064	103	46
15	·801 201	155	·912 240	258	·087 760	·888 961	103	45
16	·801 356	155	·912 498	258	·087 502	·888 858	103	44
17	·801 511	154	·912 756	258	·087 244	·888 755	104	43
18	·801 665	154	·913 014	257	·086 986	·888 651	103	42
19	·801 819	154	·913 271	258	·086 729	·888 548	104	41
20	9·801 973	155	9·913 529	258	0·086 471	9·888 444	103	40
21	·802 128	154	·913 787	257	·086 213	·888 341	104	39
22	·802 282	154	·914 044	258	·085 956	·888 237	103	38
23	·802 436	153	·914 302	258	·085 698	·888 134	104	37
24	·802 589	154	·914 560	257	·085 440	·888 030	104	36
25	·802 743	154	·914 817	258	·085 183	·887 926	104	35
26	·802 897	153	·915 075	257	·084 925	·887 822	104	34
27	·803 050	154	·915 332	258	·084 668	·887 718	104	33
28	·803 204	153	·915 590	257	·084 410	·887 614	104	32
29	·803 357	154	·915 847	257	·084 153	·887 510	104	31
30	9·803 511	153	9·916 104	258	0·083 896	9·887 406	104	30
31	·803 664	153	·916 362	257	·083 638	·887 302	104	29
32	·803 817	153	·916 619	258	·083 381	·887 198	105	28
33	·803 970	153	·916 877	257	·083 123	·887 093	104	27
34	·804 123	153	·917 134	257	·082 866	·886 989	104	26
35	·804 276	152	·917 391	257	·082 609	·886 885	105	25
36	·804 428	153	·917 648	258	·082 352	·886 780	104	24
37	·804 581	153	·917 906	257	·082 094	·886 676	105	23
38	·804 734	152	·918 163	257	·081 837	·886 571	105	22
39	·804 886	153	·918 420	257	·081 580	·886 466	104	21
40	9·805 039	152	9·918 677	257	0·081 323	9·886 362	105	20
41	·805 191	152	·918 934	257	·081 066	·886 257	105	19
42	·805 343	152	·919 191	257	·080 809	·886 152	105	18
43	·805 495	152	·919 448	257	·080 552	·886 047	105	17
44	·805 647	152	·919 705	257	·080 295	·885 942	105	16
45	·805 799	152	·919 962	257	·080 038	·885 837	105	15
46	·805 951	152	·920 219	257	·079 781	·885 732	105	14
47	·806 103	151	·920 476	257	·079 524	·885 627	105	13
48	·806 254	152	·920 733	257	·079 267	·885 522	106	12
49	·806 406	151	·920 990	257	·079 010	·885 416	105	11
50	9·806 557	152	9·921 247	256	0·078 753	9·885 311	106	10
51	·806 709	151	·921 503	257	·078 497	·885 205	105	09
52	·806 860	151	·921 760	257	·078 240	·885 100	106	08
53	·807 011	152	·922 017	257	·077 983	·884 994	105	07
54	·807 163	151	·922 274	256	·077 726	·884 889	106	06
55	·807 314	151	·922 530	257	·077 470	·884 783	106	05
56	·807 465	150	·922 787	257	·077 213	·884 677	105	04
57	·807 615	151	·923 044	256	·076 956	·884 572	106	03
58	·807 766	151	·923 300	257	·076 700	·884 466	106	02
59	·807 917	150	·923 557	257	·076 443	·884 360	106	01
60	9·808 067		9·923 814		0·076 186	9·884 254		00
	cos		cot		tan	sin		**50°** ′

P.Ps. FOR SECONDS

″	102	103	104
6	10·2	10·3	10·4
7	11·9	12·0	12·1
8	13·6	13·7	13·9
9	15·3	15·4	15·6
10	17·0	17·2	17·3
20	34·0	34·3	34·7
30	51·0	51·5	52·0
40	68·0	68·7	69·3
50	85·0	85·8	86·7

″	105	106
6	10·5	10·6
7	12·2	12·4
8	14·0	14·1
9	15·8	15·9
10	17·5	17·7
20	35·0	35·3
30	52·5	53·0
40	70·0	70·7
50	87·5	88·3

″	150	151	152
6	15·0	15·1	15·2
7	17·5	17·6	17·7
8	20·0	20·1	20·3
9	22·5	22·6	22·8
10	25·0	25·2	25·3
20	50·0	50·3	50·7
30	75·0	75·5	76·0
40	100·0	100·7	101·3
50	125·0	125·8	126·7

″	153	154	155
6	15·3	15·4	15·5
7	17·8	18·0	18·1
8	20·4	20·5	20·7
9	23·0	23·1	23·2
10	25·5	25·7	25·8
20	51·0	51·3	51·7
30	76·5	77·0	77·5
40	102·0	102·7	103·3
50	127·5	128·3	129·2

″	156	256
6	15·6	25·6
7	18·2	29·9
8	20·8	34·1
9	23·4	38·4
10	26·0	42·7
20	52·0	85·3
30	78·0	128·0
40	104·0	170·7
50	130·0	213·3

″	257	258	259
6	25·7	25·8	25·9
7	30·0	30·1	30·2
8	34·3	34·4	34·5
9	38·6	38·7	38·8
10	42·8	43·0	43·2
20	85·7	86·0	86·3
30	128·5	129·0	129·5
40	171·3	172·0	172·7
50	214·2	215·0	215·8

40°

′	sin		tan		cot	cos		′
00	9·808 067	151	9·923 814	256	0·076 186	9·884 254	106	60
01	·808 218	150	·924 070	257	·075 930	·884 148	106	59
02	·808 368	151	·924 327	256	·075 673	·884 042	106	58
03	·808 519	150	·924 583	257	·075 417	·883 936	107	57
04	·808 669	150	·924 840	256	·075 160	·883 829	106	56
05	·808 819	150	·925 096	256	·074 904	·883 723	106	55
06	·808 969	150	·925 352	257	·074 648	·883 617	107	54
07	·809 119	150	·925 609	256	·074 391	·883 510	106	53
08	·809 269	150	·925 865	257	·074 135	·883 404	107	52
09	·809 419	150	·926 122	256	·073 878	·883 297	106	51
10	9·809 569	149	9·926 378	256	0·073 622	9·883 191	107	50
11	·809 718	150	·926 634	256	·073 366	·883 084	107	49
12	·809 868	149	·926 890	257	·073 110	·882 977	106	48
13	·810 017	150	·927 147	256	·072 853	·882 871	107	47
14	·810 167	149	·927 403	256	·072 597	·882 764	107	46
15	·810 316	149	·927 659	256	·072 341	·882 657	107	45
16	·810 465	149	·927 915	256	·072 085	·882 550	107	44
17	·810 614	149	·928 171	256	·071 829	·882 443	107	43
18	·810 763	149	·928 427	257	·071 573	·882 336	107	42
19	·810 912	149	·928 684	256	·071 316	·882 229	108	41
20	9·811 061	149	9·928 940	256	0·071 060	9·882 121	107	40
21	·811 210	148	·929 196	256	·070 804	·882 014	107	39
22	·811 358	149	·929 452	256	·070 548	·881 907	108	38
23	·811 507	148	·929 708	256	·070 292	·881 799	107	37
24	·811 655	149	·929 964	256	·070 036	·881 692	108	36
25	·811 804	148	·930 220	255	·069 780	·881 584	107	35
26	·811 952	148	·930 475	256	·069 525	·881 477	108	34
27	·812 100	148	·930 731	256	·069 269	·881 369	108	33
28	·812 248	148	·930 987	256	·069 013	·881 261	108	32
29	·812 396	148	·931 243	256	·068 757	·881 153	107	31
30	9·812 544	148	9·931 499	256	0·068 501	9·881 046	108	30
31	·812 692	148	·931 755	255	·068 245	·880 938	108	29
32	·812 840	148	·932 010	256	·067 990	·880 830	108	28
33	·812 988	147	·932 266	256	·067 734	·880 722	109	27
34	·813 135	148	·932 522	256	·067 478	·880 613	108	26
35	·813 283	147	·932 778	255	·067 222	·880 505	108	25
36	·813 430	148	·933 033	256	·066 967	·880 397	108	24
37	·813 578	147	·933 289	256	·066 711	·880 289	109	23
38	·813 725	147	·933 545	255	·066 455	·880 180	108	22
39	·813 872	147	·933 800	256	·066 200	·880 072	109	21
40	9·814 019	147	9·934 056	255	0·065 944	9·879 963	108	20
41	·814 166	147	·934 311	256	·065 689	·879 855	109	19
42	·814 313	147	·934 567	255	·065 433	·879 746	109	18
43	·814 460	147	·934 822	256	·065 178	·879 637	108	17
44	·814 607	146	·935 078	255	·064 922	·879 529	109	16
45	·814 753	147	·935 333	256	·064 667	·879 420	109	15
46	·814 900	146	·935 589	255	·064 411	·879 311	109	14
47	·815 046	147	·935 844	256	·064 156	·879 202	109	13
48	·815 193	146	·936 100	255	·063 900	·879 093	109	12
49	·815 339	146	·936 355	256	·063 645	·878 984	109	11
50	9·815 485	147	9·936 611	255	0·063 389	9·878 875	109	10
51	·815 632	146	·936 866	255	·063 134	·878 766	110	09
52	·815 778	146	·937 121	256	·062 879	·878 656	109	08
53	·815 924	145	·937 377	255	·062 623	·878 547	109	07
54	·816 069	146	·937 632	255	·062 368	·878 438	110	06
55	·816 215	146	·937 887	255	·062 113	·878 328	109	05
56	·816 361	146	·938 142	256	·061 858	·878 219	110	04
57	·816 507	145	·938 398	255	·061 602	·878 109	110	03
58	·816 652	146	·938 653	255	·061 347	·877 999	109	02
59	·816 798	145	·938 908	255	·061 092	·877 890	110	01
60	9·816 943		9·939 163		0·060 837	9·877 780		00
	cos		cot		tan	sin		49°

P.Ps. for seconds

″	106	107	108
6	10·6	10·7	10·8
7	12·4	12·5	12·6
8	14·1	14·3	14·4
9	15·9	16·0	16·2
10	17·7	17·8	18·0
20	35·3	35·7	36·0
30	53·0	53·5	54·0
40	70·7	71·3	72·0
50	88·3	89·2	90·0

″	109	110
6	10·9	11·0
7	12·7	12·8
8	14·5	14·7
9	16·4	16·5
10	18·2	18·3
20	36·3	36·7
30	54·5	55·0
40	72·7	73·3
50	90·8	91·7

″	145	146	147
6	14·5	14·6	14·7
7	16·9	17·0	17·2
8	19·3	19·5	19·6
9	21·8	21·9	22·0
10	24·2	24·3	24·5
20	48·3	48·7	49·0
30	72·5	73·0	73·5
40	96·7	97·3	98·0
50	120·8	121·7	122·5

″	148	149	150
6	14·8	14·9	15·0
7	17·3	17·4	17·5
8	19·7	19·9	20·0
9	22·2	22·4	22·5
10	24·7	24·8	25·0
20	49·3	49·7	50·0
30	74·0	74·5	75·0
40	98·7	99·3	100·0
50	123·3	124·2	125·0

″	151
6	15·1
7	17·6
8	20·1
9	22·6
10	25·2
20	50·3
30	75·5
40	100·7
50	125·8

″	255	256	257
6	25·5	25·6	25·7
7	29·8	29·9	30·0
8	34·0	34·1	34·3
9	38·2	38·4	38·6
10	42·5	42·7	42·8
20	85·0	85·3	85·7
30	127·5	128·0	128·5
40	170·0	170·7	171·3
50	212·5	213·3	214·2

41°

′	sin	d	tan	d	cot	cos	d	′
00	9·816 943	145	9·939 163	255	0·060 837	9·877 780	110	60
01	·817 088	145	·939 418	255	·060 582	·877 670	110	59
02	·817 233	146	·939 673	255	·060 327	·877 560	110	58
03	·817 379	145	·939 928	255	·060 072	·877 450	110	57
04	·817 524	144	·940 183	256	·059 817	·877 340	110	56
05	·817 668	145	·940 439	255	·059 561	·877 230	110	55
06	·817 813	145	·940 694	255	·059 306	·877 120	110	54
07	·817 958	145	·940 949	255	·059 051	·877 010	111	53
08	·818 103	144	·941 204	255	·058 796	·876 899	110	52
09	·818 247	145	·941 459	254	·058 541	·876 789	111	51
10	9·818 392	144	9·941 713	255	0·058 287	9·876 678	110	50
11	·818 536	145	·941 968	255	·058 032	·876 568	111	49
12	·818 681	144	·942 223	255	·057 777	·876 457	110	48
13	·818 825	144	·942 478	255	·057 522	·876 347	111	47
14	·818 969	144	·942 733	255	·057 267	·876 236	111	46
15	·819 113	144	·942 988	255	·057 012	·876 125	111	45
16	·819 257	144	·943 243	255	·056 757	·876 014	110	44
17	·819 401	144	·943 498	254	·056 502	·875 904	111	43
18	·819 545	144	·943 752	255	·056 248	·875 793	111	42
19	·819 689	143	·944 007	255	·055 993	·875 682	111	41
20	9·819 832	144	9·944 262	255	0·055 738	9·875 571	112	40
21	·819 976	144	·944 517	254	·055 483	·875 459	111	39
22	·820 120	143	·944 771	255	·055 229	·875 348	111	38
23	·820 263	143	·945 026	255	·054 974	·875 237	111	37
24	·820 406	144	·945 281	254	·054 719	·875 126	112	36
25	·820 550	143	·945 535	255	·054 465	·875 014	111	35
26	·820 693	143	·945 790	255	·054 210	·874 903	112	34
27	·820 836	143	·946 045	254	·053 955	·874 791	111	33
28	·820 979	143	·946 299	255	·053 701	·874 680	112	32
29	·821 122	143	·946 554	254	·053 446	·874 568	112	31
30	9·821 265	142	9·946 808	255	0·053 192	9·874 456	112	30
31	·821 407	143	·947 063	255	·052 937	·874 344	112	29
32	·821 550	143	·947 318	254	·052 682	·874 232	111	28
33	·821 693	142	·947 572	255	·052 428	·874 121	112	27
34	·821 835	142	·947 827	254	·052 173	·874 009	113	26
35	·821 977	143	·948 081	254	·051 919	·873 896	112	25
36	·822 120	142	·948 335	255	·051 665	·873 784	112	24
37	·822 262	142	·948 590	254	·051 410	·873 672	112	23
38	·822 404	142	·948 844	255	·051 156	·873 560	112	22
39	·822 546	142	·949 099	254	·050 901	·873 448	113	21
40	9·822 688	142	9·949 353	255	0·050 647	9·873 335	112	20
41	·822 830	142	·949 608	254	·050 392	·873 223	113	19
42	·822 972	142	·949 862	254	·050 138	·873 110	112	18
43	·823 114	141	·950 116	255	·049 884	·872 998	113	17
44	·823 255	142	·950 371	254	·049 629	·872 885	113	16
45	·823 397	142	·950 625	254	·049 375	·872 772	113	15
46	·823 539	141	·950 879	254	·049 121	·872 659	112	14
47	·823 680	141	·951 133	255	·048 867	·872 547	113	13
48	·823 821	142	·951 388	254	·048 612	·872 434	113	12
49	·823 963	141	·951 642	254	·048 358	·872 321	113	11
50	9·824 104	141	9·951 896	254	0·048 104	9·872 208	113	10
51	·824 245	141	·952 150	255	·047 850	·872 095	114	09
52	·824 386	141	·952 405	254	·047 595	·871 981	113	08
53	·824 527	141	·952 659	254	·047 341	·871 868	113	07
54	·824 668	140	·952 913	254	·047 087	·871 755	114	06
55	·824 808	141	·953 167	254	·046 833	·871 641	113	05
56	·824 949	141	·953 421	254	·046 579	·871 528	114	04
57	·825 090	140	·953 675	254	·046 325	·871 414	113	03
58	·825 230	141	·953 929	254	·046 071	·871 301	114	02
59	·825 371	140	·954 183	254	·045 817	·871 187	114	01
60	9·825 511		9·954 437		0·045 563	9·871 073		00
	cos		cot		tan	sin		48°

P.Ps. FOR SECONDS

″	110	111	112
6	11·0	11·1	11·2
7	12·8	13·0	13·1
8	14·7	14·8	14·9
9	16·5	16·6	16·8
10	18·3	18·5	18·7
20	36·7	37·0	37·3
30	55·0	55·5	56·0
40	73·3	74·0	74·7
50	91·7	92·5	93·3

″	113	114
6	11·3	11·4
7	13·2	13·3
8	15·1	15·2
9	17·0	17·1
10	18·8	19·0
20	37·7	38·0
30	56·5	57·0
40	75·3	76·0
50	94·2	95·0

″	140	141	142
6	14·0	14·1	14·2
7	16·3	16·4	16·6
8	18·7	18·8	18·9
9	21·0	21·2	21·3
10	23·3	23·5	23·7
20	46·7	47·0	47·3
30	70·0	70·5	71·0
40	93·3	94·0	94·7
50	116·7	117·5	118·3

″	143	144	145
6	14·3	14·4	14·5
7	16·7	16·8	16·9
8	19·1	19·2	19·3
9	21·4	21·6	21·8
10	23·8	24·0	24·2
20	47·7	48·0	48·3
30	71·5	72·0	72·5
40	95·3	96·0	96·7
50	119·2	120·0	120·8

″	146
6	14·6
7	17·0
8	19·5
9	21·9
10	24·3
20	48·7
30	73·0
40	97·3
50	121·7

″	254	255	256
6	25·4	25·5	25·6
7	29·6	29·8	29·9
8	33·9	34·0	34·1
9	38·1	38·2	38·4
10	42·3	42·5	42·7
20	84·7	85·0	85·3
30	127·0	127·5	128·0
40	169·3	170·0	170·7
50	211·7	212·5	213·3

42° ′	sin		tan		cot	cos		′
00	9·825 511	140	9·954 437	254	0·045 563	9·871 073	113	60
01	·825 651	140	·954 691	255	·045 309	·870 960	114	59
02	·825 791	140	·954 946	254	·045 054	·870 846	114	58
03	·825 931	140	·955 200	254	·044 800	·870 732	114	57
04	·826 071	140	·955 454	254	·044 546	·870 618	114	56
05	·826 211	140	·955 708	253	·044 292	·870 504	114	55
06	·826 351	140	·955 961	254	·044 039	·870 390	114	54
07	·826 491	140	·956 215	254	·043 785	·870 276	115	53
08	·826 631	139	·956 469	254	·043 531	·870 161	114	52
09	·826 770	140	·956 723	254	·043 277	·870 047	114	51
10	9·826 910	139	9·956 977	254	0·043 023	9·869 933	115	50
11	·827 049	140	·957 231	254	·042 769	·869 818	114	49
12	·827 189	139	·957 485	254	·042 515	·869 704	115	48
13	·827 328	139	·957 739	254	·042 261	·869 589	115	47
14	·827 467	139	·957 993	254	·042 007	·869 474	114	46
15	·827 606	139	·958 247	253	·041 753	·869 360	115	45
16	·827 745	139	·958 500	254	·041 500	·869 245	115	44
17	·827 884	139	·958 754	254	·041 246	·869 130	115	43
18	·828 023	139	·959 008	254	·040 992	·869 015	115	42
19	·828 162	139	·959 262	254	·040 738	·868 900	115	41
20	9·828 301	138	9·959 516	253	0·040 484	9·868 785	115	40
21	·828 439	139	·959 769	254	·040 231	·868 670	115	39
22	·828 578	138	·960 023	254	·039 977	·868 555	115	38
23	·828 716	139	·960 277	253	·039 723	·868 440	116	37
24	·828 855	138	·960 530	254	·039 470	·868 324	115	36
25	·828 993	138	·960 784	254	·039 216	·868 209	116	35
26	·829 131	138	·961 038	254	·038 962	·868 093	115	34
27	·829 269	138	·961 292	253	·038 708	·867 978	116	33
28	·829 407	138	·961 545	254	·038 455	·867 862	115	32
29	·829 545	138	·961 799	253	·038 201	·867 747	116	31
30	9·829 683	138	9·962 052	254	0·037 948	9·867 631	116	30
31	·829 821	138	·962 306	254	·037 694	·867 515	116	29
32	·829 959	138	·962 560	253	·037 440	·867 399	116	28
33	·830 097	137	·962 813	254	·037 187	·867 283	116	27
34	·830 234	138	·963 067	253	·036 933	·867 167	116	26
35	·830 372	137	·963 320	254	·036 680	·867 051	116	25
36	·830 509	137	·963 574	254	·036 426	·866 935	116	24
37	·830 646	138	·963 828	253	·036 172	·866 819	116	23
38	·830 784	137	·964 081	254	·035 919	·866 703	117	22
39	·830 921	137	·964 335	253	·035 665	·866 586	116	21
40	9·831 058	137	9·964 588	254	0·035 412	9·866 470	117	20
41	·831 195	137	·964 842	253	·035 158	·866 353	116	19
42	·831 332	137	·965 095	254	·034 905	·866 237	117	18
43	·831 469	137	·965 349	253	·034 651	·866 120	116	17
44	·831 606	136	·965 602	253	·034 398	·866 004	117	16
45	·831 742	137	·965 855	254	·034 145	·865 887	117	15
46	·831 879	136	·966 109	253	·033 891	·865 770	117	14
47	·832 015	137	·966 362	254	·033 638	·865 653	117	13
48	·832 152	136	·966 616	253	·033 384	·865 536	117	12
49	·832 288	137	·966 869	254	·033 131	·865 419	117	11
50	9·832 425	136	9·967 123	253	0·032 877	9·865 302	117	10
51	·832 561	136	·967 376	253	·032 624	·865 185	117	09
52	·832 697	136	·967 629	254	·032 371	·865 068	118	08
53	·832 833	136	·967 883	253	·032 117	·864 950	117	07
54	·832 969	136	·968 136	253	·031 864	·864 833	117	06
55	·833 105	136	·968 389	254	·031 611	·864 716	118	05
56	·833 241	136	·968 643	253	·031 357	·864 598	117	04
57	·833 377	135	·968 896	253	·031 104	·864 481	118	03
58	·833 512	136	·969 149	254	·030 851	·864 363	118	02
59	·833 648	135	·969 403	253	·030 597	·864 245	118	01
60	9·833 783		9·969 656		0·030 344	9·864 127		00
	cos		cot		tan	sin		47°

P.Ps. for seconds

″	113	114	115
6	11·3	11·4	11·5
7	13·2	13·3	13·4
8	15·1	15·2	15·3
9	17·0	17·1	17·2
10	18·8	19·0	19·2
20	37·7	38·0	38·3
30	56·5	57·0	57·5
40	75·3	76·0	76·7
50	94·2	95·0	95·8

″	116	117	118
6	11·6	11·7	11·8
7	13·5	13·6	13·8
8	15·5	15·6	15·7
9	17·4	17·6	17·7
10	19·3	19·5	19·7
20	38·7	39·0	39·3
30	58·0	58·5	59·0
40	77·3	78·0	78·7
50	96·7	97·5	98·3

″	135	136	137
6	13·5	13·6	13·7
7	15·8	15·9	16·0
8	18·0	18·1	18·3
9	20·2	20·4	20·6
10	22·5	22·7	22·8
20	45·0	45·3	45·7
30	67·5	68·0	68·5
40	90·0	90·7	91·3
50	112·5	113·3	114·2

″	138	139	140
6	13·8	13·9	14·0
7	16·1	16·2	16·3
8	18·4	18·5	18·7
9	20·7	20·8	21·0
10	23·0	23·2	23·3
20	46·0	46·3	46·7
30	69·0	69·5	70·0
40	92·0	92·7	93·3
50	115·0	115·8	116·7

″	253	254	255
6	25·3	25·4	25·5
7	29·5	29·6	29·8
8	33·7	33·9	34·0
9	38·0	38·1	38·2
10	42·2	42·3	42·5
20	84·3	84·7	85·0
30	126·5	127·0	127·5
40	168·7	169·3	170·0
50	210·8	211·7	212·5

43° ′	sin	d	tan	d	cot	cos	d	′
00	9·833 783	136	9·969 656	253	0·030 344	9·864 127	117	60
01	·833 919	135	·969 909	253	·030 091	·864 010	118	59
02	·834 054	135	·970 162	254	·029 838	·863 892	118	58
03	·834 189	136	·970 416	253	·029 584	·863 774	118	57
04	·834 325	135	·970 669	253	·029 331	·863 656	118	56
05	·834 460	135	·970 922	253	·029 078	·863 538	119	55
06	·834 595	135	·971 175	254	·028 825	·863 419	118	54
07	·834 730	135	·971 429	253	·028 571	·863 301	118	53
08	·834 865	134	·971 682	253	·028 318	·863 183	119	52
09	·834 999	135	·971 935	253	·028 065	·863 064	118	51
10	9·835 134	135	9·972 188	253	0·027 812	9·862 946	119	50
11	·835 269	134	·972 441	254	·027 559	·862 827	118	49
12	·835 403	135	·972 695	253	·027 305	·862 709	119	48
13	·835 538	134	·972 948	253	·027 052	·862 590	119	47
14	·835 672	135	·973 201	253	·026 799	·862 471	118	46
15	·835 807	134	·973 454	253	·026 546	·862 353	119	45
16	·835 941	134	·973 707	253	·026 293	·862 234	119	44
17	·836 075	134	·973 960	253	·026 040	·862 115	119	43
18	·836 209	134	·974 213	253	·025 787	·861 996	119	42
19	·836 343	134	·974 466	254	·025 534	·861 877	119	41
20	9·836 477	134	9·974 720	253	0·025 280	9·861 758	120	40
21	·836 611	134	·974 973	253	·025 027	·861 638	119	39
22	·836 745	133	·975 226	253	·024 774	·861 519	119	38
23	·836 878	134	·975 479	253	·024 521	·861 400	120	37
24	·837 012	134	·975 732	253	·024 268	·861 280	119	36
25	·837 146	133	·975 985	253	·024 015	·861 161	120	35
26	·837 279	133	·976 238	253	·023 762	·861 041	119	34
27	·837 412	134	·976 491	253	·023 509	·860 922	120	33
28	·837 546	133	·976 744	253	·023 256	·860 802	120	32
29	·837 679	133	·976 997	253	·023 003	·860 682	120	31
30	9·837 812	133	9·977 250	253	0·022 750	9·860 562	120	30
31	·837 945	133	·977 503	253	·022 497	·860 442	120	29
32	·838 078	133	·977 756	253	·022 244	·860 322	120	28
33	·838 211	133	·978 009	253	·021 991	·860 202	120	27
34	·838 344	133	·978 262	253	·021 738	·860 082	120	26
35	·838 477	133	·978 515	253	·021 485	·859 962	120	25
36	·838 610	132	·978 768	253	·021 232	·859 842	121	24
37	·838 742	133	·979 021	253	·020 979	·859 721	120	23
38	·838 875	132	·979 274	253	·020 726	·859 601	121	22
39	·839 007	133	·979 527	253	·020 473	·859 480	120	21
40	9·839 140	132	9·979 780	253	0·020 220	9·859 360	121	20
41	·839 272	132	·980 033	253	·019 967	·859 239	120	19
42	·839 404	132	·980 286	252	·019 714	·859 119	121	18
43	·839 536	132	·980 538	253	·019 462	·858 998	121	17
44	·839 668	132	·980 791	253	·019 209	·858 877	121	16
45	·839 800	132	·981 044	253	·018 956	·858 756	121	15
46	·839 932	132	·981 297	253	·018 703	·858 635	121	14
47	·840 064	132	·981 550	253	·018 450	·858 514	121	13
48	·840 196	132	·981 803	253	·018 197	·858 393	121	12
49	·840 328	131	·982 056	253	·017 944	·858 272	121	11
50	9·840 459	132	9·982 309	253	0·017 691	9·858 151	122	10
51	·840 591	131	·982 562	252	·017 438	·858 029	121	09
52	·840 722	132	·982 814	253	·017 186	·857 908	122	08
53	·840 854	131	·983 067	253	·016 933	·857 786	121	07
54	·840 985	131	·983 320	253	·016 680	·857 665	122	06
55	·841 116	131	·983 573	253	·016 427	·857 543	121	05
56	·841 247	131	·983 826	253	·016 174	·857 422	122	04
57	·841 378	131	·984 079	253	·015 921	·857 300	122	03
58	·841 509	131	·984 332	252	·015 668	·857 178	122	02
59	·841 640	131	·984 584	253	·015 416	·857 056	122	01
60	9·841 771		9·984 837		0·015 163	9·856 934		00
	cos		cot		tan	sin		46°

P.Ps. for seconds

″	117	118	119
6	11·7	11·8	11·9
7	13·6	13·8	13·9
8	15·6	15·7	15·9
9	17·6	17·7	17·8
10	19·5	19·7	19·8
20	39·0	39·3	39·7
30	58·5	59·0	59·5
40	78·0	78·7	79·3
50	97·5	98·3	99·2

″	120	121	122
6	12·0	12·1	12·2
7	14·0	14·1	14·2
8	16·0	16·1	16·3
9	18·0	18·2	18·3
10	20·0	20·2	20·3
20	40·0	40·3	40·7
30	60·0	60·5	61·0
40	80·0	80·7	81·3
50	100·0	100·8	101·7

″	131	132	133
6	13·1	13·2	13·3
7	15·3	15·4	15·5
8	17·5	17·6	17·7
9	19·6	19·8	20·0
10	21·8	22·0	22·2
20	43·7	44·0	44·3
30	65·5	66·0	66·5
40	87·3	88·0	88·7
50	109·2	110·0	110·8

″	134	135	136
6	13·4	13·5	13·6
7	15·6	15·8	15·9
8	17·9	18·0	18·1
9	20·1	20·2	20·4
10	22·3	22·5	22·7
20	44·7	45·0	45·3
30	67·0	67·5	68·0
40	89·3	90·0	90·7
50	111·7	112·5	113·3

″	252	253	254
6	25·2	25·3	25·4
7	29·4	29·5	29·6
8	33·6	33·7	33·9
9	37·8	38·0	38·1
10	42·0	42·2	42·3
20	84·0	84·3	84·7
30	126·0	126·5	127·0
40	168·0	168·7	169·3
50	210·0	210·8	211·7

44°

′	sin	d	tan	d	cot	cos	d	′
00	9·841 771	131	9·984 837	253	0·015 163	9·856 934	122	60
01	·841 902	131	·985 090	253	·014 910	·856 812	122	59
02	·842 033	130	·985 343	253	·014 657	·856 690	122	58
03	·842 163	131	·985 596	252	·014 404	·856 568	122	57
04	·842 294	130	·985 848	253	·014 152	·856 446	123	56
05	·842 424	131	·986 101	253	·013 899	·856 323	122	55
06	·842 555	130	·986 354	253	·013 646	·856 201	123	54
07	·842 685	130	·986 607	253	·013 393	·856 078	122	53
08	·842 815	131	·986 860	252	·013 140	·855 956	123	52
09	·842 946	130	·987 112	253	·012 888	·855 833	122	51
10	9·843 076	130	9·987 365	253	0·012 635	9·855 711	123	50
11	·843 206	130	·987 618	253	·012 382	·855 588	123	49
12	·843 336	130	·987 871	252	·012 129	·855 465	123	48
13	·843 466	129	·988 123	253	·011 877	·855 342	123	47
14	·843 595	130	·988 376	253	·011 624	·855 219	123	46
15	·843 725	130	·988 629	253	·011 371	·855 096	123	45
16	·843 855	129	·988 882	252	·011 118	·854 973	123	44
17	·843 984	130	·989 134	253	·010 866	·854 850	123	43
18	·844 114	129	·989 387	253	·010 613	·854 727	124	42
19	·844 243	129	·989 640	253	·010 360	·854 603	123	41
20	9·844 372	130	9·989 893	252	0·010 107	9·854 480	124	40
21	·844 502	129	·990 145	253	·009 855	·854 356	123	39
22	·844 631	129	·990 398	253	·009 602	·854 233	124	38
23	·844 760	129	·990 651	252	·009 349	·854 109	123	37
24	·844 889	129	·990 903	253	·009 097	·853 986	124	36
25	·845 018	129	·991 156	253	·008 844	·853 862	124	35
26	·845 147	129	·991 409	253	·008 591	·853 738	124	34
27	·845 276	129	·991 662	252	·008 338	·853 614	124	33
28	·845 405	128	·991 914	253	·008 086	·853 490	124	32
29	·845 533	129	·992 167	253	·007 833	·853 366	124	31
30	9·845 662	128	9·992 420	252	0·007 580	9·853 242	124	30
31	·845 790	129	·992 672	253	·007 328	·853 118	124	29
32	·845 919	128	·992 925	253	·007 075	·852 994	125	28
33	·846 047	128	·993 178	253	·006 822	·852 869	124	27
34	·846 175	129	·993 431	252	·006 569	·852 745	125	26
35	·846 304	128	·993 683	253	·006 317	·852 620	124	25
36	·846 432	128	·993 936	253	·006 064	·852 496	125	24
37	·846 560	128	·994 189	252	·005 811	·852 371	124	23
38	·846 688	128	·994 441	253	·005 559	·852 247	125	22
39	·846 816	128	·994 694	253	·005 306	·852 122	125	21
40	9·846 944	127	9·994 947	252	0·005 053	9·851 997	125	20
41	·847 071	128	·995 199	253	·004 801	·851 872	125	19
42	·847 199	128	·995 452	253	·004 548	·851 747	125	18
43	·847 327	127	·995 705	252	·004 295	·851 622	125	17
44	·847 454	128	·995 957	253	·004 043	·851 497	125	16
45	·847 582	127	·996 210	253	·003 790	·851 372	126	15
46	·847 709	127	·996 463	252	·003 537	·851 246	125	14
47	·847 836	128	·996 715	253	·003 285	·851 121	125	13
48	·847 964	127	·996 968	253	·003 032	·850 996	126	12
49	·848 091	127	·997 221	252	·002 779	·850 870	125	11
50	9·848 218	127	9·997 473	253	0·002 527	9·850 745	126	10
51	·848 345	127	·997 726	253	·002 274	·850 619	126	09
52	·848 472	127	·997 979	252	·002 021	·850 493	125	08
53	·848 599	127	·998 231	253	·001 769	·850 368	126	07
54	·848 726	126	·998 484	253	·001 516	·850 242	126	06
55	·848 852	127	·998 737	252	·001 263	·850 116	126	05
56	·848 979	127	·998 989	253	·001 011	·849 990	126	04
57	·849 106	126	·999 242	253	·000 758	·849 864	126	03
58	·849 232	127	·999 495	252	·000 505	·849 738	127	02
59	·849 359	126	9·999 747	253	·000 253	·849 611	126	01
60	9·849 485		0·000 000		0·000 000	9·849 485		00
	cos		cot		tan	sin		45°

P.Ps. for seconds

″	122	123	124
6	12·2	12·3	12·4
7	14·2	14·4	14·5
8	16·3	16·4	16·5
9	18·3	18·4	18·6
10	20·3	20·5	20·7
20	40·7	41·0	41·3
30	61·0	61·5	62·0
40	81·3	82·0	82·7
50	101·7	102·5	103·3

″	125	126	127
6	12·5	12·6	12·7
7	14·6	14·7	14·8
8	16·7	16·8	16·9
9	18·8	18·9	19·0
10	20·8	21·0	21·2
20	41·7	42·0	42·3
30	62·5	63·0	63·5
40	83·3	84·0	84·7
50	104·2	105·0	105·8

″	128	129	130
6	12·8	12·9	13·0
7	14·9	15·0	15·2
8	17·1	17·2	17·3
9	19·2	19·4	19·5
10	21·3	21·5	21·7
20	42·7	43·0	43·3
30	64·0	64·5	65·0
40	85·3	86·0	86·7
50	106·7	107·5	108·3

″	131
6	13·1
7	15·3
8	17·5
9	19·6
10	21·8
20	43·7
30	65·5
40	87·3
50	109·2

″	252	253
6	25·2	25·3
7	29·4	29·5
8	33·6	33·7
9	37·8	38·0
10	42·0	42·2
20	84·0	84·3
30	126·0	126·5
40	168·0	168·7
50	210·0	210·8

TABLE IIIA—AUXILIARY FUNCTIONS FOR SMALL ANGLES

(Degrees, Minutes and Seconds)

x (″)	τ	cot x	x (″)	τ	cot x	x (″)	σ	cosec x
0		∞	5116		40·31	0		∞
	206 265			206 222			206 265	
435		473·8	5176		39·84	926		222·7
	206 264			206 221			206 266	
899		229·5	5235		39·39	1447		142·5
	206 263			206 220			206 267	
1194		172·7	5294		38·95	1825		113·0
	206 262			206 219			206 268	
1430		144·3	5352		38·53	2138		96·48
	206 261			206 218			206 269	
1632		126·4	5410		38·12	2410		85·59
	206 260			206 217			206 270	
1812		113·9	5467		37·72	2654		77·71
	206 259			206 216			206 271	
1975		104·5	5523		37·34	2878		71·67
	206 258			206 215			206 272	
2126		97·01	5579		36·97	3085		66·85
	206 257			206 214			206 273	
2267		90·98	5634		36·60	3280		62·89
	206 256			206 213			206 274	
2399		85·96	5689		36·25	3463		59·56
	206 255			206 212			206 275	
2525		81·68	5743		35·91	3637		56·71
	206 254			206 211			206 276	
2645		77·98	5796		35·58	3804		54·23
	206 253			206 210			206 277	
2759		74·75	5849		35·26	3963		52·05
	206 252			206 209			206 278	
2869		71·88	5902		34·94	4116		50·11
	206 251			206 208			206 279	
2975		69·33	5954		34·63	4264		48·38
	206 250			206 207			206 280	
3077		67·02	6006		34·34	4406		46·81
	206 249			206 206			206 281	
3176		64·93	6057		34·04	4545		45·39
	206 248			206 205			206 282	
3272		63·03	6108		33·76	4679		44·09
	206 247			206 204			206 283	
3365		61·28	6159		33·48	4809		42·89
	206 246			206 203			206 284	
3456		59·68	6209		33·21	4936		41·79
	206 245			206 202			206 285	
3544		58·19	6258		32·95	5060		40·77
	206 244			206 201			206 286	
3630		56·81	6307		32·69	5181		39·82
	206 243			206 200			206 287	
3715		55·52	6356		32·44	5299		38·93
	206 242			206 199			206 288	
3797		54·31	6405		32·20	5414		38·10
	206 241			206 198			206 289	
3878		53·18	6453		31·96	5527		37·32
	206 240			206 197			206 290	
3957		52·12	6501		31·72	5638		36·59
	206 239			206 196			206 291	
4034		51·12	6548		31·49	5747		35·90
	206 238			206 195			206 292	
4110		50·18	6595		31·27	5854		35·24
	206 237			206 194			206 293	
4185		49·28	6642		31·05	5958		34·62
	206 236			206 193			206 294	
4258		48·44	6688		30·83	6061		34·04
	206 235			206 192			206 295	
4330		47·63	6734		30·62	6162		33·48
	206 234			206 191			206 296	
4401		46·86	6780		30·41	6262		32·95
	206 233			206 190			206 297	
4471		46·13	6826		30·21	6360		32·44
	206 232			206 189			206 298	
4539		45·43	6871		30·01	6457		31·95
	206 231			206 188			206 299	
4607		44·77	6916		29·82	6552		31·49
	206 230			206 187			206 300	
4674		44·13	6960		29·63	6645		31·05
	206 229			206 186			206 301	
4739		43·52	7005		29·44	6738		30·62
	206 228			206 185			206 302	
4804		42·93	7049		29·25	6829		30·21
	206 227			206 184			206 303	
4868		42·36	7092		29·07	6919		29·82
	206 226			206 183			206 304	
4931		41·82	7136		28·90	7008		29·44
	206 225			206 182			206 305	
4994		41·30	7179		28·72	7096		29·08
	206 224			206 181			206 306	
5055		40·80	7222		28·55	7182		28·73
	206 223			206 180			206 307	
5116		40·31	7265		28·38	7268		28·39

$$\cot x = \frac{\tau}{x''} \qquad \operatorname{cosec} x = \frac{\sigma}{x''} \qquad x'' = \frac{\tau}{\cot x} = \frac{\sigma}{\operatorname{cosec} x}$$

EXAMPLE I. $x = 1^\circ\, 23'\, 45''\!\cdot\!67 = 5025''\!\cdot\!67$ $\qquad \tau = 206\ 224$

$\cot x = 206\ 224 \div 5025\!\cdot\!67 = 41\!\cdot\!0341$

EXAMPLE II. $\operatorname{cosec} x = 45\!\cdot\!6789$ $\qquad \sigma = 206\ 281$

$x'' = 206\ 281 \div 45\!\cdot\!6789 = 4515''\!\cdot\!89 = 1^\circ\, 15'\, 15''\!\cdot\!89$

EXAMPLE III. $x = 89^\circ 16' 37''\!\cdot\!28$ $\qquad 90^\circ - x = 43' 22''\!\cdot\!72 = 2602''\!\cdot\!72$ $\qquad \tau = 206\,254$

$\tan x = \cot(90^\circ - x) = 206\ 254 \div 2602\!\cdot\!72 = 79\!\cdot\!2456$

CONVERSION TO SECONDS

′	0° (″)	1° (″)
0	0	3600
1	60	3660
2	120	3720
3	180	3780
4	240	3840
5	300	3900
6	360	3960
7	420	4020
8	480	4080
9	540	4140
10	600	4200
11	660	4260
12	720	4320
13	780	4380
14	840	4440
15	900	4500
16	960	4560
17	1020	4620
18	1080	4680
19	1140	4740
20	1200	4800
21	1260	4860
22	1320	4920
23	1380	4980
24	1440	5040
25	1500	5100
26	1560	5160
27	1620	5220
28	1680	5280
29	1740	5340
30	1800	5400
31	1860	5460
32	1920	5520
33	1980	5580
34	2040	5640
35	2100	5700
36	2160	5760
37	2220	5820
38	2280	5880
39	2340	5940
40	2400	6000
41	2460	6060
42	2520	6120
43	2580	6180
44	2640	6240
45	2700	6300
46	2760	6360
47	2820	6420
48	2880	6480
49	2940	6540
50	3000	6600
51	3060	6660
52	3120	6720
53	3180	6780
54	3240	6840
55	3300	6900
56	3360	6960
57	3420	7020
58	3480	7080
59	3540	7140
60	3600	7200

TABLE IIIB—TRIGONOMETRICAL FUNCTIONS (Degrees and Minutes)

0°	sin	diff	tan	diff	sec	cosec	cot	cos	
′									′
00	0·000 000	291	0·000 000	291	1·000 000	∞	∞	1·000 000	60
01	·000 291	291	·000 291	291	·000 000	3437·7	3437·7	1·000 000	59
02	·000 582	291	·000 582	291	·000 000	1718·9	1718·9	1·000 000	58
03	·000 873	291	·000 873	291	·000 000	1145·9	1145·9	1·000 000	57
04	·001 164	290	·001 164	290	·000 001	859·44	859·44	0·999 999	56
05	·001 454	291	·001 454	291	·000 001	687·55	687·55	·999 999	55
06	·001 745	291	·001 745	291	·000 002	572·96	572·96	·999 998	54
07	·002 036	291	·002 036	291	·000 002	491·11	491·11	·999 998	53
08	·002 327	291	·002 327	291	·000 003	429·72	429·72	·999 997	52
09	·002 618	291	·002 618	291	·000 003	381·97	381·97	·999 997	51
10	0·002 909	291	0·002 909	291	1·000 004	343·78	343·77	0·999 996	50
11	·003 200	291	·003 200	291	·000 005	312·52	312·52	·999 995	49
12	·003 491	291	·003 491	291	·000 006	286·48	286·48	·999 994	48
13	·003 782	290	·003 782	290	·000 007	264·44	264·44	·999 993	47
14	·004 072	291	·004 072	291	·000 008	245·55	245·55	·999 992	46
15	·004 363	291	·004 363	291	·000 010	229·18	229·18	·999 990	45
16	·004 654	291	·004 654	291	·000 011	214·86	214·86	·999 989	44
17	·004 945	291	·004 945	291	·000 012	202·22	202·22	·999 988	43
18	·005 236	291	·005 236	291	·000 014	190·99	190·98	·999 986	42
19	·005 527	291	·005 527	291	·000 015	180·93	180·93	·999 985	41
20	0·005 818	291	0·005 818	291	1·000 017	171·89	171·89	0·999 983	40
21	·006 109	290	·006 109	291	·000 019	163·70	163·70	·999 981	39
22	·006 399	291	·006 400	291	·000 020	156·26	156·26	·999 980	38
23	·006 690	291	·006 691	290	·000 022	149·47	149·47	·999 978	37
24	·006 981	291	·006 981	291	·000 024	143·24	143·24	·999 976	36
25	·007 272	291	·007 272	291	·000 026	137·51	137·51	·999 974	35
26	·007 563	291	·007 563	291	·000 029	132·22	132·22	·999 971	34
27	·007 854	291	·007 854	291	·000 031	127·33	127·32	·999 969	33
28	·008 145	291	·008 145	291	·000 033	122·78	122·77	·999 967	32
29	·008 436	291	·008 436	291	·000 036	118·54	118·54	·999 964	31
30	0·008 727	290	0·008 727	291	1·000 038	114·59	114·59	0·999 962	30
31	·009 017	291	·009 018	291	·000 041	110·90	110·89	·999 959	29
32	·009 308	291	·009 309	291	·000 043	107·43	107·43	·999 957	28
33	·009 599	291	·009 600	291	·000 046	104·18	104·17	·999 954	27
34	·009 890	291	·009 891	290	·000 049	101·11	101·11	·999 951	26
35	·010 181	291	·010 181	291	·000 052	98·223	98·218	·999 948	25
36	·010 472	291	·010 472	291	·000 055	95·495	95·489	·999 945	24
37	·010 763	291	·010 763	291	·000 058	92·914	92·908	·999 942	23
38	·011 054	290	·011 054	291	·000 061	90·469	90·463	·999 939	22
39	·011 344	291	·011 345	291	·000 064	88·149	88·144	·999 936	21
40	0·011 635	291	0·011 636	291	1·000 068	85·946	85·940	0·999 932	20
41	·011 926	291	·011 927	291	·000 071	83·849	83·844	·999 929	19
42	·012 217	291	·012 218	291	·000 075	81·853	81·847	·999 925	18
43	·012 508	291	·012 509	291	·000 078	79·950	79·943	·999 922	17
44	·012 799	291	·012 800	291	·000 082	78·133	78·126	·999 918	16
45	·013 090	290	·013 091	291	·000 086	76·397	76·390	·999 914	15
46	·013 380	291	·013 382	291	·000 090	74·736	74·729	·999 910	14
47	·013 671	291	·013 673	291	·000 093	73·146	73·139	·999 907	13
48	·013 962	291	·013 964	290	·000 097	71·622	71·615	·999 903	12
49	·014 253	291	·014 254	291	·000 102	70·160	70·153	·999 898	11
50	0·014 544	291	0·014 545	291	1·000 106	68·757	68·750	0·999 894	10
51	·014 835	291	·014 836	291	·000 110	67·409	67·402	·999 890	09
52	·015 126	290	·015 127	291	·000 114	66·113	66·105	·999 886	08
53	·015 416	291	·015 418	291	·000 119	64·866	64·858	·999 881	07
54	·015 707	291	·015 709	291	·000 123	63·665	63·657	·999 877	06
55	·015 998	291	·016 000	291	·000 128	62·507	62·499	·999 872	05
56	·016 289	291	·016 291	291	·000 133	61·391	61·383	·999 867	04
57	·016 580	291	·016 582	291	·000 137	60·314	60·306	·999 863	03
58	·016 871	291	·016 873	291	·000 142	59·274	59·266	·999 858	02
59	·017 162	290	·017 164	291	·000 147	58·270	58·261	·999 853	01
60	0·017 452		0·017 455		1·000 152	57·299	57·290	0·999 848	00
	cos		cot		cosec	sec	tan	sin	89°

TABLE IIIb—TRIGONOMETRICAL FUNCTIONS (Degrees and Minutes)

1° ′	sin	d	tan	d	sec	d	cosec	d	cot	d	cos	d	′
00	0·017 452	291	0·017 455	291	1·000 152	5	57·299	940	57·290	939	0·999 848	5	60
01	·017 743	291	·017 746	291	·000 157	6	56·359	908	56·351	909	·999 843	6	59
02	·018 034	291	·018 037	291	·000 163	5	55·451	881	55·442	881	·999 837	5	58
03	·018 325	291	·018 328	291	·000 168	5	54·570	852	54·561	852	·999 832	5	57
04	·018 616	291	·018 619	291	·000 173	6	53·718	826	53·709	827	·999 827	6	56
05	·018 907	290	·018 910	291	·000 179	5	52·892	802	52·882	801	·999 821	5	55
06	·019 197	291	·019 201	291	·000 184	6	52·090	777	52·081	778	·999 816	6	54
07	·019 488	291	·019 492	291	·000 190	6	51·313	755	51·303	754	·999 810	6	53
08	·019 779	291	·019 783	291	·000 196	5	50·558	732	50·549	733	·999 804	5	52
09	·020 070	291	·020 074	291	·000 201	6	49·826	712	49·816	712	·999 799	6	51
10	0·020 361	291	0·020 365	291	1·000 207	6	49·114	692	49·104	692	0·999 793	6	50
11	·020 652	290	·020 656	291	·000 213	6	48·422	672	48·412	672	·999 787	6	49
12	·020 942	291	·020 947	291	·000 219	7	47·750	654	47·740	655	·999 781	6	48
13	·021 233	291	·021 238	291	·000 226	6	47·096	636	47·085	636	·999 775	7	47
14	·021 524	291	·021 529	291	·000 232	6	46·460	620	46·449	620	·999 768	6	46
15	·021 815	291	·021 820	291	·000 238	6	45·840	603	45·829	603	·999 762	6	45
16	·022 106	291	·022 111	291	·000 244	7	45·237	587	45·226	587	·999 756	7	44
17	·022 397	290	·022 402	291	·000 251	6	44·650	573	44·639	573	·999 749	6	43
18	·022 687	291	·022 693	291	·000 257	7	44·077	557	44·066	558	·999 743	7	42
19	·022 978	291	·022 984	291	·000 264	7	43·520	544	43·508	544	·999 736	7	41
20	0·023 269	291	0·023 275	291	1·000 271	7	42·976	531	42·964	531	0·999 729	7	40
21	·023 560	291	·023 566	291	·000 278	7	42·445	517	42·433	517	·999 722	6	39
22	·023 851	290	·023 857	291	·000 285	7	41·928	505	41·916	505	·999 716	7	38
23	·024 141	291	·024 148	291	·000 292	7	41·423	493	41·411	494	·999 709	8	37
24	·024 432	291	·024 439	292	·000 299	7	40·930	482	40·917	481	·999 701	7	36
25	·024 723	291	·024 731	291	·000 306	7	40·448	470	40·436	471	·999 694	7	35
26	·025 014	291	·025 022	291	·000 313	7	39·978	459	39·965	459	·999 687	7	34
27	·025 305	290	·025 313	291	·000 320	8	39·519	449	39·506	449	·999 680	8	33
28	·025 595	291	·025 604	291	·000 328	7	39·070	439	39·057	439	·999 672	7	32
29	·025 886	291	·025 895	291	·000 335	8	38·631	429	38·618	430	·999 665	8	31
30	0·026 177	291	0·026 186	291	1·000 343	7	38·202	420	38·188	419	0·999 657	7	30
31	·026 468	291	·026 477	291	·000 350	8	37·782	411	37·769	411	·999 650	8	29
32	·026 759	290	·026 768	291	·000 358	8	37·371	401	37·358	402	·999 642	8	28
33	·027 049	291	·027 059	291	·000 366	8	36·970	394	36·956	393	·999 634	8	27
34	·027 340	291	·027 350	291	·000 374	8	36·576	385	36·563	385	·999 626	8	26
35	·027 631	291	·027 641	292	·000 382	8	36·191	376	36·178	377	·999 618	8	25
36	·027 922	290	·027 933	291	·000 390	8	35·815	370	35·801	370	·999 610	8	24
37	·028 212	291	·028 224	291	·000 398	8	35·445	361	35·431	361	·999 602	8	23
38	·028 503	291	·028 515	291	·000 406	9	35·084	354	35·070	355	·999 594	9	22
39	·028 794	291	·028 806	291	·000 415	8	34·730	348	34·715	347	·999 585	8	21
40	0·029 085	290	0·029 097	291	1·000 423	9	34·382	340	34·368	341	0·999 577	9	20
41	·029 375	291	·029 388	291	·000 432	8	34·042	334	34·027	333	·999 568	8	19
42	·029 666	291	·029 679	291	·000 440	9	33·708	327	33·694	328	·999 560	9	18
43	·029 957	291	·029 970	292	·000 449	9	33·381	321	33·366	321	·999 551	9	17
44	·030 248	291	·030 262	291	·000 458	9	33·060	314	33·045	315	·999 542	8	16
45	·030 539	290	·030 553	291	·000 467	9	32·746	309	32·730	309	·999 534	9	15
46	·030 829	291	·030 844	291	·000 476	9	32·437	303	32·421	303	·999 525	9	14
47	·031 120	291	·031 135	291	·000 485	9	32·134	298	32·118	297	·999 516	9	13
48	·031 411	291	·031 426	291	·000 494	9	31·836	292	31·821	293	·999 507	10	12
49	·031 702	290	·031 717	292	·000 503	9	31·544	286	31·528	286	·999 497	9	11
50	0·031 992	291	0·032 009	291	1·000 512	10	31·258	282	31·242	282	0·999 488	9	10
51	·032 283	291	·032 300	291	·000 522	9	30·976	276	30·960	277	·999 479	10	09
52	·032 574	290	·032 591	291	·000 531	9	30·700	272	30·683	271	·999 469	9	08
53	·032 864	291	·032 882	291	·000 540	10	30·428	267	30·412	267	·999 460	10	07
54	·033 155	291	·033 173	292	·000 550	10	30·161	262	30·145	263	·999 450	9	06
55	·033 446	291	·033 465	291	·000 560	10	29·899	258	29·882	258	·999 441	10	05
56	·033 737	290	·033 756	291	·000 570	9	29·641	253	29·624	253	·999 431	10	04
57	·034 027	291	·034 047	291	·000 579	10	29·388	249	29·371	249	·999 421	10	03
58	·034 318	291	·034 338	291	·000 589	10	29·139	245	29·122	245	·999 411	10	02
59	·034 609	290	·034 630	292	·000 599	11	28·894	240	28·877	241	·999 401	10	01
60	0·034 899		0·034 921		1·000 610		28·654		28·636		0·999 391		00
	cos		cot		cosec		sec		tan		sin		88°

TABLE IIIB—TRIGONOMETRICAL FUNCTIONS (Degrees and Minutes)

2° ′	sin		tan		sec		cosec		cot		cos		
00	0·034 899	291	0·034 921	291	1·000 610	10	28·654	237	28·636	237	0·999 391	10	60′
01	·035 190	291	·035 212	291	·000 620	10	28·417	233	28·399	233	·999 381	11	59
02	·035 481	291	·035 503	292	·000 630	10	28·184	229	28·166	229	·999 370	10	58
03	·035 772	290	·035 795	291	·000 640	11	27·955	225	27·937	225	·999 360	10	57
04	·036 062	291	·036 086	291	·000 651	10	27·730	222	27·712	222	·999 350	11	56
05	·036 353	291	·036 377	291	·000 661	11	27·508	218	27·490	219	·999 339	11	55
06	·036 644	290	·036 668	292	·000 672	11	27·290	215	27·271	214	·999 328	10	54
07	·036 934	291	·036 960	291	·000 683	11	27·075	211	27·057	212	·999 318	11	53
08	·037 225	291	·037 251	291	·000 694	10	26·864	209	26·845	208	·999 307	11	52
09	·037 516	290	·037 542	292	·000 704	11	26·655	204	26·637	205	·999 296	11	51
10	0·037 806	291	0·037 834	291	1·000 715	11	26·451	202	26·432	202	0·999 285	11	50
11	·038 097	291	·038 125	291	·000 726	12	26·249	199	26·230	199	·999 274	11	49
12	·038 388	290	·038 416	291	·000 738	11	26·050	196	26·031	196	·999 263	11	48
13	·038 678	291	·038 707	292	·000 749	11	25·854	193	25·835	193	·999 252	12	47
14	·038 969	291	·038 999	291	·000 760	12	25·661	190	25·642	190	·999 240	11	46
15	·039 260	290	·039 290	291	·000 772	11	25·471	187	25·452	188	·999 229	11	45
16	·039 550	291	·039 581	292	·000 783	12	25·284	184	25·264	184	·999 218	12	44
17	·039 841	291	·039 873	291	·000 795	11	25·100	182	25·080	182	·999 206	12	43
18	·040 132	290	·040 164	292	·000 806	12	24·918	179	24·898	179	·999 194	11	42
19	·040 422	291	·040 456	291	·000 818	12	24·739	177	24·719	177	·999 183	12	41
20	0·040 713	291	0·040 747	291	1·000 830	12	24·562	174	24·542	174	0·999 171	12	40
21	·041 004	290	·041 038	292	·000 842	12	24·388	172	24·368	172	·999 159	12	39
22	·041 294	291	·041 330	291	·000 854	12	24·216	169	24·196	170	·999 147	12	38
23	·041 585	291	·041 621	291	·000 866	12	24·047	167	24·026	167	·999 135	12	37
24	·041 876	290	·041 912	292	·000 878	12	23·880	164	23·859	164	·999 123	12	36
25	·042 166	291	·042 204	291	·000 890	13	23·716	163	23·695	163	·999 111	13	35
26	·042 457	291	·042 495	292	·000 903	12	23·553	160	23·532	160	·999 098	12	34
27	·042 748	290	·042 787	291	·000 915	12	23·393	158	23·372	158	·999 086	13	33
28	·043 038	291	·043 078	292	·000 927	13	23·235	156	23·214	156	·999 073	12	32
29	·043 329	290	·043 370	291	·000 940	13	23·079	153	23·058	154	·999 061	13	31
30	0·043 619	291	0·043 661	291	1·000 953	12	22·926	152	22·904	152	0·999 048	13	30
31	·043 910	291	·043 952	292	·000 965	13	22·774	150	22·752	150	·999 035	12	29
32	·044 201	290	·044 244	291	·000 978	13	22·624	148	22·602	148	·999 023	13	28
33	·044 491	291	·044 535	292	·000 991	13	22·476	146	22·454	146	·999 010	13	27
34	·044 782	290	·044 827	291	·001 004	13	22·330	143	22·308	144	·998 997	13	26
35	·045 072	291	·045 118	292	·001 017	13	22·187	143	22·164	142	·998 984	13	25
36	·045 363	291	·045 410	291	·001 030	14	22·044	140	22·022	141	·998 971	14	24
37	·045 654	290	·045 701	292	·001 044	13	21·904	138	21·881	138	·998 957	13	23
38	·045 944	291	·045 993	291	·001 057	14	21·766	137	21·743	137	·998 944	13	22
39	·046 235	290	·046 284	292	·001 071	13	21·629	135	21·606	136	·998 931	14	21
40	0·046 525	291	0·046 576	291	1·001 084	14	21·494	134	21·470	133	0·998 917	13	20
41	·046 816	290	·046 867	292	·001 098	13	21·360	131	21·337	132	·998 904	14	19
42	·047 106	291	·047 159	291	·001 111	14	21·229	131	21·205	130	·998 890	14	18
43	·047 397	291	·047 450	292	·001 125	14	21·098	128	21·075	129	·998 876	14	17
44	·047 688	290	·047 742	291	·001 139	14	20·970	127	20·946	127	·998 862	14	16
45	·047 978	291	·048 033	292	·001 153	14	20·843	126	20·819	126	·998 848	14	15
46	·048 269	290	·048 325	292	·001 167	14	20·717	124	20·693	124	·998 834	14	14
47	·048 559	291	·048 617	291	·001 181	14	20·593	122	20·569	123	·998 820	14	13
48	·048 850	290	·048 908	292	·001 195	15	20·471	121	20·446	121	·998 806	14	12
49	·049 140	291	·049 200	291	·001 210	14	20·350	120	20·325	119	·998 792	14	11
50	0·049 431	290	0·049 491	292	1·001 224	14	20·230	118	20·206	119	0·998 778	15	10
51	·049 721	291	·049 783	292	·001 238	15	20·112	117	20·087	117	·998 763	14	09
52	·050 012	290	·050 075	291	·001 253	15	19·995	115	19·970	115	·998 749	15	08
53	·050 302	291	·050 366	292	·001 268	14	19·880	114	19·855	115	·998 734	15	07
54	·050 593	290	·050 658	291	·001 282	15	19·766	113	19·740	113	·998 719	14	06
55	·050 883	291	·050 949	292	·001 297	15	19·653	112	19·627	111	·998 705	15	05
56	·051 174	290	·051 241	292	·001 312	15	19·541	110	19·516	111	·998 690	15	04
57	·051 464	291	·051 533	291	·001 327	15	19·431	109	19·405	109	·998 675	15	03
58	·051 755	290	·051 824	292	·001 342	15	19·322	108	19·296	108	·998 660	15	02
59	·052 045	291	·052 116	292	·001 357	15	19·214	107	19·188	107	·998 645	15	01
60	0·052 336		0·052 408		1·001 372		19·107		19·081		0·998 630		00
	cos		cot		cosec		sec		tan		sin		87°

TABLE IIIB—TRIGONOMETRICAL FUNCTIONS (Degrees and Minutes)

3° ′	sin	d	tan	d	sec	d	cosec	d	cot	d	cos	d	′
00	0·052 336	290	0·052 408	291	1·001 372	16	19·1073	1054	19·0811	1056	0·998 630	16	60
01	·052 626	291	·052 699	292	·001 388	15	19·0019	1044	18·9755	1044	·998 614	15	59
02	·052 917	290	·052 991	292	·001 403	16	18·8975	1031	18·8711	1033	·998 599	16	58
03	·053 207	291	·053 283	292	·001 419	15	18·7944	1021	18·7678	1022	·998 583	15	57
04	·053 498	290	·053 575	291	·001 434	16	18·6923	1009	18·6656	1011	·998 568	16	56
05	·053 788	291	·053 866	292	·001 450	15	18·5914	999	18·5645	1000	·998 552	15	55
06	·054 079	290	·054 158	292	·001 465	16	18·4915	988	18·4645	990	·998 537	16	54
07	·054 369	291	·054 450	292	·001 481	16	18·3927	977	18·3655	978	·998 521	16	53
08	·054 660	290	·054 742	291	·001 497	16	18·2950	967	18·2677	969	·998 505	16	52
09	·054 950	291	·055 033	292	·001 513	16	18·1983	957	18·1708	958	·998 489	16	51
10	0·055 241	290	0·055 325	292	1·001 529	16	18·1026	947	18·0750	948	0·998 473	16	50
11	·055 531	291	·055 617	292	·001 545	17	18·0079	937	17·9802	939	·998 457	16	49
12	·055 822	290	·055 909	291	·001 562	16	17·9142	927	17·8863	929	·998 441	17	48
13	·056 112	290	·056 200	292	·001 578	16	17·8215	917	17·7934	919	·998 424	16	47
14	·056 402	291	·056 492	292	·001 594	17	17·7298	909	17·7015	909	·998 408	16	46
15	·056 693	290	·056 784	292	·001 611	17	17·6389	899	17·6106	901	·998 392	17	45
16	·056 983	291	·057 076	292	·001 628	16	17·5490	890	17·5205	891	·998 375	16	44
17	·057 274	290	·057 368	292	·001 644	17	17·4600	880	17·4314	882	·998 359	17	43
18	·057 564	290	·057 660	291	·001 661	17	17·3720	872	17·3432	874	·998 342	17	42
19	·057 854	291	·057 951	292	·001 678	17	17·2848	864	17·2558	865	·998 325	17	41
20	0·058 145	290	0·058 243	292	1·001 695	17	17·1984	854	17·1693	856	0·998 308	17	40
21	·058 435	291	·058 535	292	·001 712	17	17·1130	847	17·0837	847	·998 291	17	39
22	·058 726	290	·058 827	292	·001 729	17	17·0283	837	16·9990	840	·998 274	17	38
23	·059 016	290	·059 119	292	·001 746	17	16·9446	830	16·9150	831	·998 257	17	37
24	·059 306	291	·059 411	292	·001 763	18	16·8616	822	16·8319	823	·998 240	17	36
25	·059 597	290	·059 703	292	·001 781	17	16·7794	813	16·7496	815	·998 223	18	35
26	·059 887	290	·059 995	292	·001 798	18	16·6981	806	16·6681	807	·998 205	17	34
27	·060 177	291	·060 287	292	·001 816	17	16·6175	798	16·5874	799	·998 188	18	33
28	·060 468	290	·060 579	292	·001 833	18	16·5377	790	16·5075	792	·998 170	17	32
29	·060 758	291	·060 871	292	·001 851	18	16·4587	783	16·4283	784	·998 153	18	31
30	0·061 049	290	0·061 163	292	1·001 869	18	16·3804	775	16·3499	777	0·998 135	18	30
31	·061 339	290	·061 455	292	·001 887	18	16·3029	768	16·2722	770	·998 117	18	29
32	·061 629	291	·061 747	292	·001 905	18	16·2261	761	16·1952	762	·998 099	18	28
33	·061 920	290	·062 039	292	·001 923	18	16·1500	754	16·1190	755	·998 081	18	27
34	·062 210	290	·062 331	292	·001 941	18	16·0746	747	16·0435	748	·998 063	18	26
35	·062 500	291	·062 623	292	·001 959	18	15·9999	739	15·9687	742	·998 045	18	25
36	·062 791	290	·062 915	292	·001 977	19	15·9260	733	15·8945	734	·998 027	19	24
37	·063 081	290	·063 207	292	·001 996	18	15·8527	726	15·8211	728	·998 008	18	23
38	·063 371	290	·063 499	292	·002 014	19	15·7801	720	15·7483	721	·997 990	18	22
39	·063 661	291	·063 791	292	·002 033	18	15·7081	713	15·6762	714	·997 972	19	21
40	0·063 952	290	0·064 083	292	1·002 051	19	15·6368	707	15·6048	708	0·997 953	19	20
41	·064 242	290	·064 375	292	·002 070	19	15·5661	700	15·5340	702	·997 934	18	19
42	·064 532	291	·064 667	292	·002 089	19	15·4961	694	15·4638	695	·997 916	19	18
43	·064 823	290	·064 959	292	·002 108	19	15·4267	688	15·3943	689	·997 897	19	17
44	·065 113	290	·065 251	292	·002 127	19	15·3579	681	15·3254	683	·997 878	19	16
45	·065 403	290	·065 543	293	·002 146	19	15·2898	676	15·2571	678	·997 859	19	15
46	·065 693	291	·065 836	292	·002 165	19	15·2222	669	15·1893	671	·997 840	19	14
47	·065 984	290	·066 128	292	·002 184	19	15·1553	664	15·1222	665	·997 821	20	13
48	·066 274	290	·066 420	292	·002 203	20	15·0889	658	15·0557	659	·997 801	19	12
49	·066 564	290	·066 712	292	·002 223	19	15·0231	652	14·9898	654	·997 782	19	11
50	0·066 854	291	0·067 004	292	1·002 242	20	14·9579	647	14·9244	648	0·997 763	20	10
51	·067 145	290	·067 296	293	·002 262	20	14·8932	641	14·8596	642	·997 743	19	09
52	·067 435	290	·067 589	292	·002 282	19	14·8291	635	14·7954	637	·997 724	20	08
53	·067 725	290	·067 881	292	·002 301	20	14·7656	630	14·7317	632	·997 704	20	07
54	·068 015	291	·068 173	292	·002 321	20	14·7026	625	14·6685	626	·997 684	20	06
55	·068 306	290	·068 465	293	·002 341	20	14·6401	619	14·6059	621	·997 664	19	05
56	·068 596	290	·068 758	292	·002 361	20	14·5782	614	14·5438	615	·997 645	20	04
57	·068 886	290	·069 050	292	·002 381	20	14·5168	609	14·4823	611	·997 625	21	03
58	·069 176	290	·069 342	293	·002 401	21	14·4559	604	14·4212	605	·997 604	20	02
59	·069 466	290	·069 635	292	·002 422	20	14·3955	599	14·3607	600	·997 584	20	01
60	0·069 756		0·069 927		1·002 442		14·3356		14·3007		0·997 564		00
	cos		cot		cosec		sec		tan		sin		86°

4° ′	sin		tan		sec		cosec		cot		cos		′
00	0·069 756	291	0·069 927	292	1·002 442	20	14·3356	594	14·3007	596	0·997 564	20	60
01	·070 047	290	·070 219	292	·002 462	21	·2762	589	·2411	590	·997 544	21	59
02	·070 337	290	·070 511	293	·002 483	20	·2173	584	·1821	586	·997 523	20	58
03	·070 627	290	·070 804	292	·002 503	21	·1589	579	·1235	580	·997 503	21	57
04	·070 917	290	·071 096	293	·002 524	21	·1010	575	·0655	576	·997 482	20	56
05	·071 207	290	·071 389	292	·002 545	21	14·0435	570	14·0079	572	·997 462	21	55
06	·071 497	291	·071 681	292	·002 566	21	13·9865	565	13·9507	567	·997 441	21	54
07	·071 788	290	·071 973	293	·002 587	21	·9300	561	·8940	562	·997 420	21	53
08	·072 078	290	·072 266	292	·002 608	21	·8739	556	·8378	557	·997 399	21	52
09	·072 368	290	·072 558	293	·002 629	21	·8183	552	·7821	554	·997 378	21	51
10	0·072 658	290	0·072 851	292	1·002 650	21	13·7631	547	13·7267	548	0·997 357	21	50
11	·072 948	290	·073 143	292	·002 671	22	·7084	543	·6719	545	·997 336	22	49
12	·073 238	290	·073 435	293	·002 693	21	·6541	539	·6174	540	·997 314	21	48
13	·073 528	290	·073 728	292	·002 714	22	·6002	534	·5634	536	·997 293	21	47
14	·073 818	290	·074 020	293	·002 736	21	·5468	531	·5098	532	·997 272	22	46
15	·074 108	291	·074 313	292	·002 757	22	·4937	526	·4566	527	·997 250	21	45
16	·074 399	290	·074 605	293	·002 779	22	·4411	522	·4039	524	·997 229	22	44
17	·074 689	290	·074 898	292	·002 801	22	·3889	518	·3515	519	·997 207	22	43
18	·074 979	290	·075 190	293	·002 823	22	·3371	514	·2996	516	·997 185	22	42
19	·075 269	290	·075 483	292	·002 845	22	·2857	510	·2480	511	·997 163	22	41
20	0·075 559	290	0·075 775	293	1·002 867	22	13·2347	506	13·1969	508	0·997 141	22	40
21	·075 849	290	·076 068	293	·002 889	22	·1841	502	·1461	503	·997 119	22	39
22	·076 139	290	·076 361	292	·002 911	23	·1339	499	·0958	500	·997 097	22	38
23	·076 429	290	·076 653	293	·002 934	22	·0840	494	13·0458	496	·997 075	22	37
24	·076 719	290	·076 946	292	·002 956	22	13·0346	491	12·9962	493	·997 053	23	36
25	·077 009	290	·077 238	293	·002 978	23	12·9855	487	·9469	488	·997 030	22	35
26	·077 299	290	·077 531	293	·003 001	23	·9368	484	·8981	485	·997 008	23	34
27	·077 589	290	·077 824	292	·003 024	22	·8884	480	·8496	482	·996 985	22	33
28	·077 879	290	·078 116	293	·003 046	23	·8404	476	·8014	478	·996 963	23	32
29	·078 169	290	·078 409	293	·003 069	23	·7928	473	·7536	474	·996 940	23	31
30	0·078 459	290	0·078 702	292	1·003 092	23	12·7455	469	12·7062	471	0·996 917	23	30
31	·078 749	290	·078 994	293	·003 115	23	·6986	466	·6591	467	·996 894	22	29
32	·079 039	290	·079 287	293	·003 138	23	·6520	463	·6124	464	·996 872	24	28
33	·079 329	290	·079 580	293	·003 161	24	·6057	459	·5660	461	·996 848	23	27
34	·079 619	290	·079 873	292	·003 185	23	·5598	456	·5199	457	·996 825	23	26
35	·079 909	290	·080 165	293	·003 208	24	·5142	452	·4742	454	·996 802	23	25
36	·080 199	290	·080 458	293	·003 232	23	·4690	449	·4288	450	·996 779	23	24
37	·080 489	290	·080 751	293	·003 255	24	·4241	446	·3838	448	·996 756	24	23
38	·080 779	290	·081 044	292	·003 279	23	·3795	443	·3390	444	·996 732	23	22
39	·081 069	290	·081 336	293	·003 302	24	·3352	439	·2946	441	·996 709	24	21
40	0·081 359	290	0·081 629	293	1·003 326	24	12·2913	437	12·2505	438	0·996 685	24	20
41	·081 649	290	·081 922	293	·003 350	24	·2476	433	·2067	435	·996 661	24	19
42	·081 939	289	·082 215	293	·003 374	24	·2043	431	·1632	431	·996 637	23	18
43	·082 228	290	·082 508	293	·003 398	24	·1612	427	·1201	429	·996 614	24	17
44	·082 518	290	·082 801	293	·003 422	24	·1185	424	·0772	426	·996 590	24	16
45	·082 808	290	·083 094	292	·003 446	25	·0761	421	12·0346	423	·996 566	25	15
46	·083 098	290	·083 386	293	·003 471	24	12·0340	419	11·9923	419	·996 541	24	14
47	·083 388	290	·083 679	293	·003 495	24	11·9921	415	·9504	417	·996 517	24	13
48	·083 678	290	·083 972	293	·003 519	25	·9506	413	·9087	414	·996 493	25	12
49	·083 968	290	·084 265	293	·003 544	25	·9093	409	·8673	411	·996 468	24	11
50	0·084 258	289	0·084 558	293	1·003 569	24	11·8684	407	11·8262	409	0·996 444	25	10
51	·084 547	290	·084 851	293	·003 593	25	·8277	404	·7853	405	·996 419	24	09
52	·084 837	290	·085 144	293	·003 618	25	·7873	402	·7448	403	·996 395	25	08
53	·085 127	290	·085 437	293	·003 643	25	·7471	398	·7045	400	·996 370	25	07
54	·085 417	290	·085 730	293	·003 668	25	·7073	396	·6645	397	·996 345	25	06
55	·085 707	290	·086 023	293	·003 693	25	·6677	393	·6248	395	·996 320	25	05
56	·085 997	289	·086 316	293	·003 718	26	·6284	391	·5853	392	·996 295	25	04
57	·086 286	290	·086 609	293	·003 744	25	·5893	388	·5461	389	·996 270	25	03
58	·086 576	290	·086 902	294	·003 769	25	·5505	385	·5072	387	·996 245	25	02
59	·086 866	290	·087 196	293	·003 794	26	·5120	383	·4685	384	·996 220	25	01
60	0·087 156		0·087 489		1·003 820		11·4737		11·4301		0·996 195		00
	cos		cot		cosec		sec		tan		sin		85°

TABLE IIIB—TRIGONOMETRICAL FUNCTIONS (Degrees and Minutes)

5° ′	sin		tan		sec		cosec		cot		cos		′
00	0·087 156	290	0·087 489	293	1·003 820	25	11·4737	380	11·4301	382	0·996 195	26	60
01	·087 446	289	·087 782	293	·003 845	26	·4357	378	·3919	379	·996 169	25	59
02	·087 735	290	·088 075	293	·003 871	26	·3979	375	·3540	377	·996 144	26	58
03	·088 025	290	·088 368	293	·003 897	26	·3604	373	·3163	374	·996 118	25	57
04	·088 315	290	·088 661	293	·003 923	26	·3231	370	·2789	372	·996 093	26	56
05	·088 605	289	·088 954	294	·003 949	26	·2861	368	·2417	369	·996 067	26	55
06	·088 894	290	·089 248	293	·003 975	26	·2493	365	·2048	367	·996 041	26	54
07	·089 184	290	·089 541	293	·004 001	26	·2128	363	·1681	365	·996 015	26	53
08	·089 474	289	·089 834	293	·004 027	26	·1765	361	·1316	362	·995 989	26	52
09	·089 763	290	·090 127	294	·004 053	27	·1404	359	·0954	360	·995 963	26	51
10	0·090 053	290	0·090 421	293	1·004 080	26	11·1045	356	11·0594	357	0·995 937	26	50
11	·090 343	290	·090 714	293	·004 106	27	·0689	353	11·0237	355	·995 911	27	49
12	·090 633	289	·091 007	293	·004 133	26	11·0336	352	10·9882	353	·995 884	26	48
13	·090 922	290	·091 300	294	·004 159	27	10·9984	349	·9529	351	·995 858	26	47
14	·091 212	290	·091 594	293	·004 186	27	·9635	347	·9178	349	·995 832	27	46
15	·091 502	289	·091 887	293	·004 213	27	·9288	345	·8829	346	·995 805	27	45
16	·091 791	290	·092 180	294	·004 240	27	·8943	343	·8483	344	·995 778	26	44
17	·092 081	290	·092 474	293	·004 267	27	·8600	340	·8139	342	·995 752	27	43
18	·092 371	289	·092 767	294	·004 294	27	·8260	339	·7797	340	·995 725	27	42
19	·092 660	290	·093 061	293	·004 321	27	·7921	336	·7457	338	·995 698	27	41
20	0·092 950	289	0·093 354	293	1·004 348	27	10·7585	334	10·7119	336	0·995 671	27	40
21	·093 239	290	·093 647	294	·004 375	28	·7251	332	·6783	333	·995 644	27	39
22	·093 529	290	·093 941	293	·004 403	27	·6919	330	·6450	332	·995 617	28	38
23	·093 819	289	·094 234	294	·004 430	28	·6589	328	·6118	329	·995 589	27	37
24	·094 108	290	·094 528	293	·004 458	27	·6261	326	·5789	327	·995 562	27	36
25	·094 398	289	·094 821	294	·004 485	28	·5935	324	·5462	326	·995 535	28	35
26	·094 687	290	·095 115	293	·004 513	28	·5611	322	·5136	323	·995 507	28	34
27	·094 977	290	·095 408	294	·004 541	28	·5289	320	·4813	322	·995 479	27	33
28	·095 267	289	·095 702	293	·004 569	28	·4969	319	·4491	319	·995 452	28	32
29	·095 556	290	·095 995	294	·004 597	28	·4650	316	·4172	318	·995 424	28	31
30	0·095 846	289	0·096 289	294	1·004 625	28	10·4334	314	10·3854	316	0·995 396	28	30
31	·096 135	290	·096 583	293	·004 653	29	·4020	312	·3538	314	·995 368	28	29
32	·096 425	289	·096 876	294	·004 682	28	·3708	311	·3224	311	·995 340	28	28
33	·096 714	290	·097 170	294	·004 710	28	·3397	308	·2913	311	·995 312	28	27
34	·097 004	289	·097 464	293	·004 738	29	·3089	307	·2602	308	·995 284	28	26
35	·097 293	290	·097 757	294	·004 767	28	·2782	305	·2294	306	·995 256	29	25
36	·097 583	289	·098 051	294	·004 795	29	·2477	303	·1988	305	·995 227	28	24
37	·097 872	290	·098 345	293	·004 824	29	·2174	301	·1683	302	·995 199	29	23
38	·098 162	289	·098 638	294	·004 853	29	·1873	300	·1381	301	·995 170	28	22
39	·098 451	290	·098 932	294	·004 882	29	·1573	298	·1080	300	·995 142	29	21
40	0·098 741	289	0·099 226	293	1·004 911	29	10·1275	296	10·0780	297	0·995 113	29	20
41	·099 030	290	·099 519	294	·004 940	29	·0979	294	·0483	296	·995 084	28	19
42	·099 320	289	·099 813	294	·004 969	29	·0685	293	10·0187	294	·995 056	29	18
43	·099 609	290	·100 107	294	·004 998	30	·0392	291	9·9893	292	·995 027	29	17
44	·099 899	289	·100 401	294	·005 028	29	10·0101	289	·9601	291	·994 998	29	16
45	·100 188	289	·100 695	294	·005 057	29	9·9812	287	·9310	289	·994 969	30	15
46	·100 477	290	·100 989	293	·005 086	30	·9525	286	·9021	287	·994 939	29	14
47	·100 767	289	·101 282	294	·005 116	30	·9239	284	·8734	286	·994 910	29	13
48	·101 056	290	·101 576	294	·005 146	29	·8955	283	·8448	284	·994 881	30	12
49	·101 346	289	·101 870	294	·005 175	30	·8672	281	·8164	282	·994 851	29	11
50	0·101 635	289	0·102 164	294	1·005 205	30	9·8391	279	9·7882	281	0·994 822	30	10
51	·101 924	290	·102 458	294	·005 235	30	·8112	278	·7601	279	·994 792	30	09
52	·102 214	289	·102 752	294	·005 265	30	·7834	276	·7322	278	·994 762	29	08
53	·102 503	290	·103 046	294	·005 295	30	·7558	275	·7044	276	·994 733	30	07
54	·102 793	289	·103 340	294	·005 325	31	·7283	273	·6768	275	·994 703	30	06
55	·103 082	289	·103 634	294	·005 356	30	·7010	271	·6493	273	·994 673	30	05
56	·103 371	290	·103 928	294	·005 386	30	·6739	270	·6220	271	·994 643	30	04
57	·103 661	289	·104 222	294	·005 416	31	·6469	269	·5949	270	·994 613	30	03
58	·103 950	289	·104 516	294	·005 447	31	·6200	267	·5679	268	·994 583	31	02
59	·104 239	289	·104 810	294	·005 478	30	·5933	265	·5411	267	·994 552	30	01
60	0·104 528		0·105 104		1·005 508		9·5668		9·5144		0·994 522		00
	cos		cot		cosec		sec		tan		sin		84°

TABLE IIIB—TRIGONOMETRICAL FUNCTIONS (Degrees and Minutes)

6° ′	sin		tan		sec		cosec		cot		cos		′
00	0·104 528	290	0·105 104	294	1·005 508	31	9·566 77	2640	9·514 36	2655	0·994 522	31	60
01	·104 818	289	·105 398	294	·005 539	31	·540 37	2626	·487 81	2640	·994 491	30	59
02	·105 107	289	·105 692	295	·005 570	31	·514 11	2611	·461 41	2626	·994 461	31	58
03	·105 396	290	·105 987	294	·005 601	31	·488 00	2597	·435 15	2611	·994 430	30	57
04	·105 686	289	·106 281	294	·005 632	31	·462 03	2583	·409 04	2597	·994 400	31	56
05	·105 975	289	·106 575	294	·005 663	31	·436 20	2568	·383 07	2583	·994 369	31	55
06	·106 264	289	·106 869	294	·005 694	32	·410 52	2555	·357 24	2569	·994 338	31	54
07	·106 553	290	·107 163	295	·005 726	31	·384 97	2540	·331 55	2556	·994 307	31	53
08	·106 843	289	·107 458	294	·005 757	31	·359 57	2527	·305 99	2541	·994 276	31	52
09	·107 132	289	·107 752	294	·005 788	32	·334 30	2513	·280 58	2528	·994 245	31	51
10	0·107 421	289	0·108 046	294	1·005 820	32	9·309 17	2500	9·255 30	2514	0·994 214	32	50
11	·107 710	289	·108 340	295	·005 852	31	·284 17	2486	·230 16	2500	·994 182	31	49
12	·107 999	290	·108 635	294	·005 883	32	·259 31	2472	·205 16	2488	·994 151	31	48
13	·108 289	289	·108 929	294	·005 915	32	·234 59	2460	·180 28	2474	·994 120	32	47
14	·108 578	289	·109 223	295	·005 947	32	·209 99	2446	·155 54	2461	·994 088	32	46
15	·108 867	289	·109 518	294	·005 979	32	·185 53	2433	·130 93	2447	·994 056	31	45
16	·109 156	289	·109 812	295	·006 011	32	·161 20	2421	·106 46	2435	·994 025	32	44
17	·109 445	289	·110 107	294	·006 043	33	·136 99	2407	·082 11	2422	·993 993	32	43
18	·109 734	289	·110 401	294	·006 076	32	·112 92	2395	·057 89	2410	·993 961	32	42
19	·110 023	290	·110 695	295	·006 108	33	·088 97	2382	·033 79	2396	·993 929	32	41
20	0·110 313	289	0·110 990	294	1·006 141	32	9·065 15	2369	9·009 83	2385	0·993 897	32	40
21	·110 602	289	·111 284	295	·006 173	33	·041 46	2358	8·985 98	2371	·993 865	32	39
22	·110 891	289	·111 579	294	·006 206	32	9·017 88	2344	·962 27	2360	·993 833	33	38
23	·111 180	289	·111 873	295	·006 238	33	8·994 44	2333	·938 67	2347	·993 800	32	37
24	·111 469	289	·112 168	295	·006 271	33	·971 11	2320	·915 20	2335	·993 768	33	36
25	·111 758	289	·112 463	294	·006 304	33	·947 91	2309	·891 85	2323	·993 735	32	35
26	·112 047	289	·112 757	295	·006 337	33	·924 82	2296	·868 62	2311	·993 703	33	34
27	·112 336	289	·113 052	294	·006 370	33	·901 86	2285	·845 51	2299	·993 670	32	33
28	·112 625	289	·113 346	295	·006 403	33	·879 01	2273	·822 52	2288	·993 638	33	32
29	·112 914	289	·113 641	295	·006 436	34	·856 28	2261	·799 64	2275	·993 605	33	31
30	0·113 203	289	0·113 936	294	1·006 470	33	8·833 67	2249	8·776 89	2264	0·993 572	33	30
31	·113 492	289	·114 230	295	·006 503	34	·811 18	2238	·754 25	2253	·993 539	33	29
32	·113 781	289	·114 525	295	·006 537	33	·788 80	2227	·731 72	2241	·993 506	33	28
33	·114 070	289	·114 820	294	·006 570	34	·766 53	2215	·709 31	2230	·993 473	34	27
34	·114 359	289	·115 114	295	·006 604	34	·744 38	2204	·687 01	2219	·993 439	33	26
35	·114 648	289	·115 409	295	·006 638	33	·722 34	2193	·664 82	2207	·993 406	33	25
36	·114 937	289	·115 704	295	·006 671	34	·700 41	2182	·642 75	2197	·993 373	34	24
37	·115 226	289	·115 999	295	·006 705	34	·678 59	2171	·620 78	2185	·993 339	33	23
38	·115 515	289	·116 294	294	·006 739	34	·656 88	2160	·598 93	2175	·993 306	34	22
39	·115 804	289	·116 588	295	·006 773	35	·635 28	2149	·577 18	2163	·993 272	34	21
40	0·116 093	289	0·116 883	295	1·006 808	34	8·613 79	2138	8·555 55	2153	0·993 238	33	20
41	·116 382	289	·117 178	295	·006 842	34	·592 41	2128	·534 02	2143	·993 205	34	19
42	·116 671	289	·117 473	295	·006 876	35	·571 13	2117	·512 59	2131	·993 171	34	18
43	·116 960	289	·117 768	295	·006 911	34	·549 96	2107	·491 28	2121	·993 137	34	17
44	·117 249	288	·118 063	295	·006 945	35	·528 89	2096	·470 07	2111	·993 103	35	16
45	·117 537	289	·118 358	295	·006 980	35	·507 93	2086	·448 96	2101	·993 068	34	15
46	·117 826	289	·118 653	295	·007 015	34	·487 07	2075	·427 95	2090	·993 034	34	14
47	·118 115	289	·118 948	295	·007 049	35	·466 32	2066	·407 05	2080	·993 000	34	13
48	·118 404	289	·119 243	295	·007 084	35	·445 66	2055	·386 25	2070	·992 966	35	12
49	·118 693	289	·119 538	295	·007 119	35	·425 11	2045	·365 55	2059	·992 931	35	11
50	0·118 982	288	0·119 833	295	1·007 154	36	8·404 66	2035	8·344 96	2050	0·992 896	34	10
51	·119 270	289	·120 128	295	·007 190	35	·384 31	2026	·324 46	2040	·992 862	35	09
52	·119 559	289	·120 423	295	·007 225	35	·364 05	2015	·304 06	2030	·992 827	35	08
53	·119 848	289	·120 718	295	·007 260	35	·343 90	2006	·283 76	2021	·992 792	35	07
54	·120 137	289	·121 013	295	·007 295	36	·323 84	1996	·263 55	2010	·992 757	35	06
55	·120 426	288	·121 308	296	·007 331	36	·303 88	1986	·243 45	2001	·992 722	35	05
56	·120 714	289	·121 604	295	·007 367	35	·284 02	1977	·223 44	1992	·992 687	35	04
57	·121 003	289	·121 899	295	·007 402	36	·264 25	1968	·203 52	1982	·992 652	35	03
58	·121 292	289	·122 194	295	·007 438	36	·244 57	1957	·183 70	1972	·992 617	35	02
59	·121 581	288	·122 489	296	·007 474	36	·225 00	1949	·163 98	1963	·992 582	36	01
60	0·121 869		0·122 785		1·007 510		8·205 51		8·144 35		0·992 546		00
	cos		cot		cosec		sec		tan		sin		83°

7°	sin		tan		sec		cosec		cot		cos		′
00	0·121 869	289	0·122 785	295	1·007 510	36	8·205 51	1939	8·144 35	1954	0·992 546	35	60
01	·122 158	289	·123 080	295	·007 546	36	·186 12	1931	·124 81	1945	·992 511	36	59
02	·122 447	288	·123 375	295	·007 582	36	·166 81	1921	·105 36	1936	·992 475	36	58
03	·122 735	289	·123 670	296	·007 618	36	·147 60	1911	·086 00	1926	·992 439	35	57
04	·123 024	289	·123 966	295	·007 654	37	·128 49	1903	·066 74	1918	·992 404	36	56
05	·123 313	288	·124 261	296	·007 691	36	·109 46	1894	·047 56	1908	·992 368	36	55
06	·123 601	289	·124 557	295	·007 727	37	·090 52	1885	·028 48	1900	·992 332	36	54
07	·123 890	289	·124 852	295	·007 764	37	·071 67	1876	8·009 48	1890	·992 296	36	53
08	·124 179	288	·125 147	296	·007 801	36	·052 91	1868	7·990 58	1882	·992 260	36	52
09	·124 467	289	·125 443	295	·007 837	37	·034 23	1858	·971 76	1874	·992 224	37	51
10	0·124 756	289	0·125 738	296	1·007 874	37	8·015 65	1851	7·953 02	1864	0·992 187	36	50
11	·125 045	288	·126 034	295	·007 911	37	7·997 14	1841	·934 38	1856	·992 151	36	49
12	·125 333	289	·126 329	296	·007 948	37	·978 73	1833	·915 82	1848	·992 115	37	48
13	·125 622	288	·126 625	295	·007 985	37	·960 40	1824	·897 34	1839	·992 078	36	47
14	·125 910	289	·126 920	296	·008 022	37	·942 16	1817	·878 95	1831	·992 042	37	46
15	·126 199	289	·127 216	296	·008 059	38	·923 99	1807	·860 64	1822	·992 005	37	45
16	·126 488	288	·127 512	295	·008 097	37	·905 92	1800	·842 42	1814	·991 968	37	44
17	·126 776	289	·127 807	296	·008 134	38	·887 92	1791	·824 28	1806	·991 931	37	43
18	·127 065	288	·128 103	296	·008 172	37	·870 01	1783	·806 22	1797	·991 894	37	42
19	·127 353	289	·128 399	295	·008 209	38	·852 18	1775	·788 25	1790	·991 857	37	41
20	0·127 642	288	0·128 694	296	1·008 247	38	7·834 43	1766	7·770 35	1781	0·991 820	37	40
21	·127 930	289	·128 990	296	·008 285	38	·816 77	1759	·752 54	1774	·991 783	37	39
22	·128 219	288	·129 286	296	·008 323	38	·799 18	1751	·734 80	1765	·991 746	37	38
23	·128 507	289	·129 582	295	·008 361	38	·781 67	1743	·717 15	1758	·991 709	38	37
24	·128 796	288	·129 877	296	·008 399	38	·764 24	1735	·699 57	1749	·991 671	37	36
25	·129 084	289	·130 173	296	·008 437	38	·746 89	1727	·682 08	1742	·991 634	38	35
26	·129 373	288	·130 469	296	·008 475	38	·729 62	1720	·664 66	1734	·991 596	38	34
27	·129 661	288	·130 765	296	·008 513	39	·712 42	1712	·647 32	1727	·991 558	37	33
28	·129 949	289	·131 061	296	·008 552	38	·695 30	1704	·630 05	1718	·991 521	38	32
29	·130 238	288	·131 357	295	·008 590	39	·678 26	1696	·612 87	1712	·991 483	38	31
30	0·130 526	289	0·131 652	296	1·008 629	39	7·661 30	1689	7·595 75	1703	0·991 445	38	30
31	·130 815	288	·131 948	296	·008 668	38	·644 41	1682	·578 72	1696	·991 407	38	29
32	·131 103	288	·132 244	296	·008 706	39	·627 59	1674	·561 76	1689	·991 369	38	28
33	·131 391	289	·132 540	296	·008 745	39	·610 85	1667	·544 87	1681	·991 331	39	27
34	·131 680	288	·132 836	296	·008 784	39	·594 18	1659	·528 06	1674	·991 292	38	26
35	·131 968	288	·133 132	296	·008 823	39	·577 59	1652	·511 32	1667	·991 254	38	25
36	·132 256	289	·133 428	297	·008 862	40	·561 07	1645	·494 65	1659	·991 216	39	24
37	·132 545	288	·133 725	296	·008 902	39	·544 62	1637	·478 06	1652	·991 177	39	23
38	·132 833	288	·134 021	296	·008 941	39	·528 25	1631	·461 54	1645	·991 138	38	22
39	·133 121	289	·134 317	296	·008 980	40	·511 94	1623	·445 09	1638	·991 100	39	21
40	0·133 410	288	0·134 613	296	1·009 020	39	7·495 71	1616	7·428 71	1631	0·991 061	39	20
41	·133 698	288	·134 909	296	·009 059	40	·479 55	1609	·412 40	1624	·991 022	39	19
42	·133 986	288	·135 205	297	·009 099	40	·463 46	1603	·396 16	1617	·990 983	39	18
43	·134 274	289	·135 502	296	·009 139	39	·447 43	1595	·379 99	1610	·990 944	39	17
44	·134 563	288	·135 798	296	·009 178	40	·431 48	1588	·363 89	1603	·990 905	39	16
45	·134 851	288	·136 094	296	·009 218	40	·415 60	1582	·347 86	1596	·990 866	39	15
46	·135 139	288	·136 390	297	·009 258	40	·399 78	1575	·331 90	1590	·990 827	40	14
47	·135 427	289	·136 687	296	·009 298	41	·384 03	1568	·316 00	1582	·990 787	39	13
48	·135 716	288	·136 983	296	·009 339	40	·368 35	1561	·300 18	1576	·990 748	40	12
49	·136 004	288	·137 279	297	·009 379	40	·352 74	1555	·284 42	1569	·990 708	39	11
50	0·136 292	288	0·137 576	296	1·009 419	41	7·337 19	1548	7·268 73	1563	0·990 669	40	10
51	·136 580	288	·137 872	297	·009 460	40	·321 71	1541	·253 10	1556	·990 629	40	09
52	·136 868	288	·138 169	296	·009 500	41	·306 30	1535	·237 54	1550	·990 589	40	08
53	·137 156	289	·138 465	296	·009 541	40	·290 95	1529	·222 04	1543	·990 549	40	07
54	·137 445	288	·138 761	297	·009 581	41	·275 66	1522	·206 61	1536	·990 509	40	06
55	·137 733	288	·139 058	296	·009 622	41	·260 44	1515	·191 25	1531	·990 469	40	05
56	·138 021	288	·139 354	297	·009 663	41	·245 29	1510	·175 94	1523	·990 429	40	04
57	·138 309	288	·139 651	297	·009 704	41	·230 19	1502	·160 71	1518	·990 389	40	03
58	·138 597	288	·139 948	296	·009 745	41	·215 17	1497	·145 53	1511	·990 349	40	02
59	·138 885	288	·140 244	297	·009 786	42	·200 20	1490	·130 42	1505	·990 309	41	01
60	0·139 173		0·140 541		1·009 828		7·185 30		7·115 37		0·990 268		00
	cos		cot		cosec		sec		tan		sin		82°

8° ′	sin		tan		sec		cosec		cot		cos		′
00	0·139 173	288	0·140 541	296	1·009 828	41	7·185 30	1484	7·115 37	1499	0·990 268	40	60
01	·139 461	288	·140 837	297	·009 869	41	·170 46	1478	·100 38	1492	·990 228	41	59
02	·139 749	288	·141 134	297	·009 910	42	·155 68	1472	·085 46	1487	·990 187	41	58
03	·140 037	288	·141 431	297	·009 952	41	·140 96	1466	·070 59	1480	·990 146	41	57
04	·140 325	288	·141 728	296	·009 993	42	·126 30	1459	·055 79	1474	·990 105	40	56
05	·140 613	288	·142 024	297	·010 035	42	·111 71	1454	·041 05	1468	·990 065	41	55
06	·140 901	288	·142 321	297	·010 077	42	·097 17	1448	·026 37	1463	·990 024	41	54
07	·141 189	288	·142 618	297	·010 119	42	·082 69	1441	7·011 74	1456	·989 983	41	53
08	·141 477	288	·142 915	297	·010 161	42	·068 28	1436	6·997 18	1450	·989 942	42	52
09	·141 765	288	·143 212	296	·010 203	42	·053 92	1430	·982 68	1445	·989 900	41	51
10	0·142 053	288	0·143 508	297	1·010 245	42	7·039 62	1424	6·968 23	1438	0·989 859	41	50
11	·142 341	288	·143 805	297	·010 287	42	·025 38	1418	·953 85	1433	·989 818	42	49
12	·142 629	288	·144 102	297	·010 329	43	7·011 20	1412	·939 52	1427	·989 776	41	48
13	·142 917	288	·144 399	297	·010 372	42	6·997 08	1407	·925 25	1421	·989 735	42	47
14	·143 205	288	·144 696	297	·010 414	43	·983 01	1401	·911 04	1416	·989 693	42	46
15	·143 493	287	·144 993	297	·010 457	42	·969 00	1395	·896 88	1410	·989 651	41	45
16	·143 780	288	·145 290	297	·010 499	43	·955 05	1390	·882 78	1404	·989 610	42	44
17	·144 068	288	·145 587	297	·010 542	43	·941 15	1384	·868 74	1399	·989 568	42	43
18	·144 356	288	·145 884	297	·010 585	43	·927 31	1379	·854 75	1393	·989 526	42	42
19	·144 644	288	·146 181	297	·010 628	43	·913 52	1373	·840 82	1388	·989 484	42	41
20	0·144 932	288	0·146 478	298	1·010 671	43	6·899 79	1367	6·826 94	1382	0·989 442	43	40
21	·145 220	287	·146 776	297	·010 714	43	·886 12	1362	·813 12	1376	·989 399	42	39
22	·145 507	288	·147 073	297	·010 757	44	·872 50	1357	·799 36	1372	·989 357	42	38
23	·145 795	288	·147 370	297	·010 801	43	·858 93	1351	·785 64	1365	·989 315	43	37
24	·146 083	288	·147 667	297	·010 844	43	·845 42	1346	·771 99	1361	·989 272	42	36
25	·146 371	288	·147 964	298	·010 887	44	·831 96	1340	·758 38	1355	·989 230	43	35
26	·146 659	287	·148 262	297	·010 931	44	·818 56	1335	·744 83	1350	·989 187	43	34
27	·146 946	288	·148 559	297	·010 975	43	·805 21	1330	·731 33	1344	·989 144	42	33
28	·147 234	288	·148 856	298	·011 018	44	·791 91	1325	·717 89	1339	·989 102	43	32
29	·147 522	287	·149 154	297	·011 062	44	·778 66	1319	·704 50	1334	·989 059	43	31
30	0·147 809	288	0·149 451	297	1·011 106	44	6·765 47	1314	6·691 16	1329	0·989 016	43	30
31	·148 097	288	·149 748	298	·011 150	44	·752 33	1309	·677 87	1324	·988 973	43	29
32	·148 385	287	·150 046	297	·011 194	44	·739 24	1304	·664 63	1319	·988 930	44	28
33	·148 672	288	·150 343	298	·011 238	45	·726 20	1299	·651 44	1313	·988 886	43	27
34	·148 960	288	·150 641	297	·011 283	44	·713 21	1294	·638 31	1308	·988 843	43	26
35	·149 248	287	·150 938	298	·011 327	44	·700 27	1289	·625 23	1304	·988 800	44	25
36	·149 535	288	·151 236	297	·011 371	45	·687 38	1284	·612 19	1298	·988 756	43	24
37	·149 823	288	·151 533	298	·011 416	45	·674 54	1278	·599 21	1294	·988 713	44	23
38	·150 111	287	·151 831	298	·011 461	44	·661 76	1274	·586 27	1288	·988 669	43	22
39	·150 398	288	·152 129	297	·011 505	45	·649 02	1269	·573 39	1284	·988 626	44	21
40	0·150 686	287	0·152 426	298	1·011 550	45	6·636 33	1264	6·560 55	1278	0·988 582	44	20
41	·150 973	288	·152 724	298	·011 595	45	·623 69	1259	·547 77	1274	·988 538	44	19
42	·151 261	287	·153 022	297	·011 640	45	·611 10	1255	·535 03	1269	·988 494	44	18
43	·151 548	288	·153 319	298	·011 685	45	·598 55	1249	·522 34	1264	·988 450	44	17
44	·151 836	287	·153 617	298	·011 730	46	·586 06	1245	·509 70	1260	·988 406	44	16
45	·152 123	288	·153 915	298	·011 776	45	·573 61	1240	·497 10	1254	·988 362	45	15
46	·152 411	287	·154 213	297	·011 821	45	·561 21	1235	·484 56	1250	·988 317	44	14
47	·152 698	288	·154 510	298	·011 866	46	·548 86	1231	·472 06	1245	·988 273	45	13
48	·152 986	287	·154 808	298	·011 912	45	·536 55	1226	·459 61	1241	·988 228	44	12
49	·153 273	288	·155 106	298	·011 957	46	·524 29	1221	·447 20	1236	·988 184	45	11
50	0·153 561	287	0·155 404	298	1·012 003	46	6·512 08	1217	6·434 84	1231	0·988 139	45	10
51	·153 848	288	·155 702	298	·012 049	46	·499 91	1212	·422 53	1227	·988 094	44	09
52	·154 136	287	·156 000	298	·012 095	46	·487 79	1207	·410 26	1222	·988 050	45	08
53	·154 423	287	·156 298	298	·012 141	46	·475 72	1203	·398 04	1217	·988 005	45	07
54	·154 710	288	·156 596	298	·012 187	46	·463 69	1198	·385 87	1213	·987 960	45	06
55	·154 998	287	·156 894	298	·012 233	46	·451 71	1194	·373 74	1209	·987 915	45	05
56	·155 285	287	·157 192	298	·012 279	47	·439 77	1190	·361 65	1204	·987 870	46	04
57	·155 572	288	·157 490	298	·012 326	46	·427 87	1185	·349 61	1200	·987 824	45	03
58	·155 860	287	·157 788	298	·012 372	47	·416 02	1180	·337 61	1195	·987 779	45	02
59	·156 147	287	·158 086	298	·012 419	46	·404 22	1177	·325 66	1191	·987 734	46	01
60	0·156 434		0·158 384		1·012 465		6·392 45		6·313 75		0·987 688		00
	cos		cot		cosec		sec		tan		sin		81°

9° ′	sin		tan		sec		cosec		cot		cos		′
00	0·156 434	288	0·158 384	299	1·012 465	47	6·392 45	1172	6·313 75	1186	0·987 688	45	60
01	·156 722	287	·158 683	298	·012 512	47	·380 73	1167	·301 89	1182	·987 643	46	59
02	·157 009	287	·158 981	298	·012 559	46	·369 06	1163	·290 07	1178	·987 597	46	58
03	·157 296	288	·159 279	298	·012 605	47	·357 43	1159	·278 29	1174	·987 551	45	57
04	·157 584	287	·159 577	299	·012 652	47	·345 84	1155	·266 55	1169	·987 506	46	56
05	·157 871	287	·159 876	298	·012 699	48	·334 29	1150	·254 86	1165	·987 460	46	55
06	·158 158	287	·160 174	298	·012 747	47	·322 79	1146	·243 21	1161	·987 414	46	54
07	·158 445	287	·160 472	299	·012 794	47	·311 33	1142	·231 60	1157	·987 368	46	53
08	·158 732	288	·160 771	298	·012 841	48	·299 91	1138	·220 03	1152	·987 322	47	52
09	·159 020	287	·161 069	299	·012 889	47	·288 53	1134	·208 51	1148	·987 275	46	51
10	0·159 307	287	0·161 368	298	1·012 936	48	6·277 19	1129	6·197 03	1144	0·987 229	46	50
11	·159 594	287	·161 666	299	·012 984	47	·265 90	1126	·185 59	1140	·987 183	47	49
12	·159 881	287	·161 965	298	·013 031	48	·254 64	1121	·174 19	1136	·987 136	46	48
13	·160 168	287	·162 263	299	·013 079	48	·243 43	1117	·162 83	1132	·987 090	47	47
14	·160 455	288	·162 562	298	·013 127	48	·232 26	1113	·151 51	1128	·987 043	47	46
15	·160 743	287	·162 860	299	·013 175	48	·221 13	1109	·140 23	1124	·986 996	46	45
16	·161 030	287	·163 159	299	·013 223	48	·210 04	1106	·128 99	1120	·986 950	47	44
17	·161 317	287	·163 458	298	·013 271	48	·198 98	1101	·117 79	1115	·986 903	47	43
18	·161 604	287	·163 756	299	·013 319	49	·187 97	1097	·106 64	1112	·986 856	47	42
19	·161 891	287	·164 055	299	·013 368	48	·177 00	1093	·095 52	1108	·986 809	47	41
20	0·162 178	287	0·164 354	298	1·013 416	49	6·166 07	1090	6·084 44	1104	0·986 762	48	40
21	·162 465	287	·164 652	299	·013 465	48	·155 17	1085	·073 40	1100	·986 714	47	39
22	·162 752	287	·164 951	299	·013 513	49	·144 32	1082	·062 40	1097	·986 667	47	38
23	·163 039	287	·165 250	299	·013 562	49	·133 50	1077	·051 43	1092	·986 620	48	37
24	·163 326	287	·165 549	299	·013 611	48	·122 73	1074	·040 51	1089	·986 572	47	36
25	·163 613	287	·165 848	299	·013 659	49	·111 99	1070	·029 62	1084	·986 525	48	35
26	·163 900	287	·166 147	299	·013 708	49	·101 29	1067	·018 78	1081	·986 477	48	34
27	·164 187	287	·166 446	299	·013 757	50	·090 62	1062	6·007 97	1077	·986 429	48	33
28	·164 474	287	·166 745	299	·013 807	49	·080 00	1059	5·997 20	1074	·986 381	47	32
29	·164 761	287	·167 044	299	·013 856	49	·069 41	1055	·986 46	1070	·986 334	48	31
30	0·165 048	286	0·167 343	299	1·013 905	49	6·058 86	1052	5·975 76	1066	0·986 286	48	30
31	·165 334	287	·167 642	299	·013 954	50	·048 34	1047	·965 10	1062	·986 238	49	29
32	·165 621	287	·167 941	299	·014 004	50	·037 87	1044	·954 48	1058	·986 189	48	28
33	·165 908	287	·168 240	299	·014 054	49	·027 43	1041	·943 90	1055	·986 141	48	27
34	·166 195	287	·168 539	299	·014 103	50	·017 02	1036	·933 35	1052	·986 093	48	26
35	·166 482	287	·168 838	299	·014 153	50	6·006 66	1033	·922 83	1047	·986 045	49	25
36	·166 769	287	·169 137	300	·014 203	50	5·996 33	1030	·912 36	1045	·985 996	49	24
37	·167 056	286	·169 437	299	·014 253	50	·986 03	1026	·901 91	1040	·985 947	48	23
38	·167 342	287	·169 736	299	·014 303	50	·975 77	1022	·891 51	1037	·985 899	49	22
39	·167 629	287	·170 035	299	·014 353	50	·965 55	1019	·881 14	1034	·985 850	49	21
40	0·167 916	287	0·170 334	300	1·014 403	50	5·955 36	1015	5·870 80	1029	0·985 801	49	20
41	·168 203	286	·170 634	299	·014 453	51	·945 21	1012	·860 51	1027	·985 752	49	19
42	·168 489	287	·170 933	300	·014 504	50	·935 09	1008	·850 24	1023	·985 703	49	18
43	·168 776	287	·171 233	299	·014 554	51	·925 01	1005	·840 01	1019	·985 654	49	17
44	·169 063	287	·171 532	299	·014 605	51	·914 96	1001	·829 82	1016	·985 605	49	16
45	·169 350	286	·171 831	300	·014 656	50	·904 95	998	·819 66	1013	·985 556	49	15
46	·169 636	287	·172 131	299	·014 706	51	·894 97	995	·809 53	1009	·985 507	50	14
47	·169 923	286	·172 430	300	·014 757	51	·885 02	991	·799 44	1006	·985 457	49	13
48	·170 209	287	·172 730	300	·014 808	51	·875 11	987	·789 38	1002	·985 408	50	12
49	·170 496	287	·173 030	299	·014 859	51	·865 24	985	·779 36	999	·985 358	49	11
50	0·170 783	286	0·173 329	300	1·014 910	52	5·855 39	981	5·769 37	996	0·985 309	50	10
51	·171 069	287	·173 629	300	·014 962	51	·845 58	977	·759 41	992	·985 259	50	09
52	·171 356	287	·173 929	299	·015 013	51	·835 81	975	·749 49	989	·985 209	50	08
53	·171 643	286	·174 228	300	·015 064	52	·826 06	971	·739 60	986	·985 159	50	07
54	·171 929	287	·174 528	300	·015 116	51	·816 35	968	·729 74	982	·985 109	50	06
55	·172 216	286	·174 828	299	·015 167	52	·806 67	964	·719 92	979	·985 059	50	05
56	·172 502	287	·175 127	300	·015 219	52	·797 03	961	·710 13	976	·985 009	50	04
57	·172 789	286	·175 427	300	·015 271	52	·787 42	959	·700 37	973	·984 959	50	03
58	·173 075	287	·175 727	300	·015 323	52	·777 83	954	·690 64	970	·984 909	51	02
59	·173 362	286	·176 027	300	·015 375	52	·768 29	952	·680 94	966	·984 858	50	01
60	0·173 648		0·176 327		1·015 427		5·758 77		5·671 28		0·984 808		00
	cos		cot		cosec		sec		tan		sin		80°

10° ′	sin		tan		sec		cosec		cot		cos		′
00	0·173 648	287	0·176 327	300	1·015 427	52	5·758 77	948	5·671 28	963	0·984 808	51	60
01	·173 935	286	·176 627	300	·015 479	52	·749 29	946	·661 65	960	·984 757	50	59
02	·174 221	287	·176 927	300	·015 531	52	·739 83	942	·652 05	957	·984 707	51	58
03	·174 508	286	·177 227	300	·015 583	53	·730 41	939	·642 48	953	·984 656	51	57
04	·174 794	286	·177 527	300	·015 636	52	·721 02	936	·632 95	951	·984 605	51	56
05	·175 080	287	·177 827	300	·015 688	53	·711 66	932	·623 44	947	·984 554	51	55
06	·175 367	286	·178 127	300	·015 741	52	·702 34	930	·613 97	945	·984 503	51	54
07	·175 653	286	·178 427	300	·015 793	53	·693 04	927	·604 52	941	·984 452	51	53
08	·175 939	287	·178 727	301	·015 846	53	·683 77	923	·595 11	938	·984 401	51	52
09	·176 226	286	·179 028	300	·015 899	53	·674 54	921	·585 73	935	·984 350	52	51
10	0·176 512	286	0·179 328	300	1·015 952	53	5·665 33	917	5·576 38	932	0·984 298	51	50
11	·176 798	287	·179 628	300	·016 005	53	·656 16	915	·567 06	929	·984 247	51	49
12	·177 085	286	·179 928	301	·016 058	53	·647 01	911	·557 77	926	·984 196	52	48
13	·177 371	286	·180 229	300	·016 111	54	·637 90	909	·548 51	924	·984 144	52	47
14	·177 657	287	·180 529	300	·016 165	53	·628 81	905	·539 27	920	·984 092	51	46
15	·177 944	286	·180 829	301	·016 218	54	·619 76	903	·530 07	917	·984 041	52	45
16	·178 230	286	·181 130	300	·016 272	53	·610 73	899	·520 90	914	·983 989	52	44
17	·178 516	286	·181 430	301	·016 325	54	·601 74	897	·511 76	912	·983 937	52	43
18	·178 802	286	·181 731	300	·016 379	54	·592 77	894	·502 64	908	·983 885	52	42
19	·179 088	287	·182 031	301	·016 433	54	·583 83	890	·493 56	905	·983 833	52	41
20	0·179 375	286	0·182 332	300	1·016 487	54	5·574 93	888	5·484 51	903	0·983 781	52	40
21	·179 661	286	·182 632	301	·016 541	54	·566 05	885	·475 48	900	·983 729	53	39
22	·179 947	286	·182 933	301	·016 595	54	·557 20	883	·466 48	897	·983 676	52	38
23	·180 233	286	·183 234	300	·016 649	54	·548 37	879	·457 51	894	·983 624	53	37
24	·180 519	286	·183 534	301	·016 703	54	·539 58	877	·448 57	891	·983 571	52	36
25	·180 805	286	·183 835	301	·016 757	55	·530 81	873	·439 66	889	·983 519	53	35
26	·181 091	286	·184 136	301	·016 812	54	·522 08	871	·430 77	885	·983 466	52	34
27	·181 377	286	·184 437	300	·016 866	55	·513 37	869	·421 92	883	·983 414	53	33
28	·181 663	287	·184 737	301	·016 921	54	·504 68	865	·413 09	880	·983 361	53	32
29	·181 950	286	·185 038	301	·016 975	55	·496 03	863	·404 29	877	·983 308	53	31
30	0·182 236	286	0·185 339	301	1·017 030	55	5·487 40	859	5·395 52	875	0·983 255	53	30
31	·182 522	286	·185 640	301	·017 085	55	·478 81	858	·386 77	872	·983 202	53	29
32	·182 808	286	·185 941	301	·017 140	55	·470 23	854	·378 05	869	·983 149	53	28
33	·183 094	285	·186 242	301	·017 195	55	·461 69	852	·369 36	866	·983 096	54	27
34	·183 379	286	·186 543	301	·017 250	56	·453 17	849	·360 70	864	·983 042	53	26
35	·183 665	286	·186 844	301	·017 306	55	·444 68	846	·352 06	861	·982 989	54	25
36	·183 951	286	·187 145	301	·017 361	55	·436 22	844	·343 45	858	·982 935	53	24
37	·184 237	286	·187 446	301	·017 416	56	·427 78	841	·334 87	856	·982 882	54	23
38	·184 523	286	·187 747	301	·017 472	55	·419 37	838	·326 31	853	·982 828	54	22
39	·184 809	286	·188 048	301	·017 527	56	·410 99	836	·317 78	850	·982 774	53	21
40	0·185 095	286	0·188 349	302	1·017 583	56	5·402 63	833	5·309 28	848	0·982 721	54	20
41	·185 381	286	·188 651	301	·017 639	56	·394 30	830	·300 80	845	·982 667	54	19
42	·185 667	285	·188 952	301	·017 695	56	·386 00	828	·292 35	842	·982 613	54	18
43	·185 952	286	·189 253	302	·017 751	56	·377 72	825	·283 93	840	·982 559	54	17
44	·186 238	286	·189 555	301	·017 807	56	·369 47	823	·275 53	838	·982 505	55	16
45	·186 524	286	·189 856	301	·017 863	56	·361 24	820	·267 15	835	·982 450	54	15
46	·186 810	286	·190 157	302	·017 919	57	·353 04	818	·258 80	832	·982 396	54	14
47	·187 096	285	·190 459	301	·017 976	56	·344 86	815	·250 48	830	·982 342	55	13
48	·187 381	286	·190 760	302	·018 032	57	·336 71	812	·242 18	827	·982 287	54	12
49	·187 667	286	·191 062	301	·018 089	56	·328 59	810	·233 91	825	·982 233	55	11
50	0·187 953	285	0·191 363	302	1·018 145	57	5·320 49	808	5·225 66	822	0·982 178	55	10
51	·188 238	286	·191 665	301	·018 202	57	·312 41	805	·217 44	819	·982 123	54	09
52	·188 524	286	·191 966	302	·018 259	57	·304 36	802	·209 25	818	·982 069	55	08
53	·188 810	285	·192 268	302	·018 316	57	·296 34	801	·201 07	814	·982 014	55	07
54	·189 095	286	·192 570	301	·018 373	57	·288 33	797	·192 93	813	·981 959	55	06
55	·189 381	286	·192 871	302	·018 430	57	·280 36	795	·184 80	809	·981 904	55	05
56	·189 667	285	·193 173	302	·018 487	57	·272 41	793	·176 71	808	·981 849	56	04
57	·189 952	286	·193 475	302	·018 544	58	·264 48	790	·168 63	805	·981 793	55	03
58	·190 238	285	·193 777	301	·018 602	57	·256 58	788	·160 58	802	·981 738	55	02
59	·190 523	286	·194 078	302	·018 659	58	·248 70	786	·152 56	801	·981 683	56	01
60	0·190 809		0·194 380		1·018 717		5·240 84		5·144 55		0·981 627		00
	cos		cot		cosec		sec		tan		sin		79°

11°

′	sin		tan		sec		cosec		cot		cos		
00	0·190 809	286	0·194 380	302	1·018 717	57	5·240 84	783	5·144 55	797	0·981 627	55	60
01	·191 095	285	·194 682	302	·018 774	58	·233 01	780	·136 58	796	·981 572	56	59
02	·191 380	286	·194 984	302	·018 832	58	·225 21	779	·128 62	793	·981 516	56	58
03	·191 666	285	·195 286	302	·018 890	58	·217 42	776	·120 69	790	·981 460	55	57
04	·191 951	286	·195 588	302	·018 948	58	·209 66	773	·112 79	789	·981 405	56	56
05	·192 237	285	·195 890	302	·019 006	58	·201 93	772	·104 90	786	·981 349	56	55
06	·192 522	285	·196 192	302	·019 064	58	·194 21	769	·097 04	783	·981 293	56	54
07	·192 807	286	·196 494	302	·019 122	58	·186 52	766	·089 21	782	·981 237	57	53
08	·193 093	285	·196 796	303	·019 180	59	·178 86	765	·081 39	779	·981 180	56	52
09	·193 378	286	·197 099	302	·019 239	58	·171 21	762	·073 60	776	·981 124	56	51
10	0·193 664	285	0·197 401	302	1·019 297	59	5·163 59	760	5·065 84	775	0·981 068	56	50
11	·193 949	285	·197 703	302	·019 356	59	·155 99	757	·058 09	772	·981 012	57	49
12	·194 234	286	·198 005	303	·019 415	58	·148 42	755	·050 37	770	·980 955	56	48
13	·194 520	285	·198 308	302	·019 473	59	·140 87	753	·042 67	768	·980 899	57	47
14	·194 805	285	·198 610	302	·019 532	59	·133 34	751	·034 99	765	·980 842	57	46
15	·195 090	286	·198 912	303	·019 591	59	·125 83	748	·027 34	763	·980 785	57	45
16	·195 376	285	·199 215	302	·019 650	59	·118 35	747	·019 71	761	·980 728	56	44
17	·195 661	285	·199 517	303	·019 709	60	·110 88	744	·012 10	759	·980 672	57	43
18	·195 946	285	·199 820	302	·019 769	59	·103 44	742	5·004 51	756	·980 615	57	42
19	·196 231	286	·200 122	303	·019 828	59	·096 02	739	4·996 95	755	·980 558	58	41
20	0·196 517	285	0·200 425	302	1·019 887	60	5·088 63	738	4·989 40	752	0·980 500	57	40
21	·196 802	285	·200 727	303	·019 947	59	·081 25	735	·981 88	750	·980 443	57	39
22	·197 087	285	·201 030	303	·020 006	60	·073 90	733	·974 38	748	·980 386	57	38
23	·197 372	285	·201 333	302	·020 066	60	·066 57	731	·966 90	745	·980 329	58	37
24	·197 657	285	·201 635	303	·020 126	60	·059 26	729	·959 45	744	·980 271	57	36
25	·197 942	286	·201 938	303	·020 186	60	·051 97	726	·952 01	741	·980 214	58	35
26	·198 228	285	·202 241	303	·020 246	60	·044 71	725	·944 60	739	·980 156	58	34
27	·198 513	285	·202 544	303	·020 306	60	·037 46	722	·937 21	737	·980 098	57	33
28	·198 798	285	·202 847	302	·020 366	60	·030 24	721	·929 84	735	·980 041	58	32
29	·199 083	285	·203 149	303	·020 426	61	·023 03	718	·922 49	733	·979 983	58	31
30	0·199 368	285	0·203 452	303	1·020 487	60	5·015 85	716	4·915 16	731	0·979 925	58	30
31	·199 653	285	·203 755	303	·020 547	61	·008 69	714	·907 85	729	·979 867	58	29
32	·199 938	285	·204 058	303	·020 608	60	5·001 55	712	·900 56	726	·979 809	59	28
33	·200 223	285	·204 361	303	·020 668	61	4·994 43	710	·893 30	725	·979 750	58	27
34	·200 508	285	·204 664	303	·020 729	61	·987 33	708	·886 05	723	·979 692	58	26
35	·200 793	285	·204 967	304	·020 790	61	·980 25	705	·878 82	720	·979 634	59	25
36	·201 078	285	·205 271	303	·020 851	61	·973 20	704	·871 62	718	·979 575	58	24
37	·201 363	285	·205 574	303	·020 912	61	·966 16	702	·864 44	717	·979 517	59	23
38	·201 648	285	·205 877	303	·020 973	61	·959 14	699	·857 27	714	·979 458	59	22
39	·201 933	285	·206 180	303	·021 034	61	·952 15	698	·850 13	713	·979 399	58	21
40	0·202 218	284	0·206 483	304	1·021 095	62	4·945 17	696	4·843 00	710	0·979 341	59	20
41	·202 502	285	·206 787	303	·021 157	61	·938 21	693	·835 90	708	·979 282	59	19
42	·202 787	285	·207 090	303	·021 218	62	·931 28	692	·828 82	707	·979 223	59	18
43	·203 072	285	·207 393	304	·021 280	61	·924 36	690	·821 75	704	·979 164	59	17
44	·203 357	285	·207 697	303	·021 341	62	·917 46	688	·814 71	702	·979 105	60	16
45	·203 642	285	·208 000	304	·021 403	62	·910 58	685	·807 69	701	·979 045	59	15
46	·203 927	284	·208 304	303	·021 465	62	·903 73	684	·800 68	698	·978 986	59	14
47	·204 211	285	·208 607	304	·021 527	62	·896 89	682	·793 70	697	·978 927	60	13
48	·204 496	285	·208 911	303	·021 589	62	·890 07	680	·786 73	695	·978 867	59	12
49	·204 781	284	·209 214	304	·021 651	62	·883 27	678	·779 78	692	·978 808	60	11
50	0·205 065	285	0·209 518	304	1·021 713	63	4·876 49	676	4·772 86	691	0·978 748	59	10
51	·205 350	285	·209 822	304	·021 776	62	·869 73	674	·765 95	689	·978 689	60	09
52	·205 635	285	·210 126	303	·021 838	62	·862 99	672	·759 06	687	·978 629	60	08
53	·205 920	284	·210 429	304	·021 900	63	·856 27	671	·752 19	685	·978 569	60	07
54	·206 204	285	·210 733	304	·021 963	63	·849 56	668	·745 34	683	·978 509	60	06
55	·206 489	284	·211 037	304	·022 026	63	·842 88	667	·738 51	681	·978 449	60	05
56	·206 773	285	·211 341	304	·022 089	62	·836 21	665	·731 70	680	·978 389	60	04
57	·207 058	285	·211 645	304	·022 151	63	·829 56	662	·724 90	677	·978 329	61	03
58	·207 343	284	·211 949	304	·022 214	63	·822 94	661	·718 13	676	·978 268	60	02
59	·207 627	285	·212 253	304	·022 277	64	·816 33	660	·711 37	674	·978 208	60	01
60	0·207 912		0·212 557		1·022 341		4·809 73		4·704 63		0·978 148		00
	cos		cot		cosec		sec		tan		sin		78°

12°	sin		tan		sec		cosec		cot		cos		
00	0·207 912	284	0·212 557	304	1·022 341	63	4·809 73	657	4·704 63	672	0·978 148	61	60
01	·208 196	285	·212 861	304	·022 404	63	·803 16	655	·697 91	670	·978 087	61	59
02	·208 481	284	·213 165	304	·022 467	64	·796 61	654	·691 21	669	·978 026	60	58
03	·208 765	285	·213 469	304	·022 531	63	·790 07	652	·684 52	666	·977 966	61	57
04	·209 050	284	·213 773	304	·022 594	64	·783 55	650	·677 86	665	·977 905	61	56
05	·209 334	285	·214 077	304	·022 658	64	·777 05	648	·671 21	663	·977 844	61	55
06	·209 619	284	·214 381	305	·022 722	63	·770 57	646	·664 58	661	·977 783	61	54
07	·209 903	284	·214 686	304	·022 785	64	·764 11	645	·657 97	659	·977 722	61	53
08	·210 187	285	·214 990	304	·022 849	64	·757 66	643	·651 38	658	·977 661	61	52
09	·210 472	284	·215 294	305	·022 913	64	·751 23	641	·644 80	655	·977 600	61	51
10	0·210 756	284	0·215 599	304	1·022 977	65	4·744 82	639	4·638 25	654	0·977 539	62	50
11	·211 040	285	·215 903	305	·023 042	64	·738 43	638	·631 71	653	·977 477	61	49
12	·211 325	284	·216 208	304	·023 106	64	·732 05	636	·625 18	650	·977 416	62	48
13	·211 609	284	·216 512	305	·023 170	65	·725 69	634	·618 68	649	·977 354	61	47
14	·211 893	285	·216 817	304	·023 235	64	·719 35	632	·612 19	647	·977 293	62	46
15	·212 178	284	·217 121	305	·023 299	65	·713 03	630	·605 72	645	·977 231	62	45
16	·212 462	284	·217 426	305	·023 364	65	·706 73	629	·599 27	644	·977 169	61	44
17	·212 746	284	·217 731	304	·023 429	65	·700 44	627	·592 83	642	·977 108	62	43
18	·213 030	285	·218 035	305	·023 494	65	·694 17	626	·586 41	640	·977 046	62	42
19	·213 315	284	·218 340	305	·023 559	65	·687 91	624	·580 01	638	·976 984	63	41
20	0·213 599	284	0·218 645	305	1·023 624	65	4·681 67	622	4·573 63	637	0·976 921	62	40
21	·213 883	284	·218 950	304	·023 689	65	·675 45	620	·567 26	635	·976 859	62	39
22	·214 167	284	·219 254	305	·023 754	65	·669 25	618	·560 91	633	·976 797	62	38
23	·214 451	284	·219 559	305	·023 819	66	·663 07	617	·554 58	632	·976 735	63	37
24	·214 735	284	·219 864	305	·023 885	65	·656 90	616	·548 26	630	·976 672	62	36
25	·215 019	284	·220 169	305	·023 950	66	·650 74	613	·541 96	628	·976 610	63	35
26	·215 303	285	·220 474	305	·024 016	66	·644 61	612	·535 68	627	·976 547	62	34
27	·215 588	284	·220 779	305	·024 082	66	·638 49	611	·529 41	625	·976 485	63	33
28	·215 872	284	·221 084	305	·024 148	66	·632 38	608	·523 16	623	·976 422	63	32
29	·216 156	284	·221 389	306	·024 214	66	·626 30	607	·516 93	622	·976 359	63	31
30	0·216 440	284	0·221 695	305	1·024 280	66	4·620 23	606	4·510 71	620	0·976 296	63	30
31	·216 724	284	·222 000	305	·024 346	66	·614 17	604	·504 51	619	·976 233	63	29
32	·217 008	284	·222 305	305	·024 412	66	·608 13	602	·498 32	617	·976 170	63	28
33	·217 292	283	·222 610	306	·024 478	66	·602 11	600	·492 15	615	·976 107	63	27
34	·217 575	284	·222 916	305	·024 544	67	·596 11	599	·486 00	614	·976 044	64	26
35	·217 859	284	·223 221	305	·024 611	67	·590 12	598	·479 86	612	·975 980	63	25
36	·218 143	284	·223 526	306	·024 678	66	·584 14	595	·473 74	610	·975 917	64	24
37	·218 427	284	·223 832	305	·024 744	67	·578 19	595	·467 64	609	·975 853	63	23
38	·218 711	284	·224 137	306	·024 811	67	·572 24	592	·461 55	607	·975 790	64	22
39	·218 995	284	·224 443	305	·024 878	67	·566 32	591	·455 48	606	·975 726	64	21
40	0·219 279	283	0·224 748	306	1·024 945	67	4·560 41	590	4·449 42	604	0·975 662	64	20
41	·219 562	284	·225 054	306	·025 012	67	·554 51	588	·443 38	603	·975 598	63	19
42	·219 846	284	·225 360	305	·025 079	67	·548 63	586	·437 35	601	·975 535	64	18
43	·220 130	284	·225 665	306	·025 146	68	·542 77	585	·431 34	600	·975 471	65	17
44	·220 414	283	·225 971	306	·025 214	67	·536 92	583	·425 34	598	·975 406	64	16
45	·220 697	284	·226 277	306	·025 281	68	·531 09	582	·419 36	596	·975 342	64	15
46	·220 981	284	·226 583	306	·025 349	67	·525 27	580	·413 40	595	·975 278	64	14
47	·221 265	283	·226 889	305	·025 416	68	·519 47	579	·407 45	593	·975 214	65	13
48	·221 548	284	·227 194	306	·025 484	68	·513 68	577	·401 52	592	·975 149	64	12
49	·221 832	284	·227 500	306	·025 552	68	·507 91	575	·395 60	591	·975 085	65	11
50	0·222 116	283	0·227 806	306	1·025 620	68	4·502 16	574	4·389 69	588	0·975 020	64	10
51	·222 399	284	·228 112	306	·025 688	68	·496 42	573	·383 81	588	·974 956	65	09
52	·222 683	284	·228 418	306	·025 756	68	·490 69	571	·377 93	586	·974 891	65	08
53	·222 967	283	·228 724	307	·025 824	68	·484 98	570	·372 07	584	·974 826	65	07
54	·223 250	284	·229 031	306	·025 892	69	·479 28	568	·366 23	583	·974 761	65	06
55	·223 534	283	·229 337	306	·025 961	68	·473 60	567	·360 40	581	·974 696	65	05
56	·223 817	284	·229 643	306	·026 029	69	·467 93	565	·354 59	580	·974 631	65	04
57	·224 101	283	·229 949	306	·026 098	68	·462 28	564	·348 79	579	·974 566	65	03
58	·224 384	284	·230 255	307	·026 166	69	·456 64	562	·343 00	577	·974 501	66	02
59	·224 668	283	·230 562	306	·026 235	69	·451 02	561	·337 23	575	·974 435	65	01
60	0·224 951		0·230 868		1·026 304		4·445 41		4·331 48		0·974 370		00
	cos		cot		cosec		sec		tan		sin		77°

13° ′	sin	d	tan	d	sec	d	cosec	d	cot	d	cos	d	′
00	0·224 951	283	0·230 868	307	1·026 304	69	4·445 41	559	4·331 48	575	0·974 370	65	60
01	·225 234	284	·231 175	306	·026 373	69	·439 82	558	·325 73	572	·974 305	66	59
02	·225 518	283	·231 481	307	·026 442	69	·434 24	557	·320 01	571	·974 239	66	58
03	·225 801	284	·231 788	306	·026 511	70	·428 67	555	·314 30	570	·974 173	65	57
04	·226 085	283	·232 094	307	·026 581	69	·423 12	553	·308 60	569	·974 108	66	56
05	·226 368	283	·232 401	306	·026 650	69	·417 59	553	·302 91	567	·974 042	66	55
06	·226 651	284	·232 707	307	·026 719	70	·412 06	550	·297 24	565	·973 976	66	54
07	·226 935	283	·233 014	307	·026 789	70	·406 56	550	·291 59	564	·973 910	66	53
08	·227 218	283	·233 321	306	·026 859	69	·401 06	548	·285 95	563	·973 844	66	52
09	·227 501	283	·233 627	307	·026 928	70	·395 58	546	·280 32	561	·973 778	66	51
10	0·227 784	284	0·233 934	307	1·026 998	70	4·390 12	546	4·274 71	560	0·973 712	67	50
11	·228 068	283	·234 241	307	·027 068	70	·384 66	543	·269 11	559	·973 645	66	49
12	·228 351	283	·234 548	307	·027 138	70	·379 23	543	·263 52	557	·973 579	67	48
13	·228 634	283	·234 855	307	·027 208	70	·373 80	541	·257 95	556	·973 512	66	47
14	·228 917	283	·235 162	307	·027 278	71	·368 39	540	·252 39	554	·973 446	67	46
15	·229 200	284	·235 469	307	·027 349	70	·362 99	538	·246 85	553	·973 379	66	45
16	·229 484	283	·235 776	307	·027 419	71	·357 61	537	·241 32	552	·973 313	67	44
17	·229 767	283	·236 083	307	·027 490	70	·352 24	535	·235 80	550	·973 246	67	43
18	·230 050	283	·236 390	307	·027 560	71	·346 89	535	·230 30	549	·973 179	67	42
19	·230 333	283	·236 697	307	·027 631	71	·341 54	532	·224 81	548	·973 112	67	41
20	0·230 616	283	0·237 004	308	1·027 702	71	4·336 22	532	4·219 33	546	0·973 045	67	40
21	·230 899	283	·237 312	307	·027 773	71	·330 90	530	·213 87	545	·972 978	67	39
22	·231 182	283	·237 619	307	·027 844	71	·325 60	529	·208 42	544	·972 911	68	38
23	·231 465	283	·237 926	308	·027 915	71	·320 31	528	·202 98	542	·972 843	67	37
24	·231 748	283	·238 234	307	·027 986	71	·315 03	526	·197 56	541	·972 776	68	36
25	·232 031	283	·238 541	307	·028 057	72	·309 77	525	·192 15	540	·972 708	67	35
26	·232 314	283	·238 848	308	·028 129	71	·304 52	523	·186 75	538	·972 641	68	34
27	·232 597	283	·239 156	308	·028 200	72	·299 29	523	·181 37	537	·972 573	67	33
28	·232 880	283	·239 464	307	·028 272	71	·294 06	521	·176 00	536	·972 506	68	32
29	·233 163	282	·239 771	308	·028 343	72	·288 85	519	·170 64	534	·972 438	68	31
30	0·233 445	283	0·240 079	307	1·028 415	72	4·283 66	519	4·165 30	533	0·972 370	68	30
31	·233 728	283	·240 386	308	·028 487	72	·278 47	517	·159 97	532	·972 302	68	29
32	·234 011	283	·240 694	308	·028 559	72	·273 30	516	·154 65	531	·972 234	68	28
33	·234 294	283	·241 002	308	·028 631	72	·268 14	514	·149 34	529	·972 166	68	27
34	·234 577	282	·241 310	308	·028 703	73	·263 00	513	·144 05	528	·972 098	69	26
35	·234 859	283	·241 618	307	·028 776	72	·257 87	512	·138 77	527	·972 029	68	25
36	·235 142	283	·241 925	308	·028 848	72	·252 75	511	·133 50	525	·971 961	68	24
37	·235 425	283	·242 233	308	·028 920	73	·247 64	509	·128 25	524	·971 893	69	23
38	·235 708	282	·242 541	308	·028 993	73	·242 55	509	·123 01	523	·971 824	69	22
39	·235 990	283	·242 849	308	·029 066	72	·237 46	507	·117 78	522	·971 755	68	21
40	0·236 273	283	0·243 157	309	1·029 138	73	4·232 39	505	4·112 56	520	0·971 687	69	20
41	·236 556	282	·243 466	308	·029 211	73	·227 34	505	·107 36	520	·971 618	69	19
42	·236 838	283	·243 774	308	·029 284	73	·222 29	503	·102 16	517	·971 549	69	18
43	·237 121	282	·244 082	308	·029 357	73	·217 26	502	·096 99	517	·971 480	69	17
44	·237 403	283	·244 390	308	·029 430	73	·212 24	501	·091 82	516	·971 411	69	16
45	·237 686	282	·244 698	309	·029 503	74	·207 23	499	·086 66	514	·971 342	69	15
46	·237 968	283	·245 007	308	·029 577	73	·202 24	499	·081 52	513	·971 273	69	14
47	·238 251	282	·245 315	309	·029 650	74	·197 25	497	·076 39	512	·971 204	70	13
48	·238 533	283	·245 624	308	·029 724	73	·192 28	495	·071 27	511	·971 134	69	12
49	·238 816	282	·245 932	309	·029 797	74	·187 33	495	·066 16	509	·971 065	70	11
50	0·239 098	283	0·246 241	308	1·029 871	74	4·182 38	494	4·061 07	508	0·970 995	69	10
51	·239 381	282	·246 549	309	·029 945	74	·177 44	492	·055 99	507	·970 926	70	09
52	·239 663	283	·246 858	308	·030 019	74	·172 52	491	·050 92	506	·970 856	70	08
53	·239 946	282	·247 166	309	·030 093	74	·167 61	490	·045 86	505	·970 786	70	07
54	·240 228	282	·247 475	309	·030 167	74	·162 71	489	·040 81	503	·970 716	69	06
55	·240 510	283	·247 784	308	·030 241	74	·157 82	487	·035 78	502	·970 647	70	05
56	·240 793	282	·248 092	309	·030 315	75	·152 95	486	·030 76	502	·970 577	71	04
57	·241 075	282	·248 401	309	·030 390	74	·148 09	486	·025 74	500	·970 506	70	03
58	·241 357	283	·248 710	309	·030 464	75	·143 23	484	·020 74	498	·970 436	70	02
59	·241 640	282	·249 019	309	·030 539	75	·138 39	482	·015 76	498	·970 366	70	01
60	0·241 922		0·249 328		1·030 614		4·133 57		4·010 78		0·970 296		00
	cos		cot		cosec		sec		tan		sin		76°

14° ′	sin		tan		sec		cosec		cot		cos		′
00	0·241 922	282	0·249 328	309	1·030 614	74	4·133 57	482	4·010 78	496	0·970 296	71	60
01	·242 204	282	·249 637	309	·030 688	75	·128 75	481	·005 82	496	·970 225	70	59
02	·242 486	283	·249 946	309	·030 763	75	·123 94	479	4·000 86	494	·970 155	71	58
03	·242 769	282	·250 255	309	·030 838	75	·119 15	478	3·995 92	493	·970 084	70	57
04	·243 051	282	·250 564	309	·030 913	76	·114 37	477	·990 99	492	·970 014	71	56
05	·243 333	282	·250 873	310	·030 989	75	·109 60	476	·986 07	490	·969 943	71	55
06	·243 615	282	·251 183	309	·031 064	75	·104 84	475	·981 17	490	·969 872	71	54
07	·243 897	282	·251 492	309	·031 139	76	·100 09	474	·976 27	488	·969 801	71	53
08	·244 179	282	·251 801	310	·031 215	75	·095 35	472	·971 39	488	·969 730	71	52
09	·244 461	282	·252 111	309	·031 290	76	·090 63	472	·966 51	486	·969 659	71	51
10	0·244 743	282	0·252 420	309	1·031 366	76	4·085 91	470	3·961 65	485	0·969 588	71	50
11	·245 025	282	·252 729	310	·031 442	76	·081 21	469	·956 80	484	·969 517	72	49
12	·245 307	282	·253 039	309	·031 518	76	·076 52	468	·951 96	483	·969 445	71	48
13	·245 589	282	·253 348	310	·031 594	76	·071 84	467	·947 13	481	·969 374	72	47
14	·245 871	282	·253 658	310	·031 670	76	·067 17	466	·942 32	481	·969 302	71	46
15	·246 153	282	·253 968	309	·031 746	76	·062 51	465	·937 51	480	·969 231	72	45
16	·246 435	282	·254 277	310	·031 822	77	·057 86	464	·932 71	478	·969 159	71	44
17	·246 717	282	·254 587	310	·031 899	76	·053 22	462	·927 93	477	·969 088	72	43
18	·246 999	282	·254 897	310	·031 975	77	·048 60	462	·923 16	477	·969 016	72	42
19	·247 281	282	·255 207	309	·032 052	76	·043 98	460	·918 39	475	·968 944	72	41
20	0·247 563	282	0·255 516	310	1·032 128	77	4·039 38	459	3·913 64	474	0·968 872	72	40
21	·247 845	281	·255 826	310	·032 205	77	·034 79	459	·908 90	473	·968 800	72	39
22	·248 126	282	·256 136	310	·032 282	77	·030 20	457	·904 17	472	·968 728	73	38
23	·248 408	282	·256 446	310	·032 359	77	·025 63	456	·899 45	471	·968 655	72	37
24	·248 690	282	·256 756	310	·032 436	77	·021 07	455	·894 74	470	·968 583	72	36
25	·248 972	281	·257 066	311	·032 513	77	·016 52	454	·890 04	468	·968 511	73	35
26	·249 253	282	·257 377	310	·032 590	78	·011 98	453	·885 36	468	·968 438	72	34
27	·249 535	282	·257 687	310	·032 668	77	·007 45	452	·880 68	467	·968 366	73	33
28	·249 817	281	·257 997	310	·032 745	78	4·002 93	450	·876 01	465	·968 293	73	32
29	·250 098	282	·258 307	311	·032 823	77	3·998 43	450	·871 36	465	·968 220	72	31
30	0·250 380	282	0·258 618	310	1·032 900	78	3·993 93	449	3·866 71	463	0·968 148	73	30
31	·250 662	281	·258 928	310	·032 978	78	·989 44	447	·862 08	463	·968 075	73	29
32	·250 943	282	·259 238	311	·033 056	78	·984 97	447	·857 45	461	·968 002	73	28
33	·251 225	281	·259 549	310	·033 134	78	·980 50	446	·852 84	460	·967 929	73	27
34	·251 506	282	·259 859	311	·033 212	78	·976 04	444	·848 24	460	·967 856	74	26
35	·251 788	281	·260 170	310	·033 290	78	·971 60	444	·843 64	458	·967 782	73	25
36	·252 069	282	·260 480	311	·033 368	79	·967 16	442	·839 06	457	·967 709	73	24
37	·252 351	281	·260 791	311	·033 447	78	·962 74	442	·834 49	457	·967 636	74	23
38	·252 632	282	·261 102	311	·033 525	79	·958 32	440	·829 92	455	·967 562	73	22
39	·252 914	281	·261 413	310	·033 604	78	·953 92	440	·825 37	454	·967 489	74	21
40	0·253 195	282	0·261 723	311	1·033 682	79	3·949 52	438	3·820 83	453	0·967 415	73	20
41	·253 477	281	·262 034	311	·033 761	79	·945 14	438	·816 30	453	·967 342	74	19
42	·253 758	281	·262 345	311	·033 840	79	·940 76	436	·811 77	451	·967 268	74	18
43	·254 039	282	·262 656	311	·033 919	79	·936 40	436	·807 26	450	·967 194	74	17
44	·254 321	281	·262 967	311	·033 998	79	·932 04	434	·802 76	449	·967 120	74	16
45	·254 602	281	·263 278	311	·034 077	79	·927 70	433	·798 27	449	·967 046	74	15
46	·254 883	282	·263 589	311	·034 156	80	·923 37	433	·793 78	447	·966 972	74	14
47	·255 165	281	·263 900	311	·034 236	79	·919 04	431	·789 31	446	·966 898	75	13
48	·255 446	281	·264 211	312	·034 315	80	·914 73	431	·784 85	445	·966 823	74	12
49	·255 727	281	·264 523	311	·034 395	79	·910 42	429	·780 40	445	·966 749	74	11
50	0·256 008	281	0·264 834	311	1·034 474	80	3·906 13	429	3·775 95	443	0·966 675	75	10
51	·256 289	282	·265 145	312	·034 554	80	·901 84	428	·771 52	443	·966 600	74	09
52	·256 571	281	·265 457	311	·034 634	80	·897 56	426	·767 09	441	·966 526	75	08
53	·256 852	281	·265 768	311	·034 714	80	·893 30	426	·762 68	440	·966 451	75	07
54	·257 133	281	·266 079	312	·034 794	80	·889 04	425	·758 28	440	·966 376	75	06
55	·257 414	281	·266 391	311	·034 874	80	·884 79	423	·753 88	438	·966 301	75	05
56	·257 695	281	·266 702	312	·034 954	81	·880 56	423	·749 50	438	·966 226	75	04
57	·257 976	281	·267 014	312	·035 035	80	·876 33	422	·745 12	437	·966 151	75	03
58	·258 257	281	·267 326	311	·035 115	81	·872 11	421	·740 75	435	·966 076	75	02
59	·258 538	281	·267 637	312	·035 196	80	·867 90	420	·736 40	435	·966 001	75	01
60	0·258 819		0·267 949		1·035 276		3·863 70		3·732 05		0·965 926		00
	cos		cot		cosec		sec		tan		sin		75°

15°	sin		tan		sec		cosec		cot		cos		′
00	0·258 819	281	0·267 949	312	1·035 276	81	3·863 70	419	3·732 05	434	0·965 926	76	60
01	·259 100	281	·268 261	312	·035 357	81	·859 51	418	·727 71	433	·965 850	75	59
02	·259 381	281	·268 573	312	·035 438	81	·855 33	417	·723 38	431	·965 775	75	58
03	·259 662	281	·268 885	312	·035 519	81	·851 16	416	·719 07	431	·965 700	76	57
04	·259 943	281	·269 197	312	·035 600	81	·847 00	415	·714 76	430	·965 624	76	56
05	·260 224	281	·269 509	312	·035 681	81	·842 85	414	·710 46	430	·965 548	75	55
06	·260 505	280	·269 821	312	·035 762	81	·838 71	414	·706 16	428	·965 473	76	54
07	·260 785	281	·270 133	312	·035 843	82	·834 57	412	·701 88	427	·965 397	76	53
08	·261 066	281	·270 445	312	·035 925	81	·830 45	412	·697 61	426	·965 321	76	52
09	·261 347	281	·270 757	312	·036 006	82	·826 33	410	·693 35	426	·965 245	76	51
10	0·261 628	280	0·271 069	313	1·036 088	82	3·822 23	410	3·689 09	424	0·965 169	76	50
11	·261 908	281	·271 382	312	·036 170	82	·818 13	409	·684 85	424	·965 093	77	49
12	·262 189	281	·271 694	312	·036 252	82	·814 04	408	·680 61	423	·965 016	76	48
13	·262 470	281	·272 006	313	·036 334	82	·809 96	407	·676 38	421	·964 940	76	47
14	·262 751	280	·272 319	312	·036 416	82	·805 89	406	·672 17	421	·964 864	77	46
15	·263 031	281	·272 631	313	·036 498	82	·801 83	405	·667 96	420	·964 787	76	45
16	·263 312	280	·272 944	312	·036 580	82	·797 78	404	·663 76	419	·964 711	77	44
17	·263 592	281	·273 256	313	·036 662	83	·793 74	404	·659 57	419	·964 634	77	43
18	·263 873	281	·273 569	313	·036 745	82	·789 70	402	·655 38	417	·964 557	76	42
19	·264 154	280	·273 882	312	·036 827	83	·785 68	402	·651 21	416	·964 481	77	41
20	0·264 434	281	0·274 194	313	1·036 910	83	3·781 66	401	3·647 05	416	0·964 404	77	40
21	·264 715	280	·274 507	313	·036 993	83	·777 65	400	·642 89	415	·964 327	77	39
22	·264 995	281	·274 820	313	·037 076	83	·773 65	399	·638 74	413	·964 250	77	38
23	·265 276	280	·275 133	313	·037 159	83	·769 66	398	·634 61	413	·964 173	78	37
24	·265 556	281	·275 446	313	·037 242	83	·765 68	397	·630 48	412	·964 095	77	36
25	·265 837	280	·275 759	313	·037 325	83	·761 71	396	·626 36	412	·964 018	77	35
26	·266 117	280	·276 072	313	·037 408	84	·757 75	396	·622 24	410	·963 941	78	34
27	·266 397	281	·276 385	313	·037 492	83	·753 79	395	·618 14	409	·963 863	77	33
28	·266 678	280	·276 698	313	·037 575	84	·749 84	393	·614 05	409	·963 786	78	32
29	·266 958	280	·277 011	314	·037 659	83	·745 91	393	·609 96	408	·963 708	78	31
30	0·267 238	281	0·277 325	313	1·037 742	84	3·741 98	392	3·605 88	407	0·963 630	77	30
31	·267 519	280	·277 638	313	·037 826	84	·738 06	392	·601 81	406	·963 553	78	29
32	·267 799	280	·277 951	314	·037 910	84	·734 14	390	·597 75	405	·963 475	78	28
33	·268 079	280	·278 265	313	·037 994	84	·730 24	389	·593 70	404	·963 397	78	27
34	·268 359	281	·278 578	313	·038 078	84	·726 35	389	·589 66	404	·963 319	78	26
35	·268 640	280	·278 891	314	·038 162	84	·722 46	388	·585 62	402	·963 241	78	25
36	·268 920	280	·279 205	314	·038 246	85	·718 58	387	·581 60	402	·963 163	79	24
37	·269 200	280	·279 519	313	·038 331	84	·714 71	386	·577 58	401	·963 084	78	23
38	·269 480	280	·279 832	314	·038 415	85	·710 85	385	·573 57	400	·963 006	78	22
39	·269 760	280	·280 146	314	·038 500	84	·707 00	385	·569 57	400	·962 928	79	21
40	0·270 040	280	0·280 460	313	1·038 584	85	3·703 15	384	3·565 57	398	0·962 849	79	20
41	·270 320	280	·280 773	314	·038 669	85	·699 31	382	·561 59	398	·962 770	78	19
42	·270 600	280	·281 087	314	·038 754	85	·695 49	382	·557 61	397	·962 692	79	18
43	·270 880	280	·281 401	314	·038 839	85	·691 67	382	·553 64	396	·962 613	79	17
44	·271 160	280	·281 715	314	·038 924	85	·687 85	380	·549 68	395	·962 534	79	16
45	·271 440	280	·282 029	314	·039 009	86	·684 05	380	·545 73	394	·962 455	79	15
46	·271 720	280	·282 343	314	·039 095	85	·680 25	378	·541 79	394	·962 376	79	14
47	·272 000	280	·282 657	314	·039 180	86	·676 47	378	·537 85	392	·962 297	79	13
48	·272 280	280	·282 971	315	·039 266	85	·672 69	377	·533 93	392	·962 218	79	12
49	·272 560	280	·283 286	314	·039 351	86	·668 92	377	·530 01	392	·962 139	80	11
50	0·272 840	280	0·283 600	314	1·039 437	86	3·665 15	375	3·526 09	390	0·962 059	79	10
51	·273 120	280	·283 914	315	·039 523	86	·661 40	375	·522 19	390	·961 980	79	09
52	·273 400	279	·284 229	314	·039 609	86	·657 65	374	·518 29	388	·961 901	80	08
53	·273 679	280	·284 543	314	·039 695	86	·653 91	373	·514 41	388	·961 821	80	07
54	·273 959	280	·284 857	315	·039 781	86	·650 18	373	·510 53	387	·961 741	79	06
55	·274 239	280	·285 172	315	·039 867	86	·646 45	371	·506 66	387	·961 662	80	05
56	·274 519	279	·285 487	314	·039 953	87	·642 74	371	·502 79	385	·961 582	80	04
57	·274 798	280	·285 801	315	·040 040	86	·639 03	370	·498 94	385	·961 502	80	03
58	·275 078	280	·286 116	315	·040 126	87	·635 33	369	·495 09	384	·961 422	80	02
59	·275 358	279	·286 431	314	·040 213	86	·631 64	368	·491 25	384	·961 342	80	01
60	0·275 637		0·286 745		1·040 299		3·627 96		3·487 41		0·961 262		00
	cos		cot		cosec		sec		tan		sin		74°

16° ′	sin		tan		sec		cosec		cot		cos		′
00	0·275 637	280	0·286 745	315	1·040 299	87	3·627 96	368	3·487 41	382	0·961 262	81	60
01	·275 917	280	·287 060	315	·040 386	87	·624 28	367	·483 59	382	·961 181	80	59
02	·276 197	279	·287 375	315	·040 473	87	·620 61	366	·479 77	381	·961 101	80	58
03	·276 476	280	·287 690	315	·040 560	87	·616 95	365	·475 96	380	·961 021	81	57
04	·276 756	279	·288 005	315	·040 647	88	·613 30	365	·472 16	379	·960 940	80	56
05	·277 035	280	·288 320	315	·040 735	87	·609 65	364	·468 37	379	·960 860	81	55
06	·277 315	279	·288 635	315	·040 822	87	·606 01	363	·464 58	378	·960 779	81	54
07	·277 594	280	·288 950	316	·040 909	88	·602 38	362	·460 80	377	·960 698	80	53
08	·277 874	279	·289 266	315	·040 997	88	·598 76	362	·457 03	376	·960 618	81	52
09	·278 153	279	·289 581	315	·041 085	87	·595 14	360	·453 27	376	·960 537	81	51
10	0·278 432	280	0·289 896	315	1·041 172	88	3·591 54	360	3·449 51	375	0·960 456	81	50
11	·278 712	279	·290 211	316	·041 260	88	·587 94	360	·445 76	374	·960 375	81	49
12	·278 991	279	·290 527	315	·041 348	88	·584 34	358	·442 02	373	·960 294	82	48
13	·279 270	280	·290 842	316	·041 436	88	·580 76	358	·438 29	373	·960 212	81	47
14	·279 550	279	·291 158	315	·041 524	89	·577 18	357	·434 56	372	·960 131	81	46
15	·279 829	279	·291 473	316	·041 613	88	·573 61	356	·430 84	371	·960 050	82	45
16	·280 108	280	·291 789	316	·041 701	88	·570 05	356	·427 13	370	·959 968	81	44
17	·280 388	279	·292 105	315	·041 789	89	·566 49	355	·423 43	370	·959 887	82	43
18	·280 667	279	·292 420	316	·041 878	89	·562 94	354	·419 73	369	·959 805	81	42
19	·280 946	279	·292 736	316	·041 967	88	·559 40	353	·416 04	368	·959 724	82	41
20	0·281 225	279	0·293 052	316	1·042 055	89	3·555 87	353	3·412 36	367	0·959 642	82	40
21	·281 504	279	·293 368	316	·042 144	89	·552 34	351	·408 69	367	·959 560	82	39
22	·281 783	279	·293 684	316	·042 233	89	·548 83	352	·405 02	366	·959 478	82	38
23	·282 062	279	·294 000	316	·042 322	90	·545 31	350	·401 36	365	·959 396	82	37
24	·282 341	279	·294 316	316	·042 412	89	·541 81	350	·397 71	365	·959 314	82	36
25	·282 620	280	·294 632	316	·042 501	89	·538 31	349	·394 06	364	·959 232	82	35
26	·282 900	279	·294 948	317	·042 590	90	·534 82	348	·390 42	363	·959 150	83	34
27	·283 179	278	·295 265	316	·042 680	89	·531 34	347	·386 79	362	·959 067	82	33
28	·283 457	279	·295 581	316	·042 769	90	·527 87	347	·383 17	362	·958 985	83	32
29	·283 736	279	·295 897	316	·042 859	90	·524 40	346	·379 55	361	·958 902	82	31
30	0·284 015	279	0·296 213	317	1·042 949	90	3·520 94	346	3·375 94	360	0·958 820	83	30
31	·284 294	279	·296 530	316	·043 039	90	·517 48	344	·372 34	359	·958 737	83	29
32	·284 573	279	·296 846	317	·043 129	90	·514 04	344	·368 75	359	·958 654	82	28
33	·284 852	279	·297 163	317	·043 219	90	·510 60	344	·365 16	358	·958 572	83	27
34	·285 131	279	·297 480	316	·043 309	91	·507 16	342	·361 58	358	·958 489	83	26
35	·285 410	278	·297 796	317	·043 400	90	·503 74	342	·358 00	357	·958 406	83	25
36	·285 688	279	·298 113	317	·043 490	91	·500 32	341	·354 43	356	·958 323	84	24
37	·285 967	279	·298 430	317	·043 581	90	·496 91	341	·350 87	355	·958 239	83	23
38	·286 246	279	·298 747	316	·043 671	91	·493 50	340	·347 32	355	·958 156	83	22
39	·286 525	278	·299 063	317	·043 762	91	·490 10	339	·343 77	354	·958 073	83	21
40	0·286 803	279	0·299 380	317	1·043 853	91	3·486 71	338	3·340 23	353	0·957 990	84	20
41	·287 082	279	·299 697	317	·043 944	91	·483 33	338	·336 70	353	·957 906	84	19
42	·287 361	278	·300 014	317	·044 035	91	·479 95	337	·333 17	352	·957 822	83	18
43	·287 639	279	·300 331	318	·044 126	91	·476 58	337	·329 65	351	·957 739	84	17
44	·287 918	278	·300 649	317	·044 217	92	·473 21	335	·326 14	350	·957 655	84	16
45	·288 196	279	·300 966	317	·044 309	91	·469 86	335	·322 64	350	·957 571	84	15
46	·288 475	278	·301 283	317	·044 400	92	·466 51	335	·319 14	349	·957 487	83	14
47	·288 753	279	·301 600	318	·044 492	91	·463 16	333	·315 65	349	·957 404	85	13
48	·289 032	278	·301 918	317	·044 583	92	·459 83	333	·312 16	348	·957 319	84	12
49	·289 310	279	·302 235	318	·044 675	92	·456 50	333	·308 68	347	·957 235	84	11
50	0·289 589	278	0·302 553	317	1·044 767	92	3·453 17	331	3·305 21	347	0·957 151	84	10
51	·289 867	278	·302 870	318	·044 859	92	·449 86	331	·301 74	345	·957 067	84	09
52	·290 145	279	·303 188	318	·044 951	92	·446 55	331	·298 29	346	·956 983	85	08
53	·290 424	278	·303 506	317	·045 043	93	·443 24	329	·294 83	344	·956 898	84	07
54	·290 702	279	·303 823	318	·045 136	92	·439 95	329	·291 39	344	·956 814	85	06
55	·290 981	278	·304 141	318	·045 228	93	·436 66	329	·287 95	343	·956 729	85	05
56	·291 259	278	·304 459	318	·045 321	92	·433 37	327	·284 52	343	·956 644	84	04
57	·291 537	278	·304 777	318	·045 413	93	·430 10	327	·281 09	342	·956 560	85	03
58	·291 815	279	·305 095	318	·045 506	93	·426 83	327	·277 67	341	·956 475	85	02
59	·292 094	278	·305 413	318	·045 599	93	·423 56	326	·274 26	341	·956 390	85	01
60	0·292 372		0·305 731		1·045 692		3·420 30		3·270 85		0·956 305		00
	cos		cot		cosec		sec		tan		sin		73°

17° ′	sin		tan		sec		cosec		cot		cos		′
00	0·292 372	278	0·305 731	318	1·045 692	93	3·420 30	325	3·270 85	340	0·956 305	85	60
01	·292 650	278	·306 049	318	·045 785	93	·417 05	324	·267 45	339	·956 220	86	59
02	·292 928	278	·306 367	318	·045 878	93	·413 81	324	·264 06	339	·956 134	85	58
03	·293 206	278	·306 685	318	·045 971	94	·410 57	323	·260 67	338	·956 049	85	57
04	·293 484	278	·307 003	319	·046 065	93	·407 34	323	·257 29	337	·955 964	85	56
05	·293 762	278	·307 322	318	·046 158	94	·404 11	322	·253 92	337	·955 879	86	55
06	·294 040	278	·307 640	319	·046 252	93	·400 89	321	·250 55	336	·955 793	86	54
07	·294 318	278	·307 959	318	·046 345	94	·397 68	320	·247 19	336	·955 707	85	53
08	·294 596	278	·308 277	319	·046 439	94	·394 48	320	·243 83	334	·955 622	86	52
09	·294 874	278	·308 596	318	·046 533	94	·391 28	320	·240 49	335	·955 536	86	51
10	0·295 152	278	0·308 914	319	1·046 627	94	3·388 08	319	3·237 14	333	0·955 450	86	50
11	·295 430	278	·309 233	319	·046 721	94	·384 89	318	·233 81	333	·955 364	86	49
12	·295 708	278	·309 552	318	·046 815	95	·381 71	317	·230 48	333	·955 278	86	48
13	·295 986	278	·309 870	319	·046 910	94	·378 54	317	·227 15	331	·955 192	86	47
14	·296 264	278	·310 189	319	·047 004	95	·375 37	316	·223 84	331	·955 106	86	46
15	·296 542	277	·310 508	319	·047 099	94	·372 21	316	·220 53	331	·955 020	86	45
16	·296 819	278	·310 827	319	·047 193	95	·369 05	315	·217 22	330	·954 934	87	44
17	·297 097	278	·311 146	319	·047 288	95	·365 90	314	·213 92	329	·954 847	86	43
18	·297 375	278	·311 465	319	·047 383	95	·362 76	314	·210 63	329	·954 761	87	42
19	·297 653	277	·311 784	320	·047 478	95	·359 62	313	·207 34	328	·954 674	86	41
20	0·297 930	278	0·312 104	319	1·047 573	95	3·356 49	313	3·204 06	327	0·954 588	87	40
21	·298 208	278	·312 423	319	·047 668	95	·353 36	311	·200 79	327	·954 501	87	39
22	·298 486	277	·312 742	320	·047 763	96	·350 25	312	·197 52	326	·954 414	87	38
23	·298 763	278	·313 062	319	·047 859	95	·347 13	310	·194 26	326	·954 327	87	37
24	·299 041	277	·313 381	319	·047 954	96	·344 03	311	·191 00	325	·954 240	87	36
25	·299 318	278	·313 700	320	·048 050	95	·340 92	309	·187 75	324	·954 153	87	35
26	·299 596	277	·314 020	320	·048 145	96	·337 83	309	·184 51	324	·954 066	87	34
27	·299 873	278	·314 340	319	·048 241	96	·334 74	308	·181 27	323	·953 979	87	33
28	·300 151	277	·314 659	320	·048 337	96	·331 66	308	·178 04	323	·953 892	88	32
29	·300 428	278	·314 979	320	·048 433	96	·328 58	307	·174 81	322	·953 804	87	31
30	0·300 706	277	0·315 299	320	1·048 529	96	3·325 51	307	3·171 59	321	0·953 717	88	30
31	·300 983	278	·315 619	320	·048 625	97	·322 44	305	·168 38	321	·953 629	87	29
32	·301 261	277	·315 939	319	·048 722	96	·319 39	306	·165 17	320	·953 542	88	28
33	·301 538	277	·316 258	320	·048 818	97	·316 33	305	·161 97	320	·953 454	88	27
34	·301 815	278	·316 578	321	·048 915	96	·313 28	304	·158 77	319	·953 366	87	26
35	·302 093	277	·316 899	320	·049 011	97	·310 24	303	·155 58	318	·953 279	88	25
36	·302 370	277	·317 219	320	·049 108	97	·307 21	303	·152 40	318	·953 191	88	24
37	·302 647	277	·317 539	320	·049 205	97	·304 18	303	·149 22	317	·953 103	88	23
38	·302 924	278	·317 859	320	·049 302	97	·301 15	301	·146 05	317	·953 015	89	22
39	·303 202	277	·318 179	321	·049 399	97	·298 14	302	·142 88	316	·952 926	88	21
40	0·303 479	277	0·318 500	320	1·049 496	97	3·295 12	300	3·139 72	316	0·952 838	88	20
41	·303 756	277	·318 820	321	·049 593	98	·292 12	300	·136 56	315	·952 750	89	19
42	·304 033	277	·319 141	320	·049 691	97	·289 12	300	·133 41	314	·952 661	88	18
43	·304 310	277	·319 461	321	·049 788	98	·286 12	299	·130 27	314	·952 573	89	17
44	·304 587	277	·319 782	321	·049 886	98	·283 13	298	·127 13	313	·952 484	88	16
45	·304 864	277	·320 103	320	·049 984	97	·280 15	298	·124 00	313	·952 396	89	15
46	·305 141	277	·320 423	321	·050 081	98	·277 17	297	·120 87	312	·952 307	89	14
47	·305 418	277	·320 744	321	·050 179	98	·274 20	297	·117 75	311	·952 218	89	13
48	·305 695	277	·321 065	321	·050 277	99	·271 23	296	·114 64	311	·952 129	89	12
49	·305 972	277	·321 386	321	·050 376	98	·268 27	296	·111 53	311	·952 040	89	11
50	0·306 249	277	0·321 707	321	1·050 474	98	3·265 31	294	3·108 42	310	0·951 951	89	10
51	·306 526	277	·322 028	321	·050 572	99	·262 37	295	·105 32	309	·951 862	89	09
52	·306 803	277	·322 349	321	·050 671	98	·259 42	294	·102 23	309	·951 773	89	08
53	·307 080	277	·322 670	321	·050 769	99	·256 48	293	·099 14	308	·951 684	90	07
54	·307 357	276	·322 991	321	·050 868	99	·253 55	293	·096 06	308	·951 594	89	06
55	·307 633	277	·323 312	322	·050 967	99	·250 62	292	·092 98	307	·951 505	90	05
56	·307 910	277	·323 634	321	·051 066	99	·247 70	292	·089 91	306	·951 415	89	04
57	·308 187	277	·323 955	322	·051 165	99	·244 78	291	·086 85	306	·951 326	90	03
58	·308 464	276	·324 277	321	·051 264	99	·241 87	290	·083 79	306	·951 236	90	02
59	·308 740	277	·324 598	322	·051 363	99	·238 97	290	·080 73	305	·951 146	89	01
60	0·309 017		0·324 920		1·051 462		3·236 07		3·077 68		0·951 057		00
	cos		cot		cosec		sec		tan		sin		72°

18° ′	sin		tan		sec		cosec		cot		cos		′
00	0·309 017	277	0·324 920	321	1·051 462	100	3·236 07	290	3·077 68	304	0·951 057	90	60
01	·309 294	276	·325 241	322	·051 562	99	·233 17	289	·074 64	304	·950 967	90	59
02	·309 570	277	·325 563	322	·051 661	100	·230 28	288	·071 60	303	·950 877	91	58
03	·309 847	276	·325 885	322	·051 761	100	·227 40	288	·068 57	303	·950 786	90	57
04	·310 123	277	·326 207	321	·051 861	99	·224 52	287	·065 54	302	·950 696	90	56
05	·310 400	276	·326 528	322	·051 960	100	·221 65	287	·062 52	302	·950 606	90	55
06	·310 676	277	·326 850	322	·052 060	101	·218 78	286	·059 50	301	·950 516	91	54
07	·310 953	276	·327 172	322	·052 161	100	·215 92	286	·056 49	300	·950 425	90	53
08	·311 229	277	·327 494	323	·052 261	100	·213 06	285	·053 49	300	·950 335	91	52
09	·311 506	276	·327 817	322	·052 361	100	·210 21	284	·050 49	300	·950 244	90	51
10	0·311 782	277	0·328 139	322	1·052 461	101	3·207 37	284	3·047 49	299	0·950 154	91	50
11	·312 059	276	·328 461	322	·052 562	101	·204 53	284	·044 50	298	·950 063	91	49
12	·312 335	276	·328 783	323	·052 663	100	·201 69	283	·041 52	298	·949 972	91	48
13	·312 611	277	·329 106	322	·052 763	101	·198 86	282	·038 54	298	·949 881	91	47
14	·312 888	276	·329 428	323	·052 864	101	·196 04	282	·035 56	296	·949 790	91	46
15	·313 164	276	·329 751	322	·052 965	101	·193 22	282	·032 60	297	·949 699	91	45
16	·313 440	276	·330 073	323	·053 066	101	·190 40	281	·029 63	296	·949 608	91	44
17	·313 716	276	·330 396	322	·053 167	102	·187 59	280	·026 67	295	·949 517	92	43
18	·313 992	277	·330 718	323	·053 269	101	·184 79	280	·023 72	295	·949 425	91	42
19	·314 269	276	·331 041	323	·053 370	101	·181 99	279	·020 77	294	·949 334	91	41
20	0·314 545	276	0·331 364	323	1·053 471	102	3·179 20	279	3·017 83	294	0·949 243	92	40
21	·314 821	276	·331 687	323	·053 573	102	·176 41	278	·014 89	293	·949 151	92	39
22	·315 097	276	·332 010	323	·053 675	102	·173 63	278	·011 96	293	·949 059	91	38
23	·315 373	276	·332 333	323	·053 777	101	·170 85	277	·009 03	292	·948 968	92	37
24	·315 649	276	·332 656	323	·053 878	103	·168 08	277	·006 11	292	·948 876	92	36
25	·315 925	276	·332 979	323	·053 981	102	·165 31	276	·003 19	291	·948 784	92	35
26	·316 201	276	·333 302	323	·054 083	102	·162 55	276	3·000 28	290	·948 692	92	34
27	·316 477	276	·333 625	324	·054 185	102	·159 79	275	2·997 38	291	·948 600	92	33
28	·316 753	276	·333 949	323	·054 287	103	·157 04	275	·994 47	289	·948 508	92	32
29	·317 029	276	·334 272	323	·054 390	102	·154 29	274	·991 58	290	·948 416	92	31
30	0·317 305	275	0·334 595	324	1·054 492	103	3·151 55	274	2·988 68	288	0·948 324	93	30
31	·317 580	276	·334 919	323	·054 595	103	·148 81	273	·985 80	288	·948 231	92	29
32	·317 856	276	·335 242	324	·054 698	103	·146 08	273	·982 92	288	·948 139	93	28
33	·318 132	276	·335 566	324	·054 801	103	·143 35	272	·980 04	287	·948 046	92	27
34	·318 408	276	·335 890	323	·054 904	103	·140 63	272	·977 17	287	·947 954	93	26
35	·318 684	275	·336 213	324	·055 007	103	·137 91	271	·974 30	286	·947 861	93	25
36	·318 959	276	·336 537	324	·055 110	103	·135 20	271	·971 44	286	·947 768	92	24
37	·319 235	276	·336 861	324	·055 213	104	·132 49	270	·968 58	285	·947 676	93	23
38	·319 511	275	·337 185	324	·055 317	103	·129 79	270	·965 73	285	·947 583	93	22
39	·319 786	276	·337 509	324	·055 420	104	·127 09	269	·962 88	284	·947 490	93	21
40	0·320 062	275	0·337 833	324	1·055 524	104	3·124 40	269	2·960 04	283	0·947 397	93	20
41	·320 337	276	·338 157	324	·055 628	104	·121 71	268	·957 21	284	·947 304	94	19
42	·320 613	276	·338 481	325	·055 732	104	·119 03	268	·954 37	282	·947 210	93	18
43	·320 889	275	·338 806	324	·055 836	104	·116 35	268	·951 55	283	·947 117	93	17
44	·321 164	275	·339 130	324	·055 940	104	·113 67	266	·948 72	281	·947 024	94	16
45	·321 439	276	·339 454	325	·056 044	104	·111 01	267	·945 91	282	·946 930	93	15
46	·321 715	275	·339 779	324	·056 148	105	·108 34	266	·943 09	281	·946 837	94	14
47	·321 990	276	·340 103	325	·056 253	104	·105 68	265	·940 28	280	·946 743	94	13
48	·322 266	275	·340 428	324	·056 357	105	·103 03	265	·937 48	280	·946 649	94	12
49	·322 541	275	·340 752	325	·056 462	105	·100 38	264	·934 68	279	·946 555	93	11
50	0·322 816	276	0·341 077	325	1·056 567	105	3·097 74	264	2·931 89	279	0·946 462	94	10
51	·323 092	275	·341 402	325	·056 672	105	·095 10	264	·929 10	278	·946 368	94	09
52	·323 367	275	·341 727	325	·056 777	105	·092 46	263	·926 32	278	·946 274	94	08
53	·323 642	275	·342 052	325	·056 882	105	·089 83	262	·923 54	278	·946 180	95	07
54	·323 917	276	·342 377	325	·056 987	105	·087 21	262	·920 76	277	·946 085	94	06
55	·324 193	275	·342 702	325	·057 092	106	·084 59	262	·917 99	276	·945 991	94	05
56	·324 468	275	·343 027	325	·057 198	105	·081 97	261	·915 23	277	·945 897	95	04
57	·324 743	275	·343 352	325	·057 303	106	·079 36	261	·912 46	275	·945 802	94	03
58	·325 018	275	·343 677	325	·057 409	106	·076 75	260	·909 71	275	·945 708	95	02
59	·325 293	275	·344 002	326	·057 515	106	·074 15	260	·906 96	275	·945 613	94	01
60	0·325 568		0·344 328		1·057 621		3·071 55		2·904 21		0·945 519		00
	cos		cot		cosec		sec		tan		sin		71°

19° ′	sin		tan		sec		cosec		cot		cos		′
00	0·325 568	275	0·344 328	325	1·057 621	106	3·071 55	259	2·904 21	274	0·945 519	95	60
01	·325 843	275	·344 653	325	·057 727	106	·068 96	259	·901 47	274	·945 424	95	59
02	·326 118	275	·344 978	326	·057 833	106	·066 37	258	·898 73	273	·945 329	95	58
03	·326 393	275	·345 304	326	·057 939	106	·063 79	258	·896 00	273	·945 234	95	57
04	·326 668	275	·345 630	325	·058 045	107	·061 21	257	·893 27	272	·945 139	95	56
05	·326 943	275	·345 955	326	·058 152	106	·058 64	257	·890 55	272	·945 044	95	55
06	·327 218	275	·346 281	326	·058 258	107	·056 07	257	·887 83	272	·944 949	95	54
07	·327 493	275	·346 607	326	·058 365	107	·053 50	256	·885 11	271	·944 854	96	53
08	·327 768	274	·346 933	326	·058 472	107	·050 94	255	·882 40	270	·944 758	95	52
09	·328 042	275	·347 259	326	·058 579	107	·048 39	255	·879 70	270	·944 663	95	51
10	0·328 317	275	0·347 585	326	1·058 686	107	3·045 84	255	2·877 00	270	0·944 568	96	50
11	·328 592	275	·347 911	326	·058 793	107	·043 29	254	·874 30	269	·944 472	96	49
12	·328 867	274	·348 237	326	·058 900	107	·040 75	254	·871 61	269	·944 376	95	48
13	·329 141	275	·348 563	326	·059 007	108	·038 21	253	·868 92	268	·944 281	96	47
14	·329 416	275	·348 889	327	·059 115	107	·035 68	253	·866 24	268	·944 185	96	46
15	·329 691	274	·349 216	326	·059 222	108	·033 15	253	·863 56	267	·944 089	96	45
16	·329 965	275	·349 542	326	·059 330	108	·030 62	252	·860 89	267	·943 993	96	44
17	·330 240	274	·349 868	327	·059 438	107	·028 10	251	·858 22	267	·943 897	96	43
18	·330 514	275	·350 195	327	·059 545	108	·025 59	251	·855 55	266	·943 801	96	42
19	·330 789	274	·350 522	326	·059 653	109	·023 08	251	·852 89	266	·943 705	96	41
20	0·331 063	275	0·350 848	327	1·059 762	108	3·020 57	250	2·850 23	265	0·943 609	97	40
21	·331 338	274	·351 175	327	·059 870	108	·018 07	250	·847 58	264	·943 512	96	39
22	·331 612	275	·351 502	327	·059 978	109	·015 57	249	·844 94	265	·943 416	97	38
23	·331 887	274	·351 829	327	·060 087	108	·013 08	249	·842 29	264	·943 319	96	37
24	·332 161	274	·352 156	327	·060 195	109	·010 59	249	·839 65	263	·943 223	97	36
25	·332 435	275	·352 483	327	·060 304	108	·008 10	248	·837 02	263	·943 126	97	35
26	·332 710	274	·352 810	327	·060 412	109	·005 62	247	·834 39	263	·943 029	97	34
27	·332 984	274	·353 137	327	·060 521	109	·003 15	248	·831 76	262	·942 932	96	33
28	·333 258	275	·353 464	327	·060 630	109	3·000 67	246	·829 14	261	·942 836	97	32
29	·333 533	274	·353 791	328	·060 739	110	2·998 21	247	·826 53	262	·942 739	98	31
30	0·333 807	274	0·354 119	327	1·060 849	109	2·995 74	245	2·823 91	261	0·942 641	97	30
31	·334 081	274	·354 446	327	·060 958	109	·993 29	246	·821 30	260	·942 544	97	29
32	·334 355	274	·354 773	328	·061 067	110	·990 83	245	·818 70	260	·942 447	97	28
33	·334 629	274	·355 101	328	·061 177	110	·988 38	244	·816 10	260	·942 350	98	27
34	·334 903	275	·355 429	327	·061 287	109	·985 94	245	·813 50	259	·942 252	97	26
35	·335 178	274	·355 756	328	·061 396	110	·983 49	243	·810 91	258	·942 155	98	25
36	·335 452	274	·356 084	328	·061 506	110	·981 06	244	·808 33	259	·942 057	97	24
37	·335 726	274	·356 412	328	·061 616	111	·978 62	243	·805 74	258	·941 960	98	23
38	·336 000	274	·356 740	328	·061 727	110	·976 19	242	·803 16	257	·941 862	98	22
39	·336 274	273	·357 068	328	·061 837	110	·973 77	242	·800 59	257	·941 764	98	21
40	0·336 547	274	0·357 396	328	1·061 947	111	2·971 35	242	2·798 02	257	0·941 666	97	20
41	·336 821	274	·357 724	328	·062 058	110	·968 93	241	·795 45	256	·941 569	98	19
42	·337 095	274	·358 052	328	·062 168	111	·966 52	241	·792 89	256	·941 471	99	18
43	·337 369	274	·358 380	328	·062 279	111	·964 11	240	·790 33	255	·941 372	98	17
44	·337 643	274	·358 708	329	·062 390	111	·961 71	240	·787 78	255	·941 274	98	16
45	·337 917	273	·359 037	328	·062 501	111	·959 31	240	·785 23	254	·941 176	98	15
46	·338 190	274	·359 365	329	·062 612	111	·956 91	239	·782 69	255	·941 078	99	14
47	·338 464	274	·359 694	328	·062 723	111	·954 52	239	·780 14	253	·940 979	98	13
48	·338 738	274	·360 022	329	·062 834	111	·952 13	238	·777 61	254	·940 881	99	12
49	·339 012	273	·360 351	328	·062 945	112	·949 75	238	·775 07	253	·940 782	98	11
50	0·339 285	274	0·360 679	329	1·063 057	111	2·947 37	237	2·772 54	252	0·940 684	99	10
51	·339 559	273	·361 008	329	·063 168	112	·945 00	237	·770 02	252	·940 585	99	09
52	·339 832	274	·361 337	329	·063 280	112	·942 63	237	·767 50	252	·940 486	99	08
53	·340 106	274	·361 666	329	·063 392	112	·940 26	236	·764 98	251	·940 387	99	07
54	·340 380	273	·361 995	329	·063 504	112	·937 90	236	·762 47	251	·940 288	99	06
55	·340 653	274	·362 324	329	·063 616	112	·935 54	236	·759 96	250	·940 189	99	05
56	·340 927	273	·362 653	329	·063 728	112	·933 18	235	·757 46	250	·940 090	99	04
57	·341 200	273	·362 982	330	·063 840	113	·930 83	234	·754 96	250	·939 991	100	03
58	·341 473	274	·363 312	329	·063 953	112	·928 49	235	·752 46	249	·939 891	99	02
59	·341 747	273	·363 641	329	·064 065	113	·926 14	234	·749 97	249	·939 792	99	01
60	0·342 020		0·363 970		1·064 178		2·923 80		2·747 48		0·939 693		00
	cos		cot		cosec		sec		tan		sin		70°

20° ′	sin		tan		sec		cosec		cot		cos		′
00	0·342 020	273	0·363 970	330	1·064 178	112	2·923 804	2334	2·747 477	2484	0·939 693	100	60
01	·342 293	274	·364 300	329	·064 290	113	·921 470	2331	·744 993	2481	·939 593	100	59
02	·342 567	273	·364 629	330	·064 403	113	·919 139	2327	·742 512	2477	·939 493	99	58
03	·342 840	273	·364 959	329	·064 516	113	·916 812	2323	·740 035	2473	·939 394	100	57
04	·343 113	274	·365 288	330	·064 629	114	·914 489	2319	·737 562	2469	·939 294	100	56
05	·343 387	273	·365 618	330	·064 743	113	·912 170	2315	·735 093	2465	·939 194	100	55
06	·343 660	273	·365 948	330	·064 856	113	·909 855	2311	·732 628	2461	·939 094	100	54
07	·343 933	273	·366 278	330	·064 969	114	·907 544	2307	·730 167	2457	·938 994	100	53
08	·344 206	273	·366 608	330	·065 083	113	·905 237	2303	·727 710	2453	·938 894	100	52
09	·344 479	273	·366 938	330	·065 196	114	·902 934	2299	·725 257	2449	·938 794	100	51
10	0·344 752	273	0·367 268	330	1·065 310	114	2·900 635	2296	2·722 808	2446	0·938 694	101	50
11	·345 025	273	·367 598	330	·065 424	114	·898 339	2291	·720 362	2442	·938 593	100	49
12	·345 298	273	·367 928	331	·065 538	114	·896 048	2288	·717 920	2437	·938 493	100	48
13	·345 571	273	·368 259	330	·065 652	114	·893 760	2284	·715 483	2434	·938 393	101	47
14	·345 844	273	·368 589	330	·065 766	115	·891 476	2280	·713 049	2430	·938 292	101	46
15	·346 117	273	·368 919	331	·065 881	114	·889 196	2276	·710 619	2427	·938 191	100	45
16	·346 390	273	·369 250	331	·065 995	115	·886 920	2273	·708 192	2422	·938 091	101	44
17	·346 663	273	·369 581	330	·066 110	114	·884 647	2268	·705 770	2419	·937 990	101	43
18	·346 936	272	·369 911	331	·066 224	115	·882 379	2265	·703 351	2415	·937 889	101	42
19	·347 208	273	·370 242	331	·066 339	115	·880 114	2261	·700 936	2411	·937 788	101	41
20	0·347 481	273	0·370 573	331	1·066 454	115	2·877 853	2257	2·698 525	2407	0·937 687	101	40
21	·347 754	273	·370 904	331	·066 569	115	·875 596	2253	·696 118	2403	·937 586	101	39
22	·348 027	272	·371 235	331	·066 684	115	·873 343	2250	·693 715	2400	·937 485	102	38
23	·348 299	273	·371 566	331	·066 799	116	·871 093	2246	·691 315	2396	·937 383	101	37
24	·348 572	273	·371 897	331	·066 915	115	·868 847	2242	·688 919	2392	·937 282	101	36
25	·348 845	272	·372 228	331	·067 030	116	·866 605	2238	·686 527	2389	·937 181	102	35
26	·349 117	273	·372 559	331	·067 146	116	·864 367	2235	·684 138	2384	·937 079	102	34
27	·349 390	272	·372 890	332	·067 262	115	·862 132	2230	·681 754	2382	·936 977	101	33
28	·349 662	273	·373 222	331	·067 377	116	·859 902	2228	·679 372	2377	·936 876	102	32
29	·349 935	272	·373 553	332	·067 493	116	·857 674	2223	·676 995	2374	·936 774	102	31
30	0·350 207	273	0·373 885	331	1·067 609	117	2·855 451	2220	2·674 621	2369	0·936 672	102	30
31	·350 480	272	·374 216	332	·067 726	116	·853 231	2216	·672 252	2367	·936 570	102	29
32	·350 752	273	·374 548	332	·067 842	116	·851 015	2212	·669 885	2362	·936 468	102	28
33	·351 025	272	·374 880	331	·067 958	117	·848 803	2209	·667 523	2359	·936 366	102	27
34	·351 297	272	·375 211	332	·068 075	116	·846 594	2205	·665 164	2355	·936 264	102	26
35	·351 569	273	·375 543	332	·068 191	117	·844 389	2201	·662 809	2352	·936 162	102	25
36	·351 842	272	·375 875	332	·068 308	117	·842 188	2198	·660 457	2348	·936 060	103	24
37	·352 114	272	·376 207	332	·068 425	117	·839 990	2194	·658 109	2344	·935 957	102	23
38	·352 386	272	·376 539	333	·068 542	117	·837 796	2191	·655 765	2341	·935 855	103	22
39	·352 658	273	·376 872	332	·068 659	117	·835 605	2186	·653 424	2337	·935 752	102	21
40	0·352 931	272	0·377 204	332	1·068 776	118	2·833 419	2184	2·651 087	2334	0·935 650	103	20
41	·353 203	272	·377 536	333	·068 894	117	·831 235	2179	·648 753	2330	·935 547	103	19
42	·353 475	272	·377 869	332	·069 011	118	·829 056	2176	·646 423	2326	·935 444	103	18
43	·353 747	272	·378 201	333	·069 129	117	·826 880	2173	·644 097	2323	·935 341	103	17
44	·354 019	272	·378 534	332	·069 246	118	·824 707	2169	·641 774	2319	·935 238	103	16
45	·354 291	272	·378 866	333	·069 364	118	·822 538	2165	·639 455	2316	·935 135	103	15
46	·354 563	272	·379 199	333	·069 482	118	·820 373	2162	·637 139	2312	·935 032	103	14
47	·354 835	272	·379 532	332	·069 600	118	·818 211	2158	·634 827	2308	·934 929	103	13
48	·355 107	272	·379 864	333	·069 718	118	·816 053	2155	·632 519	2305	·934 826	104	12
49	·355 379	272	·380 197	333	·069 836	119	·813 898	2151	·630 214	2302	·934 722	103	11
50	0·355 651	272	0·380 530	333	1·069 955	118	2·811 747	2148	2·627 912	2298	0·934 619	104	10
51	·355 923	271	·380 863	333	·070 073	119	·809 599	2144	·625 614	2294	·934 515	103	09
52	·356 194	272	·381 196	334	·070 192	119	·807 455	2140	·623 320	2291	·934 412	104	08
53	·356 466	272	·381 530	333	·070 311	118	·805 315	2137	·621 029	2288	·934 308	104	07
54	·356 738	272	·381 863	333	·070 429	119	·803 178	2134	·618 741	2284	·934 204	103	06
55	·357 010	271	·382 196	334	·070 548	120	·801 044	2130	·616 457	2280	·934 101	104	05
56	·357 281	272	·382 530	333	·070 668	119	·798 914	2127	·614 177	2277	·933 997	104	04
57	·357 553	272	·382 863	334	·070 787	119	·796 787	2123	·611 900	2274	·933 893	104	03
58	·357 825	271	·383 197	333	·070 906	119	·794 664	2120	·609 626	2270	·933 789	104	02
59	·358 096	272	·383 530	334	·071 025	120	·792 544	2116	·607 356	2267	·933 685	105	01
60	0·358 368		0·383 864		1·071 145		2·790 428		2·605 089		0·933 580		00
	cos		cot		cosec		sec		tan		sin		69°

21° ′	sin		tan		sec		cosec		cot		cos		′
00	0·358 368	272	0·383 864	334	1·071 145	120	2·790 428	2113	2·605 089	2263	0·933 580	104	60
01	·358 640	271	·384 198	334	·071 265	119	·788 315	2109	·602 826	2260	·933 476	104	59
02	·358 911	272	·384 532	334	·071 384	120	·786 206	2106	·600 566	2257	·933 372	105	58
03	·359 183	271	·384 866	334	·071 504	120	·784 100	2103	·598 309	2253	·933 267	104	57
04	·359 454	271	·385 200	334	·071 624	120	·781 997	2099	·596 056	2249	·933 163	105	56
05	·359 725	272	·385 534	334	·071 744	121	·779 898	2096	·593 807	2246	·933 058	104	55
06	·359 997	271	·385 868	334	·071 865	120	·777 802	2092	·591 561	2243	·932 954	105	54
07	·360 268	272	·386 202	334	·071 985	121	·775 710	2089	·589 318	2240	·932 849	105	53
08	·360 540	271	·386 536	335	·072 106	120	·773 621	2086	·587 078	2236	·932 744	105	52
09	·360 811	271	·386 871	334	·072 226	121	·771 535	2082	·584 842	2233	·932 639	105	51
10	0·361 082	271	0·387 205	335	1·072 347	121	2·769 453	2079	2·582 609	2229	0·932 534	105	50
11	·361 353	272	·387 540	334	·072 468	121	·767 374	2075	·580 380	2226	·932 429	105	49
12	·361 625	271	·387 874	335	·072 589	121	·765 299	2072	·578 154	2223	·932 324	105	48
13	·361 896	271	·388 209	335	·072 710	121	·763 227	2069	·575 931	2219	·932 219	106	47
14	·362 167	271	·388 544	335	·072 831	121	·761 158	2066	·573 712	2216	·932 113	105	46
15	·362 438	271	·388 879	335	·072 952	122	·759 092	2062	·571 496	2213	·932 008	106	45
16	·362 709	271	·389 214	335	·073 074	121	·757 030	2059	·569 283	2209	·931 902	105	44
17	·362 980	271	·389 549	335	·073 195	122	·754 971	2055	·567 074	2207	·931 797	106	43
18	·363 251	271	·389 884	335	·073 317	122	·752 916	2053	·564 867	2202	·931 691	105	42
19	·363 522	271	·390 219	335	·073 439	122	·750 863	2049	·562 665	2200	·931 586	106	41
20	0·363 793	271	0·390 554	335	1·073 561	122	2·748 814	2045	2·560 465	2196	0·931 480	106	40
21	·364 064	271	·390 889	336	·073 683	122	·746 769	2043	·558 269	2193	·931 374	106	39
22	·364 335	271	·391 225	335	·073 805	122	·744 726	2039	·556 076	2190	·931 268	106	38
23	·364 606	271	·391 560	336	·073 927	122	·742 687	2036	·553 886	2187	·931 162	106	37
24	·364 877	271	·391 896	335	·074 049	123	·740 651	2032	·551 699	2183	·931 056	106	36
25	·365 148	270	·392 231	336	·074 172	123	·738 619	2030	·549 516	2180	·930 950	107	35
26	·365 418	271	·392 567	336	·074 295	122	·736 589	2026	·547 336	2177	·930 843	106	34
27	·365 689	271	·392 903	336	·074 417	123	·734 563	2023	·545 159	2174	·930 737	106	33
28	·365 960	271	·393 239	335	·074 540	123	·732 540	2020	·542 985	2170	·930 631	107	32
29	·366 231	270	·393 574	336	·074 663	123	·730 520	2016	·540 815	2167	·930 524	106	31
30	0·366 501	271	0·393 910	337	1·074 786	123	2·728 504	2013	2·538 648	2164	0·930 418	107	30
31	·366 772	270	·394 247	336	·074 909	124	·726 491	2011	·536 484	2161	·930 311	107	29
32	·367 042	271	·394 583	336	·075 033	123	·724 480	2006	·534 323	2158	·930 204	107	28
33	·367 313	271	·394 919	336	·075 156	124	·722 474	2004	·532 165	2154	·930 097	107	27
34	·367 584	270	·395 255	337	·075 280	123	·720 470	2001	·530 011	2151	·929 990	106	26
35	·367 854	271	·395 592	336	·075 403	124	·718 469	1997	·527 860	2148	·929 884	108	25
36	·368 125	270	·395 928	337	·075 527	124	·716 472	1994	·525 712	2145	·929 776	107	24
37	·368 395	270	·396 265	336	·075 651	124	·714 478	1991	·523 567	2142	·929 669	107	23
38	·368 665	271	·396 601	337	·075 775	124	·712 487	1988	·521 425	2139	·929 562	107	22
39	·368 936	270	·396 938	337	·075 899	125	·710 499	1985	·519 286	2135	·929 455	107	21
40	0·369 206	270	0·397 275	336	1·076 024	124	2·708 514	1982	2·517 151	2133	0·929 348	108	20
41	·369 476	271	·397 611	337	·076 148	125	·706 532	1978	·515 018	2129	·929 240	107	19
42	·369 747	270	·397 948	337	·076 273	124	·704 554	1976	·512 889	2126	·929 133	108	18
43	·370 017	270	·398 285	337	·076 397	125	·702 578	1972	·510 763	2123	·929 025	108	17
44	·370 287	270	·398 622	338	·076 522	125	·700 606	1969	·508 640	2120	·928 917	107	16
45	·370 557	271	·398 960	337	·076 647	125	·698 637	1966	·506 520	2117	·928 810	108	15
46	·370 828	270	·399 297	337	·076 772	125	·696 671	1963	·504 403	2114	·928 702	108	14
47	·371 098	270	·399 634	337	·076 897	125	·694 708	1960	·502 289	2111	·928 594	108	13
48	·371 368	270	·399 971	338	·077 022	126	·692 748	1957	·500 178	2107	·928 486	108	12
49	·371 638	270	·400 309	337	·077 148	125	·690 791	1954	·498 071	2105	·928 378	108	11
50	0·371 908	270	0·400 646	338	1·077 273	126	2·688 837	1950	2·495 966	2101	0·928 270	109	10
51	·372 178	270	·400 984	338	·077 399	126	·686 887	1948	·493 865	2099	·928 161	108	09
52	·372 448	270	·401 322	338	·077 525	125	·684 939	1944	·491 766	2095	·928 053	108	08
53	·372 718	270	·401 660	337	·077 650	126	·682 995	1942	·489 671	2093	·927 945	109	07
54	·372 988	270	·401 997	338	·077 776	126	·681 053	1939	·487 578	2089	·927 836	108	06
55	·373 258	270	·402 335	338	·077 902	127	·679 114	1935	·485 489	2087	·927 728	109	05
56	·373 528	269	·402 673	338	·078 029	126	·677 179	1932	·483 402	2083	·927 619	109	04
57	·373 797	270	·403 011	339	·078 155	126	·675 247	1930	·481 319	2080	·927 510	108	03
58	·374 067	270	·403 350	338	·078 281	127	·673 317	1926	·479 239	2078	·927 402	109	02
59	·374 337	270	·403 688	338	·078 408	127	·671 391	1924	·477 161	2074	·927 293	109	01
60	0·374 607		0·404 026		1·078 535		2·669 467		2·475 087		0·927 184		00
	cos		cot		cosec		sec		tan		sin		68°

22° ′	sin		tan		sec		cosec		cot		cos		′
00	0·374 607	269	0·404 026	339	1·078 535	127	2·669 467	1920	2·475 087	2072	0·927 184	109	60
01	·374 876	270	·404 365	338	·078 662	126	·667 547	1918	·473 015	2068	·927 075	109	59
02	·375 146	270	·404 703	339	·078 788	128	·665 629	1914	·470 947	2065	·926 966	109	58
03	·375 416	269	·405 042	338	·078 916	127	·663 715	1912	·468 882	2063	·926 857	110	57
04	·375 685	270	·405 380	339	·079 043	127	·661 803	1908	·466 819	2059	·926 747	109	56
05	·375 955	269	·405 719	339	·079 170	127	·659 895	1906	·464 760	2057	·926 638	109	55
06	·376 224	270	·406 058	339	·079 297	128	·657 989	1903	·462 703	2054	·926 529	110	54
07	·376 494	269	·406 397	339	·079 425	128	·656 086	1899	·460 649	2050	·926 419	109	53
08	·376 763	270	·406 736	339	·079 553	127	·654 187	1897	·458 599	2048	·926 310	110	52
09	·377 033	269	·407 075	339	·079 680	128	·652 290	1894	·456 551	2045	·926 200	110	51
10	0·377 302	269	0·407 414	339	1·079 808	128	2·650 396	1891	2·454 506	2042	0·926 090	110	50
11	·377 571	270	·407 753	339	·079 936	129	·648 505	1888	·452 464	2039	·925 980	109	49
12	·377 841	269	·408 092	340	·080 065	128	·646 617	1885	·450 425	2036	·925 871	110	48
13	·378 110	269	·408 432	339	·080 193	128	·644 732	1882	·448 389	2033	·925 761	110	47
14	·378 379	270	·408 771	340	·080 321	129	·642 850	1879	·446 356	2030	·925 651	110	46
15	·378 649	269	·409 111	339	·080 450	128	·640 971	1876	·444 326	2028	·925 541	111	45
16	·378 918	269	·409 450	340	·080 578	129	·639 095	1874	·442 298	2024	·925 430	110	44
17	·379 187	269	·409 790	340	·080 707	129	·637 221	1870	·440 274	2022	·925 320	110	43
18	·379 456	269	·410 130	340	·080 836	129	·635 351	1868	·438 252	2019	·925 210	111	42
19	·379 725	269	·410 470	340	·080 965	129	·633 483	1865	·436 233	2016	·925 099	110	41
20	0·379 994	269	0·410 810	340	1·081 094	129	2·631 618	1862	2·434 217	2013	0·924 989	111	40
21	·380 263	269	·411 150	340	·081 223	130	·629 756	1859	·432 204	2010	·924 878	110	39
22	·380 532	269	·411 490	340	·081 353	129	·627 897	1856	·430 194	2008	·924 768	111	38
23	·380 801	269	·411 830	340	·081 482	130	·626 041	1854	·428 186	2004	·924 657	111	37
24	·381 070	269	·412 170	341	·081 612	130	·624 187	1850	·426 182	2002	·924 546	111	36
25	·381 339	269	·412 511	340	·081 742	130	·622 337	1848	·424 180	1999	·924 435	111	35
26	·381 608	269	·412 851	341	·081 872	130	·620 489	1845	·422 181	1996	·924 324	111	34
27	·381 877	269	·413 192	340	·082 002	130	·618 644	1842	·420 185	1993	·924 213	111	33
28	·382 146	269	·413 532	341	·082 132	130	·616 802	1840	·418 192	1991	·924 102	111	32
29	·382 415	268	·413 873	341	·082 262	130	·614 962	1836	·416 201	1987	·923 991	111	31
30	0·382 683	269	0·414 214	340	1·082 392	131	2·613 126	1834	2·414 214	1985	0·923 880	112	30
31	·382 952	269	·414 554	341	·082 523	130	·611 292	1831	·412 229	1982	·923 768	111	29
32	·383 221	269	·414 895	341	·082 653	131	·609 461	1828	·410 247	1980	·923 657	112	28
33	·383 490	268	·415 236	341	·082 784	131	·607 633	1825	·408 267	1976	·923 545	111	27
34	·383 758	269	·415 577	342	·082 915	131	·605 808	1823	·406 291	1974	·923 434	112	26
35	·384 027	268	·415 919	341	·083 046	131	·603 985	1820	·404 317	1971	·923 322	112	25
36	·384 295	269	·416 260	341	·083 177	131	·602 165	1817	·402 346	1969	·923 210	112	24
37	·384 564	268	·416 601	342	·083 308	131	·600 348	1814	·400 377	1965	·923 098	112	23
38	·384 832	269	·416 943	341	·083 439	132	·598 534	1811	·398 412	1963	·922 986	111	22
39	·385 101	268	·417 284	342	·083 571	132	·596 723	1809	·396 449	1960	·922 875	113	21
40	0·385 369	269	0·417 626	341	1·083 703	131	2·594 914	1806	2·394 489	1957	0·922 762	112	20
41	·385 638	268	·417 967	342	·083 834	132	·593 108	1804	·392 532	1955	·922 650	112	19
42	·385 906	268	·418 309	342	·083 966	132	·591 304	1800	·390 577	1952	·922 538	112	18
43	·386 174	269	·418 651	342	·084 098	132	·589 504	1798	·388 625	1949	·922 426	113	17
44	·386 443	268	·418 993	342	·084 230	132	·587 706	1795	·386 676	1947	·922 313	112	16
45	·386 711	268	·419 335	342	·084 362	133	·585 911	1793	·384 729	1943	·922 201	113	15
46	·386 979	268	·419 677	342	·084 495	132	·584 118	1790	·382 786	1942	·922 088	112	14
47	·387 247	269	·420 019	342	·084 627	133	·582 328	1787	·380 844	1938	·921 976	113	13
48	·387 516	268	·420 361	343	·084 760	132	·580 541	1784	·378 906	1936	·921 863	113	12
49	·387 784	268	·420 704	342	·084 892	133	·578 757	1782	·376 970	1933	·921 750	112	11
50	0·388 052	268	0·421 046	343	1·085 025	133	2·576 975	1779	2·375 037	1930	0·921 638	113	10
51	·388 320	268	·421 389	342	·085 158	133	·575 196	1776	·373 107	1928	·921 525	113	09
52	·388 588	268	·421 731	343	·085 291	133	·573 420	1774	·371 179	1925	·921 412	113	08
53	·388 856	268	·422 074	343	·085 424	134	·571 646	1771	·369 254	1922	·921 299	114	07
54	·389 124	268	·422 417	342	·085 558	133	·569 875	1768	·367 332	1920	·921 185	113	06
55	·389 392	268	·422 759	343	·085 691	134	·568 107	1766	·365 412	1917	·921 072	113	05
56	·389 660	268	·423 102	343	·085 825	134	·566 341	1763	·363 495	1915	·920 959	114	04
57	·389 928	268	·423 445	343	·085 959	133	·564 578	1760	·361 580	1912	·920 845	113	03
58	·390 196	267	·423 788	344	·086 092	134	·562 818	1758	·359 668	1909	·920 732	114	02
59	·390 463	268	·424 132	343	·086 226	134	·561 060	1755	·357 759	1907	·920 618	113	01
60	0·390 731		0·424 475		1·086 360		2·559 305		2·355 852		0·920 505		00
	cos		cot		cosec		sec		tan		sin		67°

23° ′	sin		tan		sec		cosec		cot		cos		′
00	0·390 731	268	0·424 475	343	1·086 360	135	2·559 305	1753	2·355 852	1904	0·920 505	114	60
01	·390 999	268	·424 818	344	·086 495	134	·557 552	1750	·353 948	1901	·920 391	114	59
02	·391 267	267	·425 162	343	·086 629	134	·555 802	1747	·352 047	1899	·920 277	113	58
03	·391 534	268	·425 505	344	·086 763	135	·554 055	1745	·350 148	1896	·920 164	114	57
04	·391 802	268	·425 849	343	·086 898	135	·552 310	1742	·348 252	1894	·920 050	114	56
05	·392 070	267	·426 192	344	·087 033	134	·550 568	1740	·346 358	1891	·919 936	115	55
06	·392 337	268	·426 536	344	·087 167	135	·548 828	1737	·344 467	1888	·919 821	114	54
07	·392 605	267	·426 880	344	·087 302	135	·547 091	1734	·342 579	1886	·919 707	114	53
08	·392 872	268	·427 224	344	·087 437	136	·545 357	1732	·340 693	1884	·919 593	114	52
09	·393 140	267	·427 568	344	·087 573	135	·543 625	1729	·338 809	1880	·919 479	115	51
10	0·393 407	268	0·427 912	344	1·087 708	135	2·541 896	1727	2·336 929	1879	0·919 364	114	50
11	·393 675	267	·428 256	345	·087 843	136	·540 169	1724	·335 050	1875	·919 250	115	49
12	·393 942	267	·428 601	344	·087 979	136	·538 445	1721	·333 175	1873	·919 135	114	48
13	·394 209	268	·428 945	344	·088 115	136	·536 724	1719	·331 302	1871	·919 021	115	47
14	·394 477	267	·429 289	345	·088 251	136	·535 005	1717	·329 431	1868	·918 906	115	46
15	·394 744	267	·429 634	345	·088 387	136	·533 288	1714	·327 563	1865	·918 791	115	45
16	·395 011	267	·429 979	344	·088 523	136	·531 574	1711	·325 698	1863	·918 676	115	44
17	·395 278	268	·430 323	345	·088 659	136	·529 863	1709	·323 835	1861	·918 561	115	43
18	·395 546	267	·430 668	345	·088 795	137	·528 154	1706	·321 974	1858	·918 446	115	42
19	·395 813	267	·431 013	345	·088 932	136	·526 448	1704	·320 116	1855	·918 331	115	41
20	0·396 080	267	0·431 358	345	1·089 068	137	2·524 744	1701	2·318 261	1853	0·918 216	115	40
21	·396 347	267	·431 703	345	·089 205	137	·523 043	1699	·316 408	1851	·918 101	115	39
22	·396 614	267	·432 048	345	·089 342	137	·521 344	1696	·314 557	1848	·917 986	116	38
23	·396 881	267	·432 393	346	·089 479	137	·519 648	1694	·312 709	1845	·917 870	115	37
24	·397 148	267	·432 739	345	·089 616	137	·517 954	1692	·310 864	1843	·917 755	116	36
25	·397 415	267	·433 084	346	·089 753	137	·516 262	1688	·309 021	1841	·917 639	116	35
26	·397 682	267	·433 430	345	·089 890	138	·514 574	1687	·307 180	1838	·917 523	115	34
27	·397 949	266	·433 775	346	·090 028	138	·512 887	1684	·305 342	1836	·917 408	116	33
28	·398 215	267	·434 121	346	·090 166	137	·511 203	1681	·303 506	1833	·917 292	116	32
29	·398 482	267	·434 467	345	·090 303	138	·509 522	1679	·301 673	1830	·917 176	116	31
30	0·398 749	267	0·434 812	346	1·090 441	138	2·507 843	1677	2·299 843	1829	0·917 060	116	30
31	·399 016	267	·435 158	346	·090 579	138	·506 166	1674	·298 014	1826	·916 944	116	29
32	·399 283	266	·435 504	346	·090 717	138	·504 492	1671	·296 188	1823	·916 828	116	28
33	·399 549	267	·435 850	347	·090 855	139	·502 821	1670	·294 365	1821	·916 712	117	27
34	·399 816	266	·436 197	346	·090 994	138	·501 151	1666	·292 544	1818	·916 595	116	26
35	·400 082	267	·436 543	346	·091 132	139	·499 485	1665	·290 726	1816	·916 479	116	25
36	·400 349	267	·436 889	347	·091 271	139	·497 820	1661	·288 910	1814	·916 363	117	24
37	·400 616	266	·437 236	346	·091 410	139	·496 159	1660	·287 096	1811	·916 246	116	23
38	·400 882	267	·437 582	347	·091 549	139	·494 499	1657	·285 285	1809	·916 130	117	22
39	·401 149	266	·437 929	347	·091 688	139	·492 842	1655	·283 476	1807	·916 013	117	21
40	0·401 415	266	0·438 276	346	1·091 827	139	2·491 187	1652	2·281 669	1804	0·915 896	117	20
41	·401 681	267	·438 622	347	·091 966	139	·489 535	1650	·279 865	1801	·915 779	116	19
42	·401 948	266	·438 969	347	·092 105	140	·487 885	1647	·278 064	1800	·915 663	117	18
43	·402 214	266	·439 316	347	·092 245	139	·486 238	1645	·276 264	1797	·915 546	117	17
44	·402 480	267	·439 663	348	·092 384	140	·484 593	1643	·274 467	1794	·915 429	118	16
45	·402 747	266	·440 011	347	·092 524	140	·482 950	1640	·272 673	1792	·915 311	117	15
46	·403 013	266	·440 358	347	·092 664	140	·481 310	1638	·270 881	1790	·915 194	117	14
47	·403 279	266	·440 705	348	·092 804	140	·479 672	1635	·269 091	1787	·915 077	117	13
48	·403 545	266	·441 053	347	·092 944	141	·478 037	1634	·267 304	1786	·914 960	118	12
49	·403 811	267	·441 400	348	·093 085	140	·476 403	1630	·265 518	1782	·914 842	117	11
50	0·404 078	266	0·441 748	347	1·093 225	141	2·474 773	1629	2·263 736	1781	0·914 725	118	10
51	·404 344	266	·442 095	348	·093 366	140	·473 144	1626	·261 955	1778	·914 607	117	09
52	·404 610	266	·442 443	348	·093 506	141	·471 518	1624	·260 177	1775	·914 490	118	08
53	·404 876	266	·442 791	348	·093 647	141	·469 894	1621	·258 402	1774	·914 372	118	07
54	·405 142	266	·443 139	348	·093 788	141	·468 273	1619	·256 628	1771	·914 254	118	06
55	·405 408	265	·443 487	348	·093 929	141	·466 654	1617	·254 857	1768	·914 136	118	05
56	·405 673	266	·443 835	348	·094 070	142	·465 037	1614	·253 089	1767	·914 018	118	04
57	·405 939	266	·444 183	349	·094 212	141	·463 423	1612	·251 322	1764	·913 900	118	03
58	·406 205	266	·444 532	348	·094 353	142	·461 811	1610	·249 558	1762	·913 782	118	02
59	·406 471	266	·444 880	349	·094 495	141	·460 201	1608	·247 796	1759	·913 664	119	01
60	0·406 737		0·445 229		1·094 636		2·458 593		2·246 037		0·913 545		00
	cos		cot		cosec		sec		tan		sin		66°

24° ′	sin		tan		sec		cosec		cot		cos		′
00	0·406 737	265	0·445 229	348	1·094 636	142	2·458 593	1605	2·246 037	1757	0·913 545	118	60
01	·407 002	266	·445 577	349	·094 778	142	·456 988	1603	·244 280	1755	·913 427	118	59
02	·407 268	266	·445 926	349	·094 920	142	·455 385	1600	·242 525	1753	·913 309	119	58
03	·407 534	265	·446 275	349	·095 062	142	·453 785	1599	·240 772	1750	·913 190	118	57
04	·407 799	266	·446 624	349	·095 204	143	·452 186	1595	·239 022	1748	·913 072	119	56
05	·408 065	265	·446 973	349	·095 347	142	·450 591	1594	·237 274	1746	·912 953	119	55
06	·408 330	266	·447 322	349	·095 489	143	·448 997	1592	·235 528	1743	·912 834	119	54
07	·408 596	265	·447 671	349	·095 632	143	·447 405	1589	·233 785	1742	·912 715	119	53
08	·408 861	266	·448 020	349	·095 775	142	·445 816	1587	·232 043	1739	·912 596	119	52
09	·409 127	265	·448 369	350	·095 917	143	·444 229	1584	·230 304	1736	·912 477	119	51
10	0·409 392	266	0·448 719	349	1·096 060	144	2·442 645	1583	2·228 568	1735	0·912 358	119	50
11	·409 658	265	·449 068	350	·096 204	143	·441 062	1580	·226 833	1732	·912 239	119	49
12	·409 923	265	·449 418	350	·096 347	143	·439 482	1578	·225 101	1730	·912 120	119	48
13	·410 188	266	·449 768	349	·096 490	144	·437 904	1575	·223 371	1728	·912 001	120	47
14	·410 454	265	·450 117	350	·096 634	143	·436 329	1573	·221 643	1725	·911 881	119	46
15	·410 719	265	·450 467	350	·096 777	144	·434 756	1572	·219 918	1724	·911 762	119	45
16	·410 984	265	·450 817	350	·096 921	144	·433 184	1568	·218 194	1721	·911 643	120	44
17	·411 249	265	·451 167	350	·097 065	144	·431 616	1567	·216 473	1719	·911 523	120	43
18	·411 514	265	·451 517	351	·097 209	144	·430 049	1565	·214 754	1716	·911 403	119	42
19	·411 779	266	·451 868	350	·097 353	145	·428 484	1562	·213 038	1715	·911 284	120	41
20	0·412 045	265	0·452 218	350	1·097 498	144	2·426 922	1560	2·211 323	1712	0·911 164	120	40
21	·412 310	265	·452 568	351	·097 642	145	·425 362	1558	·209 611	1710	·911 044	120	39
22	·412 575	265	·452 919	350	·097 787	144	·423 804	1555	·207 901	1708	·910 924	120	38
23	·412 840	264	·453 269	351	·097 931	145	·422 249	1554	·206 193	1705	·910 804	120	37
24	·413 104	265	·453 620	351	·098 076	145	·420 695	1551	·204 488	1704	·910 684	121	36
25	·413 369	265	·453 971	351	·098 221	145	·419 144	1549	·202 784	1701	·910 563	120	35
26	·413 634	265	·454 322	351	·098 366	145	·417 595	1547	·201 083	1699	·910 443	120	34
27	·413 899	265	·454 673	351	·098 511	146	·416 048	1544	·199 384	1697	·910 323	121	33
28	·414 164	265	·455 024	351	·098 657	145	·414 504	1543	·197 687	1695	·910 202	120	32
29	·414 429	264	·455 375	351	·098 802	146	·412 961	1540	·195 992	1692	·910 082	121	31
30	0·414 693	265	0·455 726	352	1·098 948	146	2·411 421	1538	2·194 300	1691	0·909 961	120	30
31	·414 958	265	·456 078	351	·099 094	145	·409 883	1536	·192 609	1688	·909 841	121	29
32	·415 223	264	·456 429	352	·099 239	147	·408 347	1534	·190 921	1686	·909 720	121	28
33	·415 487	265	·456 781	351	·099 386	146	·406 813	1531	·189 235	1684	·909 599	121	27
34	·415 752	264	·457 132	352	·099 532	146	·405 282	1530	·187 551	1682	·909 478	121	26
35	·416 016	265	·457 484	352	·099 678	146	·403 752	1527	·185 869	1680	·909 357	121	25
36	·416 281	264	·457 836	352	·099 824	147	·402 225	1525	·184 189	1677	·909 236	121	24
37	·416 545	265	·458 188	352	·099 971	147	·400 700	1524	·182 512	1676	·909 115	121	23
38	·416 810	264	·458 540	352	·100 118	146	·399 176	1520	·180 836	1673	·908 994	122	22
39	·417 074	264	·458 892	352	·100 264	147	·397 656	1519	·179 163	1671	·908 872	121	21
40	0·417 338	265	0·459 244	352	1·100 411	147	2·396 137	1517	2·177 492	1669	0·908 751	121	20
41	·417 603	264	·459 596	353	·100 558	148	·394 620	1514	·175 823	1667	·908 630	122	19
42	·417 867	264	·459 949	352	·100 706	147	·393 106	1513	·174 156	1665	·908 508	121	18
43	·418 131	265	·460 301	353	·100 853	147	·391 593	1510	·172 491	1663	·908 387	122	17
44	·418 396	264	·460 654	352	·101 000	148	·390 083	1508	·170 828	1660	·908 265	122	16
45	·418 660	264	·461 006	353	·101 148	148	·388 575	1507	·169 168	1659	·908 143	122	15
46	·418 924	264	·461 359	353	·101 296	148	·387 068	1504	·167 509	1656	·908 021	122	14
47	·419 188	264	·461 712	353	·101 444	148	·385 564	1501	·165 853	1655	·907 899	122	13
48	·419 452	264	·462 065	353	·101 592	148	·384 063	1500	·164 198	1652	·907 777	122	12
49	·419 716	264	·462 418	353	·101 740	148	·382 563	1498	·162 546	1650	·907 655	122	11
50	0·419 980	264	0·462 771	353	1·101 888	148	2·381 065	1496	2·160 896	1648	0·907 533	122	10
51	·420 244	264	·463 124	354	·102 036	149	·379 569	1493	·159 248	1646	·907 411	122	09
52	·420 508	264	·463 478	353	·102 185	149	·378 076	1492	·157 602	1644	·907 289	123	08
53	·420 772	264	·463 831	354	·102 334	148	·376 584	1489	·155 958	1642	·907 166	122	07
54	·421 036	264	·464 185	353	·102 482	149	·375 095	1487	·154 316	1640	·907 044	122	06
55	·421 300	263	·464 538	354	·102 631	149	·373 608	1486	·152 676	1638	·906 922	123	05
56	·421 563	264	·464 892	354	·102 780	150	·372 122	1483	·151 038	1636	·906 799	123	04
57	·421 827	264	·465 246	354	·102 930	149	·370 639	1481	·149 402	1634	·906 676	122	03
58	·422 091	264	·465 600	354	·103 079	149	·369 158	1479	·147 768	1631	·906 554	123	02
59	·422 355	263	·465 954	354	·103 228	150	·367 679	1477	·146 137	1630	·906 431	123	01
60	0·422 618		0·466 308		1·103 378		2·366 202		2·144 507		0·906 308		00
	cos		cot		cosec		sec		tan		sin		65°

25° ′	sin		tan		sec		cosec		cot		cos		′
00	0·422 618	264	0·466 308	354	1·103 378	150	2·366 202	1475	2·144 507	1628	0·906 308	123	60
01	·422 882	263	·466 662	354	·103 528	150	·364 727	1473	·142 879	1625	·906 185	123	59
02	·423 145	264	·467 016	355	·103 678	150	·363 254	1471	·141 254	1624	·906 062	123	58
03	·423 409	264	·467 371	354	·103 828	150	·361 783	1469	·139 630	1621	·905 939	124	57
04	·423 673	263	·467 725	355	·103 978	150	·360 314	1467	·138 009	1620	·905 815	123	56
05	·423 936	263	·468 080	354	·104 128	150	·358 847	1465	·136 389	1618	·905 692	123	55
06	·424 199	264	·468 434	355	·104 278	151	·357 382	1463	·134 771	1615	·905 569	124	54
07	·424 463	263	·468 789	355	·104 429	151	·355 919	1461	·133 156	1614	·905 445	123	53
08	·424 726	264	·469 144	355	·104 580	150	·354 458	1459	·131 542	1611	·905 322	124	52
09	·424 990	263	·469 499	355	·104 730	151	·352 999	1457	·129 931	1610	·905 198	123	51
10	0·425 253	263	0·469 854	355	1·104 881	151	2·351 542	1454	2·128 321	1607	0·905 075	124	50
11	·425 516	263	·470 209	355	·105 032	152	·350 088	1453	·126 714	1606	·904 951	124	49
12	·425 779	263	·470 564	356	·105 184	151	·348 635	1451	·125 108	1603	·904 827	124	48
13	·426 042	264	·470 920	355	·105 335	151	·347 184	1449	·123 505	1602	·904 703	124	47
14	·426 306	263	·471 275	356	·105 486	152	·345 735	1447	·121 903	1600	·904 579	124	46
15	·426 569	263	·471 631	355	·105 638	152	·344 288	1445	·120 303	1597	·904 455	124	45
16	·426 832	263	·471 986	356	·105 790	152	·342 843	1443	·118 706	1596	·904 331	124	44
17	·427 095	263	·472 342	356	·105 942	152	·341 400	1441	·117 110	1594	·904 207	124	43
18	·427 358	263	·472 698	356	·106 094	152	·339 959	1439	·115 516	1591	·904 083	125	42
19	·427 621	263	·473 054	356	·106 246	152	·338 520	1437	·113 925	1590	·903 958	124	41
20	0·427 884	263	0·473 410	356	1·106 398	153	2·337 083	1435	2·112 335	1588	0·903 834	125	40
21	·428 147	263	·473 766	356	·106 551	152	·335 648	1433	·110 747	1586	·903 709	124	39
22	·428 410	262	·474 122	356	·106 703	153	·334 215	1431	·109 161	1584	·903 585	125	38
23	·428 672	263	·474 478	357	·106 856	153	·332 784	1429	·107 577	1582	·903 460	125	37
24	·428 935	263	·474 835	356	·107 009	153	·331 355	1427	·105 995	1580	·903 335	125	36
25	·429 198	263	·475 191	357	·107 162	153	·329 928	1426	·104 415	1578	·903 210	124	35
26	·429 461	262	·475 548	357	·107 315	153	·328 502	1423	·102 837	1576	·903 086	125	34
27	·429 723	263	·475 905	357	·107 468	153	·327 079	1421	·101 261	1575	·902 961	125	33
28	·429 986	263	·476 262	357	·107 621	154	·325 658	1420	·099 686	1572	·902 836	126	32
29	·430 249	262	·476 619	357	·107 775	154	·324 238	1418	·098 114	1570	·902 710	125	31
30	0·430 511	263	0·476 976	357	1·107 929	153	2·322 820	1415	2·096 544	1569	0·902 585	125	30
31	·430 774	262	·477 333	357	·108 082	154	·321 405	1414	·094 975	1567	·902 460	125	29
32	·431 036	263	·477 690	357	·108 236	154	·319 991	1412	·093 408	1564	·902 335	126	28
33	·431 299	262	·478 047	358	·108 390	155	·318 579	1410	·091 844	1563	·902 209	125	27
34	·431 561	262	·478 405	357	·108 545	154	·317 169	1408	·090 281	1561	·902 084	126	26
35	·431 823	263	·478 762	358	·108 699	154	·315 761	1406	·088 720	1559	·901 958	125	25
36	·432 086	262	·479 120	357	·108 853	155	·314 355	1404	·087 161	1557	·901 833	126	24
37	·432 348	262	·479 477	358	·109 008	155	·312 951	1402	·085 604	1555	·901 707	126	23
38	·432 610	263	·479 835	358	·109 163	155	·311 549	1400	·084 049	1554	·901 581	126	22
39	·432 873	262	·480 193	358	·109 318	155	·310 149	1399	·082 495	1551	·901 455	126	21
40	0·433 135	262	0·480 551	358	1·109 473	155	2·308 750	1396	2·080 944	1550	0·901 329	126	20
41	·433 397	262	·480 909	358	·109 628	155	·307 354	1395	·079 394	1547	·901 203	126	19
42	·433 659	262	·481 267	359	·109 783	155	·305 959	1393	·077 847	1546	·901 077	126	18
43	·433 921	262	·481 626	358	·109 938	156	·304 566	1391	·076 301	1544	·900 951	126	17
44	·434 183	262	·481 984	359	·110 094	156	·303 175	1389	·074 757	1542	·900 825	127	16
45	·434 445	262	·482 343	358	·110 250	156	·301 786	1387	·073 215	1541	·900 698	126	15
46	·434 707	262	·482 701	359	·110 406	156	·300 399	1386	·071 674	1538	·900 572	127	14
47	·434 969	262	·483 060	359	·110 562	156	·299 013	1383	·070 136	1537	·900 445	126	13
48	·435 231	262	·483 419	359	·110 718	156	·297 630	1382	·068 599	1534	·900 319	127	12
49	·435 493	262	·483 778	359	·110 874	156	·296 248	1379	·067 065	1533	·900 192	127	11
50	0·435 755	262	0·484 137	359	1·111 030	157	2·294 869	1378	2·065 532	1531	0·900 065	126	10
51	·436 017	261	·484 496	359	·111 187	157	·293 491	1376	·064 001	1529	·899 939	127	09
52	·436 278	262	·484 855	359	·111 344	156	·292 115	1375	·062 472	1528	·899 812	127	08
53	·436 540	262	·485 214	360	·111 500	157	·290 740	1372	·060 944	1525	·899 685	127	07
54	·436 802	261	·485 574	359	·111 657	157	·289 368	1371	·059 419	1524	·899 558	127	06
55	·437 063	262	·485 933	360	·111 814	158	·287 997	1368	·057 895	1522	·899 431	127	05
56	·437 325	262	·486 293	360	·111 972	157	·286 629	1367	·056 373	1520	·899 304	128	04
57	·437 587	261	·486 653	360	·112 129	158	·285 262	1365	·054 853	1518	·899 176	127	03
58	·437 848	262	·487 013	360	·112 287	157	·283 897	1364	·053 335	1517	·899 049	127	02
59	·438 110	261	·487 373	360	·112 444	158	·282 533	1361	·051 818	1514	·898 922	128	01
60	0·438 371		0·487 733		1·112 602		2·281 172		2·050 304		0·898 794		00
	cos		cot		cosec		sec		tan		sin		64°

26° ′	sin		tan		sec		cosec		cot		cos		′
00	0·438 371	262	0·487 733	360	1·112 602	158	2·281 172	1360	2·050 304	1513	0·898 794	128	60
01	·438 633	261	·488 093	360	·112 760	158	·279 812	1357	·048 791	1511	·898 666	127	59
02	·438 894	261	·488 453	360	·112 918	158	·278 455	1356	·047 280	1509	·898 539	128	58
03	·439 155	262	·488 813	361	·113 076	158	·277 099	1355	·045 771	1508	·898 411	128	57
04	·439 417	261	·489 174	360	·113 234	159	·275 744	1352	·044 263	1505	·898 283	127	56
05	·439 678	261	·489 534	361	·113 393	159	·274 392	1350	·042 758	1504	·898 156	128	55
06	·439 939	261	·489 895	361	·113 552	158	·273 042	1349	·041 254	1502	·898 028	128	54
07	·440 200	262	·490 256	361	·113 710	159	·271 693	1347	·039 752	1500	·897 900	129	53
08	·440 462	261	·490 617	361	·113 869	159	·270 346	1345	·038 252	1499	·897 771	128	52
09	·440 723	261	·490 978	361	·114 028	159	·269 001	1344	·036 753	1497	·897 643	128	51
10	0·440 984	261	0·491 339	361	1·114 187	160	2·267 657	1342	2·035 256	1494	0·897 515	128	50
11	·441 245	261	·491 700	361	·114 347	159	·266 315	1339	·033 762	1494	·897 387	129	49
12	·441 506	261	·492 061	361	·114 506	160	·264 976	1338	·032 268	1491	·897 258	128	48
13	·441 767	261	·492 422	362	·114 666	160	·263 638	1337	·030 777	1490	·897 130	129	47
14	·442 028	261	·492 784	361	·114 826	159	·262 301	1334	·029 287	1488	·897 001	128	46
15	·442 289	261	·493 145	362	·114 985	160	·260 967	1333	·027 799	1486	·896 873	129	45
16	·442 550	260	·493 507	362	·115 145	161	·259 634	1331	·026 313	1484	·896 744	129	44
17	·442 810	261	·493 869	362	·115 306	160	·258 303	1329	·024 829	1483	·896 615	129	43
18	·443 071	261	·494 231	362	·115 466	160	·256 974	1328	·023 346	1481	·896 486	128	42
19	·443 332	261	·494 593	362	·115 626	161	·255 646	1326	·021 865	1479	·896 358	129	41
20	0·443 593	260	0·494 955	362	1·115 787	161	2·254 320	1324	2·020 386	1477	0·896 229	130	40
21	·443 853	261	·495 317	362	·115 948	160	·252 996	1322	·018 909	1476	·896 099	129	39
22	·444 114	261	·495 679	363	·116 108	161	·251 674	1320	·017 433	1474	·895 970	129	38
23	·444 375	260	·496 042	362	·116 269	162	·250 354	1319	·015 959	1472	·895 841	129	37
24	·444 635	261	·496 404	363	·116 431	161	·249 035	1317	·014 487	1471	·895 712	130	36
25	·444 896	260	·496 767	363	·116 592	161	·247 718	1316	·013 016	1468	·895 582	129	35
26	·445 156	261	·497 130	362	·116 753	162	·246 402	1313	·011 548	1467	·895 453	130	34
27	·445 417	260	·497 492	363	·116 915	162	·245 089	1312	·010 081	1466	·895 323	129	33
28	·445 677	260	·497 855	363	·117 077	161	·243 777	1310	·008 615	1463	·895 194	130	32
29	·445 937	261	·498 218	364	·117 238	162	·242 467	1309	·007 152	1462	·895 064	130	31
30	0·446 198	260	0·498 582	363	1·117 400	163	2·241 158	1306	2·005 690	1461	0·894 934	129	30
31	·446 458	260	·498 945	363	·117 563	162	·239 852	1305	·004 229	1458	·894 805	130	29
32	·446 718	261	·499 308	364	·117 725	162	·238 547	1304	·002 771	1457	·894 675	130	28
33	·446 979	260	·499 672	363	·117 887	163	·237 243	1301	2·001 314	1455	·894 545	130	27
34	·447 239	260	·500 035	364	·118 050	162	·235 942	1300	1·999 859	1453	·894 415	131	26
35	·447 499	260	·500 399	364	·118 212	163	·234 642	1298	·998 406	1452	·894 284	130	25
36	·447 759	260	·500 763	364	·118 375	163	·233 344	1297	·996 954	1450	·894 154	130	24
37	·448 019	260	·501 127	364	·118 538	163	·232 047	1294	·995 504	1449	·894 024	130	23
38	·448 279	260	·501 491	364	·118 701	164	·230 753	1294	·994 055	1446	·893 894	131	22
39	·448 539	260	·501 855	364	·118 865	163	·229 459	1291	·992 609	1445	·893 763	130	21
40	0·448 799	260	0·502 219	364	1·119 028	164	2·228 168	1290	1·991 164	1444	0·893 633	131	20
41	·449 059	260	·502 583	365	·119 192	163	·226 878	1288	·989 720	1441	·893 502	131	19
42	·449 319	260	·502 948	364	·119 355	164	·225 590	1286	·988 279	1440	·893 371	130	18
43	·449 579	260	·503 312	365	·119 519	164	·224 304	1285	·986 839	1439	·893 241	131	17
44	·449 839	259	·503 677	364	·119 683	164	·223 019	1283	·985 400	1436	·893 110	131	16
45	·450 098	260	·504 041	365	·119 847	164	·221 736	1281	·983 964	1435	·892 979	131	15
46	·450 358	260	·504 406	365	·120 011	165	·220 455	1280	·982 529	1434	·892 848	131	14
47	·450 618	260	·504 771	365	·120 176	164	·219 175	1278	·981 095	1431	·892 717	131	13
48	·450 878	259	·505 136	366	·120 340	165	·217 897	1276	·979 664	1431	·892 586	131	12
49	·451 137	260	·505 502	365	·120 505	165	·216 621	1275	·978 233	1428	·892 455	132	11
50	0·451 397	259	0·505 867	365	1·120 670	165	2·215 346	1273	1·976 805	1427	0·892 323	131	10
51	·451 656	260	·506 232	366	·120 835	165	·214 073	1271	·975 378	1425	·892 192	131	09
52	·451 916	259	·506 598	365	·121 000	165	·212 802	1270	·973 953	1423	·892 061	132	08
53	·452 175	260	·506 963	366	·121 165	166	·211 532	1268	·972 530	1422	·891 929	131	07
54	·452 435	259	·507 329	366	·121 331	165	·210 264	1267	·971 108	1421	·891 798	132	06
55	·452 694	259	·507 695	366	·121 496	166	·208 997	1265	·969 687	1418	·891 666	132	05
56	·452 953	260	·508 061	366	·121 662	166	·207 732	1263	·968 269	1417	·891 534	132	04
57	·453 213	259	·508 427	366	·121 828	166	·206 469	1261	·966 852	1416	·891 402	132	03
58	·453 472	259	·508 793	366	·121 994	166	·205 208	1260	·965 436	1413	·891 270	131	02
59	·453 731	259	·509 159	366	·122 160	166	·203 948	1259	·964 023	1412	·891 139	132	01
60	0·453 990		0·509 525		1·122 326		2·202 689		1·962 611		0·891 007		00
	cos		cot		cosec		sec		tan		sin		63°

27° ′	sin		tan		sec		cosec		cot		cos		′
00	0·453 990	260	0·509 525	367	1·122 326	167	2·202 689	1256	1·962 611	1411	0·891 007	133	60
01	·454 250	259	·509 892	366	·122 493	166	·201 433	1256	·961 200	1409	·890 874	132	59
02	·454 509	259	·510 258	367	·122 659	167	·200 177	1253	·959 791	1407	·890 742	132	58
03	·454 768	259	·510 625	367	·122 826	167	·198 924	1252	·958 384	1406	·890 610	132	57
04	·455 027	259	·510 992	367	·122 993	167	·197 672	1250	·956 978	1404	·890 478	133	56
05	·455 286	259	·511 359	367	·123 160	167	·196 422	1249	·955 574	1403	·890 345	132	55
06	·455 545	259	·511 726	367	·123 327	167	·195 173	1247	·954 171	1401	·890 213	133	54
07	·455 804	259	·512 093	367	·123 494	168	·193 926	1245	·952 770	1399	·890 080	132	53
08	·456 063	259	·512 460	368	·123 662	167	·192 681	1244	·951 371	1398	·889 948	133	52
09	·456 322	258	·512 828	367	·123 829	168	·191 437	1242	·949 973	1396	·889 815	133	51
10	0·456 580	259	0·513 195	368	1·123 997	168	2·190 195	1241	1·948 577	1394	0·889 682	133	50
11	·456 839	259	·513 563	367	·124 165	168	·188 954	1239	·947 183	1393	·889 549	133	49
12	·457 098	259	·513 930	368	·124 333	168	·187 715	1237	·945 790	1392	·889 416	133	48
13	·457 357	258	·514 298	368	·124 501	168	·186 478	1236	·944 398	1390	·889 283	133	47
14	·457 615	259	·514 666	368	·124 669	169	·185 242	1235	·943 008	1388	·889 150	133	46
15	·457 874	259	·515 034	368	·124 838	168	·184 007	1232	·941 620	1387	·889 017	133	45
16	·458 133	258	·515 402	368	·125 006	169	·182 775	1232	·940 233	1385	·888 884	133	44
17	·458 391	259	·515 770	368	·125 175	169	·181 543	1229	·938 848	1383	·888 751	134	43
18	·458 650	258	·516 138	369	·125 344	169	·180 314	1228	·937 465	1383	·888 617	133	42
19	·458 908	258	·516 507	368	·125 513	169	·179 086	1227	·936 082	1380	·888 484	134	41
20	0·459 166	259	0·516 875	369	1·125 682	169	2·177 859	1224	1·934 702	1379	0·888 350	133	40
21	·459 425	258	·517 244	369	·125 851	170	·176 635	1224	·933 323	1377	·888 217	134	39
22	·459 683	259	·517 613	369	·126 021	170	·175 411	1222	·931 946	1376	·888 083	134	38
23	·459 942	258	·517 982	369	·126 191	169	·174 189	1220	·930 570	1374	·887 949	134	37
24	·460 200	258	·518 351	369	·126 360	170	·172 969	1218	·929 196	1373	·887 815	134	36
25	·460 458	258	·518 720	369	·126 530	170	·171 751	1217	·927 823	1371	·887 681	133	35
26	·460 716	258	·519 089	369	·126 700	170	·170 534	1216	·926 452	1370	·887 548	135	34
27	·460 974	258	·519 458	370	·126 870	171	·169 318	1214	·925 082	1368	·887 413	134	33
28	·461 232	259	·519 828	369	·127 041	170	·168 104	1212	·923 714	1367	·887 279	134	32
29	·461 491	258	·520 197	370	·127 211	171	·166 892	1211	·922 347	1365	·887 145	134	31
30	0·461 749	258	0·520 567	370	1·127 382	171	2·165 681	1210	1·920 982	1363	0·887 011	135	30
31	·462 007	258	·520 937	370	·127 553	171	·164 471	1208	·919 619	1362	·886 876	134	29
32	·462 265	258	·521 307	370	·127 724	171	·163 263	1206	·918 257	1361	·886 742	134	28
33	·462 523	257	·521 677	370	·127 895	171	·162 057	1205	·916 896	1359	·886 608	135	27
34	·462 780	258	·522 047	370	·128 066	171	·160 852	1203	·915 537	1357	·886 473	135	26
35	·463 038	258	·522 417	370	·128 237	172	·159 649	1202	·914 180	1356	·886 338	134	25
36	·463 296	258	·522 787	371	·128 409	172	·158 447	1200	·912 824	1355	·886 204	135	24
37	·463 554	258	·523 158	370	·128 581	171	·157 247	1199	·911 469	1353	·886 069	135	23
38	·463 812	257	·523 528	371	·128 752	172	·156 048	1197	·910 116	1351	·885 934	135	22
39	·464 069	258	·523 899	371	·128 924	172	·154 851	1196	·908 765	1350	·885 799	135	21
40	0·464 327	257	0·524 270	371	1·129 096	173	2·153 655	1194	1·907 415	1349	0·885 664	135	20
41	·464 584	258	·524 641	371	·129 269	172	·152 461	1193	·906 066	1347	·885 529	135	19
42	·464 842	258	·525 012	371	·129 441	173	·151 268	1191	·904 719	1345	·885 394	136	18
43	·465 100	257	·525 383	371	·129 614	172	·150 077	1189	·903 374	1344	·885 258	135	17
44	·465 357	258	·525 754	371	·129 786	173	·148 888	1189	·902 030	1343	·885 123	135	16
45	·465 615	257	·526 125	372	·129 959	173	·147 699	1186	·900 687	1341	·884 988	136	15
46	·465 872	257	·526 497	371	·130 132	173	·146 513	1186	·899 346	1339	·884 852	135	14
47	·466 129	258	·526 868	372	·130 305	174	·145 327	1183	·898 007	1338	·884 717	136	13
48	·466 387	257	·527 240	372	·130 479	173	·144 144	1182	·896 669	1337	·884 581	136	12
49	·466 644	257	·527 612	372	·130 652	174	·142 962	1181	·895 332	1335	·884 445	136	11
50	0·466 901	257	0·527 984	372	1·130 826	174	2·141 781	1179	1·893 997	1334	0·884 309	135	10
51	·467 158	258	·528 356	372	·131 000	173	·140 602	1178	·892 663	1332	·884 174	136	09
52	·467 416	257	·528 728	372	·131 173	175	·139 424	1177	·891 331	1330	·884 038	136	08
53	·467 673	257	·529 100	373	·131 348	174	·138 247	1174	·890 001	1330	·883 902	136	07
54	·467 930	257	·529 473	372	·131 522	174	·137 073	1174	·888 671	1327	·883 766	137	06
55	·468 187	257	·529 845	373	·131 696	175	·135 899	1172	·887 344	1327	·883 629	136	05
56	·468 444	257	·530 218	373	·131 871	174	·134 727	1170	·886 017	1325	·883 493	136	04
57	·468 701	257	·530 591	372	·132 045	175	·133 557	1169	·884 692	1323	·883 357	136	03
58	·468 958	257	·530 963	373	·132 220	175	·132 388	1167	·883 369	1322	·883 221	137	02
59	·469 215	257	·531 336	373	·132 395	175	·131 221	1167	·882 047	1321	·883 084	136	01
60	0·469 472		0·531 709		1·132 570		2·130 054		1·880 726		0·882 948		00
	cos		cot		cosec		sec		tan		sin		62°

28° ′	sin		tan		sec		cosec		cot		cos		′
00	0·469 472	256	0·531 709	374	1·132 570	175	2·130 054	1164	1·880 726	1319	0·882 948	137	60
01	·469 728	257	·532 083	373	·132 745	176	·128 890	1163	·879 407	1317	·882 811	137	59
02	·469 985	257	·532 456	373	·132 921	175	·127 727	1162	·878 090	1316	·882 674	136	58
03	·470 242	257	·532 829	374	·133 096	176	·126 565	1160	·876 774	1315	·882 538	137	57
04	·470 499	256	·533 203	374	·133 272	176	·125 405	1159	·875 459	1314	·882 401	137	56
05	·470 755	257	·533 577	373	·133 448	176	·124 246	1157	·874 145	1311	·882 264	137	55
06	·471 012	256	·533 950	374	·133 624	176	·123 089	1156	·872 834	1311	·882 127	137	54
07	·471 268	257	·534 324	374	·133 800	176	·121 933	1155	·871 523	1309	·881 990	137	53
08	·471 525	257	·534 698	374	·133 976	177	·120 778	1153	·870 214	1308	·881 853	138	52
09	·471 782	256	·535 072	374	·134 153	176	·119 625	1151	·868 906	1306	·881 715	137	51
10	0·472 038	256	0·535 446	375	1·134 329	177	2·118 474	1150	1·867 600	1305	0·881 578	137	50
11	·472 294	257	·535 821	374	·134 506	177	·117 324	1149	·866 295	1303	·881 441	138	49
12	·472 551	256	·536 195	375	·134 683	177	·116 175	1148	·864 992	1302	·881 303	137	48
13	·472 807	256	·536 570	375	·134 860	177	·115 027	1145	·863 690	1300	·881 166	138	47
14	·473 063	257	·536 945	374	·135 037	178	·113 882	1145	·862 390	1299	·881 028	137	46
15	·473 320	256	·537 319	375	·135 215	177	·112 737	1143	·861 091	1298	·880 891	138	45
16	·473 576	256	·537 694	375	·135 392	178	·111 594	1142	·859 793	1297	·880 753	138	44
17	·473 832	256	·538 069	376	·135 570	178	·110 452	1140	·858 496	1294	·880 615	138	43
18	·474 088	256	·538 445	375	·135 748	178	·109 312	1139	·857 202	1294	·880 477	138	42
19	·474 344	256	·538 820	375	·135 926	178	·108 173	1137	·855 908	1292	·880 339	138	41
20	0·474 600	256	0·539 195	376	1·136 104	178	2·107 036	1136	1·854 616	1291	0·880 201	138	40
21	·474 856	256	·539 571	375	·136 282	178	·105 900	1135	·853 325	1289	·880 063	138	39
22	·475 112	256	·539 946	376	·136 460	179	·104 765	1133	·852 036	1288	·879 925	138	38
23	·475 368	256	·540 322	376	·136 639	179	·103 632	1132	·850 748	1287	·879 787	138	37
24	·475 624	256	·540 698	376	·136 818	179	·102 500	1130	·849 461	1285	·879 649	139	36
25	·475 880	256	·541 074	376	·136 997	179	·101 370	1129	·848 176	1284	·879 510	138	35
26	·476 136	256	·541 450	376	·137 176	179	·100 241	1128	·846 892	1282	·879 372	139	34
27	·476 392	255	·541 826	377	·137 355	179	·099 113	1126	·845 610	1281	·879 233	138	33
28	·476 647	256	·542 203	376	·137 534	180	·097 987	1125	·844 329	1280	·879 095	139	32
29	·476 903	256	·542 579	377	·137 714	179	·096 862	1123	·843 049	1278	·878 956	139	31
30	0·477 159	255	0·542 956	376	1·137 893	180	2·095 739	1123	1·841 771	1277	0·878 817	139	30
31	·477 414	256	·543 332	377	·138 073	180	·094 616	1120	·840 494	1276	·878 678	139	29
32	·477 670	255	·543 709	377	·138 253	180	·093 496	1120	·839 218	1274	·878 539	139	28
33	·477 925	256	·544 086	377	·138 433	180	·092 376	1118	·837 944	1273	·878 400	139	27
34	·478 181	255	·544 463	377	·138 613	181	·091 258	1116	·836 671	1271	·878 261	139	26
35	·478 436	256	·544 840	378	·138 794	180	·090 142	1115	·835 400	1270	·878 122	139	25
36	·478 692	255	·545 218	377	·138 974	181	·089 027	1114	·834 130	1269	·877 983	139	24
37	·478 947	256	·545 595	378	·139 155	181	·087 913	1113	·832 861	1267	·877 844	140	23
38	·479 203	255	·545 973	377	·139 336	181	·086 800	1111	·831 594	1267	·877 704	139	22
39	·479 458	255	·546 350	378	·139 517	181	·085 689	1110	·830 327	1264	·877 565	140	21
40	0·479 713	255	0·546 728	378	1·139 698	181	2·084 579	1108	1·829 063	1264	0·877 425	139	20
41	·479 968	255	·547 106	378	·139 879	182	·083 471	1107	·827 799	1262	·877 286	140	19
42	·480 223	256	·547 484	378	·140 061	181	·082 364	1106	·826 537	1260	·877 146	140	18
43	·480 479	255	·547 862	378	·140 242	182	·081 258	1104	·825 277	1260	·877 006	139	17
44	·480 734	255	·548 240	379	·140 424	182	·080 154	1103	·824 017	1258	·876 867	140	16
45	·480 989	255	·548 619	378	·140 606	182	·079 051	1102	·822 759	1256	·876 727	140	15
46	·481 244	255	·548 997	379	·140 788	183	·077 949	1100	·821 503	1256	·876 587	140	14
47	·481 499	255	·549 376	379	·140 971	182	·076 849	1099	·820 247	1254	·876 447	140	13
48	·481 754	255	·549 755	379	·141 153	183	·075 750	1098	·818 993	1252	·876 307	140	12
49	·482 009	254	·550 134	379	·141 336	182	·074 652	1096	·817 741	1252	·876 167	141	11
50	0·482 263	255	0·550 513	379	1·141 518	183	2·073 556	1095	1·816 489	1250	0·876 026	140	10
51	·482 518	255	·550 892	379	·141 701	183	·072 461	1094	·815 239	1249	·875 886	140	09
52	·482 773	255	·551 271	379	·141 884	183	·071 367	1092	·813 990	1247	·875 746	141	08
53	·483 028	254	·551 650	380	·142 067	184	·070 275	1091	·812 743	1246	·875 605	140	07
54	·483 282	255	·552 030	379	·142 251	183	·069 184	1090	·811 497	1245	·875 465	141	06
55	·483 537	255	·552 409	380	·142 434	184	·068 094	1088	·810 252	1243	·875 324	141	05
56	·483 792	254	·552 789	380	·142 618	184	·067 006	1087	·809 009	1243	·875 183	141	04
57	·484 046	255	·553 169	380	·142 802	184	·065 919	1086	·807 766	1240	·875 042	140	03
58	·484 301	254	·553 549	380	·142 986	184	·064 833	1085	·806 526	1240	·874 902	141	02
59	·484 555	255	·553 929	380	·143 170	184	·063 748	1083	·805 286	1238	·874 761	141	01
60	0·484 810		0·554 309		1·143 354		2·062 665		1·804 048		0·874 620		00
	cos		cot		cosec		sec		tan		sin		61°

29° ′	sin		tan		sec		cosec		cot		cos		′
00	0·484 810	254	0·554 309	380	1·143 354	185	2·062 665	1081	1·804 048	1237	0·874 620	141	60
01	·485 064	254	·554 689	381	·143 539	184	·061 584	1081	·802 811	1236	·874 479	141	59
02	·485 318	255	·555 070	380	·143 723	185	·060 503	1079	·801 575	1234	·874 338	142	58
03	·485 573	254	·555 450	381	·143 908	185	·059 424	1078	·800 341	1233	·874 196	141	57
04	·485 827	254	·555 831	381	·144 093	185	·058 346	1077	·799 108	1232	·874 055	141	56
05	·486 081	254	·556 212	381	·144 278	185	·057 269	1075	·797 876	1231	·873 914	142	55
06	·486 335	255	·556 593	381	·144 463	185	·056 194	1074	·796 645	1229	·873 772	141	54
07	·486 590	254	·556 974	381	·144 648	186	·055 120	1072	·795 416	1228	·873 631	142	53
08	·486 844	254	·557 355	381	·144 834	186	·054 048	1072	·794 188	1226	·873 489	142	52
09	·487 098	254	·557 736	382	·145 020	185	·052 976	1070	·792 962	1226	·873 347	141	51
10	0·487 352	254	0·558 118	381	1·145 205	186	2·051 906	1069	1·791 736	1224	0·873 206	142	50
11	·487 606	254	·558 499	382	·145 391	187	·050 837	1067	·790 512	1223	·873 064	142	49
12	·487 860	254	·558 881	382	·145 578	186	·049 770	1066	·789 289	1221	·872 922	142	48
13	·488 114	253	·559 263	382	·145 764	186	·048 704	1065	·788 068	1221	·872 780	142	47
14	·488 367	254	·559 645	382	·145 950	187	·047 639	1064	·786 847	1219	·872 638	142	46
15	·488 621	254	·560 027	382	·146 137	187	·046 575	1062	·785 628	1217	·872 496	142	45
16	·488 875	254	·560 409	382	·146 324	187	·045 513	1062	·784 411	1217	·872 354	142	44
17	·489 129	253	·560 791	383	·146 511	187	·044 451	1059	·783 194	1215	·872 212	143	43
18	·489 382	254	·561 174	382	·146 698	187	·043 392	1059	·781 979	1214	·872 069	142	42
19	·489 636	254	·561 556	383	·146 885	188	·042 333	1057	·780 765	1213	·871 927	143	41
20	0·489 890	253	0·561 939	383	1·147 073	187	2·041 276	1056	1·779 552	1211	0·871 784	142	40
21	·490 143	254	·562 322	383	·147 260	188	·040 220	1055	·778 341	1210	·871 642	143	39
22	·490 397	253	·562 705	383	·147 448	188	·039 165	1054	·777 131	1209	·871 499	142	38
23	·490 650	254	·563 088	383	·147 636	188	·038 111	1052	·775 922	1208	·871 357	143	37
24	·490 904	253	·563 471	383	·147 824	188	·037 059	1051	·774 714	1206	·871 214	143	36
25	·491 157	254	·563 854	384	·148 012	188	·036 008	1050	·773 508	1206	·871 071	143	35
26	·491 411	253	·564 238	383	·148 200	189	·034 958	1048	·772 302	1204	·870 928	143	34
27	·491 664	253	·564 621	384	·148 389	189	·033 910	1047	·771 098	1202	·870 785	143	33
28	·491 917	253	·565 005	384	·148 578	189	·032 863	1046	·769 896	1202	·870 642	143	32
29	·492 170	254	·565 389	384	·148 767	189	·031 817	1045	·768 694	1200	·870 499	143	31
30	0·492 424	253	0·565 773	384	1·148 956	189	2·030 772	1043	1·767 494	1199	0·870 356	144	30
31	·492 677	253	·566 157	384	·149 145	189	·029 729	1043	·766 295	1198	·870 212	143	29
32	·492 930	253	·566 541	384	·149 334	190	·028 686	1041	·765 097	1196	·870 069	143	28
33	·493 183	253	·566 925	385	·149 524	189	·027 645	1039	·763 901	1196	·869 926	144	27
34	·493 436	253	·567 310	384	·149 713	190	·026 606	1039	·762 705	1194	·869 782	143	26
35	·493 689	253	·567 694	385	·149 903	190	·025 567	1037	·761 511	1193	·869 639	144	25
36	·493 942	253	·568 079	385	·150 093	190	·024 530	1036	·760 318	1191	·869 495	144	24
37	·494 195	253	·568 464	385	·150 283	190	·023 494	1035	·759 127	1191	·869 351	144	23
38	·494 448	252	·568 849	385	·150 473	191	·022 459	1034	·757 936	1189	·869 207	143	22
39	·494 700	253	·569 234	385	·150 664	190	·021 425	1032	·756 747	1188	·869 064	144	21
40	0·494 953	253	0·569 619	385	1·150 854	191	2·020 393	1031	1·755 559	1187	0·868 920	144	20
41	·495 206	253	·570 004	386	·151 045	191	·019 362	1030	·754 372	1185	·868 776	144	19
42	·495 459	252	·570 390	386	·151 236	191	·018 332	1029	·753 187	1185	·868 632	145	18
43	·495 711	253	·570 776	385	·151 427	191	·017 303	1027	·752 002	1183	·868 487	144	17
44	·495 964	253	·571 161	386	·151 618	192	·016 276	1027	·750 819	1182	·868 343	144	16
45	·496 217	252	·571 547	386	·151 810	191	·015 249	1025	·749 637	1181	·868 199	145	15
46	·496 469	253	·571 933	386	·152 001	192	·014 224	1024	·748 456	1179	·868 054	144	14
47	·496 722	252	·572 319	386	·152 193	192	·013 200	1022	·747 277	1179	·867 910	145	13
48	·496 974	252	·572 705	387	·152 385	192	·012 178	1022	·746 098	1177	·867 765	144	12
49	·497 226	253	·573 092	386	·152 577	192	·011 156	1020	·744 921	1176	·867 621	145	11
50	0·497 479	252	0·573 478	387	1·152 769	193	2·010 136	1019	1·743 745	1174	0·867 476	145	10
51	·497 731	252	·573 865	387	·152 962	192	·009 117	1018	·742 571	1174	·867 331	144	09
52	·497 983	253	·574 252	386	·153 154	193	·008 099	1016	·741 397	1172	·867 187	145	08
53	·498 236	252	·574 638	388	·153 347	193	·007 083	1016	·740 225	1172	·867 042	145	07
54	·498 488	252	·575 026	387	·153 540	193	·006 067	1014	·739 053	1170	·866 897	145	06
55	·498 740	252	·575 413	387	·153 733	193	·005 053	1013	·737 883	1169	·866 752	145	05
56	·498 992	252	·575 800	387	·153 926	193	·004 040	1012	·736 714	1167	·866 607	146	04
57	·499 244	252	·576 187	388	·154 119	194	·003 028	1010	·735 547	1167	·866 461	145	03
58	·499 496	252	·576 575	387	·154 313	194	·002 018	1010	·734 380	1165	·866 316	145	02
59	·499 748	252	·576 962	388	·154 507	194	·001 008	1008	·733 215	1164	·866 171	146	01
60	0·500 000		0·577 350		1·154 701		2·000 000		1·732 051		0·866 025		00
	cos		cot		cosec		sec		tan		sin		60°

TABLE IIIB—TRIGONOMETRICAL FUNCTIONS (Degrees and Minutes)

30° ′	sin		tan		sec		cosec		cot		cos		′
00	0·500 000	252	0·577 350	388	1·154 701	194	2·000 000	1007	1·732 051	1163	0·866 025	145	60
01	·500 252	252	·577 738	388	·154 895	194	1·998 993	1006	·730 888	1162	·865 880	146	59
02	·500 504	252	·578 126	388	·155 089	194	·997 987	1005	·729 726	1161	·865 734	145	58
03	·500 756	251	·578 514	389	·155 283	195	·996 982	1003	·728 565	1159	·865 589	146	57
04	·501 007	252	·578 903	388	·155 478	194	·995 979	1003	·727 406	1158	·865 443	146	56
05	·501 259	252	·579 291	389	·155 672	195	·994 976	1001	·726 248	1157	·865 297	146	55
06	·501 511	251	·579 680	388	·155 867	195	·993 975	1000	·725 091	1156	·865 151	145	54
07	·501 762	252	·580 068	389	·156 062	195	·992 975	999	·723 935	1155	·865 006	146	53
08	·502 014	252	·580 457	389	·156 257	195	·991 976	997	·722 780	1154	·864 860	147	52
09	·502 266	251	·580 846	389	·156 452	196	·990 979	997	·721 626	1152	·864 713	146	51
10	0·502 517	252	0·581 235	390	1·156 648	196	1·989 982	995	1·720 474	1152	0·864 567	146	50
11	·502 769	251	·581 625	389	·156 844	195	·988 987	994	·719 322	1150	·864 421	146	49
12	·503 020	251	·582 014	389	·157 039	196	·987 993	993	·718 172	1149	·864 275	147	48
13	·503 271	252	·582 403	390	·157 235	197	·987 000	992	·717 023	1148	·864 128	146	47
14	·503 523	251	·582 793	390	·157 432	196	·986 008	991	·715 875	1147	·863 982	146	46
15	·503 774	251	·583 183	390	·157 628	196	·985 017	989	·714 728	1145	·863 836	147	45
16	·504 025	251	·583 573	390	·157 824	197	·984 028	989	·713 583	1145	·863 689	147	44
17	·504 276	252	·583 963	390	·158 021	197	·983 039	987	·712 438	1143	·863 542	146	43
18	·504 528	251	·584 353	390	·158 218	197	·982 052	986	·711 295	1142	·863 396	147	42
19	·504 779	251	·584 743	391	·158 415	197	·981 066	985	·710 153	1141	·863 249	147	41
20	0·505 030	251	0·585 134	390	1·158 612	197	1·980 081	984	1·709 012	1140	0·863 102	147	40
21	·505 281	251	·585 524	391	·158 809	198	·979 097	982	·707 872	1139	·862 955	147	39
22	·505 532	251	·585 915	391	·159 007	197	·978 115	982	·706 733	1138	·862 808	147	38
23	·505 783	251	·586 306	391	·159 204	198	·977 133	980	·705 595	1136	·862 661	147	37
24	·506 034	251	·586 697	391	·159 402	198	·976 153	979	·704 459	1136	·862 514	148	36
25	·506 285	250	·587 088	391	·159 600	198	·975 174	979	·703 323	1134	·862 366	147	35
26	·506 535	251	·587 479	391	·159 798	198	·974 195	977	·702 189	1133	·862 219	147	34
27	·506 786	251	·587 870	392	·159 996	199	·973 218	975	·701 056	1132	·862 072	148	33
28	·507 037	251	·588 262	391	·160 195	198	·972 243	975	·699 924	1131	·861 924	147	32
29	·507 288	250	·588 653	392	·160 393	199	·971 268	974	·698 793	1130	·861 777	148	31
30	0·507 538	251	0·589 045	392	1·160 592	199	1·970 294	972	1·697 663	1129	0·861 629	148	30
31	·507 789	251	·589 437	392	·160 791	199	·969 322	971	·696 534	1127	·861 481	147	29
32	·508 040	250	·589 829	392	·160 990	199	·968 351	970	·695 407	1127	·861 334	148	28
33	·508 290	251	·590 221	392	·161 189	200	·967 381	970	·694 280	1125	·861 186	148	27
34	·508 541	250	·590 613	393	·161 389	200	·966 411	967	·693 155	1124	·861 038	148	26
35	·508 791	250	·591 006	392	·161 589	199	·965 444	967	·692 031	1123	·860 890	148	25
36	·509 041	251	·591 398	393	·161 788	200	·964 477	966	·690 908	1122	·860 742	148	24
37	·509 292	250	·591 791	393	·161 988	200	·963 511	965	·689 786	1121	·860 594	148	23
38	·509 542	250	·592 184	393	·162 188	201	·962 546	963	·688 665	1120	·860 446	149	22
39	·509 792	251	·592 577	393	·162 389	200	·961 583	962	·687 545	1119	·860 297	148	21
40	0·510 043	250	0·592 970	393	1·162 589	201	1·960 621	962	1·686 426	1118	0·860 149	148	20
41	·510 293	250	·593 363	394	·162 790	200	·959 659	960	·685 308	1116	·860 001	149	19
42	·510 543	250	·593 757	393	·162 990	201	·958 699	959	·684 192	1115	·859 852	148	18
43	·510 793	250	·594 150	394	·163 191	202	·957 740	958	·683 077	1115	·859 704	149	17
44	·511 043	250	·594 544	393	·163 393	201	·956 782	957	·681 962	1113	·859 555	149	16
45	·511 293	250	·594 937	394	·163 594	201	·955 825	955	·680 849	1112	·859 406	148	15
46	·511 543	250	·595 331	394	·163 795	202	·954 870	955	·679 737	1111	·859 258	149	14
47	·511 793	250	·595 725	395	·163 997	202	·953 915	954	·678 626	1110	·859 109	149	13
48	·512 043	250	·596 120	394	·164 199	202	·952 961	952	·677 516	1109	·858 960	149	12
49	·512 293	250	·596 514	394	·164 401	202	·952 009	951	·676 407	1108	·858 811	149	11
50	0·512 543	249	0·596 908	395	1·164 603	202	1·951 058	951	1·675 299	1107	0·858 662	149	10
51	·512 792	250	·597 303	395	·164 805	203	·950 107	949	·674 192	1106	·858 513	149	09
52	·513 042	250	·597 698	395	·165 008	202	·949 158	948	·673 086	1104	·858 364	150	08
53	·513 292	249	·598 093	395	·165 210	203	·948 210	947	·671 982	1104	·858 214	149	07
54	·513 541	250	·598 488	395	·165 413	203	·947 263	946	·670 878	1102	·858 065	150	06
55	·513 791	249	·598 883	395	·165 616	203	·946 317	944	·669 776	1102	·857 915	149	05
56	·514 040	250	·599 278	396	·165 819	203	·945 373	944	·668 674	1100	·857 766	150	04
57	·514 290	249	·599 674	395	·166 022	204	·944 429	943	·667 574	1099	·857 616	149	03
58	·514 539	250	·600 069	396	·166 226	204	·943 486	941	·666 475	1098	·857 467	150	02
59	·514 789	249	·600 465	396	·166 430	203	·942 545	941	·665 377	1098	·857 317	150	01
60	0·515 038		0·600 861		1·166 633		1·941 604		1·664 279		0·857 167		00
	cos		cot		cosec		sec		tan		sin		59°

31° ′	sin		tan		sec		cosec		cot		cos		′
00	0·515 038	249	0·600 861	396	1·166 633	204	1·941 604	939	1·664 279	1096	0·857 167	150	60
01	·515 287	250	·601 257	396	·166 837	205	·940 665	939	·663 183	1095	·857 017	149	59
02	·515 537	249	·601 653	396	·167 042	204	·939 726	937	·662 088	1094	·856 868	150	58
03	·515 786	249	·602 049	396	·167 246	204	·938 789	936	·660 994	1092	·856 718	151	57
04	·516 035	249	·602 445	397	·167 450	205	·937 853	935	·659 902	1092	·856 567	150	56
05	·516 284	249	·602 842	397	·167 655	205	·936 918	935	·658 810	1091	·856 417	150	55
06	·516 533	249	·603 239	396	·167 860	205	·935 983	933	·657 719	1090	·856 267	150	54
07	·516 782	249	·603 635	397	·168 065	205	·935 050	931	·656 629	1088	·856 117	151	53
08	·517 031	249	·604 032	397	·168 270	205	·934 119	931	·655 541	1088	·855 966	150	52
09	·517 280	249	·604 429	398	·168 475	206	·933 188	930	·654 453	1087	·855 816	151	51
10	0·517 529	249	0·604 827	397	1·168 681	206	1·932 258	929	1·653 366	1085	0·855 665	150	50
11	·517 778	249	·605 224	398	·168 887	206	·931 329	928	·652 281	1085	·855 515	151	49
12	·518 027	249	·605 622	397	·169 093	206	·930 401	926	·651 196	1083	·855 364	150	48
13	·518 276	249	·606 019	398	·169 299	206	·929 475	926	·650 113	1083	·855 214	151	47
14	·518 525	248	·606 417	398	·169 505	206	·928 549	925	·649 030	1081	·855 063	151	46
15	·518 773	249	·606 815	398	·169 711	207	·927 624	923	·647 949	1080	·854 912	151	45
16	·519 022	249	·607 213	398	·169 918	206	·926 701	923	·646 869	1080	·854 761	151	44
17	·519 271	248	·607 611	399	·170 124	207	·925 778	921	·645 789	1078	·854 610	151	43
18	·519 519	249	·608 010	398	·170 331	207	·924 857	920	·644 711	1077	·854 459	151	42
19	·519 768	248	·608 408	399	·170 538	208	·923 937	920	·643 634	1076	·854 308	152	41
20	0·520 016	249	0·608 807	398	1·170 746	207	1·923 017	918	1·642 558	1076	0·854 156	151	40
21	·520 265	248	·609 205	399	·170 953	208	·922 099	917	·641 482	1074	·854 005	151	39
22	·520 513	248	·609 604	399	·171 161	207	·921 182	917	·640 408	1073	·853 854	152	38
23	·520 761	249	·610 003	400	·171 368	208	·920 265	915	·639 335	1072	·853 702	151	37
24	·521 010	248	·610 403	399	·171 576	209	·919 350	914	·638 263	1071	·853 551	152	36
25	·521 258	248	·610 802	399	·171 785	208	·918 436	913	·637 192	1070	·853 399	151	35
26	·521 506	248	·611 201	400	·171 993	208	·917 523	912	·636 122	1069	·853 248	152	34
27	·521 754	248	·611 601	400	·172 201	209	·916 611	911	·635 053	1068	·853 096	152	33
28	·522 002	249	·612 001	400	·172 410	209	·915 700	910	·633 985	1067	·852 944	152	32
29	·522 251	248	·612 401	400	·172 619	209	·914 790	909	·632 918	1066	·852 792	152	31
30	0·522 499	248	0·612 801	400	1·172 828	209	1·913 881	908	1·631 852	1065	0·852 640	152	30
31	·522 747	248	·613 201	400	·173 037	209	·912 973	907	·630 787	1064	·852 488	152	29
32	·522 995	247	·613 601	401	·173 246	210	·912 066	906	·629 723	1063	·852 336	152	28
33	·523 242	248	·614 002	400	·173 456	209	·911 160	905	·628 660	1062	·852 184	152	27
34	·523 490	248	·614 402	401	·173 665	210	·910 255	904	·627 598	1061	·852 032	153	26
35	·523 738	248	·614 803	401	·173 875	210	·909 351	903	·626 537	1060	·851 879	152	25
36	·523 986	248	·615 204	401	·174 085	210	·908 448	902	·625 477	1059	·851 727	153	24
37	·524 234	247	·615 605	401	·174 295	211	·907 546	900	·624 418	1058	·851 574	152	23
38	·524 481	248	·616 006	402	·174 506	210	·906 646	900	·623 360	1057	·851 422	153	22
39	·524 729	248	·616 408	401	·174 716	211	·905 746	899	·622 303	1056	·851 269	152	21
40	0·524 977	247	0·616 809	402	1·174 927	211	1·904 847	898	1·621 247	1055	0·851 117	153	20
41	·525 224	248	·617 211	402	·175 138	211	·903 949	897	·620 192	1054	·850 964	153	19
42	·525 472	247	·617 613	402	·175 349	211	·903 052	896	·619 138	1053	·850 811	153	18
43	·525 719	248	·618 015	402	·175 560	212	·902 156	894	·618 085	1052	·850 658	153	17
44	·525 967	247	·618 417	402	·175 772	211	·901 262	894	·617 033	1051	·850 505	153	16
45	·526 214	247	·618 819	402	·175 983	212	·900 368	893	·615 982	1050	·850 352	153	15
46	·526 461	248	·619 221	403	·176 195	212	·899 475	892	·614 932	1049	·850 199	153	14
47	·526 709	247	·619 624	402	·176 407	212	·898 583	891	·613 883	1048	·850 046	153	13
48	·526 956	247	·620 026	403	·176 619	212	·897 692	889	·612 835	1047	·849 893	154	12
49	·527 203	247	·620 429	403	·176 831	213	·896 803	889	·611 788	1046	·849 739	153	11
50	0·527 450	247	0·620 832	403	1·177 044	213	1·895 914	888	1·610 742	1045	0·849 586	153	10
51	·527 697	247	·621 235	403	·177 257	212	·895 026	887	·609 697	1044	·849 433	154	09
52	·527 944	247	·621 638	404	·177 469	213	·894 139	886	·608 653	1044	·849 279	154	08
53	·528 191	247	·622 042	403	·177 682	214	·893 253	885	·607 609	1042	·849 125	153	07
54	·528 438	247	·622 445	404	·177 896	213	·892 368	883	·606 567	1041	·848 972	154	06
55	·528 685	247	·622 849	404	·178 109	213	·891 485	883	·605 526	1040	·848 818	154	05
56	·528 932	247	·623 253	404	·178 322	214	·890 602	882	·604 486	1040	·848 664	154	04
57	·529 179	247	·623 657	404	·178 536	214	·889 720	881	·603 446	1038	·848 510	154	03
58	·529 426	247	·624 061	404	·178 750	214	·888 839	880	·602 408	1037	·848 356	154	02
59	·529 673	246	·624 465	404	·178 964	214	·887 959	879	·601 371	1036	·848 202	154	01
60	0·529 919		0·624 869		1·179 178		1·887 080		1·600 335		0·848 048		00
	cos		cot		cosec		sec		tan		sin		58°

32° ′	sin		tan		sec		cosec		cot		cos		′
00	0·529 919	247	0·624 869	405	1·179 178	215	1·887 080	878	1·600 335	1036	0·848 048	154	60
01	·530 166	247	·625 274	405	·179 393	214	·886 202	877	·599 299	1034	·847 894	154	59
02	·530 413	246	·625 679	404	·179 607	215	·885 325	876	·598 265	1034	·847 740	155	58
03	·530 659	247	·626 083	405	·179 822	215	·884 449	875	·597 231	1032	·847 585	154	57
04	·530 906	246	·626 488	406	·180 037	215	·883 574	874	·596 199	1032	·847 431	155	56
05	·531 152	247	·626 894	405	·180 252	216	·882 700	873	·595 167	1030	·847 276	154	55
06	·531 399	246	·627 299	405	·180 468	215	·881 827	873	·594 137	1030	·847 122	155	54
07	·531 645	246	·627 704	406	·180 683	216	·880 954	871	·593 107	1029	·846 967	154	53
08	·531 891	247	·628 110	406	·180 899	216	·880 083	870	·592 078	1027	·846 813	155	52
09	·532 138	246	·628 516	405	·181 115	216	·879 213	869	·591 051	1027	·846 658	155	51
10	0·532 384	246	0·628 921	406	1·181 331	216	1·878 344	868	1·590 024	1026	0·846 503	155	50
11	·532 630	246	·629 327	407	·181 547	216	·877 476	868	·588 998	1025	·846 348	155	49
12	·532 876	246	·629 734	406	·181 763	217	·876 608	866	·587 973	1024	·846 193	155	48
13	·533 122	246	·630 140	406	·181 980	217	·875 742	866	·586 949	1023	·846 038	155	47
14	·533 368	247	·630 546	407	·182 197	217	·874 876	864	·585 926	1022	·845 883	155	46
15	·533 615	246	·630 953	407	·182 414	217	·874 012	864	·584 904	1021	·845 728	155	45
16	·533 861	245	·631 360	407	·182 631	217	·873 148	862	·583 883	1020	·845 573	156	44
17	·534 106	246	·631 767	407	·182 848	217	·872 286	862	·582 863	1019	·845 417	155	43
18	·534 352	246	·632 174	407	·183 065	218	·871 424	860	·581 844	1019	·845 262	156	42
19	·534 598	246	·632 581	407	·183 283	218	·870 564	860	·580 825	1017	·845 106	155	41
20	0·534 844	246	0·632 988	408	1·183 501	218	1·869 704	859	1·579 808	1016	0·844 951	156	40
21	·535 090	245	·633 396	408	·183 719	218	·868 845	858	·578 792	1016	·844 795	155	39
22	·535 335	246	·633 804	407	·183 937	218	·867 987	856	·577 776	1015	·844 640	156	38
23	·535 581	246	·634 211	408	·184 155	219	·867 131	856	·576 761	1013	·844 484	156	37
24	·535 827	245	·634 619	408	·184 374	219	·866 275	855	·575 748	1013	·844 328	156	36
25	·536 072	246	·635 027	409	·184 593	219	·865 420	854	·574 735	1012	·844 172	156	35
26	·536 318	245	·635 436	408	·184 812	219	·864 566	853	·573 723	1010	·844 016	156	34
27	·536 563	246	·635 844	409	·185 031	219	·863 713	853	·572 713	1010	·843 860	156	33
28	·536 809	245	·636 253	408	·185 250	219	·862 860	851	·571 703	1009	·843 704	156	32
29	·537 054	246	·636 661	409	·185 469	220	·862 009	850	·570 694	1008	·843 548	157	31
30	0·537 300	245	0·637 070	409	1·185 689	220	1·861 159	849	1·569 686	1008	0·843 391	156	30
31	·537 545	245	·637 479	409	·185 909	220	·860 310	849	·568 678	1006	·843 235	156	29
32	·537 790	245	·637 888	410	·186 129	220	·859 461	847	·567 672	1005	·843 079	157	28
33	·538 035	246	·638 298	409	·186 349	220	·858 614	847	·566 667	1005	·842 922	156	27
34	·538 281	245	·638 707	410	·186 569	221	·857 767	845	·565 662	1003	·842 766	157	26
35	·538 526	245	·639 117	410	·186 790	221	·856 922	845	·564 659	1003	·842 609	157	25
36	·538 771	245	·639 527	410	·187 011	221	·856 077	844	·563 656	1001	·842 452	156	24
37	·539 016	245	·639 937	410	·187 232	221	·855 233	843	·562 655	1001	·842 296	157	23
38	·539 261	245	·640 347	410	·187 453	221	·854 390	842	·561 654	1000	·842 139	157	22
39	·539 506	245	·640 757	410	·187 674	221	·853 548	841	·560 654	999	·841 982	157	21
40	0·539 751	245	0·641 167	411	1·187 895	222	1·852 707	840	1·559 655	998	0·841 825	157	20
41	·539 996	244	·641 578	411	·188 117	222	·851 867	839	·558 657	997	·841 668	157	19
42	·540 240	245	·641 989	410	·188 339	222	·851 028	838	·557 660	996	·841 511	157	18
43	·540 485	245	·642 399	411	·188 561	222	·850 190	838	·556 664	995	·841 354	158	17
44	·540 730	244	·642 810	412	·188 783	222	·849 352	836	·555 669	995	·841 196	157	16
45	·540 974	245	·643 222	411	·189 005	223	·848 516	835	·554 674	993	·841 039	157	15
46	·541 219	245	·643 633	411	·189 228	223	·847 681	835	·553 681	993	·840 882	158	14
47	·541 464	244	·644 044	412	·189 451	223	·846 846	834	·552 688	992	·840 724	157	13
48	·541 708	245	·644 456	412	·189 674	223	·846 012	833	·551 696	991	·840 567	158	12
49	·541 953	244	·644 868	412	·189 897	223	·845 179	831	·550 705	990	·840 409	158	11
50	0·542 197	245	0·645 280	412	1·190 120	224	1·844 348	831	1·549 715	989	0·840 251	157	10
51	·542 442	244	·645 692	412	·190 344	223	·843 517	830	·548 726	988	·840 094	158	09
52	·542 686	244	·646 104	412	·190 567	224	·842 687	830	·547 738	987	·839 936	158	08
53	·542 930	244	·646 516	413	·190 791	224	·841 857	828	·546 751	986	·839 778	158	07
54	·543 174	245	·646 929	413	·191 015	224	·841 029	827	·545 765	986	·839 620	158	06
55	·543 419	244	·647 342	413	·191 239	225	·840 202	827	·544 779	984	·839 462	158	05
56	·543 663	244	·647 755	413	·191 464	224	·839 375	825	·543 795	984	·839 304	158	04
57	·543 907	244	·648 168	413	·191 688	225	·838 550	825	·542 811	983	·839 146	159	03
58	·544 151	244	·648 581	413	·191 913	225	·837 725	824	·541 828	982	·838 987	158	02
59	·544 395	244	·648 994	414	·192 138	225	·836 901	823	·540 846	981	·838 829	158	01
60	0·544 639		0·649 408		1·192 363		1·836 078		1·539 865		0·838 671		00
	cos		cot		cosec		sec		tan		sin		57°

33° ′	sin		tan		sec		cosec		cot		cos		′
00	0·544 639	244	0·649 408	413	1·192 363	226	1·836 078	822	1·539 865	980	0·838 671	159	60
01	·544 883	244	·649 821	414	·192 589	225	·835 256	821	·538 885	980	·838 512	158	59
02	·545 127	244	·650 235	414	·192 814	226	·834 435	820	·537 905	978	·838 354	159	58
03	·545 371	244	·650 649	414	·193 040	226	·833 615	819	·536 927	978	·838 195	159	57
04	·545 615	243	·651 063	414	·193 266	226	·832 796	819	·535 949	976	·838 036	158	56
05	·545 858	244	·651 477	415	·193 492	226	·831 977	817	·534 973	976	·837 878	159	55
06	·546 102	244	·651 892	414	·193 718	227	·831 160	817	·533 997	975	·837 719	159	54
07	·546 346	243	·652 306	415	·193 945	226	·830 343	816	·533 022	974	·837 560	159	53
08	·546 589	244	·652 721	415	·194 171	227	·829 527	814	·532 048	973	·837 401	159	52
09	·546 833	243	·653 136	415	·194 398	227	·828 713	814	·531 075	973	·837 242	159	51
10	0·547 076	244	0·653 551	415	1·194 625	227	1·827 899	814	1·530 102	971	0·837 083	159	50
11	·547 320	243	·653 966	416	·194 852	228	·827 085	812	·529 131	971	·836 924	160	49
12	·547 563	244	·654 382	415	·195 080	227	·826 273	811	·528 160	970	·836 764	159	48
13	·547 807	243	·654 797	416	·195 307	228	·825 462	811	·527 190	968	·836 605	159	47
14	·548 050	243	·655 213	416	·195 535	228	·824 651	809	·526 222	969	·836 446	160	46
15	·548 293	243	·655 629	416	·195 763	228	·823 842	809	·525 253	967	·836 286	159	45
16	·548 536	244	·656 045	416	·195 991	228	·823 033	808	·524 286	966	·836 127	160	44
17	·548 780	243	·656 461	416	·196 219	229	·822 225	807	·523 320	965	·835 967	160	43
18	·549 023	243	·656 877	417	·196 448	229	·821 418	806	·522 355	965	·835 807	159	42
19	·549 266	243	·657 294	416	·196 677	229	·820 612	806	·521 390	964	·835 648	160	41
20	0·549 509	243	0·657 710	417	1·196 906	229	1·819 806	804	1·520 426	963	0·835 488	160	40
21	·549 752	243	·658 127	417	·197 135	229	·819 002	803	·519 463	962	·835 328	160	39
22	·549 995	243	·658 544	417	·197 364	229	·818 199	803	·518 501	961	·835 168	160	38
23	·550 238	243	·658 961	418	·197 593	230	·817 396	802	·517 540	960	·835 008	160	37
24	·550 481	243	·659 379	417	·197 823	230	·816 594	801	·516 580	960	·834 848	160	36
25	·550 724	242	·659 796	418	·198 053	230	·815 793	800	·515 620	959	·834 688	161	35
26	·550 966	243	·660 214	417	·198 283	230	·814 993	799	·514 661	957	·834 527	160	34
27	·551 209	243	·660 631	418	·198 513	231	·814 194	799	·513 704	957	·834 367	160	33
28	·551 452	242	·661 049	418	·198 744	230	·813 395	797	·512 747	957	·834 207	161	32
29	·551 694	243	·661 467	419	·198 974	231	·812 598	797	·511 790	955	·834 046	160	31
30	0·551 937	243	0·661 886	418	1·199 205	231	1·811 801	796	1·510 835	954	0·833 886	161	30
31	·552 180	242	·662 304	419	·199 436	231	·811 005	795	·509 881	954	·833 725	160	29
32	·552 422	242	·662 723	418	·199 667	231	·810 210	794	·508 927	953	·833 565	161	28
33	·552 664	243	·663 141	419	·199 898	232	·809 416	793	·507 974	952	·833 404	161	27
34	·552 907	242	·663 560	419	·200 130	232	·808 623	793	·507 022	951	·833 243	161	26
35	·553 149	243	·663 979	419	·200 362	232	·807 830	791	·506 071	950	·833 082	161	25
36	·553 392	242	·664 398	420	·200 594	232	·807 039	791	·505 121	949	·832 921	161	24
37	·553 634	242	·664 818	419	·200 826	232	·806 248	790	·504 172	949	·832 760	161	23
38	·553 876	242	·665 237	420	·201 058	233	·805 458	789	·503 223	948	·832 599	161	22
39	·554 118	242	·665 657	420	·201 291	232	·804 669	788	·502 275	947	·832 438	161	21
40	0·554 360	242	0·666 077	420	1·201 523	233	1·803 881	787	1·501 328	946	0·832 277	162	20
41	·554 602	242	·666 497	420	·201 756	233	·803 094	787	·500 382	945	·832 115	161	19
42	·554 844	242	·666 917	420	·201 989	234	·802 307	786	·499 437	945	·831 954	161	18
43	·555 086	242	·667 337	421	·202 223	233	·801 521	785	·498 492	943	·831 793	162	17
44	·555 328	242	·667 758	421	·202 456	234	·800 736	784	·497 549	943	·831 631	161	16
45	·555 570	242	·668 179	420	·202 690	234	·799 952	783	·496 606	942	·831 470	162	15
46	·555 812	242	·668 599	421	·202 924	234	·799 169	782	·495 664	941	·831 308	162	14
47	·556 054	242	·669 020	422	·203 158	234	·798 387	782	·494 723	941	·831 146	162	13
48	·556 296	241	·669 442	421	·203 392	234	·797 605	780	·493 782	939	·830 984	161	12
49	·556 537	242	·669 863	421	·203 626	235	·796 825	780	·492 843	939	·830 823	162	11
50	0·556 779	242	0·670 284	422	1·203 861	235	1·796 045	779	1·491 904	938	0·830 661	162	10
51	·557 021	241	·670 706	422	·204 096	235	·795 266	778	·490 966	937	·830 499	162	09
52	·557 262	242	·671 128	422	·204 331	235	·794 488	778	·490 029	937	·830 337	163	08
53	·557 504	241	·671 550	422	·204 566	235	·793 710	776	·489 092	935	·830 174	162	07
54	·557 745	242	·671 972	422	·204 801	236	·792 934	776	·488 157	935	·830 012	162	06
55	·557 987	241	·672 394	423	·205 037	236	·792 158	775	·487 222	934	·829 850	162	05
56	·558 228	241	·672 817	423	·205 273	236	·791 383	774	·486 288	933	·829 688	163	04
57	·558 469	241	·673 240	422	·205 509	236	·790 609	773	·485 355	932	·829 525	162	03
58	·558 710	242	·673 662	423	·205 745	236	·789 836	773	·484 423	931	·829 363	163	02
59	·558 952	241	·674 085	424	·205 981	237	·789 063	771	·483 492	931	·829 200	162	01
60	0·559 193		0·674 509		1·206 218		1·788 292		1·482 561		0·829 038		00
	cos		cot		cosec		sec		tan		sin		56°

34° ′	sin		tan		sec		cosec		cot		cos		′
00	0·559 193	241	0·674 509	423	1·206 218	237	1·788 292	771	1·482 561	930	0·829 038	163	60
01	·559 434	241	·674 932	423	·206 455	237	·787 521	770	·481 631	929	·828 875	163	59
02	·559 675	241	·675 355	424	·206 692	237	·786 751	769	·480 702	928	·828 712	163	58
03	·559 916	241	·675 779	424	·206 929	237	·785 982	769	·479 774	928	·828 549	163	57
04	·560 157	241	·676 203	424	·207 166	238	·785 213	767	·478 846	926	·828 386	163	56
05	·560 398	241	·676 627	424	·207 404	237	·784 446	767	·477 920	926	·828 223	163	55
06	·560 639	241	·677 051	424	·207 641	238	·783 679	766	·476 994	925	·828 060	163	54
07	·560 880	241	·677 475	425	·207 879	239	·782 913	765	·476 069	925	·827 897	163	53
08	·561 121	240	·677 900	424	·208 118	238	·782 148	764	·475 144	923	·827 734	163	52
09	·561 361	241	·678 324	425	·208 356	238	·781 384	764	·474 221	923	·827 571	164	51
10	0·561 602	241	0·678 749	425	1·208 594	239	1·780 620	763	1·473 298	922	0·827 407	163	50
11	·561 843	240	·679 174	425	·208 833	239	·779 857	762	·472 376	921	·827 244	163	49
12	·562 083	241	·679 599	426	·209 072	239	·779 095	761	·471 455	920	·827 081	164	48
13	·562 324	240	·680 025	425	·209 311	239	·778 334	760	·470 535	920	·826 917	164	47
14	·562 564	241	·680 450	426	·209 550	240	·777 574	759	·469 615	918	·826 753	163	46
15	·562 805	240	·680 876	426	·209 790	240	·776 815	759	·468 697	918	·826 590	164	45
16	·563 045	241	·681 302	426	·210 030	240	·776 056	758	·467 779	917	·826 426	164	44
17	·563 286	240	·681 728	426	·210 270	240	·775 298	757	·466 862	917	·826 262	164	43
18	·563 526	240	·682 154	426	·210 510	240	·774 541	756	·465 945	915	·826 098	164	42
19	·563 766	241	·682 580	427	·210 750	241	·773 785	756	·465 030	915	·825 934	164	41
20	0·564 007	240	0·683 007	426	1·210 991	240	1·773 029	755	1·464 115	914	0·825 770	164	40
21	·564 247	240	·683 433	427	·211 231	241	·772 274	754	·463 201	914	·825 606	164	39
22	·564 487	240	·683 860	427	·211 472	241	·771 520	753	·462 287	912	·825 442	164	38
23	·564 727	240	·684 287	427	·211 713	241	·770 767	752	·461 375	912	·825 278	165	37
24	·564 967	240	·684 714	428	·211 954	242	·770 015	752	·460 463	911	·825 113	164	36
25	·565 207	240	·685 142	427	·212 196	242	·769 263	750	·459 552	910	·824 949	164	35
26	·565 447	240	·685 569	428	·212 438	242	·768 513	750	·458 642	909	·824 785	165	34
27	·565 687	240	·685 997	428	·212 680	242	·767 763	750	·457 733	909	·824 620	164	33
28	·565 927	239	·686 425	428	·212 922	242	·767 013	748	·456 824	908	·824 456	165	32
29	·566 166	240	·686 853	428	·213 164	242	·766 265	748	·455 916	907	·824 291	165	31
30	0·566 406	240	0·687 281	428	1·213 406	243	1·765 517	747	1·455 009	906	0·824 126	165	30
31	·566 646	240	·687 709	429	·213 649	243	·764 770	746	·454 103	906	·823 961	164	29
32	·566 886	239	·688 138	429	·213 892	243	·764 024	745	·453 197	905	·823 797	165	28
33	·567 125	240	·688 567	428	·214 135	243	·763 279	744	·452 292	904	·823 632	165	27
34	·567 365	239	·688 995	430	·214 378	244	·762 535	744	·451 388	903	·823 467	165	26
35	·567 604	240	·689 425	429	·214 622	244	·761 791	743	·450 485	902	·823 302	166	25
36	·567 844	239	·689 854	429	·214 866	243	·761 048	742	·449 583	902	·823 136	165	24
37	·568 083	240	·690 283	430	·215 109	245	·760 306	742	·448 681	901	·822 971	165	23
38	·568 323	239	·690 713	430	·215 354	244	·759 564	740	·447 780	900	·822 806	165	22
39	·568 562	239	·691 143	429	·215 598	244	·758 824	740	·446 880	900	·822 641	166	21
40	0·568 801	239	0·691 572	431	1·215 842	245	1·758 084	739	1·445 980	899	0·822 475	165	20
41	·569 040	240	·692 003	430	·216 087	245	·757 345	739	·445 081	898	·822 310	166	19
42	·569 280	239	·692 433	430	·216 332	245	·756 606	737	·444 183	897	·822 144	166	18
43	·569 519	239	·692 863	431	·216 577	245	·755 869	737	·443 286	896	·821 978	165	17
44	·569 758	239	·693 294	431	·216 822	246	·755 132	736	·442 390	896	·821 813	166	16
45	·569 997	239	·693 725	431	·217 068	245	·754 396	735	·441 494	895	·821 647	166	15
46	·570 236	239	·694 156	431	·217 313	246	·753 661	735	·440 599	894	·821 481	166	14
47	·570 475	239	·694 587	431	·217 559	246	·752 926	734	·439 705	894	·821 315	166	13
48	·570 714	238	·695 018	432	·217 805	247	·752 192	733	·438 811	892	·821 149	166	12
49	·570 952	239	·695 450	431	·218 052	246	·751 459	732	·437 919	892	·820 983	166	11
50	0·571 191	239	0·695 881	432	1·218 298	247	1·750 727	731	1·437 027	891	0·820 817	166	10
51	·571 430	239	·696 313	432	·218 545	247	·749 996	731	·436 136	891	·820 651	166	09
52	·571 669	238	·696 745	432	·218 792	247	·749 265	730	·435 245	890	·820 485	167	08
53	·571 907	239	·697 177	433	·219 039	247	·748 535	729	·434 355	889	·820 318	166	07
54	·572 146	238	·697 610	432	·219 286	248	·747 806	728	·433 466	888	·820 152	167	06
55	·572 384	239	·698 042	433	·219 534	248	·747 078	728	·432 578	887	·819 985	166	05
56	·572 623	238	·698 475	433	·219 782	248	·746 350	727	·431 691	887	·819 819	167	04
57	·572 861	239	·698 908	433	·220 030	248	·745 623	726	·430 804	886	·819 652	166	03
58	·573 100	238	·699 341	433	·220 278	248	·744 897	726	·429 918	885	·819 486	167	02
59	·573 338	238	·699 774	434	·220 526	249	·744 171	724	·429 033	885	·819 319	167	01
60	0·573 576		0·700 208		1·220 775		1·743 447		1·428 148		0·819 152		00
	cos		cot		cosec		sec		tan		sin		55°

35°	sin		tan		sec		cosec		cot		cos		′
00	0·573 576	239	0·700 208	433	1·220 775	248	1·743 447	724	1·428 148	884	0·819 152	167	60
01	·573 815	238	·700 641	434	·221 023	249	·742 723	723	·427 264	883	·818 985	167	59
02	·574 053	238	·701 075	434	·221 272	249	·742 000	723	·426 381	882	·818 818	167	58
03	·574 291	238	·701 509	434	·221 521	250	·741 277	721	·425 499	882	·818 651	167	57
04	·574 529	238	·701 943	434	·221 771	249	·740 556	721	·424 617	881	·818 484	167	56
05	·574 767	238	·702 377	435	·222 020	250	·739 835	720	·423 736	880	·818 317	167	55
06	·575 005	238	·702 812	434	·222 270	250	·739 115	720	·422 856	879	·818 150	168	54
07	·575 243	238	·703 246	435	·222 520	250	·738 395	719	·421 977	879	·817 982	167	53
08	·575 481	238	·703 681	435	·222 770	251	·737 676	718	·421 098	878	·817 815	167	52
09	·575 719	238	·704 116	435	·223 021	250	·736 958	717	·420 220	877	·817 648	168	51
10	0·575 957	238	0·704 551	436	1·223 271	251	1·736 241	716	1·419 343	877	0·817 480	167	50
11	·576 195	237	·704 987	435	·223 522	251	·735 525	716	·418 466	876	·817 313	168	49
12	·576 432	238	·705 422	436	·223 773	251	·734 809	715	·417 590	875	·817 145	168	48
13	·576 670	238	·705 858	436	·224 024	252	·734 094	714	·416 715	874	·816 977	168	47
14	·576 908	237	·706 294	436	·224 276	251	·733 380	714	·415 841	874	·816 809	167	46
15	·577 145	238	·706 730	436	·224 527	252	·732 666	713	·414 967	873	·816 642	168	45
16	·577 383	237	·707 166	437	·224 779	252	·731 953	712	·414 094	872	·816 474	168	44
17	·577 620	238	·707 603	436	·225 031	253	·731 241	711	·413 222	871	·816 306	168	43
18	·577 858	237	·708 039	437	·225 284	252	·730 530	711	·412 351	871	·816 138	169	42
19	·578 095	237	·708 476	437	·225 536	253	·729 819	709	·411 480	870	·815 969	168	41
20	0·578 332	238	0·708 913	437	1·225 789	253	1·729 110	710	1·410 610	870	0·815 801	168	40
21	·578 570	237	·709 350	438	·226 042	253	·728 400	708	·409 740	868	·815 633	168	39
22	·578 807	237	·709 788	437	·226 295	253	·727 692	708	·408 872	868	·815 465	169	38
23	·579 044	237	·710 225	438	·226 548	253	·726 984	707	·408 004	867	·815 296	168	37
24	·579 281	237	·710 663	438	·226 801	254	·726 277	706	·407 137	867	·815 128	169	36
25	·579 518	237	·711 101	438	·227 055	254	·725 571	705	·406 270	866	·814 959	168	35
26	·579 755	237	·711 539	438	·227 309	254	·724 866	705	·405 404	865	·814 791	169	34
27	·579 992	237	·711 977	439	·227 563	255	·724 161	704	·404 539	864	·814 622	169	33
28	·580 229	237	·712 416	438	·227 818	254	·723 457	704	·403 675	864	·814 453	169	32
29	·580 466	237	·712 854	439	·228 072	255	·722 753	702	·402 811	863	·814 284	168	31
30	0·580 703	237	0·713 293	439	1·228 327	255	1·722 051	702	1·401 948	862	0·814 116	169	30
31	·580 940	236	·713 732	439	·228 582	255	·721 349	701	·401 086	862	·813 947	169	29
32	·581 176	237	·714 171	440	·228 837	255	·720 648	701	·400 224	860	·813 778	170	28
33	·581 413	237	·714 611	439	·229 092	256	·719 947	700	·399 364	861	·813 608	169	27
34	·581 650	236	·715 050	440	·229 348	256	·719 247	699	·398 503	859	·813 439	169	26
35	·581 886	237	·715 490	440	·229 604	256	·718 548	698	·397 644	859	·813 270	169	25
36	·582 123	236	·715 930	440	·229 860	256	·717 850	698	·396 785	858	·813 101	170	24
37	·582 359	237	·716 370	440	·230 116	257	·717 152	696	·395 927	857	·812 931	169	23
38	·582 596	236	·716 810	440	·230 373	256	·716 456	697	·395 070	857	·812 762	170	22
39	·582 832	237	·717 250	441	·230 629	257	·715 759	695	·394 213	856	·812 592	169	21
40	0·583 069	236	0·717 691	441	1·230 886	257	1·715 064	695	1·393 357	855	0·812 423	170	20
41	·583 305	236	·718 132	441	·231 143	257	·714 369	694	·392 502	855	·812 253	169	19
42	·583 541	236	·718 573	441	·231 400	258	·713 675	693	·391 647	854	·812 084	170	18
43	·583 777	237	·719 014	441	·231 658	258	·712 982	693	·390 793	853	·811 914	170	17
44	·584 014	236	·719 455	442	·231 916	258	·712 289	692	·389 940	852	·811 744	170	16
45	·584 250	236	·719 897	442	·232 174	258	·711 597	691	·389 088	852	·811 574	170	15
46	·584 486	236	·720 339	442	·232 432	258	·710 906	691	·388 236	851	·811 404	170	14
47	·584 722	236	·720 781	442	·232 690	259	·710 215	690	·387 385	851	·811 234	170	13
48	·584 958	236	·721 223	442	·232 949	258	·709 525	689	·386 534	850	·811 064	170	12
49	·585 194	235	·721 665	443	·233 207	259	·708 836	688	·385 684	849	·810 894	171	11
50	0·585 429	236	0·722 108	442	1·233 466	260	1·708 148	688	1·384 835	848	0·810 723	170	10
51	·585 665	236	·722 550	443	·233 726	259	·707 460	687	·383 987	848	·810 553	170	09
52	·585 901	236	·722 993	443	·233 985	260	·706 773	686	·383 139	847	·810 383	171	08
53	·586 137	235	·723 436	443	·234 245	259	·706 087	686	·382 292	846	·810 212	170	07
54	·586 372	236	·723 879	444	·234 504	260	·705 401	685	·381 446	846	·810 042	171	06
55	·586 608	236	·724 323	443	·234 764	261	·704 716	684	·380 600	845	·809 871	171	05
56	·586 844	235	·724 766	444	·235 025	260	·704 032	684	·379 755	844	·809 700	170	04
57	·587 079	235	·725 210	444	·235 285	261	·703 348	683	·378 911	844	·809 530	171	03
58	·587 314	236	·725 654	444	·235 546	261	·702 665	682	·378 067	843	·809 359	171	02
59	·587 550	235	·726 098	445	·235 807	261	·701 983	681	·377 224	842	·809 188	171	01
60	0·587 785		0·726 543		1·236 068		1·701 302		1·376 382		0·809 017		00
	cos		cot		cosec		sec		tan		sin		54°

36° ′	sin		tan		sec		cosec		cot		cos		′
00	0·587 785	236	0·726 543	444	1·236 068	261	1·701 302	681	1·376 382	842	0·809 017	171	60
01	·588 021	235	·726 987	445	·236 329	262	·700 621	680	·375 540	841	·808 846	171	59
02	·588 256	235	·727 432	445	·236 591	262	·699 941	680	·374 699	840	·808 675	171	58
03	·588 491	235	·727 877	445	·236 853	262	·699 261	679	·373 859	840	·808 504	171	57
04	·588 726	235	·728 322	445	·237 115	262	·698 582	678	·373 019	838	·808 333	172	56
05	·588 961	235	·728 767	446	·237 377	262	·697 904	677	·372 181	839	·808 161	171	55
06	·589 196	235	·729 213	445	·237 639	263	·697 227	677	·371 342	837	·807 990	172	54
07	·589 431	235	·729 658	446	·237 902	263	·696 550	676	·370 505	837	·807 818	171	53
08	·589 666	235	·730 104	446	·238 165	263	·695 874	675	·369 668	836	·807 647	172	52
09	·589 901	235	·730 550	446	·238 428	263	·695 199	675	·368 832	836	·807 475	171	51
10	0·590 136	235	0·730 996	447	1·238 691	264	1·694 524	674	1·367 996	835	0·807 304	172	50
11	·590 371	235	·731 443	446	·238 955	263	·693 850	673	·367 161	834	·807 132	172	49
12	·590 606	234	·731 889	447	·239 218	264	·693 177	672	·366 327	834	·806 960	172	48
13	·590 840	235	·732 336	447	·239 482	264	·692 505	672	·365 493	833	·806 788	171	47
14	·591 075	235	·732 783	447	·239 746	265	·691 833	672	·364 660	832	·806 617	172	46
15	·591 310	234	·733 230	448	·240 011	264	·691 161	670	·363 828	832	·806 445	172	45
16	·591 544	235	·733 678	447	·240 275	265	·690 491	670	·362 996	831	·806 273	173	44
17	·591 779	234	·734 125	448	·240 540	265	·689 821	669	·362 165	830	·806 100	172	43
18	·592 013	235	·734 573	448	·240 805	265	·689 152	669	·361 335	830	·805 928	172	42
19	·592 248	234	·735 021	448	·241 070	266	·688 483	668	·360 505	829	·805 756	172	41
20	0·592 482	234	0·735 469	448	1·241 336	266	1·687 815	667	1·359 676	828	0·805 584	173	40
21	·592 716	235	·735 917	449	·241 602	265	·687 148	667	·358 848	828	·805 411	172	39
22	·592 951	234	·736 366	449	·241 867	267	·686 481	666	·358 020	827	·805 239	173	38
23	·593 185	234	·736 815	449	·242 134	266	·685 815	665	·357 193	826	·805 066	172	37
24	·593 419	234	·737 264	449	·242 400	266	·685 150	664	·356 367	826	·804 894	173	36
25	·593 653	234	·737 713	449	·242 666	267	·684 486	664	·355 541	825	·804 721	173	35
26	·593 887	234	·738 162	449	·242 933	267	·683 822	663	·354 716	824	·804 548	172	34
27	·594 121	234	·738 611	450	·243 200	267	·683 159	663	·353 892	824	·804 376	173	33
28	·594 355	234	·739 061	450	·243 467	268	·682 496	662	·353 068	823	·804 203	173	32
29	·594 589	234	·739 511	450	·243 735	268	·681 834	661	·352 245	823	·804 030	173	31
30	0·594 823	234	0·739 961	450	1·244 003	267	1·681 173	661	1·351 422	821	0·803 857	173	30
31	·595 057	233	·740 411	451	·244 270	269	·680 512	659	·350 601	822	·803 684	173	29
32	·595 290	234	·740 862	450	·244 539	268	·679 853	660	·349 779	820	·803 511	174	28
33	·595 524	234	·741 312	451	·244 807	268	·679 193	658	·348 959	820	·803 337	173	27
34	·595 758	233	·741 763	451	·245 075	269	·678 535	658	·348 139	819	·803 164	173	26
35	·595 991	234	·742 214	452	·245 344	269	·677 877	657	·347 320	819	·802 991	174	25
36	·596 225	233	·742 666	451	·245 613	269	·677 220	657	·346 501	818	·802 817	173	24
37	·596 458	234	·743 117	452	·245 882	270	·676 563	656	·345 683	817	·802 644	174	23
38	·596 692	233	·743 569	451	·246 152	269	·675 907	655	·344 866	817	·802 470	173	22
39	·596 925	234	·744 020	452	·246 421	270	·675 252	655	·344 049	816	·802 297	174	21
40	0·597 159	233	0·744 472	453	1·246 691	270	1·674 597	654	1·343 233	815	0·802 123	174	20
41	·597 392	233	·744 925	452	·246 961	271	·673 943	653	·342 418	815	·801 949	173	19
42	·597 625	233	·745 377	453	·247 232	270	·673 290	653	·341 603	814	·801 776	174	18
43	·597 858	233	·745 830	452	·247 502	271	·672 637	652	·340 789	814	·801 602	174	17
44	·598 091	234	·746 282	453	·247 773	271	·671 985	651	·339 975	813	·801 428	174	16
45	·598 325	233	·746 735	454	·248 044	271	·671 334	651	·339 162	812	·801 254	174	15
46	·598 558	233	·747 189	453	·248 315	272	·670 683	650	·338 350	811	·801 080	174	14
47	·598 791	233	·747 642	454	·248 587	271	·670 033	650	·337 539	811	·800 906	175	13
48	·599 024	232	·748 096	453	·248 858	272	·669 383	648	·336 728	811	·800 731	174	12
49	·599 256	233	·748 549	454	·249 130	272	·668 735	649	·335 917	809	·800 557	174	11
50	0·599 489	233	0·749 003	455	1·249 402	273	1·668 086	647	1·335 108	810	0·800 383	175	10
51	·599 722	233	·749 458	454	·249 675	272	·667 439	647	·334 298	808	·800 208	174	09
52	·599 955	233	·749 912	454	·249 947	273	·666 792	646	·333 490	808	·800 034	175	08
53	·600 188	232	·750 366	455	·250 220	273	·666 146	646	·332 682	807	·799 859	174	07
54	·600 420	233	·750 821	455	·250 493	273	·665 500	645	·331 875	807	·799 685	175	06
55	·600 653	232	·751 276	455	·250 766	274	·664 855	644	·331 068	806	·799 510	175	05
56	·600 885	233	·751 731	456	·251 040	273	·664 211	644	·330 262	805	·799 335	175	04
57	·601 118	232	·752 187	455	·251 313	274	·663 567	643	·329 457	805	·799 160	175	03
58	·601 350	233	·752 642	456	·251 587	274	·662 924	642	·328 652	804	·798 985	174	02
59	·601 583	232	·753 098	456	·251 861	275	·662 282	642	·327 848	803	·798 811	175	01
60	0·601 815		0·753 554		1·252 136		1·661 640		1·327 045		0·798 636		00
	cos		cot		cosec		sec		tan		sin		53°

37° ′	sin		tan		sec		cosec		cot		cos		′
00	0·601 815	232	0·753 554	456	1·252 136	274	1·661 640	641	1·327 045	803	0·798 636	176	60
01	·602 047	233	·754 010	457	·252 410	275	·660 999	640	·326 242	802	·798 460	175	59
02	·602 280	232	·754 467	456	·252 685	275	·660 359	640	·325 440	802	·798 285	175	58
03	·602 512	232	·754 923	457	·252 960	275	·659 719	639	·324 638	801	·798 110	175	57
04	·602 744	232	·755 380	457	·253 235	276	·659 080	639	·323 837	800	·797 935	176	56
05	·602 976	232	·755 837	457	·253 511	276	·658 441	638	·323 037	800	·797 759	175	55
06	·603 208	232	·756 294	457	·253 787	275	·657 803	637	·322 237	799	·797 584	176	54
07	·603 440	232	·756 751	458	·254 062	277	·657 166	637	·321 438	799	·797 408	175	53
08	·603 672	232	·757 209	458	·254 339	276	·656 529	636	·320 639	798	·797 233	176	52
09	·603 904	232	·757 667	458	·254 615	277	·655 893	635	·319 841	797	·797 057	175	51
10	0·604 136	231	0·758 125	458	1·254 892	277	1·655 258	635	1·319 044	797	0·796 882	176	50
11	·604 367	232	·758 583	458	·255 169	277	·654 623	634	·318 247	796	·796 706	176	49
12	·604 599	232	·759 041	459	·255 446	277	·653 989	634	·317 451	795	·796 530	176	48
13	·604 831	231	·759 500	459	·255 723	277	·653 355	633	·316 656	795	·796 354	176	47
14	·605 062	232	·759 959	459	·256 000	278	·652 722	632	·315 861	794	·796 178	176	46
15	·605 294	232	·760 418	459	·256 278	278	·652 090	632	·315 067	794	·796 002	176	45
16	·605 526	231	·760 877	459	·256 556	278	·651 458	631	·314 273	793	·795 826	176	44
17	·605 757	231	·761 336	460	·256 834	279	·650 827	630	·313 480	792	·795 650	177	43
18	·605 988	232	·761 796	460	·257 113	279	·650 197	630	·312 688	792	·795 473	176	42
19	·606 220	231	·762 256	460	·257 392	279	·649 567	629	·311 896	791	·795 297	176	41
20	0·606 451	231	0·762 716	460	1·257 671	279	1·648 938	629	1·311 105	791	0·795 121	177	40
21	·606 682	232	·763 176	460	·257 950	279	·648 309	628	·310 314	790	·794 944	176	39
22	·606 914	231	·763 636	461	·258 229	280	·647 681	627	·309 524	789	·794 768	177	38
23	·607 145	231	·764 097	461	·258 509	280	·647 054	627	·308 735	789	·794 591	176	37
24	·607 376	231	·764 558	461	·258 789	280	·646 427	626	·307 946	789	·794 415	177	36
25	·607 607	231	·765 019	461	·259 069	280	·645 801	626	·307 157	787	·794 238	177	35
26	·607 838	231	·765 480	461	·259 349	280	·645 175	624	·306 370	787	·794 061	177	34
27	·608 069	231	·765 941	462	·259 629	281	·644 551	625	·305 583	787	·793 884	177	33
28	·608 300	231	·766 403	462	·259 910	281	·643 926	623	·304 796	785	·793 707	177	32
29	·608 531	230	·766 865	462	·260 191	281	·643 303	623	·304 011	786	·793 530	177	31
30	0·608 761	231	0·767 327	462	1·260 472	282	1·642 680	623	1·303 225	784	0·793 353	177	30
31	·608 992	231	·767 789	463	·260 754	282	·642 057	622	·302 441	784	·793 176	177	29
32	·609 223	231	·768 252	462	·261 036	281	·641 435	621	·301 657	784	·792 999	177	28
33	·609 454	230	·768 714	463	·261 317	283	·640 814	620	·300 873	783	·792 822	178	27
34	·609 684	231	·769 177	463	·261 600	282	·640 194	620	·300 090	782	·792 644	177	26
35	·609 915	230	·769 640	464	·261 882	283	·639 574	620	·299 308	782	·792 467	177	25
36	·610 145	231	·770 104	463	·262 165	283	·638 954	619	·298 526	781	·792 290	178	24
37	·610 376	230	·770 567	464	·262 448	283	·638 335	618	·297 745	780	·792 112	177	23
38	·610 606	230	·771 031	464	·262 731	283	·637 717	617	·296 965	780	·791 935	178	22
39	·610 836	231	·771 495	464	·263 014	284	·637 100	617	·296 185	779	·791 757	178	21
40	0·611 067	230	0·771 959	464	1·263 298	283	1·636 483	617	1·295 406	779	0·791 579	178	20
41	·611 297	230	·772 423	465	·263 581	284	·635 866	615	·294 627	778	·791 401	177	19
42	·611 527	230	·772 888	465	·263 865	285	·635 251	615	·293 849	778	·791 224	178	18
43	·611 757	230	·773 353	465	·264 150	284	·634 636	615	·293 071	777	·791 046	178	17
44	·611 987	230	·773 818	465	·264 434	285	·634 021	614	·292 294	776	·790 868	178	16
45	·612 217	230	·774 283	465	·264 719	285	·633 407	613	·291 518	776	·790 690	179	15
46	·612 447	230	·774 748	466	·265 004	285	·632 794	613	·290 742	775	·790 511	178	14
47	·612 677	230	·775 214	466	·265 289	285	·632 181	612	·289 967	775	·790 333	178	13
48	·612 907	230	·775 680	466	·265 574	286	·631 569	612	·289 192	774	·790 155	178	12
49	·613 137	230	·776 146	466	·265 860	286	·630 957	611	·288 418	773	·789 977	179	11
50	0·613 367	229	0·776 612	466	1·266 146	286	1·630 346	610	1·287 645	773	0·789 798	178	10
51	·613 596	230	·777 078	467	·266 432	287	·629 736	610	·286 872	773	·789 620	179	09
52	·613 826	230	·777 545	467	·266 719	286	·629 126	609	·286 099	771	·789 441	178	08
53	·614 056	229	·778 012	467	·267 005	287	·628 517	609	·285 328	771	·789 263	179	07
54	·614 285	230	·778 479	467	·267 292	287	·627 908	608	·284 557	771	·789 084	179	06
55	·614 515	229	·778 946	468	·267 579	287	·627 300	607	·283 786	770	·788 905	178	05
56	·614 744	230	·779 414	467	·267 866	288	·626 693	607	·283 016	769	·788 727	179	04
57	·614 974	229	·779 881	468	·268 154	288	·626 086	606	·282 247	769	·788 548	179	03
58	·615 203	229	·780 349	468	·268 442	288	·625 480	606	·281 478	769	·788 369	179	02
59	·615 432	229	·780 817	469	·268 730	288	·624 874	605	·280 709	767	·788 190	179	01
60	0·615 661		0·781 286		1·269 018		1·624 269		1·279 942		0·788 011		00
	cos		cot		cosec		sec		tan		sin		52°

38° ′	sin	d	tan	d	sec	d	cosec	d	cot	d	cos	d	′
00	0·615 661	230	0·781 286	468	1·269 018	289	1·624 269	604	1·279 942	768	0·788 011	179	60
01	·615 891	229	·781 754	469	·269 307	289	·623 665	604	·279 174	766	·787 832	180	59
02	·616 120	229	·782 223	469	·269 596	289	·623 061	603	·278 408	766	·787 652	179	58
03	·616 349	229	·782 692	469	·269 885	289	·622 458	603	·277 642	766	·787 473	179	57
04	·616 578	229	·783 161	470	·270 174	289	·621 855	602	·276 876	764	·787 294	180	56
05	·616 807	229	·783 631	469	·270 463	290	·621 253	602	·276 112	765	·787 114	179	55
06	·617 036	229	·784 100	470	·270 753	290	·620 651	601	·275 347	763	·786 935	179	54
07	·617 265	229	·784 570	470	·271 043	290	·620 050	600	·274 584	764	·786 756	180	53
08	·617 494	228	·785 040	470	·271 333	291	·619 450	600	·273 820	762	·786 576	180	52
09	·617 722	229	·785 510	471	·271 624	290	·618 850	599	·273 058	762	·786 396	179	51
10	0·617 951	229	0·785 981	470	1·271 914	291	1·618 251	599	1·272 296	762	0·786 217	180	50
11	·618 180	228	·786 451	471	·272 205	291	·617 652	598	·271 534	761	·786 037	180	49
12	·618 408	229	·786 922	472	·272 496	292	·617 054	597	·270 773	760	·785 857	180	48
13	·618 637	228	·787 394	471	·272 788	291	·616 457	597	·270 013	760	·785 677	180	47
14	·618 865	229	·787 865	471	·273 079	292	·615 860	596	·269 253	759	·785 497	180	46
15	·619 094	228	·788 336	472	·273 371	292	·615 264	596	·268 494	759	·785 317	180	45
16	·619 322	229	·788 808	472	·273 663	293	·614 668	595	·267 735	758	·785 137	180	44
17	·619 551	228	·789 280	472	·273 956	292	·614 073	595	·266 977	757	·784 957	181	43
18	·619 779	228	·789 752	473	·274 248	293	·613 478	594	·266 220	757	·784 776	180	42
19	·620 007	228	·790 225	472	·274 541	293	·612 884	593	·265 463	757	·784 596	180	41
20	0·620 235	229	0·790 697	473	1·274 834	294	1·612 291	593	1·264 706	756	0·784 416	181	40
21	·620 464	228	·791 170	473	·275 128	293	·611 698	592	·263 950	755	·784 235	180	39
22	·620 692	228	·791 643	474	·275 421	294	·611 106	592	·263 195	755	·784 055	181	38
23	·620 920	228	·792 117	473	·275 715	294	·610 514	591	·262 440	754	·783 874	181	37
24	·621 148	228	·792 590	474	·276 009	294	·609 923	591	·261 686	754	·783 693	180	36
25	·621 376	228	·793 064	474	·276 303	295	·609 332	590	·260 932	753	·783 513	181	35
26	·621 604	227	·793 538	474	·276 598	295	·608 742	589	·260 179	752	·783 332	181	34
27	·621 831	228	·794 012	474	·276 893	295	·608 153	589	·259 427	752	·783 151	181	33
28	·622 059	228	·794 486	475	·277 188	295	·607 564	588	·258 675	752	·782 970	181	32
29	·622 287	228	·794 961	475	·277 483	296	·606 976	588	·257 923	751	·782 789	181	31
30	0·622 515	227	0·795 436	475	1·277 779	295	1·606 388	587	1·257 172	750	0·782 608	181	30
31	·622 742	228	·795 911	475	·278 074	296	·605 801	587	·256 422	750	·782 427	181	29
32	·622 970	227	·796 386	476	·278 370	297	·605 214	586	·255 672	749	·782 246	181	28
33	·623 197	228	·796 862	475	·278 667	296	·604 628	585	·254 923	749	·782 065	182	27
34	·623 425	227	·797 337	476	·278 963	297	·604 043	585	·254 174	748	·781 883	181	26
35	·623 652	228	·797 813	477	·279 260	297	·603 458	585	·253 426	748	·781 702	182	25
36	·623 880	227	·798 290	476	·279 557	297	·602 873	583	·252 678	747	·781 520	181	24
37	·624 107	227	·798 766	476	·279 854	298	·602 290	584	·251 931	746	·781 339	182	23
38	·624 334	227	·799 242	477	·280 152	298	·601 706	582	·251 185	746	·781 157	181	22
39	·624 561	228	·799 719	477	·280 450	298	·601 124	582	·250 439	746	·780 976	182	21
40	0·624 789	227	0·800 196	478	1·280 748	298	1·600 542	582	1·249 693	745	0·780 794	182	20
41	·625 016	227	·800 674	477	·281 046	298	·599 960	581	·248 948	744	·780 612	182	19
42	·625 243	227	·801 151	478	·281 344	299	·599 379	580	·248 204	744	·780 430	182	18
43	·625 470	227	·801 629	478	·281 643	299	·598 799	580	·247 460	743	·780 248	181	17
44	·625 697	226	·802 107	478	·281 942	299	·598 219	580	·246 717	743	·780 067	183	16
45	·625 923	227	·802 585	478	·282 241	300	·597 639	578	·245 974	742	·779 884	182	15
46	·626 150	227	·803 063	479	·282 541	299	·597 061	579	·245 232	742	·779 702	182	14
47	·626 377	227	·803 542	479	·282 840	300	·596 482	577	·244 490	741	·779 520	182	13
48	·626 604	226	·804 021	479	·283 140	301	·595 905	577	·243 749	740	·779 338	182	12
49	·626 830	227	·804 500	479	·283 441	300	·595 328	577	·243 009	741	·779 156	183	11
50	0·627 057	227	0·804 979	479	1·283 741	301	1·594 751	576	1·242 268	739	0·778 973	182	10
51	·627 284	226	·805 458	480	·284 042	301	·594 175	575	·241 529	739	·778 791	183	09
52	·627 510	227	·805 938	480	·284 343	301	·593 600	575	·240 790	738	·778 608	182	08
53	·627 737	226	·806 418	480	·284 644	301	·593 025	575	·240 052	738	·778 426	183	07
54	·627 963	226	·806 898	481	·284 945	302	·592 450	573	·239 314	738	·778 243	183	06
55	·628 189	227	·807 379	480	·285 247	302	·591 877	574	·238 576	737	·778 060	182	05
56	·628 416	226	·807 859	481	·285 549	302	·591 303	572	·237 839	736	·777 878	183	04
57	·628 642	226	·808 340	481	·285 851	303	·590 731	573	·237 103	736	·777 695	183	03
58	·628 868	226	·808 821	482	·286 154	303	·590 158	571	·236 367	735	·777 512	183	02
59	·629 094	226	·809 303	481	·286 457	303	·589 587	571	·235 632	735	·777 329	183	01
60	0·629 320		0·809 784		1·286 760		1·589 016		1·234 897		0·777 146		00
	cos		cot		cosec		sec		tan		sin		51°

39° ′	sin	d	tan	d	sec	d	cosec	d	cot	d	cos	d	′
00	0·629 320	226	0·809 784	482	1·286 760	303	1·589 016	571	1·234 897	734	0·777 146	183	60
01	·629 546	226	·810 266	482	·287 063	303	·588 445	570	·234 163	734	·776 963	183	59
02	·629 772	226	·810 748	482	·287 366	304	·587 875	569	·233 429	733	·776 780	184	58
03	·629 998	226	·811 230	482	·287 670	304	·587 306	569	·232 696	733	·776 596	183	57
04	·630 224	226	·811 712	483	·287 974	304	·586 737	568	·231 963	732	·776 413	183	56
05	·630 450	226	·812 195	483	·288 278	305	·586 169	568	·231 231	731	·776 230	184	55
06	·630 676	226	·812 678	483	·288 583	304	·585 601	568	·230 500	731	·776 046	183	54
07	·630 902	225	·813 161	483	·288 887	305	·585 033	566	·229 769	731	·775 863	184	53
08	·631 127	226	·813 644	484	·289 192	306	·584 467	567	·229 038	730	·775 679	183	52
09	·631 353	225	·814 128	484	·289 498	305	·583 900	565	·228 308	729	·775 496	184	51
10	0·631 578	226	0·814 612	484	1·289 803	306	1·583 335	565	1·227 579	729	0·775 312	184	50
11	·631 804	225	·815 096	484	·290 109	306	·582 770	565	·226 850	729	·775 128	184	49
12	·632 029	226	·815 580	485	·290 415	306	·582 205	564	·226 121	728	·774 944	183	48
13	·632 255	225	·816 065	484	·290 721	307	·581 641	563	·225 393	727	·774 761	184	47
14	·632 480	225	·816 549	485	·291 028	307	·581 078	563	·224 666	727	·774 577	184	46
15	·632 705	226	·817 034	485	·291 335	307	·580 515	563	·223 939	727	·774 393	184	45
16	·632 931	225	·817 519	486	·291 642	307	·579 952	562	·223 212	725	·774 209	185	44
17	·633 156	225	·818 005	486	·291 949	307	·579 390	561	·222 487	726	·774 024	184	43
18	·633 381	225	·818 491	485	·292 256	308	·578 829	561	·221 761	725	·773 840	184	42
19	·633 606	225	·818 976	487	·292 564	308	·578 268	560	·221 036	724	·773 656	184	41
20	0·633 831	225	0·819 463	486	1·292 872	309	1·577 708	560	1·220 312	724	0·773 472	185	40
21	·634 056	225	·819 949	486	·293 181	308	·577 148	559	·219 588	723	·773 287	184	39
22	·634 281	225	·820 435	487	·293 489	309	·576 589	559	·218 865	723	·773 103	185	38
23	·634 506	225	·820 922	487	·293 798	309	·576 030	558	·218 142	722	·772 918	184	37
24	·634 731	224	·821 409	488	·294 107	309	·575 472	558	·217 420	722	·772 734	185	36
25	·634 955	225	·821 897	487	·294 416	310	·574 914	557	·216 698	721	·772 549	185	35
26	·635 180	225	·822 384	488	·294 726	310	·574 357	557	·215 977	721	·772 364	185	34
27	·635 405	224	·822 872	488	·295 036	310	·573 800	556	·215 256	720	·772 179	184	33
28	·635 629	225	·823 360	488	·295 346	310	·573 244	555	·214 536	720	·771 995	185	32
29	·635 854	224	·823 848	488	·295 656	311	·572 689	555	·213 816	719	·771 810	185	31
30	0·636 078	225	0·824 336	489	1·295 967	311	1·572 134	555	1·213 097	719	0·771 625	185	30
31	·636 303	224	·824 825	489	·296 278	311	·571 579	554	·212 378	718	·771 440	186	29
32	·636 527	224	·825 314	489	·296 589	311	·571 025	553	·211 660	718	·771 254	185	28
33	·636 751	225	·825 803	489	·296 900	312	·570 472	553	·210 942	717	·771 069	185	27
34	·636 976	224	·826 292	490	·297 212	312	·569 919	553	·210 225	716	·770 884	185	26
35	·637 200	224	·826 782	490	·297 524	312	·569 366	551	·209 509	717	·770 699	186	25
36	·637 424	224	·827 272	490	·297 836	313	·568 815	552	·208 792	715	·770 513	185	24
37	·637 648	224	·827 762	490	·298 149	312	·568 263	551	·208 077	715	·770 328	186	23
38	·637 872	224	·828 252	491	·298 461	313	·567 712	550	·207 362	715	·770 142	185	22
39	·638 096	224	·828 743	491	·298 774	314	·567 162	550	·206 647	714	·769 957	186	21
40	0·638 320	224	0·829 234	491	1·299 088	313	1·566 612	549	1·205 933	714	0·769 771	186	20
41	·638 544	224	·829 725	491	·299 401	314	·566 063	549	·205 219	713	·769 585	185	19
42	·638 768	224	·830 216	491	·299 715	314	·565 514	548	·204 506	713	·769 400	186	18
43	·638 992	223	·830 707	492	·300 029	314	·564 966	548	·203 793	712	·769 214	186	17
44	·639 215	224	·831 199	492	·300 343	315	·564 418	547	·203 081	712	·769 028	186	16
45	·639 439	224	·831 691	492	·300 658	314	·563 871	547	·202 369	711	·768 842	186	15
46	·639 663	223	·832 183	493	·300 972	315	·563 324	546	·201 658	711	·768 656	186	14
47	·639 886	224	·832 676	493	·301 287	316	·562 778	546	·200 947	710	·768 470	186	13
48	·640 110	223	·833 169	493	·301 603	315	·562 232	545	·200 237	709	·768 284	187	12
49	·640 333	224	·833 662	493	·301 918	316	·561 687	545	·199 528	710	·768 097	186	11
50	0·640 557	223	0·834 155	493	1·302 234	316	1·561 142	544	1·198 818	708	0·767 911	186	10
51	·640 780	223	·834 648	494	·302 550	317	·560 598	543	·198 110	708	·767 725	187	09
52	·641 003	223	·835 142	494	·302 867	316	·560 055	544	·197 402	708	·767 538	186	08
53	·641 226	224	·835 636	494	·303 183	317	·559 511	542	·196 694	707	·767 352	187	07
54	·641 450	223	·836 130	494	·303 500	317	·558 969	542	·195 987	707	·767 165	186	06
55	·641 673	223	·836 624	495	·303 817	318	·558 427	542	·195 280	706	·766 979	187	05
56	·641 896	223	·837 119	495	·304 135	318	·557 885	541	·194 574	706	·766 792	187	04
57	·642 119	223	·837 614	495	·304 453	318	·557 344	541	·193 868	705	·766 605	187	03
58	·642 342	223	·838 109	495	·304 771	318	·556 803	540	·193 163	705	·766 418	187	02
59	·642 565	223	·838 604	496	·305 089	318	·556 263	539	·192 458	704	·766 231	187	01
60	0·642 788		0·839 100		1·305 407		1·555 724		1·191 754		0·766 044		00
	cos		cot		cosec		sec		tan		sin		50°

40° ′	sin		tan		sec		cosec		cot		cos	′
00	0·642 788	222	0·839 100	495	1·305 407	319	1·555 724	539	1·191 754	704	0·766 044	187	60
01	·643 010	223	·839 595	497	·305 726	319	·555 185	539	·191 050	703	·765 857	187	59
02	·643 233	223	·840 092	496	·306 045	319	·554 646	538	·190 347	703	·765 670	187	58
03	·643 456	223	·840 588	496	·306 364	320	·554 108	537	·189 644	703	·765 483	187	57
04	·643 679	222	·841 084	497	·306 684	320	·553 571	537	·188 941	701	·765 296	187	56
05	·643 901	223	·841 581	497	·307 004	320	·553 034	537	·188 240	702	·765 109	188	55
06	·644 124	222	·842 078	497	·307 324	320	·552 497	536	·187 538	701	·764 921	187	54
07	·644 346	223	·842 575	498	·307 644	321	·551 961	536	·186 837	700	·764 734	187	53
08	·644 569	222	·843 073	498	·307 965	321	·551 425	535	·186 137	700	·764 547	188	52
09	·644 791	222	·843 571	498	·308 286	321	·550 890	534	·185 437	699	·764 359	188	51
10	0·645 013	222	0·844 069	498	1·308 607	321	1·550 356	534	1·184 738	699	0·764 171	187	50
11	·645 235	223	·844 567	499	·308 928	322	·549 822	534	·184 039	699	·763 984	188	49
12	·645 458	222	·845 066	498	·309 250	322	·549 288	533	·183 340	698	·763 796	188	48
13	·645 680	222	·845 564	499	·309 572	322	·548 755	532	·182 642	697	·763 608	188	47
14	·645 902	222	·846 063	499	·309 894	323	·548 223	532	·181 945	697	·763 420	188	46
15	·646 124	222	·846 562	500	·310 217	323	·547 691	532	·181 248	697	·763 232	188	45
16	·646 346	222	·847 062	500	·310 540	323	·547 159	531	·180 551	696	·763 044	188	44
17	·646 568	222	·847 562	500	·310 863	323	·546 628	531	·179 855	695	·762 856	188	43
18	·646 790	222	·848 062	500	·311 186	324	·546 097	530	·179 160	696	·762 668	188	42
19	·647 012	221	·848 562	500	·311 510	323	·545 567	529	·178 464	694	·762 480	188	41
20	0·647 233	222	0·849 062	501	1·311 833	325	1·545 038	529	1·177 770	694	0·762 292	188	40
21	·647 455	222	·849 563	501	·312 158	324	·544 509	529	·177 076	694	·762 104	189	39
22	·647 677	221	·850 064	501	·312 482	325	·543 980	528	·176 382	693	·761 915	188	38
23	·647 898	222	·850 565	502	·312 807	325	·543 452	528	·175 689	693	·761 727	189	37
24	·648 120	221	·851 067	501	·313 132	325	·542 924	527	·174 996	692	·761 538	188	36
25	·648 341	222	·851 568	502	·313 457	325	·542 397	526	·174 304	692	·761 350	189	35
26	·648 563	221	·852 070	503	·313 782	326	·541 871	526	·173 612	691	·761 161	189	34
27	·648 784	222	·852 573	502	·314 108	326	·541 345	526	·172 921	691	·760 972	188	33
28	·649 006	221	·853 075	503	·314 434	326	·540 819	525	·172 230	691	·760 784	189	32
29	·649 227	221	·853 578	503	·314 760	327	·540 294	525	·171 539	689	·760 595	189	31
30	0·649 448	221	0·854 081	503	1·315 087	327	1·539 769	524	1·170 850	690	0·760 406	189	30
31	·649 669	221	·854 584	503	·315 414	327	·539 245	524	·170 160	689	·760 217	189	29
32	·649 890	221	·855 087	504	·315 741	327	·538 721	523	·169 471	688	·760 028	189	28
33	·650 111	221	·855 591	504	·316 068	328	·538 198	523	·168 783	688	·759 839	189	27
34	·650 332	221	·856 095	504	·316 396	328	·537 675	522	·168 095	688	·759 650	189	26
35	·650 553	221	·856 599	505	·316 724	328	·537 153	522	·167 407	687	·759 461	190	25
36	·650 774	221	·857 104	504	·317 052	329	·536 631	521	·166 720	687	·759 271	189	24
37	·650 995	221	·857 608	505	·317 381	329	·536 110	521	·166 033	686	·759 082	189	23
38	·651 216	221	·858 113	506	·317 710	329	·535 589	520	·165 347	685	·758 893	190	22
39	·651 437	220	·858 619	505	·318 039	329	·535 069	520	·164 662	686	·758 703	189	21
40	0·651 657	221	0·859 124	506	1·318 368	330	1·534 549	519	1·163 976	684	0·758 514	190	20
41	·651 878	220	·859 630	506	·318 698	329	·534 030	519	·163 292	685	·758 324	190	19
42	·652 098	221	·860 136	506	·319 027	331	·533 511	518	·162 607	684	·758 134	189	18
43	·652 319	220	·860 642	506	·319 358	330	·532 993	518	·161 923	683	·757 945	190	17
44	·652 539	221	·861 148	507	·319 688	331	·532 475	518	·161 240	683	·757 755	190	16
45	·652 760	220	·861 655	507	·320 019	331	·531 957	517	·160 557	682	·757 565	190	15
46	·652 980	220	·862 162	507	·320 350	331	·531 440	516	·159 875	682	·757 375	190	14
47	·653 200	221	·862 669	508	·320 681	332	·530 924	516	·159 193	682	·757 185	190	13
48	·653 421	220	·863 177	508	·321 013	331	·530 408	516	·158 511	681	·756 995	190	12
49	·653 641	220	·863 685	508	·321 344	333	·529 892	515	·157 830	681	·756 805	190	11
50	0·653 861	220	0·864 193	508	1·321 677	332	1·529 377	514	1·157 149	680	0·756 615	190	10
51	·654 081	220	·864 701	508	·322 009	333	·528 863	514	·156 469	679	·756 425	191	09
52	·654 301	220	·865 209	509	·322 342	333	·528 349	514	·155 790	680	·756 234	190	08
53	·654 521	220	·865 718	509	·322 675	333	·527 835	513	·155 110	678	·756 044	191	07
54	·654 741	220	·866 227	509	·323 008	333	·527 322	513	·154 432	679	·755 853	190	06
55	·654 961	219	·866 736	510	·323 341	334	·526 809	512	·153 753	678	·755 663	191	05
56	·655 180	220	·867 246	510	·323 675	334	·526 297	512	·153 075	677	·755 472	190	04
57	·655 400	220	·867 756	510	·324 009	334	·525 785	511	·152 398	677	·755 282	191	03
58	·655 620	219	·868 266	510	·324 343	335	·525 274	511	·151 721	677	·755 091	191	02
59	·655 839	220	·868 776	511	·324 678	335	·524 763	510	·151 044	676	·754 900	190	01
60	0·656 059		0·869 287		1·325 013		1·524 253		1·150 368		0·754 710		00
	cos		cot		cosec		sec		tan		sin	49°

41° ′	sin		tan		sec		cosec		cot		cos		′
00	0·656 059	220	0·869 287	511	1·325 013	335	1·524 253	510	1·150 368	675	0·754 710	191	60
01	·656 279	219	·869 798	511	·325 348	336	·523 743	509	·149 693	675	·754 519	191	59
02	·656 498	219	·870 309	511	·325 684	335	·523 234	509	·149 018	675	·754 328	191	58
03	·656 717	220	·870 820	512	·326 019	336	·522 725	508	·148 343	674	·754 137	191	57
04	·656 937	219	·871 332	511	·326 355	337	·522 217	508	·147 669	674	·753 946	191	56
05	·657 156	219	·871 843	513	·326 692	336	·521 709	508	·146 995	673	·753 755	192	55
06	·657 375	219	·872 356	512	·327 028	337	·521 201	507	·146 322	673	·753 563	191	54
07	·657 594	220	·872 868	513	·327 365	337	·520 694	506	·145 649	673	·753 372	191	53
08	·657 814	219	·873 381	513	·327 702	338	·520 188	506	·144 976	672	·753 181	192	52
09	·658 033	219	·873 894	513	·328 040	338	·519 682	506	·144 304	671	·752 989	191	51
10	0·658 252	219	0·874 407	513	1·328 378	338	1·519 176	505	1·143 633	672	0·752 798	192	50
11	·658 471	218	·874 920	514	·328 716	338	·518 671	505	·142 961	670	·752 606	191	49
12	·658 689	219	·875 434	514	·329 054	339	·518 166	504	·142 291	670	·752 415	192	48
13	·658 908	219	·875 948	514	·329 393	338	·517 662	504	·141 621	670	·752 223	191	47
14	·659 127	219	·876 462	514	·329 731	340	·517 158	503	·140 951	670	·752 032	192	46
15	·659 346	218	·876 976	515	·330 071	339	·516 655	503	·140 281	668	·751 840	192	45
16	·659 564	219	·877 491	515	·330 410	340	·516 152	502	·139 613	669	·751 648	192	44
17	·659 783	219	·878 006	515	·330 750	340	·515 650	502	·138 944	668	·751 456	192	43
18	·660 002	218	·878 521	516	·331 090	340	·515 148	502	·138 276	667	·751 264	192	42
19	·660 220	219	·879 037	516	·331 430	341	·514 646	501	·137 609	668	·751 072	192	41
20	0·660 439	218	0·879 553	516	1·331 771	341	1·514 145	500	1·136 941	666	0·750 880	192	40
21	·660 657	218	·880 069	516	·332 112	341	·513 645	500	·136 275	666	·750 688	192	39
22	·660 875	219	·880 585	517	·332 453	341	·513 145	500	·135 609	666	·750 496	193	38
23	·661 094	218	·881 102	517	·332 794	342	·512 645	499	·134 943	666	·750 303	192	37
24	·661 312	218	·881 619	517	·333 136	342	·512 146	499	·134 277	665	·750 111	192	36
25	·661 530	218	·882 136	517	·333 478	342	·511 647	498	·133 612	664	·749 919	193	35
26	·661 748	218	·882 653	518	·333 820	343	·511 149	498	·132 948	664	·749 726	192	34
27	·661 966	218	·883 171	518	·334 163	343	·510 651	497	·132 284	664	·749 534	193	33
28	·662 184	218	·883 689	518	·334 506	343	·510 154	497	·131 620	663	·749 341	193	32
29	·662 402	218	·884 207	518	·334 849	343	·509 657	497	·130 957	663	·749 148	192	31
30	0·662 620	218	0·884 725	519	1·335 192	344	1·509 160	495	1·130 294	662	0·748 956	193	30
31	·662 838	218	·885 244	519	·335 536	344	·508 665	496	·129 632	662	·748 763	193	29
32	·663 056	217	·885 763	519	·335 880	345	·508 169	495	·128 970	661	·748 570	193	28
33	·663 273	218	·886 282	520	·336 225	344	·507 674	495	·128 309	661	·748 377	193	27
34	·663 491	218	·886 802	519	·336 569	345	·507 179	494	·127 648	661	·748 184	193	26
35	·663 709	217	·887 321	521	·336 914	345	·506 685	494	·126 987	660	·747 991	193	25
36	·663 926	218	·887 842	520	·337 259	346	·506 191	493	·126 327	660	·747 798	193	24
37	·664 144	217	·888 362	520	·337 605	346	·505 698	493	·125 667	659	·747 605	193	23
38	·664 361	218	·888 882	521	·337 951	346	·505 205	492	·125 008	659	·747 412	194	22
39	·664 579	217	·889 403	521	·338 297	346	·504 713	492	·124 349	658	·747 218	193	21
40	0·664 796	217	0·889 924	522	1·338 643	347	1·504 221	491	1·123 691	658	0·747 025	193	20
41	·665 013	217	·890 446	521	·338 990	347	·503 730	491	·123 033	658	·746 832	194	19
42	·665 230	218	·890 967	522	·339 337	347	·503 239	491	·122 375	657	·746 638	193	18
43	·665 448	217	·891 489	523	·339 684	348	·502 748	490	·121 718	656	·746 445	194	17
44	·665 665	217	·892 012	522	·340 032	347	·502 258	490	·121 062	657	·746 251	194	16
45	·665 882	217	·892 534	523	·340 379	349	·501 768	489	·120 405	655	·746 057	193	15
46	·666 099	217	·893 057	523	·340 728	348	·501 279	489	·119 750	656	·745 864	194	14
47	·666 316	216	·893 580	523	·341 076	349	·500 790	488	·119 094	655	·745 670	194	13
48	·666 532	217	·894 103	524	·341 425	349	·500 302	488	·118 439	654	·745 476	194	12
49	·666 749	217	·894 627	524	·341 774	349	·499 814	487	·117 785	655	·745 282	194	11
50	0·666 966	217	0·895 151	524	1·342 123	350	1·499 327	487	1·117 130	653	0·745 088	194	10
51	·667 183	216	·895 675	524	·342 473	350	·498 840	487	·116 477	654	·744 894	194	09
52	·667 399	217	·896 199	525	·342 823	350	·498 353	486	·115 823	652	·744 700	194	08
53	·667 616	217	·896 724	525	·343 173	350	·497 867	486	·115 171	653	·744 506	194	07
54	·667 833	216	·897 249	525	·343 523	351	·497 381	485	·114 518	652	·744 312	195	06
55	·668 049	216	·897 774	525	·343 874	351	·496 896	485	·113 866	651	·744 117	194	05
56	·668 265	217	·898 299	526	·344 225	352	·496 411	484	·113 215	652	·743 923	195	04
57	·668 482	216	·898 825	526	·344 577	351	·495 927	484	·112 563	650	·743 728	194	03
58	·668 698	216	·899 351	526	·344 928	352	·495 443	483	·111 913	651	·743 534	195	02
59	·668 914	217	·899 877	527	·345 280	353	·494 960	483	·111 262	649	·743 339	194	01
60	0·669 131		0·900 404		1·345 633		1·494 477		1·110 613		0·743 145		00
	cos		cot		cosec		sec		tan		sin		48°

42° ′	sin		tan		sec		cosec		cot		cos		′
00	0·669 131	216	0·900 404	527	1·345 633	352	1·494 477	483	1·110 613	650	0·743 145	195	60
01	·669 347	216	·900 931	527	·345 985	353	·493 994	482	·109 963	649	·742 950	195	59
02	·669 563	216	·901 458	527	·346 338	353	·493 512	482	·109 314	649	·742 755	194	58
03	·669 779	216	·901 985	528	·346 691	354	·493 030	481	·108 665	648	·742 561	195	57
04	·669 995	216	·902 513	528	·347 045	354	·492 549	481	·108 017	648	·742 366	195	56
05	·670 211	216	·903 041	528	·347 399	354	·492 068	480	·107 369	647	·742 171	195	55
06	·670 427	215	·903 569	529	·347 753	354	·491 588	480	·106 722	647	·741 976	195	54
07	·670 642	216	·904 098	529	·348 107	355	·491 108	480	·106 075	647	·741 781	195	53
08	·670 858	216	·904 627	529	·348 462	355	·490 628	479	·105 428	646	·741 586	195	52
09	·671 074	215	·905 156	529	·348 817	355	·490 149	479	·104 782	645	·741 391	196	51
10	0·671 289	216	0·905 685	530	1·349 172	356	1·489 670	478	1·104 137	646	0·741 195	195	50
11	·671 505	216	·906 215	530	·349 528	356	·489 192	478	·103 491	645	·741 000	195	49
12	·671 721	215	·906 745	530	·349 884	356	·488 714	477	·102 846	644	·740 805	196	48
13	·671 936	215	·907 275	530	·350 240	356	·488 237	477	·102 202	644	·740 609	195	47
14	·672 151	216	·907 805	531	·350 596	357	·487 760	477	·101 558	644	·740 414	196	46
15	·672 367	215	·908 336	531	·350 953	357	·487 283	476	·100 914	643	·740 218	195	45
16	·672 582	215	·908 867	531	·351 310	358	·486 807	475	·100 271	643	·740 023	196	44
17	·672 797	216	·909 398	532	·351 668	357	·486 332	476	·099 628	642	·739 827	196	43
18	·673 013	215	·909 930	532	·352 025	358	·485 856	474	·098 986	642	·739 631	196	42
19	·673 228	215	·910 462	532	·352 383	359	·485 382	475	·098 344	642	·739 435	196	41
20	0·673 443	215	0·910 994	532	1·352 742	358	1·484 907	474	1·097 702	641	0·739 239	196	40
21	·673 658	215	·911 526	533	·353 100	359	·484 433	473	·097 061	641	·739 043	195	39
22	·673 873	215	·912 059	533	·353 459	359	·483 960	473	·096 420	640	·738 848	197	38
23	·674 088	214	·912 592	533	·353 818	360	·483 487	473	·095 780	640	·738 651	196	37
24	·674 302	215	·913 125	534	·354 178	360	·483 014	472	·095 140	640	·738 455	196	36
25	·674 517	215	·913 659	534	·354 538	360	·482 542	472	·094 500	639	·738 259	196	35
26	·674 732	215	·914 193	534	·354 898	360	·482 070	471	·093 861	639	·738 063	196	34
27	·674 947	214	·914 727	534	·355 258	361	·481 599	471	·093 222	638	·737 867	197	33
28	·675 161	215	·915 261	535	·355 619	361	·481 128	471	·092 584	638	·737 670	196	32
29	·675 376	214	·915 796	535	·355 980	362	·480 657	470	·091 946	637	·737 474	197	31
30	0·675 590	215	0·916 331	535	1·356 342	361	1·480 187	469	1·091 309	638	0·737 277	196	30
31	·675 805	214	·916 866	536	·356 703	362	·479 718	470	·090 671	636	·737 081	197	29
32	·676 019	214	·917 402	536	·357 065	363	·479 248	469	·090 035	637	·736 884	197	28
33	·676 233	215	·917 938	536	·357 428	362	·478 779	468	·089 398	636	·736 687	196	27
34	·676 448	214	·918 474	536	·357 790	363	·478 311	468	·088 762	635	·736 491	197	26
35	·676 662	214	·919 010	537	·358 153	363	·477 843	467	·088 127	635	·736 294	197	25
36	·676 876	214	·919 547	537	·358 516	364	·477 376	468	·087 492	635	·736 097	197	24
37	·677 090	214	·920 084	537	·358 880	364	·476 908	466	·086 857	634	·735 900	197	23
38	·677 304	214	·920 621	538	·359 244	364	·476 442	467	·086 223	634	·735 703	197	22
39	·677 518	214	·921 159	538	·359 608	364	·475 975	466	·085 589	634	·735 506	197	21
40	0·677 732	214	0·921 697	538	1·359 972	365	1·475 509	465	1·084 955	633	0·735 309	197	20
41	·677 946	214	·922 235	538	·360 337	365	·475 044	465	·084 322	632	·735 112	197	19
42	·678 160	213	·922 773	539	·360 702	366	·474 579	465	·083 690	633	·734 915	198	18
43	·678 373	214	·923 312	539	·361 068	365	·474 114	464	·083 057	632	·734 717	197	17
44	·678 587	214	·923 851	539	·361 433	366	·473 650	464	·082 425	631	·734 520	197	16
45	·678 801	213	·924 390	540	·361 799	367	·473 186	463	·081 794	631	·734 323	198	15
46	·679 014	214	·924 930	540	·362 166	366	·472 723	463	·081 163	631	·734 125	198	14
47	·679 228	213	·925 470	540	·362 532	367	·472 260	463	·080 532	630	·733 927	197	13
48	·679 441	214	·926 010	541	·362 899	368	·471 797	462	·079 902	630	·733 730	198	12
49	·679 655	213	·926 551	540	·363 267	367	·471 335	461	·079 272	630	·733 532	198	11
50	0·679 868	213	0·927 091	541	1·363 634	368	1·470 874	462	1·078 642	629	0·733 334	197	10
51	·680 081	214	·927 632	542	·364 002	368	·470 412	461	·078 013	629	·733 137	198	09
52	·680 295	213	·928 174	541	·364 370	369	·469 951	460	·077 384	628	·732 939	198	08
53	·680 508	213	·928 715	542	·364 739	369	·469 491	460	·076 756	628	·732 741	198	07
54	·680 721	213	·929 257	543	·365 108	369	·469 031	460	·076 128	627	·732 543	198	06
55	·680 934	213	·929 800	542	·365 477	369	·468 571	459	·075 501	628	·732 345	198	05
56	·681 147	213	·930 342	543	·365 846	370	·468 112	459	·074 873	626	·732 147	198	04
57	·681 360	213	·930 885	543	·366 216	370	·467 653	458	·074 247	627	·731 949	199	03
58	·681 573	213	·931 428	543	·366 586	371	·467 195	458	·073 620	626	·731 750	198	02
59	·681 786	212	·931 971	544	·366 957	370	·466 737	458	·072 994	625	·731 552	198	01
60	0·681 998		0·932 515		1·367 327		1·466 279		1·072 369		0·731 354		00
	cos		cot		cosec		sec		tan		sin		47°

43° ′	sin		tan		sec		cosec		cot		cos		′
00	0·681 998	213	0·932 515	544	1·367 327	372	1·466 279	457	1·072 369	625	0·731 354	199	60
01	·682 211	213	·933 059	544	·367 699	371	·465 822	457	·071 744	625	·731 155	198	59
02	·682 424	212	·933 603	545	·368 070	372	·465 365	456	·071 119	625	·730 957	199	58
03	·682 636	213	·934 148	545	·368 442	372	·464 909	456	·070 494	624	·730 758	198	57
04	·682 849	212	·934 693	545	·368 814	372	·464 453	456	·069 870	623	·730 560	199	56
05	·683 061	213	·935 238	545	·369 186	373	·463 997	455	·069 247	624	·730 361	199	55
06	·683 274	212	·935 783	546	·369 559	373	·463 542	455	·068 623	623	·730 162	199	54
07	·683 486	212	·936 329	546	·369 932	373	·463 087	454	·068 000	622	·729 963	198	53
08	·683 698	213	·936 875	547	·370 305	373	·462 633	454	·067 378	622	·729 765	199	52
09	·683 911	212	·937 422	546	·370 678	374	·462 179	453	·066 756	622	·729 566	199	51
10	0·684 123	212	0·937 968	547	1·371 052	375	1·461 726	453	1·066 134	621	0·729 367	199	50
11	·684 335	212	·938 515	548	·371 427	374	·461 273	453	·065 513	621	·729 168	199	49
12	·684 547	212	·939 063	547	·371 801	375	·460 820	452	·064 892	621	·728 969	200	48
13	·684 759	212	·939 610	548	·372 176	375	·460 368	452	·064 271	620	·728 769	199	47
14	·684 971	212	·940 158	548	·372 551	376	·459 916	452	·063 651	620	·728 570	199	46
15	·685 183	212	·940 706	549	·372 927	376	·459 464	451	·063 031	619	·728 371	199	45
16	·685 395	212	·941 255	548	·373 303	376	·459 013	451	·062 412	619	·728 172	200	44
17	·685 607	211	·941 803	549	·373 679	376	·458 562	450	·061 793	619	·727 972	199	43
18	·685 818	212	·942 352	550	·374 055	377	·458 112	450	·061 174	618	·727 773	200	42
19	·686 030	212	·942 902	549	·374 432	377	·457 662	449	·060 556	618	·727 573	199	41
20	0·686 242	211	0·943 451	550	1·374 809	378	1·457 213	449	1·059 938	617	0·727 374	200	40
21	·686 453	212	·944 001	551	·375 187	377	·456 764	449	·059 321	618	·727 174	200	39
22	·686 665	211	·944 552	550	·375 564	379	·456 315	448	·058 703	616	·726 974	199	38
23	·686 876	212	·945 102	551	·375 943	378	·455 867	448	·058 087	617	·726 775	200	37
24	·687 088	211	·945 653	551	·376 321	379	·455 419	448	·057 470	616	·726 575	200	36
25	·687 299	211	·946 204	552	·376 700	379	·454 971	447	·056 854	615	·726 375	200	35
26	·687 510	211	·946 756	551	·377 079	379	·454 524	447	·056 239	615	·726 175	200	34
27	·687 721	211	·947 307	552	·377 458	380	·454 077	446	·055 624	615	·725 975	200	33
28	·687 932	212	·947 859	553	·377 838	380	·453 631	446	·055 009	615	·725 775	200	32
29	·688 144	211	·948 412	553	·378 218	380	·453 185	445	·054 394	614	·725 575	201	31
30	0·688 355	211	0·948 965	553	1·378 598	381	1·452 740	445	1·053 780	614	0·725 374	200	30
31	·688 566	210	·949 518	553	·378 979	381	·452 295	445	·053 166	613	·725 174	200	29
32	·688 776	211	·950 071	553	·379 360	382	·451 850	444	·052 553	613	·724 974	201	28
33	·688 987	211	·950 624	554	·379 742	381	·451 406	444	·051 940	612	·724 773	200	27
34	·689 198	211	·951 178	555	·380 123	382	·450 962	444	·051 328	613	·724 573	201	26
35	·689 409	211	·951 733	554	·380 505	383	·450 518	443	·050 715	612	·724 372	200	25
36	·689 620	210	·952 287	555	·380 888	382	·450 075	443	·050 103	611	·724 172	201	24
37	·689 830	211	·952 842	555	·381 270	383	·449 632	442	·049 492	611	·723 971	200	23
38	·690 041	210	·953 397	556	·381 653	384	·449 190	442	·048 881	611	·723 771	201	22
39	·690 251	211	·953 953	555	·382 037	383	·448 748	442	·048 270	610	·723 570	201	21
40	0·690 462	210	0·954 508	556	1·382 420	384	1·448 306	441	1·047 660	610	0·723 369	201	20
41	·690 672	210	·955 064	557	·382 804	385	·447 865	441	·047 050	610	·723 168	201	19
42	·690 882	211	·955 621	556	·383 189	384	·447 424	440	·046 440	609	·722 967	201	18
43	·691 093	210	·956 177	557	·383 573	385	·446 984	440	·045 831	609	·722 766	201	17
44	·691 303	210	·956 734	558	·383 958	386	·446 544	440	·045 222	608	·722 565	201	16
45	·691 513	210	·957 292	557	·384 344	385	·446 104	439	·044 614	608	·722 364	201	15
46	·691 723	210	·957 849	558	·384 729	386	·445 665	439	·044 006	608	·722 163	201	14
47	·691 933	210	·958 407	559	·385 115	387	·445 226	438	·043 398	608	·721 962	202	13
48	·692 143	210	·958 966	558	·385 502	386	·444 788	438	·042 790	607	·721 760	201	12
49	·692 353	210	·959 524	559	·385 888	387	·444 350	438	·042 183	606	·721 559	202	11
50	0·692 563	210	0·960 083	559	1·386 275	388	1·443 912	437	1·041 577	607	0·721 357	201	10
51	·692 773	210	·960 642	560	·386 663	387	·443 475	437	·040 970	606	·721 156	202	09
52	·692 983	209	·961 202	559	·387 050	388	·443 038	437	·040 364	605	·720 954	201	08
53	·693 192	210	·961 761	561	·387 438	389	·442 601	436	·039 759	605	·720 753	202	07
54	·693 402	209	·962 322	560	·387 827	388	·442 165	436	·039 154	605	·720 551	202	06
55	·693 611	210	·962 882	561	·388 215	389	·441 729	435	·038 549	605	·720 349	201	05
56	·693 821	209	·963 443	561	·388 604	390	·441 294	435	·037 944	604	·720 148	202	04
57	·694 030	210	·964 004	561	·388 994	389	·440 859	434	·037 340	603	·719 946	202	03
58	·694 240	209	·964 565	562	·389 383	390	·440 425	435	·036 737	604	·719 744	202	02
59	·694 449	209	·965 127	562	·389 773	391	·439 990	433	·036 133	603	·719 542	202	01
60	0·694 658		0·965 689		1·390 164		1·439 557		1·035 530		0·719 340		00
	cos		cot		cosec		sec		tan		sin		46°

44° ′	sin		tan		sec		cosec		cot		cos		′
00	0·694 658	210	0·965 689	562	1·390 164	390	1·439 557	434	1·035 530	602	0·719 340	202	60
01	·694 868	209	·966 251	563	·390 554	391	·439 123	433	·034 928	603	·719 138	202	59
02	·695 077	209	·966 814	563	·390 945	392	·438 690	433	·034 325	601	·718 936	203	58
03	·695 286	209	·967 377	563	·391 337	391	·438 257	432	·033 724	602	·718 733	202	57
04	·695 495	209	·967 940	564	·391 728	392	·437 825	432	·033 122	601	·718 531	202	56
05	·695 704	209	·968 504	563	·392 120	393	·437 393	431	·032 521	601	·718 329	203	55
06	·695 913	209	·969 067	565	·392 513	392	·436 962	431	·031 920	601	·718 126	202	54
07	·696 122	208	·969 632	564	·392 905	393	·436 531	431	·031 319	600	·717 924	203	53
08	·696 330	209	·970 196	565	·393 298	394	·436 100	431	·030 719	599	·717 721	202	52
09	·696 539	209	·970 761	565	·393 692	394	·435 669	430	·030 120	600	·717 519	203	51
10	0·696 748	209	0·971 326	566	1·394 086	394	1·435 239	429	1·029 520	599	0·717 316	203	50
11	·696 957	208	·971 892	566	·394 480	394	·434 810	430	·028 921	598	·717 113	202	49
12	·697 165	209	·972 458	566	·394 874	395	·434 380	428	·028 323	599	·716 911	203	48
13	·697 374	208	·973 024	566	·395 269	395	·433 952	429	·027 724	598	·716 708	203	47
14	·697 582	208	·973 590	567	·395 664	395	·433 523	428	·027 126	597	·716 505	203	46
15	·697 790	209	·974 157	567	·396 059	396	·433 095	428	·026 529	598	·716 302	203	45
16	·697 999	208	·974 724	567	·396 455	396	·432 667	427	·025 931	596	·716 099	203	44
17	·698 207	208	·975 291	568	·396 851	397	·432 240	427	·025 335	597	·715 896	203	43
18	·698 415	208	·975 859	568	·397 248	396	·431 813	427	·024 738	596	·715 693	203	42
19	·698 623	209	·976 427	569	·397 644	398	·431 386	426	·024 142	596	·715 490	204	41
20	0·698 832	208	0·976 996	568	1·398 042	397	1·430 960	426	1·023 546	595	0·715 286	203	40
21	·699 040	208	·977 564	569	·398 439	398	·430 534	425	·022 951	595	·715 083	203	39
22	·699 248	207	·978 133	570	·398 837	398	·430 109	425	·022 356	595	·714 880	204	38
23	·699 455	208	·978 703	569	·399 235	399	·429 684	425	·021 761	595	·714 676	203	37
24	·699 663	208	·979 272	570	·399 634	399	·429 259	425	·021 166	594	·714 473	204	36
25	·699 871	208	·979 842	571	·400 033	399	·428 834	424	·020 572	593	·714 269	203	35
26	·700 079	208	·980 413	570	·400 432	399	·428 410	423	·019 979	594	·714 066	204	34
27	·700 287	207	·980 983	571	·400 831	400	·427 987	423	·019 385	593	·713 862	204	33
28	·700 494	208	·981 554	572	·401 231	400	·427 564	423	·018 792	592	·713 658	204	32
29	·700 702	207	·982 126	571	·401 631	401	·427 141	423	·018 200	593	·713 454	204	31
30	0·700 909	208	0·982 697	572	1·402 032	401	1·426 718	422	1·017 607	592	0·713 250	203	30
31	·701 117	207	·983 269	573	·402 433	401	·426 296	422	·017 015	591	·713 047	204	29
32	·701 324	207	·983 842	572	·402 834	402	·425 874	421	·016 424	591	·712 843	204	28
33	·701 531	208	·984 414	573	·403 236	402	·425 453	421	·015 833	591	·712 639	205	27
34	·701 739	207	·984 987	573	·403 638	402	·425 032	421	·015 242	591	·712 434	204	26
35	·701 946	207	·985 560	574	·404 040	403	·424 611	420	·014 651	590	·712 230	204	25
36	·702 153	207	·986 134	574	·404 443	403	·424 191	420	·014 061	590	·712 026	204	24
37	·702 360	207	·986 708	574	·404 846	403	·423 771	420	·013 471	589	·711 822	205	23
38	·702 567	207	·987 282	575	·405 249	404	·423 351	419	·012 882	589	·711 617	204	22
39	·702 774	207	·987 857	575	·405 653	404	·422 932	419	·012 293	589	·711 413	204	21
40	0·702 981	207	0·988 432	575	1·406 057	405	1·422 513	418	1·011 704	589	0·711 209	205	20
41	·703 188	207	·989 007	575	·406 462	405	·422 095	418	·011 115	588	·711 004	205	19
42	·703 395	206	·989 582	576	·406 867	405	·421 677	418	·010 527	588	·710 799	204	18
43	·703 601	207	·990 158	577	·407 272	405	·421 259	417	·009 939	587	·710 595	205	17
44	·703 808	207	·990 735	576	·407 677	406	·420 842	417	·009 352	587	·710 390	205	16
45	·704 015	206	·991 311	577	·408 083	406	·420 425	417	·008 765	587	·710 185	204	15
46	·704 221	207	·991 888	577	·408 489	407	·420 008	416	·008 178	586	·709 981	205	14
47	·704 428	206	·992 465	578	·408 896	407	·419 592	416	·007 592	586	·709 776	205	13
48	·704 634	207	·993 043	578	·409 303	407	·419 176	415	·007 006	586	·709 571	205	12
49	·704 841	206	·993 621	578	·409 710	408	·418 761	416	·006 420	585	·709 366	205	11
50	0·705 047	206	0·994 199	579	1·410 118	408	1·418 345	414	1·005 835	585	0·709 161	205	10
51	·705 253	206	·994 778	579	·410 526	408	·417 931	415	·005 250	585	·708 956	206	09
52	·705 459	206	·995 357	579	·410 934	409	·417 516	414	·004 665	584	·708 750	205	08
53	·705 665	207	·995 936	579	·411 343	409	·417 102	414	·004 081	584	·708 545	205	07
54	·705 872	206	·996 515	580	·411 752	409	·416 688	413	·003 497	584	·708 340	206	06
55	·706 078	206	·997 095	581	·412 161	410	·416 275	413	·002 913	583	·708 134	205	05
56	·706 284	205	·997 676	580	·412 571	410	·415 862	413	·002 330	583	·707 929	205	04
57	·706 489	206	·998 256	581	·412 981	411	·415 449	412	·001 747	583	·707 724	206	03
58	·706 695	206	·998 837	581	·413 392	410	·415 037	412	·001 164	582	·707 518	206	02
59	·706 901	206	0·999 418	582	·413 802	412	·414 625	411	·000 582	582	·707 312	205	01
60	0·707 107		1·000 000		1·414 214		1·414 214		1·000 000		0·707 107		00
	cos		cot		cosec		sec		tan		sin		45°

TABLE IVA—CIRCULAR SINES AND COSINES

x	$\sin x$	$\cos x$	x	$\sin x$	$\cos x$	x	$\sin x$	$\cos x$
r			r			r		
0·0	+·000 000	+ Unity	**5·0**	−·958 924	+·283 662	**10·0**	−·544 021	−·839 072
0·1	·099 833	·995 004	**5·1**	·925 815	·377 978	**10·1**	·625 071	·780 568
0·2	·198 669	·980 067	**5·2**	·883 455	·468 517	**10·2**	·699 875	·714 266
0·3	·295 520	·955 336	**5·3**	·832 267	·554 374	**10·3**	·767 686	·640 826
0·4	·389 418	·921 061	**5·4**	·772 764	·634 693	**10·4**	·827 826	·560 984
0·5	+·479 426	+·877 583	**5·5**	−·705 540	+·708 670	**10**	−·544 021	−·839 072
0·6	·564 642	·825 336	**5·6**	·631 267	·775 566	**11**	−·999 990	+·004 426
0·7	·644 218	·764 842	**5·7**	·550 686	·834 713	**12**	−·536 573	+·843 854
0·8	·717 356	·696 707	**5·8**	·464 602	·885 520	**13**	+·420 167	+·907 447
0·9	·783 327	·621 610	**5·9**	·373 877	·927 478	**14**	+·990 607	+·136 737
1·0	+·841 471	+·540 302	**6·0**	−·279 415	+·960 170	**15**	+·650 288	−·759 688
1·1	·891 207	·453 596	**6·1**	·182 163	·983 268	**16**	−·287 903	−·957 659
1·2	·932 039	·362 358	**6·2**	−·083 089	·996 542	**17**	−·961 397	−·275 163
1·3	·963 558	·267 499	**6·3**	+·016 814	·999 859	**18**	−·750 987	+·660 317
1·4	·985 450	·169 967	**6·4**	·116 549	·993 185	**19**	+·149 877	+·988 705
1·5	+·997 495	+·070 737	**6·5**	+·215 120	+·976 588	**20**	+·912 945	+·408 082
1·6	·999 574	−·029 200	**6·6**	·311 541	·950 233	**21**	+·836 656	−·547 729
1·7	·991 665	·128 844	**6·7**	·404 850	·914 383	**22**	−·008 851	−·999 961
1·8	·973 848	·227 202	**6·8**	·494 113	·869 397	**23**	−·846 220	−·532 833
1·9	·946 300	·323 290	**6·9**	·578 440	·815 725	**24**	−·905 578	+·424 179
2·0	+·909 297	−·416 147	**7·0**	+·656 987	+·753 902	**25**	−·132 352	+·991 203
2·1	·863 209	·504 846	**7·1**	·728 969	·684 547	**26**	+·762 558	+·646 919
2·2	·808 496	·588 501	**7·2**	·793 668	·608 351	**27**	+·956 376	−·292 139
2·3	·745 705	·666 276	**7·3**	·850 437	·526 078	**28**	+·270 906	−·962 606
2·4	·675 463	·737 394	**7·4**	·898 708	·438 547	**29**	−·663 634	−·748 058
2·5	+·598 472	−·801 144	**7·5**	+·938 000	+·346 635	**30**	−·988 032	+·154 251
2·6	·515 501	·856 889	**7·6**	·967 920	·251 260	**31**	−·404 038	+·914 742
2·7	·427 380	·904 072	**7·7**	·988 168	·153 374	**32**	+·551 427	+·834 223
2·8	·334 988	·942 222	**7·8**	·998 543	+·053 955	**33**	+·999 912	−·013 277
2·9	·239 249	·970 958	**7·9**	·998 941	−·046 002	**34**	+·529 083	−·848 570
3·0	+·141 120	−·989 992	**8·0**	+·989 358	−·145 500	**35**	−·428 183	−·903 692
3·1	+·041 581	·999 135	**8·1**	·969 890	·243 544	**36**	−·991 779	−·127 964
3·2	−·058 374	·998 295	**8·2**	·940 731	·339 155	**37**	−·643 538	+·765 414
3·3	·157 746	·987 480	**8·3**	·902 172	·431 377	**38**	+·296 369	+·955 074
3·4	·255 541	·966 798	**8·4**	·854 599	·519 289	**39**	+·963 795	+·266 643
3·5	−·350 783	−·936 457	**8·5**	+·798 487	−·602 012	**40**	+·745 113	−·666 938
3·6	·442 520	·896 758	**8·6**	·734 397	·678 720	**41**	−·158 623	−·987 339
3·7	·529 836	·848 100	**8·7**	·662 969	·748 647	**42**	−·916 522	−·399 985
3·8	·611 858	·790 968	**8·8**	·584 917	·811 093	**43**	−·831 775	+·555 113
3·9	·687 766	·725 932	**8·9**	·501 021	·865 435	**44**	+·017 702	+·999 843
4·0	−·756 802	−·653 644	**9·0**	+·412 118	−·911 130	**45**	+·850 904	+·525 322
4·1	·818 277	·574 824	**9·1**	·319 098	·947 722	**46**	+·901 788	−·432 178
4·2	·871 576	·490 261	**9·2**	·222 890	·974 844	**47**	+·123 573	−·992 335
4·3	·916 166	·400 799	**9·3**	·124 454	·992 225	**48**	−·768 255	−·640 144
4·4	·951 602	·307 333	**9·4**	+·024 775	·999 693	**49**	−·953 753	+·300 593
4·5	−·977 530	−·210 796	**9·5**	−·075 151	−·997 172	**50**	−·262 375	+·964 966
4·6	·993 691	·112 153	**9·6**	·174 327	·984 688	**51**	+·670 229	+·742 154
4·7	·999 923	−·012 389	**9·7**	·271 761	·962 365	**52**	+·986 628	−·162 991
4·8	·996 165	+·087 499	**9·8**	·366 479	·930 426	**53**	+·395 925	−·918 283
4·9	−·982 453	+·186 512	**9·9**	−·457 536	−·889 191	**54**	−·558 789	−·829 310

Up to $x = 54$ any sine or cosine may be found by dividing x into a part x_0 that corresponds to a tabular value and a part Δx that does not exceed 1·570. Then

$\sin(x_0+\Delta x) = \sin x_0 \cos \Delta x + \cos x_0 \sin \Delta x$ $\qquad$ $\cos(x_0+\Delta x) = \cos x_0 \cos \Delta x - \sin x_0 \sin \Delta x$

If both sine and cosine are produced, it should be verified, as a partial check, that $\sin^2 x + \cos^2 x = 1$, within two (or rarely three) units of the sixth decimal.

TABLE IVB—REDUCTION TO THE FIRST QUADRANT

n	$n\frac{\pi}{2}$ r	Rev.	Deg. °	Quad.	sin	cos	tan	cot	sec	cosec
0	0·000 000	0	0	I	+sin	+cos	+tan	+cot	+sec	+cosec
1	1·570 796	0¼	90	II	+cos	−sin	−cot	−tan	−cosec	+sec
2	3·141 593	0½	180	III	−sin	−cos	+tan	+cot	−sec	−cosec
3	4·712 389	0¾	270	IV	−cos	+sin	−cot	−tan	+cosec	−sec
4	6·283 185	1	360	I	+sin	+cos	+tan	+cot	+sec	+cosec
5	7·853 982	1¼	450	II	+cos	−sin	−cot	−tan	−cosec	+sec
6	9·424 778	1½	540	III	−sin	−cos	+tan	+cot	−sec	−cosec
7	10·995 574	1¾	630	IV	−cos	+sin	−cot	−tan	+cosec	−sec
8	12·566 371	2	720	I	+sin	+cos	+tan	+cot	+sec	+cosec
9	14·137 167	2¼	810	II	+cos	−sin	−cot	−tan	−cosec	+sec
10	15·707 963	2½	900	III	−sin	−cos	+tan	+cot	−sec	−cosec
11	17·278 760	2¾	990	IV	−cos	+sin	−cot	−tan	+cosec	−sec
12	18·849 556	3	1080	I	+sin	+cos	+tan	+cot	+sec	+cosec
13	20·420 352	3¼	1170	II	+cos	−sin	−cot	−tan	−cosec	+sec
14	21·991 149	3½	1260	III	−sin	−cos	+tan	+cot	−sec	−cosec
15	23·561 945	3¾	1350	IV	−cos	+sin	−cot	−tan	+cosec	−sec
16	25·132 741	4	1440	I	+sin	+cos	+tan	+cot	+sec	+cosec
17	26·703 538	4¼	1530	II	+cos	−sin	−cot	−tan	−cosec	+sec
18	28·274 334	4½	1620	III	−sin	−cos	+tan	+cot	−sec	−cosec
19	29·845 130	4¾	1710	IV	−cos	+sin	−cot	−tan	+cosec	−sec
20	31·415 927	5	1800	I	+sin	+cos	+tan	+cot	+sec	+cosec
21	32·986 723	5¼	1890	II	+cos	−sin	−cot	−tan	−cosec	+sec
22	34·557 519	5½	1980	III	−sin	−cos	+tan	+cot	−sec	−cosec
23	36·128 316	5¾	2070	IV	−cos	+sin	−cot	−tan	+cosec	−sec
24	37·699 112	6	2160	I	+sin	+cos	+tan	+cot	+sec	+cosec
25	39·269 908	6¼	2250	II	+cos	−sin	−cot	−tan	−cosec	+sec
26	40·840 704	6½	2340	III	−sin	−cos	+tan	+cot	−sec	−cosec
27	42·411 501	6¾	2430	IV	−cos	+sin	−cot	−tan	+cosec	−sec
28	43·982 297	7	2520	I	+sin	+cos	+tan	+cot	+sec	+cosec
29	45·553 093	7¼	2610	II	+cos	−sin	−cot	−tan	−cosec	+sec
30	47·123 890	7½	2700	III	−sin	−cos	+tan	+cot	−sec	−cosec
31	48·694 686	7¾	2790	IV	−cos	+sin	−cot	−tan	+cosec	−sec
32	50·265 482	8	2880	I	+sin	+cos	+tan	+cot	+sec	+cosec
33	51·836 279	8¼	2970	II	+cos	−sin	−cot	−tan	−cosec	+sec
34	53·407 075	8½	3060	III	−sin	−cos	+tan	+cot	−sec	−cosec
35	54·977 871	8¾	3150	IV	−cos	+sin	−cot	−tan	+cosec	−sec
36	56·548 668	9	3240	I	+sin	+cos	+tan	+cot	+sec	+cosec
37	58·119 464	9¼	3330	II	+cos	−sin	−cot	−tan	−cosec	+sec
38	59·690 260	9½	3420	III	−sin	−cos	+tan	+cot	−sec	−cosec
39	61·261 057	9¾	3510	IV	−cos	+sin	−cot	−tan	+cosec	−sec
40	62·831 853	10	3600	I	+sin	+cos	+tan	+cot	+sec	+cosec
41	64·402 649	10¼	3690	II	+cos	−sin	−cot	−tan	−cosec	+sec
42	65·973 446	10½	3780	III	−sin	−cos	+tan	+cot	−sec	−cosec
43	67·544 242	10¾	3870	IV	−cos	+sin	−cot	−tan	+cosec	−sec
44	69·115 038	11	3960	I	+sin	+cos	+tan	+cot	+sec	+cosec
45	70·685 835	11¼	4050	II	+cos	−sin	−cot	−tan	−cosec	+sec
46	72·256 631	11½	4140	III	−sin	−cos	+tan	+cot	−sec	−cosec
47	73·827 427	11¾	4230	IV	−cos	+sin	−cot	−tan	+cosec	−sec
48	75·398 224	12	4320	I	+sin	+cos	+tan	+cot	+sec	+cosec
49	76·969 020	12¼	4410	II	+cos	−sin	−cot	−tan	−cosec	+sec
50	78·539 816	12½	4500	III	−sin	−cos	+tan	+cot	−sec	−cosec
51	80·110 613	12¾	4590	IV	−cos	+sin	−cot	−tan	+cosec	−sec

Multiples of 2π or one revolution

x less than 1000 radians				*x* greater than 1000 radians			
Rev.	$2\pi \times$ Rev.	Rev.	$2\pi \times$ Rev.	Rev.	$2\pi \times$ Rev.	Rev.	$2\pi \times$ Rev.
	r		r		r		r
13	81·681 409	**88**	552·920 307	**153**	961·327 352	**1045**	6565·928 646
26	163·362 818	**101**	634·601 716	**306**	1922·654 704	**1198**	7527·255 998
39	245·044 227	**114**	716·283 125	**446**	2802·300 647	**1351**	8488·583 350
52	326·725 636	**127**	797·964 534	**599**	3763·627 999	**1504**	9449·910 702
65	408·407 045	**140**	879·645 943	**752**	4724·955 351	**1657**	10411·238 054
75	471·238 898	**153**	961·327 352	**905**	5686·282 703	**1810**	11372·565 406

A table of trigonometrical functions covering the first quadrant may be used for all angles. It is first necessary to reduce the angle (if greater than $\frac{1}{2}\pi$ or $1^r{\cdot}570\ 796$ or $90°$) to the first quadrant by subtracting the largest multiple of $\frac{1}{2}\pi$ that will leave a positive remainder less than $\frac{1}{2}\pi$. Such multiples up to 51 are shown on the opposite page, both in radians and in degrees, together with the number of revolutions of 2π or $360°$ that they represent. The column Quad. shows the quadrant in which the reduced angle lies, and the remaining six columns show the function of this reduced angle that must be looked up, and the sign to be prefixed.

Example I: $\tan 1000° = -\cot(1000° - 990°) = -\cot 10°$

Example II: $\sin 12^r{\cdot}345\ 678 = -\cos(12^r{\cdot}345\ 678 - 10^r{\cdot}995\ 574) = -\cos 1^r{\cdot}350\ 104$

In order to extend the table beyond 51 multiples of $\frac{1}{2}\pi$, it is noted that the problem is unchanged if any integral number of multiples of 2π or $360°$ is discarded. The two tables at the head of this page give such multiples, so chosen that the rounding-off error of the first does not exceed 0·07 units of the sixth decimal, while that of the second does not exceed 0·003 units. The effect of this choice is to confine rounding-off errors almost entirely to the table on the opposite page. The table on the left is for reducing angles in radians lying between $81^r{\cdot}68\ldots$ and 1000^r; that on the right is for angles greater than 1000^r and up to about 12000^r.

Example III		*Example IV*	
	r		r
x	123·456 789	*x*	5000·000 000
Subtract	81·681 409	Subtract	4724·955 351
Remainder	41·775 380	Remainder	275·044 649
Subtract	40·840 704	Subtract	245·044 227
Reduced angle	0·934 676	Remainder	30·000 422
		Subtract	29·845 130
x is in the third quadrant		Reduced angle	0·155 292

If the given angle is greater than about 12000^r, a calculating machine may be used to subtract multiples of $\frac{1}{2}\pi$ or $1^r{\cdot}570\ 796\ 326\ 795$. In effect the angle is divided by $\frac{1}{2}\pi$, yielding an integral quotient that can be expressed in the form $4p + q$, and a remainder r. The angle is then in the $(q+1)$th quadrant, and the reduced angle is $\boldsymbol{r}$.

Example V: $x = 12345^r{\cdot}678\ 901$

Dividing x by 1·570 796 327 (the limit of the setting capacity of most machines), the quotient is 7859 ($= 4 \times 1964 + 3$) and the remainder, or reduced angle, is $0^r{\cdot}790\ 567$. x is in the fourth quadrant. Actually this value of x is just within the reach of the three tables, which give (correctly) a reduced angle of $0^r{\cdot}790\ 569$. The discrepancy of two units in the sixth decimal is due to the rounding-off of $\frac{1}{2}\pi$ to nine decimals, thus leading to an error of $205 \times 10^{-12} \times 7859 = 1{\cdot}6 \times 10^{-6}$ or nearly two units in the sixth decimal.

For negative angles, first add any multiple of 2π sufficiently large to render the angle positive, and then proceed as before.

TABLE IVc—FORMULÆ FOR CIRCULAR FUNCTIONS (Radians)

For similar formulæ for exponential and hyperbolic functions, see page 204.

Function	Derivative	Integral	Function	Derivative	Integral
$\sin ax$	$a\cos ax$	$-\frac{1}{a}\cos ax$	$\sin^{-1} ax$	$\frac{a}{\sqrt{1-a^2x^2}}$	$x\sin^{-1} ax+\frac{1}{a}\sqrt{1-a^2x^2}$
$\cos ax$	$-a\sin ax$	$\frac{1}{a}\sin ax$	$\cos^{-1} ax$	$-\frac{a}{\sqrt{1-a^2x^2}}$	$x\cos^{-1} ax-\frac{1}{a}\sqrt{1-a^2x^2}$
$\tan ax$	$a\sec^2 ax$	$-\frac{1}{a}\ln\cos ax$	$\tan^{-1} ax$	$\frac{a}{1+a^2x^2}$	$x\tan^{-1} ax-\frac{1}{2a}\ln(1+a^2x^2)$
$\cot ax$	$-a\operatorname{cosec}^2 ax$	$\frac{1}{a}\ln\sin ax$	$\cot^{-1} ax$	$-\frac{a}{1+a^2x^2}$	$x\cot^{-1} ax+\frac{1}{2a}\ln(1+a^2x^2)$
$\sec ax$	$a\sec ax\tan ax$	$\frac{1}{a}\operatorname{gd}^{-1} ax$	$\sec^{-1} ax$	$\frac{1}{x\sqrt{a^2x^2-1}}$	$x\sec^{-1} ax-\frac{1}{a}\cosh^{-1} ax$
$\operatorname{cosec} ax$	$-a\operatorname{cosec} ax\cot ax$	$\frac{1}{a}\ln\tan\frac{ax}{2}$	$\operatorname{cosec}^{-1} ax$	$-\frac{1}{x\sqrt{a^2x^2-1}}$	$x\operatorname{cosec}^{-1} ax+\frac{1}{a}\cosh^{-1} ax$

ax is positive in the last two lines.

$$\sin x=\frac{1}{i}\sinh ix=x-\frac{1}{3!}x^3+\frac{1}{5!}x^5-\frac{1}{7!}x^7+\frac{1}{9!}x^9-\cdots \qquad \text{(All values of } x\text{)}$$

$$\cos x=\cosh ix=1-\frac{1}{2!}x^2+\frac{1}{4!}x^4-\frac{1}{6!}x^6+\frac{1}{8!}x^8-\cdots \qquad \text{(All values of } x\text{)}$$

$$\tan x=\frac{1}{i}\tanh ix=x+\frac{1}{3}x^3+\frac{2}{15}x^5+\frac{17}{315}x^7+\frac{62}{2835}x^9+\cdots+\frac{B_n2^{2n}(2^{2n}-1)}{(2n)!}x^{2n-1}+\cdots \qquad \left(x^2<\frac{\pi^2}{4}\right)$$

$$\cot x=i\coth ix=\frac{1}{x}\left(1-\frac{1}{3}x^2-\frac{1}{45}x^4-\frac{2}{945}x^6-\frac{1}{4725}x^8-\cdots-\frac{B_n2^{2n}}{(2n)!}x^{2n}-\cdots\right)=\frac{\tau}{x} \qquad (x^2<\pi^2)$$

$$\sec x=\operatorname{sech} ix=1+\frac{1}{2}x^2+\frac{5}{24}x^4+\frac{61}{720}x^6+\frac{277}{8064}x^8+\cdots+\frac{E_n}{(2n)!}x^{2n}+\cdots=\frac{\sigma}{\tau} \qquad \left(x^2<\frac{\pi^2}{4}\right)$$

$$\operatorname{cosec} x=i\operatorname{cosech} ix=\frac{1}{x}\left(1+\frac{1}{6}x^2+\frac{7}{360}x^4+\frac{31}{15120}x^6+\cdots+\frac{B_n2(2^{2n-1}-1)}{(2n)!}x^{2n}+\cdots\right)=\frac{\sigma}{x} \qquad (x^2<\pi^2)$$

$$\log\sin x=\log x-M\left(\frac{1}{6}x^2+\frac{1}{180}x^4+\frac{1}{2835}x^6+\cdots+\frac{B_n2^{2n-1}}{n(2n)!}x^{2n}+\cdots\right)=\log x+S \qquad (x^2<\pi^2)$$

$$\log\cos x=-M\left(\frac{1}{2}x^2+\frac{1}{12}x^4+\frac{1}{45}x^6+\frac{17}{2520}x^8+\cdots+\frac{B_n2^{2n-1}(2^{2n}-1)}{n(2n)!}x^{2n}+\cdots\right)=S-T \qquad \left(x^2<\frac{\pi^2}{4}\right)$$

$$\log\tan x=\log x+M\left(\frac{1}{3}x^2+\frac{7}{90}x^4+\frac{62}{2835}x^6+\cdots+\frac{B_n2^{2n}(2^{2n-1}-1)}{n(2n)!}x^{2n}+\cdots\right)=\log x+T \qquad \left(x^2<\frac{\pi^2}{4}\right)$$

For $\ln\sin x$, etc., write ln for log and omit the modulus M.

$$x=\sin x+\frac{1}{2}\frac{\sin^3 x}{3}+\frac{1.3}{2.4}\frac{\sin^5 x}{5}+\cdots=1{\cdot}570\,796-\cos x-\frac{1}{2}\frac{\cos^3 x}{3}-\frac{1.3}{2.4}\frac{\cos^5 x}{5}-\cdots$$

$$=1{\cdot}570\,796-\sqrt{2(1-\sin x)}\left\{1+\frac{1}{12}(1-\sin x)+\frac{3}{160}(1-\sin x)^2+\frac{5}{896}(1-\sin x)^3+\cdots\right\}$$

$$=\sqrt{2(1-\cos x)}\left\{1+\frac{1}{12}(1-\cos x)+\frac{3}{160}(1-\cos x)^2+\frac{5}{896}(1-\cos x)^3+\frac{35}{18432}(1-\cos x)^4+\cdots\right\}$$

$$=\tan x-\frac{1}{3}\tan^3 x+\frac{1}{5}\tan^5 x-\frac{1}{7}\tan^7 x+\cdots=1{\cdot}570\,796-\cot x+\frac{1}{3}\cot^3 x-\frac{1}{5}\cot^5 x+\cdots$$

If x has to be found from $\sec x$ or $\operatorname{cosec} x$, first find $\cos x$ or $\sin x$.

B_n and E_n denote Bernoulli's and Euler's numbers (page 387).

REVERSAL OF SERIES

If $$y=a_0+a_1x+a_2x^2+a_3x^3+a_4x^4+a_5x^5+a_6x^6+\cdots$$

Put $B=\frac{y-a_0}{a_1}$ and $b_n=\frac{a_n}{a_1}$ Then

$$x=B[1-b_2B-(b_3-2b_2^2)B^2-(b_4-5b_2b_3+5b_2^3)B^3-\{b_5-3(2b_2b_4+b_3^2)+21b_2^2b_3-14b_2^4\}B^4$$
$$-\{b_6-7(b_2b_5+b_3b_4)+28(b_2^2b_4+b_2b_3^2)-84b_2^3b_3+42b_2^5\}B^5-\cdots]$$

TABLE IVD—AUXILIARY FUNCTIONS FOR SMALL ANGLES (Radians)

x r	τ	cot
0·000 00		∞
	1·000 000	
0·001 22		816·5
	0·999 999	
0·002 12		471·5
	0·999 998	
0·002 73		365·2
	0·999 997	
0·003 24		308·7
	0·999 996	
0·003 67		272·2
	0·999 995	
0·004 06		246·2
	0·999 994	
0·004 41		226·5
	0·999 993	
0·004 74		210·9
	0·999 992	
0·005 04		198·1
	0·999 991	
0·005 33		187·4
	0·999 990	
0·005 61		178·2
	0·999 989	
0·005 87		170·3
	0·999 988	
0·006 12		163·3
	0·999 987	
0·006 36		157·2
	0·999 986	
0·006 59		151·7
	0·999 985	
0·006 81		146·7
	0·999 984	
0·007 03		142·2
	0·999 983	
0·007 24		138·1
	0·999 982	
0·007 44		134·3
	0·999 981	
0·007 64		130·8
	0·999 980	
0·007 84		127·6
	0·999 979	
0·008 03		124·6
	0·999 978	
0·008 21		121·8
	0·999 977	
0·008 39		119·1
	0·999 976	
0·008 57		116·7
	0·999 975	
0·008 74		114·4
	0·999 974	
0·008 91		112·2
	0·999 973	
0·009 08		110·1
	0·999 972	
0·009 24		108·2
	0·999 971	
0·009 40		106·3
	0·999 970	
0·009 56		104·6
	0·999 969	
0·009 72		102·9
	0·999 968	
0·009 87		101·3

x r	σ	cosec
0·000 00		∞
	1·000 000	
0·001 73		577·4
	1·000 001	
0·002 99		333·4
	1·000 002	
0·003 87		258·2
	1·000 003	
0·004 58		218·3
	1·000 004	
0·005 19		192·5
	1·000 005	
0·005 74		174·1
	1·000 006	
0·006 24		160·2
	1·000 007	
0·006 70		149·1
	1·000 008	
0·007 14		140·1
	1·000 009	
0·007 54		132·5
	1·000 010	
0·007 93		126·0
	1·000 011	
0·008 30		120·4
	1·000 012	
0·008 66		115·5
	1·000 013	
0·008 99		111·2
	1·000 014	
0·009 32		107·3
	1·000 015	
0·009 64		103·7
	1·000 016	
0·009 94		100·6

x r	τ		σ	
0·000	1·000 000	See critical table.	1·000 000	See critical table.
·001	1·000 000		·000 000	
·002	0·999 999		·000 001	
·003	·999 997		·000 002	
·004	·999 995		·000 003	
0·005	0·999 992		1·000 004	
·006	·999 988		·000 006	
·007	·999 984		·000 008	
·008	·999 979		·000 011	
·009	·999 973		·000 014	
0·010	0·999 967	7	1·000 017	3
·011	·999 960	8	·000 020	4
·012	·999 952	8	·000 024	4
·013	·999 944	9	·000 028	5
·014	·999 935	10	·000 033	5
0·015	0·999 925	10	1·000 038	5
·016	·999 915	11	·000 043	5
·017	·999 904	12	·000 048	6
·018	·999 892	12	·000 054	6
·019	·999 880	13	·000 060	7
0·020	0·999 867	14	1·000 067	7
·021	·999 853	14	·000 074	7
·022	·999 839	15	·000 081	7
·023	·999 824	16	·000 088	8
·024	·999 808	16	·000 096	8
0·025	0·999 792	17	1·000 104	9
·026	·999 775	18	·000 113	9
·027	·999 757	18	·000 122	9
·028	·999 739	19	·000 131	9
·029	·999 720	20	·000 140	10
0·030	0·999 700	20	1·000 150	10
·031	·999 680	21	·000 160	11
·032	·999 659	22	·000 171	11
·033	·999 637	22	·000 182	11
·034	·999 615	23	·000 193	11
0·035	0·999 592	24	1·000 204	12
·036	·999 568	24	·000 216	12
·037	·999 544	25	·000 228	13
·038	·999 519	26	·000 241	13
·039	·999 493	26	·000 254	13
0·040	0·999 467	27	1·000 267	13
·041	·999 440	28	·000 280	14
·042	·999 412	28	·000 294	14
·043	·999 384	29	·000 308	15
·044	·999 355	30	·000 323	15
0·045	0·999 325	30	1·000 338	15
·046	·999 295	31	·000 353	15
·047	·999 264	32	·000 368	16
·048	·999 232	32	·000 384	16
·049	·999 200	33	·000 400	17
0·050	0·999 167		1·000 417	

x r	τ		σ	
0·050	0·999 167	34	1·000 417	17
·051	·999 133	34	·000 434	17
·052	·999 099	36	·000 451	17
·053	·999 063	35	·000 468	18
·054	·999 028	37	·000 486	18
0·055	0·998 991	37	1·000 504	19
·056	·998 954	37	·000 523	19
·057	·998 917	39	·000 542	19
·058	·998 878	39	·000 561	19
·059	·998 839	39	·000 580	20
0·060	0·998 800	41	1·000 600	20
·061	·998 759	41	·000 620	21
·062	·998 718	41	·000 641	21
·063	·998 677	43	·000 662	21
·064	·998 634	43	·000 683	22
0·065	0·998 591	43	1·000 705	21
·066	·998 548	45	·000 726	23
·067	·998 503	45	·000 749	22
·068	·998 458	46	·000 771	23
·069	·998 412	46	·000 794	23
0·070	0·998 366	47	1·000 817	24
·071	·998 319	48	·000 841	24
·072	·998 271	48	·000 865	24
·073	·998 223	49	·000 889	24
·074	·998 174	50	·000 913	25
0·075	0·998 124	50	1·000 938	25
·076	·998 074	51	·000 963	26
·077	·998 023	52	·000 989	26
·078	·997 971	52	·001 015	26
·079	·997 919	53	·001 041	26
0·080	0·997 866	54	1·001 067	27
·081	·997 812	54	·001 094	28
·082	·997 758	55	·001 122	27
·083	·997 703	56	·001 149	28
·084	·997 647	56	·001 177	28
0·085	0·997 591	58	1·001 205	29
·086	·997 533	57	·001 234	29
·087	·997 476	59	·001 263	29
·088	·997 417	59	·001 292	29
·089	·997 358	59	·001 321	30
0·090	0·997 299	61	1·001 351	31
·091	·997 238	61	·001 382	30
·092	·997 177	62	·001 412	31
·093	·997 115	62	·001 443	31
·094	·997 053	63	·001 474	32
0·095	0·996 990	64	1·001 506	32
·096	·996 926	64	·001 538	32
·097	·996 862	65	·001 570	32
·098	·996 797	66	·001 602	33
·099	·996 731	67	·001 635	34
0·100	0·996 664		1·001 669	

$$\cot x = \frac{\tau}{x^{\mathrm{r}}} \qquad \operatorname{cosec} x = \frac{\sigma}{x^{\mathrm{r}}} \qquad x^{\mathrm{r}} = \frac{\tau}{\cot x} = \frac{\sigma}{\operatorname{cosec} x}$$

TABLE IVE—CIRCULAR FUNCTIONS (Radians)

x	$\sin x$		$\cos x$		$\tan x$		$\cot x$	$\sec x$		$\operatorname{cosec} x$
r										
0·000	0·000 000	1000	1·000 000	0	0·000 000	1000	∞	1·000 000	1	∞
·001	·001 000	1000	1·000 000	2	·001 000	1000	1000·00	·000 001	1	1000·00
·002	·002 000	1000	0·999 998	2	·002 000	1000	499·999	·000 002	3	500·000
·003	·003 000	1000	·999 996	4	·003 000	1000	333·332	·000 005	3	333·334
·004	·004 000	1000	·999 992	4	·004 000	1000	249·999	·000 008	5	250·001
0·005	0·005 000	1000	0·999 988	6	0·005 000	1000	199·998	1·000 013	5	200·001
·006	·006 000	1000	·999 982	6	·006 000	1000	166·665	·000 018	7	166·668
·007	·007 000	1000	·999 976	8	·007 000	1000	142·855	·000 025	7	142·858
·008	·008 000	1000	·999 968	8	·008 000	1000	124·997	·000 032	9	125·001
·009	·009 000	1000	·999 960	10	·009 000	1000	111·108	·000 041	9	111·113
0·010	0·010 000	1000	0·999 950	10	0·010 000	1000	99·9967	1·000 050	11	100·002
·011	·011 000	1000	·999 940	12	·011 000	1001	90·9054	·000 061	11	90·9109
·012	·012 000	1000	·999 928	12	·012 001	1000	83·3293	·000 072	13	83·3353
·013	·013 000	1000	·999 916	14	·013 001	1000	76·9187	·000 085	13	76·9252
·014	·014 000	999	·999 902	14	·014 001	1000	71·4239	·000 098	15	71·4309
0·015	0·014 999	1000	0·999 888	16	0·015 001	1000	66·6617	1·000 113	15	66·6692
·016	·015 999	1000	·999 872	16	·016 001	1001	62·4947	·000 128	17	62·5027
·017	·016 999	1000	·999 856	18	·017 002	1000	58·8179	·000 145	17	58·8264
·018	·017 999	1000	·999 838	18	·018 002	1000	55·5496	·000 162	19	55·5586
·019	·018 999	1000	·999 820	20	·019 002	1001	52·6252	·000 181	19	52·6347
0·020	0·019 999	999	0·999 800	20	0·020 003	1000	49·9933	1·000 200	21	50·0033
·021	·020 998	1000	·999 780	22	·021 003	1001	47·6120	·000 221	21	47·6225
·022	·021 998	1000	·999 758	22	·022 004	1000	45·4472	·000 242	23	45·4582
·023	·022 998	1000	·999 736	24	·023 004	1001	43·4706	·000 265	23	43·4821
·024	·023 998	999	·999 712	24	·024 005	1000	41·6587	·000 288	25	41·6707
0·025	0·024 997	1000	0·999 688	26	0·025 005	1001	39·9917	1·000 313	25	40·0042
·026	·025 997	1000	·999 662	26	·026 006	1001	38·4529	·000 338	27	38·4659
·027	·026 997	999	·999 636	28	·027 007	1000	37·0280	·000 365	27	37·0415
·028	·027 996	1000	·999 608	28	·028 007	1001	35·7050	·000 392	29	35·7190
·029	·028 996	1000	·999 580	30	·029 008	1001	34·4731	·000 421	29	34·4876
0·030	0·029 996	999	0·999 550	30	0·030 009	1001	33·3233	1·000 450	31	33·3383
·031	·030 995	1000	·999 520	32	·031 010	1001	32·2477	·000 481	31	32·2632
·032	·031 995	999	·999 488	32	·032 011	1001	31·2393	·000 512	33	31·2553
·033	·032 994	999	·999 456	34	·033 012	1001	30·2920	·000 545	33	30·3085
·034	·033 993	1000	·999 422	34	·034 013	1001	29·4004	·000 578	35	29·4174
0·035	0·034 993	999	0·999 388	36	0·035 014	1002	28·5598	1·000 613	35	28·5773
·036	·035 992	1000	·999 352	36	·036 016	1001	27·7658	·000 648	37	27·7838
·037	·036 992	999	·999 316	38	·037 017	1001	27·0147	·000 685	37	27·0332
·038	·037 991	999	·999 278	38	·038 018	1002	26·3031	·000 722	39	26·3221
·039	·038 990	999	·999 240	40	·039 020	1001	25·6280	·000 761	40	25·6475
0·040	0·039 989	1000	0·999 200	40	0·040 021	1002	24·9867	1·000 801	40	25·0067
·041	·040 989	999	·999 160	42	·041 023	1002	24·3766	·000 841	42	24·3971
·042	·041 988	999	·999 118	42	·042 025	1002	23·7955	·000 883	42	23·8165
·043	·042 987	999	·999 076	44	·043 027	1001	23·2415	·000 925	44	23·2630
·044	·043 986	999	·999 032	44	·044 028	1002	22·7126	·000 969	44	22·7346
0·045	0·044 985	999	0·998 988	46	0·045 030	1002	22·2072	1·001 013	46	22·2297
·046	·045 984	999	·998 942	46	·046 032	1003	21·7238	·001 059	47	21·7468
·047	·046 983	999	·998 896	48	·047 035	1002	21·2609	·001 106	47	21·2844
·048	·047 982	998	·998 848	48	·048 037	1002	20·8173	·001 153	49	20·8413
·049	·048 980	999	·998 800	50	·049 039	1003	20·3918	·001 202	49	20·4163
0·050	0·049 979		0·998 750		0·050 042		19·9833	1·001 251		20·0083

For values of cot x and cosec x at intermediate arguments, see page 171.

TABLE IVE—CIRCULAR FUNCTIONS (Radians)

x	sin x		cos x		tan x		cot x	sec x		cosec x
r										
0·050	0·049 979	999	0·998 750	50	0·050 042	1002	19·98333	1·001 251	51	20·00834
·051	·050 978	999	·998 700	52	·051 044	1003	19·59084	·001 302	52	19·61635
·052	·051 977	998	·998 648	52	·052 047	1003	19·21343	·001 354	52	19·23944
·053	·052 975	999	·998 596	54	·053 050	1003	18·85025	·001 406	54	18·87676
·054	·053 974	998	·998 542	54	·054 053	1003	18·50052	·001 460	54	18·52752
0·055	0·054 972	999	0·998 488	56	0·055 056	1003	18·16348	1·001 514	56	18·19099
·056	·055 971	998	·998 432	56	·056 059	1003	17·83847	·001 570	57	17·86648
·057	·056 969	998	·998 376	58	·057 062	1003	17·52486	·001 627	57	17·55336
·058	·057 967	999	·998 318	58	·058 065	1004	17·22204	·001 684	59	17·25105
·059	·058 966	998	·998 260	59	·059 069	1003	16·92948	·001 743	60	16·95899
0·060	0·059 964	998	0·998 201	61	0·060 072	1004	16·64666	1·001 803	60	16·67667
·061	·060 962	998	·998 140	61	·061 076	1004	16·37310	·001 863	62	16·40361
·062	·061 960	998	·998 079	63	·062 080	1003	16·10836	·001 925	63	16·13937
·063	·062 958	998	·998 016	63	·063 083	1005	15·85201	·001 988	64	15·88352
·064	·063 956	998	·997 953	65	·064 088	1004	15·60366	·002 052	64	15·63567
0·065	0·064 954	998	0·997 888	65	0·065 092	1004	15·36294	1·002 116	66	15·39545
·066	·065 952	998	·997 823	67	·066 096	1004	15·12951	·002 182	67	15·16252
·067	·066 950	998	·997 756	67	·067 100	1005	14·90303	·002 249	67	14·93655
·068	·067 948	997	·997 689	69	·068 105	1005	14·68321	·002 316	69	14·71722
·069	·068 945	998	·997 620	69	·069 110	1005	14·46975	·002 385	70	14·50426
0·070	0·069 943	997	0·997 551	70	0·070 115	1005	14·26237	1·002 455	71	14·29739
·071	·070 940	998	·997 481	72	·071 120	1005	14·06083	·002 526	72	14·09635
·072	·071 938	997	·997 409	72	·072 125	1005	13·86488	·002 598	72	13·90090
·073	·072 935	997	·997 337	74	·073 130	1005	13·67429	·002 670	74	13·71080
·074	·073 932	998	·997 263	74	·074 135	1006	13·48884	·002 744	75	13·52585
0·075	0·074 930	997	0·997 189	76	0·075 141	1006	13·30832	1·002 819	76	13·34584
·076	·075 927	997	·997 113	76	·076 147	1006	13·13255	·002 895	77	13·17057
·077	·076 924	997	·997 037	77	·077 153	1006	12·96134	·002 972	78	12·99986
·078	·077 921	997	·996 960	79	·078 159	1006	12·79450	·003 050	79	12·83352
·079	·078 918	997	·996 881	79	·079 165	1006	12·63188	·003 129	80	12·67140
0·080	0·079 915	996	0·996 802	81	0·080 171	1007	12·47332	1·003 209	80	12·51334
·081	·080 911	997	·996 721	81	·081 178	1006	12·31867	·003 289	82	12·35919
·082	·081 908	997	·996 640	83	·082 184	1007	12·16778	·003 371	83	12·20880
·083	·082 905	996	·996 557	83	·083 191	1007	12·02051	·003 454	84	12·06204
·084	·083 901	997	·996 474	84	·084 198	1007	11·87675	·003 538	85	11·91877
0·085	0·084 898	996	0·996 390	86	0·085 205	1008	11·73636	1·003 623	86	11·77888
·086	·085 894	996	·996 304	86	·086 213	1007	11·59923	·003 709	87	11·64225
·087	·086 890	996	·996 218	88	·087 220	1008	11·46524	·003 796	89	11·50877
·088	·087 886	997	·996 130	88	·088 228	1008	11·33429	·003 885	89	11·37832
·089	·088 883	996	·996 042	89	·089 236	1008	11·20627	·003 974	90	11·25080
0·090	0·089 879	995	0·995 953	91	0·090 244	1008	11·08109	1·004 064	91	11·12613
·091	·090 874	996	·995 862	91	·091 252	1008	10·95866	·004 155	92	11·00419
·092	·091 870	996	·995 771	92	·092 260	1009	10·83888	·004 247	93	10·88491
·093	·092 866	996	·995 679	94	·093 269	1009	10·72167	·004 340	94	10·76820
·094	·093 862	995	·995 585	94	·094 278	1009	10·60695	·004 434	96	10·65398
0·095	0·094 857	996	0·995 491	95	0·095 287	1009	10·49463	1·004 530	96	10·54217
·096	·095 853	995	·995 396	97	·096 296	1009	10·38465	·004 626	97	10·43268
·097	·096 848	995	·995 299	97	·097 305	1010	10·27692	·004 723	98	10·32546
·098	·097 843	995	·995 202	99	·098 315	1010	10·17139	·004 821	100	10·22043
·099	·098 838	995	·995 103	99	·099 325	1010	10·06799	·004 921	100	10·11753
0·100	0·099 833		0·995 004		0·100 335		9·96664	1·005 021		10·01669

For values of cot x and cosec x at intermediate arguments, see page 171.

TABLE IVE—CIRCULAR FUNCTIONS (Radians)

x	sin x 0·		cos x 0·		tan x 0·		cot x		sec x 1·		cosec x	
0·100	099 833	995	995 004	100	100 335	1010	9·96664	9934	005 021	101	10·01669	9885
·101	100 828	995	994 904	101	101 345	1010	9·86730	9740	005 122	103	9·91784	9690
·102	101 823	995	994 803	103	102 355	1011	9·76990	9552	005 225	103	9·82094	9501
·103	102 818	995	994 700	103	103 366	1011	9·67438	9369	005 328	104	9·72593	9319
·104	103 813	994	994 597	104	104 377	1011	9·58069	9191	005 432	106	9·63274	9141
0·105	104 807	995	994 493	106	105 388	1011	9·48878	9018	005 538	106	9·54133	8968
·106	105 802	994	994 387	106	106 399	1011	9·39860	8850	005 644	108	9·45165	8800
·107	106 796	994	994 281	107	107 410	1012	9·31010	8687	005 752	108	9·36365	8637
·108	107 790	994	994 174	109	108 422	1012	9·22323	8528	005 860	110	9·27728	8478
·109	108 784	994	994 065	109	109 434	1012	9·13795	8374	005 970	111	9·19250	8323
0·110	109 778	994	993 956	110	110 446	1012	9·05421	8223	006 081	111	9·10927	8173
·111	110 772	994	993 846	111	111 458	1013	8·97198	8077	006 192	113	9·02754	8027
·112	111 766	994	993 735	113	112 471	1012	8·89121	7935	006 305	114	8·94727	7885
·113	112 760	993	993 622	113	113 483	1013	8·81186	7796	006 419	114	8·86842	7746
·114	113 753	994	993 509	114	114 496	1014	8·73390	7661	006 533	116	8·79096	7611
0·115	114 747	993	993 395	115	115 510	1013	8·65729	7530	006 649	117	8·71485	7480
·116	115 740	993	993 280	117	116 523	1014	8·58199	7402	006 766	118	8·64005	7351
·117	116 733	993	993 163	117	117 537	1014	8·50797	7276	006 884	119	8·56654	7227
·118	117 726	993	993 046	118	118 551	1014	8·43521	7155	007 003	120	8·49427	7104
·119	118 719	993	992 928	119	119 565	1014	8·36366	7037	007 123	120	8·42323	6986
0·120	119 712	993	992 809	121	120 579	1015	8·29329	6920	007 243	122	8·35337	6871
·121	120 705	993	992 688	121	121 594	1015	8·22409	6808	007 365	123	8·28466	6757
·122	121 698	992	992 567	122	122 609	1015	8·15601	6697	007 488	124	8·21709	6647
·123	122 690	992	992 445	123	123 624	1015	8·08904	6590	007 612	126	8·15062	6540
·124	123 682	993	992 322	124	124 639	1016	8·02314	6485	007 738	126	8·08522	6435
0·125	124 675	992	992 198	126	125 655	1016	7·95829	6383	007 864	127	8·02087	6332
·126	125 667	992	992 072	126	126 671	1016	7·89446	6282	007 991	128	7·95755	6233
·127	126 659	992	991 946	127	127 687	1017	7·83164	6185	008 119	129	7·89522	6135
·128	127 651	992	991 819	128	128 704	1016	7·76979	6090	008 248	131	7·83387	6039
·129	128 643	991	991 691	129	129 720	1017	7·70889	5996	008 379	131	7·77348	5946
0·130	129 634	992	991 562	130	130 737	1018	7·64893	5906	008 510	132	7·71402	5856
·131	130 626	991	991 432	131	131 755	1017	7·58987	5816	008 642	134	7·65546	5766
·132	131 617	991	991 301	132	132 772	1018	7·53171	5730	008 776	134	7·59780	5679
·133	132 608	991	991 169	134	133 790	1018	7·47441	5644	008 910	136	7·54101	5594
·134	133 599	991	991 035	134	134 808	1018	7·41797	5562	009 046	136	7·48507	5511
0·135	134 590	991	990 901	135	135 826	1019	7·36235	5480	009 182	138	7·42996	5430
·136	135 581	991	990 766	136	136 845	1019	7·30755	5400	009 320	138	7·37566	5351
·137	136 572	990	990 630	137	137 864	1019	7·25355	5323	009 458	140	7·32215	5272
·138	137 562	991	990 493	138	138 883	1019	7·20032	5247	009 598	141	7·26943	5197
·139	138 553	990	990 355	139	139 902	1020	7·14785	5172	009 739	142	7·21746	5122
0·140	139 543	990	990 216	140	140 922	1020	7·09613	5099	009 881	143	7·16624	5049
·141	140 533	990	990 076	141	141 942	1020	7·04514	5028	010 024	143	7·11575	4977
·142	141 523	990	989 935	142	142 962	1021	6·99486	4958	010 167	145	7·06598	4908
·143	142 513	990	989 793	143	143 983	1021	6·94528	4890	010 312	146	7·01690	4840
·144	143 503	989	989 650	144	145 004	1021	6·89638	4823	010 458	147	6·96850	4772
0·145	144 492	990	989 506	145	146 025	1021	6·84815	4757	010 605	148	6·92078	4707
·146	145 482	989	989 361	146	147 046	1022	6·80058	4693	010 753	150	6·87371	4643
·147	146 471	989	989 215	147	148 068	1022	6·75365	4630	010 903	150	6·82728	4579
·148	147 460	989	989 068	148	149 090	1023	6·70735	4568	011 053	151	6·78149	4518
·149	148 449	989	988 920	149	150 113	1022	6·66167	4508	011 204	152	6·73631	4458
0·150	149 438		988 771		151 135		6·61659		011 356		6·69173	

TABLE IVE—CIRCULAR FUNCTIONS (Radians)

x	sin *x*		cos *x*		tan *x*		cot *x*		sec *x*		cosec *x*	
	0·		0·		0·				1·			
0·150	149 438	989	988 771	150	151 135	1023	6·61659	4448	011 356	154	6·69173	4398
·151	150 427	988	988 621	151	152 158	1024	6·57211	4391	011 510	154	6·64775	4340
·152	151 415	989	988 470	152	153 182	1023	6·52820	4333	011 664	156	6·60435	4283
·153	152 404	988	988 318	153	154 205	1024	6·48487	4278	011 820	156	6·56152	4228
·154	153 392	988	988 165	153	155 229	1024	6·44209	4223	011 976	158	6·51924	4172
0·155	154 380	988	988 012	155	156 253	1025	6·39986	4169	012 134	159	6·47752	4119
·156	155 368	988	987 857	156	157 278	1025	6·35817	4116	012 293	159	6·43633	4066
·157	156 356	987	987 701	157	158 303	1025	6·31701	4065	012 452	161	6·39567	4015
·158	157 343	988	987 544	158	159 328	1026	6·27636	4014	012 613	162	6·35552	3963
·159	158 331	987	987 386	159	160 354	1025	6·23622	3964	012 775	163	6·31589	3914
0·160	159 318	987	987 227	160	161 379	1027	6·19658	3916	012 938	164	6·27675	3866
·161	160 305	987	987 067	160	162 406	1026	6·15742	3868	013 102	165	6·23809	3817
·162	161 292	987	986 907	162	163 432	1027	6·11874	3820	013 267	166	6·19992	3770
·163	162 279	987	986 745	163	164 459	1027	6·08054	3774	013 433	167	6·16222	3724
·164	163 266	986	986 582	164	165 486	1028	6·04280	3729	013 600	169	6·12498	3679
0·165	164 252	987	986 418	164	166 514	1028	6·00551	3685	013 769	169	6·08819	3634
·166	165 239	986	986 254	166	167 542	1028	5·96866	3641	013 938	170	6·05185	3590
·167	166 225	986	986 088	167	168 570	1029	5·93225	3597	014 108	172	6·01595	3548
·168	167 211	986	985 921	168	169 599	1029	5·89628	3556	014 280	172	5·98047	3505
·169	168 197	985	985 753	168	170 628	1029	5·86072	3514	014 452	174	5·94542	3464
0·170	169 182	986	985 585	170	171 657	1029	5·82558	3474	014 626	175	5·91078	3423
·171	170 168	985	985 415	171	172 686	1030	5·79084	3433	014 801	176	5·87655	3383
·172	171 153	985	985 244	171	173 716	1031	5·75651	3395	014 977	176	5·84272	3344
·173	172 138	985	985 073	173	174 747	1031	5·72256	3355	015 153	178	5·80928	3305
·174	173 123	985	984 900	173	175 778	1031	5·68901	3318	015 331	179	5·77623	3267
0·175	174 108	985	984 727	175	176 809	1031	5·65583	3280	015 510	180	5·74356	3230
·176	175 093	984	984 552	176	177 840	1032	5·62303	3244	015 690	182	5·71126	3193
·177	176 077	985	984 376	176	178 872	1032	5·59059	3207	015 872	182	5·67933	3158
·178	177 062	984	984 200	178	179 904	1033	5·55852	3172	016 054	183	5·64775	3121
·179	178 046	984	984 022	178	180 937	1033	5·52680	3137	016 237	185	5·61654	3087
0·180	179 030	983	983 844	180	181 970	1033	5·49543	3103	016 422	185	5·58567	3053
·181	180 013	984	983 664	180	183 003	1034	5·46440	3070	016 607	187	5·55514	3018
·182	180 997	983	983 484	182	184 037	1034	5·43370	3036	016 794	187	5·52496	2986
·183	181 980	984	983 302	182	185 071	1034	5·40334	3003	016 981	189	5·49510	2953
·184	182 964	983	983 120	184	186 105	1035	5·37331	2971	017 170	190	5·46557	2921
0·185	183 947	982	982 936	184	187 140	1035	5·34360	2940	017 360	191	5·43636	2889
·186	184 929	983	982 752	186	188 175	1036	5·31420	2909	017 551	192	5·40747	2858
·187	185 912	983	982 566	186	189 211	1036	5·28511	2878	017 743	193	5·37889	2828
·188	186 895	982	982 380	187	190 247	1036	5·25633	2848	017 936	194	5·35061	2797
·189	187 877	982	982 193	189	191 283	1037	5·22785	2818	018 130	196	5·32264	2768
0·190	188 859	982	982 004	189	192 320	1037	5·19967	2789	018 326	196	5·29496	2739
·191	189 841	982	981 815	190	193 357	1038	5·17178	2760	018 522	197	5·26757	2710
·192	190 823	981	981 625	192	194 395	1038	5·14418	2733	018 719	199	5·24047	2682
·193	191 804	981	981 433	192	195 433	1038	5·11685	2704	018 918	200	5·21365	2653
·194	192 785	982	981 241	193	196 471	1039	5·08981	2677	019 118	200	5·18712	2627
0·195	193 767	980	981 048	195	197 510	1039	5·06304	2650	019 318	202	5·16085	2600
·196	194 747	981	980 853	195	198 549	1040	5·03654	2624	019 520	203	5·13485	2573
·197	195 728	981	980 658	196	199 589	1040	5·01030	2597	019 723	204	5·10912	2546
·198	196 709	980	980 462	197	200 629	1040	4·98433	2571	019 927	206	5·08366	2521
·199	197 689	980	980 265	198	201 669	1041	4·95862	2547	020 133	206	5·05845	2496
0·200	198 669		980 067		202 710		4·93315		020 339		5·03349	

x	sin *x*		cos *x*		tan *x*		cot *x*		sec *x*		cosec *x*	
	0·		0·		0·				1·			
0·200	198 669	980	980 067	200	202 710	1041	4·93315	*2521*	020 339	207	5·03349	*2471*
·201	199 649	980	979 867	200	203 751	1042	4·90794	*2496*	020 546	209	5·00878	*2446*
·202	200 629	980	979 667	201	204 793	1042	4·88298	*2472*	020 755	209	4·98432	*2421*
·203	201 609	979	979 466	202	205 835	1043	4·85826	*2449*	020 964	211	4·96011	*2398*
·204	202 588	979	979 264	203	206 878	1043	4·83377	*2425*	021 175	212	4·93613	*2375*
0·205	203 567	979	979 061	204	207 921	1043	4·80952	*2401*	021 387	213	4·91238	*2351*
·206	204 546	979	978 857	205	208 964	1044	4·78551	*2379*	021 600	214	4·88887	*2328*
·207	205 525	978	978 652	206	210 008	1044	4·76172	*2356*	021 814	215	4·86559	*2306*
·208	206 503	979	978 446	207	211 052	1045	4·73816	*2334*	022 029	216	4·84253	*2283*
·209	207 482	978	978 239	208	212 097	1045	4·71482	*2312*	022 245	218	4·81970	*2261*
0·210	208 460	978	978 031	209	213 142	1046	4·69170	*2291*	022 463	218	4·79709	*2240*
·211	209 438	978	977 822	210	214 188	1046	4·66879	*2269*	022 681	220	4·77469	*2219*
·212	210 416	977	977 612	211	215 234	1047	4·64610	*2248*	022 901	220	4·75250	*2198*
·213	211 393	977	977 401	212	216 281	1047	4·62362	*2227*	023 121	222	4·73052	*2176*
·214	212 370	977	977 189	213	217 328	1047	4·60135	*2208*	023 343	223	4·70876	*2157*
0·215	213 347	977	976 976	213	218 375	1048	4·57927	*2187*	023 566	224	4·68719	*2136*
·216	214 324	977	976 763	215	219 423	1049	4·55740	*2167*	023 790	225	4·66583	*2117*
·217	215 301	976	976 548	216	220 472	1048	4·53573	*2147*	024 015	227	4·64466	*2097*
·218	216 277	977	976 332	217	221 520	1050	4·51426	*2128*	024 242	227	4·62369	*2077*
·219	217 254	976	976 115	218	222 570	1049	4·49298	*2110*	024 469	229	4·60292	*2059*
0·220	218 230	975	975 897	218	223 619	1051	4·47188	*2090*	024 698	230	4·58233	*2040*
·221	219 205	976	975 679	220	224 670	1050	4·45098	*2072*	024 928	230	4·56193	*2021*
·222	220 181	975	975 459	221	225 720	1052	4·43026	*2054*	025 158	232	4·54172	*2003*
·223	221 156	975	975 238	221	226 772	1051	4·40972	*2035*	025 390	233	4·52169	*1985*
·224	222 131	975	975 017	223	227 823	1052	4·38937	*2018*	025 623	235	4·50184	*1967*
0·225	223 106	975	974 794	223	228 875	1053	4·36919	*2000*	025 858	235	4·48217	*1950*
·226	224 081	975	974 571	225	229 928	1053	4·34919	*1983*	026 093	237	4·46267	*1932*
·227	225 056	974	974 346	226	230 981	1054	4·32936	*1966*	026 330	237	4·44335	*1915*
·228	226 030	974	974 120	226	232 035	1054	4·30970	*1949*	026 567	239	4·42420	*1899*
·229	227 004	974	973 894	228	233 089	1054	4·29021	*1932*	026 806	240	4·40521	*1881*
0·230	227 978	973	973 666	228	234 143	1055	4·27089	*1916*	027 046	241	4·38640	*1865*
·231	228 951	973	973 438	230	235 198	1056	4·25173	*1900*	027 287	242	4·36775	*1849*
·232	229 924	973	973 208	230	236 254	1056	4·23273	*1883*	027 529	243	4·34926	*1833*
·233	230 897	973	972 978	231	237 310	1057	4·21390	*1868*	027 772	245	4·33093	*1818*
·234	231 870	973	972 747	233	238 367	1057	4·19522	*1852*	028 017	245	4·31275	*1801*
0·235	232 843	972	972 514	233	239 424	1057	4·17670	*1837*	028 262	247	4·29474	*1786*
·236	233 815	973	972 281	234	240 481	1058	4·15833	*1822*	028 509	248	4·27688	*1771*
·237	234 788	971	972 047	236	241 539	1059	4·14011	*1806*	028 757	249	4·25917	*1756*
·238	235 759	972	971 811	236	242 598	1059	4·12205	*1792*	029 006	250	4·24161	*1741*
·239	236 731	972	971 575	237	243 657	1060	4·10413	*1777*	029 256	252	4·22420	*1726*
0·240	237 703	971	971 338	238	244 717	1060	4·08636	*1763*	029 508	252	4·20694	*1712*
·241	238 674	971	971 100	239	245 777	1061	4·06873	*1748*	029 760	254	4·18982	*1698*
·242	239 645	971	970 861	241	246 838	1061	4·05125	*1734*	030 014	255	4·17284	*1683*
·243	240 616	970	970 620	241	247 899	1061	4·03391	*1721*	030 269	256	4·15601	*1670*
·244	241 586	970	970 379	242	248 960	1063	4·01670	*1706*	030 525	257	4·13931	*1656*
0·245	242 556	970	970 137	243	250 023	1062	3·99964	*1693*	030 782	258	4·12275	*1642*
·246	243 526	970	969 894	244	251 085	1064	3·98271	*1680*	031 040	260	4·10633	*1629*
·247	244 496	970	969 650	245	252 149	1064	3·96591	*1666*	031 300	260	4·09004	*1615*
·248	245 466	969	969 405	246	253 213	1064	3·94925	*1653*	031 560	262	4·07389	*1602*
·249	246 435	969	969 159	247	254 277	1065	3·93272	*1640*	031 822	263	4·05787	*1590*
0·250	247 404		968 912		255 342		3·91632		032 085		4·04197	

x	sin x		cos x		tan x		cot x		sec x		cosec x	
	0·		0·		0·				1·			
r												
0·250	247 404	969	968 912	247	255 342	1065	3·91632	*1628*	032 085	264	4·04197	*1576*
·251	248 373	968	968 665	249	256 407	1066	3·90004	*1614*	032 349	265	4·02621	*1564*
·252	249 341	969	968 416	250	257 473	1067	3·88390	*1603*	032 614	267	4·01057	*1552*
·253	250 310	968	968 166	251	258 540	1067	3·86787	*1590*	032 881	268	3·99505	*1539*
·254	251 278	967	967 915	252	259 607	1068	3·85197	*1577*	033 149	268	3·97966	*1527*
0·255	252 245	968	967 663	252	260 675	1068	3·83620	*1566*	033 417	270	3·96439	*1514*
·256	253 213	967	967 411	254	261 743	1069	3·82054	*1554*	033 687	271	3·94925	*1503*
·257	254 180	967	967 157	255	262 812	1069	3·80500	*1542*	033 958	273	3·93422	*1491*
·258	255 147	967	966 902	255	263 881	1070	3·78958	*1530*	034 231	273	3·91931	*1480*
·259	256 114	967	966 647	257	264 951	1071	3·77428	*1519*	034 504	275	3·90451	*1468*
0·260	257 081	966	966 390	258	266 022	1071	3·75909	*1507*	034 779	276	3·88983	*1456*
·261	258 047	966	966 132	258	267 093	1071	3·74402	*1496*	035 055	277	3·87527	*1446*
·262	259 013	966	965 874	260	268 164	1072	3·72906	*1485*	035 332	278	3·86081	*1434*
·263	259 979	965	965 614	260	269 236	1073	3·71421	*1474*	035 610	280	3·84647	*1423*
·264	260 944	965	965 354	261	270 309	1074	3·69947	*1463*	035 890	280	3·83224	*1412*
0·265	261 909	965	965 093	263	271 383	1073	3·68484	*1453*	036 170	282	3·81812	*1402*
·266	262 874	965	964 830	263	272 456	1075	3·67031	*1442*	036 452	283	3·80410	*1391*
·267	263 839	964	964 567	265	273 531	1075	3·65589	*1431*	036 735	284	3·79019	*1380*
·268	264 803	965	964 302	265	274 606	1076	3·64158	*1421*	037 019	285	3·77639	*1370*
·269	265 768	963	964 037	266	275 682	1076	3·62737	*1411*	037 304	287	3·76269	*1360*
0·270	266 731	964	963 771	267	276 758	1077	3·61326	*1400*	037 591	288	3·74909	*1350*
·271	267 695	963	963 504	268	277 835	1078	3·59926	*1391*	037 879	289	3·73559	*1339*
·272	268 658	964	963 236	270	278 913	1078	3·58535	*1380*	038 168	290	3·72220	*1330*
·273	269 622	962	962 966	270	279 991	1078	3·57155	*1371*	038 458	291	3·70890	*1320*
·274	270 584	963	962 696	271	281 069	1080	3·55784	*1361*	038 749	293	3·69570	*1310*
0·275	271 547	962	962 425	272	282 149	1080	3·54423	*1351*	039 042	294	3·68260	*1300*
·276	272 509	962	962 153	273	283 229	1080	3·53072	*1342*	039 336	295	3·66960	*1291*
·277	273 471	962	961 880	274	284 309	1081	3·51730	*1333*	039 631	296	3·65669	*1281*
·278	274 433	961	961 606	275	285 390	1082	3·50397	*1323*	039 927	297	3·64388	*1272*
·279	275 394	962	961 331	276	286 472	1082	3·49074	*1314*	040 224	299	3·63116	*1263*
0·280	276 356	961	961 055	276	287 554	1083	3·47760	*1304*	040 523	300	3·61853	*1254*
·281	277 317	960	960 779	278	288 637	1084	3·46456	*1296*	040 823	301	3·60599	*1245*
·282	278 277	961	960 501	279	289 721	1084	3·45160	*1287*	041 124	302	3·59354	*1236*
·283	279 238	960	960 222	280	290 805	1085	3·43873	*1278*	041 426	303	3·58118	*1227*
·284	280 198	959	959 942	280	291 890	1086	3·42595	*1270*	041 729	305	3·56891	*1218*
0·285	281 157	960	959 662	282	292 976	1086	3·41325	*1260*	042 034	306	3·55673	*1210*
·286	282 117	959	959 380	283	294 062	1087	3·40065	*1253*	042 340	307	3·54463	*1201*
·287	283 076	959	959 097	283	295 149	1087	3·38812	*1243*	042 647	308	3·53262	*1193*
·288	284 035	959	958 814	285	296 236	1088	3·37569	*1236*	042 955	310	3·52069	*1184*
·289	284 994	958	958 529	285	297 324	1089	3·36333	*1227*	043 265	311	3·50885	*1176*
0·290	285 952	958	958 244	287	298 413	1089	3·35106	*1219*	043 576	312	3·49709	*1168*
·291	286 910	958	957 957	287	299 502	1090	3·33887	*1210*	043 888	313	3·48541	*1160*
·292	287 868	958	957 670	288	300 592	1091	3·32677	*1203*	044 201	314	3·47381	*1151*
·293	288 826	957	957 382	290	301 683	1091	3·31474	*1195*	044 515	316	3·46230	*1144*
·294	289 783	957	957 092	290	302 774	1092	3·30279	*1187*	044 831	317	3·45086	*1136*
0·295	290 740	957	956 802	291	303 866	1093	3·29092	*1179*	045 148	318	3·43950	*1128*
·296	291 697	956	956 511	292	304 959	1093	3·27913	*1171*	045 466	320	3·42822	*1120*
·297	292 653	956	956 219	293	306 052	1094	3·26742	*1164*	045 786	320	3·41702	*1113*
·298	293 609	956	955 926	294	307 146	1095	3·25578	*1156*	046 106	322	3·40589	*1105*
·299	294 565	955	955 632	296	308 241	1095	3·24422	*1149*	046 428	324	3·39484	*1098*
0·300	295 520		955 336		309 336		3·23273		046 752		3·38386	

TABLE IVE—CIRCULAR FUNCTIONS (Radians)

x	sin x		cos x		tan x		cot x		sec x		cosec x	
r	0·		0·		0·				1·			
0·300	295 520	955	955 336	296	309 336	1096	3·23273	*1142*	046 752	324	3·38386	*1090*
·301	296 475	955	955 040	296	310 432	1097	3·22131	*1134*	047 076	326	3·37296	*1083*
·302	297 430	955	954 744	298	311 529	1097	3·20997	*1126*	047 402	327	3·36213	*1075*
·303	298 385	954	954 446	299	312 626	1098	3·19871	*1120*	047 729	328	3·35138	*1069*
·304	299 339	954	954 147	300	313 724	1099	3·18751	*1112*	048 057	329	3·34069	*1061*
0·305	300 293	954	953 847	301	314 823	1100	3·17639	*1106*	048 386	331	3·33008	*1054*
·306	301 247	953	953 546	302	315 923	1100	3·16533	*1098*	048 717	332	3·31954	*1048*
·307	302 200	953	953 244	302	317 023	1101	3·15435	*1092*	049 049	333	3·30906	*1040*
·308	303 153	953	952 942	304	318 124	1101	3·14343	*1085*	049 382	335	3·29866	*1033*
·309	304 106	953	952 638	304	319 225	1103	3·13258	*1078*	049 717	335	3·28833	*1027*
0·310	305 059	952	952 334	306	320 328	1102	3·12180	*1071*	050 052	337	3·27806	*1020*
·311	306 011	952	952 028	306	321 430	1104	3·11109	*1064*	050 389	338	3·26786	*1014*
·312	306 963	951	951 722	308	322 534	1105	3·10045	*1058*	050 727	340	3·25772	*1006*
·313	307 914	952	951 414	308	323 639	1105	3·08987	*1052*	051 067	341	3·24766	*1000*
·314	308 866	950	951 106	310	324 744	1105	3·07935	*1045*	051 408	342	3·23766	*994*
0·315	309 816	951	950 796	310	325 849	1107	3·06890	*1038*	051 750	343	3·22772	*988*
·316	310 767	950	950 486	311	326 956	1107	3·05852	*1033*	052 093	345	3·21784	*981*
·317	311 717	950	950 175	312	328 063	1108	3·04819	*1026*	052 438	346	3·20803	*974*
·318	312 667	950	949 863	313	329 171	1109	3·03793	*1020*	052 784	347	3·19829	*969*
·319	313 617	950	949 550	315	330 280	1109	3·02773	*1013*	053 131	348	3·18860	*962*
0·320	314 567	949	949 235	315	331 389	1111	3·01760	*1008*	053 479	350	3·17898	*956*
·321	315 516	948	948 920	316	332 500	1111	3·00752	*1001*	053 829	351	3·16942	*951*
·322	316 464	949	948 604	317	333 611	1111	2·99751	*996*	054 180	353	3·15991	*944*
·323	317 413	948	948 287	317	334 722	1113	2·98755	*989*	054 533	353	3·15047	*938*
·324	318 361	948	947 970	319	335 835	1113	2·97766	*984*	054 886	355	3·14109	*933*
0·325	319 309	947	947 651	320	336 948	1114	2·96782	*978*	055 241	356	3·13176	*926*
·326	320 256	947	947 331	321	338 062	1114	2·95804	*972*	055 597	358	3·12250	*921*
·327	321 203	947	947 010	321	339 176	1116	2·94832	*967*	055 955	359	3·11329	*915*
·328	322 150	947	946 689	323	340 292	1116	2·93865	*960*	056 314	360	3·10414	*909*
·329	323 097	946	946 366	324	341 408	1117	2·92905	*955*	056 674	361	3·09505	*904*
0·330	324 043	946	946 042	324	342 525	1118	2·91950	*950*	057 035	363	3·08601	*898*
·331	324 989	945	945 718	326	343 643	1118	2·91000	*944*	057 398	364	3·07703	*893*
·332	325 934	946	945 392	326	344 761	1119	2·90056	*939*	057 762	365	3·06810	*887*
·333	326 880	945	945 066	327	345 880	1120	2·89117	*933*	058 127	367	3·05923	*882*
·334	327 825	944	944 739	329	347 000	1121	2·88184	*928*	058 494	368	3·05041	*876*
0·335	328 769	944	944 410	329	348 121	1122	2·87256	*922*	058 862	369	3·04165	*871*
·336	329 713	944	944 081	330	349 243	1122	2·86334	*917*	059 231	371	3·03294	*866*
·337	330 657	944	943 751	331	350 365	1123	2·85417	*912*	059 602	372	3·02428	*861*
·338	331 601	943	943 420	332	351 488	1124	2·84505	*907*	059 974	373	3·01567	*855*
·339	332 544	943	943 088	333	352 612	1125	2·83598	*902*	060 347	374	3·00712	*850*
0·340	333 487	943	942 755	334	353 737	1125	2·82696	*897*	060 721	376	2·99862	*845*
·341	334 430	942	942 421	335	354 862	1127	2·81799	*891*	061 097	377	2·99017	*841*
·342	335 372	942	942 086	336	355 989	1127	2·80908	*887*	061 474	379	2·98176	*835*
·343	336 314	941	941 750	337	357 116	1128	2·80021	*881*	061 853	380	2·97341	*830*
·344	337 255	942	941 413	338	358 244	1129	2·79140	*877*	062 233	381	2·96511	*825*
0·345	338 197	941	941 075	338	359 373	1129	2·78263	*872*	062 614	383	2·95686	*820*
·346	339 138	940	940 737	340	360 502	1130	2·77391	*867*	062 997	383	2·94866	*816*
·347	340 078	940	940 397	340	361 632	1132	2·76524	*862*	063 380	386	2·94050	*811*
·348	341 018	940	940 057	342	362 764	1132	2·75662	*858*	063 766	386	2·93239	*806*
·349	341 958	940	939 715	342	363 896	1132	2·74804	*853*	064 152	388	2·92433	*801*
0·350	342 898		939 373		365 028		2·73951		064 540		2·91632	

TABLE IVE—CIRCULAR FUNCTIONS (Radians)

x	sin x		cos x		tan x		cot x		sec x		cosec x	
r	0·		0·		0·		2·		1·		2·	
0·350	342 898	939	939 373	344	365 028	1134	739 512	*8482*	064 540	389	916 321	*7966*
·351	343 837	939	939 029	344	366 162	1135	731 030	*8435*	064 929	391	908 355	*7920*
·352	344 776	938	938 685	345	367 297	1135	722 595	*8390*	065 320	392	900 435	*7874*
·353	345 714	939	938 340	346	368 432	1136	714 205	*8344*	065 712	393	892 561	*7828*
·354	346 653	937	937 994	348	369 568	1137	705 861	*8299*	066 105	395	884 733	*7783*
0·355	347 590	938	937 646	348	370 705	1138	697 562	*8255*	066 500	396	876 950	*7739*
·356	348 528	937	937 298	349	371 843	1139	689 307	*8210*	066 896	397	869 211	*7694*
·357	349 465	937	936 949	350	372 982	1139	681 097	*8166*	067 293	399	861 517	*7650*
·358	350 402	936	936 599	350	374 121	1141	672 931	*8123*	067 692	400	853 867	*7606*
·359	351 338	936	936 249	352	375 262	1141	664 808	*8080*	068 092	402	846 261	*7563*
0·360	352 274	936	935 897	353	376 403	1142	656 728	*8037*	068 494	403	838 698	*7521*
·361	353 210	935	935 544	354	377 545	1143	648 691	*7994*	068 897	404	831 177	*7477*
·362	354 145	935	935 190	354	378 688	1144	640 697	*7953*	069 301	406	823 700	*7436*
·363	355 080	935	934 836	356	379 832	1144	632 744	*7910*	069 707	407	816 264	*7394*
·364	356 015	934	934 480	356	380 976	1146	624 834	*7869*	070 114	408	808 870	*7352*
0·365	356 949	934	934 124	358	382 122	1147	616 965	*7828*	070 522	410	801 518	*7311*
·366	357 883	934	933 766	358	383 269	1147	609 137	*7787*	070 932	411	794 207	*7270*
·367	358 817	933	933 408	359	384 416	1148	601 350	*7747*	071 343	412	786 937	*7229*
·368	359 750	933	933 049	361	385 564	1149	593 603	*7707*	071 755	414	779 708	*7190*
·369	360 683	932	932 688	361	386 713	1150	585 896	*7667*	072 169	416	772 518	*7149*
0·370	361 615	933	932 327	362	387 863	1151	578 229	*7628*	072 585	416	765 369	*7110*
·371	362 548	931	931 965	363	389 014	1152	570 601	*7588*	073 001	418	758 259	*7071*
·372	363 479	932	931 602	364	390 166	1153	563 013	*7550*	073 419	420	751 188	*7032*
·373	364 411	931	931 238	365	391 319	1153	555 463	*7511*	073 839	421	744 156	*6993*
·374	365 342	931	930 873	365	392 472	1155	547 952	*7473*	074 260	422	737 163	*6956*
0·375	366 273	930	930 508	367	393 627	1155	540 479	*7435*	074 682	424	730 207	*6917*
·376	367 203	930	930 141	368	394 782	1156	533 044	*7398*	075 106	425	723 290	*6879*
·377	368 133	929	929 773	368	395 938	1158	525 646	*7360*	075 531	427	716 411	*6842*
·378	369 062	930	929 405	370	397 096	1158	518 286	*7323*	075 958	428	709 569	*6805*
·379	369 992	928	929 035	370	398 254	1159	510 963	*7287*	076 386	429	702 764	*6768*
0·380	370 920	929	928 665	372	399 413	1160	503 676	*7250*	076 815	431	695 996	*6732*
·381	371 849	928	928 293	372	400 573	1161	496 426	*7214*	077 246	432	689 264	*6695*
·382	372 777	928	927 921	373	401 734	1161	489 212	*7179*	077 678	434	682 569	*6660*
·383	373 705	927	927 548	374	402 895	1163	482 033	*7143*	078 112	435	675 909	*6624*
·384	374 632	927	927 174	376	404 058	1164	474 890	*7107*	078 547	436	669 285	*6588*
0·385	375 559	927	926 798	376	405 222	1165	467 783	*7073*	078 983	438	662 697	*6554*
·386	376 486	926	926 422	377	406 387	1165	460 710	*7037*	079 421	440	656 143	*6519*
·387	377 412	926	926 045	377	407 552	1167	453 673	*7004*	079 861	440	649 624	*6484*
·388	378 338	925	925 668	379	408 719	1167	446 669	*6969*	080 301	443	643 140	*6449*
·389	379 263	925	925 289	380	409 886	1169	439 700	*6935*	080 744	443	636 691	*6416*
0·390	380 188	925	924 909	381	411 055	1169	432 765	*6902*	081 187	446	630 275	*6382*
·391	381 113	924	924 528	381	412 224	1171	425 863	*6868*	081 633	446	623 893	*6349*
·392	382 037	924	924 147	383	413 395	1171	418 995	*6835*	082 079	448	617 544	*6315*
·393	382 961	924	923 764	383	414 566	1173	412 160	*6802*	082 527	450	611 229	*6282*
·394	383 885	923	923 381	384	415 739	1173	405 358	*6769*	082 977	451	604 947	*6250*
0·395	384 808	923	922 997	386	416 912	1174	398 589	*6737*	083 428	452	598 697	*6217*
·396	385 731	922	922 611	386	418 086	1175	391 852	*6705*	083 880	454	592 480	*6184*
·397	386 653	922	922 225	387	419 261	1177	385 147	*6673*	084 334	455	586 296	*6153*
·398	387 575	922	921 838	388	420 438	1177	378 474	*6642*	084 789	457	580 143	*6121*
·399	388 497	921	921 450	389	421 615	1178	371 832	*6610*	085 246	458	574 022	*6090*
0·400	389 418		921 061		422 793		365 222		085 704		567 932	

TABLE IVE—CIRCULAR FUNCTIONS (Radians)

x	sin x		cos x		tan x		cot x		sec x		cosec x	
r	0·		0·		0·		2·		1·		2·	
0·400	389 418	921	921 061	390	422 793	1179	365 222	*6578*	085 704	460	567 932	*6058*
·401	390 339	921	920 671	391	423 972	1181	358 644	*6548*	086 164	461	561 874	*6027*
·402	391 260	920	920 280	391	425 153	1181	352 096	*6517*	086 625	463	555 847	*5996*
·403	392 180	919	919 889	393	426 334	1182	345 579	*6487*	087 088	464	549 851	*5966*
·404	393 099	920	919 496	394	427 516	1184	339 092	*6456*	087 552	466	543 885	*5935*
0·405	394 019	919	919 102	394	428 700	1184	332 636	*6426*	088 018	467	537 950	*5905*
·406	394 938	918	918 708	395	429 884	1185	326 210	*6396*	088 485	469	532 045	*5875*
·407	395 856	918	918 313	397	431 069	1186	319 814	*6367*	088 954	470	526 170	*5845*
·408	396 774	918	917 916	397	432 255	1188	313 447	*6338*	089 424	472	520 325	*5816*
·409	397 692	917	917 519	398	433 443	1188	307 109	*6308*	089 896	473	514 509	*5787*
0·410	398 609	917	917 121	399	434 631	1190	300 801	*6279*	090 369	475	508 722	*5758*
·411	399 526	917	916 722	400	435 821	1190	294 522	*6251*	090 844	476	502 964	*5728*
·412	400 443	916	916 322	401	437 011	1192	288 271	*6221*	091 320	477	497 236	*5700*
·413	401 359	916	915 921	402	438 203	1192	282 050	*6194*	091 797	480	491 536	*5672*
·414	402 275	915	915 519	403	439 395	1194	275 856	*6166*	092 277	480	485 864	*5643*
0·415	403 190	915	915 116	403	440 589	1194	269 690	*6137*	092 757	482	480 221	*5616*
·416	404 105	914	914 713	405	441 783	1196	263 553	*6110*	093 239	484	474 605	*5587*
·417	405 019	914	914 308	405	442 979	1197	257 443	*6082*	093 723	485	469 018	*5560*
·418	405 933	914	913 903	407	444 176	1198	251 361	*6055*	094 208	487	463 458	*5533*
·419	406 847	913	913 496	407	445 374	1199	245 306	*6028*	094 695	489	457 925	*5505*
0·420	407 760	913	913 089	408	446 573	1200	239 278	*6001*	095 184	489	452 420	*5478*
·421	408 673	913	912 681	409	447 773	1201	233 277	*5974*	095 673	492	446 942	*5451*
·422	409 586	912	912 272	410	448 974	1202	227 303	*5948*	096 165	493	441 491	*5425*
·423	410 498	912	911 862	411	450 176	1203	221 355	*5921*	096 658	494	436 066	*5398*
·424	411 410	911	911 451	412	451 379	1204	215 434	*5895*	097 152	496	430 668	*5372*
0·425	412 321	911	911 039	413	452 583	1206	209 539	*5869*	097 648	498	425 296	*5346*
·426	413 232	910	910 626	414	453 789	1206	203 670	*5844*	098 146	499	419 950	*5319*
·427	414 142	910	910 212	414	454 995	1208	197 826	*5817*	098 645	500	414 631	*5295*
·428	415 052	910	909 798	416	456 203	1208	192 009	*5793*	099 145	503	409 336	*5268*
·429	415 962	909	909 382	416	457 411	1210	186 216	*5766*	099 648	503	404 068	*5243*
0·430	416 871	909	908 966	418	458 621	1211	180 450	*5742*	100 151	506	398 825	*5218*
·431	417 780	908	908 548	418	459 832	1212	174 708	*5717*	100 657	507	393 607	*5193*
·432	418 688	908	908 130	419	461 044	1213	168 991	*5692*	101 164	508	388 414	*5168*
·433	419 596	907	907 711	420	462 257	1214	163 299	*5668*	101 672	510	383 246	*5143*
·434	420 503	907	907 291	421	463 471	1216	157 631	*5643*	102 182	512	378 103	*5119*
0·435	421 410	907	906 870	422	464 687	1216	151 988	*5619*	102 694	513	372 984	*5095*
·436	422 317	906	906 448	423	465 903	1218	146 369	*5595*	103 207	515	367 889	*5070*
·437	423 223	906	906 025	423	467 121	1219	140 774	*5571*	103 722	516	362 819	*5046*
·438	424 129	906	905 602	425	468 340	1219	135 203	*5547*	104 238	518	357 773	*5023*
·439	425 035	904	905 177	425	469 559	1222	129 656	*5524*	104 756	520	352 750	*4999*
0·440	425 939	905	904 752	427	470 781	1222	124 132	*5500*	105 276	521	347 751	*4975*
·441	426 844	904	904 325	427	472 003	1223	118 632	*5477*	105 797	523	342 776	*4952*
·442	427 748	904	903 898	428	473 226	1225	113 155	*5454*	106 320	524	337 824	*4928*
·443	428 652	903	903 470	429	474 451	1225	107 701	*5431*	106 844	526	332 896	*4906*
·444	429 555	903	903 041	430	475 676	1227	102 270	*5408*	107 370	527	327 990	*4882*
0·445	430 458	902	902 611	431	476 903	1228	096 862	*5386*	107 897	530	323 108	*4860*
·446	431 360	902	902 180	432	478 131	1229	091 476	*5363*	108 427	530	318 248	*4838*
·447	432 262	902	901 748	433	479 360	1231	086 113	*5341*	108 957	533	313 410	*4814*
·448	433 164	901	901 315	433	480 591	1231	080 772	*5318*	109 490	534	308 596	*4793*
·449	434 065	901	900 882	435	481 822	1233	075 454	*5297*	110 024	535	303 803	*4770*
0·450	434 966		900 447		483 055		070 157		110 559		299 033	

x	sin x		cos x		tan x		cot x		sec x		cosec x	
r	0·		0·		0·				1·		2·	
0·450	434 966	900	900 447	435	483 055	1234	2·070 157	5274	110 559	538	299 033	4749
·451	435 866	900	900 012	437	484 289	1235	·064 883	5253	111 097	539	294 284	4726
·452	436 766	899	899 575	437	485 524	1236	·059 630	5231	111 636	540	289 558	4705
·453	437 665	899	899 138	438	486 760	1238	·054 399	5210	112 176	542	284 853	4683
·454	438 564	898	898 700	439	487 998	1239	·049 189	5189	112 718	544	280 170	4662
0·455	439 462	898	898 261	440	489 237	1240	2·044 000	5167	113 262	546	275 508	4641
·456	440 360	898	897 821	441	490 477	1241	·038 833	5147	113 808	547	270 867	4619
·457	441 258	897	897 380	441	491 718	1242	·033 686	5125	114 355	549	266 248	4598
·458	442 155	897	896 939	443	492 960	1244	·028 561	5105	114 904	550	261 650	4578
·459	443 052	896	896 496	444	494 204	1245	·023 456	5084	115 454	552	257 072	4556
0·460	443 948	896	896 052	444	495 449	1246	2·018 372	5063	116 006	554	252 516	4537
·461	444 844	895	895 608	445	496 695	1247	·013 309	5044	116 560	555	247 979	4515
·462	445 739	895	895 163	446	497 942	1249	·008 265	5023	117 115	557	243 464	4496
·463	446 634	895	894 717	447	499 191	1250	2·003 242	5003	117 672	559	238 968	4475
·464	447 529	894	894 270	448	500 441	1251	1·998 239	4983	118 231	561	234 493	4455
0·465	448 423	893	893 822	449	501 692	1252	1·993 256	4963	118 792	562	230 038	4435
·466	449 316	894	893 373	450	502 944	1254	·988 293	4943	119 354	563	225 603	4415
·467	450 210	892	892 923	451	504 198	1254	·983 350	4924	119 917	566	221 188	4396
·468	451 102	892	892 472	451	505 452	1257	·978 426	4905	120 483	567	216 792	4376
·469	451 994	892	892 021	453	506 709	1257	·973 521	4885	121 050	569	212 416	4356
0·470	452 886	892	891 568	453	507 966	1259	1·968 636	4866	121 619	571	208 060	4337
·471	453 778	891	891 115	454	509 225	1260	·963 770	4847	122 190	572	203 723	4318
·472	454 669	890	890 661	455	510 485	1261	·958 923	4828	122 762	574	199 405	4299
·473	455 559	890	890 206	456	511 746	1262	·954 095	4809	123 336	576	195 106	4280
·474	456 449	889	889 750	457	513 008	1264	·949 286	4790	123 912	577	190 826	4262
0·475	457 338	890	889 293	458	514 272	1265	1·944 496	4772	124 489	579	186 564	4242
·476	458 228	888	888 835	459	515 537	1267	·939 724	4753	125 068	581	182 322	4224
·477	459 116	888	888 376	459	516 804	1267	·934 971	4735	125 649	583	178 098	4205
·478	460 004	888	887 917	461	518 071	1269	·930 236	4717	126 232	584	173 893	4187
·479	460 892	887	887 456	461	519 340	1271	·925 519	4698	126 816	586	169 706	4169
0·480	461 779	887	886 995	462	520 611	1272	1·920 821	4681	127 402	588	165 537	4150
·481	462 666	886	886 533	463	521 883	1273	·916 140	4663	127 990	590	161 387	4133
·482	463 552	886	886 070	464	523 156	1274	·911 477	4645	128 580	591	157 254	4114
·483	464 438	885	885 606	465	524 430	1276	·906 832	4627	129 171	593	153 140	4097
·484	465 323	885	885 141	466	525 706	1277	·902 205	4609	129 764	595	149 043	4079
0·485	466 208	885	884 675	467	526 983	1278	1·897 596	4593	130 359	596	144 964	4062
·486	467 093	884	884 208	467	528 261	1280	·893 003	4574	130 955	599	140 902	4044
·487	467 977	883	883 741	469	529 541	1281	·888 429	4558	131 554	600	136 858	4026
·488	468 860	883	883 272	469	530 822	1282	·883 871	4540	132 154	601	132 832	4010
·489	469 743	883	882 803	470	532 104	1284	·879 331	4524	132 755	604	128 822	3992
0·490	470 626	882	882 333	471	533 388	1285	1·874 807	4506	133 359	606	124 830	3975
·491	471 508	882	881 862	472	534 673	1287	·870 301	4490	133 965	607	120 855	3958
·492	472 390	881	881 390	473	535 960	1288	·865 811	4473	134 572	609	116 897	3942
·493	473 271	880	880 917	474	537 248	1289	·861 338	4456	135 181	610	112 955	3924
·494	474 151	881	880 443	474	538 537	1291	·856 882	4440	135 791	613	109 031	3908
0·495	475 032	879	879 969	476	539 828	1292	1·852 442	4423	136 404	614	105 123	3891
·496	475 911	880	879 493	476	541 120	1294	·848 019	4407	137 018	617	101 232	3875
·497	476 791	878	879 017	477	542 414	1294	·843 612	4391	137 635	618	097 357	3859
·498	477 669	879	878 540	478	543 708	1297	·839 221	4375	138 253	619	093 498	3842
·499	478 548	878	878 062	479	545 005	1297	·834 846	4358	138 872	622	089 656	3826
0·500	479 426		877 583		546 302		1·830 488		139 494		085 830	

TABLE IVE—CIRCULAR FUNCTIONS (Radians)

x	sin x		cos x		tan x		cot x		sec x		cosec x	
r	0·		0·		0·		1·		1·			
0·500	479 426	877	877 583	480	546 302	1300	830 488	4343	139 494	623	2·085 830	3810
·501	480 303	877	877 103	481	547 602	1300	826 145	4327	140 117	626	·082 020	3795
·502	481 180	876	876 622	482	548 902	1302	821 818	4311	140 743	627	·078 225	3778
·503	482 056	876	876 140	482	550 204	1304	817 507	4296	141 370	629	·074 447	3762
·504	482 932	875	875 658	484	551 508	1305	813 211	4280	141 999	630	·070 685	3747
0·505	483 807	875	875 174	484	552 813	1306	808 931	4264	142 629	633	2·066 938	3731
·506	484 682	875	874 690	485	554 119	1308	804 667	4249	143 262	634	·063 207	3716
·507	485 557	874	874 205	486	555 427	1309	800 418	4234	143 896	637	·059 491	3700
·508	486 431	873	873 719	487	556 736	1311	796 184	4219	144 533	638	·055 791	3685
·509	487 304	873	873 232	487	558 047	1312	791 965	4203	145 171	640	·052 106	3670
0·510	488 177	873	872 745	489	559 359	1313	787 762	4189	145 811	642	2·048 436	3654
·511	489 050	872	872 256	490	560 672	1315	783 573	4174	146 453	643	·044 782	3640
·512	489 922	871	871 766	490	561 987	1317	779 399	4159	147 096	646	·041 142	3624
·513	490 793	871	871 276	491	563 304	1318	775 240	4144	147 742	647	·037 518	3610
·514	491 664	871	870 785	492	564 622	1320	771 096	4129	148 389	650	·033 908	3595
0·515	492 535	870	870 293	493	565 942	1321	766 967	4115	149 039	651	2·030 313	3580
·516	493 405	869	869 800	494	567 263	1322	762 852	4101	149 690	653	·026 733	3565
·517	494 274	870	869 306	495	568 585	1324	758 751	4086	150 343	655	·023 168	3551
·518	495 144	868	868 811	495	569 909	1326	754 665	4071	150 998	657	·019 617	3537
·519	496 012	868	868 316	497	571 235	1327	750 594	4058	151 655	659	·016 080	3522
0·520	496 880	868	867 819	497	572 562	1328	746 536	4043	152 314	661	2·012 558	3508
·521	497 748	867	867 322	498	573 890	1331	742 493	4029	152 975	662	·009 050	3494
·522	498 615	866	866 824	499	575 221	1331	738 464	4016	153 637	665	·005 556	3479
·523	499 481	866	866 325	500	576 552	1333	734 448	4001	154 302	666	2·002 077	3466
·524	500 347	866	865 825	501	577 885	1335	730 447	3988	154 968	669	1·998 611	3451
0·525	501 213	865	865 324	502	579 220	1336	726 459	3973	155 637	670	1·995 160	3438
·526	502 078	865	864 822	502	580 556	1338	722 486	3960	156 307	672	·991 722	3424
·527	502 943	864	864 320	504	581 894	1340	718 526	3947	156 979	674	·988 298	3410
·528	503 807	863	863 816	504	583 234	1340	714 579	3933	157 653	677	·984 888	3396
·529	504 670	863	863 312	505	584 574	1343	710 646	3920	158 330	678	·981 492	3383
0·530	505 533	863	862 807	506	585 917	1344	706 726	3906	159 008	680	1·978 109	3369
·531	506 396	862	862 301	507	587 261	1346	702 820	3893	159 688	682	·974 740	3356
·532	507 258	861	861 794	507	588 607	1347	698 927	3880	160 370	684	·971 384	3343
·533	508 119	862	861 287	509	589 954	1349	695 047	3866	161 054	686	·968 041	3329
·534	508 981	860	860 778	509	591 303	1350	691 181	3854	161 740	688	·964 712	3316
0·535	509 841	860	860 269	511	592 653	1352	687 327	3840	162 428	690	1·961 396	3303
·536	510 701	860	859 758	511	594 005	1354	683 487	3828	163 118	691	·958 093	3290
·537	511 561	859	859 247	512	595 359	1355	679 659	3815	163 809	694	·954 803	3277
·538	512 420	858	858 735	513	596 714	1357	675 844	3802	164 503	696	·951 526	3264
·539	513 278	858	858 222	513	598 071	1359	672 042	3789	165 199	698	·948 262	3251
0·540	514 136	857	857 709	515	599 430	1360	668 253	3777	165 897	700	1·945 011	3239
·541	514 993	857	857 194	515	600 790	1362	664 476	3764	166 597	702	·941 772	3225
·542	515 850	857	856 679	517	602 152	1363	660 712	3752	167 299	704	·938 547	3213
·543	516 707	856	856 162	517	603 515	1365	656 960	3739	168 003	706	·935 334	3201
·544	517 563	855	855 645	518	604 880	1367	653 221	3727	168 709	708	·932 133	3188
0·545	518 418	855	855 127	519	606 247	1368	649 494	3715	169 417	710	1·928 945	3176
·546	519 273	854	854 608	519	607 615	1370	645 779	3703	170 127	712	·925 769	3163
·547	520 127	854	854 089	521	608 985	1372	642 076	3690	170 839	714	·922 606	3151
·548	520 981	853	853 568	521	610 357	1373	638 386	3678	171 553	716	·919 455	3138
·549	521 834	853	853 047	522	611 730	1375	634 708	3667	172 269	718	·916 317	3127
0·550	522 687		852 525		613 105		631 041		172 987		1·913 190	

x	sin x		cos x		tan x		cot x		sec x		cosec x	
r	0·		0·		0·		1·		1·		1·	
0·550	522 687	852	852 525	524	613 105	1377	631 041	3654	172 987	720	913 190	3114
·551	523 539	852	852 001	524	614 482	1378	627 387	3642	173 707	722	910 076	3103
·552	524 391	851	851 477	524	615 860	1381	623 745	3631	174 429	725	906 973	3090
·553	525 242	851	850 953	526	617 241	1381	620 114	3619	175 154	726	903 883	3079
·554	526 093	850	850 427	527	618 622	1384	616 495	3607	175 880	728	900 804	3067
0·555	526 943	850	849 900	527	620 006	1385	612 888	3596	176 608	731	897 737	3055
·556	527 793	849	849 373	528	621 391	1387	609 292	3584	177 339	733	894 682	3043
·557	528 642	849	848 845	529	622 778	1389	605 708	3572	178 072	734	891 639	3031
·558	529 491	848	848 316	530	624 167	1390	602 136	3562	178 806	737	888 608	3020
·559	530 339	847	847 786	531	625 557	1393	598 574	3549	179 543	739	885 588	3009
0·560	531 186	847	847 255	531	626 950	1393	595 025	3539	180 282	741	882 579	2997
·561	532 033	847	846 724	533	628 343	1396	591 486	3527	181 023	743	879 582	2986
·562	532 880	846	846 191	533	629 739	1398	587 959	3516	181 766	745	876 596	2974
·563	533 726	845	845 658	534	631 137	1399	584 443	3505	182 511	748	873 622	2963
·564	534 571	845	845 124	535	632 536	1401	580 938	3494	183 259	749	870 659	2952
0·565	535 416	844	844 589	536	633 937	1403	577 444	3483	184 008	752	867 707	2940
·566	536 260	844	844 053	537	635 340	1404	573 961	3471	184 760	754	864 767	2930
·567	537 104	843	843 516	537	636 744	1406	570 490	3461	185 514	756	861 837	2918
·568	537 947	843	842 979	539	638 150	1409	567 029	3451	186 270	758	858 919	2908
·569	538 790	842	842 440	539	639 559	1410	563 578	3439	187 028	760	856 011	2896
0·570	539 632	842	841 901	540	640 969	1411	560 139	3429	187 788	763	853 115	2886
·571	540 474	841	841 361	541	642 380	1414	556 710	3418	188 551	764	850 229	2875
·572	541 315	840	840 820	542	643 794	1415	553 292	3407	189 315	767	847 354	2864
·573	542 155	840	840 278	542	645 209	1417	549 885	3397	190 082	769	844 490	2854
·574	542 995	840	839 736	544	646 626	1419	546 488	3386	190 851	771	841 636	2842
0·575	543 835	839	839 192	544	648 045	1421	543 102	3376	191 622	773	838 794	2832
·576	544 674	838	838 648	545	649 466	1423	539 726	3366	192 395	776	835 962	2822
·577	545 512	838	838 103	546	650 889	1425	536 360	3355	193 171	778	833 140	2811
·578	546 350	837	837 557	547	652 314	1426	533 005	3345	193 949	779	830 329	2801
·579	547 187	837	837 010	547	653 740	1428	529 660	3335	194 728	783	827 528	2790
0·580	548 024	836	836 463	549	655 168	1431	526 325	3325	195 511	784	824 738	2780
·581	548 860	836	835 914	549	656 599	1432	523 000	3314	196 295	787	821 958	2770
·582	549 696	835	835 365	550	658 031	1434	519 686	3304	197 082	788	819 188	2759
·583	550 531	834	834 815	551	659 465	1435	516 382	3295	197 870	792	816 429	2750
·584	551 365	834	834 264	552	660 900	1438	513 087	3284	198 662	793	813 679	2739
0·585	552 199	834	833 712	553	662 338	1440	509 803	3275	199 455	795	810 940	2729
·586	553 033	833	833 159	553	663 778	1441	506 528	3265	200 250	798	808 211	2719
·587	553 866	832	832 606	554	665 219	1444	503 263	3255	201 048	800	805 492	2709
·588	554 698	832	832 052	555	666 663	1445	500 008	3245	201 848	803	802 783	2700
·589	555 530	831	831 497	556	668 108	1448	496 763	3235	202 651	804	800 083	2689
0·590	556 361	831	830 941	557	669 556	1449	493 528	3226	203 455	807	797 394	2679
·591	557 192	830	830 384	558	671 005	1451	490 302	3216	204 262	809	794 715	2670
·592	558 022	829	829 826	558	672 456	1453	487 086	3207	205 071	812	792 045	2660
·593	558 851	829	829 268	559	673 909	1455	483 879	3197	205 883	814	789 385	2651
·594	559 680	829	828 709	561	675 364	1458	480 682	3188	206 697	816	786 734	2641
0·595	560 509	828	828 148	560	676 822	1459	477 494	3178	207 513	818	784 093	2631
·596	561 337	827	827 588	562	678 281	1461	474 316	3169	208 331	821	781 462	2621
·597	562 164	827	827 026	563	679 742	1463	471 147	3160	209 152	823	778 841	2613
·598	562 991	826	826 463	563	681 205	1465	467 987	3150	209 975	826	776 228	2602
·599	563 817	825	825 900	564	682 670	1467	464 837	3141	210 801	827	773 626	2594
0·600	564 642		825 336		684 137		461 696		211 628		771 032	

TABLE IVE—CIRCULAR FUNCTIONS (Radians)

x	sin x		cos x		tan x		cot x		sec x		cosec x	
r	0·		0·		0·		1·		1·		1·	
0·600	564 642	826	825 336	565	684 137	1469	461 696	*3132*	211 628	830	771 032	*2584*
·601	565 468	824	824 771	566	685 606	1471	458 564	*3123*	212 458	833	768 448	*2575*
·602	566 292	824	824 205	567	687 077	1473	455 441	*3114*	213 291	835	765 873	*2565*
·603	567 116	823	823 638	568	688 550	1475	452 327	*3104*	214 126	837	763 308	*2556*
·604	567 939	823	823 070	568	690 025	1477	449 223	*3096*	214 963	839	760 752	*2548*
0·605	568 762	822	822 502	569	691 502	1479	446 127	*3087*	215 802	842	758 204	*2538*
·606	569 584	822	821 933	570	692 981	1482	443 040	*3078*	216 644	845	755 666	*2529*
·607	570 406	821	821 363	571	694 463	1483	439 962	*3069*	217 489	846	753 137	*2520*
·608	571 227	821	820 792	572	695 946	1485	436 893	*3060*	218 335	849	750 617	*2510*
·609	572 048	819	820 220	572	697 431	1488	433 833	*3052*	219 184	852	748 107	*2503*
0·610	572 867	820	819 648	573	698 919	1489	430 781	*3043*	220 036	854	745 604	*2493*
·611	573 687	819	819 075	574	700 408	1492	427 738	*3034*	220 890	856	743 111	*2484*
·612	574 506	818	818 501	575	701 900	1494	424 704	*3025*	221 746	859	740 627	*2475*
·613	575 324	817	817 926	576	703 394	1496	421 679	*3017*	222 605	861	738 152	*2467*
·614	576 141	818	817 350	577	704 890	1497	418 662	*3008*	223 466	864	735 685	*2458*
0·615	576 959	816	816 773	577	706 387	1501	415 654	*3000*	224 330	866	733 227	*2449*
·616	577 775	816	816 196	578	707 888	1502	412 654	*2992*	225 196	868	730 778	*2441*
·617	578 591	815	815 618	579	709 390	1504	409 662	*2982*	226 064	871	728 337	*2432*
·618	579 406	815	815 039	580	710 894	1506	406 680	*2975*	226 935	874	725 905	*2424*
·619	580 221	814	814 459	581	712 400	1509	403 705	*2966*	227 809	876	723 481	*2415*
0·620	581 035	814	813 878	581	713 909	1511	400 739	*2958*	228 685	878	721 066	*2406*
·621	581 849	813	813 297	582	715 420	1513	397 781	*2950*	229 563	881	718 660	*2398*
·622	582 662	812	812 715	583	716 933	1515	394 831	*2941*	230 444	883	716 262	*2390*
·623	583 474	812	812 132	584	718 448	1517	391 890	*2934*	231 327	886	713 872	*2381*
·624	584 286	811	811 548	585	719 965	1519	388 956	*2925*	232 213	889	711 491	*2373*
0·625	585 097	811	810 963	585	721 484	1522	386 031	*2917*	233 102	891	709 118	*2365*
·626	585 908	810	810 378	587	723 006	1524	383 114	*2909*	233 993	893	706 753	*2357*
·627	586 718	810	809 791	587	724 530	1526	380 205	*2901*	234 886	896	704 396	*2348*
·628	587 528	808	809 204	588	726 056	1528	377 304	*2893*	235 782	899	702 048	*2340*
·629	588 336	809	808 616	588	727 584	1531	374 411	*2885*	236 681	901	699 708	*2332*
0·630	589 145	807	808 028	590	729 115	1532	371 526	*2877*	237 582	903	697 376	*2324*
·631	589 952	808	807 438	590	730 647	1535	368 649	*2869*	238 485	906	695 052	*2316*
·632	590 760	806	806 848	592	732 182	1538	365 780	*2862*	239 391	909	692 736	*2308*
·633	591 566	806	806 256	592	733 720	1539	362 918	*2853*	240 300	911	690 428	*2300*
·634	592 372	806	805 664	592	735 259	1542	360 065	*2846*	241 211	914	688 128	*2292*
0·635	593 178	804	805 072	594	736 801	1544	357 219	*2838*	242 125	917	685 836	*2284*
·636	593 982	804	804 478	594	738 345	1546	354 381	*2831*	243 042	919	683 552	*2276*
·637	594 786	804	803 884	595	739 891	1549	351 550	*2823*	243 961	922	681 276	*2269*
·638	595 590	803	803 289	596	741 440	1551	348 727	*2815*	244 883	924	679 007	*2260*
·639	596 393	802	802 693	597	742 991	1553	345 912	*2808*	245 807	927	676 747	*2253*
0·640	597 195	802	802 096	598	744 544	1555	343 104	*2800*	246 734	929	674 494	*2245*
·641	597 997	801	801 498	598	746 099	1558	340 304	*2793*	247 663	933	672 249	*2238*
·642	598 798	801	800 900	599	747 657	1560	337 511	*2785*	248 596	935	670 011	*2230*
·643	599 599	800	800 301	600	749 217	1563	334 726	*2778*	249 531	937	667 781	*2222*
·644	600 399	799	799 701	601	750 780	1565	331 948	*2770*	250 468	940	665 559	*2215*
0·645	601 198	799	799 100	602	752 345	1567	329 178	*2763*	251 408	943	663 344	*2207*
·646	601 997	798	798 498	602	753 912	1569	326 415	*2756*	252 351	946	661 137	*2199*
·647	602 795	798	797 896	603	755 481	1572	323 659	*2748*	253 297	948	658 938	*2193*
·648	603 593	797	797 293	604	757 053	1575	320 911	*2741*	254 245	951	656 745	*2184*
·649	604 390	796	796 689	605	758 628	1576	318 170	*2734*	255 196	953	654 561	*2178*
0·650	605 186		796 084		760 204		315 436		256 149		652 383	

TABLE IVE—CIRCULAR FUNCTIONS (Radians)

x	sin x		cos x		tan x		cot x		sec x		cosec x	
r	0·		0·		0·		1·		1·		1·	
0·650	605 186	796	796 084	606	760 204	1580	315 436	*2727*	256 149	956	652 383	*2169*
·651	605 982	795	795 478	606	761 784	1581	312 709	*2720*	257 105	959	650 214	*2163*
·652	606 777	795	794 872	607	763 365	1584	309 989	*2712*	258 064	962	648 051	*2155*
·653	607 572	794	794 265	608	764 949	1586	307 277	*2706*	259 026	965	645 896	*2148*
·654	608 366	793	793 657	609	766 535	1589	304 571	*2698*	259 991	967	643 748	*2141*
0·655	609 159	793	793 048	610	768 124	1591	301 873	*2691*	260 958	970	641 607	*2134*
·656	609 952	792	792 438	610	769 715	1594	299 182	*2685*	261 928	972	639 473	*2126*
·657	610 744	792	791 828	611	771 309	1596	296 497	*2677*	262 900	976	637 347	*2119*
·658	611 536	791	791 217	612	772 905	1599	293 820	*2671*	263 876	978	635 228	*2112*
·659	612 327	790	790 605	613	774 504	1601	291 149	*2663*	264 854	981	633 116	*2106*
0·660	613 117	790	789 992	613	776 105	1603	288 486	*2657*	265 835	984	631 010	*2098*
·661	613 907	789	789 379	615	777 708	1607	285 829	*2650*	266 819	987	628 912	*2091*
·662	614 696	788	788 764	615	779 315	1608	283 179	*2643*	267 806	989	626 821	*2084*
·663	615 484	788	788 149	616	780 923	1611	280 536	*2637*	268 795	992	624 737	*2077*
·664	616 272	787	787 533	616	782 534	1614	277 899	*2629*	269 787	995	622 660	*2070*
0·665	617 059	787	786 917	618	784 148	1616	275 270	*2623*	270 782	998	620 590	*2063*
·666	617 846	786	786 299	618	785 764	1619	272 647	*2617*	271 780	1001	618 527	*2056*
·667	618 632	785	785 681	619	787 383	1621	270 030	*2609*	272 781	1004	616 471	*2050*
·668	619 417	785	785 062	620	789 004	1624	267 421	*2603*	273 785	1006	614 421	*2043*
·669	620 202	784	784 442	620	790 628	1626	264 818	*2597*	274 791	1009	612 378	*2036*
0·670	620 986	783	783 822	622	792 254	1629	262 221	*2590*	275 800	1013	610 342	*2029*
·671	621 769	783	783 200	622	793 883	1632	259 631	*2583*	276 813	1015	608 313	*2022*
·672	622 552	783	782 578	623	795 515	1634	257 048	*2577*	277 828	1018	606 291	*2016*
·673	623 335	781	781 955	624	797 149	1637	254 471	*2571*	278 846	1021	604 275	*2010*
·674	624 116	781	781 331	624	798 786	1639	251 900	*2564*	279 867	1023	602 265	*2002*
0·675	624 897	781	780 707	625	800 425	1642	249 336	*2557*	280 890	1027	600 263	*1996*
·676	625 678	779	780 082	626	802 067	1645	246 779	*2552*	281 917	1030	598 267	*1989*
·677	626 457	780	779 456	627	803 712	1647	244 227	*2544*	282 947	1032	596 278	*1983*
·678	627 237	778	778 829	628	805 359	1650	241 683	*2539*	283 979	1036	594 295	*1977*
·679	628 015	778	778 201	628	807 009	1652	239 144	*2532*	285 015	1038	592 318	*1970*
0·680	628 793	777	777 573	629	808 661	1656	236 612	*2527*	286 053	1042	590 348	*1963*
·681	629 570	777	776 944	630	810 317	1658	234 085	*2519*	287 095	1044	588 385	*1957*
·682	630 347	776	776 314	631	811 975	1660	231 566	*2514*	288 139	1048	586 428	*1951*
·683	631 123	775	775 683	632	813 635	1664	229 052	*2508*	289 187	1050	584 477	*1944*
·684	631 898	775	775 051	632	815 299	1666	226 544	*2501*	290 237	1054	582 533	*1938*
0·685	632 673	774	774 419	633	816 965	1668	224 043	*2495*	291 291	1056	580 595	*1931*
·686	633 447	774	773 786	634	818 633	1672	221 548	*2489*	292 347	1059	578 664	*1926*
·687	634 221	772	773 152	634	820 305	1674	219 059	*2483*	293 406	1063	576 738	*1919*
·688	634 993	773	772 518	636	821 979	1677	216 576	*2477*	294 469	1066	574 819	*1912*
·689	635 766	771	771 882	636	823 656	1680	214 099	*2471*	295 535	1068	572 907	*1907*
0·690	636 537	771	771 246	637	825 336	1683	211 628	*2465*	296 603	1072	571 000	*1900*
·691	637 308	770	770 609	638	827 019	1685	209 163	*2460*	297 675	1075	569 100	*1894*
·692	638 078	770	769 971	638	828 704	1688	206 703	*2453*	298 750	1077	567 206	*1889*
·693	638 848	769	769 333	639	830 392	1691	204 250	*2447*	299 827	1081	565 317	*1882*
·694	639 617	768	768 694	640	832 083	1694	201 803	*2441*	300 908	1084	563 435	*1875*
0·695	640 385	768	768 054	641	833 777	1697	199 362	*2436*	301 992	1087	561 560	*1870*
·696	641 153	767	767 413	642	835 474	1699	196 926	*2430*	303 079	1091	559 690	*1864*
·697	641 920	767	766 771	642	837 173	1702	194 496	*2424*	304 170	1093	557 826	*1858*
·698	642 687	766	766 129	643	838 875	1705	192 072	*2418*	305 263	1097	555 968	*1852*
·699	643 453	765	765 486	644	840 580	1708	189 654	*2412*	306 360	1099	554 116	*1846*
0·700	644 218		764 842		842 288		187 242		307 459		552 270	

x	$\sin x$		$\cos x$		$\tan x$		$\cot x$		$\sec x$		$\operatorname{cosec} x$	
	0·		0·		0·		1·		1·		1·	
0·700	644 218	764	764 842	644	842 288	1711	187 242	*2407*	307 459	1103	552 270	*1840*
·701	644 982	764	764 198	646	843 999	1714	184 835	*2401*	308 562	1106	550 430	*1834*
·702	645 746	763	763 552	646	845 713	1717	182 434	*2395*	309 668	1109	548 596	*1828*
·703	646 509	763	762 906	647	847 430	1719	180 039	*2390*	310 777	1113	546 768	*1822*
·704	647 272	762	762 259	647	849 149	1723	177 649	*2384*	311 890	1115	544 946	*1817*
0·705	648 034	761	761 612	649	850 872	1725	175 265	*2378*	313 005	1119	543 129	*1810*
·706	648 795	761	760 963	649	852 597	1729	172 887	*2373*	314 124	1122	541 319	*1805*
·707	649 556	760	760 314	650	854 326	1731	170 514	*2368*	315 246	1125	539 514	*1799*
·708	650 316	759	759 664	651	856 057	1734	168 146	*2361*	316 371	1129	537 715	*1794*
·709	651 075	759	759 013	651	857 791	1738	165 785	*2357*	317 500	1132	535 921	*1788*
0·710	651 834	758	758 362	652	859 529	1740	163 428	*2350*	318 632	1135	534 133	*1782*
·711	652 592	757	757 710	653	861 269	1743	161 078	*2346*	319 767	1138	532 351	*1776*
·712	653 349	757	757 057	654	863 012	1747	158 732	*2340*	320 905	1142	530 575	*1771*
·713	654 106	756	756 403	655	864 759	1749	156 392	*2334*	322 047	1145	528 804	*1765*
·714	654 862	755	755 748	655	866 508	1752	154 058	*2330*	323 192	1148	527 039	*1759*
0·715	655 617	755	755 093	656	868 260	1756	151 728	*2323*	324 340	1151	525 280	*1754*
·716	656 372	754	754 437	657	870 016	1758	149 405	*2319*	325 491	1155	523 526	*1749*
·717	657 126	754	753 780	657	871 774	1762	147 086	*2313*	326 646	1158	521 777	*1742*
·718	657 880	753	753 123	658	873 536	1764	144 773	*2308*	327 804	1162	520 035	*1738*
·719	658 633	752	752 465	659	875 300	1768	142 465	*2302*	328 966	1165	518 297	*1732*
0·720	659 385	751	751 806	660	877 068	1771	140 163	*2298*	330 131	1168	516 565	*1726*
·721	660 136	751	751 146	661	878 839	1774	137 865	*2292*	331 299	1172	514 839	*1721*
·722	660 887	750	750 485	661	880 613	1777	135 573	*2287*	332 471	1175	513 118	*1715*
·723	661 637	750	749 824	662	882 390	1780	133 286	*2282*	333 646	1178	511 403	*1711*
·724	662 387	748	749 162	663	884 170	1783	131 004	*2276*	334 824	1182	509 692	*1704*
0·725	663 135	749	748 499	663	885 953	1787	128 728	*2272*	336 006	1186	507 988	*1700*
·726	663 884	747	747 836	664	887 740	1789	126 456	*2266*	337 192	1189	506 288	*1694*
·727	664 631	747	747 172	665	889 529	1793	124 190	*2261*	338 381	1192	504 594	*1689*
·728	665 378	746	746 507	666	891 322	1796	121 929	*2256*	339 573	1196	502 905	*1683*
·729	666 124	746	745 841	667	893 118	1800	119 673	*2252*	340 769	1199	501 222	*1679*
0·730	666 870	744	745 174	667	894 918	1802	117 421	*2246*	341 968	1202	499 543	*1673*
·731	667 614	745	744 507	668	896 720	1806	115 175	*2241*	343 170	1207	497 870	*1667*
·732	668 359	743	743 839	669	898 526	1809	112 934	*2236*	344 377	1209	496 203	*1663*
·733	669 102	743	743 170	669	900 335	1812	110 698	*2231*	345 586	1214	494 540	*1657*
·734	669 845	742	742 501	670	902 147	1815	108 467	*2226*	346 800	1216	492 883	*1653*
0·735	670 587	742	741 831	671	903 962	1819	106 241	*2222*	348 016	1221	491 230	*1647*
·736	671 329	740	741 160	672	905 781	1822	104 019	*2216*	349 237	1224	489 583	*1642*
·737	672 069	741	740 488	672	907 603	1826	101 803	*2212*	350 461	1227	487 941	*1636*
·738	672 810	739	739 816	674	909 429	1828	099 591	*2206*	351 688	1231	486 305	*1632*
·739	673 549	739	739 142	673	911 257	1833	097 385	*2202*	352 919	1235	484 673	*1627*
0·740	674 288	738	738 469	675	913 090	1835	095 183	*2197*	354 154	1238	483 046	*1622*
·741	675 026	737	737 794	675	914 925	1839	092 986	*2192*	355 392	1242	481 424	*1616*
·742	675 763	737	737 119	677	916 764	1842	090 794	*2188*	356 634	1245	479 808	*1612*
·743	676 500	736	736 442	676	918 606	1845	088 606	*2183*	357 879	1250	478 196	*1607*
·744	677 236	736	735 766	678	920 451	1849	086 423	*2177*	359 129	1253	476 589	*1601*
0·745	677 972	735	735 088	678	922 300	1853	084 246	*2174*	360 382	1256	474 988	*1597*
·746	678 707	734	734 410	680	924 153	1855	082 072	*2168*	361 638	1260	473 391	*1592*
·747	679 441	733	733 730	679	926 008	1860	079 904	*2164*	362 898	1264	471 799	*1587*
·748	680 174	733	733 051	681	927 868	1862	077 740	*2159*	364 162	1268	470 212	*1582*
·749	680 907	732	732 370	681	929 730	1866	075 581	*2155*	365 430	1271	468 630	*1577*
0·750	681 639		731 689		931 596		073 426		366 701		467 053	

TABLE IVE—CIRCULAR FUNCTIONS (Radians)

x	$\sin x$		$\cos x$		$\tan x$		$\cot x$		$\sec x$		$\operatorname{cosec} x$	
	0·		0·						1·		1·	
0·750	681 639	731	731 689	682	0·931 596	1870	1·073 426	2150	366 701	1275	467 053	1573
·751	682 370	731	731 007	683	·933 466	1873	·071 276	2145	367 976	1279	465 480	1567
·752	683 101	730	730 324	683	·935 339	1877	·069 131	2141	369 255	1283	463 913	1563
·753	683 831	729	729 641	685	·937 216	1880	·066 990	2136	370 538	1286	462 350	1558
·754	684 560	729	728 956	684	·939 096	1884	·064 854	2132	371 824	1290	460 792	1553
0·755	685 289	728	728 272	686	0·940 980	1887	1·062 722	2127	373 114	1294	459 239	1548
·756	686 017	727	727 586	686	·942 867	1891	·060 595	2122	374 408	1298	457 691	1544
·757	686 744	726	726 900	688	·944 758	1894	·058 473	2119	375 706	1302	456 147	1539
·758	687 470	726	726 212	687	·946 652	1898	·056 354	2113	377 008	1305	454 608	1534
·759	688 196	725	725 525	689	·948 550	1901	·054 241	2109	378 313	1309	453 074	1530
0·760	688 921	725	724 836	689	0·950 451	1906	1·052 132	2105	379 622	1314	451 544	1525
·761	689 646	724	724 147	690	·952 357	1908	·050 027	2101	380 936	1317	450 019	1520
·762	690 370	723	723 457	691	·954 265	1913	·047 926	2095	382 253	1321	448 499	1515
·763	691 093	722	722 766	691	·956 178	1916	·045 831	2092	383 574	1325	446 984	1511
·764	691 815	722	722 075	693	·958 094	1920	·043 739	2087	384 899	1328	445 473	1507
0·765	692 537	721	721 382	693	0·960 014	1923	1·041 652	2083	386 227	1333	443 966	1502
·766	693 258	720	720 689	693	·961 937	1927	·039 569	2079	387 560	1337	442 464	1497
·767	693 978	720	719 996	694	·963 864	1931	·037 490	2074	388 897	1341	440 967	1493
·768	694 698	719	719 302	696	·965 795	1935	·035 416	2070	390 238	1344	439 474	1488
·769	695 417	718	718 606	695	·967 730	1938	·033 346	2066	391 582	1349	437 986	1484
0·770	696 135	718	717 911	697	0·969 668	1942	1·031 280	2061	392 931	1353	436 502	1479
·771	696 853	717	717 214	697	·971 610	1946	·029 219	2057	394 284	1356	435 023	1474
·772	697 570	716	716 517	698	·973 556	1950	·027 162	2053	395 640	1361	433 549	1471
·773	698 286	715	715 819	699	·975 506	1954	·025 109	2049	397 001	1365	432 078	1466
·774	699 001	715	715 120	699	·977 460	1957	·023 060	2044	398 366	1369	430 612	1461
0·775	699 716	714	714 421	700	0·979 417	1961	1·021 016	2041	399 735	1373	429 151	1457
·776	700 430	714	713 721	701	·981 378	1965	·018 975	2036	401 108	1377	427 694	1452
·777	701 144	712	713 020	701	·983 343	1969	·016 939	2032	402 485	1381	426 242	1449
·778	701 856	712	712 319	703	·985 312	1973	·014 907	2028	403 866	1385	424 793	1444
·779	702 568	711	711 616	702	·987 285	1977	·012 879	2024	405 251	1390	423 349	1439
0·780	703 279	711	710 914	704	0·989 262	1980	1·010 855	2020	406 641	1393	421 910	1435
·781	703 990	710	710 210	704	·991 242	1985	·008 835	2015	408 034	1398	420 475	1431
·782	704 700	709	709 506	705	·993 227	1988	·006 820	2012	409 432	1402	419 044	1427
·783	705 409	708	708 801	706	·995 215	1993	·004 808	2008	410 834	1406	417 617	1422
·784	706 117	708	708 095	707	·997 208	1996	·002 800	2003	412 240	1411	416 195	1418
0·785	706 825	707	707 388	707	0·999 204	2000	1·000 797	2000	413 651	1414	414 777	1414
·786	707 532	707	706 681	708	1·001 204	2005	0·998 797	1996	415 065	1419	413 363	1409
·787	708 239	705	705 973	708	·003 209	2008	·996 801	1991	416 484	1424	411 954	1406
·788	708 944	705	705 265	710	·005 217	2013	·994 810	1988	417 908	1427	410 548	1401
·789	709 649	704	704 555	710	·007 230	2016	·992 822	1984	419 335	1432	409 147	1397
0·790	710 353	704	703 845	710	1·009 246	2021	0·990 838	1979	420 767	1436	407 750	1393
·791	711 057	703	703 135	712	·011 267	2025	·988 859	1976	422 203	1440	406 357	1388
·792	711 760	702	702 423	712	·013 292	2028	·986 883	1972	423 643	1445	404 969	1385
·793	712 462	701	701 711	713	·015 320	2033	·984 911	1968	425 088	1449	403 584	1380
·794	713 163	701	700 998	713	·017 353	2037	·982 943	1965	426 537	1454	402 204	1376
0·795	713 864	700	700 285	714	1·019 390	2042	0·980 978	1960	427 991	1457	400 828	1372
·796	714 564	699	699 571	715	·021 432	2045	·979 018	1957	429 448	1463	399 456	1368
·797	715 263	698	698 856	716	·023 477	2050	·977 061	1952	430 911	1466	398 088	1364
·798	715 961	698	698 140	716	·025 527	2054	·975 109	1949	432 377	1472	396 724	1360
·799	716 659	697	697 424	717	·027 581	2058	·973 160	1945	433 849	1475	395 364	1356
0·800	717 356		696 707		1·029 639		0·971 215		435 324		394 008	

TABLE IVE—CIRCULAR FUNCTIONS (Radians)

x	sin x		cos x		tan x		cot x		sec x		cosec x	
r	0·		0·		1·		0·		1·		1·	
0·800	717 356	696	696 707	718	029 639	*2062*	971 215	1942	435 324	*1480*	394 008	1352
·801	718 052	696	695 989	718	031 701	*2066*	969 273	1937	436 804	*1485*	392 656	1348
·802	718 748	695	695 271	719	033 767	*2071*	967 336	1934	438 289	*1489*	391 308	1344
·803	719 443	694	694 552	720	035 838	*2075*	965 402	1930	439 778	*1494*	389 964	1340
·804	720 137	694	693 832	721	037 913	*2080*	963 472	1927	441 272	*1498*	388 624	1336
0·805	720 831	692	693 111	721	039 993	*2083*	961 545	1923	442 770	*1503*	387 288	1332
·806	721 523	692	692 390	722	042 076	*2089*	959 622	1919	444 273	*1507*	385 956	1328
·807	722 215	692	691 668	722	044 165	*2092*	957 703	1915	445 780	*1512*	384 628	1324
·808	722 907	690	690 946	724	046 257	*2097*	955 788	1912	447 292	*1516*	383 304	1320
·809	723 597	690	690 222	724	048 354	*2101*	953 876	1908	448 808	*1522*	381 984	1316
0·810	724 287	689	689 498	724	050 455	*2106*	951 968	1904	450 330	*1525*	380 668	1313
·811	724 976	689	688 774	726	052 561	*2110*	950 064	1901	451 855	*1531*	379 355	1308
·812	725 665	687	688 048	726	054 671	*2114*	948 163	1897	453 386	*1535*	378 047	1305
·813	726 352	687	687 322	726	056 785	*2120*	946 266	1894	454 921	*1540*	376 742	1301
·814	727 039	687	686 596	728	058 905	*2123*	944 372	1890	456 461	*1545*	375 441	1297
0·815	727 726	685	685 868	728	061 028	*2128*	942 482	1886	458 006	*1549*	374 144	1293
·816	728 411	685	685 140	728	063 156	*2133*	940 596	1883	459 555	*1554*	372 851	1289
·817	729 096	684	684 412	730	065 289	*2137*	938 713	1880	461 109	*1559*	371 562	1286
·818	729 780	683	683 682	730	067 426	*2141*	936 833	1876	462 668	*1564*	370 276	1282
·819	730 463	683	682 952	731	069 567	*2147*	934 957	1872	464 232	*1568*	368 994	1278
0·820	731 146	682	682 221	731	071 714	*2151*	933 085	1869	465 800	*1574*	367 716	1274
·821	731 828	681	681 490	732	073 865	*2155*	931 216	1865	467 374	*1578*	366 442	1271
·822	732 509	680	680 758	733	076 020	*2160*	929 351	1862	468 952	*1583*	365 171	1267
·823	733 189	680	680 025	734	078 180	*2165*	927 489	1859	470 535	*1588*	363 904	1263
·824	733 869	679	679 291	734	080 345	*2170*	925 630	1855	472 123	*1593*	362 641	1259
0·825	734 548	678	678 557	735	082 515	*2174*	923 775	1852	473 716	*1597*	361 382	1256
·826	735 226	677	677 822	735	084 689	*2179*	921 923	1848	475 313	*1603*	360 126	1252
·827	735 903	677	677 087	737	086 868	*2183*	920 075	1845	476 916	*1608*	358 874	1248
·828	736 580	676	676 350	737	089 051	*2189*	918 230	1841	478 524	*1613*	357 626	1245
·829	737 256	675	675 613	737	091 240	*2193*	916 389	1838	480 137	*1617*	356 381	1241
0·830	737 931	675	674 876	739	093 433	*2198*	914 551	1835	481 754	*1623*	355 140	1238
·831	738 606	674	674 137	738	095 631	*2203*	912 716	1831	483 377	*1628*	353 902	1234
·832	739 280	673	673 399	740	097 834	*2207*	910 885	1828	485 005	*1633*	352 668	1230
·833	739 953	672	672 659	740	100 041	*2213*	909 057	1825	486 638	*1637*	351 438	1227
·834	740 625	672	671 919	741	102 254	*2217*	907 232	1821	488 275	*1643*	350 211	1223
0·835	741 297	670	671 178	742	104 471	*2223*	905 411	1819	489 918	*1649*	348 988	1220
·836	741 967	670	670 436	742	106 694	*2227*	903 592	1814	491 567	*1653*	347 768	1216
·837	742 637	670	669 694	743	108 921	*2232*	901 778	1812	493 220	*1658*	346 552	1212
·838	743 307	668	668 951	744	111 153	*2237*	899 966	1808	494 878	*1664*	345 340	1209
·839	743 975	668	668 207	744	113 390	*2242*	898 158	1805	496 542	*1669*	344 131	1206
0·840	744 643	667	667 463	745	115 632	*2247*	896 353	1802	498 211	*1674*	342 925	1202
·841	745 310	667	666 718	746	117 879	*2253*	894 551	1799	499 885	*1679*	341 723	1198
·842	745 977	665	665 972	746	120 132	*2257*	892 752	1795	501 564	*1685*	340 525	1195
·843	746 642	665	665 226	747	122 389	*2262*	890 957	1792	503 249	*1690*	339 330	1192
·844	747 307	664	664 479	748	124 651	*2268*	889 165	1789	504 939	*1695*	338 138	1188
0·845	747 971	663	663 731	748	126 919	*2272*	887 376	1786	506 634	*1700*	336 950	1185
·846	748 634	663	662 983	749	129 191	*2278*	885 590	1783	508 334	*1706*	335 765	1181
·847	749 297	662	662 234	750	131 469	*2283*	883 807	1780	510 040	*1711*	334 584	1178
·848	749 959	661	661 484	750	133 752	*2288*	882 027	1776	511 751	*1717*	333 406	1174
·849	750 620	660	660 734	751	136 040	*2293*	880 251	1773	513 468	*1722*	332 232	1171
0·850	751 280		659 983		138 333		878 478		515 190		331 061	

x	sin x		cos x		tan x		cot x		sec x		cosec x	
r	0·		0·		1·		0·		1·		1·	
0·850	751 280	660	659 983	751	138 333	*2298*	878 478	1770	515 190	*1728*	331 061	1168
·851	751 940	659	659 232	753	140 631	*2304*	876 708	1767	516 918	*1733*	329 893	1164
·852	752 599	658	658 479	753	142 935	*2309*	874 941	1764	518 651	*1738*	328 729	1161
·853	753 257	657	657 726	753	145 244	*2314*	873 177	1761	520 389	*1744*	327 568	1157
·854	753 914	657	656 973	754	147 558	*2320*	871 416	1758	522 133	*1750*	326 411	1155
0·855	754 571	656	656 219	755	149 878	*2324*	869 658	1755	523 883	*1755*	325 256	1150
·856	755 227	655	655 464	756	152 202	*2331*	867 903	1752	525 638	*1760*	324 106	1148
·857	755 882	654	654 708	756	154 533	*2335*	866 151	1748	527 398	*1767*	322 958	1144
·858	756 536	654	653 952	757	156 868	*2341*	864 403	1746	529 165	*1771*	321 814	1141
·859	757 190	653	653 195	758	159 209	*2347*	862 657	1743	530 936	*1778*	320 673	1138
0·860	757 843	652	652 437	758	161 556	*2352*	860 914	1739	532 714	*1783*	319 535	1134
·861	758 495	651	651 679	759	163 908	*2357*	859 175	1737	534 497	*1789*	318 401	1131
·862	759 146	650	650 920	759	166 265	*2363*	857 438	1734	536 286	*1795*	317 270	1128
·863	759 796	650	650 161	760	168 628	*2369*	855 704	1731	538 081	*1800*	316 142	1125
·864	760 446	649	649 401	761	170 997	*2374*	853 973	1727	539 881	*1806*	315 017	1121
0·865	761 095	649	648 640	761	173 371	*2379*	852 246	1725	541 687	*1812*	313 896	1118
·866	761 744	647	647 879	762	175 750	*2385*	850 521	1722	543 499	*1817*	312 778	1115
·867	762 391	647	647 117	763	178 135	*2391*	848 799	1719	545 316	*1824*	311 663	1112
·868	763 038	646	646 354	763	180 526	*2397*	847 080	1716	547 140	*1829*	310 551	1108
·869	763 684	645	645 591	764	182 923	*2402*	845 364	1713	548 969	*1836*	309 443	1106
0·870	764 329	644	644 827	765	185 325	*2408*	843 651	1711	550 805	*1841*	308 337	1102
·871	764 973	644	644 062	765	187 733	*2413*	841 940	1707	552 646	*1847*	307 235	1099
·872	765 617	643	643 297	766	190 146	*2420*	840 233	1705	554 493	*1853*	306 136	1096
·873	766 260	642	642 531	767	192 566	*2425*	838 528	1701	556 346	*1859*	305 040	1093
·874	766 902	642	641 764	767	194 991	*2431*	836 827	1699	558 205	*1865*	303 947	1089
0·875	767 544	640	640 997	768	197 422	*2436*	835 128	1696	560 070	*1871*	302 858	1087
·876	768 184	640	640 229	769	199 858	*2443*	833 432	1693	561 941	*1877*	301 771	1083
·877	768 824	639	639 460	769	202 301	*2448*	831 739	1691	563 818	*1884*	300 688	1080
·878	769 463	638	638 691	769	204 749	*2455*	830 048	1687	565 702	*1889*	299 608	1078
·879	770 101	638	637 922	771	207 204	*2460*	828 361	1685	567 591	*1895*	298 530	1074
0·880	770 739	637	637 151	771	209 664	*2466*	826 676	1682	569 486	*1902*	297 456	1071
·881	771 376	636	636 380	772	212 130	*2473*	824 994	1679	571 388	*1908*	296 385	1068
·882	772 012	635	635 608	772	214 603	*2478*	823 315	1677	573 296	*1914*	295 317	1065
·883	772 647	634	634 836	773	217 081	*2484*	821 638	1674	575 210	*1920*	294 252	1062
·884	773 281	634	634 063	773	219 565	*2491*	819 964	1671	577 130	*1927*	293 190	1058
0·885	773 915	633	633 290	775	222 056	*2496*	818 293	1668	579 057	*1932*	292 132	1056
·886	774 548	632	632 515	775	224 552	*2503*	816 625	1665	580 989	*1940*	291 076	1053
·887	775 180	631	631 740	775	227 055	*2508*	814 960	1663	582 929	*1945*	290 023	1050
·888	775 811	631	630 965	776	229 563	*2515*	813 297	1660	584 874	*1952*	288 973	1047
·889	776 442	630	630 189	777	232 078	*2521*	811 637	1658	586 826	*1958*	287 926	1044
0·890	777 072	629	629 412	777	234 599	*2528*	809 979	1654	588 784	*1965*	286 882	1040
·891	777 701	628	628 635	778	237 127	*2533*	808 325	1652	590 749	*1971*	285 842	1038
·892	778 329	627	627 857	779	239 660	*2540*	806 673	1650	592 720	*1978*	284 804	1035
·893	778 956	627	627 078	779	242 200	*2547*	805 023	1647	594 698	*1984*	283 769	1032
·894	779 583	626	626 299	780	244 747	*2552*	803 376	1644	596 682	*1991*	282 737	1029
0·895	780 209	625	625 519	781	247 299	*2559*	801 732	1641	598 673	*1997*	281 708	1026
·896	780 834	625	624 738	781	249 858	*2565*	800 091	1639	600 670	*2004*	280 682	1024
·897	781 459	623	623 957	782	252 423	*2572*	798 452	1636	602 674	*2011*	279 658	1020
·898	782 082	623	623 175	782	254 995	*2578*	796 816	1634	604 685	*2017*	278 638	1017
·899	782 705	622	622 393	783	257 573	*2585*	795 182	1631	606 702	*2024*	277 621	1015
0·900	783 327		621 610		260 158		793 551		608 726		276 606	

TABLE IVE—CIRCULAR FUNCTIONS (Radians)

x	sin x		cos x		tan x		cot x		sec x		cosec x	
1	0·		0·		1·		0·		1·		1·	
0·900	783 327	621	621 610	784	260 158	2591	793 551	1628	608 726	2030	276 606	1011
·901	783 948	621	620 826	784	262 749	2598	791 923	1626	610 756	2038	275 595	1009
·902	784 569	619	620 042	785	265 347	2605	790 297	1623	612 794	2044	274 586	1006
·903	785 188	619	619 257	785	267 952	2611	788 674	1621	614 838	2051	273 580	1003
·904	785 807	618	618 472	786	270 563	2617	787 053	1618	616 889	2058	272 577	1000
0·905	786 425	617	617 686	787	273 180	2625	785 435	1616	618 947	2064	271 577	997
·906	787 042	617	616 899	788	275 805	2631	783 819	1613	621 011	2072	270 580	995
·907	787 659	616	616 111	787	278 436	2638	782 206	1611	623 083	2078	269 585	992
·908	788 275	615	615 324	789	281 074	2644	780 595	1608	625 161	2086	268 593	989
·909	788 890	614	614 535	789	283 718	2651	778 987	1605	627 247	2092	267 604	986
0·910	789 504	613	613 746	790	286 369	2659	777 382	1603	629 339	2100	266 618	983
·911	790 117	613	612 956	790	289 028	2665	775 779	1601	631 439	2106	265 635	980
·912	790 730	611	612 166	792	291 693	2671	774 178	1598	633 545	2114	264 655	978
·913	791 341	611	611 374	791	294 364	2679	772 580	1596	635 659	2120	263 677	975
·914	791 952	611	610 583	792	297 043	2686	770 984	1593	637 779	2128	262 702	972
0·915	792 563	609	609 791	793	299 729	2693	769 391	1591	639 907	2135	261 730	969
·916	793 172	609	608 998	794	302 422	2700	767 800	1588	642 042	2143	260 761	967
·917	793 781	607	608 204	794	305 122	2707	766 212	1586	644 185	2149	259 794	964
·918	794 388	607	607 410	795	307 829	2714	764 626	1583	646 334	2157	258 830	961
·919	794 995	607	606 615	795	310 543	2721	763 043	1581	648 491	2164	257 869	959
0·920	795 602	605	605 820	796	313 264	2728	761 462	1579	650 655	2171	256 910	955
·921	796 207	605	605 024	796	315 992	2735	759 883	1576	652 826	2179	255 955	953
·922	796 812	603	604 228	797	318 727	2743	758 307	1574	655 005	2186	255 002	951
·923	797 415	604	603 431	798	321 470	2750	756 733	1571	657 191	2194	254 051	947
·924	798 019	602	602 633	798	324 220	2757	755 162	1570	659 385	2201	253 104	945
0·925	798 621	601	601 835	799	326 977	2765	753 592	1566	661 586	2209	252 159	943
·926	799 222	601	601 036	800	329 742	2772	752 026	1565	663 795	2216	251 216	939
·927	799 823	600	600 236	800	332 514	2779	750 461	1562	666 011	2224	250 277	937
·928	800 423	599	599 436	801	335 293	2787	748 899	1559	668 235	2231	249 340	934
·929	801 022	598	598 635	801	338 080	2794	747 340	1558	670 466	2239	248 406	932
0·930	801 620	597	597 834	802	340 874	2802	745 782	1555	672 705	2247	247 474	929
·931	802 217	597	597 032	802	343 676	2809	744 227	1552	674 952	2254	246 545	926
·932	802 814	596	596 230	804	346 485	2817	742 675	1551	677 206	2263	245 619	924
·933	803 410	595	595 426	803	349 302	2824	741 124	1548	679 469	2270	244 695	921
·934	804 005	594	594 623	805	352 126	2832	739 576	1546	681 739	2277	243 774	919
0·935	804 599	593	593 818	804	354 958	2840	738 030	1543	684 016	2286	242 855	916
·936	805 192	593	593 014	806	357 798	2847	736 487	1542	686 302	2294	241 939	913
·937	805 785	592	592 208	806	360 645	2856	734 945	1539	688 596	2301	241 026	911
·938	806 377	591	591 402	807	363 501	2863	733 406	1536	690 897	2310	240 115	908
·939	806 968	590	590 595	807	366 364	2870	731 870	1535	693 207	2317	239 207	906
0·940	807 558	589	589 788	808	369 234	2879	730 335	1532	695 524	2326	238 301	903
·941	808 147	589	588 980	808	372 113	2887	728 803	1530	697 850	2334	237 398	901
·942	808 736	588	588 172	809	375 000	2894	727 273	1528	700 184	2342	236 497	898
·943	809 324	587	587 363	810	377 894	2903	725 745	1526	702 526	2350	235 599	895
·944	809 911	586	586 553	810	380 797	2911	724 219	1523	704 876	2358	234 704	893
0·945	810 497	585	585 743	811	383 708	2918	722 696	1521	707 234	2366	233 811	890
·946	811 082	585	584 932	811	386 626	2927	721 175	1519	709 600	2375	232 921	888
·947	811 667	584	584 121	812	389 553	2935	719 656	1517	711 975	2383	232 033	886
·948	812 251	582	583 309	813	392 488	2943	718 139	1515	714 358	2391	231 147	883
·949	812 833	583	582 496	813	395 431	2952	716 624	1512	716 749	2400	230 264	880
0·950	813 416		581 683		398 383		715 112		719 149		229 384	

TABLE IVE—CIRCULAR FUNCTIONS (Radians)

x	sin x		cos x		tan x		cot x		sec x		cosec x	
	0·		0·		1·		0·		1·		1·	
0·950	813 416	581	581 683	814	398 383	2959	715 112	1510	719 149	2408	229 384	878
·951	813 997	580	580 869	814	401 342	2968	713 602	1509	721 557	2417	228 506	875
·952	814 577	580	580 055	815	404 310	2976	712 093	1506	723 974	2425	227 631	873
·953	815 157	579	579 240	815	407 286	2985	710 587	1503	726 399	2434	226 758	871
·954	815 736	578	578 425	816	410 271	2993	709 084	1502	728 833	2443	225 887	868
0·955	816 314	577	577 609	817	413 264	3002	707 582	1500	731 276	2451	225 019	865
·956	816 891	576	576 792	817	416 266	3010	706 082	1497	733 727	2460	224 154	864
·957	817 467	576	575 975	818	419 276	3018	704 585	1496	736 187	2468	223 290	860
·958	818 043	575	575 157	818	422 294	3028	703 089	1493	738 655	2477	222 430	858
·959	818 618	574	574 339	819	425 322	3035	701 596	1491	741 132	2486	221 572	856
0·960	819 192	573	573 520	819	428 357	3045	700 105	1489	743 618	2495	220 716	854
·961	819 765	572	572 701	821	431 402	3053	698 616	1487	746 113	2504	219 862	851
·962	820 337	571	571 880	820	434 455	3062	697 129	1485	748 617	2513	219 011	848
·963	820 908	571	571 060	821	437 517	3071	695 644	1483	751 130	2522	218 163	847
·964	821 479	570	570 239	822	440 588	3080	694 161	1481	753 652	2531	217 316	843
0·965	822 049	569	569 417	822	443 668	3089	692 680	1479	756 183	2539	216 473	842
·966	822 618	568	568 595	823	446 757	3097	691 201	1476	758 722	2549	215 631	839
·967	823 186	567	567 772	824	449 854	3107	689 725	1475	761 271	2559	214 792	837
·968	823 753	567	566 948	824	452 961	3115	688 250	1473	763 830	2567	213 955	834
·969	824 320	566	566 124	824	456 076	3125	686 777	1471	766 397	2577	213 121	832
0·970	824 886	565	565 300	826	459 201	3134	685 306	1468	768 974	2586	212 289	829
·971	825 451	564	564 474	825	462 335	3143	683 838	1467	771 560	2595	211 460	828
·972	826 015	563	563 649	827	465 478	3152	682 371	1464	774 155	2605	210 632	825
·973	826 578	562	562 822	827	468 630	3162	680 907	1463	776 760	2614	209 807	822
·974	827 140	562	561 995	827	471 792	3171	679 444	1461	779 374	2623	208 985	821
0·975	827 702	561	561 168	828	474 963	3180	677 983	1458	781 997	2634	208 164	818
·976	828 263	560	560 340	828	478 143	3189	676 525	1457	784 631	2642	207 346	815
·977	828 823	559	559 512	830	481 332	3200	675 068	1455	787 273	2653	206 531	813
·978	829 382	558	558 682	829	484 532	3208	673 613	1453	789 926	2662	205 718	812
·979	829 940	557	557 853	830	487 740	3218	672 160	1450	792 588	2671	204 906	808
0·980	830 497	557	557 023	831	490 958	3228	670 710	1449	795 259	2682	204 098	807
·981	831 054	556	556 192	832	494 186	3237	669 261	1447	797 941	2691	203 291	804
·982	831 610	555	555 360	831	497 423	3248	667 814	1445	800 632	2702	202 487	802
·983	832 165	554	554 529	833	500 671	3257	666 369	1443	803 334	2711	201 685	799
·984	832 719	553	553 696	833	503 928	3266	664 926	1441	806 045	2721	200 886	798
0·985	833 272	553	552 863	833	507 194	3277	663 485	1440	808 766	2731	200 088	795
·986	833 825	551	552 030	835	510 471	3286	662 045	1437	811 497	2741	199 293	793
·987	834 376	551	551 195	834	513 757	3297	660 608	1436	814 238	2752	198 500	790
·988	834 927	550	550 361	835	517 054	3306	659 172	1433	816 990	2761	197 710	789
·989	835 477	549	549 526	836	520 360	3317	657 739	1432	819 751	2772	196 921	786
0·990	836 026	548	548 690	836	523 677	3326	656 307	1430	822 523	2782	196 135	784
·991	836 574	548	547 854	837	527 003	3337	654 877	1428	825 305	2793	195 351	782
·992	837 122	546	547 017	838	530 340	3347	653 449	1426	828 098	2803	194 569	779
·993	837 668	546	546 179	838	533 687	3358	652 023	1424	830 901	2813	193 790	777
·994	838 214	545	545 341	838	537 045	3367	650 599	1422	833 714	2824	193 013	775
0·995	838 759	544	544 503	839	540 412	3378	649 177	1421	836 538	2834	192 238	773
·996	839 303	543	543 664	840	543 790	3389	647 756	1418	839 372	2845	191 465	771
·997	839 846	543	542 824	840	547 179	3399	646 338	1417	842 217	2855	190 694	768
·998	840 389	541	541 984	840	550 578	3410	644 921	1415	845 072	2867	189 926	767
·999	840 930	541	541 144	842	553 988	3420	643 506	1413	847 939	2877	189 159	764
1·000	841 471		540 302		557 408		642 093		850 816		188 395	

TABLE IVE—CIRCULAR FUNCTIONS (Radians)

x	sin x		cos x		tan x		cot x		sec x		cosec x	
	0·		0·		1·		0·				1·	
1·000	841 471	540	540 302	841	557 408	*3431*	642 093	1412	1·850 816	*2888*	188 395	762
·001	842 011	539	539 461	843	560 839	*3441*	640 681	1409	·853 704	*2898*	187 633	760
·002	842 550	538	538 618	843	564 280	*3453*	639 272	1408	·856 602	*2910*	186 873	757
·003	843 088	537	537 775	843	567 733	*3463*	637 864	1406	·859 512	*2921*	186 116	756
·004	843 625	537	536 932	844	571 196	*3474*	636 458	1404	·862 433	*2932*	185 360	753
1·005	844 162	536	536 088	844	574 670	*3485*	635 054	1403	1·865 365	*2943*	184 607	751
·006	844 698	534	535 244	845	578 155	*3496*	633 651	1400	·868 308	*2954*	183 856	750
·007	845 232	534	534 399	846	581 651	*3507*	632 251	1399	·871 262	*2965*	183 106	746
·008	845 766	534	533 553	846	585 158	*3519*	630 852	1397	·874 227	*2977*	182 360	745
·009	846 300	532	532 707	846	588 677	*3529*	629 455	1396	·877 204	*2987*	181 615	743
1·010	846 832	531	531 861	847	592 206	*3541*	628 059	1393	1·880 191	*3000*	180 872	741
·011	847 363	531	531 014	848	595 747	*3552*	626 666	1392	·883 191	*3011*	180 131	738
·012	847 894	530	530 166	848	599 299	*3563*	625 274	1390	·886 202	*3022*	179 393	737
·013	848 424	529	529 318	849	602 862	*3575*	623 884	1388	·889 224	*3034*	178 656	734
·014	848 953	528	528 469	849	606 437	*3587*	622 496	1387	·892 258	*3046*	177 922	732
1·015	849 481	527	527 620	850	610 024	*3598*	621 109	1385	1·895 304	*3057*	177 190	730
·016	850 008	526	526 770	850	613 622	*3609*	619 724	1383	·898 361	*3069*	176 460	728
·017	850 534	526	525 920	851	617 231	*3621*	618 341	1382	·901 430	*3081*	175 732	726
·018	851 060	524	525 069	851	620 852	*3634*	616 959	1379	·904 511	*3093*	175 006	724
·019	851 584	524	524 218	852	624 486	*3644*	615 580	1379	·907 604	*3105*	174 282	722
1·020	852 108	523	523 366	852	628 130	*3657*	614 201	1376	1·910 709	*3117*	173 560	720
·021	852 631	522	522 514	853	631 787	*3669*	612 825	1375	·913 826	*3129*	172 840	717
·022	853 153	521	521 661	854	635 456	*3681*	611 450	1373	·916 955	*3141*	172 123	716
·023	853 674	521	520 807	854	639 137	*3692*	610 077	1371	·920 096	*3153*	171 407	714
·024	854 195	519	519 953	854	642 829	*3705*	608 706	1370	·923 249	*3166*	170 693	711
1·025	854 714	519	519 099	855	646 534	*3718*	607 336	1368	1·926 415	*3178*	169 982	710
·026	855 233	518	518 244	856	650 252	*3729*	605 968	1366	·929 593	*3191*	169 272	707
·027	855 751	517	517 388	856	653 981	*3742*	604 602	1365	·932 784	*3203*	168 565	706
·028	856 268	516	516 532	856	657 723	*3754*	603 237	1363	·935 987	*3216*	167 859	703
·029	856 784	515	515 676	857	661 477	*3767*	601 874	1361	·939 203	*3228*	167 156	702
1·030	857 299	514	514 819	858	665 244	*3779*	600 513	1360	1·942 431	*3241*	166 454	699
·031	857 813	514	513 961	858	669 023	*3792*	599 153	1358	·945 672	*3254*	165 755	698
·032	858 327	513	513 103	858	672 815	*3805*	597 795	1357	·948 926	*3266*	165 057	695
·033	858 840	511	512 245	859	676 620	*3817*	596 438	1355	·952 192	*3280*	164 362	694
·034	859 351	511	511 386	860	680 437	*3831*	595 083	1353	·955 472	*3292*	163 668	691
1·035	859 862	510	510 526	860	684 268	*3843*	593 730	1352	1·958 764	*3306*	162 977	690
·036	860 372	510	509 666	861	688 111	*3856*	592 378	1350	·962 070	*3319*	162 287	687
·037	860 882	508	508 805	861	691 967	*3869*	591 028	1349	·965 389	*3332*	161 600	686
·038	861 390	508	507 944	862	695 836	*3883*	589 679	1346	·968 721	*3345*	160 914	683
·039	861 898	506	507 082	862	699 719	*3896*	588 333	1346	·972 066	*3359*	160 231	682
1·040	862 404	506	506 220	862	703 615	*3909*	586 987	1344	1·975 425	*3372*	159 549	680
·041	862 910	505	505 358	864	707 524	*3922*	585 643	1342	·978 797	*3385*	158 869	677
·042	863 415	504	504 494	863	711 446	*3936*	584 301	1340	·982 182	*3400*	158 192	676
·043	863 919	503	503 631	864	715 382	*3949*	582 961	1339	·985 582	*3412*	157 516	674
·044	864 422	503	502 767	865	719 331	*3963*	581 622	1338	·988 994	*3427*	156 842	672
1·045	864 925	501	501 902	865	723 294	*3977*	580 284	1336	1·992 421	*3441*	156 170	670
·046	865 426	501	501 037	866	727 271	*3990*	578 948	1334	1·995 862	*3454*	155 500	668
·047	865 927	499	500 171	866	731 261	*4004*	577 614	1333	1·999 316	*3468*	154 832	666
·048	866 426	499	499 305	867	735 265	*4018*	576 281	1331	2·002 784	*3483*	154 166	664
·049	866 925	498	498 438	867	739 283	*4032*	574 950	1330	2·006 267	*3496*	153 502	662
1·050	867 423		497 571		743 315		573 620		2·009 763		152 840	

TABLE IVE—CIRCULAR FUNCTIONS (Radians)

x	$\sin x$		$\cos x$		$\tan x$		$\cot x$		$\sec x$		$\operatorname{cosec} x$	
r	0·		0·		1·		0·		2·		1·	
1·050	867 423	497	497 571	868	743 315	4047	573 620	1329	009 763	3511	152 840	661
·051	867 920	497	496 703	868	747 362	4060	572 291	1326	013 274	3525	152 179	658
·052	868 417	495	495 835	868	751 422	4074	570 965	1326	016 799	3540	151 521	657
·053	868 912	495	494 967	870	755 496	4089	569 639	1323	020 339	3554	150 864	654
·054	869 407	493	494 097	869	759 585	4104	568 316	1323	023 893	3568	150 210	653
1·055	869 900	493	493 228	870	763 689	4118	566 993	1320	027 461	3583	149 557	651
·056	870 393	492	492 358	871	767 807	4132	565 673	1320	031 044	3598	148 906	649
·057	870 885	491	491 487	871	771 939	4147	564 353	1317	034 642	3613	148 257	647
·058	871 376	490	490 616	872	776 086	4162	563 036	1317	038 255	3627	147 610	645
·059	871 866	489	489 744	872	780 248	4177	561 719	1314	041 882	3643	146 965	643
1·060	872 355	489	488 872	873	784 425	4191	560 405	1314	045 525	3658	146 322	642
·061	872 844	487	487 999	873	788 616	4207	559 091	1312	049 183	3672	145 680	639
·062	873 331	487	487 126	873	792 823	4222	557 779	1310	052 855	3688	145 041	638
·063	873 818	486	486 253	874	797 045	4237	556 469	1309	056 543	3704	144 403	636
·064	874 304	485	485 379	875	801 282	4252	555 160	1307	060 247	3719	143 767	634
1·065	874 789	484	484 504	875	805 534	4268	553 853	1306	063 966	3734	143 133	632
·066	875 273	483	483 629	875	809 802	4283	552 547	1305	067 700	3750	142 501	631
·067	875 756	483	482 754	876	814 085	4299	551 242	1303	071 450	3765	141 870	628
·068	876 239	481	481 878	877	818 384	4314	549 939	1302	075 215	3782	141 242	627
·069	876 720	481	481 001	877	822 698	4330	548 637	1300	078 997	3797	140 615	625
1·070	877 201	479	480 124	877	827 028	4346	547 337	1299	082 794	3814	139 990	623
·071	877 680	479	479 247	878	831 374	4362	546 038	1297	086 608	3829	139 367	621
·072	878 159	478	478 369	879	835 736	4378	544 741	1296	090 437	3846	138 746	619
·073	878 637	477	477 490	878	840 114	4394	543 445	1295	094 283	3862	138 127	618
·074	879 114	476	476 612	880	844 508	4411	542 150	1293	098 145	3878	137 509	616
1·075	879 590	475	475 732	880	848 919	4426	540 857	1292	102 023	3894	136 893	614
·076	880 065	475	474 852	880	853 345	4443	539 565	1291	105 917	3912	136 279	612
·077	880 540	473	473 972	881	857 788	4460	538 274	1289	109 829	3928	135 667	610
·078	881 013	473	473 091	881	862 248	4476	536 985	1287	113 757	3944	135 057	609
·079	881 486	472	472 210	882	866 724	4493	535 698	1287	117 701	3962	134 448	607
1·080	881 958	471	471 328	882	871 217	4510	534 411	1284	121 663	3979	133 841	605
·081	882 429	470	470 446	882	875 727	4527	533 127	1284	125 642	3995	133 236	603
·082	882 899	469	469 564	884	880 254	4544	531 843	1282	129 637	4013	132 633	602
·083	883 368	468	468 680	883	884 798	4561	530 561	1281	133 650	4030	132 031	599
·084	883 836	467	467 797	884	889 359	4578	529 280	1279	137 680	4048	131 432	598
1·085	884 303	467	466 913	885	893 937	4596	528 001	1279	141 728	4065	130 834	597
·086	884 770	465	466 028	885	898 533	4613	526 722	1276	145 793	4083	130 237	594
·087	885 235	465	465 143	885	903 146	4631	525 446	1276	149 876	4100	129 643	593
·088	885 700	464	464 258	886	907 777	4649	524 170	1274	153 976	4118	129 050	591
·089	886 164	463	463 372	887	912 426	4666	522 896	1273	158 094	4137	128 459	589
1·090	886 627	462	462 485	886	917 092	4684	521 623	1271	162 231	4154	127 870	587
·091	887 089	461	461 599	888	921 776	4702	520 352	1270	166 385	4172	127 283	586
·092	887 550	460	460 711	888	926 478	4721	519 082	1269	170 557	4191	126 697	584
·093	888 010	460	459 823	888	931 199	4738	517 813	1267	174 748	4209	126 113	582
·094	888 470	458	458 935	889	935 937	4757	516 546	1267	178 957	4228	125 531	581
1·095	888 928	458	458 046	889	940 694	4776	515 279	1264	183 185	4246	124 950	579
·096	889 386	457	457 157	889	945 470	4794	514 015	1264	187 431	4265	124 371	577
·097	889 843	455	456 268	890	950 264	4813	512 751	1262	191 696	4284	123 794	575
·098	890 298	455	455 378	891	955 077	4832	511 489	1261	195 980	4302	123 219	574
·099	890 753	454	454 487	891	959 909	4851	510 228	1260	200 282	4322	122 645	572
1·100	891 207		453 596		964 760		508 968		204 604		122 073	

TABLE IVE—CIRCULAR FUNCTIONS (Radians)

x	sin x		cos x		tan x		cot x		sec x		cosec x	
	0·		0·				0·		2·		1·	
r												
1·100	891 207	454	453 596	891	1·964 760	4870	508 968	1258	204 604	4342	122 073	570
·101	891 661	452	452 705	892	·969 630	4889	507 710	1257	208 946	4360	121 503	568
·102	892 113	451	451 813	893	·974 519	4908	506 453	1256	213 306	4380	120 935	567
·103	892 564	451	450 920	892	·979 427	4928	505 197	1255	217 686	4400	120 368	565
·104	893 015	449	450 028	894	·984 355	4947	503 942	1253	222 086	4419	119 803	564
1·105	893 464	449	449 134	893	1·989 302	4968	502 689	1252	226 505	4439	119 239	562
·106	893 913	448	448 241	894	1·994 270	4987	501 437	1251	230 944	4459	118 677	560
·107	894 361	447	447 347	895	1·999 257	5007	500 186	1250	235 403	4479	118 117	558
·108	894 808	446	446 452	895	2·004 264	5027	498 936	1248	239 882	4500	117 559	557
·109	895 254	445	445 557	895	2·009 291	5047	497 688	1247	244 382	4520	117 002	555
1·110	895 699	444	444 662	896	2·014 338	5068	496 441	1246	248 902	4540	116 447	554
·111	896 143	443	443 766	897	·019 406	5088	495 195	1244	253 442	4561	115 893	551
·112	896 586	443	442 869	897	·024 494	5109	493 951	1244	258 003	4582	115 342	550
·113	897 029	441	441 972	897	·029 603	5130	492 707	1242	262 585	4602	114 792	549
·114	897 470	441	441 075	898	·034 733	5151	491 465	1241	267 187	4624	114 243	547
1·115	897 911	440	440 177	898	2·039 884	5171	490 224	1240	271 811	4645	113 696	545
·116	898 351	438	439 279	898	·045 055	5193	488 984	1238	276 456	4666	113 151	543
·117	898 789	438	438 381	899	·050 248	5214	487 746	1237	281 122	4688	112 608	542
·118	899 227	437	437 482	900	·055 462	5236	486 509	1237	285 810	4709	112 066	540
·119	899 664	436	436 582	900	·060 698	5257	485 272	1234	290 519	4731	111 526	539
1·120	900 100	436	435 682	900	2·065 955	5279	484 038	1234	295 250	4753	110 987	537
·121	900 536	434	434 782	901	·071 234	5301	482 804	1233	300 003	4774	110 450	535
·122	900 970	433	433 881	901	·076 535	5323	481 571	1231	304 777	4797	109 915	534
·123	901 403	433	432 980	901	·081 858	5346	480 340	1230	309 574	4820	109 381	532
·124	901 836	432	432 079	902	·087 204	5367	479 110	1229	314 394	4842	108 849	530
1·125	902 268	430	431 177	903	2·092 571	5390	477 881	1228	319 236	4864	108 319	529
·126	902 698	430	430 274	903	·097 961	5413	476 653	1226	324 100	4888	107 790	527
·127	903 128	429	429 371	903	·103 374	5436	475 427	1226	328 988	4910	107 263	526
·128	903 557	428	428 468	904	·108 810	5458	474 201	1224	333 898	4933	106 737	524
·129	903 985	427	427 564	904	·114 268	5482	472 977	1223	338 831	4957	106 213	522
1·130	904 412	426	426 660	905	2·119 750	5505	471 754	1222	343 788	4980	105 691	521
·131	904 838	426	425 755	905	·125 255	5529	470 532	1221	348 768	5003	105 170	519
·132	905 264	424	424 850	905	·130 784	5552	469 311	1220	353 771	5027	104 651	518
·133	905 688	424	423 945	906	·136 336	5576	468 091	1218	358 798	5052	104 133	516
·134	906 112	422	423 039	907	·141 912	5599	466 873	1218	363 850	5075	103 617	515
1·135	906 534	422	422 132	906	2·147 511	5624	465 655	1216	368 925	5099	103 102	513
·136	906 956	421	421 226	907	·153 135	5648	464 439	1215	374 024	5124	102 589	511
·137	907 377	419	420 319	908	·158 783	5673	463 224	1214	379 148	5148	102 078	510
·138	907 796	419	419 411	908	·164 456	5697	462 010	1213	384 296	5174	101 568	508
·139	908 215	418	418 503	908	·170 153	5722	460 797	1212	389 470	5198	101 060	506
1·140	908 633	418	417 595	909	2·175 875	5747	459 585	1210	394 668	5223	100 554	505
·141	909 051	416	416 686	910	·181 622	5772	458 375	1210	399 891	5248	100 049	504
·142	909 467	415	415 776	909	·187 394	5798	457 165	1208	405 139	5274	099 545	502
·143	909 882	415	414 867	910	·193 192	5822	455 957	1208	410 413	5299	099 043	500
·144	910 297	413	413 957	911	·199 014	5849	454 749	1206	415 712	5325	098 543	499
1·145	910 710	413	413 046	911	2·204 863	5874	453 543	1205	421 037	5351	098 044	497
·146	911 123	411	412 135	911	·210 737	5901	452 338	1204	426 388	5377	097 547	496
·147	911 534	411	411 224	912	·216 638	5926	451 134	1203	431 765	5404	097 051	494
·148	911 945	410	410 312	912	·222 564	5953	449 931	1202	437 169	5430	096 557	492
·149	912 355	409	409 400	913	·228 517	5980	448 729	1201	442 599	5457	096 065	491
1·150	912 764		408 487		2·234 497		447 528		448 056		095 574	

TABLE IVE—CIRCULAR FUNCTIONS (Radians)

x	$\sin x$		$\cos x$		$\tan x$		$\cot x$		$\sec x$		$\operatorname{cosec} x$	
1	0·		0·		2·		0·		2·		1·	
1·150	912 764	408	408 487	913	234 497	*6006*	447 528	1200	448 056	*5483*	095 574	490
·151	913 172	407	407 574	913	240 503	*6034*	446 328	1198	453 539	*5511*	095 084	488
·152	913 579	406	406 661	914	246 537	*6060*	445 130	1198	459 050	*5538*	094 596	486
·153	913 985	406	405 747	914	252 597	*6088*	443 932	1197	464 588	*5566*	094 110	485
·154	914 391	404	404 833	914	258 685	*6116*	442 735	1195	470 154	*5593*	093 625	484
1·155	914 795	403	403 919	915	264 801	*6143*	441 540	1194	475 747	*5621*	093 141	482
·156	915 198	403	403 004	916	270 944	*6171*	440 346	1194	481 368	*5649*	092 659	480
·157	915 601	402	402 088	916	277 115	*6199*	439 152	1192	487 017	*5677*	092 179	479
·158	916 003	400	401 172	916	283 314	*6228*	437 960	1191	492 694	*5706*	091 700	477
·159	916 403	400	400 256	916	289 542	*6257*	436 769	1191	498 400	*5735*	091 223	476
1·160	916 803	399	399 340	917	295 799	*6285*	435 578	1189	504 135	*5763*	090 747	475
·161	917 202	398	398 423	918	302 084	*6314*	434 389	1188	509 898	*5793*	090 272	472
·162	917 600	397	397 505	918	308 398	*6343*	433 201	1187	515 691	*5822*	089 800	472
·163	917 997	396	396 587	918	314 741	*6373*	432 014	1186	521 513	*5851*	089 328	470
·164	918 393	395	395 669	918	321 114	*6402*	430 828	1185	527 364	*5881*	088 858	468
1·165	918 788	395	394 751	919	327 516	*6433*	429 643	1185	533 245	*5912*	088 390	467
·166	919 183	393	393 832	920	333 949	*6462*	428 458	1183	539 157	*5941*	087 923	465
·167	919 576	392	392 912	920	340 411	*6493*	427 275	1182	545 098	*5972*	087 458	464
·168	919 968	392	391 992	920	346 904	*6523*	426 093	1181	551 070	*6002*	086 994	463
·169	920 360	391	391 072	920	353 427	*6554*	424 912	1180	557 072	*6034*	086 531	461
1·170	920 751	389	390 152	921	359 981	*6585*	423 732	1179	563 106	*6064*	086 070	459
·171	921 140	389	389 231	922	366 566	*6616*	422 553	1178	569 170	*6096*	085 611	458
·172	921 529	388	388 309	921	373 182	*6648*	421 375	1177	575 266	*6127*	085 153	457
·173	921 917	387	387 388	922	379 830	*6680*	420 198	1176	581 393	*6160*	084 696	455
·174	922 304	386	386 466	923	386 510	*6711*	419 022	1175	587 553	*6191*	084 241	453
1·175	922 690	385	385 543	923	393 221	*6744*	417 847	1174	593 744	*6224*	083 788	452
·176	923 075	384	384 620	923	399 965	*6776*	416 673	1173	599 968	*6256*	083 336	451
·177	923 459	383	383 697	924	406 741	*6809*	415 500	1173	606 224	*6289*	082 885	449
·178	923 842	383	382 773	924	413 550	*6841*	414 327	1171	612 513	*6322*	082 436	448
·179	924 225	381	381 849	924	420 391	*6875*	413 156	1170	618 835	*6355*	081 988	446
1·180	924 606	380	380 925	925	427 266	*6909*	411 986	1169	625 190	*6389*	081 542	445
·181	924 986	380	380 000	925	434 175	*6942*	410 817	1168	631 579	*6422*	081 097	444
·182	925 366	379	379 075	926	441 117	*6976*	409 649	1168	638 001	*6457*	080 653	441
·183	925 745	377	378 149	926	448 093	*7010*	408 481	1166	644 458	*6491*	080 212	441
·184	926 122	377	377 223	926	455 103	*7045*	407 315	1166	650 949	*6526*	079 771	439
1·185	926 499	376	376 297	927	462 148	*7080*	406 149	1164	657 475	*6561*	079 332	438
·186	926 875	375	375 370	927	469 228	*7114*	404 985	1164	664 036	*6595*	078 894	436
·187	927 250	374	374 443	927	476 342	*7150*	403 821	1162	670 631	*6631*	078 458	435
·188	927 624	373	373 516	928	483 492	*7186*	402 659	1162	677 262	*6667*	078 023	433
·189	927 997	372	372 588	928	490 678	*7221*	401 497	1161	683 929	*6703*	077 590	432
1·190	928 369	371	371 660	929	497 899	*7258*	400 336	1159	690 632	*6739*	077 158	431
·191	928 740	370	370 731	929	505 157	*7294*	399 177	1159	697 371	*6776*	076 727	429
·192	929 110	370	369 802	929	512 451	*7331*	398 018	1158	704 147	*6812*	076 298	427
·193	929 480	368	368 873	930	519 782	*7368*	396 860	1157	710 959	*6850*	075 871	427
·194	929 848	368	367 943	930	527 150	*7405*	395 703	1156	717 809	*6887*	075 444	424
1·195	930 216	366	367 013	930	534 555	*7443*	394 547	1156	724 696	*6925*	075 020	424
·196	930 582	366	366 083	931	541 998	*7481*	393 391	1154	731 621	*6963*	074 596	422
·197	930 948	365	365 152	931	549 479	*7519*	392 237	1153	738 584	*7001*	074 174	421
·198	931 313	363	364 221	931	556 998	*7557*	391 084	1153	745 585	*7040*	073 753	419
·199	931 676	363	363 290	932	564 555	*7597*	389 931	1151	752 625	*7079*	073 334	418
1·200	932 039		362 358		572 152		388 780		759 704		072 916	

x	sin x		cos x		tan x		cot x		sec x		cosec x	
	0·		0·				0·				1·	
1												
1·200	932 039	362	362 358	932	2·57215	764	388 780	1151	2·75970	712	072 916	416
·201	932 401	361	361 426	933	2·57979	767	387 629	1150	2·76682	716	072 500	415
·202	932 762	360	360 493	933	2·58746	772	386 479	1149	2·77398	720	072 085	414
·203	933 122	359	359 560	933	2·59518	775	385 330	1148	2·78118	723	071 671	412
·204	933 481	358	358 627	934	2·60293	780	384 182	1147	2·78841	728	071 259	411
1·205	933 839	357	357 693	934	2·61073	783	383 035	1146	2·79569	732	070 848	409
·206	934 196	357	356 759	934	2·61856	788	381 889	1146	2·80301	736	070 439	408
·207	934 553	355	355 825	935	2·62644	792	380 743	1144	2·81037	741	070 031	407
·208	934 908	355	354 890	935	2·63436	796	379 599	1144	2·81778	744	069 624	405
·209	935 263	353	353 955	936	2·64232	800	378 455	1143	2·82522	749	069 219	404
1·210	935 616	353	353 019	935	2·65032	805	377 312	1142	2·83271	752	068 815	403
·211	935 969	351	352 084	937	2·65837	809	376 170	1141	2·84023	758	068 412	401
·212	936 320	351	351 147	936	2·66646	813	375 029	1140	2·84781	761	068 011	400
·213	936 671	350	350 211	937	2·67459	818	373 889	1139	2·85542	766	067 611	399
·214	937 021	348	349 274	937	2·68277	821	372 750	1139	2·86308	770	067 212	397
1·215	937 369	348	348 337	938	2·69098	827	371 611	1138	2·87078	775	066 815	395
·216	937 717	347	347 399	938	2·69925	831	370 473	1136	2·87853	779	066 420	395
·217	938 064	346	346 461	938	2·70756	835	369 337	1136	2·88632	784	066 025	393
·218	938 410	345	345 523	938	2·71591	840	368 201	1135	2·89416	788	065 632	392
·219	938 755	344	344 585	939	2·72431	844	367 066	1135	2·90204	793	065 240	390
1·220	939 099	344	343 646	940	2·73275	850	365 931	1133	2·90997	798	064 850	389
·221	939 443	342	342 706	939	2·74125	853	364 798	1133	2·91795	802	064 461	388
·222	939 785	341	341 767	940	2·74978	859	363 665	1132	2·92597	807	064 073	386
·223	940 126	340	340 827	940	2·75837	863	362 533	1131	2·93404	812	063 687	385
·224	940 466	340	339 887	941	2·76700	868	361 402	1130	2·94216	816	063 302	383
1·225	940 806	338	338 946	941	2·77568	873	360 272	1129	2·95032	822	062 919	383
·226	941 144	338	338 005	941	2·78441	878	359 143	1129	2·95854	826	062 536	381
·227	941 482	336	337 064	942	2·79319	882	358 014	1128	2·96680	831	062 155	379
·228	941 818	336	336 122	942	2·80201	888	356 886	1127	2·97511	836	061 776	378
·229	942 154	335	335 180	942	2·81089	893	355 759	1126	2·98347	841	061 398	377
1·230	942 489	334	334 238	943	2·81982	897	354 633	1125	2·99188	846	061 021	376
·231	942 823	332	333 295	943	2·82879	903	353 508	1125	3·00034	852	060 645	374
·232	943 155	332	332 352	943	2·83782	908	352 383	1124	3·00886	856	060 271	373
·233	943 487	331	331 409	944	2·84690	913	351 259	1123	3·01742	862	059 898	372
·234	943 818	330	330 465	944	2·85603	918	350 136	1122	3·02604	867	059 526	370
1·235	944 148	329	329 521	944	2·86521	924	349 014	1121	3·03471	872	059 156	369
·236	944 477	328	328 577	945	2·87445	929	347 893	1121	3·04343	877	058 787	368
·237	944 805	328	327 632	945	2·88374	934	346 772	1120	3·05220	883	058 419	366
·238	945 133	326	326 687	945	2·89308	940	345 652	1119	3·06103	889	058 053	365
·239	945 459	325	325 742	946	2·90248	945	344 533	1118	3·06992	893	057 688	364
1·240	945 784	324	324 796	946	2·91193	951	343 415	1118	3·07885	900	057 324	363
·241	946 108	324	323 850	946	2·92144	956	342 297	1116	3·08785	905	056 961	361
·242	946 432	322	322 904	947	2·93100	962	341 181	1116	3·09690	910	056 600	360
·243	946 754	322	321 957	946	2·94062	967	340 065	1116	3·10600	916	056 240	358
·244	947 076	320	321 011	948	2·95029	974	338 949	1114	3·11516	922	055 882	357
1·245	947 396	320	320 063	947	2·96003	979	337 835	1114	3·12438	928	055 525	356
·246	947 716	318	319 116	948	2·96982	985	336 721	1113	3·13366	933	055 169	355
·247	948 034	318	318 168	948	2·97967	990	335 608	1112	3·14299	940	054 814	353
·248	948 352	317	317 220	949	2·98957	997	334 496	1112	3·15239	945	054 461	352
·249	948 669	316	316 271	949	2·99954	1003	333 384	1111	3·16184	952	054 109	351
1·250	948 985		315 322		3·00957		332 273		3·17136		053 758	

TABLE IVE—CIRCULAR FUNCTIONS (Radians)

x	sin x		cos x		tan x		cot x		sec x		cosec x	
	0·		0·				0·				1·	
1·250	948 985	314	315 322	949	3·00957	*1009*	332 273	1110	3·17136	*957*	053 758	350
·251	949 299	314	314 373	949	3·01966	*1015*	331 163	1109	3·18093	*964*	053 408	348
·252	949 613	313	313 424	950	3·02981	*1021*	330 054	1108	3·19057	*970*	053 060	347
·253	949 926	312	312 474	950	3·04002	*1027*	328 946	1108	3·20027	*976*	052 713	345
·254	950 238	311	311 524	950	3·05029	*1034*	327 838	1107	3·21003	*982*	052 368	345
1·255	950 549	310	310 574	951	3·06063	*1040*	326 731	1107	3·21985	*989*	052 023	343
·256	950 859	310	309 623	951	3·07103	*1046*	325 624	1106	3·22974	*995*	051 680	342
·257	951 169	308	308 672	952	3·08149	*1053*	324 518	1104	3·23969	*1001*	051 338	340
·258	951 477	307	307 720	951	3·09202	*1059*	323 414	1105	3·24970	*1008*	050 998	340
·259	951 784	306	306 769	952	3·10261	*1066*	322 309	1103	3·25978	*1015*	050 658	338
1·260	952 090	306	305 817	952	3·11327	*1072*	321 206	1103	3·26993	*1021*	050 320	336
·261	952 396	304	304 865	953	3·12399	*1080*	320 103	1102	3·28014	*1028*	049 984	336
·262	952 700	304	303 912	953	3·13479	*1086*	319 001	1102	3·29042	*1035*	049 648	334
·263	953 004	302	302 959	953	3·14565	*1093*	317 899	1100	3·30077	*1042*	049 314	333
·264	953 306	302	302 006	953	3·15658	*1100*	316 799	1100	3·31119	*1049*	048 981	332
1·265	953 608	300	301 053	954	3·16758	*1107*	315 699	1100	3·32168	*1055*	048 649	330
·266	953 908	300	300 099	954	3·17865	*1113*	314 599	1098	3·33223	*1063*	048 319	329
·267	954 208	298	299 145	955	3·18978	*1122*	313 501	1098	3·34286	*1070*	047 990	328
·268	954 506	298	298 190	954	3·20100	*1128*	312 403	1097	3·35356	*1077*	047 662	327
·269	954 804	297	297 236	955	3·21228	*1135*	311 306	1097	3·36433	*1085*	047 335	325
1·270	955 101	296	296 281	955	3·22363	*1143*	310 209	1096	3·37518	*1091*	047 010	324
·271	955 397	295	295 326	956	3·23506	*1150*	309 113	1095	3·38609	*1099*	046 686	323
·272	955 692	293	294 370	956	3·24656	*1158*	308 018	1095	3·39708	*1107*	046 363	322
·273	955 985	293	293 414	956	3·25814	*1166*	306 923	1094	3·40815	*1114*	046 041	320
·274	956 278	292	292 458	956	3·26980	*1173*	305 829	1093	3·41929	*1122*	045 721	320
1·275	956 570	291	291 502	957	3·28153	*1180*	304 736	1092	3·43051	*1130*	045 401	317
·276	956 861	290	290 545	957	3·29333	*1189*	303 644	1092	3·44181	*1137*	045 084	317
·277	957 151	290	289 588	957	3·30522	*1196*	302 552	1091	3·45318	*1146*	044 767	316
·278	957 441	288	288 631	958	3·31718	*1205*	301 461	1091	3·46464	*1153*	044 451	314
·279	957 729	287	287 673	958	3·32923	*1212*	300 370	1090	3·47617	*1161*	044 137	313
1·280	958 016	286	286 715	958	3·34135	*1221*	299 280	1089	3·48778	*1170*	043 824	312
·281	958 302	285	285 757	958	3·35356	*1228*	298 191	1089	3·49948	*1177*	043 512	310
·282	958 587	285	284 799	959	3·36584	*1237*	297 102	1088	3·51125	*1186*	043 202	310
·283	958 872	283	283 840	959	3·37821	*1246*	296 014	1087	3·52311	*1195*	042 892	308
·284	959 155	282	282 881	959	3·39067	*1254*	294 927	1087	3·53506	*1203*	042 584	307
1·285	959 437	282	281 922	960	3·40321	*1262*	293 840	1086	3·54709	*1211*	042 277	305
·286	959 719	280	280 962	960	3·41583	*1271*	292 754	1085	3·55920	*1220*	041 972	305
·287	959 999	280	280 002	960	3·42854	*1280*	291 669	1085	3·57140	*1229*	041 667	303
·288	960 279	278	279 042	960	3·44134	*1289*	290 584	1084	3·58369	*1238*	041 364	302
·289	960 557	278	278 082	961	3·45423	*1298*	289 500	1083	3·59607	*1246*	041 062	301
1·290	960 835	277	277 121	961	3·46721	*1306*	288 417	1083	3·60853	*1256*	040 761	299
·291	961 112	275	276 160	961	3·48027	*1316*	287 334	1082	3·62109	*1265*	040 462	299
·292	961 387	275	275 199	962	3·49343	*1325*	286 252	1082	3·63374	*1274*	040 163	297
·293	961 662	274	274 237	962	3·50668	*1334*	285 170	1081	3·64648	*1283*	039 866	296
·294	961 936	273	273 275	962	3·52002	*1344*	284 089	1080	3·65931	*1293*	039 570	294
1·295	962 209	271	272 313	962	3·53346	*1354*	283 009	1080	3·67224	*1302*	039 276	294
·296	962 480	271	271 351	963	3·54700	*1363*	281 929	1079	3·68526	*1312*	038 982	292
·297	962 751	270	270 388	963	3·56063	*1372*	280 850	1079	3·69838	*1322*	038 690	291
·298	963 021	269	269 425	963	3·57435	*1383*	279 771	1078	3·71160	*1332*	038 399	290
·299	963 290	268	268 462	963	3·58818	*1392*	278 693	1077	3·72492	*1341*	038 109	289
1·300	963 558		267 499		3·60210		277 616		3·73833		037 820	

TABLE IVE—CIRCULAR FUNCTIONS (Radians)

x	sin x		cos x		tan x		cot x		sec x		cosec x	
	0·		0·				0·				1·	
1												
1·300	963 558	267	267 499	964	3·60210	*1403*	277 616	1077	3·73833	*1352*	037 820	287
·301	963 825	266	266 535	964	3·61613	*1413*	276 539	1076	3·75185	*1362*	037 533	287
·302	964 091	265	265 571	964	3·63026	*1423*	275 463	1076	3·76547	*1372*	037 246	285
·303	964 356	264	264 607	965	3·64449	*1433*	274 387	1075	3·77919	*1383*	036 961	284
·304	964 620	264	263 642	964	3·65882	*1444*	273 312	1074	3·79302	*1393*	036 677	283
1·305	964 884	262	262 678	965	3·67326	*1455*	272 238	1074	3·80695	*1403*	036 394	281
·306	965 146	261	261 713	966	3·68781	*1465*	271 164	1073	3·82098	*1415*	036 113	281
·307	965 407	260	260 747	965	3·70246	*1476*	270 091	1073	3·83513	*1425*	035 832	279
·308	965 667	260	259 782	966	3·71722	*1488*	269 018	1072	3·84938	*1437*	035 553	278
·309	965 927	258	258 816	966	3·73210	*1498*	267 946	1072	3·86375	*1447*	035 275	276
1·310	966 185	257	257 850	966	3·74708	*1510*	266 874	1071	3·87822	*1459*	034 999	276
·311	966 442	257	256 884	967	3·76218	*1521*	265 803	1070	3·89281	*1470*	034 723	275
·312	966 699	255	255 917	967	3·77739	*1533*	264 733	1070	3·90751	*1482*	034 448	273
·313	966 954	255	254 950	967	3·79272	*1544*	263 663	1069	3·92233	*1494*	034 175	272
·314	967 209	253	253 983	967	3·80816	*1556*	262 594	1069	3·93727	*1505*	033 903	271
1·315	967 462	253	253 016	968	3·82372	*1568*	261 525	1068	3·95232	*1517*	033 632	270
·316	967 715	251	252 048	968	3·83940	*1580*	260 457	1067	3·96749	*1530*	033 362	268
·317	967 966	251	251 080	968	3·85520	*1593*	259 390	1067	3·98279	*1541*	033 094	267
·318	968 217	249	250 112	968	3·87113	*1604*	258 323	1067	3·99820	*1554*	032 827	267
·319	968 466	249	249 144	969	3·88717	*1618*	257 256	1066	4·01374	*1567*	032 560	265
1·320	968 715	248	248 175	968	3·90335	*1630*	256 190	1065	4·02941	*1579*	032 295	264
·321	968 963	247	247 207	969	3·91965	*1643*	255 125	1065	4·04520	*1592*	032 031	262
·322	969 210	245	246 238	970	3·93608	*1655*	254 060	1064	4·06112	*1605*	031 769	262
·323	969 455	245	245 268	969	3·95263	*1669*	252 996	1064	4·07717	*1618*	031 507	260
·324	969 700	244	244 299	970	3·96932	*1683*	251 932	1063	4·09335	*1632*	031 247	259
1·325	969 944	243	243 329	970	3·98615	*1695*	250 869	1063	4·10967	*1645*	030 988	259
·326	970 187	242	242 359	971	4·00310	*1710*	249 806	1062	4·12612	*1658*	030 729	256
·327	970 429	240	241 388	970	4·02020	*1723*	248 744	1061	4·14270	*1672*	030 473	256
·328	970 669	240	240 418	971	4·03743	*1737*	247 683	1062	4·15942	*1687*	030 217	255
·329	970 909	239	239 447	971	4·05480	*1751*	246 621	1060	4·17629	*1700*	029 962	253
1·330	971 148	238	238 476	971	4·07231	*1766*	245 561	1060	4·19329	*1715*	029 709	253
·331	971 386	237	237 505	972	4·08997	*1780*	244 501	1060	4·21044	*1729*	029 456	251
·332	971 623	236	236 533	971	4·10777	*1794*	243 441	1059	4·22773	*1745*	029 205	250
·333	971 859	236	235 562	972	4·12571	*1810*	242 382	1058	4·24518	*1758*	028 955	248
·334	972 095	234	234 590	973	4·14381	*1825*	241 324	1058	4·26276	*1774*	028 707	248
1·335	972 329	233	233 617	972	4·16206	*1840*	240 266	1058	4·28050	*1790*	028 459	247
·336	972 562	232	232 645	973	4·18046	*1855*	239 208	1057	4·29840	*1804*	028 212	245
·337	972 794	231	231 672	973	4·19901	*1871*	238 151	1056	4·31644	*1821*	027 967	244
·338	973 025	230	230 699	973	4·21772	*1887*	237 095	1056	4·33465	*1836*	027 723	243
·339	973 255	230	229 726	973	4·23659	*1903*	236 039	1055	4·35301	*1852*	027 480	242
1·340	973 485	228	228 753	974	4·25562	*1919*	234 984	1055	4·37153	*1869*	027 238	241
·341	973 713	227	227 779	974	4·27481	*1936*	233 929	1055	4·39022	*1885*	026 997	240
·342	973 940	226	226 805	974	4·29417	*1952*	232 874	1054	4·40907	*1901*	026 757	238
·343	974 166	226	225 831	974	4·31369	*1969*	231 820	1053	4·42808	*1919*	026 519	238
·344	974 392	224	224 857	974	4·33338	*1987*	230 767	1053	4·44727	*1936*	026 281	236
1·345	974 616	224	223 883	975	4·35325	*2004*	229 714	1053	4·46663	*1953*	026 045	235
·346	974 840	222	222 908	975	4·37329	*2021*	228 661	1052	4·48616	*1971*	025 810	234
·347	975 062	221	221 933	975	4·39350	*2039*	227 609	1052	4·50587	*1988*	025 576	233
·348	975 283	221	220 958	976	4·41389	*2058*	226 557	1051	4·52575	*2007*	025 343	232
·349	975 504	219	219 982	975	4·43447	*2075*	225 506	1050	4·54582	*2025*	025 111	230
1·350	975 723		219 007		4·45522		224 456		4·56607		024 881	

x	sin x		cos x		tan x		cot x		sec x		cosec x	
r	0·		0·				0·				1·	
1·350	975 723	219	219 007	976	4·45522	2094	224 456	1050	4·56607	2044	024 881	230
·351	975 942	217	218 031	976	4·47616	2113	223 406	1050	4·58651	2062	024 651	228
·352	976 159	217	217 055	976	4·49729	2133	222 356	1049	4·60713	2082	024 423	227
·353	976 376	216	216 079	977	4·51862	2151	221 307	1049	4·62795	2101	024 196	226
·354	976 592	214	215 102	977	4·54013	2171	220 258	1048	4·64896	2120	023 970	225
1·355	976 806	214	214 125	977	4·56184	2191	219 210	1048	4·67016	2141	023 745	224
·356	977 020	212	213 148	977	4·58375	2212	218 162	1048	4·69157	2160	023 521	223
·357	977 232	212	212 171	977	4·60587	2231	217 114	1046	4·71317	2181	023 298	222
·358	977 444	211	211 194	978	4·62818	2253	216 068	1047	4·73498	2202	023 076	220
·359	977 655	210	210 216	977	4·65071	2273	215 021	1046	4·75700	2223	022 856	220
1·360	977 865	208	209 239	978	4·67344	2295	213 975	1045	4·77923	2244	022 636	218
·361	978 073	208	208 261	978	4·69639	2316	212 930	1046	4·80167	2266	022 418	217
·362	978 281	207	207 283	979	4·71955	2339	211 884	1044	4·82433	2288	022 201	216
·363	978 488	206	206 304	978	4·74294	2361	210 840	1044	4·84721	2310	021 985	215
·364	978 694	205	205 326	979	4·76655	2383	209 796	1044	4·87031	2333	021 770	214
1·365	978 899	203	204 347	979	4·79038	2406	208 752	1044	4·89364	2356	021 556	212
·366	979 102	203	203 368	979	4·81444	2430	207 708	1043	4·91720	2379	021 344	212
·367	979 305	202	202 389	980	4·83874	2453	206 665	1042	4·94099	2403	021 132	210
·368	979 507	201	201 409	979	4·86327	2477	205 623	1042	4·96502	2426	020 922	210
·369	979 708	200	200 430	980	4·88804	2502	204 581	1042	4·98928	2451	020 712	208
1·370	979 908	199	199 450	980	4·91306	2526	203 539	1041	5·01379	2476	020 504	207
·371	980 107	198	198 470	980	4·93832	2551	202 498	1041	5·03855	2501	020 297	206
·372	980 305	197	197 490	981	4·96383	2577	201 457	1040	5·06356	2526	020 091	205
·373	980 502	196	196 509	980	4·98960	2603	200 417	1040	5·08882	2552	019 886	204
·374	980 698	195	195 529	981	5·01563	2629	199 377	1040	5·11434	2579	019 682	203
1·375	980 893	194	194 548	981	5·04192	2655	198 337	1039	5·14013	2605	019 479	202
·376	981 087	193	193 567	981	5·06847	2683	197 298	1039	5·16618	2632	019 277	200
·377	981 280	192	192 586	982	5·09530	2710	196 259	1038	5·19250	2659	019 077	200
·378	981 472	191	191 604	981	5·12240	2737	195 221	1038	5·21909	2688	018 877	198
·379	981 663	191	190 623	982	5·14977	2767	194 183	1037	5·24597	2716	018 679	197
1·380	981 854	189	189 641	982	5·17744	2795	193 146	1037	5·27313	2744	018 482	196
·381	982 043	188	188 659	982	5·20539	2824	192 109	1037	5·30057	2774	018 286	195
·382	982 231	187	187 677	983	5·23363	2854	191 072	1036	5·32831	2804	018 091	194
·383	982 418	186	186 694	982	5·26217	2884	190 036	1036	5·35635	2833	017 897	193
·384	982 604	185	185 712	983	5·29101	2915	189 000	1036	5·38468	2865	017 704	192
1·385	982 789	185	184 729	983	5·32016	2946	187 964	1035	5·41333	2896	017 512	191
·386	982 974	183	183 746	983	5·34962	2978	186 929	1035	5·44229	2927	017 321	189
·387	983 157	182	182 763	983	5·37940	3010	185 894	1034	5·47156	2959	017 132	189
·388	983 339	182	181 780	983	5·40950	3043	184 860	1034	5·50115	2993	016 943	187
·389	983 521	180	180 797	984	5·43993	3076	183 826	1034	5·53108	3025	016 756	187
1·390	983 701	179	179 813	984	5·47069	3110	182 792	1033	5·56133	3060	016 569	185
·391	983 880	178	178 829	984	5·50179	3144	181 759	1033	5·59193	3094	016 384	184
·392	984 058	178	177 845	984	5·53323	3179	180 726	1032	5·62287	3129	016 200	183
·393	984 236	176	176 861	984	5·56502	3215	179 694	1032	5·65416	3164	016 017	182
·394	984 412	176	175 877	985	5·59717	3251	178 662	1032	5·68580	3201	015 835	181
1·395	984 588	174	174 892	984	5·62968	3288	177 630	1031	5·71781	3237	015 654	180
·396	984 762	173	173 908	985	5·66256	3325	176 599	1031	5·75018	3275	015 474	179
·397	984 935	173	172 923	985	5·69581	3364	175 568	1031	5·78293	3313	015 295	178
·398	985 108	171	171 938	985	5·72945	3402	174 537	1030	5·81606	3352	015 117	176
·399	985 279	171	170 953	986	5·76347	3441	173 507	1030	5·84958	3391	014 941	176
1·400	985 450		169 967		5·79788		172 477		5·88349		014 765	

TABLE IVE—CIRCULAR FUNCTIONS (Radians)

x	sin x		cos x		tan x		cot x		sec x		cosec x	
	0·		0·				0·				1·	
r												
1·400	985 450	169	169 967	985	5·79788	*3482*	172 477	1030	5·88349	*3431*	014 765	174
·401	985 619	169	168 982	986	5·83270	*3523*	171 447	1029	5·91780	*3473*	014 591	174
·402	985 788	167	167 996	986	5·86793	*3564*	170 418	1029	5·95253	*3513*	014 417	172
·403	985 955	167	167 010	986	5·90357	*3606*	169 389	1028	5·98766	*3557*	014 245	171
·404	986 122	165	166 024	986	5·93963	*3650*	168 361	1029	6·02323	*3599*	014 074	171
1·405	986 287	165	165 038	987	5·97613	*3693*	167 332	1027	6·05922	*3643*	013 903	169
·406	986 452	163	164 051	986	6·01306	*3739*	166 305	1028	6·09565	*3688*	013 734	168
·407	986 615	163	163 065	987	6·05045	*3783*	165 277	1027	6·13253	*3733*	013 566	167
·408	986 778	162	162 078	987	6·08828	*3830*	164 250	1027	6·16986	*3780*	013 399	166
·409	986 940	160	161 091	987	6·12658	*3878*	163 223	1026	6·20766	*3827*	013 233	165
1·410	987 100	160	160 104	987	6·16536	*3925*	162 197	1027	6·24593	*3875*	013 068	163
·411	987 260	158	159 117	987	6·20461	*3974*	161 170	1025	6·28468	*3924*	012 905	163
·412	987 418	158	158 130	988	6·24435	*4025*	160 145	1026	6·32392	*3974*	012 742	162
·413	987 576	157	157 142	987	6·28460	*4075*	159 119	1025	6·36366	*4025*	012 580	160
·414	987 733	155	156 155	988	6·32535	*4127*	158 094	1025	6·40391	*4077*	012 420	160
1·415	987 888	155	155 167	988	6·36662	*4180*	157 069	1024	6·44468	*4129*	012 260	158
·416	988 043	154	154 179	988	6·40842	*4234*	156 045	1024	6·48597	*4184*	012 102	158
·417	988 197	152	153 191	989	6·45076	*4289*	155 021	1024	6·52781	*4239*	011 944	156
·418	988 349	152	152 202	988	6·49365	*4345*	153 997	1024	6·57020	*4294*	011 788	155
·419	988 501	151	151 214	989	6·53710	*4402*	152 973	1023	6·61314	*4352*	011 633	155
1·420	988 652	149	150 225	988	6·58112	*4460*	151 950	1023	6·65666	*4410*	011 478	153
·421	988 801	149	149 237	989	6·62572	*4520*	150 927	1023	6·70076	*4470*	011 325	152
·422	988 950	148	148 248	989	6·67092	*4581*	149 904	1022	6·74546	*4530*	011 173	151
·423	989 098	147	147 259	989	6·71673	*4643*	148 882	1022	6·79076	*4593*	011 022	150
·424	989 245	146	146 270	990	6·76316	*4706*	147 860	1022	6·83669	*4655*	010 872	149
1·425	989 391	144	145 280	989	6·81022	*4770*	146 838	1021	6·88324	*4720*	010 723	148
·426	989 535	144	144 291	990	6·85792	*4836*	145 817	1021	6·93044	*4786*	010 575	146
·427	989 679	143	143 301	989	6·90628	*4904*	144 796	1021	6·97830	*4854*	010 429	146
·428	989 822	142	142 312	990	6·95532	*4972*	143 775	1021	7·02684	*4922*	010 283	145
·429	989 964	141	141 322	990	7·00504	*5042*	142 754	1020	7·07606	*4992*	010 138	144
1·430	990 105	139	140 332	991	7·05546	*5114*	141 734	1020	7·12598	*5064*	009 994	142
·431	990 244	139	139 341	990	7·10660	*5188*	140 714	1019	7·17662	*5137*	009 852	142
·432	990 383	138	138 351	990	7·15848	*5262*	139 695	1020	7·22799	*5211*	009 710	140
·433	990 521	137	137 361	991	7·21110	*5338*	138 675	1019	7·28010	*5289*	009 570	140
·434	990 658	136	136 370	991	7·26448	*5417*	137 656	1019	7·33299	*5366*	009 430	138
1·435	990 794	135	135 379	991	7·31865	*5496*	136 637	1018	7·38665	*5446*	009 292	138
·436	990 929	134	134 388	991	7·37361	*5578*	135 619	1019	7·44111	*5528*	009 154	136
·437	991 063	133	133 397	991	7·42939	*5662*	134 600	1018	7·49639	*5612*	009 018	135
·438	991 196	131	132 406	991	7·48601	*5747*	133 582	1017	7·55251	*5697*	008 883	135
·439	991 327	131	131 415	991	7·54348	*5835*	132 565	1018	7·60948	*5784*	008 748	133
1·440	991 458	130	130 424	992	7·60183	*5923*	131 547	1017	7·66732	*5873*	008 615	132
·441	991 588	129	129 432	991	7·66106	*6016*	130 530	1017	7·72605	*5965*	008 483	131
·442	991 717	128	128 441	992	7·72122	*6109*	129 513	1016	7·78570	*6059*	008 352	130
·443	991 845	127	127 449	992	7·78231	*6204*	128 497	1017	7·84629	*6155*	008 222	129
·444	991 972	126	126 457	992	7·84435	*6303*	127 480	1016	7·90784	*6252*	008 093	128
1·445	992 098	125	125 465	992	7·90738	*6403*	126 464	1016	7·97036	*6353*	007 965	127
·446	992 223	124	124 473	993	7·97141	*6507*	125 448	1015	8·03389	*6456*	007 838	126
·447	992 347	123	123 480	992	8·03648	*6611*	124 433	1016	8·09845	*6562*	007 712	125
·448	992 470	122	122 488	993	8·10259	*6720*	123 417	1015	8·16407	*6669*	007 587	124
·449	992 592	121	121 495	992	8·16979	*6830*	122 402	1015	8·23076	*6780*	007 463	123
1·450	992 713		120 503		8·23809		121 387		8·29856		007 340	

TABLE IVE—CIRCULAR FUNCTIONS (Radians)

x	sin x		cos x		tan x		cot x		sec x		cosec x	
	0·		0·				0·				1·	
1·450	992 713	120	120 503	993	8·23809	*6944*	121 387	1014	8·29856	*6894*	007 340	121
·451	992 833	119	119 510	993	8·30753	*7060*	120 373	1015	8·36750	*7010*	007 219	121
·452	992 952	118	118 517	993	8·37813	*7180*	119 358	1014	8·43760	*7129*	007 098	120
·453	993 070	117	117 524	993	8·44993	*7302*	118 344	1014	8·50889	*7252*	006 978	118
·454	993 187	116	116 531	993	8·52295	*7427*	117 330	1013	8·58141	*7377*	006 860	118
1·455	993 303	115	115 538	994	8·59722	*7556*	116 317	1014	8·65518	*7506*	006 742	117
·456	993 418	114	114 544	993	8·67278	*7688*	115 303	1013	8·73024	*7638*	006 625	115
·457	993 532	113	113 551	994	8·74966	*7825*	114 290	1013	8·80662	*7774*	006 510	115
·458	993 645	112	112 557	993	8·82791	*7963*	113 277	1013	8·88436	*7914*	006 395	113
·459	993 757	111	111 564	994	8·90754	*8107*	112 264	1012	8·96350	*8056*	006 282	113
1·460	993 868	110	110 570	994	8·98861	*8253*	111 252	1012	9·04406	*8204*	006 169	111
·461	993 978	110	109 576	994	9·07114	*8405*	110 240	1012	9·12610	*8354*	006 058	110
·462	994 088	108	108 582	994	9·15519	*8560*	109 228	1012	9·20964	*8510*	005 948	110
·463	994 196	107	107 588	995	9·24079	*8720*	108 216	1012	9·29474	*8670*	005 838	108
·464	994 303	106	106 593	994	9·32799	*8884*	107 204	1011	9·38144	*8834*	005 730	107
1·465	994 409	105	105 599	994	9·41683	*9053*	106 193	1011	9·46978	*9003*	005 623	107
·466	994 514	104	104 605	995	9·50736	*9227*	105 182	1011	9·55981	*9176*	005 516	105
·467	994 618	103	103 610	995	9·59963	*9405*	104 171	1011	9·65157	*9356*	005 411	104
·468	994 721	102	102 615	994	9·69368	*9590*	103 160	1011	9·74513	*9539*	005 307	103
·469	994 823	101	101 621	995	9·78958	*9779*	102 149	1010	9·84052	*9730*	005 204	102
1·470	994 924	100	100 626	995	9·8874	*997*	101 139	1010	9·9378	*993*	005 102	102
·471	995 024	100	099 631	995	9·9871	*1018*	100 129	1010	10·0371	*1012*	005 000	100
·472	995 124	98	098 636	995	10·0889	*1038*	099 119	1010	10·1383	*1034*	004 900	99
·473	995 222	97	097 641	996	10·1927	*1060*	098 109	1009	10·2417	*1054*	004 801	98
·474	995 319	96	096 645	995	10·2987	*1082*	097 100	1010	10·3471	*1077*	004 703	97
1·475	995 415	95	095 650	996	10·4069	*1104*	096 090	1009	10·4548	*1099*	004 606	96
·476	995 510	94	094 654	995	10·5173	*1128*	095 081	1009	10·5647	*1123*	004 510	95
·477	995 604	94	093 659	996	10·6301	*1152*	094 072	1008	10·6770	*1148*	004 415	94
·478	995 698	92	092 663	996	10·7453	*1178*	093 064	1009	10·7918	*1172*	004 321	93
·479	995 790	91	091 667	995	10·8631	*1203*	092 055	1008	10·9090	*1198*	004 228	92
1·480	995 881	90	090 672	996	10·9834	*1230*	091 047	1009	11·0288	*1225*	004 136	91
·481	995 971	89	089 676	996	11·1064	*1257*	090 038	1008	11·1513	*1252*	004 045	90
·482	996 060	88	088 680	996	11·2321	*1286*	089 030	1007	11·2765	*1281*	003 955	88
·483	996 148	88	087 684	997	11·3607	*1316*	088 023	1008	11·4046	*1311*	003 867	88
·484	996 236	86	086 687	996	11·4923	*1346*	087 015	1008	11·5357	*1341*	003 779	87
1·485	996 322	85	085 691	996	11·6269	*1378*	086 007	1007	11·6698	*1373*	003 692	86
·486	996 407	84	084 695	997	11·7647	*1411*	085 000	1007	11·8071	*1406*	003 606	85
·487	996 491	83	083 698	996	11·9058	*1444*	083 993	1007	11·9477	*1439*	003 521	84
·488	996 574	83	082 702	997	12·0502	*1480*	082 986	1007	12·0916	*1475*	003 437	82
·489	996 657	81	081 705	997	12·1982	*1517*	081 979	1006	12·2391	*1512*	003 355	82
1·490	996 738	80	080 708	996	12·3499	*1554*	080 973	1007	12·3903	*1549*	003 273	81
·491	996 818	79	079 712	997	12·5053	*1594*	079 966	1006	12·5452	*1589*	003 192	80
·492	996 897	78	078 715	997	12·6647	*1634*	078 960	1006	12·7041	*1630*	003 112	78
·493	996 975	78	077 718	997	12·8281	*1677*	077 954	1006	12·8671	*1672*	003 034	78
·494	997 053	76	076 721	997	12·9958	*1722*	076 948	1006	13·0343	*1716*	002 956	77
1·495	997 129	75	075 724	997	13·1680	*1767*	075 942	1006	13·2059	*1762*	002 879	75
·496	997 204	74	074 727	998	13·3447	*1815*	074 936	1005	13·3821	*1810*	002 804	75
·497	997 278	74	073 729	997	13·5262	*1865*	073 931	1006	13·5631	*1860*	002 729	73
·498	997 352	72	072 732	997	13·7127	*1916*	072 925	1005	13·7491	*1912*	002 656	73
·499	997 424	71	071 735	998	13·9043	*1971*	071 920	1005	13·9403	*1965*	002 583	72
1·500	997 495		070 737		14·1014		070 915		14·1368		002 511	

TABLE IVE—CIRCULAR FUNCTIONS (Radians)

x	$\sin x$		$\cos x$		$\tan x$	$\cot x$		$\sec x$	$\operatorname{cosec} x$	
1·500	0·997 495	70	0·070 737	997	14·1014	0·070 915	1005	14·1368	1·002 511	70
·501	·997 565	69	·069 740	998	14·3041	·069 910	1005	14·3390	·002 441	70
·502	·997 634	69	·068 742	998	14·5127	·068 905	1005	14·5471	·002 371	68
·503	·997 703	67	·067 744	997	14·7275	·067 900	1004	14·7614	·002 303	68
·504	·997 770	66	·066 747	998	14·9486	·066 896	1005	14·9820	·002 235	67
1·505	0·997 836	65	0·065 749	998	15·1765	0·065 891	1004	15·2094	1·002 168	65
·506	·997 901	65	·064 751	998	15·4114	·064 887	1004	15·4438	·002 103	65
·507	·997 966	63	·063 753	998	15·6536	·063 883	1004	15·6855	·002 038	63
·508	·998 029	62	·062 755	998	15·9036	·062 879	1004	15·9350	·001 975	63
·509	·998 091	61	·061 757	998	16·1616	·061 875	1004	16·1925	·001 912	61
1·510	0·998 152	61	0·060 759	998	16·4281	0·060 871	1003	16·4585	1·001 851	61
·511	·998 213	59	·059 761	999	16·7035	·059 868	1004	16·7334	·001 790	59
·512	·998 272	58	·058 762	998	16·9883	·058 864	1003	17·0177	·001 731	58
·513	·998 330	58	·057 764	998	17·2829	·057 861	1004	17·3118	·001 673	58
·514	·998 388	56	·056 766	999	17·5878	·056 857	1003	17·6162	·001 615	56
1·515	0·998 444	55	0·055 767	998	17·9037	0·055 854	1003	17·9316	1·001 559	56
·516	·998 499	54	·054 769	999	18·2311	·054 851	1003	18·2585	·001 503	54
·517	·998 553	54	·053 770	998	18·5707	·053 848	1003	18·5976	·001 449	54
·518	·998 607	52	·052 772	999	18·9231	·052 845	1002	18·9495	·001 395	52
·519	·998 659	51	·051 773	999	19·2891	·051 843	1003	19·3150	·001 343	51
1·520	0·998 710	50	0·050 774	998	19·6695	0·050 840	1002	19·6949	1·001 292	51
·521	·998 760	50	·049 776	999	20·0652	·049 838	1003	20·0901	·001 241	49
·522	·998 810	48	·048 777	999	20·4771	·048 835	1002	20·5015	·001 192	49
·523	·998 858	47	·047 778	999	20·9062	·047 833	1002	20·9301	·001 143	47
·524	·998 905	47	·046 779	999	21·3536	·046 831	1003	21·3770	·001 096	46
1·525	0·998 952	45	0·045 780	999	21·8205	0·045 828	1002	21·8434	1·001 050	46
·526	·998 997	44	·044 781	999	22·3083	·044 826	1002	22·3307	·001 004	44
·527	·999 041	43	·043 782	999	22·8184	·043 824	1002	22·8403	·000 960	44
·528	·999 084	43	·042 783	999	23·3522	·042 822	1001	23·3736	·000 916	42
·529	·999 127	41	·041 784	999	23·9116	·041 821	1002	23·9325	·000 874	41
1·530	0·999 168	40	0·040 785	999	24·4984	0·040 819	1002	24·5188	1·000 833	41
·531	·999 208	40	·039 786	999	25·1147	·039 817	1001	25·1346	·000 792	39
·532	·999 248	38	·038 787	1000	25·7627	·038 816	1002	25·7821	·000 753	38
·533	·999 286	37	·037 787	999	26·4450	·037 814	1001	26·4639	·000 715	38
·534	·999 323	36	·036 788	999	27·1644	·036 813	1001	27·1828	·000 677	36
1·535	0·999 359	36	0·035 789	1000	27·9239	0·035 812	1002	27·9418	1·000 641	35
·536	·999 395	34	·034 789	999	28·7271	·034 810	1001	28·7445	·000 606	35
·537	·999 429	33	·033 790	1000	29·5777	·033 809	1001	29·5946	·000 571	33
·538	·999 462	33	·032 790	999	30·4803	·032 808	1001	30·4967	·000 538	32
·539	·999 495	31	·031 791	1000	31·4396	·031 807	1001	31·4555	·000 506	32
1·540	0·999 526	30	0·030 791	999	32·4611	0·030 806	1001	32·4765	1·000 474	30
·541	·999 556	29	·029 792	1000	33·5513	·029 805	1001	33·5662	·000 444	29
·542	·999 585	29	·028 792	999	34·7171	·028 804	1001	34·7315	·000 415	29
·543	·999 614	27	·027 793	1000	35·9667	·027 803	1000	35·9806	·000 386	27
·544	·999 641	26	·026 793	1000	37·3096	·026 803	1001	37·3230	·000 359	26
1·545	0·999 667	26	0·025 793	999	38·7566	0·025 802	1001	38·7695	1·000 333	25
·546	·999 693	24	·024 794	1000	40·3203	·024 801	1000	40·3327	·000 308	25
·547	·999 717	23	·023 794	1000	42·0154	·023 801	1001	42·0273	·000 283	23
·548	·999 740	22	·022 794	999	43·8591	·022 800	1000	43·8705	·000 260	22
·549	·999 762	22	·021 795	1000	45·8720	·021 800	1001	45·8829	·000 238	22
1·550	0·999 784		0·020 795		48·0785	0·020 799		48·0889	1·000 216	

For values of tan x and sec x at intermediate arguments, use the formulæ on the next page.

TABLE IVE—CIRCULAR FUNCTIONS (Radians)

x	sin x		cos x		tan x	cot x		sec x	cosec x	
r										
1·550	0·999 784	20	+0·020 795	1000	+48·0785	+0·020 799	1000	+48·0889	1·000 216	20
·551	·999 804	19	·019 795	1000	50·5078	·019 799	1000	50·5177	·000 196	19
·552	·999 823	19	·018 795	1000	53·1956	·018 799	1001	53·2050	·000 177	19
·553	·999 842	17	·017 795	999	56·1854	·017 798	1000	56·1943	·000 158	17
·554	·999 859	16	·016 796	1000	59·5312	·016 798	1000	59·5396	·000 141	16
1·555	0·999 875	16	+0·015 796	1000	+63·3006	+0·015 798	1001	+63·3085	1·000 125	16
·556	·999 891	14	·014 796	1000	67·5794	·014 797	1000	67·5868	·000 109	14
·557	·999 905	13	·013 796	1000	72·4785	·013 797	1000	72·4854	·000 095	13
·558	·999 918	12	·012 796	1000	78·1432	·012 797	1000	78·1496	·000 082	12
·559	·999 930	12	·011 796	1000	84·7682	·011 797	1000	84·7741	·000 070	12
1·560	0·999 942	10	+0·010 796	1000	+92·6205	+0·010 797	1000	+92·6259	1·000 058	10
·561	·999 952	9	·009 796	1000	102·076	·009 797	1000	102·081	·000 048	9
·562	·999 961	9	·008 796	1000	113·681	·008 797	1001	113·685	·000 039	9
·563	·999 970	7	·007 796	1000	128·263	·007 796	1000	128·267	·000 030	7
·564	·999 977	6	·006 796	1000	147·136	·006 796	1000	147·139	·000 023	6
1·565	0·999 983	5	+0·005 796	1000	+172·521	+0·005 796	1000	+172·524	1·000 017	5
·566	·999 988	5	·004 796	1000	208·491	·004 796	1000	208·494	·000 012	5
·567	·999 993	3	·003 796	1000	263·411	·003 796	1000	263·413	·000 007	3
·568	·999 996	2	·002 796	1000	357·611	·002 796	1000	357·612	·000 004	2
·569	0·999 998	2	·001 796	1000	556·691	·001 796	1000	556·692	·000 002	2
1·570	1·000 000	0	+0·000 796	1000	+1255·77	+0·000 796	1000	+1255·77	1·000 000	0
·571	1·000 000	1	− ·000 204	1000	−4909·83	− ·000 204	1000	−4909·83	·000 000	1
·572	0·999 999	1	·001 204	1000	830·790	·001 204	1000	830·790	·000 001	1
·573	·999 998	3	·002 204	1000	453·787	·002 204	1000	453·788	·000 002	3
·574	·999 995	4	·003 204	1000	312·141	·003 204	1000	312·142	·000 005	4
1·575	0·999 991	5	−0·004 204	1000	−237·886	−0·004 204	1000	−237·888	1·000 009	5
·576	·999 986	5	·005 204	1000	192·170	·005 204	1000	192·173	·000 014	5
·577	·999 981	7	·006 204	1000	161·193	·006 204	1000	161·196	·000 019	7
·578	·999 974	8	·007 204	1000	138·816	·007 204	1000	138·819	·000 026	8
·579	·999 966	8	·008 204	1000	121·894	·008 204	1000	121·898	·000 034	8
1·580	0·999 958	10	−0·009 204	1000	−108·649	−0·009 204	1000	−108·654	1·000 042	10
·581	·999 948	11	·010 204	999	98·0005	·010 204	1000	98·0056	·000 052	11
·582	·999 937	11	·011 203	1000	89·2527	·011 204	1000	89·2583	·000 063	11
·583	·999 926	13	·012 203	1000	81·9385	·012 204	1000	81·9446	·000 074	13
·584	·999 913	14	·013 203	1000	75·7321	·013 204	1001	75·7387	·000 087	14
1·585	0·999 899	15	−0·014 203	1000	−70·3996	−0·014 205	1000	−70·4067	1·000 101	15
·586	·999 884	15	·015 203	1000	65·7685	·015 205	1000	65·7761	·000 116	15
·587	·999 869	17	·016 203	1000	61·7090	·016 205	1000	61·7171	·000 131	17
·588	·999 852	18	·017 203	1000	58·1214	·017 205	1001	58·1300	·000 148	18
·589	·999 834	18	·018 203	999	54·9279	·018 206	1000	54·9370	·000 166	18
1·590	0·999 816	20	−0·019 202	1000	−52·0670	−0·019 206	1000	−52·0766	1·000 184	20
·591	·999 796	21	·020 202	1000	49·4892	·020 206	1001	49·4993	·000 204	21
·592	·999 775	21	·021 202	1000	47·1546	·021 207	1000	47·1652	·000 225	22
·593	·999 754	23	·022 202	1000	45·0302	·022 207	1001	45·0413	·000 247	22
·594	·999 731	24	·023 202	999	43·0889	·023 208	1000	43·1005	·000 269	24
1·595	0·999 707	25	−0·024 201	1000	−41·3080	−0·024 208	1001	−41·3201	1·000 293	25
·596	·999 682	25	·025 201	1000	39·6684	·025 209	1001	39·6810	·000 318	25
·597	·999 657	27	·026 201	999	38·1539	·026 210	1000	38·1670	·000 343	27
·598	·999 630	28	·027 200	1000	36·7507	·027 210	1001	36·7643	·000 370	28
·599	·999 602	28	·028 200	1000	35·4470	·028 211	1001	35·4611	·000 398	29
1·600	0·999 574		−0·029 200		−34·2325	−0·029 212		−34·2471	1·000 427	

tan x = cot $(\frac{1}{2}\pi - x)$ sec x = cosec $(\frac{1}{2}\pi - x)$ $\frac{1}{2}\pi$ = 1·570 796 327 Use Table IVD (page 171).

TABLE VA—FORMULÆ FOR EXPONENTIAL AND HYPERBOLIC FUNCTIONS

For similar formulæ for circular functions and for the reversal of series, see page 170.

Function	Derivative	Integral
e^{ax}	ae^{ax}	$\frac{1}{a}e^{ax}$
$\ln ax$	$\frac{1}{x}$	$x(\ln ax - 1)$
a^{bx}	$ba^{bx}\ln a$	$\frac{a^{bx}}{b\ln a}$
$\sinh ax$	$a\cosh ax$	$\frac{1}{a}\cosh ax$
$\cosh ax$	$a\sinh ax$	$\frac{1}{a}\sinh ax$
$\tanh ax$	$a\,\mathrm{sech}^2\, ax$	$\frac{1}{a}\ln\cosh ax$
$\coth ax$	$-a\,\mathrm{cosech}^2\, ax$	$\frac{1}{a}\ln\sinh ax$
$\mathrm{sech}\, ax$	$-a\,\mathrm{sech}\, ax\tanh ax$	$\frac{1}{a}\,\mathrm{gd}\, ax$
$\mathrm{cosech}\, ax$	$-a\,\mathrm{cosech}\, ax\coth ax$	$\frac{1}{a}\ln\tanh\frac{ax}{2}$

Function	Derivative	Integral
$\sinh^{-1} ax$	$\frac{a}{\sqrt{a^2x^2+1}}$	$x\sinh^{-1} ax - \frac{1}{a}\sqrt{a^2x^2+1}$
$\cosh^{-1} ax$	$\frac{a}{\sqrt{a^2x^2-1}}$	$x\cosh^{-1} ax - \frac{1}{a}\sqrt{a^2x^2-1}$
$\tanh^{-1} ax$	$\frac{a}{1-a^2x^2}$	$x\tanh^{-1} ax + \frac{1}{2a}\ln(1-a^2x^2)$
$\coth^{-1} ax$	$-\frac{a}{a^2x^2-1}$	$x\coth^{-1} ax + \frac{1}{2a}\ln(a^2x^2-1)$
$\mathrm{sech}^{-1} ax$	$-\frac{1}{x\sqrt{1-a^2x^2}}$	$x\,\mathrm{sech}^{-1} ax + \frac{1}{a}\sin^{-1} ax$
$\mathrm{cosech}^{-1} ax$	$-\frac{1}{x\sqrt{a^2x^2+1}}$	$x\,\mathrm{cosech}^{-1} ax + \frac{1}{a}\sinh^{-1} ax$

ax is positive in the last two lines.

In the following formulæ B_n and E_n denote Bernoulli's and Euler's numbers (page 387).

$$e^x = 1 + x + \frac{x^2}{2!} + \frac{x^3}{3!} + \frac{x^4}{4!} + \cdots \qquad a^x = 1 + x\ln a + \frac{(x\ln a)^2}{2!} + \frac{(x\ln a)^3}{3!} + \cdots \qquad \text{(All values of } x\text{)}$$

$$\ln(1+x) = x - \frac{x^2}{2} + \frac{x^3}{3} - \frac{x^4}{4} + \cdots \qquad \log(1+x) = M\ln(1+x) \qquad (-1 < x \leqslant 1)$$

$$\sinh x = \frac{1}{2}(e^x - e^{-x}) = \frac{1}{i}\sin ix = x + \frac{x^3}{3!} + \frac{x^5}{5!} + \frac{x^7}{7!} + \frac{x^9}{9!} + \cdots \qquad \text{(All values of } x\text{)}$$

$$\cosh x = \frac{1}{2}(e^x + e^{-x}) = \cos ix = 1 + \frac{x^2}{2!} + \frac{x^4}{4!} + \frac{x^6}{6!} + \frac{x^8}{8!} + \cdots \qquad \text{(All values of } x\text{)}$$

$$\tanh x = \frac{e^{2x}-1}{e^{2x}+1} = \frac{1}{i}\tan ix = x - \frac{1}{3}x^3 + \frac{2}{15}x^5 - \frac{17}{315}x^7 + \cdots + \frac{(-1)^{n-1}B_n 2^{2n}(2^{2n}-1)}{(2n)!}x^{2n-1}\cdots \qquad \left(x^2 < \frac{\pi^2}{4}\right)$$

$$= 1 - 2(e^{-2x} - e^{-4x} + e^{-6x} - e^{-8x} + \cdots) \qquad \text{(Useful with large positive } x\text{)}$$

$$\coth x = i\cot ix = \frac{1}{x}\left(1 + \frac{1}{3}x^2 - \frac{1}{45}x^4 + \frac{2}{945}x^6 - \frac{1}{4725}x^8 + \cdots + \frac{(-1)^{n-1}B_n 2^{2n}}{(2n)!}x^{2n}\cdots\right) = \frac{\tau h}{x} \qquad (x^2 < \pi^2)$$

$$\mathrm{sech}\, x = \sec ix = 1 - \frac{1}{2}x^2 + \frac{5}{24}x^4 - \frac{61}{720}x^6 + \frac{277}{8064}x^8 - \cdots + \frac{(-1)^n E_n}{(2n)!}x^{2n}\cdots = \frac{\sigma h}{\tau h} \qquad \left(x^2 < \frac{\pi^2}{4}\right)$$

$$\mathrm{cosech}\, x = i\,\mathrm{cosec}\, ix = \frac{1}{x}\left(1 - \frac{1}{6}x^2 + \frac{7}{360}x^4 - \frac{31}{15120}x^6 + \cdots + \frac{(-1)^n B_n 2(2^{2n-1}-1)}{(2n)!}x^{2n}\cdots\right) = \frac{\sigma h}{x} \qquad (x^2 < \pi^2)$$

$$\log\sinh x = \log x + M\left(\frac{1}{6}x^2 - \frac{1}{180}x^4 + \frac{1}{2835}x^6 - \cdots + \frac{(-1)^{n-1}B_n 2^{2n-1}}{n(2n)!}x^{2n}\cdots\right) = \log x + Sh \qquad (x^2 < \pi^2)$$

$$\log\cosh x = M\left(\frac{1}{2}x^2 - \frac{1}{12}x^4 + \frac{1}{45}x^6 - \frac{17}{2520}x^8 + \cdots + \frac{(-1)^{n-1}B_n 2^{2n-1}(2^{2n}-1)}{n(2n)!}x^{2n}\cdots\right) = Sh - Th \qquad \left(x^2 < \frac{\pi^2}{4}\right)$$

$$\log\tanh x = \log x - M\left(\frac{1}{3}x^2 - \frac{7}{90}x^4 + \frac{62}{2835}x^6 - \cdots + \frac{(-1)^{n-1}B_n 2^{2n}(2^{2n-1}-1)}{n(2n)!}x^{2n}\cdots\right) = \log x + Th \qquad \left(x^2 < \frac{\pi^2}{4}\right)$$

$$= -2M\left(e^{-2x} + \frac{1}{3}e^{-6x} + \frac{1}{5}e^{-10x} + \cdots\right) \qquad \text{(All positive values of } x\text{)}$$

$$x = \sinh x - \frac{1}{2}\frac{\sinh^3 x}{3} + \frac{1.3}{2.4}\frac{\sinh^5 x}{5} - \frac{1.3.5}{2.4.6}\frac{\sinh^7 x}{7} + \cdots = \tanh x + \frac{1}{3}\tanh^3 x + \frac{1}{5}\tanh^5 x + \cdots$$

$$= \ln\frac{2}{\mathrm{sech}\, x} - \frac{1}{2}\frac{\mathrm{sech}^2 x}{2} - \frac{1.3}{2.4}\frac{\mathrm{sech}^4 x}{4} - \cdots = \ln\frac{2}{\mathrm{cosech}\, x} + \frac{1}{2}\frac{\mathrm{cosech}^2 x}{2} - \frac{1.3}{2.4}\frac{\mathrm{cosech}^4 x}{4} + \cdots$$

If a given function exceeds 1, use the reciprocal function of its reciprocal; thus $\sinh x = 5$ may be replaced by $\mathrm{cosech}\, x = 0{\cdot}2$. If $\cosh x$ exceeds unity by a small amount,

$$x = \sqrt{2(\cosh x - 1)}\left\{1 - \frac{1}{12}(\cosh x - 1) + \frac{3}{160}(\cosh x - 1)^2 - \frac{5}{896}(\cosh x - 1)^3 + \frac{35}{18432}(\cosh x - 1)^4 - \cdots\right\}$$

TABLE V_B—AUXILIARY FUNCTIONS FOR SMALL ARGUMENTS

(Hyperbolic Functions)

x	τh	coth
0·000 00		∞
	1·000 000	
0·001 22		816·5
	1·000 001	
0·002 12		471·5
	1·000 002	
0·002 73		365·2
	1·000 003	
0·003 24		308·7
	1·000 004	
0·003 67		272·2
	1·000 005	
0·004 06		246·2
	1·000 006	
0·004 41		226·5
	1·000 007	
0·004 74		210·9
	1·000 008	
0·005 04		198·1
	1·000 009	
0·005 33		187·4
	1·000 010	
0·005 61		178·2
	1·000 011	
0·005 87		170·3
	1·000 012	
0·006 12		163·4
	1·000 013	
0·006 36		157·2
	1·000 014	
0·006 59		151·7
	1·000 015	
0·006 81		146·7
	1·000 016	
0·007 03		142·2
	1·000 017	
0·007 24		138·1
	1·000 018	
0·007 44		134·3
	1·000 019	
0·007 64		130·8
	1·000 020	
0·007 84		127·6
	1·000 021	
0·008 03		124·6
	1·000 022	
0·008 21		121·8
	1·000 023	
0·008 39		119·2
	1·000 024	
0·008 57		116·7
	1·000 025	
0·008 74		114·4
	1·000 026	
0·008 91		112·2
	1·000 027	
0·009 08		110·1
	1·000 028	
0·009 24		108·2
	1·000 029	
0·009 40		106·4
	1·000 030	
0·009 56		104·6
	1·000 031	
0·009 72		102·9
	1·000 032	
0·009 87		101·3

x	σh	cosech
0·000 00		∞
	1·000 000	
0·001 73		577·4
	0·999 999	
0·003 00		333·4
	0·999 998	
0·003 87		258·2
	0·999 997	
0·004 58		218·3
	0·999 996	
0·005 19		192·5
	0·999 995	
0·005 74		174·1
	0·999 994	
0·006 24		160·2
	0·999 993	
0·006 70		149·1
	0·999 992	
0·007 14		140·1
	0·999 991	
0·007 54		132·5
	0·999 990	
0·007 93		126·0
	0·999 989	
0·008 30		120·4
	0·999 988	
0·008 66		115·5
	0·999 987	
0·009 00		111·2
	0·999 986	
0·009 32		107·3
	0·999 985	
0·009 64		103·7
	0·999 984	
0·009 94		100·6

x	τh		σh	
0·000	1·000 000	See critical table.	1·000 000	See critical table.
·001	·000 000		1·000 000	
·002	·000 001		0·999 999	
·003	·000 003		·999 998	
·004	·000 005		·999 997	
0·005	1·000 008		0·999 996	
·006	·000 012		·999 994	
·007	·000 016		·999 992	
·008	·000 021		·999 989	
·009	·000 027		·999 986	
0·010	1·000 033	7	0·999 983	3
·011	·000 040	8	·999 980	4
·012	·000 048	8	·999 976	4
·013	·000 056	9	·999 972	5
·014	·000 065	10	·999 967	4
0·015	1·000 075	10	0·999 963	6
·016	·000 085	11	·999 957	5
·017	·000 096	12	·999 952	6
·018	·000 108	12	·999 946	6
·019	·000 120	13	·999 940	7
0·020	1·000 133	14	0·999 933	6
·021	·000 147	14	·999 927	8
·022	·000 161	15	·999 919	7
·023	·000 176	16	·999 912	8
·024	·000 192	16	·999 904	8
0·025	1·000 208	17	0·999 896	9
·026	·000 225	18	·999 887	8
·027	·000 243	18	·999 879	10
·028	·000 261	19	·999 869	9
·029	·000 280	20	·999 860	10
0·030	1·000 300	20	0·999 850	10
·031	·000 320	21	·999 840	11
·032	·000 341	22	·999 829	10
·033	·000 363	22	·999 819	12
·034	·000 385	23	·999 807	11
0·035	1·000 408	24	0·999 796	12
·036	·000 432	24	·999 784	12
·037	·000 456	25	·999 772	13
·038	·000 481	26	·999 759	12
·039	·000 507	26	·999 747	14
0·040	1·000 533	27	0·999 733	13
·041	·000 560	28	·999 720	14
·042	·000 588	28	·999 706	14
·043	·000 616	29	·999 692	15
·044	·000 645	30	·999 677	14
0·045	1·000 675	30	0·999 663	16
·046	·000 705	31	·999 647	15
·047	·000 736	32	·999 632	16
·048	·000 768	32	·999 616	16
·049	·000 800	33	·999 600	17
0·050	1·000 833		0·999 583	

x	τh		σh	
0·050	1·000 833	34	0·999 583	16
·051	·000 867	34	·999 567	18
·052	·000 901	35	·999 549	17
·053	·000 936	36	·999 532	18
·054	·000 972	36	·999 514	18
0·055	1·001 008	37	0·999 496	18
·056	·001 045	38	·999 478	19
·057	·001 083	38	·999 459	19
·058	·001 121	39	·999 440	20
·059	·001 160	40	·999 420	20
0·060	1·001 200	40	0·999 400	20
·061	·001 240	41	·999 380	20
·062	·001 281	42	·999 360	21
·063	·001 323	42	·999 339	21
·064	·001 365	43	·999 318	22
0·065	1·001 408	44	0·999 296	22
·066	·001 452	44	·999 274	22
·067	·001 496	45	·999 252	22
·068	·001 541	45	·999 230	23
·069	·001 586	47	·999 207	23
0·070	1·001 633	47	0·999 184	24
·071	·001 680	47	·999 160	23
·072	·001 727	49	·999 137	25
·073	·001 776	49	·999 112	24
·074	·001 825	49	·999 088	25
0·075	1·001 874	51	0·999 063	25
·076	·001 925	51	·999 038	25
·077	·001 976	51	·999 013	26
·078	·002 027	52	·998 987	26
·079	·002 079	53	·998 961	27
0·080	1·002 132	54	0·998 934	27
·081	·002 186	54	·998 907	27
·082	·002 240	55	·998 880	27
·083	·002 295	56	·998 853	28
·084	·002 351	56	·998 825	28
0·085	1·002 407	57	0·998 797	29
·086	·002 464	58	·998 768	28
·087	·002 522	58	·998 740	30
·088	·002 580	59	·998 710	29
·089	·002 639	60	·998 681	30
0·090	1·002 699	60	0·998 651	30
·091	·002 759	61	·998 621	30
·092	·002 820	61	·998 591	31
·093	·002 881	63	·998 560	31
·094	·002 944	63	·998 529	32
0·095	1·003 007	63	0·998 497	31
·096	·003 070	64	·998 466	32
·097	·003 134	65	·998 434	33
·098	·003 199	66	·998 401	33
·099	·003 265	66	·998 368	33
0·100	1·003 331		0·998 335	

$$\coth x = \frac{\tau h}{x} \qquad \operatorname{cosech} x = \frac{\sigma h}{x} \qquad x = \frac{\tau h}{\coth x} = \frac{\sigma h}{\operatorname{cosech} x}$$

x	e^{x}		e^{-x}		sinh x		cosh x		tanh x		coth x
					0·				0·		
0·000	1·000 000	1001	1·000 000	1000	000 000	1000	1·000 000	1	000 000	1000	∞
·001	·001 001	1001	0·999 000	998	001 000	1000	·000 001	1	001 000	1000	1000·00
·002	·002 002	1003	·998 002	998	002 000	1000	·000 002	3	002 000	1000	500·001
·003	·003 005	1003	·997 004	996	003 000	1000	·000 005	3	003 000	1000	333·334
·004	·004 008	1005	·996 008	996	004 000	1000	·000 008	5	004 000	1000	250·001
0·005	1·005 013	1005	0·995 012	994	005 000	1000	1·000 013	5	005 000	1000	200·002
·006	·006 018	1007	·994 018	994	006 000	1000	·000 018	7	006 000	1000	166·669
·007	·007 025	1007	·993 024	992	007 000	1000	·000 025	7	007 000	1000	142·859
·008	·008 032	1009	·992 032	992	008 000	1000	·000 032	9	008 000	1000	125·003
·009	·009 041	1009	·991 040	990	009 000	1000	·000 041	9	009 000	1000	111·114
0·010	1·010 050	1011	0·990 050	990	010 000	1000	1·000 050	11	010 000	1000	100·003
·011	·011 061	1011	·989 060	988	011 000	1000	·000 061	11	011 000	999	90·9128
·012	·012 072	1013	·988 072	988	012 000	1000	·000 072	13	011 999	1000	83·3373
·013	·013 085	1013	·987 084	986	013 000	1000	·000 085	13	012 999	1000	76·9274
·014	·014 098	1015	·986 098	986	014 000	1001	·000 098	15	013 999	1000	71·4332
0·015	1·015 113	1016	0·985 112	985	015 001	1000	1·000 113	15	014 999	1000	66·6717
·016	·016 129	1016	·984 127	983	016 001	1000	·000 128	17	015 999	999	62·5053
·017	·017 145	1018	·983 144	983	017 001	1000	·000 145	17	016 998	1000	58·8292
·018	·018 163	1019	·982 161	982	018 001	1000	·000 162	19	017 998	1000	55·5616
·019	·019 182	1019	·981 179	980	019 001	1000	·000 181	19	018 998	999	52·6379
0·020	1·020 201	1021	0·980 199	980	020 001	1001	1·000 200	21	019 997	1000	50·0067
·021	·021 222	1022	·979 219	979	021 002	1000	·000 221	21	020 997	999	47·6260
·022	·022 244	1023	·978 240	978	022 002	1000	·000 242	23	021 996	1000	45·4619
·023	·023 267	1023	·977 262	976	023 002	1000	·000 265	23	022 996	999	43·4859
·024	·024 290	1025	·976 286	976	024 002	1001	·000 288	25	023 995	1000	41·6747
0·025	1·025 315	1026	0·975 310	975	025 003	1000	1·000 313	25	024 995	999	40·0083
·026	·026 341	1027	·974 335	974	026 003	1000	·000 338	27	025 994	999	38·4702
·027	·027 368	1028	·973 361	973	027 003	1001	·000 365	27	026 993	1000	37·0460
·028	·028 396	1029	·972 388	972	028 004	1000	·000 392	29	027 993	999	35·7236
·029	·029 425	1030	·971 416	970	029 004	1001	·000 421	29	028 992	999	34·4924
0·030	1·030 455	1031	0·970 446	970	030 005	1000	1·000 450	31	029 991	999	33·3433
·031	·031 486	1032	·969 476	969	031 005	1000	·000 481	31	030 990	999	32·2684
·032	·032 518	1033	·968 507	968	032 005	1001	·000 512	33	031 989	999	31·2607
·033	·033 551	1034	·967 539	967	033 006	1001	·000 545	33	032 988	999	30·3140
·034	·034 585	1035	·966 572	967	034 007	1000	·000 578	35	033 987	999	29·4231
0·035	1·035 620	1036	0·965 605	965	035 007	1001	1·000 613	35	034 986	998	28·5831
·036	·036 656	1037	·964 640	964	036 008	1000	·000 648	37	035 984	999	27·7898
·037	·037 693	1038	·963 676	963	037 008	1001	·000 685	37	036 983	999	27·0394
·038	·038 731	1039	·962 713	962	038 009	1001	·000 722	39	037 982	998	26·3285
·039	·039 770	1041	·961 751	962	039 010	1001	·000 761	39	038 980	999	25·6540
0·040	1·040 811	1041	0·960 789	960	040 011	1000	1·000 800	41	039 979	998	25·0133
·041	·041 852	1042	·959 829	959	041 011	1001	·000 841	41	040 977	998	24·4039
·042	·042 894	1044	·958 870	959	042 012	1001	·000 882	43	041 975	999	23·8235
·043	·043 938	1044	·957 911	957	043 013	1001	·000 925	43	042 974	998	23·2701
·044	·044 982	1046	·956 954	957	044 014	1001	·000 968	45	043 972	998	22·7419
0·045	1·046 028	1046	0·955 997	955	045 015	1001	1·001 013	45	044 970	998	22·2372
·046	·047 074	1048	·955 042	955	046 016	1001	·001 058	47	045 968	997	21·7545
·047	·048 122	1049	·954 087	953	047 017	1001	·001 105	47	046 965	998	21·2923
·048	·049 171	1049	·953 134	953	048 018	1002	·001 152	49	047 963	998	20·8493
·049	·050 220	1051	·952 181	952	049 020	1001	·001 201	49	048 961	997	20·4245
0·050	1·051 271		0·951 229		050 021		1·001 250		049 958		20·0167

For values of coth x at intermediate arguments, see page 205.

TABLE Vc—EXPONENTIAL AND HYPERBOLIC FUNCTIONS

x	e^x		e^{-x}		sinh x		cosh x		tanh x		coth x
			0·		0·				0·		
0·050	1·051 271	1052	951 229	950	050 021	1001	1·001 250	51	049 958	998	20·01666
·051	·052 323	1053	950 279	950	051 022	1001	·001 301	51	050 956	997	19·62484
·052	·053 376	1054	949 329	949	052 023	1002	·001 352	53	051 953	997	19·24810
·053	·054 430	1055	948 380	948	053 025	1001	·001 405	53	052 950	998	18·88559
·054	·055 485	1056	947 432	947	054 026	1002	·001 458	55	053 948	997	18·53652
0·055	1·056 541	1057	946 485	946	055 028	1001	1·001 513	55	054 945	997	18·20015
·056	·057 598	1058	945 539	945	056 029	1002	·001 568	57	055 942	996	17·87581
·057	·058 656	1059	944 594	944	057 031	1002	·001 625	57	056 938	997	17·56286
·058	·059 715	1060	943 650	943	058 033	1001	·001 682	59	057 935	997	17·26071
·059	·060 775	1062	942 707	942	059 034	1002	·001 741	60	058 932	996	16·96881
0·060	1·061 837	1062	941 765	942	060 036	1002	1·001 801	60	059 928	996	16·68666
·061	·062 899	1063	940 823	940	061 038	1002	·001 861	62	060 924	997	16·41377
·062	·063 962	1065	939 883	940	062 040	1002	·001 923	62	061 921	996	16·14969
·063	·065 027	1065	938 943	938	063 042	1002	·001 985	64	062 917	996	15·89401
·064	·066 092	1067	938 005	938	064 044	1002	·002 049	64	063 913	996	15·64633
0·065	1·067 159	1068	937 067	936	065 046	1002	1·002 113	66	064 909	995	15·40628
·066	·068 227	1068	936 131	936	066 048	1002	·002 179	66	065 904	996	15·17351
·067	·069 295	1070	935 195	935	067 050	1002	·002 245	68	066 900	995	14·94770
·068	·070 365	1071	934 260	933	068 052	1003	·002 313	68	067 895	996	14·72854
·069	·071 436	1072	933 327	933	069 055	1002	·002 381	70	068 891	995	14·51575
0·070	1·072 508	1073	932 394	932	070 057	1003	1·002 451	71	069 886	995	14·30904
·071	·073 581	1074	931 462	931	071 060	1002	·002 522	71	070 881	995	14·10817
·072	·074 655	1076	930 531	930	072 062	1003	·002 593	73	071 876	995	13·91288
·073	·075 731	1076	929 601	929	073 065	1003	·002 666	73	072 871	994	13·72295
·074	·076 807	1077	928 672	929	074 068	1002	·002 739	75	073 865	995	13·53817
0·075	1·077 884	1079	927 743	927	075 070	1003	1·002 814	75	074 860	994	13·35832
·076	·078 963	1079	926 816	926	076 073	1003	·002 889	77	075 854	994	13·18322
·077	·080 042	1081	925 890	926	077 076	1003	·002 966	78	076 848	994	13·01267
·078	·081 123	1081	924 964	924	078 079	1003	·003 044	78	077 842	994	12·84650
·079	·082 204	1083	924 040	924	079 082	1003	·003 122	80	078 836	994	12·68455
0·080	1·083 287	1084	923 116	922	080 085	1004	1·003 202	80	079 830	993	12·52666
·081	·084 371	1085	922 194	922	081 089	1003	·003 282	82	080 823	994	12·37267
·082	·085 456	1086	921 272	921	082 092	1003	·003 364	82	081 817	993	12·22244
·083	·086 542	1087	920 351	920	083 095	1004	·003 446	84	082 810	993	12·07585
·084	·087 629	1088	919 431	919	084 099	1003	·003 530	85	083 803	993	11·93275
0·085	1·088 717	1089	918 512	918	085 102	1004	1·003 615	85	084 796	993	11·79303
·086	·089 806	1091	917 594	917	086 106	1004	·003 700	87	085 789	992	11·65656
·087	·090 897	1091	916 677	916	087 110	1004	·003 787	87	086 781	993	11·52324
·088	·091 988	1093	915 761	915	088 114	1004	·003 874	89	087 774	992	11·39295
·089	·093 081	1093	914 846	915	089 118	1004	·003 963	90	088 766	992	11·26561
0·090	1·094 174	1095	913 931	913	090 122	1004	1·004 053	90	089 758	992	11·14109
·091	·095 269	1096	913 018	913	091 126	1004	·004 143	92	090 750	991	11·01933
·092	·096 365	1097	912 105	911	092 130	1004	·004 235	93	091 741	992	10·90021
·093	·097 462	1098	911 194	911	093 134	1004	·004 328	93	092 733	991	10·78367
·094	·098 560	1099	910 283	910	094 138	1005	·004 421	95	093 724	991	10·66961
0·095	1·099 659	1100	909 373	909	095 143	1005	1·004 516	96	094 715	991	10·55796
·096	·100 759	1101	908 464	908	096 148	1004	·004 612	96	095 706	991	10·44865
·097	·101 860	1103	907 556	907	097 152	1005	·004 708	98	096 697	990	10·34159
·098	·102 963	1103	906 649	906	098 157	1005	·004 806	99	097 687	991	10·23673
·099	·104 066	1105	905 743	906	099 162	1005	·004 905	99	098 678	990	10·13399
0·100	1·105 171		904 837		100 167		1·005 004		099 668		10·03331

For values of coth x at intermediate arguments, see page **205**.

x	e^x		e^{-x}		sinh x		cosh x		tanh x		coth x	
			0·		0·				0·			
0·100	1·105 171	1106	904 837	904	100 167	1005	1·005 004	101	099 668	990	10·03331	*9868*
·101	·106 277	1106	903 933	903	101 172	1005	·005 105	102	100 658	990	9·93463	*9673*
·102	·107 383	1108	903 030	903	102 177	1005	·005 207	102	101 648	989	9·83790	*9485*
·103	·108 491	1109	902 127	902	103 182	1006	·005 309	104	102 637	990	9·74305	*9302*
·104	·109 600	1111	901 225	900	104 188	1005	·005 413	105	103 627	989	9·65003	*9125*
0·105	1·110 711	1111	900 325	900	105 193	1006	1·005 518	105	104 616	989	9·55878	*8951*
·106	·111 822	1112	899 425	899	106 199	1005	·005 623	107	105 605	989	9·46927	*8784*
·107	·112 934	1114	898 526	898	107 204	1006	·005 730	108	106 594	988	9·38143	*8620*
·108	·114 048	1114	897 628	898	108 210	1006	·005 838	108	107 582	988	9·29523	*8461*
·109	·115 162	1116	896 730	896	109 216	1006	·005 946	110	108 570	988	9·21062	*8307*
0·110	1·116 278	1117	895 834	895	110 222	1006	1·006 056	111	109 558	988	9·12755	*8157*
·111	·117 395	1118	894 939	895	111 228	1006	·006 167	112	110 546	988	9·04598	*8011*
·112	·118 513	1119	894 044	893	112 234	1007	·006 279	112	111 534	987	8·96587	*7868*
·113	·119 632	1120	893 151	893	113 241	1006	·006 391	114	112 521	988	8·88719	*7729*
·114	·120 752	1121	892 258	892	114 247	1007	·006 505	115	113 509	987	8·80990	*7595*
0·115	1·121 873	1123	891 366	891	115 254	1006	1·006 620	116	114 496	986	8·73395	*7463*
·116	·122 996	1123	890 475	890	116 260	1007	·006 736	116	115 482	987	8·65932	*7335*
·117	·124 119	1125	889 585	889	117 267	1007	·006 852	118	116 469	986	8·58597	*7210*
·118	·125 244	1126	888 696	888	118 274	1007	·006 970	119	117 455	986	8·51387	*7088*
·119	·126 370	1127	887 808	888	119 281	1007	·007 089	120	118 441	986	8·44299	*6970*
0·120	1·127 497	1128	886 920	886	120 288	1007	1·007 209	120	119 427	986	8·37329	*6853*
·121	·128 625	1129	886 034	886	121 295	1008	·007 329	122	120 413	985	8·30476	*6741*
·122	·129 754	1130	885 148	884	122 303	1007	·007 451	123	121 398	985	8·23735	*6631*
·123	·130 884	1132	884 264	884	123 310	1008	·007 574	124	122 383	985	8·17104	*6523*
·124	·132 016	1132	883 380	883	124 318	1008	·007 698	125	123 368	985	8·10581	*6419*
0·125	1·133 148	1134	882 497	882	125 326	1008	1·007 823	126	124 353	984	8·04162	*6316*
·126	·134 282	1135	881 615	881	126 334	1008	·007 949	126	125 337	985	7·97846	*6216*
·127	·135 417	1136	880 734	881	127 342	1008	·008 075	128	126 322	984	7·91630	*6118*
·128	·136 553	1137	879 853	879	128 350	1008	·008 203	129	127 306	983	7·85512	*6023*
·129	·137 690	1138	878 974	879	129 358	1008	·008 332	130	128 289	984	7·79489	*5930*
0·130	1·138 828	1140	878 095	877	130 366	1009	1·008 462	131	129 273	983	7·73559	*5839*
·131	·139 968	1140	877 218	877	131 375	1009	·008 593	132	130 256	983	7·67720	*5749*
·132	·141 108	1142	876 341	876	132 384	1008	·008 725	133	131 239	982	7·61971	*5663*
·133	·142 250	1143	875 465	875	133 392	1009	·008 858	133	132 221	983	7·56308	*5578*
·134	·143 393	1144	874 590	874	134 401	1009	·008 991	135	133 204	982	7·50730	*5495*
0·135	1·144 537	1145	873 716	873	135 410	1010	1·009 126	136	134 186	982	7·45235	*5413*
·136	·145 682	1146	872 843	873	136 420	1009	·009 262	137	135 168	981	7·39822	*5334*
·137	·146 828	1148	871 970	871	137 429	1009	·009 399	138	136 149	982	7·34488	*5256*
·138	·147 976	1148	871 099	871	138 438	1010	·009 537	139	137 131	981	7·29232	*5180*
·139	·149 124	1150	870 228	870	139 448	1010	·009 676	140	138 112	980	7·24052	*5106*
0·140	1·150 274	1151	869 358	869	140 458	1010	1·009 816	141	139 092	981	7·18946	*5032*
·141	·151 425	1152	868 489	868	141 468	1010	·009 957	142	140 073	980	7·13914	*4962*
·142	·152 577	1153	867 621	867	142 478	1010	·010 099	143	141 053	980	7·08952	*4891*
·143	·153 730	1154	866 754	866	143 488	1010	·010 242	144	142 033	980	7·04061	*4823*
·144	·154 884	1156	865 888	866	144 498	1011	·010 386	145	143 013	979	6·99238	*4756*
0·145	1·156 040	1156	865 022	864	145 509	1010	1·010 531	146	143 992	979	6·94482	*4691*
·146	·157 196	1158	864 158	864	146 519	1011	·010 677	147	144 971	979	6·89791	*4626*
·147	·158 354	1159	863 294	863	147 530	1011	·010 824	148	145 950	979	6·85165	*4563*
·148	·159 513	1160	862 431	862	148 541	1011	·010 972	149	146 929	978	6·80602	*4502*
·149	·160 673	1161	861 569	861	149 552	1011	·011 121	150	147 907	978	6·76100	*4441*
0·150	1·161 834		860 708		150 563		1·011 271		148 885		6·71659	

TABLE Vc—EXPONENTIAL AND HYPERBOLIC FUNCTIONS

x	e^x		e^{-x}		sinh x		cosh x		tanh x		coth x	
			0·		0·				0·			
0·150	1·161 834	1163	860 708	860	150 563	1011	1·011 271	151	148 885	978	6·71659	*4382*
·151	·162 997	1163	859 848	860	151 574	1012	·011 422	152	149 863	977	6·67277	*4323*
·152	·164 160	1165	858 988	858	152 586	1012	·011 574	153	150 840	977	6·62954	*4267*
·153	·165 325	1166	858 130	858	153 598	1011	·011 727	154	151 817	977	6·58687	*4211*
·154	·166 491	1167	857 272	857	154 609	1012	·011 881	156	152 794	977	6·54476	*4156*
0·155	1·167 658	1168	856 415	856	155 621	1013	1·012 037	156	153 771	976	6·50320	*4103*
·156	·168 826	1170	855 559	855	156 634	1012	·012 193	157	154 747	976	6·46217	*4050*
·157	·169 996	1170	854 704	854	157 646	1012	·012 350	158	155 723	975	6·42167	*3998*
·158	·171 166	1172	853 850	854	158 658	1013	·012 508	159	156 698	976	6·38169	*3947*
·159	·172 338	1173	852 996	852	159 671	1013	·012 667	160	157 674	975	6·34222	*3898*
0·160	1·173 511	1174	852 144	852	160 684	1012	1·012 827	162	158 649	974	6·30324	*3849*
·161	·174 685	1175	851 292	851	161 696	1014	·012 989	162	159 623	975	6·26475	*3800*
·162	·175 860	1177	850 441	850	162 710	1013	·013 151	163	160 598	974	6·22675	*3754*
·163	·177 037	1177	849 591	849	163 723	1013	·013 314	164	161 572	973	6·18921	*3708*
·164	·178 214	1179	848 742	848	164 736	1014	·013 478	165	162 545	974	6·15213	*3662*
0·165	1·179 393	1180	847 894	848	165 750	1013	1·013 643	167	163 519	973	6·11551	*3618*
·166	·180 573	1181	847 046	846	166 763	1014	·013 810	167	164 492	973	6·07933	*3574*
·167	·181 754	1183	846 200	846	167 777	1014	·013 977	168	165 465	972	6·04359	*3531*
·168	·182 937	1183	845 354	845	168 791	1015	·014 145	170	166 437	972	6·00828	*3489*
·169	·184 120	1185	844 509	844	169 806	1014	·014 315	170	167 409	972	5·97339	*3448*
0·170	1·185 305	1186	843 665	843	170 820	1015	1·014 485	171	168 381	972	5·93891	*3407*
·171	·186 491	1187	842 822	843	171 835	1014	·014 656	173	169 353	971	5·90484	*3367*
·172	·187 678	1188	841 979	841	172 849	1015	·014 829	173	170 324	971	5·87117	*3327*
·173	·188 866	1190	841 138	841	173 864	1015	·015 002	174	171 295	970	5·83790	*3289*
·174	·190 056	1190	840 297	840	174 879	1016	·015 176	176	172 265	970	5·80501	*3251*
0·175	1·191 246	1192	839 457	839	175 895	1015	1·015 352	176	173 235	970	5·77250	*3214*
·176	·192 438	1193	838 618	838	176 910	1016	·015 528	177	174 205	969	5·74036	*3177*
·177	·193 631	1194	837 780	838	177 926	1015	·015 705	179	175 174	970	5·70859	*3140*
·178	·194 825	1196	836 942	836	178 941	1016	·015 884	179	176 144	968	5·67719	*3106*
·179	·196 021	1196	836 106	836	179 957	1017	·016 063	181	177 112	969	5·64613	*3070*
0·180	1·197 217	1198	835 270	835	180 974	1016	1·016 244	181	178 081	968	5·61543	*3037*
·181	·198 415	1199	834 435	834	181 990	1016	·016 425	183	179 049	968	5·58506	*3002*
·182	·199 614	1200	833 601	833	183 006	1017	·016 608	183	180 017	967	5·55504	*2969*
·183	·200 814	1202	832 768	832	184 023	1017	·016 791	185	180 984	967	5·52535	*2937*
·184	·202 016	1202	831 936	832	185 040	1017	·016 976	185	181 951	967	5·49598	*2905*
0·185	1·203 218	1204	831 104	830	186 057	1017	1·017 161	187	182 918	966	5·46693	*2873*
·186	·204 422	1205	830 274	830	187 074	1018	·017 348	188	183 884	966	5·43820	*2842*
·187	·205 627	1207	829 444	829	188 092	1017	·017 536	188	184 850	966	5·40978	*2811*
·188	·206 834	1207	828 615	828	189 109	1018	·017 724	190	185 816	965	5·38167	*2781*
·189	·208 041	1209	827 787	828	190 127	1018	·017 914	190	186 781	965	5·35386	*2752*
0·190	1·209 250	1209	826 959	826	191 145	1018	1·018 104	192	187 746	965	5·32634	*2723*
·191	·210 459	1212	826 133	826	192 163	1019	·018 296	193	188 711	964	5·29911	*2693*
·192	·211 671	1212	825 307	825	193 182	1018	·018 489	193	189 675	964	5·27218	*2666*
·193	·212 883	1213	824 482	824	194 200	1019	·018 682	195	190 639	963	5·24552	*2638*
·194	·214 096	1215	823 658	823	195 219	1019	·018 877	196	191 602	963	5·21914	*2610*
0·195	1·215 311	1216	822 835	823	196 238	1019	1·019 073	197	192 565	963	5·19304	*2583*
·196	·216 527	1217	822 012	821	197 257	1020	·019 270	197	193 528	962	5·16721	*2557*
·197	·217 744	1218	821 191	821	198 277	1019	·019 467	199	194 490	962	5·14164	*2531*
·198	·218 962	1220	820 370	820	199 296	1020	·019 666	200	195 452	962	5·11633	*2505*
·199	·220 182	1221	819 550	819	200 316	1020	·019 866	201	196 414	961	5·09128	*2479*
0·200	1·221 403		818 731		201 336		1·020 067		197 375		5·06649	

TABLE Vc—EXPONENTIAL AND HYPERBOLIC FUNCTIONS

x	e^{x}		e^{-x}		sinh x		cosh x		tanh x		coth x	
			0·		0·				0·			
0·200	1·221 403	1222	818 731	819	201 336	1020	1·020 067	202	197 375	961	5·06649	*2455*
·201	·222 625	1223	817 912	817	202 356	1021	·020 269	202	198 336	961	5·04194	*2429*
·202	·223 848	1224	817 095	817	203 377	1020	·020 471	204	199 297	960	5·01765	*2406*
·203	·225 072	1226	816 278	816	204 397	1021	·020 675	205	200 257	959	4·99359	*2382*
·204	·226 298	1227	815 462	815	205 418	1021	·020 880	206	201 216	960	4·96977	*2358*
0·205	1·227 525	1228	814 647	814	206 439	1021	1·021 086	207	202 176	959	4·94619	*2335*
·206	·228 753	1230	813 833	813	207 460	1021	·021 293	208	203 135	958	4·92284	*2312*
·207	·229 983	1230	813 020	813	208 481	1022	·021 501	209	204 093	958	4·89972	*2289*
·208	·231 213	1232	812 207	812	209 503	1022	·021 710	210	205 051	958	4·87683	*2268*
·209	·232 445	1233	811 395	811	210 525	1022	·021 920	211	206 009	957	4·85415	*2245*
0·210	1·233 678	1234	810 584	810	211 547	1022	1·022 131	212	206 966	957	4·83170	*2224*
·211	·234 912	1236	809 774	809	212 569	1023	·022 343	213	207 923	957	4·80946	*2202*
·212	·236 148	1237	808 965	809	213 592	1022	·022 556	214	208 880	956	4·78744	*2182*
·213	·237 385	1238	808 156	808	214 614	1023	·022 770	216	209 836	956	4·76562	*2161*
·214	·238 623	1239	807 348	807	215 637	1023	·022 986	216	210 792	955	4·74401	*2140*
0·215	1·239 862	1240	806 541	806	216 660	1024	1·023 202	217	211 747	955	4·72261	*2120*
·216	·241 102	1242	805 735	805	217 684	1023	·023 419	218	212 702	955	4·70141	*2101*
·217	·242 344	1243	804 930	805	218 707	1024	·023 637	219	213 657	954	4·68040	*2081*
·218	·243 587	1244	804 125	803	219 731	1024	·023 856	220	214 611	954	4·65959	*2061*
·219	·244 831	1246	803 322	803	220 755	1024	·024 076	222	215 565	953	4·63898	*2043*
0·220	1·246 077	1246	802 519	802	221 779	1024	1·024 298	222	216 518	953	4·61855	*2024*
·221	·247 323	1248	801 717	802	222 803	1025	·024 520	223	217 471	952	4·59831	*2005*
·222	·248 571	1250	800 915	800	223 828	1025	·024 743	225	218 423	953	4·57826	*1987*
·223	·249 821	1250	800 115	800	224 853	1025	·024 968	225	219 376	951	4·55839	*1969*
·224	·251 071	1252	799 315	799	225 878	1025	·025 193	226	220 327	951	4·53870	*1951*
0·225	1·252 323	1253	798 516	798	226 903	1026	1·025 419	228	221 278	951	4·51919	*1933*
·226	·253 576	1254	797 718	797	227 929	1026	·025 647	228	222 229	951	4·49986	*1917*
·227	·254 830	1255	796 921	797	228 955	1026	·025 875	230	223 180	950	4·48069	*1899*
·228	·256 085	1257	796 124	795	229 981	1026	·026 105	230	224 130	949	4·46170	*1882*
·229	·257 342	1258	795 329	795	231 007	1026	·026 335	232	225 079	949	4·44288	*1866*
0·230	1·258 600	1259	794 534	795	232 033	1027	1·026 567	232	226 028	949	4·42422	*1849*
·231	·259 859	1261	793 739	793	233 060	1027	·026 799	234	226 977	948	4·40573	*1833*
·232	·261 120	1261	792 946	792	234 087	1027	·027 033	235	227 925	948	4·38740	*1817*
·233	·262 381	1263	792 154	792	235 114	1027	·027 268	235	228 873	948	4·36923	*1801*
·234	·263 644	1265	791 362	791	236 141	1028	·027 503	237	229 821	947	4·35122	*1785*
0·235	1·264 909	1265	790 571	790	237 169	1028	1·027 740	237	230 768	946	4·33337	*1771*
·236	·266 174	1267	789 781	790	238 197	1028	·027 977	239	231 714	946	4·31566	*1754*
·237	·267 441	1268	788 991	788	239 225	1028	·028 216	240	232 660	946	4·29812	*1740*
·238	·268 709	1270	788 203	788	240 253	1029	·028 456	241	233 606	945	4·28072	*1725*
·239	·269 979	1270	787 415	787	241 282	1029	·028 697	242	234 551	945	4·26347	*1711*
0·240	1·271 249	1272	786 628	786	242 311	1029	1·028 939	242	235 496	944	4·24636	*1696*
·241	·272 521	1273	785 842	786	243 340	1029	·029 181	244	236 440	944	4·22940	*1682*
·242	·273 794	1275	785 056	784	244 369	1030	·029 425	245	237 384	943	4·21258	*1667*
·243	·275 069	1275	784 272	784	245 399	1029	·029 670	246	238 327	943	4·19591	*1654*
·244	·276 344	1277	783 488	783	246 428	1030	·029 916	247	239 270	943	4·17937	*1640*
0·245	1·277 621	1279	782 705	783	247 458	1031	1·030 163	248	240 213	942	4·16297	*1626*
·246	·278 900	1279	781 922	781	248 489	1030	·030 411	249	241 155	942	4·14671	*1613*
·247	·280 179	1281	781 141	781	249 519	1031	·030 660	250	242 097	941	4·13058	*1599*
·248	·281 460	1282	780 360	780	250 550	1031	·030 910	251	243 038	940	4·11459	*1587*
·249	·282 742	1283	779 580	779	251 581	1031	·031 161	252	243 978	941	4·09872	*1573*
0·250	1·284 025		778 801		252 612		1·031 413		244 919		4·08299	

x	e^x		e^{-x}		sinh x		cosh x		tanh x		coth x	
			0·		0·				0·			
0·250	1·284 025	1285	778 801	779	252 612	1032	1·031 413	253	244 919	939	4·08299	*1561*
·251	·285 310	1286	778 022	777	253 644	1032	·031 666	254	245 858	940	4·06738	*1548*
·252	·286 596	1287	777 245	777	254 676	1032	·031 920	256	246 798	939	4·05190	*1536*
·253	·287 883	1289	776 468	776	255 708	1032	·032 176	256	247 737	938	4·03654	*1523*
·254	·289 172	1290	775 692	776	256 740	1033	·032 432	257	248 675	938	4·02131	*1511*
0·255	1·290 462	1291	774 916	774	257 773	1032	1·032 689	258	249 613	937	4·00620	*1499*
·256	·291 753	1292	774 142	774	258 805	1033	·032 947	260	250 550	937	3·99121	*1487*
·257	·293 045	1294	773 368	773	259 838	1034	·033 207	260	251 487	937	3·97634	*1475*
·258	·294 339	1295	772 595	772	260 872	1033	·033 467	261	252 424	936	3·96159	*1464*
·259	·295 634	1296	771 823	771	261 905	1034	·033 728	263	253 360	936	3·94695	*1452*
0·260	1·296 930	1298	771 052	771	262 939	1034	1·033 991	263	254 296	935	3·93243	*1440*
·261	·298 228	1299	770 281	770	263 973	1035	·034 254	265	255 231	934	3·91803	*1430*
·262	·299 527	1300	769 511	769	265 008	1034	·034 519	265	256 165	934	3·90373	*1418*
·263	·300 827	1301	768 742	768	266 042	1035	·034 784	267	257 099	934	3·88955	*1408*
·264	·302 128	1303	767 974	768	267 077	1036	·035 051	267	258 033	933	3·87547	*1396*
0·265	1·303 431	1304	767 206	767	268 113	1035	1·035 318	269	258 966	933	3·86151	*1386*
·266	·304 735	1305	766 439	766	269 148	1036	·035 587	270	259 899	932	3·84765	*1375*
·267	·306 040	1307	765 673	765	270 184	1036	·035 857	270	260 831	932	3·83390	*1365*
·268	·307 347	1308	764 908	765	271 220	1036	·036 127	272	261 763	931	3·82025	*1354*
·269	·308 655	1309	764 143	764	272 256	1036	·036 399	273	262 694	931	3·80671	*1344*
0·270	1·309 964	1311	763 379	763	273 292	1037	1·036 672	274	263 625	930	3·79327	*1334*
·271	·311 275	1312	762 616	762	274 329	1037	·036 946	275	264 555	930	3·77993	*1324*
·272	·312 587	1313	761 854	761	275 366	1038	·037 221	276	265 485	929	3·76669	*1314*
·273	·313 900	1315	761 093	761	276 404	1037	·037 497	276	266 414	929	3·75355	*1304*
·274	·315 215	1316	760 332	760	277 441	1038	·037 773	278	267 343	928	3·74051	*1294*
0·275	1·316 531	1317	759 572	759	278 479	1038	1·038 051	279	268 271	928	3·72757	*1285*
·276	·317 848	1318	758 813	759	279 517	1039	·038 330	280	269 199	927	3·71472	*1275*
·277	·319 166	1320	758 054	757	280 556	1039	·038 610	282	270 126	927	3·70197	*1265*
·278	·320 486	1321	757 297	757	281 595	1039	·038 892	282	271 053	926	3·68932	*1257*
·279	·321 807	1323	756 540	756	282 634	1039	·039 174	283	271 979	926	3·67675	*1247*
0·280	1·323 130	1324	755 784	756	283 673	1040	1·039 457	284	272 905	925	3·66428	*1238*
·281	·324 454	1325	755 028	754	284 713	1040	·039 741	285	273 830	925	3·65190	*1230*
·282	·325 779	1326	754 274	754	285 753	1040	·040 026	286	274 755	924	3·63960	*1220*
·283	·327 105	1328	753 520	753	286 793	1040	·040 312	288	275 679	924	3·62740	*1211*
·284	·328 433	1329	752 767	753	287 833	1041	·040 600	288	276 603	923	3·61529	*1203*
0·285	1·329 762	1330	752 014	751	288 874	1041	1·040 888	290	277 526	923	3·60326	*1194*
·286	·331 092	1332	751 263	751	289 915	1041	·041 178	290	278 449	922	3·59132	*1185*
·287	·332 424	1333	750 512	750	290 956	1042	·041 468	291	279 371	922	3·57947	*1177*
·288	·333 757	1335	749 762	750	291 998	1042	·041 759	293	280 293	921	3·56770	*1169*
·289	·335 092	1335	749 012	748	293 040	1042	·042 052	294	281 214	921	3·55601	*1161*
0·290	1·336 427	1338	748 264	748	294 082	1042	1·042 346	294	282 135	920	3·54440	*1152*
·291	·337 765	1338	747 516	747	295 124	1043	·042 640	296	283 055	920	3·53288	*1144*
·292	·339 103	1340	746 769	747	296 167	1043	·042 936	296	283 975	919	3·52144	*1136*
·293	·340 443	1341	746 022	746	297 210	1044	·043 232	298	284 894	918	3·51008	*1128*
·294	·341 784	1342	745 276	744	298 254	1043	·043 530	299	285 812	918	3·49880	*1120*
0·295	1·343 126	1344	744 532	745	299 297	1044	1·043 829	300	286 730	918	3·48760	*1113*
·296	·344 470	1345	743 787	743	300 341	1045	·044 129	301	287 648	917	3·47647	*1104*
·297	·345 815	1347	743 044	743	301 386	1044	·044 430	302	288 565	916	3·46543	*1098*
·298	·347 162	1348	742 301	742	302 430	1045	·044 732	303	289 481	916	3·45445	*1089*
·299	·348 510	1349	741 559	741	303 475	1045	·045 035	304	290 397	916	3·44356	*1082*
0·300	1·349 859		740 818		304 520		1·045 339		291 313		3·43274	

TABLE Vc—EXPONENTIAL AND HYPERBOLIC FUNCTIONS

x	e^x		e^{-x}		sinh x		cosh x		tanh x		coth x	
			0·		0·				0·			
0·300	1·349 859	1350	740 818	740	304 520	1046	1·045 339	305	291 313	914	3·43274	1075
·301	·351 209	1352	740 078	740	305 566	1046	·045 644	306	292 227	915	3·42199	1067
·302	·352 561	1353	739 338	739	306 612	1046	·045 950	307	293 142	914	3·41132	1060
·303	·353 914	1355	738 599	738	307 658	1046	·046 257	308	294 056	913	3·40072	1053
·304	·355 269	1356	737 861	738	308 704	1047	·046 565	309	294 969	913	3·39019	1046
0·305	1·356 625	1357	737 123	736	309 751	1047	1·046 874	310	295 882	912	3·37973	1039
·306	·357 982	1359	736 387	736	310 798	1047	·047 184	312	296 794	911	3·36934	1031
·307	·359 341	1360	735 651	736	311 845	1048	·047 496	312	297 705	912	3·35903	1025
·308	·360 701	1361	734 915	734	312 893	1048	·047 808	314	298 617	910	3·34878	1018
·309	·362 062	1363	734 181	734	313 941	1048	·048 122	314	299 527	910	3·33860	1012
0·310	1·363 425	1364	733 447	733	314 989	1049	1·048 436	316	300 437	910	3·32848	1004
·311	·364 789	1366	732 714	732	316 038	1049	·048 752	316	301 347	908	3·31844	998
·312	·366 155	1367	731 982	732	317 087	1049	·049 068	318	302 255	909	3·30846	991
·313	·367 522	1368	731 250	731	318 136	1049	·049 386	318	303 164	908	3·29855	985
·314	·368 890	1369	730 519	730	319 185	1050	·049 704	320	304 072	907	3·28870	978
0·315	1·370 259	1371	729 789	730	320 235	1050	1·050 024	321	304 979	907	3·27892	972
·316	·371 630	1373	729 059	728	321 285	1051	·050 345	322	305 886	906	3·26920	966
·317	·373 003	1373	728 331	728	322 336	1051	·050 667	323	306 792	905	3·25954	959
·318	·374 376	1375	727 603	727	323 387	1051	·050 990	323	307 697	905	3·24995	953
·319	·375 751	1377	726 876	727	324 438	1051	·051 313	325	308 602	905	3·24042	947
0·320	1·377 128	1378	726 149	726	325 489	1052	1·051 638	326	309 507	904	3·23095	941
·321	·378 506	1379	725 423	725	326 541	1052	·051 964	327	310 411	903	3·22154	935
·322	·379 885	1380	724 698	724	327 593	1053	·052 291	329	311 314	903	3·21219	929
·323	·381 265	1382	723 974	724	328 646	1053	·052 620	329	312 217	902	3·20290	923
·324	·382 647	1384	723 250	723	329 699	1053	·052 949	330	313 119	902	3·19367	917
0·325	1·384 031	1384	722 527	722	330 752	1053	1·053 279	331	314 021	901	3·18450	911
·326	·385 415	1386	721 805	721	331 805	1054	·053 610	333	314 922	901	3·17539	906
·327	·386 801	1388	721 084	721	332 859	1054	·053 943	333	315 823	900	3·16633	899
·328	·388 189	1389	720 363	720	333 913	1054	·054 276	334	316 723	899	3·15734	894
·329	·389 578	1390	719 643	719	334 967	1055	·054 610	336	317 622	899	3·14840	889
0·330	1·390 968	1392	718 924	719	336 022	1055	1·054 946	336	318 521	898	3·13951	883
·331	·392 360	1393	718 205	718	337 077	1056	·055 282	338	319 419	898	3·13068	877
·332	·393 753	1394	717 487	717	338 133	1056	·055 620	339	320 317	897	3·12191	872
·333	·395 147	1396	716 770	716	339 189	1056	·055 959	339	321 214	896	3·11319	866
·334	·396 543	1397	716 054	716	340 245	1056	·056 298	341	322 110	896	3·10453	862
0·335	1·397 940	1399	715 338	715	341 301	1057	1·056 639	342	323 006	896	3·09591	855
·336	·399 339	1400	714 623	714	342 358	1057	·056 981	343	323 902	894	3·08736	851
·337	·400 739	1402	713 909	714	343 415	1058	·057 324	344	324 796	895	3·07885	845
·338	·402 141	1402	713 195	713	344 473	1057	·057 668	345	325 691	893	3·07040	840
·339	·403 543	1405	712 482	712	345 530	1059	·058 013	346	326 584	893	3·06200	835
0·340	1·404 948	1405	711 770	711	346 589	1058	1·058 359	347	327 477	893	3·05365	830
·341	·406 353	1407	711 059	711	347 647	1059	·058 706	348	328 370	892	3·04535	825
·342	·407 760	1409	710 348	710	348 706	1059	·059 054	349	329 262	891	3·03710	820
·343	·409 169	1410	709 638	709	349 765	1060	·059 403	351	330 153	891	3·02890	815
·344	·410 579	1411	708 929	709	350 825	1060	·059 754	351	331 044	890	3·02075	810
0·345	1·411 990	1413	708 220	708	351 885	1060	1·060 105	353	331 934	889	3·01265	805
·346	·413 403	1414	707 512	707	352 945	1061	·060 458	353	332 823	889	3·00460	801
·347	·414 817	1415	706 805	706	354 006	1061	·060 811	355	333 712	889	2·99659	795
·348	·416 232	1417	706 099	706	355 067	1061	·061 166	355	334 601	887	2·98864	791
·349	·417 649	1419	705 393	705	356 128	1062	·061 521	357	335 488	888	2·98073	786
0·350	1·419 068		704 688		357 190		1·061 878		336 376		2·97287	

x	e^{x}	Δ	e^{-x}	Δ	$\sinh x$	Δ	$\cosh x$	Δ	$\tanh x$	Δ	$\coth x$	Δ
			0·		0·				0·		2·	
0·350	1·419 068	1419	704 688	704	357 190	1062	1·061 878	358	336 376	886	972 868	7815
·351	·420 487	1422	703 984	704	358 252	1062	·062 236	358	337 262	886	965 053	7768
·352	·421 909	1422	703 280	703	359 314	1063	·062 594	360	338 148	885	957 285	7723
·353	·423 331	1424	702 577	702	360 377	1063	·062 954	361	339 033	885	949 562	7677
·354	·424 755	1426	701 875	702	361 440	1064	·063 315	362	339 918	884	941 885	7633
0·355	1·426 181	1427	701 173	700	362 504	1063	1·063 677	363	340 802	884	934 252	7587
·356	·427 608	1428	700 473	701	363 567	1065	·064 040	364	341 686	883	926 665	7544
·357	·429 036	1430	699 772	699	364 632	1064	·064 404	365	342 569	882	919 121	7499
·358	·430 466	1431	699 073	699	365 696	1065	·064 769	367	343 451	882	911 622	7456
·359	·431 897	1432	698 374	698	366 761	1066	·065 136	367	344 333	881	904 166	7412
0·360	1·433 329	1434	697 676	697	367 827	1065	1·065 503	368	345 214	881	896 754	7370
·361	·434 763	1436	696 979	697	368 892	1066	·065 871	370	346 095	879	889 384	7328
·362	·436 199	1437	696 282	696	369 958	1067	·066 241	370	346 974	880	882 056	7285
·363	·437 636	1438	695 586	695	371 025	1067	·066 611	372	347 854	878	874 771	7243
·364	·439 074	1440	694 891	694	372 092	1067	·066 983	372	348 732	879	867 528	7202
0·365	1·440 514	1441	694 197	694	373 159	1067	1·067 355	374	349 611	877	860 326	7161
·366	·441 955	1443	693 503	693	374 226	1068	·067 729	375	350 488	877	853 165	7121
·367	·443 398	1444	692 810	693	375 294	1068	·068 104	376	351 365	876	846 044	7079
·368	·444 842	1446	692 117	692	376 362	1069	·068 480	377	352 241	876	838 965	7040
·369	·446 288	1447	691 425	691	377 431	1069	·068 857	377	353 117	875	831 925	7000
0·370	1·447 735	1448	690 734	690	378 500	1070	1·069 234	380	353 992	874	824 925	6961
·371	·449 183	1450	690 044	690	379 570	1069	·069 614	380	354 866	874	817 964	6921
·372	·450 633	1451	689 354	689	380 639	1071	·069 994	381	355 740	873	811 043	6883
·373	·452 084	1453	688 665	688	381 710	1070	·070 375	382	356 613	873	804 160	6844
·374	·453 537	1454	687 977	688	382 780	1071	·070 757	383	357 486	871	797 316	6806
0·375	1·454 991	1456	687 289	687	383 851	1071	1·071 140	385	358 357	872	790 510	6768
·376	·456 447	1457	686 602	686	384 922	1072	·071 525	385	359 229	870	783 742	6730
·377	·457 904	1459	685 916	685	385 994	1072	·071 910	387	360 099	870	777 012	6693
·378	·459 363	1460	685 231	685	387 066	1073	·072 297	387	360 969	870	770 319	6657
·379	·460 823	1462	684 546	685	388 139	1073	·072 684	389	361 839	868	763 662	6619
0·380	1·462 285	1463	683 861	683	389 212	1073	1·073 073	390	362 707	869	757 043	6583
·381	·463 748	1464	683 178	683	390 285	1074	·073 463	391	363 576	867	750 460	6547
·382	·465 212	1466	682 495	682	391 359	1074	·073 854	391	364 443	867	743 913	6512
·383	·466 678	1467	681 813	682	392 433	1074	·074 245	393	365 310	866	737 401	6475
·384	·468 145	1469	681 131	680	393 507	1075	·074 638	394	366 176	866	730 926	6441
0·385	1·469 614	1471	680 451	680	394 582	1075	1·075 032	396	367 042	865	724 485	6405
·386	·471 085	1471	679 771	680	395 657	1076	·075 428	396	367 907	864	718 080	6371
·387	·472 556	1474	679 091	679	396 733	1076	·075 824	397	368 771	864	711 709	6336
·388	·474 030	1475	678 412	678	397 809	1076	·076 221	398	369 635	863	705 373	6302
·389	·475 505	1476	677 734	677	398 885	1077	·076 619	400	370 498	862	699 071	6268
0·390	1·476 981	1478	677 057	677	399 962	1077	1·077 019	400	371 360	862	692 803	6234
·391	·478 459	1479	676 380	676	401 039	1078	·077 419	402	372 222	861	686 569	6201
·392	·479 938	1480	675 704	675	402 117	1078	·077 821	403	373 083	861	680 368	6168
·393	·481 418	1483	675 029	675	403 195	1078	·078 224	403	373 944	859	674 200	6135
·394	·482 901	1483	674 354	674	404 273	1079	·078 627	405	374 803	860	668 065	6102
0·395	1·484 384	1485	673 680	673	405 352	1079	1·079 032	406	375 663	858	661 963	6070
·396	·485 869	1487	673 007	673	406 431	1080	·079 438	407	376 521	858	655 893	6038
·397	·487 356	1488	672 334	672	407 511	1080	·079 845	408	377 379	857	649 855	6006
·398	·488 844	1490	671 662	671	408 591	1080	·080 253	409	378 236	857	643 849	5974
·399	·490 334	1491	670 991	671	409 671	1081	·080 662	410	379 093	856	637 875	5943
0·400	1·491 825		670 320		410 752		1·081 072		379 949		631 932	

TABLE Vc—EXPONENTIAL AND HYPERBOLIC FUNCTIONS

x	e^x		e^{-x} 0·		sinh x 0·		cosh x		tanh x 0·		coth x 2·	
0·400	1·491 825	1492	670 320	670	410 752	1082	1·081 072	412	379 949	855	631 932	*5911*
·401	·493 317	1494	669 650	669	411 834	1081	·081 484	412	380 804	855	626 021	*5881*
·402	·494 811	1496	668 981	669	412 915	1082	·081 896	413	381 659	854	620 140	*5849*
·403	·496 307	1497	668 312	668	413 997	1083	·082 309	415	382 513	853	614 291	*5820*
·404	·497 804	1499	667 644	667	415 080	1083	·082 724	416	383 366	853	608 471	*5789*
0·405	1·499 303	1500	666 977	667	416 163	1083	1·083 140	416	384 219	852	602 682	*5759*
·406	·500 803	1501	666 310	666	417 246	1084	·083 556	418	385 071	851	596 923	*5729*
·407	·502 304	1503	665 644	665	418 330	1084	·083 974	419	385 922	851	591 194	*5699*
·408	·503 807	1505	664 979	665	419 414	1085	·084 393	420	386 773	850	585 495	*5670*
·409	·505 312	1506	664 314	664	420 499	1085	·084 813	421	387 623	850	579 825	*5641*
0·410	1·506 818	1507	663 650	663	421 584	1085	1·085 234	422	388 473	848	574 184	*5612*
·411	·508 325	1509	662 987	663	422 669	1086	·085 656	423	389 321	849	568 572	*5584*
·412	·509 834	1511	662 324	662	423 755	1086	·086 079	425	390 170	847	562 988	*5554*
·413	·511 345	1512	661 662	661	424 841	1087	·086 504	425	391 017	847	557 434	*5527*
·414	·512 857	1514	661 001	661	425 928	1087	·086 929	427	391 864	846	551 907	*5498*
0·415	1·514 371	1515	660 340	660	427 015	1088	1·087 356	427	392 710	845	546 409	*5470*
·416	·515 886	1517	659 680	659	428 103	1088	·087 783	429	393 555	845	540 939	*5443*
·417	·517 403	1518	659 021	659	429 191	1088	·088 212	429	394 400	844	535 496	*5415*
·418	·518 921	1519	658 362	658	430 279	1089	·088 641	431	395 244	844	530 081	*5387*
·419	·520 440	1522	657 704	657	431 368	1089	·089 072	432	396 088	842	524 694	*5361*
0·420	1·521 962	1522	657 047	657	432 457	1090	1·089 504	433	396 930	843	519 333	*5333*
·421	·523 484	1525	656 390	656	433 547	1090	·089 937	434	397 773	841	514 000	*5307*
·422	·525 009	1525	655 734	655	434 637	1091	·090 371	435	398 614	841	508 693	*5281*
·423	·526 534	1528	655 079	655	435 728	1091	·090 806	437	399 455	840	503 412	*5254*
·424	·528 062	1528	654 424	654	436 819	1091	·091 243	437	400 295	839	498 158	*5227*
0·425	1·529 590	1531	653 770	654	437 910	1092	1·091 680	439	401 134	839	492 931	*5202*
·426	·531 121	1532	653 116	652	439 002	1093	·092 119	439	401 973	838	487 729	*5176*
·427	·532 653	1533	652 464	653	440 095	1092	·092 558	441	402 811	838	482 553	*5150*
·428	·534 186	1535	651 811	651	441 187	1094	·092 999	441	403 649	836	477 403	*5125*
·429	·535 721	1537	651 160	651	442 281	1093	·093 440	443	404 485	836	472 278	*5100*
0·430	1·537 258	1538	650 509	650	443 374	1094	1·093 883	444	405 321	836	467 178	*5074*
·431	·538 796	1539	649 859	650	444 468	1095	·094 327	445	406 157	834	462 104	*5050*
·432	·540 335	1541	649 209	649	445 563	1095	·094 772	446	406 991	834	457 054	*5024*
·433	·541 876	1543	648 560	648	446 658	1095	·095 218	448	407 825	834	452 030	*5000*
·434	·543 419	1544	647 912	647	447 753	1096	·095 666	448	408 659	832	447 030	*4976*
0·435	1·544 963	1546	647 265	647	448 849	1097	1·096 114	449	409 491	832	442 054	*4952*
·436	·546 509	1547	646 618	647	449 946	1096	·096 563	451	410 323	832	437 102	*4927*
·437	·548 056	1549	645 971	645	451 042	1098	·097 014	451	411 155	830	432 175	*4904*
·438	·549 605	1550	645 326	645	452 140	1097	·097 465	453	411 985	830	427 271	*4880*
·439	·551 155	1552	644 681	645	453 237	1098	·097 918	454	412 815	829	422 391	*4856*
0·440	1·552 707	1554	644 036	643	454 335	1099	1·098 372	455	413 644	829	417 535	*4833*
·441	·554 261	1555	643 393	643	455 434	1099	·098 827	456	414 473	828	412 702	*4809*
·442	·555 816	1556	642 750	643	456 533	1100	·099 283	457	415 301	827	407 893	*4787*
·443	·557 372	1558	642 107	642	457 633	1100	·099 740	458	416 128	827	403 106	*4763*
·444	·558 930	1560	641 465	641	458 733	1100	·100 198	459	416 955	825	398 343	*4741*
0·445	1·560 490	1561	640 824	640	459 833	1101	1·100 657	461	417 780	825	393 602	*4718*
·446	·562 051	1563	640 184	640	460 934	1101	·101 118	461	418 605	825	388 884	*4695*
·447	·563 614	1565	639 544	639	462 035	1102	·101 579	463	419 430	824	384 189	*4673*
·448	·565 179	1566	638 905	639	463 137	1102	·102 042	463	420 254	823	379 516	*4652*
·449	·566 745	1567	638 266	638	464 239	1103	·102 505	465	421 077	822	374 864	*4628*
0·450	1·568 312		637 628		465 342		1·102 970		421 899		370 236	

TABLE Vc—EXPONENTIAL AND HYPERBOLIC FUNCTIONS

x	e^x		e^{-x}		sinh x		cosh x		tanh x		coth x	
			0·		0·				0·		2·	
0·450	1·568 312	1569	637 628	637	465 342	1103	1·102 970	466	421 899	822	370 236	*4608*
·451	·569 881	1571	636 991	637	466 445	1104	·103 436	467	422 721	821	365 628	*4585*
·452	·571 452	1572	636 354	636	467 549	1104	·103 903	468	423 542	820	361 043	*4564*
·453	·573 024	1574	635 718	635	468 653	1105	·104 371	469	424 362	819	356 479	*4542*
·454	·574 598	1575	635 083	635	469 758	1105	·104 840	471	425 181	819	351 937	*4521*
0·455	1·576 173	1577	634 448	634	470 863	1105	1·105 311	471	426 000	819	347 416	*4500*
·456	·577 750	1579	633 814	634	471 968	1106	·105 782	473	426 819	817	342 916	*4479*
·457	·579 329	1580	633 180	633	473 074	1107	·106 255	473	427 636	817	338 437	*4457*
·458	·580 909	1582	632 547	632	474 181	1107	·106 728	475	428 453	816	333 980	*4437*
·459	·582 491	1583	631 915	631	475 288	1107	·107 203	476	429 269	815	329 543	*4417*
0·460	1·584 074	1585	631 284	631	476 395	1108	1·107 679	477	430 084	815	325 126	*4396*
·461	·585 659	1586	630 653	631	477 503	1108	·108 156	478	430 899	814	320 730	*4376*
·462	·587 245	1588	630 022	629	478 611	1109	·108 634	479	431 713	813	316 354	*4355*
·463	·588 833	1590	629 393	629	479 720	1110	·109 113	480	432 526	813	311 999	*4335*
·464	·590 423	1591	628 764	629	480 830	1110	·109 593	482	433 339	812	307 664	*4316*
0·465	1·592 014	1593	628 135	628	481 940	1110	1·110 075	482	434 151	811	303 348	*4295*
·466	·593 607	1594	627 507	627	483 050	1111	·110 557	484	434 962	810	299 053	*4276*
·467	·595 201	1596	626 880	626	484 161	1111	·111 041	484	435 772	810	294 777	*4256*
·468	·596 797	1598	626 254	626	485 272	1112	·111 525	486	436 582	809	290 521	*4237*
·469	·598 395	1599	625 628	626	486 384	1112	·112 011	487	437 391	808	286 284	*4217*
0·470	1·599 994	1601	625 002	624	487 496	1113	1·112 498	488	438 199	808	282 067	*4199*
·471	·601 595	1602	624 378	624	488 609	1113	·112 986	489	439 007	807	277 868	*4179*
·472	·603 197	1604	623 754	624	489 722	1114	·113 475	491	439 814	806	273 689	*4160*
·473	·604 801	1606	623 130	623	490 836	1114	·113 966	491	440 620	806	269 529	*4141*
·474	·606 407	1607	622 507	622	491 950	1115	·114 457	493	441 426	804	265 388	*4123*
0·475	1·608 014	1609	621 885	622	493 065	1115	1·114 950	493	442 230	804	261 265	*4104*
·476	·609 623	1610	621 263	620	494 180	1115	·115 443	495	443 034	804	257 161	*4086*
·477	·611 233	1612	620 643	621	495 295	1117	·115 938	496	443 838	802	253 075	*4067*
·478	·612 845	1614	620 022	620	496 412	1116	·116 434	497	444 640	802	249 008	*4049*
·479	·614 459	1615	619 402	619	497 528	1118	·116 931	498	445 442	802	244 959	*4031*
0·480	1·616 074	1617	618 783	618	498 646	1117	1·117 429	499	446 244	800	240 928	*4012*
·481	·617 691	1619	618 165	618	499 763	1118	·117 928	500	447 044	800	236 916	*3995*
·482	·619 310	1620	617 547	617	500 881	1119	·118 428	502	447 844	799	232 921	*3977*
·483	·620 930	1622	616 930	617	502 000	1119	·118 930	502	448 643	798	228 944	*3960*
·484	·622 552	1623	616 313	616	503 119	1120	·119 432	504	449 441	798	224 984	*3941*
0·485	1·624 175	1625	615 697	615	504 239	1120	1·119 936	505	450 239	797	221 043	*3925*
·486	·625 800	1627	615 082	615	505 359	1121	·120 441	506	451 036	796	217 118	*3907*
·487	·627 427	1628	614 467	614	506 480	1121	·120 947	507	451 832	796	213 211	*3889*
·488	·629 055	1630	613 853	614	507 601	1122	·121 454	508	452 628	794	209 322	*3873*
·489	·630 685	1631	613 239	613	508 723	1122	·121 962	509	453 422	794	205 449	*3855*
0·490	1·632 316	1633	612 626	612	509 845	1123	1·122 471	511	454 216	794	201 594	*3839*
·491	·633 949	1635	612 014	612	510 968	1123	·122 982	511	455 010	792	197 755	*3822*
·492	·635 584	1637	611 402	611	512 091	1124	·123 493	513	455 802	792	193 933	*3805*
·493	·637 221	1638	610 791	610	513 215	1124	·124 006	514	456 594	791	190 128	*3788*
·494	·638 859	1639	610 181	610	514 339	1125	·124 520	515	457 385	791	186 340	*3772*
0·495	1·640 498	1642	609 571	609	515 464	1125	1·125 035	516	458 176	790	182 568	*3755*
·496	·642 140	1643	608 962	609	516 589	1126	·125 551	517	458 966	789	178 813	*3739*
·497	·643 783	1644	608 353	608	517 715	1126	·126 068	518	459 755	788	175 074	*3723*
·498	·645 427	1646	607 745	608	518 841	1127	·126 586	519	460 543	787	171 351	*3707*
·499	·647 073	1648	607 137	606	519 968	1127	·127 105	521	461 330	787	167 644	*3691*
0·500	1·648 721		606 531		521 095		1·127 626		462 117		163 953	

TABLE Vc—EXPONENTIAL AND HYPERBOLIC FUNCTIONS

x	e^x		e^{-x}		sinh x		cosh x		tanh x		coth x	
			0·		0·				0·		2·	
0·500	1·648 721	1650	606 531	607	521 095	1128	1·127 626	522	462 117	786	163 953	*3674*
·501	·650 371	1651	605 924	605	522 223	1129	·128 148	522	462 903	786	160 279	*3659*
·502	·652 022	1653	605 319	605	523 352	1129	·128 670	524	463 689	784	156 620	*3643*
·503	·653 675	1654	604 714	605	524 481	1129	·129 194	525	464 473	784	152 977	*3628*
·504	·655 329	1657	604 109	603	525 610	1130	·129 719	527	465 257	783	149 349	*3612*
0·505	1·656 986	1657	603 506	604	526 740	1130	1·130 246	527	466 040	783	145 737	*3596*
·506	·658 643	1660	602 902	602	527 870	1132	·130 773	528	466 823	781	142 141	*3581*
·507	·660 303	1661	602 300	602	529 002	1131	·131 301	530	467 604	781	138 560	*3566*
·508	·661 964	1663	601 698	602	530 133	1132	·131 831	531	468 385	781	134 994	*3551*
·509	·663 627	1664	601 096	600	531 265	1133	·132 362	531	469 166	779	131 443	*3535*
0·510	1·665 291	1666	600 496	601	532 398	1133	1·132 893	533	469 945	779	127 908	*3521*
·511	·666 957	1668	599 895	599	533 531	1134	·133 426	534	470 724	778	124 387	*3505*
·512	·668 625	1670	599 296	599	534 665	1134	·133 960	536	471 502	777	120 882	*3491*
·513	·670 295	1671	598 697	599	535 799	1135	·134 496	536	472 279	777	117 391	*3476*
·514	·671 966	1673	598 098	597	536 934	1135	·135 032	538	473 056	776	113 915	*3461*
0·515	1·673 639	1674	597 501	598	538 069	1136	1·135 570	538	473 832	775	110 454	*3447*
·516	·675 313	1676	596 903	596	539 205	1136	·136 108	540	474 607	774	107 007	*3432*
·517	·676 989	1678	596 307	596	540 341	1137	·136 648	541	475 381	774	103 575	*3418*
·518	·678 667	1679	595 711	596	541 478	1138	·137 189	542	476 155	773	100 157	*3404*
·519	·680 346	1682	595 115	594	542 616	1138	·137 731	543	476 928	772	096 753	*3389*
0·520	1·682 028	1683	594 521	595	543 754	1138	1·138 274	544	477 700	771	093 364	*3375*
·521	·683 711	1684	593 926	593	544 892	1139	·138 818	546	478 471	771	089 989	*3361*
·522	·685 395	1686	593 333	593	546 031	1140	·139 364	546	479 242	770	086 628	*3347*
·523	·687 081	1688	592 740	593	547 171	1140	·139 910	548	480 012	769	083 281	*3333*
·524	·688 769	1690	592 147	592	548 311	1141	·140 458	549	480 781	769	079 948	*3320*
0·525	1·690 459	1691	591 555	591	549 452	1141	1·141 007	550	481 550	768	076 628	*3305*
·526	·692 150	1693	590 964	591	550 593	1142	·141 557	551	482 318	767	073 323	*3292*
·527	·693 843	1695	590 373	590	551 735	1142	·142 108	553	483 085	766	070 031	*3278*
·528	·695 538	1696	589 783	589	552 877	1143	·142 661	553	483 851	765	066 753	*3265*
·529	·697 234	1698	589 194	589	554 020	1144	·143 214	555	484 616	765	063 488	*3251*
0·530	1·698 932	1700	588 605	588	555 164	1144	1·143 769	555	485 381	764	060 237	*3238*
·531	·700 632	1702	588 017	588	556 308	1144	·144 324	557	486 145	763	056 999	*3225*
·532	·702 334	1703	587 429	587	557 452	1145	·144 881	558	486 908	763	053 774	*3211*
·533	·704 037	1705	586 842	587	558 597	1146	·145 439	559	487 671	762	050 563	*3198*
·534	·705 742	1706	586 255	586	559 743	1146	·145 998	561	488 433	761	047 365	*3186*
0·535	1·707 448	1709	585 669	585	560 889	1147	1·146 559	561	489 194	760	044 179	*3172*
·536	·709 157	1710	585 084	585	562 036	1148	·147 120	563	489 954	760	041 007	*3159*
·537	·710 867	1711	584 499	584	563 184	1148	·147 683	564	490 714	759	037 848	*3146*
·538	·712 578	1714	583 915	584	564 332	1148	·148 247	565	491 473	758	034 702	*3134*
·539	·714 292	1715	583 331	583	565 480	1149	·148 812	566	492 231	757	031 568	*3121*
0·540	1·716 007	1717	582 748	582	566 629	1150	1·149 378	567	492 988	757	028 447	*3108*
·541	·717 724	1718	582 166	582	567 779	1150	·149 945	568	493 745	755	025 339	*3096*
·542	·719 442	1721	581 584	581	568 929	1151	·150 513	570	494 500	756	022 243	*3083*
·543	·721 163	1722	581 003	581	570 080	1151	·151 083	570	495 256	754	019 160	*3071*
·544	·722 885	1723	580 422	580	571 231	1152	·151 653	572	496 010	753	016 089	*3058*
0·545	1·724 608	1726	579 842	580	572 383	1153	1·152 225	573	496 763	753	013 031	*3047*
·546	·726 334	1727	579 262	579	573 536	1153	·152 798	574	497 516	752	009 984	*3034*
·547	·728 061	1729	578 683	578	574 689	1154	·153 372	575	498 268	752	006 950	*3021*
·548	·729 790	1731	578 105	578	575 843	1154	·153 947	577	499 020	750	003 929	*3010*
·549	·731 521	1732	577 527	577	576 997	1155	·154 524	577	499 770	750	000 919	*2998*
0·550	1·733 253		576 950		578 152		1·155 101		500 520		*	

* coth 0·550 = 1·997 921

x	e^x	e^{-x}	sinh x	cosh x	tanh x	coth x
		0·	0·		0·	1·
0·550	1·733 253 1734	576 950 577	578 152 1155	1·155 101 579	500 520 749	997 921 *2985*
·551	·734 987 1736	576 373 576	579 307 1156	·155 680 580	501 269 749	994 936 *2974*
·552	·736 723 1738	575 797 575	580 463 1157	·156 260 581	502 018 747	991 962 *2962*
·553	·738 461 1739	575 222 575	581 620 1157	·156 841 582	502 765 747	989 000 *2951*
·554	·740 200 1741	574 647 575	582 777 1157	·157 423 584	503 512 746	986 049 *2938*
0·555	1·741 941 1743	574 072 574	583 934 1159	1·158 007 584	504 258 746	983 111 *2927*
·556	·743 684 1744	573 498 573	585 093 1159	·158 591 586	505 004 744	980 184 *2915*
·557	·745 428 1747	572 925 572	586 252 1159	·159 177 587	505 748 744	977 269 *2904*
·558	·747 175 1748	572 353 572	587 411 1160	·159 764 588	506 492 743	974 365 *2893*
·559	·748 923 1750	571 781 572	588 571 1161	·160 352 589	507 235 742	971 472 *2881*
0·560	1·750 673 1751	571 209 571	589 732 1161	1·160 941 590	507 977 742	968 591 *2869*
·561	·752 424 1753	570 638 570	590 893 1162	·161 531 592	508 719 741	965 722 *2859*
·562	·754 177 1755	570 068 570	592 055 1162	·162 123 592	509 460 740	962 863 *2847*
·563	·755 932 1757	569 498 569	593 217 1163	·162 715 594	510 200 739	960 016 *2836*
·564	·757 689 1759	568 929 569	594 380 1164	·163 309 595	510 939 739	957 180 *2825*
0·565	1·759 448 1760	568 360 568	595 544 1164	1·163 904 596	511 678 738	954 355 *2814*
·566	·761 208 1762	567 792 567	596 708 1165	·164 500 597	512 416 737	951 541 *2803*
·567	·762 970 1764	567 225 567	597 873 1165	·165 097 599	513 153 736	948 738 *2792*
·568	·764 734 1766	566 658 567	599 038 1166	·165 696 599	513 889 735	945 946 *2782*
·569	·766 500 1767	566 091 566	600 204 1167	·166 295 601	514 624 735	943 164 *2770*
0·570	1·768 267 1769	565 525 565	601 371 1167	1·166 896 602	515 359 734	940 394 *2760*
·571	·770 036 1771	564 960 564	602 538 1168	·167 498 603	516 093 734	937 634 *2749*
·572	·771 807 1773	564 396 565	603 706 1168	·168 101 605	516 827 732	934 885 *2738*
·573	·773 580 1774	563 831 563	604 874 1169	·168 706 605	517 559 732	932 147 *2728*
·574	·775 354 1777	563 268 563	606 043 1170	·169 311 607	518 291 731	929 419 *2718*
0·575	1·777 131 1778	562 705 563	607 213 1170	1·169 918 607	519 022 730	926 701 *2707*
·576	·778 909 1779	562 142 561	608 383 1171	·170 525 609	519 752 730	923 994 *2696*
·577	·780 688 1782	561 581 562	609 554 1171	·171 134 611	520 482 728	921 298 *2687*
·578	·782 470 1783	561 019 560	610 725 1172	·171 745 611	521 210 728	918 611 *2675*
·579	·784 253 1785	560 459 561	611 897 1173	·172 356 612	521 938 727	915 936 *2666*
0·580	1·786 038 1787	559 898 559	613 070 1173	1·172 968 614	522 665 727	913 270 *2656*
·581	·787 825 1789	559 339 559	614 243 1174	·173 582 615	523 392 726	910 614 *2645*
·582	·789 614 1791	558 780 559	615 417 1175	·174 197 616	524 118 724	907 969 *2635*
·583	·791 405 1792	558 221 558	616 592 1175	·174 813 617	524 842 725	905 334 *2626*
·584	·793 197 1794	557 663 557	617 767 1176	·175 430 618	525 567 723	902 708 *2615*
0·585	1·794 991 1796	557 106 557	618 943 1176	1·176 048 620	526 290 723	900 093 *2605*
·586	·796 787 1798	556 549 556	620 119 1177	·176 668 621	527 013 722	897 488 *2596*
·587	·798 585 1799	555 993 556	621 296 1177	·177 289 622	527 735 721	894 892 *2586*
·588	·800 384 1801	555 437 555	622 473 1179	·177 911 623	528 456 720	892 306 *2576*
·589	·802 185 1803	554 882 555	623 652 1179	·178 534 624	529 176 720	889 730 *2566*
0·590	1·803 988 1805	554 327 554	624 831 1179	1·179 158 625	529 896 718	887 164 *2556*
·591	·805 793 1807	553 773 553	626 010 1180	·179 783 627	530 614 719	884 608 *2547*
·592	·807 600 1809	553 220 553	627 190 1181	·180 410 628	531 333 717	882 061 *2538*
·593	·809 409 1810	552 667 553	628 371 1181	·181 038 629	532 050 716	879 523 *2528*
·594	·811 219 1812	552 114 551	629 552 1182	·181 667 630	532 766 716	876 995 *2518*
0·595	1·813 031 1814	551 563 552	630 734 1183	1·182 297 631	533 482 715	874 477 *2509*
·596	·814 845 1816	551 011 550	631 917 1183	·182 928 633	534 197 714	871 968 *2500*
·597	·816 661 1817	550 461 551	633 100 1184	·183 561 633	534 911 714	869 468 *2490*
·598	·818 478 1820	549 910 549	634 284 1184	·184 194 635	535 625 713	866 978 *2481*
·599	·820 298 1821	549 361 549	635 468 1186	·184 829 636	536 338 712	864 497 *2471*
0·600	1·822 119	548 812	636 654	1·185 465	537 050	862 026

TABLE Vc—EXPONENTIAL AND HYPERBOLIC FUNCTIONS

x	e^x		e^{-x}		sinh x		cosh x		tanh x		coth x	
			0·		0·				0·		1·	
0·600	1·822 119	1823	548 812	549	636 654	1185	1·185 465	637	537 050	711	862 026	2463
·601	·823 942	1825	548 263	548	637 839	1187	·186 102	639	537 761	710	859 563	2453
·602	·825 767	1826	547 715	547	639 026	1187	·186 741	640	538 471	710	857 110	2445
·603	·827 593	1829	547 168	547	640 213	1188	·187 381	640	539 181	709	854 665	2435
·604	·829 422	1830	546 621	547	641 401	1188	·188 021	642	539 890	708	852 230	2426
0·605	1·831 252	1832	546 074	545	642 589	1189	1·188 663	644	540 598	707	849 804	2418
·606	·833 084	1834	545 529	546	643 778	1190	·189 307	644	541 305	707	847 386	2408
·607	·834 918	1836	544 983	544	644 968	1190	·189 951	645	542 012	706	844 978	2399
·608	·836 754	1838	544 439	545	646 158	1191	·190 596	647	542 718	705	842 579	2391
·609	·838 592	1839	543 894	543	647 349	1191	·191 243	648	543 423	704	840 188	2382
0·610	1·840 431	1842	543 351	543	648 540	1192	1·191 891	649	544 127	704	837 806	2373
·611	·842 273	1843	542 808	543	649 732	1193	·192 540	651	544 831	702	835 433	2365
·612	·844 116	1845	542 265	542	650 925	1194	·193 191	651	545 533	702	833 068	2356
·613	·845 961	1847	541 723	541	652 119	1194	·193 842	653	546 235	702	830 712	2347
·614	·847 808	1849	541 182	541	653 313	1195	·194 495	654	546 937	700	828 365	2338
0·615	1·849 657	1850	540 641	540	654 508	1195	1·195 149	655	547 637	700	826 027	2331
·616	·851 507	1853	540 101	540	655 703	1196	·195 804	656	548 337	699	823 696	2321
·617	·853 360	1854	539 561	540	656 899	1197	·196 460	658	549 036	698	821 375	2313
·618	·855 214	1856	539 021	538	658 096	1198	·197 118	658	549 734	697	819 062	2305
·619	·857 070	1858	538 483	539	659 294	1198	·197 776	660	550 431	697	816 757	2297
0·620	1·858 928	1860	537 944	537	660 492	1199	1·198 436	661	551 128	696	814 460	2288
·621	·860 788	1862	537 407	537	661 691	1199	·199 097	663	551 824	695	812 172	2280
·622	·862 650	1863	536 870	537	662 890	1200	·199 760	663	552 519	694	809 892	2271
·623	·864 513	1866	536 333	536	664 090	1201	·200 423	665	553 213	694	807 621	2264
·624	·866 379	1867	535 797	536	665 291	1201	·201 088	666	553 907	693	805 357	2255
0·625	1·868 246	1869	535 261	535	666 492	1202	1·201 754	667	554 600	692	803 102	2247
·626	·870 115	1871	534 726	534	667 694	1203	·202 421	668	555 292	691	800 855	2239
·627	·871 986	1873	534 192	534	668 897	1204	·203 089	670	555 983	691	798 616	2231
·628	·873 859	1875	533 658	533	670 101	1204	·203 759	670	556 674	689	796 385	2223
·629	·875 734	1877	533 125	533	671 305	1204	·204 429	672	557 363	689	794 162	2215
0·630	1·877 611	1878	532 592	533	672 509	1206	1·205 101	673	558 052	688	791 947	2207
·631	·879 489	1881	532 059	531	673 715	1206	·205 774	675	558 740	688	789 740	2199
·632	·881 370	1882	531 528	532	674 921	1207	·206 449	675	559 428	686	787 541	2192
·633	·883 252	1884	530 996	530	676 128	1207	·207 124	677	560 114	686	785 349	2183
·634	·885 136	1886	530 466	531	677 335	1208	·207 801	678	560 800	685	783 166	2176
0·635	1·887 022	1888	529 935	529	678 543	1209	1·208 479	679	561 485	685	780 990	2168
·636	·888 910	1890	529 406	529	679 752	1210	·209 158	680	562 170	683	778 822	2161
·637	·890 800	1892	528 877	529	680 962	1210	·209 838	682	562 853	683	776 661	2152
·638	·892 692	1893	528 348	528	682 172	1211	·210 520	683	563 536	682	774 509	2145
·639	·894 585	1896	527 820	528	683 383	1211	·211 203	684	564 218	682	772 364	2138
0·640	1·896 481	1897	527 292	527	684 594	1212	1·211 887	685	564 900	680	770 226	2130
·641	·898 378	1900	526 765	526	685 806	1213	·212 572	686	565 580	680	768 096	2122
·642	·900 278	1901	526 239	526	687 019	1214	·213 258	688	566 260	679	765 974	2115
·643	·902 179	1903	525 713	526	688 233	1214	·213 946	689	566 939	678	763 859	2108
·644	·904 082	1905	525 187	524	689 447	1215	·214 635	690	567 617	677	761 751	2100
0·645	1·905 987	1907	524 663	525	690 662	1216	1·215 325	691	568 294	677	759 651	2092
·646	·907 894	1909	524 138	524	691 878	1216	·216 016	693	568 971	676	757 559	2086
·647	·909 803	1911	523 614	523	693 094	1217	·216 709	693	569 647	675	755 473	2078
·648	·911 714	1912	523 091	523	694 311	1218	·217 402	695	570 322	674	753 395	2070
·649	·913 626	1915	522 568	522	695 529	1219	·218 097	696	570 996	674	751 325	2064
0·650	1·915 541		522 046		696 748		1·218 793		571 670		749 261	

x	e^x		e^{-x}		$\sinh x$		$\cosh x$		$\tanh x$		$\coth x$	
			0·		0·				0·		1·	
0·650	1·915 541	1916	522 046	522	696 748	1219	1·218 793	698	571 670	673	749 261	*2056*
·651	·917 457	1919	521 524	521	697 967	1220	·219 491	698	572 343	672	747 205	*2049*
·652	·919 376	1920	521 003	521	699 187	1220	·220 189	700	573 015	671	745 156	*2042*
·653	·921 296	1922	520 482	520	700 407	1221	·220 889	701	573 686	671	743 114	*2035*
·654	·923 218	1925	519 962	520	701 628	1222	·221 590	702	574 357	669	741 079	*2028*
0·655	1·925 143	1926	519 442	519	702 850	1223	1·222 292	704	575 026	669	739 051	*2021*
·656	·927 069	1928	518 923	519	704 073	1223	·222 996	704	575 695	668	737 030	*2014*
·657	·928 997	1930	518 404	518	705 296	1224	·223 700	706	576 363	668	735 016	*2007*
·658	·930 927	1932	517 886	518	706 520	1225	·224 406	707	577 031	667	733 009	*1999*
·659	·932 859	1933	517 368	517	707 745	1226	·225 113	709	577 698	665	731 010	*1993*
0·660	1·934 792	1936	516 851	516	708 971	1226	1·225 822	709	578 363	666	729 017	*1986*
·661	·936 728	1938	516 335	516	710 197	1227	·226 531	711	579 029	664	727 031	*1980*
·662	·938 666	1939	515 819	516	711 424	1227	·227 242	712	579 693	663	725 051	*1972*
·663	·940 605	1942	515 303	515	712 651	1228	·227 954	714	580 356	663	723 079	*1966*
·664	·942 547	1944	514 788	514	713 879	1229	·228 668	714	581 019	662	721 113	*1958*
0·665	1·944 491	1945	514 274	514	715 108	1230	1·229 382	716	581 681	662	719 155	*1953*
·666	·946 436	1947	513 760	514	716 338	1231	·230 098	717	582 343	660	717 202	*1945*
·667	·948 383	1950	513 246	513	717 569	1231	·230 815	718	583 003	660	715 257	*1939*
·668	·950 333	1951	512 733	512	718 800	1232	·231 533	719	583 663	659	713 318	*1932*
·669	·952 284	1953	512 221	512	720 032	1232	·232 252	721	584 322	658	711 386	*1926*
0·670	1·954 237	1956	511 709	512	721 264	1234	1·232 973	722	584 980	657	709 460	*1918*
·671	·956 193	1957	511 197	511	722 498	1234	·233 695	723	585 637	657	707 542	*1913*
·672	·958 150	1959	510 686	510	723 732	1235	·234 418	724	586 294	656	705 629	*1906*
·673	·960 109	1961	510 176	510	724 967	1235	·235 142	726	586 950	655	703 723	*1899*
·674	·962 070	1963	509 666	510	726 202	1236	·235 868	727	587 605	654	701 824	*1893*
0·675	1·964 033	1965	509 156	508	727 438	1237	1·236 595	728	588 259	654	699 931	*1887*
·676	·965 998	1967	508 648	509	728 675	1238	·237 323	729	588 913	653	698 044	*1880*
·677	·967 965	1969	508 139	508	729 913	1238	·238 052	731	589 566	652	696 164	*1874*
·678	·969 934	1971	507 631	507	731 151	1239	·238 783	731	590 218	651	694 290	*1867*
·679	·971 905	1973	507 124	507	732 390	1240	·239 514	733	590 869	650	692 423	*1861*
0·680	1·973 878	1975	506 617	506	733 630	1241	1·240 247	735	591 519	650	690 562	*1855*
·681	·975 853	1976	506 111	506	734 871	1241	·240 982	735	592 169	649	688 707	*1849*
·682	·977 829	1979	505 605	506	736 112	1242	·241 717	737	592 818	648	686 858	*1842*
·683	·979 808	1981	505 099	504	737 354	1243	·242 454	738	593 466	648	685 016	*1836*
·684	·981 789	1983	504 595	505	738 597	1244	·243 192	739	594 114	646	683 180	*1830*
0·685	1·983 772	1985	504 090	504	739 841	1244	1·243 931	740	594 760	646	681 350	*1824*
·686	·985 757	1986	503 586	503	741 085	1245	·244 671	742	595 406	645	679 526	*1818*
·687	·987 743	1989	503 083	503	742 330	1246	·245 413	743	596 051	645	677 708	*1812*
·688	·989 732	1991	502 580	502	743 576	1246	·246 156	744	596 696	643	675 896	*1805*
·689	·991 723	1993	502 078	502	744 822	1248	·246 900	746	597 339	643	674 091	*1800*
0·690	1·993 716	1994	501 576	501	746 070	1248	1·247 646	746	597 982	642	672 291	*1793*
·691	·995 710	1997	501 075	501	747 318	1249	·248 392	748	598 624	641	670 498	*1788*
·692	·997 707	1999	500 574	500	748 567	1249	·249 140	750	599 265	641	668 710	*1782*
·693	1·999 706	2000	500 074	500	749 816	1250	·249 890	750	599 906	640	666 928	*1775*
·694	2·001 706	2003	499 574	500	751 066	1251	·250 640	752	600 546	638	665 153	*1770*
0·695	2·003 709	2005	499 074	498	752 317	1252	1·251 392	753	601 184	639	663 383	*1764*
·696	·005 714	2007	498 576	499	753 569	1253	·252 145	754	601 823	637	661 619	*1758*
·697	·007 721	2008	498 077	498	754 822	1253	·252 899	755	602 460	637	659 861	*1752*
·698	·009 729	2011	497 579	497	756 075	1254	·253 654	757	603 097	636	658 109	*1747*
·699	·011 740	2013	497 082	497	757 329	1255	·254 411	758	603 733	635	656 362	*1740*
0·700	2·013 753		496 585		758 584		1·255 169		604 368		654 622	

TABLE Vc—EXPONENTIAL AND HYPERBOLIC FUNCTIONS

x	e^x		e^{-x}		$\sinh x$		$\cosh x$		$\tanh x$		$\coth x$	
			0·		0·				0·		1·	
0·700	2·013 753	2014	496 585	496	758 584	1255	1·255 169	759	604 368	634	654 622	1735
·701	·015 767	2017	496 089	496	759 839	1257	·255 928	761	605 002	634	652 887	1729
·702	·017 784	2019	495 593	495	761 096	1257	·256 689	761	605 636	633	651 158	1724
·703	·019 803	2021	495 098	495	762 353	1257	·257 450	763	606 269	632	649 434	1718
·704	·021 824	2023	494 603	494	763 610	1259	·258 213	765	606 901	631	647 716	1712
0·705	2·023 847	2025	494 109	494	764 869	1259	1·258 978	765	607 532	630	646 004	1706
·706	·025 872	2026	493 615	494	766 128	1261	·259 743	767	608 162	630	644 298	1701
·707	·027 898	2029	493 121	493	767 389	1260	·260 510	768	608 792	629	642 597	1696
·708	·029 927	2031	492 628	492	768 649	1262	·261 278	769	609 421	628	640 901	1689
·709	·031 958	2033	492 136	492	769 911	1263	·262 047	771	610 049	628	639 212	1685
0·710	2·033 991	2035	491 644	491	771 174	1263	1·262 818	772	610 677	627	637 527	1678
·711	·036 026	2037	491 153	491	772 437	1264	·263 590	773	611 304	625	635 849	1674
·712	·038 063	2039	490 662	491	773 701	1264	·264 363	774	611 929	626	634 175	1667
·713	·040 102	2042	490 171	489	774 965	1266	·265 137	776	612 555	624	632 508	1663
·714	·042 144	2043	489 682	490	776 231	1266	·265 913	776	613 179	624	630 845	1657
0·715	2·044 187	2045	489 192	489	777 497	1267	1·266 689	779	613 803	622	629 188	1651
·716	·046 232	2047	488 703	488	778 764	1268	·267 468	779	614 425	623	627 537	1647
·717	·048 279	2049	488 215	488	780 032	1269	·268 247	781	615 048	621	625 890	1640
·718	·050 328	2052	487 727	488	781 301	1269	·269 028	782	615 669	620	624 250	1636
·719	·052 380	2053	487 239	487	782 570	1270	·269 810	783	616 289	620	622 614	1630
0·720	2·054 433	2056	486 752	486	783 840	1271	1·270 593	784	616 909	619	620 984	1625
·721	·056 489	2057	486 266	486	785 111	1272	·271 377	786	617 528	619	619 359	1620
·722	·058 546	2060	485 780	486	786 383	1273	·272 163	787	618 147	617	617 739	1614
·723	·060 606	2061	485 294	485	787 656	1273	·272 950	788	618 764	617	616 125	1610
·724	·062 667	2064	484 809	484	788 929	1274	·273 738	790	619 381	616	614 515	1604
0·725	2·064 731	2066	484 325	485	790 203	1275	1·274 528	791	619 997	615	612 911	1599
·726	·066 797	2068	483 840	483	791 478	1276	·275 319	792	620 612	615	611 312	1593
·727	·068 865	2070	483 357	483	792 754	1276	·276 111	793	621 227	613	609 719	1589
·728	·070 935	2072	482 874	483	794 030	1278	·276 904	795	621 840	613	608 130	1583
·729	·073 007	2074	482 391	482	795 308	1278	·277 699	796	622 453	612	606 547	1579
0·730	2·075 081	2076	481 909	482	796 586	1279	1·278 495	797	623 065	612	604 968	1573
·731	·077 157	2078	481 427	481	797 865	1279	·279 292	799	623 677	610	603 395	1569
·732	·079 235	2080	480 946	481	799 144	1281	·280 091	799	624 287	610	601 826	1563
·733	·081 315	2083	480 465	480	800 425	1281	·280 890	801	624 897	609	600 263	1558
·734	·083 398	2084	479 985	480	801 706	1282	·281 691	803	625 506	609	598 705	1554
0·735	2·085 482	2087	479 505	479	802 988	1283	1·282 494	803	626 115	607	597 151	1548
·736	·087 569	2088	479 026	479	804 271	1284	·283 297	805	626 722	607	595 603	1544
·737	·089 657	2091	478 547	478	805 555	1284	·284 102	806	627 329	606	594 059	1538
·738	·091 748	2093	478 069	478	806 839	1286	·284 908	808	627 935	606	592 521	1534
·739	·093 841	2095	477 591	477	808 125	1286	·285 716	809	628 541	604	590 987	1529
0·740	2·095 936	2096	477 114	477	809 411	1287	1·286 525	810	629 145	604	589 458	1524
·741	·098 032	2100	476 637	476	810 698	1287	·287 335	811	629 749	603	587 934	1519
·742	·100 132	2101	476 161	476	811 985	1289	·288 146	813	630 352	602	586 415	1514
·743	·102 233	2103	475 685	476	813 274	1289	·288 959	814	630 954	602	584 901	1510
·744	·104 336	2105	475 209	475	814 563	1291	·289 773	815	631 556	601	583 391	1504
0·745	2·106 441	2108	474 734	474	815 854	1291	1·290 588	816	632 157	600	581 887	1500
·746	·108 549	2110	474 260	474	817 145	1291	·291 404	818	632 757	599	580 387	1496
·747	·110 659	2111	473 786	474	818 436	1293	·292 222	819	633 356	598	578 891	1490
·748	·112 770	2114	473 312	473	819 729	1293	·293 041	821	633 954	598	577 401	1486
·749	·114 884	2116	472 839	472	821 022	1295	·293 862	821	634 552	597	575 915	1481
0·750	2·117 000		472 367		822 317		1·294 683		635 149		574 434	

TABLE Vc—EXPONENTIAL AND HYPERBOLIC FUNCTIONS

x	e^x		e^{-x}		sinh x		cosh x		tanh x		coth x	
			0·		0·				0·		1·	
0·750	2·117 000	2118	472 367	473	822 317	1295	1·294 683	823	635 149	596	574 434	*1477*
·751	·119 118	2120	471 894	471	823 612	1296	·295 506	825	635 745	596	572 957	*1472*
·752	·121 238	2123	471 423	471	824 908	1296	·296 331	825	636 341	594	571 485	*1467*
·753	·123 361	2124	470 952	471	826 204	1298	·297 156	827	636 935	594	570 018	*1462*
·754	·125 485	2127	470 481	470	827 502	1298	·297 983	828	637 529	593	568 556	*1459*
0·755	2·127 612	2128	470 011	470	828 800	1300	1·298 811	830	638 122	593	567 097	*1453*
·756	·129 740	2131	469 541	469	830 100	1300	·299 641	830	638 715	592	565 644	*1449*
·757	·131 871	2133	469 072	469	831 400	1301	·300 471	832	639 307	590	564 195	*1444*
·758	·134 004	2135	468 603	469	832 701	1301	·301 303	834	639 897	591	562 751	*1440*
·759	·136 139	2137	468 134	468	834 002	1303	·302 137	834	640 488	589	561 311	*1436*
0·760	2·138 276	2140	467 666	467	835 305	1303	1·302 971	836	641 077	589	559 875	*1431*
·761	·140 416	2141	467 199	467	836 608	1305	·303 807	838	641 666	587	558 444	*1426*
·762	·142 557	2144	466 732	466	837 913	1305	·304 645	838	642 253	588	557 018	*1422*
·763	·144 701	2145	466 266	467	839 218	1305	·305 483	840	642 841	586	555 596	*1418*
·764	·146 846	2148	465 799	465	840 523	1307	·306 323	841	643 427	586	554 178	*1413*
0·765	2·148 994	2150	465 334	465	841 830	1308	1·307 164	843	644 013	584	552 765	*1409*
·766	·151 144	2153	464 869	465	843 138	1308	·308 007	843	644 597	585	551 356	*1405*
·767	·153 297	2154	464 404	464	844 446	1310	·308 850	846	645 182	583	549 951	*1400*
·768	·155 451	2157	463 940	464	845 756	1310	·309 696	846	645 765	583	548 551	*1396*
·769	·157 608	2158	463 476	463	847 066	1311	·310 542	848	646 348	581	547 155	*1391*
0·770	2·159 766	2161	463 013	463	848 377	1311	1·311 390	849	646 929	582	545 764	*1388*
·771	·161 927	2163	462 550	462	849 688	1313	·312 239	850	647 511	580	544 376	*1383*
·772	·164 090	2165	462 088	462	851 001	1314	·313 089	852	648 091	580	542 993	*1378*
·773	·166 255	2168	461 626	461	852 315	1314	·313 941	853	648 671	578	541 615	*1375*
·774	·168 423	2169	461 165	461	853 629	1315	·314 794	854	649 249	578	540 240	*1370*
0·775	2·170 592	2172	460 704	461	854 944	1316	1·315 648	856	649 827	578	538 870	*1366*
·776	·172 764	2174	460 243	460	856 260	1317	·316 504	856	650 405	576	537 504	*1362*
·777	·174 938	2176	459 783	459	857 577	1318	·317 360	859	650 981	576	536 142	*1357*
·778	·177 114	2178	459 324	459	858 895	1319	·318 219	859	651 557	575	534 785	*1354*
·779	·179 292	2180	458 865	459	860 214	1319	·319 078	861	652 132	575	533 431	*1349*
0·780	2·181 472	2183	458 406	458	861 533	1320	1·319 939	862	652 707	573	532 082	*1346*
·781	·183 655	2185	457 948	458	862 853	1322	·320 801	864	653 280	573	530 736	*1341*
·782	·185 840	2187	457 490	457	864 175	1322	·321 665	865	653 853	572	529 395	*1337*
·783	·188 027	2189	457 033	457	865 497	1323	·322 530	866	654 425	572	528 058	*1333*
·784	·190 216	2191	456 576	456	866 820	1324	·323 396	867	654 997	570	526 725	*1328*
0·785	2·192 407	2193	456 120	456	868 144	1324	1·324 263	869	655 567	570	525 397	*1325*
·786	·194 600	2196	455 664	456	869 468	1326	·325 132	870	656 137	569	524 072	*1321*
·787	·196 796	2198	455 208	455	870 794	1326	·326 002	872	656 706	569	522 751	*1317*
·788	·198 994	2200	454 753	454	872 120	1328	·326 874	872	657 275	567	521 434	*1313*
·789	·201 194	2202	454 299	454	873 448	1328	·327 746	875	657 842	567	520 121	*1308*
0·790	2·203 396	2205	453 845	454	874 776	1329	1·328 621	875	658 409	566	518 813	*1305*
·791	·205 601	2207	453 391	453	876 105	1330	·329 496	877	658 975	566	517 508	*1301*
·792	·207 808	2209	452 938	453	877 435	1331	·330 373	878	659 541	564	516 207	*1297*
·793	·210 017	2211	452 485	452	878 766	1331	·331 251	879	660 105	564	514 910	*1293*
·794	·212 228	2213	452 033	452	880 097	1333	·332 130	881	660 669	563	513 617	*1289*
0·795	2·214 441	2216	451 581	451	881 430	1333	1·333 011	882	661 232	563	512 328	*1285*
·796	·216 657	2217	451 130	451	882 763	1335	·333 893	884	661 795	561	511 043	*1281*
·797	·218 874	2220	450 679	450	884 098	1335	·334 777	884	662 356	561	509 762	*1278*
·798	·221 094	2222	450 229	450	885 433	1336	·335 661	887	662 917	560	508 484	*1274*
·799	·223 316	2225	449 779	450	886 769	1337	·336 548	887	663 477	560	507 210	*1269*
0·800	2·225 541		449 329		888 106		1·337 435		664 037		505 941	

x	e^x		e^{-x}		$\sinh x$		$\cosh x$		$\tanh x$		$\coth x$	
			0·		0·				0·		1·	
0·800	2·225 541	2227	449 329	449	888 106	1338	1·337 435	889	664 037	558	505 941	1266
·801	·227 768	2228	448 880	449	889 444	1339	·338 324	890	664 595	558	504 675	1262
·802	·229 996	2232	448 431	448	890 783	1339	·339 214	891	665 153	558	503 413	1259
·803	·232 228	2233	447 983	448	892 122	1341	·340 105	893	665 711	556	502 154	1254
·804	·234 461	2235	447 535	447	893 463	1341	·340 998	894	666 267	556	500 900	1251
0·805	2·236 696	2238	447 088	447	894 804	1343	1·341 892	896	666 823	555	499 649	1247
·806	·238 934	2240	446 641	446	896 147	1343	·342 788	897	667 378	554	498 402	1244
·807	·241 174	2243	446 195	446	897 490	1344	·343 685	898	667 932	553	497 158	1239
·808	·243 417	2244	445 749	446	898 834	1345	·344 583	899	668 485	553	495 919	1236
·809	·245 661	2247	445 303	445	900 179	1346	·345 482	901	669 038	552	494 683	1232
0·810	2·247 908	2249	444 858	445	901 525	1347	1·346 383	902	669 590	552	493 451	1229
·811	·250 157	2251	444 413	444	902 872	1348	·347 285	904	670 142	550	492 222	1225
·812	·252 408	2254	443 969	444	904 220	1348	·348 189	905	670 692	550	490 997	1221
·813	·254 662	2256	443 525	443	905 568	1350	·349 094	906	671 242	549	489 776	1218
·814	·256 918	2258	443 082	443	906 918	1350	·350 000	907	671 791	548	488 558	1214
0·815	2·259 176	2260	442 639	442	908 268	1352	1·350 907	909	672 339	548	487 344	1210
·816	·261 436	2263	442 197	442	909 620	1352	·351 816	911	672 887	547	486 134	1207
·817	·263 699	2264	441 755	442	910 972	1353	·352 727	911	673 434	546	484 927	1203
·818	·265 963	2267	441 313	441	912 325	1354	·353 638	913	673 980	545	483 724	1200
·819	·268 230	2270	440 872	440	913 679	1355	·354 551	915	674 525	545	482 524	1196
0·820	2·270 500	2271	440 432	441	915 034	1356	1·355 466	915	675 070	544	481 328	1192
·821	·272 771	2274	439 991	439	916 390	1357	·356 381	918	675 614	543	480 136	1189
·822	·275 045	2277	439 552	440	917 747	1358	·357 299	918	676 157	542	478 947	1186
·823	·277 322	2278	439 112	439	919 105	1358	·358 217	920	676 699	542	477 761	1182
·824	·279 600	2281	438 673	438	920 463	1360	·359 137	921	677 241	541	476 579	1179
0·825	2·281 881	2283	438 235	438	921 823	1360	1·360 058	922	677 782	540	475 400	1175
·826	·284 164	2285	437 797	438	923 183	1362	·360 980	924	678 322	540	474 225	1171
·827	·286 449	2288	437 359	437	924 545	1362	·361 904	925	678 862	539	473 054	1168
·828	·288 737	2290	436 922	436	925 907	1364	·362 829	927	679 401	538	471 886	1165
·829	·291 027	2292	436 486	437	927 271	1364	·363 756	928	679 939	537	470 721	1161
0·830	2·293 319	2294	436 049	436	928 635	1365	1·364 684	929	680 476	537	469 560	1158
·831	·295 613	2297	435 613	435	930 000	1366	·365 613	931	681 013	535	468 402	1155
·832	·297 910	2299	435 178	435	931 366	1367	·366 544	932	681 548	536	467 247	1151
·833	·300 209	2301	434 743	434	932 733	1368	·367 476	933	682 084	534	466 096	1148
·834	·302 510	2304	434 309	435	934 101	1369	·368 409	935	682 618	534	464 948	1144
0·835	2·304 814	2306	433 874	433	935 470	1370	1·369 344	936	683 152	533	463 804	1141
·836	·307 120	2308	433 441	433	936 840	1370	·370 280	938	683 685	532	462 663	1138
·837	·309 428	2311	433 008	433	938 210	1372	·371 218	939	684 217	531	461 525	1134
·838	·311 739	2313	432 575	433	939 582	1373	·372 157	940	684 748	531	460 391	1131
·839	·314 052	2315	432 142	431	940 955	1373	·373 097	942	685 279	530	459 260	1128
0·840	2·316 367	2318	431 711	432	942 328	1375	1·374 039	943	685 809	529	458 132	1125
·841	·318 685	2319	431 279	431	943 703	1375	·374 982	944	686 338	529	457 007	1121
·842	·321 004	2323	430 848	431	945 078	1377	·375 926	946	686 867	528	455 886	1118
·843	·323 327	2324	430 417	430	946 455	1377	·376 872	947	687 395	527	454 768	1115
·844	·325 651	2327	429 987	430	947 832	1378	·377 819	949	687 922	526	453 653	1111
0·845	2·327 978	2329	429 557	429	949 210	1379	1·378 768	949	688 448	526	452 542	1108
·846	·330 307	2331	429 128	429	950 589	1381	·379 717	952	688 974	525	451 434	1105
·847	·332 638	2334	428 699	428	951 970	1381	·380 669	952	689 499	524	450 329	1102
·848	·334 972	2336	428 271	428	953 351	1382	·381 621	954	690 023	524	449 227	1099
·849	·337 308	2339	427 843	428	954 733	1383	·382 575	956	690 547	522	448 128	1095
0·850	2·339 647		427 415		956 116		1·383 531		691 069		447 033	

TABLE Vc—EXPONENTIAL AND HYPERBOLIC FUNCTIONS

x	e^x		e^{-x}		sinh x		cosh x		tanh x		coth x	
			0·						0·		1·	
0·850	2·339 647	2341	427 415	427	0·956 116	1384	1·383 531	957	691 069	523	447 033	1093
·851	·341 988	2343	426 988	427	·957 500	1385	·384 488	958	691 592	521	445 940	1089
·852	·344 331	2345	426 561	426	·958 885	1386	·385 446	959	692 113	520	444 851	1086
·853	·346 676	2348	426 135	426	·960 271	1387	·386 405	961	692 633	520	443 765	1083
·854	·349 024	2350	425 709	426	·961 658	1388	·387 366	963	693 153	520	442 682	1080
0·855	2·351 374	2353	425 283	425	0·963 046	1388	1·388 329	964	693 673	518	441 602	1076
·856	·353 727	2355	424 858	425	·964 434	1390	·389 293	965	694 191	518	440 526	1074
·857	·356 082	2357	424 433	424	·965 824	1391	·390 258	966	694 709	517	439 452	1070
·858	·358 439	2360	424 009	424	·967 215	1392	·391 224	968	695 226	516	438 382	1068
·859	·360 799	2362	423 585	423	·968 607	1392	·392 192	969	695 742	516	437 314	1064
0·860	2·363 161	2364	423 162	423	0·969 999	1394	1·393 161	971	696 258	515	436 250	1061
·861	·365 525	2367	422 739	422	·971 393	1395	·394 132	972	696 773	514	435 189	1059
·862	·367 892	2369	422 317	423	·972 788	1395	·395 104	974	697 287	513	434 130	1055
·863	·370 261	2371	421 894	421	·974 183	1397	·396 078	975	697 800	513	433 075	1052
·864	·372 632	2374	421 473	421	·975 580	1397	·397 053	976	698 313	512	432 023	1049
0·865	2·375 006	2376	421 052	421	0·976 977	1399	1·398 029	977	698 825	511	430 974	1046
·866	·377 382	2379	420 631	421	·978 376	1399	·399 006	980	699 336	511	429 928	1044
·867	·379 761	2381	420 210	420	·979 775	1401	·399 986	980	699 847	510	428 884	1040
·868	·382 142	2383	419 790	419	·981 176	1401	·400 966	982	700 357	509	427 844	1037
·869	·384 525	2386	419 371	419	·982 577	1403	·401 948	983	700 866	508	426 807	1034
0·870	2·386 911	2388	418 952	419	0·983 980	1403	1·402 931	985	701 374	508	425 773	1032
·871	·389 299	2390	418 533	419	·985 383	1404	·403 916	986	701 882	507	424 741	1028
·872	·391 689	2393	418 114	417	·986 787	1406	·404 902	987	702 389	506	423 713	1026
·873	·394 082	2396	417 697	418	·988 193	1406	·405 889	989	702 895	506	422 687	1022
·874	·396 478	2397	417 279	417	·989 599	1408	·406 878	991	703 401	505	421 665	1020
0·875	2·398 875	2400	416 862	417	0·991 007	1408	1·407 869	991	703 906	504	420 645	1017
·876	·401 275	2403	416 445	416	·992 415	1409	·408 860	993	704 410	503	419 628	1014
·877	·403 678	2405	416 029	416	·993 824	1411	·409 853	995	704 913	503	418 614	1011
·878	·406 083	2407	415 613	415	·995 235	1411	·410 848	996	705 416	502	417 603	1008
·879	·408 490	2410	415 198	415	·996 646	1412	·411 844	997	705 918	501	416 595	1005
0·880	2·410 900	2412	414 783	415	0·998 058	1414	1·412 841	999	706 419	501	415 590	1003
·881	·413 312	2414	414 368	414	0·999 472	1414	·413 840	1000	706 920	500	414 587	999
·882	·415 726	2417	413 954	414	1·000 886	1415	·414 840	1002	707 420	499	413 588	997
·883	·418 143	2420	413 540	413	·002 301	1417	·415 842	1003	707 919	499	412 591	994
·884	·420 563	2421	413 127	413	·003 718	1417	·416 845	1004	708 418	497	411 597	991
0·885	2·422 984	2425	412 714	412	1·005 135	1418	1·417 849	1006	708 915	497	410 606	989
·886	·425 409	2426	412 302	412	·006 553	1420	·418 855	1007	709 412	497	409 617	985
·887	·427 835	2429	411 890	412	·007 973	1420	·419 862	1009	709 909	495	408 632	983
·888	·430 264	2432	411 478	411	·009 393	1422	·420 871	1010	710 404	495	407 649	980
·889	·432 696	2434	411 067	411	·010 815	1422	·421 881	1012	710 899	495	406 669	978
0·890	2·435 130	2436	410 656	411	1·012 237	1423	1·422 893	1013	711 394	493	405 691	974
·891	·437 566	2439	410 245	410	·013 660	1425	·423 906	1014	711 887	493	404 717	972
·892	·440 005	2441	409 835	409	·015 085	1425	·424 920	1016	712 380	492	403 745	969
·893	·442 446	2444	409 426	410	·016 510	1427	·425 936	1017	712 872	492	402 776	967
·894	·444 890	2446	409 016	408	·017 937	1427	·426 953	1019	713 364	491	401 809	963
0·895	2·447 336	2448	408 608	409	1·019 364	1429	1·427 972	1020	713 855	490	400 846	961
·896	·449 784	2451	408 199	408	·020 793	1429	·428 992	1021	714 345	489	399 885	959
·897	·452 235	2454	407 791	407	·022 222	1431	·430 013	1023	714 834	489	398 926	955
·898	·454 689	2456	407 384	408	·023 653	1431	·431 036	1025	715 323	488	397 971	953
·899	·457 145	2458	406 976	406	·025 084	1433	·432 061	1025	715 811	487	397 018	951
0·900	2·459 603		406 570		1·026 517		1·433 086		716 298		396 067	

TABLE Vc—EXPONENTIAL AND HYPERBOLIC FUNCTIONS

x	e^{x}		e^{-x}		sinh x		cosh x		tanh x		coth x	
			0·						0·		1·	
0·900	2·459 603	2461	406 570	407	1·026 517	1433	1·433 086	1028	716 298	486	396 067	947
·901	·462 064	2463	406 163	406	·027 950	1435	·434 114	1028	716 784	486	395 120	945
·902	·464 527	2466	405 757	405	·029 385	1436	·435 142	1030	717 270	485	394 175	943
·903	·466 993	2468	405 352	405	·030 821	1436	·436 172	1032	717 755	485	393 232	940
·904	·469 461	2471	404 947	405	·032 257	1438	·437 204	1033	718 240	484	392 292	937
0·905	2·471 932	2473	404 542	404	1·033 695	1439	1·438 237	1034	718 724	483	391 355	934
·906	·474 405	2476	404 138	404	·035 134	1440	·439 271	1036	719 207	482	390 421	932
·907	·476 881	2478	403 734	404	·036 574	1440	·440 307	1037	719 689	482	389 489	930
·908	·479 359	2480	403 330	403	·038 014	1442	·441 344	1039	720 171	481	388 559	927
·909	·481 839	2484	402 927	403	·039 456	1443	·442 383	1040	720 652	480	387 632	924
0·910	2·484 323	2485	402 524	402	1·040 899	1444	1·443 423	1042	721 132	480	386 708	921
·911	·486 808	2488	402 122	402	·042 343	1445	·444 465	1043	721 612	479	385 787	920
·912	·489 296	2491	401 720	402	·043 788	1446	·445 508	1045	722 091	478	384 867	916
·913	·491 787	2493	401 318	401	·045 234	1447	·446 553	1046	722 569	478	383 951	914
·914	·494 280	2495	400 917	400	·046 681	1448	·447 599	1047	723 047	476	383 037	912
0·915	2·496 775	2498	400 517	401	1·048 129	1449	1·448 646	1049	723 523	477	382 125	909
·916	·499 273	2501	400 116	400	·049 578	1451	·449 695	1050	724 000	475	381 216	906
·917	·501 774	2503	399 716	399	·051 029	1451	·450 745	1052	724 475	475	380 310	904
·918	·504 277	2505	399 317	399	·052 480	1452	·451 797	1053	724 950	474	379 406	902
·919	·506 782	2508	398 918	399	·053 932	1454	·452 850	1055	725 424	473	378 504	899
0·920	2·509 290	2511	398 519	398	1·055 386	1454	1·453 905	1056	725 897	473	377 605	896
·921	·511 801	2513	398 121	398	·056 840	1456	·454 961	1057	726 370	472	376 709	895
·922	·514 314	2516	397 723	398	·058 296	1456	·456 018	1059	726 842	472	375 814	891
·923	·516 830	2518	397 325	397	·059 752	1458	·457 077	1061	727 314	470	374 923	889
·924	·519 348	2520	396 928	397	·061 210	1458	·458 138	1062	727 784	470	374 034	887
0·925	2·521 868	2523	396 531	396	1·062 668	1460	1·459 200	1063	728 254	470	373 147	884
·926	·524 391	2526	396 135	396	·064 128	1461	·460 263	1065	728 724	468	372 263	882
·927	·526 917	2528	395 739	395	·065 589	1462	·461 328	1066	729 192	468	371 381	880
·928	·529 445	2531	395 344	396	·067 051	1463	·462 394	1068	729 660	467	370 501	877
·929	·531 976	2533	394 948	394	·068 514	1464	·463 462	1069	730 127	467	369 624	875
0·930	2·534 509	2536	394 554	395	1·069 978	1465	1·464 531	1071	730 594	466	368 749	872
·931	·537 045	2538	394 159	394	·071 443	1466	·465 602	1072	731 060	465	367 877	870
·932	·539 583	2541	393 765	393	·072 909	1467	·466 674	1074	731 525	465	367 007	867
·933	·542 124	2544	393 372	393	·074 376	1468	·467 748	1075	731 990	463	366 140	865
·934	·544 668	2545	392 979	393	·075 844	1470	·468 823	1077	732 453	464	365 275	863
0·935	2·547 213	2549	392 586	393	1·077 314	1470	1·469 900	1078	732 917	462	364 412	861
·936	·549 762	2551	392 193	392	·078 784	1472	·470 978	1079	733 379	462	363 551	858
·937	·552 313	2554	391 801	391	·080 256	1472	·472 057	1081	733 841	461	362 693	856
·938	·554 867	2556	391 410	391	·081 728	1474	·473 138	1083	734 302	460	361 837	853
·939	·557 423	2558	391 019	391	·083 202	1475	·474 221	1084	734 762	460	360 984	851
0·940	2·559 981	2562	390 628	391	1·084 677	1476	1·475 305	1085	735 222	459	360 133	849
·941	·562 543	2564	390 237	390	·086 153	1477	·476 390	1087	735 681	459	359 284	846
·942	·565 107	2566	389 847	389	·087 630	1478	·477 477	1088	736 140	458	358 438	845
·943	·567 673	2569	389 458	390	·089 108	1479	·478 565	1090	736 598	457	357 593	841
·944	·570 242	2571	389 068	388	·090 587	1480	·479 655	1091	737 055	456	356 752	840
0·945	2·572 813	2574	388 680	389	1·092 067	1481	1·480 746	1093	737 511	456	355 912	837
·946	·575 387	2577	388 291	388	·093 548	1483	·481 839	1095	737 967	455	355 075	836
·947	·577 964	2579	387 903	388	·095 031	1483	·482 934	1095	738 422	454	354 239	832
·948	·580 543	2582	387 515	387	·096 514	1485	·484 029	1098	738 876	454	353 407	831
·949	·583 125	2585	387 128	387	·097 999	1485	·485 127	1098	739 330	453	352 576	828
0·950	2·585 710		386 741		1·099 484		1·486 225		739 783		351 748	

TABLE Vc—EXPONENTIAL AND HYPERBOLIC FUNCTIONS

x	e^x		e^{-x}		sinh x		cosh x		tanh x		coth x	
			0·						0·		1·	
0·950	2·585 710	2587	386 741	387	1·099 484	1487	1·486 225	1101	739 783	452	351 748	826
·951	·588 297	2589	386 354	386	·100 971	1488	·487 326	1101	740 235	452	350 922	824
·952	·590 886	2592	385 968	385	·102 459	1489	·488 427	1103	740 687	451	350 098	822
·953	·593 478	2595	385 583	386	·103 948	1490	·489 530	1105	741 138	451	349 276	819
·954	·596 073	2598	385 197	385	·105 438	1491	·490 635	1106	741 589	449	348 457	818
0·955	2·598 671	2600	384 812	384	1·106 929	1493	1·491 741	1108	742 038	449	347 639	815
·956	·601 271	2602	384 428	385	·108 422	1493	·492 849	1109	742 487	449	346 824	813
·957	·603 873	2605	384 043	384	·109 915	1494	·493 958	1111	742 936	447	346 011	810
·958	·606 478	2608	383 659	383	·111 409	1496	·495 069	1112	743 383	447	345 201	809
·959	·609 086	2610	383 276	383	·112 905	1497	·496 181	1114	743 830	447	344 392	806
0·960	2·611 696	2613	382 893	383	1·114 402	1498	1·497 295	1115	744 277	446	343 586	804
·961	·614 309	2616	382 510	382	·115 900	1499	·498 410	1116	744 723	445	342 782	802
·962	·616 925	2618	382 128	382	·117 399	1500	·499 526	1119	745 168	444	341 980	800
·963	·619 543	2621	381 746	382	·118 899	1501	·500 645	1119	745 612	444	341 180	798
·964	·622 164	2624	381 364	381	·120 400	1502	·501 764	1121	746 056	443	340 382	795
0·965	2·624 788	2626	380 983	381	1·121 902	1504	1·502 885	1123	746 499	442	339 587	794
·966	·627 414	2628	380 602	380	·123 406	1504	·504 008	1124	746 941	442	338 793	791
·967	·630 042	2632	380 222	380	·124 910	1506	·505 132	1126	747 383	441	338 002	789
·968	·632 674	2634	379 842	380	·126 416	1507	·506 258	1127	747 824	441	337 213	787
·969	·635 308	2636	379 462	379	·127 923	1508	·507 385	1129	748 265	439	336 426	785
0·970	2·637 944	2640	379 083	379	1·129 431	1509	1·508 514	1130	748 704	439	335 641	783
·971	·640 584	2642	378 704	378	·130 940	1510	·509 644	1132	749 143	439	334 858	781
·972	·643 226	2644	378 326	379	·132 450	1511	·510 776	1133	749 582	438	334 077	779
·973	·645 870	2647	377 947	377	·133 961	1513	·511 909	1135	750 020	437	333 298	776
·974	·648 517	2650	377 570	378	·135 474	1513	·513 044	1136	750 457	436	332 522	775
0·975	2·651 167	2653	377 192	377	1·136 987	1515	1·514 180	1138	750 893	436	331 747	772
·976	·653 820	2655	376 815	376	·138 502	1516	·515 318	1139	751 329	435	330 975	771
·977	·656 475	2658	376 439	377	·140 018	1517	·516 457	1141	751 764	435	330 204	768
·978	·659 133	2660	376 062	375	·141 535	1518	·517 598	1142	752 199	434	329 436	767
·979	·661 793	2663	375 687	376	·143 053	1520	·518 740	1144	752 633	433	328 669	764
0·980	2·664 456	2666	375 311	375	1·144 573	1520	1·519 884	1145	753 066	432	327 905	762
·981	·667 122	2668	374 936	375	·146 093	1522	·521 029	1147	753 498	432	327 143	761
·982	·669 790	2672	374 561	374	·147 615	1522	·522 176	1148	753 930	432	326 382	758
·983	·672 462	2673	374 187	374	·149 137	1524	·523 324	1150	754 362	430	325 624	756
·984	·675 135	2677	373 813	374	·150 661	1525	·524 474	1152	754 792	430	324 868	754
0·985	2·677 812	2679	373 439	373	1·152 186	1527	1·525 626	1153	755 222	430	324 114	753
·986	·680 491	2682	373 066	373	·153 713	1527	·526 779	1154	755 652	428	323 361	750
·987	·683 173	2684	372 693	372	·155 240	1528	·527 933	1156	756 080	428	322 611	748
·988	·685 857	2688	372 321	373	·156 768	1530	·529 089	1158	756 508	428	321 863	747
·989	·688 545	2689	371 948	371	·158 298	1531	·530 247	1159	756 936	426	321 116	744
0·990	2·691 234	2693	371 577	372	1·159 829	1532	1·531 406	1160	757 362	426	320 372	742
·991	·693 927	2695	371 205	371	·161 361	1533	·532 566	1162	757 788	426	319 630	741
·992	·696 622	2698	370 834	370	·162 894	1534	·533 728	1164	758 214	425	318 889	738
·993	·699 320	2701	370 464	371	·164 428	1536	·534 892	1165	758 639	424	318 151	737
·994	·702 021	2703	370 093	370	·165 964	1536	·536 057	1167	759 063	423	317 414	735
0·995	2·704 724	2706	369 723	369	1·167 500	1538	1·537 224	1168	759 486	423	316 679	732
·996	·707 430	2709	369 354	369	·169 038	1539	·538 392	1170	759 909	422	315 947	731
·997	·710 139	2712	368 985	369	·170 577	1540	·539 562	1171	760 331	422	315 216	729
·998	·712 851	2714	368 616	368	·172 117	1542	·540 733	1173	760 753	421	314 487	727
·999	·715 565	2717	368 248	369	·173 659	1542	·541 906	1175	761 174	420	313 760	725
1·000	2·718 282		367 879		1·175 201		1·543 081		761 594		313 035	

TABLE Vc—EXPONENTIAL AND HYPERBOLIC FUNCTIONS

x	e^x		e^{-x}		sinh x		cosh x		tanh x		coth x	
			0·						0·		1·	
1·000	2·718 282	2719	367 879	367	1·175 201	1544	1·543 081	1176	761 594	420	313 035	723
·001	·721 001	2723	367 512	368	·176 745	1545	·544 257	1177	762 014	419	312 312	721
·002	·723 724	2725	367 144	367	·178 290	1546	·545 434	1179	762 433	418	311 591	719
·003	·726 449	2728	366 777	366	·179 836	1547	·546 613	1181	762 851	418	310 872	718
·004	·729 177	2730	366 411	366	·181 383	1548	·547 794	1182	763 269	417	310 154	715
1·005	2·731 907	2734	366 045	366	1·182 931	1550	1·548 976	1184	763 686	417	309 439	714
·006	·734 641	2736	365 679	366	·184 481	1551	·550 160	1185	764 103	415	308 725	712
·007	·737 377	2738	365 313	365	·186 032	1552	·551 345	1187	764 518	416	308 013	710
·008	·740 115	2742	364 948	365	·187 584	1553	·552 532	1188	764 934	414	307 303	708
·009	·742 857	2744	364 583	364	·189 137	1554	·553 720	1190	765 348	414	306 595	706
1·010	2·745 601	2747	364 219	364	1·190 691	1556	1·554 910	1191	765 762	413	305 889	705
·011	·748 348	2750	363 855	364	·192 247	1556	·556 101	1193	766 175	413	305 184	702
·012	·751 098	2752	363 491	363	·193 803	1558	·557 294	1195	766 588	412	304 482	701
·013	·753 850	2755	363 128	363	·195 361	1559	·558 489	1196	767 000	411	303 781	699
·014	·756 605	2758	362 765	363	·196 920	1560	·559 685	1198	767 411	411	303 082	697
1·015	2·759 363	2761	362 402	362	1·198 480	1562	1·560 883	1199	767 822	410	302 385	695
·016	·762 124	2764	362 040	362	·200 042	1563	·562 082	1201	768 232	410	301 690	694
·017	·764 888	2766	361 678	361	·201 605	1564	·563 283	1202	768 642	409	300 996	692
·018	·767 654	2769	361 317	361	·203 169	1565	·564 485	1204	769 051	408	300 304	689
·019	·770 423	2772	360 956	361	·204 734	1566	·565 689	1206	769 459	408	299 615	689
1·020	2·773 195	2774	360 595	360	1·206 300	1567	1·566 895	1207	769 867	407	298 926	686
·021	·775 969	2778	360 235	361	·207 867	1569	·568 102	1209	770 274	406	298 240	684
·022	·778 747	2780	359 874	359	·209 436	1570	·569 311	1210	770 680	406	297 556	683
·023	·781 527	2783	359 515	360	·211 006	1571	·570 521	1212	771 086	405	296 873	681
·024	·784 310	2785	359 155	359	·212 577	1572	·571 733	1213	771 491	404	296 192	679
1·025	2·787 095	2789	358 796	358	1·214 149	1574	1·572 946	1215	771 895	404	295 513	678
·026	·789 884	2791	358 438	358	·215 723	1575	·574 161	1216	772 299	403	294 835	676
·027	·792 675	2794	358 080	358	·217 298	1576	·575 377	1218	772 702	403	294 159	674
·028	·795 469	2797	357 722	358	·218 874	1577	·576 595	1220	773 105	402	293 485	672
·029	·798 266	2800	357 364	357	·220 451	1578	·577 815	1221	773 507	401	292 813	670
1·030	2·801 066	2802	357 007	357	1·222 029	1580	1·579 036	1223	773 908	401	292 143	669
·031	·803 868	2806	356 650	356	·223 609	1581	·580 259	1225	774 309	400	291 474	667
·032	·806 674	2808	356 294	356	·225 190	1582	·581 484	1226	774 709	400	290 807	665
·033	·809 482	2811	355 938	356	·226 772	1583	·582 710	1227	775 109	399	290 142	664
·034	·812 293	2813	355 582	356	·228 355	1585	·583 937	1229	775 508	398	289 478	662
1·035	2·815 106	2817	355 226	355	1·229 940	1586	1·585 166	1231	775 906	398	288 816	660
·036	·817 923	2819	354 871	354	·231 526	1587	·586 397	1232	776 304	397	288 156	659
·037	·820 742	2822	354 517	355	·233 113	1588	·587 629	1234	776 701	396	287 497	656
·038	·823 564	2825	354 162	354	·234 701	1589	·588 863	1236	777 097	396	286 841	656
·039	·826 389	2828	353 808	353	·236 290	1591	·590 099	1237	777 493	395	286 185	653
1·040	2·829 217	2831	353 455	354	1·237 881	1592	1·591 336	1239	777 888	395	285 532	652
·041	·832 048	2833	353 101	353	·239 473	1593	·592 575	1240	778 283	394	284 880	650
·042	·834 881	2836	352 748	352	·241 066	1595	·593 815	1242	778 677	393	284 230	648
·043	·837 717	2840	352 396	352	·242 661	1595	·595 057	1243	779 070	393	283 582	647
·044	·840 557	2842	352 044	352	·244 256	1597	·596 300	1245	779 463	392	282 935	645
1·045	2·843 399	2844	351 692	352	1·245 853	1599	1·597 545	1247	779 855	391	282 290	644
·046	·846 243	2848	351 340	351	·247 452	1599	·598 792	1248	780 246	391	281 646	641
·047	·849 091	2851	350 989	351	·249 051	1601	·600 040	1250	780 637	391	281 005	640
·048	·851 942	2853	350 638	350	·250 652	1602	·601 290	1251	781 028	389	280 365	639
·049	·854 795	2856	350 288	350	·252 254	1603	·602 541	1253	781 417	389	279 726	637
1·050	2·857 651		349 938		1·253 857		1·603 794		781 806		279 089	

x	e^x		e^{-x}		sinh x		cosh x		tanh x		coth x	
			0·						0·		1·	
1·050	2·857 651	2859	349 938	350	1·253 857	1604	1·603 794	1255	781 806	389	279 089	635
·051	·860 510	2862	349 588	349	·255 461	1606	·605 049	1256	782 195	388	278 454	634
·052	·863 372	2865	349 239	349	·257 067	1607	·606 305	1258	782 583	387	277 820	632
·053	·866 237	2868	348 890	349	·258 674	1608	·607 563	1260	782 970	387	277 188	630
·054	·869 105	2870	348 541	349	·260 282	1609	·608 823	1261	783 357	386	276 558	629
1·055	2·871 975	2874	348 192	348	1·261 891	1611	1·610 084	1262	783 743	385	275 929	627
·056	·874 849	2876	347 844	347	·263 502	1612	·611 346	1265	784 128	385	275 302	626
·057	·877 725	2879	347 497	348	·265 114	1613	·612 611	1266	784 513	384	274 676	624
·058	·880 604	2882	347 149	347	·266 727	1615	·613 877	1267	784 897	384	274 052	622
·059	·883 486	2885	346 802	346	·268 342	1616	·615 144	1269	785 281	383	273 430	621
1·060	2·886 371	2888	346 456	346	1·269 958	1617	1·616 413	1271	785 664	382	272 809	619
·061	·889 259	2891	346 110	346	·271 575	1618	·617 684	1273	786 046	382	272 190	618
·062	·892 150	2893	345 764	346	·273 193	1620	·618 957	1274	786 428	381	271 572	616
·063	·895 043	2897	345 418	345	·274 813	1620	·620 231	1275	786 809	381	270 956	615
·064	·897 940	2899	345 073	345	·276 433	1623	·621 506	1277	787 190	380	270 341	613
1·065	2·900 839	2902	344 728	345	1·278 056	1623	1·622 783	1279	787 570	379	269 728	611
·066	·903 741	2905	344 383	344	·279 679	1625	·624 062	1281	787 949	379	269 117	610
·067	·906 646	2909	344 039	344	·281 304	1626	·625 343	1282	788 328	379	268 507	608
·068	·909 555	2911	343 695	343	·282 930	1627	·626 625	1284	788 707	377	267 899	607
·069	·912 466	2913	343 352	343	·284 557	1628	·627 909	1285	789 084	377	267 292	605
1·070	2·915 379	2917	343 009	343	1·286 185	1630	1·629 194	1287	789 461	377	266 687	604
·071	·918 296	2920	342 666	343	·287 815	1631	·630 481	1289	789 838	376	266 083	602
·072	·921 216	2923	342 323	342	·289 446	1633	·631 770	1290	790 214	375	265 481	601
·073	·924 139	2925	341 981	342	·291 079	1634	·633 060	1292	790 589	374	264 880	599
·074	·927 064	2929	341 639	341	·292 713	1635	·634 352	1293	790 963	375	264 281	598
1·075	2·929 993	2931	341 298	341	1·294 348	1636	1·635 645	1295	791 338	373	263 683	596
·076	·932 924	2935	340 957	341	·295 984	1637	·636 940	1297	791 711	373	263 087	595
·077	·935 859	2937	340 616	341	·297 621	1639	·638 237	1299	792 084	372	262 492	593
·078	·938 796	2940	340 275	340	·299 260	1641	·639 536	1300	792 456	372	261 899	591
·079	·941 736	2944	339 935	339	·300 901	1641	·640 836	1302	792 828	371	261 308	590
1·080	2·944 680	2946	339 596	340	1·302 542	1643	1·642 138	1303	793 199	371	260 718	589
·081	·947 626	2949	339 256	339	·304 185	1644	·643 441	1305	793 570	370	260 129	587
·082	·950 575	2952	338 917	339	·305 829	1645	·644 746	1307	793 940	369	259 542	586
·083	·953 527	2955	338 578	338	·307 474	1647	·646 053	1308	794 309	369	258 956	584
·084	·956 482	2958	338 240	338	·309 121	1648	·647 361	1310	794 678	368	258 372	583
1·085	2·959 440	2961	337 902	338	1·310 769	1649	1·648 671	1311	795 046	368	257 789	581
·086	·962 401	2964	337 564	337	·312 418	1651	·649 982	1314	795 414	367	257 208	580
·087	·965 365	2966	337 227	337	·314 069	1652	·651 296	1315	795 781	366	256 628	579
·088	·968 331	2970	336 890	337	·315 721	1653	·652 611	1316	796 147	366	256 049	577
·089	·971 301	2973	336 553	337	·317 374	1655	·653 927	1318	796 513	365	255 472	575
1·090	2·974 274	2976	336 216	336	1·319 029	1656	1·655 245	1320	796 878	365	254 897	574
·091	·977 250	2979	335 880	335	·320 685	1657	·656 565	1322	797 243	364	254 323	573
·092	·980 229	2981	335 545	336	·322 342	1658	·657 887	1323	797 607	363	253 750	571
·093	·983 210	2985	335 209	335	·324 000	1660	·659 210	1325	797 970	363	253 179	570
·094	·986 195	2988	334 874	334	·325 660	1662	·660 535	1326	798 333	363	252 609	568
1·095	2·989 183	2990	334 540	335	1·327 322	1662	1·661 861	1328	798 696	362	252 041	567
·096	·992 173	2994	334 205	334	·328 984	1664	·663 189	1330	799 058	361	251 474	565
·097	·995 167	2997	333 871	334	·330 648	1665	·664 519	1332	799 419	360	250 909	564
·098	2·998 164	2999	333 537	333	·332 313	1667	·665 851	1333	799 779	361	250 345	563
·099	3·001 163	3003	333 204	333	·333 980	1667	·667 184	1335	800 140	359	249 782	561
1·100	3·004 166		332 871		1·335 647		1·668 519		800 499		249 221	

TABLE Vc—EXPONENTIAL AND HYPERBOLIC FUNCTIONS

x	e^x		e^{-x}		sinh x		cosh x		tanh x		coth x	
			0·						0·		1·	
1·100	3·004 166	3006	332 871	333	1·335 647	1670	1·668 519	1336	800 499	359	249 221	560
·101	·007 172	3008	332 538	332	·337 317	1670	·669 855	1338	800 858	358	248 661	559
·102	·010 180	3012	332 206	332	·338 987	1672	·671 193	1340	801 216	358	248 102	557
·103	·013 192	3015	331 874	332	·340 659	1673	·672 533	1342	801 574	357	247 545	555
·104	·016 207	3017	331 542	331	·342 332	1675	·673 875	1343	801 931	357	246 990	555
1·105	3·019 224	3021	331 211	331	1·344 007	1676	1·675 218	1345	802 288	356	246 435	552
·106	·022 245	3024	330 880	331	·345 683	1677	·676 563	1346	802 644	355	245 883	552
·107	·025 269	3027	330 549	330	·347 360	1679	·677 909	1348	802 999	355	245 331	550
·108	·028 296	3030	330 219	330	·349 039	1679	·679 257	1350	803 354	355	244 781	549
·109	·031 326	3032	329 889	330	·350 718	1682	·680 607	1352	803 709	353	244 232	547
1·110	3·034 358	3036	329 559	329	1·352 400	1682	1·681 959	1353	804 062	354	243 685	546
·111	·037 394	3039	329 230	329	·354 082	1684	·683 312	1355	804 416	352	243 139	545
·112	·040 433	3042	328 901	329	·355 766	1686	·684 667	1356	804 768	352	242 594	544
·113	·043 475	3045	328 572	329	·357 452	1686	·686 023	1359	805 120	352	242 050	542
·114	·046 520	3048	328 243	328	·359 138	1688	·687 382	1360	805 472	351	241 508	540
1·115	3·049 568	3051	327 915	327	1·360 826	1690	1·688 742	1361	805 823	350	240 968	540
·116	·052 619	3054	327 588	328	·362 516	1691	·690 103	1364	806 173	350	240 428	538
·117	·055 673	3058	327 260	327	·364 207	1692	·691 467	1365	806 523	349	239 890	536
·118	·058 731	3060	326 933	327	·365 899	1693	·692 832	1367	806 872	349	239 354	536
·119	·061 791	3063	326 606	326	·367 592	1695	·694 199	1368	807 221	348	238 818	534
1·120	3·064 854	3067	326 280	326	1·369 287	1696	1·695 567	1370	807 569	347	238 284	532
·121	·067 921	3069	325 954	326	·370 983	1698	·696 937	1372	807 916	347	237 752	532
·122	·070 990	3073	325 628	326	·372 681	1699	·698 309	1373	808 263	347	237 220	530
·123	·074 063	3075	325 302	325	·374 380	1700	·699 682	1376	808 610	346	236 690	528
·124	·077 138	3079	324 977	325	·376 080	1702	·701 058	1377	808 956	345	236 162	528
1·125	3·080 217	3082	324 652	324	1·377 782	1703	1·702 435	1378	809 301	345	235 634	526
·126	·083 299	3084	324 328	324	·379 485	1705	·703 813	1381	809 646	344	235 108	525
·127	·086 383	3088	324 004	324	·381 190	1706	·705 194	1382	809 990	344	234 583	523
·128	·089 471	3091	323 680	324	·382 896	1707	·706 576	1383	810 334	343	234 060	523
·129	·092 562	3095	323 356	323	·384 603	1709	·707 959	1386	810 677	342	233 537	521
1·130	3·095 657	3097	323 033	323	1·386 312	1710	1·709 345	1387	811 019	342	233 016	519
·131	·098 754	3100	322 710	322	·388 022	1711	·710 732	1389	811 361	342	232 497	519
·132	·101 854	3103	322 388	322	·389 733	1713	·712 121	1391	811 703	341	231 978	517
·133	·104 957	3107	322 066	322	·391 446	1714	·713 512	1392	812 044	340	231 461	516
·134	·108 064	3110	321 744	322	·393 160	1716	·714 904	1394	812 384	340	230 945	514
1·135	3·111 174	3112	321 422	321	1·394 876	1717	1·716 298	1396	812 724	339	230 431	514
·136	·114 286	3116	321 101	321	·396 593	1718	·717 694	1397	813 063	338	229 917	512
·137	·117 402	3119	320 780	321	·398 311	1720	·719 091	1399	813 401	339	229 405	511
·138	·120 521	3122	320 459	320	·400 031	1721	·720 490	1401	813 740	337	228 894	509
·139	·123 643	3125	320 139	320	·401 752	1723	·721 891	1403	814 077	337	228 385	508
1·140	3·126 768	3129	319 819	320	1·403 475	1724	1·723 294	1404	814 414	337	227 877	507
·141	·129 897	3131	319 499	319	·405 199	1725	·724 698	1406	814 751	335	227 370	506
·142	·133 028	3135	319 180	319	·406 924	1727	·726 104	1408	815 086	336	226 864	505
·143	·136 163	3137	318 861	319	·408 651	1728	·727 512	1409	815 422	335	226 359	503
·144	·139 300	3141	318 542	318	·410 379	1730	·728 921	1412	815 757	334	225 856	502
1·145	3·142 441	3144	318 224	318	1·412 109	1731	1·730 333	1413	816 091	334	225 354	501
·146	·145 585	3148	317 906	318	·413 840	1732	·731 746	1414	816 425	333	224 853	500
·147	·148 733	3150	317 588	317	·415 572	1734	·733 160	1417	816 758	332	224 353	498
·148	·151 883	3153	317 271	317	·417 306	1735	·734 577	1418	817 090	333	223 855	497
·149	·155 036	3157	316 954	317	·419 041	1737	·735 995	1420	817 423	331	223 358	496
1·150	3·158 193		316 637		1·420 778		1·737 415		817 754		222 862	

x	e^x		e^{-x}		sinh x		cosh x		tanh x		coth x	
			0·						0·		1·	
1·150	3·158 193	3160	316 637	317	1·420 778	1738	1·737 415	1421	817 754	331	222 862	495
·151	·161 353	3163	316 320	316	·422 516	1740	·738 836	1424	818 085	331	222 367	494
·152	·164 516	3166	316 004	316	·424 256	1741	·740 260	1425	818 416	329	221 873	492
·153	·167 682	3169	315 688	315	·425 997	1742	·741 685	1427	818 745	330	221 381	491
·154	·170 851	3172	315 373	315	·427 739	1744	·743 112	1428	819 075	329	220 890	490
1·155	3·174 023	3176	315 058	315	1·429 483	1745	1·744 540	1431	819 404	328	220 400	489
·156	·177 199	3179	314 743	315	·431 228	1747	·745 971	1432	819 732	328	219 911	488
·157	·180 378	3182	314 428	314	·432 975	1748	·747 403	1434	820 060	327	219 423	486
·158	·183 560	3185	314 114	314	·434 723	1750	·748 837	1435	820 387	327	218 937	485
·159	·186 745	3188	313 800	314	·436 473	1751	·750 272	1438	820 714	326	218 452	484
1·160	3·189 933	3192	313 486	313	1·438 224	1752	1·751 710	1439	821 040	326	217 968	483
·161	·193 125	3195	313 173	313	·439 976	1754	·753 149	1441	821 366	325	217 485	482
·162	·196 320	3197	312 860	313	·441 730	1755	·754 590	1442	821 691	324	217 003	480
·163	·199 517	3202	312 547	312	·443 485	1757	·756 032	1445	822 015	324	216 523	480
·164	·202 719	3204	312 235	312	·445 242	1758	·757 477	1446	822 339	324	216 043	478
1·165	3·205 923	3207	311 923	312	1·447 000	1760	1·758 923	1448	822 663	323	215 565	477
·166	·209 130	3211	311 611	312	·448 760	1761	·760 371	1449	822 986	322	215 088	476
·167	·212 341	3214	311 299	311	·450 521	1762	·761 820	1452	823 308	322	214 612	475
·168	·215 555	3217	310 988	311	·452 283	1764	·763 272	1453	823 630	321	214 137	473
·169	·218 772	3221	310 677	310	·454 047	1766	·764 725	1455	823 951	321	213 664	472
1·170	3·221 993	3223	310 367	310	1·455 813	1767	1·766 180	1456	824 272	320	213 192	472
·171	·225 216	3227	310 057	310	·457 580	1768	·767 636	1459	824 592	320	212 720	470
·172	·228 443	3230	309 747	310	·459 348	1770	·769 095	1460	824 912	320	212 250	469
·173	·231 673	3233	309 437	309	·461 118	1771	·770 555	1462	825 232	318	211 781	468
·174	·234 906	3237	309 128	309	·462 889	1773	·772 017	1464	825 550	318	211 313	466
1·175	3·238 143	3240	308 819	309	1·464 662	1774	1·773 481	1466	825 868	318	210 847	466
·176	·241 383	3243	308 510	308	·466 436	1776	·774 947	1467	826 186	317	210 381	464
·177	·244 626	3246	308 202	308	·468 212	1777	·776 414	1469	826 503	317	209 917	464
·178	·247 872	3249	307 894	308	·469 989	1779	·777 883	1471	826 820	316	209 453	462
·179	·251 121	3253	307 586	307	·471 768	1780	·779 354	1472	827 136	316	208 991	461
1·180	3·254 374	3256	307 279	307	1·473 548	1781	1·780 826	1475	827 452	315	208 530	460
·181	·257 630	3259	306 972	307	·475 329	1783	·782 301	1476	827 767	314	208 070	459
·182	·260 889	3263	306 665	307	·477 112	1785	·783 777	1478	828 081	314	207 611	458
·183	·264 152	3266	306 358	306	·478 897	1786	·785 255	1480	828 395	314	207 153	456
·184	·267 418	3269	306 052	306	·480 683	1787	·786 735	1482	828 709	313	206 697	456
1·185	3·270 687	3272	305 746	305	1·482 470	1789	1·788 217	1483	829 022	312	206 241	454
·186	·273 959	3276	305 441	306	·484 259	1791	·789 700	1485	829 334	312	205 787	454
·187	·277 235	3279	305 135	305	·486 050	1792	·791 185	1487	829 646	312	205 333	452
·188	·280 514	3282	304 830	304	·487 842	1793	·792 672	1489	829 958	310	204 881	451
·189	·283 796	3285	304 526	305	·489 635	1795	·794 161	1490	830 268	311	204 430	450
1·190	3·287 081	3289	304 221	304	1·491 430	1796	1·795 651	1493	830 579	310	203 980	449
·191	·290 370	3292	303 917	304	·493 226	1798	·797 144	1494	830 889	309	203 531	448
·192	·293 662	3295	303 613	303	·495 024	1800	·798 638	1496	831 198	309	203 083	447
·193	·296 957	3299	303 310	303	·496 824	1801	·800 134	1497	831 507	308	202 636	446
·194	·300 256	3302	303 007	303	·498 625	1802	·801 631	1500	831 815	308	202 190	445
1·195	3·303 558	3305	302 704	303	1·500 427	1804	1·803 131	1501	832 123	307	201 745	443
·196	·306 863	3308	302 401	302	·502 231	1805	·804 632	1503	832 430	307	201 302	443
·197	·310 171	3312	302 099	302	·504 036	1807	·806 135	1505	832 737	307	200 859	442
·198	·313 483	3315	301 797	301	·505 843	1808	·807 640	1507	833 044	305	200 417	440
·199	·316 798	3319	301 496	302	·507 651	1810	·809 147	1509	833 349	306	199 977	439
1·200	3·320 117		301 194		1·509 461		1·810 656		833 655		199 538	

x	e^x		e^{-x}		sinh x		cosh x		tanh x		coth x	
			0·						0·		1·	
1·200	3·320 117	3322	301 194	301	1·509 461	1812	1·810 656	1510	833 655	304	199 538	439
·201	·323 439	3325	300 893	301	·511 273	1813	·812 166	1512	833 959	305	199 099	437
·202	·326 764	3328	300 592	300	·513 086	1814	·813 678	1514	834 264	303	198 662	436
·203	·330 092	3332	300 292	300	·514 900	1816	·815 192	1516	834 567	304	198 226	436
·204	·333 424	3335	299 992	300	·516 716	1818	·816 708	1518	834 871	302	197 790	434
1·205	3·336 759	3339	299 692	300	1·518 534	1819	1·818 226	1519	835 173	303	197 356	433
·206	·340 098	3341	299 392	299	·520 353	1820	·819 745	1521	835 476	301	196 923	432
·207	·343 439	3345	299 093	299	·522 173	1822	·821 266	1523	835 777	302	196 491	431
·208	·346 784	3349	298 794	298	·523 995	1824	·822 789	1525	836 079	300	196 060	430
·209	·350 133	3352	298 496	299	·525 819	1825	·824 314	1527	836 379	300	195 630	429
1·210	3·353 485	3355	298 197	298	1·527 644	1826	1·825 841	1529	836 679	300	195 201	428
·211	·356 840	3358	297 899	298	·529 470	1828	·827 370	1530	836 979	299	194 773	427
·212	·360 198	3362	297 601	297	·531 298	1830	·828 900	1532	837 278	299	194 346	426
·213	·363 560	3365	297 304	297	·533 128	1831	·830 432	1534	837 577	298	193 920	425
·214	·366 925	3369	297 007	297	·534 959	1833	·831 966	1536	837 875	298	193 495	424
1·215	3·370 294	3372	296 710	297	1·536 792	1834	1·833 502	1538	838 173	297	193 071	423
·216	·373 666	3375	296 413	296	·538 626	1836	·835 040	1539	838 470	297	192 648	422
·217	·377 041	3379	296 117	296	·540 462	1837	·836 579	1542	838 767	296	192 226	421
·218	·380 420	3382	295 821	295	·542 299	1839	·838 121	1543	839 063	296	191 805	420
·219	·383 802	3386	295 526	296	·544 138	1841	·839 664	1545	839 359	295	191 385	418
1·220	3·387 188	3389	295 230	295	1·545 979	1842	1·841 209	1547	839 654	295	190 967	418
·221	·390 577	3392	294 935	295	·547 821	1843	·842 756	1549	839 949	294	190 549	417
·222	·393 969	3396	294 640	294	·549 664	1845	·844 305	1550	840 243	294	190 132	416
·223	·397 365	3399	294 346	294	·551 509	1847	·845 855	1553	840 537	293	189 716	415
·224	·400 764	3402	294 052	294	·553 356	1848	·847 408	1554	840 830	293	189 301	414
1·225	3·404 166	3406	293 758	294	1·555 204	1850	1·848 962	1556	841 123	292	188 887	413
·226	·407 572	3409	293 464	293	·557 054	1851	·850 518	1558	841 415	292	188 474	412
·227	·410 981	3413	293 171	293	·558 905	1853	·852 076	1560	841 707	291	188 062	411
·228	·414 394	3416	292 878	293	·560 758	1855	·853 636	1562	841 998	291	187 651	410
·229	·417 810	3420	292 585	292	·562 613	1855	·855 198	1563	842 289	290	187 241	409
1·230	3·421 230	3422	292 293	293	1·564 468	1858	1·856 761	1565	842 579	290	186 832	408
·231	·424 652	3427	292 000	291	·566 326	1859	·858 326	1568	842 869	289	186 424	407
·232	·428 079	3430	291 709	292	·568 185	1861	·859 894	1569	843 158	289	186 017	406
·233	·431 509	3433	291 417	291	·570 046	1862	·861 463	1571	843 447	289	185 611	406
·234	·434 942	3437	291 126	291	·571 908	1864	·863 034	1573	843 736	288	185 205	404
1·235	3·438 379	3440	290 835	291	1·573 772	1865	1·864 607	1574	844 024	287	184 801	403
·236	·441 819	3443	290 544	290	·575 637	1867	·866 181	1577	844 311	287	184 398	403
·237	·445 262	3447	290 254	290	·577 504	1869	·867 758	1578	844 598	286	183 995	401
·238	·448 709	3451	289 964	290	·579 373	1870	·869 336	1581	844 884	286	183 594	400
·239	·452 160	3453	289 674	290	·581 243	1872	·870 917	1582	845 170	286	183 194	400
1·240	3·455 613	3458	289 384	289	1·583 115	1873	1·872 499	1584	845 456	285	182 794	398
·241	·459 071	3461	289 095	289	·584 988	1875	·874 083	1586	845 741	284	182 396	398
·242	·462 532	3464	288 806	289	·586 863	1876	·875 669	1588	846 025	284	181 998	397
·243	·465 996	3468	288 517	288	·588 739	1878	·877 257	1589	846 309	284	181 601	395
·244	·469 464	3471	288 229	288	·590 617	1880	·878 846	1592	846 593	283	181 206	395
1·245	3·472 935	3474	287 941	288	1·592 497	1881	1·880 438	1593	846 876	282	180 811	394
·246	·476 409	3479	287 653	287	·594 378	1883	·882 031	1596	847 158	282	180 417	393
·247	·479 888	3481	287 366	288	·596 261	1884	·883 627	1597	847 440	282	180 024	392
·248	·483 369	3485	287 078	287	·598 145	1886	·885 224	1599	847 722	281	179 632	391
·249	·486 854	3489	286 791	286	·600 031	1888	·886 823	1601	848 003	281	179 241	390
1·250	3·490 343		286 505		1·601 919		1·888 424		848 284		178 851	

TABLE Vc—EXPONENTIAL AND HYPERBOLIC FUNCTIONS

x	e^x		e^{-x}		sinh x		cosh x		tanh x		coth x	
			0·						0·		1·	
1·250	3·490 343	3492	286 505	287	1·601 919	1889	1·888 424	1603	848 284	280	178 851	389
·251	·493 835	3496	286 218	286	·603 808	1891	·890 027	1604	848 564	280	178 462	389
·252	·497 331	3499	285 932	285	·605 699	1893	·891 631	1607	848 844	279	178 073	387
·253	·500 830	3502	285 647	286	·607 592	1894	·893 238	1609	849 123	279	177 686	386
·254	·504 332	3506	285 361	285	·609 486	1895	·894 847	1610	849 402	278	177 300	386
1·255	3·507 838	3510	285 076	285	1·611 381	1898	1·896 457	1612	849 680	278	176 914	385
·256	·511 348	3513	284 791	285	·613 279	1898	·898 069	1615	849 958	277	176 529	383
·257	·514 861	3517	284 506	284	·615 177	1901	·899 684	1616	850 235	277	176 146	383
·258	·518 378	3520	284 222	284	·617 078	1902	·901 300	1618	850 512	276	175 763	382
·259	·521 898	3523	283 938	284	·618 980	1904	·902 918	1620	850 788	276	175 381	381
1·260	3·525 421	3528	283 654	283	1·620 884	1905	1·904 538	1622	851 064	276	175 000	381
·261	·528 949	3530	283 371	284	·622 789	1907	·906 160	1623	851 340	275	174 619	379
·262	·532 479	3535	283 087	283	·624 696	1909	·907 783	1626	851 615	274	174 240	378
·263	·536 014	3537	282 804	282	·626 605	1910	·909 409	1628	851 889	274	173 862	378
·264	·539 551	3542	282 522	283	·628 515	1912	·911 037	1629	852 163	274	173 484	376
1·265	3·543 093	3545	282 239	282	1·630 427	1913	1·912 666	1631	852 437	273	173 108	376
·266	·546 638	3548	281 957	282	·632 340	1915	·914 297	1634	852 710	272	172 732	375
·267	·550 186	3552	281 675	281	·634 255	1917	·915 931	1635	852 982	273	172 357	374
·268	·553 738	3555	281 394	281	·636 172	1918	·917 566	1637	853 255	271	171 983	373
·269	·557 293	3560	281 113	281	·638 090	1920	·919 203	1639	853 526	272	171 610	372
1·270	3·560 853	3562	280 832	281	1·640 010	1922	1·920 842	1641	853 798	270	171 238	372
·271	·564 415	3566	280 551	280	·641 932	1923	·922 483	1643	854 068	271	170 866	370
·272	·567 981	3570	280 271	281	·643 855	1925	·924 126	1645	854 339	270	170 496	370
·273	·571 551	3573	279 990	279	·645 780	1927	·925 771	1647	854 609	269	170 126	369
·274	·575 124	3577	279 711	280	·647 707	1928	·927 418	1648	854 878	269	169 757	367
1·275	3·578 701	3581	279 431	279	1·649 635	1930	1·929 066	1651	855 147	269	169 390	367
·276	·582 282	3584	279 152	279	·651 565	1932	·930 717	1652	855 416	268	169 023	367
·277	·585 866	3588	278 873	279	·653 497	1933	·932 369	1655	855 684	267	168 656	365
·278	·589 454	3591	278 594	279	·655 430	1935	·934 024	1656	855 951	267	168 291	364
·279	·593 045	3595	278 315	278	·657 365	1936	·935 680	1659	856 218	267	167 927	364
1·280	3·596 640	3598	278 037	278	1·659 301	1938	1·937 339	1660	856 485	266	167 563	363
·281	·600 238	3602	277 759	277	·661 239	1940	·938 999	1662	856 751	266	167 200	362
·282	·603 840	3606	277 482	278	·663 179	1942	·940 661	1664	857 017	265	166 838	361
·283	·607 446	3609	277 204	277	·665 121	1943	·942 325	1666	857 282	265	166 477	360
·284	·611 055	3613	276 927	276	·667 064	1945	·943 991	1668	857 547	264	166 117	360
1·285	3·614 668	3616	276 651	277	1·669 009	1946	1·945 659	1670	857 811	264	165 757	358
·286	·618 284	3621	276 374	276	·670 955	1948	·947 329	1672	858 075	264	165 399	358
·287	·621 905	3623	276 098	276	·672 903	1950	·949 001	1674	858 339	263	165 041	357
·288	·625 528	3628	275 822	276	·674 853	1952	·950 675	1676	858 602	262	164 684	356
·289	·629 156	3631	275 546	275	·676 805	1953	·952 351	1678	858 864	263	164 328	355
1·290	3·632 787	3634	275 271	275	1·678 758	1955	1·954 029	1679	859 127	261	163 973	355
·291	·636 421	3638	274 996	275	·680 713	1956	·955 708	1682	859 388	261	163 618	353
·292	·640 059	3642	274 721	275	·682 669	1959	·957 390	1684	859 649	261	163 265	353
·293	·643 701	3646	274 446	274	·684 628	1959	·959 074	1685	859 910	261	162 912	352
·294	·647 347	3649	274 172	274	·686 587	1962	·960 759	1688	860 171	259	162 560	351
1·295	3·650 996	3653	273 898	274	1·688 549	1963	1·962 447	1689	860 430	260	162 209	350
·296	·654 649	3656	273 624	273	·690 512	1965	·964 136	1692	860 690	259	161 859	350
·297	·658 305	3660	273 351	274	·692 477	1967	·965 828	1693	860 949	258	161 509	349
·298	·661 965	3664	273 077	273	·694 444	1968	·967 521	1696	861 207	259	161 160	347
·299	·665 629	3668	272 804	272	·696 412	1970	·969 217	1697	861 466	257	160 813	347
1·300	3·669 297		272 532		1·698 382		1·970 914		861 723		160 466	

TABLE Vc—EXPONENTIAL AND HYPERBOLIC FUNCTIONS

x	e^x		e^{-x}		sinh x		cosh x		tanh x		coth x	
			0·						0·		1·	
1·300	3·669 297	3671	272 532	273	1·698 382	1972	1·970 914	1700	861 723	257	160 466	347
·301	·672 968	3675	272 259	272	·700 354	1974	·972 614	1701	861 980	257	160 119	345
·302	·676 643	3678	271 987	272	·702 328	1975	·974 315	1703	862 237	256	159 774	345
·303	·680 321	3682	271 715	271	·704 303	1977	·976 018	1706	862 493	256	159 429	344
·304	·684 003	3686	271 444	271	·706 280	1978	·977 724	1707	862 749	256	159 085	343
1·305	3·687 689	3690	271 173	272	1·708 258	1981	1·979 431	1709	863 005	255	158 742	342
·306	·691 379	3693	270 901	270	·710 239	1982	·981 140	1711	863 260	254	158 400	342
·307	·695 072	3697	270 631	271	·712 221	1983	·982 851	1714	863 514	254	158 058	340
·308	·698 769	3700	270 360	270	·714 204	1986	·984 565	1715	863 768	254	157 718	340
·309	·702 469	3705	270 090	270	·716 190	1987	·986 280	1717	864 022	253	157 378	339
1·310	3·706 174	3708	269 820	270	1·718 177	1989	1·987 997	1719	864 275	253	157 039	339
·311	·709 882	3711	269 550	269	·720 166	1990	·989 716	1721	864 528	253	156 700	337
·312	·713 593	3716	269 281	269	·722 156	1993	·991 437	1723	864 781	252	156 363	337
·313	·717 309	3719	269 012	269	·724 149	1994	·993 160	1726	865 033	251	156 026	336
·314	·721 028	3723	268 743	269	·726 143	1995	·994 886	1727	865 284	251	155 690	335
1·315	3·724 751	3727	268 474	268	1·728 138	1998	1·996 613	1729	865 535	251	155 355	335
·316	·728 478	3730	268 206	268	·730 136	1999	1·998 342	1731	865 786	250	155 020	333
·317	·732 208	3734	267 938	268	·732 135	2001	2·000 073	1733	866 036	250	154 687	333
·318	·735 942	3738	267 670	267	·734 136	2003	·001 806	1735	866 286	249	154 354	333
·319	·739 680	3741	267 403	268	·736 139	2004	·003 541	1737	866 535	249	154 021	331
1·320	3·743 421	3746	267 135	267	1·738 143	2006	2·005 278	1739	866 784	248	153 690	331
·321	·747 167	3749	266 868	266	·740 149	2008	·007 017	1742	867 032	248	153 359	329
·322	·750 916	3753	266 602	267	·742 157	2010	·008 759	1743	867 280	248	153 030	330
·323	·754 669	3756	266 335	266	·744 167	2011	·010 502	1745	867 528	247	152 700	328
·324	·758 425	3760	266 069	266	·746 178	2013	·012 247	1747	867 775	247	152 372	327
1·325	3·762 185	3764	265 803	266	1·748 191	2015	2·013 994	1749	868 022	246	152 045	327
·326	·765 949	3768	265 537	265	·750 206	2017	·015 743	1752	868 268	246	151 718	326
·327	·769 717	3772	265 272	265	·752 223	2018	·017 495	1753	868 514	246	151 392	326
·328	·773 489	3775	265 007	265	·754 241	2020	·019 248	1755	868 760	245	151 066	324
·329	·777 264	3779	264 742	265	·756 261	2022	·021 003	1757	869 005	244	150 742	324
1·330	3·781 043	3783	264 477	264	1·758 283	2024	2·022 760	1760	869 249	245	150 418	323
·331	·784 826	3787	264 213	264	·760 307	2025	·024 520	1761	869 494	243	150 095	323
·332	·788 613	3791	263 949	264	·762 332	2027	·026 281	1763	869 737	244	149 772	321
·333	·792 404	3794	263 685	264	·764 359	2029	·028 044	1766	869 981	243	149 451	321
·334	·796 198	3798	263 421	263	·766 388	2031	·029 810	1767	870 224	242	149 130	320
1·335	3·799 996	3802	263 158	263	1·768 419	2032	2·031 577	1769	870 466	242	148 810	320
·336	·803 798	3806	262 895	263	·770 451	2035	·033 346	1772	870 708	242	148 490	318
·337	·807 604	3809	262 632	262	·772 486	2036	·035 118	1773	870 950	241	148 172	318
·338	·811 413	3813	262 370	262	·774 522	2037	·036 891	1776	871 191	241	147 854	317
·339	·815 226	3818	262 108	262	·776 559	2040	·038 667	1778	871 432	240	147 537	317
1·340	3·819 044	3820	261 846	262	1·778 599	2041	2·040 445	1779	871 672	240	147 220	316
·341	·822 864	3825	261 584	261	·780 640	2043	·042 224	1782	871 912	240	146 904	315
·342	·826 689	3829	261 323	262	·782 683	2045	·044 006	1784	872 152	239	146 589	314
·343	·830 518	3832	261 061	261	·784 728	2047	·045 790	1785	872 391	239	146 275	313
·344	·834 350	3837	260 800	260	·786 775	2048	·047 575	1788	872 630	238	145 962	313
1·345	3·838 187	3840	260 540	261	1·788 823	2051	2·049 363	1790	872 868	238	145 649	313
·346	·842 027	3844	260 279	260	·790 874	2052	·051 153	1792	873 106	237	145 336	311
·347	·845 871	3847	260 019	260	·792 926	2054	·052 945	1794	873 343	237	145 025	311
·348	·849 718	3852	259 759	259	·794 980	2055	·054 739	1796	873 580	237	144 714	310
·349	·853 570	3856	259 500	260	·797 035	2058	·056 535	1798	873 817	236	144 404	309
1·350	3·857 426		259 240		1·799 093		2·058 333		874 053		144 095	

TABLE Vc—EXPONENTIAL AND HYPERBOLIC FUNCTIONS

x	e^x	diff.	e^{-x}	diff.	sinh x	diff.	cosh x	diff.	tanh x	diff.	coth x	diff.
			0·						0·		1·	
1·350	3·857 426	3859	259 240	259	1·799 093	2059	2·058 333	1800	874 053	236	144 095	309
·351	·861 285	3863	258 981	259	·801 152	2061	·060 133	1802	874 289	236	143 786	307
·352	·865 148	3867	258 722	258	·803 213	2063	·061 935	1804	874 525	235	143 479	308
·353	·869 015	3871	258 464	259	·805 276	2064	·063 739	1807	874 760	234	143 171	306
·354	·872 886	3875	258 205	258	·807 340	2067	·065 546	1808	874 994	234	142 865	306
1·355	3·876 761	3879	257 947	258	1·809 407	2068	2·067 354	1811	875 228	234	142 559	305
·356	·880 640	3882	257 689	257	·811 475	2070	·069 165	1812	875 462	233	142 254	304
·357	·884 522	3887	257 432	257	·813 545	2072	·070 977	1815	875 695	233	141 950	304
·358	·888 409	3890	257 175	257	·815 617	2074	·072 792	1816	875 928	233	141 646	303
·359	·892 299	3894	256 918	257	·817 691	2075	·074 608	1819	876 161	232	141 343	302
1·360	3·896 193	3898	256 661	257	1·819 766	2078	2·076 427	1821	876 393	232	141 041	302
·361	·900 091	3902	256 404	256	·821 844	2079	·078 248	1823	876 625	231	140 739	301
·362	·903 993	3906	256 148	256	·823 923	2081	·080 071	1825	876 856	231	140 438	300
·363	·907 899	3910	255 892	256	·826 004	2083	·081 896	1827	877 087	231	140 138	300
·364	·911 809	3914	255 636	255	·828 087	2084	·083 723	1829	877 318	230	139 838	299
1·365	3·915 723	3918	255 381	256	1·830 171	2087	2·085 552	1831	877 548	229	139 539	298
·366	·919 641	3921	255 125	255	·832 258	2088	·087 383	1833	877 777	230	139 241	298
·367	·923 562	3926	254 870	254	·834 346	2090	·089 216	1836	878 007	229	138 943	296
·368	·927 488	3929	254 616	255	·836 436	2092	·091 052	1837	878 236	228	138 647	297
·369	·931 417	3934	254 361	254	·838 528	2094	·092 889	1840	878 464	228	138 350	295
1·370	3·935 351	3937	254 107	254	1·840 622	2096	2·094 729	1841	878 692	228	138 055	295
·371	·939 288	3941	253 853	254	·842 718	2097	·096 570	1844	878 920	227	137 760	294
·372	·943 229	3945	253 599	253	·844 815	2099	·098 414	1846	879 147	227	137 466	294
·373	·947 174	3950	253 346	253	·846 914	2102	·100 260	1848	879 374	227	137 172	292
·374	·951 124	3953	253 093	253	·849 016	2103	·102 108	1850	879 601	226	136 880	293
1·375	3·955 077	3957	252 840	253	1·851 119	2104	2·103 958	1852	879 827	225	136 587	291
·376	·959 034	3961	252 587	253	·853 223	2107	·105 810	1855	880 052	226	136 296	291
·377	·962 995	3965	252 334	252	·855 330	2109	·107 665	1856	880 278	225	136 005	290
·378	·966 960	3969	252 082	252	·857 439	2110	·109 521	1858	880 503	224	135 715	290
·379	·970 929	3973	251 830	251	·859 549	2113	·111 379	1861	880 727	224	135 425	288
1·380	3·974 902	3977	251 579	252	1·861 662	2114	2·113 240	1863	880 951	224	135 137	289
·381	·978 879	3980	251 327	251	·863 776	2116	·115 103	1865	881 175	223	134 848	287
·382	·982 859	3985	251 076	251	·865 892	2118	·116 968	1867	881 398	223	134 561	287
·383	·986 844	3989	250 825	251	·868 010	2119	·118 835	1869	881 621	223	134 274	286
·384	·990 833	3993	250 574	250	·870 129	2122	·120 704	1871	881 844	222	133 988	286
1·385	3·994 826	3997	250 324	250	1·872 251	2124	2·122 575	1873	882 066	222	133 702	285
·386	3·998 823	4001	250 074	250	·874 375	2125	·124 448	1876	882 288	221	133 417	284
·387	4·002 824	4004	249 824	250	·876 500	2127	·126 324	1877	882 509	221	133 133	284
·388	·006 828	4009	249 574	249	·878 627	2129	·128 201	1880	882 730	221	132 849	283
·389	·010 837	4013	249 325	250	·880 756	2131	·130 081	1882	882 951	220	132 566	282
1·390	4·014 850	4017	249 075	249	1·882 887	2133	2·131 963	1884	883 171	220	132 284	282
·391	·018 867	4021	248 826	248	·885 020	2135	·133 847	1886	883 391	219	132 002	281
·392	·022 888	4025	248 578	249	·887 155	2137	·135 733	1888	883 610	219	131 721	281
·393	·026 913	4029	248 329	248	·889 292	2138	·137 621	1890	883 829	219	131 440	279
·394	·030 942	4033	248 081	248	·891 430	2141	·139 511	1893	884 048	218	131 161	280
1·395	4·034 975	4037	247 833	248	1·893 571	2142	2·141 404	1894	884 266	218	130 881	278
·396	·039 012	4041	247 585	247	·895 713	2144	·143 298	1897	884 484	217	130 603	278
·397	·043 053	4045	247 338	247	·897 857	2147	·145 195	1899	884 701	218	130 325	278
·398	·047 098	4049	247 091	247	·900 004	2148	·147 094	1901	884 919	216	130 047	276
·399	·051 147	4053	246 844	247	·902 152	2150	·148 995	1903	885 135	217	129 771	276
1·400	4·055 200		246 597		1·904 302		2·150 898		885 352		129 495	

x	e^x		e^{-x}		sinh x		cosh x		tanh x		coth x	
			0·						0·		1·	
1·400	4·055 200	4057	246 597	247	1·904 302	2151	2·150 898	1906	885 352	216	129 495	276
·401	·059 257	4061	246 350	246	·906 453	2154	·152 804	1907	885 568	215	129 219	275
·402	·063 318	4066	246 104	246	·908 607	2156	·154 711	1910	885 783	215	128 944	274
·403	·067 384	4069	245 858	245	·910 763	2157	·156 621	1912	885 998	215	128 670	273
·404	·071 453	4074	245 613	246	·912 920	2160	·158 533	1914	886 213	215	128 397	273
1·405	4·075 527	4077	245 367	245	1·915 080	2161	2·160 447	1916	886 428	214	128 124	273
·406	·079 604	4082	245 122	245	·917 241	2164	·162 363	1918	886 642	213	127 851	271
·407	·083 686	4086	244 877	245	·919 405	2165	·164 281	1921	886 855	214	127 580	272
·408	·087 772	4089	244 632	244	·921 570	2167	·166 202	1923	887 069	213	127 308	270
·409	·091 861	4094	244 388	245	·923 737	2169	·168 125	1924	887 282	212	127 038	270
1·410	4·095 955	4098	244 143	244	1·925 906	2171	2·170 049	1927	887 494	212	126 768	269
·411	·100 053	4103	243 899	244	·928 077	2173	·171 976	1929	887 706	212	126 499	269
·412	·104 156	4106	243 655	243	·930 250	2175	·173 905	1932	887 918	212	126 230	268
·413	·108 262	4110	243 412	243	·932 425	2177	·175 837	1933	888 130	211	125 962	268
·414	·112 372	4114	243 169	243	·934 602	2178	·177 770	1936	888 341	210	125 694	266
1·415	4·116 486	4119	242 926	243	1·936 780	2181	2·179 706	1938	888 551	210	125 428	267
·416	·120 605	4123	242 683	243	·938 961	2183	·181 644	1940	888 761	210	125 161	265
·417	·124 728	4126	242 440	242	·941 144	2184	·183 584	1942	888 971	210	124 896	266
·418	·128 854	4131	242 198	242	·943 328	2187	·185 526	1945	889 181	209	124 630	264
·419	·132 985	4135	241 956	242	·945 515	2188	·187 471	1946	889 390	209	124 366	264
1·420	4·137 120	4140	241 714	242	1·947 703	2191	2·189 417	1949	889 599	208	124 102	263
·421	·141 260	4143	241 472	241	·949 894	2192	·191 366	1951	889 807	208	123 839	263
·422	·145 403	4147	241 231	241	·952 086	2194	·193 317	1953	890 015	208	123 576	262
·423	·149 550	4152	240 990	241	·954 280	2196	·195 270	1956	890 223	207	123 314	262
·424	·153 702	4156	240 749	241	·956 476	2199	·197 226	1957	890 430	207	123 052	261
1·425	4·157 858	4160	240 508	240	1·958 675	2200	2·199 183	1960	890 637	207	122 791	260
·426	·162 018	4164	240 268	240	·960 875	2202	·201 143	1962	890 844	206	122 531	260
·427	·166 182	4168	240 028	240	·963 077	2204	·203 105	1964	891 050	206	122 271	259
·428	·170 350	4173	239 788	240	·965 281	2206	·205 069	1966	891 256	205	122 012	259
·429	·174 523	4176	239 548	239	·967 487	2208	·207 035	1969	891 461	206	121 753	258
1·430	4·178 699	4181	239 309	239	1·969 695	2210	2·209 004	1971	891 667	204	121 495	257
·431	·182 880	4185	239 070	239	·971 905	2212	·210 975	1973	891 871	205	121 238	257
·432	·187 065	4189	238 831	239	·974 117	2214	·212 948	1975	892 076	204	120 981	256
·433	·191 254	4193	238 592	238	·976 331	2216	·214 923	1978	892 280	203	120 725	256
·434	·195 447	4198	238 354	239	·978 547	2218	·216 901	1979	892 483	204	120 469	255
1·435	4·199 645	4202	238 115	238	1·980 765	2220	2·218 880	1982	892 687	203	120 214	255
·436	·203 847	4206	237 877	237	·982 985	2222	·220 862	1984	892 890	202	119 959	254
·437	·208 053	4210	237 640	238	·985 207	2223	·222 846	1986	893 092	202	119 705	253
·438	·212 263	4214	237 402	237	·987 430	2226	·224 832	1989	893 294	202	119 452	253
·439	·216 477	4219	237 165	237	·989 656	2228	·226 821	1991	893 496	202	119 199	252
1·440	4·220 696	4223	236 928	237	1·991 884	2230	2·228 812	1993	893 698	201	118 947	252
·441	·224 919	4227	236 691	237	·994 114	2232	·230 805	1995	893 899	201	118 695	251
·442	·229 146	4231	236 454	236	·996 346	2233	·232 800	1997	894 100	200	118 444	251
·443	·233 377	4235	236 218	236	1·998 579	2236	·234 797	2000	894 300	200	118 193	250
·444	·237 612	4240	235 982	236	2·000 815	2238	·236 797	2002	894 500	200	117 943	250
1·445	4·241 852	4244	235 746	236	2·003 053	2240	2·238 799	2004	894 700	199	117 693	249
·446	·246 096	4248	235 510	235	·005 293	2242	·240 803	2007	894 899	199	117 444	248
·447	·250 344	4253	235 275	235	·007 535	2243	·242 810	2008	895 098	199	117 196	248
·448	·254 597	4257	235 040	235	·009 778	2246	·244 818	2011	895 297	198	116 948	247
·449	·258 854	4261	234 805	235	·012 024	2248	·246 829	2013	895 495	198	116 701	247
1·450	4·263 115		234 570		2·014 272		2·248 842		895 693		116 454	

x	e^x	Δ	e^{-x}	Δ	sinh x	Δ	cosh x	Δ	tanh x	Δ	coth x	Δ
			0·						0·		1·	
1·450	4·263 115	4265	234 570	234	2·014 272	2250	2·248 842	2016	895 693	197	116 454	246
·451	·267 380	4269	234 336	234	·016 522	2252	·250 858	2017	895 890	198	116 208	246
·452	·271 649	4274	234 102	234	·018 774	2254	·252 875	2020	896 088	196	115 962	245
·453	·275 923	4278	233 868	234	·021 028	2256	·254 895	2023	896 284	197	115 717	244
·454	·280 201	4282	233 634	234	·023 284	2258	·256 918	2024	896 481	196	115 473	244
1·455	4·284 483	4287	233 400	233	2·025 542	2260	2·258 942	2027	896 677	196	115 229	244
·456	·288 770	4291	233 167	233	·027 802	2261	·260 969	2029	896 873	195	114 985	243
·457	·293 061	4295	232 934	233	·030 063	2265	·262 998	2031	897 068	195	114 742	242
·458	·297 356	4300	232 701	232	·032 328	2266	·265 029	2033	897 263	195	114 500	242
·459	·301 656	4304	232 469	233	·034 594	2268	·267 062	2036	897 458	195	114 258	241
1·460	4·305 960	4308	232 236	232	2·036 862	2270	2·269 098	2038	897 653	194	114 017	241
·461	·310 268	4312	232 004	232	·039 132	2272	·271 136	2040	897 847	193	113 776	240
·462	·314 580	4317	231 772	231	·041 404	2274	·273 176	2043	898 040	194	113 536	240
·463	·318 897	4321	231 541	232	·043 678	2276	·275 219	2045	898 234	193	113 296	239
·464	·323 218	4325	231 309	231	·045 954	2279	·277 264	2047	898 427	192	113 057	239
1·465	4·327 543	4330	231 078	231	2·048 233	2280	2·279 311	2049	898 619	193	112 818	238
·466	·331 873	4334	230 847	231	·050 513	2282	·281 360	2052	898 812	192	112 580	237
·467	·336 207	4338	230 616	230	·052 795	2285	·283 412	2054	899 004	191	112 343	237
·468	·340 545	4343	230 386	230	·055 080	2286	·285 466	2056	899 195	192	112 106	237
·469	·344 888	4347	230 156	231	·057 366	2289	·287 522	2058	899 387	190	111 869	236
1·470	4·349 235	4352	229 925	229	2·059 655	2290	2·289 580	2061	899 577	191	111 633	235
·471	·353 587	4355	229 696	230	·061 945	2293	·291 641	2063	899 768	190	111 398	235
·472	·357 942	4360	229 466	229	·064 238	2295	·293 704	2066	899 958	190	111 163	235
·473	·362 302	4365	229 237	229	·066 533	2297	·295 770	2067	900 148	190	110 928	234
·474	·366 667	4369	229 008	229	·068 830	2299	·297 837	2070	900 338	189	110 694	233
1·475	4·371 036	4373	228 779	229	2·071 129	2300	2·299 907	2073	900 527	189	110 461	233
·476	·375 409	4378	228 550	228	·073 429	2303	·301 980	2074	900 716	188	110 228	232
·477	·379 787	4382	228 322	229	·075 732	2306	·304 054	2077	900 904	189	109 996	232
·478	·384 169	4386	228 093	228	·078 038	2307	·306 131	2079	901 093	187	109 764	231
·479	·388 555	4391	227 865	227	·080 345	2309	·308 210	2082	901 280	188	109 533	231
1·480	4·392 946	4395	227 638	228	2·082 654	2311	2·310 292	2083	901 468	187	109 302	231
·481	·397 341	4399	227 410	227	·084 965	2314	·312 375	2087	901 655	187	109 071	229
·482	·401 740	4404	227 183	227	·087 279	2315	·314 462	2088	901 842	187	108 842	230
·483	·406 144	4409	226 956	227	·089 594	2318	·316 550	2091	902 029	186	108 612	228
·484	·410 553	4412	226 729	227	·091 912	2320	·318 641	2093	902 215	186	108 384	229
1·485	4·414 965	4418	226 502	226	2·094 232	2321	2·320 734	2095	902 401	185	108 155	227
·486	·419 383	4421	226 276	226	·096 553	2324	·322 829	2098	902 586	185	107 928	228
·487	·423 804	4426	226 050	226	·098 877	2326	·324 927	2100	902 771	185	107 700	226
·488	·428 230	4431	225 824	226	·101 203	2328	·327 027	2102	902 956	185	107 474	227
·489	·432 661	4435	225 598	225	·103 531	2330	·329 129	2105	903 141	184	107 247	225
1·490	4·437 096	4439	225 373	226	2·105 861	2333	2·331 234	2107	903 325	184	107 022	226
·491	·441 535	4444	225 147	225	·108 194	2334	·333 341	2109	903 509	183	106 796	224
·492	·445 979	4448	224 922	224	·110 528	2337	·335 450	2112	903 692	183	106 572	225
·493	·450 427	4452	224 698	225	·112 865	2338	·337 562	2114	903 875	183	106 347	223
·494	·454 879	4458	224 473	224	·115 203	2341	·339 676	2117	904 058	183	106 124	224
1·495	4·459 337	4461	224 249	225	2·117 544	2343	2·341 793	2118	904 241	182	105 900	222
·496	·463 798	4466	224 024	223	·119 887	2345	·343 911	2121	904 423	182	105 678	223
·497	·468 264	4471	223 801	224	·122 232	2347	·346 032	2124	904 605	181	105 455	222
·498	·472 735	4475	223 577	224	·124 579	2349	·348 156	2126	904 786	181	105 233	221
·499	·477 210	4479	223 353	223	·126 928	2351	·350 282	2128	904 967	181	105 012	221
1·500	4·481 689		223 130		2·129 279		2·352 410		905 148		104 791	

x	e^x		e^{-x}		sinh x		cosh x		tanh x		coth x	
			0·						0·		1·	
1·500	4·481 689	4484	223 130	223	2·129 279	2354	2·352 410	2130	905 148	181	104 791	220
·501	·486 173	4488	222 907	223	·131 633	2356	·354 540	2133	905 329	180	104 571	220
·502	·490 661	4493	222 684	222	·133 989	2357	·356 673	2135	905 509	180	104 351	219
·503	·495 154	4498	222 462	223	·136 346	2360	·358 808	2138	905 689	179	104 132	219
·504	·499 652	4502	222 239	222	·138 706	2362	·360 946	2139	905 868	180	103 913	218
1·505	4·504 154	4506	222 017	222	2·141 068	2364	2·363 085	2143	906 048	179	103 695	218
·506	·508 660	4511	221 795	221	·143 432	2367	·365 228	2144	906 227	178	103 477	218
·507	·513 171	4515	221 574	222	·145 799	2368	·367 372	2147	906 405	178	103 259	217
·508	·517 686	4520	221 352	221	·148 167	2371	·369 519	2150	906 583	178	103 042	216
·509	·522 206	4525	221 131	221	·150 538	2372	·371 669	2151	906 761	178	102 826	216
1·510	4·526 731	4529	220 910	221	2·152 910	2375	2·373 820	2154	906 939	177	102 610	216
·511	·531 260	4533	220 689	220	·155 285	2377	·375 974	2157	907 116	177	102 394	215
·512	·535 793	4538	220 469	221	·157 662	2380	·378 131	2159	907 293	177	102 179	214
·513	·540 331	4543	220 248	220	·160 042	2381	·380 290	2161	907 470	176	101 965	214
·514	·544 874	4547	220 028	220	·162 423	2383	·382 451	2164	907 646	176	101 751	214
1·515	4·549 421	4552	219 808	220	2·164 806	2386	2·384 615	2166	907 822	176	101 537	213
·516	·553 973	4556	219 588	219	·167 192	2388	·386 781	2168	907 998	175	101 324	213
·517	·558 529	4561	219 369	219	·169 580	2390	·388 949	2171	908 173	175	101 111	212
·518	·563 090	4565	219 150	219	·171 970	2392	·391 120	2173	908 348	175	100 899	212
·519	·567 655	4570	218 931	219	·174 362	2395	·393 293	2176	908 523	175	100 687	211
1·520	4·572 225	4575	218 712	219	2·176 757	2396	2·395 469	2177	908 698	174	100 476	211
·521	·576 800	4579	218 493	218	·179 153	2399	·397 646	2181	908 872	174	100 265	210
·522	·581 379	4583	218 275	218	·181 552	2401	·399 827	2183	909 046	173	100 055	210
·523	·585 962	4589	218 057	218	·183 953	2403	·402 010	2185	909 219	173	099 845	209
·524	·590 551	4593	217 839	218	·186 356	2405	·404 195	2187	909 392	173	099 636	209
1·525	4·595 144	4597	217 621	217	2·188 761	2408	2·406 382	2190	909 565	173	099 427	209
·526	·599 741	4602	217 404	218	·191 169	2409	·408 572	2193	909 738	172	099 218	208
·527	·604 343	4607	217 186	217	·193 578	2412	·410 765	2194	909 910	172	099 010	208
·528	·608 950	4611	216 969	217	·195 990	2414	·412 959	2198	910 082	171	098 802	207
·529	·613 561	4616	216 752	216	·198 404	2417	·415 157	2199	910 253	172	098 595	206
1·530	4·618 177	4620	216 536	217	2·200 821	2418	2·417 356	2202	910 425	171	098 389	207
·531	·622 797	4625	216 319	216	·203 239	2421	·419 558	2205	910 596	170	098 182	205
·532	·627 422	4630	216 103	216	·205 660	2423	·421 763	2207	910 766	171	097 977	206
·533	·632 052	4635	215 887	216	·208 083	2425	·423 970	2209	910 937	170	097 771	205
·534	·636 687	4639	215 671	215	·210 508	2427	·426 179	2212	911 107	169	097 566	204
1·535	4·641 326	4643	215 456	216	2·212 935	2429	2·428 391	2214	911 276	170	097 362	204
·536	·645 969	4648	215 240	215	·215 364	2432	·430 605	2216	911 446	169	097 158	204
·537	·650 617	4653	215 025	215	·217 796	2434	·432 821	2219	911 615	169	096 954	203
·538	·655 270	4658	214 810	214	·220 230	2436	·435 040	2222	911 784	168	096 751	202
·539	·659 928	4662	214 596	215	·222 666	2439	·437 262	2224	911 952	168	096 549	202
1·540	4·664 590	4667	214 381	214	2·225 105	2440	2·439 486	2226	912 120	168	096 347	202
·541	·669 257	4672	214 167	214	·227 545	2443	·441 712	2229	912 288	168	096 145	202
·542	·673 929	4676	213 953	214	·229 988	2445	·443 941	2231	912 456	167	095 943	200
·543	·678 605	4681	213 739	214	·232 433	2447	·446 172	2234	912 623	167	095 743	201
·544	·683 286	4686	213 525	213	·234 880	2450	·448 406	2236	912 790	167	095 542	200
1·545	4·687 972	4690	213 312	213	2·237 330	2452	2·450 642	2238	912 957	166	095 342	199
·546	·692 662	4695	213 099	213	·239 782	2454	·452 880	2241	913 123	166	095 143	200
·547	·697 357	4700	212 886	213	·242 236	2456	·455 121	2244	913 289	166	094 943	198
·548	·702 057	4704	212 673	213	·244 692	2458	·457 365	2246	913 455	165	094 745	198
·549	·706 761	4709	212 460	212	·247 150	2461	·459 611	2248	913 620	165	094 547	198
1·550	4·711 470		212 248		2·249 611		2·461 859		913 785		094 349	

x	e^x		e^{-x}		sinh x		cosh x		tanh x		coth x	
			0·						0·		1·	
1·550	4·711 470	4714	212 248	212	2·249 611	2463	2·461 859	2251	913 785	165	094 349	198
·551	·716 184	4719	212 036	212	·252 074	2465	·464 110	2253	913 950	165	094 151	197
·552	·720 903	4723	211 824	212	·254 539	2468	·466 363	2256	914 115	164	093 954	196
·553	·725 626	4728	211 612	211	·257 007	2470	·468 619	2258	914 279	164	093 758	196
·554	·730 354	4733	211 401	212	·259 477	2472	·470 877	2261	914 443	164	093 562	196
1·555	4·735 087	4737	211 189	211	2·261 949	2474	2·473 138	2263	914 607	163	093 366	195
·556	·739 824	4742	210 978	211	·264 423	2476	·475 401	2266	914 770	163	093 171	195
·557	·744 566	4747	210 767	210	·266 899	2479	·477 667	2268	914 933	163	092 976	194
·558	·749 313	4752	210 557	211	·269 378	2481	·479 935	2271	915 096	162	092 782	194
·559	·754 065	4756	210 346	210	·271 859	2484	·482 206	2273	915 258	162	092 588	194
1·560	4·758 821	4761	210 136	210	2·274 343	2485	2·484 479	2275	915 420	162	092 394	193
·561	·763 582	4766	209 926	210	·276 828	2488	·486 754	2278	915 582	162	092 201	193
·562	·768 348	4771	209 716	209	·279 316	2490	·489 032	2281	915 744	161	092 008	192
·563	·773 119	4776	209 507	210	·281 806	2493	·491 313	2283	915 905	161	091 816	192
·564	·777 895	4780	209 297	209	·284 299	2494	·493 596	2285	916 066	161	091 624	191
1·565	4·782 675	4785	209 088	209	2·286 793	2497	2·495 881	2289	916 227	160	091 433	191
·566	·787 460	4790	208 879	209	·289 290	2500	·498 170	2290	916 387	160	091 242	191
·567	·792 250	4795	208 670	208	·291 790	2501	·500 460	2293	916 547	160	091 051	190
·568	·797 045	4799	208 462	209	·294 291	2504	·502 753	2296	916 707	160	090 861	190
·569	·801 844	4804	208 253	208	·296 795	2507	·505 049	2298	916 867	159	090 671	189
1·570	4·806 648	4809	208 045	208	2·299 302	2508	2·507 347	2300	917 026	159	090 482	189
·571	·811 457	4814	207 837	207	·301 810	2511	·509 647	2303	917 185	158	090 293	189
·572	·816 271	4819	207 630	208	·304 321	2513	·511 950	2306	917 343	159	090 104	188
·573	·821 090	4823	207 422	207	·306 834	2515	·514 256	2308	917 502	158	089 916	187
·574	·825 913	4829	207 215	207	·309 349	2518	·516 564	2311	917 660	157	089 729	188
1·575	4·830 742	4833	207 008	207	2·311 867	2520	2·518 875	2313	917 817	158	089 541	187
·576	·835 575	4838	206 801	207	·314 387	2522	·521 188	2315	917 975	157	089 354	186
·577	·840 413	4843	206 594	207	·316 909	2525	·523 503	2319	918 132	157	089 168	186
·578	·845 256	4847	206 387	206	·319 434	2527	·525 822	2320	918 289	157	088 982	186
·579	·850 103	4853	206 181	206	·321 961	2529	·528 142	2323	918 446	156	088 796	185
1·580	4·854 956	4857	205 975	206	2·324 490	2532	2·530 465	2326	918 602	156	088 611	185
·581	·859 813	4862	205 769	205	·327 022	2534	·532 791	2328	918 758	156	088 426	184
·582	·864 675	4868	205 564	206	·329 556	2536	·535 119	2331	918 914	155	088 242	185
·583	·869 543	4872	205 358	205	·332 092	2539	·537 450	2334	919 069	155	088 057	183
·584	·874 415	4876	205 153	205	·334 631	2541	·539 784	2336	919 224	155	087 874	183
1·585	4·879 291	4882	204 948	205	2·337 172	2543	2·542 120	2338	919 379	155	087 691	183
·586	·884 173	4887	204 743	205	·339 715	2546	·544 458	2341	919 534	154	087 508	183
·587	·889 060	4891	204 538	204	·342 261	2548	·546 799	2344	919 688	154	087 325	182
·588	·893 951	4897	204 334	204	·344 809	2550	·549 143	2346	919 842	154	087 143	182
·589	·898 848	4901	204 130	204	·347 359	2553	·551 489	2348	919 996	153	086 961	181
1·590	4·903 749	4906	203 926	204	2·349 912	2555	2·553 837	2351	920 149	154	086 780	181
·591	·908 655	4911	203 722	204	·352 467	2557	·556 188	2354	920 303	152	086 599	180
·592	·913 566	4916	203 518	203	·355 024	2560	·558 542	2357	920 455	153	086 419	180
·593	·918 482	4921	203 315	203	·357 584	2562	·560 899	2358	920 608	152	086 239	180
·594	·923 403	4926	203 112	203	·360 146	2564	·563 257	2362	920 760	152	086 059	179
1·595	4·928 329	4931	202 909	203	2·362 710	2567	2·565 619	2364	920 912	152	085 880	179
·596	·933 260	4936	202 706	203	·365 277	2569	·567 983	2366	921 064	152	085 701	179
·597	·938 196	4940	202 503	202	·367 846	2572	·570 349	2369	921 216	151	085 522	178
·598	·943 136	4946	202 301	202	·370 418	2574	·572 718	2372	921 367	151	085 344	178
·599	·948 082	4950	202 099	202	·372 992	2576	·575 090	2374	921 518	151	085 166	177
1·600	4·953 032		201 897		2·375 568		2·577 464		921 669		084 989	

x	e^x		e^{-x}		$\sinh x$		$\cosh x$		$\tanh x$		$\coth x$	
			0·						0·		1·	
1·600	4·953 032	4956	201 897	202	2·375 568	2579	2·577 464	2377	921 669	150	084 989	177
·601	·957 988	4960	201 695	202	·378 147	2581	·579 841	2380	921 819	150	084 812	177
·602	·962 948	4966	201 493	201	·380 728	2583	·582 221	2382	921 969	150	084 635	176
·603	·967 914	4970	201 292	201	·383 311	2586	·584 603	2384	922 119	149	084 459	176
·604	·972 884	4976	201 091	201	·385 897	2588	·586 987	2388	922 268	150	084 283	175
1·605	4·977 860	4980	200 890	201	2·388 485	2591	2·589 375	2389	922 418	149	084 108	176
·606	·982 840	4985	200 689	201	·391 076	2593	·591 764	2393	922 567	148	083 932	174
·607	·987 825	4991	200 488	200	·393 669	2595	·594 157	2395	922 715	149	083 758	175
·608	·992 816	4995	200 288	200	·396 264	2598	·596 552	2397	922 864	148	083 583	174
·609	4·997 811	5000	200 088	200	·398 862	2600	·598 949	2400	923 012	148	083 409	173
1·610	5·002 811	5006	199 888	200	2·401 462	2602	2·601 349	2403	923 160	148	083 236	173
·611	·007 817	5010	199 688	200	·404 064	2605	·603 752	2406	923 308	147	083 063	173
·612	·012 827	5015	199 488	199	·406 669	2608	·606 158	2408	923 455	147	082 890	173
·613	·017 842	5021	199 289	199	·409 277	2609	·608 566	2410	923 602	147	082 717	172
·614	·022 863	5025	199 090	199	·411 886	2613	·610 976	2413	923 749	147	082 545	172
1·615	5·027 888	5030	198 891	199	2·414 499	2614	2·613 389	2416	923 896	146	082 373	171
·616	·032 918	5036	198 692	199	·417 113	2617	·615 805	2419	924 042	146	082 202	171
·617	·037 954	5040	198 493	198	·419 730	2620	·618 224	2421	924 188	146	082 031	170
·618	·042 994	5046	198 295	198	·422 350	2622	·620 645	2423	924 334	145	081 861	171
·619	·048 040	5050	198 097	198	·424 972	2624	·623 068	2427	924 479	145	081 690	170
1·620	5·053 090	5056	197 899	198	2·427 596	2627	2·625 495	2428	924 624	145	081 520	169
·621	·058 146	5061	197 701	198	·430 223	2629	·627 923	2432	924 769	145	081 351	169
·622	·063 207	5065	197 503	197	·432 852	2631	·630 355	2434	924 914	144	081 182	169
·623	·068 272	5071	197 306	197	·435 483	2634	·632 789	2437	925 058	144	081 013	168
·624	·073 343	5076	197 109	197	·438 117	2637	·635 226	2439	925 202	144	080 845	168
1·625	5·078 419	5081	196 912	197	2·440 754	2639	2·637 665	2442	925 346	144	080 677	168
·626	·083 500	5086	196 715	197	·443 393	2641	·640 107	2445	925 490	143	080 509	167
·627	·088 586	5091	196 518	196	·446 034	2644	·642 552	2447	925 633	143	080 342	167
·628	·093 677	5096	196 322	196	·448 678	2646	·644 999	2450	925 776	143	080 175	167
·629	·098 773	5102	196 126	196	·451 324	2649	·647 449	2453	925 919	143	080 008	166
1·630	5·103 875	5106	195 930	196	2·453 973	2651	2·649 902	2455	926 062	142	079 842	166
·631	·108 981	5112	195 734	196	·456 624	2653	·652 357	2458	926 204	142	079 676	166
·632	·114 093	5116	195 538	195	·459 277	2656	·654 815	2461	926 346	142	079 510	165
·633	·119 209	5122	195 343	196	·461 933	2659	·657 276	2463	926 488	141	079 345	165
·634	·124 331	5127	195 147	195	·464 592	2661	·659 739	2466	926 629	141	079 180	164
1·635	5·129 458	5132	194 952	194	2·467 253	2663	2·662 205	2469	926 770	141	079 016	164
·636	·134 590	5137	194 758	195	·469 916	2666	·664 674	2471	926 911	141	078 852	164
·637	·139 727	5142	194 563	195	·472 582	2669	·667 145	2474	927 052	140	078 688	163
·638	·144 869	5148	194 368	194	·475 251	2670	·669 619	2477	927 192	141	078 525	163
·639	·150 017	5153	194 174	194	·477 921	2674	·672 096	2479	927 333	140	078 362	163
1·640	5·155 170	5157	193 980	194	2·480 595	2676	2·674 575	2482	927 473	139	078 199	162
·641	·160 327	5163	193 786	194	·483 271	2678	·677 057	2484	927 612	140	078 037	162
·642	·165 490	5168	193 592	193	·485 949	2681	·679 541	2488	927 752	139	077 875	162
·643	·170 658	5173	193 399	193	·488 630	2683	·682 029	2490	927 891	139	077 713	161
·644	·175 831	5179	193 206	193	·491 313	2686	·684 519	2492	928 030	138	077 552	161
1·645	5·181 010	5184	193 013	193	2·493 999	2688	2·687 011	2496	928 168	139	077 391	161
·646	·186 194	5188	192 820	193	·496 687	2691	·689 507	2498	928 307	138	077 230	160
·647	·191 382	5194	192 627	193	·499 378	2693	·692 005	2500	928 445	138	077 070	160
·648	·196 576	5199	192 434	192	·502 071	2696	·694 505	2504	928 583	137	076 910	160
·649	·201 775	5205	192 242	192	·504 767	2698	·697 009	2506	928 720	138	076 750	159
1·650	5·206 980		192 050		2·507 465		2·699 515		928 858		076 591	

x	e^x		e^{-x}		sinh x		cosh x		tanh x		coth x	
			0·						0·		1·	
1·650	5·206 980	5209	192 050	192	2·507 465	2701	2·699 515	2509	928 858	137	076 591	159
·651	·212 189	5215	191 858	192	·510 166	2703	·702 024	2511	928 995	137	076 432	158
·652	·217 404	5220	191 666	191	·512 869	2706	·704 535	2514	929 132	136	076 274	158
·653	·222 624	5225	191 475	192	·515 575	2708	·707 049	2517	929 268	136	076 116	158
·654	·227 849	5231	191 283	191	·518 283	2711	·709 566	2520	929 404	137	075 958	158
1·655	5·233 080	5236	191 092	191	2·520 994	2713	2·712 086	2522	929 541	135	075 800	157
·656	·238 316	5241	190 901	191	·523 707	2716	·714 608	2525	929 676	136	075 643	157
·657	·243 557	5246	190 710	190	·526 423	2719	·717 133	2528	929 812	135	075 486	156
·658	·248 803	5251	190 520	191	·529 142	2720	·719 661	2531	929 947	135	075 330	156
·659	·254 054	5257	190 329	190	·531 862	2724	·722 192	2533	930 082	135	075 174	156
1·660	5·259 311	5262	190 139	190	2·534 586	2726	2·724 725	2536	930 217	135	075 018	156
·661	·264 573	5267	189 949	190	·537 312	2728	·727 261	2539	930 352	134	074 862	155
·662	·269 840	5272	189 759	190	·540 040	2732	·729 800	2541	930 486	134	074 707	155
·663	·275 112	5278	189 569	189	·542 772	2733	·732 341	2544	930 620	134	074 552	154
·664	·280 390	5283	189 380	189	·545 505	2736	·734 885	2547	930 754	134	074 398	154
1·665	5·285 673	5289	189 191	189	2·548 241	2739	2·737 432	2550	930 888	133	074 244	154
·666	·290 962	5293	189 002	189	·550 980	2741	·739 982	2552	931 021	133	074 090	154
·667	·296 255	5299	188 813	189	·553 721	2744	·742 534	2555	931 154	133	073 936	153
·668	·301 554	5304	188 624	189	·556 465	2746	·745 089	2558	931 287	132	073 783	153
·669	·306 858	5310	188 435	188	·559 211	2749	·747 647	2560	931 419	133	073 630	152
1·670	5·312 168	5315	188 247	188	2·561 960	2752	2·750 207	2564	931 552	132	073 478	152
·671	·317 483	5320	188 059	188	·564 712	2754	·752 771	2566	931 684	132	073 326	152
·672	·322 803	5325	187 871	188	·567 466	2757	·755 337	2569	931 816	131	073 174	152
·673	·328 128	5331	187 683	187	·570 223	2759	·757 906	2571	931 947	132	073 022	151
·674	·333 459	5336	187 496	188	·572 982	2761	·760 477	2575	932 079	131	072 871	151
1·675	5·338 795	5342	187 308	187	2·575 743	2765	2·763 052	2577	932 210	131	072 720	151
·676	·344 137	5346	187 121	187	·578 508	2767	·765 629	2580	932 341	130	072 569	150
·677	·349 483	5353	186 934	187	·581 275	2769	·768 209	2582	932 471	131	072 419	150
·678	·354 836	5357	186 747	187	·584 044	2772	·770 791	2586	932 602	130	072 269	149
·679	·360 193	5363	186 560	186	·586 816	2775	·773 377	2588	932 732	130	072 120	150
1·680	5·365 556	5368	186 374	186	2·589 591	2777	2·775 965	2591	932 862	129	071 970	149
·681	·370 924	5374	186 188	186	·592 368	2780	·778 556	2594	932 991	130	071 821	148
·682	·376 298	5379	186 002	186	·595 148	2783	·781 150	2596	933 121	129	071 673	148
·683	·381 677	5384	185 816	186	·597 931	2785	·783 746	2600	933 250	129	071 525	149
·684	·387 061	5390	185 630	186	·600 716	2787	·786 346	2602	933 379	128	071 376	147
1·685	5·392 451	5395	185 444	185	2·603 503	2791	2·788 948	2605	933 507	129	071 229	148
·686	·397 846	5401	185 259	185	·606 294	2792	·791 553	2607	933 636	128	071 081	147
·687	·403 247	5406	185 074	185	·609 086	2796	·794 160	2611	933 764	128	070 934	146
·688	·408 653	5411	184 889	185	·611 882	2798	·796 771	2613	933 892	128	070 788	147
·689	·414 064	5417	184 704	184	·614 680	2801	·799 384	2616	934 020	127	070 641	146
1·690	5·419 481	5422	184 520	185	2·617 481	2803	2·802 000	2619	934 147	127	070 495	146
·691	·424 903	5428	184 335	184	·620 284	2806	·804 619	2622	934 274	127	070 349	145
·692	·430 331	5433	184 151	184	·623 090	2808	·807 241	2624	934 401	127	070 204	145
·693	·435 764	5438	183 967	184	·625 898	2812	·809 865	2627	934 528	127	070 059	145
·694	·441 202	5444	183 783	184	·628 710	2813	·812 492	2631	934 655	126	069 914	145
1·695	5·446 646	5449	183 599	183	2·631 523	2817	2·815 123	2633	934 781	126	069 769	144
·696	·452 095	5455	183 416	184	·634 340	2819	·817 756	2635	934 907	126	069 625	144
·697	·457 550	5460	183 232	183	·637 159	2822	·820 391	2639	935 033	126	069 481	144
·698	·463 010	5466	183 049	183	·639 981	2824	·823 030	2641	935 159	125	069 337	143
·699	·468 476	5471	182 866	182	·642 805	2827	·825 671	2644	935 284	125	069 194	143
1·700	5·473 947		182 684		2·645 632		2·828 315		935 409		069 051	

TABLE Vc—EXPONENTIAL AND HYPERBOLIC FUNCTIONS

x	e^x	Δ	e^{-x}	Δ	sinh x	Δ	cosh x	Δ	tanh x	Δ	coth x	Δ
			0·						0·		1·	
1·700	5·473 947	5477	182 684	183	2·645 632	2830	2·828 315	2648	935 409	125	069 051	143
·701	·479 424	5482	182 501	182	·648 462	2832	·830 963	2649	935 534	125	068 908	142
·702	·484 906	5488	182 319	183	·651 294	2835	·833 612	2653	935 659	124	068 766	142
·703	·490 394	5493	182 136	182	·654 129	2837	·836 265	2656	935 783	124	068 624	142
·704	·495 887	5499	181 954	182	·656 966	2841	·838 921	2658	935 907	124	068 482	142
1·705	5·501 386	5504	181 772	181	2·659 807	2843	2·841 579	2661	936 031	124	068 340	141
·706	·506 890	5509	181 591	182	·662 650	2845	·844 240	2664	936 155	123	068 199	141
·707	·512 399	5516	181 409	181	·665 495	2848	·846 904	2667	936 278	124	068 058	140
·708	·517 915	5520	181 228	181	·668 343	2851	·849 571	2670	936 402	123	067 918	141
·709	·523 435	5526	181 047	181	·671 194	2854	·852 241	2673	936 525	123	067 777	140
1·710	5·528 961	5532	180 866	181	2·674 048	2856	2·854 914	2675	936 648	122	067 637	139
·711	·534 493	5537	180 685	181	·676 904	2859	·857 589	2678	936 770	122	067 498	140
·712	·540 030	5543	180 504	180	·679 763	2862	·860 267	2682	936 892	123	067 358	139
·713	·545 573	5549	180 324	180	·682 625	2864	·862 949	2684	937 015	121	067 219	139
·714	·551 122	5554	180 144	180	·685 489	2867	·865 633	2687	937 136	122	067 080	138
1·715	5·556 676	5559	179 964	180	2·688 356	2870	2·868 320	2689	937 258	122	066 942	138
·716	·562 235	5565	179 784	180	·691 226	2872	·871 009	2693	937 380	121	066 804	138
·717	·567 800	5571	179 604	179	·694 098	2875	·873 702	2696	937 501	121	066 666	138
·718	·573 371	5576	179 425	180	·696 973	2878	·876 398	2698	937 622	121	066 528	137
·719	·578 947	5581	179 245	179	·699 851	2880	·879 096	2701	937 743	120	066 391	137
1·720	5·584 528	5588	179 066	179	2·702 731	2883	2·881 797	2704	937 863	120	066 254	137
·721	·590 116	5593	178 887	179	·705 614	2886	·884 501	2708	937 983	120	066 117	136
·722	·595 709	5598	178 708	178	·708 500	2889	·887 209	2709	938 103	120	065 981	137
·723	·601 307	5604	178 530	179	·711 389	2891	·889 918	2713	938 223	120	065 844	135
·724	·606 911	5610	178 351	178	·714 280	2894	·892 631	2716	938 343	119	065 709	136
1·725	5·612 521	5615	178 173	178	2·717 174	2897	2·895 347	2719	938 462	119	065 573	135
·726	·618 136	5621	177 995	178	·720 071	2899	·898 066	2721	938 581	119	065 438	135
·727	·623 757	5627	177 817	178	·722 970	2902	·900 787	2725	938 700	119	065 303	135
·728	·629 384	5632	177 639	177	·725 872	2905	·903 512	2727	938 819	119	065 168	135
·729	·635 016	5638	177 462	178	·728 777	2908	·906 239	2730	938 938	118	065 033	134
1·730	5·640 654	5643	177 284	177	2·731 685	2910	2·908 969	2733	939 056	118	064 899	134
·731	·646 297	5650	177 107	177	·734 595	2913	·911 702	2736	939 174	118	064 765	133
·732	·651 947	5654	176 930	177	·737 508	2916	·914 438	2739	939 292	117	064 632	133
·733	·657 601	5661	176 753	176	·740 424	2919	·917 177	2742	939 409	118	064 499	133
·734	·663 262	5666	176 577	177	·743 343	2921	·919 919	2745	939 527	117	064 366	133
1·735	5·668 928	5672	176 400	176	2·746 264	2924	2·922 664	2748	939 644	117	064 233	133
·736	·674 600	5677	176 224	176	·749 188	2927	·925 412	2750	939 761	117	064 100	132
·737	·680 277	5683	176 048	176	·752 115	2929	·928 162	2754	939 878	116	063 968	132
·738	·685 960	5689	175 872	176	·755 044	2932	·930 916	2756	939 994	117	063 836	131
·739	·691 649	5694	175 696	176	·757 976	2936	·933 672	2760	940 111	116	063 705	132
1·740	5·697 343	5701	175 520	175	2·760 912	2937	2·936 432	2762	940 227	116	063 573	131
·741	·703 044	5706	175 345	175	·763 849	2941	·939 194	2766	940 343	115	063 442	130
·742	·708 750	5711	175 170	175	·766 790	2943	·941 960	2768	940 458	116	063 312	131
·743	·714 461	5717	174 995	175	·769 733	2946	·944 728	2771	940 574	115	063 181	130
·744	·720 178	5723	174 820	175	·772 679	2949	·947 499	2774	940 689	115	063 051	130
1·745	5·725 901	5729	174 645	175	2·775 628	2952	2·950 273	2777	940 804	115	062 921	130
·746	·731 630	5735	174 470	174	·778 580	2954	·953 050	2780	940 919	114	062 791	129
·747	·737 365	5740	174 296	174	·781 534	2958	·955 830	2783	941 033	114	062 662	129
·748	·743 105	5746	174 122	174	·784 492	2960	·958 613	2786	941 147	115	062 533	129
·749	·748 851	5752	173 948	174	·787 452	2962	·961 399	2789	941 262	114	062 404	129
1·750	5·754 603		173 774		2·790 414		2·964 188		941 376		062 275	

x	e^x		e^{-x}		sinh x		cosh x		tanh x		coth x	
			0·						0·		1·	
1·750	5·754 603	5757	173 774	174	2·790 414	2966	2·964 188	2792	941 376	113	062 275	128
·751	·760 360	5763	173 600	173	·793 380	2968	·966 980	2795	941 489	114	062 147	128
·752	·766 123	5769	173 427	174	·796 348	2972	·969 775	2798	941 603	113	062 019	128
·753	·771 892	5775	173 253	173	·799 320	2973	·972 573	2801	941 716	113	061 891	127
·754	·777 667	5781	173 080	173	·802 293	2977	·975 374	2803	941 829	113	061 764	127
1·755	5·783 448	5786	172 907	173	2·805 270	2980	2·978 177	2807	941 942	113	061 637	127
·756	·789 234	5792	172 734	172	·808 250	2982	·980 984	2810	942 055	112	061 510	127
·757	·795 026	5798	172 562	173	·811 232	2985	·983 794	2813	942 167	112	061 383	126
·758	·800 824	5804	172 389	172	·814 217	2988	·986 607	2815	942 279	112	061 257	127
·759	·806 628	5809	172 217	172	·817 205	2991	·989 422	2819	942 391	112	061 130	125
1·760	5·812 437	5816	172 045	172	2·820 196	2994	2·992 241	2822	942 503	112	061 005	126
·761	·818 253	5821	171 873	172	·823 190	2996	·995 063	2825	942 615	111	060 879	125
·762	·824 074	5827	171 701	171	·826 186	3000	2·997 888	2827	942 726	111	060 754	125
·763	·829 901	5833	171 530	172	·829 186	3002	3·000 715	2831	942 837	111	060 629	125
·764	·835 734	5838	171 358	171	·832 188	3005	·003 546	2834	942 948	111	060 504	125
1·765	5·841 572	5845	171 187	171	2·835 193	3008	3·006 380	2836	943 059	110	060 379	124
·766	·847 417	5850	171 016	171	·838 201	3010	·009 216	2840	943 169	111	060 255	124
·767	·853 267	5856	170 845	171	·841 211	3014	·012 056	2843	943 280	110	060 131	124
·768	·859 123	5862	170 674	171	·844 225	3016	·014 899	2845	943 390	110	060 007	123
·769	·864 985	5868	170 503	170	·847 241	3019	·017 744	2849	943 500	109	059 884	123
1·770	5·870 853	5874	170 333	170	2·850 260	3022	3·020 593	2852	943 609	110	059 761	123
·771	·876 727	5880	170 163	170	·853 282	3025	·023 445	2855	943 719	109	059 638	123
·772	·882 607	5885	169 993	170	·856 307	3028	·026 300	2858	943 828	109	059 515	123
·773	·888 492	5892	169 823	170	·859 335	3030	·029 158	2860	943 937	109	059 392	122
·774	·894 384	5897	169 653	170	·862 365	3034	·032 018	2864	944 046	109	059 270	122
1·775	5·900 281	5903	169 483	169	2·865 399	3036	3·034 882	2867	944 155	108	059 148	121
·776	·906 184	5910	169 314	169	·868 435	3039	·037 749	2870	944 263	109	059 027	122
·777	·912 094	5915	169 145	169	·871 474	3042	·040 619	2873	944 372	108	058 905	121
·778	·918 009	5921	168 976	169	·874 516	3045	·043 492	2876	944 480	108	058 784	121
·779	·923 930	5926	168 807	169	·877 561	3048	·046 368	2879	944 588	107	058 663	120
1·780	5·929 856	5933	168 638	168	2·880 609	3051	3·049 247	2882	944 695	108	058 543	121
·781	·935 789	5939	168 470	169	·883 660	3053	·052 129	2886	944 803	107	058 422	120
·782	·941 728	5945	168 301	168	·886 713	3057	·055 015	2888	944 910	107	058 302	120
·783	·947 673	5950	168 133	168	·889 770	3059	·057 903	2891	945 017	107	058 182	119
·784	·953 623	5957	167 965	168	·892 829	3062	·060 794	2895	945 124	106	058 063	120
1·785	5·959 580	5963	167 797	168	2·895 891	3066	3·063 689	2897	945 230	107	057 943	119
·786	·965 543	5968	167 629	167	·898 957	3068	·066 586	2900	945 337	106	057 824	119
·787	·971 511	5975	167 462	168	·902 025	3071	·069 486	2904	945 443	106	057 705	118
·788	·977 486	5980	167 294	167	·905 096	3073	·072 390	2907	945 549	106	057 587	119
·789	·983 466	5986	167 127	167	·908 169	3077	·075 297	2909	945 655	106	057 468	118
1·790	5·989 452	5993	166 960	167	2·911 246	3080	3·078 206	2913	945 761	105	057 350	118
·791	5·995 445	5998	166 793	166	·914 326	3082	·081 119	2916	945 866	105	057 232	117
·792	6·001 443	6005	166 627	167	·917 408	3086	·084 035	2919	945 971	105	057 115	118
·793	·007 448	6010	166 460	166	·920 494	3088	·086 954	2922	946 076	105	056 997	117
·794	·013 458	6017	166 294	167	·923 582	3092	·089 876	2925	946 181	105	056 880	117
1·795	6·019 475	6022	166 127	166	2·926 674	3094	3·092 801	2928	946 286	104	056 763	116
·796	·025 497	6029	165 961	165	·929 768	3097	·095 729	2932	946 390	104	056 647	117
·797	·031 526	6034	165 796	166	·932 865	3100	·098 661	2934	946 494	105	056 530	116
·798	·037 560	6041	165 630	166	·935 965	3103	·101 595	2938	946 599	103	056 414	116
·799	·043 601	6046	165 464	165	·939 068	3106	·104 533	2940	946 702	104	056 298	115
1·800	6·049 647		165 299		2·942 174		3·107 473		946 806		056 183	

TABLE Vc—EXPONENTIAL AND HYPERBOLIC FUNCTIONS

x	e^x		e^{-x}		sinh x		cosh x		tanh x		coth x	
			0·						0·		1·	
1·800	6·049 647	*6053*	165 299	165	2·942 174	3109	3·107 473	2944	946 806	103	056 183	116
·801	·055 700	*6059*	165 134	165	·945 283	3112	·110 417	2947	946 909	104	056 067	115
·802	·061 759	*6065*	164 969	165	·948 395	3115	·113 364	2950	947 013	103	055 952	115
·803	·067 824	*6071*	164 804	165	·951 510	3118	·116 314	2953	947 116	103	055 837	115
·804	·073 895	*6076*	164 639	165	·954 628	3120	·119 267	2956	947 219	102	055 722	114
1·805	6·079 971	*6083*	164 474	164	2·957 748	3124	3·122 223	2959	947 321	103	055 608	114
·806	·086 054	*6090*	164 310	164	·960 872	3127	·125 182	2963	947 424	102	055 494	114
·807	·092 144	*6095*	164 146	164	·963 999	3129	·128 145	2965	947 526	102	055 380	114
·808	·098 239	*6101*	163 982	164	·967 128	3133	·131 110	2969	947 628	102	055 266	113
·809	·104 340	*6107*	163 818	164	·970 261	3136	·134 079	2972	947 730	102	055 153	114
1·810	6·110 447	*6114*	163 654	163	2·973 397	3138	3·137 051	2975	947 832	101	055 039	113
·811	·116 561	*6120*	163 491	164	·976 535	3142	·140 026	2978	947 933	102	054 926	112
·812	·122 681	*6125*	163 327	163	·979 677	3144	·143 004	2981	948 035	101	054 814	113
·813	·128 806	*6132*	163 164	163	·982 821	3148	·145 985	2984	948 136	101	054 701	112
·814	·134 938	*6138*	163 001	163	·985 969	3150	·148 969	2988	948 237	101	054 589	112
1·815	6·141 076	*6144*	162 838	163	2·989 119	3154	3·151 957	2991	948 338	100	054 477	112
·816	·147 220	*6151*	162 675	162	·992 273	3156	·154 948	2994	948 438	100	054 365	111
·817	·153 371	*6156*	162 513	163	·995 429	3159	·157 942	2997	948 538	101	054 254	112
·818	·159 527	*6163*	162 350	162	2·998 588	3163	·160 939	3000	948 639	100	054 142	111
·819	·165 690	*6168*	162 188	162	3·001 751	3165	·163 939	3003	948 739	99	054 031	111
1·820	6·171 858	*6175*	162 026	162	3·004 916	3169	3·166 942	3007	948 838	100	053 920	110
·821	·178 033	*6182*	161 864	162	·008 085	3171	·169 949	3009	948 938	99	053 810	111
·822	·184 215	*6187*	161 702	162	·011 256	3175	·172 958	3013	949 037	100	053 699	110
·823	·190 402	*6193*	161 540	161	·014 431	3177	·175 971	3016	949 137	99	053 589	110
·824	·196 595	*6200*	161 379	161	·017 608	3181	·178 987	3019	949 236	99	053 479	110
1·825	6·202 795	*6206*	161 218	161	3·020 789	3183	3·182 006	3023	949 335	98	053 369	109
·826	·209 001	*6212*	161 057	161	·023 972	3187	·185 029	3025	949 433	99	053 260	109
·827	·215 213	*6218*	160 896	161	·027 159	3189	·188 054	3029	949 532	98	053 151	109
·828	·221 431	*6225*	160 735	161	·030 348	3193	·191 083	3032	949 630	98	053 042	109
·829	·227 656	*6231*	160 574	160	·033 541	3196	·194 115	3035	949 728	98	052 933	109
1·830	6·233 887	*6237*	160 414	161	3·036 737	3198	3·197 150	3038	949 826	98	052 824	108
·831	·240 124	*6243*	160 253	160	·039 935	3202	·200 188	3042	949 924	97	052 716	108
·832	·246 367	*6249*	160 093	160	·043 137	3205	·203 230	3045	950 021	98	052 608	108
·833	·252 616	*6256*	159 933	160	·046 342	3207	·206 275	3048	950 119	97	052 500	108
·834	·258 872	*6262*	159 773	159	·049 549	3211	·209 323	3051	950 216	97	052 392	107
1·835	6·265 134	*6268*	159 614	160	3·052 760	3214	3·212 374	3054	950 313	97	052 285	107
·836	·271 402	*6275*	159 454	159	·055 974	3217	·215 428	3058	950 410	96	052 178	107
·837	·277 677	*6281*	159 295	160	·059 191	3220	·218 486	3061	950 506	97	052 071	107
·838	·283 958	*6287*	159 135	159	·062 411	3223	·221 547	3064	950 603	96	051 964	106
·839	·290 245	*6293*	158 976	159	·065 634	3226	·224 611	3067	950 699	96	051 858	107
1·840	6·296 538	*6300*	158 817	158	3·068 860	3230	3·227 678	3070	950 795	96	051 751	106
·841	·302 838	*6306*	158 659	159	·072 090	3232	·230 748	3074	950 891	96	051 645	106
·842	·309 144	*6312*	158 500	158	·075 322	3235	·233 822	3077	950 987	95	051 539	105
·843	·315 456	*6319*	158 342	159	·078 557	3239	·236 899	3080	951 082	96	051 434	106
·844	·321 775	*6325*	158 183	158	·081 796	3241	·239 979	3084	951 178	95	051 328	105
1·845	6·328 100	*6331*	158 025	158	3·085 037	3245	3·243 063	3086	951 273	95	051 223	105
·846	·334 431	*6338*	157 867	157	·088 282	3248	·246 149	3090	951 368	95	051 118	105
·847	·340 769	*6344*	157 710	158	·091 530	3250	·249 239	3093	951 463	94	051 013	104
·848	·347 113	*6350*	157 552	158	·094 780	3254	·252 332	3097	951 557	95	050 909	104
·849	·353 463	*6357*	157 394	157	·098 034	3257	·255 429	3099	951 652	94	050 805	104
1·850	6·359 820		157 237		3·101 291		3·258 528		951 746		050 701	

TABLE Vc—EXPONENTIAL AND HYPERBOLIC FUNCTIONS

x	e^x		e^{-x}		$\sinh x$		$\cosh x$		$\tanh x$		$\coth x$	
			0·						0·		1·	
1·850	6·359 820	*6363*	157 237	157	3·101 291	3260	3·258 528	3103	951 746	94	050 701	104
·851	·366 183	*6369*	157 080	157	·104 551	3263	·261 631	3106	951 840	94	050 597	104
·852	·372 552	*6376*	156 923	157	·107 814	3267	·264 737	3110	951 934	94	050 493	103
·853	·378 928	*6382*	156 766	157	·111 081	3269	·267 847	3113	952 028	93	050 390	104
·854	·385 310	*6388*	156 609	156	·114 350	3273	·270 960	3116	952 121	94	050 286	103
1·855	6·391 698	*6395*	156 453	156	3·117 623	3275	3·274 076	3119	952 215	93	050 183	102
·856	·398 093	*6401*	156 297	157	·120 898	3279	·277 195	3122	952 308	93	050 081	103
·857	·404 494	*6408*	156 140	156	·124 177	3282	·280 317	3126	952 401	93	049 978	102
·858	·410 902	*6414*	155 984	156	·127 459	3285	·283 443	3129	952 494	92	049 876	102
·859	·417 316	*6421*	155 828	155	·130 744	3288	·286 572	3133	952 586	93	049 774	102
1·860	6·423 737	*6427*	155 673	156	3·134 032	3291	3·289 705	3135	952 679	92	049 672	102
·861	·430 164	*6433*	155 517	155	·137 323	3295	·292 840	3139	952 771	92	049 570	102
·862	·436 597	*6440*	155 362	156	·140 618	3297	·295 979	3143	952 863	92	049 468	101
·863	·443 037	*6446*	155 206	155	·143 915	3301	·299 122	3145	952 955	92	049 367	101
·864	·449 483	*6453*	155 051	155	·147 216	3304	·302 267	3149	953 047	92	049 266	101
1·865	6·455 936	*6459*	154 896	155	3·150 520	3307	3·305 416	3152	953 139	91	049 165	100
·866	·462 395	*6466*	154 741	154	·153 827	3310	·308 568	3156	953 230	91	049 065	101
·867	·468 861	*6472*	154 587	155	·157 137	3313	·311 724	3158	953 321	91	048 964	100
·868	·475 333	*6478*	154 432	154	·160 450	3317	·314 882	3163	953 412	91	048 864	100
·869	·481 811	*6485*	154 278	154	·163 767	3319	·318 045	3165	953 503	91	048 764	100
1·870	6·488 296	*6492*	154 124	154	3·167 086	3323	3·321 210	3169	953 594	91	048 664	99
·871	·494 788	*6498*	153 970	154	·170 409	3326	·324 379	3172	953 685	90	048 565	100
·872	·501 286	*6505*	153 816	154	·173 735	3329	·327 551	3175	953 775	90	048 465	99
·873	·507 791	*6511*	153 662	154	·177 064	3333	·330 726	3179	953 865	90	048 366	99
·874	·514 302	*6517*	153 508	153	·180 397	3335	·333 905	3182	953 955	90	048 267	99
1·875	6·520 819	*6524*	153 355	153	3·183 732	3339	3·337 087	3185	954 045	90	048 168	98
·876	·527 343	*6531*	153 202	153	·187 071	3342	·340 272	3189	954 135	90	048 070	99
·877	·533 874	*6537*	153 049	153	·190 413	3345	·343 461	3192	954 225	89	047 971	98
·878	·540 411	*6544*	152 896	153	·193 758	3348	·346 653	3196	954 314	89	047 873	98
·879	·546 955	*6550*	152 743	153	·197 106	3351	·349 849	3198	954 403	89	047 775	97
1·880	6·553 505	*6557*	152 590	152	3·200 457	3355	3·353 047	3203	954 492	89	047 678	98
·881	·560 062	*6563*	152 438	153	·203 812	3358	·356 250	3205	954 581	89	047 580	97
·882	·566 625	*6570*	152 285	152	·207 170	3361	·359 455	3209	954 670	88	047 483	97
·883	·573 195	*6576*	152 133	152	·210 531	3364	·362 664	3212	954 758	89	047 386	97
·884	·579 771	*6583*	151 981	152	·213 895	3368	·365 876	3216	954 847	88	047 289	97
1·885	6·586 354	*6590*	151 829	152	3·217 263	3370	3·369 092	3219	954 935	88	047 192	97
·886	·592 944	*6596*	151 677	151	·220 633	3374	·372 311	3222	955 023	88	047 095	96
·887	·599 540	*6603*	151 526	152	·224 007	3377	·375 533	3226	955 111	87	046 999	96
·888	·606 143	*6610*	151 374	151	·227 384	3381	·378 759	3229	955 198	88	046 903	96
·889	·612 753	*6616*	151 223	151	·230 765	3383	·381 988	3232	955 286	87	046 807	96
1·890	6·619 369	*6622*	151 072	151	3·234 148	3387	3·385 220	3236	955 373	87	046 711	95
·891	·625 991	*6630*	150 921	151	·237 535	3390	·388 456	3239	955 460	87	046 616	95
·892	·632 621	*6636*	150 770	151	·240 925	3394	·391 695	3243	955 547	87	046 521	95
·893	·639 257	*6642*	150 619	150	·244 319	3396	·394 938	3246	955 634	87	046 426	95
·894	·645 899	*6649*	150 469	151	·247 715	3400	·398 184	3249	955 721	86	046 331	95
1·895	6·652 548	*6656*	150 318	150	3·251 115	3403	3·401 433	3253	955 807	87	046 236	95
·896	·659 204	*6663*	150 168	150	·254 518	3406	·404 686	3256	955 894	86	046 141	94
·897	·665 867	*6669*	150 018	150	·257 924	3410	·407 942	3260	955 980	86	046 047	94
·898	·672 536	*6676*	149 868	150	·261 334	3413	·411 202	3263	956 066	86	045 953	94
·899	·679 212	*6682*	149 718	149	·264 747	3416	·414 465	3267	956 152	85	045 859	94
1·900	6·685 894		149 569		3·268 163		3·417 732		956 237		045 765	

x	e^x		e^{-x}		sinh x		cosh x		tanh x		coth x	
			0·						0·		1·	
1·900	6·685 894	6690	149 569	150	3·268 163	3419	3·417 732	3269	956 237	86	045 765	93
·901	·692 584	6696	149 419	149	·271 582	3423	·421 001	3274	956 323	85	045 672	94
·902	·699 280	6702	149 270	149	·275 005	3426	·424 275	3276	956 408	86	045 578	93
·903	·705 982	6710	149 121	149	·278 431	3429	·427 551	3281	956 494	85	045 485	93
·904	·712 692	6716	148 972	149	·281 860	3432	·430 832	3283	956 579	84	045 392	92
1·905	6·719 408	6722	148 823	149	3·285 292	3436	3·434 115	3287	956 663	85	045 300	93
·906	·726 130	6730	148 674	149	·288 728	3439	·437 402	3291	956 748	85	045 207	92
·907	·732 860	6736	148 525	148	·292 167	3443	·440 693	3293	956 833	84	045 115	92
·908	·739 596	6743	148 377	148	·295 610	3445	·443 986	3298	956 917	84	045 023	92
·909	·746 339	6750	148 229	149	·299 055	3449	·447 284	3301	957 001	84	044 931	92
1·910	6·753 089	6756	148 080	148	3·302 504	3452	3·450 585	3304	957 085	84	044 839	92
·911	·759 845	6763	147 932	147	·305 956	3456	·453 889	3308	957 169	84	044 747	91
·912	·766 608	6770	147 785	148	·309 412	3459	·457 197	3311	957 253	84	044 656	91
·913	·773 378	6777	147 637	148	·312 871	3462	·460 508	3314	957 337	83	044 565	91
·914	·780 155	6784	147 489	147	·316 333	3465	·463 822	3318	957 420	83	044 474	91
1·915	6·786 939	6790	147 342	147	3·319 798	3469	3·467 140	3322	957 503	83	044 383	91
·916	·793 729	6797	147 195	148	·323 267	3472	·470 462	3325	957 586	83	044 292	90
·917	·800 526	6804	147 047	147	·326 739	3476	·473 787	3328	957 669	83	044 202	91
·918	·807 330	6811	146 900	146	·330 215	3479	·477 115	3332	957 752	83	044 111	90
·919	·814 141	6817	146 754	147	·333 694	3482	·480 447	3336	957 835	82	044 021	90
1·920	6·820 958	6825	146 607	147	3·337 176	3485	3·483 783	3339	957 917	83	043 931	89
·921	·827 783	6831	146 460	146	·340 661	3489	·487 122	3342	958 000	82	043 842	90
·922	·834 614	6838	146 314	146	·344 150	3492	·490 464	3346	958 082	82	043 752	89
·923	·841 452	6845	146 168	146	·347 642	3496	·493 810	3349	958 164	82	043 663	89
·924	·848 297	6852	146 022	146	·351 138	3498	·497 159	3353	958 246	81	043 574	89
1·925	6·855 149	6858	145 876	146	3·354 636	3503	3·500 512	3357	958 327	82	043 485	89
·926	·862 007	6866	145 730	146	·358 139	3505	·503 869	3359	958 409	81	043 396	89
·927	·868 873	6872	145 584	145	·361 644	3509	·507 228	3364	958 490	81	043 307	88
·928	·875 745	6879	145 439	146	·365 153	3512	·510 592	3367	958 571	81	043 219	88
·929	·882 624	6886	145 293	145	·368 665	3516	·513 959	3370	958 652	81	043 131	88
1·930	6·889 510	6893	145 148	145	3·372 181	3519	3·517 329	3374	958 733	81	043 043	88
·931	·896 403	6900	145 003	145	·375 700	3522	·520 703	3378	958 814	81	042 955	88
·932	·903 303	6907	144 858	145	·379 222	3526	·524 081	3381	958 895	80	042 867	87
·933	·910 210	6913	144 713	144	·382 748	3529	·527 462	3384	958 975	80	042 780	87
·934	·917 123	6921	144 569	145	·386 277	3533	·530 846	3388	959 055	81	042 693	88
1·935	6·924 044	6928	144 424	144	3·389 810	3536	3·534 234	3392	959 136	80	042 605	87
·936	·930 972	6934	144 280	144	·393 346	3539	·537 626	3395	959 216	79	042 518	86
·937	·937 906	6941	144 136	144	·396 885	3543	·541 021	3399	959 295	80	042 432	87
·938	·944 847	6949	143 992	144	·400 428	3546	·544 420	3402	959 375	80	042 345	86
·939	·951 796	6955	143 848	144	·403 974	3550	·547 822	3405	959 455	79	042 259	86
1·940	6·958 751	6962	143 704	144	3·407 524	3552	3·551 227	3410	959 534	79	042 173	86
·941	·965 713	6969	143 560	143	·411 076	3557	·554 637	3413	959 613	79	042 087	86
·942	·972 682	6977	143 417	144	·414 633	3560	·558 050	3416	959 692	79	042 001	86
·943	·979 659	6983	143 273	143	·418 193	3563	·561 466	3420	959 771	79	041 915	86
·944	·986 642	6990	143 130	143	·421 756	3566	·564 886	3424	959 850	79	041 829	85
1·945	6·993 632	6997	142 987	143	3·425 322	3570	3·568 310	3427	959 929	78	041 744	85
·946	7·000 629	7004	142 844	142	·428 892	3574	·571 737	3430	960 007	78	041 659	85
·947	·007 633	7011	142 702	143	·432 466	3577	·575 167	3435	960 085	79	041 574	85
·948	·014 644	7018	142 559	143	·436 043	3580	·578 602	3437	960 164	78	041 489	84
·949	·021 662	7026	142 416	142	·439 623	3584	·582 039	3442	960 242	77	041 405	85
1·950	7·028 688		142 274		3·443 207		3·585 481		960 319		041 320	

TABLE Vc—EXPONENTIAL AND HYPERBOLIC FUNCTIONS

x	e^x		e^{-x}		sinh x		cosh x		tanh x		coth x	
			0·						0·		1·	
1·950	7·028 688	7032	142 274	142	3·443 207	3587	3·585 481	3445	960 319	78	041 320	84
·951	·035 720	7039	142 132	142	·446 794	3591	·588 926	3448	960 397	78	041 236	84
·952	·042 759	7046	141 990	142	·450 385	3594	·592 374	3453	960 475	77	041 152	84
·953	·049 805	7054	141 848	142	·453 979	3597	·595 827	3455	960 552	77	041 068	84
·954	·056 859	7060	141 706	142	·457 576	3601	·599 282	3460	960 629	77	040 984	83
1·955	7·063 919	7067	141 564	141	3·461 177	3605	3·602 742	3463	960 706	77	040 901	84
·956	·070 986	7075	141 423	141	·464 782	3608	·606 205	3466	960 783	77	040 817	83
·957	·078 061	7082	141 282	142	·468 390	3611	·609 671	3471	960 860	77	040 734	83
·958	·085 143	7088	141 140	141	·472 001	3615	·613 142	3473	960 937	76	040 651	83
·959	·092 231	7096	140 999	141	·475 616	3618	·616 615	3478	961 013	77	040 568	83
1·960	7·099 327	7103	140 858	140	3·479 234	3622	3·620 093	3481	961 090	76	040 485	82
·961	·106 430	7110	140 718	141	·482 856	3625	·623 574	3484	961 166	76	040 403	82
·962	·113 540	7117	140 577	141	·486 481	3629	·627 058	3489	961 242	76	040 321	83
·963	·120 657	7124	140 436	140	·490 110	3633	·630 547	3492	961 318	76	040 238	82
·964	·127 781	7132	140 296	140	·493 743	3635	·634 039	3495	961 394	76	040 156	81
1·965	7·134 913	7138	140 156	140	3·497 378	3640	3·637 534	3499	961 470	75	040 075	82
·966	·142 051	7146	140 016	140	·501 018	3642	·641 033	3503	961 545	75	039 993	82
·967	·149 197	7152	139 876	140	·504 660	3647	·644 536	3507	961 620	76	039 911	81
·968	·156 349	7160	139 736	140	·508 307	3650	·648 043	3510	961 696	75	039 830	81
·969	·163 509	7167	139 596	139	·511 957	3653	·651 553	3514	961 771	75	039 749	81
1·970	7·170 676	7175	139 457	140	3·515 610	3657	3·655 067	3517	961 846	74	039 668	81
·971	·177 851	7181	139 317	139	·519 267	3660	·658 584	3521	961 920	75	039 587	81
·972	·185 032	7189	139 178	139	·522 927	3664	·662 105	3525	961 995	75	039 506	80
·973	·192 221	7196	139 039	139	·526 591	3667	·665 630	3528	962 070	74	039 426	80
·974	·199 417	7203	138 900	139	·530 258	3671	·669 158	3532	962 144	74	039 346	81
1·975	7·206 620	7210	138 761	138	3·533 929	3675	3·672 690	3536	962 218	74	039 265	80
·976	·213 830	7217	138 623	139	·537 604	3678	·676 226	3540	962 292	74	039 185	79
·977	·221 047	7225	138 484	138	·541 282	3681	·679 766	3543	962 366	74	039 106	80
·978	·228 272	7232	138 346	139	·544 963	3685	·683 309	3547	962 440	73	039 026	80
·979	·235 504	7239	138 207	138	·548 648	3689	·686 856	3550	962 513	74	038 946	79
1·980	7·242 743	7246	138 069	138	3·552 337	3692	3·690 406	3554	962 587	73	038 867	79
·981	·249 989	7254	137 931	138	·556 029	3696	·693 960	3558	962 660	74	038 788	79
·982	·257 243	7261	137 793	137	·559 725	3699	·697 518	3562	962 734	73	038 709	79
·983	·264 504	7268	137 656	138	·563 424	3703	·701 080	3565	962 807	73	038 630	79
·984	·271 772	7275	137 518	137	·567 127	3706	·704 645	3569	962 880	72	038 551	78
1·985	7·279 047	7283	137 381	138	3·570 833	3710	3·708 214	3573	962 952	73	038 473	78
·986	·286 330	7290	137 243	137	·574 543	3714	·711 787	3576	963 025	73	038 395	79
·987	·293 620	7297	137 106	137	·578 257	3717	·715 363	3580	963 098	72	038 316	78
·988	·300 917	7305	136 969	137	·581 974	3721	·718 943	3584	963 170	72	038 238	77
·989	·308 222	7312	136 832	137	·585 695	3724	·722 527	3588	963 242	72	038 161	78
1·990	7·315 534	7319	136 695	136	3·589 419	3728	3·726 115	3591	963 314	72	038 083	78
·991	·322 853	7326	136 559	137	·593 147	3732	·729 706	3595	963 386	72	038 005	77
·992	·330 179	7334	136 422	136	·596 879	3735	·733 301	3599	963 458	72	037 928	77
·993	·337 513	7341	136 286	136	·600 614	3738	·736 900	3602	963 530	71	037 851	77
·994	·344 854	7349	136 150	136	·604 352	3743	·740 502	3606	963 601	72	037 774	77
1·995	7·352 203	7356	136 014	136	3·608 095	3746	3·744 108	3610	963 673	71	037 697	77
·996	·359 559	7363	135 878	136	·611 841	3749	·747 718	3614	963 744	71	037 620	77
·997	·366 922	7371	135 742	136	·615 590	3753	·751 332	3617	963 815	71	037 543	76
·998	·374 293	7378	135 606	135	·619 343	3757	·754 949	3622	963 886	71	037 467	76
·999	·381 671	7385	135 471	136	·623 100	3760	·758 571	3625	963 957	71	037 391	76
2·000	7·389 056		135 335		3·626 860		3·762 196		964 028		037 315	

x	e^x	Δ	e^{-x}	Δ	sinh x	Δ	cosh x	Δ	tanh x	Δ	coth x	Δ
			0·						0·		1·	
2·000	7·389 06	739	135 335	135	3·626 860	3764	3·762 196	3628	964 028	70	037 315	76
·001	·396 45	740	135 200	135	·630 624	3768	·765 824	3633	964 098	71	037 239	76
·002	·403 85	741	135 065	135	·634 392	3771	·769 457	3636	964 169	70	037 163	76
·003	·411 26	741	134 930	135	·638 163	3775	·773 093	3640	964 239	70	037 087	75
·004	·418 67	742	134 795	135	·641 938	3779	·776 733	3644	964 309	70	037 012	75
2·005	7·426 09	743	134 660	134	3·645 717	3782	3·780 377	3648	964 379	70	036 937	76
·006	·433 52	744	134 526	135	·649 499	3786	·784 025	3651	964 449	70	036 861	75
·007	·440 96	745	134 391	134	·653 285	3789	·787 676	3655	964 519	69	036 786	74
·008	·448 41	745	134 257	134	·657 074	3794	·791 331	3659	964 588	70	036 712	75
·009	·455 86	746	134 123	134	·660 868	3796	·794 990	3663	964 658	69	036 637	75
2·010	7·463 32	746	133 989	134	3·664 664	3801	3·798 653	3667	964 727	70	036 562	74
·011	·470 78	748	133 855	134	·668 465	3804	·802 320	3670	964 797	69	036 488	74
·012	·478 26	748	133 721	134	·672 269	3808	·805 990	3674	964 866	69	036 414	74
·013	·485 74	749	133 587	133	·676 077	3811	·809 664	3678	964 935	68	036 340	74
·014	·493 23	750	133 454	134	·679 888	3815	·813 342	3682	965 003	69	036 266	74
2·015	7·500 73	750	133 320	133	3·683 703	3819	3·817 024	3686	965 072	69	036 192	74
·016	·508 23	751	133 187	133	·687 522	3823	·820 710	3689	965 141	68	036 118	73
·017	·515 74	752	133 054	133	·691 345	3826	·824 399	3693	965 209	68	036 045	73
·018	·523 26	753	132 921	133	·695 171	3830	·828 092	3697	965 277	69	035 972	74
·019	·530 79	753	132 788	133	·699 001	3834	·831 789	3701	965 346	68	035 898	73
2·020	7·538 32	755	132 655	132	3·702 835	3837	3·835 490	3705	965 414	68	035 825	72
·021	·545 87	755	132 523	133	·706 672	3841	·839 195	3709	965 482	67	035 753	73
·022	·553 42	755	132 390	132	·710 513	3845	·842 904	3712	965 549	68	035 680	73
·023	·560 97	757	132 258	132	·714 358	3848	·846 616	3716	965 617	68	035 607	72
·024	·568 54	757	132 126	132	·718 206	3853	·850 332	3720	965 685	67	035 535	72
2·025	7·576 11	758	131 994	132	3·722 059	3855	3·854 052	3724	965 752	67	035 463	73
·026	·583 69	759	131 862	132	·725 914	3860	·857 776	3728	965 819	67	035 390	71
·027	·591 28	759	131 730	132	·729 774	3863	·861 504	3732	965 886	67	035 319	72
·028	·598 87	761	131 598	131	·733 637	3868	·865 236	3736	965 953	67	035 247	72
·029	·606 48	761	131 467	131	·737 505	3870	·868 972	3739	966 020	67	035 175	71
2·030	7·614 09	761	131 336	132	3·741 375	3875	3·872 711	3743	966 087	67	035 104	72
·031	·621 70	763	131 204	131	·745 250	3878	·876 454	3747	966 154	66	035 032	71
·032	·629 33	763	131 073	131	·749 128	3882	·880 201	3752	966 220	66	034 961	71
·033	·636 96	764	130 942	131	·753 010	3886	·883 953	3754	966 286	67	034 890	71
·034	·644 60	765	130 811	131	·756 896	3890	·887 707	3759	966 353	66	034 819	71
2·035	7·652 25	766	130 680	130	3·760 786	3893	3·891 466	3763	966 419	66	034 748	70
·036	·659 91	766	130 550	131	·764 679	3897	·895 229	3767	966 485	66	034 678	71
·037	·667 57	767	130 419	130	·768 576	3901	·898 996	3770	966 551	65	034 607	70
·038	·675 24	768	130 289	130	·772 477	3905	·902 766	3775	966 616	66	034 537	70
·039	·682 92	769	130 159	130	·776 382	3908	·906 541	3778	966 682	65	034 467	71
2·040	7·690 61	769	130 029	130	3·780 290	3912	3·910 319	3782	966 747	66	034 396	69
·041	·698 30	771	129 899	130	·784 202	3916	·914 101	3786	966 813	65	034 327	70
·042	·706 01	771	129 769	130	·788 118	3920	·917 887	3790	966 878	65	034 257	70
·043	·713 72	771	129 639	129	·792 038	3924	·921 677	3794	966 943	65	034 187	69
·044	·721 43	773	129 510	130	·795 962	3927	·925 471	3798	967 008	65	034 118	70
2·045	7·729 16	773	129 380	129	3·799 889	3931	3·929 269	3802	967 073	64	034 048	69
·046	·736 89	774	129 251	129	·803 820	3935	·933 071	3806	967 137	65	033 979	69
·047	·744 63	775	129 122	129	·807 755	3939	·936 877	3810	967 202	64	033 910	69
·048	·752 38	776	128 993	129	·811 694	3943	·940 687	3813	967 266	65	033 841	68
·049	·760 14	776	128 864	129	·815 637	3946	·944 500	3818	967 331	64	033 773	69
2·050	7·767 90		128 735		3·819 583		3·948 318		967 395		033 704	

TABLE Vc—EXPONENTIAL AND HYPERBOLIC FUNCTIONS

x	e^x		e^{-x}		sinh x		cosh x		tanh x		coth x	
			0·						0·		1·	
2·050	7·767 90	777	128 735	129	3·819 583	3950	3·948 318	3822	967 395	64	033 704	69
·051	·775 67	778	128 606	128	·823 533	3954	·952 140	3825	967 459	64	033 635	68
·052	·783 45	779	128 478	129	·827 487	3958	·955 965	3830	967 523	64	033 567	68
·053	·791 24	779	128 349	128	·831 445	3962	·959 795	3833	967 587	64	033 499	68
·054	·799 03	781	128 221	128	·835 407	3966	·963 628	3837	967 651	63	033 431	68
2·055	7·806 84	781	128 093	128	3·839 373	3969	3·967 465	3842	967 714	64	033 363	68
·056	·814 65	782	127 965	128	·843 342	3973	·971 307	3845	967 778	63	033 295	67
·057	·822 47	782	127 837	128	·847 315	3977	·975 152	3849	967 841	63	033 228	68
·058	·830 29	784	127 709	128	·851 292	3981	·979 001	3854	967 904	63	033 160	67
·059	·838 13	784	127 581	127	·855 273	3985	·982 855	3857	967 967	63	033 093	67
2·060	7·845 97	785	127 454	127	3·859 258	3989	3·986 712	3861	968 030	63	033 026	68
·061	·853 82	786	127 327	128	·863 247	3992	·990 573	3865	968 093	63	032 958	66
·062	·861 68	786	127 199	127	·867 239	3996	·994 438	3870	968 156	63	032 892	67
·063	·869 54	788	127 072	127	·871 235	4001	3·998 308	3873	968 219	62	032 825	67
·064	·877 42	788	126 945	127	·875 236	4004	4·002 181	3877	968 281	62	032 758	66
2·065	7·885 30	789	126 818	126	3·879 240	4008	4·006 058	3881	968 343	63	032 692	67
·066	·893 19	789	126 692	127	·883 248	4012	·009 939	3886	968 406	62	032 625	66
·067	·901 08	791	126 565	127	·887 260	4015	·013 825	3889	968 468	62	032 559	66
·068	·908 99	791	126 438	126	·891 275	4020	·017 714	3893	968 530	62	032 493	66
·069	·916 90	792	126 312	126	·895 295	4024	·021 607	3897	968 592	61	032 427	66
2·070	7·924 82	793	126 186	126	3·899 319	4027	4·025 504	3902	968 653	62	032 361	66
·071	·932 75	794	126 060	126	·903 346	4031	·029 406	3905	968 715	62	032 295	65
·072	·940 69	794	125 934	126	·907 377	4036	·033 311	3910	968 777	61	032 230	66
·073	·948 63	796	125 808	126	·911 413	4039	·037 221	3913	968 838	61	032 164	65
·074	·956 59	796	125 682	126	·915 452	4043	·041 134	3917	968 899	61	032 099	65
2·075	7·964 55	796	125 556	125	3·919 495	4047	4·045 051	3922	968 960	62	032 034	65
·076	·972 51	798	125 431	125	·923 542	4051	·048 973	3926	969 022	60	031 969	65
·077	·980 49	799	125 306	126	·927 593	4055	·052 899	3929	969 082	61	031 904	65
·078	·988 48	799	125 180	125	·931 648	4059	·056 828	3934	969 143	61	031 839	64
·079	7·996 47	800	125 055	125	·935 707	4062	·060 762	3938	969 204	61	031 775	65
2·080	8·004 47	801	124 930	125	3·939 769	4067	4·064 700	3941	969 265	60	031 710	64
·081	·012 48	801	124 805	124	·943 836	4071	·068 641	3946	969 325	60	031 646	65
·082	·020 49	803	124 681	125	·947 907	4074	·072 587	3950	969 385	61	031 581	64
·083	·028 52	803	124 556	125	·951 981	4079	·076 537	3954	969 446	60	031 517	64
·084	·036 55	804	124 431	124	·956 060	4082	·080 491	3958	969 506	60	031 453	63
2·085	8·044 59	805	124 307	124	3·960 142	4087	4·084 449	3962	969 566	60	031 390	64
·086	·052 64	806	124 183	124	·964 229	4090	·088 411	3967	969 626	59	031 326	64
·087	·060 70	806	124 059	124	·968 319	4094	·092 378	3970	969 685	60	031 262	63
·088	·068 76	807	123 935	124	·972 413	4099	·096 348	3975	969 745	60	031 199	63
·089	·076 83	809	123 811	124	·976 512	4102	·100 323	3978	969 805	59	031 136	64
2·090	8·084 92	808	123 687	123	3·980 614	4106	4·104 301	3983	969 864	59	031 072	63
·091	·093 00	810	123 564	124	·984 720	4111	·108 284	3987	969 923	60	031 009	63
·092	·101 10	811	123 440	123	·988 831	4114	·112 271	3990	969 983	59	030 946	62
·093	·109 21	811	123 317	124	·992 945	4118	·116 261	3995	970 042	59	030 884	63
·094	·117 32	812	123 193	123	3·997 063	4122	·120 256	4000	970 101	58	030 821	63
2·095	8·125 44	813	123 070	123	4·001 185	4127	4·124 256	4003	970 159	59	030 758	62
·096	·133 57	814	122 947	123	·005 312	4130	·128 259	4007	970 218	59	030 696	62
·097	·141 71	814	122 824	122	·009 442	4134	·132 266	4012	970 277	58	030 634	62
·098	·149 85	816	122 702	123	·013 576	4138	·136 278	4015	970 335	59	030 572	62
·099	·158 01	816	122 579	123	·017 714	4143	·140 293	4020	970 394	58	030 510	62
2·100	8·166 17		122 456		4·021 857		4·144 313		970 452		030 448	

TABLE Vc—EXPONENTIAL AND HYPERBOLIC FUNCTIONS

x	e^x	Δ	e^{-x} 0·	Δ	sinh x	Δ	cosh x	Δ	tanh x 0·	Δ	coth x 1·	Δ
2·100	8·166 17	817	122 456	122	4·021 857	4146	4·144 313	4024	970 452	58	030 448	62
·101	·174 34	818	122 334	122	·026 003	4150	·148 337	4028	970 510	58	030 386	62
·102	·182 52	819	122 212	122	·030 153	4155	·152 365	4032	970 568	58	030 324	61
·103	·190 71	819	122 090	122	·034 308	4158	·156 397	4037	970 626	58	030 263	62
·104	·198 90	820	121 968	122	·038 466	4163	·160 434	4040	970 684	58	030 201	61
2·105	8·207 10	821	121 846	122	4·042 629	4166	4·164 474	4045	970 742	57	030 140	61
·106	·215 31	822	121 724	122	·046 795	4171	·168 519	4049	970 799	58	030 079	61
·107	·223 53	823	121 602	121	·050 966	4174	·172 568	4053	970 857	57	030 018	61
·108	·231 76	824	121 481	122	·055 140	4179	·176 621	4057	970 914	57	029 957	61
·109	·240 00	824	121 359	121	·059 319	4183	·180 678	4062	970 971	58	029 896	60
2·110	8·248 24	825	121 238	121	4·063 502	4186	4·184 740	4065	971 029	57	029 836	61
·111	·256 49	826	121 117	121	·067 688	4191	·188 805	4070	971 086	57	029 775	60
·112	·264 75	827	120 996	121	·071 879	4195	·192 875	4074	971 143	56	029 715	60
·113	·273 02	828	120 875	121	·076 074	4199	·196 949	4078	971 199	57	029 655	60
·114	·281 30	829	120 754	121	·080 273	4203	·201 027	4083	971 256	57	029 595	60
2·115	8·289 59	829	120 633	120	4·084 476	4207	4·205 110	4086	971 313	56	029 535	60
·116	·297 88	830	120 513	121	·088 683	4212	·209 196	4091	971 369	57	029 475	60
·117	·306 18	831	120 392	120	·092 895	4215	·213 287	4095	971 426	56	029 415	60
·118	·314 49	832	120 272	120	·097 110	4219	·217 382	4099	971 482	56	029 355	59
·119	·322 81	833	120 152	120	·101 329	4224	·221 481	4104	971 538	56	029 296	60
2·120	8·331 14	833	120 032	120	4·105 553	4228	4·225 585	4107	971 594	56	029 236	59
·121	·339 47	835	119 912	120	·109 781	4231	·229 692	4112	971 650	56	029 177	59
·122	·347 82	835	119 792	120	·114 012	4236	·233 804	4116	971 706	56	029 118	59
·123	·356 17	836	119 672	120	·118 248	4240	·237 920	4121	971 762	55	029 059	59
·124	·364 53	837	119 552	119	·122 488	4244	·242 041	4124	971 817	56	029 000	59
2·125	8·372 90	837	119 433	119	4·126 732	4248	4·246 165	4129	971 873	55	028 941	58
·126	·381 27	839	119 314	120	·130 980	4253	·250 294	4133	971 928	55	028 883	59
·127	·389 66	839	119 194	119	·135 233	4256	·254 427	4138	971 983	56	028 824	58
·128	·398 05	841	119 075	119	·139 489	4261	·258 565	4141	972 039	55	028 766	59
·129	·406 46	841	118 956	119	·143 750	4265	·262 706	4146	972 094	55	028 707	58
2·130	8·414 87	842	118 837	118	4·148 015	4269	4·266 852	4150	972 149	55	028 649	58
·131	·423 29	842	118 719	119	·152 284	4273	·271 002	4155	972 204	54	028 591	58
·132	·431 71	844	118 600	119	·156 557	4277	·275 157	4158	972 258	55	028 533	58
·133	·440 15	844	118 481	118	·160 834	4281	·279 315	4163	972 313	55	028 475	57
·134	·448 59	846	118 363	118	·165 115	4286	·283 478	4168	972 368	54	028 418	58
2·135	8·457 05	846	118 245	119	4·169 401	4290	4·287 646	4171	972 422	54	028 360	57
·136	·465 51	847	118 126	118	·173 691	4294	·291 817	4176	972 476	55	028 303	58
·137	·473 98	848	118 008	118	·177 985	4298	·295 993	4180	972 531	54	028 245	57
·138	·482 46	848	117 890	117	·182 283	4302	·300 173	4184	972 585	54	028 188	57
·139	·490 94	850	117 773	118	·186 585	4306	·304 357	4189	972 639	54	028 131	57
2·140	8·499 44	850	117 655	118	4·190 891	4311	4·308 546	4193	972 693	53	028 074	57
·141	·507 94	851	117 537	117	·195 202	4315	·312 739	4198	972 746	54	028 017	57
·142	·516 45	852	117 420	118	·199 517	4319	·316 937	4201	972 800	54	027 960	56
·143	·524 97	853	117 302	117	·203 836	4323	·321 138	4206	972 854	53	027 904	57
·144	·533 50	854	117 185	117	·208 159	4328	·325 344	4211	972 907	54	027 847	56
2·145	8·542 04	855	117 068	117	4·212 487	4331	4·329 555	4214	972 961	53	027 791	57
·146	·550 59	855	116 951	117	·216 818	4336	·333 769	4219	973 014	53	027 734	56
·147	·559 14	857	116 834	117	·221 154	4340	·337 988	4224	973 067	53	027 678	56
·148	·567 71	857	116 717	116	·225 494	4345	·342 212	4227	973 120	53	027 622	56
·149	·576 28	858	116 601	117	·229 839	4348	·346 439	4232	973 173	53	027 566	56
2·150	8·584 86		116 484		4·234 187		4·350 671		973 226		027 510	

TABLE Vc—EXPONENTIAL AND HYPERBOLIC FUNCTIONS

x	e^x	Δ	e^{-x}	Δ	sinh x	Δ	cosh x	Δ	tanh x	Δ	coth x	Δ
			0·						0·		1·	
2·150	8·584 86	859	116 484	116	4·234 187	4353	4·350 671	4237	973 226	53	027 510	55
·151	·593 45	860	116 368	117	·238 540	4357	·354 908	4240	973 279	53	027 455	56
·152	·602 05	860	116 251	116	·242 897	4361	·359 148	4245	973 332	52	027 399	55
·153	·610 65	862	116 135	116	·247 258	4366	·363 393	4250	973 384	53	027 344	56
·154	·619 27	862	116 019	116	·251 624	4369	·367 643	4254	973 437	52	027 288	55
2·155	8·627 89	863	115 903	116	4·255 993	4375	4·371 897	4258	973 489	52	027 233	55
·156	·636 52	864	115 787	115	·260 368	4378	·376 155	4262	973 541	52	027 178	55
·157	·645 16	865	115 672	116	·264 746	4382	·380 417	4267	973 593	53	027 123	55
·158	·653 81	866	115 556	115	·269 128	4387	·384 684	4272	973 646	52	027 068	55
·159	·662 47	867	115 441	116	·273 515	4391	·388 956	4275	973 698	51	027 013	55
2·160	8·671 14	867	115 325	115	4·277 906	4396	4·393 231	4280	973 749	52	026 958	54
·161	·679 81	869	115 210	115	·282 302	4399	·397 511	4285	973 801	52	026 904	55
·162	·688 50	869	115 095	115	·286 701	4404	·401 796	4289	973 853	51	026 849	54
·163	·697 19	870	114 980	115	·291 105	4408	·406 085	4293	973 904	52	026 795	54
·164	·705 89	871	114 865	115	·295 513	4413	·410 378	4298	973 956	51	026 741	55
2·165	8·714 60	872	114 750	115	4·299 926	4417	4·414 676	4302	974 007	51	026 686	54
·166	·723 32	873	114 635	114	·304 343	4421	·418 978	4307	974 058	52	026 632	53
·167	·732 05	873	114 521	115	·308 764	4425	·423 285	4311	974 110	51	026 579	54
·168	·740 78	875	114 406	114	·313 189	4430	·427 596	4315	974 161	51	026 525	54
·169	·749 53	875	114 292	114	·317 619	4434	·431 911	4320	974 212	50	026 471	54
2·170	8·758 28	877	114 178	115	4·322 053	4439	4·436 231	4324	974 262	51	026 417	53
·171	·767 05	877	114 063	114	·326 492	4442	·440 555	4329	974 313	51	026 364	53
·172	·775 82	878	113 949	113	·330 934	4447	·444 884	4333	974 364	50	026 311	54
·173	·784 60	879	113 836	114	·335 381	4452	·449 217	4338	974 414	51	026 257	53
·174	·793 39	880	113 722	114	·339 833	4455	·453 555	4342	974 465	50	026 204	53
2·175	8·802 19	880	113 608	113	4·344 288	4461	4·457 897	4346	974 515	51	026 151	53
·176	·810 99	882	113 495	114	·348 749	4464	·462 243	4351	974 566	50	026 098	53
·177	·819 81	882	113 381	113	·353 213	4469	·466 594	4356	974 616	50	026 045	52
·178	·828 63	883	113 268	113	·357 682	4473	·470 950	4360	974 666	50	025 993	53
·179	·837 46	885	113 155	113	·362 155	4477	·475 310	4364	974 716	50	025 940	52
2·180	8·846 31	885	113 042	113	4·366 632	4482	4·479 674	4369	974 766	49	025 888	53
·181	·855 16	886	112 929	113	·371 114	4486	·484 043	4373	974 815	50	025 835	52
·182	·864 02	887	112 816	113	·375 600	4491	·488 416	4378	974 865	50	025 783	52
·183	·872 89	887	112 703	113	·380 091	4495	·492 794	4382	974 915	49	025 731	52
·184	·881 76	889	112 590	112	·384 586	4499	·497 176	4387	974 964	50	025 679	52
2·185	8·890 65	889	112 478	113	4·389 085	4504	4·501 563	4391	975 014	49	025 627	52
·186	·899 54	891	112 365	112	·393 589	4508	·505 954	4396	975 063	49	025 575	52
·187	·908 45	891	112 253	112	·398 097	4513	·510 350	4401	975 112	49	025 523	52
·188	·917 36	892	112 141	112	·402 610	4517	·514 751	4405	975 161	49	025 471	51
·189	·926 28	893	112 029	112	·407 127	4521	·519 156	4409	975 210	49	025 420	52
2·190	8·935 21	894	111 917	112	4·411 648	4526	4·523 565	4414	975 259	49	025 368	51
·191	·944 15	895	111 805	112	·416 174	4530	·527 979	4418	975 308	49	025 317	51
·192	·953 10	896	111 693	111	·420 704	4535	·532 397	4423	975 357	48	025 266	51
·193	·962 06	897	111 582	112	·425 239	4539	·536 820	4428	975 405	49	025 215	51
·194	·971 03	897	111 470	111	·429 778	4543	·541 248	4432	975 454	48	025 164	51
2·195	8·980 00	899	111 359	112	4·434 321	4548	4·545 680	4436	975 502	49	025 113	51
·196	·988 99	899	111 247	111	·438 869	4552	·550 116	4442	975 551	48	025 062	51
·197	8·997 98	900	111 136	111	·443 421	4557	·554 558	4445	975 599	48	025 011	50
·198	9·006 98	901	111 025	111	·447 978	4561	·559 003	4451	975 647	48	024 961	51
·199	·015 99	902	110 914	111	·452 539	4566	·563 454	4454	975 695	48	024 910	50
2·200	9·025 01		110 803		4·457 105		4·567 908		975 743		024 860	

x	e^x		e^{-x}		sinh x		cosh x		tanh x		coth x	
			0·						0·		1·	
2·200	9·025 01	903	110 803	111	4·457 105	4570	4·567 908	4460	975 743	48	024 860	50
·201	·034 04	904	110 692	110	·461 675	4575	·572 368	4464	975 791	48	024 810	51
·202	·043 08	905	110 582	111	·466 250	4579	·576 832	4468	975 839	47	024 759	50
·203	·052 13	906	110 471	110	·470 829	4584	·581 300	4473	975 886	48	024 709	50
·204	·061 19	906	110 361	110	·475 413	4588	·585 773	4478	975 934	48	024 659	50
2·205	9·070 25	908	110 251	111	4·480 001	4592	4·590 251	4482	975 982	47	024 609	49
·206	·079 33	908	110 140	110	·484 593	4597	·594 733	4487	976 029	47	024 560	50
·207	·088 41	909	110 030	110	·489 190	4601	·599 220	4492	976 076	48	024 510	50
·208	·097 50	911	109 920	110	·493 791	4606	·603 712	4496	976 124	47	024 460	49
·209	·106 61	911	109 810	109	·498 397	4611	·608 208	4501	976 171	47	024 411	49
2·210	9·115 72	912	109 701	110	4·503 008	4615	4·612 709	4505	976 218	47	024 362	50
·211	·124 84	913	109 591	110	·507 623	4619	·617 214	4510	976 265	47	024 312	49
·212	·133 97	913	109 481	109	·512 242	4624	·621 724	4514	976 312	46	024 263	49
·213	·143 10	915	109 372	109	·516 866	4629	·626 238	4520	976 358	47	024 214	49
·214	·152 25	916	109 263	109	·521 495	4633	·630 758	4523	976 405	47	024 165	49
2·215	9·161 41	917	109 154	110	4·526 128	4637	4·635 281	4529	976 452	46	024 116	48
·216	·170 58	917	109 044	109	·530 765	4642	·639 810	4533	976 498	46	024 068	49
·217	·179 75	918	108 935	108	·535 407	4647	·644 343	4538	976 544	47	024 019	49
·218	·188 93	920	108 827	109	·540 054	4651	·648 881	4542	976 591	46	023 970	48
·219	·198 13	920	108 718	109	·544 705	4656	·653 423	4547	976 637	46	023 922	48
2·220	9·207 33	921	108 609	108	4·549 361	4660	4·657 970	4552	976 683	46	023 874	49
·221	·216 54	922	108 501	109	·554 021	4665	·662 522	4556	976 729	46	023 825	48
·222	·225 76	923	108 392	108	·558 686	4669	·667 078	4561	976 775	46	023 777	48
·223	·234 99	924	108 284	108	·563 355	4674	·671 639	4566	976 821	46	023 729	48
·224	·244 23	925	108 176	109	·568 029	4679	·676 205	4570	976 867	45	023 681	48
2·225	9·253 48	926	108 067	108	4·572 708	4683	4·680 775	4575	976 912	46	023 633	48
·226	·262 74	927	107 959	108	·577 391	4687	·685 350	4580	976 958	46	023 585	47
·227	·272 01	927	107 851	107	·582 078	4693	·689 930	4584	977 004	45	023 538	48
·228	·281 28	929	107 744	108	·586 771	4696	·694 514	4589	977 049	45	023 490	47
·229	·290 57	930	107 636	108	·591 467	4702	·699 103	4594	977 094	46	023 443	48
2·230	9·299 87	930	107 528	107	4·596 169	4706	4·703 697	4599	977 140	45	023 395	47
·231	·309 17	931	107 421	107	·600 875	4710	·708 296	4603	977 185	45	023 348	47
·232	·318 48	933	107 314	108	·605 585	4716	·712 899	4608	977 230	45	023 301	47
·233	·327 81	933	107 206	107	·610 301	4719	·717 507	4613	977 275	45	023 254	47
·234	·337 14	934	107 099	107	·615 020	4725	·722 120	4617	977 320	44	023 207	47
2·235	9·346 48	935	106 992	107	4·619 745	4729	4·726 737	4622	977 364	45	023 160	47
·236	·355 83	936	106 885	107	·624 474	4734	·731 359	4627	977 409	45	023 113	47
·237	·365 19	937	106 778	106	·629 208	4738	·735 986	4632	977 454	44	023 066	46
·238	·374 56	938	106 672	107	·633 946	4743	·740 618	4636	977 498	45	023 020	47
·239	·383 94	939	106 565	106	·638 689	4747	·745 254	4641	977 543	44	022 973	46
2·240	9·393 33	940	106 459	107	4·643 436	4753	4·749 895	4646	977 587	44	022 927	47
·241	·402 73	941	106 352	106	·648 189	4756	·754 541	4650	977 631	45	022 880	46
·242	·412 14	941	106 246	106	·652 945	4762	·759 191	4656	977 676	44	022 834	46
·243	·421 55	943	106 140	106	·657 707	4766	·763 847	4660	977 720	44	022 788	46
·244	·430 98	944	106 034	106	·662 473	4771	·768 507	4665	977 764	44	022 742	46
2·245	9·440 42	944	105 928	106	4·667 244	4776	4·773 172	4669	977 808	44	022 696	46
·246	·449 86	946	105 822	106	·672 020	4780	·777 841	4675	977 852	43	022 650	46
·247	·459 32	946	105 716	106	·676 800	4785	·782 516	4679	977 895	44	022 604	45
·248	·468 78	947	105 610	105	·681 585	4789	·787 195	4684	977 939	44	022 559	46
·249	·478 25	949	105 505	106	·686 374	4794	·791 879	4689	977 983	43	022 513	45
2·250	9·487 74		105 399		4·691 168		4·796 568		978 026		022 468	

x	e^{x}		e^{-x}		$\sinh x$		$\cosh x$		$\tanh x$		$\coth x$	
			0·						0·		1·	
2·250	9·487 74	949	105 399	105	4·691 168	4799	4·796 568	4693	978 026	44	022 468	46
·251	·497 23	950	105 294	105	·695 967	4804	·801 261	4698	978 070	43	022 422	45
·252	·506 73	951	105 189	105	·700 771	4808	·805 959	4704	978 113	43	022 377	45
·253	·516 24	952	105 084	106	·705 579	4813	·810 663	4708	978 156	43	022 332	45
·254	·525 76	953	104 978	104	·710 392	4818	·815 371	4712	978 199	43	022 287	45
2·255	9·535 29	954	104 874	105	4·715 210	4822	4·820 083	4718	978 242	43	022 242	45
·256	·544 83	955	104 769	105	·720 032	4827	·824 801	4722	978 285	43	022 197	45
·257	·554 38	956	104 664	105	·724 859	4832	·829 523	4728	978 328	43	022 152	45
·258	·563 94	957	104 559	104	·729 691	4837	·834 251	4732	978 371	43	022 107	45
·259	·573 51	958	104 455	105	·734 528	4841	·838 983	4737	978 414	43	022 062	44
2·260	9·583 09	959	104 350	104	4·739 369	4846	4·843 720	4742	978 457	42	022 018	45
·261	·592 68	959	104 246	104	·744 215	4851	·848 462	4746	978 499	43	021 973	44
·262	·602 27	961	104 142	104	·749 066	4856	·853 208	4752	978 542	42	021 929	44
·263	·611 88	962	104 038	104	·753 922	4860	·857 960	4756	978 584	42	021 885	45
·264	·621 50	962	103 934	104	·758 782	4865	·862 716	4761	978 626	43	021 840	44
2·265	9·631 12	964	103 830	104	4·763 647	4870	4·867 477	4766	978 669	42	021 796	44
·266	·640 76	965	103 726	103	·768 517	4875	·872 243	4771	978 711	42	021 752	44
·267	·650 41	965	103 623	104	·773 392	4879	·877 014	4776	978 753	42	021 708	43
·268	·660 06	967	103 519	103	·778 271	4884	·881 790	4781	978 795	42	021 665	44
·269	·669 73	967	103 416	104	·783 155	4889	·886 571	4785	978 837	42	021 621	44
2·270	9·679 40	969	103 312	103	4·788 044	4894	4·891 356	4791	978 879	41	021 577	43
·271	·689 09	969	103 209	103	·792 938	4899	·896 147	4795	978 920	42	021 534	44
·272	·698 78	970	103 106	103	·797 837	4903	·900 942	4801	978 962	42	021 490	43
·273	·708 48	972	103 003	103	·802 740	4908	·905 743	4805	979 004	41	021 447	44
·274	·718 20	972	102 900	103	·807 648	4913	·910 548	4810	979 045	42	021 403	43
2·275	9·727 92	973	102 797	103	4·812 561	4918	4·915 358	4815	979 087	41	021 360	43
·276	·737 65	974	102 694	102	·817 479	4922	·920 173	4820	979 128	41	021 317	43
·277	·747 39	976	102 592	103	·822 401	4928	·924 993	4825	979 169	41	021 274	43
·278	·757 15	976	102 489	102	·827 329	4932	·929 818	4830	979 210	41	021 231	43
·279	·766 91	977	102 387	103	·832 261	4937	·934 648	4834	979 251	42	021 188	43
2·280	9·776 68	978	102 284	102	4·837 198	4942	4·939 482	4840	979 293	40	021 145	42
·281	·786 46	979	102 182	102	·842 140	4947	·944 322	4845	979 333	41	021 103	43
·282	·796 25	980	102 080	102	·847 087	4951	·949 167	4849	979 374	41	021 060	42
·283	·806 05	982	101 978	102	·852 038	4957	·954 016	4855	979 415	41	021 018	43
·284	·815 87	982	101 876	102	·856 995	4961	·958 871	4859	979 456	40	020 975	42
2·285	9·825 69	983	101 774	102	4·861 956	4966	4·963 730	4865	979 496	41	020 933	43
·286	·835 52	984	101 672	101	·866 922	4971	·968 595	4869	979 537	40	020 890	42
·287	·845 36	985	101 571	102	·871 893	4976	·973 464	4874	979 577	41	020 848	42
·288	·855 21	986	101 469	101	·876 869	4981	·978 338	4880	979 618	40	020 806	42
·289	·865 07	987	101 368	102	·881 850	4986	·983 218	4884	979 658	40	020 764	42
2·290	9·874 94	988	101 266	101	4·886 836	4990	4·988 102	4889	979 698	41	020 722	42
·291	·884 82	989	101 165	101	·891 826	4996	·992 991	4895	979 739	40	020 680	41
·292	·894 71	990	101 064	101	·896 822	5000	4·997 886	4899	979 779	40	020 639	42
·293	·904 61	991	100 963	101	·901 822	5005	5·002 785	4904	979 819	40	020 597	42
·294	·914 52	992	100 862	101	·906 827	5010	·007 689	4910	979 859	39	020 555	41
2·295	9·924 44	993	100 761	100	4·911 837	5015	5·012 599	4914	979 898	40	020 514	41
·296	·934 37	993	100 661	101	·916 852	5020	·017 513	4919	979 938	40	020 473	42
·297	·944 30	995	100 560	100	·921 872	5025	·022 432	4925	979 978	39	020 431	41
·298	·954 25	996	100 460	101	·926 897	5030	·027 357	4929	980 017	40	020 390	41
·299	·964 21	997	100 359	100	·931 927	5035	·032 286	4935	980 057	39	020 349	41
2·300	9·974 18		100 259		4·936 962		5·037 221		980 096		020 308	

TABLE Vc—EXPONENTIAL AND HYPERBOLIC FUNCTIONS

x	e^x		e^{-x}		sinh x		cosh x		tanh x		coth x	
			0·						0·		1·	
2·300	9·974 18	998	100 259	100	4·936 962	*5039*	5·037 221	*4939*	980 096	40	020 308	41
·301	·984 16	999	100 159	100	·942 001	*5045*	·042 160	*4945*	980 136	39	020 267	41
·302	9·994 15	1000	100 059	100	·947 046	*5050*	·047 105	*4949*	980 175	39	020 226	41
·303	10·004 15	1001	099 959	100	·952 096	*5054*	·052 054	*4955*	980 214	39	020 185	41
·304	·014 16	1002	099 859	100	·957 150	*5060*	·057 009	*4960*	980 253	39	020 144	40
2·305	10·024 18	1003	099 759	100	4·962 210	*5064*	5·061 969	*4964*	980 292	39	020 104	41
·306	·034 21	1004	099 659	100	·967 274	*5070*	·066 933	*4970*	980 331	39	020 063	40
·307	·044 25	1005	099 559	99	·972 344	*5074*	·071 903	*4975*	980 370	39	020 023	41
·308	·054 30	1006	099 460	99	·977 418	*5079*	·076 878	*4980*	980 409	39	019 982	40
·309	·064 36	1006	099 361	100	·982 497	*5085*	·081 858	*4985*	980 448	39	019 942	40
2·310	10·074 42	1008	099 261	99	4·987 582	*5089*	5·086 843	*4990*	980 487	38	019 902	40
·311	·084 50	1009	099 162	99	·992 671	*5094*	·091 833	*4995*	980 525	39	019 862	41
·312	·094 59	1010	099 063	99	4·997 765	*5100*	·096 828	*5001*	980 564	38	019 821	40
·313	·104 69	1011	098 964	99	5·002 865	*5104*	·101 829	*5005*	980 602	39	019 781	39
·314	·114 80	1012	098 865	99	·007 969	*5109*	·106 834	*5011*	980 641	38	019 742	40
2·315	10·124 92	1013	098 766	99	5·013 078	*5115*	5·111 845	*5015*	980 679	38	019 702	40
·316	·135 05	1014	098 667	98	·018 193	*5119*	·116 860	*5021*	980 717	38	019 662	40
·317	·145 19	1015	098 569	99	·023 312	*5124*	·121 881	*5026*	980 755	38	019 622	39
·318	·155 34	1016	098 470	98	·028 436	*5130*	·126 907	*5031*	980 793	38	019 583	40
·319	·165 50	1017	098 372	98	·033 566	*5134*	·131 938	*5036*	980 831	38	019 543	39
2·320	10·175 67	1019	098 274	99	5·038 700	*5140*	5·136 974	*5041*	980 869	38	019 504	40
·321	·185 86	1019	098 175	98	·043 840	*5144*	·142 015	*5047*	980 907	38	019 464	39
·322	·196 05	1020	098 077	98	·048 984	*5150*	·147 062	*5051*	980 945	38	019 425	39
·323	·206 25	1021	097 979	98	·054 134	*5155*	·152 113	*5057*	980 983	37	019 386	39
·324	·216 46	1022	097 881	98	·059 289	*5159*	·157 170	*5062*	981 020	38	019 347	39
2·325	10·226 68	1023	097 783	97	5·064 448	*5165*	5·162 232	*5067*	981 058	37	019 308	39
·326	·236 91	1024	097 686	98	·069 613	*5170*	·167 299	*5072*	981 095	38	019 269	39
·327	·247 15	1026	097 588	97	·074 783	*5175*	·172 371	*5077*	981 133	37	019 230	39
·328	·257 41	1026	097 491	98	·079 958	*5180*	·177 448	*5083*	981 170	37	019 191	39
·329	·267 67	1027	097 393	97	·085 138	*5185*	·182 531	*5088*	981 207	38	019 152	38
2·330	10·277 94	1028	097 296	98	5·090 323	*5190*	5·187 619	*5093*	981 245	37	019 114	39
·331	·288 22	1030	097 198	97	·095 513	*5195*	·192 712	*5098*	981 282	37	019 075	38
·332	·298 52	1030	097 101	97	·100 708	*5201*	·197 810	*5103*	981 319	37	019 037	39
·333	·308 82	1032	097 004	97	·105 909	*5205*	·202 913	*5108*	981 356	37	018 998	38
·334	·319 14	1032	096 907	97	·111 114	*5211*	·208 021	*5114*	981 393	37	018 960	38
2·335	10·329 46	1033	096 810	96	5·116 325	*5215*	5·213 135	*5119*	981 430	36	018 922	38
·336	·339 79	1035	096 714	97	·121 540	*5221*	·218 254	*5124*	981 466	37	018 884	38
·337	·350 14	1035	096 617	97	·126 761	*5226*	·223 378	*5130*	981 503	37	018 846	38
·338	·360 49	1037	096 520	96	·131 987	*5231*	·228 508	*5134*	981 540	36	018 808	38
·339	·370 86	1038	096 424	96	·137 218	*5236*	·233 642	*5140*	981 576	37	018 770	38
2·340	10·381 24	1038	096 328	97	5·142 454	*5242*	5·238 782	*5145*	981 613	36	018 732	38
·341	·391 62	1040	096 231	96	·147 696	*5246*	·243 927	*5150*	981 649	36	018 694	38
·342	·402 02	1041	096 135	96	·152 942	*5252*	·249 077	*5156*	981 685	37	018 656	37
·343	·412 43	1041	096 039	96	·158 194	*5257*	·254 233	*5161*	981 722	36	018 619	38
·344	·422 84	1043	095 943	96	·163 451	*5262*	·259 394	*5166*	981 758	36	018 581	37
2·345	10·433 27	1044	095 847	96	5·168 713	*5267*	5·264 560	*5171*	981 794	36	018 544	38
·346	·443 71	1045	095 751	95	·173 980	*5272*	·269 731	*5177*	981 830	36	018 506	37
·347	·454 16	1046	095 656	96	·179 252	*5278*	·274 908	*5182*	981 866	36	018 469	37
·348	·464 62	1047	095 560	95	·184 530	*5282*	·280 090	*5187*	981 902	36	018 432	37
·349	·475 09	1048	095 465	96	·189 812	*5288*	·285 277	*5192*	981 938	35	018 395	37
2·350	10·485 57		095 369		5·195 100		5·290 469		981 973		018 358	

TABLE Vc—EXPONENTIAL AND HYPERBOLIC FUNCTIONS

x	e^x	Δ	e^{-x}	Δ	sinh x	Δ	cosh x	Δ	tanh x	Δ	coth x	Δ
			0·						0·		1·	
2·350	10·485 57	1049	095 369	95	5·195 100	5293	5·290 469	5198	981 973	36	018 358	37
·351	·496 06	1050	095 274	95	·200 393	5299	·295 667	5203	982 009	36	018 321	37
·352	·506 56	1051	095 179	96	·205 692	5303	·300 870	5209	982 045	35	018 284	37
·353	·517 07	1053	095 083	95	·210 995	5309	·306 079	5213	982 080	36	018 247	37
·354	·527 60	1053	094 988	94	·216 304	5314	·311 292	5219	982 116	35	018 210	37
2·355	10·538 13	1054	094 894	95	5·221 618	5319	5·316 511	5224	982 151	36	018 173	36
·356	·548 67	1056	094 799	95	·226 937	5324	·321 735	5230	982 187	35	018 137	37
·357	·559 23	1056	094 704	95	·232 261	5330	·326 965	5235	982 222	35	018 100	36
·358	·569 79	1058	094 609	94	·237 591	5335	·332 200	5240	982 257	35	018 064	37
·359	·580 37	1058	094 515	95	·242 926	5340	·337 440	5246	982 292	35	018 027	36
2·360	10·590 95	1060	094 420	94	5·248 266	5345	5·342 686	5251	982 327	35	017 991	37
·361	·601 55	1060	094 326	94	·253 611	5350	·347 937	5256	982 362	35	017 954	36
·362	·612 15	1062	094 232	95	·258 961	5356	·353 193	5262	982 397	35	017 918	36
·363	·622 77	1063	094 137	94	·264 317	5361	·358 455	5267	982 432	35	017 882	36
·364	·633 40	1064	094 043	94	·269 678	5367	·363 722	5272	982 467	35	017 846	36
2·365	10·644 04	1065	093 949	94	5·275 045	5371	5·368 994	5278	982 502	34	017 810	36
·366	·654 69	1066	093 855	93	·280 416	5377	·374 272	5283	982 536	35	017 774	36
·367	·665 35	1067	093 762	94	·285 793	5383	·379 555	5288	982 571	34	017 738	35
·368	·676 02	1068	093 668	94	·291 176	5387	·384 843	5294	982 605	35	017 703	36
·369	·686 70	1069	093 574	93	·296 563	5393	·390 137	5300	982 640	34	017 667	36
2·370	10·697 39	1071	093 481	94	5·301 956	5398	5·395 437	5304	982 674	34	017 631	35
·371	·708 10	1071	093 387	93	·307 354	5403	·400 741	5310	982 708	35	017 596	36
·372	·718 81	1072	093 294	93	·312 757	5409	·406 051	5316	982 743	34	017 560	35
·373	·729 53	1074	093 201	93	·318 166	5414	·411 367	5321	982 777	34	017 525	35
·374	·740 27	1074	093 108	94	·323 580	5419	·416 688	5326	982 811	34	017 490	36
2·375	10·751 01	1076	093 014	92	5·328 999	5425	5·422 014	5332	982 845	34	017 454	35
·376	·761 77	1077	092 922	93	·334 424	5430	·427 346	5337	982 879	34	017 419	35
·377	·772 54	1077	092 829	93	·339 854	5435	·432 683	5342	982 913	34	017 384	35
·378	·783 31	1079	092 736	93	·345 289	5441	·438 025	5348	982 947	34	017 349	35
·379	·794 10	1080	092 643	92	·350 730	5446	·443 373	5354	982 981	33	017 314	35
2·380	10·804 90	1081	092 551	93	5·356 176	5452	5·448 727	5359	983 014	34	017 279	35
·381	·815 71	1082	092 458	92	·361 628	5456	·454 086	5364	983 048	34	017 244	34
·382	·826 53	1084	092 366	93	·367 084	5462	·459 450	5370	983 082	33	017 210	35
·383	·837 37	1084	092 273	92	·372 546	5468	·464 820	5375	983 115	33	017 175	35
·384	·848 21	1085	092 181	92	·378 014	5473	·470 195	5381	983 148	34	017 140	34
2·385	10·859 06	1087	092 089	92	5·383 487	5478	5·475 576	5386	983 182	33	017 106	35
·386	·869 93	1087	091 997	92	·388 965	5484	·480 962	5392	983 215	33	017 071	34
·387	·880 80	1089	091 905	92	·394 449	5489	·486 354	5397	983 248	34	017 037	34
·388	·891 69	1090	091 813	92	·399 938	5494	·491 751	5403	983 282	33	017 003	35
·389	·902 59	1090	091 721	91	·405 432	5500	·497 154	5408	983 315	33	016 968	34
2·390	10·913 49	1092	091 630	92	5·410 932	5505	5·502 562	5413	983 348	33	016 934	34
·391	·924 41	1093	091 538	91	·416 437	5511	·507 975	5420	983 381	33	016 900	34
·392	·935 34	1094	091 447	92	·421 948	5516	·513 395	5424	983 414	33	016 866	34
·393	·946 28	1096	091 355	91	·427 464	5522	·518 819	5431	983 447	32	016 832	34
·394	·957 24	1096	091 264	91	·432 986	5527	·524 250	5435	983 479	33	016 798	34
2·395	10·968 20	1097	091 173	91	5·438 513	5532	5·529 685	5442	983 512	33	016 764	34
·396	·979 17	1099	091 082	91	·444 045	5538	·535 127	5446	983 545	32	016 730	33
·397	10·990 16	1099	090 991	91	·449 583	5543	·540 573	5453	983 577	33	016 697	34
·398	11·001 15	1101	090 900	91	·455 126	5549	·546 026	5458	983 610	32	016 663	33
·399	·012 16	1102	090 809	91	·460 675	5554	·551 484	5463	983 642	33	016 630	34
2·400	11·023 18		090 718		5·466 229		5·556 947		983 675		016 596	

TABLE Vc—EXPONENTIAL AND HYPERBOLIC FUNCTIONS

x	e^x		e^{-x} 0·		sinh x		cosh x		tanh x 0·		coth x 1·	
2·400	11·023 18	1103	090 718	91	5·466 229	5560	5·556 947	5469	983 675	32	016 596	33
·401	·034 21	1103	090 627	90	·471 789	5565	·562 416	5475	983 707	32	016 563	34
·402	·045 24	1106	090 537	91	·477 354	5571	·567 891	5480	983 739	33	016 529	33
·403	·056 30	1106	090 446	90	·482 925	5576	·573 371	5486	983 772	32	016 496	33
·404	·067 36	1107	090 356	91	·488 501	5581	·578 857	5491	983 804	32	016 463	33
2·405	11·078 43	1108	090 265	90	5·494 082	5587	5·584 348	5497	983 836	32	016 430	34
·406	·089 51	1110	090 175	90	·499 669	5593	·589 845	5502	983 868	32	016 396	33
·407	·100 61	1111	090 085	90	·505 262	5598	·595 347	5508	983 900	32	016 363	32
·408	·111 72	1111	089 995	90	·510 860	5604	·600 855	5514	983 932	32	016 331	33
·409	·122 83	1113	089 905	90	·516 464	5609	·606 369	5519	983 964	32	016 298	33
2·410	11·133 96	1114	089 815	89	5·522 073	5615	5·611 888	5525	983 996	31	016 265	33
·411	·145 10	1115	089 726	90	·527 688	5620	·617 413	5531	984 027	32	016 232	33
·412	·156 25	1116	089 636	90	·533 308	5625	·622 944	5536	984 059	32	016 199	32
·413	·167 41	1118	089 546	89	·538 933	5632	·628 480	5541	984 091	31	016 167	33
·414	·178 59	1118	089 457	90	·544 565	5637	·634 021	5548	984 122	32	016 134	32
2·415	11·189 77	1120	089 367	89	5·550 202	5642	5·639 569	5553	984 154	31	016 102	33
·416	·200 97	1120	089 278	89	·555 844	5648	·645 122	5559	984 185	31	016 069	32
·417	·212 17	1122	089 189	89	·561 492	5653	·650 681	5564	984 216	32	016 037	32
·418	·223 39	1123	089 100	89	·567 145	5659	·656 245	5570	984 248	31	016 005	33
·419	·234 62	1124	089 011	89	·572 804	5665	·661 815	5575	984 279	31	015 972	32
2·420	11·245 86	1125	088 922	89	5·578 469	5670	5·667 390	5582	984 310	31	015 940	32
·421	·257 11	1126	088 833	89	·584 139	5676	·672 972	5587	984 341	31	015 908	32
·422	·268 37	1128	088 744	89	·589 815	5681	·678 559	5592	984 372	31	015 876	32
·423	·279 65	1128	088 655	88	·595 496	5687	·684 151	5599	984 403	31	015 844	32
·424	·290 93	1130	088 567	89	·601 183	5693	·689 750	5604	984 434	31	015 812	32
2·425	11·302 23	1131	088 478	88	5·606 876	5698	5·695 354	5609	984 465	31	015 780	31
·426	·313 54	1132	088 390	89	·612 574	5704	·700 963	5616	984 496	30	015 749	32
·427	·324 86	1133	088 301	88	·618 278	5709	·706 579	5621	984 526	31	015 717	32
·428	·336 19	1134	088 213	88	·623 987	5715	·712 200	5627	984 557	31	015 685	31
·429	·347 53	1135	088 125	88	·629 702	5721	·717 827	5632	984 588	30	015 654	32
2·430	11·358 88	1137	088 037	88	5·635 423	5726	5·723 459	5639	984 618	31	015 622	31
·431	·370 25	1137	087 949	88	·641 149	5732	·729 098	5644	984 649	30	015 591	32
·432	·381 62	1139	087 861	88	·646 881	5737	·734 742	5650	984 679	31	015 559	31
·433	·393 01	1140	087 773	88	·652 618	5744	·740 392	5655	984 710	30	015 528	31
·434	·404 41	1141	087 685	87	·658 362	5748	·746 047	5661	984 740	30	015 497	32
2·435	11·415 82	1142	087 598	88	5·664 110	5755	5·751 708	5667	984 770	30	015 465	31
·436	·427 24	1143	087 510	87	·669 865	5760	·757 375	5673	984 800	30	015 434	31
·437	·438 67	1145	087 423	88	·675 625	5766	·763 048	5678	984 830	31	015 403	31
·438	·450 12	1145	087 335	87	·681 391	5772	·768 726	5685	984 861	30	015 372	31
·439	·461 57	1147	087 248	87	·687 163	5777	·774 411	5690	984 891	30	015 341	31
2·440	11·473 04	1148	087 161	87	5·692 940	5783	5·780 101	5696	984 921	29	015 310	30
·441	·484 52	1149	087 074	87	·698 723	5789	·785 797	5701	984 950	30	015 280	31
·442	·496 01	1150	086 987	87	·704 512	5794	·791 498	5708	984 980	30	015 249	31
·443	·507 51	1151	086 900	87	·710 306	5800	·797 206	5713	985 010	30	015 218	31
·444	·519 02	1153	086 813	87	·716 106	5806	·802 919	5719	985 040	29	015 187	30
2·445	11·530 55	1154	086 726	87	5·721 912	5811	5·808 638	5725	985 069	30	015 157	31
·446	·542 09	1154	086 639	86	·727 723	5817	·814 363	5730	985 099	30	015 126	30
·447	·553 63	1156	086 553	87	·733 540	5823	·820 093	5737	985 129	29	015 096	31
·448	·565 19	1157	086 466	86	·739 363	5829	·825 830	5742	985 158	30	015 065	30
·449	·576 76	1159	086 380	86	·745 192	5835	·831 572	5748	985 188	29	015 035	30
2·450	11·588 35		086 294		5·751 027		5·837 320		985 217		015 005	

x	e^x		e^{-x}		sinh x		cosh x		tanh x		coth x	
			0·						0·		1·	
2·450	11·588 35	1159	086 294	87	5·751 027	5840	5·837 320	5754	985 217	29	015 005	30
·451	·599 94	1161	086 207	86	·756 867	5846	·843 074	5760	985 246	29	014 975	30
·452	·611 55	1161	086 121	86	·762 713	5851	·848 834	5766	985 275	30	014 945	31
·453	·623 16	1163	086 035	86	·768 564	5858	·854 600	5771	985 305	29	014 914	30
·454	·634 79	1164	085 949	86	·774 422	5863	·860 371	5777	985 334	29	014 884	30
2·455	11·646 43	1166	085 863	86	5·780 285	5869	5·866 148	5784	985 363	29	014 854	29
·456	·658 09	1166	085 777	85	·786 154	5875	·871 932	5789	985 392	29	014 825	30
·457	·669 75	1168	085 692	86	·792 029	5881	·877 721	5795	985 421	29	014 795	30
·458	·681 43	1168	085 606	86	·797 910	5886	·883 516	5801	985 450	29	014 765	30
·459	·693 11	1170	085 520	85	·803 796	5892	·889 317	5806	985 479	29	014 735	29
2·460	11·704 81	1171	085 435	85	5·809 688	5898	5·895 123	5813	985 508	28	014 706	30
·461	·716 52	1172	085 350	86	·815 586	5904	·900 936	5818	985 536	29	014 676	30
·462	·728 24	1174	085 264	85	·821 490	5910	·906 754	5825	985 565	29	014 646	29
·463	·739 98	1174	085 179	85	·827 400	5915	·912 579	5830	985 594	28	014 617	29
·464	·751 72	1176	085 094	85	·833 315	5922	·918 409	5836	985 622	29	014 588	30
2·465	11·763 48	1177	085 009	85	5·839 237	5927	5·924 245	5843	985 651	28	014 558	29
·466	·775 25	1178	084 924	85	·845 164	5933	·930 088	5848	985 679	29	014 529	29
·467	·787 03	1180	084 839	85	·851 097	5939	·935 936	5854	985 708	28	014 500	30
·468	·798 83	1180	084 754	85	·857 036	5944	·941 790	5860	985 736	28	014 470	29
·469	·810 63	1182	084 669	84	·862 980	5951	·947 650	5866	985 764	28	014 441	29
2·470	11·822 45	1183	084 585	85	5·868 931	5956	5·953 516	5872	985 792	29	014 412	29
·471	·834 28	1184	084 500	84	·874 887	5963	·959 388	5878	985 821	28	014 383	29
·472	·846 12	1185	084 416	85	·880 850	5968	·965 266	5883	985 849	28	014 354	29
·473	·857 97	1186	084 331	84	·886 818	5974	·971 149	5890	985 877	28	014 325	28
·474	·869 83	1188	084 247	84	·892 792	5980	·977 039	5896	985 905	28	014 297	29
2·475	11·881 71	1188	084 163	84	5·898 772	5986	5·982 935	5902	985 933	28	014 268	29
·476	·893 59	1190	084 079	84	·904 758	5992	·988 837	5908	985 961	28	014 239	28
·477	·905 49	1192	083 995	84	·910 750	5997	5·994 745	5913	985 989	27	014 211	29
·478	·917 41	1192	083 911	84	·916 747	6004	6·000 658	5920	986 016	28	014 182	29
·479	·929 33	1193	083 827	84	·922 751	6010	·006 578	5926	986 044	28	014 153	28
2·480	11·941 26	1195	083 743	83	5·928 761	6015	6·012 504	5932	986 072	27	014 125	29
·481	·953 21	1196	083 660	84	·934 776	6021	·018 436	5937	986 099	28	014 096	28
·482	·965 17	1197	083 576	84	·940 797	6028	·024 373	5944	986 127	28	014 068	28
·483	·977 14	1199	083 492	83	·946 825	6033	·030 317	5950	986 155	27	014 040	28
·484	11·989 13	1199	083 409	83	·952 858	6039	·036 267	5956	986 182	27	014 012	29
2·485	12·001 12	1201	083 326	84	5·958 897	6046	6·042 223	5962	986 209	28	013 983	28
·486	·013 13	1202	083 242	83	·964 943	6051	·048 185	5968	986 237	27	013 955	28
·487	·025 15	1203	083 159	83	·970 994	6057	·054 153	5974	986 264	27	013 927	28
·488	·037 18	1204	083 076	83	·977 051	6063	·060 127	5980	986 291	28	013 899	28
·489	·049 22	1206	082 993	83	·983 114	6069	·066 107	5986	986 319	27	013 871	28
2·490	12·061 28	1206	082 910	83	5·989 183	6075	6·072 093	5992	986 346	27	013 843	28
·491	·073 34	1208	082 827	83	5·995 258	6081	·078 085	5999	986 373	27	013 815	27
·492	·085 42	1209	082 744	82	6·001 339	6087	·084 084	6004	986 400	27	013 788	28
·493	·097 51	1211	082 662	83	·007 426	6093	·090 088	6010	986 427	27	013 760	28
·494	·109 62	1211	082 579	83	·013 519	6100	·096 098	6017	986 454	27	013 732	27
2·495	12·121 73	1213	082 496	82	6·019 619	6105	6·102 115	6023	986 481	27	013 705	28
·496	·133 86	1214	082 414	82	·025 724	6111	·108 138	6028	986 508	26	013 677	27
·497	·146 00	1215	082 332	83	·031 835	6117	·114 166	6035	986 534	27	013 650	28
·498	·158 15	1217	082 249	82	·037 952	6123	·120 201	6041	986 561	27	013 622	27
·499	·170 32	1217	082 167	82	·044 075	6129	·126 242	6047	986 588	26	013 595	28
2·500	12·182 49		082 085		6·050 204		6·132 289		986 614		013 567	

x	e^x		e^{-x}		sinh x		cosh x		tanh x		coth x	
			0·						0·		1·	
2·500	12·182 49	1219	082 085	82	6·050 20	614	6·132 29	605	986 614	27	013 567	27
·501	·194 68	1220	082 003	82	·056 34	614	·138 34	606	986 641	26	013 540	27
·502	·206 88	1222	081 921	82	·062 48	615	·144 40	607	986 667	27	013 513	27
·503	·219 10	1222	081 839	82	·068 63	615	·150 47	607	986 694	26	013 486	28
·504	·231 32	1224	081 757	81	·074 78	616	·156 54	608	986 720	27	013 458	27
2·505	12·243 56	1225	081 676	82	6·080 94	617	6·162 62	608	986 747	26	013 431	27
·506	·255 81	1226	081 594	82	·087 11	617	·168 70	609	986 773	26	013 404	27
·507	·268 07	1227	081 512	81	·093 28	618	·174 79	610	986 799	26	013 377	26
·508	·280 34	1229	081 431	81	·099 46	618	·180 89	610	986 825	27	013 351	27
·509	·292 63	1230	081 350	82	·105 64	619	·186 99	611	986 852	26	013 324	27
2·510	12·304 93	1231	081 268	81	6·111 83	620	6·193 10	611	986 878	26	013 297	27
·511	·317 24	1232	081 187	81	·118 03	620	·199 21	613	986 904	26	013 270	27
·512	·329 56	1234	081 106	81	·124 23	621	·205 34	612	986 930	26	013 243	26
·513	·341 90	1235	081 025	81	·130 44	621	·211 46	614	986 956	25	013 217	27
·514	·354 25	1236	080 944	81	·136 65	622	·217 60	614	986 981	26	013 190	26
2·515	12·366 61	1237	080 863	81	6·142 87	623	6·223 74	614	987 007	26	013 164	27
·516	·378 98	1239	080 782	81	·149 10	623	·229 88	615	987 033	26	013 137	26
·517	·391 37	1239	080 701	80	·155 33	624	·236 03	616	987 059	26	013 111	27
·518	·403 76	1241	080 621	81	·161 57	625	·242 19	617	987 085	25	013 084	26
·519	·416 17	1243	080 540	80	·167 82	625	·248 36	617	987 110	26	013 058	26
2·520	12·428 60	1243	080 460	81	6·174 07	626	6·254 53	618	987 136	25	013 032	26
·521	·441 03	1245	080 379	80	·180 33	626	·260 71	618	987 161	26	013 006	26
·522	·453 48	1246	080 299	80	·186 59	627	·266 89	619	987 187	25	012 980	27
·523	·465 94	1247	080 219	81	·192 86	628	·273 08	619	987 212	26	012 953	26
·524	·478 41	1249	080 138	80	·199 14	628	·279 27	621	987 238	25	012 927	26
2·525	12·490 90	1249	080 058	80	6·205 42	629	6·285 48	621	987 263	25	012 901	26
·526	·503 39	1251	079 978	80	·211 71	629	·291 69	621	987 288	25	012 875	25
·527	·515 90	1252	079 898	80	·218 00	630	·297 90	622	987 313	26	012 850	26
·528	·528 42	1254	079 818	79	·224 30	631	·304 12	623	987 339	25	012 824	26
·529	·540 96	1255	079 739	80	·230 61	631	·310 35	623	987 364	25	012 798	26
2·530	12·553 51	1256	079 659	80	6·236 92	632	6·316 58	624	987 389	25	012 772	26
·531	·566 07	1257	079 579	79	·243 24	633	·322 82	625	987 414	25	012 746	25
·532	·578 64	1258	079 500	80	·249 57	633	·329 07	625	987 439	25	012 721	26
·533	·591 22	1260	079 420	79	·255 90	634	·335 32	626	987 464	25	012 695	25
·534	·603 82	1261	079 341	79	·262 24	634	·341 58	627	987 489	25	012 670	26
2·535	12·616 43	1262	079 262	80	6·268 58	636	6·347 85	627	987 514	24	012 644	25
·536	·629 05	1264	079 182	79	·274 94	635	·354 12	628	987 538	25	012 619	26
·537	·641 69	1265	079 103	79	·281 29	637	·360 40	628	987 563	25	012 593	25
·538	·654 34	1266	079 024	79	·287 66	637	·366 68	629	987 588	24	012 568	25
·539	·667 00	1267	078 945	79	·294 03	637	·372 97	630	987 612	25	012 543	25
2·540	12·679 67	1269	078 866	78	6·300 40	638	6·379 27	630	987 637	25	012 518	25
·541	·692 36	1270	078 788	79	·306 78	639	·385 57	631	987 662	24	012 493	26
·542	·705 06	1271	078 709	79	·313 17	640	·391 88	632	987 686	25	012 467	25
·543	·717 77	1272	078 630	78	·319 57	640	·398 20	632	987 711	24	012 442	25
·544	·730 49	1274	078 552	79	·325 97	641	·404 52	633	987 735	24	012 417	25
2·545	12·743 23	1275	078 473	78	6·332 38	641	6·410 85	634	987 759	25	012 392	25
·546	·755 98	1276	078 395	79	·338 79	642	·417 19	634	987 784	24	012 367	24
·547	·768 74	1278	078 316	78	·345 21	643	·423 53	635	987 808	24	012 343	25
·548	·781 52	1278	078 238	78	·351 64	643	·429 88	635	987 832	24	012 318	25
·549	·794 30	1280	078 160	78	·358 07	644	·436 23	636	987 856	24	012 293	25
2·550	12·807 10		078 082		6·364 51		6·442 59		987 880		012 268	

x	e^x	Δ	e^{-x}	Δ	sinh x	Δ	cosh x	Δ	tanh x	Δ	coth x	Δ
			0·						0·		1·	
2·550	12·807 10	1282	078 082	78	6·364 51	645	6·442 59	637	987 880	24	012 268	24
·551	·819 92	1282	078 004	78	·370 96	645	·448 96	637	987 904	24	012 244	25
·552	·832 74	1284	077 926	78	·377 41	646	·455 33	639	987 928	24	012 219	25
·553	·845 58	1285	077 848	78	·383 87	646	·461 72	638	987 952	24	012 194	24
·554	·858 43	1287	077 770	78	·390 33	647	·468 10	640	987 976	24	012 170	25
2·555	12·871 30	1288	077 692	77	6·396 80	648	6·474 50	640	988 000	24	012 145	24
·556	·884 18	1289	077 615	78	·403 28	649	·480 90	640	988 024	24	012 121	24
·557	·897 07	1290	077 537	77	·409 77	649	·487 30	642	988 048	24	012 097	25
·558	·909 97	1292	077 460	78	·416 26	649	·493 72	642	988 072	23	012 072	24
·559	·922 89	1293	077 382	77	·422 75	651	·500 14	642	988 095	24	012 048	24
2·560	12·935 82	1294	077 305	78	6·429 26	651	6·506 56	643	988 119	24	012 024	24
·561	·948 76	1295	077 227	77	·435 77	651	·512 99	644	988 143	23	012 000	24
·562	·961 71	1297	077 150	77	·442 28	652	·519 43	645	988 166	24	011 976	24
·563	·974 68	1298	077 073	77	·448 80	653	·525 88	645	988 190	23	011 952	24
·564	12·987 66	1300	076 996	77	·455 33	654	·532 33	646	988 213	23	011 928	24
2·565	13·000 66	1301	076 919	77	6·461 87	654	6·538 79	646	988 236	24	011 904	24
·566	·013 67	1302	076 842	77	·468 41	655	·545 25	648	988 260	23	011 880	24
·567	·026 69	1303	076 765	76	·474 96	656	·551 73	647	988 283	23	011 856	24
·568	·039 72	1305	076 689	77	·481 52	656	·558 20	649	988 306	24	011 832	24
·569	·052 77	1305	076 612	76	·488 08	656	·564 69	649	988 330	23	011 808	24
2·570	13·065 82	1308	076 536	77	6·494 64	658	6·571 18	650	988 353	23	011 784	23
·571	·078 90	1308	076 459	76	·501 22	658	·577 68	650	988 376	23	011 761	24
·572	·091 98	1310	076 383	77	·507 80	659	·584 18	651	988 399	23	011 737	23
·573	·105 08	1311	076 306	76	·514 39	659	·590 69	652	988 422	23	011 714	24
·574	·118 19	1313	076 230	76	·520 98	660	·597 21	653	988 445	23	011 690	24
2·575	13·131 32	1314	076 154	76	6·527 58	661	6·603 74	653	988 468	23	011 666	23
·576	·144 46	1315	076 078	76	·534 19	661	·610 27	653	988 491	23	011 643	23
·577	·157 61	1316	076 002	76	·540 80	662	·616 80	655	988 514	23	011 620	24
·578	·170 77	1318	075 926	76	·547 42	663	·623 35	655	988 537	22	011 596	23
·579	·183 95	1319	075 850	76	·554 05	663	·629 90	656	988 559	23	011 573	23
2·580	13·197 14	1320	075 774	76	6·560 68	664	6·636 46	656	988 582	23	011 550	23
·581	·210 34	1322	075 698	75	·567 32	665	·643 02	657	988 605	22	011 527	24
·582	·223 56	1323	075 623	76	·573 97	665	·649 59	658	988 627	23	011 503	23
·583	·236 79	1324	075 547	75	·580 62	666	·656 17	658	988 650	23	011 480	23
·584	·250 03	1326	075 472	76	·587 28	667	·662 75	659	988 673	22	011 457	23
2·585	13·263 29	1327	075 396	75	6·593 95	667	6·669 34	660	988 695	23	011 434	23
·586	·276 56	1328	075 321	76	·600 62	668	·675 94	660	988 718	22	011 411	23
·587	·289 84	1330	075 245	75	·607 30	668	·682 54	661	988 740	22	011 388	23
·588	·303 14	1331	075 170	75	·613 98	670	·689 15	662	988 762	23	011 365	22
·589	·316 45	1332	075 095	75	·620 68	670	·695 77	663	988 785	22	011 343	23
2·590	13·329 77	1334	075 020	75	6·627 38	670	6·702 40	663	988 807	22	011 320	23
·591	·343 11	1335	074 945	75	·634 08	671	·709 03	663	988 829	22	011 297	23
·592	·356 46	1336	074 870	75	·640 79	672	·715 66	665	988 851	23	011 274	22
·593	·369 82	1338	074 795	74	·647 51	673	·722 31	665	988 874	22	011 252	23
·594	·383 20	1339	074 721	75	·654 24	673	·728 96	666	988 896	22	011 229	23
2·595	13·396 59	1340	074 646	75	6·660 97	674	6·735 62	666	988 918	22	011 206	22
·596	·409 99	1342	074 571	74	·667 71	675	·742 28	667	988 940	22	011 184	23
·597	·423 41	1343	074 497	75	·674 46	675	·748 95	668	988 962	22	011 161	22
·598	·436 84	1344	074 422	74	·681 21	676	·755 63	668	988 984	22	011 139	22
·599	·450 28	1346	074 348	74	·687 97	676	·762 31	670	989 006	21	011 117	23
2·600	13·463 74		074 274		6·694 73		6·769 01		989 027		011 094	

TABLE Vc—EXPONENTIAL AND HYPERBOLIC FUNCTIONS

x	e^x	Δ	e^{-x}	Δ	sinh x	Δ	cosh x	Δ	tanh x	Δ	coth x	Δ
			0·						0·		1·	
2·600	13·463 74	1347	074 274	75	6·694 73	677	6·769 01	669	989 027	22	011 094	22
·601	·477 21	1348	074 199	74	·701 50	678	·775 70	671	989 049	22	011 072	22
·602	·490 69	1350	074 125	74	·708 28	679	·782 41	671	989 071	22	011 050	22
·603	·504 19	1351	074 051	74	·715 07	679	·789 12	672	989 093	21	011 028	23
·604	·517 70	1353	073 977	74	·721 86	680	·795 84	672	989 114	22	011 005	22
2·605	13·531 23	1353	073 903	74	6·728 66	681	6·802 56	674	989 136	22	010 983	22
·606	·544 76	1355	073 829	74	·735 47	681	·809 30	674	989 158	21	010 961	22
·607	·558 31	1357	073 755	73	·742 28	682	·816 04	674	989 179	22	010 939	22
·608	·571 88	1358	073 682	74	·749 10	683	·822 78	675	989 201	21	010 917	22
·609	·585 46	1359	073 608	73	·755 93	683	·829 53	676	989 222	22	010 895	22
2·610	13·599 05	1361	073 535	74	6·762 76	684	6·836 29	677	989 244	21	010 873	21
·611	·612 66	1362	073 461	73	·769 60	684	·843 06	677	989 265	21	010 852	22
·612	·626 28	1363	073 388	74	·776 44	686	·849 83	678	989 286	22	010 830	22
·613	·639 91	1365	073 314	73	·783 30	686	·856 61	679	989 308	21	010 808	22
·614	·653 56	1366	073 241	73	·790 16	686	·863 40	679	989 329	21	010 786	21
2·615	13·667 22	1367	073 168	73	6·797 02	688	6·870 19	680	989 350	21	010 765	22
·616	·680 89	1369	073 095	73	·803 90	688	·876 99	681	989 371	21	010 743	22
·617	·694 58	1370	073 022	73	·810 78	689	·883 80	681	989 392	21	010 721	21
·618	·708 28	1371	072 949	73	·817 67	689	·890 61	683	989 413	21	010 700	22
·619	·721 99	1373	072 876	73	·824 56	690	·897 44	682	989 434	21	010 678	21
2·620	13·735 72	1375	072 803	73	6·831 46	691	6·904 26	684	989 455	21	010 657	21
·621	·749 47	1375	072 730	73	·838 37	691	·911 10	684	989 476	21	010 636	22
·622	·763 22	1377	072 657	72	·845 28	692	·917 94	685	989 497	21	010 614	21
·623	·776 99	1379	072 585	73	·852 20	693	·924 79	685	989 518	21	010 593	21
·624	·790 78	1379	072 512	72	·859 13	694	·931 64	687	989 539	21	010 572	22
2·625	13·804 57	1382	072 440	73	6·866 07	694	6·938 51	687	989 560	21	010 550	21
·626	·818 39	1382	072 367	72	·873 01	695	·945 38	687	989 581	20	010 529	21
·627	·832 21	1384	072 295	72	·879 96	695	·952 25	689	989 601	21	010 508	21
·628	·846 05	1385	072 223	72	·886 91	697	·959 14	689	989 622	21	010 487	21
·629	·859 90	1387	072 151	73	·893 88	697	·966 03	689	989 643	20	010 466	21
2·630	13·873 77	1388	072 078	72	6·900 85	697	6·972 92	691	989 663	21	010 445	21
·631	·887 65	1390	072 006	72	·907 82	699	·979 83	691	989 684	20	010 424	21
·632	·901 55	1390	071 934	71	·914 81	699	·986 74	692	989 704	21	010 403	21
·633	·915 45	1393	071 863	72	·921 80	699	6·993 66	692	989 725	20	010 382	21
·634	·929 38	1393	071 791	72	·928 79	701	7·000 58	694	989 745	20	010 361	21
2·635	13·943 31	1395	071 719	72	6·935 80	701	7·007 52	694	989 765	21	010 340	20
·636	·957 26	1397	071 647	71	·942 81	702	·014 46	694	989 786	20	010 320	21
·637	·971 23	1398	071 576	72	·949 83	702	·021 40	695	989 806	20	010 299	21
·638	·985 21	1399	071 504	71	·956 85	703	·028 35	697	989 826	21	010 278	20
·639	13·999 20	1400	071 433	72	·963 88	704	·035 32	696	989 847	20	010 258	21
2·640	14·013 20	1402	071 361	71	6·970 92	705	7·042 28	698	989 867	20	010 237	21
·641	·027 22	1404	071 290	71	·977 97	705	·049 26	698	989 887	20	010 216	20
·642	·041 26	1405	071 219	71	·985 02	706	·056 24	699	989 907	20	010 196	21
·643	·055 31	1406	071 148	72	·992 08	707	·063 23	699	989 927	20	010 175	20
·644	·069 37	1408	071 076	71	6·999 15	707	·070 22	701	989 947	20	010 155	20
2·645	14·083 45	1409	071 005	71	7·006 22	708	7·077 23	700	989 967	20	010 135	21
·646	·097 54	1410	070 934	71	·013 30	709	·084 23	702	989 987	20	010 114	20
·647	·111 64	1412	070 863	70	·020 39	709	·091 25	703	990 007	20	010 094	20
·648	·125 76	1413	070 793	71	·027 48	710	·098 28	703	990 027	20	010 074	21
·649	·139 89	1415	070 722	71	·034 58	711	·105 31	703	990 047	19	010 053	20
2·650	14·154 04		070 651		7·041 69		7·112 34		990 066		010 033	

TABLE Vc—EXPONENTIAL AND HYPERBOLIC FUNCTIONS

x	e^x		e^{-x}		sinh x		cosh x		tanh x		coth x	
			0·						0·		1·	
2·650	14·154 04	1416	070 651	70	7·041 69	712	7·112 34	705	990 066	20	010 033	20
·651	·168 20	1418	070 581	71	·048 81	712	·119 39	705	990 086	20	010 013	20
·652	·182 38	1418	070 510	70	·055 93	713	·126 44	706	990 106	20	009 993	20
·653	·196 56	1421	070 440	71	·063 06	714	·133 50	707	990 126	19	009 973	20
·654	·210 77	1422	070 369	70	·070 20	714	·140 57	707	990 145	20	009 953	20
2·655	14·224 99	1423	070 299	70	7·077 34	715	7·147 64	708	990 165	19	009 933	20
·656	·239 22	1424	070 229	71	·084 49	716	·154 72	709	990 184	20	009 913	20
·657	·253 46	1427	070 158	70	·091 65	717	·161 81	710	990 204	19	009 893	20
·658	·267 73	1427	070 088	70	·098 82	717	·168 91	710	990 223	20	009 873	20
·659	·282 00	1429	070 018	70	·105 99	718	·176 01	711	990 243	19	009 853	19
2·660	14·296 29	1430	069 948	70	7·113 17	719	7·183 12	712	990 262	19	009 834	20
·661	·310 59	1432	069 878	70	·120 36	719	·190 24	712	990 281	20	009 814	20
·662	·324 91	1433	069 808	69	·127 55	720	·197 36	713	990 301	19	009 794	19
·663	·339 24	1435	069 739	70	·134 75	721	·204 49	714	990 320	19	009 775	20
·664	·353 59	1436	069 669	70	·141 96	722	·211 63	714	990 339	20	009 755	20
2·665	14·367 95	1437	069 599	69	7·149 18	722	7·218 77	716	990 359	19	009 735	19
·666	·382 32	1439	069 530	70	·156 40	723	·225 93	716	990 378	19	009 716	20
·667	·396 71	1441	069 460	69	·163 63	723	·233 09	716	990 397	19	009 696	19
·668	·411 12	1442	069 391	69	·170 86	725	·240 25	718	990 416	19	009 677	20
·669	·425 54	1443	069 322	70	·178 11	725	·247 43	718	990 435	19	009 657	19
2·670	14·439 97	1445	069 252	69	7·185 36	726	7·254 61	719	990 454	19	009 638	19
·671	·454 42	1446	069 183	69	·192 62	726	·261 80	720	990 473	19	009 619	20
·672	·468 88	1447	069 114	69	·199 88	727	·269 00	720	990 492	19	009 599	19
·673	·483 35	1449	069 045	69	·207 15	728	·276 20	721	990 511	19	009 580	19
·674	·497 84	1451	068 976	69	·214 43	729	·283 41	722	990 530	19	009 561	19
2·675	14·512 35	1452	068 907	69	7·221 72	730	7·290 63	722	990 549	18	009 542	20
·676	·526 87	1453	068 838	69	·229 02	730	·297 85	724	990 567	19	009 522	19
·677	·541 40	1455	068 769	69	·236 32	731	·305 09	724	990 586	19	009 503	19
·678	·555 95	1457	068 700	68	·243 63	731	·312 33	724	990 605	19	009 484	19
·679	·570 52	1457	068 632	69	·250 94	733	·319 57	726	990 624	18	009 465	19
2·680	14·585 09	1460	068 563	68	7·258 27	733	7·326 83	726	990 642	19	009 446	19
·681	·599 69	1460	068 495	69	·265 60	733	·334 09	727	990 661	18	009 427	19
·682	·614 29	1462	068 426	68	·272 93	735	·341 36	728	990 679	19	009 408	19
·683	·628 91	1464	068 358	69	·280 28	735	·348 64	728	990 698	18	009 389	18
·684	·643 55	1465	068 289	68	·287 63	736	·355 92	729	990 716	19	009 371	19
2·685	14·658 20	1467	068 221	68	7·294 99	737	7·363 21	730	990 735	18	009 352	19
·686	·672 87	1468	068 153	68	·302 36	737	·370 51	731	990 753	19	009 333	19
·687	·687 55	1469	068 085	68	·309 73	738	·377 82	731	990 772	18	009 314	18
·688	·702 24	1471	068 017	68	·317 11	739	·385 13	732	990 790	18	009 296	19
·689	·716 95	1473	067 949	68	·324 50	740	·392 45	733	990 808	19	009 277	19
2·690	14·731 68	1473	067 881	68	7·331 90	740	7·399 78	733	990 827	18	009 258	18
·691	·746 41	1476	067 813	68	·339 30	741	·407 11	735	990 845	18	009 240	19
·692	·761 17	1477	067 745	67	·346 71	742	·414 46	735	990 863	18	009 221	18
·693	·775 94	1478	067 678	68	·354 13	743	·421 81	736	990 881	18	009 203	19
·694	·790 72	1480	067 610	68	·361 56	743	·429 17	736	990 899	18	009 184	18
2·695	14·805 52	1481	067 542	67	7·368 99	744	7·436 53	737	990 917	19	009 166	19
·696	·820 33	1483	067 475	68	·376 43	745	·443 90	738	990 936	18	009 147	18
·697	·835 16	1484	067 407	67	·383 88	745	·451 28	739	990 954	18	009 129	18
·698	·850 00	1486	067 340	67	·391 33	746	·458 67	740	990 972	18	009 111	19
·699	·864 86	1487	067 273	67	·398 79	747	·466 07	740	990 990	17	009 092	18
2·700	14·879 73		067 206		7·406 26		7·473 47		991 007		009 074	

TABLE Vc—EXPONENTIAL AND HYPERBOLIC FUNCTIONS

x	e^x		e^{-x}		sinh x		cosh x		tanh x		coth x	
			0·						0·		1·	
2·700	14·879 73	1489	067 206	68	7·406 26	748	7·473 47	741	991 007	18	009 074	18
·701	·894 62	1490	067 138	67	·413 74	748	·480 88	742	991 025	18	009 056	18
·702	·909 52	1492	067 071	67	·421 22	750	·488 30	742	991 043	18	009 038	18
·703	·924 44	1493	067 004	67	·428 72	750	·495 72	743	991 061	18	009 020	18
·704	·939 37	1495	066 937	67	·436 22	750	·503 15	744	991 079	18	009 002	19
2·705	14·954 32	1496	066 870	67	7·443 72	752	7·510 59	745	991 097	17	008 983	18
·706	·969 28	1498	066 803	66	·451 24	752	·518 04	746	991 114	18	008 965	18
·707	·984 26	1499	066 737	67	·458 76	753	·525 50	746	991 132	18	008 947	18
·708	14·999 25	1500	066 670	67	·466 29	754	·532 96	747	991 150	17	008 929	17
·709	15·014 25	1503	066 603	66	·473 83	754	·540 43	748	991 167	18	008 912	18
2·710	15·029 28	1503	066 537	67	7·481 37	755	7·547 91	748	991 185	17	008 894	18
·711	·044 31	1505	066 470	66	·488 92	756	·555 39	749	991 202	18	008 876	18
·712	·059 36	1507	066 404	67	·496 48	757	·562 88	750	991 220	17	008 858	18
·713	·074 43	1508	066 337	66	·504 05	757	·570 38	751	991 237	18	008 840	18
·714	·089 51	1510	066 271	66	·511 62	758	·577 89	752	991 255	17	008 822	17
2·715	15·104 61	1511	066 205	66	7·519 20	759	7·585 41	752	991 272	17	008 805	18
·716	·119 72	1513	066 139	66	·526 79	760	·592 93	753	991 289	18	008 787	18
·717	·134 85	1514	066 073	66	·534 39	760	·600 46	754	991 307	17	008 769	17
·718	·149 99	1516	066 007	66	·541 99	761	·608 00	755	991 324	17	008 752	18
·719	·165 15	1517	065 941	66	·549 60	762	·615 55	755	991 341	18	008 734	17
2·720	15·180 32	1519	065 875	66	7·557 22	763	7·623 10	756	991 359	17	008 717	18
·721	·195 51	1520	065 809	66	·564 85	764	·630 66	757	991 376	17	008 699	17
·722	·210 71	1522	065 743	66	·572 49	764	·638 23	757	991 393	17	008 682	18
·723	·225 93	1524	065 677	65	·580 13	765	·645 80	759	991 410	17	008 664	17
·724	·241 17	1524	065 612	66	·587 78	765	·653 39	759	991 427	17	008 647	17
2·725	15·256 41	1527	065 546	65	7·595 43	767	7·660 98	760	991 444	17	008 630	18
·726	·271 68	1528	065 481	66	·603 10	767	·668 58	761	991 461	17	008 612	17
·727	·286 96	1529	065 415	65	·610 77	768	·676 19	761	991 478	17	008 595	17
·728	·302 25	1531	065 350	65	·618 45	769	·683 80	762	991 495	17	008 578	17
·729	·317 56	1533	065 285	66	·626 14	769	·691 42	763	991 512	17	008 561	18
2·730	15·332 89	1534	065 219	65	7·633 83	771	7·699 05	764	991 529	17	008 543	17
·731	·348 23	1535	065 154	65	·641 54	771	·706 69	765	991 546	17	008 526	17
·732	·363 58	1537	065 089	65	·649 25	772	·714 34	765	991 563	16	008 509	17
·733	·378 95	1539	065 024	65	·656 97	772	·721 99	766	991 579	17	008 492	17
·734	·394 34	1540	064 959	65	·664 69	773	·729 65	767	991 596	17	008 475	17
2·735	15·409 74	1542	064 894	65	7·672 42	775	7·737 32	768	991 613	17	008 458	17
·736	·425 16	1543	064 829	65	·680 17	774	·745 00	768	991 630	16	008 441	17
·737	·440 59	1545	064 764	64	·687 91	776	·752 68	769	991 646	17	008 424	17
·738	·456 04	1547	064 700	65	·695 67	777	·760 37	770	991 663	16	008 407	17
·739	·471 51	1548	064 635	65	·703 44	777	·768 07	771	991 679	17	008 390	16
2·740	15·486 99	1549	064 570	64	7·711 21	778	7·775 78	771	991 696	16	008 374	17
·741	·502 48	1551	064 506	65	·718 99	778	·783 49	773	991 712	17	008 357	17
·742	·517 99	1553	064 441	64	·726 77	780	·791 22	773	991 729	16	008 340	17
·743	·533 52	1554	064 377	64	·734 57	780	·798 95	773	991 745	17	008 323	16
·744	·549 06	1555	064 313	65	·742 37	781	·806 68	775	991 762	16	008 307	17
2·745	15·564 61	1558	064 248	64	7·750 18	782	7·814 43	776	991 778	17	008 290	17
·746	·580 19	1558	064 184	64	·758 00	783	·822 19	776	991 795	16	008 273	16
·747	·595 77	1561	064 120	64	·765 83	783	·829 95	777	991 811	16	008 257	17
·748	·611 38	1562	064 056	64	·773 66	784	·837 72	777	991 827	16	008 240	16
·749	·627 00	1563	063 992	64	·781 50	785	·845 49	779	991 843	17	008 224	17
2·750	15·642 63		063 928		7·789 35		7·853 28		991 860		008 207	

TABLE Vc—EXPONENTIAL AND HYPERBOLIC FUNCTIONS

x	e^x	Δ	e^{-x}	Δ	sinh x	Δ	cosh x	Δ	tanh x	Δ	coth x	Δ
			0·						0·		1·	
2·750	15·642 63	1565	063 928	64	7·789 35	786	7·853 28	779	991 860	16	008 207	16
·751	·658 28	1567	063 864	64	·797 21	786	·861 07	780	991 876	16	008 191	17
·752	·673 95	1568	063 800	64	·805 07	788	·868 87	781	991 892	16	008 174	16
·753	·689 63	1570	063 736	63	·812 95	788	·876 68	782	991 908	16	008 158	17
·754	·705 33	1571	063 673	64	·820 83	789	·884 50	782	991 924	16	008 141	16
2·755	15·721 04	1573	063 609	64	7·828 72	789	7·892 32	784	991 940	16	008 125	16
·756	·736 77	1574	063 545	63	·836 61	791	·900 16	784	991 956	16	008 109	16
·757	·752 51	1576	063 482	64	·844 52	791	·908 00	785	991 972	16	008 093	17
·758	·768 27	1578	063 418	63	·852 43	792	·915 85	785	991 988	16	008 076	16
·759	·784 05	1579	063 355	63	·860 35	793	·923 70	787	992 004	16	008 060	16
2·760	15·799 84	1581	063 292	63	7·868 28	793	7·931 57	787	992 020	16	008 044	16
·761	·815 65	1582	063 229	64	·876 21	794	·939 44	788	992 036	16	008 028	16
·762	·831 47	1584	063 165	63	·884 15	796	·947 32	789	992 052	16	008 012	16
·763	·847 31	1586	063 102	63	·892 11	795	·955 21	789	992 068	16	007 996	16
·764	·863 17	1587	063 039	63	·900 06	797	·963 10	791	992 084	15	007 980	16
2·765	15·879 04	1589	062 976	63	7·908 03	798	7·971 01	791	992 099	16	007 964	16
·766	·894 93	1590	062 913	63	·916 01	798	·978 92	792	992 115	16	007 948	16
·767	·910 83	1592	062 850	63	·923 99	799	·986 84	793	992 131	15	007 932	16
·768	·926 75	1593	062 787	62	·931 98	800	7·994 77	793	992 146	16	007 916	16
·769	·942 68	1595	062 725	63	·939 98	801	8·002 70	795	992 162	16	007 900	16
2·770	15·958 63	1597	062 662	63	7·947 99	801	8·010 65	795	992 178	15	007 884	16
·771	·974 60	1598	062 599	62	·956 00	802	·018 60	796	992 193	16	007 868	16
·772	15·990 58	1600	062 537	63	·964 02	803	·026 56	797	992 209	15	007 852	15
·773	16·006 58	1602	062 474	62	·972 05	804	·034 53	797	992 224	16	007 837	16
·774	·022 60	1603	062 412	63	·980 09	805	·042 50	799	992 240	15	007 821	16
2·775	16·038 63	1604	062 349	62	7·988 14	805	8·050 49	799	992 255	16	007 805	15
·776	·054 67	1607	062 287	62	7·996 19	807	·058 48	800	992 271	15	007 790	16
·777	·070 74	1608	062 225	62	8·004 26	807	·066 48	801	992 286	15	007 774	16
·778	·086 82	1609	062 163	62	·012 33	807	·074 49	802	992 301	16	007 758	15
·779	·102 91	1611	062 101	62	·020 40	809	·082 51	802	992 317	15	007 743	16
2·780	16·119 02	1613	062 039	63	8·028 49	810	8·090 53	803	992 332	15	007 727	15
·781	·135 15	1614	061 976	61	·036 59	810	·098 56	804	992 347	15	007 712	16
·782	·151 29	1616	061 915	62	·044 69	811	·106 60	805	992 362	16	007 696	15
·783	·167 45	1618	061 853	62	·052 80	812	·114 65	806	992 378	15	007 681	16
·784	·183 63	1619	061 791	62	·060 92	812	·122 71	806	992 393	15	007 665	15
2·785	16·199 82	1621	061 729	62	8·069 04	814	8·130 77	808	992 408	15	007 650	15
·786	·216 03	1622	061 667	61	·077 18	814	·138 85	808	992 423	15	007 635	16
·787	·232 25	1624	061 606	62	·085 32	815	·146 93	809	992 438	15	007 619	15
·788	·248 49	1626	061 544	61	·093 47	816	·155 02	809	992 453	15	007 604	15
·789	·264 75	1627	061 483	62	·101 63	817	·163 11	811	992 468	15	007 589	15
2·790	16·281 02	1629	061 421	61	8·109 80	817	8·171 22	811	992 483	15	007 574	15
·791	·297 31	1630	061 360	62	·117 97	819	·179 33	813	992 498	15	007 559	16
·792	·313 61	1633	061 298	61	·126 16	819	·187 46	813	992 513	15	007 543	15
·793	·329 94	1633	061 237	61	·134 35	820	·195 59	814	992 528	15	007 528	15
·794	·346 27	1636	061 176	61	·142 55	821	·203 73	814	992 543	15	007 513	15
2·795	16·362 63	1637	061 115	61	8·150 76	821	8·211 87	816	992 558	15	007 498	15
·796	·379 00	1639	061 054	61	·158 97	823	·220 03	816	992 573	14	007 483	15
·797	·395 39	1640	060 993	61	·167 20	823	·228 19	817	992 587	15	007 468	15
·798	·411 79	1642	060 932	61	·175 43	824	·236 36	818	992 602	15	007 453	15
·799	·428 21	1644	060 871	61	·183 67	825	·244 54	819	992 617	15	007 438	15
2·800	16·444 65		060 810		8·191 92		8·252 73		992 632		007 423	

x	e^x		e^{-x}		sinh x		cosh x		tanh x		coth x	
			0·						0·		1·	
2·800	16·444 65	1645	060 810	61	8·191 92	826	8·252 73	819	992 632	14	007 423	15
·801	·461 10	1647	060 749	60	·200 18	826	·260 92	821	992 646	15	007 408	15
·802	·477 57	1648	060 689	61	·208 44	827	·269 13	821	992 661	14	007 393	14
·803	·494 05	1651	060 628	61	·216 71	828	·277 34	822	992 675	15	007 379	15
·804	·510 56	1652	060 567	60	·224 99	829	·285 56	823	992 690	15	007 364	15
2·805	16·527 08	1653	060 507	61	8·233 28	830	8·293 79	824	992 705	14	007 349	15
·806	·543 61	1655	060 446	60	·241 58	831	·302 03	824	992 719	15	007 334	14
·807	·560 16	1657	060 386	60	·249 89	831	·310 27	826	992 734	14	007 320	15
·808	·576 73	1659	060 326	61	·258 20	833	·318 53	826	992 748	14	007 305	15
·809	·593 32	1660	060 265	60	·266 53	833	·326 79	827	992 762	15	007 290	14
2·810	16·609 92	1662	060 205	60	8·274 86	834	8·335 06	828	992 777	14	007 276	15
·811	·626 54	1663	060 145	60	·283 20	834	·343 34	829	992 791	15	007 261	14
·812	·643 17	1665	060 085	60	·291 54	836	·351 63	829	992 806	14	007 247	15
·813	·659 82	1667	060 025	60	·299 90	836	·359 92	831	992 820	14	007 232	15
·814	·676 49	1669	059 965	60	·308 26	838	·368 23	831	992 834	15	007 217	14
2·815	16·693 18	1670	059 905	60	8·316 64	838	8·376 54	832	992 849	14	007 203	14
·816	·709 88	1672	059 845	60	·325 02	839	·384 86	833	992 863	14	007 189	15
·817	·726 60	1673	059 785	60	·333 41	839	·393 19	834	992 877	14	007 174	14
·818	·743 33	1675	059 725	59	·341 80	841	·401 53	834	992 891	14	007 160	15
·819	·760 08	1677	059 666	60	·350 21	841	·409 87	836	992 905	14	007 145	14
2·820	16·776 85	1679	059 606	60	8·358 62	842	8·418 23	836	992 919	15	007 131	14
·821	·793 64	1680	059 546	59	·367 04	844	·426 59	837	992 934	14	007 117	14
·822	·810 44	1682	059 487	60	·375 48	843	·434 96	838	992 948	14	007 103	15
·823	·827 26	1683	059 427	59	·383 91	845	·443 34	839	992 962	14	007 088	14
·824	·844 09	1685	059 368	59	·392 36	846	·451 73	840	992 976	14	007 074	14
2·825	16·860 94	1687	059 309	60	8·400 82	846	8·460 13	840	992 990	14	007 060	14
·826	·877 81	1689	059 249	59	·409 28	848	·468 53	842	993 004	14	007 046	14
·827	·894 70	1690	059 190	59	·417 76	848	·476 95	842	993 018	13	007 032	15
·828	·911 60	1692	059 131	59	·426 24	849	·485 37	843	993 031	14	007 017	14
·829	·928 52	1694	059 072	59	·434 73	849	·493 80	844	993 045	14	007 003	14
2·830	16·945 46	1695	059 013	59	8·443 22	851	8·502 24	844	993 059	14	006 989	14
·831	·962 41	1698	058 954	59	·451 73	852	·510 68	846	993 073	14	006 975	14
·832	·979 39	1698	058 895	59	·460 25	852	·519 14	846	993 087	14	006 961	14
·833	16·996 37	1701	058 836	59	·468 77	853	·527 60	848	993 101	13	006 947	14
·834	17·013 38	1702	058 777	58	·477 30	854	·536 08	848	993 114	14	006 933	13
2·835	17·030 40	1704	058 719	59	8·485 84	855	8·544 56	849	993 128	14	006 920	14
·836	·047 44	1706	058 660	59	·494 39	856	·553 05	850	993 142	13	006 906	14
·837	·064 50	1707	058 601	58	·502 95	856	·561 55	851	993 155	14	006 892	14
·838	·081 57	1709	058 543	59	·511 51	858	·570 06	851	993 169	14	006 878	14
·839	·098 66	1711	058 484	58	·520 09	858	·578 57	853	993 183	13	006 864	13
2·840	17·115 77	1712	058 426	59	8·528 67	859	8·587 10	853	993 196	14	006 851	14
·841	·132 89	1714	058 367	58	·537 26	860	·595 63	854	993 210	13	006 837	14
·842	·150 03	1716	058 309	58	·545 86	861	·604 17	855	993 223	14	006 823	14
·843	·167 19	1718	058 251	59	·554 47	862	·612 72	856	993 237	13	006 809	13
·844	·184 37	1719	058 192	58	·563 09	862	·621 28	857	993 250	14	006 796	14
2·845	17·201 56	1721	058 134	58	8·571 71	864	8·629 85	857	993 264	13	006 782	13
·846	·218 77	1723	058 076	58	·580 35	864	·638 42	859	993 277	13	006 769	14
·847	·236 00	1724	058 018	58	·588 99	865	·647 01	859	993 290	14	006 755	14
·848	·253 24	1726	057 960	58	·597 64	866	·655 60	860	993 304	13	006 741	13
·849	·270 50	1728	057 902	58	·606 30	867	·664 20	861	993 317	13	006 728	14
2·850	17·287 78		057 844		8·614 97		8·672 81		993 330		006 714	

x	e^x	e^{-x}	sinh x	cosh x	tanh x	coth x
		0·			0·	1·
2·850	17·287 78 1730	057 844 57	8·614 97 868	8·672 81 862	993 330 14	006 714 13
·851	·305 08 1731	057 787 58	·623 65 868	·681 43 863	993 344 13	006 701 13
·852	·322 39 1733	057 729 58	·632 33 870	·690 06 864	993 357 13	006 688 14
·853	·339 72 1735	057 671 58	·641 03 870	·698 70 864	993 370 13	006 674 13
·854	·357 07 1737	057 613 57	·649 73 871	·707 34 866	993 383 14	006 661 14
2·855	17·374 44 1738	057 556 58	8·658 44 872	8·716 00 866	993 397 13	006 647 13
·856	·391 82 1740	057 498 57	·667 16 873	·724 66 867	993 410 13	006 634 13
·857	·409 22 1742	057 441 58	·675 89 874	·733 33 868	993 423 13	006 621 14
·858	·426 64 1743	057 383 57	·684 63 874	·742 01 869	993 436 13	006 607 13
·859	·444 07 1746	057 326 57	·693 37 876	·750 70 870	993 449 13	006 594 13
2·860	17·461 53 1747	057 269 57	8·702 13 876	8·759 40 870	993 462 13	006 581 13
·861	·479 00 1748	057 212 58	·710 89 878	·768 10 872	993 475 13	006 568 13
·862	·496 48 1751	057 154 57	·719 67 878	·776 82 872	993 488 13	006 555 13
·863	·513 99 1752	057 097 57	·728 45 879	·785 54 874	993 501 13	006 542 14
·864	·531 51 1754	057 040 57	·737 24 880	·794 28 874	993 514 13	006 528 13
2·865	17·549 05 1756	056 983 57	8·746 04 880	8·803 02 875	993 527 13	006 515 13
·866	·566 61 1758	056 926 57	·754 84 882	·811 77 876	993 540 13	006 502 13
·867	·584 19 1759	056 869 57	·763 66 882	·820 53 877	993 553 12	006 489 13
·868	·601 78 1761	056 812 56	·772 48 884	·829 30 877	993 565 13	006 476 13
·869	·619 39 1763	056 756 57	·781 32 884	·838 07 879	993 578 13	006 463 13
2·870	17·637 02 1764	056 699 57	8·790 16 885	8·846 86 879	993 591 13	006 450 13
·871	·654 66 1767	056 642 56	·799 01 886	·855 65 881	993 604 13	006 437 13
·872	·672 33 1768	056 586 57	·807 87 887	·864 46 881	993 617 12	006 424 12
·873	·690 01 1770	056 529 56	·816 74 888	·873 27 882	993 629 13	006 412 13
·874	·707 71 1771	056 473 57	·825 62 888	·882 09 883	993 642 13	006 399 13
2·875	17·725 42 1774	056 416 56	8·834 50 890	8·890 92 884	993 655 12	006 386 13
·876	·743 16 1775	056 360 57	·843 40 890	·899 76 885	993 667 13	006 373 13
·877	·760 91 1777	056 303 56	·852 30 892	·908 61 885	993 680 12	006 360 12
·878	·778 68 1779	056 247 56	·861 22 892	·917 46 887	993 692 13	006 348 13
·879	·796 47 1780	056 191 56	·870 14 893	·926 33 887	993 705 13	006 335 13
2·880	17·814 27 1783	056 135 56	8·879 07 894	8·935 20 889	993 718 12	006 322 13
·881	·832 10 1784	056 079 56	·888 01 895	·944 09 889	993 730 13	006 309 12
·882	·849 94 1786	056 023 56	·896 96 895	·952 98 890	993 743 12	006 297 13
·883	·867 80 1787	055 967 56	·905 91 897	·961 88 891	993 755 12	006 284 12
·884	·885 67 1790	055 911 56	·914 88 898	·970 79 892	993 767 13	006 272 13
2·885	17·903 57 1791	055 855 56	8·923 86 898	8·979 71 893	993 780 12	006 259 13
·886	·921 48 1793	055 799 56	·932 84 899	·988 64 894	993 792 13	006 246 12
·887	·939 41 1795	055 743 56	·941 83 901	8·997 58 894	993 805 12	006 234 13
·888	·957 36 1797	055 687 55	·950 84 901	9·006 52 896	993 817 12	006 221 12
·889	·975 33 1798	055 632 56	·959 85 902	·015 48 896	993 829 13	006 209 12
2·890	17·993 31 1800	055 576 55	8·968 87 903	9·024 44 898	993 842 12	006 197 13
·891	18·011 31 1802	055 521 56	·977 90 903	·033 42 898	993 854 12	006 184 12
·892	·029 33 1804	055 465 55	·986 93 905	·042 40 899	993 866 12	006 172 13
·893	·047 37 1806	055 410 56	8·995 98 906	·051 39 900	993 878 13	006 159 12
·894	·065 43 1807	055 354 55	9·005 04 906	·060 39 901	993 891 12	006 147 12
2·895	18·083 50 1809	055 299 55	9·014 10 908	9·069 40 902	993 903 12	006 135 13
·896	·101 59 1811	055 244 55	·023 18 908	·078 42 903	993 915 12	006 122 12
·897	·119 70 1813	055 189 56	·032 26 909	·087 45 903	993 927 12	006 110 12
·898	·137 83 1815	055 133 55	·041 35 910	·096 48 905	993 939 12	006 098 12
·899	·155 98 1817	055 078 55	·050 45 911	·105 53 905	993 951 12	006 086 13
2·900	18·174 15	055 023	9·059 56	9·114 58	993 963	006 073

x	e^x		e^{-x}		sinh x		cosh x		tanh x		coth x	
			0·						0·		1·	
2·900	18·174 15	1818	055 023	55	9·059 56	912	9·114 58	907	993 963	12	006 073	12
·901	·192 33	1820	054 968	55	·068 68	913	·123 65	907	993 975	12	006 061	12
·902	·210 53	1822	054 913	55	·077 81	914	·132 72	908	993 987	12	006 049	12
·903	·228 75	1824	054 858	54	·086 95	914	·141 80	910	993 999	12	006 037	12
·904	·246 99	1825	054 804	55	·096 09	916	·150 90	910	994 011	12	006 025	12
2·905	18·265 24	1828	054 749	55	9·105 25	916	9·160 00	911	994 023	12	006 013	12
·906	·283 52	1829	054 694	55	·114 41	918	·169 11	912	994 035	12	006 001	12
·907	·301 81	1831	054 639	54	·123 59	918	·178 23	912	994 047	12	005 989	12
·908	·320 12	1833	054 585	55	·132 77	919	·187 35	914	994 059	12	005 977	12
·909	·338 45	1835	054 530	54	·141 96	920	·196 49	915	994 071	11	005 965	12
2·910	18·356 80	1836	054 476	55	9·151 16	921	9·205 64	915	994 082	12	005 953	12
·911	·375 16	1839	054 421	54	·160 37	922	·214 79	917	994 094	12	005 941	12
·912	·393 55	1840	054 367	54	·169 59	923	·223 96	917	994 106	12	005 929	12
·913	·411 95	1842	054 313	55	·178 82	924	·233 13	919	994 118	11	005 917	12
·914	·430 37	1844	054 258	54	·188 06	924	·242 32	919	994 129	12	005 905	12
2·915	18·448 81	1846	054 204	54	9·197 30	926	9·251 51	920	994 141	12	005 893	11
·916	·467 27	1848	054 150	54	·206 56	927	·260 71	921	994 153	11	005 882	12
·917	·485 75	1849	054 096	54	·215 83	927	·269 92	922	994 164	12	005 870	12
·918	·504 24	1852	054 042	54	·225 10	928	·279 14	923	994 176	12	005 858	12
·919	·522 76	1853	053 988	54	·234 38	930	·288 37	924	994 188	11	005 846	11
2·920	18·541 29	1855	053 934	54	9·243 68	930	9·297 61	925	994 199	12	005 835	12
·921	·559 84	1857	053 880	54	·252 98	931	·306 86	926	994 211	11	005 823	12
·922	·578 41	1858	053 826	54	·262 29	932	·316 12	926	994 222	12	005 811	11
·923	·596 99	1861	053 772	54	·271 61	933	·325 38	928	994 234	11	005 800	12
·924	·615 60	1863	053 718	53	·280 94	934	·334 66	929	994 245	12	005 788	12
2·925	18·634 23	1864	053 665	54	9·290 28	935	9·343 95	929	994 257	11	005 776	11
·926	·652 87	1866	053 611	54	·299 63	936	·353 24	930	994 268	12	005 765	12
·927	·671 53	1868	053 557	53	·308 99	936	·362 54	932	994 280	11	005 753	11
·928	·690 21	1870	053 504	54	·318 35	938	·371 86	932	994 291	11	005 742	12
·929	·708 91	1872	053 450	53	·327 73	939	·381 18	933	994 302	12	005 730	11
2·930	18·727 63	1874	053 397	53	9·337 12	939	9·390 51	935	994 314	11	005 719	12
·931	·746 37	1875	053 344	54	·346 51	941	·399 86	935	994 325	11	005 707	11
·932	·765 12	1878	053 290	53	·355 92	941	·409 21	936	994 336	12	005 696	12
·933	·783 90	1879	053 237	53	·365 33	942	·418 57	937	994 348	11	005 684	11
·934	·802 69	1881	053 184	53	·374 75	944	·427 94	938	994 359	11	005 673	11
2·935	18·821 50	1883	053 131	53	9·384 19	944	9·437 32	939	994 370	11	005 662	12
·936	·840 33	1885	053 078	53	·393 63	945	·446 71	939	994 381	12	005 650	11
·937	·859 18	1887	053 025	53	·403 08	946	·456 10	941	994 393	11	005 639	11
·938	·878 05	1889	052 972	53	·412 54	947	·465 51	942	994 404	11	005 628	12
·939	·896 94	1891	052 919	53	·422 01	948	·474 93	943	994 415	11	005 616	11
2·940	18·915 85	1892	052 866	53	9·431 49	949	9·484 36	943	994 426	11	005 605	11
·941	·934 77	1895	052 813	53	·440 98	950	·493 79	945	994 437	11	005 594	11
·942	·953 72	1896	052 760	53	·450 48	951	·503 24	945	994 448	11	005 583	11
·943	·972 68	1898	052 707	52	·459 99	951	·512 69	947	994 459	11	005 572	12
·944	18·991 66	1900	052 655	53	·469 50	953	·522 16	947	994 470	11	005 560	11
2·945	19·010 66	1902	052 602	53	9·479 03	954	9·531 63	949	994 481	11	005 549	11
·946	·029 68	1904	052 549	52	·488 57	954	·541 12	949	994 492	11	005 538	11
·947	·048 72	1906	052 497	53	·498 11	956	·550 61	950	994 503	11	005 527	11
·948	·067 78	1908	052 444	52	·507 67	956	·560 11	951	994 514	11	005 516	11
·949	·086 86	1909	052 392	52	·517 23	958	·569 62	953	994 525	11	005 505	11
2·950	19·105 95		052 340		9·526 81		9·579 15		994 536		005 494	

TABLE Vc—EXPONENTIAL AND HYPERBOLIC FUNCTIONS

x	e^x		e^{-x}		$\sinh x$		$\cosh x$		$\tanh x$		$\coth x$	
			0·						0·		1·	
2·950	19·105 95	1912	052 340	53	9·526 81	958	9·579 15	953	994 536	11	005 494	11
·951	·125 07	1913	052 287	52	·536 39	959	·588 68	954	994 547	11	005 483	11
·952	·144 20	1916	052 235	52	·545 98	961	·598 22	955	994 558	11	005 472	11
·953	·163 36	1917	052 183	52	·555 59	961	·607 77	956	994 569	10	005 461	11
·954	·182 53	1919	052 131	52	·565 20	962	·617 33	957	994 579	11	005 450	11
2·955	19·201 72	1921	052 079	52	9·574 82	963	9·626 90	958	994 590	11	005 439	11
·956	·220 93	1923	052 027	52	·584 45	964	·636 48	959	994 601	11	005 428	11
·957	·240 16	1925	051 975	52	·594 09	966	·646 07	960	994 612	11	005 417	10
·958	·259 41	1927	051 923	52	·603 75	966	·655 67	961	994 623	10	005 407	11
·959	·278 68	1929	051 871	52	·613 41	967	·665 28	962	994 633	11	005 396	11
2·960	19·297 97	1931	051 819	52	9·623 08	968	9·674 90	962	994 644	11	005 385	11
·961	·317 28	1933	051 767	52	·632 76	969	·684 52	964	994 655	10	005 374	11
·962	·336 61	1934	051 715	51	·642 45	969	·694 16	965	994 665	11	005 363	10
·963	·355 95	1937	051 664	52	·652 14	971	·703 81	966	994 676	11	005 353	11
·964	·375 32	1938	051 612	52	·661 85	972	·713 47	966	994 687	10	005 342	11
2·965	19·394 70	1941	051 560	51	9·671 57	973	9·723 13	968	994 697	11	005 331	11
·966	·414 11	1942	051 509	52	·681 30	974	·732 81	968	994 708	10	005 320	10
·967	·433 53	1944	051 457	51	·691 04	974	·742 49	970	994 718	11	005 310	11
·968	·452 97	1947	051 406	51	·700 78	976	·752 19	971	994 729	10	005 299	10
·969	·472 44	1948	051 355	52	·710 54	977	·761 90	971	994 739	11	005 289	11
2·970	19·491 92	1950	051 303	51	9·720 31	977	9·771 61	973	994 750	10	005 278	11
·971	·511 42	1952	051 252	51	·730 08	979	·781 34	973	994 760	11	005 267	10
·972	·530 94	1954	051 201	51	·739 87	980	·791 07	975	994 771	10	005 257	11
·973	·550 48	1956	051 150	51	·749 67	980	·800 82	975	994 781	10	005 246	10
·974	·570 04	1958	051 099	52	·759 47	982	·810 57	977	994 791	11	005 236	11
2·975	19·589 62	1960	051 047	51	9·769 29	982	9·820 34	977	994 802	10	005 225	10
·976	·609 22	1962	050 996	51	·779 11	984	·830 11	978	994 812	11	005 215	11
·977	·628 84	1964	050 945	50	·788 95	984	·839 89	980	994 823	10	005 204	10
·978	·648 48	1966	050 895	51	·798 79	986	·849 69	980	994 833	10	005 194	10
·979	·668 14	1968	050 844	51	·808 65	986	·859 49	981	994 843	10	005 184	11
2·980	19·687 82	1969	050 793	51	9·818 51	988	9·869 30	983	994 853	11	005 173	10
·981	·707 51	1972	050 742	51	·828 39	988	·879 13	983	994 864	10	005 163	11
·982	·727 23	1974	050 691	50	·838 27	989	·888 96	984	994 874	10	005 152	10
·983	·746 97	1976	050 641	51	·848 16	991	·898 80	986	994 884	10	005 142	10
·984	·766 73	1977	050 590	50	·858 07	991	·908 66	986	994 894	11	005 132	10
2·985	19·786 50	1980	050 540	51	9·867 98	992	9·918 52	987	994 905	10	005 122	11
·986	·806 30	1981	050 489	50	·877 90	994	·928 39	989	994 915	10	005 111	10
·987	·826 11	1984	050 439	51	·887 84	994	·938 28	989	994 925	10	005 101	10
·988	·845 95	1986	050 388	50	·897 78	995	·948 17	990	994 935	10	005 091	10
·989	·865 81	1987	050 338	51	·907 73	997	·958 07	991	994 945	10	005 081	11
2·990	19·885 68	1990	050 287	50	9·917 70	997	9·967 98	993	994 955	10	005 070	10
·991	·905 58	1991	050 237	50	·927 67	998	·977 91	993	994 965	10	005 060	10
·992	·925 49	1994	050 187	50	·937 65	1000	·987 84	994	994 975	10	005 050	10
·993	·945 43	1995	050 137	50	·947 65	1000	9·997 78	996	994 985	10	005 040	10
·994	·965 38	1998	050 087	50	·957 65	1001	10·007 74	996	994 995	10	005 030	10
2·995	19·985 36	2000	050 037	50	9·967 66	1002	10·017 70	997	995 005	10	005 020	10
·996	20·005 36	2001	049 987	50	·977 68	1004	·027 67	998	995 015	10	005 010	10
·997	·025 37	2004	049 937	50	·987 72	1004	·037 65	1000	995 025	10	005 000	10
·998	·045 41	2005	049 887	50	9·997 76	1005	·047 65	1000	995 035	10	004 990	10
·999	·065 46	2008	049 837	50	10·007 81	1006	·057 65	1001	995 045	10	004 980	10
3·000	20·085 54		049 787		10·017 87		10·067 66		995 055		004 970	

TABLE Vd—EXPONENTIAL AND HYPERBOLIC FUNCTIONS

x	e^x	e^{-x}	Δ	sinh x	cosh x	tanh x	Δ	coth x	Δ
3·00	20·085 54	0·049 787	495	10·017 87	10·067 66	0·995 055	97	1·004 970	99
·01	20·287 40	·049 292	491	10·119 05	10·168 35	·995 152	96	·004 871	97
·02	20·491 29	·048 801	485	10·221 25	10·270 05	·995 248	94	·004 774	94
·03	20·697 23	·048 316	481	10·324 46	10·372 77	·995 342	92	·004 680	93
·04	20·905 24	·047 835	476	10·428 70	10·476 54	·995 434	90	·004 587	91
3·05	21·115 34	0·047 359	471	10·533 99	10·581 35	0·995 524	89	1·004 496	89
·06	21·327 56	·046 888	467	10·640 33	10·687 22	·995 613	86	·004 407	88
·07	21·541 90	·046 421	462	10·747 74	10·794 16	·995 699	85	·004 319	86
·08	21·758 40	·045 959	457	10·856 22	10·902 18	·995 784	84	·004 233	84
·09	21·977 08	·045 502	453	10·965 79	11·011 29	·995 868	81	·004 149	82
3·10	22·197 95	0·045 049	448	11·076 45	11·121 50	0·995 949	80	1·004 067	81
·11	22·421 04	·044 601	444	11·188 22	11·232 82	·996 029	79	·003 986	79
·12	22·646 38	·044 157	439	11·301 11	11·345 27	·996 108	77	·003 907	77
·13	22·873 98	·043 718	435	11·415 13	11·458 85	·996 185	75	·003 830	76
·14	23·103 87	·043 283	431	11·530 29	11·573 57	·996 260	74	·003 754	75
3·15	23·336 06	0·042 852	426	11·646 61	11·689 46	0·996 334	73	1·003 679	73
·16	23·570 60	·042 426	422	11·764 09	11·806 51	·996 407	71	·003 606	71
·17	23·807 48	·042 004	418	11·882 74	11·924 74	·996 478	69	·003 535	70
·18	24·046 75	·041 586	414	12·002 58	12·044 17	·996 547	68	·003 465	69
·19	24·288 43	·041 172	410	12·123 63	12·164 80	·996 615	67	·003 396	67
3·20	24·532 53	0·040 762	405	12·245 88	12·286 65	0·996 682	66	1·003 329	66
·21	24·779 09	·040 357	402	12·369 36	12·409 72	·996 748	64	·003 263	65
·22	25·028 12	·039 955	398	12·494 08	12·534 04	·996 812	63	·003 198	64
·23	25·279 66	·039 557	393	12·620 05	12·659 61	·996 875	62	·003 134	62
·24	25·533 72	·039 164	390	12·747 28	12·786 44	·996 937	61	·003 072	61
3·25	25·790 34	0·038 774	386	12·875 78	12·914 56	0·996 998	59	1·003 011	59
·26	26·049 54	·038 388	382	13·005 57	13·043 96	·997 057	58	·002 952	59
·27	26·311 34	·038 006	378	13·136 67	13·174 67	·997 115	57	·002 893	57
·28	26·575 77	·037 628	374	13·269 07	13·306 70	·997 172	56	·002 836	56
·29	26·842 86	·037 254	371	13·402 80	13·440 06	·997 228	55	·002 780	56
3·30	27·112 64	0·036 883	367	13·537 88	13·574 76	0·997 283	54	1·002 724	54
·31	27·385 13	·036 516	363	13·674 30	13·710 82	·997 337	52	·002 670	53
·32	27·660 35	·036 153	360	13·812 10	13·848 25	·997 389	52	·002 617	51
·33	27·938 34	·035 793	356	13·951 27	13·987 07	·997 441	51	·002 566	51
·34	28·219 13	·035 437	353	14·091 84	14·127 28	·997 492	49	·002 515	50
3·35	28·502 73	0·035 084	349	14·233 82	14·268 91	0·997 541	49	1·002 465	49
·36	28·789 19	·034 735	345	14·377 23	14·411 96	·997 590	47	·002 416	48
·37	29·078 53	·034 390	343	14·522 07	14·556 46	·997 637	47	·002 368	47
·38	29·370 77	·034 047	338	14·668 36	14·702 41	·997 684	46	·002 321	46
·39	29·665 95	·033 709	336	14·816 12	14·849 83	·997 730	45	·002 275	45
3·40	29·964 10	0·033 373	332	14·965 36	14·998 74	0·997 775	44	1·002 230	44
·41	30·265 24	·033 041	329	15·116 10	15·149 14	·997 819	43	·002 186	44
·42	30·569 42	·032 712	325	15·268 35	15·301 06	·997 862	42	·002 142	42
·43	30·876 64	·032 387	322	15·422 13	15·454 51	·997 904	42	·002 100	42
·44	31·186 96	·032 065	319	15·577 45	15·609 51	·997 946	40	·002 058	40
3·45	31·500 39	0·031 746	316	15·734 32	15·766 07	0·997 986	40	1·002 018	40
·46	31·816 98	·031 430	313	15·892 77	15·924 20	·998 026	39	·001 978	40
·47	32·136 74	·031 117	310	16·052 81	16·083 93	·998 065	39	·001 938	38
·48	32·459 72	·030 807	306	16·214 46	16·245 26	·998 104	37	·001 900	38
·49	32·785 95	·030 501	304	16·377 72	16·408 22	·998 141	37	·001 862	37
3·50	33·115 45	0·030 197		16·542 63	16·572 82	0·998 178		1·001 825	

TABLE VD—EXPONENTIAL AND HYPERBOLIC FUNCTIONS

x	e^x	e^{-x}		sinh x	cosh x	tanh x		coth x	
3·50	33·115 45	0·030 197	300	16·542 63	16·572 82	0·998 178	36	1·001 825	36
·51	33·448 27	·029 897	298	16·709 19	16·739 08	·998 214	35	·001 789	35
·52	33·784 43	·029 599	294	16·877 41	16·907 01	·998 249	35	·001 754	35
·53	34·123 97	·029 305	292	17·047 33	17·076 64	·998 284	34	·001 719	34
·54	34·466 92	·029 013	288	17·218 95	17·247 97	·998 318	33	·001 685	33
3·55	34·813 32	0·028 725	286	17·392 30	17·421 02	0·998 351	33	1·001 652	33
·56	35·163 20	·028 439	283	17·567 38	17·595 82	·998 384	32	·001 619	32
·57	35·516 59	·028 156	280	17·744 22	17·772 37	·998 416	31	·001 587	32
·58	35·873 54	·027 876	278	17·922 83	17·950 71	·998 447	31	·001 555	31
·59	36·234 08	·027 598	274	18·103 24	18·130 84	·998 478	30	·001 524	30
3·60	36·598 23	0·027 324	272	18·285 46	18·312 78	0·998 508	29	1·001 494	29
·61	36·966 05	·027 052	269	18·469 50	18·496 55	·998 537	29	·001 465	29
·62	37·337 57	·026 783	267	18·655 39	18·682 18	·998 566	29	·001 436	29
·63	37·712 82	·026 516	264	18·843 15	18·869 67	·998 595	28	·001 407	28
·64	38·091 84	·026 252	261	19·032 79	19·059 04	·998 623	27	·001 379	27
3·65	38·474 67	0·025 991	258	19·224 34	19·250 33	0·998 650	27	1·001 352	27
·66	38·861 34	·025 733	257	19·417 81	19·443 54	·998 677	26	·001 325	26
·67	39·251 91	·025 476	253	19·613 21	19·638 69	·998 703	25	·001 299	26
·68	39·646 39	·025 223	251	19·810 59	19·835 81	·998 728	26	·001 273	25
·69	40·044 85	·024 972	248	20·009 94	20·034 91	·998 754	24	·001 248	25
3·70	40·447 30	0·024 724	246	20·211 29	20·236 01	0·998 778	24	1·001 223	24
·71	40·853 81	·024 478	244	20·414 66	20·439 14	·998 802	24	·001 199	24
·72	41·264 39	·024 234	241	20·620 08	20·644 31	·998 826	23	·001 175	23
·73	41·679 11	·023 993	239	20·827 56	20·851 55	·998 849	23	·001 152	23
·74	42·097 99	·023 754	236	21·037 12	21·060 87	·998 872	22	·001 129	22
3·75	42·521 08	0·023 518	234	21·248 78	21·272 30	0·998 894	22	1·001 107	22
·76	42·948 43	·023 284	232	21·462 57	21·485 85	·998 916	22	·001 085	22
·77	43·380 06	·023 052	229	21·678 51	21·701 56	·998 938	21	·001 063	21
·78	43·816 04	·022 823	227	21·896 61	21·919 43	·998 959	20	·001 042	20
·79	44·256 40	·022 596	225	22·116 90	22·139 50	·998 979	21	·001 022	21
3·80	44·701 18	0·022 371	223	22·339 41	22·361 78	0·999 000	19	1·001 001	19
·81	45·150 44	·022 148	220	22·564 15	22·586 29	·999 019	20	·000 982	20
·82	45·604 21	·021 928	218	22·791 14	22·813 07	·999 039	19	·000 962	19
·83	46·062 54	·021 710	216	23·020 41	23·042 12	·999 058	18	·000 943	19
·84	46·525 47	·021 494	214	23·251 99	23·273 48	·999 076	19	·000 924	18
3·85	46·993 06	0·021 280	212	23·485 89	23·507 17	0·999 095	18	1·000 906	18
·86	47·465 35	·021 068	210	23·722 14	23·743 21	·999 113	17	·000 888	17
·87	47·942 39	·020 858	207	23·960 76	23·981 62	·999 130	17	·000 871	18
·88	48·424 22	·020 651	206	24·201 78	24·222 43	·999 147	17	·000 853	17
·89	48·910 89	·020 445	203	24·445 22	24·465 67	·999 164	17	·000 836	16
3·90	49·402 45	0·020 242	201	24·691 10	24·711 35	0·999 181	16	1·000 820	16
·91	49·898 95	·020 041	200	24·939 46	24·959 50	·999 197	16	·000 804	16
·92	50·400 44	·019 841	197	25·190 30	25·210 14	·999 213	16	·000 788	16
·93	50·906 98	·019 644	196	25·443 67	25·463 31	·999 229	15	·000 772	15
·94	51·418 60	·019 448	193	25·699 58	25·719 02	·999 244	15	·000 757	15
3·95	51·935 37	0·019 255	192	25·958 06	25·977 31	0·999 259	14	1·000 742	15
·96	52·457 33	·019 063	190	26·219 13	26·238 19	·999 273	15	·000 727	14
·97	52·984 53	·018 873	187	26·482 83	26·501 70	·999 288	14	·000 713	14
·98	53·517 03	·018 686	186	26·749 17	26·767 86	·999 302	14	·000 699	14
·99	54·054 89	·018 500	184	27·018 19	27·036 69	·999 316	13	·000 685	14
4·00	54·598 15	0·018 316		27·289 92	27·308 23	0·999 329		1·000 671	

TABLE VD—EXPONENTIAL AND HYPERBOLIC FUNCTIONS

x	e^{x}	e^{-x}	Δ	sinh x	cosh x	tanh x	Δ	coth x	Δ
4·00	54·5982	0·018 316	183	27·2899	27·3082	0·999 329	14	1·000 671	13
·01	55·1469	·018 133	180	27·5644	27·5825	·999 343	13	·000 658	13
·02	55·7011	·017 953	179	27·8416	27·8595	·999 356	12	·000 645	13
·03	56·2609	·017 774	177	28·1216	28·1393	·999 368	13	·000 632	12
·04	56·8263	·017 597	175	28·4044	28·4220	·999 381	12	·000 620	13
4·05	57·3975	0·017 422	173	28·6900	28·7074	0·999 393	12	1·000 607	12
·06	57·9743	·017 249	172	28·9785	28·9958	·999 405	12	·000 595	12
·07	58·5570	·017 077	170	29·2699	29·2870	·999 417	11	·000 583	11
·08	59·1455	·016 907	168	29·5643	29·5812	·999 428	12	·000 572	11
·09	59·7399	·016 739	166	29·8616	29·8783	·999 440	11	·000 561	12
4·10	60·3403	0·016 573	165	30·1619	30·1784	0·999 451	11	1·000 549	10
·11	60·9467	·016 408	163	30·4652	30·4816	·999 462	10	·000 539	11
·12	61·5592	·016 245	162	30·7715	30·7877	·999 472	11	·000 528	11
·13	62·1779	·016 083	160	31·0809	31·0970	·999 483	10	·000 517	10
·14	62·8028	·015 923	159	31·3934	31·4094	·999 493	10	·000 507	10
4·15	63·4340	0·015 764	156	31·7091	31·7249	0·999 503	10	1·000 497	10
·16	64·0715	·015 608	156	32·0280	32·0436	·999 513	10	·000 487	9
·17	64·7155	·015 452	153	32·3500	32·3655	·999 523	9	·000 478	10
·18	65·3659	·015 299	153	32·6753	32·6906	·999 532	9	·000 468	9
·19	66·0228	·015 146	150	33·0038	33·0190	·999 541	9	·000 459	9
4·20	66·6863	0·014 996	150	33·3357	33·3507	0·999 550	9	1·000 450	9
·21	67·3565	·014 846	147	33·6708	33·6857	·999 559	9	·000 441	9
·22	68·0335	·014 699	147	34·0094	34·0241	·999 568	9	·000 432	8
·23	68·7172	·014 552	144	34·3513	34·3659	·999 577	8	·000 424	9
·24	69·4079	·014 408	144	34·6967	34·7111	·999 585	8	·000 415	8
4·25	70·1054	0·014 264	142	35·0456	35·0598	0·999 593	8	1·000 407	8
·26	70·8100	·014 122	140	35·3979	35·4121	·999 601	8	·000 399	8
·27	71·5216	·013 982	139	35·7538	35·7678	·999 609	8	·000 391	8
·28	72·2404	·013 843	138	36·1133	36·1271	·999 617	7	·000 383	7
·29	72·9665	·013 705	136	36·4764	36·4901	·999 624	8	·000 376	8
4·30	73·6998	0·013 569	135	36·8431	36·8567	0·999 632	7	1·000 368	7
·31	74·4405	·013 434	134	37·2135	37·2270	·999 639	7	·000 361	7
·32	75·1886	·013 300	132	37·5877	37·6010	·999 646	7	·000 354	7
·33	75·9443	·013 168	131	37·9656	37·9787	·999 653	7	·000 347	7
·34	76·7075	·013 037	130	38·3473	38·3603	·999 660	7	·000 340	7
4·35	77·4785	0·012 907	129	38·7328	38·7457	0·999 667	6	1·000 333	6
·36	78·2571	·012 778	127	39·1222	39·1350	·999 673	7	·000 327	7
·37	79·0436	·012 651	126	39·5155	39·5281	·999 680	6	·000 320	6
·38	79·8380	·012 525	124	39·9128	39·9253	·999 686	6	·000 314	6
·39	80·6404	·012 401	124	40·3140	40·3264	·999 692	7	·000 308	6
4·40	81·4509	0·012 277	122	40·7193	40·7316	0·999 699	6	1·000 302	6
·41	82·2695	·012 155	121	41·1287	41·1408	·999 705	5	·000 296	6
·42	83·0963	·012 034	120	41·5421	41·5542	·999 710	6	·000 290	6
·43	83·9314	·011 914	118	41·9598	41·9717	·999 716	6	·000 284	6
·44	84·7749	·011 796	117	42·3816	42·3934	·999 722	5	·000 278	5
4·45	85·6269	0·011 679	117	42·8076	42·8193	0·999 727	6	1·000 273	6
·46	86·4875	·011 562	115	43·2380	43·2495	·999 733	5	·000 267	5
·47	87·3567	·011 447	114	43·6726	43·6841	·999 738	5	·000 262	5
·48	88·2347	·011 333	112	44·1117	44·1230	·999 743	5	·000 257	5
·49	89·1214	·011 221	112	44·5551	44·5663	·999 748	5	·000 252	5
4·50	90·0171	0·011 109		45·0030	45·0141	0·999 753		1·000 247	

x	e^x	e^{-x}		$\sinh x$	$\cosh x$	$\tanh x$	$\coth x$
4·50	90·0171	0·011 109	111	45·0030	45·0141	0·999 753	1·000 247
·51	90·9218	·010 998	109	45·4554	45·4664	·999 758	·000 242
·52	91·8356	·010 889	108	45·9124	45·9232	·999 763	·000 237
·53	92·7586	·010 781	108	46·3739	46·3847	·999 768	·000 232
·54	93·6908	·010 673	106	46·8401	46·8507	·999 772	·000 228
4·55	94·6324	0·010 567	105	47·3109	47·3215	0·999 777	1·000 223
·56	95·5835	·010 462	104	47·7865	47·7970	·999 781	·000 219
·57	96·5441	·010 358	103	48·2669	48·2772	·999 785	·000 215
·58	97·5144	·010 255	102	48·7521	48·7623	·999 790	·000 210
·59	98·4944	·010 153	101	49·2421	49·2523	·999 794	·000 206
4·60	99·4843	0·010 052	100	49·7371	49·7472	0·999 798	1·000 202
·61	100·4841	·009 952	99	50·2371	50·2471	·999 802	·000 198
·62	101·4940	·009 853	98	50·7421	50·7519	·999 806	·000 194
·63	102·5141	·009 755	97	51·2522	51·2619	·999 810	·000 190
·64	103·5443	·009 658	96	51·7673	51·7770	·999 813	·000 187
4·65	104·5850	0·009 562	96	52·2877	52·2973	0·999 817	1·000 183
·66	105·6361	·009 466	94	52·8133	52·8228	·999 821	·000 179
·67	106·6977	·009 372	93	53·3442	53·3536	·999 824	·000 176
·68	107·7701	·009 279	92	53·8804	53·8897	·999 828	·000 172
·69	108·8532	·009 187	92	54·4220	54·4312	·999 831	·000 169
4·70	109·9472	0·009 095	90	54·9690	54·9781	0·999 835	1·000 165
·71	111·0522	·009 005	90	55·5216	55·5306	·999 838	·000 162
·72	112·1683	·008 915	89	56·0797	56·0886	·999 841	·000 159
·73	113·2956	·008 826	87	56·6434	56·6522	·999 844	·000 156
·74	114·4342	·008 739	87	57·2127	57·2215	·999 847	·000 153
4·75	115·5843	0·008 652	86	57·7878	57·7965	0·999 850	1·000 150
·76	116·7459	·008 566	86	58·3687	58·3772	·999 853	·000 147
·77	117·9192	·008 480	84	58·9554	58·9639	·999 856	·000 144
·78	119·1044	·008 396	84	59·5480	59·5564	·999 859	·000 141
·79	120·3014	·008 312	82	60·1465	60·1548	·999 862	·000 138
4·80	121·5104	0·008 230	82	60·7511	60·7593	0·999 865	1·000 135
·81	122·7316	·008 148	81	61·3617	61·3699	·999 867	·000 133
·82	123·9651	·008 067	80	61·9785	61·9866	·999 870	·000 130
·83	125·2110	·007 987	80	62·6015	62·6095	·999 872	·000 128
·84	126·4694	·007 907	79	63·2307	63·2386	·999 875	·000 125
4·85	127·7404	0·007 828	78	63·8663	63·8741	0·999 877	1·000 123
·86	129·0242	·007 750	77	64·5082	64·5160	·999 880	·000 120
·87	130·3209	·007 673	76	65·1566	65·1643	·999 882	·000 118
·88	131·6307	·007 597	76	65·8115	65·8191	·999 885	·000 115
·89	132·9536	·007 521	74	66·4730	66·4805	·999 887	·000 113
4·90	134·2898	0·007 447	75	67·1412	67·1486	0·999 889	1·000 111
·91	135·6394	·007 372	73	67·8160	67·8234	·999 891	·000 109
·92	137·0026	·007 299	72	68·4977	68·5050	·999 893	·000 107
·93	138·3795	·007 227	72	69·1861	69·1934	·999 896	·000 104
·94	139·7702	·007 155	72	69·8815	69·8887	·999 898	·000 102
4·95	141·1750	0·007 083	70	70·5839	70·5910	0·999 900	1·000 100
·96	142·5938	·007 013	70	71·2934	71·3004	·999 902	·000 098
·97	144·0269	·006 943	69	72·0100	72·0169	·999 904	·000 096
·98	145·4744	·006 874	68	72·7338	72·7406	·999 905	·000 095
·99	146·9364	·006 806	68	73·4648	73·4716	·999 907	·000 093
5·00	148·4132	0·006 738		74·2032	74·2099	0·999 909	1·000 091

x	e^{-x}		tanh x	coth x	x	e^{-x}		tanh x	coth x
5·00	0·006 738	67	0·999 909	1·000 091	**5·50**	0·004 087	41	0·999 967	1·000 033
·01	·006 671	66	911	089	**·51**	·004 046	40	967	033
·02	·006 605	66	913	087	**·52**	·004 006	40	968	032
·03	·006 539	65	914	086	**·53**	·003 966	39	969	031
·04	·006 474	65	916	084	**·54**	·003 927	40	969	031
5·05	0·006 409	63	0·999 918	1·000 082	**5·55**	0·003 887	38	0·999 970	1·000 030
·06	·006 346	64	919	081	**·56**	·003 849	39	970	030
·07	·006 282	62	921	079	**·57**	·003 810	37	971	029
·08	·006 220	62	923	077	**·58**	·003 773	38	972	028
·09	·006 158	61	924	076	**·59**	·003 735	37	972	028
5·10	0·006 097	61	0·999 926	1·000 074	**5·60**	0·003 698	37	0·999 973	1·000 027
·11	·006 036	60	927	073	**·61**	·003 661	36	973	027
·12	·005 976	59	929	071	**·62**	·003 625	36	974	026
·13	·005 917	59	930	070	**·63**	·003 589	36	974	026
·14	·005 858	59	931	069	**·64**	·003 553	35	975	025
5·15	0·005 799	57	0·999 933	1·000 067	**5·65**	0·003 518	35	0·999 975	1·000 025
·16	·005 742	57	934	066	**·66**	·003 483	35	976	024
·17	·005 685	57	935	065	**·67**	·003 448	34	976	024
·18	·005 628	56	937	063	**·68**	·003 414	34	977	023
·19	·005 572	55	938	062	**·69**	·003 380	34	977	023
5·20	0·005 517	55	0·999 939	1·000 061	**5·70**	0·003 346	33	0·999 978	1·000 022
·21	·005 462	55	940	060	**·71**	·003 313	33	978	022
·22	·005 407	53	942	058	**·72**	·003 280	33	978	022
·23	·005 354	54	943	057	**·73**	·003 247	32	979	021
·24	·005 300	52	944	056	**·74**	·003 215	32	979	021
5·25	0·005 248	53	0·999 945	1·000 055	**5·75**	0·003 183	32	0·999 980	1·000 020
·26	·005 195	51	946	054	**·76**	·003 151	31	980	020
·27	·005 144	52	947	053	**·77**	·003 120	31	981	019
·28	·005 092	50	948	052	**·78**	·003 089	31	981	019
·29	·005 042	50	949	051	**·79**	·003 058	30	981	019
5·30	0·004 992	50	0·999 950	1·000 050	**5·80**	0·003 028	31	0·999 982	1·000 018
·31	·004 942	49	951	049	**·81**	·002 997	29	982	018
·32	·004 893	49	952	048	**·82**	·002 968	30	982	018
·33	·004 844	48	953	047	**·83**	·002 938	29	983	017
·34	·004 796	48	954	046	**·84**	·002 909	29	983	017
5·35	0·004 748	47	0·999 955	1·000 045	**5·85**	0·002 880	29	0·999 983	1·000 017
·36	·004 701	47	956	044	**·86**	·002 851	28	984	016
·37	·004 654	46	957	043	**·87**	·002 823	28	984	016
·38	·004 608	46	958	042	**·88**	·002 795	28	984	016
·39	·004 562	45	958	042	**·89**	·002 767	28	985	015
5·40	0·004 517	45	0·999 959	1·000 041	**5·90**	0·002 739	27	0·999 985	1·000 015
·41	·004 472	45	960	040	**·91**	·002 712	27	985	015
·42	·004 427	44	961	039	**·92**	·002 685	27	986	014
·43	·004 383	44	962	038	**·93**	·002 658	26	986	014
·44	·004 339	43	962	038	**·94**	·002 632	26	986	014
5·45	0·004 296	42	0·999 963	1·000 037	**5·95**	0·002 606	26	0·999 986	1·000 014
·46	·004 254	43	964	036	**·96**	·002 580	26	987	013
·47	·004 211	42	965	035	**·97**	·002 554	25	987	013
·48	·004 169	41	965	035	**·98**	·002 529	25	987	013
·49	·004 128	41	966	034	**·99**	·002 504	25	987	013
5·50	0·004 087		0·999 967	1·000 033	**6·00**	0·002 479		0·999 988	1·000 012

TABLE Ve—EXPONENTIAL AND HYPERBOLIC FUNCTIONS

x	e^{-x}	
6·00	0·002 479	25
·01	·002 454	24
·02	·002 430	25
·03	·002 405	23
·04	·002 382	24
6·05	0·002 358	24
·06	·002 334	23
·07	·002 311	23
·08	·002 288	23
·09	·002 265	22
6·10	0·002 243	22
·11	·002 221	23
·12	·002 198	21
·13	·002 177	22
·14	·002 155	22
6·15	0·002 133	21
·16	·002 112	21
·17	·002 091	21
·18	·002 070	20
·19	·002 050	21
6·20	0·002 029	20
·21	·002 009	20
·22	·001 989	20
·23	·001 969	19
·24	·001 950	20
6·25	0·001 930	19
·26	·001 911	19
·27	·001 892	19
·28	·001 873	18
·29	·001 855	19
6·30	0·001 836	18
·31	·001 818	18
·32	·001 800	18
·33	·001 782	18
·34	·001 764	17
6·35	0·001 747	18
·36	·001 729	17
·37	·001 712	17
·38	·001 695	17
·39	·001 678	16
6·40	0·001 662	17
·41	·001 645	16
·42	·001 629	17
·43	·001 612	16
·44	·001 596	15
6·45	0·001 581	16
·46	·001 565	16
·47	·001 549	15
·48	·001 534	15
·49	·001 519	16
6·50	0·001 503	

x	e^{-x}	
6·50	0·001 503	15
·51	·001 488	14
·52	·001 474	15
·53	·001 459	15
·54	·001 444	14
6·55	0·001 430	14
·56	·001 416	14
·57	·001 402	14
·58	·001 388	14
·59	·001 374	14
6·60	0·001 360	13
·61	·001 347	14
·62	·001 333	13
·63	·001 320	13
·64	·001 307	13
6·65	0·001 294	13
·66	·001 281	13
·67	·001 268	12
·68	·001 256	13
·69	·001 243	12
6·70	0·001 231	12
·71	·001 219	12
·72	·001 207	12
·73	·001 195	12
·74	·001 183	12
6·75	0·001 171	12
·76	·001 159	11
·77	·001 148	12
·78	·001 136	11
·79	·001 125	11
6·80	0·001 114	11
·81	·001 103	11
·82	·001 092	11
·83	·001 081	11
·84	·001 070	11
6·85	0·001 059	10
·86	·001 049	11
·87	·001 038	10
·88	·001 028	10
·89	·001 018	10
6·90	0·001 008	10
·91	·000 998	10
·92	·000 988	10
·93	·000 978	10
·94	·000 968	9
6·95	0·000 959	10
·96	·000 949	9
·97	·000 940	10
·98	·000 930	9
·99	·000 921	9
7·00	0·000 912	

x	e^{-x}	
7·00	0·000 912	9
·01	·000 903	9
·02	·000 894	9
·03	·000 885	9
·04	·000 876	9
7·05	0·000 867	8
·06	·000 859	9
·07	·000 850	8
·08	·000 842	9
·09	·000 833	8
7·10	0·000 825	8
·11	·000 817	8
·12	·000 809	8
·13	·000 801	8
·14	·000 793	8
7·15	0·000 785	8
·16	·000 777	8
·17	·000 769	7
·18	·000 762	8
·19	·000 754	7
7·20	0·000 747	8
·21	·000 739	7
·22	·000 732	7
·23	·000 725	8
·24	·000 717	7
7·25	0·000 710	7
·26	·000 703	7
·27	·000 696	7
·28	·000 689	7
·29	·000 682	6
7·30	0·000 676	7
·31	·000 669	7
·32	·000 662	6
·33	·000 656	7
·34	·000 649	6
7·35	0·000 643	7
·36	·000 636	6
·37	·000 630	6
·38	·000 624	7
·39	·000 617	6
7·40	0·000 611	6
·41	·000 605	6
·42	·000 599	6
·43	·000 593	6
·44	·000 587	6
7·45	0·000 581	5
·46	·000 576	6
·47	·000 570	6
·48	·000 564	5
·49	·000 559	6
7·50	0·000 553	

x	e^{-x}	
7·5	0·000 553	53
7·6	·000 500	47
7·7	·000 453	43
7·8	·000 410	39
7·9	·000 371	36
8·0	0·000 335	31
8·1	·000 304	29
8·2	·000 275	26
8·3	·000 249	24
8·4	·000 225	22
8·5	0·000 203	19
8·6	·000 184	17
8·7	·000 167	16
8·8	·000 151	15
8·9	·000 136	13
9·0	0·000 123	11
9·1	·000 112	11
9·2	·000 101	10
9·3	·000 091	8
9·4	·000 083	8
9·5	0·000 075	7
9·6	·000 068	7
9·7	·000 061	6
9·8	·000 055	5
9·9	·000 050	5
10·0	0·000 045	4
10·1	·000 041	4
10·2	·000 037	3
10·3	·000 034	4
10·4	·000 030	2
10·5	0·000 028	3
10·6	·000 025	2
10·7	·000 023	3
10·8	·000 020	2
10·9	·000 018	1
11·0	0·000 017	2
11·1	·000 015	1
11·2	·000 014	2
11·3	·000 012	1
11·4	·000 011	1
11·5	0·000 010	1
11·6	·000 009	1
11·7	·000 008	0
11·8	·000 008	1
11·9	·000 007	1
12·0	0·000 006	0
12·1	·000 006	1
12·2	·000 005	0
12·3	·000 005	1
12·4	·000 004	0
12·5	0·000 004	

x	e^{-x}
12·31	
	0·000 004
12·56	
	0·000 003
12·89	
	0·000 002
13·41	
	0·000 001
14·50	
	0·000 000
∞	

x	tanh x
5·991	
	0·999 988
6·033	
	0·999 989
6·078	
	0·999 990
6·128	
	0·999 991
6·184	
	0·999 992
6·246	
	0·999 993
6·318	
	0·999 994
6·401	
	0·999 995
6·502	
	0·999 996
6·627	
	0·999 997
6·796	
	0·999 998
7·051	
	0·999 999
7·600	
	1·000 000
∞	

x	coth x
5·991	
	1·000 012
6·033	
	1·000 011
6·078	
	1·000 010
6·128	
	1·000 009
6·184	
	1·000 008
6·246	
	1·000 007
6·318	
	1·000 006
6·401	
	1·000 005
6·502	
	1·000 004
6·627	
	1·000 003
6·796	
	1·000 002
7·051	
	1·000 001
7·600	
	1·000 000
∞	

TABLE VF—EXPONENTIAL FUNCTION

x	e^x	x	e^x	x	e^x	x	e^x	x	e^x
5·00	148·4132	5·50	244·6919	6·00	403·429	6·50	665·142	7·00	1096·633
·01	149·9047	·51	247·1511	·01	407·483	·51	671·826	·01	1107·655
·02	151·4113	·52	249·6350	·02	411·579	·52	678·578	·02	1118·787
·03	152·9330	·53	252·1439	·03	415·715	·53	685·398	·03	1130·031
·04	154·4700	·54	254·6780	·04	419·893	·54	692·287	·04	1141·388
5·05	156·0225	5·55	257·2376	6·05	424·113	6·55	699·244	7·05	1152·859
·06	157·5905	·56	259·8228	·06	428·375	·56	706·272	·06	1164·445
·07	159·1743	·57	262·4341	·07	432·681	·57	713·370	·07	1176·148
·08	160·7741	·58	265·0716	·08	437·029	·58	720·539	·08	1187·969
·09	162·3899	·59	267·7356	·09	441·421	·59	727·781	·09	1199·908
5·10	164·0219	5·60	270·4264	6·10	445·858	6·60	735·095	7·10	1211·967
·11	165·6704	·61	273·1442	·11	450·339	·61	742·483	·11	1224·148
·12	167·3354	·62	275·8894	·12	454·865	·62	749·945	·12	1236·450
·13	169·0171	·63	278·6621	·13	459·436	·63	757·482	·13	1248·877
·14	170·7158	·64	281·4627	·14	464·054	·64	765·095	·14	1261·428
5·15	172·4315	5·65	284·2915	6·15	468·717	6·65	772·784	7·15	1274·106
·16	174·1645	·66	287·1486	·16	473·428	·66	780·551	·16	1286·911
·17	175·9148	·67	290·0345	·17	478·186	·67	788·396	·17	1299·845
·18	177·6828	·68	292·9494	·18	482·992	·68	796·319	·18	1312·908
·19	179·4686	·69	295·8936	·19	487·846	·69	804·322	·19	1326·103
5·20	181·2722	5·70	298·8674	6·20	492·749	6·70	812·406	7·20	1339·431
·21	183·0941	·71	301·8711	·21	497·701	·71	820·571	·21	1352·892
·22	184·9342	·72	304·9049	·22	502·703	·72	828·818	·22	1366·489
·23	186·7928	·73	307·9693	·23	507·755	·73	837·147	·23	1380·223
·24	188·6701	·74	311·0644	·24	512·859	·74	845·561	·24	1394·094
5·25	190·5663	5·75	314·1907	6·25	518·013	6·75	854·059	7·25	1408·105
·26	192·4815	·76	317·3483	·26	523·219	·76	862·642	·26	1422·257
·27	194·4160	·77	320·5377	·27	528·477	·77	871·312	·27	1436·550
·28	196·3699	·78	323·7592	·28	533·789	·78	880·069	·28	1450·988
·29	198·3434	·79	327·0130	·29	539·153	·79	888·914	·29	1465·571
5·30	200·3368	5·80	330·2996	6·30	544·572	6·80	897·847	7·30	1480·300
·31	202·3502	·81	333·6191	·31	550·045	·81	906·871	·31	1495·177
·32	204·3839	·82	336·9721	·32	555·573	·82	915·985	·32	1510·204
·33	206·4380	·83	340·3587	·33	561·157	·83	925·191	·33	1525·382
·34	208·5127	·84	343·7793	·34	566·796	·84	934·489	·34	1540·712
5·35	210·6083	5·85	347·2344	6·35	572·493	6·85	943·881	7·35	1556·197
·36	212·7249	·86	350·7241	·36	578·246	·86	953·367	·36	1571·837
·37	214·8629	·87	354·2490	·37	584·058	·87	962·949	·37	1587·634
·38	217·0223	·88	357·8092	·38	589·928	·88	972·626	·38	1603·590
·39	219·2034	·89	361·4053	·39	595·857	·89	982·401	·39	1619·706
5·40	221·4064	5·90	365·0375	6·40	601·845	6·90	992·275	7·40	1635·984
·41	223·6316	·91	368·7062	·41	607·894	·91	1002·247	·41	1652·426
·42	225·8791	·92	372·4117	·42	614·003	·92	1012·320	·42	1669·034
·43	228·1492	·93	376·1545	·43	620·174	·93	1022·494	·43	1685·808
·44	230·4422	·94	379·9349	·44	626·407	·94	1032·770	·44	1702·750
5·45	232·7582	5·95	383·7533	6·45	632·702	6·95	1043·150	7·45	1719·863
·46	235·0974	·96	387·6101	·46	639·061	·96	1053·634	·46	1737·148
·47	237·4602	·97	391·5057	·47	645·484	·97	1064·223	·47	1754·607
·48	239·8467	·98	395·4404	·48	651·971	·98	1074·918	·48	1772·241
·49	242·2572	·99	399·4146	·49	658·523	·99	1085·721	·49	1790·052
5·50	244·6919	6·00	403·4288	6·50	665·142	7·00	1096·633	7·50	1808·042

TABLE VF—EXPONENTIAL FUNCTION

x	e^x	x	e^x	x	e^x	x	e^x	x	e^x
7·50	1808·042	**8·00**	2980·958	**8·50**	4914·77	**9·00**	8103·08	**9·50**	13359·73
·51	1826·214	**·01**	3010·917	**·51**	4964·16	**·01**	8184·52	**·51**	13493·99
·52	1844·567	**·02**	3041·177	**·52**	5014·05	**·02**	8266·78	**·52**	13629·61
·53	1863·106	**·03**	3071·742	**·53**	5064·45	**·03**	8349·86	**·53**	13766·59
·54	1881·830	**·04**	3102·613	**·54**	5115·34	**·04**	8433·78	**·54**	13904·95
7·55	1900·743	**8·05**	3133·795	**8·55**	5166·75	**9·05**	8518·54	**9·55**	14044·69
·56	1919·846	**·06**	3165·290	**·56**	5218·68	**·06**	8604·15	**·56**	14185·85
·57	1939·140	**·07**	3197·102	**·57**	5271·13	**·07**	8690·62	**·57**	14328·42
·58	1958·629	**·08**	3229·233	**·58**	5324·11	**·08**	8777·97	**·58**	14472·42
·59	1978·314	**·09**	3261·688	**·59**	5377·61	**·09**	8866·19	**·59**	14617·87
7·60	1998·196	**8·10**	3294·468	**8·60**	5431·66	**9·10**	8955·29	**9·60**	14764·78
·61	2018·278	**·11**	3327·578	**·61**	5486·25	**·11**	9045·29	**·61**	14913·17
·62	2038·562	**·12**	3361·021	**·62**	5541·39	**·12**	9136·20	**·62**	15063·05
·63	2059·050	**·13**	3394·800	**·63**	5597·08	**·13**	9228·02	**·63**	15214·44
·64	2079·744	**·14**	3428·918	**·64**	5653·33	**·14**	9320·77	**·64**	15367·34
7·65	2100·646	**8·15**	3463·379	**8·65**	5710·15	**9·15**	9414·44	**9·65**	15521·79
·66	2121·757	**·16**	3498·187	**·66**	5767·53	**·16**	9509·06	**·66**	15677·78
·67	2143·081	**·17**	3533·344	**·67**	5825·50	**·17**	9604·62	**·67**	15835·35
·68	2164·620	**·18**	3568·855	**·68**	5884·05	**·18**	9701·15	**·68**	15994·50
·69	2186·375	**·19**	3604·722	**·69**	5943·18	**·19**	9798·65	**·69**	16155·24
7·70	2208·348	**8·20**	3640·950	**8·70**	6002·91	**9·20**	9897·13	**9·70**	16317·61
·71	2230·542	**·21**	3677·542	**·71**	6063·24	**·21**	9996·60	**·71**	16481·60
·72	2252·960	**·22**	3714·502	**·72**	6124·18	**·22**	10097·06	**·72**	16647·24
·73	2275·602	**·23**	3751·834	**·73**	6185·73	**·23**	10198·54	**·73**	16814·55
·74	2298·472	**·24**	3789·540	**·74**	6247·90	**·24**	10301·04	**·74**	16983·54
7·75	2321·572	**8·25**	3827·626	**8·75**	6310·69	**9·25**	10404·57	**9·75**	17154·23
·76	2344·905	**·26**	3866·094	**·76**	6374·11	**·26**	10509·13	**·76**	17326·63
·77	2368·471	**·27**	3904·949	**·77**	6438·17	**·27**	10614·75	**·77**	17500·77
·78	2392·275	**·28**	3944·194	**·78**	6502·88	**·28**	10721·43	**·78**	17676·65
·79	2416·318	**·29**	3983·834	**·79**	6568·23	**·29**	10829·18	**·79**	17854·31
7·80	2440·602	**8·30**	4023·872	**8·80**	6634·24	**9·30**	10938·02	**9·80**	18033·74
·81	2465·130	**·31**	4064·313	**·81**	6700·92	**·31**	11047·95	**·81**	18214·99
·82	2489·905	**·32**	4105·160	**·82**	6768·26	**·32**	11158·98	**·82**	18398·05
·83	2514·929	**·33**	4146·418	**·83**	6836·29	**·33**	11271·13	**·83**	18582·95
·84	2540·205	**·34**	4188·090	**·84**	6904·99	**·34**	11384·41	**·84**	18769·72
7·85	2565·734	**8·35**	4230·181	**8·85**	6974·39	**9·35**	11498·82	**9·85**	18958·35
·86	2591·520	**·36**	4272·695	**·86**	7044·48	**·36**	11614·39	**·86**	19148·89
·87	2617·566	**·37**	4315·636	**·87**	7115·28	**·37**	11731·12	**·87**	19341·34
·88	2643·873	**·38**	4359·009	**·88**	7186·79	**·38**	11849·01	**·88**	19535·72
·89	2670·444	**·39**	4402·818	**·89**	7259·02	**·39**	11968·10	**·89**	19732·06
7·90	2697·282	**8·40**	4447·067	**8·90**	7331·97	**9·40**	12088·38	**9·90**	19930·37
·91	2724·390	**·41**	4491·761	**·91**	7405·66	**·41**	12209·87	**·91**	20130·67
·92	2751·771	**·42**	4536·903	**·92**	7480·09	**·42**	12332·58	**·92**	20332·99
·93	2779·427	**·43**	4582·500	**·93**	7555·27	**·43**	12456·53	**·93**	20537·34
·94	2807·361	**·44**	4628·555	**·94**	7631·20	**·44**	12581·72	**·94**	20743·74
7·95	2835·575	**8·45**	4675·073	**8·95**	7707·89	**9·45**	12708·17	**9·95**	20952·22
·96	2864·073	**·46**	4722·058	**·96**	7785·36	**·46**	12835·88	**·96**	21162·80
·97	2892·857	**·47**	4769·515	**·97**	7863·60	**·47**	12964·89	**·97**	21375·49
·98	2921·931	**·48**	4817·450	**·98**	7942·63	**·48**	13095·19	**·98**	21590·31
·99	2951·297	**·49**	4865·866	**·99**	8022·46	**·49**	13226·80	**·99**	21807·30
8·00	2980·958	**8·50**	4914·769	**9·00**	8103·08	**9·50**	13359·73	**10·00**	22026·47

TABLE Vf—EXPONENTIAL FUNCTION

x	e^x	x	$10^{-p}.e^x$	p	x	$10^{-p}.e^x$	p	x	$10^{-p}.e^x$	p
10·0	22026·5	**15·0**	3·269 017	6	**20·0**	4·851 652	8	**25·0**	7·200 490	10
10·1	24343·0	**15·1**	3·612 823	6	**20·1**	5·361 905	8	**25·1**	7·957 772	10
10·2	26903·2	**15·2**	3·992 787	6	**20·2**	5·925 821	8	**25·2**	8·794 698	10
10·3	29732·6	**15·3**	4·412 712	6	**20·3**	6·549 045	8	**25·3**	9·719 645	10
10·4	32859·6	**15·4**	4·876 801	6	**20·4**	7·237 814	8	**25·4**	1·074 187	11
10·5	36315·5	**15·5**	5·389 698	6	**20·5**	7·999 022	8	**25·5**	1·187 160	11
10·6	40134·8	**15·6**	5·956 538	6	**20·6**	8·840 286	8	**25·6**	1·312 015	11
10·7	44355·9	**15·7**	6·582 993	6	**20·7**	9·770 027	8	**25·7**	1·450 001	11
10·8	49020·8	**15·8**	7·275 332	6	**20·8**	1·079 755	9	**25·8**	1·602 499	11
10·9	54176·4	**15·9**	8·040 485	6	**20·9**	1·193 314	9	**25·9**	1·771 035	11
11·0	59874·1	**16·0**	8·886 111	6	**21·0**	1·318 816	9	**26·0**	1·957 296	11
11·1	66171·2	**16·1**	9·820 671	6	**21·1**	1·457 517	9	**26·1**	2·163 147	11
11·2	73130·4	**16·2**	1·085 352	7	**21·2**	1·610 805	9	**26·2**	2·390 647	11
11·3	80821·6	**16·3**	1·199 499	7	**21·3**	1·780 215	9	**26·3**	2·642 073	11
11·4	89321·7	**16·4**	1·325 652	7	**21·4**	1·967 442	9	**26·4**	2·919 943	11
11·5	98 715·8	**16·5**	1·465 072	7	**21·5**	2·174 360	9	**26·5**	3·227 036	11
11·6	109 097·8	**16·6**	1·619 155	7	**21·6**	2·403 039	9	**26·6**	3·566 426	11
11·7	120 571·7	**16·7**	1·789 443	7	**21·7**	2·655 769	9	**26·7**	3·941 510	11
11·8	133 252·4	**16·8**	1·977 640	7	**21·8**	2·935 078	9	**26·8**	4·356 043	11
11·9	147 266·6	**16·9**	2·185 631	7	**21·9**	3·243 763	9	**26·9**	4·814 172	11
12·0	162 754·8	**17·0**	2·415 495	7	**22·0**	3·584 913	9	**27·0**	5·320 482	11
12·1	179 871·9	**17·1**	2·669 535	7	**22·1**	3·961 941	9	**27·1**	5·880 042	11
12·2	198 789·2	**17·2**	2·950 293	7	**22·2**	4·378 622	9	**27·2**	6·498 452	11
12·3	219 696·0	**17·3**	3·260 578	7	**22·3**	4·839 126	9	**27·3**	7·181 900	11
12·4	242 801·6	**17·4**	3·603 496	7	**22·4**	5·348 062	9	**27·4**	7·937 227	11
12·5	268 337·3	**17·5**	3·982 478	7	**22·5**	5·910 522	9	**27·5**	8·771 993	11
12·6	296 558·6	**17·6**	4·401 319	7	**22·6**	6·532 137	9	**27·6**	9·694 551	11
12·7	327 747·9	**17·7**	4·864 210	7	**22·7**	7·219 128	9	**27·7**	1·071 414	12
12·8	362 217·4	**17·8**	5·375 784	7	**22·8**	7·978 370	9	**27·8**	1·184 095	12
12·9	400 312·2	**17·9**	5·941 160	7	**22·9**	8·817 463	9	**27·9**	1·308 628	12
13·0	442 413·4	**18·0**	6·565 997	7	**23·0**	9·744 803	9	**28·0**	1·446 257	12
13·1	488 942·4	**18·1**	7·256 549	7	**23·1**	1·076 967	10	**28·1**	1·598 361	12
13·2	540 364·9	**18·2**	8·019 727	7	**23·2**	1·190 233	10	**28·2**	1·766 462	12
13·3	597 195·6	**18·3**	8·863 169	7	**23·3**	1·315 411	10	**28·3**	1·952 243	12
13·4	660 003·2	**18·4**	9·795 316	7	**23·4**	1·453 754	10	**28·4**	2·157 562	12
13·5	729 416	**18·5**	1·082 550	8	**23·5**	1·606 646	10	**28·5**	2·384 475	12
13·6	806 130	**18·6**	1·196 403	8	**23·6**	1·775 619	10	**28·6**	2·635 252	12
13·7	890 911	**18·7**	1·322 229	8	**23·7**	1·962 362	10	**28·7**	2·912 404	12
13·8	984 609	**18·8**	1·461 289	8	**23·8**	2·168 746	10	**28·8**	3·218 704	12
13·9	1088 161	**18·9**	1·614 975	8	**23·9**	2·396 835	10	**28·9**	3·557 218	12
14·0	1202 604	**19·0**	1·784 823	8	**24·0**	2·648 912	10	**29·0**	3·931 334	12
14·1	1329 083	**19·1**	1·972 534	8	**24·1**	2·927 501	10	**29·1**	4·344 796	12
14·2	1468 864	**19·2**	2·179 988	8	**24·2**	3·235 389	10	**29·2**	4·801 743	12
14·3	1623 346	**19·3**	2·409 259	8	**24·3**	3·575 657	10	**29·3**	5·306 746	12
14·4	1794 075	**19·4**	2·662 643	8	**24·4**	3·951 713	10	**29·4**	5·864 862	12
14·5	1982 759	**19·5**	2·942 676	8	**24·5**	4·367 318	10	**29·5**	6·481 674	12
14·6	2191 288	**19·6**	3·252 160	8	**24·6**	4·826 633	10	**29·6**	7·163 358	12
14·7	2421 748	**19·7**	3·594 192	8	**24·7**	5·334 254	10	**29·7**	7·916 735	12
14·8	2676 445	**19·8**	3·972 197	8	**24·8**	5·895 263	10	**29·8**	8·749 345	12
14·9	2957 929	**19·9**	4·389 956	8	**24·9**	6·515 273	10	**29·9**	9·669 522	12
15·0	3269 017	**20·0**	4·851 652	8	**25·0**	7·200 490	10	**30·0**	1·068 647	13

TABLE VF—EXPONENTIAL FUNCTION

x	$10^{-p}.e^x$	p	x	$10^{-p}.e^x$	p	x	$10^{-p}.e^x$	p	x	$10^{-p}.e^x$	p
30·0	1·068 647	13	**35·0**	1·586 013	15	**40·0**	2·353 853	17	**45·0**	3·493 427	19
30·1	1·181 038	13	**35·1**	1·752 816	15	**40·1**	2·601 410	17	**45·1**	3·860 834	19
30·2	1·305 249	13	**35·2**	1·937 161	15	**40·2**	2·875 002	17	**45·2**	4·266 882	19
30·3	1·442 523	13	**35·3**	2·140 894	15	**40·3**	3·177 369	17	**45·3**	4·715 633	19
30·4	1·594 235	13	**35·4**	2·366 054	15	**40·4**	3·511 536	17	**45·4**	5·211 581	19
30·5	1·761 902	13	**35·5**	2·614 894	15	**40·5**	3·880 847	17	**45·5**	5·759 688	19
30·6	1·947 203	13	**35·6**	2·889 905	15	**40·6**	4·288 999	17	**45·6**	6·365 439	19
30·7	2·151 992	13	**35·7**	3·193 839	15	**40·7**	4·740 077	17	**45·7**	7·034 898	19
30·8	2·378 319	13	**35·8**	3·529 738	15	**40·8**	5·238 595	17	**45·8**	7·774 765	19
30·9	2·628 449	13	**35·9**	3·900 964	15	**40·9**	5·789 543	17	**45·9**	8·592 444	19
31·0	2·904 885	13	**36·0**	4·311 232	15	**41·0**	6·398 435	17	**46·0**	9·496 119	19
31·1	3·210 394	13	**36·1**	4·764 648	15	**41·1**	7·071 364	17	**46·1**	1·049 484	20
31·2	3·548 035	13	**36·2**	5·265 750	15	**41·2**	7·815 066	17	**46·2**	1·159 859	20
31·3	3·921 185	13	**36·3**	5·819 554	15	**41·3**	8·636 984	17	**46·3**	1·281 842	20
31·4	4·333 579	13	**36·4**	6·431 602	15	**41·4**	9·545 343	17	**46·4**	1·416 655	20
31·5	4·789 346	13	**36·5**	7·108 019	15	**41·5**	1·054 924	18	**46·5**	1·565 645	20
31·6	5·293 046	13	**36·6**	7·855 576	15	**41·6**	1·165 871	18	**46·6**	1·730 306	20
31·7	5·849 720	13	**36·7**	8·681 754	15	**41·7**	1·288 487	18	**46·7**	1·912 284	20
31·8	6·464 940	13	**36·8**	9·594 822	15	**41·8**	1·423 998	18	**46·8**	2·113 400	20
31·9	7·144 864	13	**36·9**	1·060 392	16	**41·9**	1·573 761	18	**46·9**	2·335 668	20
32·0	7·896 296	13	**37·0**	1·171 914	16	**42·0**	1·739 275	18	**47·0**	2·581 313	20
32·1	8·726 757	13	**37·1**	1·295 166	16	**42·1**	1·922 196	18	**47·1**	2·852 792	20
32·2	9·644 558	13	**37·2**	1·431 379	16	**42·2**	2·124 355	18	**47·2**	3·152 823	20
32·3	1·065 888	14	**37·3**	1·581 919	16	**42·3**	2·347 776	18	**47·3**	3·484 408	20
32·4	1·177 989	14	**37·4**	1·748 291	16	**42·4**	2·594 693	18	**47·4**	3·850 866	20
32·5	1·301 879	14	**37·5**	1·932 160	16	**42·5**	2·867 580	18	**47·5**	4·255 865	20
32·6	1·438 799	14	**37·6**	2·135 367	16	**42·6**	3·169 166	18	**47·6**	4·703 459	20
32·7	1·590 119	14	**37·7**	2·359 945	16	**42·7**	3·502 470	18	**47·7**	5·198 126	20
32·8	1·757 353	14	**37·8**	2·608 143	16	**42·8**	3·870 828	18	**47·8**	5·744 817	20
32·9	1·942 175	14	**37·9**	2·882 444	16	**42·9**	4·277 926	18	**47·9**	6·349 005	20
33·0	2·146 436	14	**38·0**	3·185 593	16	**43·0**	4·727 839	18	**48·0**	7·016 736	20
33·1	2·372 178	14	**38·1**	3·520 625	16	**43·1**	5·225 071	18	**48·1**	7·754 692	20
33·2	2·621 663	14	**38·2**	3·890 892	16	**43·2**	5·774 596	18	**48·2**	8·570 261	20
33·3	2·897 385	14	**38·3**	4·300 101	16	**43·3**	6·381 916	18	**48·3**	9·471 603	20
33·4	3·202 106	14	**38·4**	4·752 347	16	**43·4**	7·053 108	18	**48·4**	1·046 774	21
33·5	3·538 874	14	**38·5**	5·252 155	16	**43·5**	7·794 889	18	**48·5**	1·156 864	21
33·6	3·911 061	14	**38·6**	5·804 529	16	**43·6**	8·614 685	18	**48·6**	1·278 533	21
33·7	4·322 391	14	**38·7**	6·414 997	16	**43·7**	9·520 700	18	**48·7**	1·412 997	21
33·8	4·776 981	14	**38·8**	7·089 668	16	**43·8**	1·052 200	19	**48·8**	1·561 603	21
33·9	5·279 380	14	**38·9**	7·835 295	16	**43·9**	1·162 861	19	**48·9**	1·725 839	21
34·0	5·834 617	14	**39·0**	8·659 340	16	**44·0**	1·285 160	19	**49·0**	1·907 347	21
34·1	6·448 249	14	**39·1**	9·570 051	16	**44·1**	1·420 321	19	**49·1**	2·107 944	21
34·2	7·126 418	14	**39·2**	1·057 654	17	**44·2**	1·569 698	19	**49·2**	2·329 638	21
34·3	7·875 910	14	**39·3**	1·168 889	17	**44·3**	1·734 785	19	**49·3**	2·574 649	21
34·4	8·704 226	14	**39·4**	1·291 822	17	**44·4**	1·917 233	19	**49·4**	2·845 427	21
34·5	9·619 658	14	**39·5**	1·427 684	17	**44·5**	2·118 871	19	**49·5**	3·144 683	21
34·6	1·063 137	15	**39·6**	1·577 835	17	**44·6**	2·341 714	19	**49·6**	3·475 412	21
34·7	1·174 948	15	**39·7**	1·743 777	17	**44·7**	2·587 994	19	**49·7**	3·840 924	21
34·8	1·298 518	15	**39·8**	1·927 172	17	**44·8**	2·860 176	19	**49·8**	4·244 878	21
34·9	1·435 084	15	**39·9**	2·129 854	17	**44·9**	3·160 984	19	**49·9**	4·691 316	21
35·0	1·586 013	15	**40·0**	2·353 853	17	**45·0**	3·493 427	19	**50·0**	5·184 706	21

	0	1	2	3	4	5	6	7	8	9	Δ
1·00	0·0 00 000	01 000	01 998	02 996	03 992	04 988	05 982	06 976	07 968	08 960	990
·01	0·0 09 950	10 940	11 929	12 916	13 903	14 889	15 873	16 857	17 840	18 822	981
·02	0·0 19 803	20 783	21 761	22 739	23 717	24 693	25 668	26 642	27 615	28 587	972
·03	0·0 29 559	30 529	31 499	32 467	33 435	34 401	35 367	36 332	37 296	38 259	962
·04	0·0 39 221	40 182	41 142	42 101	43 059	44 017	44 973	45 929	46 884	47 837	953
1·05	0·0 48 790	49 742	50 693	51 643	52 592	53 541	54 488	55 435	56 380	57 325	944
·06	0·0 58 269	59 212	60 154	61 095	62 035	62 975	63 913	64 851	65 788	66 724	935
·07	0·0 67 659	68 593	69 526	70 458	71 390	72 321	73 250	74 179	75 107	76 035	926
·08	0·0 76 961	77 887	78 811	79 735	80 658	81 580	82 501	83 422	84 341	85 260	918
·09	0·0 86 178	87 095	88 011	88 926	89 841	90 754	91 667	92 579	93 490	94 401	909
1·10	0·0 95 310	96 219	97 127	98 034	98 940	99 845	**00 750**	**01 654**	**02 557**	**03 459**	901
·11	0·1 04 360	05 261	06 160	07 059	07 957	08 854	09 751	10 647	11 541	12 435	894
·12	0·1 13 329	14 221	15 113	16 004	16 894	17 783	18 672	19 559	20 446	21 332	886
·13	0·1 22 218	23 102	23 986	24 869	25 751	26 633	27 513	28 393	29 272	30 151	877
·14	0·1 31 028	31 905	32 781	33 656	34 531	35 405	36 278	37 150	38 021	38 892	870
1·15	0·1 39 762	40 631	41 500	42 367	43 234	44 100	44 966	45 830	46 694	47 558	862
·16	0·1 48 420	49 282	50 143	51 003	51 862	52 721	53 579	54 436	55 293	56 149	855
·17	0·1 57 004	57 858	58 712	59 565	60 417	61 268	62 119	62 969	63 818	64 667	847
·18	0·1 65 514	66 362	67 208	68 054	68 899	69 743	70 586	71 429	72 271	73 113	840
·19	0·1 73 953	74 793	75 633	76 471	77 309	78 146	78 983	79 818	80 653	81 488	834
1·20	0·1 82 322	83 155	83 987	84 818	85 649	86 480	87 309	88 138	88 966	89 794	826
·21	0·1 90 620	91 446	92 272	93 097	93 921	94 744	95 567	96 389	97 210	98 031	820
·22	0·1 98 851	99 670	**00 489**	**01 307**	**02 124**	**02 941**	**03 757**	**04 572**	**05 387**	**06 201**	813
·23	0·2 07 014	07 827	08 639	09 450	10 261	11 071	11 880	12 689	13 497	14 305	806
·24	0·2 15 111	15 918	16 723	17 528	18 332	19 136	19 938	20 741	21 542	22 343	801
1·25	0·2 23 144	23 943	24 742	25 541	26 338	27 136	27 932	28 728	29 523	30 318	794
·26	0·2 31 112	31 905	32 698	33 490	34 281	35 072	35 862	36 652	37 441	38 229	788
·27	0·2 39 017	39 804	40 590	41 376	42 162	42 946	43 730	44 514	45 296	46 079	781
·28	0·2 46 860	47 641	48 421	49 201	49 980	50 759	51 537	52 314	53 091	53 867	775
·29	0·2 54 642	55 417	56 191	56 965	57 738	58 511	59 283	60 054	60 825	61 595	769
1·30	0·2 62 364	63 133	63 902	64 669	65 436	66 203	66 969	67 734	68 499	69 263	764
·31	0·2 70 027	70 790	71 553	72 315	73 076	73 837	74 597	75 356	76 115	76 874	758
·32	0·2 77 632	78 389	79 146	79 902	80 657	81 412	82 167	82 921	83 674	84 427	752
·33	0·2 85 179	85 931	86 682	87 432	88 182	88 931	89 680	90 428	91 176	91 923	747
·34	0·2 92 670	93 416	94 161	94 906	95 650	96 394	97 137	97 880	98 622	99 364	741
1·35	0·3 00 105	00 845	01 585	02 324	03 063	03 801	04 539	05 276	06 013	06 749	736
·36	0·3 07 485	08 220	08 954	09 688	10 422	11 154	11 887	12 619	13 350	14 081	730
·37	0·3 14 811	15 540	16 270	16 998	17 726	18 454	19 181	19 907	20 633	21 359	724
·38	0·3 22 083	22 808	23 532	24 255	24 978	25 700	26 422	27 143	27 864	28 584	720
·39	0·3 29 304	30 023	30 742	31 460	32 177	32 894	33 611	34 327	35 043	35 758	714
1·40	0·3 36 472	37 186	37 900	38 613	39 325	40 037	40 749	41 460	42 170	42 880	710
·41	0·3 43 590	44 299	45 007	45 715	46 423	47 130	47 836	48 542	49 247	49 952	705
·42	0·3 50 657	51 361	52 064	52 767	53 470	54 172	54 873	55 574	56 275	56 975	699
·43	0·3 57 674	58 374	59 072	59 770	60 468	61 165	61 861	62 558	63 253	63 948	695
·44	0·3 64 643	65 337	66 031	66 724	67 417	68 109	68 801	69 492	70 183	70 874	690
1·45	0·3 71 564	72 253	72 942	73 630	74 318	75 006	75 693	76 380	77 066	77 751	685
·46	0·3 78 436	79 121	79 805	80 489	81 172	81 855	82 538	83 219	83 901	84 582	680
·47	0·3 85 262	85 942	86 622	87 301	87 980	88 658	89 336	90 013	90 690	91 366	676
·48	0·3 92 042	92 718	93 393	94 067	94 741	95 415	96 088	96 761	97 433	98 105	671
·49	0·3 98 776	99 447	**00 118**	**00 788**	**01 457**	**02 126**	**02 795**	**03 463**	**04 131**	**04 798**	667
1·50	0·4 05 465	06 132	06 798	07 463	08 128	08 793	09 457	10 121	10 784	11 447	663

n	1	2	3	4	5	6	7
$\ln 10^n$	2·302 585	4·605 170	6·907 755	9·210 340	11·512 925	13·815 511	16·118 096

TABLE VI—NATURAL LOGARITHMS

		0	1	2	3	4	5	6	7	8	9	Δ
1·50	0·4	05 465	06 132	06 798	07 463	08 128	08 793	09 457	10 121	10 784	11 447	663
·51	0·4	12 110	12 772	13 433	14 094	14 755	15 415	16 075	16 735	17 394	18 052	658
·52	0·4	18 710	19 368	20 025	20 682	21 338	21 994	22 650	23 305	23 960	24 614	654
·53	0·4	25 268	25 921	26 574	27 227	27 879	28 530	29 182	29 832	30 483	31 133	649
·54	0·4	31 782	32 432	33 080	33 729	34 376	35 024	35 671	36 318	36 964	37 610	645
1·55	0·4	38 255	38 900	39 544	40 189	40 832	41 476	42 118	42 761	43 403	44 045	641
·56	0·4	44 686	45 327	45 967	46 607	47 247	47 886	48 525	49 163	49 801	50 438	638
·57	0·4	51 076	51 712	52 349	52 985	53 620	54 255	54 890	55 524	56 158	56 792	633
·58	0·4	57 425	58 058	58 690	59 322	59 953	60 584	61 215	61 845	62 475	63 105	629
·59	0·4	63 734	64 363	64 991	65 619	66 247	66 874	67 500	68 127	68 753	69 378	626
1·60	0·4	70 004	70 628	71 253	71 877	72 501	73 124	73 747	74 369	74 991	75 613	621
·61	0·4	76 234	76 855	77 476	78 096	78 716	79 335	79 954	80 573	81 191	81 809	617
·62	0·4	82 426	83 043	83 660	84 276	84 892	85 508	86 123	86 738	87 352	87 966	614
·63	0·4	88 580	89 193	89 806	90 419	91 031	91 643	92 254	92 865	93 476	94 086	610
·64	0·4	94 696	95 306	95 915	96 524	97 132	97 740	98 348	98 955	99 562	**00 169**	606
1·65	0·5	00 775	01 381	01 987	02 592	03 197	03 801	04 405	05 009	05 612	06 215	603
·66	0·5	06 818	07 420	08 022	08 623	09 224	09 825	10 426	11 026	11 625	12 225	599
·67	0·5	12 824	13 422	14 021	14 618	15 216	15 813	16 410	17 006	17 603	18 198	596
·68	0·5	18 794	19 389	19 984	20 578	21 172	21 766	22 359	22 952	23 544	24 137	592
·69	0·5	24 729	25 320	25 911	26 502	27 093	27 683	28 273	28 862	29 451	30 040	588
1·70	0·5	30 628	31 216	31 804	32 391	32 978	33 565	34 151	34 737	35 323	35 908	585
·71	0·5	36 493	37 078	37 662	38 246	38 830	39 413	39 996	40 579	41 161	41 743	581
·72	0·5	42 324	42 906	43 486	44 067	44 647	45 227	45 807	46 386	46 965	47 543	578
·73	0·5	48 121	48 699	49 277	49 854	50 431	51 007	51 584	52 159	52 735	53 310	575
·74	0·5	53 885	54 460	55 034	55 608	56 181	56 755	57 327	57 900	58 472	59 044	572
1·75	0·5	59 616	60 187	60 758	61 329	61 899	62 469	63 038	63 608	64 177	64 745	569
·76	0·5	65 314	65 882	66 450	67 017	67 584	68 151	68 717	69 283	69 849	70 414	566
·77	0·5	70 980	71 544	72 109	72 673	73 237	73 800	74 364	74 927	75 489	76 051	562
·78	0·5	76 613	77 175	77 736	78 297	78 858	79 418	79 978	80 538	81 098	81 657	559
·79	0·5	82 216	82 774	83 332	83 890	84 448	85 005	85 562	86 119	86 675	87 231	556
1·80	0·5	87 787	88 342	88 897	89 452	90 006	90 561	91 114	91 668	92 221	92 774	553
·81	0·5	93 327	93 879	94 431	94 983	95 534	96 085	96 636	97 187	97 737	98 287	550
·82	0·5	98 837	99 386	99 935	**00 483**	**01 032**	**01 580**	**02 128**	**02 675**	**03 222**	**03 769**	547
·83	0·6	04 316	04 862	05 408	05 954	06 499	07 044	07 589	08 134	08 678	09 222	544
·84	0·6	09 766	10 309	10 852	11 395	11 937	12 479	13 021	13 563	14 104	14 645	541
1·85	0·6	15 186	15 726	16 266	16 806	17 345	17 885	18 424	18 962	19 501	20 039	537
·86	0·6	20 576	21 114	21 651	22 188	22 725	23 261	23 797	24 333	24 868	25 404	534
·87	0·6	25 938	26 473	27 007	27 541	28 075	28 609	29 142	29 675	30 207	30 740	532
·88	0·6	31 272	31 804	32 335	32 866	33 397	33 928	34 458	34 988	35 518	36 048	529
·89	0·6	36 577	37 106	37 634	38 163	38 691	39 219	39 746	40 274	40 801	41 327	527
1·90	0·6	41 854	42 380	42 906	43 432	43 957	44 482	45 007	45 531	46 056	46 580	523
·91	0·6	47 103	47 627	48 150	48 673	49 195	49 718	50 240	50 761	51 283	51 804	521
·92	0·6	52 325	52 846	53 366	53 886	54 406	54 926	55 445	55 964	56 483	57 002	518
·93	0·6	57 520	58 038	58 556	59 073	59 590	60 107	60 624	61 140	61 657	62 172	516
·94	0·6	62 688	63 203	63 718	64 233	64 748	65 262	65 776	66 290	66 803	67 316	513
1·95	0·6	67 829	68 342	68 854	69 367	69 879	70 390	70 902	71 413	71 924	72 434	510
·96	0·6	72 944	73 455	73 964	74 474	74 983	75 492	76 001	76 510	77 018	77 526	508
·97	0·6	78 034	78 541	79 048	79 555	80 062	80 568	81 075	81 581	82 086	82 592	505
·98	0·6	83 097	83 602	84 106	84 611	85 115	85 619	86 123	86 626	87 129	87 632	503
·99	0·6	88 135	88 637	89 139	89 641	90 143	90 644	91 145	91 646	92 147	92 647	500
2·00	0·6	93 147	93 647	94 147	94 646	95 145	95 644	96 143	96 641	97 139	97 637	498

n	7	8	9	10	11	12
ln 10^n	16·118 096	18·420 681	20·723 266	23·025 851	25·328 436	27·631 021

TABLE VI—NATURAL LOGARITHMS

	0	1	2	3	4	5	6	7	8	9	Δ
2·00	0·6 93 147	93 647	94 147	94 646	95 145	95 644	96 143	96 641	97 139	97 637	498
·01	0·6 98 135	98 632	99 129	99 626	**00 123**	**00 619**	**01 115**	**01 611**	**02 107**	**02 602**	496
·02	0·7 03 098	03 592	04 087	04 582	05 076	05 570	06 063	06 557	07 050	07 543	493
·03	0·7 08 036	08 528	09 021	09 513	10 004	10 496	10 987	11 478	11 969	12 459	491
·04	0·7 12 950	13 440	13 930	14 419	14 909	15 398	15 887	16 375	16 864	17 352	488
2·05	0·7 17 840	18 327	18 815	19 302	19 789	20 276	20 762	21 249	21 735	22 220	486
·06	0·7 22 706	23 191	23 676	24 161	24 646	25 130	25 614	26 098	26 582	27 065	484
·07	0·7 27 549	28 032	28 514	28 997	29 479	29 961	30 443	30 925	31 406	31 887	481
·08	0·7 32 368	32 849	33 329	33 809	34 289	34 769	35 248	35 728	36 207	36 685	479
·09	0·7 37 164	37 642	38 121	38 598	39 076	39 554	40 031	40 508	40 985	41 461	476
2·10	0·7 41 937	42 413	42 889	43 365	43 840	44 315	44 790	45 265	45 740	46 214	474
·11	0·7 46 688	47 162	47 635	48 109	48 582	49 055	49 528	50 000	50 472	50 944	472
·12	0·7 51 416	51 888	52 359	52 830	53 301	53 772	54 242	54 713	55 183	55 652	470
·13	0·7 56 122	56 591	57 061	57 529	57 998	58 467	58 935	59 403	59 871	60 338	468
·14	0·7 60 806	61 273	61 740	62 207	62 673	63 140	63 606	64 072	64 537	65 003	465
2·15	0·7 65 468	65 933	66 398	66 862	67 327	67 791	68 255	68 718	69 182	69 645	463
·16	0·7 70 108	70 571	71 034	71 496	71 958	72 420	72 882	73 344	73 805	74 266	461
·17	0·7 74 727	75 188	75 648	76 109	76 569	77 029	77 488	77 948	78 407	78 866	459
·18	0·7 79 325	79 783	80 242	80 700	81 158	81 616	82 073	82 531	82 988	83 445	457
·19	0·7 83 902	84 358	84 814	85 270	85 726	86 182	86 638	87 093	87 548	88 003	454
2·20	0·7 88 457	88 912	89 366	89 820	90 274	90 728	91 181	91 634	92 087	92 540	453
·21	0·7 92 993	93 445	93 897	94 349	94 801	95 252	95 704	96 155	96 606	97 057	450
·22	0·7 97 507	97 958	98 408	98 858	99 307	99 757	**00 206**	**00 655**	**01 104**	**01 553**	449
·23	0·8 02 002	02 450	02 898	03 346	03 794	04 241	04 689	05 136	05 583	06 029	447
·24	0·8 06 476	06 922	07 368	07 814	08 260	08 706	09 151	09 596	10 041	10 486	444
2·25	0·8 10 930	11 375	11 819	12 263	12 706	13 150	13 593	14 036	14 479	14 922	443
·26	0·8 15 365	15 807	16 249	16 691	17 133	17 575	18 016	18 457	18 898	19 339	441
·27	0·8 19 780	20 220	20 661	21 101	21 540	21 980	22 420	22 859	23 298	23 737	438
·28	0·8 24 175	24 614	25 052	25 490	25 928	26 366	26 804	27 241	27 678	28 115	437
·29	0·8 28 552	28 988	29 425	29 861	30 297	30 733	31 168	31 604	32 039	32 474	435
2·30	0·8 32 909	33 344	33 778	34 213	34 647	35 081	35 514	35 948	36 381	36 815	433
·31	0·8 37 248	37 680	38 113	38 545	38 978	39 410	39 842	40 273	40 705	41 136	431
·32	0·8 41 567	41 998	42 429	42 859	43 290	43 720	44 150	44 580	45 010	45 439	429
·33	0·8 45 868	46 297	46 726	47 155	47 584	48 012	48 440	48 868	49 296	49 723	428
·34	0·8 50 151	50 578	51 005	51 432	51 859	52 285	52 712	53 138	53 564	53 990	425
2·35	0·8 54 415	54 841	55 266	55 691	56 116	56 541	56 965	57 390	57 814	58 238	424
·36	0·8 58 662	59 085	59 509	59 932	60 355	60 778	61 201	61 623	62 046	62 468	422
·37	0·8 62 890	63 312	63 733	64 155	64 576	64 997	65 418	65 839	66 260	66 680	420
·38	0·8 67 100	67 521	67 940	68 360	68 780	69 199	69 618	70 037	70 456	70 875	418
·39	0·8 71 293	71 712	72 130	72 548	72 966	73 383	73 801	74 218	74 635	75 052	417
2·40	0·8 75 469	75 885	76 302	76 718	77 134	77 550	77 966	78 381	78 797	79 212	415
·41	0·8 79 627	80 042	80 456	80 871	81 285	81 699	82 113	82 527	82 941	83 354	414
·42	0·8 83 768	84 181	84 594	85 006	85 419	85 832	86 244	86 656	87 068	87 480	411
·43	0·8 87 891	88 303	88 714	89 125	89 536	89 947	90 357	90 768	91 178	91 588	410
·44	0·8 91 998	92 408	92 817	93 227	93 636	94 045	94 454	94 863	95 271	95 680	408
2·45	0·8 96 088	96 496	96 904	97 312	97 719	98 127	98 534	98 941	99 348	99 755	406
·46	0·9 00 161	00 568	00 974	01 380	01 786	02 192	02 597	03 003	03 408	03 813	405
·47	0·9 04 218	04 623	05 028	05 432	05 836	06 240	06 644	07 048	07 452	07 855	404
·48	0·9 08 259	08 662	09 065	09 468	09 870	10 273	10 675	11 077	11 479	11 881	402
·49	0·9 12 283	12 684	13 086	13 487	13 888	14 289	14 689	15 090	15 490	15 891	400
2·50	0·9 16 291	16 691	17 090	17 490	17 889	18 289	18 688	19 087	19 486	19 884	399

n	1	2	3	4	5	6	7
$\ln 10^n$	2·302 585	4·605 170	6·907 755	9·210 340	11·512 925	13·815 511	16·118 096

TABLE VI—NATURAL LOGARITHMS

	0	1	2	3	4	5	6	7	8	9	Δ
2·50	0·9 16 291	16 691	17 090	17 490	17 889	18 289	18 688	19 087	19 486	19 884	399
·51	0·9 20 283	20 681	21 079	21 477	21 875	22 273	22 670	23 068	23 465	23 862	397
·52	0·9 24 259	24 656	25 052	25 449	25 845	26 241	26 637	27 033	27 428	27 824	395
·53	0·9 28 219	28 614	29 010	29 404	29 799	30 194	30 588	30 982	31 376	31 770	394
·54	0·9 32 164	32 558	32 951	33 344	33 738	34 131	34 524	34 916	35 309	35 701	392
2·55	0·9 36 093	36 485	36 877	37 269	37 661	38 052	38 444	38 835	39 226	39 617	390
·56	0·9 40 007	40 398	40 788	41 178	41 569	41 958	42 348	42 738	43 127	43 517	389
·57	0·9 43 906	44 295	44 684	45 073	45 461	45 850	46 238	46 626	47 014	47 402	387
·58	0·9 47 789	48 177	48 564	48 952	49 339	49 726	50 112	50 499	50 885	51 272	386
·59	0·9 51 658	52 044	52 430	52 816	53 201	53 587	53 972	54 357	54 742	55 127	384
2·60	0·9 55 511	55 896	56 280	56 665	57 049	57 433	57 816	58 200	58 584	58 967	383
·61	0·9 59 350	59 733	60 116	60 499	60 882	61 264	61 646	62 029	62 411	62 793	381
·62	0·9 63 174	63 556	63 937	64 319	64 700	65 081	65 462	65 843	66 223	66 604	380
·63	0·9 66 984	67 364	67 744	68 124	68 504	68 883	69 263	69 642	70 021	70 400	379
·64	0·9 70 779	71 158	71 536	71 915	72 293	72 671	73 049	73 427	73 805	74 182	378
2·65	0·9 74 560	74 937	75 314	75 691	76 068	76 445	76 821	77 198	77 574	77 950	376
·66	0·9 78 326	78 702	79 078	79 453	79 829	80 204	80 579	80 954	81 329	81 704	374
·67	0·9 82 078	82 453	82 827	83 201	83 575	83 949	84 323	84 697	85 070	85 444	373
·68	0·9 85 817	86 190	86 563	86 936	87 308	87 681	88 053	88 425	88 797	89 169	372
·69	0·9 89 541	89 913	90 284	90 656	91 027	91 398	91 769	92 140	92 511	92 881	371
2·70	0·9 93 252	93 622	93 992	94 362	94 732	95 102	95 472	95 841	96 210	96 580	369
·71	0·9 96 949	97 318	97 686	98 055	98 424	98 792	99 160	99 528	99 896	**00 264**	368
·72	1·0 00 632	00 999	01 367	01 734	02 101	02 468	02 835	03 202	03 569	03 935	367
·73	1·0 04 302	04 668	05 034	05 400	05 766	06 131	06 497	06 862	07 228	07 593	365
·74	1·0 07 958	08 323	08 688	09 052	09 417	09 781	10 145	10 509	10 873	11 237	364
2·75	1·0 11 601	11 964	12 328	12 691	13 054	13 417	13 780	14 143	14 506	14 868	363
·76	1·0 15 231	15 593	15 955	16 317	16 679	17 041	17 402	17 764	18 125	18 486	361
·77	1·0 18 847	19 208	19 569	19 930	20 290	20 651	21 011	21 371	21 731	22 091	360
·78	1·0 22 451	22 811	23 170	23 529	23 889	24 248	24 607	24 966	25 324	25 683	359
·79	1·0 26 042	26 400	26 758	27 116	27 474	27 832	28 190	28 547	28 905	29 262	357
2·80	1·0 29 619	29 976	30 333	30 690	31 047	31 404	31 760	32 116	32 472	32 829	355
·81	1·0 33 184	33 540	33 896	34 252	34 607	34 962	35 317	35 672	36 027	36 382	355
·82	1·0 36 737	37 091	37 446	37 800	38 154	38 508	38 862	39 216	39 570	39 923	354
·83	1·0 40 277	40 630	40 983	41 336	41 689	42 042	42 395	42 747	43 100	43 452	352
·84	1·0 43 804	44 156	44 508	44 860	45 212	45 563	45 914	46 266	46 617	46 968	351
2·85	1·0 47 319	47 670	48 021	48 371	48 722	49 072	49 422	49 772	50 122	50 472	350
·86	1·0 50 822	51 171	51 521	51 870	52 219	52 568	52 917	53 266	53 615	53 964	348
·87	1·0 54 312	54 660	55 009	55 357	55 705	56 053	56 400	56 748	57 096	57 443	347
·88	1·0 57 790	58 137	58 484	58 831	59 178	59 525	59 871	60 218	60 564	60 910	347
·89	1·0 61 257	61 602	61 948	62 294	62 640	62 985	63 330	63 676	64 021	64 366	345
2·90	1·0 64 711	65 056	65 400	65 745	66 089	66 433	66 778	67 122	67 466	67 809	344
·91	1·0 68 153	68 497	68 840	69 183	69 527	69 870	70 213	70 556	70 898	71 241	343
·92	1·0 71 584	71 926	72 268	72 610	72 953	73 294	73 636	73 978	74 320	74 661	341
·93	1·0 75 002	75 344	75 685	76 026	76 367	76 707	77 048	77 389	77 729	78 069	341
·94	1·0 78 410	78 750	79 090	79 429	79 769	80 109	80 448	80 788	81 127	81 466	339
2·95	1·0 81 805	82 144	82 483	82 822	83 160	83 499	83 837	84 175	84 513	84 851	338
·96	1·0 85 189	85 527	85 865	86 202	86 540	86 877	87 214	87 551	87 888	88 225	337
·97	1·0 88 562	88 899	89 235	89 572	89 908	90 244	90 580	90 916	91 252	91 588	335
·98	1·0 91 923	92 259	92 594	92 930	93 265	93 600	93 935	94 270	94 604	94 939	334
·99	1·0 95 273	95 608	95 942	96 276	96 610	96 944	97 278	97 612	97 945	98 279	333
3·00	1·0 98 612	98 946	99 279	99 612	99 945	**00 278**	**00 610**	**00 943**	**01 275**	**01 608**	332

n	7	8	9	10	11	12
ln 10^n	16·118 096	18·420 681	20·723 266	23·025 851	25·328 436	27·631 021

TABLE VI—NATURAL LOGARITHMS

		0	1	2	3	4	5	6	7	8	9	Δ
3·00	1·0	98 612	98 946	99 279	99 612	99 945	**00 278**	**00 610**	**00 943**	**01 275**	**01 608**	332
·01	1·1	01 940	02 272	02 604	02 936	03 268	03 600	03 931	04 263	04 594	04 926	331
·02	1·1	05 257	05 588	05 919	06 250	06 580	06 911	07 242	07 572	07 902	08 233	330
·03	1·1	08 563	08 893	09 222	09 552	09 882	10 211	10 541	10 870	11 199	11 529	329
·04	1·1	11 858	12 186	12 515	12 844	13 172	13 501	13 829	14 158	14 486	14 814	328
3·05	1·1	15 142	15 469	15 797	16 125	16 452	16 780	17 107	17 434	17 761	18 088	327
·06	1·1	18 415	18 742	19 068	19 395	19 721	20 048	20 374	20 700	21 026	21 352	326
·07	1·1	21 678	22 003	22 329	22 654	22 980	23 305	23 630	23 955	24 280	24 605	325
·08	1·1	24 930	25 254	25 579	25 903	26 227	26 552	26 876	27 200	27 524	27 847	324
·09	1·1	28 171	28 495	28 818	29 141	29 465	29 788	30 111	30 434	30 757	31 079	323
3·10	1·1	31 402	31 725	32 047	32 369	32 692	33 014	33 336	33 658	33 979	34 301	322
·11	1·1	34 623	34 944	35 266	35 587	35 908	36 229	36 550	36 871	37 192	37 512	321
·12	1·1	37 833	38 153	38 474	38 794	39 114	39 434	39 754	40 074	40 394	40 713	320
·13	1·1	41 033	41 352	41 672	41 991	42 310	42 629	42 948	43 267	43 586	43 904	319
·14	1·1	44 223	44 541	44 860	45 178	45 496	45 814	46 132	46 450	46 767	47 085	317
3·15	1·1	47 402	47 720	48 037	48 354	48 671	48 988	49 305	49 622	49 939	50 256	316
·16	1·1	50 572	50 888	51 205	51 521	51 837	52 153	52 469	52 785	53 100	53 416	316
·17	1·1	53 732	54 047	54 362	54 678	54 993	55 308	55 623	55 937	56 252	56 567	314
·18	1·1	56 881	57 196	57 510	57 824	58 138	58 452	58 766	59 080	59 394	59 707	314
·19	1·1	60 021	60 334	60 648	60 961	61 274	61 587	61 900	62 213	62 526	62 838	313
3·20	1·1	63 151	63 463	63 776	64 088	64 400	64 712	65 024	65 336	65 648	65 959	312
·21	1·1	66 271	66 582	66 894	67 205	67 516	67 827	68 138	68 449	68 760	69 071	310
·22	1·1	69 381	69 692	70 002	70 313	70 623	70 933	71 243	71 553	71 863	72 172	310
·23	1·1	72 482	72 792	73 101	73 410	73 720	74 029	74 338	74 647	74 956	75 265	308
·24	1·1	75 573	75 882	76 190	76 499	76 807	77 115	77 423	77 731	78 039	78 347	308
3·25	1·1	78 655	78 963	79 270	79 578	79 885	80 192	80 499	80 807	81 114	81 420	307
·26	1·1	81 727	82 034	82 341	82 647	82 953	83 260	83 566	83 872	84 178	84 484	306
·27	1·1	84 790	85 096	85 401	85 707	86 012	86 318	86 623	86 928	87 233	87 538	305
·28	1·1	87 843	88 148	88 453	88 758	89 062	89 367	89 671	89 975	90 279	90 584	304
·29	1·1	90 888	91 191	91 495	91 799	92 103	92 406	92 710	93 013	93 316	93 619	303
3·30	1·1	93 922	94 225	94 528	94 831	95 134	95 436	95 739	96 041	96 344	96 646	302
·31	1·1	96 948	97 250	97 552	97 854	98 156	98 458	98 759	99 061	99 362	99 664	301
·32	1·1	99 965	**00 266**	**00 567**	**00 868**	**01 169**	**01 470**	**01 770**	**02 071**	**02 372**	**02 672**	300
·33	1·2	02 972	03 273	03 573	03 873	04 173	04 473	04 772	05 072	05 372	05 671	300
·34	1·2	05 971	06 270	06 569	06 869	07 168	07 467	07 766	08 064	08 363	08 662	298
3·35	1·2	08 960	09 259	09 557	09 855	10 154	10 452	10 750	11 048	11 346	11 643	298
·36	1·2	11 941	12 239	12 536	12 833	13 131	13 428	13 725	14 022	14 319	14 616	297
·37	1·2	14 913	15 209	15 506	15 803	16 099	16 395	16 692	16 988	17 284	17 580	296
·38	1·2	17 876	18 172	18 467	18 763	19 058	19 354	19 649	19 945	20 240	20 535	295
·39	1·2	20 830	21 125	21 420	21 714	22 009	22 304	22 598	22 893	23 187	23 481	294
3·40	1·2	23 775	24 070	24 363	24 657	24 951	25 245	25 539	25 832	26 126	26 419	293
·41	1·2	26 712	27 006	27 299	27 592	27 885	28 177	28 470	28 763	29 056	29 348	293
·42	1·2	29 641	29 933	30 225	30 517	30 809	31 101	31 393	31 685	31 977	32 269	291
·43	1·2	32 560	32 852	33 143	33 435	33 726	34 017	34 308	34 599	34 890	35 181	290
·44	1·2	35 471	35 762	36 053	36 343	36 634	36 924	37 214	37 504	37 794	38 084	290
3·45	1·2	38 374	38 664	38 954	39 243	39 533	39 822	40 112	40 401	40 690	40 980	289
·46	1·2	41 269	41 558	41 846	42 135	42 424	42 713	43 001	43 290	43 578	43 866	289
·47	1·2	44 155	44 443	44 731	45 019	45 307	45 594	45 882	46 170	46 457	46 745	287
·48	1·2	47 032	47 320	47 607	47 894	48 181	48 468	48 755	49 042	49 329	49 615	287
·49	1·2	49 902	50 188	50 475	50 761	51 047	51 333	51 619	51 905	52 191	52 477	286
3·50	1·2	52 763	53 049	53 334	53 620	53 905	54 191	54 476	54 761	55 046	55 331	285

n	1	2	3	4	5	6	7
$\ln 10^n$	2·302 585	4·605 170	6·907 755	9·210 340	11·512 925	13·815 511	16·118 096

TABLE VI—NATURAL LOGARITHMS

		0	1	2	3	4	5	6	7	8	9	Δ
3·50	1·2	52 763	53 049	53 334	53 620	53 905	54 191	54 476	54 761	55 046	55 331	285
·51	1·2	55 616	55 901	56 186	56 470	56 755	57 040	57 324	57 608	57 893	58 177	284
·52	1·2	58 461	58 745	59 029	59 313	59 597	59 880	60 164	60 448	60 731	61 015	283
·53	1·2	61 298	61 581	61 864	62 147	62 430	62 713	62 996	63 279	63 562	63 844	283
·54	1·2	64 127	64 409	64 692	64 974	65 256	65 538	65 820	66 102	66 384	66 666	282
3·55	1·2	66 948	67 229	67 511	67 792	68 074	68 355	68 636	68 917	69 199	69 480	281
·56	1·2	69 761	70 041	70 322	70 603	70 884	71 164	71 445	71 725	72 005	72 285	281
·57	1·2	72 566	72 846	73 126	73 406	73 685	73 965	74 245	74 524	74 804	75 083	280
·58	1·2	75 363	75 642	75 921	76 200	76 479	76 758	77 037	77 316	77 595	77 874	278
·59	1·2	78 152	78 431	78 709	78 988	79 266	79 544	79 822	80 100	80 378	80 656	278
3·60	1·2	80 934	81 212	81 489	81 767	82 044	82 322	82 599	82 876	83 154	83 431	277
·61	1·2	83 708	83 985	84 262	84 538	84 815	85 092	85 368	85 645	85 921	86 198	276
·62	1·2	86 474	86 750	87 026	87 302	87 578	87 854	88 130	88 406	88 682	88 957	276
·63	1·2	89 233	89 508	89 783	90 059	90 334	90 609	90 884	91 159	91 434	91 709	275
·64	1·2	91 984	92 258	92 533	92 808	93 082	93 356	93 631	93 905	94 179	94 453	274
3·65	1·2	94 727	95 001	95 275	95 549	95 822	96 096	96 370	96 643	96 917	97 190	273
·66	1·2	97 463	97 736	98 009	98 282	98 555	98 828	99 101	99 374	99 647	99 919	273
·67	1·3	00 192	00 464	00 736	01 009	01 281	01 553	01 825	02 097	02 369	02 641	272
·68	1·3	02 913	03 184	03 456	03 728	03 999	04 271	04 542	04 813	05 084	05 355	271
·69	1·3	05 626	05 897	06 168	06 439	06 710	06 981	07 251	07 522	07 792	08 063	270
3·70	1·3	08 333	08 603	08 873	09 143	09 413	09 683	09 953	10 223	10 493	10 762	270
·71	1·3	11 032	11 301	11 571	11 840	12 109	12 379	12 648	12 917	13 186	13 455	269
·72	1·3	13 724	13 992	14 261	14 530	14 798	15 067	15 335	15 604	15 872	16 140	268
·73	1·3	16 408	16 676	16 944	17 212	17 480	17 748	18 016	18 283	18 551	18 818	268
·74	1·3	19 086	19 353	19 620	19 887	20 155	20 422	20 689	20 956	21 222	21 489	267
3·75	1·3	21 756	22 022	22 289	22 556	22 822	23 088	23 355	23 621	23 887	24 153	266
·76	1·3	24 419	24 685	24 951	25 217	25 482	25 748	26 013	26 279	26 544	26 810	265
·77	1·3	27 075	27 340	27 605	27 870	28 135	28 400	28 665	28 930	29 195	29 459	265
·78	1·3	29 724	29 989	30 253	30 517	30 782	31 046	31 310	31 574	31 838	32 102	264
·79	1·3	32 366	32 630	32 894	33 157	33 421	33 684	33 948	34 211	34 475	34 738	263
3·80	1·3	35 001	35 264	35 527	35 790	36 053	36 316	36 579	36 841	37 104	37 367	262
·81	1·3	37 629	37 892	38 154	38 416	38 679	38 941	39 203	39 465	39 727	39 989	261
·82	1·3	40 250	40 512	40 774	41 035	41 297	41 558	41 820	42 081	42 342	42 604	261
·83	1·3	42 865	43 126	43 387	43 648	43 909	44 169	44 430	44 691	44 951	45 212	260
·84	1·3	45 472	45 733	45 993	46 253	46 513	46 774	47 034	47 294	47 554	47 813	260
3·85	1·3	48 073	48 333	48 592	48 852	49 112	49 371	49 630	49 890	50 149	50 408	259
·86	1·3	50 667	50 926	51 185	51 444	51 703	51 962	52 220	52 479	52 738	52 996	259
·87	1·3	53 255	53 513	53 771	54 029	54 288	54 546	54 804	55 062	55 320	55 577	258
·88	1·3	55 835	56 093	56 350	56 608	56 866	57 123	57 380	57 638	57 895	58 152	257
·89	1·3	58 409	58 666	58 923	59 180	59 437	59 694	59 950	60 207	60 464	60 720	257
3·90	1·3	60 977	61 233	61 489	61 745	62 002	62 258	62 514	62 770	63 026	63 282	255
·91	1·3	63 537	63 793	64 049	64 304	64 560	64 815	65 071	65 326	65 581	65 837	255
·92	1·3	66 092	66 347	66 602	66 857	67 112	67 366	67 621	67 876	68 130	68 385	254
·93	1·3	68 639	68 894	69 148	69 402	69 657	69 911	70 165	70 419	70 673	70 927	254
·94	1·3	71 181	71 434	71 688	71 942	72 195	72 449	72 702	72 956	73 209	73 462	254
3·95	1·3	73 716	73 969	74 222	74 475	74 728	74 981	75 233	75 486	75 739	75 991	253
·96	1·3	76 244	76 497	76 749	77 001	77 254	77 506	77 758	78 010	78 262	78 514	252
·97	1·3	78 766	79 018	79 270	79 521	79 773	80 025	80 276	80 528	80 779	81 031	251
·98	1·3	81 282	81 533	81 784	82 035	82 286	82 537	82 788	83 039	83 290	83 541	250
·99	1·3	83 791	84 042	84 292	84 543	84 793	85 044	85 294	85 544	85 794	86 044	250
4·00	1·3	86 294	86 544	86 794	87 044	87 294	87 544	87 793	88 043	88 292	88 542	249

n	7	8	9	10	11	12
$\ln 10^n$	16·118 096	18·420 681	20·723 266	23·025 851	25·328 436	27·631 021

	0	1	2	3	4	5	6	7	8	9	Δ
4·00	1·3 86 294	86 544	86 794	87 044	87 294	87 544	87 793	88 043	88 292	88 542	249
·01	1·3 88 791	89 041	89 290	89 539	89 788	90 037	90 286	90 535	90 784	91 033	249
·02	1·3 91 282	91 531	91 779	92 028	92 276	92 525	92 773	93 022	93 270	93 518	248
·03	1·3 93 766	94 014	94 263	94 511	94 758	95 006	95 254	95 502	95 750	95 997	248
·04	1·3 96 245	96 492	96 740	96 987	97 234	97 482	97 729	97 976	98 223	98 470	247
4·05	1·3 98 717	98 964	99 211	99 457	99 704	99 951	**00 197**	**00 444**	**00 690**	**00 937**	246
·06	1·4 01 183	01 429	01 675	01 922	02 168	02 414	02 660	02 906	03 151	03 397	246
·07	1·4 03 643	03 889	04 134	04 380	04 625	04 871	05 116	05 361	05 607	05 852	245
·08	1·4 06 097	06 342	06 587	06 832	07 077	07 322	07 566	07 811	08 056	08 300	245
·09	1·4 08 545	08 789	09 034	09 278	09 522	09 767	10 011	10 255	10 499	10 743	244
4·10	1·4 10 987	11 231	11 475	11 718	11 962	12 206	12 449	12 693	12 936	13 180	243
·11	1·4 13 423	13 666	13 910	14 153	14 396	14 639	14 882	15 125	15 368	15 610	243
·12	1·4 15 853	16 096	16 338	16 581	16 824	17 066	17 308	17 551	17 793	18 035	242
·13	1·4 18 277	18 520	18 762	19 004	19 245	19 487	19 729	19 971	20 213	20 454	242
·14	1·4 20 696	20 937	21 179	21 420	21 662	21 903	22 144	22 385	22 626	22 867	241
4·15	1·4 23 108	23 349	23 590	23 831	24 072	24 312	24 553	24 794	25 034	25 275	240
·16	1·4 25 515	25 755	25 996	26 236	26 476	26 716	26 956	27 196	27 436	27 676	240
·17	1·4 27 916	28 156	28 396	28 635	28 875	29 114	29 354	29 593	29 833	30 072	239
·18	1·4 30 311	30 550	30 790	31 029	31 268	31 507	31 746	31 984	32 223	32 462	239
·19	1·4 32 701	32 939	33 178	33 416	33 655	33 893	34 132	34 370	34 608	34 846	239
4·20	1·4 35 085	35 323	35 561	35 799	36 036	36 274	36 512	36 750	36 987	37 225	238
·21	1·4 37 463	37 700	37 938	38 175	38 412	38 650	38 887	39 124	39 361	39 598	237
·22	1·4 39 835	40 072	40 309	40 546	40 783	41 019	41 256	41 493	41 729	41 966	236
·23	1·4 42 202	42 438	42 675	42 911	43 147	43 383	43 619	43 855	44 091	44 327	236
·24	1·4 44 563	44 799	45 035	45 271	45 506	45 742	45 977	46 213	46 448	46 684	235
4·25	1·4 46 919	47 154	47 389	47 625	47 860	48 095	48 330	48 565	48 800	49 034	235
·26	1·4 49 269	49 504	49 739	49 973	50 208	50 442	50 677	50 911	51 145	51 380	234
·27	1·4 51 614	51 848	52 082	52 316	52 550	52 784	53 018	53 252	53 486	53 719	234
·28	1·4 53 953	54 187	54 420	54 654	54 887	55 121	55 354	55 587	55 820	56 054	233
·29	1·4 56 287	56 520	56 753	56 986	57 219	57 452	57 684	57 917	58 150	58 382	233
4·30	1·4 58 615	58 848	59 080	59 312	59 545	59 777	60 009	60 242	60 474	60 706	232
·31	1·4 60 938	61 170	61 402	61 634	61 866	62 097	62 329	62 561	62 792	63 024	231
·32	1·4 63 255	63 487	63 718	63 950	64 181	64 412	64 643	64 874	65 106	65 337	231
·33	1·4 65 568	65 798	66 029	66 260	66 491	66 722	66 952	67 183	67 413	67 644	230
·34	1·4 67 874	68 105	68 335	68 565	68 796	69 026	69 256	69 486	69 716	69 946	230
4·35	1·4 70 176	70 406	70 636	70 865	71 095	71 325	71 554	71 784	72 013	72 243	229
·36	1·4 72 472	72 701	72 931	73 160	73 389	73 618	73 847	74 076	74 305	74 534	229
·37	1·4 74 763	74 992	75 221	75 449	75 678	75 907	76 135	76 364	76 592	76 820	229
·38	1·4 77 049	77 277	77 505	77 733	77 962	78 190	78 418	78 646	78 874	79 101	228
·39	1·4 79 329	79 557	79 785	80 012	80 240	80 468	80 695	80 922	81 150	81 377	228
4·40	1·4 81 605	81 832	82 059	82 286	82 513	82 740	82 967	83 194	83 421	83 648	227
·41	1·4 83 875	84 101	84 328	84 555	84 781	85 008	85 234	85 461	85 687	85 913	227
·42	1·4 86 140	86 366	86 592	86 818	87 044	87 270	87 496	87 722	87 948	88 174	226
·43	1·4 88 400	88 625	88 851	89 077	89 302	89 528	89 753	89 978	90 204	90 429	225
·44	1·4 90 654	90 880	91 105	91 330	91 555	91 780	92 005	92 230	92 455	92 679	225
4·45	1·4 92 904	93 129	93 353	93 578	93 803	94 027	94 252	94 476	94 700	94 925	224
·46	1·4 95 149	95 373	95 597	95 821	96 045	96 269	96 493	96 717	96 941	97 165	223
·47	1·4 97 388	97 612	97 836	98 059	98 283	98 506	98 730	98 953	99 177	99 400	223
·48	1·4 99 623	99 846	**00 069**	**00 292**	**00 516**	**00 738**	**00 961**	**01 184**	**01 407**	**01 630**	223
·49	1·5 01 853	02 075	02 298	02 521	02 743	02 966	03 188	03 411	03 633	03 855	222
4·50	1·5 04 077	04 300	04 522	04 744	04 966	05 188	05 410	05 632	05 854	06 075	222

n	1	2	3	4	5	6	7
$\ln 10^n$	2·302 585	4·605 170	6·907 755	9·210 340	11·512 925	13·815 511	16·118 096

TABLE VI—NATURAL LOGARITHMS

		0	1	2	3	4	5	6	7	8	9	Δ
4·50	1·5	04 077	04 300	04 522	04 744	04 966	05 188	05 410	05 632	05 854	06 075	222
·51	1·5	06 297	06 519	06 741	06 962	07 184	07 405	07 627	07 848	08 069	08 291	221
·52	1·5	08 512	08 733	08 954	09 175	09 397	09 618	09 839	10 059	10 280	10 501	221
·53	1·5	10 722	10 943	11 163	11 384	11 605	11 825	12 046	12 266	12 486	12 707	220
·54	1·5	12 927	13 147	13 367	13 588	13 808	14 028	14 248	14 468	14 688	14 907	220
4·55	1·5	15 127	15 347	15 567	15 786	16 006	16 226	16 445	16 665	16 884	17 103	220
·56	1·5	17 323	17 542	17 761	17 980	18 199	18 419	18 638	18 857	19 075	19 294	219
·57	1·5	19 513	19 732	19 951	20 169	20 388	20 607	20 825	21 044	21 262	21 481	218
·58	1·5	21 699	21 917	22 136	22 354	22 572	22 790	23 008	23 226	23 444	23 662	218
·59	1·5	23 880	24 098	24 316	24 533	24 751	24 969	25 186	25 404	25 621	25 839	217
4·60	1·5	26 056	26 274	26 491	26 708	26 925	27 143	27 360	27 577	27 794	28 011	217
·61	1·5	28 228	28 445	28 662	28 878	29 095	29 312	29 529	29 745	29 962	30 178	217
·62	1·5	30 395	30 611	30 828	31 044	31 260	31 476	31 693	31 909	32 125	32 341	216
·63	1·5	32 557	32 773	32 989	33 205	33 420	33 636	33 852	34 068	34 283	34 499	215
·64	1·5	34 714	34 930	35 145	35 361	35 576	35 791	36 007	36 222	36 437	36 652	215
4·65	1·5	36 867	37 082	37 297	37 512	37 727	37 942	38 157	38 371	38 586	38 801	214
·66	1·5	39 015	39 230	39 445	39 659	39 873	40 088	40 302	40 516	40 731	40 945	214
·67	1·5	41 159	41 373	41 587	41 801	42 015	42 229	42 443	42 657	42 871	43 084	214
·68	1·5	43 298	43 512	43 725	43 939	44 152	44 366	44 579	44 793	45 006	45 219	214
·69	1·5	45 433	45 646	45 859	46 072	46 285	46 498	46 711	46 924	47 137	47 350	213
4·70	1·5	47 563	47 775	47 988	48 201	48 413	48 626	48 838	49 051	49 263	49 476	212
·71	1·5	49 688	49 900	50 112	50 325	50 537	50 749	50 961	51 173	51 385	51 597	212
·72	1·5	51 809	52 021	52 232	52 444	52 656	52 868	53 079	53 291	53 502	53 714	211
·73	1·5	53 925	54 137	54 348	54 559	54 771	54 982	55 193	55 404	55 615	55 826	211
·74	1·5	56 037	56 248	56 459	56 670	56 881	57 091	57 302	57 513	57 723	57 934	211
4·75	1·5	58 145	58 355	58 566	58 776	58 986	59 197	59 407	59 617	59 827	60 038	210
·76	1·5	60 248	60 458	60 668	60 878	61 088	61 298	61 507	61 717	61 927	62 137	209
·77	1·5	62 346	62 556	62 766	62 975	63 185	63 394	63 603	63 813	64 022	64 231	210
·78	1·5	64 441	64 650	64 859	65 068	65 277	65 486	65 695	65 904	66 113	66 322	208
·79	1·5	66 530	66 739	66 948	67 157	67 365	67 574	67 782	67 991	68 199	68 408	208
4·80	1·5	68 616	68 824	69 032	69 241	69 449	69 657	69 865	70 073	70 281	70 489	208
·81	1·5	70 697	70 905	71 113	71 321	71 528	71 736	71 944	72 151	72 359	72 566	208
·82	1·5	72 774	72 981	73 189	73 396	73 603	73 811	74 018	74 225	74 432	74 639	207
·83	1·5	74 846	75 053	75 260	75 467	75 674	75 881	76 088	76 295	76 501	76 708	207
·84	1·5	76 915	77 121	77 328	77 534	77 741	77 947	78 154	78 360	78 566	78 772	207
4·85	1·5	78 979	79 185	79 391	79 597	79 803	80 009	80 215	80 421	80 627	80 833	205
·86	1·5	81 038	81 244	81 450	81 656	81 861	82 067	82 272	82 478	82 683	82 889	205
·87	1·5	83 094	83 299	83 505	83 710	83 915	84 120	84 325	84 530	84 735	84 940	205
·88	1·5	85 145	85 350	85 555	85 760	85 965	86 169	86 374	86 579	86 783	86 988	204
·89	1·5	87 192	87 397	87 601	87 806	88 010	88 214	88 419	88 623	88 827	89 031	204
4·90	1·5	89 235	89 439	89 643	89 847	90 051	90 255	90 459	90 663	90 867	91 070	204
·91	1·5	91 274	91 478	91 681	91 885	92 088	92 292	92 495	92 699	92 902	93 105	204
·92	1·5	93 309	93 512	93 715	93 918	94 121	94 324	94 527	94 730	94 933	95 136	203
·93	1·5	95 339	95 542	95 745	95 947	96 150	96 353	96 555	96 758	96 960	97 163	202
·94	1·5	97 365	97 568	97 770	97 972	98 175	98 377	98 579	98 781	98 983	99 186	202
4·95	1·5	99 388	99 590	99 792	99 993	**00 195**	**00 397**	**00 599**	**00 801**	**01 002**	**01 204**	202
·96	1·6	01 406	01 607	01 809	02 010	02 212	02 413	02 615	02 816	03 017	03 219	201
·97	1·6	03 420	03 621	03 822	04 023	04 224	04 425	04 626	04 827	05 028	05 229	201
·98	1·6	05 430	05 631	05 831	06 032	06 233	06 433	06 634	06 835	07 035	07 235	201
·99	1·6	07 436	07 636	07 837	08 037	08 237	08 437	08 638	08 838	09 038	09 238	200
5·00	1·6	09 438	09 638	09 838	10 038	10 238	10 437	10 637	10 837	11 037	11 236	200

n	7	8	9	10	11	12
ln 10^n	16·118 096	18·420 681	20·723 266	23·025 851	25·328 436	27·631 021

TABLE VI—NATURAL LOGARITHMS

	0	1	2	3	4	5	6	7	8	9	Δ
5·00	1·6 09 438	09 638	09 838	10 038	10 238	10 437	10 637	10 837	11 037	11 236	200
·01	1·6 11 436	11 635	11 835	12 035	12 234	12 433	12 633	12 832	13 031	13 231	199
·02	1·6 13 430	13 629	13 828	14 027	14 226	14 425	14 624	14 823	15 022	15 221	199
·03	1·6 15 420	15 619	15 818	16 016	16 215	16 414	16 612	16 811	17 009	17 208	198
·04	1·6 17 406	17 604	17 803	18 001	18 199	18 398	18 596	18 794	18 992	19 190	198
5·05	1·6 19 388	19 586	19 784	19 982	20 180	20 378	20 576	20 773	20 971	21 169	197
·06	1·6 21 366	21 564	21 762	21 959	22 157	22 354	22 552	22 749	22 946	23 144	197
·07	1·6 23 341	23 538	23 735	23 932	24 129	24 327	24 524	24 721	24 917	25 114	197
·08	1·6 25 311	25 508	25 705	25 902	26 098	26 295	26 492	26 688	26 885	27 081	197
·09	1·6 27 278	27 474	27 671	27 867	28 063	28 260	28 456	28 652	28 848	29 044	197
5·10	1·6 29 241	29 437	29 633	29 829	30 025	30 220	30 416	30 612	30 808	31 004	195
·11	1·6 31 199	31 395	31 591	31 786	31 982	32 177	32 373	32 568	32 764	32 959	195
·12	1·6 33 154	33 350	33 545	33 740	33 935	34 131	34 326	34 521	34 716	34 911	195
·13	1·6 35 106	35 301	35 495	35 690	35 885	36 080	36 275	36 469	36 664	36 859	194
·14	1·6 37 053	37 248	37 442	37 637	37 831	38 025	38 220	38 414	38 608	38 803	194
5·15	1·6 38 997	39 191	39 385	39 579	39 773	39 967	40 161	40 355	40 549	40 743	194
·16	1·6 40 937	41 130	41 324	41 518	41 711	41 905	42 099	42 292	42 486	42 679	194
·17	1·6 42 873	43 066	43 259	43 453	43 646	43 839	44 033	44 226	44 419	44 612	193
·18	1·6 44 805	44 998	45 191	45 384	45 577	45 770	45 963	46 155	46 348	46 541	193
·19	1·6 46 734	46 926	47 119	47 312	47 504	47 697	47 889	48 082	48 274	48 466	193
5·20	1·6 48 659	48 851	49 043	49 235	49 428	49 620	49 812	50 004	50 196	50 388	192
·21	1·6 50 580	50 772	50 964	51 156	51 347	51 539	51 731	51 923	52 114	52 306	191
·22	1·6 52 497	52 689	52 880	53 072	53 263	53 455	53 646	53 837	54 029	54 220	191
·23	1·6 54 411	54 602	54 794	54 985	55 176	55 367	55 558	55 749	55 940	56 131	190
·24	1·6 56 321	56 512	56 703	56 894	57 085	57 275	57 466	57 656	57 847	58 038	190
5·25	1·6 58 228	58 419	58 609	58 799	58 990	59 180	59 370	59 561	59 751	59 941	190
·26	1·6 60 131	60 321	60 511	60 701	60 891	61 081	61 271	61 461	61 651	61 841	189
·27	1·6 62 030	62 220	62 410	62 599	62 789	62 979	63 168	63 358	63 547	63 737	189
·28	1·6 63 926	64 115	64 305	64 494	64 683	64 873	65 062	65 251	65 440	65 629	189
·29	1·6 65 818	66 007	66 196	66 385	66 574	66 763	66 952	67 141	67 329	67 518	189
5·30	1·6 67 707	67 895	68 084	68 273	68 461	68 650	68 838	69 027	69 215	69 403	189
·31	1·6 69 592	69 780	69 968	70 157	70 345	70 533	70 721	70 909	71 097	71 285	188
·32	1·6 71 473	71 661	71 849	72 037	72 225	72 413	72 600	72 788	72 976	73 164	187
·33	1·6 73 351	73 539	73 726	73 914	74 101	74 289	74 476	74 664	74 851	75 038	188
·34	1·6 75 226	75 413	75 600	75 787	75 974	76 162	76 349	76 536	76 723	76 910	187
5·35	1·6 77 097	77 283	77 470	77 657	77 844	78 031	78 217	78 404	78 591	78 777	187
·36	1·6 78 964	79 151	79 337	79 524	79 710	79 896	80 083	80 269	80 455	80 642	186
·37	1·6 80 828	81 014	81 200	81 386	81 573	81 759	81 945	82 131	82 317	82 502	186
·38	1·6 82 688	82 874	83 060	83 246	83 432	83 617	83 803	83 989	84 174	84 360	185
·39	1·6 84 545	84 731	84 916	85 102	85 287	85 473	85 658	85 843	86 029	86 214	185
5·40	1·6 86 399	86 584	86 769	86 954	87 139	87 324	87 509	87 694	87 879	88 064	185
·41	1·6 88 249	88 434	88 619	88 803	88 988	89 173	89 358	89 542	89 727	89 911	185
·42	1·6 90 096	90 280	90 465	90 649	90 834	91 018	91 202	91 386	91 571	91 755	184
·43	1·6 91 939	92 123	92 307	92 491	92 676	92 860	93 043	93 227	93 411	93 595	184
·44	1·6 93 779	93 963	94 147	94 330	94 514	94 698	94 881	95 065	95 249	95 432	184
5·45	1·6 95 616	95 799	95 983	96 166	96 349	96 533	96 716	96 899	97 082	97 266	183
·46	1·6 97 449	97 632	97 815	97 998	98 181	98 364	98 547	98 730	98 913	99 096	183
·47	1·6 99 279	99 461	99 644	99 827	**00 010**	**00 192**	**00 375**	**00 558**	**00 740**	**00 923**	182
·48	1·7 01 105	01 288	01 470	01 652	01 835	02 017	02 199	02 382	02 564	02 746	182
·49	1·7 02 928	03 110	03 292	03 475	03 657	03 839	04 021	04 202	04 384	04 566	182
5·50	1·7 04 748	04 930	05 112	05 293	05 475	05 657	05 838	06 020	06 202	06 383	182

n	1	2	3	4	5	6	7
$\ln 10^n$	2·302 585	4·605 170	6·907 755	9·210 340	11·512 925	13·815 511	16·118 096

TABLE VI—NATURAL LOGARITHMS

	0	1	2	3	4	5	6	7	8	9	Δ
5·50	1·7 04 748	04 930	05 112	05 293	05 475	05 657	05 838	06 020	06 202	06 383	182
·51	1·7 06 565	06 746	06 928	07 109	07 290	07 472	07 653	07 834	08 015	08 197	181
·52	1·7 08 378	08 559	08 740	08 921	09 102	09 283	09 464	09 645	09 826	10 007	181
·53	1·7 10 188	10 369	10 549	10 730	10 911	11 092	11 272	11 453	11 633	11 814	181
·54	1·7 11 995	12 175	12 355	12 536	12 716	12 897	13 077	13 257	13 438	13 618	180
5·55	1·7 13 798	13 978	14 158	14 338	14 518	14 698	14 878	15 058	15 238	15 418	180
·56	1·7 15 598	15 778	15 958	16 138	16 317	16 497	16 677	16 856	17 036	17 216	179
·57	1·7 17 395	17 575	17 754	17 934	18 113	18 292	18 472	18 651	18 830	19 010	179
·58	1·7 19 189	19 368	19 547	19 726	19 905	20 084	20 263	20 442	20 621	20 800	179
·59	1·7 20 979	21 158	21 337	21 516	21 695	21 873	22 052	22 231	22 409	22 588	179
5·60	1·7 22 767	22 945	23 124	23 302	23 481	23 659	23 837	24 016	24 194	24 372	179
·61	1·7 24 551	24 729	24 907	25 085	25 263	25 442	25 620	25 798	25 976	26 154	178
·62	1·7 26 332	26 510	26 687	26 865	27 043	27 221	27 399	27 576	27 754	27 932	177
·63	1·7 28 109	28 287	28 465	28 642	28 820	28 997	29 175	29 352	29 529	29 707	177
·64	1·7 29 884	30 061	30 239	30 416	30 593	30 770	30 947	31 124	31 302	31 479	177
5·65	1·7 31 656	31 833	32 009	32 186	32 363	32 540	32 717	32 894	33 070	33 247	177
·66	1·7 33 424	33 601	33 777	33 954	34 130	34 307	34 483	34 660	34 836	35 013	176
·67	1·7 35 189	35 365	35 542	35 718	35 894	36 071	36 247	36 423	36 599	36 775	176
·68	1·7 36 951	37 127	37 303	37 479	37 655	37 831	38 007	38 183	38 359	38 534	176
·69	1·7 38 710	38 886	39 062	39 237	39 413	39 589	39 764	39 940	40 115	40 291	175
5·70	1·7 40 466	40 642	40 817	40 992	41 168	41 343	41 518	41 693	41 869	42 044	175
·71	1·7 42 219	42 394	42 569	42 744	42 919	43 094	43 269	43 444	43 619	43 794	175
·72	1·7 43 969	44 144	44 318	44 493	44 668	44 843	45 017	45 192	45 366	45 541	175
·73	1·7 45 716	45 890	46 065	46 239	46 413	46 588	46 762	46 936	47 111	47 285	174
·74	1·7 47 459	47 633	47 808	47 982	48 156	48 330	48 504	48 678	48 852	49 026	174
5·75	1·7 49 200	49 374	49 548	49 721	49 895	50 069	50 243	50 417	50 590	50 764	173
·76	1·7 50 937	51 111	51 285	51 458	51 632	51 805	51 979	52 152	52 325	52 499	173
·77	1·7 52 672	52 845	53 019	53 192	53 365	53 538	53 711	53 885	54 058	54 231	173
·78	1·7 54 404	54 577	54 750	54 923	55 095	55 268	55 441	55 614	55 787	55 960	172
·79	1·7 56 132	56 305	56 478	56 650	56 823	56 995	57 168	57 341	57 513	57 685	173
5·80	1·7 57 858	58 030	58 203	58 375	58 547	58 720	58 892	59 064	59 236	59 408	173
·81	1·7 59 581	59 753	59 925	60 097	60 269	60 441	60 613	60 785	60 957	61 128	172
·82	1·7 61 300	61 472	61 644	61 816	61 987	62 159	62 331	62 502	62 674	62 845	172
·83	1·7 63 017	63 189	63 360	63 531	63 703	63 874	64 046	64 217	64 388	64 560	171
·84	1·7 64 731	64 902	65 073	65 244	65 415	65 587	65 758	65 929	66 100	66 271	171
5·85	1·7 66 442	66 613	66 783	66 954	67 125	67 296	67 467	67 638	67 808	67 979	171
·86	1·7 68 150	68 320	68 491	68 661	68 832	69 002	69 173	69 343	69 514	69 684	171
·87	1·7 69 855	70 025	70 195	70 366	70 536	70 706	70 876	71 046	71 217	71 387	170
·88	1·7 71 557	71 727	71 897	72 067	72 237	72 407	72 577	72 747	72 916	73 086	170
·89	1·7 73 256	73 426	73 595	73 765	73 935	74 105	74 274	74 444	74 613	74 783	169
5·90	1·7 74 952	75 122	75 291	75 461	75 630	75 799	75 969	76 138	76 307	76 477	169
·91	1·7 76 646	76 815	76 984	77 153	77 322	77 491	77 661	77 830	77 999	78 168	168
·92	1·7 78 336	78 505	78 674	78 843	79 012	79 181	79 349	79 518	79 687	79 856	168
·93	1·7 80 024	80 193	80 361	80 530	80 699	80 867	81 036	81 204	81 372	81 541	168
·94	1·7 81 709	81 877	82 046	82 214	82 382	82 551	82 719	82 887	83 055	83 223	168
5·95	1·7 83 391	83 559	83 727	83 895	84 063	84 231	84 399	84 567	84 735	84 903	167
·96	1·7 85 070	85 238	85 406	85 574	85 741	85 909	86 077	86 244	86 412	86 579	168
·97	1·7 86 747	86 914	87 082	87 249	87 417	87 584	87 751	87 919	88 086	88 253	168
·98	1·7 88 421	88 588	88 755	88 922	89 089	89 256	89 423	89 590	89 757	89 924	167
·99	1·7 90 091	90 258	90 425	90 592	90 759	90 926	91 093	91 259	91 426	91 593	166
6·00	1·7 91 759	91 926	92 093	92 259	92 426	92 592	92 759	92 925	93 092	93 258	167

n	7	8	9	10	11	12
ln 10^n	16·118 096	18·420 681	20·723 266	23·025 851	25·328 436	27·631 021

TABLE VI—NATURAL LOGARITHMS

	0	1	2	3	4	5	6	7	8	9	Δ
6·00	1·7 91 759	91 926	92 093	92 259	92 426	92 592	92 759	92 925	93 092	93 258	167
·01	1·7 93 425	93 591	93 757	93 924	94 090	94 256	94 423	94 589	94 755	94 921	166
·02	1·7 95 087	95 253	95 419	95 585	95 751	95 917	96 083	96 249	96 415	96 581	166
·03	1·7 96 747	96 913	97 079	97 244	97 410	97 576	97 742	97 907	98 073	98 238	166
·04	1·7 98 404	98 570	98 735	98 901	99 066	99 231	99 397	99 562	99 728	99 893	165
6·05	1·8 00 058	00 224	00 389	00 554	00 719	00 884	01 050	01 215	01 380	01 545	165
·06	1·8 01 710	01 875	02 040	02 205	02 370	02 535	02 699	02 864	03 029	03 194	165
·07	1·8 03 359	03 523	03 688	03 853	04 017	04 182	04 347	04 511	04 676	04 840	165
·08	1·8 05 005	05 169	05 334	05 498	05 662	05 827	05 991	06 155	06 320	06 484	164
·09	1·8 06 648	06 812	06 976	07 141	07 305	07 469	07 633	07 797	07 961	08 125	164
6·10	1·8 08 289	08 453	08 617	08 780	08 944	09 108	09 272	09 436	09 599	09 763	164
·11	1·8 09 927	10 090	10 254	10 418	10 581	10 745	10 908	11 072	11 235	11 399	163
·12	1·8 11 562	11 725	11 889	12 052	12 215	12 379	12 542	12 705	12 868	13 032	163
·13	1·8 13 195	13 358	13 521	13 684	13 847	14 010	14 173	14 336	14 499	14 662	163
·14	1·8 14 825	14 988	15 150	15 313	15 476	15 639	15 801	15 964	16 127	16 289	163
6·15	1·8 16 452	16 615	16 777	16 940	17 102	17 265	17 427	17 590	17 752	17 914	163
·16	1·8 18 077	18 239	18 401	18 564	18 726	18 888	19 050	19 212	19 375	19 537	162
·17	1·8 19 699	19 861	20 023	20 185	20 347	20 509	20 671	20 833	20 995	21 156	162
·18	1·8 21 318	21 480	21 642	21 804	21 965	22 127	22 289	22 450	22 612	22 774	161
·19	1·8 22 935	23 097	23 258	23 420	23 581	23 743	23 904	24 065	24 227	24 388	161
6·20	1·8 24 549	24 711	24 872	25 033	25 194	25 355	25 517	25 678	25 839	26 000	161
·21	1·8 26 161	26 322	26 483	26 644	26 805	26 966	27 127	27 287	27 448	27 609	161
·22	1·8 27 770	27 931	28 091	28 252	28 413	28 573	28 734	28 895	29 055	29 216	160
·23	1·8 29 376	29 537	29 697	29 858	30 018	30 179	30 339	30 499	30 660	30 820	160
·24	1·8 30 980	31 140	31 301	31 461	31 621	31 781	31 941	32 101	32 261	32 421	160
6·25	1·8 32 581	32 741	32 901	33 061	33 221	33 381	33 541	33 701	33 861	34 020	160
·26	1·8 34 180	34 340	34 500	34 659	34 819	34 979	35 138	35 298	35 457	35 617	159
·27	1·8 35 776	35 936	36 095	36 255	36 414	36 573	36 733	36 892	37 051	37 211	159
·28	1·8 37 370	37 529	37 688	37 848	38 007	38 166	38 325	38 484	38 643	38 802	159
·29	1·8 38 961	39 120	39 279	39 438	39 597	39 756	39 915	40 073	40 232	40 391	159
6·30	1·8 40 550	40 708	40 867	41 026	41 184	41 343	41 502	41 660	41 819	41 977	159
·31	1·8 42 136	42 294	42 453	42 611	42 769	42 928	43 086	43 244	43 403	43 561	158
·32	1·8 43 719	43 877	44 036	44 194	44 352	44 510	44 668	44 826	44 984	45 142	158
·33	1·8 45 300	45 458	45 616	45 774	45 932	46 090	46 248	46 405	46 563	46 721	158
·34	1·8 46 879	47 036	47 194	47 352	47 509	47 667	47 825	47 982	48 140	48 297	158
6·35	1·8 48 455	48 612	48 770	48 927	49 085	49 242	49 399	49 557	49 714	49 871	157
·36	1·8 50 028	50 186	50 343	50 500	50 657	50 814	50 971	51 128	51 285	51 442	157
·37	1·8 51 599	51 756	51 913	52 070	52 227	52 384	52 541	52 698	52 855	53 011	157
·38	1·8 53 168	53 325	53 482	53 638	53 795	53 951	54 108	54 265	54 421	54 578	156
·39	1·8 54 734	54 891	55 047	55 204	55 360	55 516	55 673	55 829	55 985	56 142	156
6·40	1·8 56 298	56 454	56 610	56 767	56 923	57 079	57 235	57 391	57 547	57 703	156
·41	1·8 57 859	58 015	58 171	58 327	58 483	58 639	58 795	58 951	59 107	59 262	156
·42	1·8 59 418	59 574	59 730	59 885	60 041	60 197	60 352	60 508	60 663	60 819	156
·43	1·8 60 975	61 130	61 286	61 441	61 596	61 752	61 907	62 063	62 218	62 373	156
·44	1·8 62 529	62 684	62 839	62 994	63 149	63 305	63 460	63 615	63 770	63 925	155
6·45	1·8 64 080	64 235	64 390	64 545	64 700	64 855	65 010	65 165	65 320	65 475	154
·46	1·8 65 629	65 784	65 939	66 094	66 248	66 403	66 558	66 712	66 867	67 022	154
·47	1·8 67 176	67 331	67 485	67 640	67 794	67 949	68 103	68 257	68 412	68 566	155
·48	1·8 68 721	68 875	69 029	69 183	69 338	69 492	69 646	69 800	69 954	70 108	155
·49	1·8 70 263	70 417	70 571	70 725	70 879	71 033	71 187	71 341	71 494	71 648	154
6·50	1·8 71 802	71 956	72 110	72 264	72 417	72 571	72 725	72 879	73 032	73 186	153

n	1	2	3	4	5	6	7
$\ln 10^n$	2·302 585	4·605 170	6·907 755	9·210 340	11·512 925	13·815 511	16·118 096

	0	1	2	3	4	5	6	7	8	9	Δ
6·50	1·8 71 802	71 956	72 110	72 264	72 417	72 571	72 725	72 879	73 032	73 186	153
·51	1·8 73 339	73 493	73 647	73 800	73 954	74 107	74 261	74 414	74 568	74 721	153
·52	1·8 74 874	75 028	75 181	75 334	75 488	75 641	75 794	75 947	76 101	76 254	153
·53	1·8 76 407	76 560	76 713	76 866	77 019	77 172	77 325	77 478	77 631	77 784	153
·54	1·8 77 937	78 090	78 243	78 396	78 549	78 701	78 854	79 007	79 160	79 312	153
6·55	1·8 79 465	79 618	79 770	79 923	80 076	80 228	80 381	80 533	80 686	80 838	153
·56	1·8 80 991	81 143	81 295	81 448	81 600	81 753	81 905	82 057	82 209	82 362	152
·57	1·8 82 514	82 666	82 818	82 970	83 122	83 275	83 427	83 579	83 731	83 883	152
·58	1·8 84 035	84 187	84 339	84 491	84 642	84 794	84 946	85 098	85 250	85 402	151
·59	1·8 85 553	85 705	85 857	86 008	86 160	86 312	86 463	86 615	86 767	86 918	152
6·60	1·8 87 070	87 221	87 373	87 524	87 676	87 827	87 978	88 130	88 281	88 432	152
·61	1·8 88 584	88 735	88 886	89 037	89 189	89 340	89 491	89 642	89 793	89 944	151
·62	1·8 90 095	90 246	90 397	90 548	90 699	90 850	91 001	91 152	91 303	91 454	151
·63	1·8 91 605	91 756	91 906	92 057	92 208	92 359	92 509	92 660	92 811	92 961	151
·64	1·8 93 112	93 263	93 413	93 564	93 714	93 865	94 015	94 166	94 316	94 466	151
6·65	1·8 94 617	94 767	94 918	95 068	95 218	95 368	95 519	95 669	95 819	95 969	150
·66	1·8 96 119	96 270	96 420	96 570	96 720	96 870	97 020	97 170	97 320	97 470	150
·67	1·8 97 620	97 770	97 920	98 070	98 219	98 369	98 519	98 669	98 819	98 968	150
·68	1·8 99 118	99 268	99 417	99 567	99 717	99 866	**00 016**	**00 165**	**00 315**	**00 464**	150
·69	1·9 00 614	00 763	00 913	01 062	01 212	01 361	01 510	01 660	01 809	01 958	150
6·70	1·9 02 108	02 257	02 406	02 555	02 704	02 854	03 003	03 152	03 301	03 450	149
·71	1·9 03 599	03 748	03 897	04 046	04 195	04 344	04 493	04 642	04 790	04 939	149
·72	1·9 05 088	05 237	05 386	05 534	05 683	05 832	05 981	06 129	06 278	06 427	148
·73	1·9 06 575	06 724	06 872	07 021	07 169	07 318	07 466	07 615	07 763	07 912	148
·74	1·9 08 060	08 208	08 357	08 505	08 653	08 801	08 950	09 098	09 246	09 394	149
6·75	1·9 09 543	09 691	09 839	09 987	10 135	10 283	10 431	10 579	10 727	10 875	148
·76	1·9 11 023	11 171	11 319	11 467	11 614	11 762	11 910	12 058	12 206	12 353	148
·77	1·9 12 501	12 649	12 796	12 944	13 092	13 239	13 387	13 535	13 682	13 830	147
·78	1·9 13 977	14 125	14 272	14 419	14 567	14 714	14 862	15 009	15 156	15 304	147
·79	1·9 15 451	15 598	15 745	15 893	16 040	16 187	16 334	16 481	16 628	16 776	147
6·80	1·9 16 923	17 070	17 217	17 364	17 511	17 658	17 805	17 951	18 098	18 245	147
·81	1·9 18 392	18 539	18 686	18 833	18 979	19 126	19 273	19 419	19 566	19 713	146
·82	1·9 19 859	20 006	20 153	20 299	20 446	20 592	20 739	20 885	21 032	21 178	147
·83	1·9 21 325	21 471	21 617	21 764	21 910	22 056	22 203	22 349	22 495	22 642	146
·84	1·9 22 788	22 934	23 080	23 226	23 372	23 518	23 665	23 811	23 957	24 103	146
6·85	1·9 24 249	24 395	24 541	24 687	24 832	24 978	25 124	25 270	25 416	25 562	145
·86	1·9 25 707	25 853	25 999	26 145	26 290	26 436	26 582	26 727	26 873	27 019	145
·87	1·9 27 164	27 310	27 455	27 601	27 746	27 892	28 037	28 183	28 328	28 473	146
·88	1·9 28 619	28 764	28 909	29 055	29 200	29 345	29 490	29 636	29 781	29 926	145
·89	1·9 30 071	30 216	30 361	30 506	30 651	30 797	30 942	31 087	31 232	31 376	145
6·90	1·9 31 521	31 666	31 811	31 956	32 101	32 246	32 391	32 535	32 680	32 825	145
·91	1·9 32 970	33 114	33 259	33 404	33 548	33 693	33 838	33 982	34 127	34 271	145
·92	1·9 34 416	34 560	34 705	34 849	34 994	35 138	35 282	35 427	35 571	35 716	144
·93	1·9 35 860	36 004	36 148	36 293	36 437	36 581	36 725	36 869	37 014	37 158	144
·94	1·9 37 302	37 446	37 590	37 734	37 878	38 022	38 166	38 310	38 454	38 598	144
6·95	1·9 38 742	38 886	39 029	39 173	39 317	39 461	39 605	39 748	39 892	40 036	143
·96	1·9 40 179	40 323	40 467	40 610	40 754	40 898	41 041	41 185	41 328	41 472	143
·97	1·9 41 615	41 759	41 902	42 046	42 189	42 332	42 476	42 619	42 762	42 906	143
·98	1·9 43 049	43 192	43 335	43 479	43 622	43 765	43 908	44 051	44 194	44 337	144
·99	1·9 44 481	44 624	44 767	44 910	45 053	45 196	45 339	45 481	45 624	45 767	143
7·00	1·9 45 910	46 053	46 196	46 339	46 481	46 624	46 767	46 910	47 052	47 195	143

n	7	8	9	10	11	12
$\ln 10^n$	16·118 096	18·420 681	20·723 266	23·025 851	25·328 436	27·631 021

TABLE VI—NATURAL LOGARITHMS

	0	1	2	3	4	5	6	7	8	9	Δ
7·00	1·9 45 910	46 053	46 196	46 339	46 481	46 624	46 767	46 910	47 052	47 195	143
·01	1·9 47 338	47 480	47 623	47 766	47 908	48 051	48 193	48 336	48 478	48 621	142
·02	1·9 48 763	48 906	49 048	49 190	49 333	49 475	49 618	49 760	49 902	50 044	143
·03	1·9 50 187	50 329	50 471	50 613	50 756	50 898	51 040	51 182	51 324	51 466	142
·04	1·9 51 608	51 750	51 892	52 034	52 176	52 318	52 460	52 602	52 744	52 886	142
7·05	1·9 53 028	53 169	53 311	53 453	53 595	53 737	53 878	54 020	54 162	54 303	142
·06	1·9 54 445	54 587	54 728	54 870	55 011	55 153	55 295	55 436	55 578	55 719	141
·07	1·9 55 860	56 002	56 143	56 285	56 426	56 567	56 709	56 850	56 991	57 133	141
·08	1·9 57 274	57 415	57 556	57 698	57 839	57 980	58 121	58 262	58 403	58 544	141
·09	1·9 58 685	58 826	58 967	59 108	59 249	59 390	59 531	59 672	59 813	59 954	141
7·10	1·9 60 095	60 236	60 376	60 517	60 658	60 799	60 939	61 080	61 221	61 362	140
·11	1·9 61 502	61 643	61 783	61 924	62 065	62 205	62 346	62 486	62 627	62 767	141
·12	1·9 62 908	63 048	63 189	63 329	63 469	63 610	63 750	63 890	64 031	64 171	140
·13	1·9 64 311	64 451	64 592	64 732	64 872	65 012	65 152	65 293	65 433	65 573	140
·14	1·9 65 713	65 853	65 993	66 133	66 273	66 413	66 553	66 693	66 833	66 972	140
7·15	1·9 67 112	67 252	67 392	67 532	67 672	67 811	67 951	68 091	68 231	68 370	140
·16	1·9 68 510	68 650	68 789	68 929	69 068	69 208	69 348	69 487	69 627	69 766	140
·17	1·9 69 906	70 045	70 185	70 324	70 463	70 603	70 742	70 881	71 021	71 160	139
·18	1·9 71 299	71 439	71 578	71 717	71 856	71 996	72 135	72 274	72 413	72 552	139
·19	1·9 72 691	72 830	72 969	73 108	73 247	73 386	73 525	73 664	73 803	73 942	139
7·20	1·9 74 081	74 220	74 359	74 498	74 636	74 775	74 914	75 053	75 192	75 330	139
·21	1·9 75 469	75 608	75 746	75 885	76 024	76 162	76 301	76 439	76 578	76 716	139
·22	1·9 76 855	76 993	77 132	77 270	77 409	77 547	77 686	77 824	77 962	78 101	138
·23	1·9 78 239	78 377	78 516	78 654	78 792	78 930	79 069	79 207	79 345	79 483	138
·24	1·9 79 621	79 759	79 897	80 035	80 174	80 312	80 450	80 588	80 726	80 864	137
7·25	1·9 81 001	81 139	81 277	81 415	81 553	81 691	81 829	81 967	82 104	82 242	138
·26	1·9 82 380	82 518	82 655	82 793	82 931	83 068	83 206	83 344	83 481	83 619	137
·27	1·9 83 756	83 894	84 031	84 169	84 306	84 444	84 581	84 719	84 856	84 993	138
·28	1·9 85 131	85 268	85 406	85 543	85 680	85 817	85 955	86 092	86 229	86 366	138
·29	1·9 86 504	86 641	86 778	86 915	87 052	87 189	87 326	87 463	87 600	87 737	137
7·30	1·9 87 874	88 011	88 148	88 285	88 422	88 559	88 696	88 833	88 970	89 106	137
·31	1·9 89 243	89 380	89 517	89 654	89 790	89 927	90 064	90 200	90 337	90 474	136
·32	1·9 90 610	90 747	90 884	91 020	91 157	91 293	91 430	91 566	91 703	91 839	137
·33	1·9 91 976	92 112	92 248	92 385	92 521	92 657	92 794	92 930	93 066	93 203	136
·34	1·9 93 339	93 475	93 611	93 747	93 884	94 020	94 156	94 292	94 428	94 564	136
7·35	1·9 94 700	94 836	94 972	95 108	95 244	95 380	95 516	95 652	95 788	95 924	136
·36	1·9 96 060	96 196	96 332	96 467	96 603	96 739	96 875	97 011	97 146	97 282	136
·37	1·9 97 418	97 553	97 689	97 825	97 960	98 096	98 231	98 367	98 503	98 638	136
·38	1·9 98 774	98 909	99 045	99 180	99 315	99 451	99 586	99 722	99 857	99 992	136
·39	2·0 00 128	00 263	00 398	00 534	00 669	00 804	00 939	01 075	01 210	01 345	135
7·40	2·0 01 480	01 615	01 750	01 885	02 020	02 155	02 290	02 425	02 560	02 695	135
·41	2·0 02 830	02 965	03 100	03 235	03 370	03 505	03 640	03 775	03 909	04 044	135
·42	2·0 04 179	04 314	04 449	04 583	04 718	04 853	04 987	05 122	05 257	05 391	135
·43	2·0 05 526	05 660	05 795	05 930	06 064	06 199	06 333	06 468	06 602	06 736	135
·44	2·0 06 871	07 005	07 140	07 274	07 408	07 543	07 677	07 811	07 946	08 080	134
7·45	2·0 08 214	08 348	08 482	08 617	08 751	08 885	09 019	09 153	09 287	09 421	134
·46	2·0 09 555	09 689	09 823	09 957	10 091	10 225	10 359	10 493	10 627	10 761	134
·47	2·0 10 895	11 029	11 163	11 297	11 430	11 564	11 698	11 832	11 965	12 099	134
·48	2·0 12 233	12 366	12 500	12 634	12 767	12 901	13 035	13 168	13 302	13 435	134
·49	2·0 13 569	13 702	13 836	13 969	14 103	14 236	14 370	14 503	14 636	14 770	133
7·50	2·0 14 903	15 036	15 170	15 303	15 436	15 569	15 703	15 836	15 969	16 102	133

n	1	2	3	4	5	6	7
$\ln 10^n$	2·302 585	4·605 170	6·907 755	9·210 340	11·512 925	13·815 511	16·118 096

TABLE VI—NATURAL LOGARITHMS

	0	1	2	3	4	5	6	7	8	9	Δ
7·50	2·0 14 903	15 036	15 170	15 303	15 436	15 569	15 703	15 836	15 969	16 102	133
·51	2·0 16 235	16 369	16 502	16 635	16 768	16 901	17 034	17 167	17 300	17 433	133
·52	2·0 17 566	17 699	17 832	17 965	18 098	18 231	18 364	18 497	18 629	18 762	133
·53	2·0 18 895	19 028	19 161	19 293	19 426	19 559	19 692	19 824	19 957	20 090	132
·54	2·0 20 222	20 355	20 487	20 620	20 753	20 885	21 018	21 150	21 283	21 415	133
7·55	2·0 21 548	21 680	21 812	21 945	22 077	22 210	22 342	22 474	22 607	22 739	132
·56	2·0 22 871	23 003	23 136	23 268	23 400	23 532	23 665	23 797	23 929	24 061	132
·57	2·0 24 193	24 325	24 457	24 589	24 721	24 853	24 985	25 117	25 249	25 381	132
·58	2·0 25 513	25 645	25 777	25 909	26 041	26 173	26 304	26 436	26 568	26 700	132
·59	2·0 26 832	26 963	27 095	27 227	27 358	27 490	27 622	27 753	27 885	28 017	131
7·60	2·0 28 148	28 280	28 411	28 543	28 674	28 806	28 937	29 069	29 200	29 332	131
·61	2·0 29 463	29 595	29 726	29 857	29 989	30 120	30 251	30 383	30 514	30 645	131
·62	2·0 30 776	30 908	31 039	31 170	31 301	31 432	31 563	31 695	31 826	31 957	131
·63	2·0 32 088	32 219	32 350	32 481	32 612	32 743	32 874	33 005	33 136	33 267	131
·64	2·0 33 398	33 528	33 659	33 790	33 921	34 052	34 183	34 313	34 444	34 575	131
7·65	2·0 34 706	34 836	34 967	35 098	35 228	35 359	35 490	35 620	35 751	35 881	131
·66	2·0 36 012	36 143	36 273	36 404	36 534	36 665	36 795	36 925	37 056	37 186	131
·67	2·0 37 317	37 447	37 577	37 708	37 838	37 968	38 099	38 229	38 359	38 489	131
·68	2·0 38 620	38 750	38 880	39 010	39 140	39 270	39 400	39 531	39 661	39 791	130
·69	2·0 39 921	40 051	40 181	40 311	40 441	40 571	40 701	40 831	40 961	41 090	130
7·70	2·0 41 220	41 350	41 480	41 610	41 740	41 869	41 999	42 129	42 259	42 388	130
·71	2·0 42 518	42 648	42 778	42 907	43 037	43 166	43 296	43 426	43 555	43 685	129
·72	2·0 43 814	43 944	44 073	44 203	44 332	44 462	44 591	44 721	44 850	44 979	130
·73	2·0 45 109	45 238	45 368	45 497	45 626	45 755	45 885	46 014	46 143	46 272	130
·74	2·0 46 402	46 531	46 660	46 789	46 918	47 047	47 177	47 306	47 435	47 564	129
7·75	2·0 47 693	47 822	47 951	48 080	48 209	48 338	48 467	48 596	48 725	48 853	129
·76	2·0 48 982	49 111	49 240	49 369	49 498	49 626	49 755	49 884	50 013	50 141	129
·77	2·0 50 270	50 399	50 528	50 656	50 785	50 913	51 042	51 171	51 299	51 428	128
·78	2·0 51 556	51 685	51 813	51 942	52 070	52 199	52 327	52 456	52 584	52 712	129
·79	2·0 52 841	52 969	53 098	53 226	53 354	53 483	53 611	53 739	53 867	53 996	128
7·80	2·0 54 124	54 252	54 380	54 508	54 636	54 765	54 893	55 021	55 149	55 277	128
·81	2·0 55 405	55 533	55 661	55 789	55 917	56 045	56 173	56 301	56 429	56 557	128
·82	2·0 56 685	56 812	56 940	57 068	57 196	57 324	57 452	57 579	57 707	57 835	128
·83	2·0 57 963	58 090	58 218	58 346	58 473	58 601	58 729	58 856	58 984	59 111	128
·84	2·0 59 239	59 366	59 494	59 621	59 749	59 876	60 004	60 131	60 259	60 386	128
7·85	2·0 60 514	60 641	60 768	60 896	61 023	61 150	61 278	61 405	61 532	61 659	128
·86	2·0 61 787	61 914	62 041	62 168	62 295	62 423	62 550	62 677	62 804	62 931	127
·87	2·0 63 058	63 185	63 312	63 439	63 566	63 693	63 820	63 947	64 074	64 201	127
·88	2·0 64 328	64 455	64 582	64 709	64 835	64 962	65 089	65 216	65 343	65 469	127
·89	2·0 65 596	65 723	65 850	65 976	66 103	66 230	66 356	66 483	66 610	66 736	127
7·90	2·0 66 863	66 989	67 116	67 242	67 369	67 495	67 622	67 748	67 875	68 001	127
·91	2·0 68 128	68 254	68 381	68 507	68 633	68 760	68 886	69 012	69 139	69 265	126
·92	2·0 69 391	69 517	69 644	69 770	69 896	70 022	70 148	70 275	70 401	70 527	126
·93	2·0 70 653	70 779	70 905	71 031	71 157	71 283	71 409	71 535	71 661	71 787	126
·94	2·0 71 913	72 039	72 165	72 291	72 417	72 543	72 669	72 794	72 920	73 046	126
7·95	2·0 73 172	73 298	73 423	73 549	73 675	73 801	73 926	74 052	74 178	74 303	126
·96	2·0 74 429	74 555	74 680	74 806	74 931	75 057	75 182	75 308	75 434	75 559	125
·97	2·0 75 684	75 810	75 935	76 061	76 186	76 312	76 437	76 562	76 688	76 813	125
·98	2·0 76 938	77 064	77 189	77 314	77 440	77 565	77 690	77 815	77 940	78 066	125
·99	2·0 78 191	78 316	78 441	78 566	78 691	78 816	78 941	79 066	79 192	79 317	125
8·00	2·0 79 442	79 567	79 692	79 816	79 941	80 066	80 191	80 316	80 441	80 566	125

n	7	8	9	10	11	12
$\ln 10^n$	16·118 096	18·420 681	20·723 266	23·025 851	25·328 436	27·631 021

	0	1	2	3	4	5	6	7	8	9	Δ
8·00	2·0 79 442	79 567	79 692	79 816	79 941	80 066	80 191	80 316	80 441	80 566	125
·01	2·0 80 691	80 816	80 940	81 065	81 190	81 315	81 440	81 564	81 689	81 814	124
·02	2·0 81 938	82 063	82 188	82 312	82 437	82 562	82 686	82 811	82 935	83 060	125
·03	2·0 83 185	83 309	83 434	83 558	83 683	83 807	83 931	84 056	84 180	84 305	124
·04	2·0 84 429	84 553	84 678	84 802	84 926	85 051	85 175	85 299	85 424	85 548	124
8·05	2·0 85 672	85 796	85 921	86 045	86 169	86 293	86 417	86 541	86 665	86 789	125
·06	2·0 86 914	87 038	87 162	87 286	87 410	87 534	87 658	87 782	87 906	88 030	123
·07	2·0 88 153	88 277	88 401	88 525	88 649	88 773	88 897	89 021	89 144	89 268	124
·08	2·0 89 392	89 516	89 639	89 763	89 887	90 010	90 134	90 258	90 381	90 505	124
·09	2·0 90 629	90 752	90 876	90 999	91 123	91 247	91 370	91 494	91 617	91 741	123
8·10	2·0 91 864	91 988	92 111	92 234	92 358	92 481	92 605	92 728	92 851	92 975	123
·11	2·0 93 098	93 221	93 344	93 468	93 591	93 714	93 837	93 961	94 084	94 207	123
·12	2·0 94 330	94 453	94 576	94 700	94 823	94 946	95 069	95 192	95 315	95 438	123
·13	2·0 95 561	95 684	95 807	95 930	96 053	96 176	96 299	96 422	96 544	96 667	123
·14	2·0 96 790	96 913	97 036	97 159	97 281	97 404	97 527	97 650	97 772	97 895	123
8·15	2·0 98 018	98 141	98 263	98 386	98 509	98 631	98 754	98 876	98 999	99 122	122
·16	2·0 99 244	99 367	99 489	99 612	99 734	99 857	99 979	**00 102**	**00 224**	**00 347**	122
·17	2·1 00 469	00 591	00 714	00 836	00 958	01 081	01 203	01 325	01 448	01 570	122
·18	2·1 01 692	01 814	01 937	02 059	02 181	02 303	02 425	02 548	02 670	02 792	122
·19	2·1 02 914	03 036	03 158	03 280	03 402	03 524	03 646	03 768	03 890	04 012	122
8·20	2·1 04 134	04 256	04 378	04 500	04 622	04 744	04 866	04 987	05 109	05 231	122
·21	2·1 05 353	05 475	05 596	05 718	05 840	05 962	06 083	06 205	06 327	06 449	121
·22	2·1 06 570	06 692	06 813	06 935	07 057	07 178	07 300	07 421	07 543	07 665	121
·23	2·1 07 786	07 908	08 029	08 150	08 272	08 393	08 515	08 636	08 758	08 879	121
·24	2·1 09 000	09 122	09 243	09 364	09 486	09 607	09 728	09 849	09 971	10 092	121
8·25	2·1 10 213	10 334	10 456	10 577	10 698	10 819	10 940	11 061	11 182	11 304	121
·26	2·1 11 425	11 546	11 667	11 788	11 909	12 030	12 151	12 272	12 393	12 514	121
·27	2·1 12 635	12 755	12 876	12 997	13 118	13 239	13 360	13 481	13 601	13 722	121
·28	2·1 13 843	13 964	14 084	14 205	14 326	14 447	14 567	14 688	14 809	14 929	121
·29	2·1 15 050	15 171	15 291	15 412	15 532	15 653	15 773	15 894	16 015	16 135	121
8·30	2·1 16 256	16 376	16 496	16 617	16 737	16 858	16 978	17 099	17 219	17 339	121
·31	2·1 17 460	17 580	17 700	17 821	17 941	18 061	18 181	18 302	18 422	18 542	120
·32	2·1 18 662	18 782	18 903	19 023	19 143	19 263	19 383	19 503	19 623	19 743	120
·33	2·1 19 863	19 983	20 104	20 224	20 344	20 464	20 583	20 703	20 823	20 943	120
·34	2·1 21 063	21 183	21 303	21 423	21 543	21 663	21 782	21 902	22 022	22 142	120
8·35	2·1 22 262	22 381	22 501	22 621	22 740	22 860	22 980	23 100	23 219	23 339	119
·36	2·1 23 458	23 578	23 698	23 817	23 937	24 056	24 176	24 295	24 415	24 534	120
·37	2·1 24 654	24 773	24 893	25 012	25 132	25 251	25 370	25 490	25 609	25 729	119
·38	2·1 25 848	25 967	26 087	26 206	26 325	26 444	26 564	26 683	26 802	26 921	120
·39	2·1 27 041	27 160	27 279	27 398	27 517	27 636	27 755	27 874	27 994	28 113	119
8·40	2·1 28 232	28 351	28 470	28 589	28 708	28 827	28 946	29 065	29 184	29 303	118
·41	2·1 29 421	29 540	29 659	29 778	29 897	30 016	30 135	30 253	30 372	30 491	119
·42	2·1 30 610	30 729	30 847	30 966	31 085	31 203	31 322	31 441	31 559	31 678	119
·43	2·1 31 797	31 915	32 034	32 153	32 271	32 390	32 508	32 627	32 745	32 864	118
·44	2·1 32 982	33 101	33 219	33 338	33 456	33 575	33 693	33 811	33 930	34 048	118
8·45	2·1 34 166	34 285	34 403	34 521	34 640	34 758	34 876	34 995	35 113	35 231	118
·46	2·1 35 349	35 467	35 586	35 704	35 822	35 940	36 058	36 176	36 294	36 412	119
·47	2·1 36 531	36 649	36 767	36 885	37 003	37 121	37 239	37 357	37 475	37 593	117
·48	2·1 37 710	37 828	37 946	38 064	38 182	38 300	38 418	38 536	38 653	38 771	118
·49	2·1 38 889	39 007	39 125	39 242	39 360	39 478	39 595	39 713	39 831	39 949	117
8·50	2·1 40 066	40 184	40 301	40 419	40 537	40 654	40 772	40 889	41 007	41 124	118

n	1	2	3	4	5	6	7
$\ln 10^n$	2·302 585	4·605 170	6·907 755	9·210 340	11·512 925	13·815 511	16·118 096

	0	1	2	3	4	5	6	7	8	9	Δ
8·50	2·1 40 066	40 184	40 301	40 419	40 537	40 654	40 772	40 889	41 007	41 124	118
·51	2·1 41 242	41 359	41 477	41 594	41 712	41 829	41 947	42 064	42 182	42 299	117
·52	2·1 42 416	42 534	42 651	42 768	42 886	43 003	43 120	43 238	43 355	43 472	117
·53	2·1 43 589	43 707	43 824	43 941	44 058	44 175	44 293	44 410	44 527	44 644	117
·54	2·1 44 761	44 878	44 995	45 112	45 229	45 346	45 463	45 580	45 697	45 814	117
8·55	2·1 45 931	46 048	46 165	46 282	46 399	46 516	46 633	46 750	46 867	46 983	117
·56	2·1 47 100	47 217	47 334	47 451	47 567	47 684	47 801	47 918	48 034	48 151	117
·57	2·1 48 268	48 384	48 501	48 618	48 734	48 851	48 968	49 084	49 201	49 317	117
·58	2·1 49 434	49 550	49 667	49 784	49 900	50 016	50 133	50 249	50 366	50 482	117
·59	2·1 50 599	50 715	50 832	50 948	51 064	51 181	51 297	51 413	51 530	51 646	116
8·60	2·1 51 762	51 878	51 995	52 111	52 227	52 343	52 460	52 576	52 692	52 808	116
·61	2·1 52 924	53 040	53 157	53 273	53 389	53 505	53 621	53 737	53 853	53 969	116
·62	2·1 54 085	54 201	54 317	54 433	54 549	54 665	54 781	54 897	55 013	55 129	116
·63	2·1 55 245	55 360	55 476	55 592	55 708	55 824	55 940	56 055	56 171	56 287	116
·64	2·1 56 403	56 518	56 634	56 750	56 865	56 981	57 097	57 212	57 328	57 444	115
8·65	2·1 57 559	57 675	57 791	57 906	58 022	58 137	58 253	58 368	58 484	58 599	116
·66	2·1 58 715	58 830	58 946	59 061	59 177	59 292	59 407	59 523	59 638	59 753	116
·67	2·1 59 869	59 984	60 099	60 215	60 330	60 445	60 561	60 676	60 791	60 906	116
·68	2·1 61 022	61 137	61 252	61 367	61 482	61 597	61 713	61 828	61 943	62 058	115
·69	2·1 62 173	62 288	62 403	62 518	62 633	62 748	62 863	62 978	63 093	63 208	115
8·70	2·1 63 323	63 438	63 553	63 668	63 783	63 898	64 012	64 127	64 242	64 357	115
·71	2·1 64 472	64 587	64 701	64 816	64 931	65 046	65 160	65 275	65 390	65 505	114
·72	2·1 65 619	65 734	65 849	65 963	66 078	66 192	66 307	66 422	66 536	66 651	114
·73	2·1 66 765	66 880	66 994	67 109	67 223	67 338	67 452	67 567	67 681	67 796	114
·74	2·1 67 910	68 025	68 139	68 253	68 368	68 482	68 596	68 711	68 825	68 939	115
8·75	2·1 69 054	69 168	69 282	69 396	69 511	69 625	69 739	69 853	69 968	70 082	114
·76	2·1 70 196	70 310	70 424	70 538	70 652	70 767	70 881	70 995	71 109	71 223	114
·77	2·1 71 337	71 451	71 565	71 679	71 793	71 907	72 021	72 135	72 249	72 363	113
·78	2·1 72 476	72 590	72 704	72 818	72 932	73 046	73 160	73 273	73 387	73 501	114
·79	2·1 73 615	73 728	73 842	73 956	74 070	74 183	74 297	74 411	74 524	74 638	114
8·80	2·1 74 752	74 865	74 979	75 093	75 206	75 320	75 433	75 547	75 660	75 774	113
·81	2·1 75 887	76 001	76 114	76 228	76 341	76 455	76 568	76 682	76 795	76 908	114
·82	2·1 77 022	77 135	77 249	77 362	77 475	77 589	77 702	77 815	77 928	78 042	113
·83	2·1 78 155	78 268	78 381	78 495	78 608	78 721	78 834	78 947	79 061	79 174	113
·84	2·1 79 287	79 400	79 513	79 626	79 739	79 852	79 965	80 078	80 191	80 304	113
8·85	2·1 80 417	80 530	80 643	80 756	80 869	80 982	81 095	81 208	81 321	81 434	113
·86	2·1 81 547	81 660	81 772	81 885	81 998	82 111	82 224	82 337	82 449	82 562	113
·87	2·1 82 675	82 788	82 900	83 013	83 126	83 238	83 351	83 464	83 576	83 689	113
·88	2·1 83 802	83 914	84 027	84 139	84 252	84 364	84 477	84 590	84 702	84 815	112
·89	2·1 84 927	85 040	85 152	85 264	85 377	85 489	85 602	85 714	85 827	85 939	112
8·90	2·1 86 051	86 164	86 276	86 388	86 501	86 613	86 725	86 837	86 950	87 062	112
·91	2·1 87 174	87 286	87 399	87 511	87 623	87 735	87 847	87 960	88 072	88 184	112
·92	2·1 88 296	88 408	88 520	88 632	88 744	88 856	88 968	89 080	89 192	89 304	112
·93	2·1 89 416	89 528	89 640	89 752	89 864	89 976	90 088	90 200	90 312	90 424	112
·94	2·1 90 536	90 647	90 759	90 871	90 983	91 095	91 207	91 318	91 430	91 542	112
8·95	2·1 91 654	91 765	91 877	91 989	92 100	92 212	92 324	92 435	92 547	92 659	111
·96	2·1 92 770	92 882	92 993	93 105	93 217	93 328	93 440	93 551	93 663	93 774	112
·97	2·1 93 886	93 997	94 109	94 220	94 332	94 443	94 554	94 666	94 777	94 889	111
·98	2·1 95 000	95 111	95 223	95 334	95 445	95 557	95 668	95 779	95 890	96 002	111
·99	2·1 96 113	96 224	96 335	96 446	96 558	96 669	96 780	96 891	97 002	97 113	112
9·00	2·1 97 225	97 336	97 447	97 558	97 669	97 780	97 891	98 002	98 113	98 224	111

n	7	8	9	10	11	12
$\ln 10^n$	16·118 096	18·420 681	20·723 266	23·025 851	25·328 436	27·631 021

	0	1	2	3	4	5	6	7	8	9	Δ
9·00	2·1 97 225	97 336	97 447	97 558	97 669	97 780	97 891	98 002	98 113	98 224	111
·01	2·1 98 335	98 446	98 557	98 668	98 779	98 890	99 001	99 112	99 223	99 333	111
·02	2·1 99 444	99 555	99 666	99 777	99 888	99 999	**00 109**	**00 220**	**00 331**	**00 442**	110
·03	2·2 00 552	00 663	00 774	00 885	00 995	01 106	01 217	01 327	01 438	01 549	110
·04	2·2 01 659	01 770	01 880	01 991	02 102	02 212	02 323	02 433	02 544	02 654	111
9·05	2·2 02 765	02 875	02 986	03 096	03 207	03 317	03 428	03 538	03 648	03 759	110
·06	2·2 03 869	03 979	04 090	04 200	04 311	04 421	04 531	04 641	04 752	04 862	110
·07	2·2 04 972	05 083	05 193	05 303	05 413	05 523	05 634	05 744	05 854	05 964	110
·08	2·2 06 074	06 184	06 294	06 405	06 515	06 625	06 735	06 845	06 955	07 065	110
·09	2·2 07 175	07 285	07 395	07 505	07 615	07 725	07 835	07 945	08 055	08 165	109
9·10	2·2 08 274	08 384	08 494	08 604	08 714	08 824	08 934	09 043	09 153	09 263	110
·11	2·2 09 373	09 482	09 592	09 702	09 812	09 921	10 031	10 141	10 250	10 360	110
·12	2·2 10 470	10 579	10 689	10 799	10 908	11 018	11 127	11 237	11 347	11 456	110
·13	2·2 11 566	11 675	11 785	11 894	12 004	12 113	12 223	12 332	12 442	12 551	109
·14	2·2 12 660	12 770	12 879	12 989	13 098	13 207	13 317	13 426	13 535	13 645	109
9·15	2·2 13 754	13 863	13 972	14 082	14 191	14 300	14 409	14 519	14 628	14 737	109
·16	2·2 14 846	14 955	15 064	15 174	15 283	15 392	15 501	15 610	15 719	15 828	109
·17	2·2 15 937	16 046	16 155	16 264	16 373	16 482	16 591	16 700	16 809	16 918	109
·18	2·2 17 027	17 136	17 245	17 354	17 463	17 572	17 681	17 789	17 898	18 007	109
·19	2·2 18 116	18 225	18 334	18 442	18 551	18 660	18 769	18 877	18 986	19 095	108
9·20	2·2 19 203	19 312	19 421	19 530	19 638	19 747	19 855	19 964	20 073	20 181	109
·21	2·2 20 290	20 398	20 507	20 616	20 724	20 833	20 941	21 050	21 158	21 267	108
·22	2·2 21 375	21 483	21 592	21 700	21 809	21 917	22 026	22 134	22 242	22 351	108
·23	2·2 22 459	22 567	22 676	22 784	22 892	23 001	23 109	23 217	23 325	23 434	108
·24	2·2 23 542	23 650	23 758	23 867	23 975	24 083	24 191	24 299	24 407	24 515	109
9·25	2·2 24 624	24 732	24 840	24 948	25 056	25 164	25 272	25 380	25 488	25 596	108
·26	2·2 25 704	25 812	25 920	26 028	26 136	26 244	26 352	26 460	26 568	26 675	108
·27	2·2 26 783	26 891	26 999	27 107	27 215	27 323	27 430	27 538	27 646	27 754	108
·28	2·2 27 862	27 969	28 077	28 185	28 292	28 400	28 508	28 616	28 723	28 831	108
·29	2·2 28 939	29 046	29 154	29 261	29 369	29 477	29 584	29 692	29 799	29 907	107
9·30	2·2 30 014	30 122	30 229	30 337	30 444	30 552	30 659	30 767	30 874	30 982	107
·31	2·2 31 089	31 196	31 304	31 411	31 519	31 626	31 733	31 841	31 948	32 055	108
·32	2·2 32 163	32 270	32 377	32 484	32 592	32 699	32 806	32 913	33 021	33 128	107
·33	2·2 33 235	33 342	33 449	33 557	33 664	33 771	33 878	33 985	34 092	34 199	107
·34	2·2 34 306	34 413	34 520	34 627	34 734	34 841	34 948	35 055	35 162	35 269	107
9·35	2·2 35 376	35 483	35 590	35 697	35 804	35 911	36 018	36 125	36 232	36 338	107
·36	2·2 36 445	36 552	36 659	36 766	36 873	36 979	37 086	37 193	37 300	37 406	107
·37	2·2 37 513	37 620	37 727	37 833	37 940	38 047	38 153	38 260	38 367	38 473	107
·38	2·2 38 580	38 686	38 793	38 900	39 006	39 113	39 219	39 326	39 432	39 539	106
·39	2·2 39 645	39 752	39 858	39 965	40 071	40 178	40 284	40 390	40 497	40 603	107
9·40	2·2 40 710	40 816	40 922	41 029	41 135	41 241	41 348	41 454	41 560	41 667	106
·41	2·2 41 773	41 879	41 985	42 092	42 198	42 304	42 410	42 517	42 623	42 729	106
·42	2·2 42 835	42 941	43 047	43 154	43 260	43 366	43 472	43 578	43 684	43 790	106
·43	2·2 43 896	44 002	44 108	44 214	44 320	44 426	44 532	44 638	44 744	44 850	106
·44	2·2 44 956	45 062	45 168	45 274	45 380	45 486	45 591	45 697	45 803	45 909	106
9·45	2·2 46 015	46 121	46 226	46 332	46 438	46 544	46 649	46 755	46 861	46 967	105
·46	2·2 47 072	47 178	47 284	47 389	47 495	47 601	47 706	47 812	47 918	48 023	106
·47	2·2 48 129	48 234	48 340	48 446	48 551	48 657	48 762	48 868	48 973	49 079	105
·48	2·2 49 184	49 290	49 395	49 501	49 606	49 712	49 817	49 922	50 028	50 133	106
·49	2·2 50 239	50 344	50 449	50 555	50 660	50 765	50 871	50 976	51 081	51 187	105
9·50	2·2 51 292	51 397	51 502	51 608	51 713	51 818	51 923	52 028	52 134	52 239	105

n	1	2	3	4	5	6	7
$\ln 10^n$	2·302 585	4·605 170	6·907 755	9·210 340	11·512 925	13·815 511	16·118 096

TABLE VI—NATURAL LOGARITHMS

		0	1	2	3	4	5	6	7	8	9	Δ
9·50	2·2	51 292	51 397	51 502	51 608	51 713	51 818	51 923	52 028	52 134	52 239	105
·51	2·2	52 344	52 449	52 554	52 659	52 764	52 870	52 975	53 080	53 185	53 290	105
·52	2·2	53 395	53 500	53 605	53 710	53 815	53 920	54 025	54 130	54 235	54 340	105
·53	2·2	54 445	54 550	54 655	54 759	54 864	54 969	55 074	55 179	55 284	55 389	104
·54	2·2	55 493	55 598	55 703	55 808	55 913	56 017	56 122	56 227	56 332	56 436	105
9·55	2·2	56 541	56 646	56 751	56 855	56 960	57 065	57 169	57 274	57 379	57 483	105
·56	2·2	57 588	57 692	57 797	57 901	58 006	58 111	58 215	58 320	58 424	58 529	104
·57	2·2	58 633	58 738	58 842	58 947	59 051	59 156	59 260	59 364	59 469	59 573	105
·58	2·2	59 678	59 782	59 886	59 991	60 095	60 199	60 304	60 408	60 512	60 617	104
·59	2·2	60 721	60 825	60 929	61 034	61 138	61 242	61 346	61 451	61 555	61 659	104
9·60	2·2	61 763	61 867	61 971	62 076	62 180	62 284	62 388	62 492	62 596	62 700	104
·61	2·2	62 804	62 908	63 012	63 116	63 220	63 324	63 428	63 532	63 636	63 740	104
·62	2·2	63 844	63 948	64 052	64 156	64 260	64 364	64 468	64 572	64 676	64 779	104
·63	2·2	64 883	64 987	65 091	65 195	65 299	65 402	65 506	65 610	65 714	65 817	104
·64	2·2	65 921	66 025	66 129	66 232	66 336	66 440	66 543	66 647	66 751	66 854	104
9·65	2·2	66 958	67 062	67 165	67 269	67 372	67 476	67 579	67 683	67 787	67 890	104
·66	2·2	67 994	68 097	68 201	68 304	68 408	68 511	68 615	68 718	68 821	68 925	103
·67	2·2	69 028	69 132	69 235	69 338	69 442	69 545	69 649	69 752	69 855	69 959	103
·68	2·2	70 062	70 165	70 268	70 372	70 475	70 578	70 682	70 785	70 888	70 991	103
·69	2·2	71 094	71 198	71 301	71 404	71 507	71 610	71 713	71 817	71 920	72 023	103
9·70	2·2	72 126	72 229	72 332	72 435	72 538	72 641	72 744	72 847	72 950	73 053	103
·71	2·2	73 156	73 259	73 362	73 465	73 568	73 671	73 774	73 877	73 980	74 083	103
·72	2·2	74 186	74 288	74 391	74 494	74 597	74 700	74 803	74 906	75 008	75 111	103
·73	2·2	75 214	75 317	75 419	75 522	75 625	75 728	75 830	75 933	76 036	76 138	103
·74	2·2	76 241	76 344	76 446	76 549	76 652	76 754	76 857	76 960	77 062	77 165	102
9·75	2·2	77 267	77 370	77 472	77 575	77 677	77 780	77 882	77 985	78 087	78 190	102
·76	2·2	78 292	78 395	78 497	78 600	78 702	78 805	78 907	79 009	79 112	79 214	102
·77	2·2	79 316	79 419	79 521	79 623	79 726	79 828	79 930	80 033	80 135	80 237	102
·78	2·2	80 339	80 442	80 544	80 646	80 748	80 851	80 953	81 055	81 157	81 259	102
·79	2·2	81 361	81 464	81 566	81 668	81 770	81 872	81 974	82 076	82 178	82 280	102
9·80	2·2	82 382	82 484	82 586	82 688	82 790	82 892	82 994	83 096	83 198	83 300	102
·81	2·2	83 402	83 504	83 606	83 708	83 810	83 912	84 014	84 116	84 217	84 319	102
·82	2·2	84 421	84 523	84 625	84 727	84 828	84 930	85 032	85 134	85 235	85 337	102
·83	2·2	85 439	85 541	85 642	85 744	85 846	85 947	86 049	86 151	86 252	86 354	102
·84	2·2	86 456	86 557	86 659	86 761	86 862	86 964	87 065	87 167	87 268	87 370	101
9·85	2·2	87 471	87 573	87 674	87 776	87 877	87 979	88 080	88 182	88 283	88 385	101
·86	2·2	88 486	88 588	88 689	88 790	88 892	88 993	89 095	89 196	89 297	89 399	101
·87	2·2	89 500	89 601	89 702	89 804	89 905	90 006	90 108	90 209	90 310	90 411	102
·88	2·2	90 513	90 614	90 715	90 816	90 917	91 018	91 120	91 221	91 322	91 423	101
·89	2·2	91 524	91 625	91 726	91 827	91 929	92 030	92 131	92 232	92 333	92 434	101
9·90	2·2	92 535	92 636	92 737	92 838	92 939	93 040	93 141	93 242	93 343	93 443	101
·91	2·2	93 544	93 645	93 746	93 847	93 948	94 049	94 150	94 250	94 351	94 452	101
·92	2·2	94 553	94 654	94 755	94 855	94 956	95 057	95 158	95 258	95 359	95 460	100
·93	2·2	95 560	95 661	95 762	95 863	95 963	96 064	96 165	96 265	96 366	96 466	101
·94	2·2	96 567	96 668	96 768	96 869	96 969	97 070	97 170	97 271	97 372	97 472	101
9·95	2·2	97 573	97 673	97 774	97 874	97 974	98 075	98 175	98 276	98 376	98 477	100
·96	2·2	98 577	98 677	98 778	98 878	98 979	99 079	99 179	99 280	99 380	99 480	101
·97	2·2	99 581	99 681	99 781	99 881	99 982	**00 082**	**00 182**	**00 282**	**00 383**	**00 483**	100
·98	2·3	00 583	00 683	00 783	00 884	00 984	01 084	01 184	01 284	01 384	01 484	101
·99	2·3	01 585	01 685	01 785	01 885	01 985	02 085	02 185	02 285	02 385	02 485	100
10·00	2·3	02 585	02 685	02 785	02 885	02 985	03 085	03 185	03 285	03 385	03 485	100

n	7	8	9	10	11	12
ln 10^n	16·118 096	18·420 681	20·723 266	23·025 851	25·328 436	27·631 021

TABLE VII—INVERSE FUNCTIONS

x	$\sin^{-1} x$	$\cos^{-1} x$	$\tan^{-1} x$	$\cot^{-1} x$	$\sinh^{-1} x$	$\tanh^{-1} x$	$\text{sech}^{-1} x$	$\text{cosech}^{-1} x$
0·00	0·000 000	1·570 796	0·000 000	1·570 796	0·000 000	0·000 000	∞	∞
·01	·010 000	·560 796	·010 000	·560 797	·010 000	·010 000	5·298 292	5·298 342
·02	·020 001	·550 795	·019 997	·550 799	·019 999	·020 003	4·605 070	4·605 270
·03	·030 005	·540 792	·029 991	·540 805	·029 996	·030 009	4·199 480	4·199 930
·04	·040 011	·530 786	·039 979	·530 818	·039 989	·040 021	3·911 623	3·912 423
0·05	0·050 021	1·520 775	0·049 958	1·520 838	0·049 979	0·050 042	3·688 254	3·689 504
·06	·060 036	·510 760	·059 928	·510 868	·059 964	·060 072	3·505 657	3·507 457
·07	·070 057	·500 739	·069 886	·500 910	·069 943	·070 115	3·351 180	3·353 630
·08	·080 086	·490 711	·079 830	·490 966	·079 915	·080 171	3·217 272	3·220 472
·09	·090 122	·480 674	·089 758	·481 038	·089 879	·090 244	3·099 062	3·103 112
0·10	0·100 167	1·470 629	0·099 669	1·471 128	0·099 834	0·100 335	2·993 223	2·998 223
·11	·110 223	·460 573	·109 560	·461 237	·109 779	·110 447	2·897 383	2·903 433
·12	·120 290	·450 506	·119 429	·451 367	·119 714	·120 581	2·809 791	2·816 991
·13	·130 369	·440 427	·129 275	·441 521	·129 637	·130 740	2·729 116	2·737 566
·14	·140 461	·430 335	·139 096	·431 700	·139 547	·140 926	2·654 324	2·664 124
0·15	0·150 568	1·420 228	0·148 890	1·421 906	0·149 443	0·151 140	2·584 594	2·595 845
·16	·160 691	·410 106	·158 655	·412 141	·159 325	·161 387	2·519 266	2·532 068
·17	·170 830	·399 967	·168 390	·402 406	·169 192	·171 667	2·457 799	2·472 252
·18	·180 986	·389 810	·178 093	·392 703	·179 042	·181 983	2·399 745	2·415 949
·19	·191 162	·379 634	·187 762	·383 034	·188 875	·192 337	2·344 729	2·362 784
0·20	0·201 358	1·369 438	0·197 396	1·373 401	0·198 690	0·202 733	2·292 432	2·312 438
·21	·211 575	·359 221	·206 992	·363 804	·208 486	·213 171	2·242 583	2·264 642
·22	·221 814	·348 982	·216 550	·354 246	·218 263	·223 656	2·194 949	2·219 161
·23	·232 078	·338 719	·226 068	·344 728	·228 019	·234 189	2·149 328	2·175 793
·24	·242 366	·328 430	·235 545	·335 251	·237 754	·244 774	2·105 542	2·134 362
0·25	0·252 680	1·318 116	0·244 979	1·325 818	0·247 466	0·255 413	2·063 437	2·094 713
·26	·263 022	·307 774	·254 368	·316 428	·257 156	·266 108	2·022 876	2·056 708
·27	·273 393	·297 403	·263 712	·307 084	·266 823	·276 864	1·983 736	2·020 227
·28	·283 794	·287 002	·273 009	·297 788	·276 465	·287 682	1·945 910	1·985 160
·29	·294 227	·276 569	·282 257	·288 539	·286 082	·298 566	1·909 301	1·951 413
0·30	0·304 693	1·266 104	0·291 457	1·279 340	0·295 673	0·309 520	1·873 820	1·918 896
·31	·315 193	·255 603	·300 606	·270 191	·305 238	·320 545	1·839 390	1·887 533
·32	·325 729	·245 067	·309 703	·261 093	·314 776	·331 647	1·805 938	1·857 251
·33	·336 304	·234 493	·318 748	·252 049	·324 286	·342 828	1·773 401	1·827 986
·34	·346 917	·223 879	·327 739	·243 058	·333 768	·354 093	1·741 717	1·799 679
0·35	0·357 571	1·213 225	0·336 675	1·234 122	0·343 222	0·365 444	1·710 833	1·772 276
·36	·368 268	·202 528	·345 556	·225 241	·352 645	·376 886	1·680 700	1·745 728
·37	·379 009	·191 787	·354 380	·216 416	·362 039	·388 423	1·651 270	1·719 990
·38	·389 796	·181 000	·363 147	·207 649	·371 402	·400 060	1·622 503	1·695 020
·39	·400 632	·170 165	·371 856	·198 940	·380 735	·411 800	1·594 358	1·670 779
0·40	0·411 517	1·159 279	0·380 506	1·190 290	0·390 035	0·423 649	1·566 799	1·647 231
·41	·422 454	·148 342	·389 097	·181 699	·399 304	·435 611	1·539 793	1·624 344
·42	·433 445	·137 351	·397 628	·173 168	·408 540	·447 692	1·513 307	1·602 087
·43	·444 493	·126 304	·406 098	·164 698	·417 743	·459 897	1·487 312	1·580 431
·44	·455 599	·115 198	·414 507	·156 289	·426 913	·472 231	1·461 780	1·559 350
0·45	0·466 765	1·104 031	0·422 854	1·147 942	0·436 050	0·484 700	1·436 686	1·538 818
·46	·477 995	·092 801	·431 139	·139 658	·445 152	·497 311	1·412 004	1·518 812
·47	·489 291	·081 506	·439 361	·131 435	·454 219	·510 070	1·387 712	1·499 311
·48	·500 655	·070 142	·447 520	·123 276	·463 252	·522 984	1·363 787	1·480 294
·49	·512 090	·058 707	·455 616	·115 181	·472 250	·536 060	1·340 209	1·461 741
0·50	0·523 599	1·047 198	0·463 648	1·107 149	0·481 212	0·549 306	1·316 958	1·443 635

TABLE VII—INVERSE FUNCTIONS

x	$\sin^{-1} x$	$\cos^{-1} x$	$\tan^{-1} x$	$\cot^{-1} x$	$\sinh^{-1} x$	$\tanh^{-1} x$	$\text{sech}^{-1} x$	$\text{cosech}^{-1} x$
0·50	0·523 599	1·047 198	0·463 648	1·107 149	0·481 212	0·549 306	1·316 958	1·443 635
·51	·535 185	·035 612	·471 616	·099 181	·490 138	·562 730	·294 015	·425 959
·52	·546 851	·023 945	·479 519	·091 277	·499 028	·576 340	·271 362	·408 696
·53	·558 601	·012 196	·487 359	·083 438	·507 882	·590 145	·248 981	·391 830
·54	·570 437	1·000 359	·495 133	·075 663	·516 700	·604 156	·226 856	·375 348
0·55	0·582 364	0·988 432	0·502 843	1·067 953	0·525 480	0·618 381	1·204 971	1·359 237
·56	·594 386	·976 411	·510 488	·060 308	·534 224	·632 833	·183 310	·343 482
·57	·606 506	·964 290	·518 069	·052 728	·542 931	·647 523	·161 859	·328 072
·58	·618 729	·952 068	·525 584	·045 213	·551 600	·662 463	·140 601	·312 995
·59	·631 059	·939 737	·533 034	·037 762	·560 231	·677 666	·119 524	·298 239
0·60	0·643 501	0·927 295	0·540 420	1·030 377	0·568 825	0·693 147	1·098 612	1·283 796
·61	·656 061	·914 736	·547 740	·023 056	·577 381	·708 921	·077 853	·269 653
·62	·668 743	·902 054	·554 996	·015 801	·585 899	·725 005	·057 231	·255 802
·63	·681 553	·889 243	·562 187	·008 610	·594 379	·741 416	·036 734	·242 234
·64	·694 498	·876 298	·569 313	1·001 483	·602 821	·758 174	1·016 348	·228 939
0·65	0·707 584	0·863 212	0·576 375	0·994 421	0·611 224	0·775 299	0·996 059	1·215 910
·66	·720 819	·849 978	·583 373	·987 423	·619 590	·792 814	·975 854	·203 138
·67	·734 209	·836 588	·590 307	·980 490	·627 916	·810 743	·955 719	·190 617
·68	·747 763	·823 034	·597 177	·973 620	·636 205	·829 114	·935 639	·178 337
·69	·761 489	·809 307	·603 983	·966 813	·644 455	·847 956	·915 600	·166 293
0·70	0·775 397	0·795 399	0·610 726	0·960 070	0·652 667	0·867 301	0·895 588	1·154 477
·71	·789 498	·781 298	·617 406	·953 390	·660 840	·887 184	·875 587	·142 884
·72	·803 802	·766 994	·624 023	·946 773	·668 974	·907 645	·855 581	·131 507
·73	·818 322	·752 474	·630 578	·940 219	·677 070	·928 727	·835 554	·120 340
·74	·833 070	·737 726	·637 070	·933 726	·685 128	·950 479	·815 489	·109 377
0·75	0·848 062	0·722 734	0·643 501	0·927 295	0·693 147	0·972 955	0·795 365	1·098 612
·76	·863 313	·707 483	·649 870	·920 926	·701 128	0·996 215	·775 166	·088 041
·77	·878 841	·691 955	·656 179	·914 618	·709 070	1·020 328	·754 868	·077 659
·78	·894 666	·676 131	·662 426	·908 370	·716 975	·045 371	·734 449	·067 460
·79	·910 809	·659 987	·668 614	·902 183	·724 841	·071 432	·713 884	·057 439
0·80	0·927 295	0·643 501	0·674 741	0·896 055	0·732 668	1·098 612	0·693 147	1·047 593
·81	·944 152	·626 644	·680 809	·889 987	·740 458	·127 029	·672 207	·037 916
·82	·961 411	·609 385	·686 818	·883 979	·748 210	·156 817	·651 031	·028 405
·83	·979 108	·591 689	·692 768	·878 028	·755 923	·188 136	·629 581	·019 055
·84	0·997 283	·573 513	·698 660	·872 137	·763 599	·221 174	·607 814	·009 862
0·85	1·015 985	0·554 811	0·704 494	0·866 302	0·771 237	1·256 153	0·585 682	1·000 822
·86	·035 270	·535 527	·710 271	·860 525	·778 838	·293 345	·563 127	0·991 933
·87	·055 202	·515 594	·715 991	·854 805	·786 401	·333 080	·540 084	·983 189
·88	·075 862	·494 934	·721 655	·849 141	·793 927	·375 768	·516 474	·974 588
·89	·097 345	·473 451	·727 263	·843 534	·801 415	·421 926	·492 200	·966 126
0·90	1·119 770	0·451 027	0·732 815	0·837 981	0·808 867	1·472 219	0·467 145	0·957 800
·91	·143 284	·427 512	·738 313	·832 484	·816 281	·527 524	·441 163	·949 608
·92	·168 080	·402 716	·743 756	·827 041	·823 659	·589 027	·414 065	·941 544
·93	·194 413	·376 383	·749 145	·821 652	·831 000	·658 390	·385 598	·933 608
·94	·222 630	·348 166	·754 480	·816 316	·838 305	·738 049	·355 421	·925 796
0·95	1·253 236	0·317 560	0·759 763	0·811 034	0·845 573	1·831 781	0·323 036	0·918 104
·96	·287 002	·283 794	·764 993	·805 803	·852 805	1·945 910	·287 682	·910 532
·97	·325 231	·245 566	·770 171	·800 625	·860 000	2·092 296	·248 071	·903 075
·98	·370 461	·200 335	·775 297	·795 499	·867 161	2·297 560	·201 688	·895 731
·99	·429 257	·141 539	·780 373	·790 423	·874 285	2·646 652	·142 014	·888 498
1·00	1·570 796	0·000 000	0·785 398	0·785 398	0·881 374	∞	0·000 000	0·881 374

TABLE VII—INVERSE FUNCTIONS

x	$\sin^{-1} x$	$\cos^{-1} x$	$\tanh^{-1} x$
0·900	1·119 770	0·451 027	1·472 219
·901	·122 069	·448 727	·477 508
·902	·124 380	·446 417	·482 847
·903	·126 702	·444 095	·488 238
·904	·129 035	·441 761	·493 682
0·905	1·131 380	0·439 417	1·499 180
·906	·133 736	·437 060	·504 734
·907	·136 105	·434 692	·510 344
·908	·138 485	·432 311	·516 011
·909	·140 878	·429 918	·521 738
0·910	1·143 284	0·427 512	1·527 524
·911	·145 702	·425 094	·533 373
·912	·148 134	·422 663	·539 284
·913	·150 578	·420 218	·545 260
·914	·153 036	·417 760	·551 302
0·915	1·155 508	0·415 288	1·557 411
·916	·157 994	·412 803	·563 589
·917	·160 493	·410 303	·569 838
·918	·163 008	·407 789	·576 160
·919	·165 537	·405 260	·582 555
0·920	1·168 080	0·402 716	1·589 027
·921	·170 640	·400 157	·595 577
·922	·173 215	·397 582	·602 206
·923	·175 805	·394 991	·608 918
·924	·178 412	·392 384	·615 714
0·925	1·181 036	0·389 761	1·622 597
·926	·183 676	·387 120	·629 568
·927	·186 333	·384 463	·636 630
·928	·189 008	·381 788	·643 786
·929	·191 701	·379 095	·651 039
0·930	1·194 413	0·376 383	1·658 390
·931	·197 143	·373 653	·665 843
·932	·199 892	·370 904	·673 402
·933	·202 661	·368 135	·681 068
·934	·205 450	·365 347	·688 845
0·935	1·208 259	0·362 537	1·696 738
·936	·211 089	·359 707	·704 748
·937	·213 941	·356 855	·712 880
·938	·216 815	·353 982	·721 139
·939	·219 711	·351 085	·729 527
0·940	1·222 630	0·348 166	1·738 049
·941	·225 573	·345 223	·746 711
·942	·228 541	·342 256	·755 515
·943	·231 533	·339 264	·764 469
·944	·234 551	·336 246	·773 576
0·945	1·237 595	0·333 202	1·782 842
·946	·240 666	·330 131	·792 274
·947	·243 765	·327 032	·801 877
·948	·246 892	·323 904	·811 657
·949	·250 049	·320 748	·821 623
0·950	1·253 236	0·317 560	1·831 781

x	$\sin^{-1} x$	$\cos^{-1} x$	$\tanh^{-1} x$	$\text{sech}^{-1} x$
0·950	1·253 236	0·317 560	1·831 78	0·323 036
·951	·256 454	·314 342	·842 14	·319 651
·952	·259 705	·311 092	·852 70	·316 234
·953	·262 988	·307 808	·863 49	·312 787
·954	·266 306	·304 490	·874 50	·309 307
0·955	1·269 660	0·301 137	1·885 74	0·305 794
·956	·273 050	·297 747	·897 23	·302 246
·957	·276 478	·294 319	·908 98	·298 662
·958	·279 945	·290 852	·921 00	·295 041
·959	·283 452	·287 344	·933 31	·291 382
0·960	1·287 002	0·283 794	1·945 91	0·287 682
·961	·290 596	·280 201	·958 82	·283 941
·962	·294 235	·276 562	·972 07	·280 156
·963	·297 921	·272 875	·985 66	·276 326
·964	·301 657	·269 140	1·999 61	·272 449
0·965	1·305 443	0·265 353	2·013 95	0·268 523
·966	·309 284	·261 513	·028 70	·264 545
·967	·313 180	·257 616	·043 88	·260 514
·968	·317 135	·253 662	·059 52	·256 427
·969	·321 151	·249 646	·075 65	·252 280
0·970	1·325 231	0·245 566	2·092 30	0·248 071
·971	·329 379	·241 418	·109 50	·243 798
·972	·333 597	·237 199	·127 30	·239 455
·973	·337 891	·232 905	·145 74	·235 040
·974	·342 264	·228 532	·164 86	·230 548
0·975	1·346 721	0·224 075	2·184 72	0·225 974
·976	·351 267	·219 530	·205 39	·221 314
·977	·355 907	·214 889	·226 92	·216 563
·978	·360 648	·210 148	·249 40	·211 712
·979	·365 497	·205 299	·272 91	·206 757
0·980	1·370 461	0·200 335	2·297 56	0·201 688
·981	·375 550	·195 246	·323 46	·196 498
·982	·380 774	·190 022	·350 74	·191 176
·983	·386 143	·184 653	·379 58	·185 711
·984	·391 672	·179 125	·410 14	·180 090
0·985	1·397 374	0·173 422	2·442 66	0·174 298
·986	·403 268	·167 528	·477 41	·168 317
·987	·409 376	·161 420	·514 72	·162 126
·988	·415 722	·155 075	·554 99	·155 700
·989	·422 336	·148 460	·598 75	·149 009
0·990	1·429 257	0·141 539	2·646 65	0·142 014
·991	·436 531	·134 265	·699 58	·134 670
·992	·444 221	·126 576	·758 73	·126 915
·993	·452 406	·118 391	·825 74	·118 668
·994	·461 197	·109 599	·903 07	·109 819
0·995	1·470 755	0·100 042	2·994 48	0·100 209
·996	·481 324	·089 473	3·106 30	·089 592
·997	·493 317	·077 479	3·250 39	·077 557
·998	·507 540	·063 256	3·453 38	·063 298
·999	·526 071	·044 725	3·800 20	·044 740
1·000	1·570 796	0·000 000	∞	0·000 000

TABLE VII—INVERSE FUNCTIONS

x	$\sec^{-1} x$	$\operatorname{cosec}^{-1} x$	$\cosh^{-1} x$	$\coth^{-1} x$
1·000	0·000 000	1·570 796	0·000 000	∞
·001	·044 703	·526 094	·044 718	3·800 70
·002	·063 193	·507 603	·063 235	3·454 38
·003	·077 363	·493 433	·077 440	3·251 89
·004	·089 294	·481 502	·089 413	3·108 30
1·005	0·099 792	1·471 004	0·099 958	2·996 98
·006	·109 272	·461 525	·109 490	·906 07
·007	·117 978	·452 818	·118 253	·829 24
·008	·126 072	·444 725	·126 407	·762 73
·009	·133 664	·437 132	·134 064	·704 08
1·010	0·140 836	1·429 960	0·141 304	2·651 65
·011	·147 649	·423 147	·148 188	·604 25
·012	·154 151	·416 646	·154 765	·560 99
·013	·160 379	·410 417	·161 071	·521 22
·014	·166 365	·404 432	·167 137	·484 41
1·015	0·172 133	1·398 663	0·172 989	2·450 16
·016	·177 705	·393 091	·178 648	·418 14
·017	·183 099	·387 697	·184 131	·388 08
·018	·188 330	·382 466	·189 453	·359 75
·019	·193 411	·377 385	·194 629	·332 96
1·020	0·198 355	1·372 442	0·199 668	2·307 56
·021	·203 170	·367 627	·204 582	·283 41
·022	·207 866	·362 931	·209 379	·260 40
·023	·212 451	·358 346	·214 067	·238 42
·024	·216 931	·353 865	·218 653	·217 39
1·025	0·221 314	1·349 482	0·223 144	2·197 22
·026	·225 605	·345 191	·227 544	·177 86
·027	·229 809	·340 987	·231 859	·159 24
·028	·233 931	·336 865	·236 094	·141 30
·029	·237 975	·332 821	·240 254	·124 00
1·030	0·241 945	1·328 851	0·244 341	2·107 30
·031	·245 845	·324 952	·248 359	·091 15
·032	·249 677	·321 119	·252 312	·075 52
·033	·253 446	·317 351	·256 203	·060 38
·034	·257 153	·313 643	·260 035	·045 70
1·035	0·260 802	1·309 995	0·263 809	2·031 45
·036	·264 394	·306 402	·267 530	2·017 61
·037	·267 933	·302 863	·271 198	2·004 16
·038	·271 420	·299 376	·274 815	1·991 07
·039	·274 857	·295 939	·278 385	1·978 33
1·040	0·278 247	1·292 550	0·281 908	1·965 91
·041	·281 590	·289 206	·285 387	·953 81
·042	·284 889	·285 907	·288 823	·942 01
·043	·288 145	·282 652	·292 217	·930 49
·044	·291 359	·279 437	·295 571	·919 24
1·045	0·294 533	1·276 263	0·298 886	1·908 25
·046	·297 668	·273 128	·302 164	·897 50
·047	·300 766	·270 030	·305 406	·886 99
·048	·303 827	·266 969	·308 613	·876 71
·049	·306 853	·263 943	·311 785	·866 64
1·050	0·309 845	1·260 952	0·314 925	1·856 79

x	$\sec^{-1} x$	$\operatorname{cosec}^{-1} x$	$\cosh^{-1} x$
1·050	0·309 845	1·260 952	0·314 925
·051	·312 803	·257 993	·318 032
·052	·315 729	·255 068	·321 109
·053	·318 623	·252 173	·324 155
·054	·321 487	·249 309	·327 172
1·055	0·324 321	1·246 475	0·330 161
·056	·327 126	·243 670	·333 122
·057	·329 903	·240 893	·336 055
·058	·332 652	·238 144	·338 963
·059	·335 375	·235 422	·341 844
1·060	0·338 071	1·232 725	0·344 701
·061	·340 742	·230 055	·347 533
·062	·343 387	·227 409	·350 342
·063	·346 009	·224 787	·353 127
·064	·348 607	·222 190	·355 890
1·065	0·351 181	1·219 615	0·358 630
·066	·353 733	·217 064	·361 349
·067	·356 262	·214 534	·364 046
·068	·358 770	·212 027	·366 723
·069	·361 256	·209 541	·369 380
1·070	0·363 721	1·207 075	0·372 017
·071	·366 166	·204 630	·374 634
·072	·368 591	·202 205	·377 233
·073	·370 997	·199 800	·379 812
·074	·373 383	·197 413	·382 374
1·075	0·375 750	1·195 046	0·384 918
·076	·378 099	·192 697	·387 444
·077	·380 430	·190 366	·389 953
·078	·382 743	·188 054	·392 445
·079	·385 038	·185 758	·394 921
1·080	0·387 317	1·183 480	0·397 380
·081	·389 578	·181 218	·399 824
·082	·391 823	·178 973	·402 252
·083	·394 052	·176 745	·404 664
·084	·396 265	·174 532	·407 062
1·085	0·398 462	1·172 335	0·409 445
·086	·400 643	·170 153	·411 813
·087	·402 810	·167 987	·414 167
·088	·404 961	·165 835	·416 506
·089	·407 098	·163 698	·418 832
1·090	0·409 221	1·161 576	0·421 145
·091	·411 329	·159 467	·423 444
·092	·413 423	·157 373	·425 730
·093	·415 504	·155 293	·428 003
·094	·417 571	·153 226	·430 263
1·095	0·419 624	1·151 172	0·432 511
·096	·421 665	·149 131	·434 746
·097	·423 693	·147 104	·436 970
·098	·425 707	·145 089	·439 181
·099	·427 710	·143 087	·441 380
1·100	0·429 700	1·141 097	0·443 568

TABLE VII—INVERSE FUNCTIONS

x	$\tan^{-1} x$	$\cot^{-1} x$	$\sec^{-1} x$	$\operatorname{cosec}^{-1} x$	$\sinh^{-1} x$	$\cosh^{-1} x$	$\coth^{-1} x$	$\operatorname{cosech}^{-1} x$
1·00	0·785 398	0·785 398	0·000 000	1·570 796	0·881 374	0·000 000	∞	0·881 374
·01	·790 373	·780 423	·140 836	·429 960	·888 427	·141 304	2·651 652	·874 355
·02	·795 299	·775 497	·198 355	·372 442	·895 445	·199 668	2·307 560	·867 441
·03	·800 175	·770 621	·241 945	·328 851	·902 428	·244 341	2·107 297	·860 628
·04	·805 003	·765 793	·278 247	·292 550	·909 377	·281 908	1·965 913	·853 914
1·05	0·809 784	0·761 013	0·309 845	1·260 952	0·916 291	0·314 925	1·856 786	0·847 298
·06	·814 516	·756 280	·338 071	·232 725	·923 170	·344 701	·768 058	·840 777
·07	·819 202	·751 595	·363 721	·207 075	·930 015	·372 017	·693 404	·834 350
·08	·823 841	·746 956	·387 317	·183 480	·936 826	·397 380	·629 048	·828 014
·09	·828 434	·742 363	·409 221	·161 576	·943 603	·421 145	·572 555	·821 767
1·10	0·832 981	0·737 815	0·429 700	1·141 097	0·950 347	0·443 568	1·522 261	0·815 609
·11	·837 484	·733 313	·448 956	·121 841	·957 057	·464 845	·476 981	·809 536
·12	·841 942	·728 855	·467 146	·103 650	·963 734	·485 127	·435 840	·803 548
·13	·846 355	·724 441	·484 397	·086 399	·970 377	·504 534	·398 171	·797 643
·14	·850 726	·720 071	·500 812	·069 984	·976 988	·523 164	·363 459	·791 818
1·15	0·855 053	0·715 744	0·516 475	1·054 321	0·983 566	0·541 097	1·331 294	0·786 073
·16	·859 337	·711 459	·531 458	·039 338	·990 112	·558 402	·301 345	·780 406
·17	·863 579	·707 217	·545 822	·024 974	0·996 625	·575 136	·273 342	·774 815
·18	·867 780	·703 016	·559 619	1·011 178	1·003 106	·591 346	·247 062	·769 299
·19	·871 939	·698 857	·572 893	0·997 903	·009 556	·607 076	·222 316	·763 857
1·20	0·876 058	0·694 738	0·585 686	0·985 111	1·015 973	0·622 363	1·198 948	0·758 486
·21	·880 136	·690 660	·598 030	·972 766	·022 359	·637 237	·176 820	·753 186
·22	·884 175	·686 622	·609 958	·960 838	·028 714	·651 729	·155 817	·747 956
·23	·888 174	·682 623	·621 496	·949 300	·035 038	·665 864	·135 839	·742 794
·24	·892 134	·678 662	·632 670	·938 126	·041 331	·679 663	·116 796	·737 698
1·25	0·896 055	0·674 741	0·643 501	0·927 295	1·047 593	0·693 147	1·098 612	0·732 668
·26	·899 939	·670 857	·654 010	·916 787	·053 825	·706 335	·081 219	·727 703
·27	·903 785	·667 012	·664 214	·906 582	·060 026	·719 243	·064 557	·722 800
·28	·907 593	·663 203	·674 131	·896 666	·066 198	·731 887	·048 571	·717 960
·29	·911 365	·659 431	·683 775	·887 022	·072 339	·744 279	·033 213	·713 180
1·30	0·915 101	0·655 696	0·693 160	0·877 636	1·078 451	0·756 433	1·018 441	0·708 461
·31	·918 800	·651 996	·702 300	·868 497	·084 533	·768 360	1·004 215	·703 800
·32	·922 464	·648 332	·711 205	·859 591	·090 587	·780 071	0·990 501	·699 197
·33	·926 093	·644 703	·719 888	·850 909	·096 611	·791 575	·977 265	·694 650
·34	·929 688	·641 109	·728 358	·842 439	·102 606	·802 882	·964 480	·690 159
1·35	0·933 248	0·637 549	0·736 624	0·834 172	1·108 572	0·814 000	0·952 119	0·685 723
·36	·936 774	·634 023	·744 696	·826 101	·114 511	·824 937	·940 156	·681 341
·37	·940 266	·630 530	·752 581	·818 215	·120 420	·835 701	·928 571	·677 011
·38	·943 726	·627 071	·760 288	·810 508	·126 302	·846 297	·917 342	·672 734
·39	·947 152	·623 644	·767 823	·802 973	·132 156	·856 733	·906 451	·668 507
1·40	0·950 547	0·620 249	0·775 193	0·795 603	1·137 982	0·867 015	0·895 880	0·664 331
·41	·953 909	·616 887	·782 405	·788 391	·143 781	·877 147	·885 612	·660 203
·42	·957 240	·613 556	·789 465	·781 331	·149 552	·887 137	·875 634	·656 125
·43	·960 540	·610 256	·796 378	·774 419	·155 296	·896 987	·865 931	·652 094
·44	·963 809	·606 988	·803 149	·767 648	·161 014	·906 704	·856 489	·648 109
1·45	0·967 047	0·603 749	0·809 784	0·761 013	1·166 704	0·916 291	0·847 298	0·644 171
·46	·970 255	·600 541	·816 287	·754 510	·172 368	·925 753	·838 345	·640 278
·47	·973 434	·597 363	·822 663	·748 134	·178 006	·935 093	·829 620	·636 430
·48	·976 583	·594 214	·828 915	·741 881	·183 618	·944 316	·821 114	·632 625
·49	·979 703	·591 094	·835 050	·735 747	·189 203	·953 425	·812 816	·628 864
1·50	0·982 794	0·588 003	0·841 069	0·729 728	1·194 763	0·962 424	0·804 719	0·625 145

TABLE VII—INVERSE FUNCTIONS

x	$\tan^{-1} x$	$\cot^{-1} x$	$\sec^{-1} x$	$\text{cosec}^{-1} x$	$\sinh^{-1} x$	$\cosh^{-1} x$	$\coth^{-1} x$	$\text{cosech}^{-1} x$
1·50	0·982 794	0·588 003	0·841 069	0·729 728	1·194 763	0·962 424	0·804 719	0·625 145
·51	·985 857	·584 940	·846 976	·723 820	·200 297	·971 315	·796 814	·621 468
·52	·988 891	·581 905	·852 776	·718 020	·205 806	·980 102	·789 093	·617 832
·53	·991 898	·578 898	·858 472	·712 324	·211 290	·988 787	·781 549	·614 236
·54	·994 878	·575 919	·864 066	·706 730	·216 748	0·997 374	·774 175	·610 680
1·55	0·997 830	0·572 966	0·869 562	0·701 234	1·222 182	1·005 865	0·766 965	0·607 163
·56	1·000 756	·570 040	·874 962	·695 834	·227 591	·014 263	·759 913	·603 684
·57	·003 655	·567 141	·880 270	·690 526	·232 975	·022 570	·753 012	·600 244
·58	·006 528	·564 268	·885 488	·685 308	·238 335	·030 788	·746 258	·596 841
·59	·009 375	·561 421	·890 619	·680 177	·243 671	·038 920	·739 645	·593 474
1·60	1·012 197	0·558 599	0·895 665	0·675 132	1·248 983	1·046 968	0·733 169	0·590 144
·61	·014 993	·555 803	·900 628	·670 168	·254 271	·054 934	·726 823	·586 849
·62	·017 765	·553 031	·905 511	·665 286	·259 536	·062 819	·720 605	·583 589
·63	·020 512	·550 285	·910 315	·660 481	·264 777	·070 626	·714 510	·580 364
·64	·023 234	·547 562	·915 044	·655 753	·269 995	·078 357	·708 533	·577 173
1·65	1·025 932	0·544 864	0·919 698	0·651 099	1·275 189	1·086 013	0·702 671	0·574 015
·66	·028 607	·542 189	·924 280	·646 517	·280 361	·093 597	·696 921	·570 890
·67	·031 258	·539 538	·928 791	·642 005	·285 509	·101 108	·691 278	·567 798
·68	·033 886	·536 911	·933 234	·637 562	·290 635	·108 550	·685 740	·564 737
·69	·036 490	·534 306	·937 611	·633 186	·295 739	·115 924	·680 302	·561 708
1·70	1·039 072	0·531 724	0·941 921	0·628 875	1·300 820	1·123 231	0·674 963	0·558 711
·71	·041 632	·529 165	·946 169	·624 628	·305 880	·130 472	·669 719	·555 743
·72	·044 169	·526 627	·950 354	·620 443	·310 917	·137 650	·664 568	·552 806
·73	·046 684	·524 112	·954 478	·616 318	·315 932	·144 764	·659 506	·549 899
·74	·049 178	·521 618	·958 543	·612 253	·320 926	·151 818	·654 532	·547 021
1·75	1·051 650	0·519 146	0·962 551	0·608 246	1·325 898	1·158 810	0·649 641	0·544 171
·76	·054 101	·516 695	·966 502	·604 295	·330 848	·165 744	·644 834	·541 350
·77	·056 531	·514 265	·970 397	·600 399	·335 778	·172 620	·640 106	·538 557
·78	·058 940	·511 856	·974 239	·596 557	·340 686	·179 439	·635 456	·535 792
·79	·061 329	·509 467	·978 028	·592 768	·345 574	·186 202	·630 882	·533 054
1·80	1·063 698	0·507 099	0·981 765	0·589 031	1·350 441	1·192 911	0·626 381	0·530 343
·81	·066 046	·504 750	·985 452	·585 344	·355 287	·199 566	·621 953	·527 658
·82	·068 375	·502 421	·989 090	·581 706	·360 113	·206 168	·617 594	·524 999
·83	·070 684	·500 112	·992 679	·578 117	·364 918	·212 718	·613 303	·522 366
·84	·072 974	·497 822	·996 221	·574 575	·369 703	·219 218	·609 079	·519 758
1·85	1·075 245	0·495 552	0·999 717	0·571 079	1·374 468	1·225 667	0·604 919	0·517 175
·86	·077 496	·493 300	1·003 167	·567 629	·379 213	·232 068	·600 822	·514 617
·87	·079 729	·491 067	·006 573	·564 223	·383 939	·238 420	·596 787	·512 084
·88	·081 944	·488 852	·009 936	·560 860	·388 645	·244 725	·592 812	·509 574
·89	·084 140	·486 656	·013 256	·557 540	·393 331	·250 983	·588 895	·507 088
1·90	1·086 318	0·484 478	1·016 534	0·554 262	1·397 998	1·257 196	0·585 036	0·504 625
·91	·088 479	·482 318	·019 772	·551 024	·402 646	·263 363	·581 232	·502 185
·92	·090 621	·480 175	·022 969	·547 827	·407 275	·269 486	·577 483	·499 768
·93	·092 746	·478 050	·026 128	·544 669	·411 885	·275 566	·573 787	·497 373
·94	·094 854	·475 942	·029 247	·541 549	·416 476	·281 602	·570 142	·495 000
1·95	1·096 945	0·473 851	1·032 329	0·538 467	1·421 049	1·287 597	0·566 549	0·492 649
·96	·099 019	·471 778	·035 374	·535 422	·425 602	·293 550	·563 006	·490 320
·97	·101 076	·469 720	·038 383	·532 413	·430 138	·299 462	·559 511	·488 012
·98	·103 116	·467 680	·041 356	·529 440	·434 655	·305 333	·556 063	·485 725
·99	·105 141	·465 656	·044 294	·526 502	·439 154	·311 165	·552 662	·483 458
2·00	1·107 149	0·463 648	1·047 198	0·523 599	1·443 635	1·316 958	0·549 306	0·481 212

TABLE VII—INVERSE FUNCTIONS

x	$\tan^{-1} x$	$\cot^{-1} x$	$\sec^{-1} x$	$\mathrm{cosec}^{-1} x$	$\sinh^{-1} x$	$\cosh^{-1} x$	$\coth^{-1} x$	$\mathrm{cosech}^{-1} x$
2·00	1·107 149	0·463 648	1·047 198	0·523 599	1·443 635	1·316 958	0·549 306	0·481 212
·01	·109 141	·461 656	·050 068	·520 729	·448 099	·322 712	·545 995	·478 986
·02	·111 117	·459 679	·052 905	·517 892	·452 544	·328 429	·542 727	·476 780
·03	·113 078	·457 719	·055 709	·515 087	·456 972	·334 108	·539 502	·474 593
·04	·115 023	·455 774	·058 482	·512 315	·461 382	·339 750	·536 318	·472 426
2·05	1·116 952	0·453 844	1·061 223	0·509 573	1·465 775	1·345 356	0·533 176	0·470 278
·06	·118 867	·451 929	·063 934	·506 863	·470 151	·350 926	·530 073	·468 148
·07	·120 767	·450 030	·066 614	·504 182	·474 509	·356 461	·527 009	·466 038
·08	·122 651	·448 145	·069 265	·501 532	·478 851	·361 961	·523 984	·463 945
·09	·124 521	·446 275	·071 886	·498 910	·483 175	·367 427	·520 997	·461 871
2·10	1·126 377	0·444 419	1·074 479	0·496 317	1·487 483	1·372 859	0·518 046	0·459 815
·11	·128 218	·442 578	·077 044	·493 753	·491 774	·378 258	·515 131	·457 777
·12	·130 045	·440 751	·079 581	·491 216	·496 048	·383 624	·512 252	·455 756
·13	·131 859	·438 938	·082 091	·488 706	·500 306	·388 957	·509 408	·453 752
·14	·133 658	·437 139	·084 574	·486 223	·504 548	·394 258	·506 597	·451 765
2·15	1·135 443	0·435 353	1·087 030	0·483 766	1·508 773	1·399 528	0·503 820	0·449 795
·16	·137 215	·433 581	·089 461	·481 335	·512 983	·404 767	·501 076	·447 842
·17	·138 973	·431 823	·091 867	·478 930	·517 176	·409 974	·498 364	·445 905
·18	·140 718	·430 078	·094 247	·476 549	·521 353	·415 152	·495 683	·443 985
·19	·142 450	·428 346	·096 603	·474 193	·525 515	·420 299	·493 034	·442 080
2·20	1·144 169	0·426 627	1·098 934	0·471 862	1·529 660	1·425 417	0·490 415	0·440 191
·21	·145 875	·424 922	·101 242	·469 554	·533 791	·430 505	·487 825	·438 318
·22	·147 568	·423 228	·103 527	·467 270	·537 906	·435 565	·485 265	·436 460
·23	·149 248	·421 548	·105 788	·465 009	·542 005	·440 596	·482 734	·434 618
·24	·150 916	·419 880	·108 026	·462 770	·546 089	·445 599	·480 231	·432 791
2·25	1·152 572	0·418 224	1·110 242	0·460 554	1·550 158	1·450 575	0·477 756	0·430 978
·26	·154 215	·416 581	·112 436	·458 360	·554 212	·455 522	·475 308	·429 180
·27	·155 847	·414 950	·114 609	·456 187	·558 251	·460 443	·472 887	·427 397
·28	·157 466	·413 330	·116 760	·454 036	·562 275	·465 337	·470 492	·425 628
·29	·159 073	·411 723	·118 890	·451 906	·566 284	·470 204	·468 123	·423 874
2·30	1·160 669	0·410 127	1·120 999	0·449 797	1·570 279	1·475 045	0·465 779	0·422 133
·31	·162 253	·408 543	·123 088	·447 708	·574 259	·479 860	·463 461	·420 407
·32	·163 826	·406 971	·125 157	·445 639	·578 224	·484 650	·461 167	·418 694
·33	·165 387	·405 410	·127 207	·443 590	·582 175	·489 414	·458 897	·416 994
·34	·166 937	·403 860	·129 236	·441 560	·586 112	·494 153	·456 651	·415 308
2·35	1·168 475	0·402 321	1·131 247	0·439 550	1·590 035	1·498 868	0·454 428	0·413 635
·36	·170 003	·400 793	·133 238	·437 558	·593 943	·503 558	·452 228	·411 976
·37	·171 520	·399 277	·135 211	·435 585	·597 838	·508 224	·450 051	·410 329
·38	·173 025	·397 771	·137 166	·433 631	·601 718	·512 866	·447 896	·408 695
·39	·174 521	·396 276	·139 102	·431 694	·605 585	·517 484	·445 763	·407 074
2·40	1·176 005	0·394 791	1·141 021	0·429 775	1·609 438	1·522 079	0·443 652	0·405 465
·41	·177 479	·393 317	·142 922	·427 874	·613 277	·526 651	·441 561	·403 869
·42	·178 943	·391 853	·144 806	·425 991	·617 103	·531 200	·439 492	·402 285
·43	·180 396	·390 400	·146 672	·424 124	·620 915	·535 727	·437 443	·400 712
·44	·181 839	·388 957	·148 522	·422 274	·624 714	·540 231	·435 414	·399 152
2·45	1·183 273	0·387 524	1·150 355	0·420 441	1·628 500	1·544 713	0·433 405	0·397 604
·46	·184 696	·386 101	·152 172	·418 624	·632 272	·549 173	·431 416	·396 067
·47	·186 109	·384 688	·153 972	·416 824	·636 031	·553 612	·429 446	·394 542
·48	·187 512	·383 284	·155 757	·415 039	·639 778	·558 029	·427 495	·393 029
·49	·188 906	·381 890	·157 526	·413 270	·643 511	·562 424	·425 563	·391 526
2·50	1·190 290	0·380 506	1·159 279	0·411 517	1·647 231	1·566 799	0·423 649	0·390 035

TABLE VII—INVERSE FUNCTIONS

x	$\tan^{-1} x$	$\cot^{-1} x$	$\sec^{-1} x$	$\mathrm{cosec}^{-1} x$	$\sinh^{-1} x$	$\cosh^{-1} x$	$\coth^{-1} x$	$\mathrm{cosech}^{-1} x$
2·50	1·190 290	0·380 506	1·159 279	0·411 517	1·647 231	1·566 799	0·423 649	0·390 035
·51	·191 665	·379 132	·161 018	·409 779	·650 939	·571 153	·421 753	·388 555
·52	·193 030	·377 767	·162 741	·408 056	·654 633	·575 487	·419 875	·387 086
·53	·194 385	·376 411	·164 449	·406 348	·658 316	·579 800	·418 015	·385 628
·54	·195 732	·375 064	·166 142	·404 654	·661 985	·584 093	·416 172	·384 180
2·55	1·197 070	0·373 727	1·167 821	0·402 975	1·665 642	1·588 366	0·414 346	0·382 743
·56	·198 398	·372 398	·169 486	·401 310	·669 287	·592 619	·412 537	·381 317
·57	·199 717	·371 079	·171 136	·399 660	·672 919	·596 852	·410 745	·379 901
·58	·201 028	·369 769	·172 773	·398 023	·676 539	·601 067	·408 969	·378 495
·59	·202 330	·368 467	·174 396	·396 400	·680 147	·605 262	·407 209	·377 099
2·60	1·203 622	0·367 174	1·176 005	0·394 791	1·683 743	1·609 438	0·405 465	0·375 713
·61	·204 907	·365 889	·177 601	·393 195	·687 327	·613 595	·403 737	·374 338
·62	·206 183	·364 614	·179 184	·391 613	·690 899	·617 734	·402 024	·372 972
·63	·207 450	·363 346	·180 753	·390 043	·694 459	·621 854	·400 326	·371 616
·64	·208 709	·362 087	·182 310	·388 486	·698 007	·625 956	·398 644	·370 269
2·65	1·209 960	0·360 837	1·183 854	0·386 942	1·701 543	1·630 040	0·396 976	0·368 932
·66	·211 202	·359 594	·185 386	·385 411	·705 068	·634 106	·395 323	·367 604
·67	·212 436	·358 360	·186 905	·383 892	·708 581	·638 154	·393 684	·366 286
·68	·213 662	·357 134	·188 411	·382 385	·712 083	·642 184	·392 059	·364 977
·69	·214 880	·355 916	·189 906	·380 890	·715 573	·646 198	·390 449	·363 677
2·70	1·216 091	0·354 706	1·191 389	0·379 408	1·719 052	1·650 193	0·388 852	0·362 386
·71	·217 293	·353 503	·192 859	·377 937	·722 519	·654 172	·387 269	·361 105
·72	·218 488	·352 309	·194 319	·376 478	·725 976	·658 134	·385 700	·359 832
·73	·219 674	·351 122	·195 766	·375 030	·729 421	·662 079	·384 143	·358 567
·74	·220 854	·349 943	·197 203	·373 594	·732 855	·666 007	·382 600	·357 312
2·75	1·222 025	0·348 771	1·198 628	0·372 169	1·736 278	1·669 919	0·381 070	0·356 065
·76	·223 189	·347 607	·200 042	·370 755	·739 690	·673 814	·379 553	·354 826
·77	·224 346	·346 450	·201 445	·369 352	·743 091	·677 694	·378 048	·353 596
·78	·225 496	·345 301	·202 837	·367 959	·746 481	·681 557	·376 555	·352 375
·79	·226 638	·344 159	·204 218	·366 578	·749 860	·685 404	·375 075	·351 161
2·80	1·227 772	0·343 024	1·205 589	0·365 207	1·753 229	1·689 236	0·373 607	0·349 956
·81	·228 900	·341 896	·206 949	·363 847	·756 587	·693 051	·372 151	·348 759
·82	·230 021	·340 776	·208 299	·362 497	·759 934	·696 852	·370 707	·347 570
·83	·231 134	·339 662	·209 639	·361 157	·763 271	·700 637	·369 274	·346 388
·84	·232 241	·338 556	·210 969	·359 827	·766 598	·704 406	·367 853	·345 215
2·85	1·233 340	0·337 456	1·212 289	0·358 508	1·769 914	1·708 161	0·366 444	0·344 049
·86	·234 433	·336 363	·213 598	·357 198	·773 220	·711 900	·365 045	·342 892
·87	·235 519	·335 277	·214 899	·355 898	·776 515	·715 625	·363 658	·341 741
·88	·236 598	·334 198	·216 189	·354 607	·779 800	·719 335	·362 282	·340 599
·89	·237 671	·333 125	·217 470	·353 327	·783 075	·723 030	·360 916	·339 463
2·90	1·238 737	0·332 059	1·218 741	0·352 055	1·786 340	1·726 711	0·359 561	0·338 336
·91	·239 796	·331 000	·220 003	·350 793	·789 595	·730 377	·358 217	·337 215
·92	·240 849	·329 947	·221 256	·349 540	·792 840	·734 029	·356 883	·336 102
·93	·241 896	·328 901	·222 500	·348 296	·796 075	·737 667	·355 560	·334 996
·94	·242 936	·327 860	·223 735	·347 062	·799 300	·741 291	·354 246	·333 897
2·95	1·243 970	0·326 827	1·224 961	0·345 836	1·802 515	1·744 902	0·352 943	0·332 805
·96	·244 997	·325 799	·226 178	·344 619	·805 721	·748 498	·351 650	·331 721
·97	·246 019	·324 778	·227 386	·343 410	·808 917	·752 080	·350 366	·330 643
·98	·247 034	·323 763	·228 586	·342 211	·812 103	·755 649	·349 092	·329 572
·99	·248 043	·322 754	·229 777	·341 020	·815 279	·759 205	·347 828	·328 508
3·00	1·249 046	0·321 751	1·230 959	0·339 837	1·818 446	1·762 747	0·346 574	0·327 450

TABLE VII—INVERSE FUNCTIONS

x	$\tan^{-1} x$	$\cot^{-1} x$	$\sec^{-1} x$	$\mathrm{cosec}^{-1} x$	$\sinh^{-1} x$	$\cosh^{-1} x$	$\coth^{-1} x$	$\mathrm{cosech}^{-1} x$
3·0	1·249 046	0·321 751	1·230 959	0·339 837	1·818 446	1·762 747	0·346 574	0·327 450
3·1	·258 754	·312 042	·242 342	·328 455	·849 604	·797 457	·334 525	·317 233
3·2	·267 911	·302 885	·252 973	·317 824	·879 864	·830 938	·323 314	·307 625
3·3	·276 562	·294 235	·262 925	·307 871	·909 274	·863 279	·312 853	·298 574
3·4	·284 745	·286 051	·272 264	·298 532	·937 879	·894 559	·303 068	·290 034
3·5	1·292 497	0·278 300	1·281 045	0·289 752	1·965 720	1·924 847	0·293 893	0·281 963
3·6	·299 849	·270 947	·289 316	·281 480	1·992 836	1·954 208	·285 272	·274 324
3·7	·306 833	·263 964	·297 123	·273 674	2·019 261	1·982 697	·277 155	·267 084
3·8	·313 473	·257 324	·304 502	·266 294	·045 028	2·010 367	·269 498	·260 211
3·9	·319 794	·251 003	·311 490	·259 306	·070 169	2·037 266	·262 262	·253 681
4·0	1·325 818	0·244 979	1·318 116	0·252 680	2·094 713	2·063 437	0·255 413	0·247 466
4·1	·331 565	·239 232	·324 409	·246 388	·118 685	·088 919	·248 919	·241 547
4·2	·337 053	·233 743	·330 392	·240 404	·142 112	·113 748	·242 754	·235 901
4·3	·342 300	·228 497	·336 089	·234 707	·165 017	·137 959	·236 892	·230 511
4·4	·347 320	·223 477	·341 520	·229 276	·187 422	·161 581	·231 312	·225 360
4·5	1·352 127	0·218 669	1·346 703	0·224 093	2·209 348	2·184 644	0·225 993	0·220 433
4·6	·356 736	·214 061	·351 655	·219 141	·230 814	·207 174	·220 916	·215 714
4·7	·361 156	·209 640	·356 391	·214 405	·251 840	·229 195	·216 067	·211 193
4·8	·365 401	·205 395	·360 926	·209 871	·272 441	·250 731	·211 428	·206 855
4·9	·369 479	·201 317	·365 271	·205 526	·292 636	·271 804	·206 988	·202 691
5·0	1·373 401	0·197 396	1·369 438	0·201 358	2·312 438	2·292 432	0·202 733	0·198 690
5·1	·377 174	·193 622	·373 439	·197 357	·331 864	·312 634	·198 651	·194 843
5·2	·380 808	·189 988	·377 283	·193 513	·350 926	·332 429	·194 732	·191 142
5·3	·384 309	·186 487	·380 979	·189 817	·369 637	·351 833	·190 967	·187 577
5·4	·387 686	·183 111	·384 536	·186 260	·388 011	·370 860	·187 347	·184 143
5·5	1·390 943	0·179 853	1·387 961	0·182 835	2·406 059	2·389 526	0·183 862	0·180 831
5·6	·394 087	·176 709	·391 262	·179 534	·423 792	·407 845	·180 507	·177 636
5·7	·397 125	·173 671	·394 445	·176 351	·441 221	·425 828	·177 273	·174 551
5·8	·400 061	·170 735	·397 517	·173 280	·458 355	·443 489	·174 153	·171 571
5·9	·402 900	·167 896	·400 483	·170 314	·475 205	·460 839	·171 143	·168 690
6·0	1·405 648	0·165 149	1·403 348	0·167 448	2·491 780	2·477 889	0·168 236	0·165 905
6·1	·408 307	·162 489	·406 119	·164 678	·508 088	·494 649	·165 427	·163 209
6·2	·410 883	·159 913	·408 798	·161 998	·524 138	·511 128	·162 711	·160 599
6·3	·413 379	·157 417	·411 392	·159 404	·539 937	·527 338	·160 084	·158 071
6·4	·415 800	·154 997	·413 903	·156 893	·555 494	·543 285	·157 541	·155 621
6·5	1·418 147	0·152 649	1·416 337	0·154 460	2·570 815	2·558 979	0·155 077	0·153 246
6·6	·420 425	·150 371	·418 695	·152 101	·585 907	·574 428	·152 691	·150 941
6·7	·422 636	·148 160	·420 983	·149 814	·600 778	·589 638	·150 377	·148 705
6·8	·424 784	·146 012	·423 202	·147 594	·615 433	·604 619	·148 133	·146 534
6·9	·426 871	·143 925	·425 357	·145 440	·629 879	·619 376	·145 955	·144 425
7·0	1·428 899	0·141 897	1·427 449	0·143 348	2·644 121	2·633 916	0·143 841	0·142 376
7·1	·430 872	·139 925	·429 481	·141 315	·658 165	·648 245	·141 788	·140 384
7·2	·432 790	·138 006	·431 457	·139 339	·672 016	·662 370	·139 792	·138 446
7·3	·434 657	·136 139	·433 378	·137 418	·685 680	·676 297	·137 853	·136 561
7·4	·436 475	·134 321	·435 246	·135 550	·699 162	·690 030	·135 967	·134 727
7·5	1·438 245	0·132 552	1·437 065	0·133 732	2·712 465	2·703 576	0·134 132	0·132 941
7·6	·439 969	·130 827	·438 835	·131 962	·725 596	·716 939	·132 346	·131 202
7·7	·441 649	·129 147	·440 558	·130 238	·738 558	·730 124	·130 608	·129 508
7·8	·443 287	·127 510	·442 237	·128 559	·751 355	·743 136	·128 915	·127 856
7·9	·444 884	·125 913	·443 874	·126 923	·763 992	·755 980	·127 265	·126 247
8·0	1·446 441	0·124 355	1·445 469	0·125 328	2·776 472	2·768 659	0·125 657	0·124 677

x	$\tan^{-1} x$	$\cot^{-1} x$	$\sec^{-1} x$	$\operatorname{cosec}^{-1} x$	$\sinh^{-1} x$	$\cosh^{-1} x$	$\coth^{-1} x$	$\operatorname{cosech}^{-1} x$
8·0	1·446 441	0·124 355	1·445 469	0·125 328	2·776 472	2·768 659	0·125 657	0·124 677
8·1	·447 961	·122 835	·447 024	·123 773	·788 800	·781 179	·124 090	·123 145
8·2	·449 444	·121 352	·448 541	·122 256	·800 979	·793 542	·122 561	·121 651
8·3	·450 892	·119 904	·450 021	·120 775	·813 012	·805 754	·121 070	·120 192
8·4	·452 306	·118 490	·451 466	·119 331	·824 903	·817 817	·119 615	·118 768
8·5	1·453 688	0·117 109	1·452 876	0·117 920	2·836 656	2·829 735	0·118 194	0·117 377
8·6	·455 037	·115 759	·454 254	·116 543	·848 273	·841 512	·116 807	·116 019
8·7	·456 356	·114 440	·455 599	·115 197	·859 757	·853 151	·115 453	·114 691
8·8	·457 645	·113 151	·456 914	·113 882	·871 112	·864 655	·114 129	·113 393
8·9	·458 906	·111 890	·458 199	·112 597	·882 340	·876 027	·112 836	·112 124
9·0	1·460 139	0·110 657	1·459 455	0·111 341	2·893 444	2·887 271	0·111 572	0·110 884
9·1	·461 345	·109 451	·460 684	·110 112	·904 427	·898 389	·110 336	·109 670
9·2	·462 526	·108 271	·461 885	·108 911	·915 291	·909 384	·109 127	·108 483
9·3	·463 681	·107 115	·463 061	·107 735	·926 040	·920 258	·107 944	·107 321
9·4	·464 812	·105 984	·464 212	·106 585	·936 674	·931 015	·106 787	·106 183
9·5	1·465 919	0·104 877	1·465 338	0·105 459	2·947 198	2·941 657	0·105 655	0·105 070
9·6	·467 004	·103 792	·466 440	·104 356	·957 612	·952 187	·104 546	·103 979
9·7	·468 066	·102 730	·467 520	·103 276	·967 920	·962 605	·103 460	·102 911
9·8	·469 107	·101 689	·468 578	·102 219	·978 123	·972 916	·102 397	·101 865
9·9	·470 128	·100 669	·469 614	·101 183	·988 223	·983 121	·101 356	·100 839
10·0	1·471 128	0·099 669	1·470 629	0·100 167	2·998 223	2·993 223	0·100 335	0·099 834
10·1	·472 108	·098 688	·471 624	·099 172	3·008 124	3·003 223	·099 335	·098 849
10·2	·473 069	·097 727	·472 599	·098 197	·017 929	·013 123	·098 355	·097 883
10·3	·474 012	·096 784	·473 556	·097 241	·027 639	·022 926	·097 394	·096 935
10·4	·474 937	·095 859	·474 494	·096 303	·037 256	·032 634	·096 452	·096 006
10·5	1·475 845	0·094 952	1·475 414	0·095 383	3·046 782	3·042 247	0·095 528	0·095 095
10·6	·476 735	·094 061	·476 316	·094 480	·056 219	·051 769	·094 621	·094 200
10·7	·477 609	·093 187	·477 202	·093 595	·065 567	·061 200	·093 731	·093 322
10·8	·478 467	·092 329	·478 071	·092 725	·074 830	·070 543	·092 859	·092 461
10·9	·479 309	·091 487	·478 924	·091 872	·084 008	·079 799	·092 002	·091 615
11·0	1·480 136	0·090 660	1·479 762	0·091 035	3·093 102	3·088 970	0·091 161	0·090 784
11·1	·480 949	·089 848	·480 584	·090 212	·102 115	·098 057	·090 335	·089 969
11·2	·481 747	·089 050	·481 392	·089 405	·111 048	·107 062	·089 524	·089 168
11·3	·482 531	·088 266	·482 185	·088 611	·119 902	·115 986	·088 728	·088 380
11·4	·483 301	·087 495	·482 964	·087 832	·128 679	·124 831	·087 945	·087 607
11·5	1·484 058	0·086 738	1·483 730	0·087 066	3·137 379	3·133 598	0·087 177	0·086 847
11·6	·484 802	·085 994	·484 482	·086 314	·146 005	·142 289	·086 421	·086 100
11·7	·485 533	·085 263	·485 222	·085 574	·154 557	·150 905	·085 679	·085 366
11·8	·486 253	·084 544	·485 949	·084 848	·163 037	·159 446	·084 950	·084 645
11·9	·486 960	·083 837	·486 663	·084 133	·171 446	·167 915	·084 232	·083 935
12·0	1·487 655	0·083 141	1·487 366	0·083 430	3·179 785	3·176 313	0·083 527	0·083 237
12·1	·488 339	·082 457	·488 057	·082 739	·188 056	·184 641	·082 834	·082 551
12·2	·489 012	·081 784	·488 737	·082 059	·196 259	·192 899	·082 152	·081 876
12·3	·489 674	·081 122	·489 406	·081 391	·204 395	·201 090	·081 481	·081 212
12·4	·490 325	·080 471	·490 063	·080 733	·212 466	·209 214	·080 821	·080 558
12·5	1·490 966	0·079 830	1·490 711	0·080 086	3·220 472	3·217 272	0·080 171	0·079 915
12·6	·491 597	·079 199	·491 348	·079 449	·228 415	·225 266	·079 532	·079 282
12·7	·492 218	·078 578	·491 975	·078 822	·236 296	·233 196	·078 903	·078 659
12·8	·492 830	·077 967	·492 592	·078 205	·244 115	·241 063	·078 285	·078 046
12·9	·493 432	·077 365	·493 199	·077 597	·251 873	·248 869	·077 675	·077 442
13·0	1·494 024	0·076 772	1·493 797	0·076 999	3·259 573	3·256 614	0·077 075	0·076 847

TABLE VII—INVERSE FUNCTIONS

x	$\tan^{-1} x$	$\cot^{-1} x$	$\sec^{-1} x$	$\operatorname{cosec}^{-1} x$	$\sinh^{-1} x$	$\cosh^{-1} x$	$\coth^{-1} x$	$\operatorname{cosech}^{-1} x$
10	1·471 128	0·099 669	1·470 629	0·100 167	2·998 22	2·993 22	0·100 335	0·099 834
11	·480 136	·090 660	·479 762	·091 035	3·093 10	3·088 97	·091 161	·090 784
12	·487 655	·083 141	·487 366	·083 430	·179 79	·176 31	·083 527	·083 237
13	·494 024	·076 772	·493 797	·076 999	·259 57	·256 61	·077 075	·076 847
14	·499 489	·071 307	·499 307	·071 489	·333 48	·330 93	·071 550	·071 368
15	1·504 228	0·066 568	1·504 080	0·066 716	3·402 31	3·400 08	0·066 766	0·066 617
16	·508 378	·062 419	·508 256	·062 541	·466 71	·464 76	·062 582	·062 459
17	·512 041	·058 756	·511 939	·058 858	·527 22	·525 49	·058 892	·058 790
18	·515 298	·055 499	·515 212	·055 584	·584 29	·582 75	·055 613	·055 527
19	·518 213	·052 583	·518 140	·052 656	·638 28	·636 89	·052 680	·052 607
20	1·520 838	0·049 958	1·520 775	0·050 021	3·689 50	3·688 25	0·050 042	0·049 979
21	·523 213	·047 583	·523 159	·047 637	·738 24	·737 10	·047 655	·047 601
22	·525 373	·045 423	·525 326	·045 470	·784 71	·783 67	·045 486	·045 439
23	·527 345	·043 451	·527 304	·043 492	·829 11	·828 17	·043 506	·043 465
24	·529 154	·041 643	·529 118	·041 679	·871 63	·870 77	·041 691	·041 655
25	1·530 818	0·039 979	1·530 786	0·040 011	3·912 42	3·911 62	0·040 021	0·039 989
26	·532 354	·038 443	·532 325	·038 471	3·951 61	3·950 87	·038 481	·038 452
27	·533 776	·037 020	·533 751	·037 046	3·989 33	3·988 64	·037 054	·037 029
28	·535 097	·035 699	·535 074	·035 722	4·025 67	4·025 03	·035 729	·035 707
29	·536 327	·034 469	·536 307	·034 490	4·060 74	4·060 15	·034 496	·034 476
30	1·537 475	0·033 321	1·537 457	0·033 340	4·094 62	4·094 07	0·033 346	0·033 327
31	·538 549	·032 247	·538 533	·032 264	·127 39	·126 87	·032 269	·032 252
32	·539 556	·031 240	·539 541	·031 255	·159 13	·158 64	·031 260	·031 245
33	·540 503	·030 294	·540 489	·030 308	·189 88	·189 43	·030 312	·030 298
34	·541 393	·029 403	·541 380	·029 416	·219 72	·219 29	·029 420	·029 408
35	1·542 233	0·028 564	1·542 221	0·028 575	4·248 70	4·248 29	0·028 579	0·028 568
36	·543 026	·027 771	·543 015	·027 781	·276 86	·276 47	·027 785	·027 774
37	·543 776	·027 020	·543 766	·027 030	·304 25	·303 88	·027 034	·027 024
38	·544 487	·026 310	·544 477	·026 319	·330 91	·330 56	·026 322	·026 313
39	·545 161	·025 635	·545 152	·025 644	·356 87	·356 54	·025 647	·025 638
40	1·545 802	0·024 995	1·545 794	0·025 003	4·382 18	4·381 87	0·025 005	0·024 997
41	·546 411	·024 385	·546 404	·024 393	·406 87	·406 57	·024 395	·024 388
42	·546 991	·023 805	·546 985	·023 812	·430 96	·430 68	·023 814	·023 807
43	·547 545	·023 252	·547 538	·023 258	·454 48	·454 21	·023 260	·023 254
44	·548 073	·022 723	·548 067	·022 729	·477 47	·477 21	·022 731	·022 725
45	1·548 578	0·022 219	1·548 572	0·022 224	4·499 93	4·499 69	0·022 226	0·022 220
46	·549 061	·021 736	·549 055	·021 741	·521 91	·521 67	·021 743	·021 737
47	·549 523	·021 273	·549 518	·021 278	·543 41	·543 18	·021 280	·021 275
48	·549 966	·020 830	·549 961	·020 835	·564 46	·564 24	·020 836	·020 832
49	·550 391	·020 405	·550 387	·020 410	·585 07	·584 86	·020 411	·020 407
50	1·550 799	0·019 997	1·550 795	0·020 001	4·605 27	4·605 07	0·020 003	0·019 999
51	·551 191	·019 605	·551 187	·019 609	·625 07	·624 88	·019 610	·019 607
52	·551 568	·019 228	·551 564	·019 232	·644 48	·644 30	·019 233	·019 230
53	·551 931	·018 866	·551 927	·018 869	·663 53	·663 35	·018 870	·018 867
54	·552 280	·018 516	·552 277	·018 520	·682 22	·682 05	·018 521	·018 517
55	1·552 617	0·018 180	1·552 614	0·018 183	4·700 56	4·700 40	0·018 184	0·018 181
56	·552 941	·017 855	·552 938	·017 858	·718 58	·718 42	·017 859	·017 856
57	·553 254	·017 542	·553 252	·017 545	·736 28	·736 12	·017 546	·017 543
58	·553 557	·017 240	·553 554	·017 242	·753 66	·753 52	·017 243	·017 241
59	·553 849	·016 948	·553 846	·016 950	·770 76	·770 61	·016 951	·016 948
60	1·554 131	0·016 665	1·554 129	0·016 667	4·787 56	4·787 42	0·016 668	0·016 666

x	$\tan^{-1} x$	$\cot^{-1} x$	$\sec^{-1} x$	$\operatorname{cosec}^{-1} x$	$\sinh^{-1} x$	$\cosh^{-1} x$	$\coth^{-1} x$	$\operatorname{cosech}^{-1} x$
60	1·554 131	0·016 665	1·554 129	0·016 667	4·787 56	4·787 42	0·016 668	0·016 666
61	·554 404	·016 392	·554 402	·016 394	·804 09	·803 95	·016 395	·016 393
62	·554 669	·016 128	·554 667	·016 130	·820 35	·820 22	·016 130	·016 128
63	·554 925	·015 872	·554 923	·015 874	·836 34	·836 22	·015 874	·015 872
64	·555 173	·015 624	·555 171	·015 626	·852 09	·851 97	·015 626	·015 624
65	1·555 413	0·015 383	1·555 411	0·015 385	4·867 59	4·867 48	0·015 386	0·015 384
66	·555 646	·015 150	·555 644	·015 152	·882 86	·882 74	·015 153	·015 151
67	·555 872	·014 924	·555 870	·014 926	·897 90	·897 78	·014 926	·014 925
68	·556 092	·014 705	·556 090	·014 706	·912 71	·912 60	·014 707	·014 705
69	·556 305	·014 492	·556 303	·014 493	·927 31	·927 20	·014 494	·014 492
70	1·556 512	0·014 285	1·556 510	0·014 286	4·941 69	4·941 59	0·014 287	0·014 285
71	·556 713	·014 084	·556 711	·014 085	·955 88	·955 78	·014 085	·014 084
72	·556 908	·013 888	·556 907	·013 889	·969 86	·969 77	·013 890	·013 888
73	·557 099	·013 698	·557 097	·013 699	·983 65	·983 56	·013 699	·013 698
74	·557 284	·013 513	·557 282	·013 514	4·997 26	4·997 17	·013 514	·013 513
75	1·557 464	0·013 333	1·557 463	0·013 334	5·010 68	5·010 59	0·013 334	0·013 333
76	·557 639	·013 157	·557 638	·013 158	·023 92	·023 84	·013 159	·013 158
77	·557 810	·012 986	·557 809	·012 987	·036 99	·036 91	·012 988	·012 987
78	·557 977	·012 820	·557 975	·012 821	·049 90	·049 81	·012 821	·012 820
79	·558 139	·012 658	·558 138	·012 659	·062 64	·062 55	·012 659	·012 658
80	1·558 297	0·012 499	1·558 296	0·012 500	5·075 21	5·075 13	0·012 501	0·012 500
81	·558 451	·012 345	·558 450	·012 346	·087 63	·087 56	·012 346	·012 345
82	·558 602	·012 195	·558 601	·012 195	·099 90	·099 83	·012 196	·012 195
83	·558 749	·012 048	·558 748	·012 048	·112 02	·111 95	·012 049	·012 048
84	·558 892	·011 904	·558 891	·011 905	·124 00	·123 93	·011 905	·011 904
85	1·559 032	0·011 764	1·559 031	0·011 765	5·135 83	5·135 76	0·011 765	0·011 764
86	·559 169	·011 627	·559 168	·011 628	·147 53	·147 46	·011 628	·011 628
87	·559 303	·011 494	·559 302	·011 495	·159 09	·159 02	·011 495	·011 494
88	·559 433	·011 363	·559 432	·011 364	·170 52	·170 45	·011 364	·011 363
89	·559 561	·011 235	·559 560	·011 236	·181 82	·181 75	·011 236	·011 236
90	1·559 686	0·011 111	1·559 685	0·011 111	5·192 99	5·192 93	0·011 112	0·011 111
91	·559 808	·010 989	·559 807	·010 989	·204 04	·203 98	·010 989	·010 989
92	·559 927	·010 869	·559 927	·010 870	·214 97	·214 91	·010 870	·010 869
93	·560 044	·010 752	·560 043	·010 753	·225 78	·225 72	·010 753	·010 752
94	·560 158	·010 638	·560 158	·010 638	·236 47	·236 41	·010 639	·010 638
95	1·560 270	0·010 526	1·560 270	0·010 527	5·247 05	5·247 00	0·010 527	0·010 526
96	·560 380	·010 416	·560 379	·010 417	·257 52	·257 47	·010 417	·010 416
97	·560 487	·010 309	·560 487	·010 309	·267 88	·267 83	·010 310	·010 309
98	·560 593	·010 204	·560 592	·010 204	·278 14	·278 09	·010 204	·010 204
99	·560 696	·010 101	·560 695	·010 101	·288 29	·288 24	·010 101	·010 101
100	1·560 797	0·010 000	1·560 796	0·010 000	5·298 34	5·298 29	0·010 000	0·010 000

If x is greater than 100,

$$\tan^{-1} x = \sec^{-1} x = 1{\cdot}570\ 796 - \frac{1}{x}$$

$$\cot^{-1} x = \operatorname{cosec}^{-1} x = \coth^{-1} x = \operatorname{cosech}^{-1} x = \frac{1}{x}$$

x	A
74·5	
	4
84·5	
	3
100·0	
	2
129·1	
	1
223·6	
	0
∞	

If x is greater than 75,

$$\sinh^{-1} x = \ln 2x + A$$

$$\cosh^{-1} x = \ln 2x - A$$

where A, in units of the fifth decimal, is tabulated alongside in a critical table.

Example: Find $\sec^{-1}$ 123·45.

$\frac{1}{2}\pi$	1·570 796
1/123·45	0·008 100
$\sec^{-1}$ 123·45	1·562 696

Example: Find $\cosh^{-1}$ 123·45.

ln 2·4690 (page 278)	0·903 81
ln 100 (page 278)	4·605 17
A (above)	− 2
Sum = $\cosh^{-1}$ 123·45	5·508 96

TABLE VIII—POWERS AND SQUARE ROOTS

No.	Square	Cube	Fourth Power	Fifth Power	Square Root	Root of Reciprocal	Reciprocal of Square
x	x^2	x^3	x^4	x^5	$\sqrt{x} = x^{\frac{1}{2}}$	$\frac{1}{\sqrt{x}} = x^{-\frac{1}{2}}$	$\frac{1}{x^2} = x^{-2}$
0	0	0	0	0	0·000 000	∞	∞
1	1	1	1	1	1·000 000	1·000 000	1·000 000 000
2	4	8	16	32	1·414 214	0·707 107	0·250 000 000
3	9	27	81	243	1·732 051	·577 350	·111 111 111
4	16	64	256	1024	2·000 000	·500 000	·062 500 000
5	25	125	625	3125	2·236 068	0·447 214	0·040 000 000
6	36	216	1296	7776	2·449 490	·408 248	·027 777 778
7	49	343	2401	16807	2·645 751	·377 964	·020 408 163
8	64	512	4096	32768	2·828 427	·353 553	·015 625 000
9	81	729	6561	59049	3·000 000	·333 333	·012 345 679
10	1 00	1 000	1 0000	1 00000	3·162 278	0·316 228	0·010 000 000
11	1 21	1 331	1 4641	1 61051	3·316 625	·301 511	·008 264 463
12	1 44	1 728	2 0736	2 48832	3·464 102	·288 675	·006 944 444
13	1 69	2 197	2 8561	3 71293	3·605 551	·277 350	·005 917 160
14	1 96	2 744	3 8416	5 37824	3·741 657	·267 261	·005 102 041
15	2 25	3 375	5 0625	7 59375	3·872 983	0·258 199	0·004 444 444
16	2 56	4 096	6 5536	10 48576	4·000 000	·250 000	·003 906 250
17	2 89	4 913	8 3521	14 19857	4·123 106	·242 536	·003 460 208
18	3 24	5 832	10 4976	18 89568	4·242 641	·235 702	·003 086 420
19	3 61	6 859	13 0321	24 76099	4·358 899	·229 416	·002 770 083
20	4 00	8 000	16 0000	32 00000	4·472 136	0·223 607	0·002 500 000
21	4 41	9 261	19 4481	40 84101	4·582 576	·218 218	·002 267 574
22	4 84	10 648	23 4256	51 53632	4·690 416	·213 201	·002 066 116
23	5 29	12 167	27 9841	64 36343	4·795 832	·208 514	·001 890 359
24	5 76	13 824	33 1776	79 62624	4·898 979	·204 124	·001 736 111
25	6 25	15 625	39 0625	97 65625	5·000 000	0·200 000	0·001 600 000
26	6 76	17 576	45 6976	118 81376	5·099 020	·196 116	·001 479 290
27	7 29	19 683	53 1441	143 48907	5·196 152	·192 450	·001 371 742
28	7 84	21 952	61 4656	172 10368	5·291 503	·188 982	·001 275 510
29	8 41	24 389	70 7281	205 11149	5·385 165	·185 695	·001 189 061
30	9 00	27 000	81 0000	243 00000	5·477 226	0·182 574	0·001 111 111
31	9 61	29 791	92 3521	286 29151	5·567 764	·179 605	·001 040 583
32	10 24	32 768	104 8576	335 54432	5·656 854	·176 777	·000 976 562
33	10 89	35 937	118 5921	391 35393	5·744 563	·174 078	·000 918 274
34	11 56	39 304	133 6336	454 35424	5·830 952	·171 499	·000 865 052
35	12 25	42 875	150 0625	525 21875	5·916 080	0·169 031	0·000 816 327
36	12 96	46 656	167 9616	604 66176	6·000 000	·166 667	·000 771 605
37	13 69	50 653	187 4161	693 43957	6·082 763	·164 399	·000 730 460
38	14 44	54 872	208 5136	792 35168	6·164 414	·162 221	·000 692 521
39	15 21	59 319	231 3441	902 24199	6·244 998	·160 128	·000 657 462
40	16 00	64 000	256 0000	1024 00000	6·324 555	0·158 114	0·000 625 000
41	16 81	68 921	282 5761	1158 56201	6·403 124	·156 174	·000 594 884
42	17 64	74 088	311 1696	1306 91232	6·480 741	·154 303	·000 566 893
43	18 49	79 507	341 8801	1470 08443	6·557 439	·152 499	·000 540 833
44	19 36	85 184	374 8096	1649 16224	6·633 250	·150 756	·000 516 529
45	20 25	91 125	410 0625	1845 28125	6·708 204	0·149 071	0·000 493 827
46	21 16	97 336	447 7456	2059 62976	6·782 330	·147 442	·000 472 590
47	22 09	103 823	487 9681	2293 45007	6·855 655	·145 865	·000 452 694
48	23 04	110 592	530 8416	2548 03968	6·928 203	·144 338	·000 434 028
49	24 01	117 649	576 4801	2824 75249	7·000 000	·142 857	·000 416 493
50	25 00	125 000	625 0000	3125 00000	7·071 068	0·141 421	0·000 400 000

TABLE VIII—RECIPROCALS, ROOTS, FACTORIALS AND FACTORS

No. x	Reciprocal $1/x = x^{-1}$	Cube Root $\sqrt[3]{x}$	Fourth Root $\sqrt[4]{x}$	Fifth Root $\sqrt[5]{x}$	Factorials * $x!$	Factorials $\log x!$	Prime Factors
0	∞	0·000 000	0·000 000	0·000 000	1	0·000 000	...
1	1·0000 000	1·000 000	1·000 000	1·000 000	1	0·000 000	1
2	0·5000 000	1·259 921	1·189 207	1·148 698	2	0·301 030	2
3	·3333 333	1·442 250	1·316 074	1·245 731	6	0·778 151	3
4	·2500 000	1·587 401	1·414 214	1·319 508	24	1·380 211	2^2
5	0·2000 000	1·709 976	1·495 349	1·379 730	120	2·079 181	5
6	·1666 667	1·817 121	1·565 085	1·430 969	720	2·857 332	2.3
7	·1428 571	1·912 931	1·626 577	1·475 773	5 040	3·702 431	7
8	·1250 000	2·000 000	1·681 793	1·515 717	40 320	4·605 521	2^3
9	·1111 111	2·080 084	1·732 051	1·551 846	362 880	5·559 763	3^2
10	0·1000 000	2·154 435	1·778 279	1·584 893	3·628 800	6·559 763	2.5
11	·0909 091	2·223 980	1·821 160	1·615 394	3·991 680	7·601 156	11
12	·0833 333	2·289 428	1·861 210	1·643 752	4·790 016	8·680 337	$2^2.3$
13	·0769 231	2·351 335	1·898 829	1·670 278	6·227 021	9·794 280	13
14	·0714 286	2·410 142	1·934 336	1·695 218	8·717 829	10·940 408	2.7
15	0·0666 667	2·466 212	1·967 990	1·718 772	1·307 674	12·116 500	3.5
16	·0625 000	2·519 842	2·000 000	1·741 101	2·092 279	13·320 620	2^4
17	·0588 235	2·571 282	2·030 543	1·762 340	3·556 874	14·551 069	17
18	·0555 556	2·620 741	2·059 767	1·782 602	6·402 374	15·806 341	2.3^2
19	·0526 316	2·668 402	2·087 798	1·801 983	1·216 451	17·085 095	19
20	0·0500 000	2·714 418	2·114 743	1·820 564	2·432 902	18·386 125	$2^2.5$
21	·0476 190	2·758 924	2·140 695	1·838 416	5·109 094	19·708 344	3.7
22	·0454 545	2·802 039	2·165 737	1·855 601	1·124 001	21·050 767	2.11
23	·0434 783	2·843 867	2·189 939	1·872 171	2·585 202	22·412 494	23
24	·0416 667	2·884 499	2·213 364	1·888 175	6·204 484	23·792 706	$2^3.3$
25	0·0400 000	2·924 018	2·236 068	1·903 654	1·551 121	25·190 646	5^2
26	·0384 615	2·962 496	2·258 101	1·918 645	4·032 915	26·605 619	2.13
27	·0370 370	3·000 000	2·279 507	1·933 182	1·088 887	28·036 983	3^3
28	·0357 143	3·036 589	2·300 327	1·947 294	3·048 883	29·484 141	$2^2.7$
29	·0344 828	3·072 317	2·320 596	1·961 009	8·841 762	30·946 539	29
30	0·0333 333	3·107 233	2·340 347	1·974 350	2·652 529	32·423 660	2.3.5
31	·0322 581	3·141 381	2·359 611	1·987 341	8·222 839	33·915 022	31
32	·0312 500	3·174 802	2·378 414	2·000 000	2·631 308	35·420 172	2^5
33	·0303 030	3·207 534	2·396 782	2·012 347	8·683 318	36·938 686	3.11
34	·0294 118	3·239 612	2·414 736	2·024 397	2·952 328	38·470 165	2.17
35	0·0285 714	3·271 066	2·432 299	2·036 168	1·033 315	40·014 233	5.7
36	·0277 778	3·301 927	2·449 490	2·047 673	3·719 933	41·570 535	$2^2.3^2$
37	·0270 270	3·332 222	2·466 326	2·058 924	1·376 375	43·138 737	37
38	·0263 158	3·361 975	2·482 824	2·069 935	5·230 226	44·718 520	2.19
39	·0256 410	3·391 211	2·498 999	2·080 717	2·039 788	46·309 585	3.13
40	0·0250 000	3·419 952	2·514 867	2·091 279	8·159 153	47·911 645	$2^3.5$
41	·0243 902	3·448 217	2·530 440	2·101 632	3·345 253	49·524 429	41
42	·0238 095	3·476 027	2·545 730	2·111 786	1·405 006	51·147 678	2.3.7
43	·0232 558	3·503 398	2·560 750	2·121 747	6·041 526	52·781 147	43
44	·0227 273	3·530 348	2·575 510	2·131 526	2·658 272	54·424 599	$2^2.11$
45	0·0222 222	3·556 893	2·590 020	2·141 127	1·196 222	56·077 812	$3^2.5$
46	·0217 391	3·583 048	2·604 291	2·150 560	5·502 622	57·740 570	2.23
47	·0212 766	3·608 826	2·618 330	2·159 830	2·586 232	59·412 668	47
48	·0208 333	3·634 241	2·632 148	2·168 944	1·241 392	61·093 909	$2^4.3$
49	·0204 082	3·659 306	2·645 751	2·177 906	6·082 819	62·784 105	7^2
50	0·0200 000	3·684 031	2·659 148	2·186 724	3·041 409	64·483 075	2.5^2

* If x is greater than 9, multiply by 10^c, where c is the characteristic of $\log x!$

$\sqrt[4]{10} = 1·778\ 279$ $\sqrt[4]{100} = 3·162\ 278$ $\sqrt[4]{1000} = 5·623\ 413$

$\sqrt[5]{10} = 1·584\ 893$ $\sqrt[5]{100} = 2·511\ 886$ $\sqrt[5]{1000} = 3·981\ 072$ $\sqrt[5]{10{,}000} = 6·309\ 573$

TABLE VIII—POWERS AND SQUARE ROOTS

No.	Square	Cube	Fourth Power	Fifth Power	Square Root	Root of Reciprocal	Reciprocal of Square
x	x^2	x^3	x^4	x^5	$\sqrt{x}=x^{\frac{1}{2}}$	$\frac{1}{\sqrt{x}}=x^{-\frac{1}{2}}$	$\frac{1}{x^2}=x^{-2}$
							0·000
50	25 00	125 000	625 0000	3125 00000	7·071 068	0·141 421	400 000
51	26 01	132 651	676 5201	3450 25251	7·141 428	·140 028	384 468
52	27 04	140 608	731 1616	3802 04032	7·211 103	·138 675	369 822
53	28 09	148 877	789 0481	4181 95493	7·280 110	·137 361	355 999
54	29 16	157 464	850 3056	4591 65024	7·348 469	·136 083	342 936
55	30 25	166 375	915 0625	5032 84375	7·416 198	0·134 840	330 579
56	31 36	175 616	983 4496	5507 31776	7·483 315	·133 631	318 878
57	32 49	185 193	1055 6001	6016 92057	7·549 834	·132 453	307 787
58	33 64	195 112	1131 6496	6563 56768	7·615 773	·131 306	297 265
59	34 81	205 379	1211 7361	7149 24299	7·681 146	·130 189	287 274
60	36 00	216 000	1296 0000	7776 00000	7·745 967	0·129 099	277 778
61	37 21	226 981	1384 5841	8445 96301	7·810 250	·128 037	268 745
62	38 44	238 328	1477 6336	9161 32832	7·874 008	·127 000	260 146
63	39 69	250 047	1575 2961	9924 36543	7·937 254	·125 988	251 953
64	40 96	262 144	1677 7216	10737 41824	8·000 000	·125 000	244 141
65	42 25	274 625	1785 0625	11602 90625	8·062 258	0·124 035	236 686
66	43 56	287 496	1897 4736	12523 32576	8·124 038	·123 091	229 568
67	44 89	300 763	2015 1121	13501 25107	8·185 353	·122 169	222 767
68	46 24	314 432	2138 1376	14539 33568	8·246 211	·121 268	216 263
69	47 61	328 509	2266 7121	15640 31349	8·306 624	·120 386	210 040
70	49 00	343 000	2401 0000	16807 00000	8·366 600	0·119 523	204 082
71	50 41	357 911	2541 1681	18042 29351	8·426 150	·118 678	198 373
72	51 84	373 248	2687 3856	19349 17632	8·485 281	·117 851	192 901
73	53 29	389 017	2839 8241	20730 71593	8·544 004	·117 041	187 652
74	54 76	405 224	2998 6576	22190 06624	8·602 325	·116 248	182 615
75	56 25	421 875	3164 0625	23730 46875	8·660 254	0·115 470	177 778
76	57 76	438 976	3336 2176	25355 25376	8·717 798	·114 708	173 130
77	59 29	456 533	3515 3041	27067 84157	8·774 964	·113 961	168 663
78	60 84	474 552	3701 5056	28871 74368	8·831 761	·113 228	164 366
79	62 41	493 039	3895 0081	30770 56399	8·888 194	·112 509	160 231
80	64 00	512 000	4096 0000	32768 00000	8·944 272	0·111 803	156 250
81	65 61	531 441	4304 6721	34867 84401	9·000 000	·111 111	152 416
82	67 24	551 368	4521 2176	37073 98432	9·055 385	·110 432	148 721
83	68 89	571 787	4745 8321	39390 40643	9·110 434	·109 764	145 159
84	70 56	592 704	4978 7136	41821 19424	9·165 151	·109 109	141 723
85	72 25	614 125	5220 0625	44370 53125	9·219 544	0·108 465	138 408
86	73 96	636 056	5470 0816	47042 70176	9·273 618	·107 833	135 208
87	75 69	658 503	5728 9761	49842 09207	9·327 379	·107 211	132 118
88	77 44	681 472	5996 9536	52773 19168	9·380 832	·106 600	129 132
89	79 21	704 969	6274 2241	55840 59449	9·433 981	·106 000	126 247
90	81 00	729 000	6561 0000	59049 00000	9·486 833	0·105 409	123 457
91	82 81	753 571	6857 4961	62403 21451	9·539 392	·104 828	120 758
92	84 64	778 688	7163 9296	65908 15232	9·591 663	·104 257	118 147
93	86 49	804 357	7480 5201	69568 83693	9·643 651	·103 695	115 620
94	88 36	830 584	7807 4896	73390 40224	9·695 360	·103 142	113 173
95	90 25	857 375	8145 0625	77378 09375	9·746 794	0·102 598	110 803
96	92 16	884 736	8493 4656	81537 26976	9·797 959	·102 062	108 507
97	94 09	912 673	8852 9281	85873 40257	9·848 858	·101 535	106 281
98	96 04	941 192	9223 6816	90392 07968	9·899 495	·101 015	104 123
99	98 01	970 299	9605 9601	95099 00499	9·949 874	·100 504	102 030
100	100 00	1000 000	10000 0000	100000 00000	10·000 000	0·100 000	100 000

TABLE VIII—RECIPROCALS, ROOTS, FACTORIALS AND FACTORS

No. x	Reciprocal $1/x = x^{-1}$	Cube Root $\sqrt[3]{x}$	Fourth Root $\sqrt[4]{x}$	Fifth Root $\sqrt[5]{x}$	Factorials * $x!$	Factorials $\log x!$	Prime Factors
50	0·0200 000	3·684 031	2·659 148	2·186 724	3·041 409	64·483 075	2.5^2
51	·0196 078	3·708 430	2·672 345	2·195 402	1·551 119	66·190 645	3.17
52	·0192 308	3·732 511	2·685 350	2·203 945	8·065 818	67·906 648	$2^2.13$
53	·0188 679	3·756 286	2·698 168	2·212 357	4·274 883	69·630 924	**53**
54	·0185 185	3·779 763	2·710 806	2·220 643	2·308 437	71·363 318	2.3^3
55	0·0181 818	3·802 952	2·723 270	2·228 807	1·269 640	73·103 681	5.11
56	·0178 571	3·825 862	2·735 565	2·236 854	7·109 986	74·851 869	$2^3.7$
57	·0175 439	3·848 501	2·747 696	2·244 786	4·052 692	76·607 744	3.19
58	·0172 414	3·870 877	2·759 669	2·252 608	2·350 561	78·371 172	2.29
59	·0169 492	3·892 996	2·771 488	2·260 322	1·386 831	80·142 024	**59**
60	0·0166 667	3·914 868	2·783 158	2·267 933	8·320 987	81·920 175	$2^2.3.5$
61	·0163 934	3·936 497	2·794 682	2·275 443	5·075 802	83·705 505	**61**
62	·0161 290	3·957 892	2·806 066	2·282 855	3·146 997	85·497 896	2.31
63	·0158 730	3·979 057	2·817 313	2·290 172	1·982 608	87·297 237	$3^2.7$
64	·0156 250	4·000 000	2·828 427	2·297 397	1·268 869	89·103 417	2^6
65	0·0153 846	4·020 726	2·839 412	2·304 532	8·247 651	90·916 330	5.13
66	·0151 515	4·041 240	2·850 270	2·311 579	5·443 449	92·735 874	2.3.11
67	·0149 254	4·061 548	2·861 006	2·318 542	3·647 111	94·561 949	**67**
68	·0147 059	4·081 655	2·871 622	2·325 422	2·480 036	96·394 458	$2^2.17$
69	·0144 928	4·101 566	2·882 121	2·332 222	1·711 225	98·233 307	3.23
70	0·0142 857	4·121 285	2·892 508	2·338 943	1·197 857	100·078 405	2.5.7
71	·0140 845	4·140 818	2·902 783	2·345 588	8·504 786	101·929 663	**71**
72	·0138 889	4·160 168	2·912 951	2·352 158	6·123 446	103·786 996	$2^3.3^2$
73	·0136 986	4·179 339	2·923 013	2·358 656	4·470 115	105·650 319	**73**
74	·0135 135	4·198 336	2·932 972	2·365 083	3·307 885	107·519 550	2.37
75	0·0133 333	4·217 163	2·942 831	2·371 441	2·480 914	109·394 612	3.5^2
76	·0131 579	4·235 824	2·952 592	2·377 731	1·885 495	111·275 425	$2^2.19$
77	·0129 870	4·254 321	2·962 257	2·383 956	1·451 831	113·161 916	7.11
78	·0128 205	4·272 659	2·971 828	2·390 116	1·132 428	115·054 011	2.3.13
79	·0126 582	4·290 840	2·981 308	2·396 213	8·946 182	116·951 638	**79**
80	0·0125 000	4·308 869	2·990 698	2·402 249	7·156 946	118·854 728	$2^4.5$
81	·0123 457	4·326 749	3·000 000	2·408 225	5·797 126	120·763 213	3^4
82	·0121 951	4·344 481	3·009 217	2·414 142	4·753 643	122·677 027	2.41
83	·0120 482	4·362 071	3·018 349	2·420 001	3·945 524	124·596 105	**83**
84	·0119 048	4·379 519	3·027 400	2·425 805	3·314 240	126·520 384	$2^2.3.7$
85	0·0117 647	4·396 830	3·036 370	2·431 553	2·817 104	128·449 803	5.17
86	·0116 279	4·414 005	3·045 262	2·437 248	2·422 710	130·384 301	2.43
87	·0114 943	4·431 048	3·054 076	2·442 890	2·107 757	132·323 821	3.29
88	·0113 636	4·447 960	3·062 814	2·448 480	1·854 826	134·268 303	$2^3.11$
89	·0112 360	4·464 745	3·071 479	2·454 019	1·650 796	136·217 693	**89**
90	0·0111 111	4·481 405	3·080 070	2·459 509	1·485 716	138·171 936	$2.3^2.5$
91	·0109 890	4·497 941	3·088 591	2·464 951	1·352 002	140·130 977	7.13
92	·0108 696	4·514 357	3·097 041	2·470 345	1·243 841	142·094 765	$2^2.23$
93	·0107 527	4·530 655	3·105 423	2·475 692	1·156 773	144·063 248	3.31
94	·0106 383	4·546 836	3·113 737	2·480 993	1·087 366	146·036 376	2.47
95	0·0105 263	4·562 903	3·121 986	2·486 250	1·032 998	148·014 099	5.19
96	·0104 167	4·578 857	3·130 169	2·491 462	9·916 779	149·996 371	$2^5.3$
97	·0103 093	4·594 701	3·138 289	2·496 631	9·619 276	151·983 142	**97**
98	·0102 041	4·610 436	3·146 346	2·501 758	9·426 890	153·974 368	2.7^2
99	·0101 010	4·626 065	3·154 342	2·506 842	9·332 622	155·970 004	$3^2.11$
100	0·0100 000	4·641 589	3·162 278	2·511 886	9·332 622	157·970 004	$2^2.5^2$

* Multiply by 10^c, where c is the characteristic of $\log x!$

$\sqrt[4]{10} = 1·778\ 279$ $\sqrt[4]{100} = 3·162\ 278$ $\sqrt[4]{1000} = 5·623\ 413$

$\sqrt[5]{10} = 1·584\ 893$ $\sqrt[5]{100} = 2·511\ 886$ $\sqrt[5]{1000} = 3·981\ 072$ $\sqrt[5]{10,000} = 6·309\ 573$

TABLE VIII—POWERS AND SQUARE ROOTS

No.	Square	Cube	Fourth Power	Square Roots		Reciprocals of Square Roots	
x	x^2	x^3	x^4	$\sqrt{x}=x^{\frac{1}{2}}$	$\sqrt{10x}$	$\frac{1}{\sqrt{x}}=x^{-\frac{1}{2}}$	$\frac{1}{\sqrt{10x}}$
							0·0
100	1 00 00	1 000 000	1 0000 0000	10·000 000	31·622 777	0·1000 000	316 228
101	1 02 01	1 030 301	1 0406 0401	10·049 876	31·780 497	·0995 037	314 658
102	1 04 04	1 061 208	1 0824 3216	10·099 505	31·937 439	·0990 148	313 112
103	1 06 09	1 092 727	1 1255 0881	10·148 892	32·093 613	·0985 329	311 588
104	1 08 16	1 124 864	1 1698 5856	10·198 039	32·249 031	·0980 581	310 087
105	1 10 25	1 157 625	1 2155 0625	10·246 951	32·403 703	0·0975 900	308 607
106	1 12 36	1 191 016	1 2624 7696	10·295 630	32·557 641	·0971 286	307 148
107	1 14 49	1 225 043	1 3107 9601	10·344 080	32·710 854	·0966 736	305 709
108	1 16 64	1 259 712	1 3604 8896	10·392 305	32·863 353	·0962 250	304 290
109	1 18 81	1 295 029	1 4115 8161	10·440 307	33·015 148	·0957 826	302 891
110	1 21 00	1 331 000	1 4641 0000	10·488 088	33·166 248	0·0953 463	301 511
111	1 23 21	1 367 631	1 5180 7041	10·535 654	33·316 662	·0949 158	300 150
112	1 25 44	1 404 928	1 5735 1936	10·583 005	33·466 401	·0944 911	298 807
113	1 27 69	1 442 897	1 6304 7361	10·630 146	33·615 473	·0940 721	297 482
114	1 29 96	1 481 544	1 6889 6016	10·677 078	33·763 886	·0936 586	296 174
115	1 32 25	1 520 875	1 7490 0625	10·723 805	33·911 650	0·0932 505	294 884
116	1 34 56	1 560 896	1 8106 3936	10·770 330	34·058 773	·0928 477	293 610
117	1 36 89	1 601 613	1 8738 8721	10·816 654	34·205 263	·0924 500	292 353
118	1 39 24	1 643 032	1 9387 7776	10·862 780	34·351 128	·0920 575	291 111
119	1 41 61	1 685 159	2 0053 3921	10·908 712	34·496 377	·0916 698	289 886
120	1 44 00	1 728 000	2 0736 0000	10·954 451	34·641 016	0·0912 871	288 675
121	1 46 41	1 771 561	2 1435 8881	11·000 000	34·785 054	·0909 091	287 480
122	1 48 84	1 815 848	2 2153 3456	11·045 361	34·928 498	·0905 357	286 299
123	1 51 29	1 860 867	2 2888 6641	11·090 537	35·071 356	·0901 670	285 133
124	1 53 76	1 906 624	2 3642 1376	11·135 529	35·213 634	·0898 027	283 981
125	1 56 25	1 953 125	2 4414 0625	11·180 340	35·355 339	0·0894 427	282 843
126	1 58 76	2 000 376	2 5204 7376	11·224 972	35·496 479	·0890 871	281 718
127	1 61 29	2 048 383	2 6014 4641	11·269 428	35·637 059	·0887 357	280 607
128	1 63 84	2 097 152	2 6843 5456	11·313 708	35·777 088	·0883 883	279 508
129	1 66 41	2 146 689	2 7692 2881	11·357 817	35·916 570	·0880 451	278 423
130	1 69 00	2 197 000	2 8561 0000	11·401 754	36·055 513	0·0877 058	277 350
131	1 71 61	2 248 091	2 9449 9921	11·445 523	36·193 922	·0873 704	276 289
132	1 74 24	2 299 968	3 0359 5776	11·489 125	36·331 804	·0870 388	275 241
133	1 76 89	2 352 637	3 1290 0721	11·532 563	36·469 165	·0867 110	274 204
134	1 79 56	2 406 104	3 2241 7936	11·575 837	36·606 010	·0863 868	273 179
135	1 82 25	2 460 375	3 3215 0625	11·618 950	36·742 346	0·0860 663	272 166
136	1 84 96	2 515 456	3 4210 2016	11·661 904	36·878 178	·0857 493	271 163
137	1 87 69	2 571 353	3 5227 5361	11·704 700	37·013 511	·0854 358	270 172
138	1 90 44	2 628 072	3 6267 3936	11·747 340	37·148 351	·0851 257	269 191
139	1 93 21	2 685 619	3 7330 1041	11·789 826	37·282 704	·0848 189	268 221
140	1 96 00	2 744 000	3 8416 0000	11·832 160	37·416 574	0·0845 154	267 261
141	1 98 81	2 803 221	3 9525 4161	11·874 342	37·549 967	·0842 152	266 312
142	2 01 64	2 863 288	4 0658 6896	11·916 375	37·682 887	·0839 181	265 372
143	2 04 49	2 924 207	4 1816 1601	11·958 261	37·815 341	·0836 242	264 443
144	2 07 36	2 985 984	4 2998 1696	12·000 000	37·947 332	·0833 333	263 523
145	2 10 25	3 048 625	4 4205 0625	12·041 595	38·078 866	0·0830 455	262 613
146	2 13 16	3 112 136	4 5437 1856	12·083 046	38·209 946	·0827 606	261 712
147	2 16 09	3 176 523	4 6694 8881	12·124 356	38·340 579	·0824 786	260 820
148	2 19 04	3 241 792	4 7978 5216	12·165 525	38·470 768	·0821 995	259 938
149	2 22 01	3 307 949	4 9288 4401	12·206 556	38·600 518	·0819 232	259 064
150	2 25 00	3 375 000	5 0625 0000	12·247 449	38·729 833	0·0816 497	258 199

TABLE VIII—RECIPROCALS, CUBE ROOTS, FACTORIALS AND FACTORS

No. x	Reciprocal $\frac{1}{x} = x^{-1}$	Cube Roots $\sqrt[3]{x}$	$\sqrt[3]{10x}$	$\sqrt[3]{100x}$	Factorials * $x!$	$\log x!$	Prime Factors
	0·00						
100	...	4·641 589	10·000 000	21·544 347	9·332 622	157·970 004	$2^2.5^2$
101	990 099	·657 010	·033 223	21·615 923	9·425 948	159·974 325	**101**
102	980 392	·672 329	·066 227	21·687 029	9·614 467	161·982 925	2.3.17
103	970 874	·687 548	·099 016	21·757 671	9·902 901	163·995 762	**103**
104	961 538	·702 669	·131 594	21·827 858	1·029 902	166·012 796	$2^3.13$
105	952 381	4·717 694	10·163 964	21·897 596	1·081 397	168·033 985	3.5.7
106	943 396	·732 623	·196 128	21·966 892	1·146 281	170·059 291	2.53
107	934 579	·747 459	·228 091	22·035 755	1·226 520	172·088 675	**107**
108	925 926	·762 203	·259 856	22·104 189	1·324 642	174·122 098	$2^2.3^3$
109	917 431	·776 856	·291 425	22·172 202	1·443 860	176·159 525	**109**
110	909 091	4·791 420	10·322 801	22·239 801	1·588 246	178·200 918	2.5.11
111	900 901	·805 896	·353 988	22·306 991	1·762 953	180·246 241	3.37
112	892 857	·820 285	·384 988	22·373 779	1·974 507	182·295 459	$2^4.7$
113	884 956	·834 588	·415 804	22·440 170	2·231 193	184·348 537	**113**
114	877 193	·848 808	·446 439	22·506 171	2·543 560	186·405 442	2.3.19
115	869 565	4·862 944	10·476 896	22·571 787	2·925 094	188·466 140	5.23
116	862 069	·876 999	·507 176	22·637 024	3·393 109	190·530 598	$2^2.29$
117	854 701	·890 973	·537 282	22·701 887	3·969 937	192·598 784	$3^2.13$
118	847 458	·904 868	·567 218	22·766 381	4·684 526	194·670 666	2.59
119	840 336	·918 685	·596 985	22·830 512	5·574 586	196·746 213	7.17
120	833 333	4·932 424	10·626 586	22·894 285	6·689 503	198·825 394	$2^3.3.5$
121	826 446	·946 087	·656 022	22·957 704	8·094 299	200·908 179	11^2
122	819 672	·959 676	·685 297	23·020 775	9·875 044	202·994 539	2.61
123	813 008	·973 190	·714 413	23·083 502	1·214 630	205·084 444	3.41
124	806 452	4·986 631	·743 371	23·145 891	1·506 142	207·177 866	$2^2.31$
125	800 000	5·000 000	10·772 173	23·207 944	1·882 677	209·274 776	5^3
126	793 651	·013 298	·800 823	23·269 668	2·372 173	211·375 146	$2.3^2.7$
127	787 402	·026 526	·829 321	23·331 066	3·012 660	213·478 950	**127**
128	781 250	·039 684	·857 670	23·392 142	3·856 205	215·586 160	2^7
129	775 194	·052 774	·885 872	23·452 901	4·974 504	217·696 750	3.43
130	769 231	5·065 797	10·913 929	23·513 347	6·466 855	219·810 693	2.5.13
131	763 359	·078 753	·941 842	23·573 484	8·471 581	221·927 964	**131**
132	757 576	·091 643	·969 613	23·633 315	1·118 249	224·048 538	$2^2.3.11$
133	751 880	·104 469	10·997 244	23·692 845	1·487 271	226·172 390	7.19
134	746 269	·117 230	11·024 738	23·752 077	1·992 943	228·299 495	2.67
135	740 741	5·129 928	11·052 094	23·811 016	2·690 473	230·429 829	$3^3.5$
136	735 294	·142 563	·079 317	23·869 664	3·659 043	232·563 367	$2^3.17$
137	729 927	·155 137	·106 405	23·928 025	5·012 889	234·700 088	**137**
138	724 638	·167 649	·133 363	23·986 103	6·917 786	236·839 967	2.3.23
139	719 424	·180 101	·160 190	24·043 901	9·615 723	238·982 982	**139**
140	714 286	5·192 494	11·186 889	24·101 423	1·346 201	241·129 110	$2^2.5.7$
141	709 220	·204 828	·213 462	24·158 671	1·898 144	243·278 329	3.47
142	704 225	·217 103	·239 909	24·215 649	2·695 364	245·430 617	2.71
143	699 301	·229 322	·266 232	24·272 360	3·854 371	247·585 953	11.13
144	694 444	·241 483	·292 432	24·328 808	5·550 294	249·744 316	$2^4.3^2$
145	689 655	5·253 588	11·318 512	24·384 995	8·047 926	251·905 684	5.29
146	684 932	·265 637	·344 472	24·440 924	1·174 997	254·070 037	2.73
147	680 272	·277 632	·370 314	24·496 598	1·727 246	256·237 354	3.7^2
148	675 676	·289 572	·396 038	24·552 021	2·556 324	258·407 616	$2^2.37$
149	671 141	·301 459	·421 648	24·607 194	3·808 923	260·580 802	**149**
150	666 667	5·313 293	11·447 142	24·662 121	5·713 384	262·756 893	$2.3.5^2$

* Multiply by 10^c, where c is the characteristic of $\log x!$

TABLE VIII—POWERS AND SQUARE ROOTS

No.	Square	Cube	Fourth Power	Square Roots		Reciprocals of Square Roots	
x	x^2	x^3	x^4	$\sqrt{x}=x^{\frac{1}{2}}$	$\sqrt{10x}$	$\frac{1}{\sqrt{x}}=x^{-\frac{1}{2}}$	$\frac{1}{\sqrt{10x}}$
						0·0	0·0
150	2 25 00	3 375 000	5 0625 0000	12·247 449	38·729 833	816 497	258 199
151	2 28 01	3 442 951	5 1988 5601	12·288 206	38·858 718	813 788	257 343
152	2 31 04	3 511 808	5 3379 4816	12·328 828	38·987 177	811 107	256 495
153	2 34 09	3 581 577	5 4798 1281	12·369 317	39·115 214	808 452	255 655
154	2 37 16	3 652 264	5 6244 8656	12·409 674	39·242 834	805 823	254 824
155	2 40 25	3 723 875	5 7720 0625	12·449 900	39·370 039	803 219	254 000
156	2 43 36	3 796 416	5 9224 0896	12·489 996	39·496 835	800 641	253 185
157	2 46 49	3 869 893	6 0757 3201	12·529 964	39·623 226	798 087	252 377
158	2 49 64	3 944 312	6 2320 1296	12·569 805	39·749 214	795 557	251 577
159	2 52 81	4 019 679	6 3912 8961	12·609 520	39·874 804	793 052	250 785
160	2 56 00	4 096 000	6 5536 0000	12·649 111	40·000 000	790 569	250 000
161	2 59 21	4 173 281	6 7189 8241	12·688 578	40·124 805	788 110	249 222
162	2 62 44	4 251 528	6 8874 7536	12·727 922	40·249 224	785 674	248 452
163	2 65 69	4 330 747	7 0591 1761	12·767 145	40·373 258	783 260	247 689
164	2 68 96	4 410 944	7 2339 4816	12·806 248	40·496 913	780 869	246 932
165	2 72 25	4 492 125	7 4120 0625	12·845 233	40·620 192	778 499	246 183
166	2 75 56	4 574 296	7 5933 3136	12·884 099	40·743 098	776 151	245 440
167	2 78 89	4 657 463	7 7779 6321	12·922 848	40·865 633	773 823	244 704
168	2 82 24	4 741 632	7 9659 4176	12·961 481	40·987 803	771 517	243 975
169	2 85 61	4 826 809	8 1573 0721	13·000 000	41·109 610	769 231	243 252
170	2 89 00	4 913 000	8 3521 0000	13·038 405	41·231 056	766 965	242 536
171	2 92 41	5 000 211	8 5503 6081	13·076 697	41·352 146	764 719	241 825
172	2 95 84	5 088 448	8 7521 3056	13·114 877	41·472 883	762 493	241 121
173	2 99 29	5 177 717	8 9574 5041	13·152 946	41·593 269	760 286	240 424
174	3 02 76	5 268 024	9 1663 6176	13·190 906	41·713 307	758 098	239 732
175	3 06 25	5 359 375	9 3789 0625	13·228 757	41·833 001	755 929	239 046
176	3 09 76	5 451 776	9 5951 2576	13·266 499	41·952 354	753 778	238 366
177	3 13 29	5 545 233	9 8150 6241	13·304 135	42·071 368	751 646	237 691
178	3 16 84	5 639 752	10 0387 5856	13·341 664	42·190 046	749 532	237 023
179	3 20 41	5 735 339	10 2662 5681	13·379 088	42·308 392	747 435	236 360
180	3 24 00	5 832 000	10 4976 0000	13·416 408	42·426 407	745 356	235 702
181	3 27 61	5 929 741	10 7328 3121	13·453 624	42·544 095	743 294	235 050
182	3 31 24	6 028 568	10 9719 9376	13·490 738	42·661 458	741 249	234 404
183	3 34 89	6 128 487	11 2151 3121	13·527 749	42·778 499	739 221	233 762
184	3 38 56	6 229 504	11 4622 8736	13·564 660	42·895 221	737 210	233 126
185	3 42 25	6 331 625	11 7135 0625	13·601 471	43·011 626	735 215	232 495
186	3 45 96	6 434 856	11 9688 3216	13·638 182	43·127 717	733 236	231 869
187	3 49 69	6 539 203	12 2283 0961	13·674 794	43·243 497	731 272	231 249
188	3 53 44	6 644 672	12 4919 8336	13·711 309	43·358 967	729 325	230 633
189	3 57 21	6 751 269	12 7598 9841	13·747 727	43·474 130	727 393	230 022
190	3 61 00	6 859 000	13 0321 0000	13·784 049	43·588 989	725 476	229 416
191	3 64 81	6 967 871	13 3086 3361	13·820 275	43·703 547	723 575	228 814
192	3 68 64	7 077 888	13 5895 4496	13·856 406	43·817 805	721 688	228 218
193	3 72 49	7 189 057	13 8748 8001	13·892 444	43·931 765	719 816	227 626
194	3 76 36	7 301 384	14 1646 8496	13·928 388	44·045 431	717 958	227 038
195	3 80 25	7 414 875	14 4590 0625	13·964 240	44·158 804	716 115	226 455
196	3 84 16	7 529 536	14 7578 9056	14·000 000	44·271 887	714 286	225 877
197	3 88 09	7 645 373	15 0613 8481	14·035 669	44·384 682	712 470	225 303
198	3 92 04	7 762 392	15 3695 3616	14·071 247	44·497 191	710 669	224 733
199	3 96 01	7 880 599	15 6823 9201	14·106 736	44·609 416	708 881	224 168
200	4 00 00	8 000 000	16 0000 0000	14·142 136	44·721 360	707 107	223 607

No.	Reciprocal	Cube Roots			Factorials		Prime Factors
x	$\frac{1}{x}=x^{-1}$	$\sqrt[3]{x}$	$\sqrt[3]{10x}$	$\sqrt[3]{100x}$	* $x!$	$\log x!$	
	0·00						
150	666 667	5·313 293	11·447 142	24·662 121	5·713 384	262·756 893	$2.3.5^2$
151	662 252	·325 074	·472 524	24·716 804	8·627 210	264·935 870	**151**
152	657 895	·336 803	·497 794	24·771 247	1·311 336	267·117 714	$2^3.19$
153	653 595	·348 481	·522 954	24·825 451	2·006 344	269·302 405	$3^2.17$
154	649 351	·360 108	·548 004	24·879 419	3·089 770	271·489 926	2.7.11
155	645 161	5·371 685	11·572 945	24·933 155	4·789 143	273·680 258	5.31
156	641 026	·383 213	·597 780	24·986 660	7·471 063	275·873 382	$2^2.3.13$
157	636 943	·394 691	·622 509	25·039 936	1·172 957	278·069 282	**157**
158	632 911	·406 120	·647 133	25·092 987	1·853 272	280·267 939	2.79
159	628 931	·417 502	·671 653	25·145 815	2·946 702	282·469 336	3.53
160	625 000	5·428 835	11·696 071	25·198 421	4·714 724	284·673 456	$2^5.5$
161	621 118	·440 122	·720 387	25·250 809	7·590 705	286·880 282	7.23
162	617 284	·451 362	·744 603	25·302 980	1·229 694	289·089 797	2.3^4
163	613 497	·462 556	·768 719	25·354 937	2·004 402	291·301 985	**163**
164	609 756	·473 704	·792 737	25·406 682	3·287 219	293·516 829	$2^2.41$
165	606 061	5·484 807	11·816 658	25·458 217	5·423 911	295·734 313	3.5.11
166	602 410	·495 865	·840 481	25·509 544	9·003 692	297·954 421	2.83
167	598 802	·506 878	·864 210	25·560 666	1·503 617	300·177 137	**167**
168	595 238	·517 848	·887 844	25·611 583	2·526 076	302·402 446	$2^3.3.7$
169	591 716	·528 775	·911 384	25·662 299	4·269 068	304·630 333	13^2
170	588 235	5·539 658	11·934 832	25·712 816	7·257 416	306·860 782	2.5.17
171	584 795	·550 499	·958 188	25·763 135	1·241 018	309·093 778	$3^2.19$
172	581 395	·561 298	11·981 453	25·813 258	2·134 551	311·329 307	$2^2.43$
173	578 035	·572 055	12·004 628	25·863 187	3·692 773	313·567 353	**173**
174	574 713	·582 770	·027 714	25·912 924	6·425 426	315·807 902	2.3.29
175	571 429	5·593 445	12·050 711	25·962 471	1·124 449	318·050 940	$5^2.7$
176	568 182	·604 079	·073 621	26·011 829	1·979 031	320·296 453	$2^4.11$
177	564 972	·614 672	·096 445	26·061 001	3·502 885	322·544 426	3.59
178	561 798	·625 226	·119 183	26·109 988	6·235 135	324·794 846	2.89
179	558 659	·635 741	·141 835	26·158 792	1·116 089	327·047 699	**179**
180	555 556	5·646 216	12·164 404	26·207 414	2·008 961	329·302 971	$2^2.3^2.5$
181	552 486	·656 653	·186 889	26·255 857	3·636 219	331·560 650	**181**
182	549 451	·667 051	·209 291	26·304 121	6·617 918	333·820 721	2.7.13
183	546 448	·677 411	·231 612	26·352 209	1·211 079	336·083 172	3.61
184	543 478	·687 734	·253 851	26·400 122	2·228 385	338·347 990	$2^3.23$
185	540 541	5·698 019	12·276 010	26·447 862	4·122 513	340·615 162	5.37
186	537 634	·708 267	·298 089	26·495 431	7·667 874	342·884 675	2.3.31
187	534 759	·718 479	·320 090	26·542 829	1·433 892	345·156 517	11.17
188	531 915	·728 654	·342 012	26·590 058	2·695 718	347·430 674	$2^2.47$
189	529 101	·738 794	·363 856	26·637 120	5·094 907	349·707 136	$3^3.7$
190	526 316	5·748 897	12·385 623	26·684 016	9·680 323	351·985 890	2.5.19
191	523 560	·758 965	·407 314	26·730 749	1·848 942	354·266 923	**191**
192	520 833	·768 998	·428 930	26·777 318	3·549 968	356·550 224	$2^6.3$
193	518 135	·778 997	·450 471	26·823 726	6·851 438	358·835 782	**193**
194	515 464	·788 960	·471 937	26·869 974	1·329 179	361·123 583	2.97
195	512 821	5·798 890	12·493 330	26·916 063	2·591 899	363·413 618	3.5.13
196	510 204	·808 786	·514 649	26·961 995	5·080 122	365·705 874	$2^2.7^2$
197	507 614	·818 648	·535 897	27·007 771	1·000 784	368·000 340	**197**
198	505 051	·828 477	·557 072	27·053 392	1·981 552	370·297 006	$2.3^2.11$
199	502 513	·838 272	·578 177	27·098 860	3·943 289	372·595 859	**199**
200	500 000	5·848 035	12·599 210	27·144 176	7·886 579	374·896 889	$2^3.5^2$

*** Multiply by 10^c, where c is the characteristic of $\log x!$**

TABLE VIII—POWERS AND SQUARE ROOTS

No.	Square	Cube	Fourth Power	Square Roots		Reciprocals of Square Roots	
x	x^2	x^3	x^4	$\sqrt{x} = x^{\frac{1}{2}}$	$\sqrt{10x}$	$\frac{1}{\sqrt{x}} = x^{-\frac{1}{2}}$	$\frac{1}{\sqrt{10x}}$
						0·0	0·0
200	4 00 00	8 000 000	16 0000 0000	14·142 136	44·721 360	707 107	223 607
201	4 04 01	8 120 601	16 3224 0801	14·177 447	44·833 024	705 346	223 050
202	4 08 04	8 242 408	16 6496 6416	14·212 670	44·944 410	703 598	222 497
203	4 12 09	8 365 427	16 9818 1681	14·247 807	45·055 521	701 862	221 948
204	4 16 16	8 489 664	17 3189 1456	14·282 857	45·166 359	700 140	221 404
205	4 20 25	8 615 125	17 6610 0625	14·317 821	45·276 926	698 430	220 863
206	4 24 36	8 741 816	18 0081 4096	14·352 700	45·387 223	696 733	220 326
207	4 28 49	8 869 743	18 3603 6801	14·387 495	45·497 253	695 048	219 793
208	4 32 64	8 998 912	18 7177 3696	14·422 205	45·607 017	693 375	219 265
209	4 36 81	9 129 329	19 0802 9761	14·456 832	45·716 518	691 714	218 739
210	4 41 00	9 261 000	19 4481 0000	14·491 377	45·825 757	690 066	218 218
211	4 45 21	9 393 931	19 8211 9441	14·525 839	45·934 736	688 428	217 700
212	4 49 44	9 528 128	20 1996 3136	14·560 220	46·043 458	686 803	217 186
213	4 53 69	9 663 597	20 5834 6161	14·594 520	46·151 923	685 189	216 676
214	4 57 96	9 800 344	20 9727 3616	14·628 739	46·260 134	683 586	216 169
215	4 62 25	9 938 375	21 3675 0625	14·662 878	46·368 092	681 994	215 666
216	4 66 56	10 077 696	21 7678 2336	14·696 938	46·475 800	680 414	215 166
217	4 70 89	10 218 313	22 1737 3921	14·730 920	46·583 259	678 844	214 669
218	4 75 24	10 360 232	22 5853 0576	14·764 823	46·690 470	677 285	214 176
219	4 79 61	10 503 459	23 0025 7521	14·798 649	46·797 436	675 737	213 687
220	4 84 00	10 648 000	23 4256 0000	14·832 397	46·904 158	674 200	213 201
221	4 88 41	10 793 861	23 8544 3281	14·866 069	47·010 637	672 673	212 718
222	4 92 84	10 941 048	24 2891 2656	14·899 664	47·116 876	671 156	212 238
223	4 97 29	11 089 567	24 7297 3441	14·933 185	47·222 876	669 650	211 762
224	5 01 76	11 239 424	25 1763 0976	14·966 630	47·328 638	668 153	211 289
225	5 06 25	11 390 625	25 6289 0625	15·000 000	47·434 165	666 667	210 819
226	5 10 76	11 543 176	26 0875 7776	15·033 296	47·539 457	665 190	210 352
227	5 15 29	11 697 083	26 5523 7841	15·066 519	47·644 517	663 723	209 888
228	5 19 84	11 852 352	27 0233 6256	15·099 669	47·749 346	662 266	209 427
229	5 24 41	12 008 989	27 5005 8481	15·132 746	47·853 944	660 819	208 969
230	5 29 00	12 167 000	27 9841 0000	15·165 751	47·958 315	659 380	208 514
231	5 33 61	12 326 391	28 4739 6321	15·198 684	48·062 459	657 952	208 063
232	5 38 24	12 487 168	28 9702 2976	15·231 546	48·166 378	656 532	207 614
233	5 42 89	12 649 337	29 4729 5521	15·264 338	48·270 074	655 122	207 168
234	5 47 56	12 812 904	29 9821 9536	15·297 059	48·373 546	653 720	206 725
235	5 52 25	12 977 875	30 4980 0625	15·329 710	48·476 799	652 328	206 284
236	5 56 96	13 144 256	31 0204 4416	15·362 291	48·579 831	650 945	205 847
237	5 61 69	13 312 053	31 5495 6561	15·394 804	48·682 646	649 570	205 412
238	5 66 44	13 481 272	32 0854 2736	15·427 249	48·785 244	648 204	204 980
239	5 71 21	13 651 919	32 6280 8641	15·459 625	48·887 626	646 846	204 551
240	5 76 00	13 824 000	33 1776 0000	15·491 933	48·989 795	645 497	204 124
241	5 80 81	13 997 521	33 7340 2561	15·524 175	49·091 751	644 157	203 700
242	5 85 64	14 172 488	34 2974 2096	15·556 349	49·193 496	642 824	203 279
243	5 90 49	14 348 907	34 8678 4401	15·588 457	49·295 030	641 500	202 860
244	5 95 36	14 526 784	35 4453 5296	15·620 499	49·396 356	640 184	202 444
245	6 00 25	14 706 125	36 0300 0625	15·652 476	49·497 475	638 877	202 031
246	6 05 16	14 886 936	36 6218 6256	15·684 387	49·598 387	637 577	201 619
247	6 10 09	15 069 223	37 2209 8081	15·716 234	49·699 095	636 285	201 211
248	6 15 04	15 252 992	37 8274 2016	15·748 016	49·799 598	635 001	200 805
249	6 20 01	15 438 249	38 4412 4001	15·779 734	49·899 900	633 724	200 401
250	6 25 00	15 625 000	39 0625 0000	15·811 388	50·000 000	632 456	200 000

TABLE VIII—RECIPROCALS, CUBE ROOTS, FACTORIALS AND FACTORS

No. x	Reciprocal $\frac{1}{x} = x^{-1}$	Cube Roots $\sqrt[3]{x}$	$\sqrt[3]{10x}$	$\sqrt[3]{100x}$	Factorials * $x!$	$\log x!$	Prime Factors
	0·00						
200	500 000	5·848 035	12·599 210	27·144 176	7·886 579	374·896 889	$2^3.5^2$
201	497 512	·857 766	·620 174	27·189 341	1·585 202	377·200 085	3.67
202	495 050	·867 464	·641 069	27·234 357	3·202 109	379·505 436	2.101
203	492 611	·877 131	·661 894	27·279 224	6·500 281	381·812 932	7.29
204	490 196	·886 765	·682 651	27·323 944	1·326 057	384·122 562	$2^2.3.17$
205	487 805	5·896 369	12·703 341	27·368 518	2·718 417	386·434 316	5.41
206	485 437	·905 941	·723 963	27·412 948	5·599 940	388·748 183	2.103
207	483 092	·915 482	·744 519	27·457 234	1·159 188	391·064 154	$3^2.23$
208	480 769	·924 992	·765 009	27·501 377	2·411 110	393·382 217	$2^4.13$
209	478 469	·934 472	·785 433	27·545 380	5·039 220	395·702 363	11.19
210	476 190	5·943 922	12·805 792	27·589 242	1·058 236	398·024 583	2.3.5.7
211	473 934	·953 342	·826 086	27·632 965	2·232 878	400·348 865	**211**
212	471 698	·962 732	·846 317	27·676 550	4·733 702	402·675 201	$2^2.53$
213	469 484	·972 093	·866 484	27·719 998	1·008 279	405·003 581	3.71
214	467 290	·981 424	·886 587	27·763 311	2·157 716	407·333 994	2.107
215	465 116	5·990 726	12·906 629	27·806 489	4·639 090	409·666 433	5.43
216	462 963	6·000 000	·926 608	27·849 533	1·002 043	412·000 887	$2^3.3^3$
217	460 829	·009 245	·946 526	27·892 445	2·174 434	414·337 346	7.31
218	458 716	·018 462	·966 383	27·935 224	4·740 266	416·675 803	2.109
219	456 621	·027 650	12·986 179	27·977 874	1·038 118	419·016 247	3.73
220	454 545	6·036 811	13·005 914	28·020 393	2·283 860	421·358 670	$2^2.5.11$
221	452 489	·045 944	·025 591	28·062 784	5·047 331	423·703 062	13.17
222	450 450	·055 049	·045 208	28·105 048	1·120 508	426·049 415	2.3.37
223	448 430	·064 127	·064 766	28·147 184	2·498 732	428·397 720	**223**
224	446 429	·073 178	·084 265	28·189 195	5·597 159	430·747 968	$2^5.7$
225	444 444	6·082 202	13·103 707	28·231 081	1·259 361	433·100 150	$3^2.5^2$
226	442 478	·091 199	·123 091	28·272 843	2·846 156	435·454 259	2.113
227	440 529	·100 170	·142 418	28·314 482	6·460 773	437·810 284	**227**
228	438 596	·109 115	·161 689	28·355 999	1·473 056	440·168 219	$2^2.3.19$
229	436 681	·118 033	·180 903	28·397 394	3·373 299	442·528 055	**229**
230	434 783	6·126 926	13·200 061	28·438 670	7·758 587	444·889 783	2.5.23
231	432 900	·135 792	·219 164	28·479 826	1·792 234	447·253 395	3.7.11
232	431 034	·144 634	·238 212	28·520 863	4·157 982	449·618 883	$2^3.29$
233	429 185	·153 449	·257 205	28·561 782	9·688 098	451·986 239	**233**
234	427 350	·162 240	·276 144	28·602 585	2·267 015	454·355 454	$2.3^2.13$
235	425 532	6·171 006	13·295 029	28·643 272	5·327 485	456·726 522	5.47
236	423 729	·179 747	·313 860	28·683 843	1·257 287	459·099 434	$2^2.59$
237	421 941	·188 463	·332 639	28·724 300	2·979 769	461·474 183	3.79
238	420 168	·197 154	·351 364	28·764 643	7·091 850	463·850 760	2.7.17
239	418 410	·205 822	·370 038	28·804 873	1·694 952	466·229 157	**239**
240	416 667	6·214 465	13·388 659	28·844 991	4·067 885	468·609 369	$2^4.3.5$
241	414 938	·223 084	·407 229	28·884 998	9·803 604	470·991 386	**241**
242	413 223	·231 680	·425 747	28·924 895	2·372 472	473·375 201	2.11^2
243	411 523	·240 251	·444 214	28·964 682	5·765 107	475·760 807	3^5
244	409 836	·248 800	·462 631	29·004 359	1·406 686	478·148 197	$2^2.61$
245	408 163	6·257 325	13·480 997	29·043 929	3·446 381	480·537 363	5.7^2
246	406 504	·265 827	·499 314	29·083 391	8·478 097	482·928 298	2.3.41
247	404 858	·274 305	·517 581	29·122 746	2·094 090	485·320 995	13.19
248	403 226	·282 761	·535 799	29·161 995	5·193 343	487·715 447	$2^3.31$
249	401 606	·291 195	·553 968	29·201 138	1·293 143	490·111 646	3.83
250	400 000	6·299 605	13·572 088	29·240 177	3·232 856	492·509 586	2.5^3

* Multiply by 10^c, where c is the characteristic of $\log x!$

TABLE VIII—POWERS AND SQUARE ROOTS

No.	Square	Cube	Fourth Power	Square Roots		Reciprocals of Square Roots	
x	x^2	x^3	x^4	$\sqrt{x} = x^{\frac{1}{2}}$	$\sqrt{10x}$	$\frac{1}{\sqrt{x}} = x^{-\frac{1}{2}}$	$\frac{1}{\sqrt{10x}}$
						0·0	0·0
250	6 25 00	15 625 000	39 0625 0000	15·811 388	50·000 000	632 456	200 000
251	6 30 01	15 813 251	39 6912 6001	15·842 980	50·099 900	631 194	199 601
252	6 35 04	16 003 008	40 3275 8016	15·874 508	50·199 602	629 941	199 205
253	6 40 09	16 194 277	40 9715 2081	15·905 974	50·299 105	628 695	198 811
254	6 45 16	16 387 064	41 6231 4256	15·937 377	50·398 413	627 456	198 419
255	6 50 25	16 581 375	42 2825 0625	15·968 719	50·497 525	626 224	198 030
256	6 55 36	16 777 216	42 9496 7296	16·000 000	50·596 443	625 000	197 642
257	6 60 49	16 974 593	43 6247 0401	16·031 220	50·695 167	623 783	197 257
258	6 65 64	17 173 512	44 3076 6096	16·062 378	50·793 700	622 573	196 875
259	6 70 81	17 373 979	44 9986 0561	16·093 477	50·892 043	621 370	196 494
260	6 76 00	17 576 000	45 6976 0000	16·124 515	50·990 195	620 174	196 116
261	6 81 21	17 779 581	46 4047 0641	16·155 494	51·088 159	618 984	195 740
262	6 86 44	17 984 728	47 1199 8736	16·186 414	51·185 936	617 802	195 366
263	6 91 69	18 191 447	47 8435 0561	16·217 275	51·283 526	616 626	194 994
264	6 96 96	18 399 744	48 5753 2416	16·248 077	51·380 930	615 457	194 625
265	7 02 25	18 609 625	49 3155 0625	16·278 821	51·478 151	614 295	194 257
266	7 07 56	18 821 096	50 0641 1536	16·309 506	51·575 188	613 139	193 892
267	7 12 89	19 034 163	50 8212 1521	16·340 135	51·672 043	611 990	193 528
268	7 18 24	19 248 832	51 5868 6976	16·370 706	51·768 716	610 847	193 167
269	7 23 61	19 465 109	52 3611 4321	16·401 219	51·865 210	609 711	192 807
270	7 29 00	19 683 000	53 1441 0000	16·431 677	51·961 524	608 581	192 450
271	7 34 41	19 902 511	53 9358 0481	16·462 078	52·057 660	607 457	192 095
272	7 39 84	20 123 648	54 7363 2256	16·492 423	52·153 619	606 339	191 741
273	7 45 29	20 346 417	55 5457 1841	16·522 712	52·249 402	605 228	191 390
274	7 50 76	20 570 824	56 3640 5776	16·552 945	52·345 009	604 122	191 040
275	7 56 25	20 796 875	57 1914 0625	16·583 124	52·440 442	603 023	190 693
276	7 61 76	21 024 576	58 0278 2976	16·613 248	52·535 702	601 929	190 347
277	7 67 29	21 253 933	58 8733 9441	16·643 317	52·630 789	600 842	190 003
278	7 72 84	21 484 952	59 7281 6656	16·673 332	52·725 705	599 760	189 661
279	7 78 41	21 717 639	60 5922 1281	16·703 293	52·820 451	598 684	189 321
280	7 84 00	21 952 000	61 4656 0000	16·733 201	52·915 026	597 614	188 982
281	7 89 61	22 188 041	62 3483 9521	16·763 055	53·009 433	596 550	188 646
282	7 95 24	22 425 768	63 2406 6576	16·792 856	53·103 672	595 491	188 311
283	8 00 89	22 665 187	64 1424 7921	16·822 604	53·197 744	594 438	187 978
284	8 06 56	22 906 304	65 0539 0336	16·852 300	53·291 650	593 391	187 647
285	8 12 25	23 149 125	65 9750 0625	16·881 943	53·385 391	592 349	187 317
286	8 17 96	23 393 656	66 9058 5616	16·911 535	53·478 968	591 312	186 989
287	8 23 69	23 639 903	67 8465 2161	16·941 074	53·572 381	590 281	186 663
288	8 29 44	23 887 872	68 7970 7136	16·970 563	53·665 631	589 256	186 339
289	8 35 21	24 137 569	69 7575 7441	17·000 000	53·758 720	588 235	186 016
290	8 41 00	24 389 000	70 7281 0000	17·029 386	53·851 648	587 220	185 695
291	8 46 81	24 642 171	71 7087 1761	17·058 722	53·944 416	586 210	185 376
292	8 52 64	24 897 088	72 6994 9696	17·088 007	54·037 024	585 206	185 058
293	8 58 49	25 153 757	73 7005 0801	17·117 243	54·129 474	584 206	184 742
294	8 64 36	25 412 184	74 7118 2096	17·146 428	54·221 767	583 212	184 428
295	8 70 25	25 672 375	75 7335 0625	17·175 564	54·313 902	582 223	184 115
296	8 76 16	25 934 336	76 7656 3456	17·204 651	54·405 882	581 238	183 804
297	8 82 09	26 198 073	77 8082 7681	17·233 688	54·497 706	580 259	183 494
298	8 88 04	26 463 592	78 8615 0416	17·262 677	54·589 376	579 284	183 186
299	8 94 01	26 730 899	79 9253 8801	17·291 616	54·680 892	578 315	182 879
300	9 00 00	27 000 000	81 0000 0000	17·320 508	54·772 256	577 350	182 574

No.	Reciprocal	Cube Roots			Factorials		Prime
x	$\frac{1}{x} = x^{-1}$	$\sqrt[3]{x}$	$\sqrt[3]{10x}$	$\sqrt[3]{100x}$	* $x!$	$\log x!$	Factors
	0·00						
250	400 000	6·299 605	13·572 088	29·240 177	3·232 856	492·509 586	2.5^3
251	398 406	·307 994	·590 160	29·279 112	8·114 469	494·909 260	**251**
252	396 825	·316 360	·608 184	29·317 944	2·044 846	497·310 661	$2^2.3^2.7$
253	395 257	·324 704	·626 161	29·356 673	5·173 461	499·713 781	11.23
254	393 701	·333 026	·644 090	29·395 301	1·314 059	502·118 615	2.127
255	392 157	6·341 326	13·661 972	29·433 827	3·350 851	504·525 155	3.5.17
256	390 625	·349 604	·679 808	29·472 252	8·578 178	506·933 395	2^8
257	389 105	·357 861	·697 597	29·510 577	2·204 592	509·343 328	**257**
258	387 597	·366 097	·715 340	29·548 804	5·687 847	511·754 948	2.3.43
259	386 100	·374 311	·733 037	29·586 931	1·473 152	514·168 248	7.37
260	384 615	6·382 504	13·750 689	29·624 961	3·830 196	516·583 221	$2^2.5.13$
261	383 142	·390 677	·768 295	29·662 893	9·996 811	518·999 861	$3^2.29$
262	381 679	·398 828	·785 857	29·700 728	2·619 165	521·418 163	2.131
263	380 228	·406 959	·803 374	29·738 467	6·888 403	523·838 119	**263**
264	378 788	·415 069	·820 846	29·776 111	1·818 538	526·259 722	$2^3.3.11$
265	377 358	6·423 158	13·838 275	29·813 660	4·819 127	528·682 968	5.53
266	375 940	·431 228	·855 660	29·851 114	1·281 888	531·107 850	2.7.19
267	374 532	·439 277	·873 001	29·888 475	3·422 640	533·534 361	3.89
268	373 134	·447 306	·890 299	29·925 742	9·172 675	535·962 496	$2^2.67$
269	371 747	·455 315	·907 554	29·962 917	2·467 450	538·392 248	**269**
270	370 370	6·463 304	13·924 767	30·000 000	6·662 114	540·823 612	$2.3^3.5$
271	369 004	·471 274	·941 936	30·036 991	1·805 433	543·256 581	**271**
272	367 647	·479 224	·959 064	30·073 892	4·910 778	545·691 150	$2^4.17$
273	366 300	·487 154	·976 150	30·110 702	1·340 642	548·127 313	3.7.13
274	364 964	·495 065	13·993 194	30·147 423	3·673 360	550·565 063	2.137
275	363 636	6·502 957	14·010 197	30·184 054	1·010 174	553·004 396	$5^2.11$
276	362 319	·510 830	·027 158	30·220 596	2·788 080	555·445 305	$2^2.3.23$
277	361 011	·518 684	·044 079	30·257 050	7·722 982	557·887 785	**277**
278	359 712	·526 519	·060 959	30·293 417	2·146 989	560·331 830	2.139
279	358 423	·534 335	·077 798	30·329 697	5·990 099	562·777 434	$3^2.31$
280	357 143	6·542 133	14·094 597	30·365 890	1·677 228	565·224 592	$2^3.5.7$
281	355 872	·549 912	·111 357	30·401 997	4·713 010	567·673 298	**281**
282	354 610	·557 672	·128 076	30·438 018	1·329 069	570·123 547	2.3.47
283	353 357	·565 414	·144 757	30·473 954	3·761 265	572·575 334	**283**
284	352 113	·573 138	·161 398	30·509 806	1·068 199	575·028 652	$2^2.71$
285	350 877	6·580 844	14·177 999	30·545 574	3·044 368	577·483 497	3.5.19
286	349 650	·588 532	·194 562	30·581 258	8·706 892	579·939 863	2.11.13
287	348 432	·596 202	·211 087	30·616 859	2·498 878	582·397 745	7.41
288	347 222	·603 854	·227 573	30·652 377	7·196 768	584·857 138	$2^5.3^2$
289	346 021	·611 489	·244 021	30·687 814	2·079 866	587·318 035	17^2
290	344 828	6·619 106	14·260 431	30·723 168	6·031 612	589·780 433	2.5.29
291	343 643	·626 705	·276 804	30·758 442	1·755 199	592·244 326	3.97
292	342 466	·634 287	·293 139	30·793 634	5·125 181	594·709 709	$2^2.73$
293	341 297	·641 852	·309 437	30·828 747	1·501 678	597·176 577	**293**
294	340 136	·649 400	·325 698	30·863 780	4·414 933	599·644 924	$2.3.7^2$
295	338 983	6·656 930	14·341 921	30·898 733	1·302 405	602·114 746	5.59
296	337 838	·664 444	·358 109	30·933 607	3·855 120	604·586 038	$2^3.37$
297	336 700	·671 940	·374 260	30·968 403	1·144 971	607·058 794	$3^3.11$
298	335 570	·679 420	·390 374	31·003 121	3·412 012	609·533 011	2.149
299	334 448	·686 883	·406 453	31·037 762	1·020 192	612·008 682	13.23
300	333 333	6·694 330	14·422 496	31·072 325	3·060 575	614·485 803	$2^2.3.5^2$

* Multiply by 10^c, where c is the characteristic of $\log x!$

No.	Square	Cube	Fourth Power	Square Roots		Reciprocals of Square Roots	
x	x^2	x^3	x^4	$\sqrt{x}=x^{\frac{1}{2}}$	$\sqrt{10x}$	$\frac{1}{\sqrt{x}}=x^{-\frac{1}{2}}$	$\frac{1}{\sqrt{10x}}$
						0·0	0·0
300	9 00 00	27 000 000	81 0000 0000	17·320 508	54·772 256	577 350	182 574
301	9 06 01	27 270 901	82 0854 1201	17·349 352	54·863 467	576 390	182 271
302	9 12 04	27 543 608	83 1816 9616	17·378 147	54·954 527	575 435	181 969
303	9 18 09	27 818 127	84 2889 2481	17·406 895	55·045 436	574 485	181 668
304	9 24 16	28 094 464	85 4071 7056	17·435 596	55·136 195	573 539	181 369
305	9 30 25	28 372 625	86 5365 0625	17·464 249	55·226 805	572 598	181 071
306	9 36 36	28 652 616	87 6770 0496	17·492 856	55·317 267	571 662	180 775
307	9 42 49	28 934 443	88 8287 4001	17·521 415	55·407 581	570 730	180 481
308	9 48 64	29 218 112	89 9917 8496	17·549 929	55·497 748	569 803	180 187
309	9 54 81	29 503 629	91 1662 1361	17·578 396	55·587 768	568 880	179 896
310	9 61 00	29 791 000	92 3521 0000	17·606 817	55·677 644	567 962	179 605
311	9 67 21	30 080 231	93 5495 1841	17·635 192	55·767 374	567 048	179 316
312	9 73 44	30 371 328	94 7585 4336	17·663 522	55·856 960	566 139	179 029
313	9 79 69	30 664 297	95 9792 4961	17·691 806	55·946 403	565 233	178 743
314	9 85 96	30 959 144	97 2117 1216	17·720 045	56·035 703	564 333	178 458
315	9 92 25	31 255 875	98 4560 0625	17·748 239	56·124 861	563 436	178 174
316	9 98 56	31 554 496	99 7122 0736	17·776 389	56·213 877	562 544	177 892
317	10 04 89	31 855 013	100 9803 9121	17·804 494	56·302 753	561 656	177 611
318	10 11 24	32 157 432	102 2606 3376	17·832 555	56·391 489	560 772	177 332
319	10 17 61	32 461 759	103 5530 1121	17·860 571	56·480 085	559 893	177 054
320	10 24 00	32 768 000	104 8576 0000	17·888 544	56·568 542	559 017	176 777
321	10 30 41	33 076 161	106 1744 7681	17·916 473	56·656 862	558 146	176 501
322	10 36 84	33 386 248	107 5037 1856	17·944 358	56·745 044	557 278	176 227
323	10 43 29	33 698 267	108 8454 0241	17·972 201	56·833 089	556 415	175 954
324	10 49 76	34 012 224	110 1996 0576	18·000 000	56·920 998	555 556	175 682
325	10 56 25	34 328 125	111 5664 0625	18·027 756	57·008 771	554 700	175 412
326	10 62 76	34 645 976	112 9458 8176	18·055 470	57·096 410	553 849	175 142
327	10 69 29	34 965 783	114 3381 1041	18·083 141	57·183 914	553 001	174 874
328	10 75 84	35 287 552	115 7431 7056	18·110 770	57·271 284	552 158	174 608
329	10 82 41	35 611 289	117 1611 4081	18·138 357	57·358 522	551 318	174 342
330	10 89 00	35 937 000	118 5921 0000	18·165 902	57·445 626	550 482	174 078
331	10 95 61	36 264 691	120 0361 2721	18·193 405	57·532 599	549 650	173 814
332	11 02 24	36 594 368	121 4933 0176	18·220 867	57·619 441	548 821	173 553
333	11 08 89	36 926 037	122 9637 0321	18·248 288	57·706 152	547 997	173 292
334	11 15 56	37 259 704	124 4474 1136	18·275 667	57·792 733	547 176	173 032
335	11 22 25	37 595 375	125 9445 0625	18·303 005	57·879 185	546 358	172 774
336	11 28 96	37 933 056	127 4550 6816	18·330 303	57·965 507	545 545	172 516
337	11 35 69	38 272 753	128 9791 7761	18·357 560	58·051 701	544 735	172 260
338	11 42 44	38 614 472	130 5169 1536	18·384 776	58·137 767	543 928	172 005
339	11 49 21	38 958 219	132 0683 6241	18·411 953	58·223 707	543 125	171 751
340	11 56 00	39 304 000	133 6336 0000	18·439 089	58·309 519	542 326	171 499
341	11 62 81	39 651 821	135 2127 0961	18·466 185	58·395 205	541 530	171 247
342	11 69 64	40 001 688	136 8057 7296	18·493 242	58·480 766	540 738	170 996
343	11 76 49	40 353 607	138 4128 7201	18·520 259	58·566 202	539 949	170 747
344	11 83 36	40 707 584	140 0340 8896	18·547 237	58·651 513	539 164	170 499
345	11 90 25	41 063 625	141 6695 0625	18·574 176	58·736 701	538 382	170 251
346	11 97 16	41 421 736	143 3192 0656	18·601 075	58·821 765	537 603	170 005
347	12 04 09	41 781 923	144 9832 7281	18·627 936	58·906 706	536 828	169 760
348	12 11 04	42 144 192	146 6617 8816	18·654 758	58·991 525	536 056	169 516
349	12 18 01	42 508 549	148 3548 3601	18·681 542	59·076 222	535 288	169 273
350	12 25 00	42 875 000	150 0625 0000	18·708 287	59·160 798	534 522	169 031

No.	Reciprocal	Cube Roots			Factorials		Prime Factors
x	$\frac{1}{x}=x^{-1}$	$\sqrt[3]{x}$	$\sqrt[3]{10x}$	$\sqrt[3]{100x}$	* $x!$	$\log x!$	
	0·00						
300	333 333	6·694 330	14·422 496	31·072 325	3·060 575	614·485 803	$2^2.3.5^2$
301	332 226	·701 759	·438 503	31·106 812	9·212 331	616·964 370	7.43
302	331 126	·709 173	·454 475	31·141 222	2·782 124	619·444 376	2.151
303	330 033	·716 570	·470 411	31·175 556	8·429 836	621·925 819	3.101
304	328 947	·723 951	·486 313	31·209 815	2·562 670	624·408 693	$2^4.19$
305	327 869	6·731 315	14·502 180	31·243 999	7·816 144	626·892 993	5.61
306	326 797	·738 664	·518 012	31·278 108	2·391 740	629·378 714	$2.3^2.17$
307	325 733	·745 997	·533 809	31·312 143	7·342 642	631·865 852	**307**
308	324 675	·753 313	·549 573	31·346 104	2·261 534	634·354 403	$2^2.7.11$
309	323 625	·760 614	·565 302	31·379 992	6·988 139	636·844 362	3.103
310	322 581	6·767 899	14·580 997	31·413 807	2·166 323	639·335 723	2.5.31
311	321 543	·775 169	·596 659	31·447 549	6·737 265	641·828 484	**311**
312	320 513	·782 423	·612 287	31·481 218	2·102 027	644·322 638	$2^3.3.13$
313	319 489	·789 661	·627 882	31·514 816	6·579 343	646·818 183	**313**
314	318 471	·796 884	·643 444	31·548 343	2·065 914	649·315 112	2.157
315	317 460	6·804 092	14·658 972	31·581 798	6·507 628	651·813 423	$3^2.5.7$
316	316 456	·811 285	·674 468	31·615 183	2·056 411	654·313 110	$2^2.79$
317	315 457	·818 462	·689 931	31·648 497	6·518 822	656·814 169	**317**
318	314 465	·825 624	·705 362	31·681 741	2·072 985	659·316 596	2.3.53
319	313 480	·832 771	·720 760	31·714 916	6·612 823	661·820 387	11.29
320	312 500	6·839 904	14·736 126	31·748 021	2·116 103	664·325 537	$2^6.5$
321	311 526	·847 021	·751 460	31·781 058	6·792 692	666·832 042	3.107
322	310 559	·854 124	·766 763	31·814 025	2·187 247	669·339 898	2.7.23
323	309 598	·861 212	·782 033	31·846 925	7·064 807	671·849 100	17.19
324	308 642	·868 285	·797 272	31·879 757	2·288 997	674·359 645	$2^2.3^4$
325	307 692	6·875 344	14·812 480	31·912 522	7·439 242	676·871 529	$5^2.13$
326	306 748	·882 389	·827 657	31·945 219	2·425 193	679·384 746	2.163
327	305 810	·889 419	·842 803	31·977 849	7·930 380	681·899 294	3.109
328	304 878	·896 434	·857 918	32·010 413	2·601 165	684·415 168	$2^3.41$
329	303 951	·903 436	·873 002	32·042 911	8·557 832	686·932 364	7.47
330	303 030	6·910 423	14·888 056	32·075 343	2·824 085	689·450 878	2.3.5.11
331	302 115	·917 396	·903 079	32·107 710	9·347 720	691·970 706	**331**
332	301 205	·924 356	·918 072	32·140 012	3·103 443	694·491 844	$2^2.83$
333	300 300	·931 301	·933 035	32·172 248	1·033 447	697·014 288	$3^2.37$
334	299 401	·938 232	·947 968	32·204 421	3·451 711	699·538 034	2.167
335	298 507	6·945 150	14·962 871	32·236 529	1·156 323	702·063 079	5.67
336	297 619	·952 053	·977 745	32·268 573	3·885 246	704·589 419	$2^4.3.7$
337	296 736	·958 943	14·992 589	32·300 554	1·309 328	707·117 048	**337**
338	295 858	·965 820	15·007 404	32·332 471	4·425 529	709·645 965	2.13^2
339	294 985	·972 683	·022 189	32·364 326	1·500 254	712·176 165	3.113
340	294 118	6·979 532	15·036 946	32·396 118	5·100 864	714·707 644	$2^2.5.17$
341	293 255	·986 368	·051 674	32·427 848	1·739 395	717·240 398	11.31
342	292 398	6·993 191	·066 373	32·459 516	5·948 730	719·774 424	$2.3^2.19$
343	291 545	7·000 000	·081 043	32·491 122	2·040 414	722·309 718	7^3
344	290 698	·006 796	·095 685	32·522 667	7·019 026	724·846 277	$2^3.43$
345	289 855	7·013 579	15·110 298	32·554 150	2·421 564	727·384 096	3.5.23
346	289 017	·020 349	·124 883	32·585 573	8·378 611	729·923 172	2.173
347	288 184	·027 106	·139 440	32·616 936	2·907 378	732·463 502	**347**
348	287 356	·033 850	·153 970	32·648 238	1·011 768	735·005 081	$2^2.3.29$
349	286 533	·040 581	·168 471	32·679 480	3·531 069	737·547 906	**349**
350	285 714	7·047 299	15·182 945	32·710 663	1·235 874	740·091 974	$2.5^2.7$

* Multiply by 10^c, where c is the characteristic of $\log x!$

No.	Square	Cube	Fourth Power	Square Roots		Reciprocals of Square Roots	
x	x^2	x^3	x^4	$\sqrt{x}=x^{\frac{1}{2}}$	$\sqrt{10x}$	$\frac{1}{\sqrt{x}}=x^{-\frac{1}{2}}$	$\frac{1}{\sqrt{10x}}$
						0·0	0·0
350	12 25 00	42 875 000	150 0625 0000	18·708 287	59·160 798	534 522	169 031
351	12 32 01	43 243 551	151 7848 6401	18·734 994	59·245 253	533 761	168 790
352	12 39 04	43 614 208	153 5220 1216	18·761 663	59·329 588	533 002	168 550
353	12 46 09	43 986 977	155 2740 2881	18·788 294	59·413 803	532 246	168 311
354	12 53 16	44 361 864	157 0409 9856	18·814 888	59·497 899	531 494	168 073
355	12 60 25	44 738 875	158 8230 0625	18·841 444	59·581 876	530 745	167 836
356	12 67 36	45 118 016	160 6201 3696	18·867 962	59·665 736	529 999	167 600
357	12 74 49	45 499 293	162 4324 7601	18·894 444	59·749 477	529 256	167 365
358	12 81 64	45 882 712	164 2601 0896	18·920 888	59·833 101	528 516	167 132
359	12 88 81	46 268 279	166 1031 2161	18·947 295	59·916 609	527 780	166 899
360	12 96 00	46 656 000	167 9616 0000	18·973 666	60·000 000	527 046	166 667
361	13 03 21	47 045 881	169 8356 3041	19·000 000	60·083 276	526 316	166 436
362	13 10 44	47 437 928	171 7252 9936	19·026 298	60·166 436	525 588	166 206
363	13 17 69	47 832 147	173 6306 9361	19·052 559	60·249 481	524 864	165 977
364	13 24 96	48 228 544	175 5519 0016	19·078 784	60·332 413	524 142	165 748
365	13 32 25	48 627 125	177 4890 0625	19·104 973	60·415 230	523 424	165 521
366	13 39 56	49 027 896	179 4420 9936	19·131 126	60·497 934	522 708	165 295
367	13 46 89	49 430 863	181 4112 6721	19·157 244	60·580 525	521 996	165 070
368	13 54 24	49 836 032	183 3965 9776	19·183 326	60·663 004	521 286	164 845
369	13 61 61	50 243 409	185 3981 7921	19·209 373	60·745 370	520 579	164 622
370	13 69 00	50 653 000	187 4161 0000	19·235 384	60·827 625	519 875	164 399
371	13 76 41	51 064 811	189 4504 4881	19·261 360	60·909 769	519 174	164 177
372	13 83 84	51 478 848	191 5013 1456	19·287 302	60·991 803	518 476	163 956
373	13 91 29	51 895 117	193 5687 8641	19·313 208	61·073 726	517 780	163 737
374	13 98 76	52 313 624	195 6529 5376	19·339 080	61·155 539	517 088	163 517
375	14 06 25	52 734 375	197 7539 0625	19·364 917	61·237 244	516 398	163 299
376	14 13 76	53 157 376	199 8717 3376	19·390 719	61·318 839	515 711	163 082
377	14 21 29	53 582 633	202 0065 2641	19·416 488	61·400 326	515 026	162 866
378	14 28 84	54 010 152	204 1583 7456	19·442 222	61·481 705	514 344	162 650
379	14 36 41	54 439 939	206 3273 6881	19·467 922	61·562 976	513 665	162 435
380	14 44 00	54 872 000	208 5136 0000	19·493 589	61·644 140	512 989	162 221
381	14 51 61	55 306 341	210 7171 5921	19·519 221	61·725 197	512 316	162 008
382	14 59 24	55 742 968	212 9381 3776	19·544 820	61·806 149	511 645	161 796
383	14 66 89	56 181 887	215 1766 2721	19·570 386	61·886 994	510 976	161 585
384	14 74 56	56 623 104	217 4327 1936	19·595 918	61·967 734	510 310	161 374
385	14 82 25	57 066 625	219 7065 0625	19·621 417	62·048 368	509 647	161 165
386	14 89 96	57 512 456	221 9980 8016	19·646 883	62·128 898	508 987	160 956
387	14 97 69	57 960 603	224 3075 3361	19·672 316	62·209 324	508 329	160 748
388	15 05 44	58 411 072	226 6349 5936	19·697 716	62·289 646	507 673	160 540
389	15 13 21	58 863 869	228 9804 5041	19·723 083	62·369 865	507 020	160 334
390	15 21 00	59 319 000	231 3441 0000	19·748 418	62·449 980	506 370	160 128
391	15 28 81	59 776 471	233 7260 0161	19·773 720	62·529 993	505 722	159 923
392	15 36 64	60 236 288	236 1262 4896	19·798 990	62·609 903	505 076	159 719
393	15 44 49	60 698 457	238 5449 3601	19·824 228	62·689 712	504 433	159 516
394	15 52 36	61 162 984	240 9821 5696	19·849 433	62·769 419	503 793	159 313
395	15 60 25	61 629 875	243 4380 0625	19·874 607	62·849 025	503 155	159 111
396	15 68 16	62 099 136	245 9125 7856	19·899 749	62·928 531	502 519	158 910
397	15 76 09	62 570 773	248 4059 6881	19·924 859	63·007 936	501 886	158 710
398	15 84 04	63 044 792	250 9182 7216	19·949 937	63·087 241	501 255	158 511
399	15 92 01	63 521 199	253 4495 8401	19·974 984	63·166 447	500 626	158 312
400	16 00 00	64 000 000	256 0000 0000	20·000 000	63·245 553	500 000	158 114

No. x	Reciprocal $\frac{1}{x}=x^{-1}$	Cube Roots $\sqrt[3]{x}$	$\sqrt[3]{10x}$	$\sqrt[3]{100x}$	Factorials * $x!$	$\log x!$	Prime Factors
	0·00						
350	285 714	7·047 299	15·182 945	32·710 663	1·235 874	740·091 974	$2.5^2.7$
351	284 900	·054 004	·197 391	32·741 786	4·337 918	742·637 281	$3^3.13$
352	284 091	·060 697	·211 810	32·772 851	1·526 947	745·183 824	$2^5.11$
353	283 286	·067 377	·226 201	32·803 856	5·390 123	747·731 599	**353**
354	282 486	·074 044	·240 566	32·834 803	1·908 104	750·280 602	2.3.59
355	281 690	7·080 699	15·254 903	32·865 692	6·773 768	752·830 830	5.71
356	280 899	·087 341	·269 213	32·896 523	2·411 461	755·382 280	$2^2.89$
357	280 112	·093 971	·283 497	32·927 296	8·608 917	757·934 949	3.7.17
358	279 330	·100 588	·297 754	32·958 012	3·081 992	760·488 832	2.179
359	278 552	·107 194	·311 985	32·988 671	1·106 435	763·043 926	**359**
360	277 778	7·113 787	15·326 189	33·019 272	3·983 167	765·600 229	$2^3.3^2.5$
361	277 008	·120 367	·340 366	33·049 818	1·437 923	768·157 736	19^2
362	276 243	·126 936	·354 518	33·080 306	5·205 282	770·716 444	2.181
363	275 482	·133 492	·368 644	33·110 739	1·889 517	773·276 351	3.11^2
364	274 725	·140 037	·382 743	33·141 116	6·877 843	775·837 452	$2^2.7.13$
365	273 973	7·146 569	15·396 817	33·171 437	2·510 413	778·399 745	5.73
366	273 224	·153 090	·410 865	33·201 703	9·188 111	780·963 226	2.3.61
367	272 480	·159 599	·424 888	33·231 914	3·372 037	783·527 892	**367**
368	271 739	·166 096	·438 885	33·262 070	1·240 910	786·093 740	$2^4.23$
369	271 003	·172 581	·452 857	33·292 171	4·578 956	788·660 766	$3^2.41$
370	270 270	7·179 054	15·466 804	33·322 219	1·694 214	791·228 968	2.5.37
371	269 542	·185 516	·480 725	33·352 212	6·285 533	793·798 342	7.53
372	268 817	·191 966	·494 622	33·382 151	2·338 218	796·368 885	$2^2.3.31$
373	268 097	·198 405	·508 493	33·412 036	8·721 554	798·940 594	**373**
374	267 380	·204 832	·522 340	33·441 868	3·261 861	801·513 465	2.11.17
375	266 667	7·211 248	15·536 163	33·471 648	1·223 198	804·087 497	3.5^3
376	265 957	·217 652	·549 960	33·501 374	4·599 224	806·662 685	$2^3.47$
377	265 252	·224 045	·563 733	33·531 047	1·733 908	809·239 026	13.29
378	264 550	·230 427	·577 482	33·560 668	6·554 171	811·816 518	$2.3^3.7$
379	263 852	·236 797	·591 207	33·590 237	2·484 031	814·395 157	**379**
380	263 158	7·243 156	15·604 908	33·619 754	9·439 317	816·974 941	$2^2.5.19$
381	262 467	·249 505	·618 584	33·649 219	3·596 380	819·555 866	3.127
382	261 780	·255 842	·632 237	33·678 633	1·373 817	822·137 929	2.191
383	261 097	·262 167	·645 865	33·707 995	5·261 719	824·721 128	**383**
384	260 417	·268 482	·659 471	33·737 307	2·020 500	827·305 459	$2^7.3$
385	259 740	7·274 786	15·673 052	33·766 567	7·778 926	829·890 920	5.7.11
386	259 067	·281 079	·686 610	33·795 777	3·002 665	832·477 507	2.193
387	258 398	·287 362	·700 145	33·824 936	1·162 031	835·065 218	$3^2.43$
388	257 732	·293 633	·713 656	33·854 046	4·508 682	837·654 050	$2^2.97$
389	257 069	·299 894	·727 144	33·883 105	1·753 877	840·243 999	**389**
390	256 410	7·306 144	15·740 609	33·912 114	6·840 122	842·835 064	2.3.5.13
391	255 754	·312 383	·754 051	33·941 074	2·674 488	845·427 241	17.23
392	255 102	·318 611	·767 470	33·969 985	1·048 399	848·020 527	$2^3.7^2$
393	254 453	·324 829	·780 867	33·998 847	4·120 209	850·614 919	3.131
394	253 807	·331 037	·794 240	34·027 659	1·623 362	853·210 415	2.197
395	253 165	7·337 234	15·807 591	34·056 423	6·412 281	855·807 013	5.79
396	252 525	·343 420	·820 920	34·085 138	2·539 263	858·404 708	$2^2.3^2.11$
397	251 889	·349 597	·834 226	34·113 805	1·008 087	861·003 498	**397**
398	251 256	·355 762	·847 510	34·142 424	4·012 188	863·603 381	2.199
399	250 627	·361 918	·860 771	34·170 996	1·600 863	866·204 354	3.7.19
400	250 000	7·368 063	15·874 011	34·199 519	6·403 452	868·806 414	$2^4.5^2$

* Multiply by 10^c, where c is the characteristic of $\log x!$

No.	Square	Cube	Fourth Power	Square Roots		Reciprocals of Square Roots	
x	x^2	x^3	x^4	$\sqrt{x} = x^{\frac{1}{2}}$	$\sqrt{10x}$	$\frac{1}{\sqrt{x}} = x^{-\frac{1}{2}}$	$\frac{1}{\sqrt{10x}}$
						0·0	**0·0**
400	16 00 00	64 000 000	256 0000 0000	20·000 000	63·245 553	500 000	158 114
401	16 08 01	64 481 201	258 5696 1601	20·024 984	63·324 561	499 376	157 917
402	16 16 04	64 964 808	261 1585 2816	20·049 938	63·403 470	498 755	157 720
403	16 24 09	65 450 827	263 7668 3281	20·074 860	63·482 281	498 135	157 524
404	16 32 16	65 939 264	266 3946 2656	20·099 751	63·560 994	497 519	157 329
405	16 40 25	66 430 125	269 0420 0625	20·124 612	63·639 610	496 904	157 135
406	16 48 36	66 923 416	271 7090 6896	20·149 442	63·718 129	496 292	156 941
407	16 56 49	67 419 143	274 3959 1201	20·174 241	63·796 552	495 682	156 748
408	16 64 64	67 917 312	277 1026 3296	20·199 010	63·874 878	495 074	156 556
409	16 72 81	68 417 929	279 8293 2961	20·223 748	63·953 108	494 468	156 365
410	16 81 00	68 921 000	282 5761 0000	20·248 457	64·031 242	493 865	156 174
411	16 89 21	69 426 531	285 3430 4241	20·273 135	64·109 282	493 264	155 984
412	16 97 44	69 934 528	288 1302 5536	20·297 783	64·187 226	492 665	155 794
413	17 05 69	70 444 997	290 9378 3761	20·322 401	64·265 076	492 068	155 606
414	17 13 96	70 957 944	293 7658 8816	20·346 990	64·342 832	491 473	155 417
415	17 22 25	71 473 375	296 6145 0625	20·371 549	64·420 494	490 881	155 230
416	17 30 56	71 991 296	299 4837 9136	20·396 078	64·498 062	490 290	155 043
417	17 38 89	72 511 713	302 3738 4321	20·420 578	64·575 537	489 702	154 857
418	17 47 24	73 034 632	305 2847 6176	20·445 048	64·652 920	489 116	154 672
419	17 55 61	73 560 059	308 2166 4721	20·469 489	64·730 209	488 532	154 487
420	17 64 00	74 088 000	311 1696 0000	20·493 902	64·807 407	487 950	154 303
421	17 72 41	74 618 461	314 1437 2081	20·518 285	64·884 513	487 370	154 120
422	17 80 84	75 151 448	317 1391 1056	20·542 639	64·961 527	486 792	153 937
423	17 89 29	75 686 967	320 1558 7041	20·566 964	65·038 450	486 217	153 755
424	17 97 76	76 225 024	323 1941 0176	20·591 260	65·115 282	485 643	153 574
425	18 06 25	76 765 625	326 2539 0625	20·615 528	65·192 024	485 071	153 393
426	18 14 76	77 308 776	329 3353 8576	20·639 767	65·268 675	484 502	153 213
427	18 23 29	77 854 483	332 4386 4241	20·663 978	65·345 237	483 934	153 033
428	18 31 84	78 402 752	335 5637 7856	20·688 161	65·421 709	483 368	152 854
429	18 40 41	78 953 589	338 7108 9681	20·712 315	65·498 092	482 805	152 676
430	18 49 00	79 507 000	341 8801 0000	20·736 441	65·574 385	482 243	152 499
431	18 57 61	80 062 991	345 0714 9121	20·760 539	65·650 590	481 683	152 322
432	18 66 24	80 621 568	348 2851 7376	20·784 610	65·726 707	481 125	152 145
433	18 74 89	81 182 737	351 5212 5121	20·808 652	65·802 736	480 569	151 969
434	18 83 56	81 746 504	354 7798 2736	20·832 667	65·878 676	480 015	151 794
435	18 92 25	82 312 875	358 0610 0625	20·856 654	65·954 530	479 463	151 620
436	19 00 96	82 881 856	361 3648 9216	20·880 613	66·030 296	478 913	151 446
437	19 09 69	83 453 453	364 6915 8961	20·904 545	66·105 976	478 365	151 272
438	19 18 44	84 027 672	368 0412 0336	20·928 450	66·181 568	477 818	151 099
439	19 27 21	84 604 519	371 4138 3841	20·952 327	66·257 075	477 274	150 927
440	19 36 00	85 184 000	374 8096 0000	20·976 177	66·332 496	476 731	150 756
441	19 44 81	85 766 121	378 2285 9361	21·000 000	66·407 831	476 190	150 585
442	19 53 64	86 350 888	381 6709 2496	21·023 796	66·483 081	475 651	150 414
443	19 62 49	86 938 307	385 1367 0001	21·047 565	66·558 245	475 114	150 244
444	19 71 36	87 528 384	388 6260 2496	21·071 308	66·633 325	474 579	150 075
445	19 80 25	88 121 125	392 1390 0625	21·095 023	66·708 320	474 045	149 906
446	19 89 16	88 716 536	395 6757 5056	21·118 712	66·783 231	473 514	149 738
447	19 98 09	89 314 623	399 2363 6481	21·142 375	66·858 059	472 984	149 571
448	20 07 04	89 915 392	402 8209 5616	21·166 010	66·932 802	472 456	149 404
449	20 16 01	90 518 849	406 4296 3201	21·189 620	67·007 462	471 929	149 237
450	20 25 00	91 125 000	410 0625 0000	21·213 203	67·082 039	471 405	149 071

TABLE VIII—RECIPROCALS, CUBE ROOTS, FACTORIALS AND FACTORS

No. x	Reciprocal $\frac{1}{x} = x^{-1}$	Cube Roots $\sqrt[3]{x}$	$\sqrt[3]{10x}$	$\sqrt[3]{100x}$	Factorials $* x!$	$\log x!$	Prime Factors
	0·00						
400	250 000	7·368 063	15·874 011	34·199 519	6·403 452	868·806 414	$2^4.5^2$
401	249 377	·374 198	·887 228	34·227 995	2·567 784	871·409 559	**401**
402	248 756	·380 323	·900 423	34·256 423	1·032 249	874·013 785	2.3.67
403	248 139	·386 437	·913 597	34·284 805	4·159 965	876·619 090	13.31
404	247 525	·392 542	·926 748	34·313 139	1·680 626	879·225 471	$2^2.101$
405	246 914	7·398 636	15·939 879	34·341 427	6·806 534	881·832 926	$3^4.5$
406	246 305	·404 721	·952 987	34·369 669	2·763 453	884·441 452	2.7.29
407	245 700	·410 795	·966 074	34·397 864	1·124 725	887·051 046	11.37
408	245 098	·416 860	·979 139	34·426 012	4·588 879	889·661 707	$2^3.3.17$
409	244 499	·422 914	15·992 184	34·454 115	1·876 852	892·273 430	**409**
410	243 902	7·428 959	16·005 207	34·482 172	7·695 092	894·886 214	2.5.41
411	243 309	·434 994	·018 208	34·510 184	3·162 683	897·500 056	3.137
412	242 718	·441 019	·031 189	34·538 150	1·303 025	900·114 953	$2^2.103$
413	242 131	·447 034	·044 149	34·566 071	5·381 494	902·730 903	7.59
414	241 546	·453 040	·057 088	34·593 947	2·227 939	905·347 903	$2.3^2.23$
415	240 964	7·459 036	16·070 006	34·621 778	9·245 946	907·965 951	5.83
416	240 385	·465 022	·082 903	34·649 564	3·846 313	910·585 045	$2^5.13$
417	239 808	·470 999	·095 780	34·677 306	1·603 913	913·205 181	3.139
418	239 234	·476 966	·108 636	34·705 004	6·704 355	915·826 357	2.11.19
419	238 663	·482 924	·121 471	34·732 657	2·809 125	918·448 571	**419**
420	238 095	7·488 872	16·134 286	34·760 266	1·179 832	921·071 820	$2^2.3.5.7$
421	237 530	·494 811	·147 081	34·787 832	4·967 094	923·696 102	**421**
422	236 967	·500 741	·159 856	34·815 354	2·096 114	926·321 415	2.211
423	236 407	·506 661	·172 610	34·842 833	8·866 562	928·947 755	$3^2.47$
424	235 849	·512 572	·185 345	34·870 268	3·759 422	931·575 121	$2^3.53$
425	235 294	7·518 473	16·198 059	34·897 660	1·597 754	934·203 510	$5^2.17$
426	234 742	·524 365	·210 753	34·925 010	6·806 434	936·832 920	2.3.71
427	234 192	·530 248	·223 428	34·952 316	2·906 347	939·463 347	7.61
428	233 645	·536 122	·236 083	34·979 580	1·243 917	942·094 791	$2^2.107$
429	233 100	·541 987	·248 718	35·006 801	5·336 402	944·727 249	3.11.13
430	232 558	7·547 842	16·261 333	35·033 981	2·294 653	947·360 717	2.5.43
431	232 019	·553 689	·273 929	35·061 118	9·889 954	949·995 194	**431**
432	231 481	·559 526	·286 506	35·088 213	4·272 460	952·630 678	$2^4.3^3$
433	230 947	·565 355	·299 063	35·115 266	1·849 975	955·267 166	**433**
434	230 415	·571 174	·311 601	35·142 278	8·028 893	957·904 656	2.7.31
435	229 885	7·576 985	16·324 119	35·169 248	3·492 568	960·543 145	3.5.29
436	229 358	·582 787	·336 618	35·196 177	1·522 760	963·182 631	$2^2.109$
437	228 833	·588 579	·349 099	35·223 065	6·654 460	965·823 113	19.23
438	228 311	·594 363	·361 560	35·249 912	2·914 654	968·464 587	2.3.73
439	227 790	·600 139	·374 002	35·276 718	1·279 533	971·107 051	**439**
440	227 273	7·605 905	16·386 425	35·303 483	5·629 945	973·750 504	$2^3.5.11$
441	226 757	·611 663	·398 830	35·330 208	2·482 806	976·394 943	$3^2.7^2$
442	226 244	·617 412	·411 216	35·356 893	1·097 400	979·040 365	2.13.17
443	225 734	·623 152	·423 583	35·383 537	4·861 482	981·686 769	**443**
444	225 225	·628 884	·435 932	35·410 141	2·158 498	984·334 152	$2^2.3.37$
445	224 719	7·634 607	16·448 262	35·436 705	9·605 317	986·982 512	5.89
446	224 215	·640 321	·460 573	35·463 230	4·283 971	989·631 847	2.223
447	223 714	·646 027	·472 866	35·489 715	1·914 935	992·282 154	3.149
448	223 214	·651 725	·485 141	35·516 160	8·578 910	994·933 432	$2^6.7$
449	222 717	·657 414	·497 398	35·542 566	3·851 931	997·585 678	**449**
450	222 222	7·663 094	16·509 636	35·568 933	1·733 369	1000·238 891	$2.3^2.5^2$

* Multiply by 10^c, where c is the characteristic of $\log x!$

No.	Square	Cube	Fourth Power	Square Roots		Reciprocals of Square Roots	
x	x^2	x^3	x^4	$\sqrt{x}=x^{\frac{1}{2}}$	$\sqrt{10x}$	$\frac{1}{\sqrt{x}}=x^{-\frac{1}{2}}$	$\frac{1}{\sqrt{10x}}$
						0·0	0·0
450	20 25 00	91 125 000	410 0625 0000	21·213 203	67·082 039	471 405	149 071
451	20 34 01	91 733 851	413 7196 6801	21·236 761	67·156 534	470 882	148 906
452	20 43 04	92 345 408	417 4012 4416	21·260 292	67·230 945	470 360	148 741
453	20 52 09	92 959 677	421 1073 3681	21·283 797	67·305 275	469 841	148 577
454	20 61 16	93 576 664	424 8380 5456	21·307 276	67·379 522	469 323	148 413
455	20 70 25	94 196 375	428 5935 0625	21·330 729	67·453 688	468 807	148 250
456	20 79 36	94 818 816	432 3738 0096	21·354 157	67·527 772	468 293	148 087
457	20 88 49	95 443 993	436 1790 4801	21·377 558	67·601 775	467 780	147 925
458	20 97 64	96 071 912	440 0093 5696	21·400 935	67·675 697	467 269	147 764
459	21 06 81	96 702 579	443 8648 3761	21·424 285	67·749 539	466 760	147 602
460	21 16 00	97 336 000	447 7456 0000	21·447 611	67·823 300	466 252	147 442
461	21 25 21	97 972 181	451 6517 5441	21·470 911	67·896 981	465 746	147 282
462	21 34 44	98 611 128	455 5834 1136	21·494 185	67·970 582	465 242	147 122
463	21 43 69	99 252 847	459 5406 8161	21·517 435	68·044 103	464 739	146 964
464	21 52 96	99 897 344	463 5236 7616	21·540 659	68·117 545	464 238	146 805
465	21 62 25	100 544 625	467 5325 0625	21·563 859	68·190 908	463 739	146 647
466	21 71 56	101 194 696	471 5672 8336	21·587 033	68·264 193	463 241	146 490
467	21 80 89	101 847 563	475 6281 1921	21·610 183	68·337 398	462 745	146 333
468	21 90 24	102 503 232	479 7151 2576	21·633 308	68·410 526	462 250	146 176
469	21 99 61	103 161 709	483 8284 1521	21·656 408	68·483 575	461 757	146 020
470	22 09 00	103 823 000	487 9681 0000	21·679 483	68·556 546	461 266	145 865
471	22 18 41	104 487 111	492 1342 9281	21·702 534	68·629 440	460 776	145 710
472	22 27 84	105 154 048	496 3271 0656	21·725 561	68·702 256	460 287	145 556
473	22 37 29	105 823 817	500 5466 5441	21·748 563	68·774 995	459 800	145 402
474	22 46 76	106 496 424	504 7930 4976	21·771 541	68·847 658	459 315	145 248
475	22 56 25	107 171 875	509 0664 0625	21·794 495	68·920 244	458 831	145 095
476	22 65 76	107 850 176	513 3668 3776	21·817 424	68·992 753	458 349	144 943
477	22 75 29	108 531 333	517 6944 5841	21·840 330	69·065 187	457 869	144 791
478	22 84 84	109 215 352	522 0493 8256	21·863 211	69·137 544	457 389	144 639
479	22 94 41	109 902 239	526 4317 2481	21·886 069	69·209 826	456 912	144 488
480	23 04 00	110 592 000	530 8416 0000	21·908 902	69·282 032	456 435	144 338
481	23 13 61	111 284 641	535 2791 2321	21·931 712	69·354 164	455 961	144 187
482	23 23 24	111 980 168	539 7444 0976	21·954 498	69·426 220	455 488	144 038
483	23 32 89	112 678 587	544 2375 7521	21·977 261	69·498 201	455 016	143 889
484	23 42 56	113 379 904	548 7587 3536	22·000 000	69·570 109	454 545	143 740
485	23 52 25	114 084 125	553 3080 0625	22·022 716	69·641 941	454 077	143 592
486	23 61 96	114 791 256	557 8855 0416	22·045 408	69·713 700	453 609	143 444
487	23 71 69	115 501 303	562 4913 4561	22·068 076	69·785 385	453 143	143 296
488	23 81 44	116 214 272	567 1256 4736	22·090 722	69·856 997	452 679	143 150
489	23 91 21	116 930 169	571 7885 2641	22·113 344	69·928 535	452 216	143 003
490	24 01 00	117 649 000	576 4801 0000	22·135 944	70·000 000	451 754	142 857
491	24 10 81	118 370 771	581 2004 8561	22·158 520	70·071 392	451 294	142 712
492	24 20 64	119 095 488	585 9498 0096	22·181 073	70·142 712	450 835	142 566
493	24 30 49	119 823 157	590 7281 6401	22·203 603	70·213 959	450 377	142 422
494	24 40 36	120 553 784	595 5356 9296	22·226 111	70·285 134	449 921	142 278
495	24 50 25	121 287 375	600 3725 0625	22·248 595	70·356 236	449 467	142 134
496	24 60 16	122 023 936	605 2387 2256	22·271 057	70·427 267	449 013	141 990
497	24 70 09	122 763 473	610 1344 6081	22·293 497	70·498 227	448 561	141 848
498	24 80 04	123 505 992	615 0598 4016	22·315 914	70·569 115	448 111	141 705
499	24 90 01	124 251 499	620 0149 8001	22·338 308	70·639 932	447 661	141 563
500	25 00 00	125 000 000	625 0000 0000	22·360 680	70·710 678	447 214	141 421

No.	Reciprocal	Cube Roots			Factorials		Prime Factors
x	$\frac{1}{x} = x^{-1}$	$\sqrt[3]{x}$	$\sqrt[3]{10x}$	$\sqrt[3]{100x}$	* $x!$	log $x!$	
	0·00						
450	222 222	7·663 094	16·509 636	35·568 933	1·733 369	1000·238 891	$2.3^2.5^2$
451	221 729	·668 766	·521 857	35·595 261	7·817 493	1002·893 068	11.41
452	221 239	·674 430	·534 059	35·621 550	3·533 507	1005·548 206	$2^2.113$
453	220 751	·680 086	·546 243	35·647 800	1·600 679	1008·204 304	3.151
454	220 264	·685 733	·558 409	35·674 012	7·267 081	1010·861 360	2.227
455	219 780	7·691 372	16·570 558	35·700 185	3·306 522	1013·519 371	5.7.13
456	219 298	·697 002	·582 689	35·726 320	1·507 774	1016·178 336	$2^3.3.19$
457	218 818	·702 625	·594 802	35·752 416	6·890 527	1018·838 252	**457**
458	218 341	·708 239	·606 897	35·778 475	3·155 861	1021·499 118	2.229
459	217 865	·713 845	·618 975	35·804 496	1·448 540	1024·160 931	$3^3.17$
460	217 391	7·719 443	16·631 035	35·830 479	6·663 286	1026·823 688	$2^2.5.23$
461	216 920	·725 032	·643 078	35·856 424	3·071 775	1029·487 389	**461**
462	216 450	·730 614	·655 103	35·882 332	1·419 160	1032·152 031	2.3.7.11
463	215 983	·736 188	·667 111	35·908 202	6·570 710	1034·817 612	**463**
464	215 517	·741 753	·679 102	35·934 036	3·048 810	1037·484 130	$2^4.29$
465	215 054	7·747 311	16·691 075	35·959 832	1·417 696	1040·151 583	3.5.31
466	214 592	·752 861	·703 032	35·985 591	6·606 465	1042·819 969	2.233
467	214 133	·758 402	·714 971	36·011 313	3·085 219	1045·489 286	**467**
468	213 675	·763 936	·726 893	36·036 999	1·443 883	1048·159 532	$2^2.3^2.13$
469	213 220	·769 462	·738 798	36·062 648	6·771 810	1050·830 705	7.67
470	212 766	7·774 980	16·750 687	36·088 261	3·182 751	1053·502 803	2.5.47
471	212 314	·780 490	·762 558	36·113 837	1·499 076	1056·175 824	3.157
472	211 864	·785 993	·774 413	36·139 377	7·075 636	1058·849 766	$2^3.59$
473	211 416	·791 488	·786 251	36·164 882	3·346 776	1061·524 627	11.43
474	210 970	·796 975	·798 072	36·190 350	1·586 372	1064·200 405	2.3.79
475	210 526	7·802 454	16·809 877	36·215 782	7·535 266	1066·877 099	$5^2.19$
476	210 084	·807 925	·821 665	36·241 179	3·586 787	1069·554 706	$2^2.7.17$
477	209 644	·813 389	·833 437	36·266 540	1·710 897	1072·233 224	$3^2.53$
478	209 205	·818 846	·845 192	36·291 866	8·178 089	1074·912 652	2.239
479	208 768	·824 294	·856 931	36·317 157	3·917 305	1077·592 987	**479**
480	208 333	7·829 735	16·868 653	36·342 412	1·880 306	1080·274 229	$2^5.3.5$
481	207 900	·835 169	·880 360	36·367 632	9·044 273	1082·956 374	13.37
482	207 469	·840 595	·892 050	36·392 817	4·359 339	1085·639 421	2.241
483	207 039	·846 013	·903 723	36·417 968	2·105 561	1088·323 368	3.7.23
484	206 612	·851 424	·915 381	36·443 084	1·019 092	1091·008 213	$2^2.11^2$
485	206 186	7·856 828	16·927 023	36·468 165	4·942 594	1093·693 955	5.97
486	205 761	·862 224	·938 649	36·493 212	2·402 101	1096·380 591	2.3^5
487	205 339	·867 613	·950 258	36·518 224	1·169 823	1099·068 120	**487**
488	204 918	·872 994	·961 852	36·543 203	5·708 736	1101·756 540	$2^3.61$
489	204 499	·878 368	·973 430	36·568 147	2·791 572	1104·445 849	3.163
490	204 082	7·883 735	16·984 993	36·593 057	1·367 870	1107·136 045	$2.5.7^2$
491	203 666	·889 095	16·996 539	36·617 933	6·716 243	1109·827 126	**491**
492	203 252	·894 447	17·008 070	36·642 776	3·304 392	1112·519 092	$2^2.3.41$
493	202 840	·899 792	·019 585	36·667 585	1·629 065	1115·211 938	17.29
494	202 429	·905 129	·031 085	36·692 360	8·047 581	1117·905 665	2.13.19
495	202 020	7·910 460	17·042 569	36·717 102	3·983 553	1120·600 271	$3^2.5.11$
496	201 613	·915 783	·054 038	36·741 811	1·975 842	1123·295 752	$2^4.31$
497	201 207	·921 099	·065 491	36·766 487	9·819 936	1125·992 109	7.71
498	200 803	·926 408	·076 929	36·791 129	4·890 328	1128·689 338	2.3.83
499	200 401	·931 710	·088 352	36·815 738	2·440 274	1131·387 439	**499**
500	200 000	7·937 005	17·099 759	36·840 315	1·220 137	1134·086 409	$2^2.5^3$

* Multiply by 10^c, where c is the characteristic of log $x!$

TABLE VIII—POWERS AND SQUARE ROOTS

No.	Square	Cube	Fourth Power	Square Roots		Reciprocals of Square Roots	
x	x^2	x^3	x^4	$\sqrt{x} = x^{\frac{1}{2}}$	$\sqrt{10x}$	$\frac{1}{\sqrt{x}} = x^{-\frac{1}{2}}$	$\frac{1}{\sqrt{10x}}$
						0·0	0·0
500	25 00 00	125 000 000	625 0000 0000	22·360 680	70·710 678	447 214	141 421
501	25 10 01	125 751 501	630 0150 2001	22·383 029	70·781 353	446 767	141 280
502	25 20 04	126 506 008	635 0601 6016	22·405 357	70·851 958	446 322	141 139
503	25 30 09	127 263 527	640 1355 4081	22·427 661	70·922 493	445 878	140 999
504	25 40 16	128 024 064	645 2412 8256	22·449 944	70·992 957	445 435	140 859
505	25 50 25	128 787 625	650 3775 0625	22·472 205	71·063 352	444 994	140 720
506	25 60 36	129 554 216	655 5443 3296	22·494 444	71·133 677	444 554	140 580
507	25 70 49	130 323 843	660 7418 8401	22·516 660	71·203 932	444 116	140 442
508	25 80 64	131 096 512	665 9702 8096	22·538 855	71·274 119	443 678	140 303
509	25 90 81	131 872 229	671 2296 4561	22·561 028	71·344 236	443 242	140 165
510	26 01 00	132 651 000	676 5201 0000	22·583 180	71·414 284	442 807	140 028
511	26 11 21	133 432 831	681 8417 6641	22·605 309	71·484 264	442 374	139 891
512	26 21 44	134 217 728	687 1947 6736	22·627 417	71·554 175	441 942	139 754
513	26 31 69	135 005 697	692 5792 2561	22·649 503	71·624 018	441 511	139 618
514	26 41 96	135 796 744	697 9952 6416	22·671 568	71·693 793	441 081	139 482
515	26 52 25	136 590 875	703 4430 0625	22·693 611	71·763 500	440 653	139 347
516	26 62 56	137 388 096	708 9225 7536	22·715 633	71·833 140	440 225	139 212
517	26 72 89	138 188 413	714 4340 9521	22·737 634	71·902 712	439 799	139 077
518	26 83 24	138 991 832	719 9776 8976	22·759 613	71·972 217	439 375	138 943
519	26 93 61	139 798 359	725 5534 8321	22·781 571	72·041 655	438 951	138 809
520	27 04 00	140 608 000	731 1616 0000	22·803 509	72·111 026	438 529	138 675
521	27 14 41	141 420 761	736 8021 6481	22·825 424	72·180 330	438 108	138 542
522	27 24 84	142 236 648	742 4753 0256	22·847 319	72·249 567	437 688	138 409
523	27 35 29	143 055 667	748 1811 3841	22·869 193	72·318 739	437 269	138 277
524	27 45 76	143 877 824	753 9197 9776	22·891 046	72·387 844	436 852	138 145
525	27 56 25	144 703 125	759 6914 0625	22·912 878	72·456 884	436 436	138 013
526	27 66 76	145 531 576	765 4960 8976	22·934 690	72·525 857	436 021	137 882
527	27 77 29	146 363 183	771 3339 7441	22·956 481	72·594 766	435 607	137 751
528	27 87 84	147 197 952	777 2051 8656	22·978 251	72·663 608	435 194	137 620
529	27 98 41	148 035 889	783 1098 5281	23·000 000	72·732 386	434 783	137 490
530	28 09 00	148 877 000	789 0481 0000	23·021 729	72·801 099	434 372	137 361
531	28 19 61	149 721 291	795 0200 5521	23·043 437	72·869 747	433 963	137 231
532	28 30 24	150 568 768	801 0258 4576	23·065 125	72·938 330	433 555	137 102
533	28 40 89	151 419 437	807 0655 9921	23·086 793	73·006 849	433 148	136 973
534	28 51 56	152 273 304	813 1394 4336	23·108 440	73·075 304	432 742	136 845
535	28 62 25	153 130 375	819 2475 0625	23·130 067	73·143 694	432 338	136 717
536	28 72 96	153 990 656	825 3899 1616	23·151 674	73·212 021	431 934	136 590
537	28 83 69	154 854 153	831 5668 0161	23·173 260	73·280 284	431 532	136 462
538	28 94 44	155 720 872	837 7782 9136	23·194 827	73·348 483	431 131	136 335
539	29 05 21	156 590 819	844 0245 1441	23·216 374	73·416 619	430 730	136 209
540	29 16 00	157 464 000	850 3056 0000	23·237 900	73·484 692	430 331	136 083
541	29 26 81	158 340 421	856 6216 7761	23·259 407	73·552 702	429 934	135 957
542	29 37 64	159 220 088	862 9728 7696	23·280 893	73·620 649	429 537	135 831
543	29 48 49	160 103 007	869 3593 2801	23·302 360	73·688 534	429 141	135 706
544	29 59 36	160 989 184	875 7811 6096	23·323 808	73·756 356	428 746	135 582
545	29 70 25	161 878 625	882 2385 0625	23·345 235	73·824 115	428 353	135 457
546	29 81 16	162 771 336	888 7314 9456	23·366 643	73·891 813	427 960	135 333
547	29 92 09	163 667 323	895 2602 5681	23·388 031	73·959 448	427 569	135 209
548	30 03 04	164 566 592	901 8249 2416	23·409 400	74·027 022	427 179	135 086
549	30 14 01	165 469 149	908 4256 2801	23·430 749	74·094 534	426 790	134 963
550	30 25 00	166 375 000	915 0625 0000	23·452 079	74·161 985	426 401	134 840

TABLE VIII—RECIPROCALS, CUBE ROOTS, FACTORIALS AND FACTORS

No. x	Reciprocal $\frac{1}{x}=x^{-1}$	Cube Roots $\sqrt[3]{x}$	$\sqrt[3]{10x}$	$\sqrt[3]{100x}$	Factorials * $x!$	$\log x!$	Prime Factors
	0·00						
500	200 000	7·937 005	17·099 759	36·840 315	1·220 137	1134·086 409	$2^2.5^3$
501	199 601	·942 293	·111 152	36·864 859	6·112 885	1136·786 246	3.167
502	199 203	·947 574	·122 529	36·889 370	3·068 669	1139·486 950	2.251
503	198 807	·952 848	·133 891	36·913 849	1·543 540	1142·188 518	**503**
504	198 413	·958 114	·145 238	36·938 295	7·779 443	1144·890 948	$2^3.3^2.7$
505	198 020	7·963 374	17·156 570	36·962 709	3·928 619	1147·594 240	5.101
506	197 628	·968 627	·167 887	36·987 091	1·987 881	1150·298 390	2.11.23
507	197 239	·973 873	·179 189	37·011 440	1·007 856	1153·003 398	3.13^2
508	196 850	·979 112	·190 476	37·035 758	5·119 907	1155·709 262	$2^2.127$
509	196 464	·984 344	·201 749	37·060 044	2·606 033	1158·415 980	**509**
510	196 078	7·989 570	17·213 006	37·084 298	1·329 077	1161·123 550	2.3.5.17
511	195 695	7·994 788	·224 249	37·108 520	6·791 582	1163·831 971	7.73
512	195 312	8·000 000	·235 478	37·132 711	3·477 290	1166·541 241	2^9
513	194 932	·005 205	·246 691	37·156 870	1·783 850	1169·251 358	$3^3.19$
514	194 553	·010 403	·257 890	37·180 998	9·168 987	1171·962 321	2.257
515	194 175	8·015 595	17·269 075	37·205 094	4·722 028	1174·674 129	5.103
516	193 798	·020 779	·280 245	37·229 160	2·436 567	1177·386 778	$2^2.3.43$
517	193 424	·025 957	·291 401	37·253 194	1·259 705	1180·100 269	11.47
518	193 050	·031 129	·302 542	37·277 197	6·525 272	1182·814 599	2.7.37
519	192 678	·036 293	·313 669	37·301 170	3·386 616	1185·529 766	3.173
520	192 308	8·041 452	17·324 782	37·325 112	1·761 040	1188·245 769	$2^3.5.13$
521	191 939	·046 603	·335 881	37·349 023	9·175 020	1190·962 607	**521**
522	191 571	·051 748	·346 965	37·372 903	4·789 361	1193·680 278	$2.3^2.29$
523	191 205	·056 886	·358 035	37·396 753	2·504 836	1196·398 779	**523**
524	190 840	·062 018	·369 091	37·420 573	1·312 534	1199·118 111	$2^2.131$
525	190 476	8·067 143	17·380 133	37·444 362	6·890 803	1201·838 270	$3.5^2.7$
526	190 114	·072 262	·391 161	37·468 121	3·624 562	1204·559 256	2.263
527	189 753	·077 374	·402 175	37·491 850	1·910 144	1207·281 066	17.31
528	189 394	·082 480	·413 175	37·515 549	1·008 556	1210·003 700	$2^4.3.11$
529	189 036	·087 579	·424 162	37·539 218	5·335 262	1212·727 156	23^2
530	188 679	8·092 672	17·435 134	37·562 858	2·827 689	1215·451 432	2.5.53
531	188 324	·097 759	·446 093	37·586 467	1·501 503	1218·176 526	$3^2.59$
532	187 970	·102 839	·457 037	37·610 047	7·987 995	1220·902 438	$2^2.7.19$
533	187 617	·107 913	·467 969	37·633 598	4·257 601	1223·629 165	13.41
534	187 266	·112 980	·478 886	37·657 119	2·273 559	1226·356 706	2.3.89
535	186 916	8·118 041	17·489 790	37·680 610	1·216 354	1229·085 060	5.107
536	186 567	·123 096	·500 680	37·704 073	6·519 658	1231·814 225	$2^3.67$
537	186 220	·128 145	·511 557	37·727 506	3·501 056	1234·544 199	3.179
538	185 874	·133 187	·522 420	37·750 910	1·883 568	1237·274 981	2.269
539	185 529	·138 223	·533 270	37·774 285	1·015 243	1240·006 570	$7^2.11$
540	185 185	8·143 253	17·544 106	37·797 631	5·482 314	1242·738 964	$2^2.3^3.5$
541	184 843	·148 276	·554 929	37·820 949	2·965 932	1245·472 161	**541**
542	184 502	·153 294	·565 739	37·844 238	1·607 535	1248·206 160	2.271
543	184 162	·158 305	·576 536	37·867 498	8·728 916	1250·940 960	3.181
544	183 824	·163 310	·587 319	37·890 729	4·748 530	1253·676 559	$2^5.17$
545	183 486	8·168 309	17·598 089	37·913 933	2·587 949	1256·412 956	5.109
546	183 150	·173 302	·608 845	37·937 107	1·413 020	1259·150 148	2.3.7.13
547	182 815	·178 289	·619 589	37·960 254	7·729 220	1261·888 136	**547**
548	182 482	·183 269	·630 320	37·983 372	4·235 613	1264·626 916	$2^2.137$
549	182 149	·188 244	·641 037	38·006 462	2·325 351	1267·366 489	$3^2.61$
550	181 818	8·193 213	17·651 742	38·029 525	1·278 943	1270·106 851	$2.5^2.11$

* Multiply by 10^c, where c is the characteristic of $\log x!$

No.	Square	Cube	Fourth Power	Square Roots		Reciprocals of Square Roots	
x	x^2	x^3	x^4	$\sqrt{x}=x^{\frac{1}{2}}$	$\sqrt{10x}$	$\frac{1}{\sqrt{x}}=x^{-\frac{1}{2}}$	$\frac{1}{\sqrt{10x}}$
						0·0	0·0
550	30 25 00	166 375 000	915 0625 0000	23·452 079	74·161 985	426 401	134 840
551	30 36 01	167 284 151	921 7356 7201	23·473 389	74·229 374	426 014	134 718
552	30 47 04	168 196 608	928 4452 7616	23·494 680	74·296 702	425 628	134 595
553	30 58 09	169 112 377	935 1914 4481	23·515 952	74·363 970	425 243	134 474
554	30 69 16	170 031 464	941 9743 1056	23·537 205	74·431 176	424 859	134 352
555	30 80 25	170 953 875	948 7940 0625	23·558 438	74·498 322	424 476	134 231
556	30 91 36	171 879 616	955 6506 6496	23·579 652	74·565 408	424 094	134 110
557	31 02 49	172 808 693	962 5444 2001	23·600 847	74·632 433	423 714	133 990
558	31 13 64	173 741 112	969 4754 0496	23·622 024	74·699 398	423 334	133 870
559	31 24 81	174 676 879	976 4437 5361	23·643 181	74·766 303	422 955	133 750
560	31 36 00	175 616 000	983 4496 0000	23·664 319	74·833 148	422 577	133 631
561	31 47 21	176 558 481	990 4930 7841	23·685 439	74·899 933	422 200	133 511
562	31 58 44	177 504 328	997 5743 2336	23·706 539	74·966 659	421 825	133 393
563	31 69 69	178 453 547	1004 6934 6961	23·727 621	75·033 326	421 450	133 274
564	31 80 96	179 406 144	1011 8506 5216	23·748 684	75·099 933	421 076	133 156
565	31 92 25	180 362 125	1019 0460 0625	23·769 729	75·166 482	420 703	133 038
566	32 03 56	181 321 496	1026 2796 6736	23·790 755	75·232 971	420 331	132 920
567	32 14 89	182 284 263	1033 5517 7121	23·811 762	75·299 402	419 961	132 803
568	32 26 24	183 250 432	1040 8624 5376	23·832 751	75·365 775	419 591	132 686
569	32 37 61	184 220 009	1048 2118 5121	23·853 721	75·432 089	419 222	132 570
570	32 49 00	185 193 000	1055 6001 0000	23·874 673	75·498 344	418 854	132 453
571	32 60 41	186 169 411	1063 0273 3681	23·895 606	75·564 542	418 487	132 337
572	32 71 84	187 149 248	1070 4936 9856	23·916 521	75·630 682	418 121	132 221
573	32 83 29	188 132 517	1077 9993 2241	23·937 418	75·696 763	417 756	132 106
574	32 94 76	189 119 224	1085 5443 4576	23·958 297	75·762 788	417 392	131 991
575	33 06 25	190 109 375	1093 1289 0625	23·979 158	75·828 754	417 029	131 876
576	33 17 76	191 102 976	1100 7531 4176	24·000 000	75·894 664	416 667	131 762
577	33 29 29	192 100 033	1108 4171 9041	24·020 824	75·960 516	416 305	131 647
578	33 40 84	193 100 552	1116 1211 9056	24·041 631	76·026 311	415 945	131 533
579	33 52 41	194 104 539	1123 8652 8081	24·062 419	76·092 050	415 586	131 420
580	33 64 00	195 112 000	1131 6496 0000	24·083 189	76·157 731	415 227	131 306
581	33 75 61	196 122 941	1139 4742 8721	24·103 942	76·223 356	414 870	131 193
582	33 87 24	197 137 368	1147 3394 8176	24·124 676	76·288 924	414 513	131 081
583	33 98 89	198 155 287	1155 2453 2321	24·145 393	76·354 437	414 158	130 968
584	34 10 56	199 176 704	1163 1919 5136	24·166 092	76·419 893	413 803	130 856
585	34 22 25	200 201 625	1171 1795 0625	24·186 773	76·485 293	413 449	130 744
586	34 33 96	201 230 056	1179 2081 2816	24·207 437	76·550 637	413 096	130 632
587	34 45 69	202 262 003	1187 2779 5761	24·228 083	76·615 925	412 744	130 521
588	34 57 44	203 297 472	1195 3891 3536	24·248 711	76·681 158	412 393	130 410
589	34 69 21	204 336 469	1203 5418 0241	24·269 322	76·746 335	412 043	130 299
590	34 81 00	205 379 000	1211 7361 0000	24·289 916	76·811 457	411 693	130 189
591	34 92 81	206 425 071	1219 9721 6961	24·310 492	76·876 524	411 345	130 079
592	35 04 64	207 474 688	1228 2501 5296	24·331 050	76·941 536	410 997	129 969
593	35 16 49	208 527 857	1236 5701 9201	24·351 591	77·006 493	410 651	129 859
594	35 28 36	209 584 584	1244 9324 2896	24·372 115	77·071 395	410 305	129 750
595	35 40 25	210 644 875	1253 3370 0625	24·392 622	77·136 243	409 960	129 641
596	35 52 16	211 708 736	1261 7840 6656	24·413 111	77·201 036	409 616	129 532
597	35 64 09	212 776 173	1270 2737 5281	24·433 583	77·265 775	409 273	129 423
598	35 76 04	213 847 192	1278 8062 0816	24·454 039	77·330 460	408 930	129 315
599	35 88 01	214 921 799	1287 3815 7601	24·474 477	77·395 090	408 589	129 207
600	36 00 00	216 000 000	1296 0000 0000	24·494 897	77·459 667	408 248	129 099

No.	Reciprocal	Cube Roots			Factorials		Prime Factors
x	$\frac{1}{x} = x^{-1}$	$\sqrt[3]{x}$	$\sqrt[3]{10x}$	$\sqrt[3]{100x}$	$* x!$	$\log x!$	
	0·00						
550	181 818	8·193 213	17·651 742	38·029 525	1·278 943	1270·106 851	$2.5^{2}.11$
551	181 488	·198 175	·662 433	38·052 559	7·046 977	1272·848 003	19.29
552	181 159	·203 132	·673 112	38·075 565	3·889 931	1275·589 942	$2^{3}.3.23$
553	180 832	·208 082	·683 778	38·098 544	2·151 132	1278·332 667	7.79
554	180 505	·213 027	·694 430	38·121 495	1·191 727	1281·076 177	2.277
555	180 180	8·217 966	17·705 071	38·144 418	6·614 086	1283·820 470	3.5.37
556	179 856	·222 899	·715 698	38·167 314	3·677 432	1286·565 545	$2^{2}.139$
557	179 533	·227 825	·726 312	38·190 182	2·048 329	1289·311 400	**557**
558	179 211	·232 746	·736 914	38·213 023	1·142 968	1292·058 034	$2.3^{2}.31$
559	178 891	·237 661	·747 503	38·235 837	6·389 190	1294·805 446	13.43
560	178 571	8·242 571	17·758 080	38·258 624	3·577 946	1297·553 634	$2^{4}.5.7$
561	178 253	·247 474	·768 644	38·281 383	2·007 228	1300·302 597	3.11.17
562	177 936	·252 372	·779 195	38·304 116	1·128 062	1303·052 333	2.281
563	177 620	·257 263	·789 734	38·326 821	6·350 990	1305·802 841	**563**
564	177 305	·262 149	·800 261	38·349 500	3·581 958	1308·554 121	$2^{2}.3.47$
565	176 991	8·267 029	17·810 775	38·372 151	2·023 806	1311·306 169	5.113
566	176 678	·271 904	·821 277	38·394 776	1·145 474	1314·058 985	2.283
567	176 367	·276 773	·831 766	38·417 375	6·494 840	1316·812 568	$3^{4}.7$
568	176 056	·281 635	·842 243	38·439 947	3·689 069	1319·566 917	$2^{3}.71$
569	175 747	·286 493	·852 707	38·462 492	2·099 080	1322·322 029	**569**
570	175 439	8·291 344	17·863 160	38·485 011	1·196 476	1325·077 904	2.3.5.19
571	175 131	·296 190	·873 600	38·507 504	6·831 877	1327·834 540	**571**
572	174 825	·301 031	·884 028	38·529 970	3·907 833	1330·591 936	$2^{2}.11.13$
573	174 520	·305 865	·894 444	38·552 411	2·239 189	1333·350 091	3.191
574	174 216	·310 694	·904 848	38·574 825	1·285 294	1336·109 003	2.7.41
575	173 913	8·315 517	17·915 239	38·597 213	7·390 442	1338·868 670	$5^{2}.23$
576	173 611	·320 335	·925 619	38·619 575	4·256 894	1341·629 093	$2^{6}.3^{2}$
577	173 310	·325 148	·935 987	38·641 912	2·456 228	1344·390 269	**577**
578	173 010	·329 954	·946 342	38·664 222	1·419 700	1347·152 197	2.17^{2}
579	172 712	·334 755	·956 686	38·686 507	8·220 062	1349·914 875	3.193
580	172 414	8·339 551	17·967 018	38·708 766	4·767 636	1352·678 303	$2^{2}.5.29$
581	172 117	·344 341	·977 338	38·731 000	2·769 997	1355·442 479	7.83
582	171 821	·349 126	·987 646	38·753 208	1·612 138	1358·207 402	2.3.97
583	171 527	·353 905	17·997 942	38·775 391	9·398 764	1360·973 071	11.53
584	171 233	·358 678	18·008 227	38·797 548	5·488 878	1363·739 484	$2^{3}.73$
585	170 940	8·363 447	18·018 499	38·819 680	3·210 994	1366·506 639	$3^{2}.5.13$
586	170 648	·368 209	·028 761	38·841 787	1·881 642	1369·274 537	2.293
587	170 358	·372 967	·039 010	38·863 869	1·104 524	1372·043 175	**587**
588	170 068	·377 719	·049 248	38·885 926	6·494 602	1374·812 553	$2^{2}.3.7^{2}$
589	169 779	·382 465	·059 474	38·907 957	3·825 320	1377·582 668	19.31
590	169 492	8·387 207	18·069 689	38·929 964	2·256 939	1380·353 520	2.5.59
591	169 205	·391 942	·079 892	38·951 946	1·333 851	1383·125 107	3.197
592	168 919	·396 673	·090 083	38·973 903	7·896 398	1385·897 429	$2^{4}.37$
593	168 634	·401 398	·100 264	38·995 836	4·682 564	1388·670 484	**593**
594	168 350	·406 118	·110 432	39·017 743	2·781 443	1391·444 270	$2.3^{3}.11$
595	168 067	8·410 833	18·120 589	39·039 627	1·654 959	1394·218 787	5.7.17
596	167 785	·415 542	·130 735	39·061 485	9·863 553	1396·994 033	$2^{2}.149$
597	167 504	·420 246	·140 870	39·083 320	5·888 541	1399·770 008	3.199
598	167 224	·424 945	·150 993	39·105 129	3·521 348	1402·54[illegible] 709	2.13.23
599	166 945	·429 638	·161 105	39·126 915	2·109 287	1405·324 136	**599**
600	166 667	8·434 327	18·171 206	39·148 676	1·265 572	1408·102 287	$2^{3}.3.5^{2}$

* Multiply by 10^{c}, where c is the characteristic of log x!

No. x	Square x^2	Cube x^3	Fourth Power x^4	Square Roots $\sqrt{x} = x^{\frac{1}{2}}$	$\sqrt{10x}$	Reciprocals of Square Roots $\frac{1}{\sqrt{x}} = x^{-\frac{1}{2}}$	$\frac{1}{\sqrt{10x}}$
						0·0	0·0
600	36 00 00	216 000 000	1296 0000 0000	24·494 897	77·459 667	408 248	129 099
601	36 12 01	217 081 801	1304 6616 2401	24·515 301	77·524 190	407 909	128 992
602	36 24 04	218 167 208	1313 3665 9216	24·535 688	77·588 659	407 570	128 885
603	36 36 09	219 256 227	1322 1150 4881	24·556 058	77·653 075	407 231	128 778
604	36 48 16	220 348 864	1330 9071 3856	24·576 411	77·717 437	406 894	128 671
605	36 60 25	221 445 125	1339 7430 0625	24·596 748	77·781 746	406 558	128 565
606	36 72 36	222 545 016	1348 6227 9696	24·617 067	77·846 002	406 222	128 459
607	36 84 49	223 648 543	1357 5466 5601	24·637 370	77·910 205	405 887	128 353
608	36 96 64	224 755 712	1366 5147 2896	24·657 656	77·974 355	405 554	128 247
609	37 08 81	225 866 529	1375 5271 6161	24·677 925	78·038 452	405 220	128 142
610	37 21 00	226 981 000	1384 5841 0000	24·698 178	78·102 497	404 888	128 037
611	37 33 21	228 099 131	1393 6856 9041	24·718 414	78·166 489	404 557	127 932
612	37 45 44	229 220 928	1402 8320 7936	24·738 634	78·230 429	404 226	127 827
613	37 57 69	230 346 397	1412 0234 1361	24·758 837	78·294 317	403 896	127 723
614	37 69 96	231 475 544	1421 2598 4016	24·779 023	78·358 152	403 567	127 619
615	37 82 25	232 608 375	1430 5415 0625	24·799 194	78·421 936	403 239	127 515
616	37 94 56	233 744 896	1439 8685 5936	24·819 347	78·485 667	402 911	127 412
617	38 06 89	234 885 113	1449 2411 4721	24·839 485	78·549 348	402 585	127 309
618	38 19 24	236 029 032	1458 6594 1776	24·859 606	78·612 976	402 259	127 205
619	38 31 61	237 176 659	1468 1235 1921	24·879 711	78·676 553	401 934	127 103
620	38 44 00	238 328 000	1477 6336 0000	24·899 799	78·740 079	401 610	127 000
621	38 56 41	239 483 061	1487 1898 0881	24·919 872	78·803 553	401 286	126 898
622	38 68 84	240 641 848	1496 7922 9456	24·939 928	78·866 977	400 963	126 796
623	38 81 29	241 804 367	1506 4412 0641	24·959 968	78·930 349	400 642	126 694
624	38 93 76	242 970 624	1516 1366 9376	24·979 992	78·993 671	400 320	126 592
625	39 06 25	244 140 625	1525 8789 0625	25·000 000	79·056 942	400 000	126 491
626	39 18 76	245 314 376	1535 6679 9376	25·019 992	79·120 162	399 680	126 390
627	39 31 29	246 491 883	1545 5041 0641	25·039 968	79·183 332	399 362	126 289
628	39 43 84	247 673 152	1555 3873 9456	25·059 928	79·246 451	399 043	126 189
629	39 56 41	248 858 189	1565 3180 0881	25·079 872	79·309 520	398 726	126 088
630	39 69 00	250 047 000	1575 2961 0000	25·099 801	79·372 539	398 410	125 988
631	39 81 61	251 239 591	1585 3218 1921	25·119 713	79·435 508	398 094	125 888
632	39 94 24	252 435 968	1595 3953 1776	25·139 610	79·498 428	397 779	125 789
633	40 06 89	253 636 137	1605 5167 4721	25·159 491	79·561 297	397 464	125 689
634	40 19 56	254 840 104	1615 6862 5936	25·179 357	79·624 117	397 151	125 590
635	40 32 25	256 047 875	1625 9040 0625	25·199 206	79·686 887	396 838	125 491
636	40 44 96	257 259 456	1636 1701 4016	25·219 040	79·749 608	396 526	125 392
637	40 57 69	258 474 853	1646 4848 1361	25·238 859	79·812 280	396 214	125 294
638	40 70 44	259 694 072	1656 8481 7936	25·258 662	79·874 902	395 904	125 196
639	40 83 21	260 917 119	1667 2603 9041	25·278 449	79·937 476	395 594	125 098
640	40 96 00	262 144 000	1677 7216 0000	25·298 221	80·000 000	395 285	125 000
641	41 08 81	263 374 721	1688 2319 6161	25·317 978	80·062 476	394 976	124 902
642	41 21 64	264 609 288	1698 7916 2896	25·337 719	80·124 902	394 669	124 805
643	41 34 49	265 847 707	1709 4007 5601	25·357 445	80·187 281	394 362	124 708
644	41 47 36	267 089 984	1720 0594 9696	25·377 155	80·249 611	394 055	124 611
645	41 60 25	268 336 125	1730 7680 0625	25·396 850	80·311 892	393 750	124 515
646	41 73 16	269 586 136	1741 5264 3856	25·416 530	80·374 125	393 445	124 418
647	41 86 09	270 840 023	1752 3349 4881	25·436 195	80·436 310	393 141	124 322
648	41 99 04	272 097 792	1763 1936 9216	25·455 844	80·498 447	392 837	124 226
649	42 12 01	273 359 449	1774 1028 2401	25·475 478	80·560 536	392 534	124 130
650	42 25 00	274 625 000	1785 0625 0000	25·495 098	80·622 577	392 232	124 035

TABLE VIII—RECIPROCALS, CUBE ROOTS, FACTORIALS AND FACTORS

No.	Reciprocal	Cube Roots			Factorials		Prime
x	$\frac{1}{x} = x^{-1}$	$\sqrt[3]{x}$	$\sqrt[3]{10x}$	$\sqrt[3]{100x}$	* $x!$	$\log x!$	Factors
	0·00						
600	166 667	8·434 327	18·171 206	39·148 676	1·265 572	1408·102 287	$2^3.3.5^2$
601	166 389	·439 010	·181 295	39·170 414	7·606 090	1410·881 161	**601**
602	166 113	·443 688	·191 374	39·192 127	4·578 866	1413·660 758	2.7.43
603	165 837	·448 361	·201 441	39·213 816	2·761 056	1416·441 075	$3^2.67$
604	165 563	·453 028	·211 497	39·235 481	1·667 678	1419·222 112	$2^2.151$
605	165 289	8·457 691	18·221 542	39·257 122	1·008 945	1422·003 868	5.11^2
606	165 017	·462 348	·231 576	39·278 739	6·114 208	1424·786 340	2.3.101
607	164 745	·467 000	·241 599	39·300 333	3·711 324	1427·569 529	**607**
608	164 474	·471 647	·251 611	39·321 903	2·256 485	1430·353 432	$2^5.19$
609	164 204	·476 289	·261 611	39·343 449	1·374 199	1433·138 050	3.7.29
610	163 934	8·480 926	18·271 601	39·364 972	8·382 616	1435·923 380	2.5.61
611	163 666	·485 558	·281 580	39·386 471	5·121 778	1438·709 421	13.47
612	163 399	·490 185	·291 549	39·407 947	3·134 528	1441·496 172	$2^2.3^2.17$
613	163 132	·494 807	·301 506	39·429 399	1·921 466	1444·283 633	**613**
614	162 866	·499 423	·311 452	39·450 828	1·179 780	1447·071 801	2.307
615	162 602	8·504 035	18·321 388	39·472 234	7·255 647	1449·860 676	3.5.41
616	162 338	·508 642	·331 313	39·493 616	4·469 479	1452·650 257	$2^3.7.11$
617	162 075	·513 243	·341 227	39·514 976	2·757 668	1455·440 542	**617**
618	161 812	·517 840	·351 131	39·536 312	1·704 239	1458·231 531	2.3.103
619	161 551	·522 432	·361 023	39·557 626	1·054 924	1461·023 221	**619**
620	161 290	8·527 019	18·370 906	39·578 916	6·540 529	1463·815 613	$2^2.5.31$
621	161 031	·531 601	·380 777	39·600 184	4·061 668	1466·608 704	$3^3.23$
622	160 772	·536 178	·390 638	39·621 428	2·526 358	1469·402 495	2.311
623	160 514	·540 750	·400 488	39·642 650	1·573 921	1472·196 983	7.89
624	160 256	·545 317	·410 328	39·663 850	9·821 266	1474·992 167	$2^4.3.13$
625	160 000	8·549 880	18·420 157	39·685 026	6·138 291	1477·788 047	5^4
626	159 744	·554 437	·429 976	39·706 180	3·842 570	1480·584 622	2.313
627	159 490	·558 990	·439 785	39·727 312	2·409 292	1483·381 889	3.11.19
628	159 236	·563 538	·449 583	39·748 421	1·513 035	1486·179 849	$2^2.157$
629	158 983	·568 081	·459 370	39·769 508	9·516 991	1488·978 500	17.37
630	158 730	8·572 619	18·469 148	39·790 572	5·995 704	1491·777 840	$2.3^2.5.7$
631	158 479	·577 152	·478 914	39·811 614	3·783 289	1494·577 870	**631**
632	158 228	·581 681	·488 671	39·832 634	2·391 039	1497·378 587	$2^3.79$
633	157 978	·586 205	·498 417	39·853 632	1·513 528	1500·179 990	3.211
634	157 729	·590 724	·508 153	39·874 607	9·595 765	1502·982 080	2.317
635	157 480	8·595 238	18·517 879	39·895 561	6·093 311	1505·784 853	5.127
636	157 233	·599 748	·527 595	39·916 492	3·875 346	1508·588 310	$2^2.3.53$
637	156 986	·604 252	·537 300	39·937 402	2·468 595	1511·392 450	$7^2.13$
638	156 740	·608 753	·546 995	39·958 290	1·574 964	1514·197 271	2.11.29
639	156 495	·613 248	·556 680	39·979 156	1·006 402	1517·002 771	$3^2.71$
640	156 250	8·617 739	18·566 355	40·000 000	6·440 972	1519·808 951	$2^7.5$
641	156 006	·622 225	·576 020	40·020 822	4·128 663	1522·615 809	**641**
642	155 763	·626 706	·585 675	40·041 623	2·650 602	1525·423 344	2.3.107
643	155 521	·631 183	·595 320	40·062 403	1·704 337	1528·231 555	**643**
644	155 280	·635 655	·604 955	40·083 160	1·097 593	1531·040 441	$2^2.7.23$
645	155 039	8·640 123	18·614 580	40·103 897	7·079 474	1533·850 001	3.5.43
646	154 799	·644 585	·624 195	40·124 611	4·573 340	1536·660 234	2.17.19
647	154 560	·649 044	·633 800	40·145 305	2·958 951	1539·471 138	**647**
648	154 321	·653 497	·643 395	40·165 977	1·917 400	1542·282 713	$2^3.3^4$
649	154 083	·657 947	·652 980	40·186 628	1·244 393	1545·094 958	11.59
650	153 846	8·662 391	18·662 556	40·207 258	8·088 554	1547·907 871	$2.5^2.13$

* Multiply by 10^c, where c is the characteristic of $\log x!$

No.	Square	Cube	Fourth Power	Square Roots		Reciprocals of Square Roots	
x	x^2	x^3	x^4	$\sqrt{x} = x^{\frac{1}{2}}$	$\sqrt{10x}$	$\frac{1}{\sqrt{x}} = x^{-\frac{1}{2}}$	$\frac{1}{\sqrt{10x}}$
						0·0	0·0
650	42 25 00	274 625 000	1785 0625 0000	25·495 098	80·622 577	392 232	124 035
651	42 38 01	275 894 451	1796 0728 7601	25·514 702	80·684 571	391 931	123 939
652	42 51 04	277 167 808	1807 1341 0816	25·534 291	80·746 517	391 630	123 844
653	42 64 09	278 445 077	1818 2463 5281	25·553 865	80·808 415	391 330	123 749
654	42 77 16	279 726 264	1829 4097 6656	25·573 424	80·870 266	391 031	123 655
655	42 90 25	281 011 375	1840 6245 0625	25·592 968	80·932 070	390 732	123 560
656	43 03 36	282 300 416	1851 8907 2896	25·612 497	80·993 827	390 434	123 466
657	43 16 49	283 593 393	1863 2085 9201	25·632 011	81·055 537	390 137	123 372
658	43 29 64	284 890 312	1874 5782 5296	25·651 511	81·117 199	389 841	123 278
659	43 42 81	286 191 179	1885 9998 6961	25·670 995	81·178 815	389 545	123 185
660	43 56 00	287 496 000	1897 4736 0000	25·690 465	81·240 384	389 249	123 091
661	43 69 21	288 804 781	1908 9996 0241	25·709 920	81·301 906	388 955	122 998
662	43 82 44	290 117 528	1920 5780 3536	25·729 361	81·363 382	388 661	122 905
663	43 95 69	291 434 247	1932 2090 5761	25·748 786	81·424 812	388 368	122 813
664	44 08 96	292 754 944	1943 8928 2816	25·768 197	81·486 195	388 075	122 720
665	44 22 25	294 079 625	1955 6295 0625	25·787 594	81·547 532	387 783	122 628
666	44 35 56	295 408 296	1967 4192 5136	25·806 976	81·608 823	387 492	122 536
667	44 48 89	296 740 963	1979 2622 2321	25·826 343	81·670 068	387 202	122 444
668	44 62 24	298 077 632	1991 1585 8176	25·845 696	81·731 267	386 912	122 352
669	44 75 61	299 418 309	2003 1084 8721	25·865 034	81·792 420	386 622	122 261
670	44 89 00	300 763 000	2015 1121 0000	25·884 358	81·853 528	386 334	122 169
671	45 02 41	302 111 711	2027 1695 8081	25·903 668	81·914 590	386 046	122 078
672	45 15 84	303 464 448	2039 2810 9056	25·922 963	81·975 606	385 758	121 988
673	45 29 29	304 821 217	2051 4467 9041	25·942 244	82·036 577	385 472	121 897
674	45 42 76	306 182 024	2063 6668 4176	25·961 510	82·097 503	385 186	121 806
675	45 56 25	307 546 875	2075 9414 0625	25·980 762	82·158 384	384 900	121 716
676	45 69 76	308 915 776	2088 2706 4576	26·000 000	82·219 219	384 615	121 626
677	45 83 29	310 288 733	2100 6547 2241	26·019 224	82·280 010	384 331	121 536
678	45 96 84	311 665 752	2113 0937 9856	26·038 433	82·340 755	384 048	121 447
679	46 10 41	313 046 839	2125 5880 3681	26·057 628	82·401 456	383 765	121 357
680	46 24 00	314 432 000	2138 1376 0000	26·076 810	82·462 113	383 482	121 268
681	46 37 61	315 821 241	2150 7426 5121	26·095 977	82·522 724	383 201	121 179
682	46 51 24	317 214 568	2163 4033 5376	26·115 130	82·583 291	382 920	121 090
683	46 64 89	318 611 987	2176 1198 7121	26·134 269	82·643 814	382 639	121 001
684	46 78 56	320 013 504	2188 8923 6736	26·153 394	82·704 293	382 360	120 913
685	46 92 25	321 419 125	2201 7210 0625	26·172 505	82·764 727	382 080	120 824
686	47 05 96	322 828 856	2214 6059 5216	26·191 602	82·825 117	381 802	120 736
687	47 19 69	324 242 703	2227 5473 6961	26·210 685	82·885 463	381 524	120 648
688	47 33 44	325 660 672	2240 5454 2336	26·229 754	82·945 765	381 246	120 561
689	47 47 21	327 082 769	2253 6002 7841	26·248 809	83·006 024	380 970	120 473
690	47 61 00	328 509 000	2266 7121 0000	26·267 851	83·066 239	380 693	120 386
691	47 74 81	329 939 371	2279 8810 5361	26·286 879	83·126 410	380 418	120 299
692	47 88 64	331 373 888	2293 1073 0496	26·305 893	83·186 537	380 143	120 212
693	48 02 49	332 812 557	2306 3910 2001	26·324 893	83·246 622	379 869	120 125
694	48 16 36	334 255 384	2319 7323 6496	26·343 880	83·306 662	379 595	120 038
695	48 30 25	335 702 375	2333 1315 0625	26·362 853	83·366 660	379 322	119 952
696	48 44 16	337 153 536	2346 5886 1056	26·381 812	83·426 614	379 049	119 866
697	48 58 09	338 608 873	2360 1038 4481	26·400 758	83·486 526	378 777	119 780
698	48 72 04	340 068 392	2373 6773 7616	26·419 690	83·546 394	378 506	119 694
699	48 86 01	341 532 099	2387 3093 7201	26·438 608	83·606 220	378 235	119 608
700	49 00 00	343 000 000	2401 0000 0000	26·457 513	83·666 003	377 964	119 523

No. x	Reciprocal $\frac{1}{x}=x^{-1}$ 0·00	Cube Roots $\sqrt[3]{x}$	$\sqrt[3]{10x}$	$\sqrt[3]{100x}$	Factorials * $x!$	$\log x!$	Prime Factors
650	153 846	8·662 391	18·662 556	40·207 258	8·088 554	1547·907 871	$2.5^{2}.13$
651	153 610	·666 831	·672 121	40·227 866	5·265 648	1550·721 452	3.7.31
652	153 374	·671 266	·681 677	40·248 454	3·433 203	1553·535 699	$2^{2}.163$
653	153 139	·675 697	·691 223	40·269 020	2·241 881	1556·350 613	**653**
654	152 905	·680 124	·700 760	40·289 565	1·466 190	1559·166 190	2.3.109
655	152 672	8·684 546	18·710 286	40·310 090	9·603 547	1561·982 432	5.131
656	152 439	·688 963	·719 803	40·330 594	6·299 927	1564·799 336	$2^{4}.41$
657	152 207	·693 376	·729 311	40·351 076	4·139 052	1567·616 901	$3^{2}.73$
658	151 976	·697 784	·738 808	40·371 538	2·723 496	1570·435 127	2.7.47
659	151 745	·702 188	·748 296	40·391 980	1·794 784	1573·254 012	**659**
660	151 515	8·706 588	18·757 775	40·412 400	1·184 557	1576·073 556	$2^{2}.3.5.11$
661	151 286	·710 983	·767 243	40·432 800	7·829 925	1578·893 758	**661**
662	151 057	·715 373	·776 703	40·453 180	5·183 410	1581·714 616	2.331
663	150 830	·719 760	·786 152	40·473 539	3·436 601	1584·536 129	3.13.17
664	150 602	·724 141	·795 593	40·493 877	2·281 903	1587·358 297	$2^{3}.83$
665	150 376	8·728 519	18·805 024	40·514 195	1·517 466	1590·181 119	5.7.19
666	150 150	·732 892	·814 445	40·534 493	1·010 632	1593·004 593	$2.3^{2}.37$
667	149 925	·737 260	·823 857	40·554 770	6·740 916	1595·828 719	23.29
668	149 701	·741 625	·833 259	40·575 027	4·502 932	1598·653 495	$2^{2}.167$
669	149 477	·745 985	·842 653	40·595 264	3·012 461	1601·478 921	3.223
670	149 254	8·750 340	18·852 036	40·615 481	2·018 349	1604·304 996	2.5.67
671	149 031	·754 691	·861 411	40·635 678	1·354 312	1607·131 719	11.61
672	148 810	·759 038	·870 776	40·655 854	9·100 978	1609·959 088	$2^{5}.3.7$
673	148 588	·763 381	·880 132	40·676 011	6·124 958	1612·787 103	**673**
674	148 368	·767 719	·889 478	40·696 148	4·128 222	1615·615 763	2.337
675	148 148	8·772 053	18·898 816	40·716 264	2·786 550	1618·445 067	$3^{3}.5^{2}$
676	147 929	·776 383	·908 144	40·736 361	1·883 708	1621·275 014	$2^{2}.13^{2}$
677	147 710	·780 708	·917 463	40·756 438	1·275 270	1624·105 602	**677**
678	147 493	·785 030	·926 773	40·776 495	8·646 331	1626·936 832	2.3.113
679	147 275	·789 347	·936 073	40·796 533	5·870 859	1629·768 702	7.97
680	147 059	8·793 659	18·945 365	40·816 551	3·992 184	1632·601 211	$2^{3}.5.17$
681	146 843	·797 968	·954 647	40·836 549	2·718 677	1635·434 358	3.227
682	146 628	·802 272	·963 920	40·856 528	1·854 138	1638·268 142	2.11.31
683	146 413	·806 572	·973 185	40·876 487	1·266 376	1641·102 563	**683**
684	146 199	·810 868	·982 440	40·896 427	8·662 013	1643·937 619	$2^{2}.3^{2}.19$
685	145 985	8·815 160	18·991 686	40·916 347	5·933 479	1646·773 309	5.137
686	145 773	·819 447	19·000 923	40·936 248	4·070 367	1649·609 634	2.7^{3}
687	145 560	·823 731	·010 152	40·956 130	2·796 342	1652·446 590	3.229
688	145 349	·828 010	·019 371	40·975 992	1·923 883	1655·284 179	$2^{4}.43$
689	145 138	·832 285	·028 581	40·995 835	1·325 556	1658·122 398	13.53
690	144 928	8·836 556	19·037 783	41·015 659	9·146 333	1660·961 247	2.3.5.23
691	144 718	·840 823	·046 975	41·035 464	6·320 116	1663·800 725	**691**
692	144 509	·845 085	·056 159	41·055 250	4·373 521	1666·640 831	$2^{2}.173$
693	144 300	·849 344	·065 334	41·075 016	3·030 850	1669·481 564	$3^{2}.7.11$
694	144 092	·853 599	·074 500	41·094 764	2·103 410	1672·322 924	2.347
695	143 885	8·857 849	19·083 657	41·114 493	1·461 870	1675·164 909	5.139
696	143 678	·862 095	·092 805	41·134 202	1·017 461	1678·007 518	$2^{3}.3.29$
697	143 472	·866 338	·101 945	41·153 893	7·091 706	1680·850 751	17.41
698	143 266	·870 576	·111 076	41·173 565	4·950 010	1683·694 606	2.349
699	143 062	·874 810	·120 198	41·193 218	3·460 057	1686·539 083	3.233
700	142 857	8·879 040	19·129 312	41·212 853	2·422 040	1689·384 181	$2^{2}.5^{2}.7$

* Multiply by 10^{c}, where c is the characteristic of $\log x!$

No. x	Square x^2	Cube x^3	Fourth Power x^4	Square Roots $\sqrt{x} = x^{\frac{1}{2}}$	$\sqrt{10x}$	Reciprocals of Square Roots $\frac{1}{\sqrt{x}} = x^{-\frac{1}{2}}$	$\frac{1}{\sqrt{10x}}$
						0·0	0·0
700	49 00 00	343 000 000	2401 0000 0000	26·457 513	83·666 003	377 964	119 523
701	49 14 01	344 472 101	2414 7494 2801	26·476 405	83·725 743	377 695	119 438
702	49 28 04	345 948 408	2428 5578 2416	26·495 283	83·785 440	377 426	119 352
703	49 42 09	347 428 927	2442 4253 5681	26·514 147	83·845 095	377 157	119 268
704	49 56 16	348 913 664	2456 3521 9456	26·532 998	83·904 708	376 889	119 183
705	49 70 25	350 402 625	2470 3385 0625	26·551 836	83·964 278	376 622	119 098
706	49 84 36	351 895 816	2484 3844 6096	26·570 661	84·023 806	376 355	119 014
707	49 98 49	353 393 243	2498 4902 2801	26·589 472	84·083 292	376 089	118 930
708	50 12 64	354 894 912	2512 6559 7696	26·608 269	84·142 736	375 823	118 846
709	50 26 81	356 400 829	2526 8818 7761	26·627 054	84·202 138	375 558	118 762
710	50 41 00	357 911 000	2541 1681 0000	26·645 825	84·261 498	375 293	118 678
711	50 55 21	359 425 431	2555 5148 1441	26·664 583	84·320 816	375 029	118 595
712	50 69 44	360 944 128	2569 9221 9136	26·683 328	84·380 092	374 766	118 511
713	50 83 69	362 467 097	2584 3904 0161	26·702 060	84·439 327	374 503	118 428
714	50 97 96	363 994 344	2598 9196 1616	26·720 778	84·498 521	374 241	118 345
715	51 12 25	365 525 875	2613 5100 0625	26·739 484	84·557 673	373 979	118 262
716	51 26 56	367 061 696	2628 1617 4336	26·758 176	84·616 783	373 718	118 180
717	51 40 89	368 601 813	2642 8749 9921	26·776 856	84·675 853	373 457	118 097
718	51 55 24	370 146 232	2657 6499 4576	26·795 522	84·734 881	373 197	118 015
719	51 69 61	371 694 959	2672 4867 5521	26·814 175	84·793 868	372 937	117 933
720	51 84 00	373 248 000	2687 3856 0000	26·832 816	84·852 814	372 678	117 851
721	51 98 41	374 805 361	2702 3466 5281	26·851 443	84·911 719	372 419	117 769
722	52 12 84	376 367 048	2717 3700 8656	26·870 058	84·970 583	372 161	117 688
723	52 27 29	377 933 067	2732 4560 7441	26·888 659	85·029 407	371 904	117 606
724	52 41 76	379 503 424	2747 6047 8976	26·907 248	85·088 190	371 647	117 525
725	52 56 25	381 078 125	2762 8164 0625	26·925 824	85·146 932	371 391	117 444
726	52 70 76	382 657 176	2778 0910 9776	26·944 387	85·205 634	371 135	117 363
727	52 85 29	384 240 583	2793 4290 3841	26·962 938	85·264 295	370 879	117 282
728	52 99 84	385 828 352	2808 8304 0256	26·981 475	85·322 916	370 625	117 202
729	53 14 41	387 420 489	2824 2953 6481	27·000 000	85·381 497	370 370	117 121
730	53 29 00	389 017 000	2839 8241 0000	27·018 512	85·440 037	370 117	117 041
731	53 43 61	390 617 891	2855 4167 8321	27·037 012	85·498 538	369 863	116 961
732	53 58 24	392 223 168	2871 0735 8976	27·055 499	85·556 999	369 611	116 881
733	53 72 89	393 832 837	2886 7946 9521	27·073 973	85·615 419	369 358	116 801
734	53 87 56	395 446 904	2902 5802 7536	27·092 434	85·673 800	369 107	116 722
735	54 02 25	397 065 375	2918 4305 0625	27·110 883	85·732 141	368 856	116 642
736	54 16 96	398 688 256	2934 3455 6416	27·129 320	85·790 442	368 605	116 563
737	54 31 69	400 315 553	2950 3256 2561	27·147 744	85·848 704	368 355	116 484
738	54 46 44	401 947 272	2966 3708 6736	27·166 155	85·906 926	368 105	116 405
739	54 61 21	403 583 419	2982 4814 6641	27·184 554	85·965 109	367 856	116 326
740	54 76 00	405 224 000	2998 6576 0000	27·202 941	86·023 253	367 607	116 248
741	54 90 81	406 869 021	3014 8994 4561	27·221 315	86·081 357	367 359	116 169
742	55 05 64	408 518 488	3031 2071 8096	27·239 677	86·139 422	367 112	116 091
743	55 20 49	410 172 407	3047 5809 8401	27·258 026	86·197 448	366 864	116 013
744	55 35 36	411 830 784	3064 0210 3296	27·276 363	86·255 435	366 618	115 935
745	55 50 25	413 493 625	3080 5275 0625	27·294 688	86·313 383	366 372	115 857
746	55 65 16	415 160 936	3097 1005 8256	27·313 001	86·371 292	366 126	115 779
747	55 80 09	416 832 723	3113 7404 4081	27·331 301	86·429 162	365 881	115 702
748	55 95 04	418 508 992	3130 4472 6016	27·349 589	86·486 993	365 636	115 624
749	56 10 01	420 189 749	3147 2212 2001	27·367 864	86·544 786	365 392	115 547
750	56 25 00	421 875 000	3164 0625 0000	27·386 128	86·602 540	365 148	115 470

No. x	Reciprocal $\frac{1}{x}=x^{-1}$ 0·00	Cube Roots $\sqrt[3]{x}$	$\sqrt[3]{10x}$	$\sqrt[3]{100x}$	Factorials * $x!$	$\log x!$	Prime Factors
700	142 857	8·879 040	19·129 312	41·212 853	2·422 040	1689·384 181	$2^2.5^2.7$
701	142 653	·883 266	·138 417	41·232 469	1·697 850	1692·229 899	**701**
702	142 450	·887 488	·147 513	41·252 066	1·191 891	1695·076 236	$2.3^3.13$
703	142 248	·891 706	·156 600	41·271 645	8·378 992	1697·923 192	19.37
704	142 045	·895 920	·165 679	41·291 205	5·898 811	1700·770 764	$2^6.11$
705	141 844	8·900 130	19·174 750	41·310 746	4·158 661	1703·618 954	3.5.47
706	141 643	·904 337	·183 812	41·330 269	2·936 015	1706·467 758	2.353
707	141 443	·908 539	·192 865	41·349 774	2·075 763	1709·317 178	7.101
708	141 243	·912 737	·201 910	41·369 260	1·469 640	1712·167 211	$2^2.3.59$
709	141 044	·916 931	·210 946	41·388 728	1·041 975	1715·017 857	**709**
710	140 845	8·921 121	19·219 973	41·408 177	7·398 020	1717·869 116	2.5.71
711	140 647	·925 308	·228 993	41·427 609	5·259 992	1720·720 985	$3^2.79$
712	140 449	·929 490	·238 003	41·447 022	3·745 115	1723·573 465	$2^3.89$
713	140 252	·933 669	·247 006	41·466 417	2·670 267	1726·426 555	23.31
714	140 056	·937 843	·256 000	41·485 794	1·906 570	1729·280 253	2.3.7.17
715	139 860	8·942 014	19·264 985	41·505 153	1·363 198	1732·134 559	5.11.13
716	139 665	·946 181	·273 962	41·524 493	9·760 497	1734·989 472	$2^2.179$
717	139 470	·950 344	·282 931	41·543 816	6·998 276	1737·844 991	3.239
718	139 276	·954 503	·291 892	41·563 121	5·024 762	1740·701 116	2.359
719	139 082	·958 658	·300 844	41·582 408	3·612 804	1743·557 844	**719**
720	138 889	8·962 809	19·309 788	41·601 676	2·601 219	1746·415 177	$2^4.3^2.5$
721	138 696	·966 957	·318 723	41·620 928	1·875 479	1749·273 112	7.103
722	138 504	·971 101	·327 651	41·640 161	1·354 096	1752·131 649	2.19^2
723	138 313	·975 241	·336 570	41·659 377	9·790 112	1754·990 788	3.241
724	138 122	·979 377	·345 481	41·678 574	7·088 041	1757·850 526	$2^2.181$
725	137 931	8·983 509	19·354 383	41·697 755	5·138 830	1760·710 864	$5^2.29$
726	137 741	·987 637	·363 278	41·716 917	3·730 790	1763·571 801	$2.3.11^2$
727	137 552	·991 762	·372 164	41·736 062	2·712 285	1766·433 335	**727**
728	137 363	8·995 883	·381 042	41·755 190	1·974 543	1769·295 467	$2^3.7.13$
729	137 174	9·000 000	·389 912	41·774 300	1·439 442	1772·158 194	3^6
730	136 986	9·004 113	19·398 774	41·793 392	1·050 793	1775·021 517	2.5.73
731	136 799	·008 223	·407 628	41·812 467	7·681 295	1777·885 434	17.43
732	136 612	·012 329	·416 474	41·831 525	5·622 708	1780·749 945	$2^2.3.61$
733	136 426	·016 431	·425 311	41·850 565	4·121 445	1783·615 049	**733**
734	136 240	·020 529	·434 141	41·869 588	3·025 140	1786·480 746	2.367
735	136 054	9·024 624	19·442 963	41·888 594	2·223 478	1789·347 033	$3.5.7^2$
736	135 870	·028 715	·451 777	41·907 582	1·636 480	1792·213 911	$2^5.23$
737	135 685	·032 802	·460 582	41·926 553	1·206 086	1795·081 378	11.67
738	135 501	·036 886	·469 380	41·945 508	8·900 913	1797·949 435	$2.3^2.41$
739	135 318	·040 966	·478 170	41·964 445	6·577 774	1800·818 079	**739**
740	135 135	9·045 042	19·486 952	41·983 365	4·867 553	1803·687 311	$2^2.5.37$
741	134 953	·049 114	·495 726	42·002 267	3·606 857	1806·557 129	3.13.19
742	134 771	·053 183	·504 492	42·021 153	2·676 288	1809·427 533	2.7.53
743	134 590	·057 248	·513 250	42·040 022	1·988 482	1812·298 522	**743**
744	134 409	·061 310	·522 000	42·058 874	1·479 430	1815·170 095	$2^3.3.31$
745	134 228	9·065 368	19·530 743	42·077 709	1·102 176	1818·042 251	5.149
746	134 048	·069 422	·539 477	42·096 528	8·222 231	1820·914 990	2.373
747	133 869	·073 473	·548 204	42·115 329	6·142 006	1823·788 310	$3^2.83$
748	133 690	·077 520	·556 923	42·134 114	4·594 221	1826·662 212	$2^2.11.17$
749	133 511	·081 563	·565 635	42·152 882	3·441 071	1829·536 694	7.107
750	133 333	9·085 603	19·574 338	42·171 633	2·580 803	1832·411 755	$2.3.5^3$

* Multiply by 10^c, where c is the characteristic of $\log x!$

No.	Square	Cube	Fourth Power	Square Roots		Reciprocals of Square Roots	
x	x^2	x^3	x^4	$\sqrt{x}=x^{\frac{1}{2}}$	$\sqrt{10x}$	$\frac{1}{\sqrt{x}}=x^{-\frac{1}{2}}$	$\frac{1}{\sqrt{10x}}$
						0·0	0·0
750	56 25 00	421 875 000	3164 0625 0000	27·386 128	86·602 540	365 148	115 470
751	56 40 01	423 564 751	3180 9712 8001	27·404 379	86·660 256	364 905	115 393
752	56 55 04	425 259 008	3197 9477 4016	27·422 618	86·717 934	364 662	115 316
753	56 70 09	426 957 777	3214 9920 6081	27·440 845	86·775 573	364 420	115 240
754	56 85 16	428 661 064	3232 1044 2256	27·459 060	86·833 173	364 179	115 163
755	57 00 25	430 368 875	3249 2850 0625	27·477 263	86·890 736	363 937	115 087
756	57 15 36	432 081 216	3266 5339 9296	27·495 454	86·948 260	363 696	115 011
757	57 30 49	433 798 093	3283 8515 6401	27·513 633	87·005 747	363 456	114 935
758	57 45 64	435 519 512	3301 2379 0096	27·531 800	87·063 195	363 216	114 859
759	57 60 81	437 245 479	3318 6931 8561	27·549 955	87·120 606	362 977	114 783
760	57 76 00	438 976 000	3336 2176 0000	27·568 098	87·177 979	362 738	114 708
761	57 91 21	440 711 081	3353 8113 2641	27·586 228	87·235 314	362 500	114 632
762	58 06 44	442 450 728	3371 4745 4736	27·604 347	87·292 611	362 262	114 557
763	58 21 69	444 194 947	3389 2074 4561	27·622 455	87·349 871	362 024	114 482
764	58 36 96	445 943 744	3407 0102 0416	27·640 550	87·407 094	361 787	114 407
765	58 52 25	447 697 125	3424 8830 0625	27·658 633	87·464 278	361 551	114 332
766	58 67 56	449 455 096	3442 8260 3536	27·676 705	87·521 426	361 315	114 258
767	58 82 89	451 217 663	3460 8394 7521	27·694 765	87·578 536	361 079	114 183
768	58 98 24	452 984 832	3478 9235 0976	27·712 813	87·635 609	360 844	114 109
769	59 13 61	454 756 609	3497 0783 2321	27·730 849	87·692 645	360 609	114 035
770	59 29 00	456 533 000	3515 3041 0000	27·748 874	87·749 644	360 375	113 961
771	59 44 41	458 314 011	3533 6010 2481	27·766 887	87·806 606	360 141	113 887
772	59 59 84	460 099 648	3551 9692 8256	27·784 888	87·863 531	359 908	113 813
773	59 75 29	461 889 917	3570 4090 5841	27·802 878	87·920 419	359 675	113 739
774	59 90 76	463 684 824	3588 9205 3776	27·820 855	87·977 270	359 443	113 666
775	60 06 25	465 484 375	3607 5039 0625	27·838 822	88·034 084	359 211	113 592
776	60 21 76	467 288 576	3626 1593 4976	27·856 777	88·090 862	358 979	113 519
777	60 37 29	469 097 433	3644 8870 5441	27·874 720	88·147 603	358 748	113 446
778	60 52 84	470 910 952	3663 6872 0656	27·892 651	88·204 308	358 517	113 373
779	60 68 41	472 729 139	3682 5599 9281	27·910 571	88·260 977	358 287	113 300
780	60 84 00	474 552 000	3701 5056 0000	27·928 480	88·317 609	358 057	113 228
781	60 99 61	476 379 541	3720 5242 1521	27·946 377	88·374 204	357 828	113 155
782	61 15 24	478 211 768	3739 6160 2576	27·964 263	88·430 764	357 599	113 083
783	61 30 89	480 048 687	3758 7812 1921	27·982 137	88·487 287	357 371	113 011
784	61 46 56	481 890 304	3778 0199 8336	28·000 000	88·543 774	357 143	112 938
785	61 62 25	483 736 625	3797 3325 0625	28·017 851	88·600 226	356 915	112 867
786	61 77 96	485 587 656	3816 7189 7616	28·035 692	88·656 641	356 688	112 795
787	61 93 69	487 443 403	3836 1795 8161	28·053 520	88·713 020	356 462	112 723
788	62 09 44	489 303 872	3855 7145 1136	28·071 338	88·769 364	356 235	112 651
789	62 25 21	491 169 069	3875 3239 5441	28·089 144	88·825 672	356 009	112 580
790	62 41 00	493 039 000	3895 0081 0000	28·106 939	88·881 944	355 784	112 509
791	62 56 81	494 913 671	3914 7671 3761	28·124 722	88·938 181	355 559	112 438
792	62 72 64	496 793 088	3934 6012 5696	28·142 495	88·994 382	355 335	112 367
793	62 88 49	498 677 257	3954 5106 4801	28·160 256	89·050 547	355 110	112 296
794	63 04 36	500 566 184	3974 4955 0096	28·178 006	89·106 678	354 887	112 225
795	63 20 25	502 459 875	3994 5560 0625	28·195 744	89·162 773	354 663	112 154
796	63 36 16	504 358 336	4014 6923 5456	28·213 472	89·218 832	354 441	112 084
797	63 52 09	506 261 573	4034 9047 3681	28·231 188	89·274 856	354 218	112 014
798	63 68 04	508 169 592	4055 1933 4416	28·248 894	89·330 846	353 996	111 943
799	63 84 01	510 082 399	4075 5583 6801	28·266 588	89·386 800	353 775	111 873
800	64 00 00	512 000 000	4096 0000 0000	28·284 271	89·442 719	353 553	111 803

TABLE VIII—RECIPROCALS, CUBE ROOTS, FACTORIALS AND FACTORS

No. x	Reciprocal $\frac{1}{x}=x^{-1}$ 0·00	Cube Roots $\sqrt[3]{x}$	$\sqrt[3]{10x}$	$\sqrt[3]{100x}$	Factorials * $x!$	$\log x!$	Prime Factors
750	133 333	9·085 603	19·574 338	42·171 633	2·580 803	1832·411 755	$2.3.5^{3}$
751	133 156	·089 639	·583 034	42·190 368	1·938 183	1835·287 395	**751**
752	132 979	·093 672	·591 722	42·209 086	1·457 514	1838·163 613	$2^{4}.47$
753	132 802	·097 701	·600 403	42·227 787	1·097 508	1841·040 408	3.251
754	132 626	·101 727	·609 075	42·246 472	8·275 210	1843·917 779	2.13.29
755	132 450	9·105 748	19·617 740	42·265 141	6·247 784	1846·795 726	5.151
756	132 275	·109 767	·626 398	42·283 792	4·723 325	1849·674 248	$2^{2}.3^{3}.7$
757	132 100	·113 782	·635 048	42·302 428	3·575 557	1852·553 344	**757**
758	131 926	·117 793	·643 690	42·321 047	2·710 272	1855·433 013	2.379
759	131 752	·121 801	·652 324	42·339 650	2·057 096	1858·313 255	3.11.23
760	131 579	9·125 805	19·660 951	42·358 236	1·563 393	1861·194 068	$2^{3}.5.19$
761	131 406	·129 806	·669 571	42·376 806	1·189 742	1864·075 453	**761**
762	131 234	·133 803	·678 183	42·395 360	9·065 836	1866·957 408	2.3.127
763	131 062	·137 797	·686 787	42·413 897	6·917 233	1869·839 932	7.109
764	130 890	·141 787	·695 384	42·432 419	5·284 766	1872·723 026	$2^{2}.191$
765	130 719	9·145 774	19·703 973	42·450 924	4·042 846	1875·606 687	$3^{2}.5.17$
766	130 548	·149 758	·712 555	42·469 413	3·096 820	1878·490 916	2.383
767	130 378	·153 738	·721 130	42·487 886	2·375 261	1881·375 711	13.59
768	130 208	·157 714	·729 697	42·506 343	1·824 200	1884·261 073	$2^{8}.3$
769	130 039	·161 687	·738 256	42·524 784	1·402 810	1887·146 999	**769**
770	129 870	9·165 656	19·746 808	42·543 209	1·080 164	1890·033 490	2.5.7.11
771	129 702	·169 623	·755 353	42·561 618	8·328 063	1892·920 544	3.257
772	129 534	·173 585	·763 890	42·580 011	6·429 265	1895·808 161	$2^{2}.193$
773	129 366	·177 544	·772 420	42·598 388	4·969 821	1898·696 341	**773**
774	129 199	·181 500	·780 943	42·616 749	3·846 642	1901·585 082	$2.3^{2}.43$
775	129 032	9·185 453	19·789 458	42·635 095	2·981 147	1904·474 383	$5^{2}.31$
776	128 866	·189 402	·797 966	42·653 425	2·313 370	1907·364 245	$2^{3}.97$
777	128 700	·193 347	·806 467	42·671 739	1·797 489	1910·254 666	3.7.37
778	128 535	·197 290	·814 960	42·690 037	1·398 446	1913·145 646	2.389
779	128 370	·201 229	·823 446	42·708 320	1·089 390	1916·037 183	19.41
780	128 205	9·205 164	19·831 925	42·726 587	8·497 239	1918·929 278	$2^{2}.3.5.13$
781	128 041	·209 096	·840 396	42·744 838	6·636 344	1921·821 929	11.71
782	127 877	·213 025	·848 861	42·763 074	5·189 621	1924·715 136	2.17.23
783	127 714	·216 950	·857 318	42·781 294	4·063 473	1927·608 897	$3^{3}.29$
784	127 551	·220 873	·865 768	42·799 499	3·185 763	1930·503 213	$2^{4}.7^{2}$
785	127 389	9·224 791	19·874 211	42·817 689	2·500 824	1933·398 083	5.157
786	127 226	·228 707	·882 646	42·835 862	1·965 648	1936·293 506	2.3.131
787	127 065	·232 619	·891 075	42·854 021	1·546 965	1939·189 480	**787**
788	126 904	·236 528	·899 496	42·872 164	1·219 008	1942·086 007	$2^{2}.197$
789	126 743	·240 433	·907 910	42·890 292	9·617 974	1944·983 084	3.263
790	126 582	9·244 335	19·916 317	42·908 404	7·598 200	1947·880 711	2.5.79
791	126 422	·248 234	·924 717	42·926 501	6·010 176	1950·778 887	7.113
792	126 263	·252 130	·933 110	42·944 583	4·760 059	1953·677 612	$2^{3}.3^{2}.11$
793	126 103	·256 022	·941 496	42·962 650	3·774 727	1956·576 886	13.61
794	125 945	·259 911	·949 874	42·980 702	2·997 133	1959·476 706	2.397
795	125 786	9·263 797	19·958 246	42·998 738	2·382 721	1962·377 073	3.5.53
796	125 628	·267 680	·966 611	43·016 759	1·896 646	1965·277 986	$2^{2}.199$
797	125 471	·271 559	·974 969	43·034 765	1·511 627	1968·179 445	**797**
798	125 313	·275 435	·983 319	43·052 757	1·206 278	1971·081 447	2.3.7.19
799	125 156	·279 308	19·991 663	43·070 733	9·638 163	1973·983 994	17.47
800	125 000	9·283 178	20·000 000	43·088 694	7·710 530	1976·887 084	$2^{5}.5^{2}$

* Multiply by 10^{c}, where c is the characteristic of $\log x!$

TABLE VIII—POWERS AND SQUARE ROOTS

No.	Square	Cube	Fourth Power	Square Roots		Reciprocals of Square Roots	
x	x^2	x^3	x^4	$\sqrt{x} = x^{\frac{1}{2}}$	$\sqrt{10x}$	$\frac{1}{\sqrt{x}} = x^{-\frac{1}{2}}$	$\frac{1}{\sqrt{10x}}$
						0·0	0·0
800	64 00 00	512 000 000	4096 0000 0000	28·284 271	89·442 719	353 553	111 803
801	64 16 01	513 922 401	4116 5184 3201	28·301 943	89·498 603	353 333	111 734
802	64 32 04	515 849 608	4137 1138 5616	28·319 605	89·554 453	353 112	111 664
803	64 48 09	517 781 627	4157 7864 6481	28·337 255	89·610 267	352 892	111 594
804	64 64 16	519 718 464	4178 5364 5056	28·354 894	89·666 047	352 673	111 525
805	64 80 25	521 660 125	4199 3640 0625	28·372 522	89·721 792	352 454	111 456
806	64 96 36	523 606 616	4220 2693 2496	28·390 139	89·777 503	352 235	111 386
807	65 12 49	525 557 943	4241 2526 0001	28·407 745	89·833 179	352 017	111 317
808	65 28 64	527 514 112	4262 3140 2496	28·425 341	89·888 820	351 799	111 249
809	65 44 81	529 475 129	4283 4537 9361	28·442 925	89·944 427	351 581	111 180
810	65 61 00	531 441 000	4304 6721 0000	28·460 499	90·000 000	351 364	111 111
811	65 77 21	533 411 731	4325 9691 3841	28·478 062	90·055 538	351 147	111 043
812	65 93 44	535 387 328	4347 3451 0336	28·495 614	90·111 043	350 931	110 974
813	66 09 69	537 367 797	4368 8001 8961	28·513 155	90·166 513	350 715	110 906
814	66 25 96	539 353 144	4390 3345 9216	28·530 685	90·221 949	350 500	110 838
815	66 42 25	541 343 375	4411 9485 0625	28·548 205	90·277 350	350 285	110 770
816	66 58 56	543 338 496	4433 6421 2736	28·565 714	90·332 718	350 070	110 702
817	66 74 89	545 338 513	4455 4156 5121	28·583 212	90·388 052	349 856	110 634
818	66 91 24	547 343 432	4477 2692 7376	28·600 699	90·443 352	349 642	110 566
819	67 07 61	549 353 259	4499 2031 9121	28·618 176	90·498 619	349 428	110 499
820	67 24 00	551 368 000	4521 2176 0000	28·635 642	90·553 851	349 215	110 432
821	67 40 41	553 387 661	4543 3126 9681	28·653 098	90·609 050	349 002	110 364
822	67 56 84	555 412 248	4565 4886 7856	28·670 542	90·664 216	348 790	110 297
823	67 73 29	557 441 767	4587 7457 4241	28·687 977	90·719 347	348 578	110 230
824	67 89 76	559 476 224	4610 0840 8576	28·705 400	90·774 446	348 367	110 163
825	68 06 25	561 515 625	4632 5039 0625	28·722 813	90·829 511	348 155	110 096
826	68 22 76	563 559 976	4655 0054 0176	28·740 216	90·884 542	347 945	110 030
827	68 39 29	565 609 283	4677 5887 7041	28·757 608	90·939 540	347 734	109 963
828	68 55 84	567 663 552	4700 2542 1056	28·774 989	90·994 505	347 524	109 897
829	68 72 41	569 722 789	4723 0019 2081	28·792 360	91·049 437	347 314	109 830
830	68 89 00	571 787 000	4745 8321 0000	28·809 721	91·104 336	347 105	109 764
831	69 05 61	573 856 191	4768 7449 4721	28·827 071	91·159 201	346 896	109 698
832	69 22 24	575 930 368	4791 7406 6176	28·844 410	91·214 034	346 688	109 632
833	69 38 89	578 009 537	4814 8194 4321	28·861 739	91·268 834	346 479	109 566
834	69 55 56	580 093 704	4837 9814 9136	28·879 058	91·323 600	346 272	109 501
835	69 72 25	582 182 875	4861 2270 0625	28·896 367	91·378 334	346 064	109 435
836	69 88 96	584 277 056	4884 5561 8816	28·913 665	91·433 036	345 857	109 370
837	70 05 69	586 376 253	4907 9692 3761	28·930 952	91·487 704	345 651	109 304
838	70 22 44	588 480 472	4931 4663 5536	28·948 230	91·542 340	345 444	109 239
839	70 39 21	590 589 719	4955 0477 4241	28·965 497	91·596 943	345 238	109 174
840	70 56 00	592 704 000	4978 7136 0000	28·982 753	91·651 514	345 033	109 109
841	70 72 81	594 823 321	5002 4641 2961	29·000 000	91·706 052	344 828	109 044
842	70 89 64	596 947 688	5026 2995 3296	29·017 236	91·760 558	344 623	108 979
843	71 06 49	599 077 107	5050 2200 1201	29·034 462	91·815 031	344 418	108 915
844	71 23 36	601 211 584	5074 2257 6896	29·051 678	91·869 473	344 214	108 850
845	71 40 25	603 351 125	5098 3170 0625	29·068 884	91·923 882	344 010	108 786
846	71 57 16	605 495 736	5122 4939 2656	29·086 079	91·978 258	343 807	108 721
847	71 74 09	607 645 423	5146 7567 3281	29·103 264	92·032 603	343 604	108 657
848	71 91 04	609 800 192	5171 1056 2816	29·120 440	92·086 915	343 401	108 593
849	72 08 01	611 960 049	5195 5408 1601	29·137 605	92·141 196	343 199	108 529
850	72 25 00	614 125 000	5220 0625 0000	29·154 759	92·195 445	342 997	108 465

No.	Reciprocal	Cube Roots			Factorials		Prime Factors
x	$\frac{1}{x} = x^{-1}$	$\sqrt[3]{x}$	$\sqrt[3]{10x}$	$\sqrt[3]{100x}$	* $x!$	$\log x!$	
	0·00						
800	125 000	9·283 178	20·000 000	43·088 694	7·710 530	1976·887 084	$2^5.5^2$
801	124 844	·287 044	·008 330	43·106 640	6·176 135	1979·790 717	$3^2.89$
802	124 688	·290 907	·016 653	43·124 571	4·953 260	1982·694 891	2.401
803	124 533	·294 767	·024 969	43·142 487	3·977 468	1985·599 607	11.73
804	124 378	·298 624	·033 278	43·160 389	3·197 884	1988·504 863	$2^2.3.67$
805	124 224	9·302 477	20·041 580	43·178 276	2·574 297	1991·410 659	5.7.23
806	124 069	·306 328	·049 876	43·196 147	2·074 883	1994·316 994	2.13.31
807	123 916	·310 175	·058 164	43·214 004	1·674 431	1997·223 867	3.269
808	123 762	·314 019	·066 446	43·231 847	1·352 940	2000·131 279	$2^3.101$
809	123 609	·317 860	·074 720	43·249 674	1·094 528	2003·039 227	**809**
810	123 457	9·321 698	20·082 989	43·267 487	8·865 680	2005·947 712	$2.3^4.5$
811	123 305	·325 532	·091 250	43·285 285	7·190 067	2008·856 733	**811**
812	123 153	·329 363	·099 504	43·303 069	5·838 334	2011·766 289	$2^2.7.29$
813	123 001	·333 192	·107 752	43·320 838	4·746 566	2014·676 380	3.271
814	122 850	·337 017	·115 993	43·338 592	3·863 705	2017·587 004	2.11.37
815	122 699	9·340 839	20·124 227	43·356 332	3·148 919	2020·498 162	5.163
816	122 549	·344 657	·132 454	43·374 058	2·569 518	2023·409 852	$2^4.3.17$
817	122 399	·348 473	·140 675	43·391 769	2·099 296	2026·322 074	19.43
818	122 249	·352 286	·148 889	43·409 465	1·717 224	2029·234 827	2.409
819	122 100	·356 095	·157 096	43·427 147	1·406 407	2032·148 111	$3^2.7.13$
820	121 951	9·359 902	20·165 297	43·444 815	1·153 254	2035·061 925	$2^2.5.41$
821	121 803	·363 705	·173 491	43·462 468	9·468 211	2037·976 268	**821**
822	121 655	·367 505	·181 678	43·480 107	7·782 870	2040·891 140	2.3.137
823	121 507	·371 302	·189 859	43·497 732	6·405 302	2043·806 540	**823**
824	121 359	·375 096	·198 033	43·515 342	5·277 969	2046·722 467	$2^3.103$
825	121 212	9·378 887	20·206 200	43·532 938	4·354 324	2049·638 921	$3.5^2.11$
826	121 065	·382 675	·214 361	43·550 520	3·596 672	2052·555 901	2.7.59
827	120 919	·386 460	·222 515	43·568 088	2·974 448	2055·473 406	**827**
828	120 773	·390 242	·230 663	43·585 642	2·462 843	2058·391 437	$2^2.3^2.23$
829	120 627	·394 021	·238 804	43·603 181	2·041 697	2061·309 991	**829**
830	120 482	9·397 796	20·246 939	43·620 707	1·694 608	2064·229 069	2.5.83
831	120 337	·401 569	·255 067	43·638 218	1·408 219	2067·148 670	3.277
832	120 192	·405 339	·263 188	43·655 715	1·171 638	2070·068 794	$2^6.13$
833	120 048	·409 105	·271 303	43·673 199	9·759 749	2072·989 439	$7^2.17$
834	119 904	·412 869	·279 412	43·690 668	8·139 630	2075·910 605	2.3.139
835	119 760	9·416 630	20·287 514	43·708 123	6·796 591	2078·832 291	5.167
836	119 617	·420 387	·295 609	43·725 565	5·681 950	2081·754 497	$2^2.11.19$
837	119 474	·424 142	·303 698	43·742 992	4·755 792	2084·677 223	$3^3.31$
838	119 332	·427 894	·311 781	43·760 406	3·985 354	2087·600 467	2.419
839	119 190	·431 642	·319 857	43·777 805	3·343 712	2090·524 229	**839**
840	119 048	9·435 388	20·327 927	43·795 191	2·808 718	2093·448 508	$2^3.3.5.7$
841	118 906	·439 131	·335 991	43·812 564	2·362 132	2096·373 304	29^2
842	118 765	·442 870	·344 048	43·829 922	1·988 915	2099·298 616	2.421
843	118 624	·446 607	·352 098	43·847 267	1·676 655	2102·224 444	3.281
844	118 483	·450 341	·360 143	43·864 598	1·415 097	2105·150 786	$2^2.211$
845	118 343	9·454 072	20·368 181	43·881 915	1·195 757	2108·077 643	5.13^2
846	118 203	·457 800	·376 212	43·899 218	1·011 611	2111·005 013	$2.3^2.47$
847	118 064	·461 525	·384 237	43·916 508	8·568 341	2113·932 897	7.11^2
848	117 925	·465 247	·392 256	43·933 785	7·265 953	2116·861 293	$2^4.53$
849	117 786	·468 966	·400 269	43·951 047	6·168 794	2119·790 200	3.283
850	117 647	9·472 682	20·408 276	43·968 297	5·243 475	2122·719 619	$2.5^2.17$

* Multiply by 10^c, where c is the characteristic of $\log x!$

No.	Square	Cube	Fourth Power	Square Roots		Reciprocals of Square Roots	
x	x^2	x^3	x^4	$\sqrt{x} = x^{\frac{1}{2}}$	$\sqrt{10x}$	$\frac{1}{\sqrt{x}} = x^{-\frac{1}{2}}$	$\frac{1}{\sqrt{10x}}$
						0·0	0·0
850	72 25 00	614 125 000	5220 0625 0000	29·154 759	92·195 445	342 997	108 465
851	72 42 01	616 295 051	5244 6708 8401	29·171 904	92·249 661	342 796	108 401
852	72 59 04	618 470 208	5269 3661 7216	29·189 039	92·303 846	342 594	108 338
853	72 76 09	620 650 477	5294 1485 6881	29·206 164	92·357 999	342 393	108 274
854	72 93 16	622 835 864	5319 0182 7856	29·223 278	92·412 120	342 193	108 211
855	73 10 25	625 026 375	5343 9755 0625	29·240 383	92·466 210	341 993	108 148
856	73 27 36	627 222 016	5369 0204 5696	29·257 478	92·520 268	341 793	108 084
857	73 44 49	629 422 793	5394 1533 3601	29·274 562	92·574 294	341 593	108 021
858	73 61 64	631 628 712	5419 3743 4896	29·291 637	92·628 289	341 394	107 958
859	73 78 81	633 839 779	5444 6837 0161	29·308 702	92·682 253	341 196	107 896
860	73 96 00	636 056 000	5470 0816 0000	29·325 757	92·736 185	340 997	107 833
861	74 13 21	638 277 381	5495 5682 5041	29·342 802	92·790 086	340 799	107 770
862	74 30 44	640 503 928	5521 1438 5936	29·359 837	92·843 955	340 601	107 708
863	74 47 69	642 735 647	5546 8086 3361	29·376 862	92·897 793	340 404	107 645
864	74 64 96	644 972 544	5572 5627 8016	29·393 877	92·951 600	340 207	107 583
865	74 82 25	647 214 625	5598 4065 0625	29·410 882	93·005 376	340 010	107 521
866	74 99 56	649 461 896	5624 3400 1936	29·427 878	93·059 121	339 814	107 459
867	75 16 89	651 714 363	5650 3635 2721	29·444 864	93·112 835	339 618	107 397
868	75 34 24	653 972 032	5676 4772 3776	29·461 840	93·166 518	339 422	107 335
869	75 51 61	656 234 909	5702 6813 5921	29·478 806	93·220 169	339 227	107 273
870	75 69 00	658 503 000	5728 9761 0000	29·495 762	93·273 791	339 032	107 211
871	75 86 41	660 776 311	5755 3616 6881	29·512 709	93·327 381	338 837	107 150
872	76 03 84	663 054 848	5781 8382 7456	29·529 646	93·380 940	338 643	107 088
873	76 21 29	665 338 617	5808 4061 2641	29·546 573	93·434 469	338 449	107 027
874	76 38 76	667 627 624	5835 0654 3376	29·563 491	93·487 967	338 255	106 966
875	76 56 25	669 921 875	5861 8164 0625	29·580 399	93·541 435	338 062	106 904
876	76 73 76	672 221 376	5888 6592 5376	29·597 297	93·594 872	337 869	106 843
877	76 91 29	674 526 133	5915 5941 8641	29·614 186	93·648 278	337 676	106 783
878	77 08 84	676 836 152	5942 6214 1456	29·631 065	93·701 654	337 484	106 722
879	77 26 41	679 151 439	5969 7411 4881	29·647 934	93·755 000	337 292	106 661
880	77 44 00	681 472 000	5996 9536 0000	29·664 794	93·808 315	337 100	106 600
881	77 61 61	683 797 841	6024 2589 7921	29·681 644	93·861 600	336 909	106 540
882	77 79 24	686 128 968	6051 6574 9776	29·698 485	93·914 855	336 718	106 479
883	77 96 89	688 465 387	6079 1493 6721	29·715 316	93·968 080	336 527	106 419
884	78 14 56	690 807 104	6106 7347 9936	29·732 137	94·021 274	336 336	106 359
885	78 32 25	693 154 125	6134 4140 0625	29·748 950	94·074 439	336 146	106 299
886	78 49 96	695 506 456	6162 1872 0016	29·765 752	94·127 573	335 957	106 239
887	78 67 69	697 864 103	6190 0545 9361	29·782 545	94·180 677	335 767	106 179
888	78 85 44	700 227 072	6218 0163 9936	29·799 329	94·233 752	335 578	106 119
889	79 03 21	702 595 369	6246 0728 3041	29·816 103	94·286 797	335 389	106 059
890	79 21 00	704 969 000	6274 2241 0000	29·832 868	94·339 811	335 201	106 000
891	79 38 81	707 347 971	6302 4704 2161	29·849 623	94·392 796	335 013	105 940
892	79 56 64	709 732 288	6330 8120 0896	29·866 369	94·445 752	334 825	105 881
893	79 74 49	712 121 957	6359 2490 7601	29·883 106	94·498 677	334 637	105 822
894	79 92 36	714 516 984	6387 7818 3696	29·899 833	94·551 573	334 450	105 762
895	80 10 25	716 917 375	6416 4105 0625	29·916 551	94·604 440	334 263	105 703
896	80 28 16	719 323 136	6445 1352 9856	29·933 259	94·657 277	334 077	105 644
897	80 46 09	721 734 273	6473 9564 2881	29·949 958	94·710 084	333 890	105 585
898	80 64 04	724 150 792	6502 8741 1216	29·966 648	94·762 862	333 704	105 527
899	80 82 01	726 572 699	6531 8885 6401	29·983 329	94·815 611	333 519	105 468
900	81 00 00	729 000 000	6561 0000 0000	30·000 000	94·868 330	333 333	105 409

No. x	Reciprocal $\frac{1}{x}=x^{-1}$	Cube Roots $\sqrt[3]{x}$	$\sqrt[3]{10x}$	$\sqrt[3]{100x}$	Factorials * $x!$	$\log x!$	Prime Factors
	0·00						
850	117 647	9·472 682	20·408 276	43·968 297	5·243 475	2122·719 619	$2.5^2.17$
851	117 509	·476 396	·416 276	43·985 532	4·462 197	2125·649 549	23.37
852	117 371	·480 106	·424 269	44·002 755	3·801 792	2128·579 988	$2^2.3.71$
853	117 233	·483 814	·432 257	44·019 963	3·242 929	2131·510 937	**853**
854	117 096	·487 518	·440 238	44·037 159	2·769 461	2134·442 395	2.7.61
855	116 959	9·491 220	20·448 214	44·054 341	2·367 889	2137·374 361	$3^2.5.19$
856	116 822	·494 919	·456 182	44·071 509	2·026 913	2140·306 835	$2^3.107$
857	116 686	·498 615	·464 145	44·088 664	1·737 065	2143·239 816	**857**
858	116 550	·502 308	·472 102	44·105 806	1·490 401	2146·173 303	2.3.11.13
859	116 414	·505 998	·480 052	44·122 934	1·280 255	2149·107 296	**859**
860	116 279	9·509 685	20·487 996	44·140 050	1·101 019	2152·041 795	$2^2.5.43$
861	116 144	·513 370	·495 934	44·157 152	9·479 775	2154·976 798	3.7.41
862	116 009	·517 052	·503 866	44·174 240	8·171 566	2157·912 305	2.431
863	115 875	·520 730	·511 792	44·191 316	7·052 061	2160·848 316	**863**
864	115 741	·524 406	·519 711	44·208 378	6·092 981	2163·784 830	$2^5.3^3$
865	115 607	9·528 079	20·527 625	44·225 427	5·270 429	2166·721 846	5.173
866	115 473	·531 750	·535 532	44·242 463	4·564 191	2169·659 364	2.433
867	115 340	·535 417	·543 434	44·259 486	3·957 154	2172·597 383	3.17^2
868	115 207	·539 082	·551 329	44·276 496	3·434 809	2175·535 903	$2^2.7.31$
869	115 075	·542 744	·559 218	44·293 493	2·984 849	2178·474 922	11.79
870	114 943	9·546 403	20·567 101	44·310 476	2·596 819	2181·414 442	2.3.5.29
871	114 811	·550 059	·574 978	44·327 447	2·261 829	2184·354 460	13.67
872	114 679	·553 712	·582 849	44·344 405	1·972 315	2187·294 976	$2^3.109$
873	114 548	·557 363	·590 714	44·361 349	1·721 831	2190·235 991	$3^2.97$
874	114 416	·561 011	·598 573	44·378 281	1·504 880	2193·177 502	2.19.23
875	114 286	9·564 656	20·606 426	44·395 200	1·316 770	2196·119 510	$5^3.7$
876	114 155	·568 298	·614 274	44·412 106	1·153 491	2199·062 014	$2^2.3.73$
877	114 025	·571 938	·622 115	44·428 999	1·011 611	2202·005 014	**877**
878	113 895	·575 574	·629 950	44·445 880	8·881 949	2204·948 508	2.439
879	113 766	·579 208	·637 779	44·462 747	7·807 233	2207·892 497	3.293
880	113 636	9·582 840	20·645 602	44·479 602	6·870 365	2210·836 980	$2^4.5.11$
881	113 507	·586 468	·653 420	44·496 444	6·052 792	2213·781 956	**881**
882	113 379	·590 094	·661 231	44·513 273	5·338 562	2216·727 424	$2.3^2.7^2$
883	113 250	·593 717	·669 037	44·530 089	4·713 950	2219·673 385	**883**
884	113 122	·597 337	·676 836	44·546 893	4·167 132	2222·619 837	$2^2.13.17$
885	112 994	9·600 955	20·684 630	44·563 684	3·687 912	2225·566 781	3.5.59
886	112 867	·604 570	·692 418	44·580 463	3·267 490	2228·514 214	2.443
887	112 740	·608 182	·700 200	44·597 229	2·898 264	2231·462 138	**887**
888	112 613	·611 791	·707 976	44·613 982	2·573 658	2234·410 551	$2^3.3.37$
889	112 486	·615 398	·715 746	44·630 723	2·287 982	2237·359 453	7.127
890	112 360	9·619 002	20·723 511	44·647 451	2·036 304	2240·308 843	2.5.89
891	112 233	·622 603	·731 270	44·664 167	1·814 347	2243·258 720	$3^4.11$
892	112 108	·626 202	·739 023	44·680 870	1·618 397	2246·209 085	$2^2.223$
893	111 982	·629 797	·746 770	44·697 560	1·445 229	2249·159 937	19.47
894	111 857	·633 391	·754 511	44·714 239	1·292 035	2252·111 274	2.3.149
895	111 732	9·636 981	20·762 247	44·730 904	1·156 371	2255·063 097	5.179
896	111 607	·640 569	·769 976	44·747 558	1·036 108	2258·015 405	$2^7.7$
897	111 483	·644 154	·777 700	44·764 199	9·293 893	2260·968 198	3.13.23
898	111 359	·647 737	·785 419	44·780 827	8·345 915	2263·921 474	2.449
899	111 235	·651 317	·793 131	44·797 444	7·502 978	2266·875 234	29.31
900	111 111	9·654 894	20·800 838	44·814 047	6·752 680	2269·829 476	$2^2.3^2.5^2$

* Multiply by 10^c, where c is the characteristic of $\log x!$

No.	Square	Cube	Fourth Power	Square Roots		Reciprocals of Square Roots	
x	x^2	x^3	x^4	$\sqrt{x} = x^{\frac{1}{2}}$	$\sqrt{10x}$	$\frac{1}{\sqrt{x}} = x^{-\frac{1}{2}}$	$\frac{1}{\sqrt{10x}}$
						0·0	0·0
900	81 00 00	729 000 000	6561 0000 0000	30·000 000	94·868 330	333 333	105 409
901	81 18 01	731 432 701	6590 2086 3601	30·016 662	94·921 020	333 148	105 351
902	81 36 04	733 870 808	6619 5146 8816	30·033 315	94·973 681	332 964	105 292
903	81 54 09	736 314 327	6648 9183 7281	30·049 958	95·026 312	332 779	105 234
904	81 72 16	738 763 264	6678 4199 0656	30·066 593	95·078 915	332 595	105 176
905	81 90 25	741 217 625	6708 0195 0625	30·083 218	95·131 488	332 411	105 118
906	82 08 36	743 677 416	6737 7173 8896	30·099 834	95·184 032	332 228	105 060
907	82 26 49	746 142 643	6767 5137 7201	30·116 441	95·236 548	332 045	105 002
908	82 44 64	748 613 312	6797 4088 7296	30·133 038	95·289 034	331 862	104 944
909	82 62 81	751 089 429	6827 4029 0961	30·149 627	95·341 491	331 679	104 886
910	82 81 00	753 571 000	6857 4961 0000	30·166 206	95·393 920	331 497	104 828
911	82 99 21	756 058 031	6887 6886 6241	30·182 777	95·446 320	331 315	104 771
912	83 17 44	758 550 528	6917 9808 1536	30·199 338	95·498 691	331 133	104 713
913	83 35 69	761 048 497	6948 3727 7761	30·215 890	95·551 033	330 952	104 656
914	83 53 96	763 551 944	6978 8647 6816	30·232 433	95·603 347	330 771	104 599
915	83 72 25	766 060 875	7009 4570 0625	30·248 967	95·655 632	330 590	104 542
916	83 90 56	768 575 296	7040 1497 1136	30·265 492	95·707 889	330 409	104 485
917	84 08 89	771 095 213	7070 9431 0321	30·282 008	95·760 117	330 229	104 428
918	84 27 24	773 620 632	7101 8374 0176	30·298 515	95·812 317	330 049	104 371
919	84 45 61	776 151 559	7132 8328 2721	30·315 013	95·864 488	329 870	104 314
920	84 64 00	778 688 000	7163 9296 0000	30·331 502	95·916 630	329 690	104 257
921	84 82 41	781 229 961	7195 1279 4081	30·347 982	95·968 745	329 511	104 201
922	85 00 84	783 777 448	7226 4280 7056	30·364 453	96·020 831	329 332	104 144
923	85 19 29	786 330 467	7257 8302 1041	30·380 915	96·072 889	329 154	104 088
924	85 37 76	788 889 024	7289 3345 8176	30·397 368	96·124 919	328 976	104 031
925	85 56 25	791 453 125	7320 9414 0625	30·413 813	96·176 920	328 798	103 975
926	85 74 76	794 022 776	7352 6509 0576	30·430 248	96·228 894	328 620	103 919
927	85 93 29	796 597 983	7384 4633 0241	30·446 675	96·280 839	328 443	103 863
928	86 11 84	799 178 752	7416 3788 1856	30·463 092	96·332 757	328 266	103 807
929	86 30 41	801 765 089	7448 3976 7681	30·479 501	96·384 646	328 089	103 751
930	86 49 00	804 357 000	7480 5201 0000	30·495 901	96·436 508	327 913	103 695
931	86 67 61	806 954 491	7512 7463 1121	30·512 293	96·488 341	327 737	103 639
932	86 86 24	809 557 568	7545 0765 3376	30·528 675	96·540 147	327 561	103 584
933	87 04 89	812 166 237	7577 5109 9121	30·545 049	96·591 925	327 385	103 528
934	87 23 56	814 780 504	7610 0499 0736	30·561 414	96·643 675	327 210	103 473
935	87 42 25	817 400 375	7642 6935 0625	30·577 770	96·695 398	327 035	103 418
936	87 60 96	820 025 856	7675 4420 1216	30·594 117	96·747 093	326 860	103 362
937	87 79 69	822 656 953	7708 2956 4961	30·610 456	96·798 760	326 686	103 307
938	87 98 44	825 293 672	7741 2546 4336	30·626 786	96·850 400	326 512	103 252
939	88 17 21	827 936 019	7774 3192 1841	30·643 107	96·902 012	326 338	103 197
940	88 36 00	830 584 000	7807 4896 0000	30·659 419	96·953 597	326 164	103 142
941	88 54 81	833 237 621	7840 7660 1361	30·675 723	97·005 155	325 991	103 087
942	88 73 64	835 896 888	7874 1486 8496	30·692 019	97·056 684	325 818	103 033
943	88 92 49	838 561 807	7907 6378 4001	30·708 305	97·108 187	325 645	102 978
944	89 11 36	841 232 384	7941 2337 0496	30·724 583	97·159 662	325 472	102 923
945	89 30 25	843 908 625	7974 9365 0625	30·740 852	97·211 110	325 300	102 869
946	89 49 16	846 590 536	8008 7464 7056	30·757 113	97·262 531	325 128	102 815
947	89 68 09	849 278 123	8042 6638 2481	30·773 365	97·313 925	324 956	102 760
948	89 87 04	851 971 392	8076 6887 9616	30·789 609	97·365 292	324 785	102 706
949	90 06 01	854 670 349	8110 8216 1201	30·805 844	97·416 631	324 614	102 652
950	90 25 00	857 375 000	8145 0625 0000	30·822 070	97·467 943	324 443	102 598

TABLE VIII—RECIPROCALS, CUBE ROOTS, FACTORIALS AND FACTORS

No. x	Reciprocal $\frac{1}{x}=x^{-1}$	Cube Roots $\sqrt[3]{x}$	$\sqrt[3]{10x}$	$\sqrt[3]{100x}$	Factorials * $x!$	$\log x!$	Prime Factors
	0·00						
900	111 111	9·654 894	20·800 838	44·814 047	6·752 680	2269·829 476	$2^2.3^2.5^2$
901	110 988	·658 468	·808 539	44·830 639	6·084 165	2272·784 201	17.53
902	110 865	·662 040	·816 235	44·847 218	5·487 917	2275·739 408	2.11.41
903	110 742	·665 610	·823 925	44·863 786	4·955 589	2278·695 095	3.7.43
904	110 619	·669 176	·831 609	44·880 341	4·479 852	2281·651 264	$2^3.113$
905	110 497	9·672 740	20·839 287	44·896 883	4·054 266	2284·607 912	5.181
906	110 375	·676 302	·846 960	44·913 414	3·673 165	2287·565 040	2.3.151
907	110 254	·679 860	·854 627	44·929 932	3·331 561	2290·522 648	**907**
908	110 132	·683 417	·862 289	44·946 438	3·025 057	2293·480 734	$2^2.227$
909	110 011	·686 970	·869 945	44·962 932	2·749 777	2296·439 297	$3^2.101$
910	109 890	9·690 521	20·877 595	44·979 414	2·502 297	2299·398 339	2.5.7.13
911	109 769	·694 069	·885 239	44·995 884	2·279 593	2302·357 857	**911**
912	109 649	·697 615	·892 879	45·012 342	2·078 989	2305·317 852	$2^4.3.19$
913	109 529	·701 158	·900 512	45·028 788	1·898 117	2308·278 323	11.83
914	109 409	·704 699	·908 140	45·045 222	1·734 879	2311·239 269	2.457
915	109 290	9·708 237	20·915 762	45·061 644	1·587 414	2314·200 690	3.5.61
916	109 170	·711 772	·923 379	45·078 054	1·454 071	2317·162 586	$2^2.229$
917	109 051	·715 305	·930 990	45·094 452	1·333 383	2320·124 955	7.131
918	108 932	·718 835	·938 596	45·110 838	1·224 046	2323·087 798	$2.3^3.17$
919	108 814	·722 363	·946 196	45·127 212	1·124 898	2326·051 113	**919**
920	108 696	9·725 888	20·953 791	45·143 574	1·034 906	2329·014 901	$2^3.5.23$
921	108 578	·729 411	·961 380	45·159 925	9·531 486	2331·979 161	3.307
922	108 460	·732 931	·968 964	45·176 263	8·788 030	2334·943 892	2.461
923	108 342	·736 448	·976 542	45·192 590	8·111 352	2337·909 093	13.71
924	108 225	·739 963	·984 115	45·208 905	7·494 889	2340·874 765	$2^2.3.7.11$
925	108 108	9·743 476	20·991 682	45·225 208	6·932 773	2343·840 907	$5^2.37$
926	107 991	·746 986	20·999 244	45·241 500	6·419 747	2346·807 518	2.463
927	107 875	·750 493	21·006 801	45·257 780	5·951 106	2349·774 598	$3^2.103$
928	107 759	·753 998	·014 351	45·274 048	5·522 626	2352·742 146	$2^5.29$
929	107 643	·757 500	·021 897	45·290 304	5·130 520	2355·710 161	**929**
930	107 527	9·761 000	21·029 437	45·306 549	4·771 383	2358·678 644	2.3.5.31
931	107 411	·764 497	·036 972	45·322 782	4·442 158	2361·647 594	$7^2.19$
932	107 296	·767 992	·044 501	45·339 004	4·140 091	2364·617 010	$2^2.233$
933	107 181	·771 485	·052 025	45·355 213	3·862 705	2367·586 892	3.311
934	107 066	·774 974	·059 544	45·371 412	3·607 767	2370·557 238	2.467
935	106 952	9·778 462	21·067 057	45·387 598	3·373 262	2373·528 050	5.11.17
936	106 838	·781 946	·074 565	45·403 774	3·157 373	2376·499 326	$2^3.3^2.13$
937	106 724	·785 429	·082 067	45·419 937	2·958 458	2379·471 065	**937**
938	106 610	·788 909	·089 565	45·436 089	2·775 034	2382·443 268	2.7.67
939	106 496	·792 386	·097 056	45·452 230	2·605 757	2385·415 934	3.313
940	106 383	9·795 861	21·104 543	45·468 359	2·449 412	2388·389 062	$2^2.5.47$
941	106 270	·799 334	·112 024	45·484 477	2·304 896	2391·362 651	**941**
942	106 157	·802 804	·119 500	45·500 584	2·171 212	2394·336 702	2.3.157
943	106 045	·806 271	·126 971	45·516 679	2·047 453	2397·311 214	23.41
944	105 932	·809 736	·134 436	45·532 762	1·932 796	2400·286 186	$2^4.59$
945	105 820	9·813 199	21·141 896	45·548 835	1·826 492	2403·261 618	$3^3.5.7$
946	105 708	·816 659	·149 351	45·564 896	1·727 861	2406·237 509	2.11.43
947	105 597	·820 117	·156 801	45·580 945	1·636 285	2409·213 859	**947**
948	105 485	·823 572	·164 245	45·596 983	1·551 198	2412·190 667	$2^2.3.79$
949	105 374	·827 025	·171 684	45·613 011	1·472 087	2415·167 933	13.73
950	105 263	9·830 476	21·179 118	45·629 026	1·398 483	2418·145 657	$2.5^2.19$

* Multiply by 10^c, where c is the characteristic of $\log x!$

No.	Square	Cube	Fourth Power	Square Roots		Reciprocals of Square Roots	
x	x^2	x^3	x^4	$\sqrt{x}=x^{\frac{1}{2}}$	$\sqrt{10x}$	$\frac{1}{\sqrt{x}}=x^{-\frac{1}{2}}$	$\frac{1}{\sqrt{10x}}$
						0·0	0·0
950	90 25 00	857 375 000	8145 0625 0000	30·822 070	97·467 943	324 443	102 598
951	90 44 01	860 085 351	8179 4116 8801	30·838 288	97·519 229	324 272	102 544
952	90 63 04	862 801 408	8213 8694 0416	30·854 497	97·570 487	324 102	102 490
953	90 82 09	865 523 177	8248 4358 7681	30·870 698	97·621 719	323 932	102 436
954	91 01 16	868 250 664	8283 1113 3456	30·886 890	97·672 924	323 762	102 383
955	91 20 25	870 983 875	8317 8960 0625	30·903 074	97·724 101	323 592	102 329
956	91 39 36	873 722 816	8352 7901 2096	30·919 250	97·775 252	323 423	102 275
957	91 58 49	876 467 493	8387 7939 0801	30·935 417	97·826 377	323 254	102 222
958	91 77 64	879 217 912	8422 9075 9696	30·951 575	97·877 474	323 085	102 169
959	91 96 81	881 974 079	8458 1314 1761	30·967 725	97·928 545	322 917	102 115
960	92 16 00	884 736 000	8493 4656 0000	30·983 867	97·979 590	322 749	102 062
961	92 35 21	887 503 681	8528 9103 7441	31·000 000	98·030 607	322 581	102 009
962	92 54 44	890 277 128	8564 4659 7136	31·016 125	98·081 599	322 413	101 956
963	92 73 69	893 056 347	8600 1326 2161	31·032 241	98·132 563	322 245	101 903
964	92 92 96	895 841 344	8635 9105 5616	31·048 349	98·183 502	322 078	101 850
965	93 12 25	898 632 125	8671 8000 0625	31·064 449	98·234 414	321 911	101 797
966	93 31 56	901 428 696	8707 8012 0336	31·080 541	98·285 299	321 745	101 745
967	93 50 89	904 231 063	8743 9143 7921	31·096 624	98·336 158	321 578	101 692
968	93 70 24	907 039 232	8780 1397 6576	31·112 698	98·386 991	321 412	101 639
969	93 89 61	909 853 209	8816 4775 9521	31·128 765	98·437 798	321 246	101 587
970	94 09 00	912 673 000	8852 9281 0000	31·144 823	98·488 578	321 081	101 535
971	94 28 41	915 498 611	8889 4915 1281	31·160 873	98·539 332	320 915	101 482
972	94 47 84	918 330 048	8926 1680 6656	31·176 915	98·590 060	320 750	101 430
973	94 67 29	921 167 317	8962 9579 9441	31·192 948	98·640 762	320 585	101 378
974	94 86 76	924 010 424	8999 8615 2976	31·208 973	98·691 438	320 421	101 326
975	95 06 25	926 859 375	9036 8789 0625	31·224 990	98·742 088	320 256	101 274
976	95 25 76	929 714 176	9074 0103 5776	31·240 999	98·792 712	320 092	101 222
977	95 45 29	932 574 833	9111 2561 1841	31·256 999	98·843 310	319 928	101 170
978	95 64 84	935 441 352	9148 6164 2256	31·272 992	98·893 883	319 765	101 118
979	95 84 41	938 313 739	9186 0915 0481	31·288 976	98·944 429	319 601	101 067
980	96 04 00	941 192 000	9223 6816 0000	31·304 952	98·994 949	319 438	101 015
981	96 23 61	944 076 141	9261 3869 4321	31·320 920	99·045 444	319 275	100 964
982	96 43 24	946 966 168	9299 2077 6976	31·336 879	99·095 913	319 113	100 912
983	96 62 89	949 862 087	9337 1443 1521	31·352 831	99·146 356	318 950	100 861
984	96 82 56	952 763 904	9375 1968 1536	31·368 774	99·196 774	318 788	100 810
985	97 02 25	955 671 625	9413 3655 0625	31·384 710	99·247 166	318 626	100 759
986	97 21 96	958 585 256	9451 6506 2416	31·400 637	99·297 533	318 465	100 707
987	97 41 69	961 504 803	9490 0524 0561	31·416 556	99·347 874	318 304	100 656
988	97 61 44	964 430 272	9528 5710 8736	31·432 467	99·398 189	318 142	100 605
989	97 81 21	967 361 669	9567 2069 0641	31·448 370	99·448 479	317 982	100 555
990	98 01 00	970 299 000	9605 9601 0000	31·464 265	99·498 744	317 821	100 504
991	98 20 81	973 242 271	9644 8309 0561	31·480 152	99·548 983	317 660	100 453
992	98 40 64	976 191 488	9683 8195 6096	31·496 031	99·599 197	317 500	100 402
993	98 60 49	979 146 657	9722 9263 0401	31·511 903	99·649 385	317 340	100 352
994	98 80 36	982 107 784	9762 1513 7296	31·527 766	99·699 549	317 181	100 301
995	99 00 25	985 074 875	9801 4950 0625	31·543 621	99·749 687	317 021	100 251
996	99 20 16	988 047 936	9840 9574 4256	31·559 468	99·799 800	316 862	100 201
997	99 40 09	991 026 973	9880 5389 2081	31·575 307	99·849 887	316 703	100 150
998	99 60 04	994 011 992	9920 2396 8016	31·591 138	99·899 950	316 544	100 100
999	99 80 01	997 002 999	9960 0599 6001	31·606 961	99·949 987	316 386	100 050
1000	100 00 00	1000 000 000	10000 0000 0000	31·622 777	100·000 000	316 228	100 000

No. x	Reciprocal $\frac{1}{x}=x^{-1}$	Cube Roots $\sqrt[3]{x}$	$\sqrt[3]{10x}$	$\sqrt[3]{100x}$	Factorials * $x!$	$\log x!$	Prime Factors
	0·00						
950	105 263	9·830 476	21·179 118	45·629 026	1·398 483	2418·145 657	$2.5^{2}.19$
951	105 152	·833 924	·186 547	45·645 031	1·329 957	2421·123 838	3.317
952	105 042	·837 369	·193 970	45·661 024	1·266 119	2424·102 475	$2^{3}.7.17$
953	104 932	·840 813	·201 388	45·677 006	1·206 611	2427·081 567	**953**
954	104 822	·844 254	·208 801	45·692 977	1·151 107	2430·061 116	$2.3^{2}.53$
955	104 712	9·847 692	21·216 209	45·708 937	1·099 307	2433·041 119	5.191
956	104 603	·851 128	·223 612	45·724 886	1·050 938	2436·021 577	$2^{2}.239$
957	104 493	·854 562	·231 010	45·740 824	1·005 748	2439·002 489	3.11.29
958	104 384	·857 993	·238 402	45·756 750	9·635 062	2441·983 855	2.479
959	104 275	·861 422	·245 789	45·772 665	9·240 024	2444·965 673	7.137
960	104 167	9·864 848	21·253 171	45·788 570	8·870 423	2447·947 944	$2^{6}.3.5$
961	104 058	·868 272	·260 548	45·804 463	8·524 477	2450·930 668	31^{2}
962	103 950	·871 694	·267 920	45·820 345	8·200 547	2453·913 843	2.13.37
963	103 842	·875 113	·275 287	45·836 217	7·897 126	2456·897 469	$3^{2}.107$
964	103 734	·878 530	·282 649	45·852 077	7·612 830	2459·881 546	$2^{2}.241$
965	103 627	9·881 945	21·290 005	45·867 926	7·346 381	2462·866 073	5.193
966	103 520	·885 357	·297 357	45·883 765	7·096 604	2465·851 051	2.3.7.23
967	103 413	·888 767	·304 703	45·899 592	6·862 416	2468·836 477	**967**
968	103 306	·892 175	·312 045	45·915 408	6·642 819	2471·822 352	$2^{3}.11^{2}$
969	103 199	·895 580	·319 381	45·931 214	6·436 891	2474·808 676	3.17.19
970	103 093	9·898 983	21·326 712	45·947 009	6·243 784	2477·795 448	2.5.97
971	102 987	·902 384	·334 039	45·962 793	6·062 715	2480·782 667	**971**
972	102 881	·905 782	·341 360	45·978 566	5·892 959	2483·770 333	$2^{2}.3^{5}$
973	102 775	·909 178	·348 676	45·994 328	5·733 849	2486·758 446	7.139
974	102 669	·912 571	·355 987	46·010 080	5·584 769	2489·747 005	2.487
975	102 564	9·915 962	21·363 293	46·025 820	5·445 150	2492·736 010	$3.5^{2}.13$
976	102 459	·919 351	·370 595	46·041 550	5·314 466	2495·725 460	$2^{4}.61$
977	102 354	·922 738	·377 891	46·057 270	5·192 233	2498·715 354	**977**
978	102 249	·926 122	·385 182	46·072 978	5·078 004	2501·705 693	2.3.163
979	102 145	·929 504	·392 468	46·088 676	4·971 366	2504·696 476	11.89
980	102 041	9·932 884	21·399 750	46·104 363	4·871 939	2507·687 702	$2^{2}.5.7^{2}$
981	101 937	·936 261	·407 026	46·120 039	4·779 372	2510·679 371	$3^{2}.109$
982	101 833	·939 636	·414 297	46·135 705	4·693 343	2513·671 482	2.491
983	101 729	·943 009	·421 564	46·151 360	4·613 556	2516·664 036	**983**
984	101 626	·946 380	·428 825	46·167 005	4·539 739	2519·657 031	$2^{3}.3.41$
985	101 523	9·949 748	21·436 082	46·182 639	4·471 643	2522·650 467	5.197
986	101 420	·953 114	·443 334	46·198 262	4·409 040	2525·644 344	2.17.29
987	101 317	·956 478	·450 581	46·213 875	4·351 723	2528·638 661	3.7.47
988	101 215	·959 839	·457 822	46·229 477	4·299 502	2531·633 418	$2^{2}.13.19$
989	101 112	·963 198	·465 060	46·245 069	4·252 208	2534·628 614	23.43
990	101 010	9·966 555	21·472 292	46·260 650	4·209 686	2537·624 250	$2.3^{2}.5.11$
991	100 908	·969 910	·479 519	46·276 221	4·171 798	2540·620 323	**991**
992	100 806	·973 262	·486 741	46·291 781	4·138 424	2543·616 835	$2^{5}.31$
993	100 705	·976 612	·493 959	46·307 331	4·109 455	2546·613 784	3.331
994	100 604	·979 960	·501 172	46·322 870	4·084 798	2549·611 171	2.7.71
995	100 503	9·983 305	21·508 380	46·338 399	4·064 374	2552·608 994	5.199
996	100 402	·986 649	·515 583	46·353 918	4·048 117	2555·607 253	$2^{2}.3.83$
997	100 301	·989 990	·522 781	46·369 426	4·035 972	2558·605 948	**997**
998	100 200	·993 329	·529 974	46·384 924	4·027 901	2561·605 079	2.499
999	100 100	9·996 666	·537 163	46·400 411	4·023 873	2564·604 644	$3^{3}.37$
1000	100 000	10·000 000	21·544 347	46·415 888	4·023 873	2567·604 644	$2^{3}.5^{3}$

* Multiply by 10^{c}, where c is the characteristic of $\log x!$

TABLE IX—PRIME NUMBERS

1	257	601	977	1373	1777	2213	2659	3079	3539	3989	4451	4937	5417	5857
2	263	607	983	1381	1783	2221	2663	3083	3541	4001	4457	4943	5419	5861
3	269	613	991	1399	1787	2237	2671	3089	3547	4003	4463	4951	5431	5867
5	271	617	997	1409	1789	2239	2677	3109	3557	4007	4481	4957	5437	5869
7	277	619	1009	1423	1801	2243	2683	3119	3559	4013	4483	4967	5441	5879
11	281	631	1013	1427	1811	2251	2687	3121	3571	4019	4493	4969	5443	5881
13	283	641	1019	1429	1823	2267	2689	3137	3581	4021	4507	4973	5449	5897
17	293	643	1021	1433	1831	2269	2693	3163	3583	4027	4513	4987	5471	5903
19	307	647	1031	1439	1847	2273	2699	3167	3593	4049	4517	4993	5477	5923
23	311	653	1033	1447	1861	2281	2707	3169	3607	4051	4519	4999	5479	5927
29	313	659	1039	1451	1867	2287	2711	3181	3613	4057	4523	5003	5483	5939
31	317	661	1049	1453	1871	2293	2713	3187	3617	4073	4547	5009	5501	5953
37	331	673	1051	1459	1873	2297	2719	3191	3623	4079	4549	5011	5503	5981
41	337	677	1061	1471	1877	2309	2729	3203	3631	4091	4561	5021	5507	5987
43	347	683	1063	1481	1879	2311	2731	3209	3637	4093	4567	5023	5519	6007
47	349	691	1069	1483	1889	2333	2741	3217	3643	4099	4583	5039	5521	6011
53	353	701	1087	1487	1901	2339	2749	3221	3659	4111	4591	5051	5527	6029
59	359	709	1091	1489	1907	2341	2753	3229	3671	4127	4597	5059	5531	6037
61	367	719	1093	1493	1913	2347	2767	3251	3673	4129	4603	5077	5557	6043
67	373	727	1097	1499	1931	2351	2777	3253	3677	4133	4621	5081	5563	6047
71	379	733	1103	1511	1933	2357	2789	3257	3691	4139	4637	5087	5569	6053
73	383	739	1109	1523	1949	2371	2791	3259	3697	4153	4639	5099	5573	6067
79	389	743	1117	1531	1951	2377	2797	3271	3701	4157	4643	5101	5581	6073
83	397	751	1123	1543	1973	2381	2801	3299	3709	4159	4649	5107	5591	6079
89	401	757	1129	1549	1979	2383	2803	3301	3719	4177	4651	5113	5623	6089
97	409	761	1151	1553	1987	2389	2819	3307	3727	4201	4657	5119	5639	6091
101	419	769	1153	1559	1993	2393	2833	3313	3733	4211	4663	5147	5641	6101
103	421	773	1163	1567	1997	2399	2837	3319	3739	4217	4673	5153	5647	6113
107	431	787	1171	1571	1999	2411	2843	3323	3761	4219	4679	5167	5651	6121
109	433	797	1181	1579	2003	2417	2851	3329	3767	4229	4691	5171	5653	6131
113	439	809	1187	1583	2011	2423	2857	3331	3769	4231	4703	5179	5657	6133
127	443	811	1193	1597	2017	2437	2861	3343	3779	4241	4721	5189	5659	6143
131	449	821	1201	1601	2027	2441	2879	3347	3793	4243	4723	5197	5669	6151
137	457	823	1213	1607	2029	2447	2887	3359	3797	4253	4729	5209	5683	6163
139	461	827	1217	1609	2039	2459	2897	3361	3803	4259	4733	5227	5689	6173
149	463	829	1223	1613	2053	2467	2903	3371	3821	4261	4751	5231	5693	6197
151	467	839	1229	1619	2063	2473	2909	3373	3823	4271	4759	5233	5701	6199
157	479	853	1231	1621	2069	2477	2917	3389	3833	4273	4783	5237	5711	6203
163	487	857	1237	1627	2081	2503	2927	3391	3847	4283	4787	5261	5717	6211
167	491	859	1249	1637	2083	2521	2939	3407	3851	4289	4789	5273	5737	6217
173	499	863	1259	1657	2087	2531	2953	3413	3853	4297	4793	5279	5741	6221
179	503	877	1277	1663	2089	2539	2957	3433	3863	4327	4799	5281	5743	6229
181	509	881	1279	1667	2099	2543	2963	3449	3877	4337	4801	5297	5749	6247
191	521	883	1283	1669	2111	2549	2969	3457	3881	4339	4813	5303	5779	6257
193	523	887	1289	1693	2113	2551	2971	3461	3889	4349	4817	5309	5783	6263
197	541	907	1291	1697	2129	2557	2999	3463	3907	4357	4831	5323	5791	6269
199	547	911	1297	1699	2131	2579	3001	3467	3911	4363	4861	5333	5801	6271
211	557	919	1301	1709	2137	2591	3011	3469	3917	4373	4871	5347	5807	6277
223	563	929	1303	1721	2141	2593	3019	3491	3919	4391	4877	5351	5813	6287
227	569	937	1307	1723	2143	2609	3023	3499	3923	4397	4889	5381	5821	6299
229	571	941	1319	1733	2153	2617	3037	3511	3929	4409	4903	5387	5827	6301
233	577	947	1321	1741	2161	2621	3041	3517	3931	4421	4909	5393	5839	6311
239	587	953	1327	1747	2179	2633	3049	3527	3943	4423	4919	5399	5843	6317
241	593	967	1361	1753	2203	2647	3061	3529	3947	4441	4931	5407	5849	6323
251	599	971	1367	1759	2207	2657	3067	3533	3967	4447	4933	5413	5851	6329

6337	6833	7333	7829	8329	8831	9337	9811	10313	10859	11393	11933	12437
6343	6841	7349	7841	8353	8837	9341	9817	10321	10861	11399	11939	12451
6353	6857	7351	7853	8363	8839	9343	9829	10331	10867	11411	11941	12457
6359	6863	7369	7867	8369	8849	9349	9833	10333	10883	11423	11953	12473
6361	6869	7393	7873	8377	8861	9371	9839	10337	10889	11437	11959	12479
6367	6871	7411	7877	8387	8863	9377	9851	10343	10891	11443	11969	12487
6373	6883	7417	7879	8389	8867	9391	9857	10357	10903	11447	11971	12491
6379	6899	7433	7883	8419	8887	9397	9859	10369	10909	11467	11981	12497
6389	6907	7451	7901	8423	8893	9403	9871	10391	10937	11471	11987	12503
6397	6911	7457	7907	8429	8923	9413	9883	10399	10939	11483	12007	12511
6421	6917	7459	7919	8431	8929	9419	9887	10427	10949	11489	12011	12517
6427	6947	7477	7927	8443	8933	9421	9901	10429	10957	11491	12037	12527
6449	6949	7481	7933	8447	8941	9431	9907	10433	10973	11497	12041	12539
6451	6959	7487	7937	8461	8951	9433	9923	10453	10979	11503	12043	12541
6469	6961	7489	7949	8467	8963	9437	9929	10457	10987	11519	12049	12547
6473	6967	7499	7951	8501	8969	9439	9931	10459	10993	11527	12071	12553
6481	6971	7507	7963	8513	8971	9461	9941	10463	11003	11549	12073	12569
6491	6977	7517	7993	8521	8999	9463	9949	10477	11027	11551	12097	12577
6521	6983	7523	8009	8527	9001	9467	9967	10487	11047	11579	12101	12583
6529	6991	7529	8011	8537	9007	9473	9973	10499	11057	11587	12107	12589
6547	6997	7537	8017	8539	9011	9479	10007	10501	11059	11593	12109	12601
6551	7001	7541	8039	8543	9013	9491	10009	10513	11069	11597	12113	12611
6553	7013	7547	8053	8563	9029	9497	10037	10529	11071	11617	12119	12613
6563	7019	7549	8059	8573	9041	9511	10039	10531	11083	11621	12143	12619
6569	7027	7559	8069	8581	9043	9521	10061	10559	11087	11633	12149	12637
6571	7039	7561	8081	8597	9049	9533	10067	10567	11093	11657	12157	12641
6577	7043	7573	8087	8599	9059	9539	10069	10589	11113	11677	12161	12647
6581	7057	7577	8089	8609	9067	9547	10079	10597	11117	11681	12163	12653
6599	7069	7583	8093	8623	9091	9551	10091	10601	11119	11689	12197	12659
6607	7079	7589	8101	8627	9103	9587	10093	10607	11131	11699	12203	12671
6619	7103	7591	8111	8629	9109	9601	10099	10613	11149	11701	12211	12689
6637	7109	7603	8117	8641	9127	9613	10103	10627	11159	11717	12227	12697
6653	7121	7607	8123	8647	9133	9619	10111	10631	11161	11719	12239	12703
6659	7127	7621	8147	8663	9137	9623	10133	10639	11171	11731	12241	12713
6661	7129	7639	8161	8669	9151	9629	10139	10651	11173	11743	12251	12721
6673	7151	7643	8167	8677	9157	9631	10141	10657	11177	11777	12253	12739
6679	7159	7649	8171	8681	9161	9643	10151	10663	11197	11779	12263	12743
6689	7177	7669	8179	8689	9173	9649	10159	10667	11213	11783	12269	12757
6691	7187	7673	8191	8693	9181	9661	10163	10687	11239	11789	12277	12763
6701	7193	7681	8209	8699	9187	9677	10169	10691	11243	11801	12281	12781
6703	7207	7687	8219	8707	9199	9679	10177	10709	11251	11807	12289	12791
6709	7211	7691	8221	8713	9203	9689	10181	10711	11257	11813	12301	12799
6719	7213	7699	8231	8719	9209	9697	10193	10723	11261	11821	12323	12809
6733	7219	7703	8233	8731	9221	9719	10211	10729	11273	11827	12329	12821
6737	7229	7717	8237	8737	9227	9721	10223	10733	11279	11831	12343	12823
6761	7237	7723	8243	8741	9239	9733	10243	10739	11287	11833	12347	12829
6763	7243	7727	8263	8747	9241	9739	10247	10753	11299	11839	12373	12841
6779	7247	7741	8269	8753	9257	9743	10253	10771	11311	11863	12377	12853
6781	7253	7753	8273	8761	9277	9749	10259	10781	11317	11867	12379	12889
6791	7283	7757	8287	8779	9281	9767	10267	10789	11321	11887	12391	12893
6793	7297	7759	8291	8783	9283	9769	10271	10799	11329	11897	12401	12899
6803	7307	7789	8293	8803	9293	9781	10273	10831	11351	11903	12409	12907
6823	7309	7793	8297	8807	9311	9787	10289	10837	11353	11909	12413	12911
6827	7321	7817	8311	8819	9319	9791	10301	10847	11369	11923	12421	12917
6829	7331	7823	8317	8821	9323	9803	10303	10853	11383	11927	12433	12919

TABLE Xa—BESSELIAN INTERPOLATION COEFFICIENTS

n	B'' (–)
·000	
	·00
·020	
	·01
·064	
	·02
·112	
	·03
·168	
	·04
·235	
	·05
·326	
	·06
·673	
	·05
·764	
	·04
·831	
	·03
·887	
	·02
·935	
	·01
·979	
	·00
·999	

n	B'''
·000	
	·000
·006	
	+·001
·019	
	+·002
·033	
	+·003
·048	
	+·004
·066	
	+·005
·087	
	+·006
·114	
	+·007
·153	
	+·008
·273	
	+·007
·321	
	+·006
·356	
	+·005
·386	
	+·004
·413	
	+·003
·439	
	+·002
·463	
	+·001
·487	
	·000
·512	
	−·001
·536	
	−·002
·560	
	−·003
·586	
	−·004
·613	
	−·005
·643	
	−·006
·678	
	−·007
·726	
	−·008
·846	
	−·007
·885	
	−·006
·912	
	−·005
·933	
	−·004
·951	
	−·003
·966	
	−·002
·980	
	−·001
·993	
	·000
·999	

In critical cases ascend.

n	B'' (–)	n	B'' (–)	n	B'' (–)	n	B'' (–)	n	B'' (–)	n	B^{iv} (+)
·000		·110		·271		·728		·889		·000	
	·000		·025		·050		·049		·024		·000
·002		·115		·280		·736		·894		·012	
	·001		·026		·051		·048		·023		·001
·006		·120		·290		·744		·900		·036	
	·002		·027		·052		·047		·022		·002
·010		·125		·300		·752		·904		·062	
	·003		·028		·053		·046		·021		·003
·014		·131		·310		·760		·909		·088	
	·004		·029		·054		·045		·020		·004
·018		·136		·321		·768		·914		·116	
	·005		·030		·055		·044		·019		·005
·022		·142		·332		·775		·919		·145	
	·006		·031		·056		·043		·018		·006
·026		·147		·345		·782		·924		·176	
	·007		·032		·057		·042		·017		·007
·030		·153		·358		·789		·928		·210	
	·008		·033		·058		·041		·016		·008
·035		·159		·373		·796		·933		·248	
	·009		·034		·059		·040		·015		·009
·039		·165		·390		·803		·938		·291	
	·010		·035		·060		·039		·014		·010
·043		·171		·410		·809		·942		·346	
	·011		·036		·061		·038		·013		·011
·048		·177		·436		·816		·947		·435	
	·012		·037		·062		·037		·012		·012
·052		·183		·563		·822		·951		·564	
	·013		·038		·061		·036		·011		·011
·057		·190		·589		·828		·956		·653	
	·014		·039		·060		·035		·010		·010
·061		·196		·609		·834		·960		·708	
	·015		·040		·059		·034		·009		·009
·066		·203		·626		·840		·964		·751	
	·016		·041		·058		·033		·008		·008
·071		·210		·641		·846		·969		·789	
	·017		·042		·057		·032		·007		·007
·075		·217		·654		·852		·973		·823	
	·018		·043		·056		·031		·006		·006
·080		·224		·667		·857		·977		·854	
	·019		·044		·055		·030		·005		·005
·085		·231		·678		·863		·981		·883	
	·020		·045		·054		·029		·004		·004
·090		·239		·689		·868		·985		·911	
	·021		·046		·053		·028		·003		·003
·095		·247		·700		·874		·989		·937	
	·022		·047		·052		·027		·002		·002
·100		·255		·709		·879		·993		·963	
	·023		·048		·051		·026		·001		·001
·105		·263		·719		·884		·997		·987	
	·024		·049		·050		·025		·000		·000
·110		·271		·728		·889		·999		·999	

Note that B'' is always negative.

The notation used is:

Tabular values of function	Differences: First	Second	Third	Fourth
f_{-2}				
	$\Delta'_{-1\frac{1}{2}}$			
f_{-1}		Δ''_{-1}		
	$\Delta'_{-\frac{1}{2}}$		$\Delta'''_{-\frac{1}{2}}$	
f_0		Δ''_0		Δ^{iv}_0
	$\Delta'_{\frac{1}{2}}$		$\Delta'''_{\frac{1}{2}}$	
f_1		Δ''_1		
	$\Delta'_{1\frac{1}{2}}$			
f_2				

Note that

$$\Delta''_0 + \Delta''_1 = \Delta'_{1\frac{1}{2}} - \Delta'_{-\frac{1}{2}}$$

$$\Delta^{iv}_0 + \Delta^{iv}_1 = \Delta'''_{1\frac{1}{2}} - \Delta'''_{-\frac{1}{2}}$$

$$M'' = \Delta'' - 0{\cdot}184\Delta^{iv}$$

$$f_0 + n\Delta'_{\frac{1}{2}} = (1-n)f_0 + nf_1$$

FORMULÆ

If n is a fraction of the interval between the arguments of f_0 and f_1,

$$f_n = f_0 + n\Delta'_{\frac{1}{2}} + B''(\Delta''_0 + \Delta''_1) + B'''\Delta'''_{\frac{1}{2}} + B^{iv}(\Delta^{iv}_0 + \Delta^{iv}_1)$$

$$= f_0 + n\Delta'_{\frac{1}{2}} + B''(M''_0 + M''_1) + B'''\Delta'''_{\frac{1}{2}}$$

$f_{\frac{1}{2}} = \frac{1}{16}(-f_{-1} + 9f_0 + 9f_1 - f_2)$ if mean Δ^{iv} is less than 20.

$= \frac{1}{256}(3f_{-2} - 25f_{-1} + 150f_0 + 150f_1 - 25f_2 + 3f_3)$ if mean Δ^{vi} is less than 100.

With Bessel's formula, third differences less than 60, fourth differences less than 20, fifth differences less than 500 or sixth differences less than 100 are negligible.

TABLE Xb—NUMERICAL DIFFERENTIATION

The derivatives of a function tabulated at equal intervals of the independent variable may be computed from the differences of that function. Unfortunately, in order to obtain sufficient significant figures in the derivative it is often necessary to use a wide interval and high orders of differences.

The notation used is given on page xxii and summarised below.

Tabular values of function	Differences					
	First	Second	Third	Fourth	Fifth	Sixth
f_{-3}						
	$\Delta'_{-2\frac{1}{2}}$					
f_{-2}		Δ''_{-2}				
	$\Delta'_{-1\frac{1}{2}}$		$\Delta'''_{-1\frac{1}{2}}$			
f_{-1}		Δ''_{-1}		Δ^{iv}_{-1}		
	$\Delta'_{-\frac{1}{2}}$		$\Delta'''_{-\frac{1}{2}}$		$\Delta^{v}_{-\frac{1}{2}}$	
f_0		Δ''_0		Δ^{iv}_0		Δ^{vi}_0
	$\Delta'_{\frac{1}{2}}$		$\Delta'''_{\frac{1}{2}}$		$\Delta^{v}_{\frac{1}{2}}$	
f_1		Δ''_1		Δ^{iv}_1		
	$\Delta'_{1\frac{1}{2}}$		$\Delta'''_{1\frac{1}{2}}$			
f_2		Δ''_2				
	$\Delta'_{2\frac{1}{2}}$					
f_3						

In particular $\quad \Delta'_0 = \frac{1}{2}(\Delta'_{-\frac{1}{2}} + \Delta'_{\frac{1}{2}}) \qquad \Delta''_{\frac{1}{2}} = \frac{1}{2}(\Delta''_0 + \Delta''_1)$ etc.

Derivatives at Tabular Points

$$wf'_0 = \Delta'_0 - \frac{1}{6}\Delta'''_0 + \frac{1}{30}\Delta^{v}_0 - \frac{1}{140}\Delta^{vii}_0 + \frac{1}{630}\Delta^{ix}_0 - \frac{1}{2772}\Delta^{xi}_0 + \cdots$$

$$w^2f''_0 = \Delta''_0 - \frac{1}{12}\Delta^{iv}_0 + \frac{1}{90}\Delta^{vi}_0 - \frac{1}{560}\Delta^{viii}_0 + \frac{1}{3150}\Delta^{x}_0 - \frac{1}{16632}\Delta^{xii}_0 + \cdots$$

$$w^3f'''_0 = \Delta'''_0 - \frac{1}{4}\Delta^{v}_0 + \frac{7}{120}\Delta^{vii}_0 - \frac{41}{3024}\Delta^{ix}_0 + \frac{479}{151200}\Delta^{xi}_0 - \cdots$$

$$w^4f^{iv}_0 = \Delta^{iv}_0 - \frac{1}{6}\Delta^{vi}_0 + \frac{7}{240}\Delta^{viii}_0 - \frac{41}{7560}\Delta^{x}_0 + \frac{479}{453600}\Delta^{xii}_0 - \cdots$$

Derivatives at Half-way Points

$$wf'_{\frac{1}{2}} = \Delta'_{\frac{1}{2}} - \frac{1}{24}\Delta'''_{\frac{1}{2}} + \frac{3}{640}\Delta^{v}_{\frac{1}{2}} - \frac{5}{7168}\Delta^{vii}_{\frac{1}{2}} + \frac{35}{294912}\Delta^{ix}_{\frac{1}{2}} - \frac{63}{2883584}\Delta^{xi}_{\frac{1}{2}} + \cdots$$

$$w^2f''_{\frac{1}{2}} = \Delta''_{\frac{1}{2}} - \frac{5}{24}\Delta^{iv}_{\frac{1}{2}} + \frac{259}{5760}\Delta^{vi}_{\frac{1}{2}} - \frac{3229}{322560}\Delta^{viii}_{\frac{1}{2}} + \frac{117469}{51609600}\Delta^{x}_{\frac{1}{2}} - \frac{7156487}{13624934400}\Delta^{xii}_{\frac{1}{2}} + \cdots$$

$$w^3f'''_{\frac{1}{2}} = \Delta'''_{\frac{1}{2}} - \frac{1}{8}\Delta^{v}_{\frac{1}{2}} + \frac{37}{1920}\Delta^{vii}_{\frac{1}{2}} - \frac{3229}{967680}\Delta^{ix}_{\frac{1}{2}} + \frac{10679}{17203200}\Delta^{xi}_{\frac{1}{2}} - \cdots$$

It will be noted that even derivatives are best computed at tabular points and odd derivatives at half-way points, these being the cases that do not involve mean differences.

Derivatives at Intermediate Points

$$wf'_n = \Delta'_0 - \frac{1}{6}(1-3n^2)\Delta'''_0 + \frac{1}{120}(4-15n^2+5n^4)\Delta^{v}_0 + \cdots + n\left\{\Delta''_0 - \frac{1}{12}(1-2n^2)\Delta^{iv}_0 + \frac{1}{360}(4-10n^2+3n^4)\Delta^{vi}_0 + \cdots\right\}$$

$$w^2f''_n = \Delta''_0 - \frac{1}{12}(1-6n^2)\Delta^{iv}_0 + \frac{1}{360}(4-30n^2+15n^4)\Delta^{vi}_0 + \cdots + n\left\{\Delta'''_0 - \frac{1}{12}(3-2n^2)\Delta^{v}_0 + \cdots\right\}$$

Derivatives from Tabular Points

$wf'_0 = \frac{1}{12}(f_{-2} - 8f_{-1} + 8f_1 - f_2)$ if five points are available.

$= \frac{1}{60}(-f_{-3} + 9f_{-2} - 45f_{-1} + 45f_1 - 9f_2 + f_3)$ if seven points are available.

$w^2f''_0 = \frac{1}{12}(-f_{-2} + 16f_{-1} - 30f_0 + 16f_1 - f_2)$ (five points).

$= \frac{1}{180}(2f_{-3} - 27f_{-2} + 270f_{-1} - 490f_0 + 270f_1 - 27f_2 + 2f_3)$ (seven points).

TABLE Xc—SECOND DIFFERENCE CORRECTION $B''(\Delta_0'' + \Delta_1'')$

9	
·000	0
·333	1
·666	0
·999	

12	
·000	0
·211	1
·788	0
·999	

15	
·000	0
·158	1
·841	0
·999	

18	
·000	0
·127	1
·872	0
·999	

21	
·000	0
·106	1
·893	0
·999	

24	
·000	0
·091	1
·908	0
·999	

27	
·000	0
·080	1
·333	2
·666	1
·919	0
·999	

30	
·000	0
·071	1
·276	2
·723	1
·928	0
·999	

33	
·000	0
·064	1
·238	2
·761	1
·935	0
·999	

36	
·000	0
·059	1
·211	2
·788	1
·940	0
·999	

39	
·000	0
·054	1
·189	2
·810	1
·945	0
·999	

42	
·000	0
·050	1
·172	2
·390	3
·609	2
·827	1
·949	0
·999	

45	
·000	0
·046	1
·158	2
·333	3
·666	2
·841	1
·953	0
·999	

48	
·000	0
·043	1
·146	2
·295	3
·704	2
·853	1
·956	0
·999	

51	
·000	0
·040	1
·136	2
·267	3
·732	2
·863	1
·959	0
·999	

54	
·000	0
·038	1
·127	2
·245	3
·754	2
·872	1
·961	0
·999	

57	
·000	0
·036	1
·119	2
·226	3
·433	4
·566	3
·773	2
·880	1
·963	0
·999	

60	
·000	0
·034	1
·112	2
·211	3
·370	4
·629	3
·788	2
·887	1
·965	0
·999	

63	
·000	0
·032	1
·106	2
·197	3
·333	4
·666	3
·802	2
·893	1
·967	0
·999	

66	
·000	0
·031	1
·101	2
·186	3
·305	4
·694	3
·813	2
·898	1
·968	0
·999	

69	
·000	0
·029	1
·096	2
·175	3
·282	4
·717	3
·824	2
·903	1
·970	0
·999	

72	
·000	0
·028	1
·091	2
·166	3
·264	4
·735	3
·833	2
·908	1
·971	0
·999	

75	
·000	0
·027	1
·087	2
·158	3
·248	4
·399	5
·600	4
·751	3
·841	2
·912	1
·972	0
·999	

78	
·000	0
·026	1
·083	2
·151	3
·234	4
·361	5
·638	4
·765	3
·848	2
·916	1
·973	0
·999	

81	
·000	0
·025	1
·080	2
·144	3
·222	4
·333	5
·666	4
·777	3
·855	2
·919	1
·974	0
·999	

84	
·000	0
·024	1
·077	2
·138	3
·211	4
·311	5
·688	4
·788	3
·861	2
·922	1
·975	0
·999	

87	
·000	0
·023	1
·074	2
·132	3
·201	4
·292	5
·707	4
·798	3
·867	2
·925	1
·976	0
·999	

90	
·000	0
·022	1
·071	2
·127	3
·192	4
·276	5
·425	6
·574	5
·723	4
·807	3
·872	2
·928	1
·977	0
·999	

93	
·000	0
·021	1
·069	2
·122	3
·184	4
·262	5
·384	6
·615	5
·737	4
·815	3
·877	2
·930	1
·978	0
·999	

96	
·000	0
·021	1
·066	2
·118	3
·177	4
·250	5
·355	6
·644	5
·749	4
·822	3
·881	2
·933	1
·978	0
·999	

99	
·000	0
·020	1
·064	2
·114	3
·170	4
·238	5
·333	6
·666	5
·761	4
·829	3
·885	2
·935	1
·979	0
·999	

102	
·000	0
·020	1
·062	2
·110	3
·164	4
·228	5
·314	6
·685	5
·771	4
·835	3
·889	2
·937	1
·979	0
·999	

105	
·000	0
·019	1
·060	2
·106	3
·158	4
·219	5
·298	6
·451	7
·548	6
·701	5
·780	4
·841	3
·893	2
·939	1
·980	0
·999	

108	
·000	0
·018	1
·059	2
·103	3
·153	4
·211	5
·284	6
·403	7
·596	6
·715	5
·788	4
·846	3
·896	2
·940	1
·981	0
·999	

111	
·000	0
·018	1
·057	2
·100	3
·148	4
·203	5
·272	6
·374	7
·625	6
·727	5
·796	4
·851	3
·899	2
·942	1
·981	0
·999	

114	
·000	0
·017	1
·055	2
·097	3
·143	4
·196	5
·261	6
·351	7
·648	6
·738	5
·803	4
·856	3
·902	2
·944	1
·982	0
·999	

117	
·000	0
·017	1
·054	2
·094	3
·138	4
·189	5
·251	6
·333	7
·666	6
·748	5
·810	4
·861	3
·905	2
·945	1
·982	0
·999	

120	
·000	0
·016	1
·052	2
·091	3
·134	4
·183	5
·241	6
·317	7
·682	6
·758	5
·816	4
·865	3
·908	2
·947	1
·983	0
·999	

123	
·000	0
·016	1
·051	2
·089	3
·130	4
·178	5
·233	6
·303	7
·421	8
·578	7
·696	6
·766	5
·821	4
·869	3
·910	2
·948	1
·983	0
·999	

126	
·000	0
·016	1
·050	2
·086	3
·127	4
·172	5
·225	6
·291	7
·390	8
·609	7
·708	6
·774	5
·827	4
·872	3
·913	2
·949	1
·983	0
·999	

129	
·000	0
·015	1
·048	2
·084	3
·123	4
·167	5
·218	6
·279	7
·367	8
·632	7
·720	6
·781	5
·832	4
·876	3
·915	2
·951	1
·984	0
·999	

132	
·000	0
·015	1
·047	2
·082	3
·120	4
·162	5
·211	6
·269	7
·349	8
·650	

132	
·349	8
·650	7
·730	6
·788	5
·837	4
·879	3
·917	2
·952	1
·984	0
·999	

135	
·000	0
·015	1
·046	2
·080	3
·117	4
·158	5
·204	6
·260	7
·333	8
·666	7
·739	6
·795	5
·841	4
·882	3
·919	2
·953	1
·984	0
·999	

138	
·000	0
·014	1
·045	2
·078	3
·114	4
·154	5
·199	6
·251	7
·319	8
·439	9
·560	8
·680	7
·748	6
·800	5
·845	4
·885	3
·921	2
·954	1
·985	0
·999	

141	
·000	0
·014	1
·044	2
·076	3
·111	4
·150	5
·193	6
·243	7
·307	8
·405	9
·594	8
·692	7
·756	6
·806	5
·849	4
·888	3
·923	2
·955	1
·985	0
·999	

144	
·000	0
·014	1
·043	2
·075	3
·109	4
·146	5
·188	6
·236	7
·295	8
·382	9
·617	8
·704	7
·763	6
·811	5
·853	4
·890	3
·924	2
·956	1
·985	0
·999	

147

·000	
	0
·013	
	1
·042	
	2
·073	
	3
·106	
	4
·142	
	5
·183	
	6
·229	
	7
·285	
	8
·363	
	9
·636	
	8
·714	
	7
·770	
	6
·816	
	5
·857	
	4
·893	
	3
·926	
	2
·957	
	1
·986	
	0
·999	

150

·000	
	0
·013	
	1
·041	
	2
·071	
	3
·104	
	4
·139	
	5
·178	
	6
·223	
	7
·276	
	8
·347	
	9
·652	
	8
·723	
	7
·776	
	6
·821	
	5
·860	
	4
·895	
	3
·928	
	2
·958	
	1
·986	
	0
·999	

153

·000	
	0
·013	
	1
·040	
	2
·070	
	3
·101	
	4
·136	
	5
·174	
	6
·217	
	7
·267	
	8
·333	
	9
·459	
	10
·540	
	9
·666	
	8
·732	
	7
·782	
	6
·825	
	5
·863	
	4
·898	
	3
·929	
	2
·959	
	1
·986	
	0
·999	

156

·000	
	0
·012	
	1
·040	
	2
·068	
	3
·099	
	4
·133	
	5
·169	
	6
·211	
	7
·259	
	8
·320	
	9
·419	
	10
·580	
	9
·679	
	8
·740	
	7
·788	
	6
·830	
	5
·866	
	4
·900	
	3
·931	
	2
·959	
	1
·987	
	0
·999	

159

·000	
	0
·012	
	1
·039	
	2
·067	
	3
·097	
	4
·130	
	5
·165	
	6
·205	
	7
·252	
	8
·309	
	9
·395	
	10
·604	
	9
·690	
	8
·747	
	7
·794	
	6
·834	
	5
·869	
	4
·902	
	3
·932	
	2
·960	
	1
·987	
	0
·999	

162

·000	
	0
·012	
	1
·038	
	2
·066	
	3
·095	
	4
·127	
	5
·162	
	6
·200	
	7
·245	
	8
·299	
	9
·375	
	10
·624	
	9
·700	
	8
·754	
	7
·799	
	6
·837	
	5
·872	
	4
·904	
	3
·933	
	2
·961	
	1
·987	
	0
·999	

165

·000	
	0
·012	
	1
·037	
	2
·064	
	3
·093	
	4
·124	
	5
·158	
	6
·195	
	7
·238	
	8
·290	
	9
·359	
	10
·640	
	9
·709	
	8
·761	
	7
·804	
	6
·841	
	5
·875	
	4
·906	
	3
·935	
	2
·962	
	1
·987	
	0
·999	

168

·000	
	0
·012	
	1
·037	
	2
·063	
	3
·091	
	4
·122	
	5
·154	
	6
·191	
	7
·232	
	8
·281	
	9
·345	
	10
·654	
	9
·718	
	8
·767	
	7
·808	
	6
·845	
	5
·877	
	4
·908	
	3
·936	
	2
·962	
	1
·987	
	0
·999	

171

·000	
	0
·011	
	1
·036	
	2
·062	
	3
·089	
	4
·119	
	5
·151	
	6
·187	
	7
·226	
	8
·273	
	9
·333	
	10
·433	
	11
·566	
	10
·666	
	9
·726	
	8
·773	
	7
·812	
	6
·848	
	5
·880	
	4
·910	
	3
·937	
	2
·963	
	1
·988	
	0
·999	

174

·000	
	0
·011	
	1
·035	
	2
·061	
	3
·088	
	4
·117	
	5
·148	
	6
·182	
	7
·221	
	8
·266	
	9
·322	
	10
·407	
	11
·592	
	10
·677	
	9
·733	
	8
·778	
	7
·817	
	6
·851	
	5
·882	
	4
·911	
	3
·938	
	2
·964	
	1
·988	
	0
·999	

177

·000	
	0
·011	
	1
·035	
	2
·060	
	3
·086	
	4
·114	
	5
·145	
	6
·178	
	7
·216	
	8
·259	
	9
·312	
	10
·387	
	11
·612	
	10
·687	
	9
·740	
	8
·783	
	7
·821	
	6
·854	
	5
·885	
	4
·913	
	3
·939	
	2
·964	
	1
·988	
	0
·999	

180

·000	
	0
·011	
	1
·034	
	2
·059	
	3
·085	
	4
·112	
	5
·142	
	6
·175	
	7
·211	
	8
·252	
	9
·302	
	10
·370	
	11
·629	
	10
·697	
	9
·747	
	8
·788	
	7
·824	
	6
·857	
	5
·887	
	4
·914	
	3
·940	
	2
·965	
	1
·988	
	0
·999	

183

·000	
	0
·011	
	1
·033	
	2
·058	
	3
·083	
	4
·110	
	5
·139	
	6
·171	
	7
·206	
	8
·246	
	9
·294	
	10
·356	
	11
·643	
	10
·705	
	9
·753	
	8
·793	
	7
·828	
	6
·860	
	5
·889	
	4
·916	
	3
·941	
	2
·966	
	1
·988	
	0
·999	

186

·000	
	0
·010	
	1
·033	
	2
·057	
	3
·081	
	4
·108	
	5
·137	
	6
·168	
	7
·202	
	8
·240	
	9
·286	
	10
·344	
	11
·448	
	12
·551	
	11
·655	
	10
·713	
	9
·759	
	8
·797	
	7
·831	
	6
·862	
	5
·891	
	4
·918	
	3
·942	
	2
·966	
	1
·989	
	0
·999	

189

·000	
	0
·010	
	1
·032	
	2
·056	
	3
·080	
	4
·106	
	5
·134	
	6
·164	
	7
·197	
	8
·235	
	9
·278	
	10
·333	
	11
·418	
	12
·581	
	11
·666	
	10
·721	
	9
·764	
	8
·802	
	7
·835	
	6
·865	
	5
·893	
	4
·919	
	3
·943	
	2
·967	
	1
·989	
	0
·999	

192

·000	
	0
·010	
	1
·032	
	2
·055	
	3
·079	
	4
·104	
	5
·132	
	6
·161	
	7
·193	
	8
·229	
	9
·271	
	10
·323	
	11
·397	
	12
·602	
	11
·676	
	10
·728	
	9
·770	
	8
·806	
	7
·838	
	6
·867	
	5
·895	
	4
·920	
	3
·944	
	2
·967	
	1
·989	
	0
·999	

195

·000	
	0
·010	
	1
·031	
	2
·054	
	3
·077	
	4
·102	
	5
·129	
	6
·158	
	7
·189	
	8
·224	
	9
·265	
	10
·313	
	11
·381	
	12
·618	
	11
·686	
	10
·734	
	9
·775	
	8
·810	
	7
·841	
	6
·870	
	5
·897	
	4
·922	
	3
·945	
	2
·968	
	1
·989	
	0
·999	

198

·000	
	0
·010	
	1
·031	
	2
·053	
	3
·076	
	4
·101	
	5
·127	
	6
·155	
	7
·186	
	8
·220	
	9
·259	
	10
·305	
	11
·367	
	12
·632	
	11
·694	
	10
·740	
	9
·779	
	8
·813	
	7
·844	
	6
·872	
	5
·898	
	4
·923	
	3
·946	
	2
·968	
	1
·989	
	0
·999	

201

·000	
	0
·010	
	1
·030	
	2
·052	
	3
·075	
	4
·099	
	5
·125	
	6
·152	
	7
·182	
	8
·215	
	9
·253	
	10
·297	
	11
·354	
	12
·464	
	13
·535	
	12
·645	
	11
·702	
	10
·746	
	9
·784	
	8
·817	
	7
·847	
	6
·874	
	5
·900	
	4
·924	
	3
·947	
	2
·969	
	1
·989	
	0
·999	

If a function cannot be integrated analytically, or is known by certain numerical values only, the value of the definite integral

$$I=\int_a^b y\,dx$$

can be found by numerical integration, otherwise known as quadrature. This subject is treated and illustrated more fully in *Chambers's Six-Figure Mathematical Tables*, Volume II, which also gives the coefficients required when solving differential equations. The method now presented is a summary of those particularly applicable when a single definite integral is to be evaluated.

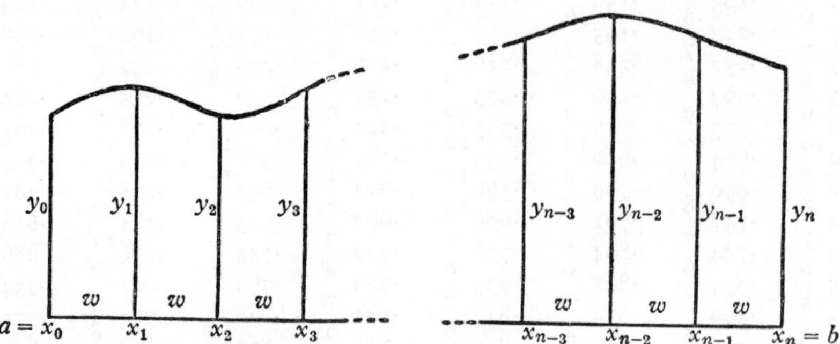

The range of integration, i.e. the interval between a and b, is divided into a convenient number n of equal parts, and for each point of this subdivision the value of the integrand y must be known. The desired integral is then the sum of the areas of n strips, each bounded on the sides by two adjacent ordinates, at the base by the constant interval w in x, and at the top by the portion of the curve cut off by the two ordinates.

As a first approximation the curve can be replaced by straight lines joining the tops of adjacent ordinates. Each strip then becomes a trapezium whose area is $\frac{1}{2}w(y_r+y_{r+1})$ and, since $nw=b-a$, the approximate integral becomes

$$I=\frac{b-a}{n}(\tfrac{1}{2}y_0+y_1+y_2+y_3+\cdots+y_{n-2}+y_{n-1}+\tfrac{1}{2}y_n)$$

This is known as the trapezoidal rule. It is easily seen that it is, at best, only a rough approximation and that each strip is treated independently.

Better results are obtained by fitting curves of the type $y=c_0+c_1x+c_2x^2+c_3x^3+\cdots$ to the tops of sufficient adjacent equidistant ordinates and then integrating this curve analytically. The fitting and integration do not actually have to be done, because the entire effect of this process can be expressed in a series of simple formulæ.

If a parabola is fitted to the three ordinates defining two strips, i.e. if $n=2$, the area or integral becomes

$$I=\frac{b-a}{6}(y_0+4y_1+y_2) \qquad \text{(Simpson's rule)}$$

This well-known formula is a favourite among engineers and physicists, because of its simplicity and effectiveness if the strips are not too wide.

The use of formulæ involving three or five strips is not generally recommended, so we quote the three-eighths rule only for purposes of record. Formulæ involving a greater number of ordinates are as follows:

$$I=\frac{b-a}{8}(y_0+3y_1+3y_2+y_3) \qquad \text{(Three-eighths rule)}$$

$$I=\frac{b-a}{90}(7y_0+32y_1+12y_2+32y_3+7y_4) \qquad \text{(Four-strip)}$$

$$I=\frac{b-a}{840}(41y_0+216y_1+27y_2+272y_3+27y_4+216y_5+41y_6) \qquad \text{(Six-strip)}$$

$$I=\frac{b-a}{20}(y_0+5y_1+y_2+6y_3+y_4+5y_5+y_6) \qquad \text{(Weddle)}$$

Weddle's formula differs only from the six-strip formula in that it requires a further term (in the bracket) of $-\frac{1}{42}\Delta^{vi}$, which is often negligible. The great virtue of this excellent formula is the extraordinary simplicity of its coefficients. A good computer usually checks the smoothness of his ordinates by differencing them; if this is done (as is strongly recommended), the simple Weddle result can easily be perfected (within the limits imposed by having only six strips) by adding the equally simple correction $-\frac{w}{140}\Delta^{vi}$ (since $b-a=6w$).

It will be observed that the complexity of the formulæ increases with increasing number of strips. For this reason, if only two or three orders of differences are significant, it is customary, in a long integration, to use repetitions of one of the simpler formulæ. If n is even, and if Δ''' is approximately constant, we may use

$$I=\frac{b-a}{3n}(y_0+4y_1+2y_2+4y_3+2y_4+\cdots+2y_{n-2}+4y_{n-1}+y_n) \qquad \text{(Repeated Simpson)}$$

When the repeated Simpson formula fails because of the neglected higher-order differences, the recommended procedure is to use the four-strip formula or even to jump to the very simple and highly accurate Weddle formula (coupled, if necessary, with its correcting term). For a long range of integration, if high-order differences are to be expected, it is advisable to make n a multiple of 6. Alternatively, we may use the general formula of Gregory or preferably that of Gauss, as described below.

Interesting variants, in which the labour of differencing the ordinates offers two rewards—a check on the smoothness of the ordinates and a simpler formula for the integral, are

$$I=(b-a)\left(y_1+\frac{1}{6}\Delta_1''\right) \qquad \text{(Simpson)}$$

$$I=(b-a)\left(y_2+\frac{2}{3}\Delta_2''+\frac{7}{90}\Delta_2^{iv}\right) \qquad \text{(Four-strip)}$$

The physical meaning of these two variants is simply that the integral of an area is the product of its base by its average height. The latter is expressed as the middle ordinate, together with the necessary correction, which is derived from the even differences at the level of the middle ordinate. In this form the four-strip formula becomes attractive for those numerous cases in which fifth differences are approximately constant.

The general formula for covering any range of integration is that of Gregory, namely

$$I=\frac{b-a}{n}\Big\{\frac{1}{2}y_0+y_1+y_2+y_3+\cdots+y_{n-2}+y_{n-1}+\frac{1}{2}y_n-\frac{1}{12}(\Delta'_{n-\frac{1}{2}}-\Delta'_{\frac{1}{2}})-\frac{1}{24}(\Delta''_{n-1}+\Delta''_1)$$
$$-\frac{19}{720}(\Delta'''_{n-1\frac{1}{2}}-\Delta'''_{1\frac{1}{2}})-\frac{3}{160}(\Delta^{iv}_{n-2}+\Delta^{iv}_2)-\frac{863}{60480}(\Delta^{v}_{n-2\frac{1}{2}}-\Delta^{v}_{2\frac{1}{2}})-\frac{275}{24192}(\Delta^{vi}_{n-3}+\Delta^{vi}_3)-\cdots\Big\}$$

The formulæ already given (except that of Weddle) may be derived from that of Gregory by expressing the differences in terms of the y's.

If sufficient values of the integrand beyond the range of integration are known, Gregory's formula may with advantage be replaced by the more rapidly convergent formula of Gauss, which uses central differences rather than the forward and backward differences that must otherwise be used.

$$I=\frac{b-a}{n}\Big\{\frac{1}{2}y_0+y_1+y_2+y_3+\cdots+y_{n-2}+y_{n-1}+\frac{1}{2}y_n-\frac{1}{12}(\Delta'_n-\Delta'_0)$$
$$+\frac{11}{720}(\Delta'''_n-\Delta'''_0)-\frac{191}{60480}(\Delta^{v}_n-\Delta^{v}_0)+\frac{2497}{3628800}(\Delta^{vii}_n-\Delta^{vii}_0)-\cdots\Big\}$$

where $\Delta'_i=\frac{1}{2}(\Delta'_{i-\frac{1}{2}}+\Delta'_{i+\frac{1}{2}})$ and similarly for higher odd differences.

TABLE XI—CONVERSION OF DEGREES, MINUTES AND SECONDS TO RADIANS

Deg.	Radians	x	x°	x'	x''	x	x°	x'	x''
°	r			$0^{r}\cdot 0$	$0^{r}\cdot 000$		r	$0^{r}\cdot 0$	$0^{r}\cdot 000$
100	1·745 3293	**00**	0·000 0000	00 0000	0000	**50**	0·872 6646	14 5444	2424
200	3·490 6585	**01**	·017 4533	00 2909	0048	**51**	·890 1179	14 8353	2473
300	5·235 9878	**02**	·034 9066	00 5818	0097	**52**	·907 5712	15 1262	2521
400	6·981 3170	**03**	·052 3599	00 8727	0145	**53**	·925 0245	15 4171	2570
500	8·726 6463	**04**	·069 8132	01 1636	0194	**54**	·942 4778	15 7080	2618
600	10·471 9755	**05**	0·087 2665	01 4544	0242	**55**	0·959 9311	15 9989	2666
700	12·217 3048	**06**	·104 7198	01 7453	0291	**56**	0·977 3844	16 2897	2715
800	13·962 6340	**07**	·122 1730	02 0362	0339	**57**	0·994 8377	16 5806	2763
900	15·707 9633	**08**	·139 6263	02 3271	0388	**58**	1·012 2910	16 8715	2812
1000	17·453 2925	**09**	·157 0796	02 6180	0436	**59**	1·029 7443	17 1624	2860
1100	19·198 6218	**10**	0·174 5329	02 9089	0485	**60**	1·047 1976	17 4533	2909
1200	20·943 9510	**11**	·191 9862	03 1998	0533	**61**	·064 6508	17 7442	2957
1300	22·689 2803	**12**	·209 4395	03 4907	0582	**62**	·082 1041	18 0351	3006
1400	24·434 6095	**13**	·226 8928	03 7815	0630	**63**	·099 5574	18 3260	3054
1500	26·179 9388	**14**	·244 3461	04 0724	0679	**64**	·117 0107	18 6168	3103
1600	27·925 2680	**15**	0·261 7994	04 3633	0727	**65**	1·134 4640	18 9077	3151
1700	29·670 5973	**16**	·279 2527	04 6542	0776	**66**	·151 9173	19 1986	3200
1800	31·415 9265	**17**	·296 7060	04 9451	0824	**67**	·169 3706	19 4895	3248
1900	33·161 2558	**18**	·314 1593	05 2360	0873	**68**	·186 8239	19 7804	3297
2000	34·906 5850	**19**	·331 6126	05 5269	0921	**69**	·204 2772	20 0713	3345
2100	36·651 9143	**20**	0·349 0659	05 8178	0970	**70**	1·221 7305	20 3622	3394
2200	38·397 2435	**21**	·366 5191	06 1087	1018	**71**	·239 1838	20 6531	3442
2300	40·142 5728	**22**	·383 9724	06 3995	1067	**72**	·256 6371	20 9440	3491
2400	41·887 9020	**23**	·401 4257	06 6904	1115	**73**	·274 0904	21 2348	3539
2500	43·633 2313	**24**	·418 8790	06 9813	1164	**74**	·291 5436	21 5257	3588
2600	45·378 5606	**25**	0·436 3323	07 2722	1212	**75**	1·308 9969	21 8166	3636
2700	47·123 8898	**26**	·453 7856	07 5631	1261	**76**	·326 4502	22 1075	3685
2800	48·869 2191	**27**	·471 2389	07 8540	1309	**77**	·343 9035	22 3984	3733
2900	50·614 5483	**28**	·488 6922	08 1449	1357	**78**	·361 3568	22 6893	3782
3000	52·359 8776	**29**	·506 1455	08 4358	1406	**79**	·378 8101	22 9802	3830
		30	0·523 5988	08 7266	1454	**80**	1·396 2634	23 2711	3879
		31	·541 0521	09 0175	1503	**81**	·413 7167	23 5619	3927
3008	52·499 5039	**32**	·558 5054	09 3084	1551	**82**	·431 1700	23 8528	3975
4639	80·965 8240	**33**	·575 9587	09 5993	1600	**83**	·448 6233	24 1437	4024
7647	133·465 3279	**34**	·593 4119	09 8902	1648	**84**	·466 0766	24 4346	4072
10655	185·964 8318	**35**	0·610 8652	10 1811	1697	**85**	1·483 5299	24 7255	4121
13663	238·464 3357	**36**	·628 3185	10 4720	1745	**86**	·500 9832	25 0164	4169
15294	266·930 6558	**37**	·645 7718	10 7629	1794	**87**	·518 4364	25 3073	4218
18302	319·430 1597	**38**	·663 2251	11 0538	1842	**88**	·535 8897	25 5982	4266
21310	371·929 6636	**39**	·680 6784	11 3446	1891	**89**	·553 3430	25 8891	4315
24318	424·429 1675	**40**	0·698 1317	11 6355	1939	**90**	1·570 7963	26 1799	4363
25949	452·895 4876	**41**	·715 5850	11 9264	1988	**91**	·588 2496	26 4708	4412
28957	505·394 9915	**42**	·733 0383	12 2173	2036	**92**	·605 7029	26 7617	4460
31965	557·894 4954	**43**	·750 4916	12 5082	2085	**93**	·623 1562	27 0526	4509
33596	586·360 8155	**44**	·767 9449	12 7991	2133	**94**	·640 6095	27 3435	4557
36604	638·860 3194	**45**	0·785 3982	13 0900	2182	**95**	1·658 0628	27 6344	4606
39612	691·359 8233	**46**	·802 8515	13 3809	2230	**96**	·675 5161	27 9253	4654
42620	743·859 3272	**47**	·820 3047	13 6717	2279	**97**	·692 9694	28 2162	4703
44251	772·325 6473	**48**	·837 7580	13 9626	2327	**98**	·710 4227	28 5070	4751
47259	824·825 1512	**49**	0·855 2113	14 2535	2376	**99**	1·727 8760	28 7979	4800

$1^{\circ} = 0^{r}\cdot 017\ 453\ 292\ 519\ 943$ $\quad 1' = 0^{r}\cdot 000\ 290\ 888\ 208\ 666$ $\quad 1'' = 0^{r}\cdot 000\ 004\ 848\ 136\ 811\ 095$

For the first pair of decimals of x, drop the second half of the last group of four decimals in the radian equivalent; for the second pair, drop the whole of this group.

TABLE XI—CONVERSION OF RADIANS TO DEGREES, MINUTES AND SECONDS

Integral Radians		Decimals of a Radian				Decimals of a Radian			
			First pair	Second pair	Third pair		First pair	Second pair	Third pair
r	° ′ ″		° ′ ″	′ ″	″		° ′ ″	′ ″	″
0	0·0	00	0·0	0·0	0·0	50	28 38 52·4	17 11·3	10·3
1	57 17 44·8	01	34 22·6	20·6	0·2	51	29 13 15·1	17 32·0	10·5
2	114 35 29·6	02	1 08 45·3	41·3	0·4	52	29 47 37·7	17 52·6	10·7
3	171 53 14·4	03	1 43 07·9	1 01·9	0·6	53	30 22 00·3	18 13·2	10·9
4	229 10 59·2	04	2 17 30·6	1 22·5	0·8	54	30 56 23·0	18 33·8	11·1
5	286 28 44·0	05	2 51 53·2	1 43·1	1·0	55	31 30 45·6	18 54·5	11·3
6	343 46 28·8	06	3 26 15·9	2 03·8	1·2	56	32 05 08·3	19 15·1	11·6
7	401 04 13·6	07	4 00 38·5	2 24·4	1·4	57	32 39 30·9	19 35·7	11·8
8	458 21 58·4	08	4 35 01·2	2 45·0	1·7	58	33 13 53·6	19 56·3	12·0
9	515 39 43·3	09	5 09 23·8	3 05·6	1·9	59	33 48 16·2	20 17·0	12·2
10	572 57 28·1	10	5 43 46·5	3 26·3	2·1	60	34 22 38·9	20 37·6	12·4
11	630 15 12·9	11	6 18 09·1	3 46·9	2·3	61	34 57 01·5	20 58·2	12·6
12	687 32 57·7	12	6 52 31·8	4 07·5	2·5	62	35 31 24·2	21 18·8	12·8
13	744 50 42·5	13	7 26 54·4	4 28·1	2·7	63	36 05 46·8	21 39·5	13·0
14	802 08 27·3	14	8 01 17·1	4 48·8	2·9	64	36 40 09·5	22 00·1	13·2
15	859 26 12·1	15	8 35 39·7	5 09·4	3·1	65	37 14 32·1	22 20·7	13·4
16	916 43 56·9	16	9 10 02·4	5 30·0	3·3	66	37 48 54·8	22 41·3	13·6
17	974 01 41·7	17	9 44 25·0	5 50·7	3·5	67	38 23 17·4	23 02·0	13·8
18	1031 19 26·5	18	10 18 47·7	6 11·3	3·7	68	38 57 40·1	23 22·6	14·0
19	1088 37 11·3	19	10 53 10·3	6 31·9	3·9	69	39 32 02·7	23 43·2	14·2
20	1145 54 56·1	20	11 27 33·0	6 52·5	4·1	70	40 06 25·4	24 03·9	14·4
21	1203 12 40·9	21	12 01 55·6	7 13·2	4·3	71	40 40 48·0	24 24·5	14·6
22	1260 30 25·7	22	12 36 18·3	7 33·8	4·5	72	41 15 10·7	24 45·1	14·9
23	1317 48 10·5	23	13 10 40·9	7 54·4	4·7	73	41 49 33·3	25 05·7	15·1
24	1375 05 55·3	24	13 45 03·6	8 15·0	5·0	74	42 23 56·0	25 26·4	15·3
25	1432 23 40·2	25	14 19 26·2	8 35·7	5·2	75	42 58 18·6	25 47·0	15·5
26	1489 41 25·0	26	14 53 48·8	8 56·3	5·4	76	43 32 41·3	26 07·6	15·7
27	1546 59 09·8	27	15 28 11·5	9 16·9	5·6	77	44 07 03·9	26 28·2	15·9
28	1604 16 54·6	28	16 02 34·1	9 37·5	5·8	78	44 41 26·5	26 48·9	16·1
29	1661 34 39·4	29	16 36 56·8	9 58·2	6·0	79	45 15 49·2	27 09·5	16·3
30	1718 52 24·2	30	17 11 19·4	10 18·8	6·2	80	45 50 11·8	27 30·1	16·5
31	1776 10 09·0	31	17 45 42·1	10 39·4	6·4	81	46 24 34·5	27 50·7	16·7
32	1833 27 53·8	32	18 20 04·7	11 00·0	6·6	82	46 58 57·1	28 11·4	16·9
33	1890 45 38·6	33	18 54 27·4	11 20·7	6·8	83	47 33 19·8	28 32·0	17·1
34	1948 03 23·4	34	19 28 50·0	11 41·3	7·0	84	48 07 42·4	28 52·6	17·3
35	2005 21 08·2	35	20 03 12·7	12 01·9	7·2	85	48 42 05·1	29 13·3	17·5
36	2062 38 53·0	36	20 37 35·3	12 22·6	7·4	86	49 16 27·7	29 33·9	17·7
37	2119 56 37·8	37	21 11 58·0	12 43·2	7·6	87	49 50 50·4	29 54·5	17·9
38	2177 14 22·6	38	21 46 20·6	13 03·8	7·8	88	50 25 13·0	30 15·1	18·2
39	2234 32 07·4	39	22 20 43·3	13 24·4	8·0	89	50 59 35·7	30 35·8	18·4
40	2291 49 52·2	40	22 55 05·9	13 45·1	8·3	90	51 33 58·3	30 56·4	18·6
41	2349 07 37·1	41	23 29 28·6	14 05·7	8·5	91	52 08 21·0	31 17·0	18·8
42	2406 25 21·9	42	24 03 51·2	14 26·3	8·7	92	52 42 43·6	31 37·6	19·0
43	2463 43 06·7	43	24 38 13·9	14 46·9	8·9	93	53 17 06·3	31 58·3	19·2
44	2521 00 51·5	44	25 12 36·5	15 07·6	9·1	94	53 51 28·9	32 18·9	19·4
45	2578 18 36·3	45	25 46 59·2	15 28·2	9·3	95	54 25 51·6	32 39·5	19·6
46	2635 36 21·1	46	26 21 21·8	15 48·8	9·5	96	55 00 14·2	33 00·1	19·8
47	2692 54 05·9	47	26 55 44·5	16 09·4	9·7	97	55 34 36·9	33 20·8	20·0
48	2750 11 50·7	48	27 30 07·1	16 30·1	9·9	98	56 08 59·5	33 41·4	20·2
49	2807 29 35·5	49	28 04 29·8	16 50·7	10·1	99	56 43 22·2	34 02·0	20·4

$50^{r} = 2864^{\circ}\ 47'\ 20''{\cdot}3$ $100^{r} = 5729^{\circ}\ 34'\ 40''{\cdot}6$ $200^{r} = 11459^{\circ}\ 09'\ 21''{\cdot}2$ $1000^{r} = 57295^{\circ}\ 46'\ 46''{\cdot}2$

TABLE XI—CONVERSION OF TIME TO ARC

ALSO CONVERSION OF MEAN TIME TO SIDEREAL TIME

Time h	Arc °	Acc. m	Acc. s
1	15		9·9
2	30		19·7
3	45		29·6
4	60		39·4
5	75		49·3
6	90		59·1
7	105	1	09·0
8	120	1	18·9
9	135	1	28·7
10	150	1	38·6
11	165	1	48·4
12	180	1	58·3
13	195	2	08·1
14	210	2	18·0
15	225	2	27·8
16	240	2	37·7
17	255	2	47·6
18	270	2	57·4
19	285	3	07·3
20	300	3	17·1
21	315	3	27·0
22	330	3	36·8
23	345	3	46·7
24	360	3	56·6

Time m	Arc °	Arc ′
1		15
2		30
3		45
4	1	00
5	1	15
6	1	30
7	1	45
8	2	00
9	2	15
10	2	30
11	2	45
12	3	00
13	3	15
14	3	30
15	3	45
16	4	00
17	4	15
18	4	30
19	4	45
20	5	00
21	5	15
22	5	30
23	5	45
24	6	00
25	6	15
26	6	30
27	6	45
28	7	00
29	7	15
30	7	30
31	7	45
32	8	00
33	8	15
34	8	30
35	8	45
36	9	00
37	9	15
38	9	30
39	9	45
40	10	00
41	10	15
42	10	30
43	10	45
44	11	00
45	11	15
46	11	30
47	11	45
48	12	00
49	12	15
50	12	30
51	12	45
52	13	00
53	13	15
54	13	30
55	13	45
56	14	00
57	14	15
58	14	30
59	14	45

Time s	Arc ′	Arc ″
1		15
2		30
3		45
4	1	00
5	1	15
6	1	30
7	1	45
8	2	00
9	2	15
10	2	30
11	2	45
12	3	00
13	3	15
14	3	30
15	3	45
16	4	00
17	4	15
18	4	30
19	4	45
20	5	00
21	5	15
22	5	30
23	5	45
24	6	00
25	6	15
26	6	30
27	6	45
28	7	00
29	7	15
30	7	30
31	7	45
32	8	00
33	8	15
34	8	30
35	8	45
36	9	00
37	9	15
38	9	30
39	9	45
40	10	00
41	10	15
42	10	30
43	10	45
44	11	00
45	11	15
46	11	30
47	11	45
48	12	00
49	12	15
50	12	30
51	12	45
52	13	00
53	13	15
54	13	30
55	13	45
56	14	00
57	14	15
58	14	30
59	14	45

Time s	Arc ″	Time s	Arc ″
0·00	0·0	0·50	7·5
·01	0·2	·51	7·6
·02	0·3	·52	7·8
·03	0·4	·53	8·0
·04	0·6	·54	8·1
0·05	0·8	0·55	8·2
·06	0·9	·56	8·4
·07	1·0	·57	8·6
·08	1·2	·58	8·7
·09	1·4	·59	8·8
0·10	1·5	0·60	9·0
·11	1·6	·61	9·2
·12	1·8	·62	9·3
·13	2·0	·63	9·4
·14	2·1	·64	9·6
0·15	2·2	0·65	9·8
·16	2·4	·66	9·9
·17	2·6	·67	10·0
·18	2·7	·68	10·2
·19	2·8	·69	10·4
0·20	3·0	0·70	10·5
·21	3·2	·71	10·6
·22	3·3	·72	10·8
·23	3·4	·73	11·0
·24	3·6	·74	11·1
0·25	3·8	0·75	11·2
·26	3·9	·76	11·4
·27	4·0	·77	11·6
·28	4·2	·78	11·7
·29	4·4	·79	11·8
0·30	4·5	0·80	12·0
·31	4·6	·81	12·2
·32	4·8	·82	12·3
·33	5·0	·83	12·4
·34	5·1	·84	12·6
0·35	5·2	0·85	12·8
·36	5·4	·86	12·9
·37	5·6	·87	13·0
·38	5·7	·88	13·2
·39	5·8	·89	13·4
0·40	6·0	0·90	13·5
·41	6·2	·91	13·6
·42	6·3	·92	13·8
·43	6·4	·93	14·0
·44	6·6	·94	14·1
0·45	6·8	0·95	14·2
·46	6·9	·96	14·4
·47	7·0	·97	14·6
·48	7·2	·98	14·7
·49	7·4	·99	14·8
0·50	7·5	1·00	15·0

Mean Time m s	Acc. s	Mean Time m s	Acc. s
0 00·0		30 07·9	
	0·0		5·0
0 18·2		30 44·4	
	0·1		5·1
0 54·7		31 20·9	
	0·2		5·2
1 31·3		31 57·5	
	0·3		5·3
2 07·8		32 34·0	
	0·4		5·4
2 44·3		33 10·5	
	0·5		5·5
3 20·8		33 47·0	
	0·6		5·6
3 57·4		34 23·6	
	0·7		5·7
4 33·9		35 00·1	
	0·8		5·8
5 10·4		35 36·6	
	0·9		5·9
5 46·9		36 13·1	
	1·0		6·0
6 23·5		36 49·7	
	1·1		6·1
7 00·0		37 26·2	
	1·2		6·2
7 36·5		38 02·7	
	1·3		6·3
8 13·0		38 39·2	
	1·4		6·4
8 49·6		39 15·8	
	1·5		6·5
9 26·1		39 52·3	
	1·6		6·6
10 02·6		40 28·8	
	1·7		6·7
10 39·1		41 05·3	
	1·8		6·8
11 15·6		41 41·9	
	1·9		6·9
11 52·2		42 18·4	
	2·0		7·0
12 28·7		42 54·9	
	2·1		7·1
13 05·2		43 31·4	
	2·2		7·2
13 41·7		44 08·0	
	2·3		7·3
14 18·3		44 44·5	
	2·4		7·4
14 54·8		45 21·0	
	2·5		7·5
15 31·3		45 57·5	
	2·6		7·6
16 07·8		46 34·1	
	2·7		7·7
16 44·4		47 10·6	
	2·8		7·8
17 20·9		47 47·1	
	2·9		7·9
17 57·4		48 23·6	
	3·0		8·0
18 33·9		49 00·1	
	3·1		8·1
19 10·5		49 36·7	
	3·2		8·2
19 47·0		50 13·2	
	3·3		8·3
20 23·5		50 49·7	
	3·4		8·4
21 00·0		51 26·2	
	3·5		8·5
21 36·6		52 02·8	
	3·6		8·6
22 13·1		52 39·3	
	3·7		8·7
22 49·6		53 15·8	
	3·8		8·8
23 26·1		53 52·3	
	3·9		8·9
24 02·7		54 28·9	
	4·0		9·0
24 39·2		55 05·4	
	4·1		9·1
25 15·7		55 41·9	
	4·2		9·2
25 52·2		56 18·4	
	4·3		9·3
26 28·8		56 55·0	
	4·4		9·4
27 05·3		57 31·5	
	4·5		9·5
27 41·8		58 08·0	
	4·6		9·6
28 18·3		58 44·5	
	4·7		9·7
28 54·9		59 21·1	
	4·8		9·8
29 31·4		59 57·6	
	4·9		9·9
30 07·9		59 59·9	

The last column above gives the acceleration of sidereal time on mean time for hours.

The critical table on the right of this page caters for minutes and seconds.

An example of the conversion of mean time to sidereal time is given on the next page.

Example: Convert $12^h\ 34^m\ 56^s{\cdot}78$ to arc.

	°	′	″
12^h	180		
34^m	8	30	
56^s		14	00
$0^s{\cdot}78$			11·7
Sum	188	44	11·7

TABLE XI—CONVERSION OF ARC TO TIME

Arc °	Time h m	Arc °	Time h m	Arc °	Time h m	Arc °	Time h m
0	0	50	3 20	100	6 40	140	9 20
1	4	51	3 24	101	6 44	141	9 24
2	8	52	3 28	102	6 48	142	9 28
3	12	53	3 32	103	6 52	143	9 32
4	16	54	3 36	104	6 56	144	9 36
5	20	55	3 40	105	7 00	145	9 40
6	24	56	3 44	106	7 04	146	9 44
7	28	57	3 48	107	7 08	147	9 48
8	32	58	3 52	108	7 12	148	9 52
9	36	59	3 56	109	7 16	149	9 56
10	40	60	4 00	110	7 20	150	10 00
11	44	61	4 04	111	7 24	151	10 04
12	48	62	4 08	112	7 28	152	10 08
13	52	63	4 12	113	7 32	153	10 12
14	56	64	4 16	114	7 36	154	10 16
15	1 00	65	4 20	115	7 40	155	10 20
16	1 04	66	4 24	116	7 44	156	10 24
17	1 08	67	4 28	117	7 48	157	10 28
18	1 12	68	4 32	118	7 52	158	10 32
19	1 16	69	4 36	119	7 56	159	10 36
20	1 20	70	4 40	120	8 00	160	10 40
21	1 24	71	4 44	121	8 04	161	10 44
22	1 28	72	4 48	122	8 08	162	10 48
23	1 32	73	4 52	123	8 12	163	10 52
24	1 36	74	4 56	124	8 16	164	10 56
25	1 40	75	5 00	125	8 20	165	11 00
26	1 44	76	5 04	126	8 24	166	11 04
27	1 48	77	5 08	127	8 28	167	11 08
28	1 52	78	5 12	128	8 32	168	11 12
29	1 56	79	5 16	129	8 36	169	11 16
30	2 00	80	5 20	130	8 40	170	11 20
31	2 04	81	5 24	131	8 44	171	11 24
32	2 08	82	5 28	132	8 48	172	11 28
33	2 12	83	5 32	133	8 52	173	11 32
34	2 16	84	5 36	134	8 56	174	11 36
35	2 20	85	5 40	135	9 00	175	11 40
36	2 24	86	5 44	136	9 04	176	11 44
37	2 28	87	5 48	137	9 08	177	11 48
38	2 32	88	5 52	138	9 12	178	11 52
39	2 36	89	5 56	139	9 16	179	11 56
40	2 40	90	6 00	140	9 20	180	12 00
41	2 44	91	6 04				
42	2 48	92	6 08				
43	2 52	93	6 12				
44	2 56	94	6 16				
45	3 00	95	6 20				
46	3 04	96	6 24				
47	3 08	97	6 28				
48	3 12	98	6 32				
49	3 16	99	6 36				
50	3 20	100	6 40				

Example: What is the interval of sidereal time corresponding to a mean time interval of $8^h\ 47^m\ 28^s{\cdot}3$?

		h m s
Mean time	.	8 47 28·3
Acceleration	.	1 18·9
		7·8
Sidereal time	.	8 48 55·0

Arc ′	Time m s	Arc ″	Time s
1	4	1	0·07
2	8	2	0·13
3	12	3	0·20
4	16	4	0·27
5	20	5	0·33
6	24	6	0·40
7	28	7	0·47
8	32	8	0·53
9	36	9	0·60
10	40	10	0·67
11	44	11	0·73
12	48	12	0·80
13	52	13	0·87
14	56	14	0·93
15	1 0	15	1·00
16	1 4	16	1·07
17	1 8	17	1·13
18	1 12	18	1·20
19	1 16	19	1·27
20	1 20	20	1·33
21	1 24	21	1·40
22	1 28	22	1·47
23	1 32	23	1·53
24	1 36	24	1·60
25	1 40	25	1·67
26	1 44	26	1·73
27	1 48	27	1·80
28	1 52	28	1·87
29	1 56	29	1·93
30	2 0	30	2·00
31	2 4	31	2·07
32	2 8	32	2·13
33	2 12	33	2·20
34	2 16	34	2·27
35	2 20	35	2·33
36	2 24	36	2·40
37	2 28	37	2·47
38	2 32	38	2·53
39	2 36	39	2·60
40	2 40	40	2·67
41	2 44	41	2·73
42	2 48	42	2·80
43	2 52	43	2·87
44	2 56	44	2·93
45	3 0	45	3·00
46	3 4	46	3·07
47	3 8	47	3·13
48	3 12	48	3·20
49	3 16	49	3·27
50	3 20	50	3·33
51	3 24	51	3·40
52	3 28	52	3·47
53	3 32	53	3·53
54	3 36	54	3·60
55	3 40	55	3·67
56	3 44	56	3·73
57	3 48	57	3·80
58	3 52	58	3·87
59	3 56	59	3·93

Arc ″	Time s
3	0·2
6	0·4
9	0·6
12	0·8
15	1·0
18	1·2
21	1·4
24	1·6
27	1·8
30	2·0
33	2·2
36	2·4
39	2·6
42	2·8
45	3·0
48	3·2
51	3·4
54	3·6
57	3·8

″	s
0·00	
	·00
0·07	
	·01
0·22	
	·02
0·37	
	·03
0·52	
	·04
0·67	
	·05
0·82	
	·06
0·97	
	·07
1·12	
	·08
1·27	
	·09
1·42	
	·10
1·57	
	·11
1·72	
	·12
1·87	
	·13
2·02	
	·14
2·17	
	·15
2·32	
	·16
2·47	
	·17
2·62	
	·18
2·77	
	·19
2·92	
	·20
2·99	

Arc ″	Time s
0·0	
	0·0
0·7	
	0·1
2·2	
	0·2
3·7	
	0·3
5·2	
	0·4
6·7	
	0·5
8·2	
	0·6
9·7	
	0·7
11·2	
	0·8
12·7	
	0·9
14·2	
	1·0
15·7	
	1·1
17·2	
	1·2
18·7	
	1·3
20·2	
	1·4
21·7	
	1·5
23·2	
	1·6
24·7	
	1·7
26·2	
	1·8
27·7	
	1·9
29·2	
	2·0
30·7	
	2·1
32·2	
	2·2
33·7	
	2·3
35·2	
	2·4
36·7	
	2·5
38·2	
	2·6
39·7	
	2·7
41·2	
	2·8
42·7	
	2·9
44·2	
	3·0
45·7	
	3·1
47·2	
	3·2
48·7	
	3·3
50·2	
	3·4
51·7	
	3·5
53·2	
	3·6
54·7	
	3·7
56·2	
	3·8
57·7	
	3·9
59·2	
	4·0
59·9	

In critical cases ascend.

Examples:

$23''{\cdot}45 = 1^s{\cdot}40 + 0^s{\cdot}16$

$= 1^s{\cdot}56$

$34''{\cdot}5 = 2^s{\cdot}3$

TABLE XII—PROPORTIONAL PARTS

	1	2	3	4	5	6	7	8	9
1	0·1	0·2	0·3	0·4	0·5	0·6	0·7	0·8	0·9
2	0·2	0·4	0·6	0·8	1·0	1·2	1·4	1·6	1·8
3	0·3	0·6	0·9	1·2	1·5	1·8	2·1	2·4	2·7
4	0·4	0·8	1·2	1·6	2·0	2·4	2·8	3·2	3·6
5	0·5	1·0	1·5	2·0	2·5	3·0	3·5	4·0	4·5
6	0·6	1·2	1·8	2·4	3·0	3·6	4·2	4·8	5·4
7	0·7	1·4	2·1	2·8	3·5	4·2	4·9	5·6	6·3
8	0·8	1·6	2·4	3·2	4·0	4·8	5·6	6·4	7·2
9	0·9	1·8	2·7	3·6	4·5	5·4	6·3	7·2	8·1

	10	11	12	13	14	15	16	17	18	19
1	1·0	1·1	1·2	1·3	1·4	1·5	1·6	1·7	1·8	1·9
2	2·0	2·2	2·4	2·6	2·8	3·0	3·2	3·4	3·6	3·8
3	3·0	3·3	3·6	3·9	4·2	4·5	4·8	5·1	5·4	5·7
4	4·0	4·4	4·8	5·2	5·6	6·0	6·4	6·8	7·2	7·6
5	5·0	5·5	6·0	6·5	7·0	7·5	8·0	8·5	9·0	9·5
6	6·0	6·6	7·2	7·8	8·4	9·0	9·6	10·2	10·8	11·4
7	7·0	7·7	8·4	9·1	9·8	10·5	11·2	11·9	12·6	13·3
8	8·0	8·8	9·6	10·4	11·2	12·0	12·8	13·6	14·4	15·2
9	9·0	9·9	10·8	11·7	12·6	13·5	14·4	15·3	16·2	17·1

	20	21	22	23	24	25	26	27	28	29
1	2·0	2·1	2·2	2·3	2·4	2·5	2·6	2·7	2·8	2·9
2	4·0	4·2	4·4	4·6	4·8	5·0	5·2	5·4	5·6	5·8
3	6·0	6·3	6·6	6·9	7·2	7·5	7·8	8·1	8·4	8·7
4	8·0	8·4	8·8	9·2	9·6	10·0	10·4	10·8	11·2	11·6
5	10·0	10·5	11·0	11·5	12·0	12·5	13·0	13·5	14·0	14·5
6	12·0	12·6	13·2	13·8	14·4	15·0	15·6	16·2	16·8	17·4
7	14·0	14·7	15·4	16·1	16·8	17·5	18·2	18·9	19·6	20·3
8	16·0	16·8	17·6	18·4	19·2	20·0	20·8	21·6	22·4	23·2
9	18·0	18·9	19·8	20·7	21·6	22·5	23·4	24·3	25·2	26·1

	30	31	32	33	34	35	36	37	38	39
1	3·0	3·1	3·2	3·3	3·4	3·5	3·6	3·7	3·8	3·9
2	6·0	6·2	6·4	6·6	6·8	7·0	7·2	7·4	7·6	7·8
3	9·0	9·3	9·6	9·9	10·2	10·5	10·8	11·1	11·4	11·7
4	12·0	12·4	12·8	13·2	13·6	14·0	14·4	14·8	15·2	15·6
5	15·0	15·5	16·0	16·5	17·0	17·5	18·0	18·5	19·0	19·5
6	18·0	18·6	19·2	19·8	20·4	21·0	21·6	22·2	22·8	23·4
7	21·0	21·7	22·4	23·1	23·8	24·5	25·2	25·9	26·6	27·3
8	24·0	24·8	25·6	26·4	27·2	28·0	28·8	29·6	30·4	31·2
9	27·0	27·9	28·8	29·7	30·6	31·5	32·4	33·3	34·2	35·1

	40	41	42	43	44	45	46	47	48	49
1	4·0	4·1	4·2	4·3	4·4	4·5	4·6	4·7	4·8	4·9
2	8·0	8·2	8·4	8·6	8·8	9·0	9·2	9·4	9·6	9·8
3	12·0	12·3	12·6	12·9	13·2	13·5	13·8	14·1	14·4	14·7
4	16·0	16·4	16·8	17·2	17·6	18·0	18·4	18·8	19·2	19·6
5	20·0	20·5	21·0	21·5	22·0	22·5	23·0	23·5	24·0	24·5
6	24·0	24·6	25·2	25·8	26·4	27·0	27·6	28·2	28·8	29·4
7	28·0	28·7	29·4	30·1	30·8	31·5	32·2	32·9	33·6	34·3
8	32·0	32·8	33·6	34·4	35·2	36·0	36·8	37·6	38·4	39·2
9	36·0	36·9	37·8	38·7	39·6	40·5	41·4	42·3	43·2	44·1

TABLE XII—PROPORTIONAL PARTS

	50	51	52	53	54	55	56	57	58	59
1	5·0	5·1	5·2	5·3	5·4	5·5	5·6	5·7	5·8	5·9
2	10·0	10·2	10·4	10·6	10·8	11·0	11·2	11·4	11·6	11·8
3	15·0	15·3	15·6	15·9	16·2	16·5	16·8	17·1	17·4	17·7
4	20·0	20·4	20·8	21·2	21·6	22·0	22·4	22·8	23·2	23·6
5	25·0	25·5	26·0	26·5	27·0	27·5	28·0	28·5	29·0	29·5
6	30·0	30·6	31·2	31·8	32·4	33·0	33·6	34·2	34·8	35·4
7	35·0	35·7	36·4	37·1	37·8	38·5	39·2	39·9	40·6	41·3
8	40·0	40·8	41·6	42·4	43·2	44·0	44·8	45·6	46·4	47·2
9	45·0	45·9	46·8	47·7	48·6	49·5	50·4	51·3	52·2	53·1

	60	61	62	63	64	65	66	67	68	69
1	6·0	6·1	6·2	6·3	6·4	6·5	6·6	6·7	6·8	6·9
2	12·0	12·2	12·4	12·6	12·8	13·0	13·2	13·4	13·6	13·8
3	18·0	18·3	18·6	18·9	19·2	19·5	19·8	20·1	20·4	20·7
4	24·0	24·4	24·8	25·2	25·6	26·0	26·4	26·8	27·2	27·6
5	30·0	30·5	31·0	31·5	32·0	32·5	33·0	33·5	34·0	34·5
6	36·0	36·6	37·2	37·8	38·4	39·0	39·6	40·2	40·8	41·4
7	42·0	42·7	43·4	44·1	44·8	45·5	46·2	46·9	47·6	48·3
8	48·0	48·8	49·6	50·4	51·2	52·0	52·8	53·6	54·4	55·2
9	54·0	54·9	55·8	56·7	57·6	58·5	59·4	60·3	61·2	62·1

	70	71	72	73	74	75	76	77	78	79
1	7·0	7·1	7·2	7·3	7·4	7·5	7·6	7·7	7·8	7·9
2	14·0	14·2	14·4	14·6	14·8	15·0	15·2	15·4	15·6	15·8
3	21·0	21·3	21·6	21·9	22·2	22·5	22·8	23·1	23·4	23·7
4	28·0	28·4	28·8	29·2	29·6	30·0	30·4	30·8	31·2	31·6
5	35·0	35·5	36·0	36·5	37·0	37·5	38·0	38·5	39·0	39·5
6	42·0	42·6	43·2	43·8	44·4	45·0	45·6	46·2	46·8	47·4
7	49·0	49·7	50·4	51·1	51·8	52·5	53·2	53·9	54·6	55·3
8	56·0	56·8	57·6	58·4	59·2	60·0	60·8	61·6	62·4	63·2
9	63·0	63·9	64·8	65·7	66·6	67·5	68·4	69·3	70·2	71·1

	80	81	82	83	84	85	86	87	88	89
1	8·0	8·1	8·2	8·3	8·4	8·5	8·6	8·7	8·8	8·9
2	16·0	16·2	16·4	16·6	16·8	17·0	17·2	17·4	17·6	17·8
3	24·0	24·3	24·6	24·9	25·2	25·5	25·8	26·1	26·4	26·7
4	32·0	32·4	32·8	33·2	33·6	34·0	34·4	34·8	35·2	35·6
5	40·0	40·5	41·0	41·5	42·0	42·5	43·0	43·5	44·0	44·5
6	48·0	48·6	49·2	49·8	50·4	51·0	51·6	52·2	52·8	53·4
7	56·0	56·7	57·4	58·1	58·8	59·5	60·2	60·9	61·6	62·3
8	64·0	64·8	65·6	66·4	67·2	68·0	68·8	69·6	70·4	71·2
9	72·0	72·9	73·8	74·7	75·6	76·5	77·4	78·3	79·2	80·1

	90	91	92	93	94	95	96	97	98	99
1	9·0	9·1	9·2	9·3	9·4	9·5	9·6	9·7	9·8	9·9
2	18·0	18·2	18·4	18·6	18·8	19·0	19·2	19·4	19·6	19·8
3	27·0	27·3	27·6	27·9	28·2	28·5	28·8	29·1	29·4	29·7
4	36·0	36·4	36·8	37·2	37·6	38·0	38·4	38·8	39·2	39·6
5	45·0	45·5	46·0	46·5	47·0	47·5	48·0	48·5	49·0	49·5
6	54·0	54·6	55·2	55·8	56·4	57·0	57·6	58·2	58·8	59·4
7	63·0	63·7	64·4	65·1	65·8	66·5	67·2	67·9	68·6	69·3
8	72·0	72·8	73·6	74·4	75·2	76·0	76·8	77·6	78·4	79·2
9	81·0	81·9	82·8	83·7	84·6	85·5	86·4	87·3	88·2	89·1

TABLE XII—PROPORTIONAL PARTS

	100	101	102	103	104	105	106	107	108	109
1	10·0	10·1	10·2	10·3	10·4	10·5	10·6	10·7	10·8	10·9
2	20·0	20·2	20·4	20·6	20·8	21·0	21·2	21·4	21·6	21·8
3	30·0	30·3	30·6	30·9	31·2	31·5	31·8	32·1	32·4	32·7
4	40·0	40·4	40·8	41·2	41·6	42·0	42·4	42·8	43·2	43·6
5	50·0	50·5	51·0	51·5	52·0	52·5	53·0	53·5	54·0	54·5
6	60·0	60·6	61·2	61·8	62·4	63·0	63·6	64·2	64·8	65·4
7	70·0	70·7	71·4	72·1	72·8	73·5	74·2	74·9	75·6	76·3
8	80·0	80·8	81·6	82·4	83·2	84·0	84·8	85·6	86·4	87·2
9	90·0	90·9	91·8	92·7	93·6	94·5	95·4	96·3	97·2	98·1

	110	111	112	113	114	115	116	117	118	119
1	11·0	11·1	11·2	11·3	11·4	11·5	11·6	11·7	11·8	11·9
2	22·0	22·2	22·4	22·6	22·8	23·0	23·2	23·4	23·6	23·8
3	33·0	33·3	33·6	33·9	34·2	34·5	34·8	35·1	35·4	35·7
4	44·0	44·4	44·8	45·2	45·6	46·0	46·4	46·8	47·2	47·6
5	55·0	55·5	56·0	56·5	57·0	57·5	58·0	58·5	59·0	59·5
6	66·0	66·6	67·2	67·8	68·4	69·0	69·6	70·2	70·8	71·4
7	77·0	77·7	78·4	79·1	79·8	80·5	81·2	81·9	82·6	83·3
8	88·0	88·8	89·6	90·4	91·2	92·0	92·8	93·6	94·4	95·2
9	99·0	99·9	100·8	101·7	102·6	103·5	104·4	105·3	106·2	107·1

	120	121	122	123	124	125	126	127	128	129
1	12·0	12·1	12·2	12·3	12·4	12·5	12·6	12·7	12·8	12·9
2	24·0	24·2	24·4	24·6	24·8	25·0	25·2	25·4	25·6	25·8
3	36·0	36·3	36·6	36·9	37·2	37·5	37·8	38·1	38·4	38·7
4	48·0	48·4	48·8	49·2	49·6	50·0	50·4	50·8	51·2	51·6
5	60·0	60·5	61·0	61·5	62·0	62·5	63·0	63·5	64·0	64·5
6	72·0	72·6	73·2	73·8	74·4	75·0	75·6	76·2	76·8	77·4
7	84·0	84·7	85·4	86·1	86·8	87·5	88·2	88·9	89·6	90·3
8	96·0	96·8	97·6	98·4	99·2	100·0	100·8	101·6	102·4	103·2
9	108·0	108·9	109·8	110·7	111·6	112·5	113·4	114·3	115·2	116·1

	130	131	132	133	134	135	136	137	138	139
1	13·0	13·1	13·2	13·3	13·4	13·5	13·6	13·7	13·8	13·9
2	26·0	26·2	26·4	26·6	26·8	27·0	27·2	27·4	27·6	27·8
3	39·0	39·3	39·6	39·9	40·2	40·5	40·8	41·1	41·4	41·7
4	52·0	52·4	52·8	53·2	53·6	54·0	54·4	54·8	55·2	55·6
5	65·0	65·5	66·0	66·5	67·0	67·5	68·0	68·5	69·0	69·5
6	78·0	78·6	79·2	79·8	80·4	81·0	81·6	82·2	82·8	83·4
7	91·0	91·7	92·4	93·1	93·8	94·5	95·2	95·9	96·6	97·3
8	104·0	104·8	105·6	106·4	107·2	108·0	108·8	109·6	110·4	111·2
9	117·0	117·9	118·8	119·7	120·6	121·5	122·4	123·3	124·2	125·1

	140	141	142	143	144	145	146	147	148	149
1	14·0	14·1	14·2	14·3	14·4	14·5	14·6	14·7	14·8	14·9
2	28·0	28·2	28·4	28·6	28·8	29·0	29·2	29·4	29·6	29·8
3	42·0	42·3	42·6	42·9	43·2	43·5	43·8	44·1	44·4	44·7
4	56·0	56·4	56·8	57·2	57·6	58·0	58·4	58·8	59·2	59·6
5	70·0	70·5	71·0	71·5	72·0	72·5	73·0	73·5	74·0	74·5
6	84·0	84·6	85·2	85·8	86·4	87·0	87·6	88·2	88·8	89·4
7	98·0	98·7	99·4	100·1	100·8	101·5	102·2	102·9	103·6	104·3
8	112·0	112·8	113·6	114·4	115·2	116·0	116·8	117·6	118·4	119·2
9	126·0	126·9	127·8	128·7	129·6	130·5	131·4	132·3	133·2	134·1

	150	151	152	153	154	155	156	157	158	159
1	15·0	15·1	15·2	15·3	15·4	15·5	15·6	15·7	15·8	15·9
2	30·0	30·2	30·4	30·6	30·8	31·0	31·2	31·4	31·6	31·8
3	45·0	45·3	45·6	45·9	46·2	46·5	46·8	47·1	47·4	47·7
4	60·0	60·4	60·8	61·2	61·6	62·0	62·4	62·8	63·2	63·6
5	75·0	75·5	76·0	76·5	77·0	77·5	78·0	78·5	79·0	79·5
6	90·0	90·6	91·2	91·8	92·4	93·0	93·6	94·2	94·8	95·4
7	105·0	105·7	106·4	107·1	107·8	108·5	109·2	109·9	110·6	111·3
8	120·0	120·8	121·6	122·4	123·2	124·0	124·8	125·6	126·4	127·2
9	135·0	135·9	136·8	137·7	138·6	139·5	140·4	141·3	142·2	143·1

	160	161	162	163	164	165	166	167	168	169
1	16·0	16·1	16·2	16·3	16·4	16·5	16·6	16·7	16·8	16·9
2	32·0	32·2	32·4	32·6	32·8	33·0	33·2	33·4	33·6	33·8
3	48·0	48·3	48·6	48·9	49·2	49·5	49·8	50·1	50·4	50·7
4	64·0	64·4	64·8	65·2	65·6	66·0	66·4	66·8	67·2	67·6
5	80·0	80·5	81·0	81·5	82·0	82·5	83·0	83·5	84·0	84·5
6	96·0	96·6	97·2	97·8	98·4	99·0	99·6	100·2	100·8	101·4
7	112·0	112·7	113·4	114·1	114·8	115·5	116·2	116·9	117·6	118·3
8	128·0	128·8	129·6	130·4	131·2	132·0	132·8	133·6	134·4	135·2
9	144·0	144·9	145·8	146·7	147·6	148·5	149·4	150·3	151·2	152·1

	170	171	172	173	174	175	176	177	178	179
1	17·0	17·1	17·2	17·3	17·4	17·5	17·6	17·7	17·8	17·9
2	34·0	34·2	34·4	34·6	34·8	35·0	35·2	35·4	35·6	35·8
3	51·0	51·3	51·6	51·9	52·2	52·5	52·8	53·1	53·4	53·7
4	68·0	68·4	68·8	69·2	69·6	70·0	70·4	70·8	71·2	71·6
5	85·0	85·5	86·0	86·5	87·0	87·5	88·0	88·5	89·0	89·5
6	102·0	102·6	103·2	103·8	104·4	105·0	105·6	106·2	106·8	107·4
7	119·0	119·7	120·4	121·1	121·8	122·5	123·2	123·9	124·6	125·3
8	136·0	136·8	137·6	138·4	139·2	140·0	140·8	141·6	142·4	143·2
9	153·0	153·9	154·8	155·7	156·6	157·5	158·4	159·3	160·2	161·1

	180	181	182	183	184	185	186	187	188	189
1	18·0	18·1	18·2	18·3	18·4	18·5	18·6	18·7	18·8	18·9
2	36·0	36·2	36·4	36·6	36·8	37·0	37·2	37·4	37·6	37·8
3	54·0	54·3	54·6	54·9	55·2	55·5	55·8	56·1	56·4	56·7
4	72·0	72·4	72·8	73·2	73·6	74·0	74·4	74·8	75·2	75·6
5	90·0	90·5	91·0	91·5	92·0	92·5	93·0	93·5	94·0	94·5
6	108·0	108·6	109·2	109·8	110·4	111·0	111·6	112·2	112·8	113·4
7	126·0	126·7	127·4	128·1	128·8	129·5	130·2	130·9	131·6	132·3
8	144·0	144·8	145·6	146·4	147·2	148·0	148·8	149·6	150·4	151·2
9	162·0	162·9	163·8	164·7	165·6	166·5	167·4	168·3	169·2	170·1

	190	191	192	193	194	195	196	197	198	199
1	19·0	19·1	19·2	19·3	19·4	19·5	19·6	19·7	19·8	19·9
2	38·0	38·2	38·4	38·6	38·8	39·0	39·2	39·4	39·6	39·8
3	57·0	57·3	57·6	57·9	58·2	58·5	58·8	59·1	59·4	59·7
4	76·0	76·4	76·8	77·2	77·6	78·0	78·4	78·8	79·2	79·6
5	95·0	95·5	96·0	96·5	97·0	97·5	98·0	98·5	99·0	99·5
6	114·0	114·6	115·2	115·8	116·4	117·0	117·6	118·2	118·8	119·4
7	133·0	133·7	134·4	135·1	135·8	136·5	137·2	137·9	138·6	139·3
8	152·0	152·8	153·6	154·4	155·2	156·0	156·8	157·6	158·4	159·2
9	171·0	171·9	172·8	173·7	174·6	175·5	176·4	177·3	178·2	179·1

	200	201	202	203	204
1	20·0	20·1	20·2	20·3	20·4
2	40·0	40·2	40·4	40·6	40·8
3	60·0	60·3	60·6	60·9	61·2
4	80·0	80·4	80·8	81·2	81·6
5	100·0	100·5	101·0	101·5	102·0
6	120·0	120·6	121·2	121·8	122·4
7	140·0	140·7	141·4	142·1	142·8
8	160·0	160·8	161·6	162·4	163·2
9	180·0	180·9	181·8	182·7	183·6

	205	206	207	208	209
1	20·5	20·6	20·7	20·8	20·9
2	41·0	41·2	41·4	41·6	41·8
3	61·5	61·8	62·1	62·4	62·7
4	82·0	82·4	82·8	83·2	83·6
5	102·5	103·0	103·5	104·0	104·5
6	123·0	123·6	124·2	124·8	125·4
7	143·5	144·2	144·9	145·6	146·3
8	164·0	164·8	165·6	166·4	167·2
9	184·5	185·4	186·3	187·2	188·1

	210	211	212	213	214
1	21·0	21·1	21·2	21·3	21·4
2	42·0	42·2	42·4	42·6	42·8
3	63·0	63·3	63·6	63·9	64·2
4	84·0	84·4	84·8	85·2	85·6
5	105·0	105·5	106·0	106·5	107·0
6	126·0	126·6	127·2	127·8	128·4
7	147·0	147·7	148·4	149·1	149·8
8	168·0	168·8	169·6	170·4	171·2
9	189·0	189·9	190·8	191·7	192·6

	215	216	217	218	219
1	21·5	21·6	21·7	21·8	21·9
2	43·0	43·2	43·4	43·6	43·8
3	64·5	64·8	65·1	65·4	65·7
4	86·0	86·4	86·8	87·2	87·6
5	107·5	108·0	108·5	109·0	109·5
6	129·0	129·6	130·2	130·8	131·4
7	150·5	151·2	151·9	152·6	153·3
8	172·0	172·8	173·6	174·4	175·2
9	193·5	194·4	195·3	196·2	197·1

	220	221	222	223	224
1	22·0	22·1	22·2	22·3	22·4
2	44·0	44·2	44·4	44·6	44·8
3	66·0	66·3	66·6	66·9	67·2
4	88·0	88·4	88·8	89·2	89·6
5	110·0	110·5	111·0	111·5	112·0
6	132·0	132·6	133·2	133·8	134·4
7	154·0	154·7	155·4	156·1	156·8
8	176·0	176·8	177·6	178·4	179·2
9	198·0	198·9	199·8	200·7	201·6

	225	226	227	228	229
1	22·5	22·6	22·7	22·8	22·9
2	45·0	45·2	45·4	45·6	45·8
3	67·5	67·8	68·1	68·4	68·7
4	90·0	90·4	90·8	91·2	91·6
5	112·5	113·0	113·5	114·0	114·5
6	135·0	135·6	136·2	136·8	137·4
7	157·5	158·2	158·9	159·6	160·3
8	180·0	180·8	181·6	182·4	183·2
9	202·5	203·4	204·3	205·2	206·1

	230	231	232	233	234
1	23·0	23·1	23·2	23·3	23·4
2	46·0	46·2	46·4	46·6	46·8
3	69·0	69·3	69·6	69·9	70·2
4	92·0	92·4	92·8	93·2	93·6
5	115·0	115·5	116·0	116·5	117·0
6	138·0	138·6	139·2	139·8	140·4
7	161·0	161·7	162·4	163·1	163·8
8	184·0	184·8	185·6	186·4	187·2
9	207·0	207·9	208·8	209·7	210·6

	235	236	237	238	239
1	23·5	23·6	23·7	23·8	23·9
2	47·0	47·2	47·4	47·6	47·8
3	70·5	70·8	71·1	71·4	71·7
4	94·0	94·4	94·8	95·2	95·6
5	117·5	118·0	118·5	119·0	119·5
6	141·0	141·6	142·2	142·8	143·4
7	164·5	165·2	165·9	166·6	167·3
8	188·0	188·8	189·6	190·4	191·2
9	211·5	212·4	213·3	214·2	215·1

	240	241	242	243	244
1	24·0	24·1	24·2	24·3	24·4
2	48·0	48·2	48·4	48·6	48·8
3	72·0	72·3	72·6	72·9	73·2
4	96·0	96·4	96·8	97·2	97·6
5	120·0	120·5	121·0	121·5	122·0
6	144·0	144·6	145·2	145·8	146·4
7	168·0	168·7	169·4	170·1	170·8
8	192·0	192·8	193·6	194·4	195·2
9	216·0	216·9	217·8	218·7	219·6

	245	246	247	248	249
1	24·5	24·6	24·7	24·8	24·9
2	49·0	49·2	49·4	49·6	49·8
3	73·5	73·8	74·1	74·4	74·7
4	98·0	98·4	98·8	99·2	99·6
5	122·5	123·0	123·5	124·0	124·5
6	147·0	147·6	148·2	148·8	149·4
7	171·5	172·2	172·9	173·6	174·3
8	196·0	196·8	197·6	198·4	199·2
9	220·5	221·4	222·3	223·2	224·1

	250	251	252	253	254	255	256	257	258	259
1	25·0	25·1	25·2	25·3	25·4	25·5	25·6	25·7	25·8	25·9
2	50·0	50·2	50·4	50·6	50·8	51·0	51·2	51·4	51·6	51·8
3	75·0	75·3	75·6	75·9	76·2	76·5	76·8	77·1	77·4	77·7
4	100·0	100·4	100·8	101·2	101·6	102·0	102·4	102·8	103·2	103·6
5	125·0	125·5	126·0	126·5	127·0	127·5	128·0	128·5	129·0	129·5
6	150·0	150·6	151·2	151·8	152·4	153·0	153·6	154·2	154·8	155·4
7	175·0	175·7	176·4	177·1	177·8	178·5	179·2	179·9	180·6	181·3
8	200·0	200·8	201·6	202·4	203·2	204·0	204·8	205·6	206·4	207·2
9	225·0	225·9	226·8	227·7	228·6	229·5	230·4	231·3	232·2	233·1

	260	261	262	263	264	265	266	267	268	269
1	26·0	26·1	26·2	26·3	26·4	26·5	26·6	26·7	26·8	26·9
2	52·0	52·2	52·4	52·6	52·8	53·0	53·2	53·4	53·6	53·8
3	78·0	78·3	78·6	78·9	79·2	79·5	79·8	80·1	80·4	80·7
4	104·0	104·4	104·8	105·2	105·6	106·0	106·4	106·8	107·2	107·6
5	130·0	130·5	131·0	131·5	132·0	132·5	133·0	133·5	134·0	134·5
6	156·0	156·6	157·2	157·8	158·4	159·0	159·6	160·2	160·8	161·4
7	182·0	182·7	183·4	184·1	184·8	185·5	186·2	186·9	187·6	188·3
8	208·0	208·8	209·6	210·4	211·2	212·0	212·8	213·6	214·4	215·2
9	234·0	234·9	235·8	236·7	237·6	238·5	239·4	240·3	241·2	242·1

	270	271	272	273	274	275	276	277	278	279
1	27·0	27·1	27·2	27·3	27·4	27·5	27·6	27·7	27·8	27·9
2	54·0	54·2	54·4	54·6	54·8	55·0	55·2	55·4	55·6	55·8
3	81·0	81·3	81·6	81·9	82·2	82·5	82·8	83·1	83·4	83·7
4	108·0	108·4	108·8	109·2	109·6	110·0	110·4	110·8	111·2	111·6
5	135·0	135·5	136·0	136·5	137·0	137·5	138·0	138·5	139·0	139·5
6	162·0	162·6	163·2	163·8	164·4	165·0	165·6	166·2	166·8	167·4
7	189·0	189·7	190·4	191·1	191·8	192·5	193·2	193·9	194·6	195·3
8	216·0	216·8	217·6	218·4	219·2	220·0	220·8	221·6	222·4	223·2
9	243·0	243·9	244·8	245·7	246·6	247·5	248·4	249·3	250·2	251·1

	280	281	282	283	284	285	286	287	288	289
1	28·0	28·1	28·2	28·3	28·4	28·5	28·6	28·7	28·8	28·9
2	56·0	56·2	56·4	56·6	56·8	57·0	57·2	57·4	57·6	57·8
3	84·0	84·3	84·6	84·9	85·2	85·5	85·8	86·1	86·4	86·7
4	112·0	112·4	112·8	113·2	113·6	114·0	114·4	114·8	115·2	115·6
5	140·0	140·5	141·0	141·5	142·0	142·5	143·0	143·5	144·0	144·5
6	168·0	168·6	169·2	169·8	170·4	171·0	171·6	172·2	172·8	173·4
7	196·0	196·7	197·4	198·1	198·8	199·5	200·2	200·9	201·6	202·3
8	224·0	224·8	225·6	226·4	227·2	228·0	228·8	229·6	230·4	231·2
9	252·0	252·9	253·8	254·7	255·6	256·5	257·4	258·3	259·2	260·1

	290	291	292	293	294	295	296	297	298	299
1	29·0	29·1	29·2	29·3	29·4	29·5	29·6	29·7	29·8	29·9
2	58·0	58·2	58·4	58·6	58·8	59·0	59·2	59·4	59·6	59·8
3	87·0	87·3	87·6	87·9	88·2	88·5	88·8	89·1	89·4	89·7
4	116·0	116·4	116·8	117·2	117·6	118·0	118·4	118·8	119·2	119·6
5	145·0	145·5	146·0	146·5	147·0	147·5	148·0	148·5	149·0	149·5
6	174·0	174·6	175·2	175·8	176·4	177·0	177·6	178·2	178·8	179·4
7	203·0	203·7	204·4	205·1	205·8	206·5	207·2	207·9	208·6	209·3
8	232·0	232·8	233·6	234·4	235·2	236·0	236·8	237·6	238·4	239·2
9	261·0	261·9	262·8	263·7	264·6	265·5	266·4	267·3	268·2	269·1

	300	301	302	303	304		305	306	307	308	309
1	30·0	30·1	30·2	30·3	30·4	1	30·5	30·6	30·7	30·8	30·9
2	60·0	60·2	60·4	60·6	60·8	2	61·0	61·2	61·4	61·6	61·8
3	90·0	90·3	90·6	90·9	91·2	3	91·5	91·8	92·1	92·4	92·7
4	120·0	120·4	120·8	121·2	121·6	4	122·0	122·4	122·8	123·2	123·6
5	150·0	150·5	151·0	151·5	152·0	5	152·5	153·0	153·5	154·0	154·5
6	180·0	180·6	181·2	181·8	182·4	6	183·0	183·6	184·2	184·8	185·4
7	210·0	210·7	211·4	212·1	212·8	7	213·5	214·2	214·9	215·6	216·3
8	240·0	240·8	241·6	242·4	243·2	8	244·0	244·8	245·6	246·4	247·2
9	270·0	270·9	271·8	272·7	273·6	9	274·5	275·4	276·3	277·2	278·1

	310	311	312	313	314		315	316	317	318	319
1	31·0	31·1	31·2	31·3	31·4	1	31·5	31·6	31·7	31·8	31·9
2	62·0	62·2	62·4	62·6	62·8	2	63·0	63·2	63·4	63·6	63·8
3	93·0	93·3	93·6	93·9	94·2	3	94·5	94·8	95·1	95·4	95·7
4	124·0	124·4	124·8	125·2	125·6	4	126·0	126·4	126·8	127·2	127·6
5	155·0	155·5	156·0	156·5	157·0	5	157·5	158·0	158·5	159·0	159·5
6	186·0	186·6	187·2	187·8	188·4	6	189·0	189·6	190·2	190·8	191·4
7	217·0	217·7	218·4	219·1	219·8	7	220·5	221·2	221·9	222·6	223·3
8	248·0	248·8	249·6	250·4	251·2	8	252·0	252·8	253·6	254·4	255·2
9	279·0	279·9	280·8	281·7	282·6	9	283·5	284·4	285·3	286·2	287·1

	320	321	322	323	324		325	326	327	328	329
1	32·0	32·1	32·2	32·3	32·4	1	32·5	32·6	32·7	32·8	32·9
2	64·0	64·2	64·4	64·6	64·8	2	65·0	65·2	65·4	65·6	65·8
3	96·0	96·3	96·6	96·9	97·2	3	97·5	97·8	98·1	98·4	98·7
4	128·0	128·4	128·8	129·2	129·6	4	130·0	130·4	130·8	131·2	131·6
5	160·0	160·5	161·0	161·5	162·0	5	162·5	163·0	163·5	164·0	164·5
6	192·0	192·6	193·2	193·8	194·4	6	195·0	195·6	196·2	196·8	197·4
7	224·0	224·7	225·4	226·1	226·8	7	227·5	228·2	228·9	229·6	230·3
8	256·0	256·8	257·6	258·4	259·2	8	260·0	260·8	261·6	262·4	263·2
9	288·0	288·9	289·8	290·7	291·6	9	292·5	293·4	294·3	295·2	296·1

	330	331	332	333	334		335	336	337	338	339
1	33·0	33·1	33·2	33·3	33·4	1	33·5	33·6	33·7	33·8	33·9
2	66·0	66·2	66·4	66·6	66·8	2	67·0	67·2	67·4	67·6	67·8
3	99·0	99·3	99·6	99·9	100·2	3	100·5	100·8	101·1	101·4	101·7
4	132·0	132·4	132·8	133·2	133·6	4	134·0	134·4	134·8	135·2	135·6
5	165·0	165·5	166·0	166·5	167·0	5	167·5	168·0	168·5	169·0	169·5
6	198·0	198·6	199·2	199·8	200·4	6	201·0	201·6	202·2	202·8	203·4
7	231·0	231·7	232·4	233·1	233·8	7	234·5	235·2	235·9	236·6	237·3
8	264·0	264·8	265·6	266·4	267·2	8	268·0	268·8	269·6	270·4	271·2
9	297·0	297·9	298·8	299·7	300·6	9	301·5	302·4	303·3	304·2	305·1

	340	341	342	343	344		345	346	347	348	349
1	34·0	34·1	34·2	34·3	34·4	1	34·5	34·6	34·7	34·8	34·9
2	68·0	68·2	68·4	68·6	68·8	2	69·0	69·2	69·4	69·6	69·8
3	102·0	102·3	102·6	102·9	103·2	3	103·5	103·8	104·1	104·4	104·7
4	136·0	136·4	136·8	137·2	137·6	4	138·0	138·4	138·8	139·2	139·6
5	170·0	170·5	171·0	171·5	172·0	5	172·5	173·0	173·5	174·0	174·5
6	204·0	204·6	205·2	205·8	206·4	6	207·0	207·6	208·2	208·8	209·4
7	238·0	238·7	239·4	240·1	240·8	7	241·5	242·2	242·9	243·6	244·3
8	272·0	272·8	273·6	274·4	275·2	8	276·0	276·8	277·6	278·4	279·2
9	306·0	306·9	307·8	308·7	309·6	9	310·5	311·4	312·3	313·2	314·1

	350	351	352	353	354
1	35·0	35·1	35·2	35·3	35·4
2	70·0	70·2	70·4	70·6	70·8
3	105·0	105·3	105·6	105·9	106·2
4	140·0	140·4	140·8	141·2	141·6
5	175·0	175·5	176·0	176·5	177·0
6	210·0	210·6	211·2	211·8	212·4
7	245·0	245·7	246·4	247·1	247·8
8	280·0	280·8	281·6	282·4	283·2
9	315·0	315·9	316·8	317·7	318·6

	355	356	357	358	359
1	35·5	35·6	35·7	35·8	35·9
2	71·0	71·2	71·4	71·6	71·8
3	106·5	106·8	107·1	107·4	107·7
4	142·0	142·4	142·8	143·2	143·6
5	177·5	178·0	178·5	179·0	179·5
6	213·0	213·6	214·2	214·8	215·4
7	248·5	249·2	249·9	250·6	251·3
8	284·0	284·8	285·6	286·4	287·2
9	319·5	320·4	321·3	322·2	323·1

	360	361	362	363	364
1	36·0	36·1	36·2	36·3	36·4
2	72·0	72·2	72·4	72·6	72·8
3	108·0	108·3	108·6	108·9	109·2
4	144·0	144·4	144·8	145·2	145·6
5	180·0	180·5	181·0	181·5	182·0
6	216·0	216·6	217·2	217·8	218·4
7	252·0	252·7	253·4	254·1	254·8
8	288·0	288·8	289·6	290·4	291·2
9	324·0	324·9	325·8	326·7	327·6

	365	366	367	368	369
1	36·5	36·6	36·7	36·8	36·9
2	73·0	73·2	73·4	73·6	73·8
3	109·5	109·8	110·1	110·4	110·7
4	146·0	146·4	146·8	147·2	147·6
5	182·5	183·0	183·5	184·0	184·5
6	219·0	219·6	220·2	220·8	221·4
7	255·5	256·2	256·9	257·6	258·3
8	292·0	292·8	293·6	294·4	295·2
9	328·5	329·4	330·3	331·2	332·1

	370	371	372	373	374
1	37·0	37·1	37·2	37·3	37·4
2	74·0	74·2	74·4	74·6	74·8
3	111·0	111·3	111·6	111·9	112·2
4	148·0	148·4	148·8	149·2	149·6
5	185·0	185·5	186·0	186·5	187·0
6	222·0	222·6	223·2	223·8	224·4
7	259·0	259·7	260·4	261·1	261·8
8	296·0	296·8	297·6	298·4	299·2
9	333·0	333·9	334·8	335·7	336·6

	375	376	377	378	379
1	37·5	37·6	37·7	37·8	37·9
2	75·0	75·2	75·4	75·6	75·8
3	112·5	112·8	113·1	113·4	113·7
4	150·0	150·4	150·8	151·2	151·6
5	187·5	188·0	188·5	189·0	189·5
6	225·0	225·6	226·2	226·8	227·4
7	262·5	263·2	263·9	264·6	265·3
8	300·0	300·8	301·6	302·4	303·2
9	337·5	338·4	339·3	340·2	341·1

	380	381	382	383	384
1	38·0	38·1	38·2	38·3	38·4
2	76·0	76·2	76·4	76·6	76·8
3	114·0	114·3	114·6	114·9	115·2
4	152·0	152·4	152·8	153·2	153·6
5	190·0	190·5	191·0	191·5	192·0
6	228·0	228·6	229·2	229·8	230·4
7	266·0	266·7	267·4	268·1	268·8
8	304·0	304·8	305·6	306·4	307·2
9	342·0	342·9	343·8	344·7	345·6

	385	386	387	388	389
1	38·5	38·6	38·7	38·8	38·9
2	77·0	77·2	77·4	77·6	77·8
3	115·5	115·8	116·1	116·4	116·7
4	154·0	154·4	154·8	155·2	155·6
5	192·5	193·0	193·5	194·0	194·5
6	231·0	231·6	232·2	232·8	233·4
7	269·5	270·2	270·9	271·6	272·3
8	308·0	308·8	309·6	310·4	311·2
9	346·5	347·4	348·3	349·2	350·1

	390	391	392	393	394
1	39·0	39·1	39·2	39·3	39·4
2	78·0	78·2	78·4	78·6	78·8
3	117·0	117·3	117·6	117·9	118·2
4	156·0	156·4	156·8	157·2	157·6
5	195·0	195·5	196·0	196·5	197·0
6	234·0	234·6	235·2	235·8	236·4
7	273·0	273·7	274·4	275·1	275·8
8	312·0	312·8	313·6	314·4	315·2
9	351·0	351·9	352·8	353·7	354·6

	395	396	397	398	399
1	39·5	39·6	39·7	39·8	39·9
2	79·0	79·2	79·4	79·6	79·8
3	118·5	118·8	119·1	119·4	119·7
4	158·0	158·4	158·8	159·2	159·6
5	197·5	198·0	198·5	199·0	199·5
6	237·0	237·6	238·2	238·8	239·4
7	276·5	277·2	277·9	278·6	279·3
8	316·0	316·8	317·6	318·4	319·2
9	355·5	356·4	357·3	358·2	359·1

	400	401	402	403	404		405	406	407	408	409
1	40·0	40·1	40·2	40·3	40·4	1	40·5	40·6	40·7	40·8	40·9
2	80·0	80·2	80·4	80·6	80·8	2	81·0	81·2	81·4	81·6	81·8
3	120·0	120·3	120·6	120·9	121·2	3	121·5	121·8	122·1	122·4	122·7
4	160·0	160·4	160·8	161·2	161·6	4	162·0	162·4	162·8	163·2	163·6
5	200·0	200·5	201·0	201·5	202·0	5	202·5	203·0	203·5	204·0	204·5
6	240·0	240·6	241·2	241·8	242·4	6	243·0	243·6	244·2	244·8	245·4
7	280·0	280·7	281·4	282·1	282·8	7	283·5	284·2	284·9	285·6	286·3
8	320·0	320·8	321·6	322·4	323·2	8	324·0	324·8	325·6	326·4	327·2
9	360·0	360·9	361·8	362·7	363·6	9	364·5	365·4	366·3	367·2	368·1

	410	411	412	413	414		415	416	417	418	419
1	41·0	41·1	41·2	41·3	41·4	1	41·5	41·6	41·7	41·8	41·9
2	82·0	82·2	82·4	82·6	82·8	2	83·0	83·2	83·4	83·6	83·8
3	123·0	123·3	123·6	123·9	124·2	3	124·5	124·8	125·1	125·4	125·7
4	164·0	164·4	164·8	165·2	165·6	4	166·0	166·4	166·8	167·2	167·6
5	205·0	205·5	206·0	206·5	207·0	5	207·5	208·0	208·5	209·0	209·5
6	246·0	246·6	247·2	247·8	248·4	6	249·0	249·6	250·2	250·8	251·4
7	287·0	287·7	288·4	289·1	289·8	7	290·5	291·2	291·9	292·6	293·3
8	328·0	328·8	329·6	330·4	331·2	8	332·0	332·8	333·6	334·4	335·2
9	369·0	369·9	370·8	371·7	372·6	9	373·5	374·4	375·3	376·2	377·1

	420	421	422	423	424		425	426	427	428	429
1	42·0	42·1	42·2	42·3	42·4	1	42·5	42·6	42·7	42·8	42·9
2	84·0	84·2	84·4	84·6	84·8	2	85·0	85·2	85·4	85·6	85·8
3	126·0	126·3	126·6	126·9	127·2	3	127·5	127·8	128·1	128·4	128·7
4	168·0	168·4	168·8	169·2	169·6	4	170·0	170·4	170·8	171·2	171·6
5	210·0	210·5	211·0	211·5	212·0	5	212·5	213·0	213·5	214·0	214·5
6	252·0	252·6	253·2	253·8	254·4	6	255·0	255·6	256·2	256·8	257·4
7	294·0	294·7	295·4	296·1	296·8	7	297·5	298·2	298·9	299·6	300·3
8	336·0	336·8	337·6	338·4	339·2	8	340·0	340·8	341·6	342·4	343·2
9	378·0	378·9	379·8	380·7	381·6	9	382·5	383·4	384·3	385·2	386·1

	430	431	432	433	434		435	436	437	438	439
1	43·0	43·1	43·2	43·3	43·4	1	43·5	43·6	43·7	43·8	43·9
2	86·0	86·2	86·4	86·6	86·8	2	87·0	87·2	87·4	87·6	87·8
3	129·0	129·3	129·6	129·9	130·2	3	130·5	130·8	131·1	131·4	131·7
4	172·0	172·4	172·8	173·2	173·6	4	174·0	174·4	174·8	175·2	175·6
5	215·0	215·5	216·0	216·5	217·0	5	217·5	218·0	218·5	219·0	219·5
6	258·0	258·6	259·2	259·8	260·4	6	261·0	261·6	262·2	262·8	263·4
7	301·0	301·7	302·4	303·1	303·8	7	304·5	305·2	305·9	306·6	307·3
8	344·0	344·8	345·6	346·4	347·2	8	348·0	348·8	349·6	350·4	351·2
9	387·0	387·9	388·8	389·7	390·6	9	391·5	392·4	393·3	394·2	395·1

	440	441	442	443	444		445	446	447	448	449
1	44·0	44·1	44·2	44·3	44·4	1	44·5	44·6	44·7	44·8	44·9
2	88·0	88·2	88·4	88·6	88·8	2	89·0	89·2	89·4	89·6	89·8
3	132·0	132·3	132·6	132·9	133·2	3	133·5	133·8	134·1	134·4	134·7
4	176·0	176·4	176·8	177·2	177·6	4	178·0	178·4	178·8	179·2	179·6
5	220·0	220·5	221·0	221·5	222·0	5	222·5	223·0	223·5	224·0	224·5
6	264·0	264·6	265·2	265·8	266·4	6	267·0	267·6	268·2	268·8	269·4
7	308·0	308·7	309·4	310·1	310·8	7	311·5	312·2	312·9	313·6	314·3
8	352·0	352·8	353·6	354·4	355·2	8	356·0	356·8	357·6	358·4	359·2
9	396·0	396·9	397·8	398·7	399·6	9	400·5	401·4	402·3	403·2	404·1

	450	451	452	453	454
1	45·0	45·1	45·2	45·3	45·4
2	90·0	90·2	90·4	90·6	90·8
3	135·0	135·3	135·6	135·9	136·2
4	180·0	180·4	180·8	181·2	181·6
5	225·0	225·5	226·0	226·5	227·0
6	270·0	270·6	271·2	271·8	272·4
7	315·0	315·7	316·4	317·1	317·8
8	360·0	360·8	361·6	362·4	363·2
9	405·0	405·9	406·8	407·7	408·6

	455	456	457	458	459
1	45·5	45·6	45·7	45·8	45·9
2	91·0	91·2	91·4	91·6	91·8
3	136·5	136·8	137·1	137·4	137·7
4	182·0	182·4	182·8	183·2	183·6
5	227·5	228·0	228·5	229·0	229·5
6	273·0	273·6	274·2	274·8	275·4
7	318·5	319·2	319·9	320·6	321·3
8	364·0	364·8	365·6	366·4	367·2
9	409·5	410·4	411·3	412·2	413·1

	460	461	462	463	464
1	46·0	46·1	46·2	46·3	46·4
2	92·0	92·2	92·4	92·6	92·8
3	138·0	138·3	138·6	138·9	139·2
4	184·0	184·4	184·8	185·2	185·6
5	230·0	230·5	231·0	231·5	232·0
6	276·0	276·6	277·2	277·8	278·4
7	322·0	322·7	323·4	324·1	324·8
8	368·0	368·8	369·6	370·4	371·2
9	414·0	414·9	415·8	416·7	417·6

	465	466	467	468	469
1	46·5	46·6	46·7	46·8	46·9
2	93·0	93·2	93·4	93·6	93·8
3	139·5	139·8	140·1	140·4	140·7
4	186·0	186·4	186·8	187·2	187·6
5	232·5	233·0	233·5	234·0	234·5
6	279·0	279·6	280·2	280·8	281·4
7	325·5	326·2	326·9	327·6	328·3
8	372·0	372·8	373·6	374·4	375·2
9	418·5	419·4	420·3	421·2	422·1

	470	471	472	473	474
1	47·0	47·1	47·2	47·3	47·4
2	94·0	94·2	94·4	94·6	94·8
3	141·0	141·3	141·6	141·9	142·2
4	188·0	188·4	188·8	189·2	189·6
5	235·0	235·5	236·0	236·5	237·0
6	282·0	282·6	283·2	283·8	284·4
7	329·0	329·7	330·4	331·1	331·8
8	376·0	376·8	377·6	378·4	379·2
9	423·0	423·9	424·8	425·7	426·6

	475	476	477	478	479
1	47·5	47·6	47·7	47·8	47·9
2	95·0	95·2	95·4	95·6	95·8
3	142·5	142·8	143·1	143·4	143·7
4	190·0	190·4	190·8	191·2	191·6
5	237·5	238·0	238·5	239·0	239·5
6	285·0	285·6	286·2	286·8	287·4
7	332·5	333·2	333·9	334·6	335·3
8	380·0	380·8	381·6	382·4	383·2
9	427·5	428·4	429·3	430·2	431·1

	480	481	482	483	484
1	48·0	48·1	48·2	48·3	48·4
2	96·0	96·2	96·4	96·6	96·8
3	144·0	144·3	144·6	144·9	145·2
4	192·0	192·4	192·8	193·2	193·6
5	240·0	240·5	241·0	241·5	242·0
6	288·0	288·6	289·2	289·8	290·4
7	336·0	336·7	337·4	338·1	338·8
8	384·0	384·8	385·6	386·4	387·2
9	432·0	432·9	433·8	434·7	435·6

	485	486	487	488	489
1	48·5	48·6	48·7	48·8	48·9
2	97·0	97·2	97·4	97·6	97·8
3	145·5	145·8	146·1	146·4	146·7
4	194·0	194·4	194·8	195·2	195·6
5	242·5	243·0	243·5	244·0	244·5
6	291·0	291·6	292·2	292·8	293·4
7	339·5	340·2	340·9	341·6	342·3
8	388·0	388·8	389·6	390·4	391·2
9	436·5	437·4	438·3	439·2	440·1

	490	491	492	493	494
1	49·0	49·1	49·2	49·3	49·4
2	98·0	98·2	98·4	98·6	98·8
3	147·0	147·3	147·6	147·9	148·2
4	196·0	196·4	196·8	197·2	197·6
5	245·0	245·5	246·0	246·5	247·0
6	294·0	294·6	295·2	295·8	296·4
7	343·0	343·7	344·4	345·1	345·8
8	392·0	392·8	393·6	394·4	395·2
9	441·0	441·9	442·8	443·7	444·6

	495	496	497	498	499
1	49·5	49·6	49·7	49·8	49·9
2	99·0	99·2	99·4	99·6	99·8
3	148·5	148·8	149·1	149·4	149·7
4	198·0	198·4	198·8	199·2	199·6
5	247·5	248·0	248·5	249·0	249·5
6	297·0	297·6	298·2	298·8	299·4
7	346·5	347·2	347·9	348·6	349·3
8	396·0	396·8	397·6	398·4	399·2
9	445·5	446·4	447·3	448·2	449·1

	500	501	502	503	504
1	50·0	50·1	50·2	50·3	50·4
2	100·0	100·2	100·4	100·6	100·8
3	150·0	150·3	150·6	150·9	151·2
4	200·0	200·4	200·8	201·2	201·6
5	250·0	250·5	251·0	251·5	252·0
6	300·0	300·6	301·2	301·8	302·4
7	350·0	350·7	351·4	352·1	352·8
8	400·0	400·8	401·6	402·4	403·2
9	450·0	450·9	451·8	452·7	453·6

	505	506	507	508	509
1	50·5	50·6	50·7	50·8	50·9
2	101·0	101·2	101·4	101·6	101·8
3	151·5	151·8	152·1	152·4	152·7
4	202·0	202·4	202·8	203·2	203·6
5	252·5	253·0	253·5	254·0	254·5
6	303·0	303·6	304·2	304·8	305·4
7	353·5	354·2	354·9	355·6	356·3
8	404·0	404·8	405·6	406·4	407·2
9	454·5	455·4	456·3	457·2	458·1

	510	511	512	513	514
1	51·0	51·1	51·2	51·3	51·4
2	102·0	102·2	102·4	102·6	102·8
3	153·0	153·3	153·6	153·9	154·2
4	204·0	204·4	204·8	205·2	205·6
5	255·0	255·5	256·0	256·5	257·0
6	306·0	306·6	307·2	307·8	308·4
7	357·0	357·7	358·4	359·1	359·8
8	408·0	408·8	409·6	410·4	411·2
9	459·0	459·9	460·8	461·7	462·6

	515	516	517	518	519
1	51·5	51·6	51·7	51·8	51·9
2	103·0	103·2	103·4	103·6	103·8
3	154·5	154·8	155·1	155·4	155·7
4	206·0	206·4	206·8	207·2	207·6
5	257·5	258·0	258·5	259·0	259·5
6	309·0	309·6	310·2	310·8	311·4
7	360·5	361·2	361·9	362·6	363·3
8	412·0	412·8	413·6	414·4	415·2
9	463·5	464·4	465·3	466·2	467·1

	520	521	522	523	524
1	52·0	52·1	52·2	52·3	52·4
2	104·0	104·2	104·4	104·6	104·8
3	156·0	156·3	156·6	156·9	157·2
4	208·0	208·4	208·8	209·2	209·6
5	260·0	260·5	261·0	261·5	262·0
6	312·0	312·6	313·2	313·8	314·4
7	364·0	364·7	365·4	366·1	366·8
8	416·0	416·8	417·6	418·4	419·2
9	468·0	468·9	469·8	470·7	471·6

	525	526	527	528	529
1	52·5	52·6	52·7	52·8	52·9
2	105·0	105·2	105·4	105·6	105·8
3	157·5	157·8	158·1	158·4	158·7
4	210·0	210·4	210·8	211·2	211·6
5	262·5	263·0	263·5	264·0	264·5
6	315·0	315·6	316·2	316·8	317·4
7	367·5	368·2	368·9	369·6	370·3
8	420·0	420·8	421·6	422·4	423·2
9	472·5	473·4	474·3	475·2	476·1

	530	531	532	533	534
1	53·0	53·1	53·2	53·3	53·4
2	106·0	106·2	106·4	106·6	106·8
3	159·0	159·3	159·6	159·9	160·2
4	212·0	212·4	212·8	213·2	213·6
5	265·0	265·5	266·0	266·5	267·0
6	318·0	318·6	319·2	319·8	320·4
7	371·0	371·7	372·4	373·1	373·8
8	424·0	424·8	425·6	426·4	427·2
9	477·0	477·9	478·8	479·7	480·6

	535	536	537	538	539
1	53·5	53·6	53·7	53·8	53·9
2	107·0	107·2	107·4	107·6	107·8
3	160·5	160·8	161·1	161·4	161·7
4	214·0	214·4	214·8	215·2	215·6
5	267·5	268·0	268·5	269·0	269·5
6	321·0	321·6	322·2	322·8	323·4
7	374·5	375·2	375·9	376·6	377·3
8	428·0	428·8	429·6	430·4	431·2
9	481·5	482·4	483·3	484·2	485·1

	540	541	542	543	544
1	54·0	54·1	54·2	54·3	54·4
2	108·0	108·2	108·4	108·6	108·8
3	162·0	162·3	162·6	162·9	163·2
4	216·0	216·4	216·8	217·2	217·6
5	270·0	270·5	271·0	271·5	272·0
6	324·0	324·6	325·2	325·8	326·4
7	378·0	378·7	379·4	380·1	380·8
8	432·0	432·8	433·6	434·4	435·2
9	486·0	486·9	487·8	488·7	489·6

	545	546	547	548	549
1	54·5	54·6	54·7	54·8	54·9
2	109·0	109·2	109·4	109·6	109·8
3	163·5	163·8	164·1	164·4	164·7
4	218·0	218·4	218·8	219·2	219·6
5	272·5	273·0	273·5	274·0	274·5
6	327·0	327·6	328·2	328·8	329·4
7	381·5	382·2	382·9	383·6	384·3
8	436·0	436·8	437·6	438·4	439·2
9	490·5	491·4	492·3	493·2	494·1

TABLE XII—PROPORTIONAL PARTS

	550	551	552	553	554		555	556	557	558	559
1	55·0	55·1	55·2	55·3	55·4	1	55·5	55·6	55·7	55·8	55·9
2	110·0	110·2	110·4	110·6	110·8	2	111·0	111·2	111·4	111·6	111·8
3	165·0	165·3	165·6	165·9	166·2	3	166·5	166·8	167·1	167·4	167·7
4	220·0	220·4	220·8	221·2	221·6	4	222·0	222·4	222·8	223·2	223·6
5	275·0	275·5	276·0	276·5	277·0	5	277·5	278·0	278·5	279·0	279·5
6	330·0	330·6	331·2	331·8	332·4	6	333·0	333·6	334·2	334·8	335·4
7	385·0	385·7	386·4	387·1	387·8	7	388·5	389·2	389·9	390·6	391·3
8	440·0	440·8	441·6	442·4	443·2	8	444·0	444·8	445·6	446·4	447·2
9	495·0	495·9	496·8	497·7	498·6	9	499·5	500·4	501·3	502·2	503·1

	560	561	562	563	564		565	566	567	568	569
1	56·0	56·1	56·2	56·3	56·4	1	56·5	56·6	56·7	56·8	56·9
2	112·0	112·2	112·4	112·6	112·8	2	113·0	113·2	113·4	113·6	113·8
3	168·0	168·3	168·6	168·9	169·2	3	169·5	169·8	170·1	170·4	170·7
4	224·0	224·4	224·8	225·2	225·6	4	226·0	226·4	226·8	227·2	227·6
5	280·0	280·5	281·0	281·5	282·0	5	282·5	283·0	283·5	284·0	284·5
6	336·0	336·6	337·2	337·8	338·4	6	339·0	339·6	340·2	340·8	341·4
7	392·0	392·7	393·4	394·1	394·8	7	395·5	396·2	396·9	397·6	398·3
8	448·0	448·8	449·6	450·4	451·2	8	452·0	452·8	453·6	454·4	455·2
9	504·0	504·9	505·8	506·7	507·6	9	508·5	509·4	510·3	511·2	512·1

	570	571	572	573	574		575	576	577	578	579
1	57·0	57·1	57·2	57·3	57·4	1	57·5	57·6	57·7	57·8	57·9
2	114·0	114·2	114·4	114·6	114·8	2	115·0	115·2	115·4	115·6	115·8
3	171·0	171·3	171·6	171·9	172·2	3	172·5	172·8	173·1	173·4	173·7
4	228·0	228·4	228·8	229·2	229·6	4	230·0	230·4	230·8	231·2	231·6
5	285·0	285·5	286·0	286·5	287·0	5	287·5	288·0	288·5	289·0	289·5
6	342·0	342·6	343·2	343·8	344·4	6	345·0	345·6	346·2	346·8	347·4
7	399·0	399·7	400·4	401·1	401·8	7	402·5	403·2	403·9	404·6	405·3
8	456·0	456·8	457·6	458·4	459·2	8	460·0	460·8	461·6	462·4	463·2
9	513·0	513·9	514·8	515·7	516·6	9	517·5	518·4	519·3	520·2	521·1

	580	581	582	583	584		585	586	587	588	589
1	58·0	58·1	58·2	58·3	58·4	1	58·5	58·6	58·7	58·8	58·9
2	116·0	116·2	116·4	116·6	116·8	2	117·0	117·2	117·4	117·6	117·8
3	174·0	174·3	174·6	174·9	175·2	3	175·5	175·8	176·1	176·4	176·7
4	232·0	232·4	232·8	233·2	233·6	4	234·0	234·4	234·8	235·2	235·6
5	290·0	290·5	291·0	291·5	292·0	5	292·5	293·0	293·5	294·0	294·5
6	348·0	348·6	349·2	349·8	350·4	6	351·0	351·6	352·2	352·8	353·4
7	406·0	406·7	407·4	408·1	408·8	7	409·5	410·2	410·9	411·6	412·3
8	464·0	464·8	465·6	466·4	467·2	8	468·0	468·8	469·6	470·4	471·2
9	522·0	522·9	523·8	524·7	525·6	9	526·5	527·4	528·3	529·2	530·1

	590	591	592	593	594		595	596	597	598	599
1	59·0	59·1	59·2	59·3	59·4	1	59·5	59·6	59·7	59·8	59·9
2	118·0	118·2	118·4	118·6	118·8	2	119·0	119·2	119·4	119·6	119·8
3	177·0	177·3	177·6	177·9	178·2	3	178·5	178·8	179·1	179·4	179·7
4	236·0	236·4	236·8	237·2	237·6	4	238·0	238·4	238·8	239·2	239·6
5	295·0	295·5	296·0	296·5	297·0	5	297·5	298·0	298·5	299·0	299·5
6	354·0	354·6	355·2	355·8	356·4	6	357·0	357·6	358·2	358·8	359·4
7	413·0	413·7	414·4	415·1	415·8	7	416·5	417·2	417·9	418·6	419·3
8	472·0	472·8	473·6	474·4	475·2	8	476·0	476·8	477·6	478·4	479·2
9	531·0	531·9	532·8	533·7	534·6	9	535·5	536·4	537·3	538·2	539·1

	600	601	602	603	604
1	60·0	60·1	60·2	60·3	60·4
2	120·0	120·2	120·4	120·6	120·8
3	180·0	180·3	180·6	180·9	181·2
4	240·0	240·4	240·8	241·2	241·6
5	300·0	300·5	301·0	301·5	302·0
6	360·0	360·6	361·2	361·8	362·4
7	420·0	420·7	421·4	422·1	422·8
8	480·0	480·8	481·6	482·4	483·2
9	540·0	540·9	541·8	542·7	543·6

	605	606	607	608	609
1	60·5	60·6	60·7	60·8	60·9
2	121·0	121·2	121·4	121·6	121·8
3	181·5	181·8	182·1	182·4	182·7
4	242·0	242·4	242·8	243·2	243·6
5	302·5	303·0	303·5	304·0	304·5
6	363·0	363·6	364·2	364·8	365·4
7	423·5	424·2	424·9	425·6	426·3
8	484·0	484·8	485·6	486·4	487·2
9	544·5	545·4	546·3	547·2	548·1

	610	611	612	613	614
1	61·0	61·1	61·2	61·3	61·4
2	122·0	122·2	122·4	122·6	122·8
3	183·0	183·3	183·6	183·9	184·2
4	244·0	244·4	244·8	245·2	245·6
5	305·0	305·5	306·0	306·5	307·0
6	366·0	366·6	367·2	367·8	368·4
7	427·0	427·7	428·4	429·1	429·8
8	488·0	488·8	489·6	490·4	491·2
9	549·0	549·9	550·8	551·7	552·6

	615	616	617	618	619
1	61·5	61·6	61·7	61·8	61·9
2	123·0	123·2	123·4	123·6	123·8
3	184·5	184·8	185·1	185·4	185·7
4	246·0	246·4	246·8	247·2	247·6
5	307·5	308·0	308·5	309·0	309·5
6	369·0	369·6	370·2	370·8	371·4
7	430·5	431·2	431·9	432·6	433·3
8	492·0	492·8	493·6	494·4	495·2
9	553·5	554·4	555·3	556·2	557·1

	620	621	622	623	624
1	62·0	62·1	62·2	62·3	62·4
2	124·0	124·2	124·4	124·6	124·8
3	186·0	186·3	186·6	186·9	187·2
4	248·0	248·4	248·8	249·2	249·6
5	310·0	310·5	311·0	311·5	312·0
6	372·0	372·6	373·2	373·8	374·4
7	434·0	434·7	435·4	436·1	436·8
8	496·0	496·8	497·6	498·4	499·2
9	558·0	558·9	559·8	560·7	561·6

	625	626	627	628	629
1	62·5	62·6	62·7	62·8	62·9
2	125·0	125·2	125·4	125·6	125·8
3	187·5	187·8	188·1	188·4	188·7
4	250·0	250·4	250·8	251·2	251·6
5	312·5	313·0	313·5	314·0	314·5
6	375·0	375·6	376·2	376·8	377·4
7	437·5	438·2	438·9	439·6	440·3
8	500·0	500·8	501·6	502·4	503·2
9	562·5	563·4	564·3	565·2	566·1

	630	631	632	633	634
1	63·0	63·1	63·2	63·3	63·4
2	126·0	126·2	126·4	126·6	126·8
3	189·0	189·3	189·6	189·9	190·2
4	252·0	252·4	252·8	253·2	253·6
5	315·0	315·5	316·0	316·5	317·0
6	378·0	378·6	379·2	379·8	380·4
7	441·0	441·7	442·4	443·1	443·8
8	504·0	504·8	505·6	506·4	507·2
9	567·0	567·9	568·8	569·7	570·6

	635	636	637	638	639
1	63·5	63·6	63·7	63·8	63·9
2	127·0	127·2	127·4	127·6	127·8
3	190·5	190·8	191·1	191·4	191·7
4	254·0	254·4	254·8	255·2	255·6
5	317·5	318·0	318·5	319·0	319·5
6	381·0	381·6	382·2	382·8	383·4
7	444·5	445·2	445·9	446·6	447·3
8	508·0	508·8	509·6	510·4	511·2
9	571·5	572·4	573·3	574·2	575·1

	640	641	642	643	644
1	64·0	64·1	64·2	64·3	64·4
2	128·0	128·2	128·4	128·6	128·8
3	192·0	192·3	192·6	192·9	193·2
4	256·0	256·4	256·8	257·2	257·6
5	320·0	320·5	321·0	321·5	322·0
6	384·0	384·6	385·2	385·8	386·4
7	448·0	448·7	449·4	450·1	450·8
8	512·0	512·8	513·6	514·4	515·2
9	576·0	576·9	577·8	578·7	579·6

	645	646	647	648	649
1	64·5	64·6	64·7	64·8	64·9
2	129·0	129·2	129·4	129·6	129·8
3	193·5	193·8	194·1	194·4	194·7
4	258·0	258·4	258·8	259·2	259·6
5	322·5	323·0	323·5	324·0	324·5
6	387·0	387·6	388·2	388·8	389·4
7	451·5	452·2	452·9	453·6	454·3
8	516·0	516·8	517·6	518·4	519·2
9	580·5	581·4	582·3	583·2	584·1

	650	**651**	**652**	**653**	**654**	**655**	**656**	**657**	**658**	**659**
1	65·0	65·1	65·2	65·3	65·4	65·5	65·6	65·7	65·8	65·9
2	130·0	130·2	130·4	130·6	130·8	131·0	131·2	131·4	131·6	131·8
3	195·0	195·3	195·6	195·9	196·2	196·5	196·8	197·1	197·4	197·7
4	260·0	260·4	260·8	261·2	261·6	262·0	262·4	262·8	263·2	263·6
5	325·0	325·5	326·0	326·5	327·0	327·5	328·0	328·5	329·0	329·5
6	390·0	390·6	391·2	391·8	392·4	393·0	393·6	394·2	394·8	395·4
7	455·0	455·7	456·4	457·1	457·8	458·5	459·2	459·9	460·6	461·3
8	520·0	520·8	521·6	522·4	523·2	524·0	524·8	525·6	526·4	527·2
9	585·0	585·9	586·8	587·7	588·6	589·5	590·4	591·3	592·2	593·1

	660	**661**	**662**	**663**	**664**	**665**	**666**	**667**	**668**	**669**
1	66·0	66·1	66·2	66·3	66·4	66·5	66·6	66·7	66·8	66·9
2	132·0	132·2	132·4	132·6	132·8	133·0	133·2	133·4	133·6	133·8
3	198·0	198·3	198·6	198·9	199·2	199·5	199·8	200·1	200·4	200·7
4	264·0	264·4	264·8	265·2	265·6	266·0	266·4	266·8	267·2	267·6
5	330·0	330·5	331·0	331·5	332·0	332·5	333·0	333·5	334·0	334·5
6	396·0	396·6	397·2	397·8	398·4	399·0	399·6	400·2	400·8	401·4
7	462·0	462·7	463·4	464·1	464·8	465·5	466·2	466·9	467·6	468·3
8	528·0	528·8	529·6	530·4	531·2	532·0	532·8	533·6	534·4	535·2
9	594·0	594·9	595·8	596·7	597·6	598·5	599·4	600·3	601·2	602·1

	670	**671**	**672**	**673**	**674**	**675**	**676**	**677**	**678**	**679**
1	67·0	67·1	67·2	67·3	67·4	67·5	67·6	67·7	67·8	67·9
2	134·0	134·2	134·4	134·6	134·8	135·0	135·2	135·4	135·6	135·8
3	201·0	201·3	201·6	201·9	202·2	202·5	202·8	203·1	203·4	203·7
4	268·0	268·4	268·8	269·2	269·6	270·0	270·4	270·8	271·2	271·6
5	335·0	335·5	336·0	336·5	337·0	337·5	338·0	338·5	339·0	339·5
6	402·0	402·6	403·2	403·8	404·4	405·0	405·6	406·2	406·8	407·4
7	469·0	469·7	470·4	471·1	471·8	472·5	473·2	473·9	474·6	475·3
8	536·0	536·8	537·6	538·4	539·2	540·0	540·8	541·6	542·4	543·2
9	603·0	603·9	604·8	605·7	606·6	607·5	608·4	609·3	610·2	611·1

	680	**681**	**682**	**683**	**684**	**685**	**686**	**687**	**688**	**689**
1	68·0	68·1	68·2	68·3	68·4	68·5	68·6	68·7	68·8	68·9
2	136·0	136·2	136·4	136·6	136·8	137·0	137·2	137·4	137·6	137·8
3	204·0	204·3	204·6	204·9	205·2	205·5	205·8	206·1	206·4	206·7
4	272·0	272·4	272·8	273·2	273·6	274·0	274·4	274·8	275·2	275·6
5	340·0	340·5	341·0	341·5	342·0	342·5	343·0	343·5	344·0	344·5
6	408·0	408·6	409·2	409·8	410·4	411·0	411·6	412·2	412·8	413·4
7	476·0	476·7	477·4	478·1	478·8	479·5	480·2	480·9	481·6	482·3
8	544·0	544·8	545·6	546·4	547·2	548·0	548·8	549·6	550·4	551·2
9	612·0	612·9	613·8	614·7	615·6	616·5	617·4	618·3	619·2	620·1

	690	**691**	**692**	**693**	**694**	**695**	**696**	**697**	**698**	**699**
1	69·0	69·1	69·2	69·3	69·4	69·5	69·6	69·7	69·8	69·9
2	138·0	138·2	138·4	138·6	138·8	139·0	139·2	139·4	139·6	139·8
3	207·0	207·3	207·6	207·9	208·2	208·5	208·8	209·1	209·4	209·7
4	276·0	276·4	276·8	277·2	277·6	278·0	278·4	278·8	279·2	279·6
5	345·0	345·5	346·0	346·5	347·0	347·5	348·0	348·5	349·0	349·5
6	414·0	414·6	415·2	415·8	416·4	417·0	417·6	418·2	418·8	419·4
7	483·0	483·7	484·4	485·1	485·8	486·5	487·2	487·9	488·6	489·3
8	552·0	552·8	553·6	554·4	555·2	556·0	556·8	557·6	558·4	559·2
9	621·0	621·9	622·8	623·7	624·6	625·5	626·4	627·3	628·2	629·1

TABLE XII—PROPORTIONAL PARTS

	700	701	702	703	704	705	706	707	708	709
1	70·0	70·1	70·2	70·3	70·4	70·5	70·6	70·7	70·8	70·9
2	140·0	140·2	140·4	140·6	140·8	141·0	141·2	141·4	141·6	141·8
3	210·0	210·3	210·6	210·9	211·2	211·5	211·8	212·1	212·4	212·7
4	280·0	280·4	280·8	281·2	281·6	282·0	282·4	282·8	283·2	283·6
5	350·0	350·5	351·0	351·5	352·0	352·5	353·0	353·5	354·0	354·5
6	420·0	420·6	421·2	421·8	422·4	423·0	423·6	424·2	424·8	425·4
7	490·0	490·7	491·4	492·1	492·8	493·5	494·2	494·9	495·6	496·3
8	560·0	560·8	561·6	562·4	563·2	564·0	564·8	565·6	566·4	567·2
9	630·0	630·9	631·8	632·7	633·6	634·5	635·4	636·3	637·2	638·1

	710	711	712	713	714	715	716	717	718	719
1	71·0	71·1	71·2	71·3	71·4	71·5	71·6	71·7	71·8	71·9
2	142·0	142·2	142·4	142·6	142·8	143·0	143·2	143·4	143·6	143·8
3	213·0	213·3	213·6	213·9	214·2	214·5	214·8	215·1	215·4	215·7
4	284·0	284·4	284·8	285·2	285·6	286·0	286·4	286·8	287·2	287·6
5	355·0	355·5	356·0	356·5	357·0	357·5	358·0	358·5	359·0	359·5
6	426·0	426·6	427·2	427·8	428·4	429·0	429·6	430·2	430·8	431·4
7	497·0	497·7	498·4	499·1	499·8	500·5	501·2	501·9	502·6	503·3
8	568·0	568·8	569·6	570·4	571·2	572·0	572·8	573·6	574·4	575·2
9	639·0	639·9	640·8	641·7	642·6	643·5	644·4	645·3	646·2	647·1

	720	721	722	723	724	725	726	727	728	729
1	72·0	72·1	72·2	72·3	72·4	72·5	72·6	72·7	72·8	72·9
2	144·0	144·2	144·4	144·6	144·8	145·0	145·2	145·4	145·6	145·8
3	216·0	216·3	216·6	216·9	217·2	217·5	217·8	218·1	218·4	218·7
4	288·0	288·4	288·8	289·2	289·6	290·0	290·4	290·8	291·2	291·6
5	360·0	360·5	361·0	361·5	362·0	362·5	363·0	363·5	364·0	364·5
6	432·0	432·6	433·2	433·8	434·4	435·0	435·6	436·2	436·8	437·4
7	504·0	504·7	505·4	506·1	506·8	507·5	508·2	508·9	509·6	510·3
8	576·0	576·8	577·6	578·4	579·2	580·0	580·8	581·6	582·4	583·2
9	648·0	648·9	649·8	650·7	651·6	652·5	653·4	654·3	655·2	656·1

	730	731	732	733	734	735	736	737	738	739
1	73·0	73·1	73·2	73·3	73·4	73·5	73·6	73·7	73·8	73·9
2	146·0	146·2	146·4	146·6	146·8	147·0	147·2	147·4	147·6	147·8
3	219·0	219·3	219·6	219·9	220·2	220·5	220·8	221·1	221·4	221·7
4	292·0	292·4	292·8	293·2	293·6	294·0	294·4	294·8	295·2	295·6
5	365·0	365·5	366·0	366·5	367·0	367·5	368·0	368·5	369·0	369·5
6	438·0	438·6	439·2	439·8	440·4	441·0	441·6	442·2	442·8	443·4
7	511·0	511·7	512·4	513·1	513·8	514·5	515·2	515·9	516·6	517·3
8	584·0	584·8	585·6	586·4	587·2	588·0	588·8	589·6	590·4	591·2
9	657·0	657·9	658·8	659·7	660·6	661·5	662·4	663·3	664·2	665·1

	740	741	742	743	744	745	746	747	748	749
1	74·0	74·1	74·2	74·3	74·4	74·5	74·6	74·7	74·8	74·9
2	148·0	148·2	148·4	148·6	148·8	149·0	149·2	149·4	149·6	149·8
3	222·0	222·3	222·6	222·9	223·2	223·5	223·8	224·1	224·4	224·7
4	296·0	296·4	296·8	297·2	297·6	298·0	298·4	298·8	299·2	299·6
5	370·0	370·5	371·0	371·5	372·0	372·5	373·0	373·5	374·0	374·5
6	444·0	444·6	445·2	445·8	446·4	447·0	447·6	448·2	448·8	449·4
7	518·0	518·7	519·4	520·1	520·8	521·5	522·2	522·9	523·6	524·3
8	592·0	592·8	593·6	594·4	595·2	596·0	596·8	597·6	598·4	599·2
9	666·0	666·9	667·8	668·7	669·6	670·5	671·4	672·3	673·2	674·1

	750	751	752	753	754		755	756	757	758	759
1	75·0	75·1	75·2	75·3	75·4	1	75·5	75·6	75·7	75·8	75·9
2	150·0	150·2	150·4	150·6	150·8	2	151·0	151·2	151·4	151·6	151·8
3	225·0	225·3	225·6	225·9	226·2	3	226·5	226·8	227·1	227·4	227·7
4	300·0	300·4	300·8	301·2	301·6	4	302·0	302·4	302·8	303·2	303·6
5	375·0	375·5	376·0	376·5	377·0	5	377·5	378·0	378·5	379·0	379·5
6	450·0	450·6	451·2	451·8	452·4	6	453·0	453·6	454·2	454·8	455·4
7	525·0	525·7	526·4	527·1	527·8	7	528·5	529·2	529·9	530·6	531·3
8	600·0	600·8	601·6	602·4	603·2	8	604·0	604·8	605·6	606·4	607·2
9	675·0	675·9	676·8	677·7	678·6	9	679·5	680·4	681·3	682·2	683·1

	760	761	762	763	764		765	766	767	768	769
1	76·0	76·1	76·2	76·3	76·4	1	76·5	76·6	76·7	76·8	76·9
2	152·0	152·2	152·4	152·6	152·8	2	153·0	153·2	153·4	153·6	153·8
3	228·0	228·3	228·6	228·9	229·2	3	229·5	229·8	230·1	230·4	230·7
4	304·0	304·4	304·8	305·2	305·6	4	306·0	306·4	306·8	307·2	307·6
5	380·0	380·5	381·0	381·5	382·0	5	382·5	383·0	383·5	384·0	384·5
6	456·0	456·6	457·2	457·8	458·4	6	459·0	459·6	460·2	460·8	461·4
7	532·0	532·7	533·4	534·1	534·8	7	535·5	536·2	536·9	537·6	538·3
8	608·0	608·8	609·6	610·4	611·2	8	612·0	612·8	613·6	614·4	615·2
9	684·0	684·9	685·8	686·7	687·6	9	688·5	689·4	690·3	691·2	692·1

	770	771	772	773	774		775	776	777	778	779
1	77·0	77·1	77·2	77·3	77·4	1	77·5	77·6	77·7	77·8	77·9
2	154·0	154·2	154·4	154·6	154·8	2	155·0	155·2	155·4	155·6	155·8
3	231·0	231·3	231·6	231·9	232·2	3	232·5	232·8	233·1	233·4	233·7
4	308·0	308·4	308·8	309·2	309·6	4	310·0	310·4	310·8	311·2	311·6
5	385·0	385·5	386·0	386·5	387·0	5	387·5	388·0	388·5	389·0	389·5
6	462·0	462·6	463·2	463·8	464·4	6	465·0	465·6	466·2	466·8	467·4
7	539·0	539·7	540·4	541·1	541·8	7	542·5	543·2	543·9	544·6	545·3
8	616·0	616·8	617·6	618·4	619·2	8	620·0	620·8	621·6	622·4	623·2
9	693·0	693·9	694·8	695·7	696·6	9	697·5	698·4	699·3	700·2	701·1

	780	781	782	783	784		785	786	787	788	789
1	78·0	78·1	78·2	78·3	78·4	1	78·5	78·6	78·7	78·8	78·9
2	156·0	156·2	156·4	156·6	156·8	2	157·0	157·2	157·4	157·6	157·8
3	234·0	234·3	234·6	234·9	235·2	3	235·5	235·8	236·1	236·4	236·7
4	312·0	312·4	312·8	313·2	313·6	4	314·0	314·4	314·8	315·2	315·6
5	390·0	390·5	391·0	391·5	392·0	5	392·5	393·0	393·5	394·0	394·5
6	468·0	468·6	469·2	469·8	470·4	6	471·0	471·6	472·2	472·8	473·4
7	546·0	546·7	547·4	548·1	548·8	7	549·5	550·2	550·9	551·6	552·3
8	624·0	624·8	625·6	626·4	627·2	8	628·0	628·8	629·6	630·4	631·2
9	702·0	702·9	703·8	704·7	705·6	9	706·5	707·4	708·3	709·2	710·1

	790	791	792	793	794		795	796	797	798	799
1	79·0	79·1	79·2	79·3	79·4	1	79·5	79·6	79·7	79·8	79·9
2	158·0	158·2	158·4	158·6	158·8	2	159·0	159·2	159·4	159·6	159·8
3	237·0	237·3	237·6	237·9	238·2	3	238·5	238·8	239·1	239·4	239·7
4	316·0	316·4	316·8	317·2	317·6	4	318·0	318·4	318·8	319·2	319·6
5	395·0	395·5	396·0	396·5	397·0	5	397·5	398·0	398·5	399·0	399·5
6	474·0	474·6	475·2	475·8	476·4	6	477·0	477·6	478·2	478·8	479·4
7	553·0	553·7	554·4	555·1	555·8	7	556·5	557·2	557·9	558·6	559·3
8	632·0	632·8	633·6	634·4	635·2	8	636·0	636·8	637·6	638·4	639·2
9	711·0	711·9	712·8	713·7	714·6	9	715·5	716·4	717·3	718·2	719·1

TABLE XII—PROPORTIONAL PARTS

	800	801	802	803	804		805	806	807	808	809
1	80·0	80·1	80·2	80·3	80·4	1	80·5	80·6	80·7	80·8	80·9
2	160·0	160·2	160·4	160·6	160·8	2	161·0	161·2	161·4	161·6	161·8
3	240·0	240·3	240·6	240·9	241·2	3	241·5	241·8	242·1	242·4	242·7
4	320·0	320·4	320·8	321·2	321·6	4	322·0	322·4	322·8	323·2	323·6
5	400·0	400·5	401·0	401·5	402·0	5	402·5	403·0	403·5	404·0	404·5
6	480·0	480·6	481·2	481·8	482·4	6	483·0	483·6	484·2	484·8	485·4
7	560·0	560·7	561·4	562·1	562·8	7	563·5	564·2	564·9	565·6	566·3
8	640·0	640·8	641·6	642·4	643·2	8	644·0	644·8	645·6	646·4	647·2
9	720·0	720·9	721·8	722·7	723·6	9	724·5	725·4	726·3	727·2	728·1

	810	811	812	813	814		815	816	817	818	819
1	81·0	81·1	81·2	81·3	81·4	1	81·5	81·6	81·7	81·8	81·9
2	162·0	162·2	162·4	162·6	162·8	2	163·0	163·2	163·4	163·6	163·8
3	243·0	243·3	243·6	243·9	244·2	3	244·5	244·8	245·1	245·4	245·7
4	324·0	324·4	324·8	325·2	325·6	4	326·0	326·4	326·8	327·2	327·6
5	405·0	405·5	406·0	406·5	407·0	5	407·5	408·0	408·5	409·0	409·5
6	486·0	486·6	487·2	487·8	488·4	6	489·0	489·6	490·2	490·8	491·4
7	567·0	567·7	568·4	569·1	569·8	7	570·5	571·2	571·9	572·6	573·3
8	648·0	648·8	649·6	650·4	651·2	8	652·0	652·8	653·6	654·4	655·2
9	729·0	729·9	730·8	731·7	732·6	9	733·5	734·4	735·3	736·2	737·1

	820	821	822	823	824		825	826	827	828	829
1	82·0	82·1	82·2	82·3	82·4	1	82·5	82·6	82·7	82·8	82·9
2	164·0	164·2	164·4	164·6	164·8	2	165·0	165·2	165·4	165·6	165·8
3	246·0	246·3	246·6	246·9	247·2	3	247·5	247·8	248·1	248·4	248·7
4	328·0	328·4	328·8	329·2	329·6	4	330·0	330·4	330·8	331·2	331·6
5	410·0	410·5	411·0	411·5	412·0	5	412·5	413·0	413·5	414·0	414·5
6	492·0	492·6	493·2	493·8	494·4	6	495·0	495·6	496·2	496·8	497·4
7	574·0	574·7	575·4	576·1	576·8	7	577·5	578·2	578·9	579·6	580·3
8	656·0	656·8	657·6	658·4	659·2	8	660·0	660·8	661·6	662·4	663·2
9	738·0	738·9	739·8	740·7	741·6	9	742·5	743·4	744·3	745·2	746·1

	830	831	832	833	834		835	836	837	838	839
1	83·0	83·1	83·2	83·3	83·4	1	83·5	83·6	83·7	83·8	83·9
2	166·0	166·2	166·4	166·6	166·8	2	167·0	167·2	167·4	167·6	167·8
3	249·0	249·3	249·6	249·9	250·2	3	250·5	250·8	251·1	251·4	251·7
4	332·0	332·4	332·8	333·2	333·6	4	334·0	334·4	334·8	335·2	335·6
5	415·0	415·5	416·0	416·5	417·0	5	417·5	418·0	418·5	419·0	419·5
6	498·0	498·6	499·2	499·8	500·4	6	501·0	501·6	502·2	502·8	503·4
7	581·0	581·7	582·4	583·1	583·8	7	584·5	585·2	585·9	586·6	587·3
8	664·0	664·8	665·6	666·4	667·2	8	668·0	668·8	669·6	670·4	671·2
9	747·0	747·9	748·8	749·7	750·6	9	751·5	752·4	753·3	754·2	755·1

	840	841	842	843	844		845	846	847	848	849
1	84·0	84·1	84·2	84·3	84·4	1	84·5	84·6	84·7	84·8	84·9
2	168·0	168·2	168·4	168·6	168·8	2	169·0	169·2	169·4	169·6	169·8
3	252·0	252·3	252·6	252·9	253·2	3	253·5	253·8	254·1	254·4	254·7
4	336·0	336·4	336·8	337·2	337·6	4	338·0	338·4	338·8	339·2	339·6
5	420·0	420·5	421·0	421·5	422·0	5	422·5	423·0	423·5	424·0	424·5
6	504·0	504·6	505·2	505·8	506·4	6	507·0	507·6	508·2	508·8	509·4
7	588·0	588·7	589·4	590·1	590·8	7	591·5	592·2	592·9	593·6	594·3
8	672·0	672·8	673·6	674·4	675·2	8	676·0	676·8	677·6	678·4	679·2
9	756·0	756·9	757·8	758·7	759·6	9	760·5	761·4	762·3	763·2	764·1

	850	851	852	853	854	855	856	857	858	859
1	85·0	85·1	85·2	85·3	85·4	85·5	85·6	85·7	85·8	85·9
2	170·0	170·2	170·4	170·6	170·8	171·0	171·2	171·4	171·6	171·8
3	255·0	255·3	255·6	255·9	256·2	256·5	256·8	257·1	257·4	257·7
4	340·0	340·4	340·8	341·2	341·6	342·0	342·4	342·8	343·2	343·6
5	425·0	425·5	426·0	426·5	427·0	427·5	428·0	428·5	429·0	429·5
6	510·0	510·6	511·2	511·8	512·4	513·0	513·6	514·2	514·8	515·4
7	595·0	595·7	596·4	597·1	597·8	598·5	599·2	599·9	600·6	601·3
8	680·0	680·8	681·6	682·4	683·2	684·0	684·8	685·6	686·4	687·2
9	765·0	765·9	766·8	767·7	768·6	769·5	770·4	771·3	772·2	773·1

	860	861	862	863	864	865	866	867	868	869
1	86·0	86·1	86·2	86·3	86·4	86·5	86·6	86·7	86·8	86·9
2	172·0	172·2	172·4	172·6	172·8	173·0	173·2	173·4	173·6	173·8
3	258·0	258·3	258·6	258·9	259·2	259·5	259·8	260·1	260·4	260·7
4	344·0	344·4	344·8	345·2	345·6	346·0	346·4	346·8	347·2	347·6
5	430·0	430·5	431·0	431·5	432·0	432·5	433·0	433·5	434·0	434·5
6	516·0	516·6	517·2	517·8	518·4	519·0	519·6	520·2	520·8	521·4
7	602·0	602·7	603·4	604·1	604·8	605·5	606·2	606·9	607·6	608·3
8	688·0	688·8	689·6	690·4	691·2	692·0	692·8	693·6	694·4	695·2
9	774·0	774·9	775·8	776·7	777·6	778·5	779·4	780·3	781·2	782·1

	870	871	872	873	874	875	876	877	878	879
1	87·0	87·1	87·2	87·3	87·4	87·5	87·6	87·7	87·8	87·9
2	174·0	174·2	174·4	174·6	174·8	175·0	175·2	175·4	175·6	175·8
3	261·0	261·3	261·6	261·9	262·2	262·5	262·8	263·1	263·4	263·7
4	348·0	348·4	348·8	349·2	349·6	350·0	350·4	350·8	351·2	351·6
5	435·0	435·5	436·0	436·5	437·0	437·5	438·0	438·5	439·0	439·5
6	522·0	522·6	523·2	523·8	524·4	525·0	525·6	526·2	526·8	527·4
7	609·0	609·7	610·4	611·1	611·8	612·5	613·2	613·9	614·6	615·3
8	696·0	696·8	697·6	698·4	699·2	700·0	700·8	701·6	702·4	703·2
9	783·0	783·9	784·8	785·7	786·6	787·5	788·4	789·3	790·2	791·1

	880	881	882	883	884	885	886	887	888	889
1	88·0	88·1	88·2	88·3	88·4	88·5	88·6	88·7	88·8	88·9
2	176·0	176·2	176·4	176·6	176·8	177·0	177·2	177·4	177·6	177·8
3	264·0	264·3	264·6	264·9	265·2	265·5	265·8	266·1	266·4	266·7
4	352·0	352·4	352·8	353·2	353·6	354·0	354·4	354·8	355·2	355·6
5	440·0	440·5	441·0	441·5	442·0	442·5	443·0	443·5	444·0	444·5
6	528·0	528·6	529·2	529·8	530·4	531·0	531·6	532·2	532·8	533·4
7	616·0	616·7	617·4	618·1	618·8	619·5	620·2	620·9	621·6	622·3
8	704·0	704·8	705·6	706·4	707·2	708·0	708·8	709·6	710·4	711·2
9	792·0	792·9	793·8	794·7	795·6	796·5	797·4	798·3	799·2	800·1

	890	891	892	893	894	895	896	897	898	899
1	89·0	89·1	89·2	89·3	89·4	89·5	89·6	89·7	89·8	89·9
2	178·0	178·2	178·4	178·6	178·8	179·0	179·2	179·4	179·6	179·8
3	267·0	267·3	267·6	267·9	268·2	268·5	268·8	269·1	269·4	269·7
4	356·0	356·4	356·8	357·2	357·6	358·0	358·4	358·8	359·2	359·6
5	445·0	445·5	446·0	446·5	447·0	447·5	448·0	448·5	449·0	449·5
6	534·0	534·6	535·2	535·8	536·4	537·0	537·6	538·2	538·8	539·4
7	623·0	623·7	624·4	625·1	625·8	626·5	627·2	627·9	628·6	629·3
8	712·0	712·8	713·6	714·4	715·2	716·0	716·8	717·6	718·4	719·2
9	801·0	801·9	802·8	803·7	804·6	805·5	806·4	807·3	808·2	809·1

	900	901	902	903	904	905	906	907	908	909
1	90·0	90·1	90·2	90·3	90·4	90·5	90·6	90·7	90·8	90·9
2	180·0	180·2	180·4	180·6	180·8	181·0	181·2	181·4	181·6	181·8
3	270·0	270·3	270·6	270·9	271·2	271·5	271·8	272·1	272·4	272·7
4	360·0	360·4	360·8	361·2	361·6	362·0	362·4	362·8	363·2	363·6
5	450·0	450·5	451·0	451·5	452·0	452·5	453·0	453·5	454·0	454·5
6	540·0	540·6	541·2	541·8	542·4	543·0	543·6	544·2	544·8	545·4
7	630·0	630·7	631·4	632·1	632·8	633·5	634·2	634·9	635·6	636·3
8	720·0	720·8	721·6	722·4	723·2	724·0	724·8	725·6	726·4	727·2
9	810·0	810·9	811·8	812·7	813·6	814·5	815·4	816·3	817·2	818·1

	910	911	912	913	914	915	916	917	918	919
1	91·0	91·1	91·2	91·3	91·4	91·5	91·6	91·7	91·8	91·9
2	182·0	182·2	182·4	182·6	182·8	183·0	183·2	183·4	183·6	183·8
3	273·0	273·3	273·6	273·9	274·2	274·5	274·8	275·1	275·4	275·7
4	364·0	364·4	364·8	365·2	365·6	366·0	366·4	366·8	367·2	367·6
5	455·0	455·5	456·0	456·5	457·0	457·5	458·0	458·5	459·0	459·5
6	546·0	546·6	547·2	547·8	548·4	549·0	549·6	550·2	550·8	551·4
7	637·0	637·7	638·4	639·1	639·8	640·5	641·2	641·9	642·6	643·3
8	728·0	728·8	729·6	730·4	731·2	732·0	732·8	733·6	734·4	735·2
9	819·0	819·9	820·8	821·7	822·6	823·5	824·4	825·3	826·2	827·1

	920	921	922	923	924	925	926	927	928	929
1	92·0	92·1	92·2	92·3	92·4	92·5	92·6	92·7	92·8	92·9
2	184·0	184·2	184·4	184·6	184·8	185·0	185·2	185·4	185·6	185·8
3	276·0	276·3	276·6	276·9	277·2	277·5	277·8	278·1	278·4	278·7
4	368·0	368·4	368·8	369·2	369·6	370·0	370·4	370·8	371·2	371·6
5	460·0	460·5	461·0	461·5	462·0	462·5	463·0	463·5	464·0	464·5
6	552·0	552·6	553·2	553·8	554·4	555·0	555·6	556·2	556·8	557·4
7	644·0	644·7	645·4	646·1	646·8	647·5	648·2	648·9	649·6	650·3
8	736·0	736·8	737·6	738·4	739·2	740·0	740·8	741·6	742·4	743·2
9	828·0	828·9	829·8	830·7	831·6	832·5	833·4	834·3	835·2	836·1

	930	931	932	933	934	935	936	937	938	939
1	93·0	93·1	93·2	93·3	93·4	93·5	93·6	93·7	93·8	93·9
2	186·0	186·2	186·4	186·6	186·8	187·0	187·2	187·4	187·6	187·8
3	279·0	279·3	279·6	279·9	280·2	280·5	280·8	281·1	281·4	281·7
4	372·0	372·4	372·8	373·2	373·6	374·0	374·4	374·8	375·2	375·6
5	465·0	465·5	466·0	466·5	467·0	467·5	468·0	468·5	469·0	469·5
6	558·0	558·6	559·2	559·8	560·4	561·0	561·6	562·2	562·8	563·4
7	651·0	651·7	652·4	653·1	653·8	654·5	655·2	655·9	656·6	657·3
8	744·0	744·8	745·6	746·4	747·2	748·0	748·8	749·6	750·4	751·2
9	837·0	837·9	838·8	839·7	840·6	841·5	842·4	843·3	844·2	845·1

	940	941	942	943	944	945	946	947	948	949
1	94·0	94·1	94·2	94·3	94·4	94·5	94·6	94·7	94·8	94·9
2	188·0	188·2	188·4	188·6	188·8	189·0	189·2	189·4	189·6	189·8
3	282·0	282·3	282·6	282·9	283·2	283·5	283·8	284·1	284·4	284·7
4	376·0	376·4	376·8	377·2	377·6	378·0	378·4	378·8	379·2	379·6
5	470·0	470·5	471·0	471·5	472·0	472·5	473·0	473·5	474·0	474·5
6	564·0	564·6	565·2	565·8	566·4	567·0	567·6	568·2	568·8	569·4
7	658·0	658·7	659·4	660·1	660·8	661·5	662·2	662·9	663·6	664·3
8	752·0	752·8	753·6	754·4	755·2	756·0	756·8	757·6	758·4	759·2
9	846·0	846·9	847·8	848·7	849·6	850·5	851·4	852·3	853·2	854·1

	950	951	952	953	954	955	956	957	958	959
1	95·0	95·1	95·2	95·3	95·4	95·5	95·6	95·7	95·8	95·9
2	190·0	190·2	190·4	190·6	190·8	191·0	191·2	191·4	191·6	191·8
3	285·0	285·3	285·6	285·9	286·2	286·5	286·8	287·1	287·4	287·7
4	380·0	380·4	380·8	381·2	381·6	382·0	382·4	382·8	383·2	383·6
5	475·0	475·5	476·0	476·5	477·0	477·5	478·0	478·5	479·0	479·5
6	570·0	570·6	571·2	571·8	572·4	573·0	573·6	574·2	574·8	575·4
7	665·0	665·7	666·4	667·1	667·8	668·5	669·2	669·9	670·6	671·3
8	760·0	760·8	761·6	762·4	763·2	764·0	764·8	765·6	766·4	767·2
9	855·0	855·9	856·8	857·7	858·6	859·5	860·4	861·3	862·2	863·1

	960	961	962	963	964	965	966	967	968	969
1	96·0	96·1	96·2	96·3	96·4	96·5	96·6	96·7	96·8	96·9
2	192·0	192·2	192·4	192·6	192·8	193·0	193·2	193·4	193·6	193·8
3	288·0	288·3	288·6	288·9	289·2	289·5	289·8	290·1	290·4	290·7
4	384·0	384·4	384·8	385·2	385·6	386·0	386·4	386·8	387·2	387·6
5	480·0	480·5	481·0	481·5	482·0	482·5	483·0	483·5	484·0	484·5
6	576·0	576·6	577·2	577·8	578·4	579·0	579·6	580·2	580·8	581·4
7	672·0	672·7	673·4	674·1	674·8	675·5	676·2	676·9	677·6	678·3
8	768·0	768·8	769·6	770·4	771·2	772·0	772·8	773·6	774·4	775·2
9	864·0	864·9	865·8	866·7	867·6	868·5	869·4	870·3	871·2	872·1

	970	971	972	973	974	975	976	977	978	979
1	97·0	97·1	97·2	97·3	97·4	97·5	97·6	97·7	97·8	97·9
2	194·0	194·2	194·4	194·6	194·8	195·0	195·2	195·4	195·6	195·8
3	291·0	291·3	291·6	291·9	292·2	292·5	292·8	293·1	293·4	293·7
4	388·0	388·4	388·8	389·2	389·6	390·0	390·4	390·8	391·2	391·6
5	485·0	485·5	486·0	486·5	487·0	487·5	488·0	488·5	489·0	489·5
6	582·0	582·6	583·2	583·8	584·4	585·0	585·6	586·2	586·8	587·4
7	679·0	679·7	680·4	681·1	681·8	682·5	683·2	683·9	684·6	685·3
8	776·0	776·8	777·6	778·4	779·2	780·0	780·8	781·6	782·4	783·2
9	873·0	873·9	874·8	875·7	876·6	877·5	878·4	879·3	880·2	881·1

	980	981	982	983	984	985	986	987	988	989
1	98·0	98·1	98·2	98·3	98·4	98·5	98·6	98·7	98·8	98·9
2	196·0	196·2	196·4	196·6	196·8	197·0	197·2	197·4	197·6	197·8
3	294·0	294·3	294·6	294·9	295·2	295·5	295·8	296·1	296·4	296·7
4	392·0	392·4	392·8	393·2	393·6	394·0	394·4	394·8	395·2	395·6
5	490·0	490·5	491·0	491·5	492·0	492·5	493·0	493·5	494·0	494·5
6	588·0	588·6	589·2	589·8	590·4	591·0	591·6	592·2	592·8	593·4
7	686·0	686·7	687·4	688·1	688·8	689·5	690·2	690·9	691·6	692·3
8	784·0	784·8	785·6	786·4	787·2	788·0	788·8	789·6	790·4	791·2
9	882·0	882·9	883·8	884·7	885·6	886·5	887·4	888·3	889·2	890·1

	990	991	992	993	994	995	996	997	998	999
1	99·0	99·1	99·2	99·3	99·4	99·5	99·6	99·7	99·8	99·9
2	198·0	198·2	198·4	198·6	198·8	199·0	199·2	199·4	199·6	199·8
3	297·0	297·3	297·6	297·9	298·2	298·5	298·8	299·1	299·4	299·7
4	396·0	396·4	396·8	397·2	397·6	398·0	398·4	398·8	399·2	399·6
5	495·0	495·5	496·0	496·5	497·0	497·5	498·0	498·5	499·0	499·5
6	594·0	594·6	595·2	595·8	596·4	597·0	597·6	598·2	598·8	599·4
7	693·0	693·7	694·4	695·1	695·8	696·5	697·2	697·9	698·6	699·3
8	792·0	792·8	793·6	794·4	795·2	796·0	796·8	797·6	798·4	799·2
9	891·0	891·9	892·8	893·7	894·6	895·5	896·4	897·3	898·2	899·1

CHAMBERS'S SIX-FIGURE MATHEMATICAL TABLES

Contents of Volume I

This 600-page volume is designed for those who do not have access to a calculating machine but want the best facilities that can be provided with logarithms. The principal contents are:

Table I—Logarithms of Numbers, 10,000(1)100,000. This table is ten times as extensive as the one in this volume. The differences range from 44 down to 4, whereas here they range from 434 to 43. As there is a 5-figure argument, interpolation is required for one decimal only (or perhaps two if the number begins with 1) and the proportional part never exceeds two figures. Critical tables have been used where appropriate for the proportional parts. In the inverse use of this table interpolation for one decimal is required, whereas an antilogarithm table with only 10,000 values demands interpolation for the last two decimals. As a working table for regular logarithmic use, both directly and inversely, nothing better could be offered to the computer.

Tables II and III—8-Figure Logarithms, 1(1)1000 and 1·00000(0·00001)1·10000. These enable 8-figure logarithms (e.g. for interest problems) to be found by a simple factorising method, which is explained and illustrated. Multiples of the modulus and of its reciprocal for $x = 0(0{\cdot}001)1$ for converting natural to common logarithms, and vice versa.

Table IV—Antilogarithms, 0(0·0001)1.

Table V—Logarithms of Trigonometrical Functions of Angles in Degrees, Minutes and Seconds. Log sin and log tan, $0°(1'')1°\ 20'$; logarithms of the four principal functions, $0°(10'')10°(1')45°$. The table at interval $1''$ does not appear in this volume, but the others do.

Table VI—Logarithms of Trigonometrical Functions of Angles in Degrees and Decimals and in Radians. Log sin and log tan for $0°(0°{\cdot}001)5°$ and $0^{r}(0^{r}{\cdot}000\ 0175)0^{r}{\cdot}087$; logarithms of the four principal functions $0°(0°{\cdot}01)45°$ and at interval $0^{r}{\cdot}000\ 175$ throughout the quadrant. The argument for decimals of a degree is exact, so interpolation is done by ordinary proportional parts; special proportional parts are provided for the radian argument.

Table VII—*S* and *T* Functions. Given to 3° for seconds of arc, and to $1\frac{1}{2}°$ for minutes of arc, degrees and decimals, seconds of time, minutes of time, radians and hyperbolic functions.

Table VIII—Logarithms of Hyperbolic Functions. Log sinh, log cosh and log tanh for $x = 0(0{\cdot}001)3(0{\cdot}01)5(\text{various})\infty$; also *Sh* and *Th* for $x = 0(0{\cdot}001)0{\cdot}4$. These logarithms are now more readily accessible than in any previous publication.

Table IX—Logarithms of the Gamma or Factorial Function, 1(0·001)2.

Table X—Conversion Tables. Tables for the conversion of degrees, minutes, seconds and decimals of any one of these units to radians. Also for the conversion of radians to degrees and decimals, degrees, minutes and decimals, or degrees, minutes and seconds. Provision is also made for expressing large angles in revolutions and degrees. Also tables for converting time to arc and vice versa.

Table XI—Proportional Parts. Decimal proportional parts of all 3-figure numbers, as in this volume.

A 3-page bibliography of multiplication tables (for proportional parts) and of logarithmic tables to more than six decimals.

Physical and mathematical constants, as in this volume.

Contents of Volume II

This 612-page volume of tables is mainly—but not exclusively—for those who use calculating machines. The principal contents are:

Table I—Trigonometrical Functions of Angles in Degrees, Minutes and Seconds. Six functions, $0°(1')45°$; cotangents and cosecants, $0°(1'')0°\ 24'(10'')4°$; τ and σ, 0°-2°, for seconds and for minutes. Mostly reproduced here.

Table II—Trigonometrical Functions of Angles in Degrees and Decimals. Six functions, $0°(0°{\cdot}01)45°$; cotangents and cosecants, $0°(0°{\cdot}001)1°$; τ and σ, 0°-2°·5.

TABLE III—CIRCULAR OR TRIGONOMETRICAL FUNCTIONS WITH THE ARGUMENT IN RADIANS. Six functions, $0^r(0^r\cdot001)1^r\cdot6$; cotangents and cosecants, $0^r(0^r\cdot0001)0^r\cdot1$; sines and cosines, $0^r(0^r\cdot1)50^r(1^r)100^r(100^r)5000^r$; τ and σ; tables for reduction to the first quadrant; formulæ. Mostly reproduced here.

TABLE IV—EXPONENTIAL AND HYPERBOLIC FUNCTIONS. e^x, sinh and cosh, 0(0·001)6(various)100; e^{-x}, tanh and coth, 0(0·001)3(0·01)6(various)∞; coth and cosech, 0(0·0001)0·1; τh and σh; formulæ. Mostly reproduced here.

TABLE V—NATURAL LOGARITHMS, 1(0·001)10. Reproduced here.

TABLE VI—INVERSE CIRCULAR AND HYPERBOLIC FUNCTIONS. 66 pages as compared with the skeleton table of 12 pages given in this volume.

TABLE VII—THE GUDERMANNIAN AND ITS INVERSE. Gd x for x = 0(0·001)4·5(0·01)10(critical)∞; the inverse gudermannian or $\text{gd}^{-1}\, x$ for x = 0(0·001)1·4(0·0001)1·57(0·00001)1·5705(0·000001)1·570 796; formulæ.

TABLE VIII—POWERS, ROOTS, RECIPROCALS, FACTORS AND FACTORIALS, 1-1000, as in this volume. Squares, cubes and prime factors, 1000-3400.

TABLE IX—PRIME NUMBERS, 1-12919.

TABLE X—CONVERSION FROM RECTANGULAR TO POLAR CO-ORDINATES. If tan θ is the (positive) ratio of the smaller rectangular co-ordinate to the larger, this table gives sec θ, as well as θ in degrees and decimals and in radians. The opportunity has been taken of including a column, with argument 0(0·001)1 and expansion near 1, for the direct conversion of sine to cosine and vice versa.

TABLE XI—THE GAMMA OR FACTORIAL FUNCTION, 1(0·001)2.

TABLE XII—THE PROBABILITY INTEGRAL IN DIFFERENT FORMS. $\text{Erf}\, x = H(x) = \frac{2}{\sqrt{\pi}}\int_0^x e^{-t^2}\, dt$ for x = 0(0·01)4, with an increasing number of decimals, and an auxiliary table for extension to $x = 8$. The probability ordinate $z = \frac{1}{\sqrt{2\pi}} e^{-\frac{1}{2}x^2}$ for x = 0(0·01)5 and integral $\alpha = \int_{-x}^{x} z(t)\, dt = \sqrt{\frac{2}{\pi}}\int_0^x e^{-\frac{1}{2}t^2}\, dt$ for x = 0(0·01)6, with an increasing number of decimals in α; also an auxiliary table to give three significant figures up to x = 12·5. The probability abscissa x and ordinate z, with argument $\frac{1}{2}(1+\alpha)$, 0·5(0·001)0·99(0·0001)1, with extensions and auxiliary tables.

TABLE XIII—INTERPOLATION FORMULÆ AND COEFFICIENTS. More extended interpolation tables than are given here, including a critical table of B'' to four decimals, as well as 4-point and 6-point Lagrangian coefficients for n = 0(0·01)1, with means for avoiding interpolation when n has more than two decimals.

TABLE XIV—A section on numerical differentiation and integration and the numerical solution of differential equations, of which an abbreviated version is reproduced here.

TABLE XV—PROPORTIONAL PARTS FOR SECONDS. The proportional parts for 6″(1″)10″(10″)50″ are given for differences up to 1015.

TABLE XVI—CONVERSION TABLES. As in Volume I.

A 6-page bibliography of multiplication tables and of natural tables with more than six decimals.

Physical and mathematical constants, as in this volume.

CIRCULAR AND HYPERBOLIC FUNCTIONS

See also pages 170 and 204

$$\cos^2 x + \sin^2 x = 1 \qquad \cosh^2 x - \sinh^2 x = 1$$
$$\sec^2 x - \tan^2 x = 1 \qquad \operatorname{sech}^2 x + \tanh^2 x = 1$$
$$\operatorname{cosec}^2 x - \cot^2 x = 1 \qquad \operatorname{cosech}^2 x - \coth^2 x = -1$$

$$\sin x = \frac{2\tan\frac{1}{2}x}{1+\tan^2\frac{1}{2}x} \qquad \sinh x = \frac{2\tanh\frac{1}{2}x}{1-\tanh^2\frac{1}{2}x}$$
$$\cos x = \frac{1-\tan^2\frac{1}{2}x}{1+\tan^2\frac{1}{2}x} \qquad \cosh x = \frac{1+\tanh^2\frac{1}{2}x}{1-\tanh^2\frac{1}{2}x}$$
$$\tan x = \frac{2\tan\frac{1}{2}x}{1-\tan^2\frac{1}{2}x} \qquad \tanh x = \frac{2\tanh\frac{1}{2}x}{1+\tanh^2\frac{1}{2}x}$$

$$2\cot 2x = \cot x - \tan x$$
$$2\operatorname{cosec} 2x = \cot x + \tan x$$
$$\tan\tfrac{1}{2}x = \operatorname{cosec} x - \cot x$$
$$\cot\tfrac{1}{2}x = \operatorname{cosec} x + \cot x$$

$$2\coth 2x = \coth x + \tanh x$$
$$2\operatorname{cosech} 2x = \coth x - \tanh x$$
$$\tanh\tfrac{1}{2}x = \coth x - \operatorname{cosech} x$$
$$\coth\tfrac{1}{2}x = \coth x + \operatorname{cosech} x$$

$$2\sin^2\tfrac{1}{2}x = 1 - \cos x \qquad 2\sinh^2\tfrac{1}{2}x = \cosh x - 1$$
$$2\cos^2\tfrac{1}{2}x = 1 + \cos x \qquad 2\cosh^2\tfrac{1}{2}x = \cosh x + 1$$

$$\tan\tfrac{1}{2}x = \frac{1-\cos x}{\sin x} = \frac{\sin x}{1+\cos x}$$
$$\tanh\tfrac{1}{2}x = \frac{\cosh x - 1}{\sinh x} = \frac{\sinh x}{\cosh x + 1}$$

$$\sin 2x = 2\sin x\cos x \qquad \sinh 2x = 2\sinh x\cosh x$$
$$\cos 2x = \cos^2 x - \sin^2 x = 2\cos^2 x - 1 = 1 - 2\sin^2 x$$
$$\cosh 2x = \cosh^2 x + \sinh^2 x = 2\cosh^2 x - 1 = 1 + 2\sinh^2 x$$

$$\tan 2x = \frac{2\tan x}{1-\tan^2 x} \qquad \tanh 2x = \frac{2\tanh x}{1+\tanh^2 x}$$

$$\sin 3x = 3\sin x - 4\sin^3 x$$
$$\cos 3x = 4\cos^3 x - 3\cos x$$
$$\tan 3x = \frac{3\tan x - \tan^3 x}{1 - 3\tan^2 x}$$

$$\sinh 3x = 3\sinh x + 4\sinh^3 x$$
$$\cosh 3x = 4\cosh^3 x - 3\cosh x$$
$$\tanh 3x = \frac{3\tanh x + \tanh^3 x}{1 + 3\tanh^2 x}$$

$$\sin(A+B) = \sin A\cos B + \cos A\sin B$$
$$\sin(A-B) = \sin A\cos B - \cos A\sin B$$
$$\cos(A+B) = \cos A\cos B - \sin A\sin B$$
$$\cos(A-B) = \cos A\cos B + \sin A\sin B$$

$$\tan(A+B) = \frac{\tan A + \tan B}{1 - \tan A\tan B}$$
$$\tan(A-B) = \frac{\tan A - \tan B}{1 + \tan A\tan B}$$

$$2\sin A\cos B = \sin(A+B) + \sin(A-B)$$
$$2\cos A\sin B = \sin(A+B) - \sin(A-B)$$
$$2\cos A\cos B = \cos(A+B) + \cos(A-B)$$
$$2\sin A\sin B = -\cos(A+B) + \cos(A-B)$$
$$\sin A + \sin B = 2\sin\tfrac{1}{2}(A+B)\cos\tfrac{1}{2}(A-B)$$
$$\sin A - \sin B = 2\cos\tfrac{1}{2}(A+B)\sin\tfrac{1}{2}(A-B)$$
$$\cos A + \cos B = 2\cos\tfrac{1}{2}(A+B)\cos\tfrac{1}{2}(A-B)$$
$$\cos A - \cos B = -2\sin\tfrac{1}{2}(A+B)\sin\tfrac{1}{2}(A-B)$$

$$\sinh(A+B) = \sinh A\cosh B + \cosh A\sinh B$$
$$\sinh(A-B) = \sinh A\cosh B - \cosh A\sinh B$$
$$\cosh(A+B) = \cosh A\cosh B + \sinh A\sinh B$$
$$\cosh(A-B) = \cosh A\cosh B - \sinh A\sinh B$$

$$\tanh(A+B) = \frac{\tanh A + \tanh B}{1 + \tanh A\tanh B}$$
$$\tanh(A-B) = \frac{\tanh A - \tanh B}{1 - \tanh A\tanh B}$$

$$2\sinh A\cosh B = \sinh(A+B) + \sinh(A-B)$$
$$2\cosh A\sinh B = \sinh(A+B) - \sinh(A-B)$$
$$2\cosh A\cosh B = \cosh(A+B) + \cosh(A-B)$$
$$2\sinh A\sinh B = \cosh(A+B) - \cosh(A-B)$$
$$\sinh A + \sinh B = 2\sinh\tfrac{1}{2}(A+B)\cosh\tfrac{1}{2}(A-B)$$
$$\sinh A - \sinh B = 2\cosh\tfrac{1}{2}(A+B)\sinh\tfrac{1}{2}(A-B)$$
$$\cosh A + \cosh B = 2\cosh\tfrac{1}{2}(A+B)\cosh\tfrac{1}{2}(A-B)$$
$$\cosh A - \cosh B = 2\sinh\tfrac{1}{2}(A+B)\sinh\tfrac{1}{2}(A-B)$$

Inverse Functions.

$\sin^{-1} x = \frac{1}{2}\pi - \cos^{-1} x \qquad x^2 \leqslant 1$

$\tan^{-1} x = \frac{1}{2}\pi - \cot^{-1} x \qquad$ Any x

$\operatorname{cosec}^{-1} x = \frac{1}{2}\pi - \sec^{-1} x \qquad x^2 \geqslant 1$

Principal values of the inverse functions may be chosen so that each of the *expressions* given above lies in the range $-\frac{1}{2}\pi \leqslant y \leqslant +\frac{1}{2}\pi$. This gives $0 \leqslant \cot^{-1} x \leqslant \pi$; the range $-\frac{1}{2}\pi$ to $\frac{1}{2}\pi$ is, however, sometimes used for $\cot^{-1} x$ also, although it clearly cannot be used for $\cos^{-1} x$ and $\sec^{-1} x$.

$\sinh^{-1} x = \ln(x + \sqrt{x^2+1}) = -\ln(-x + \sqrt{x^2+1}) \qquad$ Any x

$\cosh^{-1} x = \pm\ln(x + \sqrt{x^2-1}) = \mp\ln(x - \sqrt{x^2-1}) \qquad x \geqslant 1$

$\tanh^{-1} x = \coth^{-1}\frac{1}{x} = \frac{1}{2}\ln\frac{1+x}{1-x} \qquad x^2 < 1$

$\coth^{-1} x = \tanh^{-1}\frac{1}{x} = \frac{1}{2}\ln\frac{x+1}{x-1} \qquad x^2 > 1$

$\operatorname{cosech}^{-1} x = \sinh^{-1}\frac{1}{x} = \ln\frac{1+\sqrt{1+x^2}}{x} = \ln\frac{x}{\sqrt{1+x^2}-1} \qquad$ Any x

$\operatorname{sech}^{-1} x = \cosh^{-1}\frac{1}{x} = \pm\ln\frac{1+\sqrt{1-x^2}}{x} = \mp\ln\frac{1-\sqrt{1-x^2}}{x} \qquad \frac{1}{x} \geqslant 1$

The positive sign is understood with all radicals. The principal values of $\cosh^{-1} x$ and $\operatorname{sech}^{-1} x$ are taken as positive, corresponding to the upper signs.

General Formulæ.

$$\frac{dy}{dx}=\frac{dy}{du}\times\frac{du}{dx}$$

$$\frac{d^2y}{dx^2}=\frac{dy}{du}\times\frac{d^2u}{dx^2}+\frac{d^2y}{du^2}\times\left(\frac{du}{dx}\right)^2$$

$$\frac{d(uv)}{dx}=u\frac{dv}{dx}+v\frac{du}{dx}$$

$$\frac{d}{dx}\left(\frac{u}{v}\right)=\frac{1}{v^2}\left(v\frac{du}{dx}-u\frac{dv}{dx}\right)$$

$$\frac{d}{dx}u^v=u^v\left(\ln u\times\frac{dv}{dx}+\frac{v}{u}\frac{du}{dx}\right)$$

Transformation of Differential Equations.

If $y=uz$, $x=f(t)$, $dx/dt=\dot{x}$, etc., then

$$\frac{dy}{dx}=\frac{u}{\dot{x}}\frac{dz}{dt}+\frac{\dot{u}}{\dot{x}}z$$

$$\frac{d^2y}{dx^2}=\frac{u}{\dot{x}^2}\frac{d^2z}{dt^2}+\left(2\frac{\dot{u}}{\dot{x}^2}-\frac{u\ddot{x}}{\dot{x}^3}\right)\frac{dz}{dt}+\left(\frac{\ddot{u}}{\dot{x}^2}-\frac{\dot{u}\ddot{x}}{\dot{x}^3}\right)z$$

$$\frac{dy}{dx}=\frac{1}{\dot{x}}\frac{dy}{dt}$$

$$\frac{d^2y}{dx^2}=\frac{1}{\dot{x}^2}\frac{d^2y}{dt^2}-\frac{\ddot{x}}{\dot{x}^3}\frac{dy}{dt}$$

If $y=\dfrac{u_1u_2\cdots}{v_1v_2\cdots}$, then $\dfrac{1}{y}\dfrac{dy}{dx}=\dfrac{1}{u_1}\dfrac{du_1}{dx}+\dfrac{1}{u_2}\dfrac{du_2}{dx}+\cdots-\dfrac{1}{v_1}\dfrac{dv_1}{dx}-\dfrac{1}{v_2}\dfrac{dv_2}{dx}-\cdots$

$$\frac{d^n(uv)}{dx^n}=v\times\frac{d^nu}{dx^n}+n\frac{dv}{dx}\times\frac{d^{n-1}u}{dx^{n-1}}+\cdots+\binom{n}{r}\frac{d^rv}{dx^r}\times\frac{d^{n-r}u}{dx^{n-r}}+\cdots+\frac{d^nv}{dx^n}\times u$$

$$\int y\,dx=\int y\frac{dx}{dt}\,dt \qquad \int u\,dv=uv-\int v\,du \qquad \int uv\,dx=u\int v\,dx-\int\frac{du}{dx}\left(\int v\,dx\right)dx$$

$$\int_a^c y\,dx=\int_a^b y\,dx+\int_b^c y\,dx \qquad \int_a^b y\,dx=-\int_b^a y\,dx \qquad \frac{d}{dx}\int_a^x f(t)\,dt=-\frac{d}{dx}\int_x^a f(t)\,dt=f(x)$$

If f is a function of u and v, each a function of independent variables x and y

then $$\frac{\partial f}{\partial x}=\frac{\partial f}{\partial u}\frac{\partial u}{\partial x}+\frac{\partial f}{\partial v}\frac{\partial v}{\partial x} \qquad \frac{\partial f}{\partial y}=\frac{\partial f}{\partial u}\frac{\partial u}{\partial y}+\frac{\partial f}{\partial v}\frac{\partial v}{\partial y}$$

If f is a homogeneous function of degree n in the variables $x, y, z, \cdots$

then $$x\frac{\partial f}{\partial x}+y\frac{\partial f}{\partial y}+z\frac{\partial f}{\partial z}+\cdots=nf.$$

Standard Forms. See pages 170 and 204.

Higher Derivatives.

$$\frac{d^p}{dx^p}x^n=\frac{n!}{(n-p)!}x^{n-p}\quad(n>p) \qquad \frac{d^n}{dx^n}x^n=n! \qquad \frac{d^p}{dx^p}x^n=0\quad(p-n>0\text{ an integer})$$

$$\frac{d^p}{dx^p}x^{-n}=\frac{(-1)^n(n+p-1)!}{(n-1)!}x^{-n-p} \qquad \frac{d^p}{dx^p}\frac{1}{1-x^2}=\frac{1}{2}\frac{p!}{(1-x)^{p+1}}+\frac{1}{2}\frac{(-1)^p\,p!}{(1+x)^{p+1}}$$

$$\frac{d^p}{dx^p}\sin(ax+b)=a^p\sin(ax+b+\tfrac{1}{2}p\pi) \qquad \frac{d^p}{dx^p}\cos(ax+b)=a^p\cos(ax+b+\tfrac{1}{2}p\pi)$$

$$\frac{d^p}{dx^p}e^{ax}=a^pe^{ax} \qquad \frac{d^p}{dx^p}e^{ax}\cos(bx+c)=(a^2+b^2)^{\frac{1}{2}p}e^{ax}\cos(bx+c+p\epsilon)\quad\text{where }\tan\epsilon=b/a$$

$$\frac{d^{2p}}{dx^{2p}}\sinh ax=a^{2p}\sinh ax \qquad \frac{d^{2p+1}}{dx^{2p+1}}\sinh ax=a^{2p+1}\cosh ax$$

$$\frac{d^{2p}}{dx^{2p}}\cosh ax=a^{2p}\cosh ax \qquad \frac{d^{2p+1}}{dx^{2p+1}}\cosh ax=a^{2p+1}\sinh ax$$

Integrals. Constants of integration are omitted, but may be important.

$$\int ae^{ax}\,dx=e^{ax}$$

$$\int a^2x\,e^{ax}\,dx=e^{ax}(ax-1)$$

$$\int ax^n\,e^{ax}\,dx=x^n\,e^{ax}-n\int x^{n-1}\,e^{ax}\,dx$$

$$\int\ln ax\,dx=x(\ln ax-1)$$

$$\int\frac{dx}{x\ln x}=\ln\ln x$$

$$\left.\begin{aligned}&\int_0^\infty ae^{-ax}\,dx=1\\&\int_0^\infty a^2x\,e^{-ax}\,dx=1\\&\int_0^\infty(ax)^n\,e^{-ax}\,dx=n\int_0^\infty(ax)^{n-1}\,e^{-ax}\,dx=n!/a\end{aligned}\right\}a>0$$

$$\int x^p\ln x\,dx=\frac{x^{p+1}}{p+1}\left(\ln x-\frac{1}{p+1}\right),\ p\neq-1$$

$$\int\frac{\ln x}{x}\,dx=\tfrac{1}{2}(\ln x)^2$$

Circular and Hyperbolic Functions.

$$\int a \sin ax\, dx = -\cos ax \qquad \int a \sinh ax\, dx = \cosh ax$$

$$\int a \sin^2 ax\, dx = \tfrac{1}{2}ax - \tfrac{1}{2}\sin ax \cos ax \qquad \int a \sinh^2 ax\, dx = -\tfrac{1}{2}ax + \tfrac{1}{2}\sinh ax \cosh ax$$

$$n\int a \sin^n ax\, dx = -\cos ax \sin^{n-1} ax + (n-1)\int a \sin^{n-2} ax\, dx$$

$$n\int a \sinh^n ax\, dx = \cosh ax \sinh^{n-1} ax - (n-1)\int a \sinh^{n-2} ax\, dx$$

$$\int a \cos ax\, dx = \sin ax \qquad \int a \cosh ax\, dx = \sinh ax$$

$$\int a \cos^2 ax\, dx = \tfrac{1}{2}ax + \tfrac{1}{2}\sin ax \cos ax \qquad \int a \cosh^2 ax\, dx = \tfrac{1}{2}ax + \tfrac{1}{2}\sinh ax \cosh ax$$

$$n\int a \cos^n ax\, dx = \sin ax \cos^{n-1} ax + (n-1)\int a \cos^{n-2} ax\, dx$$

$$n\int a \cosh^n ax\, dx = \sinh ax \cosh^{n-1} ax + (n-1)\int a \cosh^{n-2} ax\, dx$$

$$\int a \tan ax\, dx = -\ln \cos ax \qquad \int a \tanh ax\, dx = \ln \cosh ax$$

$$\int a \tan^2 ax\, dx = -ax + \tan ax \qquad \int a \tanh^2 ax\, dx = ax - \tanh ax$$

$$\int \tan^n ax\, dx = \frac{\tan^{n-1} ax}{a(n-1)} - \int \tan^{n-2} ax\, dx \qquad \int \tanh^n ax\, dx = -\frac{\tanh^{n-1} ax}{a(n-1)} + \int \tanh^{n-2} ax\, dx$$

$$\int a \cot ax\, dx = \ln \sin ax \qquad \int a \coth ax\, dx = \ln \sinh ax$$

$$\int a \cot^2 ax\, dx = -ax - \cot ax \qquad \int a \coth^2 ax\, dx = ax - \coth ax$$

$$\int a \sin^2 ax \cos^2 ax\, dx = \tfrac{1}{8}(ax - \tfrac{1}{4}\sin 4ax) \qquad \int a \sinh^2 ax \cosh^2 ax\, dx = \tfrac{1}{8}(-ax + \tfrac{1}{4}\sinh 4ax)$$

$$\int a^2 x \sin ax\, dx = \sin ax - ax \cos ax \qquad \int a^2 x \sinh ax\, dx = -\sinh ax + ax \cosh ax$$

$$\int a^2 x \cos ax\, dx = \cos ax + ax \sin ax \qquad \int a^2 x \cosh ax\, dx = -\cosh ax + ax \sinh ax$$

$$(a^2+b^2)\int e^{ax} \sin(bx+c)\, dx = e^{ax}\{a \sin(bx+c) - b\cos(bx+c)\}$$

The Gudermannian.

$$\operatorname{gd} x = \int_0^x \operatorname{sech} x\, dx = 2\tan^{-1} e^x - \tfrac{1}{2}\pi = \tfrac{1}{2}\pi - 2\tan^{-1} e^{-x} = 2\tan^{-1}\tanh \tfrac{1}{2}x$$

$$\operatorname{gd}^{-1}\theta = \int_0^\theta \sec\theta\, d\theta = \ln\tan(\tfrac{1}{4}\pi + \tfrac{1}{2}\theta) = \ln\cot(\tfrac{1}{4}\pi - \tfrac{1}{2}\theta) = 2\tanh^{-1}\tan\tfrac{1}{2}\theta$$
$$= \ln(\sec\theta + \tan\theta) = -\ln(\sec\theta - \tan\theta)$$

$$\sin\theta = \tanh x \qquad \cos\theta = \operatorname{sech} x \qquad \tan\theta = \sinh x$$
$$\operatorname{cosec}\theta = \coth x \qquad \sec\theta = \cosh x \qquad \cot\theta = \operatorname{cosech} x$$

Algebraic Functions.

$$\int \frac{dx}{ax+b} = \frac{1}{a}\ln(ax+b) \qquad \int \frac{dx}{\sqrt{b^2-a^2x^2}} = \frac{1}{a}\sin^{-1}\frac{ax}{b}$$

$$\int \frac{dx}{(x-a)(x-b)} = \frac{1}{a-b}\ln\frac{x-a}{x-b},\ a \neq b \qquad \int \frac{dx}{\sqrt{a^2x^2+b^2}} = \frac{1}{a}\sinh^{-1}\frac{ax}{b}$$

$$\int \frac{dx}{(x-a)^2} = -\frac{1}{x-a} \qquad \int \frac{dx}{\sqrt{ax^2-b^2}} = \frac{1}{a}\cosh^{-1}\frac{ax}{b}$$

$$\int \frac{dx}{a^2x^2+b^2} = \frac{1}{ab}\tan^{-1}\frac{ax}{b} \qquad \int \frac{dx}{a^2x^2-b^2} = -\frac{1}{ab}\coth^{-1}\frac{ax}{b} \quad ax>b$$

$$\int \frac{x\,dx}{cx^2+d} = \frac{1}{2c}\ln(cx^2+d) \qquad \int \frac{dx}{b^2-a^2x^2} = \frac{1}{ab}\tanh^{-1}\frac{ax}{b} \quad ax<b$$

For $\int \frac{dx}{ax^2+bx+c}$ or for $\int \frac{dx}{\sqrt{ax^2+bx+c}}$ write ax^2+bx+c in one of the forms $\pm(\alpha x+\beta)^2 \pm \gamma^2$ and use the results given above. Also

$$\int \frac{px+q}{ax^2+bx+c}\,dx = \frac{p}{2a}\ln(ax^2+bx+c) + \left(q-\frac{bp}{2a}\right)\int \frac{dx}{ax^2+bx+c}$$

$$\int \frac{px+q}{\sqrt{ax^2+bx+c}}\,dx = \frac{p}{a}\sqrt{ax^2+bx+c} + \left(q-\frac{bp}{2a}\right)\int \frac{dx}{\sqrt{ax^2+bx+c}}$$

$$\int \sqrt{ax^2+bx+c}\,dx = \frac{2ax+b}{4a}\sqrt{ax^2+bx+c} - \frac{b^2-4ac}{8a}\int \frac{dx}{\sqrt{ax^2+bx+c}}$$

Definite Integrals.

$$\int_0^{\frac{1}{2}\pi} \sin^{2p} x\,dx = \int_0^{\frac{1}{2}\pi} \cos^{2p} x\,dx = \frac{2p-1}{2p}\,\frac{2p-3}{2p-2}\cdots\frac{3}{4}\,\frac{1}{2}\,\frac{\pi}{2}$$

$$\int_0^{\frac{1}{2}\pi} \sin^{2p+1} x\,dx = \int_0^{\frac{1}{2}\pi} \cos^{2p+1} x\,dx = \frac{2p}{2p+1}\,\frac{2p-2}{2p-1}\cdots\frac{4}{5}\,\frac{2}{3}\,1$$

$$I_{m,n} = \int_0^{\frac{1}{2}\pi} \sin^m x\cos^n x\,dx = \frac{m-1}{m+n}\int_0^{\frac{1}{2}\pi} \sin^{m-2} x\cos^n x\,dx = \frac{n-1}{m+n}\int_0^{\frac{1}{2}\pi} \sin^m x\cos^{n-2} x\,dx$$

whence $I_{m,n}$ can be expressed in terms of one of

$$I_{0,0} = \tfrac{1}{2}\pi, \qquad I_{0,1} = I_{1,0} = 1, \qquad I_{1,1} = \tfrac{1}{2}$$

$$\int_0^\infty x^n e^{-x}\,dx = \Gamma(n+1) = n! \qquad \int_0^\infty x^n e^{-x^2}\,dx = \tfrac{1}{2}\Gamma(\tfrac{1}{2}n+\tfrac{1}{2}) = \tfrac{1}{2}(\tfrac{1}{2}n-\tfrac{1}{2})!$$

$$\int_0^\infty e^{-a^2x^2}\,dx = \frac{\sqrt{\pi}}{2a} \qquad \int_0^\infty x\,e^{-a^2x^2}\,dx = \frac{1}{2a^2}$$

$$(a^2+b^2)\int_0^\infty e^{-ax}\sin bx\,dx = b \qquad (a^2+b^2)\int_0^\infty e^{-ax}\cos bx\,dx = a \qquad (a>0)$$

SERIES

See pages 170 and 204 for other series connected with circular, hyperbolic, exponential and logarithmic functions. Suffixes x and y below denote partial differentiation.

Maclaurin's Series.

$$f(x) = f(0) + xf'(0) + \frac{x^2}{2!}f''(0) + \frac{x^3}{3!}f'''(0) + \cdots + \frac{x^n}{n!}f^{(n)}(\theta x) \qquad (0<\theta<1)$$

$$f(x,y) = f(0,0) + \{xf_x(0,0)+yf_y(0,0)\} + \frac{1}{2!}\{x^2f_{xx}(0,0)+2xyf_{xy}(0,0)+y^2f_{yy}(0,0)\} + \cdots$$

Taylor's Series.

$$f(x+h) = f(x) + hf'(x) + \frac{h^2}{2!}f''(x) + \frac{h^3}{3!}f'''(x) + \cdots + \frac{h^n}{n!}f^{(n)}(x+\theta h) \qquad (0<\theta<1)$$

$$f(x+h,y+k) = f(x,y) + \{hf_x(x,y)+kf_y(x,y)\} + \frac{1}{2!}\{h^2f_{xx}(x,y)+2hkf_{xy}(x,y)+k^2f_{yy}(x,y)\} + \cdots$$

Lagrange's Theorem.

If $y = x + h\phi(y)$

$$f(y) = f(x) + h\phi(x)f'(x) + \frac{h^2}{2!}\frac{d}{dx}[\{\phi(x)\}^2f'(x)] + \cdots + \frac{h^n}{n!}\frac{d^{n-1}}{dx^{n-1}}[\{\phi(x)\}^nf'(x)] + \cdots$$

Fourier's Series.

Full-range series; a to $a+c$ corresponds to 0 to 2π.

$$f(x) = \tfrac{1}{2}a_0 + a_1 \cos\frac{2\pi x}{c} + a_2 \cos\frac{4\pi x}{c} + a_3 \cos\frac{6\pi x}{c} + \cdots$$

$$+ b_1 \sin\frac{2\pi x}{c} + b_2 \sin\frac{4\pi x}{c} + b_3 \sin\frac{6\pi x}{c} + \cdots$$

where $a_n = \frac{2}{c}\int_a^{a+c} f(x) \cos\frac{2n\pi x}{c}dx,\ n = 0 \text{ to } \infty$ $\qquad b_n = \frac{2}{c}\int_a^{a+c} f(x) \sin\frac{2n\pi x}{c}dx,\ n = 1 \text{ to } \infty$

Fourier's Series.

Half-range series; a to $a+c'$ corresponds to 0 to π

$$f(x) = \tfrac{1}{2}a_0 + a_1 \cos\frac{\pi x}{c'} + a_2 \cos\frac{2\pi x}{c'} + a_3 \cos\frac{3\pi x}{c'} + \cdots \quad \text{where } a_n = \frac{2}{c'}\int_a^{a+c'} f(x) \cos\frac{n\pi x}{c'}\,dx,\ n = 0 \text{ to } \infty$$

$$f(x) = b_1 \sin\frac{\pi x}{c'} + b_2 \sin\frac{2\pi x}{c'} + b_3 \sin\frac{3\pi x}{c'} + \cdots \quad \text{where } b_n = \frac{2}{c'}\int_a^{a+c'} f(x) \sin\frac{n\pi x}{c'}\,dx$$

Binomial Series.

Ranges of convergence are given in brackets.

$$(a+b)^n = a^n + na^{n-1}b + \frac{n(n-1)}{2!}a^{n-2}b^2 + \frac{n(n-1)(n-2)}{3!}a^{n-3}b^3 + \cdots + \binom{n}{r}a^{n-r}b^r + \cdots \quad (\text{Integer } n \geqslant 0)$$

$$= a^n\left\{1 + n\left(\frac{b}{a}\right) + \frac{n(n-1)}{2!}\left(\frac{b}{a}\right)^2 + \frac{n(n-1)(n-2)}{3!}\left(\frac{b}{a}\right)^3 + \cdots + \binom{n}{r}\left(\frac{b}{a}\right)^r + \cdots\right\} \quad (b^2 < a^2)$$

$$= b^n\left\{1 + n\left(\frac{a}{b}\right) + \frac{n(n-1)}{2!}\left(\frac{a}{b}\right)^2 + \frac{n(n-1)(n-2)}{3!}\left(\frac{a}{b}\right)^3 + \cdots + \binom{n}{r}\left(\frac{a}{b}\right)^r + \cdots\right\} \quad (a^2 < b^2)$$

$$(1+x)^n = 1 + nx + \frac{n(n-1)}{2!}x^2 + \frac{n(n-1)(n-2)}{3!}x^3 + \cdots + \binom{n}{r}x^r + \cdots \quad (x^2 < 1 \text{ or integer } n \geqslant 0)$$

$$(1+x)^{-1} = 1 - x + x^2 - x^3 + x^4 - \cdots \qquad (1-x)^{-1} = 1 + x + x^2 + x^3 + x^4 + \cdots \quad (x^2 < 1)$$

$$(1+x)^{-2} = 1 - 2x + 3x^2 - 4x^3 + 5x^4 - \cdots \qquad (1-x)^{-2} = 1 + 2x + 3x^2 + 4x^3 + 5x^4 + \cdots \quad (x^2 < 1)$$

$$(1+x)^{\frac{1}{2}} = 1 + \frac{1}{2}x - \frac{1}{8}x^2 + \frac{1}{16}x^3 - \frac{5}{128}x^4 + \frac{7}{256}x^5 - \frac{21}{1024}x^6 + \frac{33}{2048}x^7 - \cdots \quad (x^2 < 1)$$

$$(1+x)^{-\frac{1}{2}} = 1 - \frac{1}{2}x + \frac{3}{8}x^2 - \frac{5}{16}x^3 + \frac{35}{128}x^4 - \frac{63}{256}x^5 + \frac{231}{1024}x^6 - \frac{429}{2048}x^7 + \cdots \quad (x^2 < 1)$$

Arithmetical Progression.

$$a + (a+d) + (a+2d) + \cdots + \{a + (n-1)d\} = \tfrac{1}{2}n\{2a + (n-1)d\}$$

Geometrical Progression.

$$a + ar + ar^2 + \cdots + ar^{n-1} = a\frac{1-r^n}{1-r}$$

$$a + ar + ar^2 + \cdots \text{ to infinity} = \frac{a}{1-r} \quad (r^2 < 1)$$

Natural Numbers.

$$1 + 2 + 3 + \cdots + n = \tfrac{1}{2}n(n+1)$$

$$1^2 + 2^2 + 3^2 + \cdots + n^2 = \tfrac{1}{6}n(n+1)(2n+1)$$

$$1^3 + 2^3 + 3^3 + \cdots + n^3 = \tfrac{1}{4}n^2(n+1)^2$$

CONVERSION FACTORS AND PHYSICAL CONSTANTS AND DEFINITIONS IN SI

For the 1970 edition the physical concepts and conversion tables have been revised by Mr K. L. Morphew of Southampton University, who wishes to acknowledge among his sources E. R. Cohen and J. W. M. DuMond, "Fundamental Constants in 1965", *Reviews of Modern Physics*, 37, 537-594 (1965).

The "Système International d'Unités", called "SI" by international agreement, is now coming into general use. The system derives from the metric system and so is continuous with the basis of scientific measurement, although many anomalies have been eradicated. The conversion tables below, for Metric-British and British-Metric, will become obsolete when the U.K. adopts SI units completely. The table on p. 387 gives definitions of physical quantities in SI, including the basic SI units of length (metre), mass (kilogramme), time (second), electric current (ampere), thermodynamic temperature (degree Kelvin) and luminous intensity (candela). The mole is also defined as a basic unit for chemistry. Precise information about SI, the Metric Units and Letters, Signs and Symbols are contained in BS 3763: 1964; P.D. 5686: 1969 and BS 1991 respectively. Details of Conversion Factors are in BS 350: 1959. Complete international agreement on all points has yet to be reached, but of those given here the seemingly most common usage has dictated the choice where uncertainty exists, with guidance from the B.S.I. specifications.

K. L. Morphew.

METRIC TO BRITISH

Length

1 km = 0·621 371 mile
1 m = 1·093 61 yd
1 cm = 0·393 701 in
1 μm = 39·370 1 μin

Area

1 km^2 = 0·386 103 mile^2
1 m^2 = 1·195 99 yd^2
1 cm^2 = 0·155 000 in^2
1 hectare = 2·471 05 acres

Volume

1 m^3 = 1·307 95 yd^3
1 cm^3 = 0·061 023 7 in^3
1 l (litre) = 1·759 80 U.K. pt

Velocity

1 km/h = 0·621 371 mile/h
1 m/s = 3·280 84 ft/s
1 international knot (kn) = 0·999 361 U.K. knot

Mass

1 tonne = 1000 kg = 0·984 205 U.K. ton
1 kg = 2·204 62 lb
1 g = 0·035 2740 oz

Force

1 N = 0·224 809 lbf
1 N = 7·233 01 pdl
1 kgf = 2·204 62 lbf

Energy (work)

1 J = 0·737 562 ft lbf
1 J = 23·730 4 ft pdl

BRITISH TO METRIC

Length

1 mile = 1·609 34 km
1 yd = **0·914 4** m
1 in = **2·54** cm
1 μin = **0·025 4** μm

Area

1 mile^2 = 2·589 99 km^2
1 yd^2 = 0·836 127 m^2
1 in^2 = **6·451 6** cm^2
1 acre = 0·404 686 hectare

Volume

1 yd^3 = 0·764 555 m^3
1 in^3 = 16·387 1 cm^3
1 U.K. pt = 0·568 245 l (litre)

Velocity

1 mile/h = 1·609 34 km/h
1 ft/s = 0·304 8 m/s
1 U.K. knot = 1·000 64 international knot

Mass

1 U.K. ton = 1016·05 kg = 1·016 05 tonne
1 lb = **0·453 592 37** kg
1 oz = 28·349 5 g

Force

1 lbf = 4·448 22 N
1 pdl = 0·138 255 N
1 lbf = 0·453 592 kgf

Energy (work)

1 ft lbf = 1·355 82 J
1 ft pdl = 0·042 140 1 J

Exact values in **bold** type

OTHER CONVERSION FACTORS

1 micron (μ) = 10^{-6} m
1 angstrom (Å) = 10^{-10} m
1 internat. nautical mile = 1852 m
1 U.K. nautical mile = 1853·18 m = 6080 ft
1 fathom = 1·828 8 m

1 chain (Gunter's) = 22 yd
1 chain (engineer's) = 100 ft
1 bar = 10^5 N/m^2
1 millibar = 0·750 1 torr
1 light year = 5·880 × 10^{12} miles

1 parsec = 3·263 light years
1 U.K. gallon = 4·546 09 l
1 U.S. gallon = 3·785 41 l
1 torr = 1/760 atmosphere (= 1 mm Hg to 1 in 7 × 10^6)

PHYSICAL CONSTANTS, STANDARD VALUES AND EQUIVALENTS IN SI UNITS

(Figures in parentheses are the standard deviation in last digit(s))

Velocity of light in vacuo

$c = 2{\cdot}997\,925(1) \times 10^{8}$ m s^{-1}

Gravitational constant

$G = 6{\cdot}670(5) \times 10^{-11}$ N m^{2} kg^{-2}

Standard acceleration of gravity

$g = 9{\cdot}806\,65$ m s^{-2} ($= 32{\cdot}1740$ ft s^{-2})

Acceleration of gravity at Greenwich $g = 9{\cdot}818\,83$ m s^{-2}

Standard atmosphere ($\equiv 760$ mm Hg to 1 in 7×10^{6})

1 atm $= 101\,325$ N m^{-2}

Electron charge

$e = 1{\cdot}602\,10(2) \times 10^{-19}$ C

Avogadro constant

$N_A = 6{\cdot}022\,52(9) \times 10^{26}$ kmole^{-1}

Mass unit

$u = 1{\cdot}660\,43(2) \times 10^{-27}$ kg

Electron rest mass

$m_e = 9{\cdot}109\,08(13) \times 10^{-31}$ kg

$= 5{\cdot}485\,97(3) \times 10^{-4}$ u

Proton rest mass

$m_p = 1{\cdot}672\,52(3) \times 10^{-27}$ kg

$= 1{\cdot}007\,276\,63(8) \times$ u

Neutron rest mass

$m_n = 1{\cdot}674\,82(3) \times 10^{-27}$ kg

$= 1{\cdot}008\,665\,4(4) \times$ u

Charge/mass ratio for electron

$\dfrac{e}{m_e} = 1{\cdot}758\,796(6) \times 10^{11}$ C kg^{-1}

Faraday constant

$F = 9{\cdot}648\,70(5) \times 10^{4}$ C mole^{-1}

Planck constant

$h = 6{\cdot}625\,59(16) \times 10^{-34}$ J s

Fine structure constant

$\alpha = 7{\cdot}297\,20(3) \times 10^{-3}$

$\dfrac{1}{\alpha} = 137{\cdot}038\,8(6)$

Rydberg constant

$R_\infty = 1{\cdot}097\,373\,1(1) \times 10^{7}$ m^{-1}

Bohr radius

$a_0 = 5{\cdot}291\,67(2) \times 10^{-11}$ m

Compton wavelength of electron

$\lambda_{Ce} = 2{\cdot}426\,21(2) \times 10^{-12}$ m

Electron radius

$r_e = 2{\cdot}817\,77(4) \times 10^{-15}$ m

Compton wavelength of proton

$\lambda_{Cp} = 1{\cdot}321\,398(13) \times 10^{-15}$ m

Gyromagnetic ratio of proton

$\gamma = 2{\cdot}675\,192(7) \times 10^{8}$ rad s^{-1} T^{-1}

Bohr magneton

$\mu_B = 9{\cdot}273\,2(2) \times 10^{-24}$ J T^{-1}

Gas constant

$R_0 = 8{\cdot}314\,34(35)$ J deg^{-1} mole^{-1}

Standard volume of ideal gas

$V_0 = 2{\cdot}241\,36 \times 10^{-2}$ m^{3} mole^{-1}

Boltzmann constant

$k = 1{\cdot}380\,54(6) \times 10^{-23}$ J deg^{-1}

First radiation constant

$c_1 = 3{\cdot}741\,50(9) \times 10^{-16}$ W m^{2}

Second radiation constant

$c_2 = 1{\cdot}438\,79(6) \times 10^{-2}$ m deg

Stefan-Boltzmann constant

$\sigma = 5{\cdot}669\,7(10) \times 10^{-8}$ W m^{-2} deg^{-4}

Solar year = 365 d 5 hr 48 min 45·5 s

Sidereal year contains 365·256 360 42 mean solar days.

Mean Solar second = 1/86400 mean solar day.

International Temperature Scale of 1948. (All at a pressure of one standard atmosphere)

b.p. Oxygen	−182·970	°C
m.p. Ice	0	°C
b.p. Water	100	°C
b.p. Sulphur	444·600	°C
f.p. Silver	960·8	°C
f.p. Gold	1063·0	°C

m.p. Ice on Kelvin Scale. 0 °C = 273·16 K.

Triple point of water = 0·0100 °C.

International Table calorie 1 $\text{cal}_{IT} = 4{\cdot}186\,8$ J.

15°C calorie 1 $\text{cal}_{15} = 4{\cdot}1855$ J.

British thermal unit 1 Btu = 778·169 ft lbf = 1055·06 J.

Therm 1 therm = 10^{5} Btu.

Horsepower 1 hp = 550 ft lbf s^{-1} = 745·700 W.

c.g.s. units: Force 1 dyn = 10^{-5} N. Work 1 erg = 10^{-7} J.

Electron volt 1 eV = $1{\cdot}602\,1 \times 10^{-19}$ J.

e.m.f. of Normal Weston Cadmium cell at 20° C.

$E = (1{\cdot}018\,74 \pm 0{\cdot}000\,003)$ V

Standard transmitted radio frequencies. (All better than 1 in 10^{9}) Droitwich (carrier) 200 kc/s. Rugby 60 kc/s; 2·5 Mc/s; 5 Mc/s; 10 Mc/s.

Velocity of sound at sea level at 0 °C. $V = 1088$ ft s^{-1} $= 331{\cdot}7$ m s^{-1}

Standard concert pitch A = 440 c/s.

Scientific pitch C = 256 c/s.

SI PREFIXES

The following prefixes are used to indicate decimal multiples and sub-multiples of SI units.

Factor	Prefix	Symbol
10^{12}	tera	T
10^{9}	giga	G
10^{6}	mega	M
10^{3}	kilo	k
10^{2}	hecto	h
10	deca	da
10^{-1}	deci	d
10^{-2}	centi	c
10^{-3}	milli	m
10^{-6}	micro	μ
10^{-9}	nano	n
10^{-12}	pico	p
10^{-15}	femto	f
10^{-18}	atto	a

THERMOMETRIC SCALES: CONVERSIONS

The following relationships are suggested "for most practical purposes" in *Changing to the Metric System* (HMSO, 1967). The symbols t, r, T and θ represent respectively the same temperature on the Fahrenheit, Rankine (absolute Fahrenheit), Kelvin and Celsius (Centigrade) scales.

Celsius $\theta = \frac{5}{9}(t - 32)$

Kelvin $T = \frac{5}{9}(t + 459{\cdot}67)$

Rankine $r = t + 459{\cdot}67$

PHYSICAL CONCEPTS IN RATIONALISED M.K.S. UNITS

Concept	Symbol	Name of Unit	Abbreviation of Unit Name	Definition or Defining Equation	Explanations; Equivalent Units; Alternative Definitions; etc.
Length	l	metre	m	1 m = 1 650 763·73 wavelengths of radiation ($2p_{10}-5d_5$) of Kr 86.	
Mass	m	kilogramme	kg	International Prototype Kilogramme.	
Time	t	second	s	1/31 556 925·974 7 of the tropical year for 1900 January 0 at 12h ephemeris time.	
Electric current	I	ampere	A	An ampere in each of two infinitely long parallel conductors of negligible cross-section in vacuo will produce on each a force of 2×10^{-7} N/m on each.	
Thermodynamic temperature	K	kelvin		The kelvin is 1/273·16 of the thermo-dynamic temperature of the triple point of water.	
Luminous intensity	cd	candela	cd	The luminous intensity of a black body radiator at the temperature of freezing Pt at a pressure of 1 std. atm. viewed normal to the surface is 6×10^5 cd/m².	
Amount of substance	mol	mole	mol	The amount of substance of a system which contains as many elementary units as there are carbon atoms in 0·012 kg of ^{12}C. The elementary unit must be specified (atom, molecule, ion, etc.).	

All the above are internationally agreed basic units except the mole, which, though recommended, awaits international acceptance.

Concept	Symbol	Name of Unit	Abbreviation of Unit Name	Definition or Defining Equation	Explanations; Equivalent Units; Alternative Definitions; etc.
Plane angle		radian	rad	A radian is equal to the angle subtended at the centre of a circle by an arc equal in length to the radius.	
Solid angle		steradian	sr	A steradian is equal to the angle in three dimensions subtended at the centre of a sphere by an area on the surface equal to the radius squared.	
Area	A, a	square metre	m²	$a=l^2$	
Volume	V, v	cubic metre	m³	$V=l^3$	
Velocity	v, u	metre/second	m s⁻¹	$v=\mathrm{d}l/\mathrm{d}t$	
Acceleration	a	metre/second²	m s⁻²	$a=\mathrm{d}^2l/\mathrm{d}t^2$	
Density	ρ	kilogramme/metre³	kg m⁻³	$\rho=m/V$	
Mass rate of flow		kilogramme/sec	kg s⁻¹	$\mathrm{d}m/\mathrm{d}t$	
Volume rate of flow		cubic metre/sec	m³ s⁻¹	$\mathrm{d}V/\mathrm{d}b$	
Moment of inertia	I	kilogramme metre²	kg m²	$I=Mk^2$	
Momentum	p	kilogramme metre/sec	kg m s⁻¹	$p=mv$	
Angular momentum	$I\omega$	kilogramme metre²/sec	kg m² s⁻¹	$\mathrm{d}I/\mathrm{d}t$	
Kinetic energy	T, (W)	kilogramme metre²/sec²	kg m² s⁻²	$T=\frac{1}{2}mv^2$	Newton metre (N m)
Force	F	newton	N	$F=ma$	kg m s⁻²
Torque (Moment of Force)	T, (M)	newton metre	N m	$T=Fl$	$T=P/2\pi n$ n in rev/sec
Potential energy	V, (w)	newton metre	N m	$V=\int F\mathrm{d}l$	kg m² s⁻²
Work (Energy, Heat)	W, (U)	joule	J	$W=\int F\mathrm{d}l$	Nm 1 J = 1 N m by definition
Heat (Enthalpy)	Q, (H)	joule	J	$H=U+p\,v$	Definition for a fluid U = Internal energy
Power	P	watt	W	$P=\mathrm{d}W/\mathrm{d}t$	1 W = 1 J s⁻¹ by definition
Pressure (Stress)	p (σ, f)	newton/metre²	N m⁻²	$p=F/A$	Usually pressure in fluid; stress in solids
Surface tension	γ (σ)	newton/metre	N m⁻¹	$\gamma=F/l$	Free surface energy
Viscosity, dynamic	η, μ	poise	P	$\frac{P}{A}=\eta dv/dl$	10⁻¹ N s m⁻² Defined in c.g.s. units
Viscosity, kinematic	ν	stokes	S, St	$\nu=\eta/\rho$	10⁻⁴ m² s⁻¹ Defined in c.g.s. units
Temperature	θ, T	degree C, degree K	°C, K	$T\,\mathrm{K}=(\theta+273{\cdot}16)°\mathrm{C}$	International Temperature Scale
Velocity of light	c	metre/second	m s⁻¹	Fundamental, measured, constant	
Permeability of vacuum	μ_0	henry/metre	H m⁻¹	$\mu_0=4\pi\times10^{-7}$ H/m	Defined value to give coherent rationalised electrical units
Permittivity of vacuum	ε_0	farad/metre	F m⁻¹	$\varepsilon_0=1/\mu_0c^2$	Derived in Maxwell's theory of e.m. radiation
Electric charge	Q	coulomb	C	$F=(Q_1Q_5)/(4\pi\varepsilon_0r^2)$	A s Coulomb's Law Also $Q=\int Idt$
Electric potential (Potential difference)	V	volt	V	$V_r=\int_\infty Fdl(V_{ab}=\int_a Fdl)$	N m C⁻¹ or J C⁻¹
Electric field strength (Electric force)	E	volt/metre	V m⁻¹	$E=dV/dl$	N C⁻¹ E = Force on unit point charge

PHYSICAL CONCEPTS IN RATIONALISED M.K.S. UNITS

Concept	Symbol	Name of Unit	Abbreviation of Unit Name	Definition or Defining Equation	Explanations; Equivalent Units; Alternative Definitions; etc.
Electric resistance	R	ohm	Ω	$R=V/I$	J/A
Conductance	G	siemens	S	$G=\frac{1}{R}$ or $\frac{R}{Z^2}$	℧
Electric flux	Ψ	coulomb	Ψ=Q		
Electric flux density (Displacement)	D	coulomb/metre²	C m⁻²	$D=d\Psi/dA$	
Frequency	f	hertz	Hz		s^{-1}
Permittivity	ε	farad/metre	F m⁻¹	$\varepsilon=D/E$	
Relative permittivity	ε_r			$\varepsilon_r=\varepsilon/\varepsilon_0$	a numeric
Magnetic field strength (Magnetic force)	H	amp.turn/metre	AT m⁻¹	$dH=Idl\sin\theta/4\pi r^2$	The *turn* is a numeric not a unit
Magnetic flux	Φ	weber	Wb	$\Phi=-\int e\,dt$	V s Faraday Lenz Law
Magnetic flux density	B	tesla	T	$B=d\Phi/dA$	V s m⁻¹ Wb m⁻²
Permeability	μ	henry/metre	H/m	$\mu=B/H$	
Relative permeability	μ_r			$\mu_r=\mu/\mu_0$	a numeric
Coefficient of mutual induction	M	henry	H	$e_2=-M dI_1/dt$	Wb A⁻¹
Coefficient of self induction	L	henry	H	$e=-L dI/dt$	Wb A⁻¹
Capacitance	C	farad	F	$C=Q/V$	C V⁻¹
Reactance	X	equivalent ohm	Ω	$X=\omega L$ or $\frac{1}{\omega C}$	Sinusoidal a.c.
Impedance	Z	equivalent ohm	Ω	$Z=\sqrt{R^2+X^2}$	Sinusoidal a.c.
Susceptance	B	siemens	S	$B=\frac{X}{Z^2}$	℧ (Sinusoidal a.c.)
Admittance	Y	siemens	S	$Y=\frac{1}{Z}$	℧ (Sinusoidal a.c.)
Apparent power	VA	volt amp			
Reactive power	VAr	volt amp reactive			
Power factor	ϕ		p.f.	$\cos\phi=\frac{\text{power}}{\text{apparent power}}$	
Luminous flux	lm	lumen	lm	$lm=cd\ sr$	
Illumination	lx	lux	lx	$lx=lm\ m^{-2}$	

Names in parentheses with upper case initial, e.g. (Energy), are alternatives. Words in parentheses with l.c. initial, e.g. (dynamic), are adjectival.

Symbols and names are in accordance with B.S. 1991, B.S. 350, and B.S. 1637.

MATHEMATICAL CONSTANTS

Constant	Number	Log	Constant	Number	Log
π	3·14159 26535 89793	0·49714 98727	$\frac{1}{\pi}$	0·31830 98861 83791	9·50285 01273
2π	6·28318 53071 79586	0·79817 98684	$\frac{1}{2\pi}$	0·15915 49430 91895	9·20182 01316
$\frac{\pi}{2}$	1·57079 63267 94897	0·19611 98770	$\frac{2}{\pi}$	0·63661 97723 67581	9·80388 01230
$\frac{\pi}{4}$	0·78539 81633 97448	9·89508 98814	$\frac{4}{\pi}$	1·27323 95447 35163	0·10491 01186
$\frac{4}{3}\pi$	4·18879 02047 86391	0·62208 86093	$\sqrt[3]{\frac{3}{4\pi}}$	0·62035 04908 99400	9·79263 71302
π^2	9·86960 44010 89359	0·99429 97454	$\frac{1}{\pi^2}$	0·10132 11836 42338	9·00570 02546
$\sqrt{\pi}$	1·77245 38509 05516	0·24857 49363	$\frac{1}{\sqrt{\pi}}$	0·56418 95835 47756	9·75142 50637
$\sqrt{2\pi}$	2·50662 82746 31001	0·39908 99342	$\frac{1}{\sqrt{2\pi}}$	0·39894 22804 01433	9·60091 00658
$\frac{\sqrt{\pi}}{2}$	0·88622 69254 52758	9·94754 49407	$\frac{2}{\sqrt{\pi}}$	1·12837 91670 95513	0·05245 50593
$\sqrt{\frac{\pi}{2}}$	1·25331 41373 15500	0·09805 99385	$\sqrt{\frac{2}{\pi}}$	0·79788 45608 02865	9·90194 00615
e	2·71828 18284 59045	0·43429 44819	e^{-1}	0·36787 94411 71442	9·56570 55181
e^2	7·38905 60989 30650	0·86858 89638	e^{-2}	0·13533 52832 36613	9·13141 10362
$\sqrt{e}$	1·64872 12707 00128	0·21714 72410	$e^{-\frac{1}{2}}$	0·60653 06597 12633	9·78285 27590
e^{π}	23·14069 26327 79269	1·36437 63538	$e^{-\pi}$	0·04321 39182 63772	8·63562 36462
$e^{\frac{\pi}{2}}$	4·81047 73809 65352	0·68218 81769	$e^{-\frac{\pi}{2}}$	0·20787 95763 50762	9·31781 18231
$\sqrt{2}$	1·41421 35623 73095	0·15051 49978	$\sqrt{\frac{1}{2}}$	0·70710 67811 86548	9·84948 50022
$\sqrt{3}$	1·73205 08075 68877	0·23856 06274	$\sqrt[3]{10}$	2·15443 46900 31884	0·33333 33333
$\sqrt{10}$	3·16227 76601 68379	0·50000 00000	$\sqrt[3]{100}$	4·64158 88336 12779	0·66666 66667
$M = \log e$	0·43429 44819 03252	9·63778 43113	$\frac{1}{M} = \ln 10$	2·30258 50929 94046	0·36221 56887
γ	0·57721 56649 01533	9·76133 81088	$\ln \pi$	1·14472 98858 49400	...
ρ	0·47693 62762 04470	9·67846 03565	$\sqrt{2}\rho$	0·67448 97501 96081	9·82897 53544
1 radian	57°·29577 95131	1·75812 26324	1°	0^r·01745 32925 19943	8·24187 73676
,,	3437′·74677 07849	3·53627 38828	1′	0^r·00029 08882 08666	6·46372 61172
,,	206264″·80624 70964	5·31442 51332	1″	0^r·00000 48481 36811	4·68557 48668

If B_n and E_n denote Bernoulli's and Euler's numbers,

$$B_1 = \frac{1}{6} \quad B_2 = \frac{1}{30} \quad B_3 = \frac{1}{42} \quad B_4 = \frac{1}{30} \quad B_5 = \frac{5}{66} \quad B_6 = \frac{691}{2730} \quad B_7 = \frac{7}{6} \quad B_8 = \frac{3617}{510} \quad B_9 = \frac{43867}{798} \quad B_{10} = \frac{174611}{330}$$

$$E_1 = 1 \quad E_2 = 5 \quad E_3 = 61 \quad E_4 = 1385 \quad E_5 = 50521 \quad E_6 = 27\,02765 \quad E_7 = 1993\,60981 \quad E_8 = 1\,93915\,12145$$

BINOMIAL COEFFICIENTS

The binomial coefficients are the coefficients of powers of x in the binomial expansion

$$(1+x)^n = 1 + nx + \frac{n(n-1)}{2!}x^2 + \frac{n(n-1)(n-2)}{3!}x^3 + \cdots$$

If x^2 is less than 1,

$$(1+x)^{\frac{1}{2}} = 1 + \frac{1}{2}x - \frac{1}{8}x^2 + \frac{1}{16}x^3 - \frac{5}{128}x^4 + \frac{7}{256}x^5 - \cdots$$

$$(1+x)^{-\frac{1}{2}} = 1 - \frac{1}{2}x + \frac{3}{8}x^2 - \frac{5}{16}x^3 + \frac{35}{128}x^4 - \cdots$$

$$(1+x)^{-1} = 1 - x + x^2 - x^3 + x^4 - \cdots$$

$$(1+x)^{-2} = 1 - 2x + 3x^2 - \cdots$$

n													
1	1	1											
2	1	2	1										
3	1	3	3	1									
4	1	4	6	4	1								
5	1	5	10	10	5	1							
6	1	6	15	20	15	6	1						
7	1	7	21	35	35	21	7	1					
8	1	8	28	56	70	56	28	8	1				
9	1	9	36	84	126	126	84	36	9	1			
10	1	10	45	120	210	252	210	120	45	10	1		
11	1	11	55	165	330	462	462	330	165	55	11	1	
12	1	12	66	220	495	792	924	792	495	220	66	12	1